ENCYCLOPEDIA OF HUMAN BIOLOGY

VOLUME 2 Br–De

ENCYCLOPEDIA OF HUMAN BIOLOGY

VOLUME 2 Br–De

Editor–in–Chief
Renato Dulbecco
The Salk Institute
La Jolla, California

ACADEMIC PRESS, INC. *Harcourt Brace Jovanovich, Publishers*
San Diego New York Boston London Sydney Tokyo Toronto

This book is printed on acid-free paper. ∞

Academic Press, Inc.
San Diego, California 92101

United Kingdom Edition published by
Academic Press Limited
24–28 Oval Road, London NW1 7DX

Library of Congress Cataloging-in-Publication Data

Encyclopedia of human biology / [edited by] Renato Dulbecco.
p. cm.
Includes index.
ISBN 0-12-226751-6 (v. 1). -- ISBN 0-12-226752-4 (v. 2). -- ISBN 0-12-226753-2 (v. 3). -- ISBN 0-12-226754-0 (v. 4). -- ISBN 0-12-226755-9 (v. 5). -- ISBN 0-12-226756-7 (v. 6). -- ISBN 0-12-226757-5 (v. 7). -- ISBN 0-12-226758-3 (v. 8)
1. Human biology--Encyclopedias. I. Dulbecco, Renato, 1914-
[DNLM: 1. Biology--encyclopedias. 2. Physiology--Encyclopedias.
QH 302.5 E56]
QP11.E53 1991
612'.003--dc20
DNLM/DLC
for Library of Congress 91-45538
CIP

PRINTED IN THE UNITED STATES OF AMERICA
91 92 93 94 9 8 7 6 5 4 3 2 1

CONTENTS OF VOLUME 2

HOW TO USE THE ENCYCLOPEDIA

We have organized this encyclopedia in a manner that we believe will be the most useful to you and would like to acquaint you with some of its features.

The volumes are organized alphabetically as you would expect to find them in, for example, magazine articles. Thus, "Food Toxicology" is listed as such and would *not* be found under "Toxicology, Food." If the first words in a title are *not* the primary subject matter contained in an article, the main subject of the title is listed first: (e.g., "Sex Differences, Biocultural," "Sex Differences, Psychological," "Aging, Psychiatric Aspects," "Bone, Embryonic Development.") This is also true if the primary word of a title is too general (e.g., "Coenzymes, Biochemistry.") Here, the word "coenzymes" is listed first as "biochemistry" is a very broad topic. Titles are alphabetized letter-by-letter so that "Gangliosides" is followed by "Gangliosides and Neuronal Differentiation" and then "GangliosideTransport."

Each article contains a brief introductory Glossary wherein terms that may be unfamiliar to you are defined *in the context of their use in the article*. Thus, a term may appear in another article defined in a slightly different manner or with a subtle pedagogic nuance that is specific to that particular article. For clarity, we have allowed these differences in definition to remain so that the terms are defined relative to the context of each article.

Articles about closely related subjects are identified in the Index of Related Articles at the end of the last volume (Volume 8.) The article titles that are cross-referenced within each article may be found in this index, along with other articles on related topics.

The Subject Index contains specific, detailed information about any subject discussed in the *Encyclopedia*. Entries appear with the source volume number in boldface followed by a colon and the page number in that volume where the information occurs (e.g., "Diuretics, **3:** 93"). Each article is also indexed by its title (or a shortened version thereof) and the page ranges of the article appear in boldface (e.g., "Abortion, **1: 1–10**" means that the primary coverage of the topic of abortion occurs on pages 1–10 of Volume 1).

If a topic is covered primarily under one heading but is occasionally referred to in a slightly different manner or by a related term, it is indexed under the term that is most commonly used and a cross-reference is given to the minor usage. For example, "B Lymphocytes" would contain all page entries where relevant information occurs, followed by "*see also* B Cells." In addition, "B Cells, *see* B Lymphocytes" would lead the reader to the primary usages of the more general term. Similarly, "*see under*" would mean that the subject is covered under a subheading of the more common term.

An additional feature of the Subject Index is the identification of Glossary terms. These appear in the index where the word "defined" (or the words "definition of") follows an entry. As we noted earlier, there may be more than one definition for a particular term and, as when using a dictionary, you will be able to choose among several different usages to find the particular meaning that is specifically of interest to you.

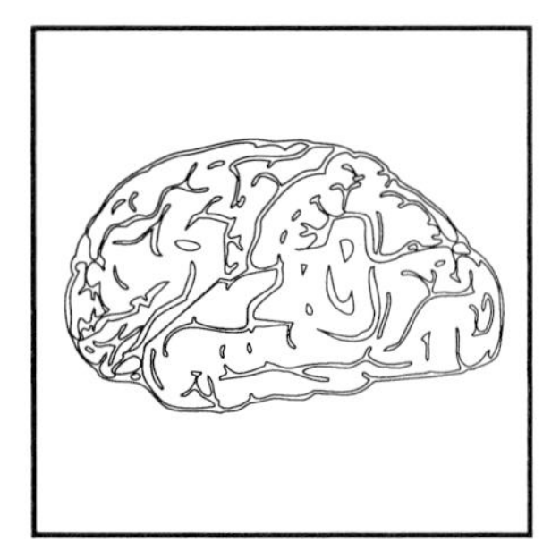

Brain

BRYAN KOLB AND IAN Q. WHISHAW, *University of Lethbridge, Canada*

Glossary

Action potential Brief electrical impulse by which information is conducted along an axon. It results from brief changes in the membrane's permeability to sodium

Amnesia Partial or total loss of memory

Aphasia Defect or loss of power of expression by speech, writing, or signs of comprehending spoken or written language caused by injury or disease of the brain

Cerebral cortex Layer of gray matter on the surface of the cerebral hemispheres composed of neurons and their synaptic connections that form 4–6 sublayers

Hippocampus Primitive cortical structure lying in the medial region of the temporal lobe; named after its shape, which is similar to a sea horse, or hippocampus

Neuron Basic unit of the nervous system; the nerve cell. Its function is to transmit and store information; includes the cell body (*soma*), many processes called *dendrites,* and an *axon*

Neurotransmitter Chemical released from a synapse in response to an action potential and acting on postsynaptic receptors to change the resting potential of the receiving cell; chemically transmits information from one neuron to another

Synapse Functional junction between one neuron and another

THE BRAIN is that part of the central nervous system that is contained in the skull. It weighs approximately 1,450 g at maturity and is composed of brain cells (neurons) and their processes, as well as support cells, that are organized into hundreds of functionally distinct regions. Neurons communicate both chemically and electrically so that different brain regions from functional systems to control behavior. Measurement of brain structure, activity, and behavior has allowed neuroscientists to reach inferences regarding the mechanisms of the basic functions, which include (1) the body's interactions with the environment through the sensory systems (e.g., vision, audition, touch) and motor systems, (2) internal activities of the body (e.g., breathing, temperature, blood pressure), and (3) mental activities (e.g., thought, language, affect). By studying people with brain injuries it is possible to propose brain circuits that underly human behavior.

I. Anatomical and Physiological Organization of the Human Brain

A. Cellular Composition

The brain is composed of two general classes of cells: neurons and glial cells. Neurons are the functional units of the nervous system, whereas glial cells are support cells. Estimates of the numbers of cells in the human brain usually run around 10^{10} neurons and 10^{12} glial cells, although the numbers could be even higher. Only about 2–3 million cells (motor neurons) send their connections out of the brain to animate muscle fibers, leaving an enormous number of cells with other functions.

There are numerous types of neurons (e.g., pyramidal, granule, Purkinje, Golgi I, motoneurons),

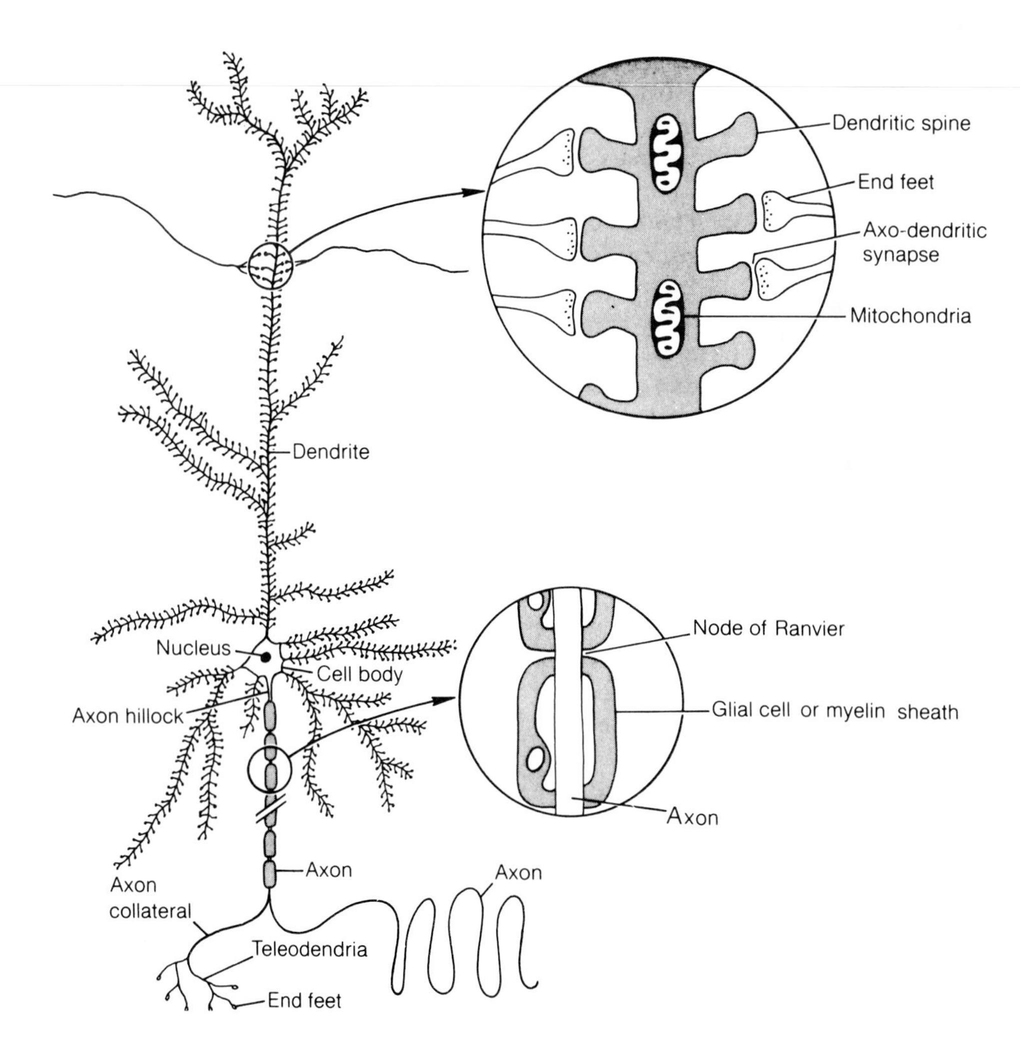

FIGURE 1 Summary of the major parts of a stylized neuron. Enlargement at the *top right* shows the gross structure of the synapse between axons and the spines on dendrites. Enlargement on the *bottom right* shows the myelin sheath that surrounds the axon and acts as insulation. [From Kolb, B., and Whishaw, I. Q. (1990). "Fundamentals of Human Neuropsychology," 3rd ed. New York: W. H. Freeman.]

but they share several features in common. First, they have a cell body, which like most cells contains a variety of substances that determines the function of the cell; processes called *dendrites,* which function primarily to increase the surface area on which a cell can receive information from other cells; and a process called an *axon,* which normally originates in the cell body and transmits information to other cells (Fig. 1). Different types of neurons are morphologically distinct, reflecting differences in function. The different types are distributed differentially to different regions of the brain reflecting regional differences in brain function.

Neurons are connected with one another via their axons; any given neuron may have as many as 15,000 connections with other neurons. These connections are highly organized so that certain regions of the brain are more closely connected to one another than they are to others. As a result, these closely associated regions form functional systems in the brain, which control certain types of behavior.

B. Gross Anatomical Organization of the Brain

The most obvious feature of the human brain is that there are two large hemispheres, which sit on a stem, known as the brainstem. Both structures are composed of hundreds of regions; nearly all of them are found bilaterally. Traditionally, the brain is described by the gross divisions observed phylogenetically and embryologically as summarized in Table I. The most primitive region is the hindbrain, whose

TABLE I Divisions of the Central Nervous System

Primitive divisions	Mammalian divisions	Major structures
Prosencephalon (forebrain)	Telencephalon (endbrain)	Neocortex
		Basal ganglia
		Limbic system
		Olfactory bulb
		Lateral ventricles
	Diencephalon (between brain)	Thalamus
		Hypothalamus
		Epithalamus
		Third ventricle
Mesencephalon (midbrain)	Mesencephalon (midbrain)	Tectum
		Tegmentum
		Cerebral aqueduct
Rhombencephalon (hindbrain)	Metencephalon (acrossbrain)	Cerebellum
		Pons
		Fourth ventricle
	Myelencephalon (spinal brain)	Medulla oblongata
		Fourth ventricle

principal structures include the cerebellum, pons, and medulla. The cerebellum was originally specialized for sensory-motor coordination, which remains its major function. The pons and medulla also contribute to equilibrium, balance, and the control of gross movements (including breathing). The midbrain consists of two main structures, the tectum and the tegmentum. The tectum consists primarily of two sets of nuclei, the superior and inferior colliculi, which mediate whole body movements to visual and auditory stimuli, respectively. The tegmentum contains various structures including regions associated with (1) the nerves of the head (so-called cranial nerves), (2) sensory nerves from the body, (3) connections from higher structures that function to control movement, and (4) a number of structures involved in movement (substantia nigra, red nucleus), as well as a region known as the reticular formation. The latter system plays a major role in the control of sleep and waking.

The forebrain is conventionally divided into five anatomical areas: (1) the neocortex, (2) the basal ganglia, (3) the limbic system, (4) the thalamus, and (5) the olfactory bulbs and tract. Each of these regions can be dissociated into numerous smaller regions on the basis of neuronal type, physiological and chemical properties, and connections with other brain regions.

The neocortex, which is usually called the *cortex,* is composed of approximately six layers, each of which have distinct neuronal populations, comprises 80% of the human forebrain by volume, is grossly divided into four regions, which are named by the cranial bones lying above them (Fig. 2). It is wrinkled, which is nature's solution to the problem of confining a large surface area into a shell that is still small enough to pass through the birth canal. The cortex has a thickness of only 1.5–3.0 mm but has a total area of about 2,500 cm^2. The cortex can be subdivided into dozens of subregions on the basis of the distribution of neuron types, their chemical and physiological characteristics, and their connections. These subregions can be shown to be functionally distinct. [*See* NEOCORTEX.]

The basal ganglia are a collection of nuclei lying beneath the neocortex. They include the putamen, caudate nucleus, globus pallidus, and amygdala. These nuclei have intimate connections with the neocortex as well as having major connections with midbrain structures. The basal ganglia have principally a motor function, as damage to different regions can produce changes in posture or muscle tone and abnormal movements such as twitches, jerks, and tremors.

The limbic system is not really a unitary system but refers to a number of structures that were once believed to function together to produce emotion. These include the hippocampus, septum, cingulate cortex, and hypothalamus, each of which have different functions (Fig. 3). [*See* HIPPOCAMPAL FORMATION; HYPOTHALAMUS; LIMBIC MOTOR SYSTEM.]

The thalamus provides the major route of information to the neocortex, and different neocortical regions are associated with inputs from distinct thalamic regions connected with the neocortex. The different thalamic areas receive information from sensory and motor regions in the brainstem as well as the limbic system. [*See* THALAMUS.]

C. Physiological Organization of the Brain

Like other cells in the body, the neuron has an electrical voltage (potential) across its membrane, which results from the differential distribution of different ions on the two sides of the membrane. In contrast to other body cells, however, this electrical potential is used to transmit information from one neuron to another in the nervous system, which is accomplished in the following way. Neurons have a resting potential across the membrane of the dendrites, cell body, and axon, which remains relatively constant at about −70 mV. If the membrane permeability for different ions changes, the electrical potential will also change. If it becomes more

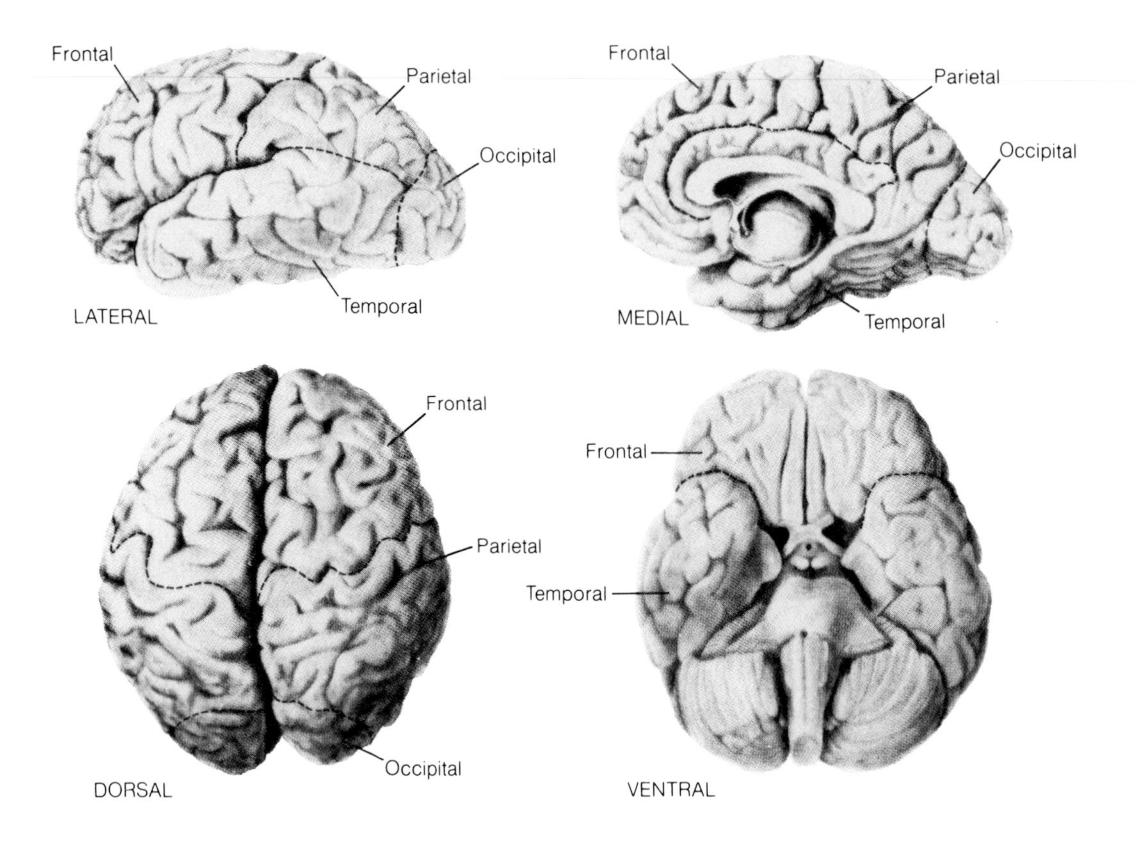

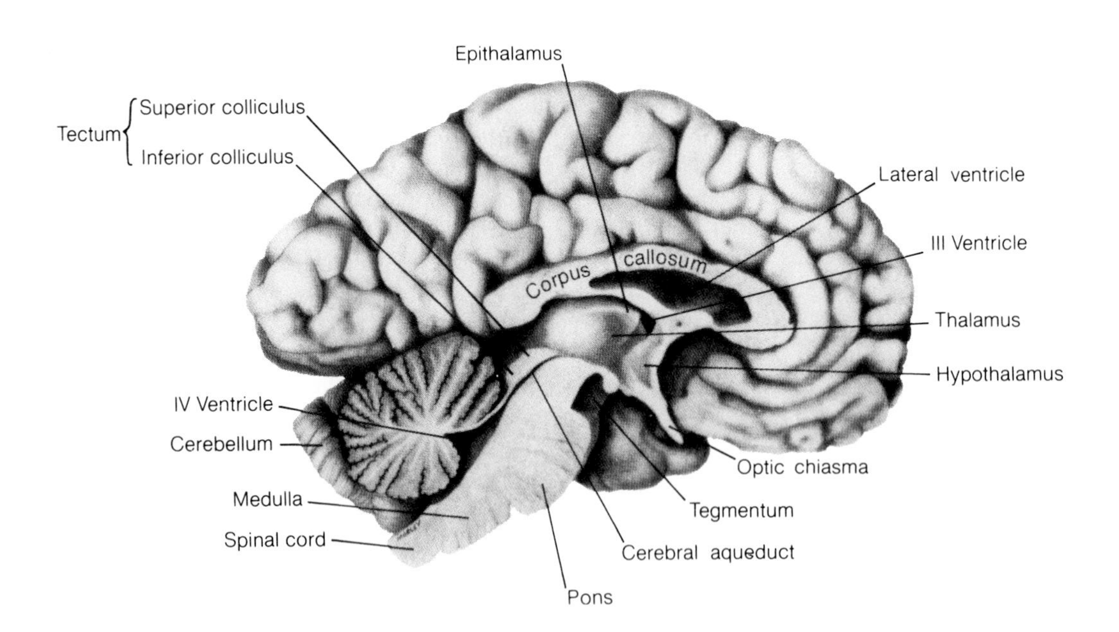

FIGURE 2 *Top:* Summary of gross regions of the neocortex of the human brain. *Bottom:* View of major structures of the brain. [From Kolb, B., and Whishaw, I. Q. (1990). ''Fundamentals of Human Neuropsychology,'' 3rd ed. New York: W. H. Freeman.]

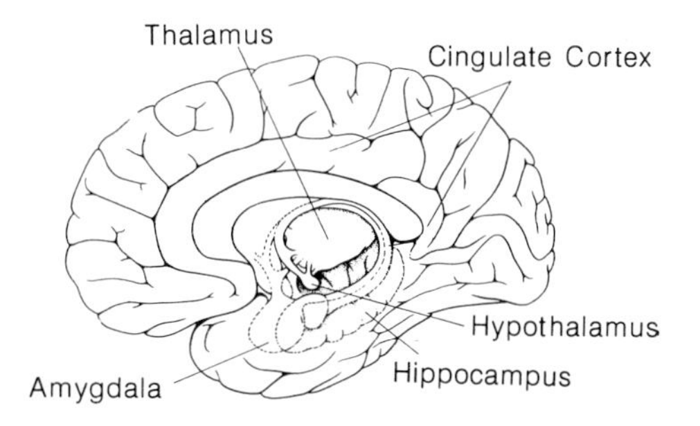

FIGURE 3 Medial view of the right hemisphere illustrating positions of limbic structures. Anterior is to the *left*.

negative the cell is said to be hyperpolarized, and if it becomes less negative (i.e., more positive) it is said to be depolarized. When the membrane of a neuron is perturbed by the signals coming from other neurons or by certain external agents (e.g., chemicals), the voltage across the membrane changes, becoming either hyperpolarized or depolarized. These changes are normally restricted to the area of membrane stimulated, but if there are numerous signals through many synapses to the same cell, they will summate, altering the membrane potential of a larger region of the cell. If the excitation is sufficient to reduce the membrane potential to about −50 mV, the membrane permeability for positive Na^+ ions changes. The influence of ions raises the potential until it becomes positive (e.g., +40 mV). This change in membrane potential spreads across the cell and if it reaches the axon, it travels down the axon, producing a propagating signal. This change in permeability is quickly reversed by the cell in about 0.5 msec, allowing the cell to send repeated signals during a short period of time. The signal that travels down the axon is known as an action potential (or nerve impulse), and when it occurs, the cell is said to have fired. The rate at which the impulse travels along the axon varies from 1 to 100 m/sec and can occur as frequently as 1,000 times/sec, depending on the diameter of the axon; the most common rate is about 100/sec.

D. Chemical Organization

Once the nerve impulse reaches the end of the axon (the axon terminal), it initiates biochemical changes that result in the release of a chemical known as a neurotransmitter into the synapse. Although the action of transmitters is complex, their effect is either to raise or to lower the membrane potential of the postsynaptic cell, with the effect of making it more or less likely to transmit a nerve impulse.

Dozens of chemicals are known to be neurotransmitters, including a variety of amino acids [e.g., glutamic acid, glycine, aspartate, and gamma-aminobutyric acid (GABA)], monoamines (e.g., dopamine, norepinephrine, serotonin), and peptides [e.g., substance P, β-endorphin, corticotrophin (ACTH)]. Any neuron can receive signals from neurotransmitters through different synapses. The distribution of different transmitters is not homogeneous in the brain as different regions are dominated by different types. Because drugs that affect the brain act by either mimicking certain transmitters or interfering with the normal function of particular transmitters, different drugs alter different regions of the brain and subsequently have different behavioral effects. [*See* NEUROTRANSMITTER AND NEUROPEPTIDE RECEPTORS IN THE BRAIN.]

II. Functional Organization of the Brain

A. Principles of Brain Organization

The fundamental principle of brain organization is that it is organized hierarchically such that the same behavior is represented at several levels in the nervous system. The function of each level can be inferred from studies in which the outer levels have been removed, as is summarized in Fig. 4. The principal idea is that the basic units of behavior are produced by the lowest level, the spinal cord, and at each successive level there is the addition of greater control over these simple behavioral units. At the highest level the neocortex allows the addition of flexibility to the relatively stereotyped movement sequences generated by lower levels, as well as allowing greater control of behavior by complex concepts such as space and time. Although new abilities are added at each level in the hierarchy, it remains difficult to localize any process to a particular level because any behavior requires the activity of all regions for its successful execution. Nonetheless, brain injury at different levels will produce different symptoms, depending on what functions are added at that level. [*See* SPINAL CORD.]

B. Principles of Neocortical Organization

The cortex can be divided into three general types of areas: (1) sensory areas, (2) motor areas, and (3) association areas. The sensory areas are the regions

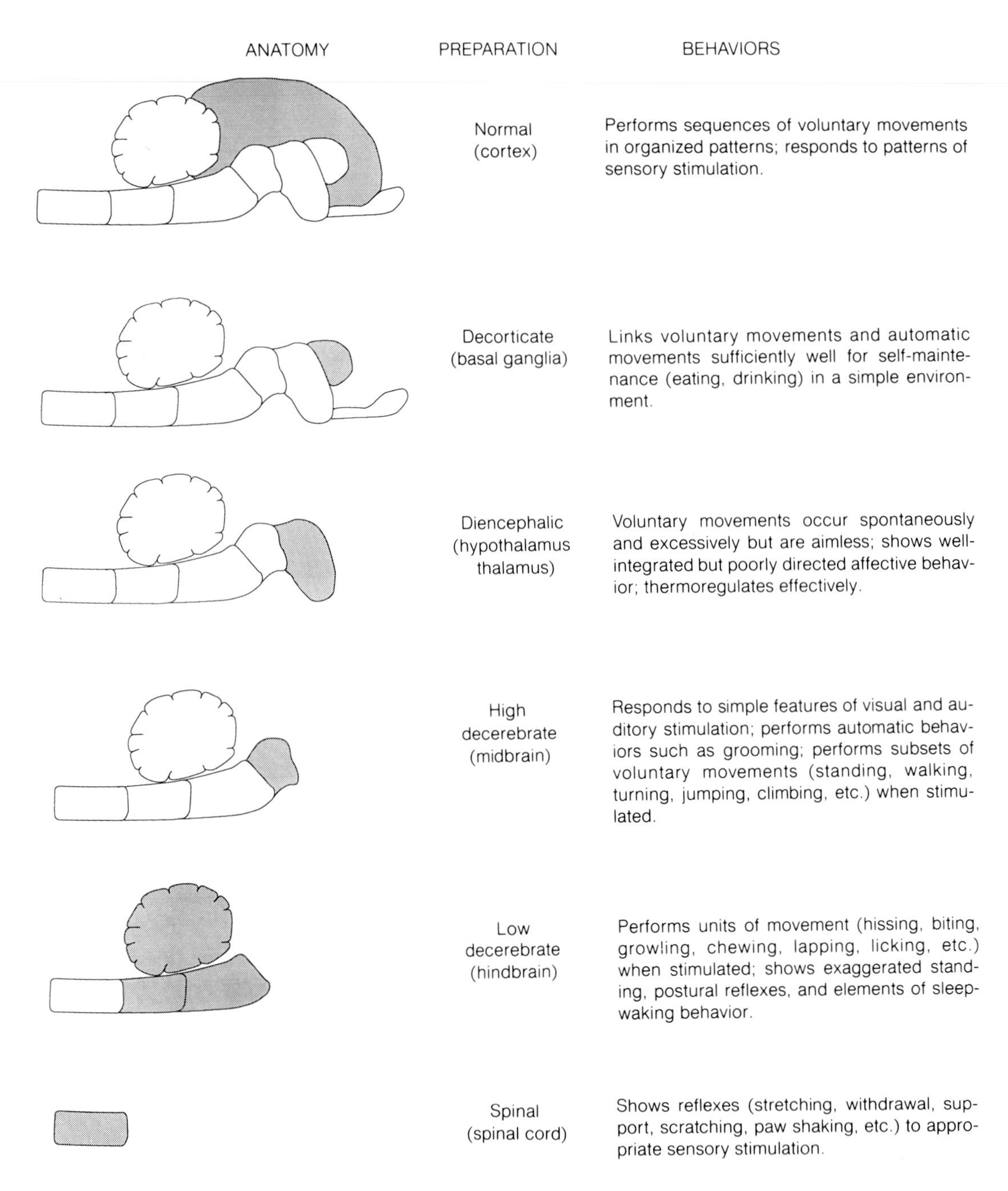

FIGURE 4 Summary of the behavior that can be supported by different levels of the nervous system. *Shading* indicates highest remaining functional area in each preparation.

that function to identify and to interpret information coming from the receptor structures in the eyes, ears, nose, mouth, and skin. The motor complex is the region involved in the direct control of movement. The association cortex is the cortex that is not ascribed specific sensory or motor functions.

Distinct sensory cortex is associated with each of these *sensory systems,* and each is made up of numerous subregions, each of which functions to process a distinct type of sensory information. For example, in the visual system separate regions are devoted to the analysis of form, color, size, movement, etc. Damage to each of these regions will produce a distinct loss of sensory experience. In each system there is a region of sensory cortex that produces an apparent inability to detect sensory information such that a person will, for instance, appear to be blind, unable to taste, or will have numbness of the skin, etc. In each case, however, it can be shown that other aspects of sensory function are intact. Thus, a person who is unable to "see" an object may be able to indicate its color and position! Similarly, a person may be able to locate a place on the body where they were touched while being unable to "feel" the touch. Regions of sensory cortex producing such symptoms are referred to as primary sensory cortex. The other regions are known as unimodal association regions, and damage to them is associated with various other symptoms. Thus, in every sensory system there are regions of cortex that, when damaged, result in an inability to understand the significance of sensory events. Such symptoms are known as agnosias. For example, although able to perceive an object (e.g., toothbrush) and to pick it up, a person may be unable to name it or to identify its use. Similarly, a person may be able to perceive a sound (e.g., that of an insect) but be unable to indicate what the sound is from. Some agnosias are relatively specific (e.g., an inability to recognize faces or colors). [*See* VISUAL SYSTEM.]

The motor cortex represents a relatively small region of the cortex that controls all voluntary movements and is specialized to produce fine movements such as independent finger movements and complex tongue movements. Damage to this region prevents certain movements (e.g., of the fingers), although others (e.g., arm and body movements) may be relatively normal. Although the motor cortex is a relatively small region of the cortex, a much larger region contributes to motor functions, including some of the sensory regions as well as the association cortex. For example, a person may be unable to organize behaviors such as those required for dressing or for using objects. Such disorders are known as apraxias, which refers to the inability to make voluntary movements in the absence of any damage to the motor cortex. [*See* MOTOR CONTROL.]

The regions of the neocortex not specialized as sensory or motor regions are referred to as association cortex. This cortex receives information from one or more of the sensory systems and functions to organize complex behaviors such as the three-dimensional control of movement, the comprehension of written language, and the making of plans of action. The principal regions of association cortex include the prefrontal cortex in the frontal lobe, the posterior parietal cortex, and regions of the temporal cortex. Although not neocortex, it is convenient to consider the medial temporal structures, including the hippocampus, amygdala, and associated cortical regions, as a type of association cortex. Taken together, damage to the association regions causes a puzzling array of behavioral symptoms that include changes in affect and personality, memory, and language. [*See* CORTEX.]

III. Cerebral Asymmetry

One of the most distinctive features of the human brain is that the two cerebral hemispheres are both anatomically and functionally different, a property referred to as cerebral asymmetry.

A. Functional Asymmetry

The clearest functional difference between the two sides of the brain is that structures of the left hemisphere are involved in language functions and those of the right hemisphere are involved in nonlanguage functions such as the control of spatial abilities. This asymmetry can be demonstrated in each of the sensory systems, with differences in left-handed and right-handed persons. We will first consider the common case, in the right-handed individual. In the visual system the left hemisphere is specialized to recognize printed words or numbers, whereas the right hemisphere is specialized to process complex nonverbal material such as is seen in geometric figures, faces, route maps, etc. Similarly, in the auditory system the left hemisphere analyzes words, whereas the right hemisphere analyzes tone of voice (prosody) and certain aspects of music. Asymmetrical functions go beyond sensation, however, to include memory, and affect. In the control of movement, the left hemisphere is specialized for the production of certain types of complex movement sequences, as in meaningful gestures (e.g., salute, wave) or writing. The right hemisphere has a complementary role in the production of other movements such as in drawing, dressing, or constructing objects. Similarly, the left hemisphere has a favored role in memory functions related to language (e.g., written and spoken words), whereas the right hemisphere plays a major role in the memory of places and nonverbal information such as music and faces.

B. Anatomical Asymmetry

The functional asymmetry of the human brain is correlated with various asymmetries in gross brain morphology, cell structure, neurochemical distribution, and blood flow. Differences in gross morphology and cell structure are most easily seen in the regions specialized for language, including the anterior (Broca's) and posterior speech areas (Fig. 5). For example, one region in the posterior speech area, the planum temporale, is twice as large on the left hemisphere in most brains, whereas a region involved in the processing of musical notes, Heschl's gyrus, is larger in the right temporal lobe than in the left. [*See* LANGUAGE.]

C. Variations in Cerebral Asymmetry: Handedness and Sex

There is considerable variation in the details of both functional and anatomical asymmetry in different people. Two factors, handedness and sex, appear to account for much of this variation. First, left-handers have a different pattern of anatomical organization than do right-handers. For example, they appear to have a larger bundle of fibers connecting the two cerebral hemispheres, the corpus callosum, which implies that the nature of hemispheric interaction differs in left- and right-handers. Similarly, left-handers are less likely to show the large asymmetries in the structure of the language-related areas than are right-handers. Functionally, the organization of the left-handed brain shows considerable variation: Language is located primarily in the left hemisphere in about $\frac{2}{3}$ of left-handers, in the right hemisphere in about $\frac{1}{6}$, and in both hemispheres in about $\frac{1}{6}$. Left-handers with different speech organization than right-handers do not simply have a reversal of brain organization, however, although the nature of their cerebral organization is still poorly understood. Second, males and females also differ in functional and structural organization. For example, the corpus callosum of females is larger relative to brain size than that of males, and females are less likely to show gross asymmetries or to have reversed asymmetries. Animal studies have shown a clear relation between anatomical organization and the presence of the perinatal gonadal hormones present at about the time of birth, suggesting that these hormones differentially organize the brain of males and females. Functionally, the effect of brain damage in males and females differs as well, although in complex ways that are poorly understood. It does appear, however, that frontal lobe injuries in both human and nonhuman subjects have differential effects in the two sexes, with larger behavioral effects of frontal lube injury observed in females. Other factors are also believed to influence the nature of cerebral asymmetry, especially experience, interacting with both sex and handedness. [*See* HEMISPHERIC INTERACTIONS.]

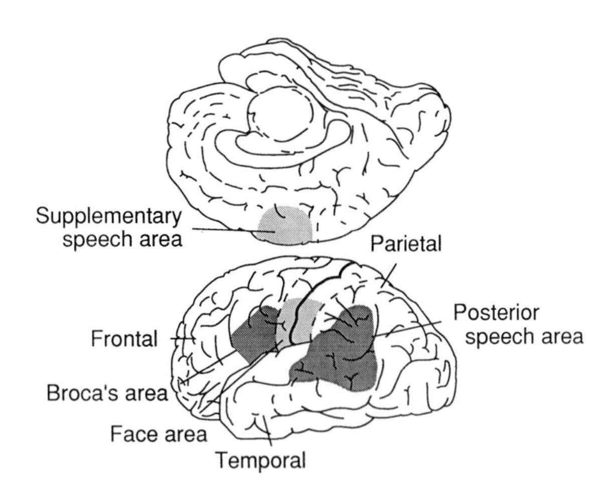

FIGURE 5 Schematic diagram showing the anterior speech area (Broca's area) in the frontal lobe, the posterior speech area (including Wernicke's area) in posterior temporal and parietal regions. *Shaded region* between speech areas is the motor and sensory cortex controlling the face and tongue. [From Kolb, B., and Whishaw, I. Q. (1990). "Fundamentals of Human Neuropsychology," 3rd ed. New York: W. H. Freeman.]

IV. Organization of Higher Functions

Complex functions such as memory or emotion are not easily localized in the brain, as the circuits involved include vast areas of both the cerebral hemispheres and other forebrain structures. Part of the difficulty in localizing such functions is that they are not unitary things but are inferred from behavior, which in turn results from numerous processes. Nonetheless, it is possible to reach some generalizations regarding such functions.

A. Memory

Memory is an inferred process that results in a relatively permanent change in behavior, which presumably results from a change in the brain. Psychologists distinguish many types of memory, each of which may have a distinct neural basis. These include, among others, (1) long-term memory, which is the recall of information over hours, days, weeks, years, etc.; (2) short-term memory, which is the recall of information over seconds or minutes; (3)

declarative memory, which is the recall of facts that are accessible to conscious recollection; (4) procedural memory, which is the ability to perform skills that are "automatic" and that are not stored with respect to specific times or places (e.g., the movements required to drive); (5) verbal memory, which is memory of language-related material; and (6) spatial memory, which is the recall of places or locations.

The neural basis of human memory can be considered at two levels: cellular and neural location. Thus, changes in cell activity and structure are associated with processes like memory, which may occur extremely rapidly in the brain, possibly in the order of seconds or at least minutes. Further, there is a variety of candidate regions for memory processes, the region varying with the nature of memory process. One structure that plays a major role in various forms of memory is the hippocampus. Bilateral damage to this structure leads to a condition of anterograde amnesia, which refers to the inability to recall, after a few minutes, any new material that is experienced after the damage. There is only a brief period of retrograde amnesia, which refers to the inability to recall material before the injury. The relatively selective effects of hippocampal injuries in producing anterograde but not retrograde amnesia suggests that different brain regions are involved in the initial learning of information and their later retrieval from memory. As a generalization, it appears that the temporal lobe is involved in various types of long-term memory processes, whereas the frontal and parietal lobes play a role in certain short-term memory processes. Damage to these regions thus produces different forms of memory loss, which are further complicated by whether the injury is to the left or right hemisphere. [*See* LEARNING AND MEMORY.]

B. Language

Damage to either of the major speech areas (Broca's or the posterior speech zone, which is sometimes referred to as Wernicke's area) will produce a variety of dissociable syndromes, including aphasia, an inability to comprehend language, alexia, an inability to read; and agraphia, an inability to write. There are various forms of each of these syndromes that relate to the precise details of the brain injury. Various other forms of language disturbance result from damage outside the speech areas, including changes in speech fluency [i.e., the ability to generate words according to certain criteria (e.g., write down words starting with "D"; give the name of objects)], spontaneous talking in conversation, the ability to categorize words (e.g., apple and banana are fruits), and so on. It appears that nearly any left hemisphere injury will affect some aspect of these language functions, as does damage to some regions of the right hemisphere. [*See* SPEECH AND LANGUAGE PATHOLOGY.]

C. Emotional Processes

Like memory processes, emotional processes are inferred from behavior and include many different functions including autonomic nervous system activity, "feelings," facial expression, and tone of voice. Certain subcortical regions (hypothalamus, amygdala) play a major role in the generation of affective behaviors, especially the autonomic components such as blood pressure, respiration, and heart rate. In addition, damage nearly anywhere in the cortex will alter some aspect of cognitive function, which in turn will alter personality and emotional behavior, but damage to the right hemisphere produces a greater effect on emotional behavior than similar damage to the left. Moreover, the frontal lobe plays a special role as well, possibly because it has direct control of autonomic function as well as of spontaneous facial expression and other nonverbal aspects of personality. Thus, damage to the right hemisphere, or the frontal lobe of either hemisphere, is likely to lead to complaints from relatives regarding a change in "personality" or "affect." The control of emotional behavior may not only be relatively localized to different regions of the brain, but also related to specific neurotransmitter systems in the brain. For example, one dominant theory of the cause of schizophrenia proposes that there is an overactivity of the dopaminergic neurons (i.e., neurons that use dopamine as neurotransmitter) in the forebrain, likely in the frontal cortex (hence, the dopamine hypothesis of schizophrenia). Similarly, the dominant theory of depression is that it is related to low activity in systems that employ norepinephrine or serotonin as neurotransmitters. Because there are asymmetries in the cerebral distribution of these transmitters, it is reasonable to expect that depression may be more related to the right than left hemisphere, which is consistent with the dominant role of the right hemisphere in emotion. [*See* DEPRESSION, NEURO-

TRANSMITTERS AND RECEPTORS; SCHIZOPHRENIC DISORDERS.]

D. Space

The concept of space has many different interpretations, which are not equivalent. Objects (and bodies) occupy space, move through space, and interact with other things in space; we can form mental representations of space, and we have memories for the location of things. It is difficult therefore to define space or to know how the brain codes spatial information. Damage to different parts of the cerebral hemispheres can produce a wide variety of spatial disturbances including the inability to appreciate the location of one's body, or even the location of one's body parts relative to one another. It is generally accepted that the right hemisphere plays the major role in spatial behavior, but there can also be spatial disruptions from left hemisphere damage, especially if the behavior involves verbalizing space. The major region in the control of spatial behavior is the parietal cortex, although the hippocampus is also involved. [*See* BINOCULAR VISION AND SPACE PERCEPTION.]

Bibliography

Kolb, B., and Whishaw, I. Q. (1990). "Fundamentals of Human Neuropsychology." 3rd ed. New York: W. H. Freeman.

Nauta, W. J. H., and Feirtage, M. (1986). "Fundamental Neuroanatomy." New York: W. H. Freeman.

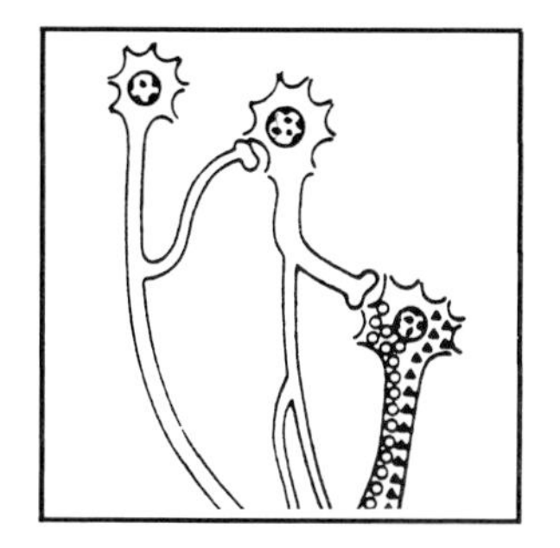

Brain Messengers and the Pituitary Gland

EUGENIO E. MÜLLER, *University of Milan*

Glossary

Hypophysiotropic regulatory hormone Low–medium-molecular-weight compound, especially synthesized in mediobasal hypothalamic nuclei, released episodically from nerve terminals into the hypophyseal portal capillaries, and transported to the anterior pituitary cells where it interacts with specific receptor sites

Median eminence Neurohemal structure consisting of nerve cells and blood vessels located at the base of the hypothalamus where hypophysiotropic regulatory hormones are released into portal capillaries

Neuroactive drug Acts by mimicking or opposing via different modalities the action of one or more neurotransmitters, thus affecting the secretion of anterior pituitary hormones

Neuropeptide Peptide generated from a large pre-pro-peptide molecule in the rough endoplasmic reticulum, packaged as pre-peptide in the Golgi complex, released by exocytosis at the nerve terminal into the synaptic cleft, and inactivated by proteolytic enzymes

Neurotransmitter Low-molecular-weight water-soluble compound, mostly monoamine or amino acid, that exerts localized, short-lived responses at the synapse; it is rapidly inactivated following the completion of the signal

ONCE THOUGHT TO BE the master gland of the body, it is now clear that the anterior pituitary gland (AP) is under the influence of hypothalamic and extrahypothalamic structures. A host of chemical messenger substances are released from neurons located in the hypothalamus and conveyed to the AP via a portal system of capillaries. The functional activity of hypothalamic neurosecretory neurons, which elaborate and deliver specific hypophysiotropic regulatory hormones into the portal system, is, in turn, regulated by a host of neurotransmitters and neuropeptides. These substances via typical or atypical synaptic connections relay to the hypophysiotropic neurosecretory neurons of the hypothalamus neural or neurohormonal influences, which are translated into hormonal responses to be conveyed to the AP. As a corollary, pharmacologic-induced suppression or activation of this neurotransmitter–neuropeptide system of control induces profound changes in the secretion of AP hormones, and regulatory hormones and central nervous system (CNS)-acting compounds can be used as probes of pituitary or hypothalamic function, respectively, and in humans in the diagnosis and therapy of neuroendocrine disorders.

I. Brain Neurotransmitters and Neuropeptides

Neurotransmitter and peptidergic systems in the endocrine hypothalamus and extrahypothalamic-related areas interact functionally to ensure the proper control of AP function. The widely held distinction between the two systems is now blurred. It is now generally recognized that hypothalamic peptidergic neurons are widely distributed to extrahypothalamic CNS areas, which is compatible with their additional behavioral role, that few classical neurotransmitters (e.g., dopamine [DA] and norepinephrine [NE]) mediate a neurohormonal type (e.g., diffuse and slow) of synaptic transmission and in addition neurotransmitters (e.g., DA, γ-aminobutyric acid [GABA], adrenaline) may be delivered into the portal capillaries and vehicled to the AP where they act as neurohormones.

Another reason for breakdown of demarcations between these messenger substances is the recognition that they may coexist in the same neuron in different CNS areas as well as in the mediobasal hypothalamus. The functional role of these costored neurotransmitters and neuropeptides is unsettled because demonstrating costorage also within nerve terminals of these neuronal systems at the median eminence level has not yet been possible. However, evidence for interaction between release mechanisms for the transmission lines (corelease), between decoding mechanisms (codecoding), and between transduction mechanisms (cotransduction) does exist. Figure 1 is a diagram illustrating the principal neural afferents involved in pituitary regulation.

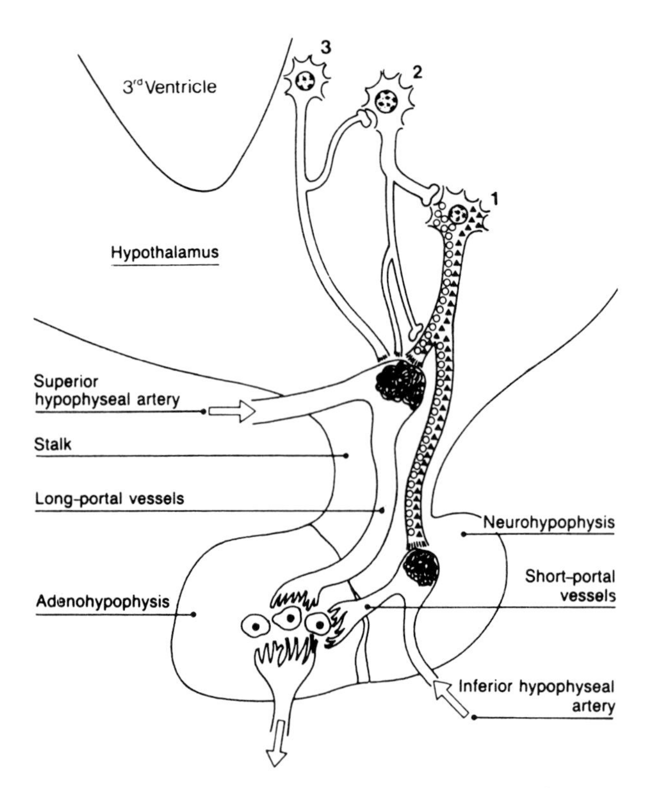

FIGURE 1 Neurotransmitter-neurohormonal control of AP secretion. The diagram illustrates the principal neural afferents involved in pituitary regulation. Neuron 1 denotes a tubero-hypophyseal peptidergic neuron that manufactures hypothalamic hormones, which are then released into the hypophyseal portal vessels and are relayed to the AP. Different symbols in the neuron (▲, ○) refer to the possibility of peptide–amine, peptide–peptide colocalization. Neuron 2 denotes a neurotransmitter nerve ending in relation to a peptidergic neuron (axosomatic or axoaxonic contact). The possibility is also depicted that a neurotransmitter neuron (DA, epinephrine, GABA, etc.) ends directly in relation with the primary plexus of blood capillaries at the median eminence level. Neuron 3 is a peptidergic neuron that modulates the activity of a neurotransmitter neuron (axosomatic or axoaxonic contact). This neuron may end in direct contact with hypophyseal portal vessels. For the sake of clarity, peptidergic neurons of the supraoptic and paraventricular-hypophyseal systems, with cell bodies in the hypothalamus and nerve endings in the neurohypophysis, have been omitted. [Reproduced, with permission, from E. E. Müller, 1986, Brian neurotransmitters and the secretion of growth hormone, *Growth and Growth Factors* **1**, 65–74.]

Description of the influences exerted by brain messengers on the neuroendocrine control of AP secretion requires a sketch of the mechanisms of biosynthesis, release, metabolic disposal of principal neurotransmitters, nonhypophysiotropic and hypophysiotropic neuropeptides of the CNS-acting compounds capable of interfering with the different metabolic steps or with pre-postsynaptic receptors, and, finally, mention of the regional distribution of the main neuronal systems in the CNS. [*See* NEUROENDOCRINOLOGY.]

A. Principal Neurotransmitters

1. Catecholamines

Synthesis of catecholamines (DA, NE, and epinephrine) occurs *in vivo* from precursor amino acids phenylalanine and tyrosine. The first limiting step in catecholamine biosynthesis is the transformation of L-tyrosine to L-dopa, a reaction catalyzed by the enzyme tyrosine hydroxylase. Dopamine formed from L-dopa by L-aromatic amino acid decarboxylase is subsequently oxidized to NE. There are neurons in the CNS that take up NE and methylate it to epinephrine.

Catecholamines synthesized and stored in specific granules in nerve terminals are released into the extracellular space following neuronal depolarization and interact with specific receptor sites located on the postsynaptic or presynaptic membranes. Following release, most of the neurotransmitter is captured into the presynaptic nerve terminal where it undergoes deamination by monoamine oxidases or is stored again in secretory granules. The uptake process is the principal mechanism for termination of catecholamine effects at postsynaptic sites.

Many drugs are capable of inhibiting the functional activity of catecholamine neurons. They comprise biosynthesis inhibitors at the level of different enzymatic steps: reserpine, which depletes the granular pool, and neurotoxic agents, such as 6-hydroxydopamine, which destroy catecholamine terminals. Conversely, catecholamine precursors or inhibitors of metabolic degradation or uptake potentiate catecholamine mechanisms.

Adrenergic receptors can be subdivided into four subtypes: α_1, α_2, β_1, and β_2. α-adrenergic receptors comprise α_1- and α_2-adrenoceptors: α_1-receptors are located postsynaptically and are functionally related to phosphoinositide metabolism; α_2-receptors are located pre- and postsynaptically and are functionally coupled to inhibition of adenylate cyclase. DA receptors can be subdivided into D_1 and D_2 sites, linked, respectively, to activation or inhibition of adenylate cyclase. There are now specific agonists and antagonists for D_1- and D_2-receptors.

2. Serotonin

The synthesis of serotonin (5-HT) involves several mechanisms: uptake of the precursor amino acid L-tryptophan (Trp) into the terminals, hydroxylation of Trp to 5-hydroxytryptophan (5-HTP), and decarboxylation to 5-HT, by the same enzyme that forms DA from L-dopa. Administration of either Trp or 5-HTP increases 5-HT neurotransmission, although the specificity of 5-HTP is poor. At high doses, in fact, this compound leads to substantial 5-HT accumulation in cells that do not ordinarily contain this indoleamine and also interferes with catecholamines and their precursors for transport, storage, and metabolism within the CNS.

5-HT synthesis can also be controlled pharmacologically, either via inhibition of Trp-hydroxylase or by destroying 5-HT nerve terminals with specific neurotoxic agents.

Based on the actions of 5-HT receptor agonists and antagonists, it has been suggested that three subgroups of 5-HT receptors exist: 5-HT_1-like, 5-HT_2, and 5-HT_3. To date, at least four different subtypes of 5-HT_1 receptors have been described: 5-HT_{1A}, 5-HT_{1B}, 5-HT_{1C}, and 5-HT_{1D}.

Many of the 5-HT receptor agonists and antagonists have poor neuropharmacological specificity. Drugs that seem to be most effective in influencing 5-HT neurotransmission comprise a series of amine-uptake inhibitors (chloroimipramine, fluoxetine, fluvoxamine) and 5-HT releasers (fenfluramine).

3. Acetylcholine

In general, the biochemical mechanisms on which cholinergic (acetylcholine [Ach]-mediated) neurotransmission depends are very similar to those of catecholamines and 5-HT. The biosynthesis and metabolic degradation of ACh are controlled by two enzymatic activities, i.e., choline acetyltransferase (CAT) and acetylcholinesterase (ACh-ase). Choline, the percursor of ACh, is taken up by the extracellular fluid into the axoplasm. ACh is stored in synaptic vesicles and, following its release, is hydrolyzed almost immediately to choline and acetic acid. Activation of cholinergic neurotransmission ensues inhibition of ACh-ase by drugs. There are two classes of ACh receptors: nicotinic and muscarinic, subdivided into M_1- M_2- and M_3-receptors. They have distinct anatomical distributions and physiological functions. Muscarinic receptors also are present in the AP, whereas both receptor types are present in the CNS.

4. Histamine

Histamine formation is made by decarboxylation of histidine. The existence of three types of histamine receptors (H_1, H_2, and H_3) is well established.

5. GABA

GABA, the main inhibitory amino acid neurotransmitter, is formed in the nervous tissue by the decarboxylation of L-glutamate. The catabolism of GABA involves a transamination. Several drugs have been found capable of inhibiting transamination, thus increasing brain GABA concentrations. GABA, once released into the synaptic cleft, acts on specific receptors to produce hyperpolarization of the postsynaptic membrane a selective increase in Cl^- permeability. GABA receptors are part of a supramolecular assembly in which there is a GABA recognition site operationally coupled with a Cl-ionophore and a benzodiazepine receptor, whose functional activation leads to an enhancement of GABA binding to its high-affinity binding sites.

Multiple sites exist for GABA action: $GABA_A$-receptors, which are bicuculline-sensitive, and $GABA_B$-receptors, which are bicuculline-insensitive, present in both the CNS and the periphery, and mainly at the presynaptic level, at least peripherally.

Figure 2 reports the principal steps involved in the formation of classical neurotransmitters here reviewed, and Table I a series of drugs that act as agonists or antagonists at neurotransmitter receptors or alter different aspects of neurotransmitter function.

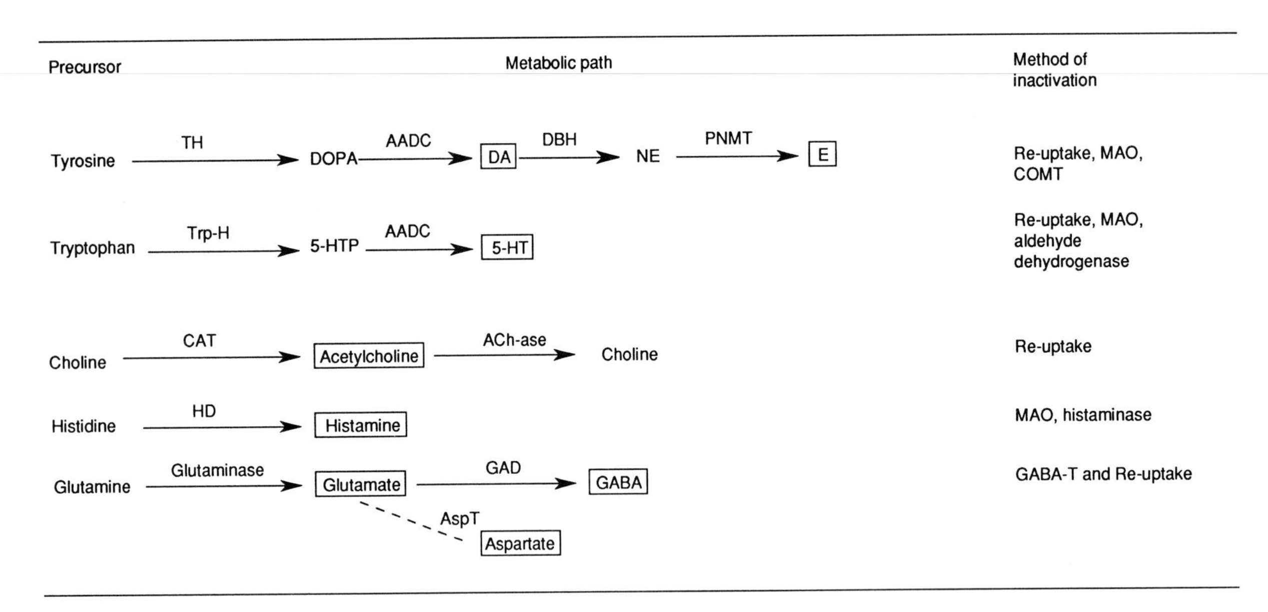

FIGURE 2 Steps in the formation of classical neurotransmitters. AAAD, aromatic amino acid decarboxylase; ACh, acetylcholine-sterase, AspT, mitochondrial aspartate transaminase; CAT, choline acetyltransferase; COMT, catechol-*O*-methyltransferase; DA, dopamine; DβH, dopamine β-hydroxylase; DOPA, dihydroxyphenylalanine; E, epinephrine; GABA-T, γ-aminobutyric acid transaminase; GAD, glutamic acid decarboxylase; HD, histidine decarboxylase; 5-HT, 5-hydroxytryptamine; 5-HTP, 5-hydroxytryptophan; MAO, monoamine oxidase; NE, norepinephrine; PNMT, phenylethanolamine-*N*-methyltransferase; TH, tyrosine hydroxylase; Trp-H, tryptophan hydroxylase.

B. Topographical Localization of Neurotransmitter Systems of Neuroendocrine Relevance in the CNS

Progress in the topographical localization of principal neurotransmitter systems has been made possible by many technical advances, especially immunohistochemical methodology, which is suitable for analyzing multiple antigens in one and the same neuron.

Most studies of catecholamine neurons have been carried out on the rat brain, but tissues from other species including primates have also been analyzed. There are about 50,000 catecholamine neurons in the rat brainstem, of which DA neurons constitute 80% and NE neurons 20%. The number of epinephrine neurons has so far not been calculated. About 70% of all catecholamine neurons are found in the mesencephalon. The ascending NE pathway comprises (1) a dorsal bundle, which originates in the locus coeruleus and innervates the cortex, the hippocampus, and the cerebellum, and (2) a ventral bundle, which originates in the pons and medulla and gives rise to pathways innervating the lower brainstem, the hypothalamus, and the limbic sys-

TABLE I Drugs Altering Neurotransmitter Function by Various Mechanisms

Drug	Mechanisms of Action	Observations
Apomorphine	Direct stimulation of pre- and postsynaptic receptors	Increase in brain 5-HT and 5-HIAA
Bromocriptine (2-Br-α-ergocriptine)	Direct stimulation of pre- and postsynaptic DA receptors	Action depending in part on brain CA stores
Lisuride	Direct stimulation of pre- and postsynaptic DA receptors	Peripheral antiserotoninergic activity
Piribedil	Direct and indirect stimulation of DA receptors	—
Amantadine	Direct and indirect stimulation of DA receptors	Blockade of DA re-uptake
Nomifensine	Blockade of DA and NE re-uptake	—
Clonidine	Direct stimulation of central and peripheral α_2-adrenoceptors	Stimulation of central histamine H_2-receptors

TABLE I Drugs Altering Neurotransmitter Function by Various Mechanisms[a]

Drug	Mechanisms of Action	Observations
L-Dopa	Increase in synthesis of DA and NE	Displacement of 5-HT by DA formed in serotoninergic neurons
α-MpT (α-methyl-paratyrosine)	Blockade of TH and of DA, NE, and E synthesis	Conversion to α-methylDA and α-methyl NE endowed with intrinsic receptor stimulating activity
FLA-63 bis (4-methyl-1-homo-piperazinyl-thiocarbonyl) disulphide	Blockade of the conversion of DA to NE	—
6-OHDA (6-hydroxydopamine)	Neurotoxic for CA neurons	Does not cross the BBB
Reserpine	Release of intragranular pool of CAs and inhibition of granular uptake; same effect on 5-HT neurons	—
Carbidopa	Selective inhibition of peripheral aromatic L-amino acid decarboxylase	—
Chlorpromazine	Blockade of DA and NE receptors	Hypersensitivity to adrenergic stimulation
Haloperidol	Blockade of DA receptors	—
Pimozide	Blockade of DA receptors	At high doses blockade of NE receptors
Phentolamine	Blockade of α_1- and α_2-adrenoceptors	Long-lasting action
Phenoxybenzamine	Blockade of α_1- and α_2-adrenoceptors	Long-lasting action; blockade also of 5-HT, H, and ACh receptors
Prazosin	Blockade of α_1-adrenoceptors	—
Yohimbine	Blockade of α_2-adrenoceptors	—
Propranolol	Blockade of β_1- and β_2-adrenoceptors	—
L-tryptophan	Selective increase in 5-HT synthesis, storage, and metabolism	—
5-Hydroxytryptophan	Increase in 5-HT synthesis	Decrease (by displacement) of brain DA and NE levels
pCPA (*p*-chlorophenyl-alanine)	Blockade of Trp-H and of 5-HT synthesis	Release of CAs
Methysergide	Blockade of 5-HT receptors	Potential dopaminergic, antidopaminergic, and antiserotoninergic activity
Cyproheptadine	Blockade of 5-HT receptors	Antihistamine, anticholinergic, and anti-catecholaminergic activity
Metergoline	Blockade of 5-HT receptors	Potential dopaminergic activity
5,6-DHT (5,6-dihydroxytryptamine)	Neurotoxic for serotoninergic neurons	Release of CAs
5,7-DHT (5,7-dihydroxytryptamine)	Neurotoxic for serotoninergic neurons	Release of CAs
Fluoxetine	Selective blockade of 5-HT re-uptake	—
Acetylcholine	Direct stimulation of muscarinic and nicotinic receptors	—
Pyridostigmine	Reversible inhibition of acetylcholin-esterase	Does not cross the BBB
Atropine	Antagonism at muscarinic M_1- and M_2-receptors	—
Pirenzepine	Antagonism at muscarinic M_1-receptors	Poor penetration of the BBB
Aminooxyacetic acid	Inhibition of GABA catabolism	—
Sodium valproate	Inhibition of GABA catabolism	—
Muscimol	Direct stimulation of $GABA_A$-receptors	Poor penetration of the BBB
Baclofen	Direct stimulation of $GABA_B$-receptors	Action not antagonized by bicuculline
Bicuculline	Antagonism at GABA receptors	—
2-Methylhistamine	H_1-receptor agonist	—
Dimaprit	H_2-receptor agonist	—
Diphenhydramine	H_1-receptor antagonist	Anticholinergic activity
Meclastine	H_1-receptor antagonist	No anticholinergic activity
Cimetidine	H_2-receptor antagonist	Poor penetration of the BBB
Ranitidine	H_2-receptor antagonist	Poor penetration of the BBB

[a] CA, catecholamine; E, epinephrine; H, histamine; 5-HIAA, 5-hydroxy-indoleacetic acid; Trp-H, tryptophan hydroxylase; BBB, blood–brain barrier; for other abbreviations see text.

[Reproduced with permission, from E. E. Müller, 1986, Brain neurotransmitters and the secretion of growth hormone, *Growth and Growth Factors* **1,** 65–74.]

tem. The DA pathways comprise three main systems of long DA neurons: the nigrostriatal, the mesolimbic, and the mesocortical systems and two systems composed of short intrahypothalamic DA axons, which originate in the arcuate nucleus and project to the external layer of the median eminence (tuberoinfundibular DA [TIDA] neurons) and in the rostral periventricular hypothalamus and project to anterior and dorsal hypothalamic areas and to the zona incerta (incertohypothalamic system). Of particular relevance for neuroendocrine control is the TIDA system, which innervates the median eminence, where regulatory hormones, neurotransmitters, and neuropeptides are released into the portal capillaries to be conveyed to the AP.

The 5-HT system arises from cell bodies situated in the mesencephalic and pontine raphe and send axons especially to the hypothalamus, median eminence, preoptic area, limbic system, septal area, striatum, and cerebral cortex. There are also 5-HT neurons in the hypothalamus.

Availability of antibodies specific for CAT has made it possible to outline cholinergic neurons and to study their projections. The enzymatic activity is present in most hypothalamic nuclei, including the arcuate nucleus and the median eminence. Since only small changes have been detected in the concentrations of CAT in the mediobasal hypothalamus following mechanical separation of this area, the existence of a TI-cholinergic pathway, similar to the TIDA pathway, has been envisaged. This may play an important role in mediating most of the neuroendocrine effects of cholinergic drugs.

The regional localization of histamine in the brain and also in individual nuclei of the hypothalamus has been studied in rodents and primates. In the monkey hypothalamus, the highest concentrations are found in the mammillary bodies, the supraoptic nucleus, the ventromedial nucleus, the ventrolateral nucleus, and the median eminence. A similar pattern of distribution is present for histamine in the human brain; the highest levels are found in the mammillary bodies and the mid-hypothalamus. After deafferentation of the mediobasal hypothalamus in rats, levels of histamine do not decrease significantly from the control value in the arcuate nucleus, ventromedial nucleus, dorsomedial nucleus, and median eminence, which suggests that histamine is present in the posterior two-thirds of the hypothalamus in cells that are intrinsic to this area.

Specific antibodies raised against glutamate decarboxylase (GAD-I) have made precise localization of GABAergic neurons in the hypothalamus possible. A dense network of GAD-positive nerve fibers is present in different hypothalamic nuclei of rodent and cat brains. In the median eminence, a dense immunofluorescent plexus is found in the external layer, extending across the entire mediolateral axis from the rostral part to the pituitary gland. In the median eminence, GAD activity remains unaffected by a total deafferentation of the mediobasal hypothalamus but decreases about 50% following neurochemical lesioning of the arcuate nucleus, by monosodium glutamate. This suggests the existence of a TI-GABAergic system, a hypothesis confirmed by the presence of GAD-positive immunoreactive cell bodies in the arcuate nucleus and periventricular nuclei.

C. Neuropeptides

During the last 15 years, it has become increasingly apparent that large numbers of neuropeptides are present in CNS neurons (Table II). Availability of synthetic neuropeptides to be used for the production of specific antibodies for radioimmunoassay and immunocytochemistry has served as a powerful tool for localization; application of rDNA or rRNA technologies has permitted the study of neuropeptide biosynthesis and its regulation. Detailed analysis of biosynthesis, metabolism, and localization in the CNS of principal neuropeptides is beyond the scope of this chapter. Only some generalities will be considered here.

Neurosecretory neurons synthesize, transport, process, and secrete neuropeptides by mechanisms similar to those in peripheral hormone-producing tissues. Peptides are generated from large precursor molecules produced in the rough endoplasmic reticulum and packaged in secretory granules or vesicles in the Golgi complex. The granules are transported out from cell bodies to the terminals (axonal transport), where they release their content by exocytosis upon neuronal depolarization. Neuropeptide release can also be stimulated or inhibited by application of neurotransmitters or other neuropeptides, indicating the existence of neurotransmitter–neuropeptide and neuropeptide–neuropeptide interactions, an important step in the control of neurohormonal or neuromodulator function of the peptide. [*See* PEPTIDES.]

Proteolytic enzymes are not only important in the processing of prohormones to active component forms but also in terminating the action of active neuropeptides upon their release. In general, neuropeptides are present in the brain at concentrations

TABLE II Neuropeptides[a]

Pituitary peptides	Opioid peptides
Adrenocorticotropin hormone	Enkephalins
α-melanocyte-stimulating hormone	β-endorphin
Growth hormone	Dynorphin
Luteinizing hormone	Kyotorphin
Prolactin	Dermorphin
Thyroid-stimulating hormone	Hypothalamic regulatory hormones
Oxytocin	Corticotropin-releasing hormone
Vasopressin	Gonadotropin-releasing hormone
Circulating hormones	Thyrotropin-releasing hormone
Angiotensin	Growth hormone-releasing hormone
Calcitonin	Somatostatin
Glucagon	Miscellaneous peptides
Insulin	Bombesin
IGF-I; IGF-II	Gastrin-releasing peptide
Atrial natriuretic factor	Bradykinin
Gut hormones	Neuropeptide Y
Avian pancreatic polypeptide	Histidine-isoleucine peptide
Pancreatic polypeptide	Neurotensin
Cholecystokinin	Carnosine
Gastrin	Proctolin
Motilin	
Galanin	
Secretin	
Substance P	
Vasoactive intestinal peptide	

[a] These peptides have been described in mammalian CNS neurons and nerve terminals other than those related to endocrine or neuroendocrine functions.

much lower than those of the classical amine neurotransmitters, and their concentrations are even lower when compared with those of amino acid neurotransmitters such as GABA and glycine. Once released into the synaptic cleft or transported via the extracellular fluid for a short or long distance (volume transmission), neuropeptides act on specific receptors to produce an alteration in the level of one or another intracellular second messenger.

Detailed maps of the distribution of various types of neuropeptides are available, and some generalities can be expressed on the basis of these distribution studies. Thus, there are some brain areas that are rich in both peptide-immunoreactive cell bodies and terminals. Other areas, such as the cerebellum, have low levels of most neuropeptides; the thalamus is also poor in neuropeptides. Cortical areas are particularly rich in some neuropeptides, such as vasoactive intestinal peptide, colecystokinin, somatostatin, and corticotropin-releasing hormone (CRH). Peptides are present mainly in small interneurons, a notable exception being the long projections of endorphinergic neurons whose cell bodies in the arcuate nucleus send terminals to innervate various nuclei of the brainstem.

In many peptidergic neuronal systems, a neuropeptide-receptor mismatch exists, (e.g., neuropeptide stores) but no receptors are present (topological mismatch), or the amount of neuropeptide stores does not correlate across several brain regions with the density of the receptors (functional mismatch). Factors contributing to the mismatch phenomenon include the existence of volume transmission (see Section I,C), breaking down of the parent peptide (see Section I,C) after release into active fragments which are then recognized by specific receptors, nonfunctional or spare receptors, unrecognized low-affinity or occupied receptors, which impede proper evaluation of the density of the peptide receptor (i.e., functional mismatch). Figure 3 depicts schematically the topography of some neurotransmitters and neuropeptides in the rat brain.

II. Regulatory Hormones for AP Control

In mammals, the existence of at least six hypothalamic regulators of the AP is now reasonably well established: five of them have been chemically char-

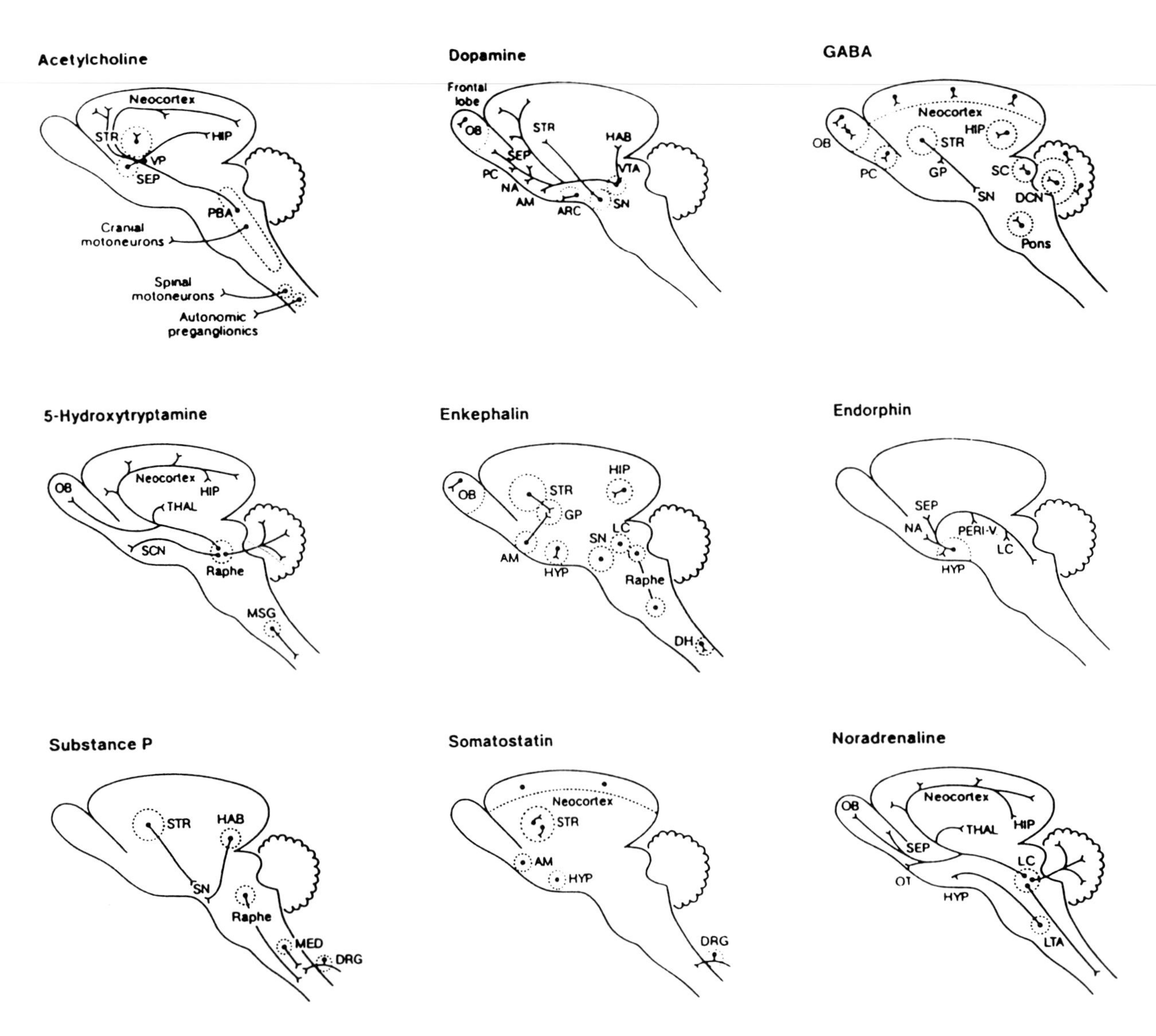

acterized (Table III) and therefore conform to the designation of regulatory hormones. Although this designation, in view of the expanding list of neuropeptides, which are also endowed with intrinsic hypophysiotropic actions, now appear to be a misnomer, it has historical and heuristic value and will be used here.

A. Generalities

Regulatory hormones are low–medium-molecular-weight compounds of about 800–4,000 daltons (Table III), which are especially synthesized in the mediobasal hypothalamus nuclei and in the anterior hypothalamus, although they can be elaborated ubiquitously in the CNS. Each peptide is synthesized as a pre-pro-hormone in the ribosomes; the pre-pro-hormone enters the rough endoplasmic reticulum where the leader sequence is removed; and

FIGURE 3 Principal locations of neuronal pathways. AM, amygdala; ARC, arcuate nucleus; DCN, deep cerebellar nuclei; DH, dorsal horn; DRG, dorsal root ganglion; GP, globus pallidus; HAB, habenula; HIP, hippocampus; HYP, hypothalamus; LC, locus coeruleus; LTA, lateral tegmental area; MED, medulla; MSG, medullary serotonin group; NA, nucleus accumbens; OB, olfactory bulb; OT, olfactory tubercle; PBA, parabrachial area; PC, pyriform cortex; PERI-V, periventricular gray; SC, superior colliculus; SCN, suprachiasmatic nucleus; SEP, septum; SN, substantia nigra; STR, striatum; THAL, thalamus; VP, ventral pallidum; VTA, ventral tegmental area. [Reproduced, with permission, from J. K. McQueen, 1987, Classical transmitters and neuromodulators, *in* "Basic and Clinical Aspects of Neuroscience," Vol. 2 (E. Flückiger, E. E. Müller, and M. O. Thorner, eds.), pp. 7–16. Springer Sandoz Advanced Texts, Springer-Verlag, Heidelberg.]

TABLE III Hypothalamic Regulatory Hormones

Hypothalamic hormone or factor	Structure or candidate
Corticotropin-releasing hormone (CRH or CRF)[a]	H-Ser-Glu-Glu-Pro-Pro-Ile-Ser-Leu-Asp-Leu-Thr-Phe-His-Leu-Leu-Arg-Glu-Val-Leu-Glu-Met-Ala-Arg-Ala-Glu-Gln-Leu-Ala-Gin-Gin-Ala-His-Ser-Asn-Arg-Lys-Leu-Met-Glu-Ile-Ile-NH_2
Thyrotropin-releasing hormone (TRH)	pGlu-His-Pro-NH_2
Luteinizing hormone–follicle-stimulating hormone-releasing hormone (LHRH, or GnRH)	pGlu-His-Trp-Ser-Tyr-Gly-Leu-Arg-Pro-Gly-NH_2
Growth hormone-releasing hormone (GHRH, or GRF)[b]	H-Tyr-Ala-Asp-Ala-Ile-Phe-Thr-Asn-Ser-Tyr-Arg-Lys-Val-Leu-Gly-Gln-Leu-Ser-Ala-Arg-Lys-Leu-Leu-Gln-Asp-Ile-Met-Ser-Arg-Gln-Gln-Gly-Glu-Ser-Asn-Gln-Glu-Arg-Gly-Ala-Arg-Ala-Arg-Leu-NH_2
Growth hormone-releasing–inhibiting hormone (GH-RIH, or somatostatin)[c]	H-Al-Gly-Cys-Lys-Asn-Phe-Phe-Trp-Lys-Lys-Thr-Phe-Thr-ser-Cys-OH (S-S bridge between Cys residues)
Prolactin-inhibiting factor (PIF)[d]	Dopamine
Prolactin-releasing factor (PRF)[e]	

[a] The sequence of human CRF is shown.
[b] The 44-amino acid structure.
[c] The 14-amino acid molecular form of somatostatin is indicated.
[d] Dopamine is the most likely candidate to be the PIF, although the sequence of a 56-amino acid peptide named GAP (GnRH-associated peptide) and endowed with PIF activity has also been reported.
[e] Vasoactive intestinal peptide may be a PRF.
[Reproduced, with permission, from E. E. Müller and G. Nisticò, 1989, "Brain Messengers and the Pituitary," Academic Press, San Diego.]

the pre-hormone accumulates in secretory granules of the Golgi apparatus, which are then transported by axoplasmic flow to the nerve terminals abuting on the median eminence and released into the hypophyseal portal system. Regulatory hormone secretion is not continuous but occurs in episodic fashion (one major pulse each 1–3 hr), likely due to the rhythmic, spontaneous discharge of neurosecretory neurons. The pulsatile release of regulatory hormones into the portal capillaries is a *sine qua non* condition to allow proper sensitivity of pituitary receptor sites, which otherwise will undergo a process of downregulation (diminution of receptor number). The interaction between regulatory hormones and their receptors prompts the formation or inhibits the functional activation of one or more intracellular messengers (cAMP, Ca^{2+}, hydrolysis products of phosphoinositides, e.g., diacylglycerol, PI_3), which mediate the hormonal response of the AP.

A peculiar feature of regulatory hormones is provided by their abundant extrahypothalamic distribution, particularly in the most ancient phylogenetic species (some even lacking the AP), which suggests a primitive role as "regulators" of CNS function and more recent cooptation during the phylogenesis to act as controllers of the pituitary function. The ubiquitous distribution in the CNS, and its ability to evoke neurochemical and behavioral effects in animals and humans, even in the absence of the pituitary, denote the neurotransmitter or neuromodulator role of these substances. It is also noteworthy that, contrary to previous belief, regulatory hormones do not present with a rigid neurohormonal specificity; in fact, some of them affect the release of more than one AP hormone. Thyrotropin-releasing hormone (TRH), for instance, in addition to thyroid-stimulating hormone (TSH), affects also prolactin secretion and, in some pathologic conditions, growth hormone secretion; somatostatin inhibits growth hormone but also TSH and adrenocorticotropin hormone (ACTH) secretion.

1. Corticotropin-Releasing Hormone

It is now evident that the secretion of ACTH is under regulation by a specific 41-residue peptide (CRF-41) first identified in sheep hypothalami. CRF-41 residue peptides with identical biological activities were isolated from the rat and the pig, and the sequence of the human peptide was deduced from cloned human genomic DNA. The human and rat CRF-41 molecules proved to be identical in sequence, differing in seven residues from the sheep peptide. CRF alone does not account for the

ACTH-releasing activity of the hypothalamus; the bioactivity of CRF may be modulated in fact by other hypothalamic substances, which show a synergistic interaction with CRF (e.g., vasopressin, oxytocin, angiotensin). Within the hypothalamus, CRF is localized in the parvocellular neurons of the paraventricular nucleus with the axons projecting to the external palisade zone of the median eminence, in both rats and humans. From the same source arise vasopressin-containing fibers present in the median eminence in about 50% of which CRH and vasopressin coexist in the same neurosecretory granules.

Acute administration of ovine CRH triggers in normal healthy volunteers a prompt, consistent, and long-lasting rise in circulating levels of ACTH, followed by a gradual increase of cortisol levels; after CRH doses of 0.3–30 μg kg, i.v., plasma ACTH reaches a peak at 10–15 min, declines until 90 min, and rises to a second peak at 3 hr. Plasma ACTH and cortisol levels are elevated 7 and 10 hr, respectively, after the highest dose. The hypercortisolemia that follows CRH injection blunts the responsiveness of the pituitary gland to subsequent stimuli, and with high CRH doses this effect may last for >12 hr. In the doses commonly used (approximately 100 μg), the only side effects observed are facial flushing, which occurs in about one-third of subjects, and an occasional transient tachycardia. CRH, although of no therapeutic use, has diagnostic power in conditions of hyperdysfunction of the hypothalamo–pituitary adrenal axis and in differentiating primary from secondary insufficiency of the adrenal gland.

2. Thyrotrophin-Releasing Hormone

TRH, the first regulatory hormone to be isolated and characterized, is the dominant regulator of the secretion of TSH through a tonic stimulating action. In mammals, TRH is synthesized in the thyrotropic area of the hypothalamus, principally the parvocellular division of the paraventricular nucleus, and then transported to the AP via portal capillaries. TRH interacts with high-affinity receptors on the pituitary thyrotrophs, whose number is regulated in part by thyroid hormones, and activates the phosphatidylinositol pathway. TRH stimulates the synthesis of TSH, in addition to its release.

In humans, i.v. doses of as little as 15 μg TRH induce significant increases in plasma TSH; the effect is dose-dependent up to a concentration of 400 μg, is higher in females than in males (a reflection of estrogen's ability to increase pituitary TRH-binding sites), and decreases in aged subjects. Administration of TRH causes not only a prompt rise in plasma TSH concentration, 15–30 min later, but also a significant increase in circulating plasma triiodothyronine (T_3) (the thyroid hormone), which occurs 20 min after the elevation in TSH. Administration of TRH provides a valuable tool for the assessment of pituitary or thyroid ability to respond to respective therapies and for distinguishing whether idiopathic hypopituitarism is due to a pituitary disease (impaired TSH response) or hypothalamic disease (normal or exaggerated response). In primary hypothyroidism, the response of TSH to TRH is exaggerated in >70% of cases; conversely, in hyperthyroidism the response is usually absent, to indicate that TSH secretion is controlled by the negative feedback of thyroid hormones on the AP.

TRH is also a potent prolactin releaser *in vivo* and from *in vitro* APs in both animals and humans. In humans, the minimum effective dose that releases prolactin is similar to the dose that releases TRH.

3. Gonadotropin-Releasing Hormone

Unequivocal demonstration for the existence of a neurohormonal control of gonadotropin secretion was given with the isolation and chemical identification of a peptide that had both luteinizing hormone (LH)- and follicle-stimulating hormone (FSH)-releasing activity, termed LHRH or, more recently, GnRH (gonadotropin-releasing hormone). GnRH induces FSH and LH release in many animal species, including primates.

GnRH is a linear decapeptide (Table III), which in the mammalian hypothalamus and the placenta is synthesized as part of a 92-amino acid precursor protein.

The structure of the GnRH precursor comprises the decapeptide preceded by a signal sequence of 23 amino acids and followed by a Gly-Lys-Arg sequence necessary for enzymatic cleavage of the precursor and C-terminal amidation of GnRH. A sequence of 56 amino acid residues occupies the C-terminal region of the precursor and constitutes the GnRH-associated peptide, GAP. GAP is also capable of stimulating gonadotropin secretion and inhibiting prolactin secretion from rat AP cells in culture; although GAP has been suggested to be a candidate prolactin-inhibiting factor, unequivocal evidence for this role is still lacking. GnRH acts in the pituitary on specific receptors to foster release of LH and also FSH from the gland. Peripheral factors (e.g., gonadal steroids and inhibin, a peptide produced by the gonads, which selectively suppresses FSH release) probably exert a crucial role in dictating preferential FSH or LH release. Also, the modalities of pulsatile GnRH secretion contribute to the differential secretion of gonadotropins.

In mammals, the hypothalamo-infundibular GnRH tract constitutes the major axonal route for GnRH neurons. GnRH is present in high concentrations in the median eminence region (lateral wings of the external layer) but also in the suprachiasmatic nucleus, in the preoptic area and the vascular organ of the *lamina terminalis,* and in other circumventricular organs. In primates, GnRH cell bodies are mainly localized in the mediobasal hypothalamus, and GnRH neurons present in the preoptic area exert likely extrapituitary actions. In the mediobasal hypothalamus, the arcuate nucleus is an oscillator that generates signals (period 1/hr in the monkey or 1/1–2 hr in the human) that result in the release of an amount of GnRH into the stalk blood and, consequently, a pulse of LH and FSH from the gonadotrophs. Continuous infusion of or frequently repeated GnRH amounts or, conversely, reduction in the discharge frequency inhibits the secretion of both or only one (LH) gonadotropin, respectively.

During each menstrual cycle, the maturation of the ovarian follicle is largely due to FSH secretion, while secretion of estrogen from the maturing follicle is FSH- and LH-dependent. Persistence of plasma estradiol levels exceeding a threshold of approximately 200 pg/ml for at least 2 days at midcycle stops the negative feedback action of the steroid and triggers the preovulatory gonadotropin surge (positive feedback).

The secretion of FSH and LH is maximal after 15–30 min from i.v. or subcutaneous (s.c.) administration of a GnRH amount (50–150 μg), while chronic, pulsatile delivery of GnRH via minipumps effects successful pituitary and gonadal stimulation in men and women with hypothalamic hypogonadotropic hypogonadism. Superactive analogues of GnRH, after an initial phase of pituitary stimulation, lead to an impaired gonadotropin secretion by persistently occupying and down-regulating pituitary receptors. The use of these compounds allows to effect a selective chemical castration therapy in conditions requiring temporary, reversible suppression of gonadotropin secretion (precocious puberty, endometriosis, prostate cancer, breast cancer, etc.).

4. Growth Hormone-Releasing Hormone and Somatostatin

The secretion of growth hormone is regulated by the CNS via two specific regulatory hormones, a stimulatory growth-hormone-releasing hormone (GHRH) and an inhibitory somatostatin, both of which have been isolated, chemically identified, and synthesized.

GHRH-containing neurons are located mainly in the arcuate nucleus, in the medial perifornical region of the lateral hypothalamus, paraventricular nucleus, dorsomedial nucleus, and lateral and medial borders of the ventromedial nucleus. Somatostatin-containing neurons are, instead, mainly localized in the periventricular-anterior region, from where they send axons directed caudally through the hypothalamus to enter the median eminence at the level of the ventromedial nucleus.

GHRH and somatostatin are rhythmically secreted from median eminence nerve terminals into the portal circulation with a periodicity of about 3–4 hr but 180° out of phase. The asynchronous secretion of GHRH and somatostatin allows a pulsatile pattern of growth hormone secretion, which is crucial for maintaining optimal sensitivity of growth hormone receptors in target tissues.

The action of GHRH appears to be very specific because both *in vivo* and *in vitro* it stimulates the pituitary to secrete only growth hormone. At the dose commonly used to study pituitary function (1 μg/kg, i.v.), the peptide does not induce side effects, with the exception sometimes of facial flushing. GHRH, at high doses and in subjects affected by a GH-secreting adenoma (acromegaly), may trigger the release of prolactin.

Somatostatin decreases either basal or stimulated growth hormone secretion (i.e., that occurring after insulin hypoglycemia, arginine, L-dopa, physical exercise, or sleep) and supresses growth hormone secretion also in acromegalics, although its effect is evanescent. The peptide also inhibits basal and stimulated TSH secretion, prolactin secretion under certain conditions, and ACTH secretion from tumor tissues.

Growth hormone autoregulates its own secretion through a feedback mechanism operating either at hypothalamic or AP level. Intracerebroventricular administration of growth hormone in the rat increases somatostatin concentrations in the portal capillaries, suppresses pulsatile growth hormone release, and the growth hormone-releasing effect of many secretagogues, including GHRH.

Growth hormone also may influence its own secretion via the formation of peripheral peptides, i.e., somatomedins secreted by many tissues, particularly liver. Somatomedins, and especially somatomedin-C, induce, like growth hormone, release of somatostatin from the hypothalamus but, in addition, are potent inhibitors of growth hormone synthesis and GHRH-stimulated release. GHRH and somatostatin interact functionally: GHRH increases somatostatin secretion and gene expression in the hypothalamus; conversely, somatostatin inhibits release of GHRH acting on specific receptors located in the arcuate nucleus.

5. Prolactin-Inhibiting and -Releasing Factors

The secretion of prolactin, an AP hormone with multiple roles and sites of action, is under dual hypothalamic control, with a predominant dopaminergic inhibition and sometimes overlapping stimulatory input. DA, a classical neurotransmitter produced by neurons with cell bodies in the arcuate nucleus, is secreted as a neurohormone and transported via the portal circulation to the AP, where it interacts with high-affinity, specific DA D_2-type receptors situated in the cell membrane of the lactotroph. DA inhibits the secretion of prolactin so that, under most circumstances, blockade of DA receptors results in an elevation of prolactin levels. Fluctuations in DA levels in portal blood occur, and this may contribute to the changing levels of prolactin in animals and humans of both sexes during different physiological situations. To date, DA remains the single most important inhibitory neural signal in the regulation of prolactin secretion, although other candidates appear as potential prolactin-inhibiting factors; they include GABA and the previously alluded to GAP.

It is also apparent that the neural lobe of the pituitary provides both inhibitory and stimulatory signals for prolactin release, although the exact nature of these factors is, as yet, unknown. In addition to the inhibitory dopaminergic tone, certain aspects of prolactin secretion are mediated by substances with prolactin-releasing factor activity. Factors that have been found to participate in the stimulation of prolactin release under different physiological conditions include TRH, vasoactive intestinal peptide, peptide histidine isoleucine amide, and oxytocin. These substances are selectively involved in the stimulation of prolactin secretion in specific situations but do not entirely fulfill the criteria necessary to identify them as a physiological prolactin-releasing factor, because they do not play the same role as other regulatory hormones when regulating specific pituitary hormones.

III. Neurotransmitter Regulation-AP

A. ACTH

1. Catecholamines

The effect of catecholamines on the hypothalamo–pituitary adrenal axis in experimental animals and in humans has received much attention. There seems to be a central α_1-adrenergic mechanism that exerts a stimulant effect on ACTH secretion in humans. Methoxamine (Methox), a highly selective α_1-adrenergic agonist, stimulates ACTH secretion in a dose-dependent manner, when infused into normal subjects. The site of action of Methox is within the BBB, presumably on the paraventricular nucleus, and not at the pituitary. Administration of β_1- and β_2-adrenergic agonists had no effect on the secretion of ACTH and cortisol, suggesting that in humans, in contrast to rodents, circulating catecholamines do not play an important physiological role on ACTH and have ready access to the pituitary level, an area lying outside the blood-brain barrier. [*See* CATECHOLAMINES AND BEHAVIOR.]

2. Serotonin

Available results on the effect of the 5-HT system on the hypothalamo–pituitary adrenal axis are sometimes confusing, and many points still need elucidation. Nonspecificity of effects of some of the drugs at the doses used, existence of multiple 5-HT receptor subtypes, uncertainty about the site(s) of action, and the likelihood of multiple sites of action, differences in species, or experimental design, etc. are among the contributing factors.

5-HT precursors, direct 5-HT receptor signals, 5-HT uptake inhibitors, or releasing drugs induce an elevation in plasma corticosteroids when injected into rodents or members of other species, including humans. In particular, 5-HT stimulates bioactive CRH secretion from rat hypothalamic fragments *in vitro*. Oral administration of quipazine, or other direct 5-HT agonists, to normal volunteers reliably increases plasma cortisol levels. Administration of 5-HTP was found to be followed by an increase of plasma ACTH and cortisol, but the ambiguous pharmacologic effects of this compound dictate caution in interpreting these effects.

Studies in rats on the nature of the 5-HT receptors involved in the activation of the hypothalamo–pituitary adrenal axis indicate that they may be of the 5-HT_{1A}- and 5-HT_2-receptor subtypes. Also, animal studies indicate that 5-HT plays a modulatory role in the regulation of the time of ACTH secretion. Lesions of 5-HT pathways or blockade of 5-HT synthesis abolished circadian rhythm of plasma corticosterone in rats.

Like ACTH, β-endorphin and related peptides are under a stimulatory serotoninergic control in both animals and humans.

3. ACh

It is now reasonably well established that in several subprimate species an increase in cholinergic transmission is often accompanied by hypothalamo–pituitary adrenal axis activation. The mechanism underlying this effect appears to be direct stimulation of CRH release from the hypothalamus, likely the paraventricular nucleus, as suggested by *in vitro* experiments. Nicotinic receptors would be involved in this effect. Cholinergic agonists do not have a direct stimulant effect on ACTH secretion by the AP *in vivo* or *in vitro*.

In humans, stimulation of ACTH secretion is observed after administration of the ACh-ase inhibitor physostigmine an effect which is more evident in older subjects. Physostigmine administered to normal adults also induces an escape from suppression of the hypothalamo–pituitary adrenal axis induced by dexamethasone.

4. Histamine

Histamine, given systemically, enhances plasma ACTH and corticosteroids in animals and humans. It is unclear whether a peripheral or a central site is involved. Failure of histamine to stimulate ACTH release from rat hypothalamus *in vitro* suggests an indirect mechanism, perhaps mediated by vasopressin or catecholamine. Similar to 5-HT, histamine seems to participate in the circadian rhythmic regulation of hypothalamo–pituitary adrenal axis function. The role of histamine in regulation of ACTH release in response to stress has not yet been clarified. In humans, meclastine, a rather specific H_1 antagonist, causes a decrease of plasma ACTH concentrations in hypoadrenal patients and blunts the ACTH response to insulin hypoglycemia and metyrapone.

5. GABA

It is now sufficiently well established that GABA exerts an inhibitory role on ACTH secretion, although some *in vivo* studies also point to a stimulatory function. These data conclude the involvement of both $GABA_A$- and $GABA_B$-receptors, which can be shown as present on corticotrophs. Similar to GABA, benzodiagepines supress *in vitro* 5-HT-stimulated CRH-like immunoreactivity secretion, whereas potent inverse agonists of the benzodiagepine receptors induce a significant release of CRH-like immunoreactivity.

In humans, some results, in part consistent with those of animal studies, have been presented. In healthy subjects, baclofen or y-vinyl-GABA, an irreversible inhibitor of GABA-T, decreases baseline plasma cortisol levels and the response to insulin hypoglycemia. These findings agree with the ability of sodium valproate to lower high plasma ACTH concentrations in some subjects with Nelson's syndrome (see Section IV, C).

B. TSH

1. Catecholamines

A series of studies in rodents demonstrate that central NE transmission increases secretion of TSH by acting on the α_2-adrenoceptors. The experimental evidence suggests that a facilitatory effect of α_2-adrenergic stimulation on TSH secretion is exerted through modifications in the activity of TRH-containing neurons within the paraventricular nucleus and dorsomedial nucleus areas.

In contrast to the facilitatory effect exerted on TSH secretion by NE transmission is the inhibitory influence that the dopaminergic system exerts in both animals and humans. In subjects with primary hypothyroidism, a single L-dopa dose lowers the elevated TSH levels, although it does not alter the response to TRH. Bromocriptine and DA are instead effective in lowering baseline levels and inhibiting the TSH response to TRH in hypothyroid subjects, and DA also in euthyroid subjects. The site of action of DA is unknown, although a median eminence or pituitary site of action is suggested by many findings.

2. Histamine

Evidence from few *in vitro* studies indicates that H_2-receptors may be involved in the stimulatory control of TSH secretion. Specific H_2-antagonists counteracted TRH release from rat mediobasal hypothalamus slices or hypothalamic synaptosomes induced by histamine.

C. Gonadotropins

1. Catecholamines

Based on studies in rodents, central NE neurons clearly exert a facilitatory role for allowing GnRH neurons to produce and discharge their products on the afternoon of proestrus, the phase of the estrous cycle during which ovulation occurs. This activation, however, is not an absolute requirement, because in rats the estrous cycle is reestablished a few days after severance of the ascending NE bundle. In ovariectomized rats or monkeys, pulsatile secretion is inhibited by the use of blockers of NE synthesis or of α-adrenergic receptors, suggesting that NE is an essential neurotransmitter in the regulation of pulsatile LH secretion in both species.

In humans, the role played by catecholamines seems less prominent. DA might exert an inhibitory action on GnRH-producing neurons, as suggested by the existence of significant overlapping between GnRH and DA nerve terminals in the lateral wings of the external layer of the median eminence. Inhibition of pulsatile and phasic gonadotropin secretion by DA might be one of the mechanisms through which reproductive function is impaired in hyperprolactinemic states.

2. ACh

Under certain circumstances, the cholinergic system may exert a braking action on LH release. No evidence has been given for a role of ACh on gonadotropin secretion in humans.

3. Histamine

Numerous observations suggest that histamine may be involved in the control of gonadotropin release, although the sites and mechanism of action are uncertain. The observations indicate that the effect of histamine on gonadotropins is sex-related both in humans and in rodents.

4. GABA

The predominant effect of GABA appears to inhibit LH release. Data presented from animal studies suggest a direct action of GABA at the gonadotrophs and are consistent with some of the human findings.

D. Growth Hormone

1. Catecholamines

α-adrenergic mechanisms are important in the regulation of growth hormone secretion in both subprimate and primate species. Blockade of catecholamine synthesis, or depletion of catecholamine storage granules by drugs, almost completely suppresses episodic growth hormone secretion in conscious rats. Selective inhibition of NE and epinephrine synthesis also suppress spontaneous bursts of growth hormone secretion; human growth hormone antagonizes this effect. Apparently, α_2-adrenergic influences act by stimulating GHRH release from the α_2-hypothalamus, while α_1-adrenergic stimulation, which is inhibitory to growth hormone release in rats and dogs, occurs via stimulation of somatostatin release. Dopaminergic pathways are also involved in the control of growth hormone secretion, although their role appears to be largely ancillary. It is generally recognized that the DA system may stimulate growth hormone release in humans: both directly and indirectly acting DA agonists elevate human growth hormone levels; however, DA and its agonists inhibit human growth hormone release in patients with acromegaly and also inhibit stimulated human growth hormone release. These effects are due to direct activation of DA receptors located on the tumoral somatotrophs (acromegaly) and the demonstration that DA agonists trigger not only GHRH but also somatostatin release from the hypothalamus, respectively.

2. ACh

Cholinergic neurotransmission is an important modulator of growth hormone secretion. Studies in rats and dogs have shown that the central or peripheral administration of muscarinic agonists or antagonists stimulates or suppresses, respectively, growth hormone release. In humans, administration of ACh precursors or agonists is invariably associated with a small, but unequivocal, rise in plasma growth hormone. Conversely, atropine or its congeners, regardless of whether or not they cross the BBB, proves effective in blocking most of the physiological or neurotransmitter stimuli to growth hormone release, except insulin hypoglycemia. Cholinergic modulation appears to affect growth hormone release via stimulation or inhibition of hypothalamic somatostatin release.

3. Histamine

Studies in both dogs and humans point to a facilitatory role of brain histamine system on growth hormone release.

4. GABA

Activation of GABAergic function induces in rats a dual effect on growth hormone secretion. Both central administration of GABA and systemic administration of a GABA-T inhibitor that freely crosses the BBB, amino-oxyacetic acid, elicit a dose-related, prompt increase in serum growth hormone, an effect blocked by bicuculline. Opposing, growth-hormone lowering effects are obtained instead by systemic administration of GABAergic drugs unable to cross the BBB. These findings are best explained by the ability of GABAergic neurotransmission to inhibit somatostatin neurons located within the BBB, and to act similarly on terminals of GHRH neurons located outside the BBB (median eminence). In humans, studies point to a stimulatory role for GABAergic neurotransmission on baseline growth hormone secretion.

E. Prolactin

1. Catecholamines

The role of DA in the inhibitory control of prolactin secretion already has been mentioned (Section II.B.5). Drugs mimicking at the receptor level the action of the amine (directly or indirectly acting DA agonists) are potent inhibitors of prolactin secretion and can be used on different paradigms of prolactin hypersecretion. Conversely, blockers of DA biosynthesis or receptor antagonists do increase plasma prolactin levels. Both NE and epinephrine exert only an ancillary role in the control of prolactin secretion.

2. Serotonin

The 5-HT system exerts in rodents and monkeys an important facilitatory role on the stimulated prolactin secretion (stress or suckling in lactating dams). The action of 5-HT is not direct on the pituitary but is mediated by a prolactin-releasing factor, most likely vasoactive intestinal peptide. Baseline prolactin secretion is instead barely affected in the rat by functional alterations of 5-HT neurons. In humans, the evidence for a facilitatory role of 5-HT is not so clear-cut.

3. ACh

ACh, via muscarinic receptors, may play an inhibitory role on phasic prolactin secretion. Cholinergic nicotinic receptors located in the hypothalamus also seem to be involved in the inhibitory control.

4. GABA

GABA exerts in rodents an inhibitory control over prolactin secretion. Neurons of the TI-GABA system directly secrete the amino acid into the hypophyseal capillaries from where it is transported to the lactotroph cells to interact with specific binding sites. Pituitary GABA receptors are not a homogeneous population, and both high-affinity, low-capacity and low-affinity, high-capacity sites have been detected. GABA receptors are also present in human APs. The high-affinity GABA-binding site would be the one responsible for the GABA-induced inhibition of prolactin secretion in the rat.

In addition to this inhibitory component of GABA action on prolactin secretion, which is preferentially activated by systemically administered GABA or GABA mimetic drugs, a central, stimulatory component exerts its action via inhibition of TIDA neuronal function. This stimulatory component is also evident from studies in humans, following intracisternal administration of GABAergic compounds to psychiatric patients. Intravenous administration of GABA to normal subjects elicits a biphasic response, because a transient increase in plasma prolactin levels is then followed by a sustained inhibition of prolactin secretion.

Table IV lists the stimulatory or inhibitory effects on AP secretion of brain neurotransmitters in humans, as derived from the reviewed experimental evidence.

F. Brain Peptides and AP Hormone Secretion

In addition to the known regulatory hormones, a host of CNS neuropeptides (Section I.C) exerts profound neuroendocrine effects. The presence of some of these compounds in specific hypothalamic nuclei in relatively high concentrations, their secretion from hypothalamic nerve endings, and their detection in hypophyseal portal blood suggest a neurohormonal role and a hypophysiotropic function. However, these peptides can alter endocrine function not only as hormones but also as neurotransmitters or neuromodulators; moreover, their physiological role in the control of AP hormone secretion

TABLE IV Neurotransmitters and AP Hormone Secretion[a,b]

Hypothalamic pituitary axis	E	α_1	α_2	β_1	β_2	NE	DA	5-HT	ACh	H_1	H	H_2
CRH–ACTH	—	(↑)	(→)[b]	(→)	(→)	→	↑	↑	↑ ?	(↑)	—	—
GHRH-somatostatin–growth hormone	↑[c]	—	(↑)	(↓)	—	↑	↑[d]	↑ ↓ ?	↑	(↑ ↓)	↑	?
GnRH–FSH-LH	—	(→)	(→)	—	—	—	↓→	—	↑ ↓	—	—	?
PIF-PRF–prolactin	—	—	(→ ↓)[f]	—	—	→	↓	↑	—	—	—	(→)
TRH–TSH	—	—	(→)	(→)	(→)	—	↓[g]	→ ↓	—	(→)	—	(→)

[a] Key to symbols: →, no effect; ↑, stimulation; ↓, inhibition; —, action no ascertained; ?, action still questionable. The effect of activation of receptor subtypes is indicated with parentheses.
[b] Inhibition in depressed patients.
[c] In combination with propranolol.
[d] Inhibitory in acromegaly and *in vitro*.
[e] Inhibition of the GnRH-induced LH rise.
[f] Inhibitory in children
[g] TRH-induced rise; hypothyroid subjects.

has yet to be unequivocally established. A separate, succinct description of the neuroendocrine effects of some of these compounds, therefore, seems proper. Discussion of the neuroendocrine effects of products of the immune system or growth factors is beyond the scope of this chapter.

1. Opioid Peptides

Among neuropeptides, a major role in the neural mechanisms underlying the control of pituitary function must be credited to opioid peptides. These peptides used so far stimulate the secretion of growth hormone in either subprimates and primates acting on the hypothalamus to release GHRH. The physiological significance of opioid peptides for growth hormone secretion, however, is not clear, because naloxone, the antagonist of opioid peptide receptors, fails to alter basal growth hormone secretion. Opioid peptides also stimulate prolactin secretion via a suprapituitary site of action, i.e., by inhibiting TIDA neuronal function, but, at least in humans, do not exert a tonic stimulatory action on basal prolactin secretion. There is unequivocal evidence that opioid peptides exert an important role in the control of gonadotropin secretion and, hence, reproductive function in both animals and humans. They act mainly by inhibiting GnRH release from the hypothalamus via decrease of excitatory adrenergic influences and also exert an important role in the inhibitory feedback effects of gonadal steroids on LH release, making hypothalamic neurons hyperresponsive to the steroids. Concerning their role on the hypothalamo–pituitary adrenal axis, their major effect is to tonically inhibit the secretion of ACTH. Naloxone, although at high doses, increases ACTH and corticosterone secretion in rats and cortisol secretion in humans, and chronic administration of morphine inhibits stress-induced activation of the hypothalamo–pituitary adrenal axis.

2. Other Neuropeptides

An interesting feature of the neuroendocrine activity of CNS neuropeptides is the diversity of effects they can exert on the secretion of distinct hormones, according to their target site of action, a fact that compounds interpretation of their actual, physiological role. Thus, experimental evidence suggests for substance P in the rat a dual stimulatory and inhibitory role on the secretion of gonadotropins, exerted at the hypothalamic and pituitary levels, respectively.

It is also apparent that different organismic variables may greatly influence the effects of neuropeptides. For instance, neurotensin injected into the medial preoptic area of ovariectomized, estrogen-primed rats significantly facilitates the circadian afternoon rise of LH secretion, although in unprimed rats it does not affect the existing LH secretion; vasoactive intestinal peptides, essentially ineffective in eliciting growth hormone release from superfused rat pituitary cell reaggregates, strongly stimulates growth hormone release in the presence of dexamethasone in the culture medium. In addition, vasoactive intestinal peptide effects direct stimulation of growth hormone release from human somatotropinomas *in vitro* and is an active prolactin releaser in healthy humans but not in subjects bearing prolactin-secreting adenomas. Also, species-related differences are present.

To summarize from these and other data, the interactions either within different brain messengers or between them and pituitary and target gland hormones, whose ultimate result is that of assuring proper functioning of the pituitary and the endocrine system, are still poorly understood. The effects on the hypothalamus and/or the pituitary of some neuropeptides are reported in Tables V and VI.

TABLE V Neuropeptides and Pituitary Hormone Release: Action on the CNS

	Hormone[a]					
Peptide (dosage)	ACTH	Prolactin	Growth Hormone	TSH	FSH	LH
Substance P (μg)	NT	+?	−?	0	0	+
Neurotensin (μg)	NT	−, +	+	0	0	+?
Vasoactive intestinal peptide (ng)	NT	+	+	0	0	+
Gastric inhibitory polypeptide (μg)	NT	0	+	0	−	0
Motilin (μg)	NT	NT	−[b]	NT	NT	NT
Galanin (ng)	0[c]	+[c]	+	0[c]	0[c]	0[c]
Cholecystokinin (ng)	+	+	+	−	0	−
Angiotensin II (μg)	+	−	−	0	NT	+
Neuropeptide Y (ng)	+	0[d]	−	0[d]	0	−?
Bombesin (ng)	NT	+[e]	+	0	0[c]	0[c]
Calcitonin (νg)	+[c]	−?	−	NT	NT	NT

[a] Key to symbols: NT, not tested; +, stimulation; −, inhibition; 0, no effect; ?, controversial findings.
[b] Given intracerebroventricularly.
[c] Human data.
[d] Data derived from the effect of bovine and avian pancreatic polypeptides.
[e] Blockade of stress-induced Prl release.
[Reproduced, with permission, from E. E. Müller and G. Nisticò, 1989, "Brain Messengers and the Pituitary," Academic Press, San Diego.]

TABLE VI Neuropeptides and Pituitary Hormone Release: Action on the Pituitary

	Hormone[a]					
Peptide (dosage)	ACTH	Prolactin	Growth Hormone	TSH	FSH	LH
Substance P (ng)	NT	+	0	0	−[b]	−[b]
Neurotensin (ng)	NT	+	0	+?	0	0
Vasoactive intestinal peptide (μg)	+[c]	+	+[d,e]	0	0	0
Peptide histidine isoleucine amide (μg)	NT	+	+[d,e]	NT	NT	NT
Gastric inhibitory polypeptide (μg)	NT	NT	−	NT	+	+
Motilin (μg)	NT	NT	+	NT	NT	NT
Galanin (μg)	NT	0	0	NT	NT	NT
Colecystokinin (μg)	0	+[f]	0	0	0	0
Angiotensin II (ng)	+	+	0	0	NT	0
Neuropeptide Y (μg)	NT	NT	−[e]	NT	+	+
Bombesin (ng)	NT	+[c]	+[c]	NT	NT	NT
Calcitonin (μg)	NT	+	0	−	NT	−

[a] Same symbols as in Table V.
[b] Inhibition of GnRH-stimulated release.
[c] Only on tumor cells.
[d] In the presence of dexamethasone.
[e] On human somatotropinomas.
[f] At huge doses on the rat AP.
[Reproduced, with permission, from E. E. Müller and G. Nisticò, 1989, "Brain Messengers and the Pituitary," Academic Press, San Diego.]

IV. Neuroactive Compounds in the Diagnosis and Treatment of Neuroendocrine Disorders

The notion gained from the preceding sections that in the CNS a host of neuropeptides and neurotransmitters interact functionally to ensure the physiological secretion of AP hormones leads to the ultimate conclusion that their dysfunction may be the trigger for specific neuroendocrine disorders. The principal diagnostic and therapeutic uses of CNS-acting compounds will be succinctly reviewed here. Mention of the clinical applications of regulatory hormones has already been made in the specific subsections.

A. Growth Hormone Deficiency and Excess

The ability of GHRH to affect somatotroph function cannot be used as a test for identifying patients with inadequate spontaneous growth hormone secretion, owing to the poor inter- and intraindividual reproducibility of the growth hormone responses to GHRH. Because the major factors that plague evaluation of the growth hormone response to GHRH are fluctuations in the hypothalamic function and release of somatostatin, compounds such as pyridostigmine, which deprive the pituitary from inhibitory somatostatin inputs, given in advance to GHRH, appear to be useful for a full evaluation of pituitary somatotroph function and, thus, differentiation from a primary hypothalamic origin of the disease.

The finding that most adults and children with growth hormone deficiency show variable but unequivocal rises in plasma growth hormone levels after administration of GHRH suggests that growth hormone deficiency is rarely due to a functional impairment of somatotrophs but is more likely to be due to hypothalamic dysfunction. Pituitary growth hormone is present but not secreted, probably due to lack of GHRH synthesis and/or release. The presence of low but detectable GHRH levels in the cerebrospinal fluid of children with idiopathic growth hormone deficiency is suggestive of dysfunction of neurons regulating the release of GHRH from GHRH-containing neurons. In this context, recent attempts to stimulate growth hormone release in children with idiopathic growth hormone deficiency, intrauterine growth retardation, constitutional growth delay by treatment with DA agonists (L-dopa, bromocriptine), or α_2-adrenergic agonists (Clo) must be considered. In all, the results obtained seem to be promising, although broadening and confirmation of these findings are awaited. Finally, administration of pyridostigmine alone or combined with GHRH has been considered for the treatment of short stature, but results obtained so far are elusive.

Evidence that in acromegalic patients direct DA agonists induce a consistent suppression of the elevated growth hormone levels in about 60% of patients for a direct action on DA receptors located on the somatotrophs was the rationale for the use of ergot drugs, bromocriptine, lisuride, pergolide, and cabergoline. Apart from lowering plasma human growth hormone levels, clinical and metabolic improvements have been reported following the institution of medical therapy, although apparently patients benefit from but are not cured by chronic treatment with ergot drugs. The introduction of potent analogues of somatostatin capable of long-lasting reductions in plasma growth hormone levels will limit the therapeutic use of ergot derivatives in acromegaly.

B. Hypogonadotropic Hypogonadism

Secondary amenorrhea is by far the most common symptom attributable to pituitary function in women. It is usually transient and unaccompanied by structural abnormalities of hypothalamus, pituitary, or ovary (hypothalamic amenorrhea). Suptary, or ovary (hypothalamic amenorrhea). Šupporting evidence for a role of opioid peptides in the etiology of amenorrhea is derived from studies of amenorrheic women treated with naloxone. A clear increment in LH levels was observed in women with amenorrhea and/or hyperprolactinemia, suggesting that the acyclicity was due, at least in part, to the effect of an increased opioid peptide tone on GnRH and gonadotropin secretion. Thus, opioid antagonists may represent a useful therapeutic approach in these cases.

C. Cushing's Disease

The awareness that a host of neurotransmitters and neuropeptides are involved in the regulation of ACTH secretion, and the evidence that some neuroactive drugs initially thought to act on the hypothalamus to decrease CRH activity actually may act at the pituitary level account for a medical approach to therapy in Cushing's disease. Thus far, drugs used in this context encompass cyproheptadine, whose use relied on the known stimulatory action of the 5-HT neuronal system on the hypothalamo–pituitary adrenal axis, bromocriptine and sodium valproate, ultimately capable to stimulate DA and GABA receptors located on the corticotrophs. Although on distinct cases some clinical and biochemical remission is evident, the role of the medical therapy in Cushing's disease is ancillary to transsphenoidal microsurgery.

D. Prolactinomas

Prolactin-secreting tumors, either ≤10 mm diameter (microprolactinomas) or higher (macroprolactinomas), are the most frequently occurring neoplasms in the human pituitary. Clinically, hyperprolactinemia is associated with amenorrhea, galactorrhea, infertility, decreased libido, impotence, and, in macroprolactinomas, visual disturbances. In patients with microprolactinomas, DA agonist–ergot-related drugs represent a primary medical therapy. Administration of these drugs causes immediate and sustained prolactin suppression with restoration of fertility in women and normalization of hyperprolactinemic hypogonadism in men. Usually within 2 mo of the return of menstruation, ovulation and adequacy of the luteal phase is achieved and galactorrhea disappears. In men, libido and potency return to normal and, if reduced, the seminal volume also is normalized. When treatment is started, bromocriptine, the drug most commonly used, and other ergolines may cause different neurovegetative symptoms due to activation of central and peripheral DA receptors. Thus, low doses of drugs that are increased slowly and taken during a meal rather than after food are mandatory to minimize the side effects. Two long-acting injectable preparations of bromocriptine are available whose injection in patients with prolactinoma is followed by a prompt and steep prolactin decrease lasting for weeks or months. In many cases, shrinkage of the pituitary tumor can also be documented. Only transient and mild to moderate side effects are noticed.

The effects of medical therapy are particularly noteworthy in macroprolactinomas because these tumors are rarely cured by surgery and with radiotherapy subsequent hypopituitarism is common. It is now evident that reduction of tumor size as documented by tomographic scan and amelioration of visual disturbances can be anticipated in about 75% of the patients. Interestingly, a tumor shrinks only when prolactin secretion is inhibited by dopaminergic stimulation, but tumor size may remain unaltered despite the reduction of prolactin secretion. From the foregoing evidence, it appears that medical treatment of macroprolactinomas is more appropriate than neurosurgical transsphenoidal exploration for the primary treatment of the disease. The critical issue is whether or not this therapy effects a real cure of the disease. Overall, it would seem that only in a minority of patients does long-term medical treatment result in a persisting correction of the underlying cause of the adenoma. Possibly, more prolonged drug regimens, different drug doses, or newer DA agonists will prove more effective in this context.

Bibliography

Hökfelt, T., Meister, B., Everitt, B., Staines, W., Melander, T., Schalling, M., Mutt, V., Hulting, A.-N., Werner, S., Bartfai, T., Nordström, O., Fahrenkrug, J., and Goldstein, M. (1986). Chemical neuroanatomy of the hypothalamo-pituitary axis: Focus on multimessengers systems. *In* "Integrative Neuroendocrinology: Molecular, Cellular and Clinical Aspects" (S. M. McCann and R. I. Weiner, eds.), pp. 1–34. Karger, Basel.

Martin, J. B., and Reichlin, S. (1987). "Clinical Neuroendocrinology." F. A. Davis Company, Philadelphia.

Müller, E. E., and Nisticò, G. (1989). "Brain Messengers and the Pituitary." Academic Press, San Diego.

Reichlin, S. (1985). Neuroendocrinology. *in* "Williams Textbook of Endocrinology" (J. D. Wilson, and D. W. Foster, eds.) pp. 492–567. W. B. Saunders Co., Philadelphia.

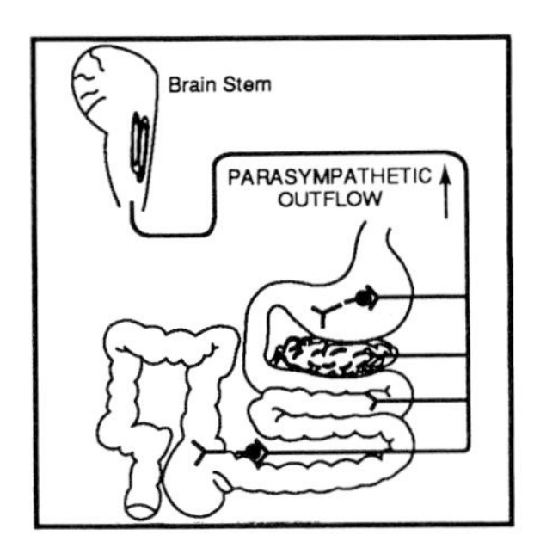

Brain Regulation of Gastrointestinal Function

YVETTE TACHÉ, University of California Los Angeles

ERIK BARQUIST, University of California Los Angeles

Glossary

Dorsal vagal complex Association of two medullary nuclei: the dorsal motor nucleus of the vagus (which contains neurons projecting the gut through the vagus), and the nucleus tractus solitarius (which contains terminals of afferent vagal neurons from the gut)

ENS Enteric nervous system: neuronal network embedded within the gut wall that serves as relay for signals from and to the brain or spinal cord but also can, independently from the brain, receive information from various kind of sensory receptors and generate neural outflow

Hypothalamus Nuclei in the forebrain that are involved in the regulation of pituitary hormone secretion and visceral function and are subdivided in lateral, ventromedial, and paraventricular parts

MMC Migrating myoelectrical complex refers to a cyclically occurring phenomenon that begins in the stomach and duodenum and is propagated to the ileum. It is composed of three well-defined phases: phase I: noncontractile activity; II: intermittent and irregular contractions; III, or activity front: period of intense spikes and contractile activity

Monosynaptic vago-vagal reflex Transmission of the information along vagal sensory neurons directly to vagal motoneurons in the dorsal motor nucleus of the vagus

Peptide Molecules formed of a small number (below 100) of amino acids

MAMMALIAN GASTROINTESTINAL FUNCTIONS are subjected to a diversity of regulatory controls exerted at multiple levels including the brain, spinal cord, peripheral autonomic ganglia, and enteric plexuses (Fig. 1) and by hormones acting through endocrine or paracrine mechanisms. The term *brain–gut axis* refers to the control of gut functions including secretion, absorption, and motility exerted by the brain or spinal cord through the autonomous nervous system. The extent of the central nervous system (CNS) influence on the gut ranges from the esophagus to the colon and includes the liver, pancreas, and gallbladder. Brain–gut interactions encompass knowledge on localization of the specific brain nuclei involved in modulating gut function, the anatomical, chemical, and electrophysiological characterization of the connections between the brain and the gut (efferent pathways) and from the gut to the brain (afferent pathways), the identification of chemicals in the brain, and the gut coding the neuronal transmission and physiological stimuli that use these regulatory mechanisms.

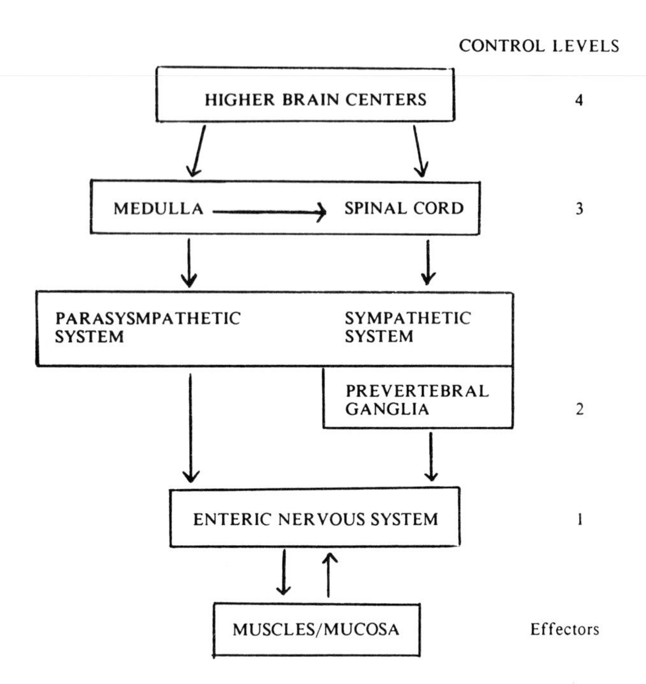

FIGURE 1 Various levels of neural control of gastrointestinal function.

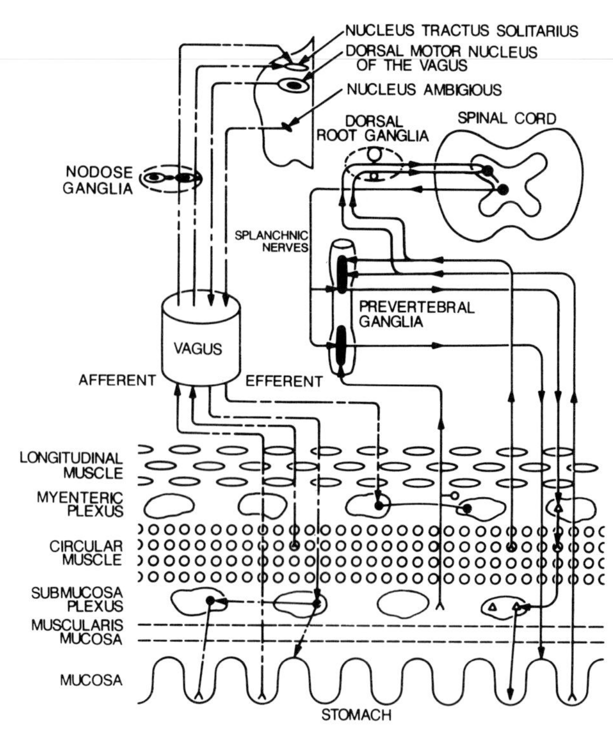

FIGURE 2 Schematic diagram illustrating medullary afferent and efferent pathways to the stomach.

I. Neuroanatomical Basis for Brain–Gut Interactions

A. Brain Sites Influencing Gastrointestinal Function

1. Medullary Nuclei

Primary nuclei in the brainstem involved in the control of gastrointestinal function are the dorsal motor nucleus of the vagus and the nucleus tractus solitarius, referred to as the dorsal motor vagal complex. Other medullary nuclei eliciting a gastrointestinal response are the nucleus ambiguus, raphe nucleus, reticular substance, and medial longitudinal fascicle (Fig. 2). Preganglionic gastric neurons located in the dorsal motor nucleus of the vagus have a dendritic arborization with direct synaptic contacts on axon terminals from vagal gastric sensory neurons arising from the nucleus tractus solitarius. These axodendritic contacts provide the anatomical basis for monosynaptic gastric vago-vagal reflexes in the brainstem.

2. Forebrain Nuclei

In the forebrain, several nuclei can influence gastrointestinal function: the cingulate cortex, hypothalamus (lateral, ventromedial, posterior and paraventricular nuclei), locus ceruleus, bed nucleus of the stria terminalis, and central nucleus of the amygdala. These nuclei, the hypothalamus in particular, can receive and process interoceptive and exteroceptive afferent information through poly- or monosynaptic reciprocal connections with medullary nuclei and provide a major input to the brainstem circuitry responsible for coordinating autonomic reflexes. The influence of these higher centers can be exerted either on the efferents of the dorsal motor nucleus of the vagus and/or the afferents to the nucleus tractus solitarius before the sensory signals are relayed to neurons in the dorsal motor nucleus of the vagus. [*See* HYPOTHALAMUS.]

B. Extrinsic Innervation of the Gut

The gastrointestinal tract is innervated by the parasympathetic and sympathetic divisions of the autonomic nervous system. The efferent autonomic innervation is involved in transmitting the information from the brain to the gut, whereas the autonomic afferent innervation allows peripheral information to reach the brain. The afferent fibers outnumber the efferent fibers by a ratio of 9 : 1 in the vagus and 3 : 1 in the splanchnic nerve.

1. Efferent Innervation

The autonomic efferent pathways convey input from the CNS to an intrinsic neuronal circuitry present in the gastrointestinal wall [enteric nervous system (ENS)], which then relays the information to gastrointestinal effectors such as the mucosal cells, smooth muscles, or blood vessels. The ENS serves not only as a relay station to the autonomic nervous system but also acts as a local integrative system particularly in the small intestine.

a. Parasympathetic The efferent parasympathetic preganglionic neurons originate in the medial portion of the dorsal motor nucleus of the vagus and to a smaller extent in the nucleus ambiguus (Fig. 2). These neurons send axons in the vagus nerve of the same side and end in the gastrointestinal tract from the esophagus to proximal colon. The parasympathetic innervation to the distal colon and rectum comes from the pelvic nerves. Cell bodies arise in the spinal cord at the sacral levels 2–4, and terminals synapse with the pelvis plexus. The vagal efferent terminals are proportionally more numerous in the stomach than other bowel segments. Electrophysiological studies demonstrated that vagal efferent fibers display continuous spontaneous discharges generated centrally independently from vagal input to the brainstem. This ongoing activity provides a background vagal tone. In the gastrointestinal wall, terminals of preganglionic vagal fibers synapse with neurons in the intrinsic plexuses. These postganglionic neurons innervate smooth muscles, mucosal secretory cells, and blood vessels. Neuroeffector transmission of excitatory vagal pathways use acetylcholine as transmitter and inhibitory vagal pathways use peptidergic [vasoactive intestinal peptide (VIP)] or purinergic transmitter (serotonin). There is a large disparity between the numbers of vagal efferent fibers and ganglia in the gastrointestinal tract. This has led to the postulate that vagal efferents activate "command" postganglionic neurons in the ENS, which amplify and distribute the signal through intramural networks.

b. Sympathetic The sympathetic efferent pathways are organized hierarchically between spinal and supraspinal levels. The preganglionic neurons of the efferent sympathetic nerve fibers are cholinergic. They are located in the intermediolateral column of the spinal cord mainly between the 5th to the 11th thoracic and lumbar 1–3 segments, although species variation exists. They send processes, which synapse with the postganglionic fibers arising in the prevertebral ganglia (celiac and mesenteric); these postganglionic noradrenergic fibers project to the gut where they further synapse within the intrinsic plexuses (Fig. 2). The noradrenergic terminals are densely represented particularly at the levels of sphincters and blood vessels and to a lower extent in the mucosa and muscular layers.

2. Afferent Pathways

a. Parasympathetic The majority of gastric vagal afferents are composed of unmyelinated fibers, which arise from the nodose ganglia. Axons project centrally almost exclusively to the medial subnucleus of the caudal areas of the nucleus tractus solitarius and peripherally to the vagus nerve, where terminals innervate the various mucosal and muscular components of the gastrointestinal wall (Fig. 2). Vagal sensory fibers transmit nonconsciously perceived signals generated by food ingestion and digestions, which are encoded by mechano-, chemo-, osmo-, and thermoreceptors located in the gastrointestinal tract. There is evidence that vagal afferent inputs arising from the stomach and liver can converge on to the same neurons in the nucleus tractus solitarius. However, there is also a somatotopic organization of vagal afferents innervating the gut because they project along the rostro-caudal axis of the nucleus tractus solitarius.

b. Sympathetic The sympathetic afferents have their cell bodies in the thoracolumbar and sacral dorsal root ganglia. They terminate centrally in the dorsal horn of the spinal cord at the level of the laminae I, II, V–VII, and X. Hence, terminals synapse on to ascending spinal pathways. These spinal pathways play an important role to convey nociceptive signals and CNS representation of visceral pain. Sympathetic afferents also initiate enterogastric, intestino-intestino and intestino-colonic reflexes in response to physiological stimulation of enteroreceptors. These reflexes are mediated through projections of the afferents to preganglionic sympathetic neurons located in intermediolateral column of the spinal cord or to postganglionic neurons located in the prevertebral ganglia. Chemical coding of afferent projections from the gut to the prevertebral ganglia are cholinergic and contains a variety of peptides including cholecystokinin, enkephalin, dynorphin, gastrin-releasing peptide, and VIP.

II. Brain Regulation of Gastric Function

The main function of the stomach can be divided into motor and secretory, either exocrine (acid, pepsin, bicarbonate, mucus) or endocrine (gastrin, somatostatin, serotonin, histamine). It is well-documented that the CNS can modulate gastric function and that the vagus plays an important role in mediating this influence. Yet, as we will see for other areas of the gastrointestinal tract, only recent developments in research techniques have allowed elucidation of the chemical messengers that make up this control.

A. Gastric Secretion

1. Centrally Mediated Stimulation of Gastric Secretion

Gastric exocrine and endocrine secretion are well-established to be under a vagal stimulatory control expressed through peripheral muscarinic receptors. Experimental evidence indicates that vagal activation can be triggered centrally by hypothalamic, limbic, and/or medullary input to preganglionic neurons in the dorsal motor nucleus of the vagus leading to increase in parasympathetic outflow and stimulation of gastric secretion. Convergent neuropharmacological, electrophysiological, and neuroanatomical data indicate that the tripeptide thyrotropin-releasing hormone (TRH) and TRH receptors located in the dorsal vagal complex and raphe nuclei are involved in initiating the activation of the parasympathetic outflow, leading to the stimulation of gastric secretion (acid, pepsin, bicarbonate, serotonin, and histamine). Other transmitters, gamma-aminobutyric acid (GABA) acting on $GABA_B$ receptors, acetylcholine acting on muscarinic receptors, or peptides such as oxytocin or somatostatin can also stimulate gastric secretion through central activation of vagal pathways. However, less substantial evidence has accumulated to assign them a physiological role.

Several physiological stimuli increase gastric acid and bicarbonate secretion through an action on preganglionic neurons in the dorsal motor nucleus of the vagus. Stimuli can originate in the brain [e.g., the sight, smell, or chewing of tasty food and the suggestion or anticipation of eating (cephalic phase of gastric secretion)]. The exact neural circuitries involved in the various afferent components of the cephalic phase are not completely elucidated. Processing of gustatory signals takes place in the hypothalamus where the sensory information projects by ways of the parabranchial nucleus. Then hypothalamic efferents converge on the brainstem nuclei responsible for coordinating vagal outflow. The cephalic phase of acid secretion accounts for one-third of the total acid response to eating in healthy subjects and is prolonged by gastric distention (Fig. 3). Other stimuli can originate in the periphery (e.g., gastric distention or cold exposure). The acid response to gastric distention is initiated through mechanoreceptors located in the gastric wall. They trigger a flow of vagal afferent impulses to discrete neurons in the hypothalamus and brainstem, which feed back to excite preganglionic neurons in the dorsal motor nucleus of the vagus. Hypoglycemia induced by injection of nonmetabolized glucose analogues or insulin has been established for several decades to be potent vagal stimulants of gastric secretion through an hypothalamic site of action. However, the physiologic importance of this activation is uncertain, because the levels of hypoglycemia required to stimulate gastric secretion are not commonly seen even under a fasting state. The centrally mediated gastric secretory response to all these stimuli does not return to baseline immedi-

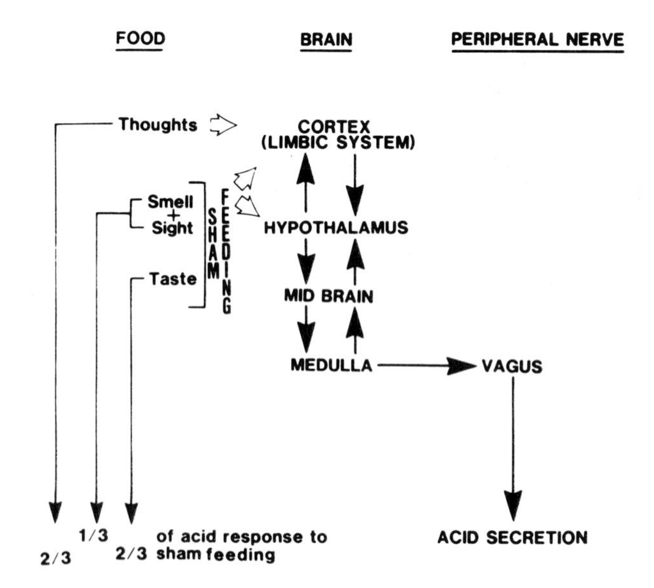

FIGURE 3 Schematic representation of pathways involved in shamfeeding–induced gastric acid secretion.

ately after withdrawing the stimuli. For instance, acid secretion induced by sham feeding outlasts the duration of chewing by 60–90 min. This sustained effect required the integrity of the vagus; however, its underlying central mechanisms are unknown. The chemicals encoding the afferent and efferent transmission evoked by these stimuli are also yet to be established. However, circumstantial evidence indicates that TRH located in the dorsal vagal complex, nucleus ambiguus, and raphe nucleus is a likely candidate involved in initiating vagal stimulation.

2. CNS-Mediated Inhibition of Gastric Secretion

Knowledge has accumulated recently on the inhibitory influence exerted by the brain on gastric secretion. Many peptides, specifically bombesin, calcitonin, calcitonin gene-related peptide (CGRP), corticotropin-releasing factor (CRF), interleukin 1-β, and opioid peptides, have been shown to act in the brain to inhibit gastric acid and pepsin secretion. Their sites of action are in the hypothalamus (lateral, ventromedial, or paraventricular nuclei), nucleus ambiguus, and/or the dorsal vagal complex. Thus, it seems that the same nuclei can be involved in both stimulation or inhibition of gastric secretion, depending on the type of receptors activated. Pharmacological studies further demonstrated that an interaction exists in the dorsal vagal complex between centrally acting stimulatory (TRH) and inhibitory (bombesin, CRF) peptides.

The centrally mediated inhibition of gastric acid secretion can be achieved either by inhibiting the vagal stimulatory pathways or by stimulating sympathetic noradrenergic inhibitory pathways, or by both. For instance, the central inhibition of gastric acid secretion induced by CGRP or opioid peptides is vagally dependent, whereas that of bombesin is mediated by sympathetic activation. CRF involves both pathways.

Stimuli inhibiting gastric acid secretion through cephalic influence are mostly related to stress exposure. CRF, which is centrally released during stress, and central CRF receptors play a physiological role in the inhibition of gastric acid secretion induced by various stressors in experimental animals. Removal of sympathetic pathways stimulates gastric acid secretion. To what extent this represents a CNS tonic inhibitory control exerted by these peptides is still to be addressed.

B. Gastric Motility

The motor function of the stomach varies strikingly in relation to the prandial state. During the interdigestive period, the motility in the corpus and antrum is characterized by a cyclic recurring period of high-amplitude contractions known as phase III of the interdigestive migrating motor complex (MMC). Phase III usually is initiated in the antrum and propagated to the duodenum. On food ingestion, the fasted pattern is disrupted and replaced by a postprandial pattern consisting of continuous irregular contractions. Also the fundus must accommodate to incoming nutrients by the process of receptive relaxation while antral and fundic motor activity are increased. The stomach also regulates the amount of food entering the intestine by modulating gastric emptying. Neural mechanisms, either local or long reflexes as well as hormonal factors, regulate these processes.

1. Central Vagal Regulation of Gastric Motility

Vagal activation conveys both excitatory and inhibitory inputs to gastric intrinsic plexus projecting to smooth muscle effectors. This is unlike the effect of vagal activation on gastric secretion, which is always stimulatory. The anatomical substrate of such a dual response is the existence in the vagus of two kinds of preganglionic fibers: excitatory and inhibitory. The ones that convey excitatory response on motility have a lower threshold of excitability than the inhibitory fibers. They are composed of preganglionic cholinergic fibers, which synapse with postganglionic cholinergic neurons. Their stimulation leads to an increase in the amplitude of corpus, antral, and pyloric phasic contractions and, to a lesser extent, their frequency.

Based on experimental studies, brain nuclei that can induce a vagally mediated excitation of gastric contractility are the dorsal vagal complex, nucleus ambiguus, raphe nucleus, and hypothalamus (lateral and paraventricular parts). TRH or acetylcholine in these medullary nuclei serve as mediators to activate the vagal excitatory pathways. Preliminary evidence, still to be further substantiated, indicates that somatostatin acting in the dorsal vagal complex may also play a role in such a process. Physiological stimuli influencing gastric motor function through the vagal excitatory pathways are sham feeding or distention of the corpus and antrum. The latter increases antral motility through vago-vagal reflexes well-characterized electrophysiologically.

The vagal inhibitory pathway contains vagal efferent fibers that fire spontaneously at low frequency. The preganglionic vagal fibers synapse directly or through a nicotinic intermediate on intramural inhibitory neurons. These postganglionic inhibitory neurons are nonadrenergic and noncholinergic and most likely use VIP as their neurotransmitter. Excitation of vagal inhibitory pathways leads to the relaxation of the gastric wall predominantly at the proximal stomach (fundus). Preliminary evidence suggests that oxytocin in the dorsal motor nucleus of the vagus may be the chemical signal activating this efferent inhibitory pathway particularly during conditions that provoke satiety and nausea. Several physiological stimuli (e.g., swallowing, distention of the esophagus or antrum, or strong antral contractions) induce relaxation of the fundus through the long vago-vagal reflex, using the inhibitory pathways as the efferent limb of the reflex. Sham feeding has also been reported to inhibit gastric motility. This may be related to activation of vagal inhibitory pathways and/or subsequent inhibitory reflexes caused by the increased acid secretion and arrival of acid in the proximal duodenum.

The role of the vagus in the control of interdigestive patterns of gastric motility remains controversial. There are two schools of thoughts as to whether the extrinsic neural control is required to initiate gastric MMC. Some studies indicate that the cyclic interdigestive motor patterns of the stomach can occur in the absence of extrinsic neural control. Other more recent data, based on total extrinsic denervation or vagal cooling, favor a vagal tonic modulation of the gastric MMC. This is a further supported by electrophysiological studies demonstrating that the majority of vagal efferent fibers display continuous activity. Moreover, the spontaneous discharge rate of excitatory fibers fluctuates in relation with the various phases of the MMC.

Similarly, the part that central vagal input plays in the initiation of the fed motility pattern is not clearly established. In favor of such a role are the facts that as soon as food is offered, there is an increase in the discharge rate of vagal excitatory fibers, which is maintained for several hours if the food is ingested. Moreover, an increase in parasympathetic activity triggers a postprandial pattern of motor activity. Peptides such as cholecystokinin (CCK) and neurotension have been proposed to be putative peptides acting in the ventromedial hypothalamus to initiate (neurotensin) and maintain (CCK) the fed pattern of motility.

2. Central Sympathetic Regulation of Gastric Motility

Activation of the splanchnic sympathetic activity lead to decreased tone in the proximal stomach and peristaltic activity in the antrum. This inhibitory effect is adrenergic in nature and is exerted at the myenteric nervous plexus by inhibiting acetylcholine release from intramural cholinergic neurons through an action on α-adrenergic receptors. Some studies also indicate a direct effect on the smooth muscles exerted by β-adrenergic receptors. Increased sympathetic outflow caused by removal of the inhibitory tone exerted by supraspinal centers or by blockade of vagal input prevents the initiation of phase III of the interdigestive MMC.

3. Central Regulation of Gastric Emptying

Various psychological, physical, or chemical stressors alter gastric contractility and emptying, producing mainly an inhibitory effect on motor function. In addition, delayed gastric emptying has been described in a variety of neurological disorders including brain tumors, bulbar poliomyelitis, diabetic gastroparesis, paraplegia, and high cord transection. Neural pathways initiating changes in the rate of gastric transit in response to these physiological or pathological conditions are not fully characterized, although it has been ascribed to an increased sympathetic outflow. Regarding the central chemical coding, growing experimental evidence indicates that CRF plays a physiological role in mediating the delay in gastric emptying related to stress exposure. Other peptides (e.g., bombesin, calcitonin, CGRP, neurotensin, neuropeptide Y, and μ- or δ-opioid peptides) act centrally to delay gastric emptying of a nonnutritive solution. The inhibitory effect of these peptides, except calcitonin, is also exerted through the vagus. However, their physiological role in the control of gastric transit is not clearly established because of the lack of specific antagonists for most of them and the inability to monitor their release in the brain in response to centrally acting stimuli.

III. Brain Regulation of Intestinal Function

After nutrients leave the stomach, they are further digested in the duodenum. It is here that gastric acid must be neutralized and nutrients mixed with pancreatic enzymes. The duodenum, as the recipient of gastric acid, has extensive feedback pathways to the stomach at both local, hormonal, and neural levels to limit the amount of liquid entering the duodenum. Also local protective mechanisms exist (e.g., a thin layer of bicarbonate rich in mucus, which titrates incoming acid). Proper control of bicarbonate secretion and mucus production is essential to good duodenal function.

The autonomic nervous system whether through central or spinal input, while playing a major role in the control of gastric, colonic, and pancreatic function, appears less involved in the overall control of small intestinal function. In the small intestine, the enteric nervous system is reciprocally connected with the prevertebral ganglia. For instance, in the prevertebral ganglia, cell bodies of sympathetic secretomotor inhibitory and motility inhibitory neurons received direct input from sensory pathways originating from the intestine. These connections allow an array of reflex activity initiated by enteroreceptors and processed in the prevertebral ganglions or the ENS with a large degree of autonomy from the CNS.

A. Duodenal Bicarbonate Secretion

1. Vagal Control

It has long been held that vagal stimulation causes increased duodenal bicarbonate secretion. The peripheral transmitters involved in this effect are unknown, although pharmacological studies indicate a nicotinic receptor intermediate and a nonmuscarinic postganglionic transmission. Knowledge on central control of the vagally mediated alkaline secretion is still fragmentary. Several peptides (e.g., bombesin, CRF, somatostatin, TRH) cause a centrally mediated increase in duodenal bicarbonate secretion through vagal efferent pathways. The end neurotransmitter is most likely VIP and not cholinergic, as atropine does not block either TRH- or sham feeding–induced duodenal bicarbonate secretion increases.

2. Splanchnic Control

The local splanchnic nerves, normally quiescent, inhibit bicarbonate secretion when activated by stressful stimuli. For instance, hypovolemic stress induced by controlled blood loss is known to cause decreased bicarboante secretion through splanchnic adrenergic pathways. Because the response is blocked by α_2-adrenergic antagonists, it implies an effect mediated by endogenous α_2-adrenergic receptors. Stimulation of the medial hypothalamus can trigger inhibition of duodenal alkaline secretion through spinal adrenergic pathways.

B. Intestinal Absorption/Secretion

Although CNS control of the duodenum is primarily directed toward protecting against gastric acid and enzyme effect and, to a lesser extent, motility, the more distal duodenum and remaining small bowel require coordination of nutrient absorption, fluid and electrolyte absorption/secretion, and motility. Intestinal absorption takes place simultaneously with intestinal secretion, making studies of either mechanism difficult. Most studies have measured either net absorption or net secretion.

1. Influence of the Vagus

Central activation of vagal efferent outflow causes net secretion. Neurotransmitters involved in this effect are both cholinergic and peptidergic. Atropine blocked the centrally stimulated effect of increased secretion. More distal stimulation of the vagus in atropinized rats still caused net secretion. A likely candidate for the more distal transmitter is VIP, which is released during electrical stimulation and causes net secretion when administered locally. However, in two studies in which vagal stimulation was produced by sham feeding, the increased net secretion in response to this stimulus was inconclusive. More recent studies performed in dogs with isolated jejunal segments concluded to the absence of a cephalic phase in the intestinal absorption/secretion.

2. Influence of the Sympathetic Nerve

Direct stimulation of sympathetic outflow generally causes intestinal absorption. This effect is blocked by the peripheral use of adrenergic antagonists. One center in particular, the lateral nucleus of the hypothalamus, causes absorption when stimulated with microelectrodes in rats. Several centrally acting agents (e.g., δ- and μ-opioid peptides, as well as angiotensin II and III) cause centrally mediated increases in intestinal secretion. Part of this effect may be mediated in hypothalamic centers.

The tonic nature of sympathetic control of intestinal absorption can be shown by the profuse hypersecretion and diarrhea that results from chemical or surgical sympathectomy. With the removal of the sympathetic efferent arm, the bowel tends to secrete more fluid and electrolytes than it absorbs, leading to diarrhea. Inputs from higher centers probably play a role in this effect but are not mandatory for its tonic functioning. Norepinephrine and somatostatin act as neurotransmitters in this system, and fibers containing these compounds project to the villi and crypts and originate in the prevertebral ganglia.

C. CNS Control of Intestinal Motility

In the fasting state, the small intestine exhibits a cyclic burst of contractions, which lasts a few minutes, and propagates from the proximal duodenum to ileocolonic junction with a velocity of 2–8 cm/min. The intestinal MMC is controlled by neural and hormonal mechanisms. There is clear evidence that the initiation and the propagation of the MMC takes place at the level of the ENS and does not require the CNS. However, the brain can modulate the characteristics of the MMC by reducing their time intervals, disrupting the pattern, or initiating MMC in postprandial state. For instance, cephalic influence produced by sham feeding can increase duodenal motor activity. Experimental studies demonstrated that several peptides present in the brain (e.g., calcitonin, CGRP, neurotensin, neuropeptide Y, and μ-opioid peptides) act centrally to induce a fasted MMC pattern of intestinal motility in fed animals, whereas growth hormone–releasing factor (GRF) and substance P shorten the duration of the fed pattern. TRH acts centrally to stimulate intestinal motility and, microinjected into the medial septum, medial, or lateral hypothalamus, increases intestinal transit in fasted animals. Central injection of CRF bombesin, calcitonin, CGRP, CRF, neurotensin, and μ-opioid peptides inhibit intestinal transit through vagal (CGRP, neurotensin, CRF) or nonvagal pathways. The periaqueductal gray is a site of action for neurotensin and opioid peptide antitransit effect.

The sympathetic nervous system is well known to inhibit intestinal motility. In the normal, unstressed state, however, these neurons are silent and not required to assume normal digestive function. When activated, they cause contraction of sphincters, inhibit intestinal motility, and delay the incidence of MMC. The action is exerted presynaptically through inhibition of neural signals within the myenteric ganglia and not by a direct effect on the smooth muscles. The activation of this sympathetic efferent arm can be caused either by stimulation of intestinal receptors (e.g., by intestinal distention, intraperitoneal administration of irritants, or peritonitis) or by nonenteric stimuli. Nonenteric stimuli include those associated with severe pain or stress (e.g., surgery to nonintraperitoneal structures, systemic hypotension, or psychological stressors). It appears that this sympathetic effect is mediated at both the CNS and spinal levels.

IV. Brain Regulation of Exocrine Pancreatic Secretion

The CNS control of the pancreas is less well-studied than that of the stomach, yet the two organs are closely interrelated to allow integrated functioning of the foregut. Both organs mix their outputs in the proximal duodenum, where alkalization of gastric content allows further enzymatic degradation of nutrients. Information from the duodenum is extensively used as feedback on pancreatic function, through both vagal and sympathetic pathways.

A. Central Vagal Stimulation of Pancreatic Secretion

As with acid secretion, the vagus represents the main pathway that stimulates pancreatic exocrine secretion. This is achieved by activation of cholinergic and noncholinergic secretomotor pathways using serotonin as well as peptides, VIP, bombesin, GRP, CCK, substance P, enkephalins, neuropeptide Y, and CGRP acting as neurotransmitters or hormones. The central transmitters implicated in modulating vagal pancreatic secretion are less well known. Based on pharmacological studies, TRH is a likely candidate because the peptide acts centrally to increase pancreatic secretion. Its action is expressed through noncholinergic, VIP postganglionic neurons. [*See* PEPTIDE HORMONES OF THE GUT; PEPTIDES.]

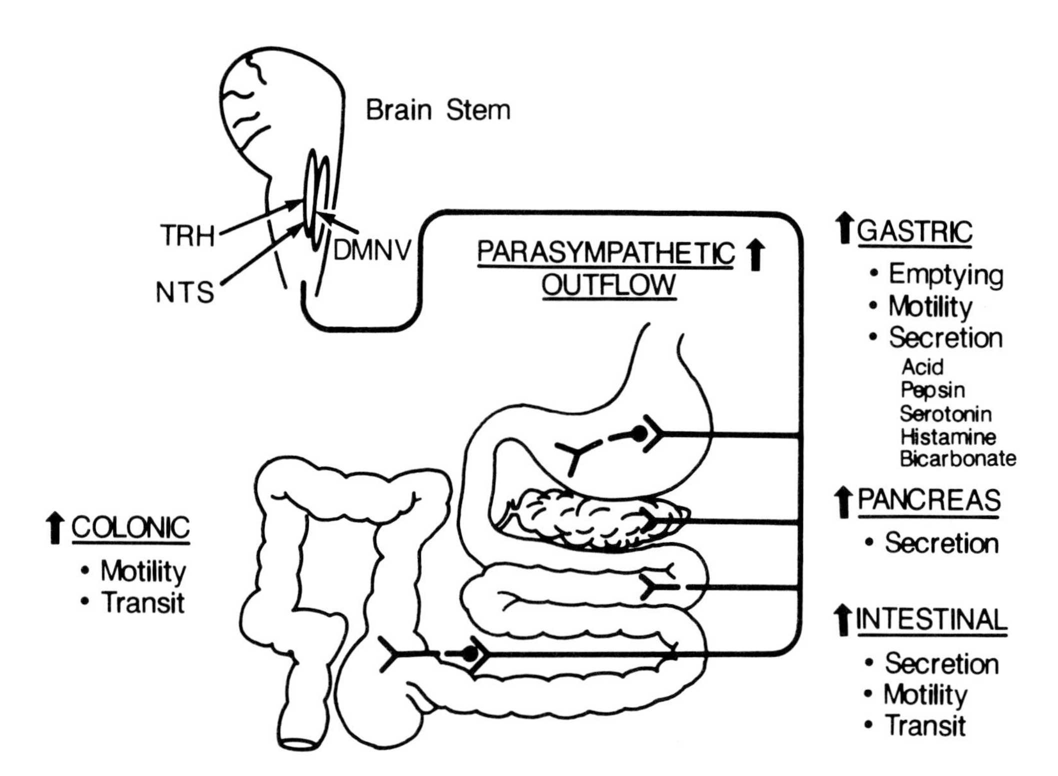

FIGURE 4 Summary of centrally mediated effect of tripeptide thyrotropin-releasing hormone (TRH) on gastrointestinal secretory and motor function exerted through activation of the vagal outflow in the dorsal vagal complex (DVC).

A physiological stimuli activating this central vagal pathway is sham feeding. In humans, this response occurs within 1 min of food ingestion. This rapid response is unlikely to be related to gastric acid contamination of the duodenum, but rather represents a direct cephalic influence. In dogs, a sham-feeding pancreatic response may approximate 60% to 70% of a maximal response to CCK and is similar to the effect of chemical vagal stimulation produced by glucose analogues (e.g., 2-deoxy glucose or 5-thio glucose) or electrical stimulation of the vagus. Stimuli originating peripherally such as stimulation of intestinal chemoreceptors by amino acids, acid, or fatty acids induced marked pancreatic secretory response, 50% of which is mediated by enteropancreatic long vago-vagal cholinergic reflexes.

Pancreatic secretion can be influenced by nuclei in the brainstem, hypothalamus, and medial amygdala. Stimulation of the anterior hypothalamus tends to exert parasympathetic effect and increases basal pancreatic output without effecting periodicity.

B. Central Sympathetic Regulation of Pancreatic Secretion

The influence of sympathetic stimulation on pancreatic exocrine secretion are more controversial. Stimulation of the posterior hypothalamus has a sympathetic effect to increase the periodicity of basal pancreatic output. Much less is known about centrally acting agents that might stimulate the sympathetic arm of pancreatic innervation. Opioid peptides acting on μ receptors or α_2 agonists such as clonidine decrease pancreatic outflow when administered in the brain.

V. Conclusions

In the past years there has been a resurgence of interest in investigating the interactions between the CNS and the gastrointestinal tract. Positive impacts came from the characterization of better tools to probe these interactions, such as selective peptides able to influence gastrointestinal secretory and motor functions through the CNS. The development of sophisticated electrophysiological and neuroanatomical techniques for recording neural activity and tracing neural circuitry has also contributed to this knowledge. This has lead to a recent increase in the understanding of CNS control of gastrointestinal motility and secretion, particularly in relation to neuroanatomy and neurochemistry. Functional studies have established that vagal efferents and afferents play an important tonic or phasic role in the overall regulation of gastrointestinal secretory and motor function, particularly to initiate the cephalic phase or food-induced receptive fundic relaxation. Vagal control is exerted through the dorsal vagal complex, along with input to these nuclei from higher hypothalamic or limbic centers. Convincing experimental evidence suggests that the tripeptide TRH and its receptors, localized in medullary nuclei, play a physiological role in mediating the increased parasympathetic outflow to the gut leading to increased gastrointestinal secretory and motor activity (Fig. 4). The sympathetic nervous system through modulation of parasympathetic outflow and spinal or prevertebral reflexes can also markedly influence gastrointestinal function. Sympathetic influence can be tonic (intestinal absorption) but is mostly recruited in relation with stress situation. Growing experimental data support a role of central CRF in mediating the inhibitory effect of stress on gastric function and stimulation of colonic activity.

Bibliography

Buéno, L., Collins, S., and Junien, J. L., (1989). "Stress and Digestive Motility." John Libbey Eurotext, Montrouge, France.

Grundy, D. (1988). Speculation on the structure/function relationship for vagal and splanchnic afferent endings supplying the gastrointestinal tract. *J. Auton. Nerv. Syst.* **22,** 175.

Mayer, A. E., and Raybould, H. (1989). Role of neural control in gastrointestinal motility and visceral pain. *In* "Pathogenesis of Functional Bowel Disease" (W. J. Snape, ed.). Plenum Publishing, New York.

Roman, C., and Gonella, J. (1987). Extrinsic control of digestive tract motility. *In* "Physiology of the Gastrointestinal Tract" (L. R. Johnson, J. Christensen, M. Jackson, E. D. Jacobson, and J. H. Walsh, eds.). Raven Press, New York.

Singer, M. V., and Goebell, H., Eds. (1989). "Nerves and the Gastrointestinal Tract" MTP Press, Dordrecht, The Netherlands.

Taché, Y. (1987). Central regulation of gastric acid secretion. *In* "Physiology of the Gastrointestinal Tract" (L. R. Johnson, J. Christensen, M. Jackson, E. D. Jacobson, and J. H. Walsh, eds.). Raven Press, New York.

Taché, Y. (1989). Central control of gastrointestinal transit and motility by brain-gut peptides. *In* "Pathogenesis of Functional Bowel Disease" (W. J. Snape, ed.), pp. 55–78. Plenum Press, New York.

Taché, Y., Stephens, R. L., and Ishikawa, T. (1989). Central nervous system action of TRH to influence gastrointestinal function and ulceration. *Ann. N.Y. Acad. Sci.* **553,** 269–285.

Taché, Y., Stephens, R. L., and Ishikawa, T. (1989). Stress-induced alterations of gastrointestinal function: Involvement of brain CRF and TRH. *In* "IV. New Frontiers of Stress Research" (H. Weiner, I. Florin, D. Hellhammer, and M. Murison, eds.). Hans Huber Publishers, Lewiston, New York.

Brain Spectrin

WARREN E. ZIMMER, STEVEN R. GOODMAN, *University of South Alabama College of Medicine*

Glossary

Actin 43-kDa protein found in eukaryotic cells that has the capability to form a thin helical filaments

Amelin Spectrin binding protein found in the soma and dendrites of neurons

Ankyrin Family of proteins that binds with spectrin and anchors the spectrin membrane skeleton to the membrane via association with integral membrane proteins

Calmodulin Highly conserved calcium binding protein that appears to be ubiquitious among eukaryotes, and thought to be involved in the transmission of intracellular calcium signals by it's calcium-dependent interaction with proteins

Complementary DNA Synthesized from a messenger RNA

Fodrin Original name given to a brain protein (Grk. *fodros,* lining) that has been demonstrated to be a structural analogue of erythrocyte spectrin and is now referred to as brain spectrin (240/235).

Indirect immunofluorescence Procedure to find the location of a protein–antibody complex within a cell by binding a secondary antibody containing a fluorescence dye to the complex, and detection of the fluorescent antibody using a fluorescence microscope

Peptide mapping Technique for examining the structure of a protein by partial digestion of the protein with proteases and analysis of the cleavage products

Spectrin Class of related proteins that contains two high-molecular-weight subunits and forms the structural framework of the spectrin membrane skeleton

Spectrin membrane skeleton Protein complex that contains spectrin as its major component and forms a three-dimensional meshwork lining the cytoplasmic surface of eukaryotic cell membranes

SPECTRIN IS THE MAJOR CONSTITUENT of the skeletal protein meshwork that is closely associated with the cortical cytoplasm of most, if not all, eukaryotic cells. This membrane protein structure was first identified in red blood cells, and it was considered to be a cellular structure peculiar to these cells until spectrin and spectrin-associated proteins were detected in many cell types. The complex has been termed the spectrin membrane skeleton because spectrin appears to form the structural backbone of the complex, and this nomenclature distinguishes the membrane skeletal structure from cytoskeletal structures, which transverse the cytoplasm of nucleated cells. The best-characterized nonerythroid spectrin molecules are those found in the brain. While brain spectrin appears to be similar to its erythroid counterpart with respect to physical and morphological parameters, structural analyses of isolated brain spectrin molecules demonstrate that they represent a diverse class of proteins that are related but clearly not identical to red blood cell spectrin. This diversity of brain spectrin is manifested not only at the structural level but includes expression of multiple spectrin isoforms exhibiting distinct localization within a single neuronal or glial cell. The diversity of brain spectrin molecules must be taken into account when assigning functional roles to the spectrin membrane skeleton in brain cells. The close association of brain spectrin and spectrin-binding proteins with the plasma membrane and membranes of various organelles suggests that the membrane skeleton may play a role in the maintenance and modifications of membrane behavior. The best-characterized examples of spectrin's involvement in neural processes are axonal transport, control of synaptic transmission, and regulation of the lateral mobility of the 180-kDa neural cell adhesion molecule (N-CAM$_{180}$) in the plasma membrane. [*See* BRAIN.]

I. Nonerythrocyte Spectrin

A. Discovery

Spectrin was originally described as the major constituent detected in low-ionic-strength extracts of human erythrocyte membranes. The structure and function of erythrocyte spectrin has been the subject of intensive investigations beginning with its description in the late 1960s. During the course of these studies, the prevailing thought was that spectrin existed only in erythrocyte cells.

The first demonstration of spectrin or spectrinlike molecules in nonerythroid cells came in the early 1980s by Goodman and colleagues. Using an antibody that reacted with human erythrocyte spectrin, the presence of spectrin was detected in a diverse set of nonerythroid cells including embryonic chicken heart cells, mouse fibroblast cells, and rat hepatoma (liver) cells by indirect immunofluorescence techniques. While these experiments suggested that these very different cell types contained proteins that shared antigenetic determinants with erythroid spectrin, they did not address whether or not these molecules were close relatives of the erythrocyte protein. Thus, further experiments were performed, demonstrating that the erythrocyte spectrin antibodies precipitated two polypeptides from lysates of the chicken heart cells that not only represented spectrin in size (240 and 230 kDa) but were present in equal numbers (i.e., at 1 : 1 mol/mol stoichiometry). Comparison of the immunoprecipitated polypeptides with spectrin isolated from chicken erythrocytes, by proteolytic digestion with the enzyme chymotrypsin, documented that the proteins immunoprecipitated from chicken heart cells were similar in structure to the human erythrocyte spectrin α and β subunits. These experiments established that nonerythroid cells contain polypeptides that are related to erythrocyte spectrin subunits, not only in size and stoichiometry but also antigenically and structurally. Although these results were originally viewed with skepticism by some investigators, these experiments opened the door for investigations of spectrin molecules and the spectrin membrane skeleton in nonerythroid cells.

B. Early Experiments with Brain Spectrin

Studies appeared in the literature about the same time as the discovery of nonerythroid spectrins, which, in retrospect, provided insights about the spectrin molecules found in brain cells. A high-molecular-weight protein, which bound the calcium-binding protein calmodulin, contained subunits of 235 kDa and 230 kDa, and had the ability to bind f-actin, was purified from bovine brain tissues. Although this complex resembled spectrin in subunit molecular weight, it did not react with erythrocyte spectrin antibodies and was termed calmodulin-binding protein I (CBP-I). Additionally, a 240-kDa polypeptide, which was able to bind f-actin and stimulate actomyosin Mg^{+2}-adenosine triphosphatase (ATPase) activity, was isolated from pig brain tissue by calmodulin-affinity chromatography and was called brain actin-binding protein (BABP). Finally, two axonally transported actin-binding proteins of 250- and 240-kDa molecular weight were found to be concentrated at the internal periphery of neurons, Schwann cells, and a variety of nonneural cells. This protein complex was termed fodrin because of its location lining the cell membrane. Subsequent to these results, each of these proteins were shown to be related to erythrocyte spectrin. Thus, these experiments provided the knowledge that brain spectrin bound calmodulin to its 240-kDa α subunit, associated with f-actin and was axonally transported, and that related molecules were present in a variety of nonneural cells. These experiments also provided some confusion because of the various names that had been given to the same protein. Although the name fodrin is sometimes still used to mean brain spectrin, emerging practice favors the term spectrin, prefaced by the tissue (and often the species) of origin followed by the molecular weights (in kilodaltons) of the α and β subunits in naming the spectrin molecules.

II. Brain Spectrin

A. Structure

Spectrin has been isolated from brain tissue of a variety of species including chicken, pig, mouse, rat, bovine, and human. In general, the brain spectrin from this variety of species contains high molecular subunits of 240 kDa (α) and 235 kDa (β), which are present in a 1 : 1 mol/mol stoichiometry. Examination of the physical properties of brain spectrin demonstrate that it is a highly asymmetrical molecule with a calculated molecular weight of ~970,000. These experiments indicate that, similar to erythrocyte spectrin, the spectrin isolated from brain tissue exhibits properties of an $(\alpha\beta)_2$ tetrameric unit.

Isolated brain spectrin takes the shape of a long, flexible rod of ~200 nm contour length when visualized by electron microscopy. The morphology of the brain spectrin revealed by these studies is that of two $\alpha\beta$ heterodimers found in a head-to-head interaction with the two strands of the spectrin interwoven into a tight double-helix structure containing few gaps. In this regard, brain spectrin is nearly identical to the erythrocyte spectrin, although the latter forms a helical structure that is more loosely woven than that of the brain molecule. The head-to-head alignment of the heterodimers is indicated by the bilateral symmetry of binding sites for monoclonal antibodies against brain spectrin, f-actin, ankyrin, calmodulin, and synapsin I. These studies showed that when two copies of any one of these molecules are bound to brain spectrin, the binding sites appear on opposite strands, equidistant from the center of spectrin tetramer. Thus, this bivalent, symmetrical binding of proteins along the brain spectrin tetramers demonstrates the head-to-head association of the $\alpha\beta$ spectrin heterodimers. A model of the brain spectrin tetramer summarizing its structure and protein-binding sites is presented in Fig. 1.

Isolated α and β subunits of brain spectrin will reassociate to form $(\alpha\beta)_2$ tetramers. Hybrid molecules can be formed by mixing erythrocyte β subunits with brain α subunits. These hybrid molecules form tetramers that retain the helical morphology, actin-binding capacity, and ankyrin-binding capacity of the parent molecules. This observation is a clear indication of the relatedness of the erythrocyte and nonerythroid molecules.

Peptide mapping analyses of brain spectrin from mammalian species have shown that both α and β subunits are substantially different from the α and β subunits of the erythrocyte spectrin. In contrast, in avian species homologies are limited among the β subunits, whereas the α subunit of spectrin is nearly identical in all cells. This observation may indicate that a single α spectrin gene is expressed in this species. Although the α subunit of mammalian brain spectrin is more homologous to the avian α spectrins than to mammalian erythrocyte spectrin, there is clearly a difference between mammals and birds in that mammalian brain cells appear to express α and β spectrin subunits that are different from their erythroid counterparts.

Examination of the primary sequence of brain spectrin has relied on analyses of nucleic acid sequence from cloned DNA molecules representative of the messenger RNA (mRNA) (complementary DNAs [cDNAs]) encoding spectrin. Only partial sequences of the α subunit of mammalian brain spectrins are known, whereas the complete sequence for the α subunit of the avian (chicken) brain spectrin has been elucidated. The deduced amino acid sequences from the cloned DNAs, regardless of species, have demonstrated a 106-amino acid repeating motif, which is characteristic of the internal structure described for erythrocyte spectrin. These 106-amino acid repeats may each form a triple helical structure, and the spectrin molecules perhaps contain multiples of these internal structures. Comparison of the α subunit sequences has revealed a striking homology between nonerythroid spectrins from diverse species (chicken and human are ~95% identical in protein sequence). This is contrasted with the limited degree of homology seen between the nonerythroid and erythroid spectrin subunits within the same species, which is on the order of 50–60%. The high degree of conservation among the nonerythroid spectrin α subunits from diverse species suggests that most regions of spectrin are required for its function, possibly because it must maintain the ability to bind with a variety of proteins including multiple β subunits. Notably, among the many partial sequences for human α spectrin, a single human brain cDNA encoding the α subunit has shown a 60-base pair (equal to 20 amino acids) insert in the region corresponding to the center of the protein, which may be the result of an alternate splicing event. However, this result remains to be rigorously examined.

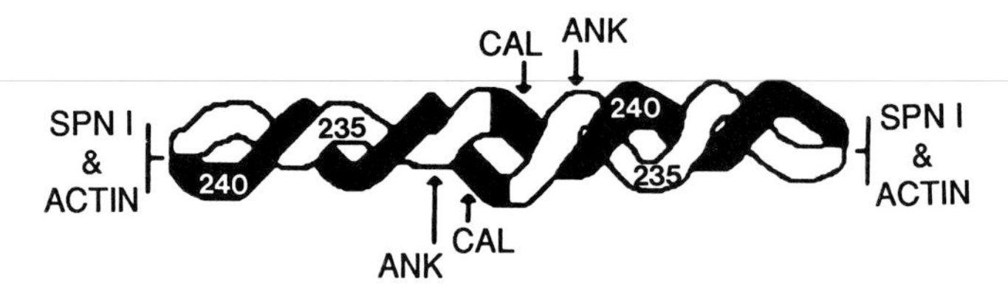

FIGURE 1 Model of the brain spectrin molecule. The morphology of the 200-nm brain spectrin tetramer is shown, along with the bivalent binding sites for calmodulin (CAL), ankyrin (ANK), actin, and synapsin I (SPN I). [Reproduced, with permission, from S. R. Goodman, B. M. Riederer, and I. S. Zagon, 1986, *Bioessays* **5**, 25.]

Inspection of the deduced sequence for the avian brain α subunit revealed that this molecule is composed of 22 distinct domains numbered $\alpha1$–$\alpha22$. The domains numbered $\alpha1$–$\alpha9$ and $\alpha11$–$\alpha19$ demonstrate compliance with the 106-amino acid internal repeat units. There is a nonhomologous region of sequence at each terminus, presumably representing functional regions for actin-binding and spectrin self-association. Domain $\alpha10$ is shortened and domain $\alpha11$ is lengthened, relative to the 106-amino acid motif. These nonhomologous domains may also represent specialized sequences, conferring functional capacities, since neutral calcium-dependent proteases have been demonstrated to specifically cleave the molecule within this region. Additionally, the calcium-binding protein calmodulin has been demonstrated to bind within the region of the molecule between $\alpha11$ and $\alpha12$ in mammalian α subunits; however, a region of sequence near the carboxyl-terminal region of the avian α subunit appears to resemble closely a calmodulin-binding domain in structure. Comparisons of the α subunit sequence with databases of protein sequences has revealed that a region of sequence homology with phospholipase C and the noncatalytic regions of the src tyrosine kinases lies within domain $\alpha10$. These studies also revealed that both dystrophin and α-actinin display regions of repetitive structure similar to the spectrin internal repeat units. While the significance of these findings is not yet clear, it is intriguing that these three proteins, each having the capacity to bind f-actin, share apparent structural conformations. They may represent a distinct class of actin-binding proteins related to a special function.

Molecular cloning and sequencing experiments directly examine the structure of the 235-kDa β subunit of mammalian brain spectrin. cDNA clones representing two distinct β spectrin molecules have been characterized from mouse brain tissue. Amino acid sequence deduced from clones representing one class of the brain β subunits exhibits ~50–60% homology with the human red blood cell β spectrin sequence. This class β spectrin is encoded by a 9-kilobase (Kb) mRNA transcript in brain tissue. In contrast, clones representing a second class of brain β subunits are encoded by an 11 Kb mRNA transcript and demonstrate strong homology (~90–95% identity) with erythroid β spectrin sequences. However, this class of brain spectrin molecules is clearly different from its erythroid counterpart because of the difference in size between the two polypeptides (235 kDa compared with 220-kDa for red blood cell β spectrin) and the difference in size of the mRNA's encoding these molecules (11 Kb in brain compared with 8 Kb mRNA transcript in erythroid tissues). Although, the exact nature of diversity among the classes of 235-kDa β subunits of brain spectrin will not be totally clear until the complete structures for the different molecules are elucidated these observations are consistent with studies (see Section II,b) demonstrating that brain cells contain spectrin isoforms showing variation in the β subunits.

Similar to its erythroid counterpart, the β subunit (235 kDa) of brain spectrin can be phosphorylated *in vitro* and *in vivo* by cyclic adenosine monophosphate (cAMP)-independent and calcium–calmodulin-dependent protein kinases. It has also been suggested that this subunit can be phosphorylated by the tyrosine kinase activity of the insulin receptor. The significance of β subunit phosphorylation is not clear, although the addition of phosphate may disrupt the molecular interactions of this polypeptide, such as its association with the α subunit, or regulation of f-actin-binding.

B. Isoforms

The discovery of mammalian brain spectrin isoforms actually resulted from a reconciliation of seemingly contradictory data. On the one hand, immunohistochemistry experiments used antibodies, which recognized axonally transported spectrin (the antibody reacting primarily with the 240-kDa α subunit of the 240/235-kDa doublet), and localized the protein in the cortical cytoplasm of guinea pig neuronal cell bodies, dendrites, and axons in the peripheral nervous system. On the other hand, an antibody made against mouse red blood cell spectrin that reacted with the 240- and 235-kDa spectrin subunits among total mouse brain proteins localized brain spectrin to mouse cell neuronal cell bodies and dendrites, but no reactivity was detected in axons. These data presented the rather interesting question of how two antibodies, both of which had been demonstrated to react specifically with the brain spectrin subunits, could demonstrate very different localization of spectrin in the brain cells. The answer to this question came from careful preparation of antibodies against isolated mouse brain spectrin from a fraction enriched in synaptic–axonal membranes and mouse erythrocyte spectrin, and use of these antibodies in parallel studies to localize spectrin within brain tissue. After cleaning these antibodies through an affinity column to which spectrin from the opposite tissue was attached, the antibody made from the synaptic–axonal spectrin reacted with the 240-kDa α subunit of the synaptic–axonal spectrin but did not cross-react with erythrocyte spectrin. In contrast, the erythrocyte spectrin antibody did not show reaction with the synaptic–axonal spectrin while retaining strong reaction with the erythrocyte spectrin. These two antibodies revealed two different localizations of spectrin when used to stain mouse brain tissue in parallel immunohistochemistry experiments. Brain spectrin (240/235), reacting with the synaptic–axonal spectrin, was found to be enriched in mouse neuronal axons and to a lesser extent in cell bodies, but did not stain glial cells. Conversely, brain spectrin (240/235E), so named because the erythrocyte antibody reacted with the 235-kDa subunit and was thus erythroidlike, was present in mouse neuronal cell bodies and dendrites and in glial cell types. These findings demonstrated that at least two-immunologically distinct spectrin isoforms exist in mouse brain. Recently, using similar techniques, a third isoform named brain spectrin (240/235A) has been found exclusively in astrocyte soma and processes. Significantly, these isoforms exhibit the same localizations in all mammalian brain tissues examined to date, including human. A summary of the location of brain spectrin (240/235) and (240/235E) within a single neuron is presented in Fig. 2.

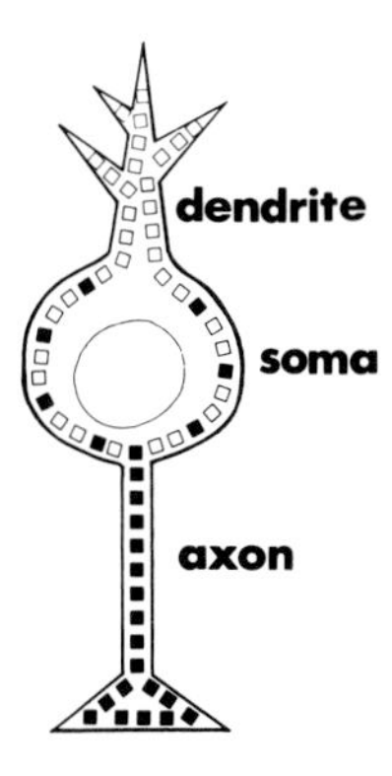

FIGURE 2 Summary of the distribution of spectrin isoforms within the mammalian neuronal cell. Brain spectrin (240/235E), detected with a red blood cell spectrin antibody, is localized throughout the dendrites and cell body (□). Brain spectrin (240/235) is found primarily in the axon and to a lesser extent in the cell body (■). [Reprinted, with permission, from S. R. Goodman, B. M. Riederer, and I. S. Zagon, 1986, *Bioessays* **5,** 25.]

A detailed examination of the spectrin isoform distribution in brain tissue by immunohistochemistry techniques using electron microscopy revealed that brain spectrin (240/235E) was associated with the cytoplasmic surface of the plasma membrane and with organelle membranes, including mitochondria, endoplasmic reticulum, and the nuclear envelope. This isoform was located in the synaptic spines of dendrites exhibiting strong association with postsynaptic densities. Brain spectrin (240/235) was found to be in axons and presynaptic elements, where it was associated with the cytoplasmic surface of the plasma membrane, organelle membranes, synaptic vesicles, and cytoskeletal structures. These findings are consistent with the observation that this isoform is transported down the axon, with specific neural cell structures. Interestingly, both of the brain spectrin isoforms were found to be associated with microtubules, neurofilaments, and actin filaments throughout the neuronal cytoplasm indicating that although these isoforms exhibit very different localizations within the cell, there is a continuum of contacts between the spectrin membrane skeletal apparatus and the cytoplasmic organization within the cell. Moreover, the discovery that two spectrin isoforms exhibit specific compartmentalization within a single neuronal cell would suggest that brain spectrin and perhaps the skeletal membrane complex is more structurally and functionally versatile than that ascribed for erythroid cells. The location and associations of brain spectrin revealed by electronmicroscopy are summarized in Fig. 3.

There is a differential expression of the spectrin

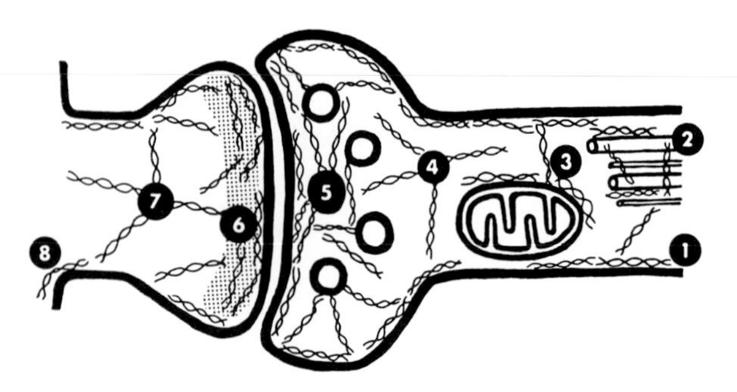

FIGURE 3 Summary of spectrin isoform location at an axo-dendritic synapse. This model summarizes the location and interactions of spectrin isoforms at an axo-dendrite synapse based on immuelectron microscopy. The locations of brain spectrin (240/235) are demonstrated at positions 1–5, and brain spectrin (240/235E) are shown at positions 6–8. 1, axonal plasma membrane; 2, neurofilaments and microtubules; 3, mitochondria; 4, cytoplasmic cytoskeleton; 5, vesicles and presynaptic membrane; 6, postsynaptic densities and postsynaptic membrane; 7, dendritic spine shaft; 8, dendritic plasma membrane. [Reprinted, with permission, from S. R. Goodman, K. E. Krebs, C. F. Witfield, B. M. Riederer, and I. S. Zagon, 1988, *CRC Crit. Rev. Biochem.* **23,** 171. Copyright CRC Press, Inc., Boca Raton, Florida.]

isoforms during brain development. Using specific antibodies to quantitate the amount of spectrin isoform, brain spectrin (240/235) was detected in fetal mouse brain tissues, and increased twofold in content to the levels measured in adult tissues. This isoform was enriched in the cortical cytoplasm of primary and secondary germinative neural cells and was detected within fibers resembling axons in fetal tissues. The erythroidlike brain spectrin (240/235E) was not detected in mouse fetal and neonatal brain tissues but exhibited a rapid increase in concentration during the second postnatal week. This isoform was detected in the cell body and dendrites of differentiating neurons and glial cells but was not found in mitotic cells. A similar pattern of spectrin expression has been demonstrated during avian brain development. In these experiments, the erythroid-related chicken brain spectrin isoform (240/230) was detected only at the stage of synaptogenesis of chicken brain development. The differential expression of these brain spectrins might indicate that early neural differentiation events rely on the functional attributes of the axonal brain spectrin (240/235), which is present at the origin of the neuronal tissues. Additionally, the observation that the erythroidlike brain spectrin isoform is expressed at later stages of brain development, at a time of cell specialization within the neural tissue, implies that this isoform may impart a specialized role of the associated membrane skeleton within the cell.

Taken together, these results have demonstrated the presence of spectrin isoforms within brain cells. These isoforms not only exhibit discrete localizations within a brain cell but are differentially expressed during development and specialization of the brain. Although we have concentrated on evidence concerning differences in β subunits of the brain spectrin, emerging evidence suggests that there is also a variation of α subunits within brain cells. Clearly, the organization of brain spectrin and perhaps the spectrin-based membrane skeleton is much more diverse, or complicated, than that of the erythrocyte. Whether this diversity of the spectrin in brain tissue is due to a requirement for specialized or regional functions within a single neuronal cell is, at present, not well understood. However, it is evident from these observations that experiments aimed at understanding the role of "brain spectrin" must take the diverse nature of the spectrin isoforms into consideration in assigning the functional capabilities of spectrin within brain tissues and cells.

III. Protein Interactions

In this section, we review the current knowledge of proteins that interact directly and indirectly with brain spectrin. These proteins include calmodulin, ankyrin, adducin, brain 4.1 (amelin and synapsin I), and actin. We will discuss experiments that were conducted with comixtures of brain spectrin isoforms, thus the affinity of a spectrin-binding protein to a specific isoform has not been addressed in general. Additionally, the list of brain spectrin-binding proteins is probably not exhaustive, as additional proteins are probably associated with specific neuronal structures, or specific spectrin isoforms will be characterized as examination of brain spectrins progresses.

A. Calmodulin

As described previously, the interaction of brain spectrin with calmodulin was first demonstrated by the isolation of the 240-kDa subunit by calmodulin-affinity chromatography. Subsequent studies have demonstrated that spectrin is a major calmodulin-binding protein in brain cells and that calmodulin binds with the 240-kDa α subunit in the presence of physiological concentrations of calcium. Analysis of calmodulin-binding by electron microscopy revealed that the calmodulin-binding sites on the α subunits occur at a distance of ~15 nm from the junction of the heterodimers. The exact placement of calmodulin-binding along the α subunit has been defined as within the region of sequence between $\alpha 11$ and $\alpha 12$ domains of the molecule by examination of proteins derived from bacteria, which had been induced to express this region of the α subunit from isolated cDNA sequences. These studies localized the calmodulin-binding domain to an extra arm of 36 amino acids that do not exhibit the 106-repeat consensus motif of the spectrin molecule, which is located just after the $\alpha 11$ repeat unit. Interestingly, inspection of the deduced sequence from cDNA clones isolated from chicken brain tissue implicated a sequence near the carboxyl-terminal region of the molecule as a calmodulin-binding site by homology with calmodulin-binding sites found in other proteins; however, the ability of this sequence to bind calmodulin has not been investigated. In mammals, erythrocyte spectrin does not contain a high-affinity calmodulin-binding site, although a weak interaction between calmodulin and erythrocyte β subunit has been reported. The exact role for the calmodulin–spectrin interaction remains unclear.

B. Ankyrin

The presence of an immunoreactive analogue of erythrocyte ankyrin in brain tissue was demonstrated before the discovery of brain spectrin. There are three sequence-related proteins with apparent molecular weights of 220, 210, and 150 kDa, which exhibit reaction with an antibody prepared to erythrocyte ankyrin. Recently it has been demonstrated that there are ankyrin molecules present in nonerythroid cells which do not share antigenetic sites with the erythrocyte ankyrin molecules. Thus, similar to the spectrins, ankyrin appears to represent a diverse class of related, but not identical, proteins. Brain ankyrin binds to the 235-kDa β subunit of brain spectrin with high affinity at sites that are ~20 nm from the junction of the spectrin heterodimers. Similar to the erythrocyte ankyrin, brain ankyrin can be divided into a 72-kDa spectrin-binding domain and a 93-kDa membrane-bound domain by digestion with the proteolytic enzyme chymotrypsin. Although the exact role of ankyrin–spectrin interaction in brain is not well defined, brain spectrin may be associated with the neuronal plasma membrane through a direct or indirect interaction (via ankyrin association) with N-CAM_{180}.

C. Actin

In erythrocytes, spectrin tetramers are capable of cross-linking f-actin filaments (protofilaments of ~13 actin monomers) into a two-dimensional meshwork. The direct spectrin–actin interaction is relatively weak, however, this interaction is strengthened in the presence of a second protein found in erythrocytes called protein 4.1. Similar to erythrocyte spectrin, brain spectrin binds end-on to f-actin, cross-linking the actin filaments. These studies indicate that the brain spectrin isoforms are involved in linking f-actin to the plasma membrane and organelle membranes.

D. Amelin

Initial characterizations of brain spectrin included experiments that demonstrated that this protein could bind with f-actin and that f-actin-binding was stimulated or strengthened with the addition of erythrocyte protein 4.1. These experiments suggested that brain cells might contain a protein that could stimulate the f-actin–spectrin association analogous to the erythroid protein 4.1. Using antibodies that reacted with the erythrocyte protein 4.1, an immunoreactive and structural analogue of the erythrocyte protein has been identified in mammalian brain tissue. The brain 4.1 analogue, termed amelin (Grk. *amelew,* overlook), is a single polypeptide exhibiting a molecular weight of 93 kDa as determined by polyacrylamide gel electrophoresis in the presence of sodium dodecyl sulfate (SDS-PAGE). This protein has been demonstrated to be structurally related to the erythrocyte protein 4.1 by peptide mapping studies. Amelin is localized in the cell bodies and dendrites of neurons as well as in certain glial cells, which is the same localization observed for the spectrin (240/235E) isoform. This colocalization of amelin and brain spectrin (240/235E) might reflect a difference in the affinity of specific brain isoforms for amelin and may imply that there are multiple forms of brain 4.1 proteins, which provide a specialization for the spectrin–f-actin interactions. Although amelin has been demonstrated to bind brain spectrin, these experiments have been performed with total brain spectrin, and whether or not amelin exhibits a stronger affinity to the (240/235E) isoform over the axonal (240/235) spectrin isoform remains to be determined.

E. Synapsin I

Synapsin I is a neuron-specific phosphoprotein that is associated with the cytoplasmic surface of small synaptic vesicles. This protein has the capacity to bind spectrin, and this binding stimulates the f-actin–brain spectrin (240/235) interaction. Originally, synapsin I was suggested as a structural analogue in brain of erythrocyte protein 4.1; however, subsequent analyses of this protein demonstrated that, although synapsin I is a functional analogue of the erythrocyte protein 4.1, it bears little structural resemblance to the erythroid 4.1.

Synapsin I is composed of two polypeptides of 76 and 70 kDA (termed Ia and Ib) molecular weight, which give nearly identical peptide maps independent of which mammalian species is the source of the protein. Examination of synapsin I by molecular cloning techniques has demonstrated that the synapsin I proteins belong to a family of at least four distinct proteins (termed synapsin Ia, Ib, IIa, and IIb), all of which exhibit localization to synaptic nerve terminals. Whether or not all of the synapsins mediate their function through an interaction with brain spectrin is not yet clear.

As indicated above, synapsin I is a phosphoprotein, and it has been demonstrated that the protein is differentially phosphorylated by calcium–calmodulin-dependent protein kinase II and cAMP-dependent protein kinase. Synapsin I, which has been phosphorylated by calcium–calmodulin-dependent protein kinase II, is not able to stimulate the interaction of f-actin and brain spectrin (240/235), whereas dephosphorylated synapsin I strongly stimulated this interaction. These observations suggest that regulation of the synapsin I–spectrin interaction may provide a means for modulation of movement of synaptic vesicles to the plasma membrane and, thus, a regulation of synaptic transmission.

F. Adducin

Adducin is a recently characterized calmodulin-binding phosphoprotein, which in erythrocytes is a 205-kDa heterodimer complex. Adducin stimulates the spectrin–f-actin interaction and has been demonstrated to cause f-actin bundling *in vitro*. Although these functions suggest that this protein is similar to protein 4.1, adducin and protein 4.1 appear to be independent and do not compete for spectrin–actin binding but are additive in their activities. Whether or not adducin cross-links spectrin with f-actin or creates new (altered) binding sites on the actin filament is unknown; however, the activities of adducin are downregulated by calmodulin and calcium. It has been recently demonstrated that three sequence-related polypeptides purified from brain tissue share properties with erythrocyte adducin, including the ability to bind calmodulin, to stimulate brain spectrin–actin interactions, and to react with antibodies to the erythrocyte adducin molecule. It remains to be demonstrated whether or not these molecules exhibit a localization similar to that of the brain spectrin isoforms. However, these observations suggest that calmodulin may regulate several aspects of the spectrin membrane skeletal complex.

IV. Functions

We are in the early stages of experimentation aimed at elucidating the function(s) of spectrin in brain cells. The static views of brain spectrin isoforms within individual neuronal and glial cells supplied by immunoelectron microscopy as well as the studies of its protein and membrane interactions *in vitro* have led to an initial understanding of brain spectrin function. Based on the detailed knowledge of the erythrocyte spectrin membrane skeleton, we can predict that brain spectrin isoforms, which are bound to the cytoplasmic surface of the plasma membrane and to organelle membranes, will exhibit functions of (1) giving stability to these membranes, (2) regulating their contour, (3) controlling the flip-flop of phospholipids across the membrane bilayer, and (4) limiting the lateral mobility of integral membrane proteins through the bilayer. While it is clear that in erythrocytes a single spectrin species manifests multiple functions, the finding of multiple spectrins within a single brain cell might indicate that the individual spectrins each impart some specialized function(s) within the cell. This suggestion is strengthened by the observations that the brain spectrin isoforms show distinct compartmentalization within neural cells and that certain spectrin-binding proteins exhibit a colocalization with a specific spectrin isoform. Thus, as the examinations of the functions of spectrin in brain cells proceed, it will certainly be interesting to analyze whether an individual spectrin isoform can adequately perform all the tasks required by the brain cells, or if individual spectrin isoforms have evolved to provide a specific functional capability to the cells. Although the preponderance of data supports the latter, the former cannot be totally excluded until experiments are designed to assay the individual brain spectrin isoforms.

A. Axonal Transport

Brain spectrin is synthesized in the neuronal cell body, and, thus, brain spectrin, which is found in axons and presynaptic terminals, must be transported from the cell body to the synapse. Experiments examining the movement of proteins through the axon have shown that five different populations of proteins, referred to as I–V, travel through the axon at different velocities. Brain spectrin (presumably brain spectrin [240/235]) travels down the axon at various rates, suggesting that distinct populations of brain spectrin are associated with neuronal structures. The brain spectrin traveling down the axon with the group of proteins exhibiting the slowest rate of migration, termed group V (at a rate of ~1 mm/day), has been demonstrated to be associated with large, complex structures. The predominant protein components of group V include the neurofilament proteins and tubulin, indicating that this group may include movement of the cytoskeleton. Recent studies have demonstrated that a crude particulate fraction of brain tissue when added to purified polymerized tubulin caused ATP-dependent gelation-contraction *in vitro*. This particulate fraction had microtubule-stimulated ATPase activity and moved slowly (~1 μm/min) along microtubule walls in the presence of ATP. This fraction was essential for the gelation-contraction, which is thought to be the equivalent of slow axonal transport (group V), and contained brain spectrin as a major protein component. These observations suggest an important role of brain spectrin (240/235) in slow axonal transport.

B. Synaptic Transmission

The transmission of information through the neural system occurs via the specific release of neurotransmitters from the axon of one cell, which stimulates a second cell. This neurotransmitter release occurs at a region where the two cells are in close apposition, termed the synaptic cleft, which has a width of 10–20 nm. A large number of vesicles ranging in size from 10 to 140 nm in diameter are present in presynaptic cytoplasm, closely associated with the presynaptic membrane. These vesicles are thought to contain the neurotransmitters that are released from the presynaptic terminal in response to nerve stimulation, with the resultant stimulation of the postsynaptic cell. At the cellular level, the first steps in the release of neurotransmitter would include (1) release of the vesicle, (2) vesicle translocation, (3) vesicle attachment to the presynaptic plasma membrane, and (4) fusion of the vesicle membrane with the presynaptic membrane releasing the neurotransmitter into the synaptic cleft. The demonstration that brain spectrin is in contact with the axonal plasma membrane as well as vesicle membranes suggests that spectrin may play a role in the neurotransmitter release from the synaptic vesicles. [*See* SYNAPTIC PHYSIOLOGY OF THE BRAIN.]

Characterizing the role of brain spectrin in synaptic transmission has advanced with the identification and characterization of proteins associated with the membrane of the synaptic vesicle and the characterization of the neural spectrin cytoskeleton in the presynaptic terminal of the neuron. The neuron-specific phosphoprotein synapsin I is associated with the cytoplasmic surface of small (40–60 nm in diameter) synaptic vesicles. The phosphorylation of synapsin I in response to neural stimulation is correlated with the translocation of synaptic vesicles to the presynaptic membrane and release of neurotransmitters. Synapsin I is phosphorylated *in vitro* and *in vivo* by calcium–calmodulin-dependent protein kinases and by a cAMP-dependent protein kinase, and the role of synapsin I in synaptic transmission is thought to be mediated through changes in intracellular calcium and cAMP concentrations. As previously described, synapsin I has been demonstrated as a functional analogue of the erythrocytes protein 4.1, in that it can bind ends of a brain spectrin molecule, and this binding stimulates the f-actin–brain spectrin (240/235) interaction. Additionally, electron microscopy experiments have shown that purified synaptic vesicles will bind to the terminal ends of brain spectrin (240/235) molecules. These observations indicate a direct role for brain spectrin in regulating the release of synaptic vesicles following the stimulation of the neuron.

A model for the regulation of synaptic transmission would be that in the resting state neuron, where synapsin I is thought to be in the dephosphorylated state, the small synaptic vesicles are crosslinked via synapsin I in a three-dimensional meshwork of brain spectrin (240/235) and f-actin. Under these conditions, the cytoplasm at the synaptic terminal would be a viscous gel, effectively preventing diffusion of the vesicles to the presynaptic membrane. A stimulation of the neuron causing an increase in the intracellular concentration of calcium would activate the calcium–calmodulin-dependent protein kinase at the synapse, resulting in the phosphorylation of synapsin I, which, in turn, would cause (1) the disassociation of synapsin I and synaptic vesicles from the brain spectrin–f-actin network and (2) a transient decrease in the viscosity of the synaptic cytoplasm that would facilitate the diffusion and/or active movement of the released vesicles toward the presynaptic membrane. The key points of this model, the cytoskeletal-mediated release hypothesis, are outlined in Fig. 4.

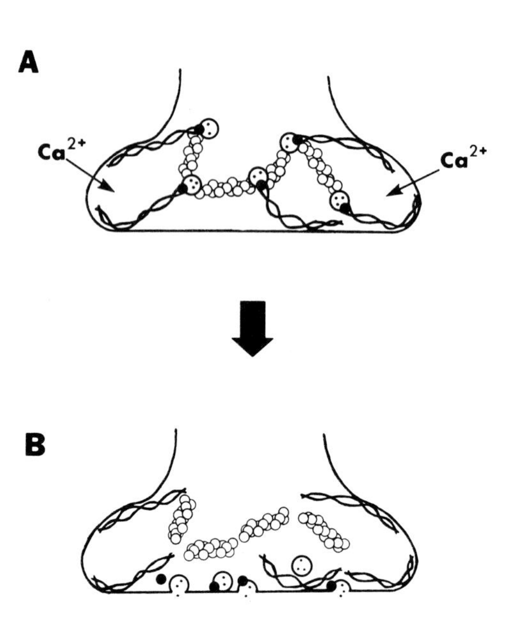

FIGURE 4 A model for the role of brain spectrin (240/235) in synaptic transmission, the cytoskeletal-mediated release hypothesis. In the resting state neuron (A) brain spectrin (240/235) (shown by the twisted lines) and f-actin (small circles) anchor small synaptic vesicles (large circles), through interaction with dephosphorylated synapsin I (filled circles), to the neuronal cytoskeleton. Upon stimulation of the neuron, calcium enters the presynaptic terminal and stimulates Ca^{+2}–calmodulin-dependent protein kinase II, which phosphorylates synapsin I. Spectrin (240/235) and f-actin become dissociated from phosphorylated synapsin I, resulting in the release of vesicles from the spectrin–actin network, and a decrease in viscosity of the synaptic cytosol, allowing the now free vesicles to diffuse toward the active zone of the presynaptic plasma membrane and fuse with the presynaptic membrane releasing neurotransmitter depicted as stars in this diagram (B). [Reprinted, with permission, from S. R. Goodman, K. E. Krebs, C. F. Whitfield, B. M. Riederer, and I. S. Zagon, 1988, *CRC Crit. Rev. Biochem.* **23,** 171. Copyright CRC Press, Inc., Boca Raton, Florida.]

C. Regulation of the Lateral Mobility of N-CAM$_{180}$

The neural cell adhesion molecule (N-CAM) has been implicated in morphogenesis of neural and nonneural tissues. In the mouse, N-CAM consists of three integral membrane proteins of 180, 140, and 120 kDa molecular weight. The extracellular amino-terminal domain of these molecules share common sequence, whereas the cytoplasmic carboxyl-terminal regions exhibit differences in length and sequence. N-CAM$_{180}$ contains the largest cytoplasmic domain and is accumulated at sites of cell–cell contact. This molecule has a restricted lateral mobility within the neuronal plasma membrane compared with the other N-CAMs, suggesting that the cytoplasmic domain may play a role in the restrictive movement of the molecule. The observation that ankyrin and brain spectrin coisolates with N-CAM$_{180}$, but not N-CAM$_{140}$, suggests that the membrane skeleton may play a role in regulating the lateral mobility of this molecule in the neuronal plasma membrane. Brain spectrin binds directly with N-CAM$_{180}$ and does not bind with N-CAM$_{140}$ or N-CAM$_{120}$. This interaction is presumably mediated by the cytoplasmic domain of the N-CAM$_{180}$ and may be specific for the brain spectrin (240/235E) isoform based on colocalization studies. Taken together, these results suggest that brain spectrin (240/235E) plays a role in limiting the lateral mobility of this essential adhesion molecule, restricting it to regions of cell–cell contact. Although the role of ankyrin in this process is unclear, these studies demonstrate a possible function of brain spectrin that is analogous to the role of spectrin on the erythrocyte membrane.

It is anticipated that in the near future, many more functions of the neural spectrin membrane skeleton will be identified, and its roles in axonal transport, synaptic transmission, and cell–cell contact will be further clarified.

Bibliography

Bennett, V. (1985). The membrane skeleton of human erythrocytes and its implications for more complex cells. *Annu. Rev. Biochem.* **54,** 273.

Goodman, S. R., Krebs, K. E., Whitfield, C. F., Riederer, B. M., and Zagon, I. S. (1988). Spectrin and related molecules. *CRC Crit. Rev. Biochem.* **23,** 171.

Goodman, S. R., Zagon, I. S., and Kulikowski, R. R. (1981). Identification of a spectrin-like molecule in non-erythroid cells. *Proc. Nat. Acad. Sci. USA* **78,** 7570.

Riederer, B. M., Lopresti, L. L., Krebs, K. E., Zagon, I. S., and Goodman, S. R. (1988). Brain spectrin (240/235) and brain spectrin (240/235E): conservation of structure and location within mammalian neural tissue. *Brain Res. Bull.* **21,** 607.

Riederer, B. M., Zagon, I. S., and Goodman, S. R. (1986). Brain spectrin (240/235) and brain spectrin (240/235E): Two distinct spectrin subtypes with different locations within mammalian neural cells. *J. Cell Biol.* **102,** 2088.

Speicher, D. W. (1986). The present status of erythrocyte spectrin structure: The 106-residue repetitive structure is a basic feature of an entire class of proteins. *J. Cell. Biochem.* **30,** 245.

Breast Cancer Biology

SAM C. BROOKS *Wayne State University School of Medicine*

ROBERT J. PAULEY *Michigan Cancer Foundation*

Glossary

Allelic loss or loss of heterozygosity Absence of one of two distinguishable alleles at a heterozygous locus in tumor DNA as compared to nontumor DNA

Antioncogene, recessive oncogene, or tumor-suppressor gene Gene whose functional product diminishes the likelihood of transformation of normal cells to tumor cells

cDNA Complementary DNA synthesized *in vitro* from an mRNA template rather than a DNA template

Epigenetic Describes changes in the genome that do not alter the structure of the genome per se, but affect its expression

Growth factor Molecule, usually a polypeptide, that influences cellular proliferation and differentiation through binding to a specific high-affinity cell membrane receptor

Hormone-responsive element DNA element which is recognized by the receptor DNA binding domain and is a 15-bp palindrome comprising two 6-bp arms separated by a 3-bp spacer

Oncogene Gene whose product potentiates the transformation of a normal cell to a tumor cell

Protooncogene Normal gene which, by alteration, can become an oncogene

Zinc finger The DNA binding domain of steroid receptors is rich in cysteines and basic amino acids. Four cysteine residues are coordinated with a zinc ion, forming a positively charged loop (i.e., finger) with the intervening basic amino acids. Two such fingers are formed in the DNA binding domain.

BREAST CANCER is a tumor of the mammary gland epithelium occurring almost exclusively in women. The tumor arises from a hyperplastic growth of the epithelium comprising the ducts, terminal ducts, or ductules. Although the exact causes of the human disease remain unknown, breast cancer has been shown to be related to endocrine factors, environmental agents, genetic anomalies, and socioeconomic discrepancies. These neoplasias are complex, displaying heterogeneity in cellular composition, hormone dependence, karyotype, genotype, metastatic potential, and prognosis. Unfortunately, the rate of death due to this disease in populations of the Western civilizations has been unchanged over the past three decades.

I. Introduction

Breast cancer is a major disease of women in the Western hemisphere. More than 130,000 women in the United States develop this disease each year, which represents approximately 30% of all cancer in women. Most disturbing is the fact that the incidence of this disease has not decreased in the past 30 years.

Several risk factors associated with increased incidence of primary breast cancer appear to be endocrine related, including (1) first full-term pregnancy after the age of 30, (2) early menarche or late menopause, and (3) nulliparity (i.e., the condition of having never given birth to a viable infant). On the other hand, oophorectomy before the age of 35 has been associated with a lower incidence of breast cancer. In addition, a genetic factor might be involved in a small percentage of these patients, as evidenced by the existence of high-risk families with a history of breast cancer. Furthermore, rates of breast cancer vary strikingly among countries, and migrants moving from low- to high-risk countries adopt the rates of the new habitat. Thus, the environment appears to have a role in this disease. Neither the use of birth control pills nor the use of estrogens in postmenopausal women has been reliably associated with increased risk for the development of breast cancer.

The hormones that regulate breast tissue have been identified. In humans it is believed that estrogen is essential for the growth of the ductal epithelium of the breast and that progesterone is required for the development of acini. Maintenance of human mammary glands also involves growth hormone and prolactin. Additionally, tissue culture investigations have implicated cortisol, insulin, and certain growth factors in the proliferation of human breast epithelium.

It would appear that the hormonal milieu or, possibly, the past exposure of breast tissue to hormonal influences could play a role in the initiation of breast tumors. While the exact nature of hormonal involvement in the human disease is unknown, certain extrapolations might be made from experiments with laboratory animals and the culture of breast cancer cells. In the Sprague–Dawley rat, for example, estrogens must be present for the initiation and growth of mammary tumors induced by chemical carcinogens. More direct evidence has been gained from experiments with cell cultures of neoplastic human breast epithelium (MCF-7 and ZR75-1). The growth of these cells *in vitro* is enhanced by estrogen, and their *in vivo* growth, when transplanted into athymic mice, depends on the presence of estrogens, as well as prolactin.

Like carcinomas of other endocrine target tissues, the growth and progression of breast tumors to more malignant stages can be influenced by hormones and antihormones. Moreover, there is direct and indirect evidence that hormones have a role in the etiology of breast cancer. Finally, recent evidence implicates dietary animal fat consumption in the incidence of breast cancer. While not conclusive, these studies show a relationship between the occurrence of breast tumors and the level of saturated fatty acids in the diet, as well as the total calories consumed.

II. Differentiation of the Mammary Gland and Susceptibility to Carcinogenesis

A. Differentiation

The human mammary gland is a complex organ. Under the influence of body growth and hormonal stimulation, its glandular structure undergoes continuous changes from birth to senescence. At birth the mammary gland contains primitive lobular structures composed of ducts and ductules lined by a single layer each of epithelial and myoepithelial cells. Throughout childhood there is little change in these structures other than growth commensurate with general body growth.

Puberty initiates the most dramatic changes in the mammary gland (Fig. 1), beginning with the growth of glandular tissue and the surrounding stroma. At this time the glandular tissue forms bundles of primary and secondary ducts undergoing repeated bifurcation. Ducts ultimately grow into terminal end buds, which proceed to form alveolar buds, a primitive form of the mature resting acinus. The alveolar buds cluster around the terminal duct in a structure termed a type 1 lobule, which contains approximately 11 alveolar buds. These buds are lined with two layers of epithelial cells, whereas the terminal end buds are lined with four layers of cells. [*See* PUBERTY.]

In the adult woman normal breast epithelium undergoes cyclic variations in proliferation, which can be measured by DNA labeling with tritiated thymidine [termed the DNA-labeling index (DNA-LI)]. Diminished DNA-LI has been observed during the follicular phase of the menstrual cycle, followed by a significant increase in the luteal phase. Cellular proliferation and cell death appear to balance in the resting breast. However, the mammary development induced by ovarian hormones throughout the menstrual cycle never fully returns to the level of the preceding cycle. Accordingly, each cycle slightly fosters mammary development, with new budding of structures occurring continuously until about age 35. This development is expressed in the appearance of two additional types of lobules in the

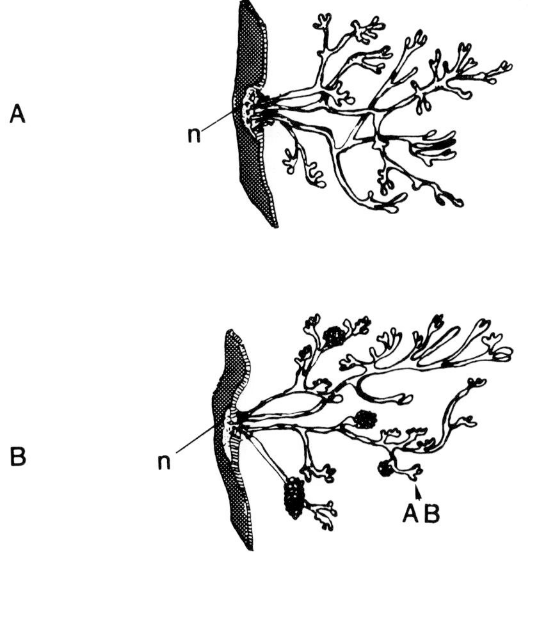

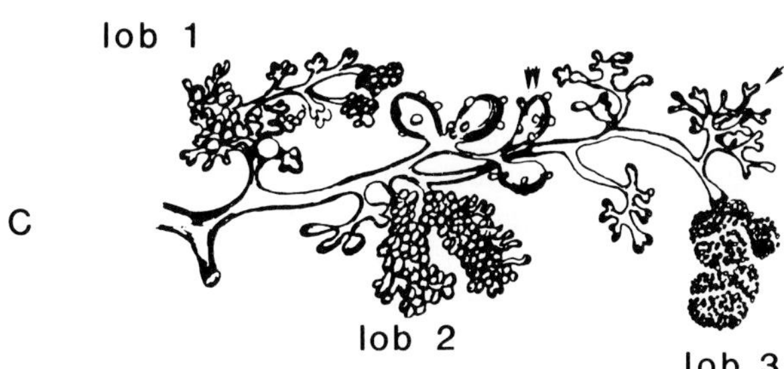

FIGURE 1 Breast development. (A) At puberty or during its onset the ducts grow and divide in a dichotomous sympodial basis, ending in terminal end buds. (B) After the first menstruation the initial lobular structures appear (lobule type 1); these are composed of alveolar buds (AB). Some branches end in terminal end buds or terminal ducts. (C) The number of lobules increases with age, and in the adult nulliparous female breast three types of lobules can be found (lobule types 1, 2, and 3). n, Nipple; lob, lobule. [From M. C. Neville and C. W. Daniel (eds.), "The Mammary Gland," by permission of Plenum Publishing Corporation.]

breast of adult women. These are designated types 2 and 3 (Fig. 1).

The gradual development of type 1 lobules into type 2, followed by further differentiation into type 3, is a process of sprouting of new alveolar buds which reach approximately 47 in number (type 2 lobules) and as many as 80 in type 3 lobules. Type 1 lobules are predominantly found in the breasts of nulliparous young women, whereas types 2 and 3 lobules are more frequent in the breasts of parous women. Measurement of the DNA-LI of these structures indicates that the proliferative activity of the breast epithelium decreases appreciably from the most active terminal end bud through each lobule type.

The systemic hormonal patterns that accompany pregnancy stimulate the breast to attain its maximum development. Growth is initiated which is characterized by proliferation of the distal elements of the ductal tree, resulting in the formation of new ductules and bringing about the development of type 3 lobules to a level of budding and degree of lobule formation beyond that seen in a virginal breast. Following pregnancy the parous organ contains more glandular tissue than if pregnancy had never occurred. At menopause involution of the glandular tissue occurs.

B. Susceptibility to Carcinogenesis

The development of experimental systems, principally the rat 7,12-dimethylbenz[*a*]anthracene (DMBA)-induced mammary tumor model, has enabled investigators to pinpoint the site of tumor induction in the breast. The similarity of mammary gland growth and differentiation between the rat and human mammary glands is significant; therefore, it is reasonable to assume that a comparable state exists during tumor initiation in these species. Furthermore, as has been demonstrated in the human disease, there is an endocrine (i.e., estrogen, progesterone, and prolactin), dietary (i.e., fat), parity, and genetic influence on the initiation of mammary neoplasia in the rat model.

Treatment of virgin female rats with the carcinogen DMBA at the age when the terminal end buds begin to differentiate into alveolar buds (i.e., 35–42 days) ultimately results in the greatest number of mammary tumors. The observation that mammary carcinomas arise from undifferentiated structures such as the terminal end bud indicates that the carcinogen requires an adequate structural target for the induction of neoplastic lesions. Benign lesions have been associated with the more differentiated mammary structures (e.g., the alveolar buds) following carcinogen administration. The higher susceptibility of terminal end buds to neoplastic transformation is attribted to the fact that this structure is composed of actively proliferating epithelium. Furthermore, autoradiographic studies show that the greatest uptake of tritiated DMBA occurs in the nucleus of epithelial cells of the terminal end buds, indicating that the highest DMBA–DNA interaction is associated with the structure with the highest proliferative rate. This observation has been corroborated by *in vitro* experiments using human breast tissue.

Terminal end buds have also been demonstrated to metabolically produce more polar metabolites of the carcinogen than the differentiated lobular cells. Such polar metabolites are required for the carcinogenic action, since they are associated with the formation of DNA adducts during carcinogenesis. Furthermore, the removal of these adducts (i.e., DNA repair) is less efficient in the terminal end buds, thereby facilitating tumor induction.

III. Steroid Hormone Metabolism

A. Androgens

More than three decades ago investigators were attempting to relate the steroid content of a breast cancer patient's urine to the prognosis of her disease. The outcome of this extensive effort was the discovery that, as a consequence of nonspecific illness or exposure to therapeutic drugs, the metabolic fate of endogeneous steroids in many patients is altered, resulting in diminished adrenal androgen excretion. A prospective study of 5000 apparently normal women on the island of Guernsey in the English Channel, has showed the prediagnosis excretion of low levels of the urinary metabolite of adrenal androgens, etiocholanolone, to be correlated with the patient's eventual presentation with breast cancer. This well-controlled comprehensive investigation, which spanned some 20 years, established that urinary androgens are indeed decreased in the breast cancer population, both before and after diagnosis. This finding might reflect a dimunition of the urinary 17-ketosteroid (etiocholanolone) levels via an increased hepatic steroid hydroxylation, principally at position 16α [*See* STEROIDS.]

This specific metabolic pathway could result from the induction of certain hepatic mixed-function oxidases (e.g., cytochrome *P*-450) by stress or drugs, bringing to mind the earlier studies of chronically ill patients. Nevertheless, the discovery, in laboratory animals, that the inducibility of certain cytochrome *P*-450 systems is genetically determined allows speculation that certain women destined to develop breast cancer are predisposed to elevated levels of hepatic mixed-function oxidase activity over an extended period. The continuous bathing of breast tissue with an altered pattern of hepatically influenced plasma steroid metabolites might be causally related to breast cancer. [*See* CYTOCHROME *P*-450.]

Examination of the plasma from women with a high risk of breast cancer (e.g., those displaying low urinary etiocholanolone 5 years before detection of disease, those who have experienced early menarche or a first full-term pregnancy above the age of 30, or those with a family history of breast cancer) has shown low androgen concentrations to be characteristic of the individuals who later present with this disease. This interesting observation, carried out on white women from western Europe or North America has not been seen in African or Asian women. Thus, the reliability of this prognostic discriminant for breast cancer could vary among populations.

B. Estrogens

Of all of the steroid hormones examined for a possible role in breast cancer, the estrogens have received the most attention. The obvious relationship between estrogens and mammary gland growth and function has prompted numerous laboratories to investigate the influence of this active steroid in neoplastic breast disease. Initially, these studies were limited to the examination of the urinary levels of estrone, estradiol, or estriol in breast cancer patients. The results from these investigations were offset by the discovery of a number of other estrogen metabolites. For the most part, these metabolites were composed of estrogens that had been hydroxylated at various positions on the molecule (e.g., 2- or 4-hydroxyestrogens and 16α-hydroxyestrone). With a more complete knowledge of the metabolic fate of estrogens in the women, it has recently become possible to attempt to relate estrogen metabolism to the disease process.

As discussed in the previous section with respect to the urinary androgen metabolites, hepatic hydroxylating enzymes might influence the pattern of estrogens in the urine and the plasma, a pattern that has been related to breast cancer occurrence. Increased 2-hydroxylation, whether of estrone, estradial, or estriol, has been associated with a lowered risk of breast cancer. On the other hand, enhanced 16α-hydroxylation of estrone has been linked to the propensity of certain women to develop breast cancer and is elevated in the urine of breast cancer patients. In addition, this reactive metabolite of estrone is formed in mice infected with the mammary tumor virus, which induces breast cancer in these animals. It has been proposed that heightened 16α-hydroxylation of estrogens at the expense of 2-hydroxylation might be a metabolic pattern of women with breast cancer or at high risk for breast cancer.

Although the levels of numerous other estrogen metabolites have been examined in the urine and the plasma, only the elevation of 16α-hydroxylated estrogens in systemic fluids remain related to neoplasia of the breast. It is known that the reactive α-hydroxyketone structure on carbons 16 and 17 of 16α-hydroxyestrone will form covalent linkages with amines, sulfhydryl groups, and the guanine moiety. In view of the fact that this metabolite binds efficiently to the nuclear estrogen receptor, it is postulated that 16α-hydroxyestrone might react with informational macromolecules within the nucleus, resulting in transformation of the breast epithelial target cell.

Steroids are also metabolized by the tissues of breast tumors. For the most part this metabolism serves to deactivate the hormones or their precursors which enter the tumor from the plasma. Once within the neoplastic target tissue, estradiol might be bound by its specific nuclear receptor, oxidized to estrone, or esterified to sulfate at the 3-phenolic hydroxyl. Both of these metabolic products are considered to be attenuated or inactive estrogens, since they do not bind to receptor at physiological concentrations. Possibly of greater importance is the observation that certain breast tumors contain the enzyme aromatase and are therefore capable of converting common plasma steroids such as dehydroepiandrosterone and androestenediol into estradiol. This *in situ* synthesis of the active estrogen is carried out to such a small degree that the contribution of the hormonal product to tumor growth is believed to be negligible.

C. Steroid Hormones as Carcinogens

To date, extensive investigations of a possible role of steroid hormones in the etiology of breast cancer have produced considerable data regarding the association of these hormones in a process which ultimately results in neoplasia. However, the available data do not support steroid hormones as causative. Present theories of carcinogenesis stipulate that the responsible agents form DNA adducts or strand breaks in order to promote transformation. Steroid hormones or their metabolites have not been shown to carry out these prerequisite functions. Furthermore, epidemiological data compiled from populations of women using oral contraceptives do not indicate a direct association between estrogens and the appearance of breast tumors. Finally, estrogen's stimulation of the pituitary gland to secrete prolactin, although implicated in rodent mammary tumorigenesis, has not been shown to be involved in the human disease.

IV. Steroid Receptors

A. Receptor Theory

Breast epithelium, like all target tissues for steroid hormones, contains nuclear proteins which bind specific steroids with finite capacity and high affinity. These proteins, termed receptors, are essential to the proliferative activity or the promotion of differentiation brought about by steroid hormones in responsive cells. Distinct receptors have been characterized, each of which is bound with high affinity by a particular steroid hormone. The structural characteristics of each hormone are recognized by the binding site of its receptor. Therefore, a cell that is responsive to a given hormone must contain the receptor specific for that steroid hormone.

The cDNA of each receptor has been cloned, and the make-up of these important gene-regulatory proteins has been determined. A superfamily of nuclear receptors has thus been identified, each containing a ligand binding domain at the carboxyl end (domain E in Fig. 2), a highly conserved DNA binding domain made up of two zinc fingers near the center (domain C in Fig. 2), and a more diverse immunogenic amino end (domain B in Fig. 2).

It is presently envisioned that steroid hormones act on normal or neoplastic breast epithelium by spontaneously diffusing into the epithelial cell from the plasma (Fig. 2). Upon their appearance in the nucleus, these hormones bind to the steroid binding domain of their specific receptors, thereby derepressing the DNA binding domain, which in turn binds to precise areas of the DNA known as the hormone-responsive elements. This interaction of the receptor complex with DNA results in the initiation of transcription of the responsive genes. The new proteins induced by this process are responsible for proliferation and differentiation of the target breast epithelial cell. [*See* DNA AND GENE TRANSCRIPTION.]

Peptide hormones function in breast epithelial cells by binding to the extracellular portion of

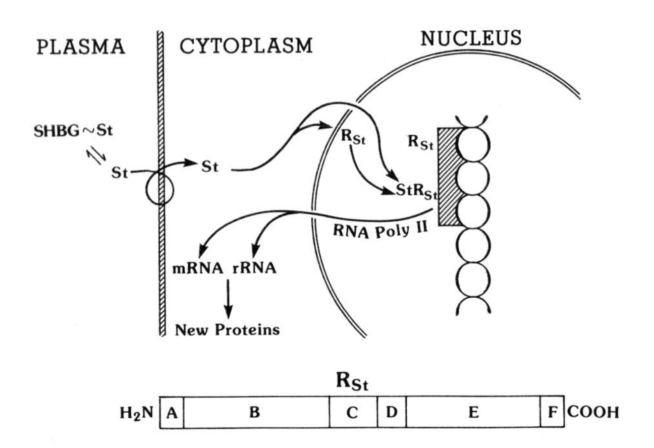

FIGURE 2 The mechanism of steroid hormone action in target tissues. Plasma free sex steroids (S_t) which are in equilibrium with those bound to sex hormone-binding globulin (SHBG) enter the target cell by diffusion. The specific nuclear binding protein of each steroid is represented by R_{st}, chromatin is indicated by a helix, and specific genes are shown by boxed material. RNA polymerase II (RNA poly II), mRNA, and rRNA are also indicated. Two of the "new proteins" specifically induced by the interaction of the estradiol-receptor complex with chromatin are estrogen receptor and progesterone receptor. The structural domains of the R_{st} are indicated at the bottom.

plasma membrane receptors and initiating the production of secondary messengers such as cAMP and the tyrosine kinases. Prolactin is an example of a hormone which causes production of cAMP. Although the role of prolactin in human breast cancer is not clear, there is ample evidence that the peptide growth factors [i.e., epidermal growth factor (EGF), transforming growth factor α (TGF-α), and insulinlike growth factor I (IGF-I)] do contribute to the process of neoplastic transformation. Many of these factors, or their receptors, are the products of protooncogenes, which are growth-promoting genes (discussed in Section VII). At this point it will suffice to mention that estrogens have been shown to induce certain of these growth factors (e.g., TGF-α) during their stimulation of breast epithelium, particularly in neoplastic tissue. [*See* POLYPEPTIDE HORMONES; TRANSFORMING GROWTH FACTOR-α.]

B. Steroid Receptors in Therapy and Prognosis

1. Therapy

The knowledge of the precise mechanisms by which particular agents regulate the growth of breast epithelium has been important to recent developments in the understanding of breast tumor growth, resulting in improved diagnosis and treatment of breast cancer. Examination of a large number of breast tumor biopsies has shown that 75% of the patients bear neoplasias which contain significant levels of the estrogen receptor (Table I). Clinical experience has demonstrated that approximately one-half of the breast tumors which contain a level of estrogen receptor greater than 10 fmol/mg of soluble protein are hormone (i.e., estrogen) dependent for growth. Patients with higher concentrations of estrogen receptors in their tumors (some are greater than 1000 fmol/mg of protein) more often respond favorably to hormonal therapy.

Although the reasons underlying these variations are unknown, there is evidence that each breast tumor is made up of both estrogen receptor-positive and -negative cells. Thus, a higher receptor value could represent a greater percentage of estrogen receptor-containing cells in the biopsy. The heterogeneity of the cellular make-up of breast neoplasias is discussed in Section V.

Tumors devoid of estrogen receptor do not respond to estrogens or the various protocols for hormone therapy and are considered to be hormone-independent neoplasias. Soon after the clinical importance of the level of estrogen receptors in breast tumor biopsies was recognized, it became accepted practice to initiate treatment of patients with estrogen receptor-negative tumors by administering chemotherapeutic agents or radiation.

The empirical experience composing decades of patient observation has ascertained that one-third of breast cancer patients respond to hormonal therapy. Furthermore, as pointed out above, only one-half of the estrogen receptor-positive tumors respond when the patient is treated by hormonal

TABLE I Estrogen and Progesterone Receptor Status of Primary Breast Tumors

	Receptor status[a]			
Biopsies	ER^+, PgR^+	ER^+, PgR^-	ER^-, PgR^+	ER^-, PgR^-
Number	112	111	6	74
Percentage	37%	37%	2%	24%

[a] Estrogen receptor-positive (ER^+) tumors contained more than 3 fmol of estrogen-binding capacity per 10 mg of tumor tissue. Progesterone receptor-positive (PgR^+) tumors contained more than 10 fmol of progesterone-binding capacity per 10 mg of tumor tissue. ER^- and PgR^-, breast tumors with binding capacity below the defined amounts.

manipulations. The most likely explanation for this phenomenon is that the neoplastic process in certain estrogen receptor-positive tumors prevents the estrogen receptor complex from its normal interaction with the genes. Indeed, mutations have been discovered in the estrogen receptor of certain unresponsive receptor-positive breast tumors. It is clinically important to determine which estrogen receptor-containing tumors will respond to hormonal therapy.

Such a determination can be achieved by recognizing that estrogen target tissues display, as one parameter of response, the induction of progesterone receptor (Fig. 2). In fact, this finding probably explains the requirement for estrogen priming to render uterine endometrium capable of responding to progesterone. With this in mind, numerous breast tumor biopsies were examined for progesterone receptor, showing that approximately one-half of the estrogen receptor-positive tumors are progesterone receptor positive (Table I). Furthermore, the clinical experience has been that breast cancers which contain both estrogen and progesterone receptors respond more often (i.e., 75%) to hormonal therapy. Patients with tumors devoid of both receptors usually do not respond to hormone therapy.

While this dual-receptor assay is helpful in predicting the response of breast tumors, it must be kept in mind that in many postmenopausal women breast tumors might contain estrogen receptor but no progesterone receptor, because endogenous estrogen is insufficient in amount to bring about induction of the progesterone receptor within the tumor. Such patients, when primed with physiological levels of estradiol, display the induction of progesterone receptors in tumors previously devoid of this protein. Finally, a small fraction (i.e., 2%) of breast cancer has frequently been shown to contain progesterone receptor in the absence of estrogen receptors (Table I). This anomaly appears to be an unnatural condition resulting from the neoplastic process.

2. Prognosis

The retention of estrogen receptor in breast tumor cells suggests a more differentiated state of the neoplasia, better prognosis, and longer survival. This, indeed, proves to be true. In fact, the prognosis is even more favorable when estrogen's capability to induce progesterone receptor is retained. More important prognostically is the ability of breast tumor cells to invade regional lymph nodes, which can be determined when the tumor is removed by surgery. Consideration of this metastatic potential of a breast cancer, together with its receptor status, yields a discriminant with the greatest prognostic value (Fig. 3).

The presence of steroid receptors might, however, simply indicate slower growth of these more differentiated tumors. Indeed, retrospective studies and studies carried out for longer periods (i.e., longer than 54 months) indicate that the prognostic advantage of patients with estrogen receptor-positive tumors eventually disappears. These findings could reflect the tendency of breast tumor cells to dedifferentiate and hence lose their capacity to generate the estrogen and progesterone receptors. Such tumors usually display a higher growth rate.

3. Metastasis

The receptor and lymph node status of primary breast tumors has proved to be of considerable value to the clinician in designing protocols for the treatment of recurrent disease. Regardless of receptor content, lymph node involvement indicates dissemination of the tumor and calls for systemic hormonal therapy and/or chemotherapy. Virtually all primary neoplasias that are initially steroid receptor negative remain negative upon recurrence. Receptor-positive cancers usually, although not always, recur as receptor-positive metastatic lesions.

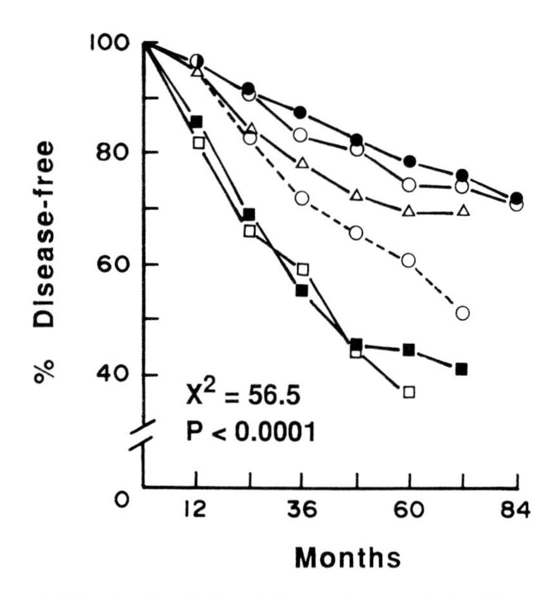

FIGURE 3 Survival and disease-free survival of breast cancer patients based on progesterone receptor (PgR), estrogen receptor (ER), and lymph node (LN) status. —○—, PgR^+, ER^+, LN^-; —●—, PgR^-, ER^+, LN^-; —△—, PgR^-, ER^-, LN^-; --○--, PgR^+, ER^+, LN^+; —■—, PgR^-, ER^+, LN^+; —□—, PgR^-, ER^-, LN^+.

Different patterns of metastatic sites are associated with the estrogen receptor status of neoplastic cells. Estrogen receptor-positive breast tumors predominantly metastasize to bone, whereas recurrent receptor-negative breast tumors most commonly involve the viscera. Metastasis to other soft tissues is the same for receptor-positive or -negative tumors. [*See* METASTASIS.]

4. Antiestrogens

Knowledge of the receptor-mediated mechanism of activity of steroid hormones has stimulated the design of compounds capable of interfering with the binding of natural hormones to receptor. Once bound, these agents form a complex which is ineffective in gene activation. Such "antihormones" have been used successfully in the treatment of a number of neoplasias of the endocrine system. The most effective antiestrogens have the structure of triphenylethylenes. They are relatively nontoxic. They have less affinity for the estrogen receptor than natural estrogens, but, at high doses, they are capable of binding to the receptor and preventing the binding of estrogen. Typical of the antiestrogens are clomiphene and tamoxifen. The latter compound has been used most extensively, particularly in the treatment of hormone-dependent breast cancer. Tamoxifen is metabolically converted *in vivo* to an even more potent antiestrogen via hepatic hydroxylation at position 4.

Following the binding of 4-hydroxytamoxifen by the estrogen receptor, the cells are unable to initiate growth or differentiation, which are induced by the natural hormone. Under the influence of this antihormone, estrogen-dependent cells remain in a dormant or nondividing state for long periods. Hormone-dependent tumors will actually regress. Laboratory experiments have shown hormone-dependent neoplastic mammary cells, which have been implanted in mice treated with tamoxifen, to reenter an active state of growth after removal of the antiestrogen. While being growth inhibitory, antihormones do not appear to be lethal.

V. Cellular Heterogeneity and Breast Cancer Progression

A. Cellular Heterogeneity

A salient property of breast cancer is the extensive phenotypic heterogeneity of the cells which compose a single tumor. Distinct subpopulations of neoplastic cells can be observed in rodent tumors induced by hormones, viruses, or chemicals, and also in human breast cancers. Variations in the cellular characteristics within a single tumor span a wide range of histological and biochemical properties (e.g., ultrastructural features) and biochemical markers (e.g., the expression of casein, keratin, type IV collagen, and a variety of tumor-associated antigens). Of importance to therapy is the heterogeneity in the cellular distribution of hormone receptors. Cells composing a single tumor might also show differences in ploidy (i.e., the number of chromosomes).

Diversity in the cellular subpopulations of breast cancers can be reflected in the behavior of the neoplasia. Thus, a single rodent tumor can yield cellular clones that differ in their ability to metastasize. Tumor subpopulations can also differ in growth rate, immunogenicity, and, as mentioned, hormone dependency. Of particular clinical significance are the reported differences among tumor subpopulations in response to chemotherapy and irradiation.

Curiously, breast tumors do not usually display the extreme behavior seen among the individual subpopulations isolated from biopsies, suggesting that characteristics of subpopulations can be modified by the other cell types which make up the environment within the tumor. Indeed, subpopulation interactions that modify growth rate, genetic stability, drug sensitivity, and ability to metastasize have been demonstrated directly in experiments with mixed tumor subpopulations. The mechanisms of such interactions include host-mediated events such as the immune response, drug metabolism, as well as tumor-mediated events such as the secretion by one subpopulation of growth factors that affect other subpopulations. The behavior of a neoplasm, then, is not just the sum of the component parts, but rather a reflection of an interlocking network of cellular subpopulations.

B. Tumor Progression

It has been hypothesized that cancer cells are "genetically instable" in comparison to their normal counterparts. It is therefore conceivable that, within a breast tumor, genetic instability might underlie the emergence of the cellular heterogeneity seen, even if the tumor was of clonal origin. Over time this evolution of new phenotypic characteristics, or tumor progression, is thought to be responsible for a cancer's becoming refractory to a therapeutic protocol or exhibiting a new property (e.g., metastasis).

Understanding the underlying mechanisms of genetic instability is clearly important to the eventual control of tumor growth and progression. Possibly contributing to this phenomenon are genomic alterations such as point mutations, base pair shifts, gene amplification, chromosomal rearrangements, and ploidy alterations. The appearance of new cellular variants might also arise from epigenetic changes, including normal differentiational processes. Regardless of the mechanism, tumor heterogeneity forces one to view breast cancer as a dynamic interacting tissue, the behavior of which is the result of a multiplicity of factors, rather than a series of linear responses or a hierarchy of cause–effect relationships.

VI. Growth Factors

Elucidation of the role of growth factors in normal and neoplastic breast cells has evolved from a number of observations. These are (1) demonstration of the presence of growth factors and their receptors in the cells, (2) assessment of the effect of growth factor supplementation on growth, (3) measurement of the effect of mitogenic steroid hormones and their antagonists on growth factors and their receptors, (4) evaluation of oncogene-transformed mammary cells for growth factor dependence, and (5) studies of the growth properties of cells following introduction of a growth factor gene. The importance of growth factors and their receptors in breast cancer resides not only in their control of cell growth, but in their potential as targets for therapy. Numerous growth factors have been described in normal and neoplastic breast tissue and cells. Only the best understood and most widely studied are reviewed here.

A. EGF and TGF-α

Although EGF and TGF-α are distinct protein molecules, they are discussed together since both bind to the EGF receptor. EGF is a biologically active 53-amino-acid polypeptide produced by a variety of normal mouse tissues. TGF-α is 50-amino-acid polypeptide originally described in retrovirus-transformed rodent cells, which, like EGF, stimulates the growth of cultured cells. TGF-α and EGF are antigenically distinct and have significant, but not complete, sequence homology. TGF-α could be a form of EGF that is inappropriately expressed in the neoplastic state.

EGF and TGF-α bind to the EGF receptor with the same affinity. EGF receptor, a product of the *EGFR/ERBB1/HER1* gene, is a 180-kDa transmembrane protein characterized by an extracellular ligand binding domain, as well as transmembrane and cytoplasmic tyrosine kinase domains. The cytoplasmic domain contains sites for tyrosine phosphorylation. The tyrosine kinase activity of EGF receptor is initiated by ligand binding and brings about a postreceptor signaling pathway that is initiated by autophosphorylation of the EGF receptor. Following internalization and degradation of the ligand–receptor complex, the phosphorylation of other cellular proteins takes place, resulting in the enhanced expression of nuclear protein genes (*MYC* and *FOS*) that influence cellular proliferation.

EGF and TGF-α both have important roles in normal and neoplastic mammary gland development in rodents and humans. Short-term cultures and established cell lines of normal breast cells, when cultured in a defined medium, require EGF for cellular growth. Biologically active TGF-α has been detected in chemical carcinogen-induced rat mammary tumors, in most primary human breast tumors, and in several human breast cancer cell lines. Murine mammary cells transfected with the TGF-α gene acquire a transformed phenotype *in vitro* (colony formation in soft agar) and tumor formation *in vivo,* indicating that TGF-α is an important determinant of the neoplastic phenotype.

The concept that EGF or TGF-α and the EGF receptor have a role in breast cancer is further supported by observations of EGF production and an increased quantity of EGF receptor in some human breast cancer cell lines. In addition, elevated levels of the EGF receptor in human breast tumors has been correlated with increased rates of proliferation and with a clinically more aggressive breast tumor phenotype. Recently, it has been postulated that TGF-α is an autocrine modulator of human breast cancer cell growth. [*See* TRANSFORMING GROWTH FACTOR-α.]

B. TGF-β

TGF-β can either stimulate or inhibit proliferation, depending on the cell type. Originally described in retrovirus-transformed rodent cells, TGF-β is distinct from TGF-α in its structure, the biological response it elicits, and the receptor to which it binds.

TGF-β protein is composed of two 112-amino-acid subunits, which can be identical or different, linked by disulfide bonds. TGF-β is synthesized and released from certain cells in association with other polypeptides, which render it inactive. A possible primary regulatory mechanism for this growth factor is the requirement for its conversion to an active form, probably by proteolysis. A unique high-affinity receptor for TGF-β is expressed on nearly all cell types, including epithelial cells. Limited characterization of the TGF-β receptor indicates that this receptor lacks a tyrosine kinase domain, implying the necessity for a different postreceptor signaling pathway.

Expression of TGF-β has been demonstrated in normal as well as malignant human breast cells. It has been postulated to have a role in hormone-dependent breast cancer growth because antiestrogens that inhibit the growth of MCF-7 cultures also increase TGF-β secretion. No correlation, however, has been demonstrated among TGF-β expression and hormone receptor status, cellular growth rate, or tumorigenicity. [*See* TRANSFORMING GROWTH FACTOR-β.]

C. IGF-I and -II

IGF-I and -II correspond to human somatomedins C and A, respectively. The mature 70-amino-acid IGF-I and 67-amino-acid IGF-II are encoded by distinct structural genes. Both IGF-I and -II have been postulated to stimulate cellular growth and metabolism in an autocrine fashion.

The receptors for IGF-I and -II are structurally and functionally different from each other and from the insulin receptor. IGF-I receptor, of approximately 350 kDa, is composed of four subunits. The two extracellular α subunits bind, in order of decreasing affinity, IGF-I, IGF-II, and insulin. Disulfide bonds link the α subunits to two transmembrane β subunits, which contain intracellular tyrosine kinase domains. IGF-II receptor is a 250-kDa single polypeptide composed of a transmembrane protein with high, low, and undetectable binding for IGF-II, IGF-I, and insulin, respectively. The IGF-II receptor is devoid of tyrosine kinase activity, has high sequence homology to the mannose 6-phosphate receptor, and might not mediate the biological activity of IGF-II. The biological significance of the IGF-II receptor remains to be established. In addition, there are extracellular IGF-binding proteins with high ligand affinity. These binding proteins could complicate the identification of IGFs and IGF receptors in tissues and cells and also could influence the ability of IGFs to be biologically active.

The insulin requirement for the growth of normal murine and human mammary cells *in vitro* is replaced by IGF-I, indicating that IGF-I is acting through the IGF-I receptor. IGF-I receptors are elevated in breast cancer as compared to normal or benign breast tissue. Human breast cancer cell lines have IGF-I or -II receptor, or both. Furthermore, these cell lines are capable of producing IGF-I, as well as IGF-II, mRNA and the secreted protein. It is likely, therefore, that the growth characterization and the response of normal and neoplastic breast cells to these growth factors is dependent on the amount of each of the interacting components, IGFs, receptors, and IGF-binding proteins, in the cell itself and in its environment.

D. Other Growth Factors

Although breast carcinoma is derived from epithelial cells, these tumors also contain stromal tissue, which is composed of fibroblasts, endothelial cells, and infiltrating lymphoid cells. The excessive proliferation of stromal fibroblasts, resulting in collagen formation and desmoplasia, might be the result of growth factor production by the transformed epithelial component of the tumor. In fact, breast cancer cell lines produce a growth factor with mitogenic activity that is similar, if not identical, to the platelet-derived growth factor. This is a potent growth enhancer for connective tissue-derived cells, which possess high-affinity PDGF receptors. It is possible that fibroblast growth factors (FGFs) or heparin-binding growth factors are important in breast cancer, because a member of the FGF gene family, *INT2,* is a murine and a putative human breast cancer oncogene. There is no direct evidence to support a role for FGFs in breast cancer growth, but FGF action on tumor stroma might, in turn, influence breast tumor growth.

VII. Oncogenes and Antioncogenes

A. *HRAS* Gene

Several lines of evidence suggest that alteration of the *HRAS* protooncogene to an oncogene contributes to the neoplastic progression of mammary cells. In carcinogen-induced murine mammary tumors mutational activation to a transforming form of *HRAS* is frequently observed. Also, mouse mammary cells are transformed when infected with the activated viral *H-ras* oncogene, and in transgenic mice containing either the v-H-*ras* or a mutationally activated *HRAS1* oncogene. However, this transformation probably requires cooperating phenotypic changes (e.g., *MYC* expression). [*See* ONCOGENE AMPLIFICATION IN HUMAN CANCER.]

Human breast cancer cell lines and human breast tumors have been examined for mutationally activated members of the *RAS* gene family (e.g., *HRAS, KRAS2,* and *NRAS*), for amplification of the protooncogenes, and for overexpression of mRNA and p21 protein. Mutational activation has been rarely observed in cell lines and primary breast tumors, and it is thought not to be involved in human breast tumorigenesis. Conflicting reports concerning overexpression of the *RAS* protooncogene at the level of either mRNA or p21 antigen do not allow any conclusion regarding expression of *HRAS1,* and *KRAS2* or *NRAS,* and the development of human breast tumors or their phenotypic properties. The observation that an immortalized human breast cell line is only weakly transformed by the viral (i.e., activated) *H-ras* suggests this oncogene might require other concurring alterations for manifestation of the transformed phenotype.

B. *MYC* Gene

A role for the *MYC* oncogene in mammary neoplasia is implied from the development of mammary tumors in transgenic mice containing activated *MYC* constructs. Approximately 30% of human breast tumors contain amplified *MYC* protooncogenes, whereas about 15% have *MYC* gene rearrangements. Neither of these markers has consistently correlated with the behavior of the primary tumor in terms of short-term prognosis for the patient. A high proportion, about 75% of human breast tumors, have higher levels of *MYC* RNA expression than benign breast tissue, and in many cases gene amplification and rearrangement correlated with enhanced expression. The precise role for *MYC* expression during breast tumorigenesis has not been elucidated. Elevated *MYC* expression could reflect the proliferative status of tumor cells and the presence of lymphoid cell infiltrates.

C. FGF-Related Genes *INT2* and *HSTF1*

Mammary tumors occurring in mouse mammary tumor virus (MMTV)-infected cells contain acquired MMTV proviral DNA that acts as an insertional mutagen activating adjacent *int2* and *int1* protooncogenes. An activated oncogene identified in human stomach tumor DNA transfected into NIH/3T3 cells, designated *HSTF1,* is proximal to, by about 1 megabase, *INT2* on human chromosome 11. These genes are members of the basic FGF gene family. Whether the *INT2* and *HSTF1* gene products display the mitogenic and angiogenic activities of basic FGF, as well as the tumorigenic activity of a basic FGF-transforming gene construct, remains to be elucidated.

Human *INT2* and *HSTF1* genes are components of an amplified region of contiguous DNA in approximately 20% of breast tumors. The significance of DNA amplification remains to be determined, because there is only limited evidence for *INT2* or *HSTF1* RNA expression. Some evidence suggests that *INT2* amplification correlates with poor prognosis in terms of local recurrence or distal metastasis.

D. *ERB*B2/HER2/*neu NGL* Gene

The transforming v-*erb*B oncogene, the human EGF receptor gene, and the rat *neu* oncogene are closely homologous. The *neu* oncogene encodes a 185-kDa transmembrane domain that differs from the nontransforming gene c-*erb*B2 by a single mutation. The observation that transgenic mice with the rat *neu* oncogene have a high incidence of early-appearing mammary tumors has implicated *neu* and its protooncogene counterpart, *erbB2,* in the initiation and progression of mammary tumors. The *erbB2/neu* receptor does not bind EGF, and the ligand for this receptor has not been identified.

Several studies have demonstrated amplification of the chromosome 17q21 *ERBB2* locus in primary tumors and in numerous human breast cancer cell lines. The amplified segment generally includes the *ERBA1/THRA* locus at chromosome 17q22, which is homologous to the viral *erb*A oncogene and related to the thyroid hormone receptor gene THR. Amplification and increased *ERBB2* mRNA and cell surface antigen expression occur in approximately 20% of primary breast tumors. *ERBB2* amplification might be correlated with shorter disease-free survival in a patient population, but the evidence remains controversial.

E. Antioncogenes

The concept that antioncogenes have a role in human breast cancer stems from several lines of indirect evidence. For example, mothers of children with osteosarcomas, a second primary tumor observed in individuals with bilateral retinoblastoma and caused by alteration of the *RB1* antioncogene, are at increased risk for breast cancer. There is an increased risk for breast cancer development in first-degree relatives of women with breast cancer, particularly in families with at least two individuals with premenopausal breast cancer. Furthermore, diploid primary breast tumors have a variety of chromosomal alterations involving specific chromosomes. The strategy to identify antioncogenes has involved identification of genes for which there is loss of an allele in tumor DNA as compared to normal cell DNA. The allelic loss is usually detected by the loss of a region on a chromosome, which can be mapped using multiple markers. [*See* TUMOR SUPPRESSOR GENES.]

Recent studies have shown, in breast tumors and breast cancer cell lines, gene loss at several loci; the *RB1* locus on chromosome 13q14, the p arm of chromosome 11 proximal to the HRAS locus and the β-globin locus at p15.5, the q arm of chromosome 1 in the region q23–q32, and the p arm of chromosome 17 in the region of p13.3. The low frequency of loss of heterozygosity reported, for example, on chromosomes 1 and 11, might, in part, be due to a long separation between the markers and the antioncogene and to the tumors' being mixed populations of cells.

Analysis of a few primary breast tumors and breast cancer cell lines indicates specific chromosome 13 alterations to the *RB1* locus in approximately 15% of the samples examined, based on analysis for *Rb* locus expression. Many human breast cancer cell lines (e.g., MCF-7, BT20, T47D, and ZR-75) lacked *Rb* locus alteration by restriction mapping. In several breast cancer cell lines (e.g., MDA-MB-468, MDA-MB-436, BT-549, and DU4475) there were homozygous internal or 3′ deletions, homozygous total deletions, or duplication of a portion of the *RB1* gene that altered transcript and protein expression. Together, these observations suggest that mutational alteration of the *RB1* antioncogene is implicated in breast tumorigenesis.

VIII. Chromosomal Abnormalities

Karyotypic characterization of most human breast cancers has been limited to enumeration of the chromosome complement. In general, the modal chromosome number is abnormal (i.e., aneuploid), with few diploid cells. Identification of chromosomal alterations by chromosomal banding has been limited generally to established breast cancer cell lines, because direct karyotyping of tumor biopsies is technically difficult. Karyotypes observed in established cell lines might differ from the primary tumor karyotype. Nevertheless, recurring sites of chromosomal changes in breast cancer have been reported. Chromosome 1 has frequently been observed to be overrepresented, and the q arm has been observed to undergo translocation. Similarly, chromosome 11 is structurally altered in many human breast cancer cells. However, the translocation breakpoints and the chromosomal partner are variable. Chromosomes 6 and 7 and, to a lesser extent, chromosomes 3 and 9 also show frequent structural alterations in human breast cancer cells. In contrast to chromosomal translocations in cancers such as Burkitt's lymphoma and retinoblastoma, in which karyotypic alterations have contributed to the demonstration of specific genetic changes to protooncogenes and antioncogenes, karyotypic changes in breast cancer cells have not been correlated with specific gene alterations. [*See* CHROMOSOME PATTERNS IN HUMAN CANCER AND LEUKEMIA.]

Bibliography

Bradshaw, R. A., and Prentis, S. (eds.) (1987). "Oncogenes and Growth Factors." Elsevier, New York.

Brooks, S. C., and Singhakowinta, A. (1982). Steroid hormones in breast cancer. *In* "Special Topics in Endocrinology and Metabolism" (M. Cohen and P. Foa, eds.), Vol. 4, p. 29. Liss, New York.

Brunner, N., Zugmaier, G., Bano, M., Ennis, B. W., Clarke, R., Cullen, K. J., Kern, F. G., Dickson, R. B., and Lippmann, M. E. (1989). Endocrine therapy of human breast cancer cells: The role of secreted polypeptide growth factors. *Cancer Cells* **1**, 81.

Callahan, R., and Campbell, G. (1989). Mutations in human breast cancer: An overview. *JNCI* **81**, 1780.

Ethier, S. P., and Heppner, G. H. (1987). Biology of breast cancer in vivo and in vitro. *In* "Breast Diseases" (J. R. Harris, S. Hellman, I. C. Henderson, and D. W. Kime, eds.), p. 135. Lippincott, Philadelphia, Pennsylvania.

Kidwell, W. R., Monaham, S., and Salomon, D. S. (1987). Growth factor production by mammary tumor cells. *In* "Cellular and Molecular Biology of Mammary Cancer" (D. Medina, W. Kidwell, G. Heppner, and E. Anderson, eds.), p. 239. Plenum, New York.

Russo, J., and Russo, I. (1987). Development of the human mammary gland. *In* "The Mammary Gland" (M. C. Neville and C. W. Daniel, eds.), p. 67. Plenum, New York.

Russo, J., and Russo, I. (1987). Biological and molecular basis of mammary carcinogenesis. *Lab. Invest.* **57,** 112.

Slamon, D. J., Godolphin, W., Jones, L. A., Hoh, J. A., Wong, S. G., Keith, D. E., Levin, W. J., Stuart, S. G., Udove, J., Ullrich, A., and Press, M. F. (1989). Studies of the HER-2/*neu* proto-oncogene in human breast and ovarian cancer. *Science* **244,** 707.

Slocum, H. K., Heppner, G. H., and Rustum, Y. M. (1985). Cellular heterogeneity of human tumors. *In* "Biological Responses in Cancer" (E. Mihich, ed.), Vol. 4, p. 183. Plenum, New York.

Weinberg, R. A. (1989). Oncogenes, antioncogenes, and the molecular basis of multistep carcinogenesis. *Cancer Res.* **49,** 3713.

Wolman, S. R. (1987). Chromosomes in breast cancer. *In* "Cellular and Molecular Biology of Mammary Cancer" (D. Medina, W. Kidwell, G. Heppner, and E. Anderson, eds.), p. 47. Plenum, New York.

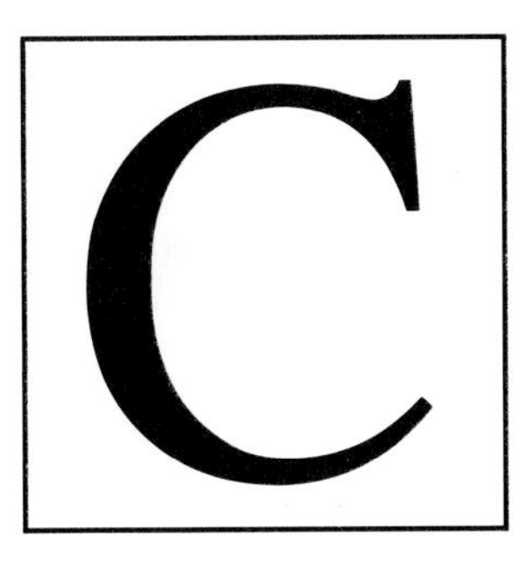

Caffeine

JOSEPH P. BLOUNT, *Widener University*

W. MILES COX, *North Chicago Veterans Affairs Medical Center, University of Health Sciences/The Chicago Medical School*

Glossary

Double blind Type of research design in which neither the subject nor the experimenter (but only some third party) knows whether the subject has received a drug (e.g., caffeinated coffee) or an inert substance (e.g., decaffeinated coffee); the advantage of this design is that it allows the investigator to isolate the pharmacological effects of the drug from the psychological effects

Epidemiological study Study that deals with the incidence, distribution, and control of a disease in a population

Ergogenic aid Substance that helps an athlete generate increased force or endurance

Etiologic science Science that deals with the causes of a disease or other abnormal condition

Mutagenic risk Factor that increases the chances that a mutation (a permanent change in chromosomes or genes) will occur

Protopathic bias Bias in an epidemiological study resulting from analyses based on nonrepresentative early measures of factors contributing to a disease without recognizing that early manifestations of the disease might have triggered the patient or physician to alter those very factors

State-dependent learning Drug-induced effect on memory that occurs when information that is learned in one drug state is later better recalled when the individual is in the same drug state rather than a different one

Statistical significance Assurance, through a statistical test, that an observed data pattern is genuine rather than having occurred by chance

Teratogen Exogenous agent, such as a drug ingested by the mother, that produces structural changes in the fetus

Theobromine Xanthine closely related to caffeine that is found especially in cacao bean and chocolate

Theophylline Xanthine closely related to caffeine that is found in tea leaves

Withdrawal symptoms Symptoms that occur when a person suddenly stops taking a drug that he or she is accustomed to taking; headache is the most common symptom associated with withdrawal from caffeine

CAFFEINE IS A PSYCHOACTIVE drug, a methylated xanthine, that occurs naturally in many foods and beverages (e.g., coffee, tea, cocoa) and is added to many commercial products (e.g., soft drinks, analgesics). It is used widely, often in high doses, and it has been the subject of much research. Four major questions have been raised regarding caffeine and its effects: (1) What are its physical and mental benefits? (2) What are the possible harmful effects of caffeine from single doses or chronic use? (3) Is caffeine consumption a social problem? (4) What are the short- and long-term mechanisms that cause its physiological and behavioral effects? Scientific studies have given us a definitive understanding of the chemical reactions, byproducts, and biological cycles involved when caffeine is metabolized by the body. Although other research results are less conclusive, they have greatly increased our understanding of the complex ways in which caffeine can produce beneficial effects in one individual but harmful effects in another, or different effects in the same individual at different times. Researchers have also shown the apparent absence of certain suspected health risks and the seriousness of others. Given the current state of knowledge, mild caution about excessive use of caffeine is prudent for the general population, reduced intake is indicated for certain clinical populations, and further research is needed in particular areas.

I. Introduction

Caffeine is almost certainly the most widely used psychoactive substance. It is present in coffee, tea, chocolate, soft drinks, analgesics, and other medications. Despite recent concerns about health risks, more than 100 billion doses of caffeine are consumed annually in the United States. This popularity is probably due to the stimulant action of caffeine. Caffeine has a long history: Written records mention a legendary Chinese emperor drinking tea in 2737 B.C. Because it occurs naturally in more than 60 plant species, many of which have been used as food sources, people may have consumed it as early as the Paleolithic period. So prevalent a substance deserves the attention that caffeine has received. In the last 20 yr, caffeine research has increased, perhaps stimulated by media attention to possible health risks. The purpose of this chapter is to present some basic facts about caffeine as they relate to human biology, to survey the major methodologies that have been used in caffeine research, and to discuss the implications of this research for health and social policy. However, because empirical knowledge about caffeine is constantly expanding, the summaries of research that we present here should be taken as the current state of the field and should be interpreted with caution.

Caffeine intake can be calculated as the product of a person's consumption of food, beverages, or other products containing caffeine × the caffeine content of these products. Such data can be obtained by having people keep diaries of what they consume or retrospectively report what they remember consuming or by having a third party record what they consume. There is substantial agreement among these different methods of collecting data. Furthermore, there is close agreement between these methods and certain measures of caffeine production, such as the amount of coffee imported per capita. These intake data reveal interesting patterns. For instance, the most prominent source of caffeine in Britain is tea, but in the United States, it is coffee. The shift in preference from tea to coffee in the United States has been attributed to the British tax on tea in the 1700s. The U.S. peak of more than 3 cups of coffee per day per capita in 1962 fell to less than 2 cups in 1983, a shift that has been variously attributed to poor advertising and/or a decline in the quality of coffees in the 1960s. Currently in the United States, females drink more coffee than males, and the younger generation gets its caffeine primarily from soft drinks. Caffeine-consuming habits also change with age. For example, 35–64-year-olds drink more coffee than younger or older age groups; on the other hand, tea consumption shows no difference across ages, and soft drink consumption declines.

The second factor necessary to calculate caffeine intake is the amount of caffeine in a source. The amount of caffeine depends on both methods of commercial production and personal methods of preparation. Tea is especially variable, ranging from 8 to 91 mg per serving. Due to such variation and to differences in the definitions of serving size (and sometimes definitions are not reported), various authors report different data on caffeine content. Nonetheless, representative values for the caffeine content of major dietary sources are useful (Tables I and II). Many people believe that the dark color of soft drinks indicates which ones contain caffeine. In actuality, caffeine forms a white powder, a yellow residue (caffeine is a xanthine [Grk. for yellow]), or a clear solution. Note that root beer is dark and contains no caffeine, whereas Mountain Dew is clear and contains caffeine.

Despite this variability, a number of authors have used the figures for caffeine content to calculate the average person's daily caffeine intake. For American adults (consumers and nonconsumers), a representative average is 200 mg (3 mg/kg body weight) per day. Adults who would be considered heavy consumers ingest 500 mg (7 mg/kg) per day or more. Some individuals who consumed 2,000 mg/day have sought professional help to reduce their intake. Note that several segments of society consume less than these figures: Pregnant women appear to consume about 2.1 mg/kg, although the data are limited and pregnancy-related weight gains have not always been taken into account. Children <18 yr old (both consumers and nonconsumers) ingest about 37 mg (1 mg/kg) per day. The health code of the Mormons, as well as that of certain other religious groups, prohibits all use of stimulants.

TABLE I Levels of Caffeine in Common Sources: Beverages and Foods

Beverage/food	Serving size (oz.)	Approximate mg caffeine/serving
Coffee		
Drip	5	150
Percolated	5	110
Instant, regular	5	50–100
Instant, flavored mix	5	25–75
Decaffeinated coffee	5	1–6
Tea		
Black		
1-min brew	5	20–35
3-min brew	5	35–45
5-min brew	5	40–50
Green		
1-min brew	5	10–20
3-min brew	5	20–35
5-min brew	5	25–35
Instant	5	30–60
Cocoa beverage	5	2–20
Soft drinks		
Jolt	12	70
Caffeinated cola drinks	12	30–65
Mountain Dew®, Mello Yello®, Sunkist Orange®	12	40–50
7-Up®, Sprite®, RC-100®, Fanta Orange®, Hires Root Beer®	12	0
Chocolate		
Cake	$\frac{1}{16}$ of 9-in. cake	14
Ice cream	$\frac{2}{3}$ cup	5
Mr. Goodbar®	1.65	6
Special dark, Hershey	1.02	23

II. Biochemistry

This section addresses the chemistry of caffeine, the processes by which caffeine is taken into the body, how it is distributed in fluids and tissues, its immediate effects on body systems, how tolerance to caffeine develops, and how it is cleared from the body.

A. Chemistry

Caffeine is an alkaline compound in the family of naturally occurring derivatives of xanthine. The three primary xanthines are caffeine, theophylline, and theobromine. All three are structurally similar with only minor variations in the number and position of methylated sites. Thus, they have similar effects on the body. Specifically, caffeine is 1,3,7-trimethylxanthine. Because it was one of the first pharmacological agents to be chemically isolated (in Germany in 1820), it is the most extensively studied xanthine.

B. Distribution in the Body

Although most other alkaloids are insoluble in water, caffeine is slightly soluble and becomes still more so in the form of complex double salts. Of greatest importance to humans is that caffeine is readily absorbed by the gastrointestinal tract after oral ingestion (99% complete within 45 min), and it is distributed into all body fluids and tissues, includ-

TABLE II Levels of Caffeine in Common Sources: Drugs

Drugs	Standard adult dose	Approximate mg caffeine/standard dose
Prescription painkillers		
Darvon® compound capsule	1	32
Cafergot® tablet (migraine)	1	100
Nonprescription (over-the-counter)		
Painkillers		
Anacin®, Midol®, Vanquish®	2	65
Plain aspirin	2	0
Cold/allergy		
Dristan®	2	30
Coryban-D®, Sinarest®, Triaminicin®	1	30
Stimulants		
No-Doz®	2	200
Vivarin®	1	200

ing the brain, testes, fetus, and mother's milk. Rates of caffeine absorption have been observed to be slower for soft drinks than for coffee, a surprising finding because the rate of absorption of alcohol is increased by carbonation. Peak blood plasma levels of caffeine are reached within 15–45 min after ingestion.

C. Effects

Caffeine has a number of effects on the nervous system, cardiovascular system, and other body systems. It is a powerful central nervous system stimulant. Moderate doses of 200 mg (about 2 cups of coffee taken close together) activate the cortex of the brain enough to show changes in a person's electroencephalogram. However, considerably higher doses or injection are required to show effects in the medulla, spinal cord, or autonomic nervous system. All such effects begin about 0.5 hr after intake and maximal effects are reached about 2 hr after intake.

Caffeine has several effects on the cardiovascular system, although the effects of caffeine are weaker and not as clinically useful as the effects of theophylline. The major effects of caffeine are on heart rate, blood pressure, blood flow, and serum cholesterol. Caffeine stimulates the heart directly by acting on the myocardium to increase the force of muscle contraction, heart rate, and cardiac output (stroke index). Such effects may be masked at low dosage levels because caffeine also stimulates the medullary vagal nuclei, which in turn tends to produce a decrease in heart rate. Single caffeine doses decrease heart rate during the first hour after administration and increase it during the following 2 hr. Chronic use appears to elevate basal heart rate so that clinically significant reductions in heart rate are obtained upon abstention from caffeine. Previous contradictory reports about effects on blood pressure have now been reconciled. Single doses can increase blood pressure among both users and nonusers; however, the size of the increase is significantly larger for nonusers. Doses from 100 to 300 mg of caffeine are enough to cause noticeable increases in nonusers, but higher doses are needed to cause such increases in users. Caffeine dilates systemic blood vessels but constricts cerebral blood vessels. The accompanying change in peripheral blood flow is too short-lived and of too small a magnitude to be of therapeutic value. On the other hand, the reduced cerebral blood flow has been used in the treatment of headache. Both acute and chronic caffeine intake seems partially responsible for increased blood levels of cholesterol. Some researchers have disagreed; however, repeated replications and demonstrations that discontinuing coffee reduces serum cholesterol and at least one finding of a dose–response relationship tend to outweigh such reservations. [*See* CHOLESTEROL.]

In addition to effects on the nervous system and circulatory system, caffeine has a number of other effects. Caffeine causes a slight increase in basal metabolic rate (10%). The increase occurs as a sharp rise in rate during the first several hours after ingestion. Caffeine increases the secretion of stomach acids, increases the respiratory rate, and slightly increases the production of urine. It acts as a bronchodilator by relaxing the smooth muscles. On the other hand, caffeine strengthens the contraction of skeletal muscles.

It has been suggested that the basis for the effects of caffeine is multifactorial. The primary mechanism seems to be that caffeine blocks receptors for adenosine. Caffeine is also supposed to block receptors for benzodiazepines and to alter the movement of calcium within cells via the cyclic nucleotides. The molecular basis for the effects of caffeine remains unclear.

D. Tolerance and Individual Sensitivity

When an individual has developed tolerance to a drug, larger dosages are required than previously to achieve a given effect. This pattern of decreasing effects can cause drug users to increase dosages to compensate for the decrease. With chronic caffeine use, it takes larger dosages to achieve the same (mild) diuretic and salivary effects. As mentioned above, caffeine users show smaller blood pressure increases than nonusers for the same caffeine dosage. Some evidence suggests that these different levels of tolerance develop after only a few days of chronic use and are lost within a day, although one would like to see this research replicated. On the other hand, less tolerance seems to develop for the stimulating effects of caffeine on the central nervous system. Furthermore, for many individuals, increases in caffeine dosage would be self-limiting because higher doses exacerbate undesirable symptoms (nervousness, anxiety, restlessness, insomnia, tremors, gastrointestinal disturbances, and feelings of uneasiness). (Withdrawal symptoms will be discussed in Section VII.)

E. Toxicity

In contrast to the high toxicity of theophylline, caffeine is not very toxic. Nevertheless, caffeine can produce symptoms that require medical consultation. For example, sudden increases in consumption have been associated with a variety of adverse effects, such as delirium, abdominal cramps, vomiting, high anxiety and hostility, and psychosis. More gradual increases may not show toxicity because tolerance develops. Higher doses of caffeine can cause convulsions and still higher doses can cause death from respiratory failure. A lethal dose in adults appears to be 5–10 g, the amount of caffeine in approximately 200 colas. There is little concern that death could occur from beverage consumption because gastric distress and vomiting would prevent concentrations from reaching life-threatening levels. Although similar principles would seem to apply to over-the-counter caffeinated drugs, at least seven deaths from ingested caffeinated medications were reported between 1959 and 1980. One death after injection of 3.2 g has also been reported.

F. Clearance

About 95% of a dose of caffeine is eliminated from the body by being metabolized to other products in the liver; the remainder is excreted unchanged by the kidneys. The exact metabolites may vary among races. For example, Orientals have different metabolites than Caucasians. Rates of clearance vary greatly. For example, clearance is much slower in infants because they lack certain enzymes. Even among healthy adults of the same race, there is wide variation. For example, it takes between 2.5 and 7.5 hr to remove 50% of a dose of caffeine from the body, and 97% is eliminated in 15–30 hr. Many factors may cause the rate of clearance to vary. For example, smoking, liver disease, pregnancy, and the use of oral contraceptives have all been found to decrease clearance. Other medications can increase or decrease clearance. When a chronic user abruptly discontinues all use of caffeine, complete removal of caffeine from the body can take up to 7 days.

In summary, a comprehensive metabolic pathway has been established for caffeine. Future work may produce minor modifications in our understanding of this pathway, but it is more likely to focus on individual differences and on the mechanisms by which caffeine achieves its various effects.

III. Research Methodology

Much of the research on caffeine has produced seemingly contradictory findings. In point of fact, the differences can be attributed to insufficient attention to methodological details. The most prominent methodological issues involve measurement, biased self-reports, confounding factors, and improper interpretation of results.

At the most basic level of measurement of quantity, there is a problem with caffeine research. Is a "cup" of coffee 5, 6, or 8 oz.? Often researchers do not report their reference volumes. Furthermore, when researchers ask members of the general public to report their consumption habits, those people may not realize that their "mug" of coffee contains two of the researcher's reference volumes—and thus they may be consuming twice the caffeine they think they are.

There are also problems at the chemical level: Measured caffeine content of beverages or foods can vary due to the analytical method employed. Caffeine has no recommended clinical chemistry because it is not routinely measured. However, a number of methods could be used, including ultraviolet spectrophotometry, liquid chromatography, thin-layer chromatography, immunoassays, and gas chromatography, as well as other methods if they too provide appropriate sensitivity, specificity, and feasibility. Such techniques unanimously show that plant variety, growing conditions, and method of preparation (for coffee and tea) dramatically affect caffeine content.

Even when measured quantities of caffeine are administered during a laboratory experiment, specifying the pharmacologically active dose is difficult because many factors affect responses to caffeine. In other words, what is a relatively small dose for one person can be a relatively large dose for another person. As a dramatic example, consider that, because of individual sensitivity and body size, a single candy bar could have the same effect on a young child as 5 cups of instant coffee would have on an adult. The important factors to take into consideration are differences in body weights, different rates of stomach emptying and intestinal absorption, and inherited caffeine-sensitivities. Furthermore, differences in chronic, occasional, or no-caffeine use; recent intake; length of abstention; and blood caffeine levels can cause differences in response to caffeine. Psychological factors, such as stress or being told (deceptively) that one was administered caffeine when one was not, also cause differences in response. A common way to rule out psychological factors and study pharmacological effects in isolation is to use the "double-blind" design. In this design, caffeinated and uncaffeinated sources are identical in appearance, and a third party schedules which source will be given to each participant. Neither the participant nor the researcher, who has direct contact with the participant, knows whether the caffeinated or uncaffeinated treatment is being used. They are both blind to the kind of treatment administered. In some studies, it has been found that participants who are "blind" and receive caffeine have different responses from those who know that they are receiving caffeine. Keeping participants blind to the nature of their treatment condition is difficult in some situations because people can taste and feel differences in treatments of 200 mg of caffeine or larger. Differences of 50 mg or smaller are not detectable by most individuals. The balanced-placebo design, a procedure that is still more refined, allows the experimenter to separate the psychological and pharmacological effects of caffeine.

Among all these difficulties in accuracy of measurement, there is a reassuring fact: A number of biochemical and other validations have shown that self-reports of caffeine habits are, for the most part, reliable. Such findings allay concerns about recall bias in retrospective studies.

Important research that cannot be done in the laboratory is, for example, studying whether caffeine contributes to the occurrence or nonoccurrence of certain diseases in human populations. No such epidemiological studies have been conducted with caffeine itself, but there are many with coffee drinking. Caffeine is only one of many active ingredients in coffee. It would be remarkable if caffeine proved to be the only substance of significance. Furthermore, those who are coffee drinkers bring confounding characteristics with them: they smoke more, consume more alcohol, are more extraverted, and have different levels of education, dietary habits, and life-styles (e.g., they exercise more). Controlling for these factors has often made apparent links between coffee consumption and health disappear.

The critical reader of an epidemiological study should watch for a number of potential difficulties. One problem is insufficient objectivity of recorded data when the researchers did not use "blinding" (explained above) or did not seek corroboration of self-reports through spouses or other sources. Another common problem is insufficient attention to protopathic bias; i.e., analyses are based on nonrepresentative early measures of factors contributing to a disease without recognizing that early manifestations of the disease triggered the patient or physician to make alterations in those very factors. Before accepting the results of these studies, the critical reader will require the researcher to have appropriately tested the statistical significance of the findings. Chance factors might always cause a pattern to appear in data that would not be observed if the study were repeated. Thus, researchers should take statistical precautions not to read meanings into patterns that are not really there. To do so would be analogous to reading meaning into tea leaves or "seeing" objects in the random shapes of clouds or a man on the moon. For researchers to say the pattern of data is compatible with their theory is not enough; they must also demonstrate that the results are statistically significant, i.e., that they are very unlikely to have occurred only by chance. Moreover, although a simple test of statistical significance might be appropriate for research that tests only one hypothesis, performing the same statistical test repeatedly on different portions of the data is improper. Instead, the researcher must use a different statistical technique designed for so-called "multiple comparisons."

When a statistically significant association is found between coffee intake and some health risk, this result by itself is not sufficient to conclude that coffee intake causes the health effect. Other explanations are possible. For example, the disease symptoms may cause the increase in coffee intake, or a third factor may cause both. Consequently, when researchers find an association, they should look further to see if the size of a dose is related to the size of a response. The lack of a dose–response relationship is usually taken as lack of causality and as a clue that other factors are involved. Another possibility is that caffeine has no direct effect but modulates the effects of other agents, either antagonistically or synergistically. While epidemiological studies must take into account one set of factors, laboratory studies of performance effects must take into account a different set of factors. For instance, caffeine has been shown to have different effects on the performance of introverted and extraverted people. Factors affecting a person's level of arousal are also important. For example, caffeine may have a different effect on the same person earlier in the day rather than later, because of the person's diurnal rhythm of bodily arousal. Finally, caffeine may have a different effect depending on whether or not a person is under stress.

In summary, caffeine research involves many complexities, and there are rigorous scientific methods for handling them.

IV. Physical Performance

People report that caffeine increases their physiological arousal. Assuming these subjective impressions reflect underlying physiological changes, one would expect effects on fine motor coordination and athletic endurance. In fact, hand steadiness has been shown to be about 25% worse after consuming 200 mg of caffeine (about 2 cups of strong coffee). Furthermore, numerous empirical reports show that caffeine impairs motor skills that involve delicate muscular coordination and accurate timing. The few studies that have failed to find effects have had small numbers of subjects or poor laboratory controls. Given that caffeine impairs fine coordination, one would expect caffeine to adversely affect skilled automobile driving, but no studies have been reported on this topic.

Many people believe that caffeine is an ergogenic aid, a substance that helps an athlete generate increased force or endurance. Witness the coffee-drinking rituals preceding marathons or the use of coffee to get through the daily grind of training. Empirical studies have shown improved work production in trained cyclists, runners, and cross-country skiers, for example, extending mean cycling time to exhaustion by 20% when cycling at 80% of maximal capacity. Such ergogenic effects occur only during prolonged work and not during short-term work episodes. When the work conditions have been varied or when dosage or caffeine habits are not sufficiently accounted for, effects have been unclear and equivocal (although none have been in the reverse direction).

Caffeine seems to delay deterioration in performance due to fatigue through both psychological and physical effects. It decreases perceived exertion, perception of fatigue, and drowsiness; it increases self-reported alertness and motivation. The mechanisms for these psychological effects may involve reduced neuronal thresholds in the central nervous system, influences on catecholamine receptors, and/or direct effects on the adrenal medulla. On the other hand, caffeine may produce its physical effects through increased fuel availability or increased contractile activity of skeletal muscles. In turn, several mechanisms have been suggested for the increase in contractile activity: enhanced transmission of nerve impulses or potentiated twitch responses in both rested and fatigued muscles. Similarly, several mechanisms have been suggested for the increase in fuel availability. For example, one proposal involves enhanced fat metabolism: Caffeine may stimulate adipocyte lipolysis via the activation of lipase. Unfortunately, research has failed to clearly demonstrate such enhanced fat metabolism and competitive athletes should have plenty of epinephrine for lipolysis, even if they do not ingest caffeine. This example demonstrates the intricacies needed in any viable explanation of the effects of caffeine.

An athlete concerned about the use of caffeine should realize that responses to caffeine vary greatly. For example, in contrast to others, more sensitive individuals may become overstimulated and show performance decrements. To obtain a beneficial effect while avoiding acquiring tolerance, the athlete might consider abstaining from caffeine for several days before a major event and then consume a moderate serving of coffee (e.g., 2 cups) 1 hr before competition. For some activities, the diuretic effects of caffeine may be a problem if maintaining hydration is important and difficult. Regular, heavy use throughout training may reduce benefits during competitive performance and may increase blood cholesterol and the risk of heart attack or other medical problems (see below). Finally, the athlete should realize that the International Olympic Committee has banned caffeine when its values are greater than 15 μg/ml in a urine test.

V. Perceptual and Cognitive Performance

In addition to physical arousal, people report that caffeine increases their mental arousal. This arousal might affect perceptual processes, attention and speed of reaction, memory, and/or the flow of thoughts. At the perceptual level, we can ask whether or not the taste of caffeine is noticeable. It has been claimed to enhance the taste of colas. However, humans in experimental settings have been unable to detect its presence or absence at common cola levels of concentration. Caffeine intensifies the taste of certain sweeteners, lowers visual luminance threshold, and improves auditory vigilance. More definitive studies that showed effects of caffeine on vision, hearing, and skin conductance would be desirable. Small to moderate amounts of caffeine (i.e., 32–200 mg) help speed reactions to simple, routinized tasks, such as indicating whether an even or odd digit was presented, pressing buttons corresponding to bulbs lit in a circular pattern, or watching for strings of three even numbers. Conversely, when habitual caffeine users abstain for 2 days, their reactions are slowed and their attention impaired. (The physical indicator of impaired attention is less anticipatory heart-rate deceleration.) For novel or slightly more complex tasks, whether caffeine will improve or impair performance is difficult to predict. An example of such a task is watching a random sequence of letters and responding each time the letter A is preceded by the letter X. Because to a large extent driving a car is routinized, the improvements in auditory vigilance and visual reaction time would seem to imply benefits for late-night driving—if not counteracted by loss of fine motor coordination. No researcher seems to have carefully tested the net effects. It is noteworthy, though, that several researchers have tested the common belief that caffeine counteracts the effects of alcohol. Surprisingly, coffee further impairs rather than improves performance. For example, a person who has consumed enough alcohol to be close to the legal level of intoxication and then drinks a cup and a half of coffee (150 mg caffeine) has even slower reaction times than if only the alcohol has been consumed. In short, the extra coffee may make one more prone to accidents rather than less so.

Many people believe that arousal produced by caffeine is beneficial to learning and retention. Some research has, in fact, shown beneficial effects; however, many studies involving short-term memory have shown no effect or impaired performance due to caffeine. It has been proposed and partially demonstrated that this lack of consistency can be resolved by disentangling the complex interactions among personality type (extravert–introvert and/or high–low impulsivity), diurnal rhythm of arousal, task requirements (e.g., sustained information transfer, short-term memory), dosages matched to body weights, and the curvilinear relationship with arousal level (moderate arousal enhances performance, whereas excessive arousal hinders it). For example, coffee may improve an extravert's performance on a particular task in the morning, but impair the same person's performance on the same task in the afternoon. Conversely, coffee may hinder an introvert in the morning, but facilitate the introvert in the afternoon.

Another kind of drug-induced effect on memory is state-dependent learning, which occurs when information learned in one drug state is later recalled better when the individual is in the same drug state rather than a different one. An example of state-dependent learning is the alcohol drinker, who while sober forgets what he or she did while intoxicated but recalls it again when next intoxicated. In the experimental laboratory, state-dependent learning with alcohol has been demonstrated with social drinkers; however, several attempts to demonstrate state-dependent learning in the laboratory with caffeine have been unsuccessful. One possible cause for the latter negative results is that the "drug" and "nondrug" states that were supposed to be different actually were not. The experimenters assumed that the drug state involved a high level of arousal because caffeine was consumed, and that the other state involved a low-level of arousal because a placebo (i.e., no caffeine) was consumed. However, subjects in the latter condition may have also been aroused, because, for example, they had been challenged to perform well on the experimental task, or they may have been excited by being in an unfamiliar setting and in the presence of a stranger. The first explanation is especially plausible because many people find that having to take any kind of a test causes them to be anxious. Assuming that alcohol reduces test anxiety would account for the state-dependent learning that was found in one study that, from the learning to the test phase of the experiment, shifted subjects from a combined alcohol–caffeine state to either (1) the same state (which produced no decrement in performance), (2) an alcohol-only state (which produced a performance decrement), (3) a caffeine-only state (which produced small decrement), or (4) a no-drug state (which produced maximal detriment). In short, at the present time, whether caffeine does or does not produce state-dependent learning is not entirely clear. There seems to be a good chance that caffeine does produce state-dependent learning, but that this effect will be difficult to isolate. [*See* LEARNING AND MEMORY.]

Many people report that caffeine helps them think more clearly and creatively. Whether or not people's actual cognitive performance improves while they are under the influence of caffeine has been tested using a variety of tasks, ranging from simple subtraction, to identifying errors in written passages, to taking the Graduate Record Examination (among many others). A few studies have reported that caffeine did not improve performance, but their negative findings could be due to the fact that they did not control for variables such as the personality characteristics of their subjects or the time of day that they were tested. These variables are important because other research has demonstrated that they modulate the effects of caffeine. For example, several studies that controlled for personality characteristics and time of day found that caffeine both improved performance and showed dose–response effects. On the other hand, one study reported that high caffeine intake among college students was associated with low academic grades; however, it is impossible to conclude that caffeine consumption is the cause of the low grades. Because this study is frequently cited, it should be replicated and cause and effect relationships should be investigated. Furthermore, it would be important to have more studies of other "real-world" forms of thinking processes, such as reading comprehension, creativity, and problem solving. Studies that compared fatigued or bored subjects with alert ones would help resolve the issue of whether caffeine actually improves performance generally or is largely confined to restorative effects. Do the benefits of caffeine occur at the expense of impaired performance later in the day? Is the period of increased stimulation–metabolism followed by a restorative period of decreased stimulation–metabolism? Researchers do not seem to have addressed these simple, practical questions.

VI. Physical Health

Physicians have long been interested in caffeine for its therapeutic effects. Caffeine has been beneficially used as a cardiac stimulant, to reduce bronchial asthma, therapeutically with infant apnea, to treat acne and other skin disorders, to reduce migraine headaches, to enhance the effects of analgesics, and to counteract the depressant side effects of various medications. There are two reports of anticancer effects of caffeine, but unfortunately they are not convincing.

There have been sporadic reports of evidence and counter-evidence for health risks due to caffeine. As a result, public interest in caffeine has mushroomed in the last 20 yr. The major debates have involved birth-related problems, cancer, ailments of the gastrointestinal system, and diseases of the cardiovascular system. In the sections below, we will evaluate the evidence related to each of these possible effects.

A. Birth-Related Risks

There is reason to be concerned about whether or not caffeine increases birth-related risks. In general, fetal growth is a time of great risk. With regard to caffeine in particular, animal studies have demonstrated that caffeine is a teratogen among mammals. As a result of these animal studies, the Food and Drug Administration removed caffeine from its list of drugs "generally regarded as safe" (GRAS) and in 1978 released a warning concerning the ingestion of caffeine during pregnancy. Moreover, three studies with humans have shown a relationship between coffee consumption and miscarriage or spontaneous abortion. Another study, however, found negative results, and it would be important to have further studies to clarify the exact nature of the relationship, if one actually exists. Research has shown no association between coffee consumption and preterm birth, preterm labor, Caesarean sections, breech births, or premature births. Whether or not coffee consumption affects ease of conception is an open question that is difficult to study because many additional life-style variables must be taken into account. [*See* ABORTION, SPONTANEOUS; TERATOGENIC DEFECTS.]

In addition to effects on the mother and her child's birth as discussed above, a pregnant woman's caffeine consumption may have effects both on her fetus and the infant after it is born. Certainly, many fetuses are exposed to caffeine. In fact, more than 80% of pregnant women report consuming caffeine during their gestational period. Although many report reducing the quantity of their caffeine intake, few report that they totally abstain. In fact, pregnancy is a time of low energy, and women desiring a "lift" may actually increase their intake of coffee or other beverages containing caffeine. Furthermore, some women may unknowingly ingest caffeine through medications. These possibilities are of concern because caffeine is known to cross the placenta and diffuse into breast milk. Moreover, the concentrations of caffeine in the fetus or newborn may be considerably higher than in the mother because the former lacks the enzymes necessary to metabolize caffeine. In fact, the slower clearance of caffeine in young infants has been empirically verified.

When the possibility of an association between coffee drinking and congenital malformation was first suggested, suspecting detrimental effects of caffeine was partially justified, because studies with microorganisms had shown that caffeine causes mutations. However, closer examination of the original rationale and of the studies with humans currently indicates that there is no mutagenic risk. The extrapolation of the results from microorganisms to humans does not hold because of biological differences between these organisms and because the experiments used caffeine concentrations 40–4,000× those achieved by heavy coffee drinkers. Most large-scale epidemiological studies with humans have found no association between coffee intake and congenital malformations, although a few recent studies with small numbers of subjects have found associations. Based on the large-scale studies, mutagenic risk, then, is not presently a major concern, but further epidemiological research is needed.

The relationship between caffeine intake and low birthweight is closely related to concerns about mutagenic risk. With regard to birthweight, the evidence is mixed. Although some early studies found that increased caffeine consumption by pregnant women was associated with slightly greater frequency of low birthweights, the magnitude of the associations was small and, because of a lack of statistical significance, they were discounted. Another early study did find a significant association, but it too was discounted because the study had not controlled for smoking and gestational age (and researchers assumed that controlling for such extraneous factors would give a nonsignificant pattern of results). At the time that this study was conducted, the cumulative pattern of evidence seemed to indicate no association between coffee intake and low birthweights; however, several more recent studies, including one carefully controlled one, have found a significant association. The import of the association is strengthened by two additional statistically significantly results: Maternal caffeine consumption has been found to be associated with smaller head circumferences and neuromuscular immaturity of the newborn. This shift in the balance of evidence led researchers to reanalyze the early study whose significant results had been discounted. They found that when appropriate corrections were made for smoking and gestational age, a significant association was still present, indicating that the original results should not have been discounted. In summary, caffeine intake does seem to be related to low birthweights.

The research summarized thus far involved chronic coffee use. However, short-term intake of caffeine during pregnancy has also been associated with detrimental effects, albeit behavioral rather than physical. For example, maternal caffeine consumption 3 days before delivery has been shown to increase general arousal level, depress muscle tone, and make it harder to console the infant when it was upset. There is also some suggestion that spontaneous sleep states are affected.

In short, many of the suspected birth-related risks of caffeine have been disproved, and evidence regarding other effects is mixed. However, the current evidence regarding spontaneous abortion, low birthweight, and behavioral effects is convincing enough to advise pregnant mothers to limit, or discontinue entirely, their caffeine consumption, especially because safe levels of caffeine use during pregnancy have not been established. In fact, pregnant mothers should avoid taking any unnecessary drugs during pregnancy, or should take them only under the supervision of a physician.

B. Caffeine and Cancer

As just described, caffeine has been suspected of being a mutagen. Many mutagens also cause cancer. Caffeine has been suspected of both directly initiating cancer and interacting with other carcinogenic agents. In studies with animal tissue, caffeine has been shown to either increase or decrease the growth of tumors at the cellular level, depending on which carcinogen was used. Epidemiological studies in humans have focused on coffee rather than caffeine specifically and have been fraught with all the potential difficulties of etiologic science (as discussed in Section II). In fact, initial reports of a link between coffee consumption and pancreatic cancer have been retracted by the original authors (as well as criticized by others). Many epidemiological studies show no link with cancer of the urinary bladder, although one study attributes 25% of all bladder tumors to drinking more than 1 cup of coffee a day. Many, but not all, reports show no link between coffee drinking and cancer of the breast through fibrocystic breast disease. Only a small amount of research exists on associations between coffee drinking and cancer of the ovary, and the results of these studies are mixed. On the other hand, there have been isolated reports of beneficial effects of caffeine: Higher coffee consumption has been linked to lower carcinoma of the renal parenchyma, the skin, and the colon. In summary, the evidence does not allow us to conclude that coffee intake causes cancer. At the same time, more research is needed with all the rigorous controls of etiologic science (see above).

C. Gastrointestinal Risks

There are several ways in which caffeine could lead to gastrointestinal problems. Caffeinated beverages can increase the secretion of stomach acids and could thereby exacerbate an ulcer that is already present or could contribute to the formation of a new one. Caffeinated beverages also lower esophageal sphincter pressure, perhaps allowing or adding to gastric acid reflux. One piece of evidence that tends to support this reasoning is the fact that a reduction in caffeine intake has been associated with reduced abdominal complaints. In particular, patients taking the drug cimetidine for gastrointestinal ulcers may find that it slows caffeine clearance by the body. In summary, coffee drinkers with ulcer problems may want to reduce their intake of caffeine, but switching to decaffeinated coffee is not a solution because it also stimulates the secretion of stomach acids.

D. Cardiovascular Risks

The effects of caffeine on the cardiovascular system have been extensively studied, yet remain unresolved. This research is especially difficult to interpret because of numerous interactions of caffeine with other life-style habits. Because of its effects on heart rate, blood pressure, and serum cholesterol, caffeine is implicated in arrhythmias, hypertension, and ischemic heart disease. Caffeine is contraindicated for individuals with cardiac arrhythmias. Perhaps because this dictum is so widely believed, there is disappointingly little published evidence of a link between caffeine intake and arrhythmias. More research is needed for an understanding of the real risks and the mechanisms through which caffeine operates. For example, it has been suggested that caffeine causes irregularity by affecting the heart's contractility through alterations in the movement of calcium in and out of the cells.

The relation of caffeine to hypertension has also been questioned. Although the effects of caffeine may be small, they may be clinically significant if a patient is exposed to stress or other factors associated with cardiovascular disease. Clinically significant reductions in blood pressure have been observed in chronic users who abstain. Future research should also investigate whether or not caffeine interacts with antihypertensive medication and negates the effect of the medication. Until more evidence has been gathered, individuals worried about blood pressure may consider reducing or eliminating caffeine from their diet. [*See* HYPERTENSION.]

During the 1970s, sporadic correlations between coffee intake and heart attacks were reported, but researchers tended to believe that these results did not indicate that caffeine consumption was one of the causal factors in ischemic heart disease. However, two reports since 1985 have shown a link between heavy coffee use and heart attacks, thus raising the issue again. Actually, two issues may be distinguished: Does chronic use contribute to the development of the disease? Does a single dose produce the stimulation that triggers a heart attack? Future research should distinguish these issues and should include careful controls for smoking, diet, exercise, and stress. Researchers should attempt to obtain coffee consumption habits both immediately preceding the coronary event and distant in time from it.

E. Other Health Problems

Research on sleep disturbances has shown that coffee consumed shortly before bedtime increases time to fall asleep and the number of spontaneous awakenings during the night. It also decreases total sleep time, amount of deep sleep (time in stage 3 sleep and stage 4 sleep), and perceived quality of sleep. Habitual caffeine use has been shown to be correlated with habitual sleep duration. Single doses >300 mg can produce temporary insomnia. Reducing caffeine intake has been shown to reduce sleep disturbances, but with wide individual variations. For example, eliminating an evening soft drink may help some children with problems getting to sleep, but not others. [*See* SLEEP DISORDERS.]

Several other health problems have been suggested: Because it affects calcium flow, caffeine may accelerate the development of osteoporosis. Habitual use of caffeine has been associated with subclinical (i.e., premorbid) symptoms of poor somatic and psychological health. However, because this research was based on self-reports, one must be cautious about whether or not the symptoms are more imagined than real.

VII. Mental Health

Many years ago, the well-known psychologist Harry Stack Sullivan observed "incipient depression and neurasthenic states" in a client after "unwitting denial of the accustomed caffeine dosage." He surmised that "there might be times when a cup of coffee would delay the outcropping of a mental disorder." Recent concerns reflect the opposite point of view, namely, that consumption of caffeine might lead to caffeine intoxication, that it might exacerbate other psychological disorders, or that its symptoms might be misdiagnosed as another disorder.

A. Addiction and Withdrawal

Is caffeine really a drug of addiction? The cardinal properties of a drug of addiction are that it have psychoactive properties, that it have reinforcing properties, and that its removal result in withdrawal symptoms. Caffeine is psychoactive: Witness the stimulating effects people report. Caffeine is reinforcing according to people's reports as well as carefully controlled experimental studies. Does removing caffeine produce withdrawal symptoms? When caffeine users abstain, they experience such symptoms as dysphoria, drowsiness, yawning, poor concentration, disinterest in work, runny nose, facial flushing, headache, fatigue, irritability, and anxiety. The symptoms typically begin between 12 and 24 hr after the person discontinues use of caffeine. The symptoms vary from individual to individual; they can be mild to extreme, peak within 20–48 hr and can last for a week. Headache, for example, is reported by about one-quarter of the heavy users who abstain. In a few cases, the symptoms have been reported to appear when caffeine intake was gradually reduced over several weeks rather than abruptly. Even someone who is a relatively light user, habitually consuming as little as 200 mg of caffeine per day, may experience withdrawal symptoms. Nonusers can become quickly addicted—within as little as 6–15 days, if high doses are consumed. Drugs of addiction tend to upset the homeostasis of the body. Addicts would, in fact, be in constant disequilibrium except for the fact that they develop compensatory responses, physiological changes opposite of those induced by the drug. Furthermore, these compensatory responses become conditioned to the cues that precede drug use. In short, the body prepares itself for the drug assault. Caffeine users, like users of other addicting drugs, develop compensatory responses, which become conditioned to the stimulus cues associated with caffeine consumption. For example, the sight of coffee inhibits salivation in chronic users of caffeine, a response that compensates for the increase in salivation produced by caffeine. Note that decaffeinated coffee provides the same visual and gustatory cues as caffeinated coffee, thereby also inhibiting salivation. Clearly, then, caffeine is a drug of addiction in spite of the fact that the general public does not generally regard it as such. [*See* NON-NARCOTIC DRUG USE AND ABUSE.]

B. Anxiety and Panic Disorders

The relationship between caffeine and emotional health has been most often studied in terms of anxiety and panic disorders. At low doses, some people interpret the stimulation provided by caffeine as a pleasant, general elevation in mood, whereas others find it unpleasant. After consuming moderate amounts of caffeine (200 mg), many people report increased feelings of restlessness, tension, and anxiety. Larger doses (300 mg) can lead to further anxiety, hostility, and depression. Although the data indicating these effects were obtained from self-reports, they have been corroborated by objective observers. In double-blind experiments, observers were able to reliably see the increased restlessness and "drug effect" of caffeine on users. In even larger doses, the symptoms may be indistinguishable from anxiety disorders. Reductions in daily caffeine level can reduce anxiety, although sudden withdrawal can increase it. Patients with panic disorders are particularly sensitive to the anxiogenic (anxiety-producing) effects of caffeine. In addition to the usual symptoms, they show palpitations, nervousness, fear, nausea, and tremors. They show clear dose–response effects. A 480-mg dose is enough to create a panic attack in these patients, although a much larger dose would be required to create panic in a "normal" person. The mechanisms that give rise to anxiogenic effects are unknown, but it has been proposed that the effects are mediated by autonomic nervous system activity, plasma adenosine levels, blocking the actions of adenosine, or increased lactate. Additional studies are needed to establish the true cause. In the meantime, all the available data reinforce the clinical wisdom that people with anxiety disorders should avoid caffeine-containing foods and beverages. [*See* MENTAL DISORDERS.]

C. Depression

Caffeine increases feelings of depression and exacerbates manic-depressive symptoms. Although reducing daily caffeine intake can improve mood, there is a note of caution. In two cases of bipolar affective disorder patients who were on lithium treatment, reduction of caffeine intake increased lithium tremors. In spite of its links to depression, caffeine does not seem to play a major role in dysfunction of the hypothalamic-pituitary-adrenal axis in major depression. [*See* DEPRESSION; MOOD DISORDERS.]

D. Other Issues

Two issues are important for their practical implications, although there has not been a lot of research related to them. Clinicians working with anorexics may want to monitor their patients' caffeine intake. In their striving to be thin, anorexics have been observed to consume large quantities of diet colas or coffee, apparently because these beverages have few calories and suppress the appetite. Clinicians working with patients who are taking psychotropic medication should note that diazapam has antagonistic and synergistic interactions with caffeine, although the exact nature of these interactions is controversial. Presumably other benzodiazapines have similar interactions.

VIII. Conclusions

The research to date has provided a definitive metabolic pathway for caffeine in humans and a number of well-established immediate effects of caffeine on the bodily systems. Caffeine has both beneficial and adverse effects on physical and mental performance, but these are less well-established because of the intricacies of the research methodology. The data suggest, but do not conclusively prove, that caffeine has adverse effects on physical and mental health. Specifically, there appear to be links between caffeine intake and hypercholesterolemia, miscarriages, low birthweights, hypertension, heart disease, anxiety, and psychiatric disorders. There is much more to be learned about the mechanisms by which caffeine produces its effects. Researchers have come to opposite conclusions on many of the issues and, in the process, have revealed extraneous variables that must be taken into account. It is important that there be future research to resolve these issues, and that the studies be well-designed and well-controlled, taking these new variables into account.

Some organizations have used the research on caffeine to advocate that caffeine consumption be considered a social problem. This level of alarm is inappropriate; however, more education of the general public regarding caffeine is warranted. Educational efforts should cover known and probable health risks, misconceptions that caffeine counteracts the effects of alcohol, misconceptions about sources of caffeine, becoming aware of one's actual intake, and ways to reduce one's intake. Those who want to reduce or eliminate caffeine intake may do so on their own by reducing the concentration of caffeine in the foods or beverages that they consume (e.g., by boiling tea 1 min instead of 5 min, mixing caffeinated and decaffeinated coffee), substituting noncaffeinated products for caffeinated ones (e.g., carob for chocolate, fruit juice during "coffee" breaks, caffeine-free for caffeinated over-the-counter medications), gradually eliminating occasions on which caffeine is consumed (e.g., coffee with the evening meal), and by organizing a support group of friends or coworkers. Some individuals may find it hard to reduce their intake of caffeine because they lack motivation, because of social pressure to consume, or because they do not want to give up the stimulatory effects of caffeine. Such individuals may want to seek the help of health-care professionals who use systematic multicomponent interventions proven successful. In short, for individuals to decide to continue or change their caffeine habits, they must be informed.

Bibliography

Ashton, C. H. (1987). Caffeine and health. *Br. Med. J.* **295,** 1293–1294.

Blount, J. P., and Cox, W. M. (1985). Perception of caffeine and its effects: Laboratory and everyday abilities. *Percep. Psychophys.* **38,** 55–62.

Bruce, M. S., and Lader, M. H. (1986). Caffeine: Clinical and experimental effects in humans. *Hum. Psychopharmacol.* **1,** 63–82.

Dews, P. B. (ed.) (1984). "Caffeine: Perspectives from Recent Research." Springer-Verlag, Berlin.

Griffiths, R. R., and Woodson, P. P. (1988). Caffeine physical dependence: A review of human and laboratory animal studies. *Psychopharmacology* **94,** 437–451.

Heller, J. (1987). What do we know about the risks of caffeine consumption in pregnancy? *Br. J. Addict.* **82,** 885–889.

Humphreys, M. S., and Revelle, W. (1984). Personality, motivation, and performance: A theory of the relationship between individual differences and information processing. *Psychol. Rev.* **91,** 153–184.

Revelle, W., Humphreys, M. S., Simon, L., and Gilliland, K. (1980). The interactive effect of personality, time of day, and caffeine: A test of the arousal model. *J. Exp. Psychol. General* **109,** 1–31.

Troyer, R. J., and Markle, G. E. (1984). Coffee drinking: An emerging social problem? *Social Problems* **31,** 403–416.

Watson, R. R. (1988). Caffeine: Is it dangerous to health? *Am. J. Health Prom.* **2(4),** 13–22.

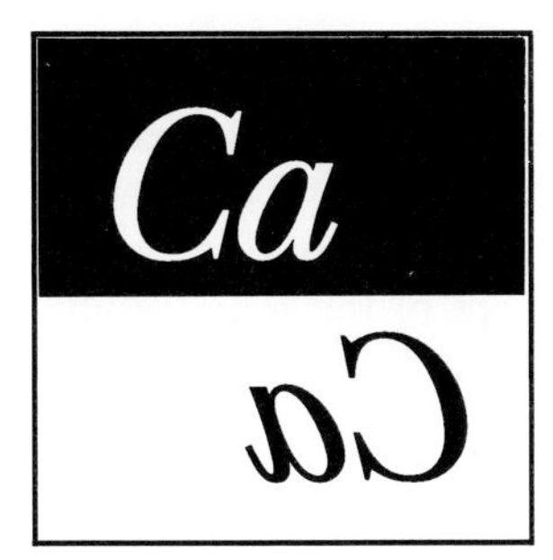

Calcium Antagonists

WINIFRED G. NAYLER, *University of Melbourne*

Glossary

Ca^{2+} Calcium in its ionized form
Extracellular fluid Fluid bathing the outer surface of the cells
Myocytes Single muscle cells
Na^{+} Sodium in its ionized form

THE CALCIUM ANTAGONISTS are a newly discovered group of drugs that are now being used in the management of patients with a wide variety of disorders, including angina and hypertension (high blood pressure). The prototypes of the group were initially developed because of their ability to dilate the coronary blood vessels of the heart, but almost by chance it was discovered that these drugs also depress the pumping activity of the heart. Moreover, it was noted that this depressant effect on the heart could be counteracted or reversed simply by adding more calcium. This discovery, coupled with the fact that calcium ions (Ca^{2+}) are needed for muscle contraction, resulted in the conclusion that these drugs limit the availability of calcium to the muscle cells. Hence the term "calcium antagonist" was introduced to define them, and although subsequent investigations have shown that these drugs interact with only one of the possible ways in which calcium ions can be made available for muscle contraction, the term remains in common usage. Alternatives include "calcium channel blockers" and "calcium entry blockers."

1. Definition of Calcium Antagonism

A. Background

This chapter deals mainly with the synthetic calcium antagonists that are now available. The discussion will concentrate on the chemistry of these drugs, how they modulate calcium transport, and the consequences of that modulation in terms of their current and prospective clinical use. It may be useful to know, however, that calcium antagonists are not the perogative of Western medicine. They form an active ingredient in some of the traditional Chinese medicines used in the treatment of cardiac disorders. One such compound is Tanshinone; another is Tetrandrine. As far as Western medicine is concerned, however, a wide variety of calcium antagonists is now available, the enormity of the range reflecting the need to have drugs to target a particular end organ. For example, one of the calcium antagonists, nimodipine, has a high degree of specificity for the cerebral blood vessels. Another, nisoldipine, has a high degree of specificity for the blood vessels that regulate the supply of blood to the heart (the coronary vasculature). Others, including the prototype of the group, verapamil, are most effective in modulating calcium availability in the tissues that are responsible for regulating the distribution of the excitatory stimulus from its site of origin in the "pacemaker" area of the heart to the other chambers of the heart. This makes this particular calcium antagonist useful for treating patients with irregular heart beats (arrhythmias) caused by excessive or irregular pacemaker activity.

B. Site of Action

All muscle and excitable cells, including those of the brain and nervous tissue, are surrounded by a complex membrane that contains thousands of "pore-like" channels. When closed (inactive), these channels are impermeable to ions, but when open (active), they allow millions of ions to traverse the membrane each second (Fig. 1). The direction of movement of the ions through these channels is governed by the concentration gradients of the ions across the membrane. This means that sodium ions (Na^+) move in an inward direction into the cells, whereas potassium ions (K^+) move in the opposite direction, to accumulate in the fluid (extracellular) that surrounds the cells. Calcium ions (Ca^{2+}) resemble Na^+ ions in moving from the extracellular fluid into the cell. The movement of these ions (Ca^{2+}, Na^+, and K^+) does not occur in a haphazard fashion but rather through ion-specific channels. Thus, there are channels that selectively admit Na^+ into the cell. Other channels provide the route of K^+ exit, whereas others selectively admit Ca^{2+}. The calcium antagonists exert their effect by modulating the movement of Ca^{2+} through these Ca^{2+}-selective channels. [*See* ION PUMPS.]

In most excitable tissues the Na^+-carrying channels open within milliseconds of the cells being stimulated, whereas the Ca^{2+}-conducting channels open more slowly. It was for this reason that the terminology of "*fast*" Na^+ and "*slow*" Ca^{2+} channels was introduced. The currents that are generated by the inward movement of the ions through their respective channels accordingly are referred to as "fast inward Na^+" and "slow inward Ca^{2+}" currents.

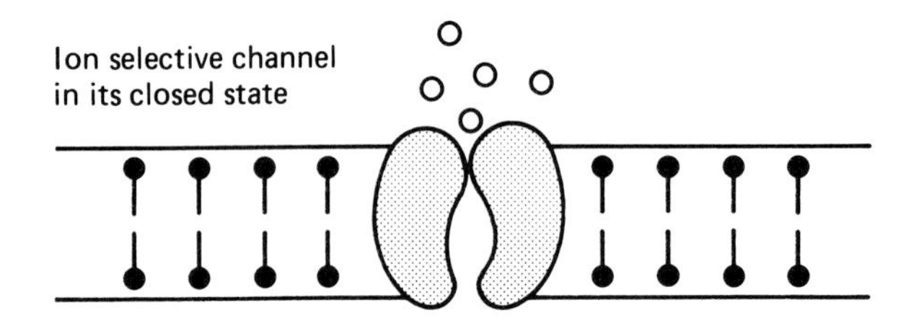

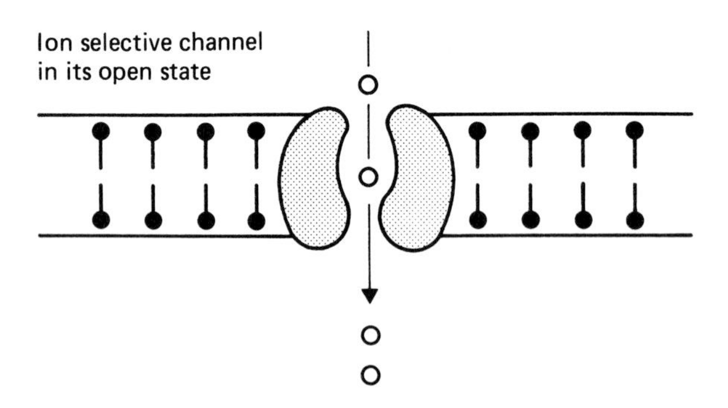

FIGURE 1 Schematic diagram of an ion-selective channel. The opening of each channel has a diameter large enough to admit Ca^{2+}. Larger ions are excluded.

When the Na^+ channels are open, Na^+ ions file through them at a rate of approximately 50 million ions every 10 sec. The Ca^{2+} conducting channels selectively admit Ca^{2+} ions at about the same rate. Their ability to exclude Na^+ ions has nothing to do with the actual size of the channels, because the diameter of a Na^+ ion is smaller than that of a Ca^{2+} ion. Instead the specificity for Ca^{2+} revolves around the chemical rejection of Na^+ from the lumen of the Ca^{2+} channel.

The importance of the Ca^{2+} conducting channels and of the Ca^{2+} ions that move through them cannot be underestimated, because these Ca^{2+} ions play a critical role in a wide variety of physiological processes, including muscle contraction, impulse propagation in the heart, and neuronal activity. It is these channels that provide the primary site of action of the calcium antagonist drugs.

The overall effect of the calcium antagonists is to restrict Ca^{2+} ion entry, so that over a given period of time fewer Ca^{2+} ions become available for participation in the various intracellular events with which they are involved. Hence the alternative terminology: "calcium channel blockers," or "slow calcium channel blockers," or "Ca^{2+}-entry blockers."

II. Mode of Action of Calcium Antagonists

A. Techniques Used to Study the Functioning of Ca^{2+} Channels

The techniques that have been used to study the electrical activity associated with the opening and closing of Ca^{2+} channels include

(1) Insertion of a conducting electrode into a single cell to record the electrical activity associated with the opening and closing of the channels in the membrane. By using appropriate bathing solutions it is possible to preset the potential difference across the membrane to

a point at which only the Ca^{2+}-conducting channels are operative.

(2) The electrical activity associated with the opening and closing of a single Ca^{2+}-conducting channel can be recorded from a small "patch" of the membrane that has been sucked into the orifice of a pipette. This is often called "patch-clamping."

(3) In addition, with modern technology it is now possible to harvest a single Ca^{2+} channel and insert it into an artificial lipid bilayer so that its properties can be studied without possible interference from surrounding tissue.

As soon as investigators started using these last two techniques it became apparent that the membranes of many excitable tissues, including those of the heart and the blood vessels, contain several different types of Ca^{2+} channels.

B. Heterogeneity of the Ca^{2+} Channels

There are at least three different types of Ca^{2+} channels, and by convention they are designated as L, T, and N types. These channels can be differentiated from one another in terms of

(1) the transmembrane potential difference at which they become operational,
(2) their Ca^{2+} ion carrying capacity, or conductance,
(3) their stability in isolated membrane patches,
(4) their sensitivity to calcium antagonists,
(5) the rate at which they open and close, and
(6) their sensitivity to naturally occurring toxins.

The L-type channels were originally given this designation because once activated they remain open for a relatively long period of time and have a large Ca^{2+} ion–carrying capacity. By contrast the T channels were so designated because of their brief opening time. The N channels were originally called N because their characteristics were neither of the L nor T type. However, N now is commonly assumed to mean neuronal, because these channels assume prevalence in neuronal tissue. Only the L-type channels are sensitive to the drugs that are now classed as calcium antagonists. The Ca^{2+} conducting T-type channels, however, can be blocked by certain inorganic divalent cations, including cobalt and nickel. The sensitivity of the L-type channels to the organic-based calcium antagonist drugs can be explained simply in terms of the fact that sites with which these drugs interact actually form part of the L-type calcium channel.

C. Size and Structure of the Calcium Antagonist Binding Sites

Biochemists who are interested in establishing how the calcium antagonists modulate Ca^{2+} ion entry through the L-type Ca^{2+} ion–conducting channels have recently identified the complex that contains the binding sites for these drugs. The molecular weight is around 170 kDa, and as such, it accounts for approximately half of the calcium channel. The binding site can be subdivided into four polypeptide subunits designed as α_1, α_2, β, and γ. It is the α_1 subunit that contains the binding sites for the calcium antagonist drugs.

Until recently the calcium antagonist binding site was thought to have the same structure and composition, irrespective of its tissue location. Quite recently, however, Schwartz and colleagues presented evidence that there are tissue-dependent differences in these binding sites. Thus it might help to explain why some of the antagonists act preferentially on certain tissues.

D. Effect of Calcium Antagonists on the Functioning of the Ca^{2+} Channels

Each Ca^{2+} conducting channels has a half-life of about 40 hr, and it admits as many as 4 million Ca^{2+} ions per second when it is in its open or "activated" state. The calcium antagonists modulate the Ca^{2+} carrying capacity of these channels not by providing a physical block, or "plug" at the channel orifice, but rather by reducing the likelihood of the channels being open. Each channel can exist in three modes: open, shut (and hence unavailable for activation), or resting (and hence available for activation). The calcium antagonist drugs either prolong the closed time for each channel or reduce the likelihood of its being available for activation. Only those tissues that contain L-type channels are sensitive to the calcium antagonists (organic). They include the heart, the blood vessels, the electrical conducting pathways within the heart, the uterus, and the gastrointestinal tract.

III. Chemistry of the Calcium Antagonists

The calcium antagonists can be subdivided into two major groups, according to their chemistry: inorganic and organic. Cobalt (Co^{2+}), nickel (Ni^{2+}), manganese (Mn^{2+}), and lanthanum (La^{3+}) belong to the inorganic group, and although they are useful laboratory tools, they cannot be used clinically because of their toxicity and because they do not discriminate between the L- and T-types of calcium channels. By contrast the organic calcium antagonists are relatively specific for the L-type channel. Nevertheless, they exhibit a large amount of chemical heterogeneity. Thus (Table I) some are phenylalkylamines, some are dihydropyridines, and others are either benzothiazepines, or piperazines.

They differ from one another in terms of their potency, tissue selectivity, and duration of action.

A. Phenylalkylamines

The prototype of the calcium antagonists is the phenylalkylamine verapamil (Fig. 2). Compounds that have a similar chemistry and that are now classed as calcium antagonists include (Fig. 2) gallopamil, verapamil, and anipamil.

TABLE I Calcium Antagonists Subgroups

Subgroup			Molecular weights
Phenylalkylamines			
	Prototype:	Verapamil	454.5
	Derivatives:	Gallopamil (D600)	485.6
		Desmethoxyverapmil (D888)	461.0
		Animamil	520.8
		Ronipamil	460.7
		Devapamil	424.6
		Tiapamil	592.1
		Fendiline	315.5
Dihydropyridines			
	Prototype:	Nifedipine	346.3
	Derivatives:	Nisoldipine	388.4
		Nitrendipine	490.6
		Nimodipine	418.5
		Niludipine	490.5
		Nicardipine	388.4
		Felodipine	384.3
		Isradipine	371.4
		Amoldipine	408.9
Benzothiazepines			
	Prototype:	Diltiazem	414.5
Piperazines			
	Prototype:	Lidoflazine	490.6
	Derivatives:	Cinnarizine	360.5
		Flunarizine	406.5
Others			
		Bepridil	366.6
		Perhexiline	277.5
		Prenylamine	329.5

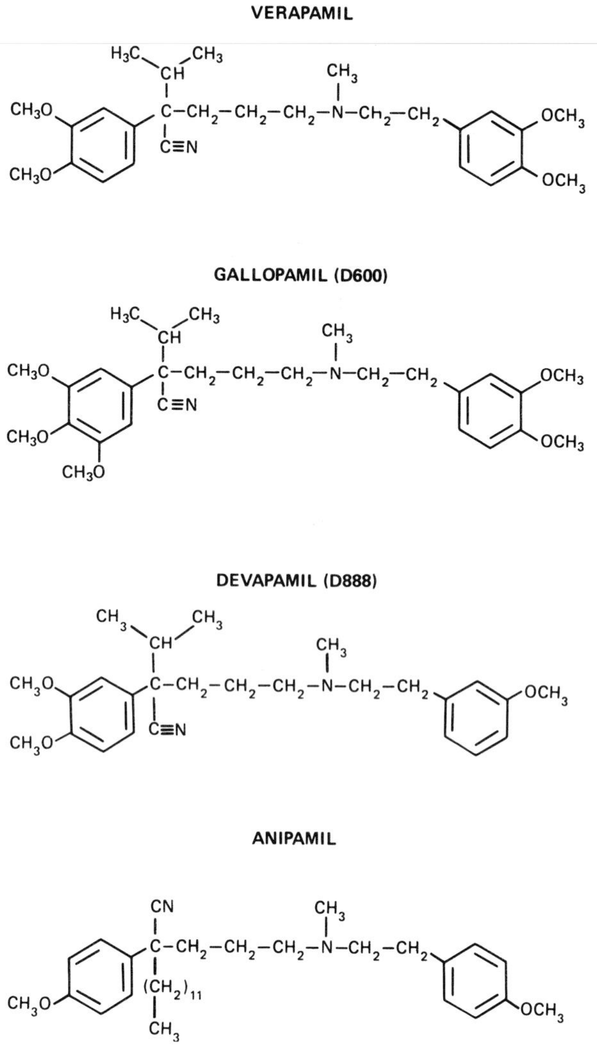

FIGURE 2 Chemical structure of verapamil and its derivatives: gallopamil, devapamil, anipamil.

Verapamil was originally selected for development as a therapeutic agent because of its potency as a coronary blood vessel dilator, but relatively early studies on this compound showed that it also slowed the heart rate and depressed the force of cardiac contractions (negative inotropy). When the drug reached clinical trial status, it became apparent that it would be useful in the management of patients with hypertension, because it has a blood pressure–lowering effect. Gallopamil resembles verapamil in its pharmacology, but its potency as a calcium antagonist exceeds that of the parent compound by a factor of 10. Anipamil matches gallopamil in terms of potency, but because it only slowly dissociates from its binding sites, it has a long duration of action.

The essential features of the phenylalkylamine-based calcium antagonists are:

(1) the presence of the two benzene rings, and

(2) a tertiary-amino nitrogen in the chain linking the two benzene rings (Fig. 2). As a rule the drugs exist in either the d or l isomeric state, but it is the l isomer that has the greatest calcium antagonist activity.

B. Dihydropyridines

Nifedipine is the prototype of this group. Structurally it is quite dissimilar from verapamil (Fig. 3),

VERAPAMIL

mol wt 454.59

NIFEDIPINE

mol wt 346.34

DILTIAZEM

mol wt 414.52

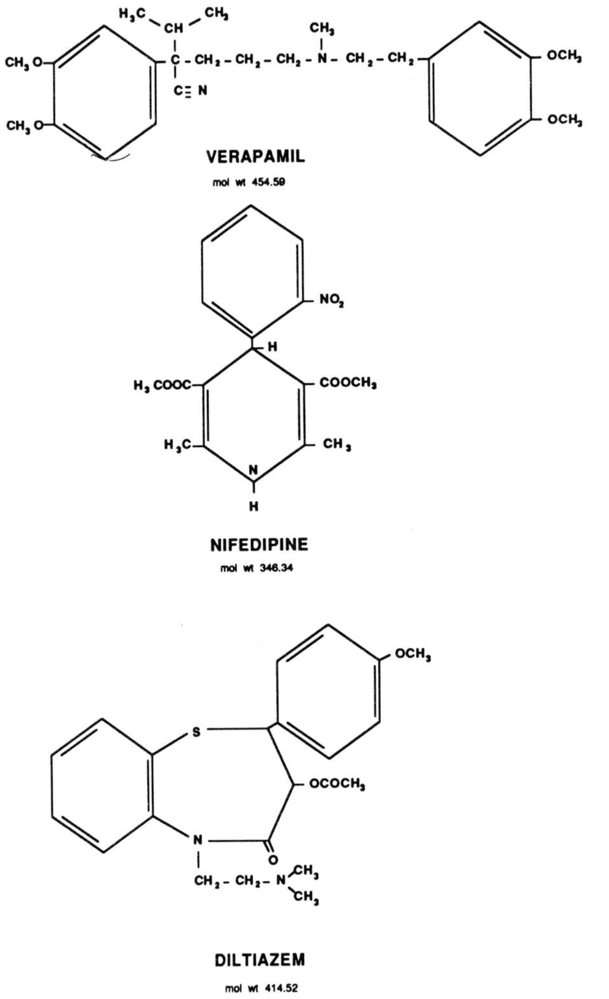

FIGURE 3 Structural formulas of the three calcium antagonist prototypes: verapamil (a phenylalkylamine), nifedipine (a dehydropyridine), and diltiazem (a benzothiazepine).

and unlike verapamil, it is rapidly degraded and inactivated when exposed to either daylight or ultraviolet light.

Various substitutions within the two rings of nifedipine have provided a wide range of dihydropyridine-based calcium antagonists, some of which are listed in Table I. The compounds differ from one another not only in terms of their potency and duration of action, but also in terms of their tissue specificity. For example, amlodipine is a long-acting calcium antagonist, relative to nifedipine, but like nifedipine it does not discriminate between the various vascular beds. By contrast nisodipine, another dihydropyridine-based calcium antagonist, acts preferentially on the coronary blood vessels, whereas nimodipine is highly selective for the cerebral blood vessels.

C. Benzothiazepines

Diltiazem (Fig. 3) is the prototype of this group. It resembles the phenylalkylamines in lacking vascular selectivity, but within the vasculature, it has a powerful dilator effect on the coronary arteries. It is not light sensitive, and like verapamil but in contrast to nifedipine, it is water-soluble. Derivatives of diltiazem are being developed, but they are not yet available for clinical use.

D. Piperazines

Lidoflazine, cinnarizine, and flunnarizine are examples of this group. However, because these drugs also inhibit Na^+ ion entry through the Na^+-selective channels, some investigators argue that they should not be classed simply as calcium antagonists. Nevertheless they exert a potent inhibitory effect on the slow calcium channels in the vasculature.

IV. Criterion for Classifying a Drug as a Calcium Antagonist

Two criteria at least must be satisfied:

(1) The drug must exert a dose-dependent inhibitory effect on the slow inward Ca^{2+} current (reflecting Ca^{2+} ion entry through the Ca^{2+}-selective channels); and

(2) the drug must interact specifically with the known binding sites associated with the Ca^{2+} channels.

V. Tissue Selectivity of Calcium Antagonists

There are many examples of the tissue selectivity of these drugs. For example, although all the calcium antagonists have a dose-dependent inhibitory effect on cardiac and vascular smooth muscle, they have little effect on skeletal muscle. In addition they have relatively little effect on antigen- or spasmogen-induced contractions in tracheal and bronchial smooth muscle, although they relax vascular smooth muscle. As a rule they do not interfere with excitation–secretion process, although many of these are Ca^{2+}-dependent.

Various factors contribute to the tissue selectivity of these drugs. They include:

(1) differences in the source of Ca^{2+} ions needed for the physiological response;
(2) differences in the chemical profiles of the drugs, which will determine how closely they can approach the receptors in a particular tissue;
(3) differences in the density, distribution, and type (L, T, or N) of Ca^{2+}-selective channel;
(4) differences in the requirements of the binding sites, and other ancillary properties of the drugs, including their interaction with other receptors, their effect on other ion-conducting channels, and in some cases, possible intracellular effects.

The insensitivity of skeletal muscle is readily explicable because in that tissue the Ca^{2+} ions required to activate the contractile proteins originate from within the muscle cell. This contrasts with the situation that exists in cardiac and smooth muscle, where some extracellular Ca^{2+} must move inward across the cell membrane before contraction occurs.

Some calcium antagonists exhibit "use-dependence." This is probably associated with the fact that the drug only interacts with the relevant receptor when the channel is in the "open" state. This applies particularly to the calcium antagonists that are ionized when in solution (verapamil and diltiazem). As a corollary, these drugs will be particularly effective in those tissues in which the Ca^{2+}-conducting channels are being repeatedly activated, as at the atrioventricular node. This property of "use"-dependence probably explains why drugs like verapamil, and to a lesser extent diltiazem, are particularly useful in controlling certain cardiac arrhythmias.

As far as the vascular selectivity of the dihydropyridines is concerned (Table II), its most likely explanation is that the transmembrane potential difference across vascular smooth muscle cells, which differs from that across the membranes of cardiac muscle cells, favors the binding of the dihydropyridine-based molecule. Different sensitivities within the vascular system involve a host of possibilities, including

(1) differences in T-, L-, and N-type channel distribution,
(2) relevance of reflex control mechanisms,
(3) different stimuli for contraction; and
(4) regional differences in membrane excitability and lipid composition, the latter being of importance because the lipid-soluble dihydropyridines may approach their receptors by way of the lipid bilayer of the membrane.

The relative insensitivity of neuronal tissue is not altogether surprising, because although the Ca^{2+}-selective channels that are of importance for transmitter release do occur in the dendrites, most of them are N-type channels, which are insensitive to the calcium antagonists.

The fact that the calcium antagonists have little or no effect on neuronal tissues is clinically important, and it explains why neuronally mediated processes are not affected by these drugs, which therefore lack mental side effects. In general, therefore, the calcium antagonists will only affect the functioning of those tissues in which

(1) the L-type Ca^{2+} channels predominate,
(2) the Ca^{2+} ions that mediate the functional response originate largely from the extracellular fluid, and
(3) the Ca^{2+}-selective L-type channels are in an operational state that favors the binding of the drugs to the receptors associated with the Ca^{2+} channels.

TABLE II. Tissue Selectivity of the Calcium Antagonists[a]

Drug	Myocardium	Vasculature	Conducting tissue	Skeletal muscle
Verapamil	+	+	+	–
Gallopamil	+	+	+	–
Diltiazem	+	+	+	–
Nifedipine	+	++	–	–
Nisoldipine	+	++++	–	–
Amlodipine	+	++++	–	–
Felodipine	=	++++	–	–

[a] + denotes the presence and – the absence of an effect on the activity of that tissue at therapeutically achievable concentrations.

VI. Clinical Relevance

When first introduced into clinical practice, the calcium antagonists were considered to be potentially useful for the management of patients with ischemic (inadequate blood supply to the wall of the heart) heart disease, including angina pectoris. Experience has shown, however, that their usefulness includes a wide variety of cardiovascular disorders (Table III). Hypertension, certain forms of arrhythmias, and even, under certain circumstances, heart failure can provide areas of use. The drugs also have other potential uses, including the management of premature labor and Raynaud's disease (i.e., a disease of the peripheral blood vessels that is caused by their intense constriction so that the fingers and toes receive an inadequate supply of blood). The drugs are also being used to treat patients with esophageal achalasia (a disorder characterized by sustained constriction of the esophagus, which makes swallowing difficult or impossible). Other areas include the protection of muscle and brain tissue from secondary damage caused by inadequate perfusion, such as may occur if a thrombus (blood clot) disturbs blood flow or the vessels suddenly develop a sustained contraction (spasm).

TABLE III Clinical Uses of Calcium Antagonists

Cardiovascular
Arrhythmias:
atrial flutter
paroxysmal supraventricular tachycardia
Angina:
chronic, stable, and vasospastic
Hypertension
Peripheral vascular disease
Raynaud's disease
Cerebral spasm
Congestive heart failure
Hypertrophic cardiomyopathy
Stroke
Migraine headache
Others
Achalasia
Premature labor
Adjunct to immunosuppressive therapy
Potentiation on oncotic drug activity
Tissue protection:
heart, kidney, brain
Cocaine intoxication:
prevention of cardiac toxicity
Morphine withdrawal symptoms
Ethanol withdrawal symptoms
Atherosclerosis

There is even evidence that suggests that these drugs might provide a useful way of slowing the growth of atherogenic plaques (calcified fatty deposits), which form in the lumen of blood vessels, including those of the heart).

A. Calcium Antagonists and Angina

The pathology of angina is complex. The condition can arise for a variety of reasons, including

(1) a sudden reduction in the lumen of a major coronary artery caused by contraction of the muscle in the wall of the artery;

(2) progressive reduction in blood flow through the coronary arteries because of platelet aggregations (blood clots) or atherosclerotic lesions; or

(3) a sudden increase in the oxygen requirements of the heart, caused by an increase in workload on the heart.

The calcium antagonists, particularly the vascular selective calcium antagonists, can be useful for the management of these conditions, because by restricting the entry of Ca^{2+} ions into the muscle cells of the coronary vasculature, they cause the vessels to dilate, thereby increasing the effective blood flow. However, if the inadequate flow of blood is due to a fixed lesion (e.g., an adherent clot or large atherosclerotic plaque), the drugs will be of little benefit.

B. Calcium Antagonists and Arrhythmias

Some calcium antagonists can be used to manage certain arrhythmias caused by excessive or irregular conduction of impulses through the conducting pathways in the heart. The dihydropyridines (nifedipine), however, cannot be used for this purpose because of their potency as vasodilators, but verapamil and, to a lesser extent, diltiazem are effective. Their use depends on their ability

(1) to depress the electrical activity of the sinus node (i.e., the region of the heart where the "heart beat" actually originates), and

(2) to slow down conduction through the conducting tissues, so that even if the sinus node does trigger a rapid volley of excitatory stimuli, the excess pulses are prevented from spreading to the other areas of the heart.

C. Calcium Antagonists and Hypertension

An essential characteristic of hypertension is excessive constriction of the peripheral vasculature. This excessive constriction is caused by abnormalities in the mechanisms that are responsible for maintaining homeostasis with respect to Ca^{2+}. Because the force of muscle contraction is directly proportional to the amount of Ca^{2+} that is available, it is logical to use drugs that limit Ca^{2+} availability to treat a disorder that reflects either an excess availability of or excessive sensitivity to Ca^{2+} ions. The calcium antagonists are particularly useful for this purpose, and in contrast to many other blood pressure lowering agents, they neither alter the plasma lipid profile (ratio of high-density to low-density lipoproteins), nor do they cause the retention of sodium ions (Na^+). To the contrary, because of their effect on the glomerular filtration rate in the kidney, they cause Na^+ and water loss, thereby preventing the development of edema (water retention). [*See* HYPERTENSION.]

D. Calcium Antagonists and the Slowing of Atherogenesis

Laboratory studies on cholesterol-fed rabbits and monkeys and one study in humans that extended over a 3-year period indicate that the calcium antagonists may slow the rate of growth of atherogenic plaques. There is some evidence that suggests that this effect is due to the ability of these drugs to slow the rate of growth of smooth muscle cells, an event that is Ca^{2+}-dependent and is fundamental to the development of atherogenic lesions. [*See* ATHEROSCLEROSIS.]

E. Calcium Antagonists and Cell Protection

In most organs the immediate cause of cell death appears to be an uncontrolled entry to Ca^{2+} ions. This applies to the heart that has been deprived of blood flow through its coronary vessels for a prolonged period of time and, likewise, to the brain. There is laboratory evidence that shows that the calcium antagonists can have a beneficial effect under these conditions, but only if they are present before blood flow to the affected tissue is restored. In clinical practice, this presents a dilemma, because of the difficulty of predicting which patients are at risk.

F. Other Uses and Future Developments

There are other conditions is which the calcium antagonists have been found to be of use, including the management of patients receiving immunosuppressive therapy associated with transplantation procedures.

Research on new drugs is continuing, with the aim of producing drugs with greater tissue specificity, prolonged duration of action, and reduced side effects. The newly developed drugs, however, like their prototypes, must retain the basic property of modulating Ca^{2+} ion movement through the Ca^{2+}-selective channels.

Bibliography

Finkel, M. S., Patterson, R. E., Roberts, W. G., Smith, T. D., and Keiser, H. R. (1988). Calcium channel binding characteristics in the human heart. *Am. J. Cardiol.* **62,** 1281–1284.

Fleckenstein, A., ed. (1983). "Calcium Antagonists in the Heart and Smooth Muscle." J. Wiley, New York.

Mikam, A., Imotoa, K., Tanabe, T., Niidome, T., Mori, Y., Takeshima, H., Narumiya, S., and Numa, S. (1989). Primary structure and functional expression of the dihydropyridine-sensitive calcium channel. *Nature* **340,** 230–236.

Nayler, W. G., ed. (1988). "Calcium Antagonists." Academic Press, London.

Stone, P. H., and Antman, E. M., eds. (1983). "Calcium Channel Blocking Agents in the Treatment of Cardiovascular Disorders." Futura Publishing, New York.

Vaghy, P. L., Itagaki, K., Miwa, K., McKenna, E., and Schwartz, A. (1988). Mechanism of action of calcium channel modulator drugs. Identification of a unique, labile, drug-binding polypeptide in a purified calcium channel preparation. *Ann. N. Y. Acad. Sci.* **522,** 176–186.

Vaghy, P. L., McKenna, E., Itagki, K. and Schwartz, A. (1988). Resolution of the identity of the Ca^{2+}-antagonist receptor in skeletal muscle. *Trends Pharmacol. Sci.* **9,** 398–402.

Weiss, W. F., and Simic, M. G. (1988). Calcium antagonists in tissue protection. *Pharmacol. Ther.* **39,** 385–388.

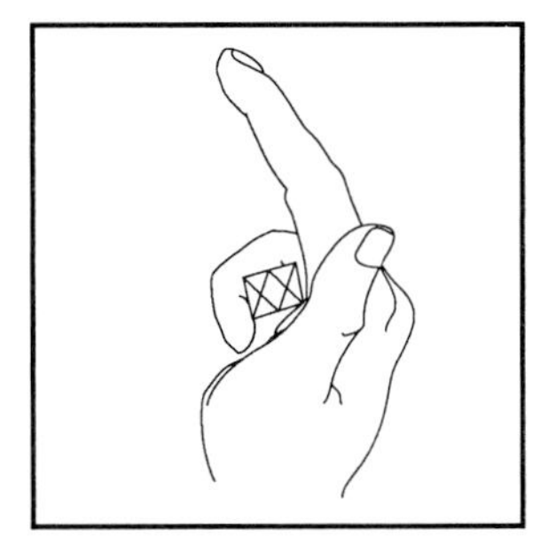

Calcium, Biochemistry

JOACHIM KREBS, *Swiss Federal Institute of Technology*

Glossary

Calmodulin Intracellular Ca^{2+}-binding modulator protein involved in triggering Ca^{2+} signals in the cell. It is a highly conserved, acidic protein containing four specific Ca^{2+}-binding domains (EF-hands). It belongs to a superfamily of highly homologous Ca^{2+}-binding proteins

Channel Transmembrane proteins responsible for ionspecific transport across the membrane down the ion's gradient

Coordination number Definition of the number of ligands bound to a central atom. It attributes it to a geometric configuration (e.g., tetraeder, octaeder)

EF-hand motif Expression coined by R. H. Kretsinger to describe Ca^{2+}-binding domains of specific proteins. It is composed of two helices and the Ca^{2+}-binding loop enclosed by the former; therefore it is also called the ''helix-loop-helix'' motif. The motif can be resembled by the forefinger and the thumb of the right hand (representing the two helices) and by the bent midfinger (representing the loop)

EF-hand type proteins Ca^{2+}-binding proteins containing ''EF-hand'' or ''helix-loop-helix'' motifs (e.g., calmodulin, parvalbumin and troponin C)

Electrogenic transport Net-transport of charges across a membrane, usually facilitated by an integral membrane protein

G proteins Specific GTP-binding proteins, consisting of three subunits (α, β, γ), involved in the signal-transducing process, transferring the message from a specific, hormonally activated receptor to the enzyme producing the second messenger

Growth factors Small peptide hormones [e.g., epidermal growth factor (EGF), platelet-derived growth factor (PDGF), insulin-like growth factor (IGF)] involved in cell proliferation. By binding to specific receptors they induce cell proliferation as a result of specific signal-transducing pathways

Ion exchanger Transmembrane protein involved in ion transport using the downhill gradient of one ion as an energy source to transport the other against its gradient (e.g., Na^{+}/Ca^{2+} exchanger)

Ion pump Ion-transporting enzyme, transporting the ion (e.g., Ca^{2+}) across the membrane against its gradient by using ATP as an energy source, thereby forming an acylphosphate intermediate

Reticulum Intracellular organellar system (e.g., sarcoplasmic reticulum, endoplasmic reticulum), important for the control of Ca^{2+} homeostasis in the cell

Second messenger Either small molecules (e.g., cyclic nucleotides or inositolpolyphosphates) or ions (e.g., Ca^{2+}) that are pivotal for the transduction of extracellular stimuli (e.g., hormonal signals = primary messengers) into intracellular events

CALCIUM PLAYS a pivotal role in biological systems. It occurs predominantly in a complexed form, fulfilling either a static, structure-stabilizing function or a dynamic, signal-transducing role. Because of large gradients of Ca^{2+} across cellular membranes, the cell can use even small changes of its membrane permeability for the ion to cause significant changes in the intracellular free-Ca^{2+} concentration, thereby transmitting and amplifying a signal from the extracellular space. Therefore the concentration of ionized calcium within the cell has to be carefully controlled involving a number of different systems to use Ca^{2+} as an intracellular second messenger. This also involves specific intracellular Ca^{2+}-binding proteins with characteristic "helix-loop-helix" motifs as Ca^{2+}-binding domains important for the modulation of the Ca^{2+} message.

I. Introduction

Calcium is one of the most common elements found on earth, making up about 3% of the earth's crust composition. It was discovered by Humphry Davy as a chemical element in 1808. The biological importance of calcium minerals and their stabilizing function of shells, bones, and teeth became soon apparent, and today it is known that calcium is an old component of organisms, the record (blue-green algae) being about 2×10^9 years. Therefore, it has long been recognized that calcium plays an important role in biology. It occurs predominantly in a complex form, both in minerals and in solution. In most higher organisms, more than 99% of the total calcium is precipitated as hydroxyapatite [$Ca_{10}(PO_4)_6(OH)_2$] in the skeleton, thereby fulfilling a rather static, structure-stabilizing function.

Because of the experiments carried out by Sidney Ringer at the end of the 19th century concerning muscle contraction, it was first recognized that calcium also played an important dynamic function in the regulation of cellular events. However, first the discovery of cyclic AMP by Earl Sutherland as an intracellular regulatory constituent and the subsequent development of the "second-messenger" concept prepared the ground to recognize the pivotal role of Ca^{2+} as an intracellular regulator.

Compared with the amount of calcium precipitated in the skeleton, the Ca^{2+} found in the extracellular fluid and intracellularly in the cytosol and in other intracellular compartments is rather minute. The concentration of calcium in the extracellular fluid or in the intracellular reticular system is in the millimolar range (2–5 mM, of which about 50% is ionized, i.e., unbound), whereas the free intracellular Ca^{2+} concentration in the cytosol of a resting cell is in the submicromolar range (200–500 nM). This results in a 10,000-fold concentration difference of ionized Ca^{2+} across the membrane. The resulting large electrochemical force makes it possible for minor changes of the free-Ca^{2+} concentration, due to membrane permeability changes as the result of an external signal, to cause significant oscillations of its concentration in the cytosol. These fluctuations provide the possibility of transmitting signals for a large variety of different biochemical activities (e.g., intracellular metabolic events, muscle contraction, neurotransmitter release, cell growth, proliferation and fertilization, stimulus-secretion coupling, and mineralization). Many of these activities are brought about by the interaction of Ca^{2+} with specific proteins, resulting in conformational changes and subsequent specific modulations of protein–protein interactions.

The dynamic role of Ca^{2+} as a signal transducer in the form of a so-called second messenger became the center of intense research activities in the past two decades. In this article, the different properties of calcium as a structure-stabilizing and signal-transducing factor will be described in detail, and the interrelation of the different roles and functions will be discussed.

II. Ca^{2+} Ligation

Ca^{2+} ligation in complexes usually occurs via carboxylates (mono- or bidentate) or neutral oxygen donors. The number of donor centers and the geometry of their arrangement has a great influence on the binding strength (e.g., can be in the range between 10^3 and 10^{12}). The superior binding properties of Ca^{2+} over Mg^{2+} or other cations such as Na^+ or K^+, which are present at much higher concentrations, enable Ca^{2+} to fulfill its function as a signal transducer. Furthermore, next to its ability to choose oxygen donors as ligands is its great flexibility in coordination (coordination numbers 6–8) with a largely irregular geometry both in bond length and bond angles. This geometry is suitable for binding to proteins, which usually provides irregularly shaped cavities. Because Mg^{2+} (0.64 Å) is smaller in diameter than Ca^{2+} (0.97 Å), it requires a regular octahedron with six coordinating oxygen atoms, whereas Ca^{2+} puts much less constraint on the complexing protein; thus it has greater versatility in coordinating ligands and has a much higher rate of exchanging them. The latter, which is about three orders of magnitude faster for Ca^{2+} than in the case of Mg^{2+} because of a significant slower dehydration rate of the latter, makes Ca^{2+} ideally suited for a signal-transducing factor. Since for Ca^{2+} the exchange rate with water in the inner coordination sphere is close to the diffusion limit, the on-rate in Ca^{2+} binding to proteins is often diffusion-limited, whereas the off-rates are dependent on the binding strength of the proteins. These advantages of Ca^{2+} over other cations make it possible for Ca^{2+} to bind to cavities of proteins accepting a greater variation of interatomic distances. This ability in turn permits selective binding because of a favorable charge-to-size ratio.

III. Calcium in the Extracellular Space

It has long been recognized that one of the important roles of Ca^{2+} in the extracellular space is to stabilize the structure of proteins. Several such proteins have been crystallized, and the Ca^{2+}-binding sites have been determined. By comparing their features with those of the intracellular trigger proteins (e.g., calmodulin) discussed in detail later, several important differences could be noted. Usually the Ca^{2+}-binding site of extracellular proteins, unlike those of the intracellular proteins, is preformed and relatively fixed. Therefore the on-rate for Ca^{2+} can be relatively slow (i.e., much slower than the diffusion limit). Since the off-rate often is fairly fast the affinity of Ca^{2+} for the protein is relatively low. However, because the Ca^{2+} concentration of the extracellular fluid is adequate for this affinity, these proteins usually occur in their Ca^{2+}-bound form, which is necessary to protect against proteolytic cleavage. The binding of Ca^{2+} to the protein reduces the capability of unfolding, which would increase the probability of proteolytic cleavage.

The preformed cavity of Ca^{2+}-binding sites of extracellular proteins and their relative high degree of rigidity is reflected in the arrangement of the chemical groups contributing to the binding site. In contrast to the intracellular Ca^{2+}-binding proteins (see below), the amino acids participating in Ca^{2+} binding usually are located at distant positions in the amino acid sequence and are not sequential. Therefore, the Ca^{2+}-binding sites in those proteins are preformed and Ca^{2+}-binding does not induce a significant conformational change of the protein. On the other hand, the Ca^{2+}-binding sites of the intracellular EF-hand-type proteins permit a significant conformational change due to the sequential arrangement of the liganding residues, thereby fulfilling the triggering function of these proteins. In addition, the extracellular Ca^{2+}-binding proteins are not homologous to each other nor to one of the intracellular, so-called EF-hand-type proteins (see Glossary).

An interesting exception to the rule that EF-type Ca^{2+}-binding proteins belong to the family of *intracellular* trigger proteins is found in osteonectin, the extracellular Ca^{2+}-binding protein involved in bone formation, or in fibronectin, an extracellular protein involved in blood clotting. Both proteins contain a single EF-type Ca^{2+}-binding domain.

IV. Signal Transduction Principles

Extracellular signals recognized by cell surface receptors are transmitted into a limited number of intracellular signal transducers, so-called "second messengers" to multiply the incoming signals using a cascade of intracellular events such as phosphorylation/dephosphorylation of a variety of different proteins or enzymes. It has long been known that the action of hormones such as epinephrine might be accompanied by changes of intracellular Ca^{2+} concentration, but it was not until the discovery of the intracellular formation of cyclic nucleotides on an extracellular stimulus (e.g., β-adrenergic receptor stimulation) and the subsequent development of the second-messenger hypothesis that the role of Ca^{2+} as an intracellular second messenger was recognized. To date, three classes of intracellular messengers are known:

(1) cyclic nucleotides [cyclic adenosine 3′-5′ monophosphate (cyclic AMP); cyclic guanosine 3′-5′ monophosphate (cyclic GMP)];
(2) derivatives of phosphatidylinositol [inositolpolyphosphates, e.g., inositol-1,4,5-triphosphate (IP_3), and diacylglycerol (DAG), stemming from the same precursor phosphatidylinositol-4,5-diphosphate (PIP_2)]; and
(3) free Ca^{2+} ions.

Common to all signal-transducing pathways is that a primary, external signal is received by a specific receptor located in the cellular membrane, which then will activate a chain of reactions, resulting finally in an intracellular response. This indicates that the cellular or plasma membrane is the main barrier that information coming from outside has to overcome to trigger intracellular events.

Thus the extracellular primary messenger (e.g., a hormone) has to transduce its message using special pathways to create intracellular second messengers, which distribute the signal throughout the cell.

Several steps are common to most signal-transducing pathways. After the receptor in the plasma membrane receives the primary signal, the information is transferred either indirectly, via so-called G proteins (proteins that are modulated in their function because of the binding to GTP or GDP), or directly to signal-transducing enzymes responsible for the production of the second messengers. These usually are small phosphorylated molecules which originate either by converting energy-rich nucleotides (i.e., ATP, GTP) into cyclic nucleotides with the use of specific enzymes (e.g., adenylate- or guanylatecyclases) or by the cleavage of phosphorylated forms of phosphatidylinositol located in the plasma membrane. Thus PIP_2 is cleaved into IP_3 and DAG by a specific, membrane-bound phospholipase C. IP_3 is closely connected to Ca^{2+} as a second messenger, because it can release Ca^{2+} from intracellular stores, probably the endoplasmatic reticulum. DAG amplifies its signal because of the activation of protein kinase C, a Ca^{2+}- and phospholipid-dependent enzyme. [SEE G PROTEINS.]

The way calcium performs its second-messenger function is quite different. The main difference is that for Ca^{2+}, changes in the distribution at the two sides of the cell membrane modulate its messenger function, instead of metabolic synthesis and degradation. On receiving a signal from outside the cell (e.g., a nerve pulse), the voltage across the cell membrane changes (depolarization), opening voltage-sensitive Ca^{2+} channels, which permits Ca^{2+} to enter the cell down its concentration gradient. The increased intracellular Ca^{2+} binds to specific proteins causing in them a conformational change. These proteins provide the triggering device to multiply the signal. The occurrence of those specific Ca^{2+}-binding intracellular trigger proteins were first described in muscle tissues (i.e., troponin C), paving the way for the detection of a whole class of homologous proteins, as it will be discussed in Section V.

The intracellular interplay between the different second messengers can be manifold and complex. From the diagram of Fig. 1, it becomes clear that the different second messengers not only are central components of intracellular control mechanisms but that they are connected in their action by a complex network of feedback relations. Signal transduction pathways are also connected with the interaction of polypeptide growth factors and their specific, high-affinity receptors on target cells. One of the earliest responses (within seconds) on binding of growth factors (e.g., EGF, PDGF) to their receptors is an increase in ion fluxes (Na^+, K^+, H^+) across plasma membranes and the release of Ca^{2+} from intracellular stores leading to a rapid increase of cytosolic free Ca^{2+}. It is proposed that on binding of the growth factor to its receptor and subsequent autophosphorylation of the latter because of its Tyr-specific kinase activity, the receptor activates the membrane-bound phospholipase C, which specifically cleaves PIP_2 to release IP_3 and DAG. IP_3 in turn releases Ca^{2+} from intracellular stores, whereas DAG activates the Ca^{2+}- and phospholipid-dependent protein kinase C.

V. Intracellular Calcium-Binding Proteins

In contrast to extracellular Ca^{2+}-binding proteins, which display a rather low affinity for calcium, intracellular Ca^{2+}-binding proteins bind Ca^{2+} with high affinity. In these proteins the amino acids that bind Ca^{2+} usually are arranged in a sequential manner, often following a characteristic pattern. Here Ca^{2+} has a triggering rather than a structure-stabilizing function. There may exist an additional class of Ca^{2+}-binding proteins (i.e., the exchange proteins), which transport Ca^{2+} across membranes, using either ATP or an appropriate ion gradient as an energy source. Little is known, however, about the structural properties of these proteins.

Intracellular Ca^{2+}-binding proteins (e.g., calmodulin, troponin C) are nonenzymatic effector molecules transducing the Ca^{2+} signal into cellular responses. Calmodulin is the pivotal intracellular Ca^{2+}-dependent modulator protein, which controls a variety of cellular events, including phosphorylation and dephosphorylation of proteins, synthesis and degradation of cyclic nucleotides, or cell growth and differentiation. It is ubiquitous among eucaryotic organisms. By contrast, troponin C is specific in its function as the Ca^{2+}-dependent regulator of the troponin–tropomyosin system in muscle cells and is limited in its distribution. This difference in their functional specificity is also reflected in the conservation of the primary structure of these proteins. Calmodulin is one of the most conserved

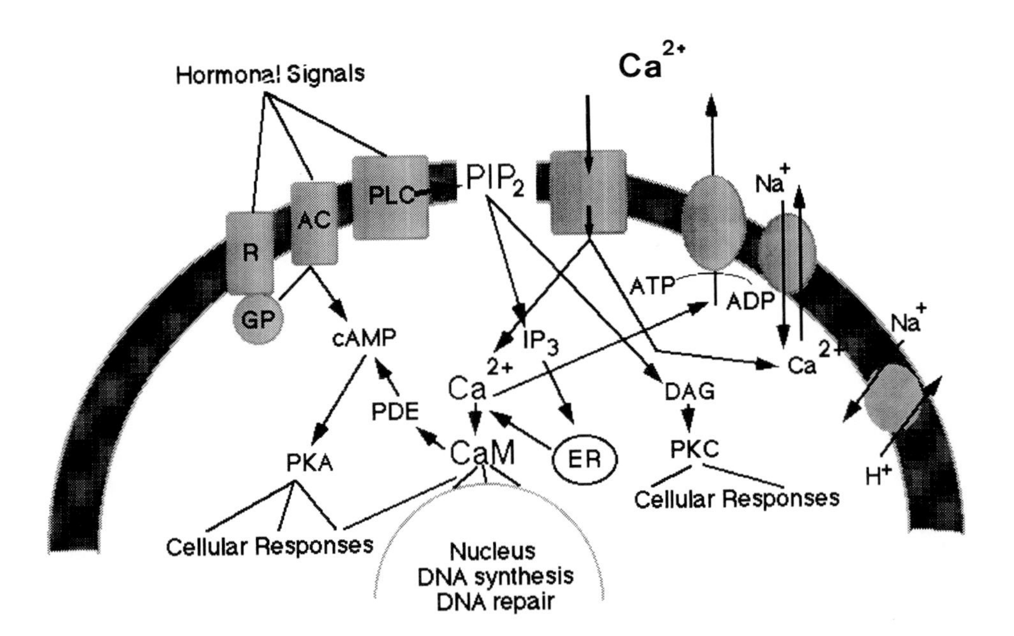

FIGURE 1 Schematic view of signal transduction pathways. AC, adenylate cyclase; CaM, calmodulin; DAG, diacylglycerol; ER, endoplasmic reticulum; GP, GTP-binding protein; IP_3, inositol-1,4,5-triphosphate; PDE, cyclic nucleotide phosphodiesterase; PIP_2, phosphatidylinositol-4,5-diphosphate; PKA, cyclic AMP–dependent protein kinase (protein kinase A); PKC, protein kinase C; PLC, phospholipase C; R, receptor.

proteins known to date (e.g., sequences known from different vertebrates, including humans, are identical). By contrast, troponin C displays a much greater degree of variation in its amino acid sequences, including tissue-specific differences.

In 1973 Kretsinger and Nockolds reported the first crystal structure of an intracellular Ca^{2+}-binding protein, parvalbumin. It consisted of three homologous domains, each of which contained two α-helices perpendicular to each other, enclosing a loop responsible for the specific Ca^{2+} binding. In parvalbumin the six helices were named A–F. The protein contains only two Ca^{2+}-binding sites, named CD and EF sites (see Fig. 2). Kretsinger coined the name ''EF-hand'' to describe the binding domain of a Ca^{2+}-binding protein because it can be resembled by the forefinger (helix E), the thumb (helix F), and the bent middle finger (Ca^{2+}-binding loop) of the right hand. It is now often also called the ''helix-loop-helix'' model. The helices of the homologous Ca^{2+}-binding proteins usually are 10 amino acids in length, often start with glutamic acid, and have hydrophobic amino acids in positions 2, 5, 6, and 9. The loop domain usually contains 12 amino acids and has carboxylate containing amino acids in positions 1, 3, 5, and 12. Because Ca^{2+} usually is complexed in a octahedral fashion, two further ligands are in positions 7 and 9. The ligands complexing Ca^{2+} can be vertexed according to a cartesian system with X nominating the first position. On the basis of the similarity of ''EF-hand'' domains of different Ca^{2+}-binding proteins, the following general sequence principle of ''helix-loop-helix'' motifs of these proteins emerged:

X Y Z −Y −X −Z
Eh**hh**hD*D*DG*hD**Eh**hh**h

(D, aspartic acid; E, glutamic acid; G, glycine; h, hydrophobic; *, variable.)

The original ''EF-hand'' model (Fig. 2), designed on the basis of the parvalbumin crystal structure, appears to be valid also for other Ca^{2+}-binding proteins like troponin C or calmodulin, which have a high degree of sequence homology. This generalization is supported by the determination of the crystal structure of the intestinal calcium-binding protein (ICaBP), of troponin C, and of calmodulin. But in contrast to parvalbumin and ICaBP, which consist only of two Ca^{2+}-binding domains as pairs of EF-hands, calmodulin, as well as troponin C (from skeletal muscle), display Ca^{2+}-binding characteristics compatible with a ''pair of pairs''-model of EF-hands (i.e., both proteins bind 4 moles of Ca^{2+}/mole of protein with high affinity). The crystal structures of troponin C and of calmodulin provided evidence that the two EF-hand pairs are connected by a long helix, separating the two domains. This provided the two proteins with an unusual dumbbell-shaped

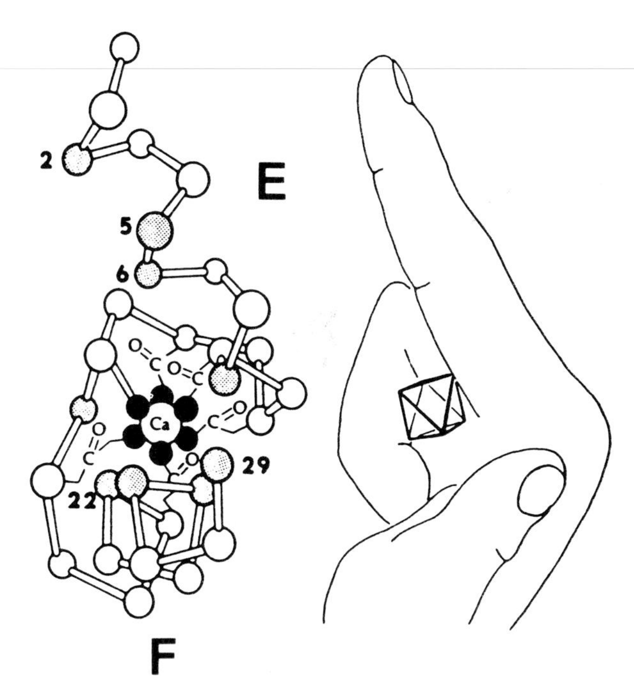

FIGURE 2 EF-hand model is built according to suggestions made by Kretsinger. [From Kretsinger, R. H., et al. (1988). *In* "The Calcium Channel: Structure, Function and Implications" (M. Morad, W. E. Nayler, S. Kazda, M. Schramm, eds.), pp. 16–35. Springer-Verlag, Berlin, with permission.]

appearance, suggesting the possibility that the two domains bind ions independently, as was indeed found using nuclear magnetic resonance (NMR) techniques. In addition, it is interesting to note that the central part of this long helix has a high degree of conformational mobility. This is probably important in the interaction of these proteins with their targets, which can involve bending of the central helix.

In contrast to the Ca^{2+}-binding domains of calmodulin or troponin C, the sites of interaction with target proteins are much less well-characterized. It is known that the Ca^{2+}-induced conformational change of calmodulin or of troponin C causes exposure of hydrophobic sites that are thought to be important for the interaction with targets.

The crystal structure suggests that a hydrophobic cleft visible in each half of the protein could be responsible for this interaction. These clefts are made mainly from ring clusters of four aromatic amino acids, as confirmed by NMR studies. Some polypeptides (e.g., melittin or mastoparan), which somewhat resemble the calmodulin-binding domains of different target proteins, appear also to interact with these hydrophobic pockets. This is in line with results obtained from NMR studies of peptides corresponding to the calmodulin-binding domain of myosin light-chain kinase and the plasma membrane Ca^{2+} pump.

To date, more than 150 different intracellular Ca^{2+}-binding proteins of the EF-hand type have been identified by sequence determination. These can be grouped into about 10 subfamilies. In Fig. 3 an ancestral tree is shown, indicating the possible evolutionary relation of these subfamilies. Proteins within a subfamily of this tree are more related to each other than to members of another subfamily. How close such a relation between two subfamilies can be is indicated by the high degree of sequence homology between calmodulin and skeletal troponin C, which share about 45% of identity and more than 70% homology. This high degree of homology in the primary structure is somewhat reflected also functionally because calmodulin and troponin C can replace each other to a certain extent.

The so-called S-100 proteins (named by Moore according to their peculiar property to be soluble in 100% ammonium sulfate solution) are closely related to ICaBP. No specific function has yet been

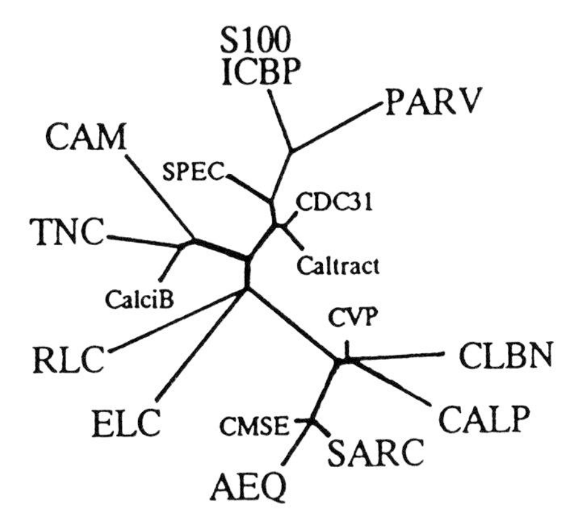

FIGURE 3 Evolutionary tree of the relation among the 10 subfamilies of EF-hand-type proteins. Caltract, caltractin from *Chlamydomonas reinhardtii*; CDC31, the cell division cycle gene 31 product from *Saccharomyces cerevisiae*: SPEC, the group of ectodermal proteins from *Strongylocentrotus purpuratus*; PARV, parvalbumin; S100/ICBP, several two EF-hand domain families including the intestinal calcium-binding protein 9K; CAM, calmodulin; TNC, troponin C; CalciB, B subunit of calcineurin; RLC, regulatory light chain of myosin; ELC, enzymatic light chain of myosin; CMSE, calcium-binding protein from *Streptomyces erythraeus*: AEQ, aequorin; SARC, sarcoplasmic calcium-binding protein; CALP, calpain; CLBN, calbindin and calretinin; CVP, calcium vector protein from *Branchiostoma lanceolatum*. [From Kretsinger, R. H., et al. (1988). *In* "The Calcium Channel: Structure, Function and Implications" (M. Morad, W. G. Nayler, S. Kazda, and M. Schramm, eds.), pp. 16–35. Springer-Verlag, Berlin, with permission.]

ascribed to these proteins, which are mainly localized in the nervous tissue of vertebrates and invertebrates. The protein usually exists as a dimer, which can bind four moles of Ca^{2+}.

Parvalbumin, the subfamily of proteins that have been crystallized first, has originally been found only in fish and in reptiles but was subsequently also identified in higher organisms. No specific function has been assigned to this class of Ca^{2+}-binding proteins, but it is interesting to note that the highest concentration occurs in muscle cells of fast twitch (i.e., fast relaxing fibers). In brain, parvalbumin is found in certain types of neurons.

VI. Systems Controlling Intracellular Ca^{2+} Concentration: Structural and Functional Properties

As it has been discussed in the previous sections, the ionized Ca^{2+} is kept in a narrow concentration range (100–500 nM) in the resting cell because this is crucial for its signaling function. Several transmembrane Ca^{2+}-transporting systems participate in controlling the free-Ca^{2+} concentration in the cell. These systems are either located in the plasma membrane (i.e., Ca^{2+} channel, ATP-dependent transporting system, Na^+/Ca^{2+} exchanger) or in the sarco(endo)plasmic reticular system (i.e., ATPase, Ca^{2+}-release channel) in mitochondria (i.e., an electrophoretic uptake system, a Na^+-dependent Ca^{2+} exchanger). Most probable also, the Golgi membranes and lysosomes contain ATP-driven Ca^{2+} pumps, which have not yet been characterized. Some important structural features of these systems will be summarized in this section.

A. The Calcium Channel

As indicated before transient, changes of free Ca^{2+} inside cells trigger many different cellular functions. These Ca^{2+} ions can either derive from intracellular stores or from the extracellular fluid by passing through the plasma membrane down their electrochemical gradient, using specifically regulated (i.e., gated) channels. The Ca^{2+} channels located in the plasma membranes in cells of different tissues mediate calcium influx during depolarization. They play an important role in excitation–contraction coupling in muscle tissues or couple changes in membrane potential at nerve terminals to the release of neurotransmitters. However, in this context we should emphasize that Ca^{2+} channels do also exist in nonexcitable cells, because no cell can afford diffusion of Ca^{2+} into the cell down its concentration gradient without a tight control, which only selectively gated channels (i.e., proteins) can provide. The selectivity of these gates gives rise to different permeability rates for various cations decreasing in the sequence Ba^{2+}, Sr^{2+}, Ca^{2+}, Mg^{2+}, whereas monovalent cations are much less permeable. However, cations such as La^{3+}, Cd^{2+}, Co^{2+}, Ni^{2+}, or Mn^{2+} can bind to the Ca^{2+}-binding sites with high affinity without being transported (i.e., these cations block the channel). Excitable cells provide great advantages in their way to study the properties of Ca^{2+} channels, and therefore most of the studies have been performed with those cell types. [*See* Ion Pumps.]

On the basis of their physiological characteristics to date, we can distinguish between three different types of channels: L, T, and N. Using the now well-established patch-clamp technique for studying single channels, it can be shown that these channels differ by their opening kinetics and their conductance. Channels of the T type produce a *transient* inward current with relatively small conductance whereas the L type can be characterized by its *long-lasting* inward current and about threefold stronger conductance. The third channel, the N type, has been found so far only in sensory neurons and has properties found neither in L nor in T channels. Characteristic for the N channel is an activation by rather strong depolarizations and an intermediate conductance. Ion flux through Ca^{2+} channels is about 10^6 Ca^{2+} ions/sec. The channel density in membranes can be as much as $1/\mu m^2$ in heart cells but is well above this value in membranes like the T tubules of skeletal muscle. The three channel types differ in their tissue distribution and also in their response to neurotransmitters or pharmacological agents. It is now well-established that the sensitivity of Ca^{2+} channels to adrenergic neurotransmitters is due to cyclic AMP–dependent increase in channel opening probability (i.e., increase in inward calcium current). The same effect can be obtained by intracellular injection of the catalytic subunit of the cyclic AMP–dependent protein kinase, indicating that protein phosphorylation mediates this effect. [*See* Protein Phosphorylation.]

Pharmacological agents can block the channel in a well-defined conformational state [e.g., either in the open (agonist) or closed conformation (antagonist)]. A number of different compounds interacting with Ca^{2+} channels have been described in recent years, among them tertiary amines (e.g., verapamil, D600), benzothiazepines (e.g., diltiazem) or 1,4-dihydropyridines (e.g., nifedipine, felodipine). The latter block selectively the L-type channel, which seems to be the most common type of Ca^{2+} channels.

The L-type channel is a voltage-dependent Ca^{2+} channel with distinct modes of gating (i.e., control of opening). The open-state probability increases with increasing depolarization of the membrane. It can be modulated by β-adrenergic agonists. As a consequence of this stimulation, one of the subunits of the channel proteins can be phosphorylated in a cyclic AMP–dependent fashion, increasing its opening probability. The selective interaction of 1,4-dihydropyridines and their derivatives (e.g., nifedipine) with the L-type channel allowed their identification and isolation in a pure form from brain cells, from rabbit skeletal muscles, and from guinea pig skeletal muscles. According to these studies, the L-type channel is a glycoprotein of 170–220 kDa with a subunit composition of either 140 (α), 50 (β), and (γ), or 140 and 30 kDa. Subunits α and β can be phosphorylated in a cyclic AMP–dependent manner, but only phosphorylation of the α subunit seems to be of a functional relevance.

Cloning of the cDNA for the larger subunit of the dihydropyridine receptor from skeletal muscle showed that it contains 1873 amino acids (212 kDA) and suggested that it is similar to two other channel proteins, a voltage-sensitive Na^{+} channel and a K^{+} channel from *Drosophila*. The similarity is especially significant in several internal repeat units, which are thought to serve as voltage sensors.

B. Ca^{2+} Pump of Plasma Membranes

The ATP-dependent Ca^{2+} transport system of plasma membranes is a protein of low abundance (e.g., 0.1% of the total membrane proteins of human erythrocytes). It has a high affinity and specificity for Ca^{2+} and extrudes the ions from the cell against its concentration gradient, using ATP as an energy source. Therefore it belongs to the class of the so-called E_1-E_2 ATPases (i.e., it has at least two different conformational states, E_1 and E_2, during the reaction cycle). The energy provided by ATP is conserved intramolecularly during the cycle by forming an aspartylphosphate intermediate. [*See* ADENOSINE TRIPHOSPHATE (ATP).]

The pump can be activated by calmodulin. As a result, the Ca^{2+} affinity and its transport rate are increased. The interaction between the two proteins has been used to isolate the Ca^{2+} pump from solubilized erythrocyte membranes by applying calmodulin affinity chromatography. The purified enzyme consists of a single polypeptide of 1,220 amino acids (135 kDa). A number of functional domains highly conserved among ion-transporting pumps could be identified: Asp 475 is the residue of the aforementioned acylphosphate, Lys 601 is part of the ATP-binding site. These two residues are well-separated within the sequence but are thought to be brought close to each other in space during the reaction cycle with the help of another highly conserved domain, the so-called hinge-region. Residues 1100–1127 comprise the calmodulin-binding domain. The Ca^{2+} pump can also be regulated by cyclic AMP-dependent phosphorylation (Ser 1178), which increases the affinity of the enzyme for Ca^{2+}. The pump is suggested to have 10 membrane spanning domains, which appear in comparable areas of the sequence of other ion-transporting pumps. The enzyme has a stoichiometry of transported Ca^{2+}/hydrolyzed ATP close to 1 and is an electroneutral Ca^{2+}/H^{+} antiporter (i.e., it transports Ca^{2+} and H^{+} in opposite directions and equivalent amounts).

The stimulation of the Ca^{2+}-ATPase by calmodulin has been studied using calmodulin fragments obtained by controlled proteolysis. The structural properties of some of these fragments are similar to those of the native protein. These studies showed that the C-terminal half of CaM (i.e., fragment 78-148) can fully activate the ATPase. Short fragments at the N-terminus do not stimulate the ATPase, but fragment 1-106 does, pointing to the importance of the third Ca^{2+}-binding site of CaM as the appropriate recognition site. however, the N-terminal half of CaM, albeit unable to activate the ATPase, confers to the third Ca^{2+}-binding site the ability to interact with the enzyme, because fragment 78-106 (i.e., the binding site proper) does not have this ability. There is also evidence indicating cooperativity between the first and second half of CaM and that it interacts differently with different targets.

Activation of the plasma membrane Ca^{2+} pump without calmodulin can be achieved by either exposure of the enzyme to negatively charged phospholipids or to fatty acids or by controlled proteolysis. These observations were important for the understanding of structure–function relations of the enzyme. So it could be shown that about the first 300 amino acids of the pump (including also the first two proposed transmembrane segments) can be cleaved off by trypsin, resulting in a fragment of about 90 kDa without apparently affecting any function of the enzyme including calmodulin-sensitive Ca^{2+} transport in a reconstituted system. However, further proteolysis converts this fragment into a peptide of 85 kDa, which still can bind calmodulin but has lost its sensitivity to be stimulated by it. Further cleavage produces a 81-kDa and finally a 76-kDa fragment, which both have full Ca^{2+}-dependent ATPase activity but have completely lost their calmodulin sensitivity. These findings indicated that the CaM-binding domain of the ATPase is located near the C-terminus of the protein.

DNA cloning techniques and peptide sequencing identified at least four different gene families of the plasma membrane Ca^{2+} pump in both humans and in rats. Due to alternative splicing which mainly occurs in the C-terminal region of the protein where most of the regulatory domains are located, each of these gene products may subdivide into several isoforms. The latter may then give rise to the possibility of expressing different forms with distinct properties (differently regulated?) in different tissues.

C. Sarco (Endo) Plasmic Reticulum

Skeletal, heart, and smooth muscle cells contain a reticular system, which is important for the control of the free-Ca^{2+} concentration in the cell. Most studies have been carried out on the sarcoplasmic reticulum (SR) of striated muscles, but recently the interest in studying the endoplasmic reticular (ER) system increased rapidly because it became evident that Ca^{2+} stored in this system can be released into the cytosol by inositol-1,4,5-phosphate. The corresponding receptor has been purified (M_r 260 kDa) and localized in the ER membrane.

The SR is an important element concerning the excitation-contraction coupling of muscles. In striated muscles, the SR forms longitudinal tubules. In the vicinity of the transverse tubules (T tubules), which are periodic inflections from the plasma membranes at the level of the Z lines, the SR forms compartments called the *terminal cisternae*, which contain "feet-like" projections connecting the SR to the plasma membrane and to the T tubules (diadic or triadic junctions).

Recently, one of these "feet-like" structures (i.e., of the triadic junction) has been identified as the calcium-release channel of SR. It was identified as the ryanodine receptor (ryanodine is a plant alkaloid, which potentiates Ca^{2+} release in SR) in skeletal as well as in heart muscle cells. The functional unit of the channel is a tetramer. The amino acid sequence of the monomer, deduced from the cDNA, consists of 5,037 amino acids (molecular mass, 565,223). Only four potential transmembrane domains are situated right at the C-terminal end, suggesting that 90% of this protein protrudes into the cytosol. This is in line with the morphological observation that the ryanodine receptor spans the 150-Å gap between the SR and the T tubules, forming the basis for the attractive hypothesis that there exists a physical interaction between the Ca^{2+}-release channel and the voltage-dependent Ca^{2+}-channel of the T tubules. This would provide a rational explanation for a direct triggering of Ca^{2+} release by the Ca^{2+} channel located in the T tubules, as observed by a number of laboratories.

A major component of the SR (less abundant in ER) is an ATPase that pumps Ca^{2+} out of the cytosol into the lumen of the reticular system. It can represent as much as 90% of the membrane protein (in SR of skeletal muscles), but even in less favorable cases (e.g., heart cells), it is still 50%. The SR Ca^{2+} pump also belongs to the E_1-E_2 type of ion-transporting ATPases, similar to the plasma membrane Ca^{2+} pump. The SR enzyme is smaller than its counterpart from the plasma membrane (1,001 amino acids as compared with 1,220 amino acids of the plasma membrane Ca^{2+} pump). The difference reflects a major difference in the regulation of the two enzymes: the SR Ca^{2+}-ATPase does *not* directly interact with calmodulin and therefore lacks the corresponding regulatory domain.

The Ca^{2+} pump of cardiac or smooth muscles SR (but not of fast skeletal muscles) is regulated via a highly hydrophobic, phosphorylatable protein called *phospholamban*, a pentamer of a subunit with an M_r of 5–6 kDa. The protein serves as a substrate of several protein kinases (e.g., cyclic AMP–dependent protein kinase, calmodulin-dependent kinase, and protein kinase C). The recently determined primary structure of the monomer (52 amino acids = 6,080 Da) showed that the amino acids phosphorylated by the various kinases are located next to the N-terminus. They are a serine residue, which becomes phosphorylated by a protein kinase A, and next to it a threonine residue, which can be phosphorylated by a calmodulin-dependent kinase. A direct interaction between phospholamban and the SR Ca^{2+} pump exists and influences the Ca^{2+} transport rate across the membrane. Transport is strongly influenced by the phosphorylation state of the phospholamban: Phosphorylation induces an increase in the enzyme's affinity of Ca^{2+}. The action of phospholamban on the SR Ca^{2+} pump could be compared with an inhibitor of the enzyme, which becomes ineffective by phosphorylation and thereby permits the pump to express its full activity.

D. Na^+/Ca^{2+} Exchanger of Plasma Membranes

It has been pointed out before that the fine and rapid tuning of the Ca^{2+} signal is performed by a membrane Ca^{2+} pump because only this enzyme possesses a high Ca^{2+} affinity. However, these enzymes have low capacity and are not able to transport Ca^{2+} in bulk quantities across the membrane. In contrast to these, there exists in the plasma membrane a Na^+/Ca^{2+} exchanger, which has a rather low affinity but high transport capacity for Ca^{2+}. It is an electrogenic system (i.e., it transports $3Na^+$ for $1Ca^{2+}$ in directions that depend on the magnitude and direction of the transmembrane gradients of Na^+ and Ca^{2+} or the transmembrane electrical potential).

The Na^+/Ca^{2+} exchanger is particularly active in plasma membranes of cells from excitable tissues. It may exist as a tetramer with a subunit of M_r 30–35 kDa. An interesting observation concerns the ATP-dependent activation of the Na^+/Ca^{2+} exchange process, suggesting the possibility that the exchanger is regulated by a phosphorylation/dephosphorylation process. Indeed it was observed that the Ca^{2+} affinity of the Na^+/Ca^{2+} exchanger can be influenced by a phosphorylation process under the control of a CaM-dependent protein kinase, which can be reversed by a CaM-dependent phosphatase.

E. Ca^{2+}-Transporting Systems in Mitochondria

Over the years the Ca^{2+} transport systems of isolated mitochondria have been studied most extensively because for a long time these systems have been regarded as extremely important in buffering the free cytosolic Ca^{2+} concentration. However, more detailed kinetic studies revealed that the affinity of mitochondria for Ca^{2+} is not sufficiently sensitive to play a role in the regulation of reactions in the sub-micromolar range of Ca^{2+}. In this context it is relevant that the mitochondrial Ca^{2+} uptake rate is about 10-fold slower than that of the SR. However, mitochondria possesses by far the largest buffering capacity within the cell, which might be especially significant when the Ca^{2+} concentration in the cytosol increases pathologically. Mitochondria are able to accumulate Ca^{2+} at a high rate in the presence of the permeant anion inorganic phosphate, by producing insoluble calcium phosphate precipitation, probably as amorphous hydroxyapatite.

It has been extensively documented that under the influence of toxic agents interfering with the Ca^{2+} permeability of plasma membranes, mitochondria accumulate substantial amounts of Ca^{2+}, which become visible under the electron microscope as electron-opaque granules. Similar observations have been made with hormonally induced hypercalcemias, resulting in similar dense mitochondrial granules, especially in tissues where the hormones specifically increase Ca^{2+} transport (kidney, bone). This safety device is of utmost importance for the cells because such an excess of Ca^{2+} entry into cells is an early and frequent observable phenomenon in cell injury (''calcium overload'') and can be secured because of significant Ca^{2+}-storing capacity of mitochondria.

The electrogenic Ca^{2+} uptake system of mitochondria operates under normal physiological conditions only at a marginal rate. Because of their large membrane potential (negative inside), mitochondria would continuously take up Ca^{2+} if there were not a mechanism for exporting Ca^{2+} against the large membrane potential. Indeed, such a system has been identified as a electroneutral Na^+/Ca^{2+} exchanger, which operates independently of the electrophoretic uptake route.

Besides the Na^+-promoted Ca^{2+} release route, there exists also a Na^+-independent route, which has highest activity in mitochondria of nonexcitable tissues (liver, kidney) where the Na^+/Ca^{2+} exchanger has its lowest activity. Whether this route is a Ca^{2+}/H^+ antiporter or caused by changes of the permeability properties of the inner mitochondrial membrane is still controversial. It is possible that a Ca^{2+} release route is linked to the hydrolysis of pyridine nucleotides concomitant with the ADP-ribosylation of a protein from the inner mitochondrial membrane.

VII. Conclusions

The following general points can be made on the basis of the material presented here:

1. Calcium plays a pivotal role in biological systems. It predominantly occurs in a complexed form fulfilling either a static, structure-stabilizing function or a dynamic, signal-transducing role.

2. Cells are exposed to a large gradient of Ca^{2+} across their membranes. This is the condition that permits the cell to use even small changes of membrane permeability to cause significant changes in the free-Ca^{2+} concentration inside the cell, which can be used to transmit and amplify a signal transduced from the extracellular space. However, it is clear that the concentration of ionized calcium within the cell has to be carefully controlled to use Ca^{2+} as an intracellular second messenger.

3. Intracellular Ca^{2+}-binding proteins have developed during evolution, which complex Ca^{2+} with high affinity and specificity. Their characteristic structural principle is a common motif (i.e., the helix-loop-helix or EF-hand principle), which permits these proteins to bind Ca^{2+} with high selectivity in the presence of high concentrations of other ions. By contrast, Ca^{2+}-binding properties of extracellular proteins normally are of lesser specificity and affinity. In the latter proteins, calcium often plays a structure stabilizing function.

4. Signal transformation from outside the cell, changing the free-Ca^{2+} concentration in the cytosol, can occur either directly by opening calcium channels in the plasma membrane or indirectly by opening Ca^{2+} pathways from intracellular stores using other second messengers (i.e., inositolpolyphosphates). In this way the signal is transferred from the extracellular space to intracellular receptors and amplified using different second-messenger pathways.

5. The intracellular interplay between different second-messenger systems can be manifold and complex, as outlined in Fig. 1. In this respect it is interesting to note that under many conditions signal transduction is not only transferred from the plasma membrane to the cytosol but also to the nucleus. Thus the synthesis of several calmodulin-binding proteins is induced in the nucleus. Future research will dissect in more detail the complexity of signal transduction pathways and especially will identify not only cytosolic but also nuclear responses of the cell on receiving extracellular signals.

Bibliography

Carafoli, E. (1987). Intracellular calcium homeostasis. *Annu. Rev. Biochem.* **56,** 395.

Carafoli, E., Krebs, J., and Chiesi, M. (1988). Calmodulin in the transport of calcium across biomembranes. *In* "Calmodulin" (C. B. Klee and P. Cohen, eds.). Elsevier, Amsterdam.

Cheung, W. Y., ed. (1980). "Calcium and Cell Function," a series of monographs. Academic Press, New York.

Kretsinger, R. H. (1987). Calcium coordination and the calmodulin fold: Divergent versus convergent evolution. *Cold Spring Harbor Symp. Quant. Bio.* **52,** 499.

Strynadka, N. C. J., and James, M. N. G. (1989). Crystal structures of the helix-loop-helix calcium-binding proteins. *Annu. Rev. Biochem.* **58,** 85.

Carbohydrates (Nutrition)

CAROL N. MEREDITH, *University of California, Davis*

Glossary

Cariogenicity Capacity of a food to produce dental caries

Complex carbohydrates Polysaccharides such as starch, cellulose, and hemicellulose; only starch can be digested to glucose in the gut, whereas the others make up dietary fiber

Corn sweeteners Sweet mixture of fructose, glucose, and other carbohydrates made by the enzymatic breakdown of corn starch

Glycemic index Ratio between the increase in blood sugar after consuming a test carbohydrate and the increase obtained after consuming an equivalent amount of carbohydrate from a standard food such as white bread

Invert sugar Liquid mixture of glucose and fructose made by chemical hydrolysis of sucrose

Nonsugar sweeteners Natural and manufactured simple sugars such as fructose, sorbitol, and xylitol that are absorbed less efficiently than sugar or glucose

NUTRITIONAL CARBOHYDRATES ARE organic compounds of carbon, hydrogen, and oxygen, which can be digested, absorbed, and broken down in the tissues to water and carbon dioxide. They make up the bulk of human diets throughout the world, starch-rich foods being the least expensive to produce and store. The main function of carbohydrates is to provide energy. They also contribute to the taste and texture of food, with sucrose and fructose used as sweeteners. Diets rich in carbohydrates, especially complex carbohydrates, are healthier than high-fat diets and are not necessarily linked to obesity, diabetes, atherosclerosis, or hyperactivity.

I. Consumption Patterns in the United States and the World

Primitive nomadic peoples subsisted on meat and milk-based diets providing protein and fat as the main sources of energy. The domestication of plants with starch-rich seeds, such as wheat, rice, and corn, allowed humans to settle communities with an assured supply of food, which was the basis of civilization. A diet where about 80% of the energy came from starch was the norm before the twentieth century in the United States and is still typical in poor countries. Today, in the United States and other western nations, only 50% of the daily energy intake comes from carbohydrates.

The types of carbohydrates in the diet have also changed over the years. The intake of starch in the United States from foods such as potatoes, bread, and legumes has declined about 50% over this century, while the intake of sugar and other sweeteners has increased. Over the past 30 yr, the consumption of sweeteners has further increased, due to a greater use of sugar in manufactured foods (e.g., baked goods, breakfast cereals, ketchup, canned fruit) and the consumption of sweetened beverages (e.g., sodas, fruit juices) (Fig. 1). New sweeteners other than sugar have emerged, such as fructose and corn sweeteners, both manufactured from corn

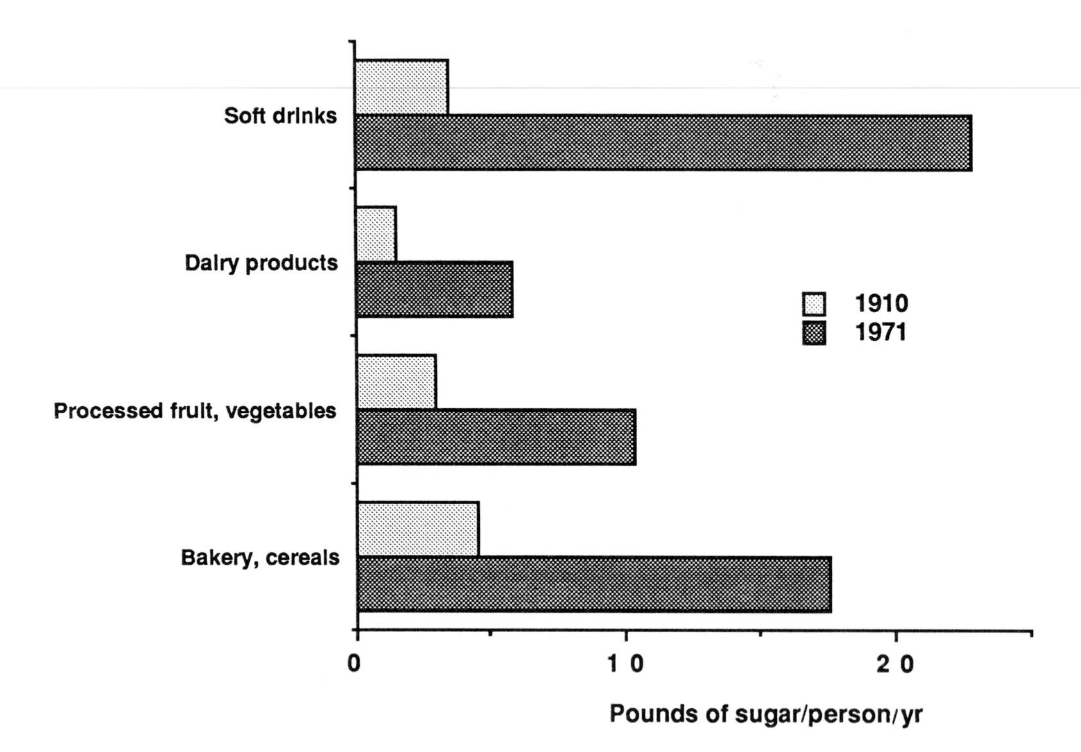

FIGURE 1 Increased yearly intake of sugar in prepared foods in the United States, comparing the year 1910 with 1971. Household use of sugar has declined.

starch, and the nonsugar sweeteners such as xylitol and sorbitol, used in chewing gum and candy. These new sweeteners may be digested and absorbed slightly differently from other carbohydrates, but they provide about the same amount of calories per unit of weight.

Carbohydrates are not essential nutrients because they can be made in the body from other compounds. However, a diet with little or no carbohydrates is perceived by most people as extremely unappetizing, and adaptation to such a diet is associated with fatigue, headaches, and lack of well-being. National guidelines for a healthier diet promote an increased intake of carbohydrates, especially as complex carbohydrates, and a lower intake of sugar.

II. Chemical Composition of Nutritional Carbohydrates in Foods

Carbohydrates are made up of one or more units of simple sugars. According to the number of sugar units, they are classified as monosaccharides (1 sugar), disaccharides (2 sugars), oligosaccharides (3–10 sugars), and polysaccharides (10 to several thousand sugars). The nutritional carbohydrates are mostly converted to glucose in the body and can be enzymatically broken down to carbon dioxide and water, producing an average 4 kcal/g carbohydrate oxidized.

A. Monosaccharides

The simple sugars, or monosaccharides, that are nutritionally important are six-carbon compounds with alcohol (−OH) groups in different positions and spatial arrangements. They are highly soluble in water and have a sweet taste (Table I). Glucose is the most important, because it is the carbohydrate that enters the tissues and is converted to energy or other products. Glucose is also called dextrose, because in solution it bends a beam of polarized light to the right. Fructose is also known as levulose, because it bends polarized light to the left. Fructose is sweeter than sugar and enhances the taste of fruit products. The simple sugar galactose is found only as a component of the disaccharide lactose and does not exist in free form in foods.

Polysaccharides and disaccharides must be hydrolyzed to simple sugars before being absorbed in the gut. Patients who must be fed with intravenous solutions can only utilize simple sugars (usually only glucose), because these are the only carbohy-

TABLE I Sweetness of Different Sugars Found Naturally in Food or Added to Foods

Name	Composition	Relative sweetness[a]	Origin
Starch	[glucose]n	—	cereals, tubers
Sucrose	fructose + glucose	100%	sugar cane, beet
Lactose	galactose + glucose	30%	milk
Maltose	glucose + glucose	40%	germinating seeds
Glucose	$C_6H_{12}O_6$	70%	fruit
Fructose	$C_6H_{12}O_6$	120%	fruit
Galactose	$C_6H_{12}O_6$	—	not found free
Sorbitol	$C_6H_{14}O_6$	54%	artificial sweetener
Xylitol	$C_5H_{12}O_5$	134%	artificial sweetener

[a] Compared to sucrose.

drates that can cross cell membranes to enter the tissues.

B. Disaccharides

The carbohydrates made up of two similar or different sugar units are the disaccharides. Lactose, the carbohydrate in milk, is made up of glucose + galactose. Maltose, a product of starch breakdown during germination of seeds or during digestion in the gut, is made up of two glucose units. Sucrose, or table sugar, is made up of glucose + fructose. The disaccharides are soluble in water.

C. Polysaccharides

The main polysaccharide in the diet is starch, made up of 300 to many thousand glucose units bound to each other in linkages denoted α-glucosidic bonds. While the polysaccharide cellulose is also a polymer of glucose, they are linked in a way that resists hydrolysis by the enzymes of the digestive system of humans or other mammals. Herbivores can feed off grass and leaves not through the action of their own digestive enzymes but because bacteria in the gut produce enzymes that can break the glucose–glucose bonds in cellulose.

The glucose chains in starch are about 25% straight chains forming spirals, called amylose, and about 75% branched chains, called amylopectin. Starch is a stable product, almost insoluble in cold water, as suggested by its use in cosmetic dusting powders. It is an important source of stored energy in plants. The starch in seeds, tubers, and other plant parts is found in granules. The starch in unbroken granules is different to digest, but during moist cooking, the granules swell and burst, facilitating digestion.

Starches that are modified chemically or enzymatically are used industrially to improve the texture, appearance, and keeping qualities of processed foods (Table II).

The animal polysaccharide is glycogen, similar in structure to amylopectin. Although the main store of energy in humans is as fat, the body of an adult man contains about 300 g glycogen in liver and muscle. Liver glycogen is a reserve of energy, but more importantly it is a store of easily released glucose units that can nourish tissues that are totally glu-

TABLE II Examples of the Use of Modified Starches and Starch-Derived Sweeteners in the Food Industry

Processed food	Effect of adding various starch-derived products
Canned food	Improved resistance to heat and acid; lower freezing temperature
Frozen foods	Improved stability
Ice cream	Decreased formation of large ice crystals
Jams and jellies	Decreased formation of sugar crystals; enhanced fruit flavor
Confections	Changes in viscosity, sweetness, shine, and transparency
Fruit juice	Enhanced fruit flavor; lower freezing temperature

cose-dependent (such as the brain) during times of fasting. Muscle glycogen can be broken down to lactate and other products, producing energy that can be used for physical work, but muscle cells lack the enzymes for regenerating glucose from stored glycogen. In meats such as steak, liver, or fish, the amounts of glycogen are insignificant, as these stores disappear before the food is eaten.

The carbohydrates containing 3–10 different simple sugars are known as oligosaccharides and are found in beans and other legumes. They are considered undesirable products because they are not digestible and lead to flatulence.

Carbohydrates in foods are often not measured directly but, rather, as the amount remaining after accounting for water, protein, fat, and minerals. In older food tables, the term "carbohydrate" may include starch as well as the nonabsorbable cellulose, sugar as well as nonabsorbable oligosaccharides, other types of plant fiber, organic acids and minor compounds. The accurate analysis of each type of carbohydrate in a typical food is a major undertaking given the variety of compounds involved.

III. Dietary Sources of Carbohydrates

Traditional staple foods (bread, rice, tortillas, corn porridge) usually provide starch as the main nutrient, but in our diet about 25% of the carbohydrates are sweeteners (sugar, glucose, fructose). Refined sugar foods are the most concentrated form of carbohydrate and are termed "empty calories" because they provide no other useful nutrients. In hard candies, artificially flavored fruit drinks, and sodas, >95% of the energy is provided as sugar or high-fructose syrup, with trivial amounts of essential nutrients or nonnutritive components such as fiber.

Honey is a natural sweetener that is a mixture of free fructose and glucose. It is not superior to sugar, although it provides some minerals and vitamins. Honey contains 70% more calories per spoonful than sugar, and both sweeteners enter the circulation as fructose and glucose.

Foods with a high content of sugar are jams, jellies, and dried fruits. Fresh fruits and vegetables provide a wide range of sucrose, glucose, starch, and nondigestible carbohydrates, depending on the species, variety, and state of ripeness.

Dried cereal grains such as wheat, rice, and corn contain 33–37% carbohydrates by weight. These foods are generally the cheapest source of calories. Because they make up such a large part of the diet, they provide substantial amounts of protein, fiber, minerals, and vitamins to the daily ration in most of the world.

IV. Digestion of Carbohydrates

All carbohydrates made up of more than one sugar unit must be hydrolyzed before being absorbed or utilized by cells. The breakdown of carbohydrates to smaller units or to simple sugars sometimes beings with food-processing and cooking.

When a food is eaten, digestion of carbohydrates beings in the mouth, where the salivary enzyme amylase splits starch into smaller units. In the stomach, the acid environment inhibits amylase but favors some chemical breakdown of sugar and starch. In the small intestine, the well-mixed liquid products of gastric digestion encounter the enzymes of pancreatic juice and of the intestinal cells. Pancreatic juice contains amylase that breaks down starch to short, straight chains of glucose units (maltose, maltotriose) and to short, branched chains of 5–10 glucose units, known as α-dextrins. The enzymes of the brush border of the cells lining the intestine complete the breakdown of the disaccharides, α-dextrins, and maltotriose. The brush-border enzymes include maltase, sucrase, isomaltase, and lactase, which hydrolyze maltose and maltotriose, sucrose, α-dextrins, and lactose, respectively; the simple sugars released (glucose, fructose, and galactose) are readily absorbed.

Carbohydrate digestion is generally efficient. Impaired digestion and absorption due to low activity of one or more digestive enzymes allows the carbohydrate to pass to the large bowel, where it becomes a nutrient for bacteria. The fermentation of the CHO by the bacteria leads to the production of gas, acid products, and toxins, which irritate the gut and lead to pain and diarrhea. The activity of lactase declines during childhood in about 80% of the population, except for Caucasians of Northern European origin, producing lactose intolerance. Among older children and adults of non-Caucasian ancestry, a dose of >10 g lactose, equivalent to a glass of cow's milk, quickly produces gastric pain, flatulence, and diarrhea. The only dairy products containing <10 g lactose per serving are butter, cream, cheese, cottage cheese, and some ice creams. Even in persons or infants who normally tolerate lactose, an episode of gastrointestinal infection and diarrhea can produce a transient decline in lactase activity and an intolerance for milk. [*See* LACTOSE MALABSORPTION AND INTOLERANCE.]

Young infants should not be fed starch, because amylase activity is low in pancreatic juice until 1 or 2 mo of age, leading to starch intolerance.

Some digestive enzymes adapt to the dietary supply of their substrate. Diets rich in sugar increase sucrase and maltase activity, but feeding more lactose does not increase lactase activity or improve lactose digestibility.

The passage of simple sugars from the gut mucosa to the blood is by diffusion and, for some sugars, also by active transport. The actively absorbed sugars are glucose and galactose, while fructose is absorbed more slowly by facilitated diffusion. A dose of >100 g pure fructose is only slowly absorbed, and the excess forms a concentrated solution in the gut that draws water from its surroundings by osmotic pressure and leads to diarrhea. The nonsugar sweeteners such as xylitol and sorbitol can produce a similar effect when consumed in large amounts.

V. Glucose Uptake and Utilization by Tissues

After the digestion and absorption of a carbohydrate-containing meal, the liver takes up the glucose and other simple sugars transported from the gut. The main end product of carbohydrate digestion is glucose, as it makes up 70–90% of all the carbohydrates we eat. Of all the sugars, glucose has the tightest metabolic control, by hormones and other regulators. The liver can use fructose and galactose as fuel or convert them to glucose and other products. Glucose is oxidized for energy. It can be stored as the polysaccharide glycogen, the excess can be converted into fat, and a carefully regulated amount is released into the circulation to be transported to other tissues. Most of the tissues obtain energy from the oxidation of fat and its breakdown products plus glucose, but the brain, red blood cells, and developing fetus are almost entirely dependent on glucose for fuel.

After a meal, in response to blood glucose levels >100 mg/100 ml, insulin is released from the pancreas. This hormone has a key role in regulating blood glucose and the metabolism of many tissues. Insulin increases the uptake of glucose by muscle and adipose tissue, promoting the storage of glycogen, protein, and fat. If no food is consumed for several days, the tissues gradually adapt to utilizing the fuels obtained from the breakdown of fat stores and tissue proteins, but the body still needs about 130 g glucose per day. Initially, glucose is obtained by breaking down glycogen stores in the liver, mediated by the hormone glucagon or the stress hormones. After 1 or 2 days, liver glycogen stores are exhausted and new glucose is made in the liver from the carbon chains produced in the partial breakdown of fats and proteins. Glucose must also be manufactured from other substrates in the liver when the diet provides less than about 130 g glucose per day. The provision of enough carbohydrate in the diet is protein-sparing because it prevents the use of amino acids for the synthesis of glucose. [*See* INSULIN AND GLUCAGON.]

Under stress conditions, hormones that increase blood glucose levels are released. Glucose is the most important fuel for muscles during rapid and intense exercise. The effect of a surge of anger and fear is to fuel the muscle for "flight or fight." Unfortunately, in a civilized context where persons do not immediately respond to stress by violent movement or running away, the increased glucose is channeled into the synthesis of triglycerides.

Glucose absorbed into the cells of a tissue can follow several pathways. It can be used to produce energy, by hydrolysis to lactate in the absence of oxygen, as occurs in muscles during intense contractions, or by complete oxidation to carbon dioxide and water. Glucose can also be used for making other substrates or products for energy storage. In muscle and liver, it can be stored as the polysaccharide glycogen. In adipose tissue and the liver, it can be converted into fat and stored. It is directly or indirectly a substrate for making dispensable amino acids, cholesterol, nucleic acids, mucous secretions, and components of cartilage and for detoxifying certain drugs in the liver.

VI. Utilization of Dietary Carbohydrates

A. Characteristics of the Consumer

1. Age

The digestion of carbohydrates and the utilization of glucose by the tissues are affected by age. Lactose digestibility is greatest in infants, declines sharply in childhood in most ethnic groups, and declines further in old age. The sensitivity of tissues to insulin declines in old age, especially in persons who are obese and inactive. A drink containing 75 g glucose can produce high levels of blood glucose (120–180 mg/100 ml) for at least 2 hr. Although this is not diabetes, it is defined as glucose intolerance.

2. Physical Activity

The energy for intense, brief exercise is produced by the anaerobic breakdown of glucose to lactate, while prolonged low-intensity exercise is fueled by the oxidation of glucose and fat. The muscles of endurance-trained athletes contain more glycogen and have a greater capacity for taking up glucose from the blood, yet during aerobic exercise they produce energy from a fuel mix that is richer in fat, compared with the muscles of sedentary persons. The increased ability of trained muscles to take up glucose is a transient effect of individual bouts of exercise, lasting for 2–3 days.

3. Genetic Background

Digestion and absorption of certain carbohydrates can be impaired because of a genetic deficiency in specific digestive enzymes. The problem of impaired utilization of glucose and inadequate regulation of blood glucose has a strong genetic component. In diabetes mellitus, high-circulating glucose coexists with poor nutrition of tissues, because the lack of insulin prevents adequate entry of glucose into cells. Juvenile diabetes is a serious disease requiring insulin administration. The more common form of diabetes is a milder disease that emerges during middle age, especially in persons who are excessively fat and whose body fat is accumulated in the trunk rather than the hips and thighs. Obesity is found in 80% of patients with adult diabetes. In the United States, persons of Native American ancestry, including most Mexican Americans, have a 3–4× greater risk of developing diabetes than Caucasians or Blacks. Maintaining appropriate body weight by dieting and exercise decreases the risk of diabetes in susceptible persons. Chronic high blood glucose in diabetics produces accelerated atherosclerosis, nerve damage, cataracts, and capillary damage that can lead to kidney disease and retinal damage.

Abnormally low blood glucose, or hypoglycemia, can occur with an overdose of insulin, with very prolonged exercise or, more rarely, due to an inherent inability to increase blood glucose during fasting. True hypoglycemia is a rare disease. Its symptoms are headache, fatigue, confusion, seizures, and even loss of consciousness brought on by fasting 8–14 hr. These symptoms can be avoided by eating small frequent meals.

B. Characteristics of the Diet

Overnutrition leading to fatness impairs glucose utilization, whereas weight loss in obese subjects improves glucose utilization. The composition of the diet has immediate and long-term effects. The increase in blood glucose and accompanying insulin is different for the various carbohydrates and is affected by the physical composition of food, the presence of other nutrients, and the presence of different sorts of dietary fiber. After consuming glucose or sugar, blood glucose immediately increases. The response to eating complex carbohydrates is slower but varies for different starches. The glucose and insulin response to carbohydrate consumed as potatoes is greater than the same amount of carbohydrate consumed as rice. The glucose response to apple juice is greater than for applesauce, which in turn is greater than for a raw, unpeeled apple, reflecting the difference in rate of digestion and absorption due to fiber.

Changing to a high-carbohydrate diet improves glucose utilization by the tissues. Some scientists attribute this effect to the increased fiber in diets that are rich in complex carbohydrates. Conversely, eating a high-fat diet impairs glucose utilization, probably because it involves lowering carbohydrate intake. Excess sugar is not a cause of diabetes, but it is harmful for persons who have the disease. The easiest way to increase the carbohydrate content of the diet is to choose low-fat foods, as shown in the examples of Table III.

The micronutrients affect the efficiency of glucose utilization in cells, because many enzymes required for glucose metabolism derive from the B vitamins and require mineral cofactors such as magnesium, zinc, and chromium.

VII. Effects of a Very High-Carbohydrate Diet

A. Athletic Performance

Endurance-trained persons who compete in events lasting several hours have found that an increased store of muscle glycogen can delay fatigue by about

TABLE III Examples of Daily Meals for a High-Carbohydrate or a Low-Carbohydrate Diet

High-carbohydrate diet	Low-carbohydrate diet
Breakfast	
English muffin (large)	Whole-grain bread (thin slice)
Jelly	Butter
Cornflakes	Fried egg and bacon
Coffee and skim milk	Coffee and cream
Sugar	Artificial sweetener
Lunch	
Grilled chicken (no skin)	Tuna in oil
Steamed rice	Crackers
Steamed zucchini	Avocado and lettuce
Bread roll	Creamy salad dressing
Lemonade	Diet soda
Grapes	Chocolate chip cookies
Dinner	
Pea soup	Hamburger on a roll
Spaghetti with tomato sauce	French fried potatoes
Roasted apple with syrup	Ice cream
Soda	Diet soda
Snacks	
Popcorn	Peanuts
Nonfat yogurt with sugar	Full-fat yogurt, no sugar
Crackers	Doughnut
Skim milk	Whole milk

23% or allow a greater speed toward the end of a race. Experimentally, athletes can increase muscle glycogen by eating a diet where about 70% of the calories come from carbohydrates, during about 3 days before competition, while reducing training intensity. A typical precompetition diet is shown by the high-carbohydrate meals in Table III. However, a large intake of carbohydrate immediately before exercising is not advisable, especially in the form of rapidly absorbed glucose or sugar. This will produce a surge of insulin at a time when exercise itself is causing the rapid removal of glucose from the blood to the muscles. The combination of an insulin peak together with exercise can drive down blood glucose to levels that cause fatigue, discomfort, and reduced endurance. At the end of an exercise bout, consuming dilute solutions of sugar or other easily assimilated carbohydrates helps rapidly restore body water and muscle glycogen. This is important for sports involving repeated events, such as triathlons or cross-country cycling.

B. Caries

There is a linear relationship between a country's per capita sugar intake and the average number of dental caries in its inhabitants. The bacteria that cause caries reproduce on any food particles adhering to teeth, and starch as well as sugar can support the formation of plaque and bacterial growth. The most cariogenic foods, however, are rich in sugar and sticky, such as chewy candy, cookies, or raisins, while the dissolved sugar in sodas or milk-moistened breakfast cereals is less harmful. [*See* DENTAL CARIES.]

C. Sugar and Hyperactivity in Children

In recent years, sugar has been blamed for the hyperactivity of some children, despite little experimental evidence. Properly designed double-blind studies, in which neither the children nor the investigators could tell if sugar or a fake dose of sugar was given before examining behavior, have not shown that sugar has any effect on hyperactivity.

D. Atherosclerosis and Cardiovascular Disease

High-carbohydrate, low-fat diets reduce plasma cholesterol and blood lipids only if the diet is also rich in fiber. Diets rich in complex carbohydrates are made up of a variety of plant foods, which include high amounts of different types of fiber. [*See* ATHEROSCLEROSIS.]

Bibliography

Dobbing, J. (1989). "Dietary Starches and Sugars in Man: A Comparison." Springer-Verlag, London.

MacDonald, I. (1988). Carbohydrates. *In* "Modern Nutrition in Health and Disease" (M. E. Shils and V. R. Young, eds.). Lea & Febiger, New York.

National Research Council. (1989). "Recommended Dietary Allowances," 10th ed. National Academy of Sciences, Washington, D.C.

Sipple, H. L., and McNutt, K. W. (1974). "Sugars in Nutrition." Academic Press, New York.

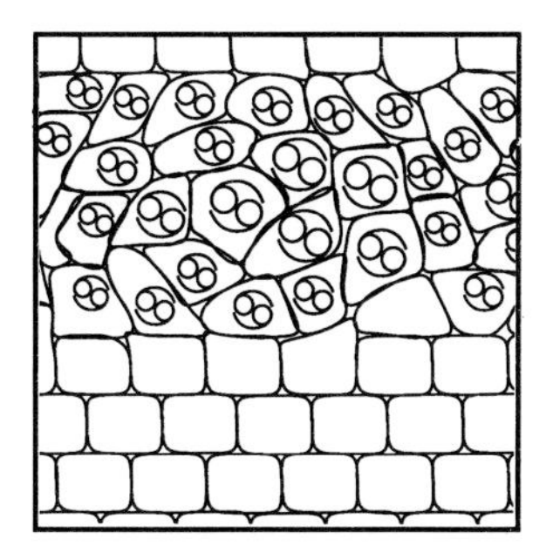

Carcinogenic Chemicals

RICHARD L. CARTER, *Institute of Cancer Research and Royal Marsden Hospital*

Glossary

Carcinogenesis Origin and development of tumors; applies to all forms of tumors and not only to carcinomas (cf. tumorigenesis in American usage)

Carcinogens Causal factors of tumors including exogenous factors (chemicals, physical agents, viruses), endogenous factors (hormones) and also general factors (e.g. nutrition, reproductive activities). A broad distinction may be drawn between genotoxic carcinogens which react directly with and mutate DNA, and nongenotoxic carcinogens which appear to act through other (non-genetic) mechanisms

Mutation Permanent change in the base sequence of the DNA constituting the genetic material (the genome); alterations may involve a single gene, a block of genes, or a whole chromosome; effects involving single genes may be a consequence of modifications of single DNA bases (point mutations) or of larger changes, including deletions, within the gene, whereas effects on whole chromosomes may involve changes in number and/or structure; chemicals that produce such changes are mutagens and mutagenic carcinogens are often described as genotoxic

Toxicokinetics Mathematical description of rates of absorption, distribution, and elimination of toxic chemicals

Tumor (syn. neoplasm) Mass of abnormal, disorganized tissue characterized by excessive and uncoordinated cell proliferation and by impaired differentiation; benign tumors generally show a close morphological resemblance to their tissue of origin, grow slowly, and do not disseminate—they are rarely fatal; malignant tumors resemble the parent tissue less closely, are composed of increasingly abnormal pleomorphic cells, grow rapidly, and disseminate (metastasize) to distant sites—untreated, they are fatal; the nomenclature of tumors is based on the tissue of origin and the presence of benign or malignant features, judged by morphology; cancer is a general synonym for malignant tumors

CARCINOGENIC CHEMICALS ARE substances that are causally associated with an increased risk of tumors in humans and/or experimental animals. Their activity, which can in part be predicted from chemical structure, may reside in the parent compound or (more commonly) in activated metabolites derived from it in the body. Human exposures to carcinogenic chemicals occur in occupational, environmental, and social or cultural contexts. Most chemical carcinogens exert their effects after prolonged exposure, show a dose–response relationship, and tend to act on a limited number of susceptible target tissues. There is a long latent period between the first encounter with a carcinogenic chemical and the appearance of a tumor. The mechanisms of chemical carcinogenesis are complex, and at least three phases can be identified: (1) initiation: cells in the target tissue sustain one or more mutations following covalent binding of the reactive carcinogen to DNA; (2) promotion: the initiated (mutant) cells give rise to progressively expanding populations (clones) to form gross tumors; and (3) progression: the later phases of tumor growth involving benign → malignant transformation and dissemination. Additional mutations occur during promotion and progression, resulting in activation of normal proto-oncogenes and mutation or deletion of suppressor genes. Overall, there is cumulative ge-

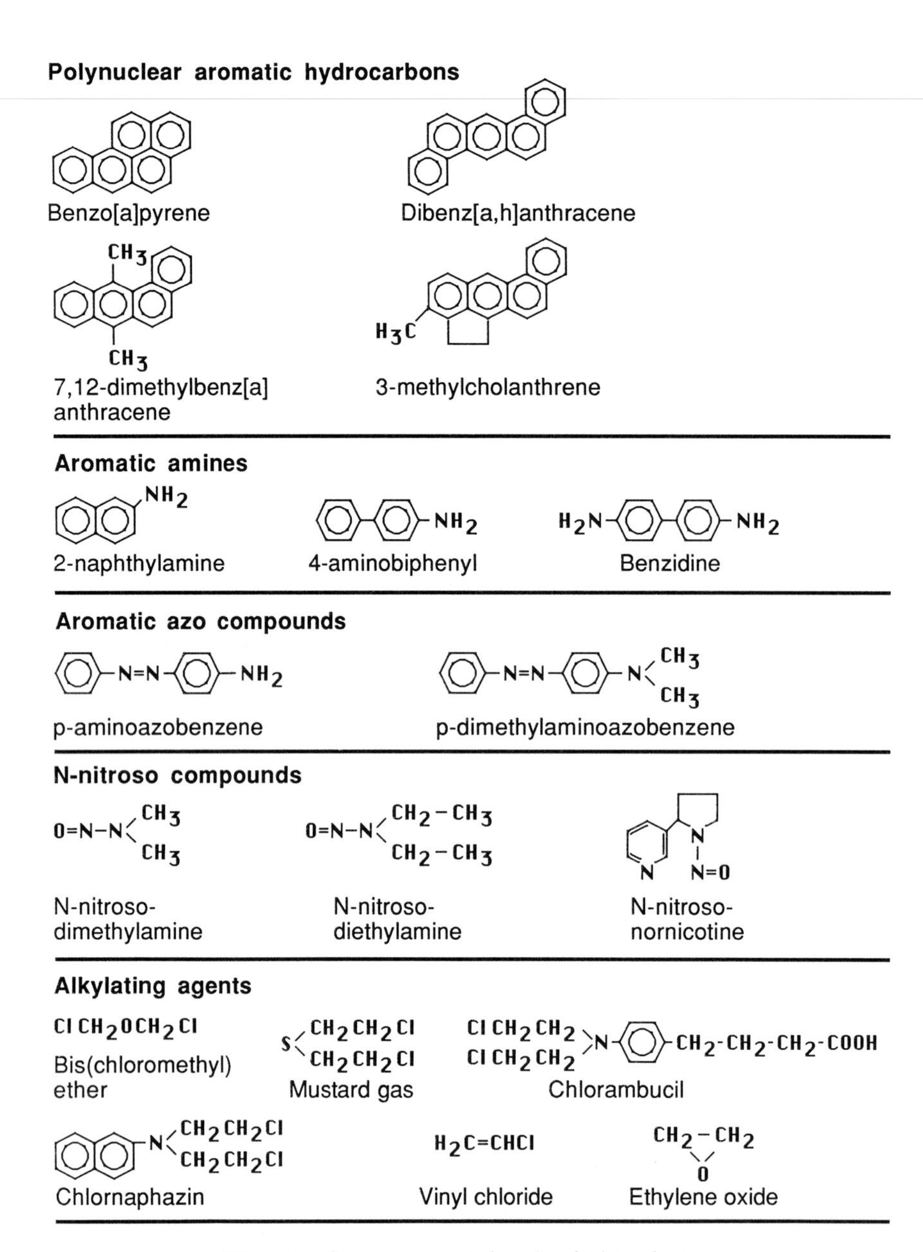

FIGURE 1 Some representative chemical carcinogens.

netic damage, which is enhanced by nongenetic (epigenetic) cellular and tissue damage. Chemical carcinogens appear to act through a combination of genotoxic and nongenotoxic effects.

I. Chemicals as Carcinogens

It is widely accepted that the majority of human cancers are caused by extrinsic factors, a broad group that includes specific chemical carcinogens. Such chemicals, which induce tumors in humans and/or experimental animals, are structurally diverse and comprise naturally occurring substances and hormones as well as synthetic chemicals (some examples are illustrated in Fig. 1). Several human carcinogens occur as mixtures (pyrolysis products, ores) in which it is difficult or impossible to attribute effects to any one constituent. The mode of action of chemical carcinogens is determined by the initial route(s) of exposure and the processes of absorption, distribution within the body, and metabolism. Toxic effects, separate from carcinogenic activity, may also contribute. Chemical structure provides

Aflatoxin B_1 Cycasin Safrole

Steroid hormones and related compounds

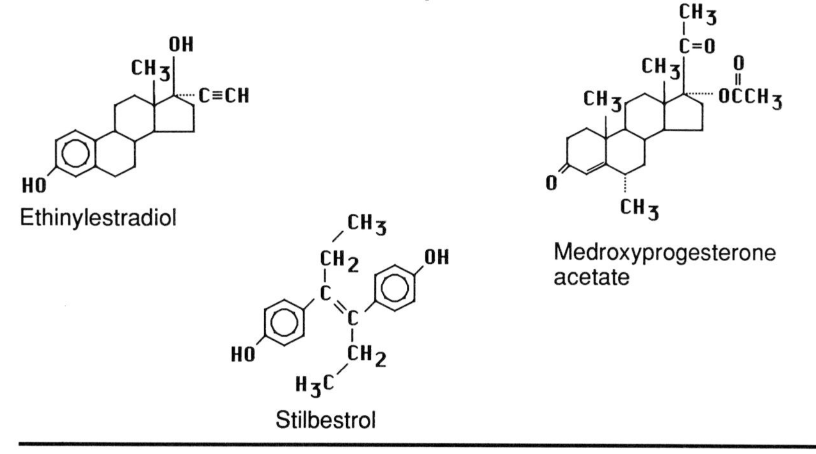

Miscellaneous

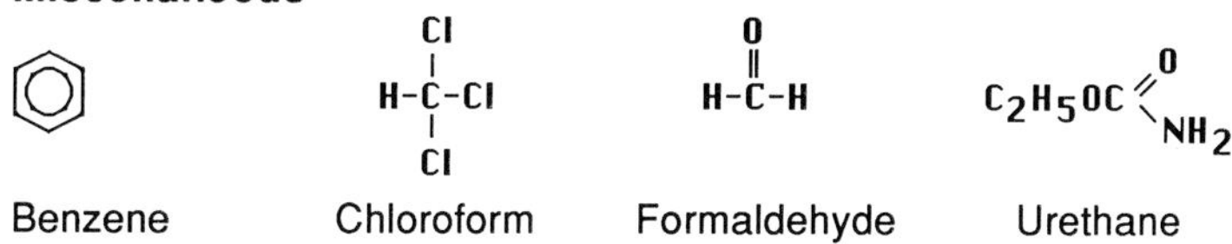

Metals and organometallic compounds: arsenic, asbestos, chromium, nickel, cadmium, beryllium

important clues to carcinogenic activity, particularly for genotoxic carcinogens; and, in the case of *N*-nitroso compounds, it may indicate likely target organ susceptibility. But such "structural alerts" tend to be oversensitive and cannot be viewed in isolation—a suspicious chemical may prove innocuous because it is not absorbed or is rapidly detoxified or excreted. Many carcinogenic chemicals undergo metabolic changes and are converted from inactive procarcinogens to proximate carcinogens and then to ultimate carcinogens which bind to, and irreversibly alter, DNA in one or more target sites. Some of these genotoxic chemicals react directly with DNA without previous metabolic activation, but an important group of carcinogens (designated as nongenotoxic) do not act directly with the genome at all (see Sections III and IV).

Although this account is concerned with chemical carcinogens, such substances cannot be viewed as causal agents in isolation. Several different chemicals may interact, and chemical carcinogens may operate with other exogenous factors such as physical agents (ionizing radiation, ultraviolet light) and viruses (Epstein–Barr, hepatitis B). Constitutional factors in the host (genetic susceptibility, hormonal status) may also contribute, emphasizing the multifactorial nature of the carcinogenic process.

II. Chemically Induced Tumors in Humans

Direct epidemiological evidence indicates that certain chemicals are carcinogenic in humans. Information is available for industrial chemicals and processes and for chemicals encountered in more general cultural and environmental contexts. Data are obtained principally from case-control and cohort studies, which are concerned with individual exposure and estimates of risk ratios, and from case reports. (Cohort studies are investigations in which information is collected prospectively from a group of individuals (a cohort) identified by a demographic characteristic such as an age group or an occupation.) Details of methodology fall outside the scope of this account, but the choice of appropriate protocols and statistical techniques and, in particular, the exclusion of bias and confounding factors is of prime importance. Evidence for carcinogenic activity will appear as a large relative risk between exposed and unexposed populations and a dose–response with respect to increasing risk with

TABLE I Industrial Chemicals and Processes Identified as Carcinogenic for Humans

	Routes of exposure[a]	Main target sites for tumors	Reproduced in experimental animals[b]
I. Industrial chemicals			
Aromatic amines			
4-Aminobiphenyl	inh, o, sk	bladder	d, rab
Benzidine	inh, o, sk	bladder	d
2-Naphthylamine	inh, o, sk?	bladder	d, ham, mon
Metals and metallic compounds			
Arsenic and certain arsenic compounds	inh, o, sk	skin, lung	ham (lung) (limited data)
Chromium compounds, hexavalent	inh	lung	rat
Nickel and certain nickel compounds	inh, o	nose, sinuses, lung	mo, rat (lung)
Mineral fibers, naturally occurring			
Asbestos	inh, o	lung, pleura	rat (lung) rat, ham (mesothelioma)
Talc-containing asbestiform fibers	inh		inadequate data
Miscellaneous organic chemicals			
BCME and CMME bis(chloromethyl) ether Chloromethyl methyl ether	inh	lung	mo, rat
Benzene	inh, o, sk	bone marrow	mo (lymphohaemopoietic)
Mustard gas	inh	lung	mo
Vinyl chloride	inh, sk	liver	mo, rat, ham
Polynuclear aromatic hydrocarbons			
Coal tar	inh, sk	skin, lung	mo, rab (skin)
Coal tar pitches	inh, sk	skin, lung	mo (skin)
Mineral oils, untreated and lightly treated	inh, sk, o	skin	mo
Shale oils	inh, sk	skin, lung	mo (skin), rat (lung)
Soots	inh, sk	skin, lung	mo (skin)
II. Industrial Processes			
Aluminum production	inh	lung	
Auramine production	inh, o, sk?	bladder	
Boot and shoe manufacture and repair	inh	nose, sinuses	
Coal gasification	inh, o, sk	skin, bladder, lung	
Coke production	inh	lung	
Furniture and cabinet making	inh	nose, sinuses	
Iron and steel founding	inh	lung	
Isopropyl alcohol production (strong acid process)	inh	sinuses	
Magenta manufacture	inh, o, sk?	bladder	
Rubber industry	inh, o, sk	bladder, lymphoid system, etc.	

[a] inh, inhalation; o, oral; sk, skin.
[b] d, dog; ham, hamster; mo, mouse; mon, monkey; rab, rabbit.

increasing exposure, the latter involving considerations of duration of exposure and concentration of the suspected agent or agents. Identification of a carcinogenic chemical is aided if the tumors occurring among the exposed group are unusual in terms of their site, morphology or the clinical setting in which they are encountered (e.g., a particular tumor type developing at an uncharacteristically early age). Epidemiological findings are, however, often more problematic. Circumstances where people come into contact with a single chemical are unusual and occupational exposures, in particular, generally involve complex and inconstant mixtures encountered under variable conditions. Duration of exposure and concentration of the agents concerned is often difficult to determine. Unrecognized confounding factors, particularly cigarette smoking, will distort the results. A small increase in the inci-

TABLE I Extended

General comments
Aromatic amines: major carcinogens in dyestuff and rubber industries, first suspected at end of 19th century. Early examples of carcinogens acting at site distant from initial contact. Pronounced species variation in target sites for tumors—bladder in dogs compared with liver in rats and mice—reflecting different patterns of metabolism (dog unable to acetylate aromatic amines). Potent genotoxins.
Metals and metallic compounds: occur as complex ores, frequently contaminated with other metallic substances, and as byproducts of refining processes. Bio-availability depends on local persistence, solubility, and valency. Animal data for arsenic still very limited. Mode of carcinogenic action unclear. Little evidence for genotoxicity except for hexavalent chromium.
Asbestos: carcinogenic effects best documented for crocidolite, amosite, chrysotile, and anthophyllite; exposures usually mixed. Tumors of mesothelium (pleural and peritoneal mesotheliomas) very rare in general population and their occurrence serves as valuable marker for probable exposure to asbestos or related fibers. Strong multiplicative effects with smoking for lung cancer but not for mesothelioma. Carcinogenic activity related to physical dimensions of fiber (<2.5 μm thick and 10–80 μm long) rather than chemical structure. Mode of action still unknown. Limited evidence of genotoxicity, consisting mainly of production of aneuploidy.
Mustard gas: carcinogenic effects described in production workers but not in troops exposed on the battlefields in World War I. An alkylating agent, predictably genotoxic.
Vinyl chloride: Associated with very rare form of liver tumor, angiosarcoma. Not pathognomonic of previous exposure to vinyl chloride but a strong pointer, comparable to mesotheliomas in individuals exposed to asbestos. Metabolized to form an alkylating agent and genotoxin.
Polynuclear aromatic hydrocarbons: human exposure is to complex mixtures and not to single compounds, at leat 15 of which are carcinogenic for experimental animals. The earliest recognized occupational carcinogens: scrotal/skin cancers described among "climbing boys" (chimney sweeps) in 18th century London by Percival Pott and, later, among workers exposed to shale oils, mineral oils, and coal tars. Repeated applications of coal tar shown to produce local skin cancers in rabbits and mice in 1915 and 1918. First pure carcinogen—benzo(a) pyrene—isolated from coal tar by Kennaway in 1932 at Institute of Cancer Research, London.
A group of processes that have been shown, on epidemiological grounds, to carry a clear carcinogenic risk, although the chemicals responsible are not identified. Animal data are, for the most part, unhelpful. Polynuclear aromatic hydrocarbons (particularly coal tar pitch volatiles) may be involved in aluminum production, coal gasification, coke production, and iron and steel founding. The manufacture of auramine and magenta involves exposure to other chemicals, including aromatic amines. Carcinogenic risks associated with isopropyl alcohol manufacture may be associated with intermediates (di-isopropyl sulfate), byproducts (isopropyl oils), or other factors. Carcinogenic wood dusts associated with furniture and cabinet making appear to be mostly hardwoods such as oak and beech. The tumors (adenocarcinomas of the nasal sinuses) are usually very rare and they may provide important clues for previous exposure: see also mesotheliomas (asbestos) and hepatic angiosarcomas (vinyl chloride).

TABLE II Other (Nonindustrial) Chemicals and Products Identified as Carcinogenic for Humans

	Routes of exposure[a]	Main target sites for tumors	Reproduced in experimental animals[b]
I. Medicinal			
Alkylating agents			
Chlorambucil	oral and/or intravenous	bone marrow	mo, rat (lymphohaemopoietic)
Chlornaphazine	oral and/or intravenous	bladder	inadequate data
Cyclophosphamide	oral and/or intravenous	bladder, bone marrow	rat (bladder)
Melphalan	oral and/or intravenous	bone marrow	mo (lymphohaemopoietic)
Methyl-CCNU	oral and/or intravenous	bone marrow	inadequate data
Myleran	oral and/or intravenous	bone marrow	mo (lymphohaemopoietic)
Certain combined chemotherapy regimes	oral and/or intravenous	bone marrow, lymphoid system	—
Anabolic steroids (oxymethalone)	o	liver	no data
Oestrogens			
Steroidal (oestrogen replacement)	o	endometrium	
Nonsteroidal (diethylstilbestrol)	transplancental	cervix, vagina	mo, ham
Estrogen–progestin combinations, oral contraceptives			
Combined	o	liver	
Sequential	o	endometrium	inadequate data
Miscellaneous			
Analgesic mixtures containing: Phenacetin	o	urothelium	—
Azathioprine	o, intravenous	lymphoid system	mo, rat
8-Methoxypsoralen + ultraviolet light	o + sk	skin	mo
Arsenic	o	skin	—
Coal tar	sk	skin	mo, rab
II. Environmental			
Aflatoxins	o	liver	mo, rat, ham, mon, etc.
Erionite	inh	mesothelium (pleura)	mo, rat
Arsenic	o	skin	—
Asbestos	inh	mesothelium	rat, ham
III. Social/cultural			
Alcoholic beverages		mouth, pharynx, larynx	inadequate data
Betel quid + tobacco		mouth	ham (cheek pouch) (limited data)
Tobacco smoke		lung; also mouth, pharynx, larynx, oesophagus, bladder, pancreas	mo, rat, d (lung) rat (mouth) ham (larynx)
Smokeless tobacco products		mouth	rat (mouth) (limited data)

[a] inh, inhalation; o, oral; sk, skin.
[b] d, dog; ham, hamster; mo, mouse; mon, monkey; rab, rabbit.

TABLE II Extended

General comments
Alkylating agents: nucleophilic substances that alkylate purine and pyrimidine bases. Some (e.g., cyclophosphamide) require metabolic activation. Effects on DNA include single-base substitution, single- and double-strand breaks, inter- and intrastrand cross-linkages. Potent genotoxic effects. Predictably carcinogenic, but valuable in chemotherapy of potentially fatal malignant disease. Chlornaphazine no longer in clinical use. Best-documented combined regime is MOPP—nitrogen mustard, vincristine, procarbazine, and prednisone; almost certainly associated with additional increased risk of non-Hodgkin lymphoma and perhaps solid tumors (lung, breast). Effects extremely complex, with interaction of several drugs and—usually—irradiation.
Evidence incriminating oxymethalone based on descriptive studies; no analytical studies reported but drug almost certainly carcinogenic.
Steroidal estrogens and estrogen–progestin combinations: difficult to test in realistic protocols in laboratory animals. Appear to act mainly as tumor promoters for liver, breast, and other sites. No consistent evidence for genotoxic effects. Combined oral contraceptives protect against ovarian and endometrial cancer. Altered relative risks for breast cancer remain controversial; ? significant trend in rise of breast cancer with total duration of use among young women below age 36 yr.
Phenacetin, given alone, induces urothelial tumors in rats and mice; the mixtures appear to be negative. Azathioprine: evidence derived mainly from patients with renal transplants; carcinogenic effects less clear in patients with other disorders. Drug is genotoxic, but general effects of sustained immune suppression on the carcinogenic process are complex. Arsenic: an archaic medicine used as 1% aqueous solution of potassium arsenite (Fowler's solution). Coal tar preparations: used in chronic eczema and psoriasis.
Aflatoxins: mycotoxins produced by *Aspergillus flavus* spp. Dose-related association between intake of aflatoxin-contaminated foodstuffs (some cereals, ground nuts) and hepatocellular carcinoma in South and East Africa, China, and Southeast Asia. Problems of confounding with pre-existing hepatitis B infection adequately addressed. Aflatoxin B_1 induces liver tumors in many laboratory animals, although species susceptibility varies; strongly genotoxic. Erionite: a widely distributed silicate, usually associated with other zeolite minerals. Carcinogenic effects so far identified in remote villages in Turkey, where heavy long-term exposure (soil, road dust, building materials) is associated with high incidence of mesothelioma. Arsenic: high concentrations in drinking water associated with skin cancer in parts of Chile, Argentina, Silecia, and Taiwan. Asbestos: environmental exposure seen in household contacts of asbestos workers and, in some studies, in people living near asbestos factories and mines.
Alcohol: no consistent differences between carcinogenic effects of commercial beers, wines, or spirits. No reliable data from animals for pure ethanol although main metabolite, acetaldehyde, is carcinogenic in rats and hamsters. Tobacco smoke: the single most important chemical carcinogen, now responsible for ~30% of all cancer deaths in developed countries. Strongly genotoxic. Effects of passive smoking remain controversial; probably slightly enhanced risk. Smokeless tobacco products: long established use in many different parts of the world, particularly India, Pakistan, parts of Southeast Asia, Soviet Central Asia, and parts of the Middle East, with consumption up to 5 kg/yr. Wide variations in composition. Some products contain tobacco-specific *N*-nitrosamines at much higher levels (mg/kg) than tobacco smoke. Note general difficulties in testing "social/cultural" carcinogens in laboratory animals in a realistic manner.

dence of a common tumor such as lung cancer may be missed because of its high prevalence in the unexposed group and among the population as a whole. An inherent limitation of the epidemiological approach is imposed by the long latent period (20, 30, or more yr) that elapses between first exposure to carcinogenic chemicals and the appearance of a tumor—the current epidemic of lung cancer in developed countries is the consequence of social trends that started in the 1930s and 1940s.

These comments should be borne in mind when examining Tables I and II which summarize data on chemicals and chemical processes generally accepted as carcinogenic for humans. The chemicals are structurally diverse and they act on a range of target tissues although most of them induce tumors at or near the points of initial contact—skin, mouth, respiratory tract. There is close congruence between findings in humans and in laboratory animals exposed by comparable routes to the same materials, albeit at proportionately higher doses for longer times. The contribution made by the various categories to overall mortality from cancer are difficult to quantitate but the series of "best estimates" published by Doll and Peto in 1981, based on standardized cancer death certification rates in the United States, provide a valuable guide.

The authors calculate that industrial cancers account for ~4% of all deaths from malignant disease, occupational bladder and lung cancer making up ~15% and ~20%, respectively, of all deaths from these two tumors. Asbestos is responsible for ~5% of all fatal lung cancers. Among nonoccupational chemical exposures, Doll and Peto conclude that therapeutic drugs are associated with ~0.5% of cancer deaths and alcohol with ~3%. Cigarette smoke is the single most important carcinogenic agent, causally associated with ~30% of all cancer deaths and, worldwide, calculated to be responsible for more than 1 million deaths from lung cancer each year. Increased risk ratios associated with passive smoking appear so far to be fairly small. The contribution made by smokeless tobacco products to cancers of the mouth and pharynx cannot be reliably quantitated, but these materials are important etiological agents in high-incidence areas. (Tumors of the mouth and pharynx are uncommon in developed countries but, on a global scale, they rank high in the list of major fatal cancers—fourth among males, sixth among females, and sixth overall). [*See* ASBESTOS; TOBACCO SMOKING, IMPACT ON HEALTH.]

Approximately 2% of cancer deaths are attributed to pollution of air, water and food and—very speculatively—~35% to dietary factors. Both categories are complex. The diet, for example, may contain carcinogenic chemicals in the form of plant products (pyrrolizidine alkaloids, flavenoids), mycotoxins (aflatoxins) and additives, contaminants, and constituents acquired during the processing and preparation of foodstuffs (nitrosamines, products of combustion); but for people living in developed societies, general factors such as total fat content and excessive calorie intake (overnutrition, obesity) are far more important. Certain naturally occurring dietary substances (vitamin C, tocopherols, phenols) may exert anticarcinogenic effects by preventing the formation of carcinogens from precursors, and others (vitamin A derivatives, selenium, isothiocyanates, glutathione) may block access of carcinogens to their target tissues. Useful leads may come from studying the relatively simple traditional diets eaten by people who live in rural communities in nonindustrial countries. It is likely, for example, that dietary factors are important etiological agents in esophageal cancer—a tumor that shows exceptionally wide differences in incidence over small areas and is particularly common in parts of South and East Africa and in contiguous areas of Iran, Soviet Central Asia and northwest China. In Western countries, dietary factors are also implicated in cancers of the stomach, large intestine, pancreas, and breast—common sites that are conspicuously absent from Tables I and II. A problem common to both dietary factors and to environmental pollution is the occurrence of putative carcinogens (nitroarenes in diesel exhaust particulates, halogenated hydrocarbons in drinking water) at very low concentrations but widely distributed, so that large segments of the population are likely to be exposed for much or all of their lives.

III. Chemically Induced Tumors in Laboratory Animals

The chemical industry in the United States and other developed countries has grown exponentially since the late 1930s, with production of synthetic chemicals virtually doubling in each decade. Increasing numbers of new chemicals—therapeutic drugs, food additives, agrochemicals, veterinary and consumer products, industrial chemicals—are introduced whose potential carcinogenic effects in

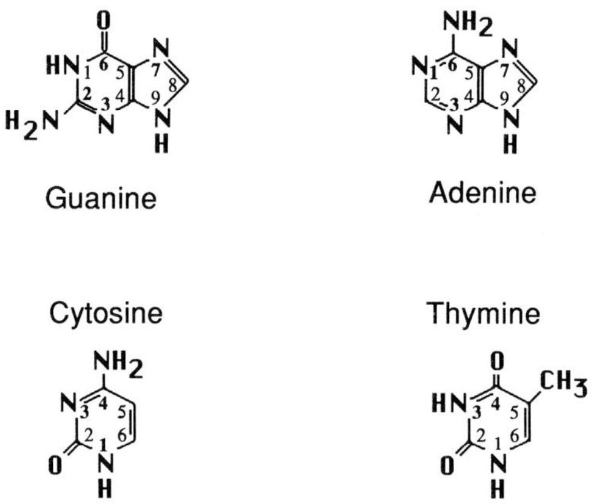

Guanine Adenine

Cytosine Thymine

FIGURE 2 Binding sites for ultimate reactive carcinogens in purine and pyrimidine bases:

- PURINES
  - *Guanine* (G) N-7, O-6, N-3 alkylating agents; NH_2 polynuclear aromatic hydrocarbons; C-8 2-acetylaminofluorene, ? azo-dyes.
  - *Adenine* (A) N-1, N-3, N-7 alkylating agents; NH_2 polynuclear aromatic hydrocarbons.
- PYRIMIDINES
  - *Cytosine* (C) N-1, N-3 alkylating agents; NH_2 ? polycyclic aromatic hydrocarbons.
  - *Thymine* (T) N-3, O-4 alkylating agents.

humans are unknown. Considerable reliance is therefore placed on predictive studies in laboratory animals, and two questions arise: How effectively do animals function as human surrogates in terms of their response to carcinogenic chemicals? And to what extent do studies in animals clarify some of the more fundamental aspects of chemical carcinogenesis that cannot be investigated in humans?

The design, conduct, and interpretation of carcinogenicity tests in rats and mice fall outside the scope of this article. They present several controversial issues such as the choice of dose regimens for the test chemicals and the assessment of certain neoplasms that occur as "spontaneous" lesions among control animals as well as in the treated groups; prime examples are the liver cell tumors commonly found in several strains of mice. Extrapolation of information from animals to humans does, however, merit a brief comment. It is prudent to regard a chemical that is clearly carcinogenic in rats and mice as a potential human hazard. The analysis involves two stages—extrapolation *within* the test species, from the high doses used in the bioassay down to "no adverse effect levels" and on to the low doses proposed for human exposures; and then *across* species from rodents to humans. The first extrapolation is based on dose–response relationships which, at low concentration, may not be linear although they are often assumed to be so. Further assumptions are made on the existence (or otherwise) of thresholds for carcinogenic effects. Both impinge on fundamental aspects of carcinogenic mechanisms (see Section IV). The extrapolation from animals to humans is even more problematic, particularly with respect to the use of bioassays as a basis for quantitative assessments of human risk. Although such calculations are widely used in radiation biology, they are more difficult to apply to chemicals whose carcinogenic activity is often critically dependent on toxicokinetics and metabolism. Large biological assumptions are made in the calculations, and different mathematical models may give widely varying estimates of risk for the same compound.

It is fair to conclude that, despite important differences, the responses of laboratory animals and humans to carcinogenic chemicals are more often similar than divergent. Seven of the chemicals listed in Tables I and II—4-aminobiphenyl, bis(chloromethyl) ether, diethylstilbestrol, melphalan, 8-methoxypsoralen + UV-A, mustard gas, vinyl chloride—were predicted to be human carcinogens on the basis of effects first described in rodents. Furthermore, there is enough common ground for animals to serve as plausible models for studying some mechanistic aspects of chemical carcinogenesis in greater detail.

The toxicokinetics and metabolism of carcinogenic chemicals are major determinants for their biological activity. Two main groups of enzymes are involved in metabolizing foreign compounds (xenobiotics) such as chemical carcinogens: the cytochrome *P*-450 mixed-function mono-oxygenases and various conjugating enzymes which catalyze the formation of glucuronides, sulphate esters and mercapturic acids. Most of these metabolizing enzymes are located in the endoplasmic reticulum (which is disrupted by homogenization to form "microsomes"—hence the generic term "microsomal enzymes"). The *P*-450 mono-oxygenase system is particularly important because it has a broad substrate specificity, catalyzes a wide variety of oxidative reactions, and can be induced in response to exposure to xenobiotics. Metabolic patterns, often under genetic control, provide at least a partial explanation for species-, sex-, and strain-dependent differences in response to chemical carcinogens.

The general importance of choosing appropriate levels of chemicals in bioassays, noted earlier, will now be apparent because chronic overdosing is likely to have profound toxicokinetic and metabolic consequences: altered patterns of absorption, modification of microsomal and other biotransforming enzymes, metabolic overloading, and switching to alternative metabolic pathways. Other, more specific, examples may be cited. The common occurrence of hepatic tumors in mice is (at least in part) the consequence of very high levels of the cytochrome *P*-448 system in liver cells, which potentiates activation rather than detoxification of foreign substances, and high levels of other metabolizing enzymes such as epoxide hydratase and glutathione transferase. The overall rate of metabolic oxidation of chemicals is rapid in the murine liver and the turnover time of hepatocytes is short. The susceptibility of the dog to develop bladder tumors after exposure to aromatic amines is related to an absence of tissue acetyl transferase and a consequent inability to detoxify this group of compounds by deacetylation. Rodents respond to agents such as lipid-lowering drugs by intense proliferation of peroxisomes (cytoplasmic organelles) in the liver and a corresponding increase in the peroxisome-associated enzymes which catalyze the β-oxidation of long-chain fatty acids. The response is species-related and is negligible or absent in the dog, primates, and humans. Male rats (only) synthesize α_2-microglobulin in the liver, which is partly reabsorbed by the kidney where it accumulates in the proximal convoluted tubules. Various chemicals including 1,4-dinitrobenzene bind to the absorbed protein and render it less susceptible to destruction by local proteases. Lysosomal enzymes accumulate, the tubular cells die, and a chronic cycle of cell regeneration and cell death is set up which results eventually in the development of renal cell tumors—an example of indirect, nongenotoxic carcinogenic mechanisms which are discussed further in Section IV. [*See* PHARMACOKINETICS.]

The most fundamental consequence of metabolism is, however, the conversion of inactive procarcinogens into proximate carcinogens and then to ultimate carcinogens—strongly electrophile reactants that bind covalently to certain nucleophile sites in DNA in target tissues to form carcinogen-DNA adducts. The terms electrophile and nucleophile refer to agents that, respectively, acquire or donate electrons in the course of chemical reactions; the identification of potential electrophilic, DNA-reactive groups within chemical structures forms the basis for structure-activity relationships with respect to carcinogenic action for genotoxic carcinogens (cf. Section I). Although the generation of electrophiles is an important unifying concept for carcinogenic chemicals that react directly with DNA, the structural variety of the different chemical categories of carcinogens (Fig. 1) is still apparent in their final reactive forms. The activating process takes place mainly in the liver and, to a lesser extent, in the kidneys, lungs, and gastrointestinal tract where bacterial enzymes may also contribute. Details of the steps involved in individual activation pathways fall beyond the scope of this account, but examples include formation of epoxides (polynuclear aromatic hydrocarbons, aflatoxins, vinyl chloride) and *N*-hydroxylation and esterification (aromatic amines, azo-dyes, urethane) with generation of carbonium or nitrenium ions. Certain alkylating agents function as direct acting ultimate carcinogens without previous metabolic activation. Examples include mustard gas, nitrogen mustard, melphalan and bis(chloromethyl) ether. Such compounds are, however, susceptible to metabolic inactivation under appropriate conditions.

The process whereby ultimate carcinogens bind covalently to DNA can only be noted briefly here. All four nitrogenous bases of DNA may act as targets, particularly the reactive (i.e. nucleophilic) purines. Some examples of binding sites are shown in Fig. 2. There are extensive data for alkylating agents, indicating covalent binding to N and O atoms in all four bases. N^7-alkylguanine accounts for at least 70% of bound product but alkylation at other, subsidiary sites—O^6-guanine, N^2-adenine—may be particularly important in terms of subsequent carcinogenic activity. Binding sites for several chemical carcinogens are still unknown.

The immediate consequences of carcinogens binding covalently to DNA include single-base substitution, depurination, single- or double-strand breaks, and (in the case of bifunctional alkylating agents such as mustard gas) inter- or intra-strand cross-linkages. The long-term outcome depends on the extent and nature of the altered DNA. The damage may kill the cell because, for example, a cross-linking adduct may prevent replication of DNA or synthesis of an essential protein by blocking transcription of its own RNA. Second, the damage may be completely repaired and the cell restored to its former intact state. Third, the damage may be repaired imperfectly such that the cell survives with a permanently altered genome. "Error-prone" and other forms of inadequate DNA repair are highly complex and fall outside the scope of this account but their consequence—the development of one or more somatic mutation—forms the molecular basis for tumor development. [*See* REPAIR OF DAMAGED DNA.]

IV. Mechanisms of Chemical Carcinogenesis

Carcinogenesis is a multistage process in the course of which there is progressive emergence of the aberrant phenotype which constitutes the cancer cell. Separate phases of initiation, promotion and progression are recognized. Initiation consists of one or more mutation developing in the target cell genome after the formation of adducts between DNA and a reactive (electrophile) carcinogen. Such changes are, by definition, irreversible and heritable; they are also dose-related and, theoretically, show no threshold. The short phase of initiation is succeeded by a long period of promotion when there is progressive clonal expansion of the initiated cells to form gross tumors. The mechanisms of promotion are diverse, impinging on cell proliferation and differentiation directly and also through a variety of indirect biological effects such as tissue damage, cycles of cell death and regeneration, immune suppression and hormonal imbalance. The mechanisms are dose-related and, in contrast to initiation, often have a threshold and are partly reversible. Initiation and promotion may be mediated by the same compound or its metabolites, or by separate initiating and promoting agents. A number of promoters have now been chemically characterized, notably a series of phorbol esters derived from croton oil; the best known is TPA or 12-*O*-tetradecanoylphorbol-13-acetate (Table III). The biochemical effects of TPA are numerous and include induction of ornithine decarboxylase and several other enzymes, alterations in membrane function and intercellular communications, and inhibition of terminal differentiation in a number of cell types. Many of these effects are linked to the capacity of TPA to activate a calcium-dependent protein kinase known as protein kinase C. Promoting agents such as TPA may indirectly lead to additional genotoxic damage in target tissues by releasing free oxygen radicals. Progression covers the later stages of tumor growth when various subclones of neoplastic cells have emerged. It includes transformation from benign to malignant neoplasms, and the process of metastasis whereby invasive malignant cells disseminate from the primary tumor to distant sites

TABLE III Promoting Agents and Initiation—Promotion Systems

Promoting Agents

Phorbol esters: isolated from croton oil. Promoters for skin carcinogenesis in mice. General structure shown in inset.

TPA (12-O-tetradecanoylphorbol-13-acetate): the most potent, in which R_1 (position 12) is $CO\text{-}(CH_2)_{12}\text{-}CH_3$ and R_2 (position 13) is $CO\text{-}CH_3$. TPA is not directly genotoxic and does not undergo metabolic activation.

Other promoting agents: skin (mouse)—indole alkaloids, anthralin, retinoic acid, benzo(a)pyrene; bladder (rat)—saccharin, cyclamate; thyroid (rat)—goitrogens

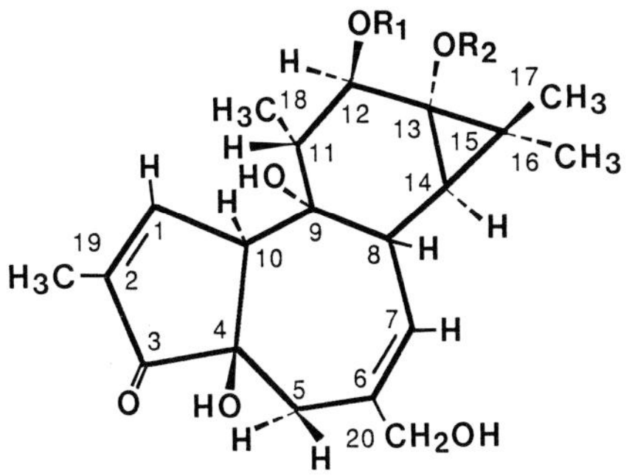

Example of Initiation–Promotion System[a]

1	i	no tumors
2	i p p p p p p . . . . up to 20 wk	tumors
	i [delay of up to 1 yr] p p p p p p . . . . up to 20 weeks	tumors
3	p p p p p p . . . . up to 20 wk	no tumors
	p p p p p p . . . . up to 20 wk i	no tumors

[a] i (initiator) is given as single subcarcinogenic dose; p (promoter) is given as a series of doses, varying according to test system.

and form secondary (metastatic) deposits. More mutational changes occur at this stage, reinforcing the view that the tumor cell genome is modified in a progressive and cumulative fashion throughout the carcinogenic process. [*See* METASTASIS; TUMOR CLONALITY.]

There is an increasing tendency to describe carcinogens operationally in terms of their likely modes of action in the carcinogenic process. Genotoxic carcinogens are responsible for initiation and for other genetic changes that develop subsequently. Nongenotoxic carcinogens function through other, mainly unknown, mechanisms and probably act in conjunction with promoting agents. Further comments on these two groups follow.

Genotoxic carcinogens are mutagens whose activity can be demonstrated in a variety of short-term tests with end-points as diverse as gene mutations, chromosome damage, sister chromatid exchanges, aneuploidy and formation of micronuclei. Details of these tests are not relevant here, but the increasing emphasis placed on sequential *in vitro* → *in vivo* testing is worth stressing. Clear advantages are evident in subjecting a mutagen, demonstrated initially *in vitro,* to further investigation in the whole animal where toxicokinetic and metabolic factors can be taken into consideration; analogies with bioassays for carcinogens will be apparent. The demonstration that some chemical carcinogens are mutagens has had far-reaching consequences. Short-term testing procedures carry predictive information for genotoxic carcinogens which is valuable for industrial concerns in developing new chemicals, and for regulatory authorities who control their introduction and use. They may be useful in the future for screening high-risk populations and augmenting traditional epidemiology (cf. Section II), and they point to fundamental mechanisms of action.

It is now possible to investigate these mechanisms in detail as a result of conceptual and technical advances in molecular biology. Studies of oncogenic retroviruses with a relatively simple genetic structure have identified specific viral oncogenes, which are responsible for the neoplastic properties of cells transformed by RNA viruses. These retroviral oncogenes are homologues of normal genes—proto-oncogenes—which are present in the genome of normal cells. They are highly conserved, occurring in living forms from yeasts to mammals. Their gene products are involved in the normal control of cell growth and differentiation, acting as growth factors, growth factor receptors, signal transducers, and transcriptional activators. It has become clear that proto-oncogenes can be activated or mutated by chemical carcinogens as well as oncogenic viruses. Chemicals may, for example, change gene structure as a result of point-mutations and chromosomal rearrangements; or copies of the normal gene may be increased ~5–50-fold as a result of amplification. [*See* ONCOGENE AMPLIFICATION IN HUMAN CANCER.]

Some examples of oncogenes in human and rodent tumors are listed in Table IV. The functional significance of oncogenes in human neoplasms is difficult to assess, given their rather low overall occurrence in tumors of generally obscure etiology. More information (particularly for the *ras* oncogenes) has come from chemically induced tumors in rodents where oncogene levels are generally higher. Certain segments of the *ras* oncogene appear to be particularly susceptible to point-mutations inflicted by chemical carcinogens (codons 12, 13, 61), and the mutations produced may be remarkably consistent: the A → T transversion of the second nucleotide of codon 61 of *H-ras* induced by DMBA in mouse skin and the G → A transversion of the second nucleotide of codon 12 of *H-ras* induced by NMU in rat breast (see Table IV). The subsequent stages whereby the activated *ras* oncogene initiates neoplastic changes, or modifies a pre-existing neoplastic phenotype, are obscure. Little at present is known of the functions of normal *ras* P21 proteins, and still less about the corresponding altered *ras* gene products. Activated oncogenes of the *ras* type have now been implicated in several models of tumor initiation and in tumor progression, and it is almost certain that additional oncogenic activation occurs during promotion as well.

Oncogenes illustrate situations where somatic mutations (usually dominant) have resulted in a gain of gene function. It is now clear that recessive mutations resulting in a loss of gene function are also involved in the carcinogenic process. The loss involves both copies of normal genes (anti-oncogenes, tumor suppressor genes) that control normal cell proliferation and differentiation. The two copies may either be lost as a result of deletion of part of a chromosome or inactivated by point mutation or methylation. The clearest evidence for deletion or inactivation of specific suppressor genes comes from studies in human tumors—particularly certain rare neoplasms in children (i.e., retinoblastomas, Wilms' tumors) but also increasingly in common cancers in adults such as those arising in the breast,

TABLE IV Examples of Oncogenes in Human and Experimental Tumors

A. Human tumors

Present in one or a few tumor types

abl chronic myeloid leukemia	c-*myc* lung
N-myc neuroblastoma	*neu*/*erb B2* breast

Present in several tumor types at varying incidence

H-, *K*-, *N-ras* genes present in DNA from 10–20% of unselected series of fresh tumor biopsies and in tumor cell lines

Distribution in biopsies from specific tumor types

colon 11/27 *K-ras*	lung 5/39 *N-ras*	liver 3/10 *N-ras*	head & neck 2/37 *H-ras*
26/66 *K-ras*			
bladder 2/38 *H-ras*	breast 0/16	stomach 0/26	cervix uteri 0/30
1/15 *H-ras*			

Increased evidence of *ras* oncogenes in tumor cell lines

B. Experimental tumors

Carcinogen	Species	Tumor	Oncogene	
7,12-Dimethylbenz(a)anthracene	rat	breast	*H-ras*	6/6
7,12-Dimethylbenz(a)anthracene + TPA	mouse	skin	*H-ras*	33/33
1,8-Dinitropyrene	rat	soft tissues	*K-ras*	
3-Methylcholanthrene	mouse	soft tissues	*K-ras*	2/2
Methyl(methoxymethyl)nitrosamine	rat	kidney	*K-ras*	
			N-ras	
Dimethylnitrosamine	rat	kidney	*K-ras*	10/11
Diethylnitrosamine	mouse	liver	*H-ras*	7/7
N-nitroso-*N*-methylurea	rat	breast	*H-ras*	16/61
	mouse	thymus	*N-ras*	5/5
			K-ras	5/5
Vinyl carbamate	mouse	liver	*H-ras*	7/7
N-hydroxy-2-acetylaminofluorene	mouse	liver	*H-ras*	7/7
Aflatoxin	rat	liver	*K-ras*	2/8
Tetra-aminomethane	rat	lung	*K-ras*	
	mouse			
[Transplacental]				
N-nitroso-*N*-methylurea	rat	nervous system	*neu*	4/5
N-nitroso-*N*-ethylurea	rat	nervous system	*neu*	2/3
[Spontaneous]	mouse	liver	*H-ras*	11/13

Nomenclature: Oncogenes given three-letter codes based on animal or tumor from which they were first derived: *abl*, Abelson mouse leukemia virus; *erb B*, avian erythroblastosis virus; *myc*, avian myelocytoma virus; *N-myc*, human neuroblastoma; *neu*, rat neuroglioblastoma; *ras: H*-, (Harvey) rat sarcoma virus; *K*-, (Kirsten) rat sarcoma virus; *N*-, human neuroblastoma cell line.

lung, and large intestine. Suppressor genes have been cloned in retinoblastoma (Rb-1 on the long arm of chromosome 13) and in colorectal cancer (p53 on the short arm of chromosome 17), and DCC on the long arm of chromosome 18). P53, as a tumor suppressor gene, appears to be a frequent target for mutation in several of the common human cancers; but no association with specific genotoxic chemicals has yet been established. Remarkable advances have been made in our understanding of the molecular genetics of colorectal cancer where it is now possible to identify four separate genomic alterations: activation of the *k-ras* oncogene and loss or inactivation of the suppressor genes apc, p53 and DCC. Loss of the apc gene from the long arm of chromosome 5 is particularly interesting as this is also a consistent feature of familial adenomatous polyposis (FAP)—an inherited precancerous condition in which very large numbers of adenomas develop in the large intestine and other parts of the gastrointestinal tract and almost invariably progress to multiple carcinomas. The combination of both gain and loss of gene function is notable and points to interactions between oncogenes and tumor suppressor genes. Indeed, recent evidence suggests that certain oncogenes may function by producing proteins which bind to gene products of tumor suppressor genes, thereby inactivating them. [*See* CHROMOSOME PATTERNS IN HUMAN CANCER AND LEUKEMIA.]

The association between carcinogens and mutagens has been described as "useful but imperfect," and it is now clear that some chemical carcinogens have no mutagenic or clastogenic activity and that genetic aberrations, however diverse, do not provide a complete explanation of the carcinogenic process. Nongenotoxic carcinogens are diverse in terms of their chemical structure and biological effects and include enzyme inducers, endogenous and synthetic hormones, and hormone-modifying substances. Nongenotoxic carcinogens which induce the cytochrome *P*-450 mono-oxygenase system act primarily in the liver of rats and mice (see Section III) and include various halogenated hydrocarbons, polychlorinated biphenyls, phenobarbital, dioxin and peroxisome proliferators. Hormones and hormone-modifying compounds induce tumors in the appropriate endocrine glands and, in some circumstances, in the liver which is the prime site for their metabolism. Their mode of action usually involves effective loss of one or more of the corresponding endogenous hormones, either by inhibiting their synthesis or enhancing their destruction or removal. In the case of endocrine glands which are regulated by trophic hormones released from the hypothalamic–pituitary axis, these changes will break the normal feedback control mechanisms and excessive stimulation of the target gland by one or more trophic hormones will ensue. The glandular tissue becomes hyperplastic and neoplasms eventually develop. Benign tumors (adenomas) appear first, a variable proportion of which will evolve into carcinomas.

The modes of action of most nongenotoxic carcinogens at cellular and molecular levels are ill-understood. They may, for example, bind to proteins which either regulate cell-surface receptors for growth factors or control feedback regulation or cell proliferation. Alternatively, they may interact with tRNA, mRNA, and regulatory proteins and disturb normal protein synthesis. Interplay between nongenotoxic and genotoxic carcinogens is close. The latter bind to RNA and protein in addition to DNA, while enhanced tissue proliferation, evoked by nongenotoxic chemicals, will facilitate further genomic damage. More DNA adducts are likely to form under these circumstances because, with increased cell turnover, DNA is less protected by chromosomal proteins and chromatin structure; there is also less time for DNA repair.

Brief mention should finally be made of certain plastics and naturally occurring and synthetic fibers which are carcinogenic in humans and in experimental animals. The prime example is asbestos (see Section II). Such substances are chemically inert and their mode of action as carcinogens appears to be related to their physical properties. Asbestos fibers do, however, induce chromosome damage and aneuploidy, illustrating that the distinction between genotoxic and nongenotoxic carcinogens is sometimes indistinct.

Many aspects of the carcinogenic process are still obscure, but it is not unrealistic to hope that a clearer understanding will emerge of the etiology of human cancers and the mechanisms whereby chemicals and other extrinsic carcinogenic agents exert their effects—an understanding that will lead to more effective prevention of more tumor types and, perhaps, amelioration of some of their devastating effects. [*See* NEOPLASMS, ETIOLOGY.]

Bibliography[1]

Ames, B. N., Magaw, R., and Gold, L. S. (1987). Ranking possible carcinogenic hazards. *Science* **236,** 271.

Balmain, A., and Brown, K. (1988). Oncogenic activation and chemical carcinogenesis. *Adv. Cancer Res.* **51,** 147.

Bishop, J. M. (1987). The molecular genetics of cancer. *Science* **235,** 305.

Doll, R., and Peto, R. (1981). "The Causes of Cancer." Oxford University Press, Oxford, New York.

Graham, J. D., Green, L. C., and Roberts, M. J. (1988). "In Search of Safety: Chemicals and Cancer Risk." Harvard University Press, London, Boston.

International Agency for Research on Cancer (1972–1990). "Monographs on the Evaluation of Carcinogenic

1. More detailed reviews and original papers may be found in specialist journals such as *Nature, Science, Proceedings of the National Academy of Science (USA), Journal of the National Cancer Institute, Cancer, International Journal of Cancer, British Journal of Cancer, Carcinogenesis, Mutagenesis, Mutation Research, and Cell.*

Risks to Humans," Vols. 1–50 (a continuing series). IARC, Lyon.

Miller, E. C., and Miller, J. A. (1981). Mechanisms of chemical carcinogenesis. *Cancer* **47,** 1055.

Parkin, S. M., Läärä, E., and Muir, C. S. (1988). Estimates of the worldwide frequency of sixteen major cancers in 1980. *Int. J. Cancer* **41,** 184.

Sager, R. Tumor suppressor genes: The puzzle and the promise (1989). *Science* **246,** 1406.

Weinberg, R. A. Oncogenes, anti-oncogenes, and the molecular bases of multistep carcinogenesis (1989). *Cancer Research* **49,** 3713.

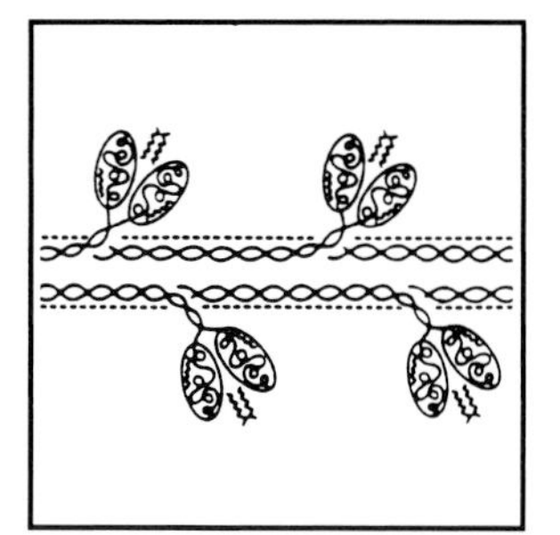

Cardiac Muscle

C. E. CHALLICE, *The University of Calgary*

Glossary

Action potential Sequence of changes in transmembrane potential which are responsible for triggering contraction of the myocardial cells

Myofibril Component of cardiac cells responsible for providing contraction

Sarcomere Regular repeating element along the length of a myofibril consisting of interdigitating actin and myosin filaments. In a relaxed state it is approximately 2.5 μm in length and in a contracted state it is approximately 1.5 μm.

Sarcoplasmic reticulum Tubular structure generally surrounding the myofibrillar elements; responsible for controlling the concentration of calcium ions, which, in turn, modulates contractile activity and contractile force

CARDIAC MUSCLE is the muscle which is unique to the heart. Although it is similar in many ways to skeletal muscle, it differs from it in a number of important respects, and because of this it is designated as a third type of muscle, the others being smooth muscle and skeletal muscle (Table I). Cellular structure and properties vary with their position in the heart. While the majority of heart muscle cells have as their purpose the generation of a contractile force and the resultant production of a periodic contraction, all cardiac muscle cells conduct the electrical event which triggers contraction, and some cardiac muscle cells have conduction as their main purpose. Such muscle is often termed "specialized" muscle, and this includes the muscle cells responsible for the initiation of the contractile impulse—the so-called pacemaker.

I. Introduction

A. Historical Background

Older light-microscopic studies showed a great deal of structural similarity between cardiac muscle and skeletal muscle, but these studies were unable to distinguish individual cellular elements or fibers. Thus, it was thought that the heart consisted of a single multinucleated cell or syncytium. The advent of electron microscopy made it possible to distinguish individual cellular elements, in which cellular junctions crossed the direction of contraction, and where the junction included intercalated disks. Later it was shown that these junctions included regions in which cellular contact was so intimate as to permit the passage not only of the electrical impulse, but also of molecules of substantial size (through so-called intercellular gap junctions). Thus, although later work demonstrated the multicellular nature of the heart, it showed that in many respects it could be regarded as a functional syncytium. [*See* ELECTRON MICROSCOPY.]

B. Origin of Information

The available information on the cardiac cell comes from microscopic and experimental studies. For both types of analysis, it is necessary to have healthy tissue, removed from the host and processed within seconds of its normal *in vivo* existence. These requirements make it impossible to

TABLE I Principal Differences between Skeletal and Cardiac Muscle

Skeletal Muscle	Cardiac Muscle
Fibrillar elements continuous in single fiber cell from end to end	Fibrillar continuity is interrupted by the intercellular abuttments at intercalated disks
Fibrils discrete along length	Fibrillar structures branch and interconnect, such that a single fibril cannot be precisely defined
Sarcoplasmic reticulum forms well defined sheath around fibrils, forming triads at Z line level with T system	Sarcoplasmic reticulum also well developed but less abundant. Partially encloses contractile material. Junctions with T system (called "couplings") also less abundant
Activation neurogenic; controlled by central nervous system	Activation myogenic; production and conduction of activating action potential occur by the muscle cells themselves
Can maintain sustained contraction or tetanus	Cannot maintain sustained tension
Contraction irregular and widely variable in force, usually produced by voluntary action. Force of contraction modulated by increase in recruitment of fibers	Contraction regularly periodic, with frequency and force modulated in response to demands of body metabolism

have more than a very occasional observation on human tissue. Thus, most of our information is derived from studies of laboratory animals. However, occasional studies on primate hearts have shown that the information obtained from laboratory mammals can reliably be taken as closely indicative of the structure and function of the human cardiac cell.

C. Function of Cardiac Muscle

The essential function of the heart is that of a pump, operating by a regular periodic decrease of volume, which, in combination with unidirectional valves, generates a regular unidirectional flow of blood through the cavity, or lumen. Thus, the essential function of the cells of the heart is to produce a regular periodic contraction, and in describing the cardiac cell it is perhaps simplest to focus on this function and describe first the element responsible for the contraction. The other elements can then be seen to serve this function. Following this description the intercellular relationships important in the electrical activation of the cardiac cells are examined.

II. Structure of Myocardial Cells

A. Contractile Fibrils

The contractile element is, as with skeletal muscle, made up of fibrillar elements which, in turn, consist of sarcomeres (Fig. 1). These vary in length between about 2.5 μm when relaxed and about 1.5 μm when contracted. In cross-section the sarcomeres are seen to consist of interdigitating thin (actin) and thick (myosin) filaments. Electron microscopy of longitudinal sections reveals them to interdigitate in a regular symmetrical fashion, producing bands of width dependent on the degree of contraction (Fig. 1). Dense (by electron microscopy) Z bands join adjacent sarcomeres by joining their respective abutting actin filament ends. The interdigitating part of the sarcomeres (i.e., the middle section) behaves anisotropically in examination by polarized light (i.e., it causes polarization), and is thus known as the anisotropic, or A, band and the remainder—the region between the A bands, which has the Z band in the middle—behaves isotropically in examination by polarized light, and is thus the isotropic, or I, band. A fine structure can often be seen at the middle of the A band in the form of a thin dark band (the M line), each side of which is a thin pale band.

Whereas in skeletal muscle the fibrils are longitudinally continuous and transversely discrete, in cardiac muscle they branch and link neighboring elements, making the fibrillar contents of a cell essentially a single, branching, interconnected contractile element (Fig. 1).

B. Mechanism of Contraction

In the sarcomeres of the fibrils, energy is used (by hydrolysis of ATP to ADP) to generate a mechanical force and consequent contraction. The following represents a simplified summary of the process and of the microstructures involved.

Detailed analyses have shown that the thick myosin filament is made up of a bundle of molecules. Each molecule is a long double-helical coil, one end of which terminates abruptly, while at the other each strand terminates in a globular head (Fig. 2). The bundle (i.e., the thick filament) is assembled symmetrically about the middle, with the molecular

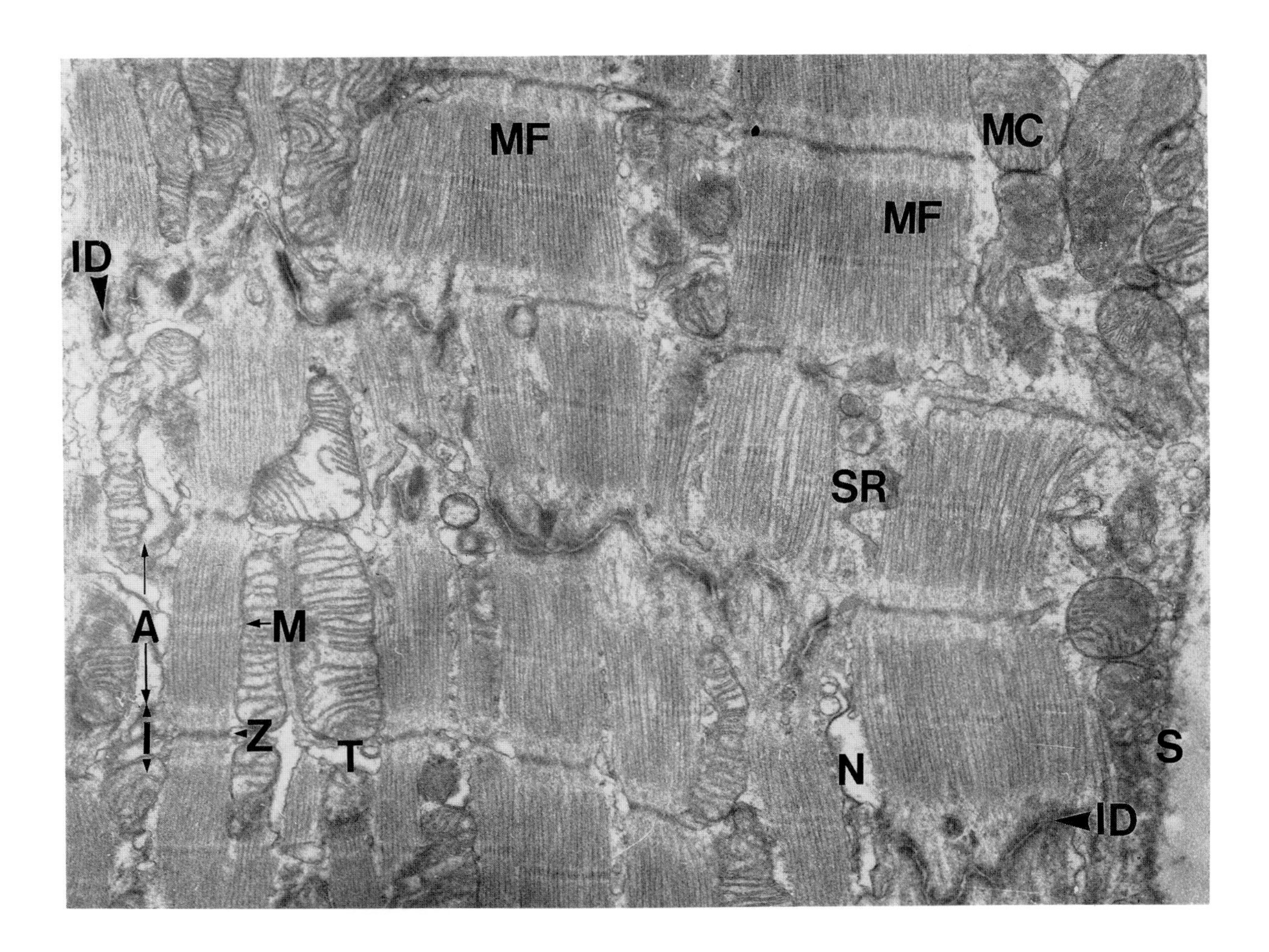

FIGURE 1 Transmission electron micrograph of a section from the ventricle of a rat heart in which the myofibrils (MF) are cut longitudinally. The bands of the sarcomeres are labeled (lower left): A, anisotropic band; I, isotropic band; Z, Z disk or line; M, M line. An intercellular junction traverses the micrograph from top left to lower right, showing the structure of the intercalated disks (ID). A nexus, or gap junction (N), is present. The way in which the myofibrillar elements divide and link up can be seen by following the two central elements from the top of the micrograph to the bottom. The discretenes of the elements seen at the top is not maintained. SR, Sarcoplasmic reticulum; T, transverse tubule; S, plasmalemma. Original magnification ×117,000.

heads farthest from the middle, equal numbers in each direction, and regularly distributed along the length (Fig. 3). As they interdigitate in the fibril with the actin filaments (each of which is a double strand of connected globular actin molecules), the myosin heads produce regularly arrayed cross-bridges, approximately at right angles with the filaments, connecting them with the actin filaments (Fig. 4).

Under specific conditions notably of Ca^{2+} and ATP concentration, the myosin heads go through a motion similar to rowing. The myosin heads (the "oars") contact the actin, bend in such a manner as to produce a force on the filaments (and thus movement in the direction of contraction), disconnect, return to their original condition, and then repeat the whole process. The control factor in this mechanism is the concentration of Ca^{2+} adjacent to the fibrils, which determines the degree of contractile activity and, hence, the degree of contraction.

C. Sarcoplasmic Reticulum

The control of the Ca^{2+} concentration in the immediate vicinity of the myofibrils is effected by the sarcotubular system, or sarcoplasmic reticulum (SR). Because of a more symmetrical arrangement, the SR can be seen in skeletal muscle to form a fenestrated tubular sheath around each sarcomere

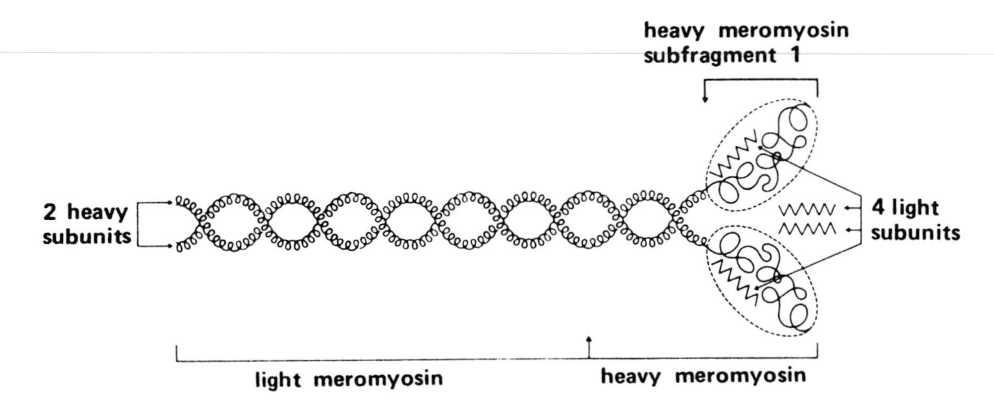

FIGURE 2 Structure of a myosin molecule showing the long double-helical tail and the double-globular head. The length of the tail is about 134 nm. It consists of two identical heavy chains, each of about 200,000 Da. One end of the helical coil ends abruptly (i.e., the carboxy terminus), while at the other globular end (i.e., the amino terminus) each molecule has two light chains (each about 20,000 Da) attached. [From Katz, A. M. (1977). "Physiology of the Heart." Raven Press. Reproduced by permission.]

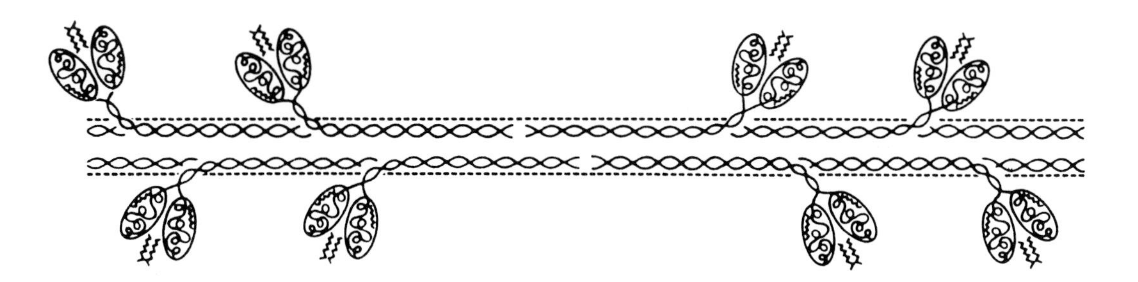

FIGURE 3 Myosin molecules associating to form a myosin filament. The long chains associate end to end at the carboxy termini in an overlapping manner to create a "bare" zone in the middle and a length at each end with regularly distributed globular heads. [From Katz, A. M. (1977). "Physiology of the Heart." Raven Press. Reproduced by permission.]

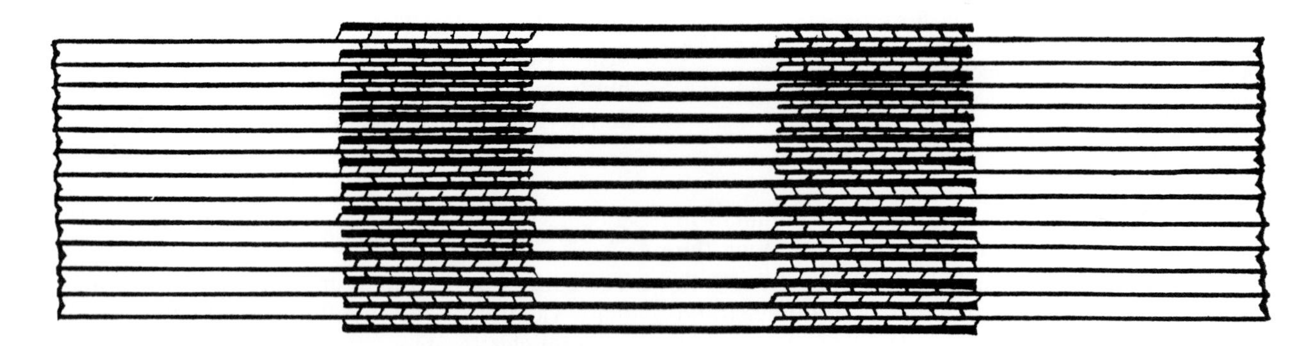

FIGURE 4 A single sarcomere showing the manner in which the thin (actin) and thick (myosin) filaments interdigitate in a sarcomere, indicating the way in which the myosin heads form the cross-bridges between the actin and the myosin.

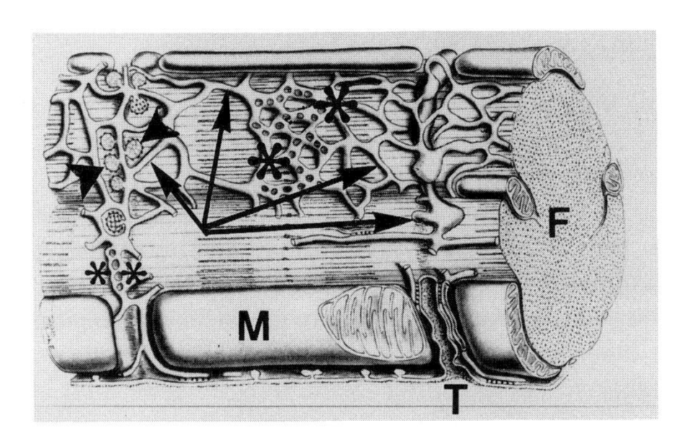

FIGURE 5 A sarcomere in a cardiac ventricular cell. The invagination of the sarcolemma forming a T tubule (T) is seen (lower right). The detailed structure of the sarcoplasmic reticulum (SR) is seen as it relates to the sarcomere. Blebs on the SR (''corbular'' SR) are seen at Z line level (arrowheads). SR consists generally of narrow tubules (arrows) with a tendency to form flattened lamellae with fenestrations at the M line level (large asterisk) and the same occurs to a lesser extent at the Z line level (small asterisk). F, Myofibril; M, mitochondrion. (Modified from Bossen, E., Sommer, J. R., and Waugh, R. A. (1978). *Tissue Cell* **10** 773–784. Reproduced by permission.)

and each fibril. In addition to being less abundant in cardiac cells, the less ordered fibrillar arrangement reflects itself in a less ordered arrangement of SR, but in general principle the arrangements are similar. The elements of the SR are tubular, in places flattened round the fibril. It is more abundant around the middle of the sarcomeres, where the tubules tend to be flattened with fenestrations, and also to a lesser extent around the Z bands, where there are also some small bulbous appendages to the SR (Fig. 5, and Color Plate 1).

The full details of the mechanism by which Ca^{2+} are released and sequestered by the SR have yet to be determined, but it is established that Ca^{2+} is stored in the SR, that release from the SR activates contraction, and that Ca^{2+} sequestration by the SR (and hence a decrease in Ca^{2+} concentration around the fibrils) leads to relaxation.

D. Transverse Tubules

Another set of tubules, running generally at right angles to the fibrillar direction, is present in the sarcoplasm. These are the transverse, or T, tubules (i.e., the T system) and are formed by invaginations of the outer plasma membrane of the cardiac cell, principally at the level of the Z bands of the sarcomeres (Fig. 1). They penetrate throughout the cross-section of the cell, making close contact with the fibrils, primarily close to the Z bands. They vary in diameter between 30 and 200 nm. Their function (analogous with skeletal muscle) is still debated, but is generally believed to be the propagation of the action potential across the cell (see Section V,E).

E. Cytoskeleton

As in other cells, filaments are present which have the functions of both stabilizing the distribution of cell components and controlling the size and shape of the cells. A measure of rigidity is important in many types of cells, but particularly in contractile cells where the action of contraction introduces a strong tendency to change the shape of the cell.

While various filaments serve as the cytoskeleton, the principal element in adult cardiac muscle is in the so-called ''intermediate'' class (approximately 10 nm in diameter) and is primarily desmin (Fig. 6). It is to be found aligned both parallel to and at right angles with the myofibrils, attaching to many types of organelles. They are sometimes referred to as ''anchor fibers,'' present in the intercalated disks (see Section III,B) also attached to the invaginations of the T system and connecting Z discs to each other and to the plasmalemma.

F. Other Structures

Mitochondria, the site of ATP production in cells, are abundant (i.e., approximately 30% of the total volume of cardiac muscle cells), reflecting an available high level of energy conversion. They assume a location and shape largely dictated by the arrangement of the myofibrils, but on average are distributed uniformly within the sarcoplasm, again reflecting the use of energy throughout the cell (Fig. 1). The Golgi apparatus or complex is usually prominent, again reflecting the consistently high level of metabolic activity. The nucleus is located centrally, and frequently more than one can be seen, although it has not been possible to establish whether or not there exists a consistent number of nuclei per cell. [*See* GOLGI APPARATUS; MITOCHONDRIAL RESPIRATORY CHAIN.]

Other cytoplasmic components common to most cells are found, including liposomes, multivesicular bodies, autophagic vacuoles, lipofuchsin granules, peroxisomes, glycogen granules, and lipid droplets.

III. Cell Membrane and Intercellular Relationships

As with all living cells, the outer cell membrane functions as a chemical control barrier between the cytoplasm (i.e., sarcoplasm) and the environment. However, the cardiac cell membrane also fulfills an essential electrical function involving the transmembrane movement of ions and, hence, electrical charge. The cycle of ionic movement which generates an action potential is described later (see Section V,B).

A. Membrane Structure

The cell membrane (i.e., plasma membrane, or plasmalemma) has the regular "unit" membrane structure and has a number of specialized regions described later (see Sections III,B,C, and D). It is covered by a feltlike coating approximately 50 nm thick (called the laminar coat, or glycocalyx) on the outside surface of which are some collagen fibers. The total structure (i.e., plasma membrane, glycocalyx, and collagen) is referred to as the sarcolemma. The plasmalemma is held in place by the intermediate (i.e., desmin) fibers, particularly at the Z line level. Unlike skeletal muscle, the glycocalyx penetrates into the T tubules.

The basic structure of the membrane is the standard bilayer, each layer made up of phospholipids in which the hydophobic lipids from the apposing layers face each other. Associated with this are various so-called membrane proteins which sometimes sit on one surface leaving the bilayer undisturbed, sometimes penetrate just one of the layers, and sometimes completely traverse the membrane. These proteins give the membrane its specific properties. Some form the structural support for the membrane, while others include antigens, receptors, enzymes, ion channels, and pumps. The membrane structure has a high degree of fluidity, permitting the protein molecules to move around in a manner that has been compared with "protein icebergs floating in a lipid sea" (A. Katz, 1986).

The cells of the myocardial walls are, in most places, separated by a space of varying width. However, there are a number of specialized regions of intercellular abutment.

B. Intercalated Disks

The intercalated disk is the structure present where the continuity of the myofibrillar structure crosses from one cell to another (Figs. 6 and 7). It is usually found to be convoluted and is rarely at right angles to the myofibrillar orientation. It interdigitates, forming almost a two-dimensional dovetail junction, which arises presumably by virtue of the intercellular adhesion necessary to hold the cells together when subjected to the tension generated by the myofibrillar contraction. Within the intercalated disk are also found the specific specialized junctions referred to in the next section.

C. Nexuses, or Gap Junctions

In gap junctions the two apposing membranes contact each other to form essentially a single structure (Fig. 7). Freeze-fracture electron-microscopic studies have shown pores to exist in these junctions which communicate between the respective sarcoplasms. The pores (i.e., connexons) are formed by a hexagonal array of six parallel protein molecules (i.e., a connexin) which form a tube. It has been suggested that the opening and closing are effected by a rotation of one end of the tube with respect to the other. Gap junctions permit the intercellular passage of not only ions but molecules, and are the structures which render cardiac muscle a functional syncytium.

FIGURE 6 Electron micrograph of rat myocardium cut approximately at right angles to the myofibrils, at the level of an intercalated disk. The convoluted geometry is readily seen; the two cells interdigitate extensively. (Lower left) The cells are cut almost transversely where the fairly regular array of thick (myosin) and thin (actin) filaments can be seen, while at the right the myofibril is cut obliquely, showing that in these regions the myofibrillar orientation is not entirely constant. D, Desmo somes; IJ, intermediate junction; IF, intermediate filaments; MF, myofibrillar elements. Original magnification ×60,000.

FIGURE 7 Electron micrograph of rat myocardium cut parallel with the myofibrils (MF) at the level of an intercalated disk. A desmosome (D) and intermediate junctions (IJ) are seen. A long nexus (Ns) is present, whose orientation with respect to the section varies along its length, showing the general irregularity of arrangement. M, Mitochondrion. Original magnification ×60,000.

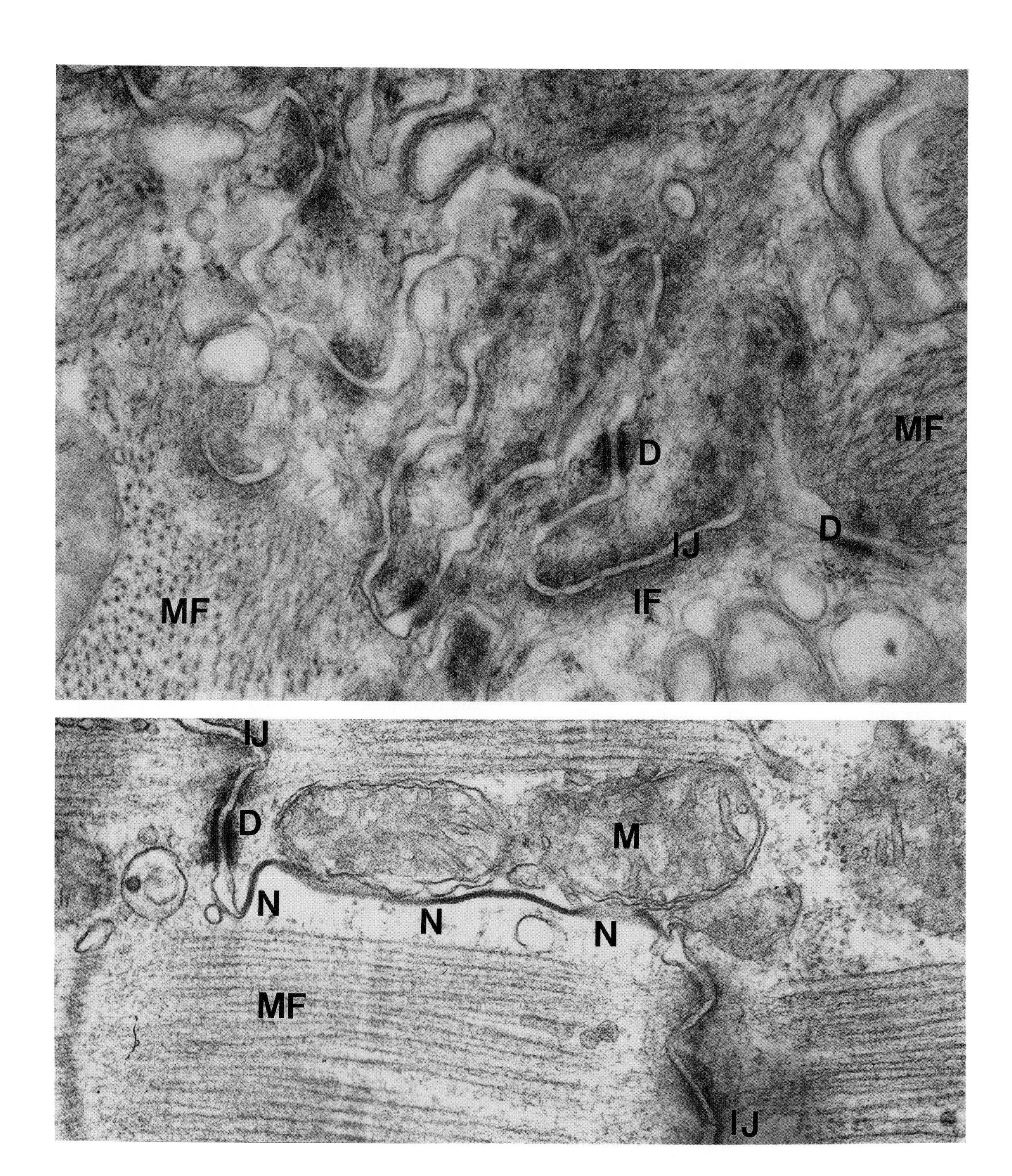

D. Desmosomes and Intermediate Junctions

Desmosomes and intermediate junctions are believed to serve mainly to prevent the cells being pulled apart. Desmosomes have a discrete substructure (Fig. 6). Intermediate junctions contain vinculin and α actin, which are both absent in desmosomes. Desmin is associated with desmosomes, but not actually within the structures.

IV. Other Muscle Cells

A. Atrial Muscle Cells

During the embryonic developmental process in the ventricle, contractile trabeculae consolidate to form the ventricular wall, which, at maturity, consists of layers within which the fibrillar alignment is parallel. The atrial wall is much thinner, made up of smaller cells (i.e., 6–8 μm in diameter), and even at maturity is formed of bundles rather than layers. Historically, these bundles have been given names and proposed as specific conduction pathways, but this is no longer accepted. The smaller cells often lack transverse tubules, and atrial cells commonly contain dense "specific atrial granules," which have been shown to contain a natriuretic factor. [*See* ATRIAL NATRIURETIC FACTOR.]

B. "Specialized" Tissues

Skeletal muscle is activated by nervous stimulation, the nerves receiving their signals from the central nervous system. Although nerves are present in the heart, their function is only to modify the contractile functioning of the heart cells, not to initiate it. Both the initiation and conduction of the electrical stimulus which leads to contraction in the heart are performed by cardiac myocytes, many of which are specialized in their development to perform that function.

The heart beat is initiated by a small group of cells located in the right atrial wall close to its junction with the sinus venosus. These form the pacemaker, or sinoatrial node. From here the impulse spreads over the atrium, generating atrial contraction, and then arrives at another small group of cells situated on the dorsal side of the junction of the atria and the ventricles. Here the impulse is carried for a short distance quite slowly (about one-hundredth of the speed elsewhere) through the atrioventricular node to the ventricular septum, to connect with the trunk (i.e., common, or His, bundle) of a treelike conduction system (i.e., the Purkinje system) which is insulated from the surrounding myocardium by blood vessels and connective tissue. This system conveys the impulse to the various parts of the ventricular musculature, causing it to contract in a systematic manner beginning at the periphery and moving toward the aorta.

The most obvious feature of these cells is that the contractile material is sparse and largely unaligned, with a substantial part of the sarcoplasm containing apparently undifferentiated material. Nodal cells are small (Figs. 8 and 9), while Purkinje cells are often large. Specialized junctions (i.e., nexuses, or gap junctions) between cells are numerous. As one progresses along the Purkinje arborizations from the His bundle toward the myocardial wall, the cell structure is observed to change steadily toward that of the ordinary contractile myocardial cells.

V. Electrical Activity of Cardiac Muscle

A. Resting Membrane Potential

By penetrating the sarcolemma with a very fine glass micropipette (i.e., approximately 0.5 μm in diameter) which is filled with a 3 *M* KCl solution; using this as a electrode, the transmembrane electrical potential of a cardiac cell can be measured. In the resting state it is about −85 mV with respect to its surroundings. This resting membrane potential reflects the difference in composition between the intracellular and extracellular fluids, which includes a difference in ionic content. This difference arises as a result of the selective permeability of the membrane combined with an active transmembrane ion pumping process. A steady state is reached in which transmembrane ionic diffusion is prevented from producing equal ionic concentrations on either side of the membrane, or equal charge distribution, and hence a transmembrane potential difference is produced. In the resting state the K^+ concentration is greater within the cell than outside it, whereas Na^+ is more concentrated outside than inside, as is

FIGURE 8 Comparatively low-magnification transmission electron micrograph of a section through the sinoatrial (i.e., pacemaker) region of a ferret heart, showing the small size of the nodal cells and both wide and very narrow spacings, between them. Nervous elements (NE) and capillaries (C) are present. Original magnification ×4,250.

FIGURE 9 Higher magnification electron micrograph from a region similar to that from which Fig. 8 was made. The nucleus (NU) is rounded, and the sarcoplasm has comparatively large areas without definable organelles. A Golgi complex (G) is present. Myofibrils (MF) are sparse and inconsistently oriented. Nervous elements (NE) are also present. M, Mitochondrion. Original magnification ×12,750.

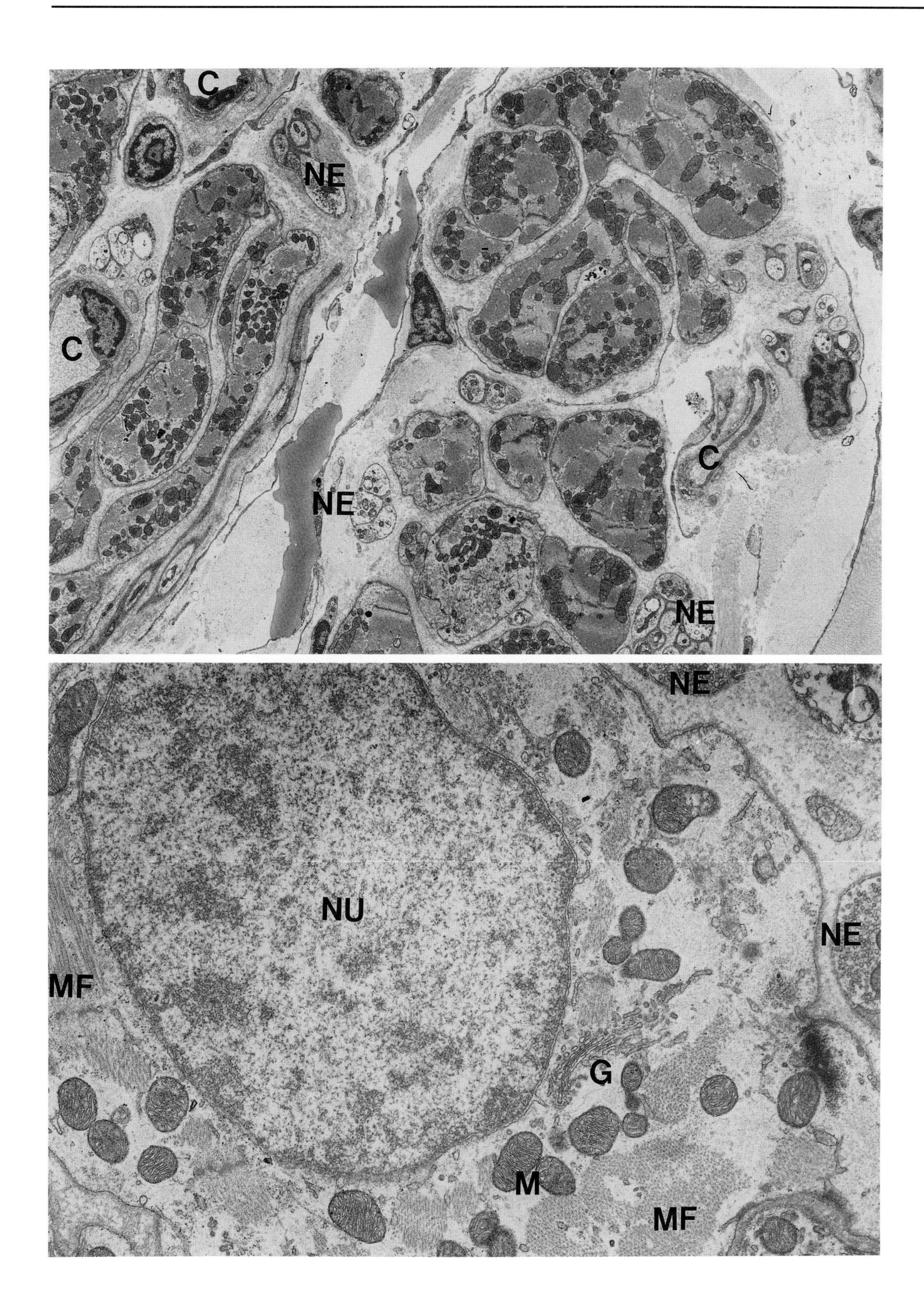
C
NE
C
NE
C
NE
NE
NU
NE
MF
G
M
MF

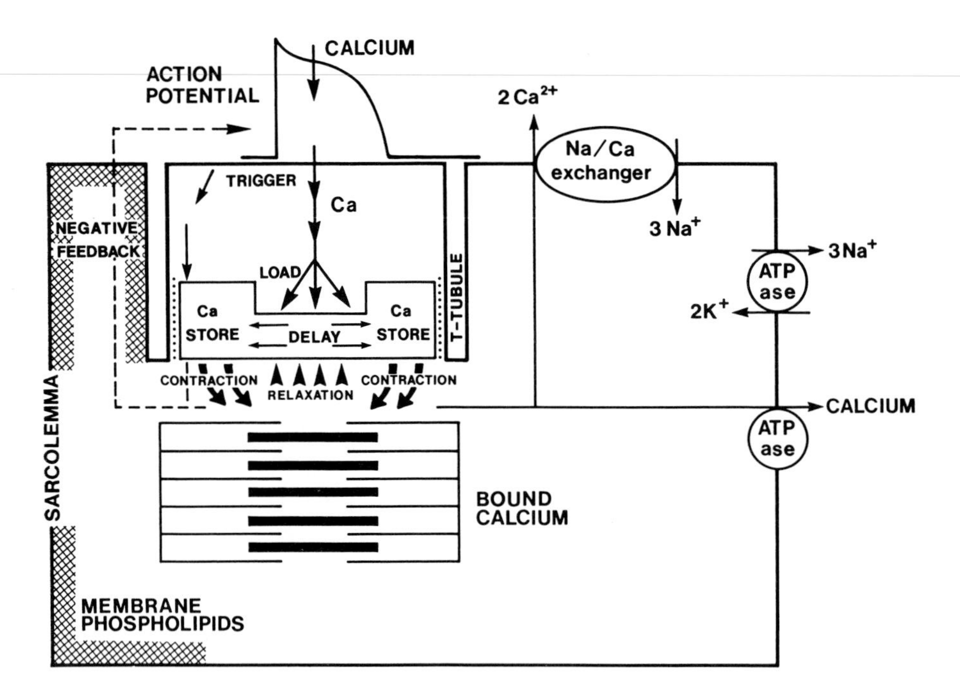

FIGURE 10 Ionic movement in the cardiac cell. The triggering cyclic action of calcium ions is indicated by the heavy arrows. During an action potential (*top left*), together with transmembrane movement of sodium and potassium ions (see Fig. 12), calcium ions enter the cell. This, in turn, stimulates the release of more calcium ions from the lateral (or junctional) regions of the sarcoplasmic reticulum (SR) by the mechanism of "calcium-induced calcium release" enunciated by Fabiato, and also produces an increase in cytoplasmic calcium, permitting the SR to be "reloaded." The sharply increased Ca^{2+} concentration in the immediate proximity of the contractile fibers triggers the contraction mechanism, and the subsequent drop in Ca^{2+} concentration is brought about by Ca^{2+} uptake by the central (or longitudinal) region of the SR, which in turn brings about relaxation. This Ca^{2+} release–sequestration sequence is part of the regular cycle of events associated with the contraction–relaxation periodicity in the heart. There is a delay before the sequestered calcium is available for release by the SR. It has been speculated that this delay represents the control process preventing sustained contraction by cardiac muscle and thereby contributing to stabilization of the heartbeat. The thinner arrows indicate the method by which the overall increase in cytoplasmic Ca^{2+} is remedied—by a sarcolemmal Na/Ca exchange mechanism and by ATPase activity. ATPase activity also brings about the resting potential sarcolemmal balance of Na^+ and K^+ ions. (Modified from H. Banijamali, M.Sc. Thesis, The University of Calgary, 1989, with permission).

Ca^{2+}. Figure 10 is a schematic representation of the way the resting potential is maintained, along with the way an action potential leads to contraction.

B. Action Potential

While all living cells generate a measure of resting potential, nerve and muscle are peculiar in developing an action potential. Reduction of the resting potential to about −60 mV (i.e., depolarization) causes the transmembrane potential to go through a spontaneous cycle, quickly becoming approximately +20 mV, then returning more slowly to its resting potential. For cardiac muscle the cycle is shown in Fig. 11; this cycle and the associated transmembrane ionic currents trigger the contractile process (Fig. 10).

Following depolarization of the membrane to −60 mV (the threshold potential), ionic channels in the membrane open, permitting the inward flow of Na^+ and Ca^{2+} and the outward flow of K^+. The sharp reversal of the potential is produced by a fast inward flow of Na^+, resulting in a closure of the channel and reestablishment of the resting transmembrane balance of Na^+. The same process occurs (in the opposite direction) with K^+, but the mechanism is slower. Peculiar to the muscle cell is the inward transmembrane current which carries Ca^{2+}. This is a slower process than the Na^+ current and is responsible for the overall transmembrane potential's demonstrating a "plateau" (Figs. 11 and 12).

When this process takes place locally in a cardiac cell, the depolarization then causes depolarization of the neighboring area to the threshold potential,

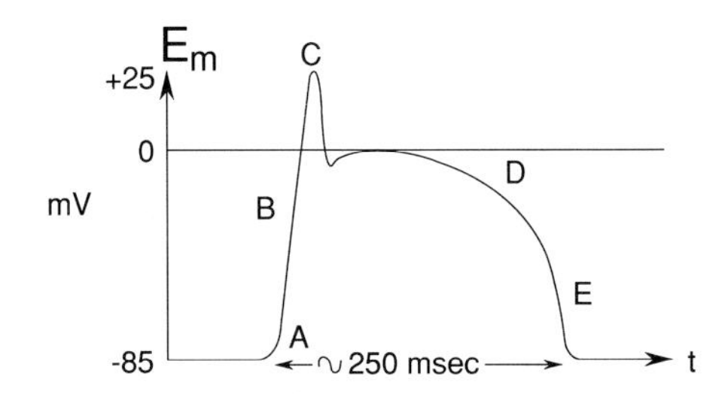

FIGURE 11 Sequence of transmembrane potential (E_m) as a function of time (t) in an action potential of a cardiac cell. (A) The point at which the depolarization is applied; (B) the period of rapid inward flow of Na^+ (fast inward current); (C) the point at which sodium current is "switched off"; (D) the plateau; (E) the period when inward current ceases.

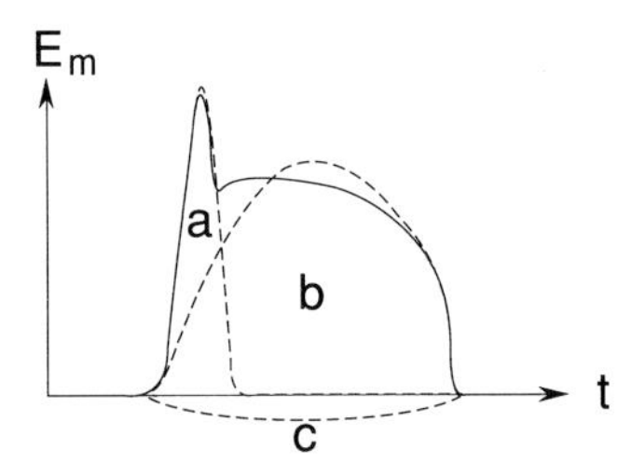

FIGURE 12 The principal current ingredients which produce the transmembrane potential changes. (a) The inward fast sodium current; (b) the slow inward calcium-containing current; (c) the outward potassium current. The dotted lines refer to the individual ionic currents; the solid line to the total electrical effect of the sum of these currents.

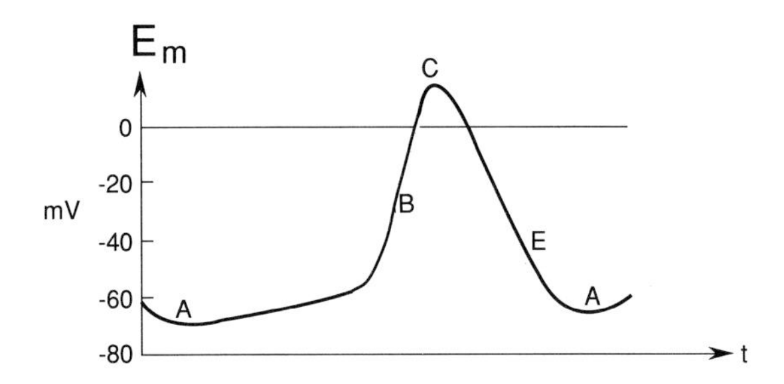

FIGURE 13 Spontaneous action potential recorded from a pacemaker cell in the sinoatrial region of the heart. (A) The greatest (negative) transmembrane potential recorded, but this is not stable and rises steadily until the threshold is reached, whereupon the fast depolarization (B—cf. Fig. 11) occurs. The maximum depolarization peak (C) is not as sharp as for a ventricular cell, and the plateau (D—cf. Fig. 11) does not exist. Repolarization takes place (E), following which the sequence repeats.

with repetition of the described events. Thus, the action potential travels along the length of the cell to be conveyed also to neighboring cells.

An important feature of the cardiac action potential is that immediately following an action potential there is a brief refractory period during which it cannot be triggered, and this helps secure the regular periodicity of the heart. Under some circumstances (e.g., electric shock) this property can fail, notably in the atrium, leading to a very fast periodicity (i.e., fibrillation), which results in a kind of seizure and thus failure of the whole organ.

C. Origin of the Action Potential

In the heart some cells spontaneously self-depolarize, thereby initiating the contraction of the whole heart (i.e., pacemaker). In such cells, notably those of the sinus node (Figs. 8 and 9), the transmembrane potential spontaneously decreases until it reaches the threshold, when the action potential occurs (Fig. 13). While some other cells of the heart (notably in the atrium and parts of the conduction system) have their property, the periodicity is slower, and as a result they are triggered by those with the faster periodicity, thus giving the whole organ its synchronous contraction.

VI. Nervous Elements

While the action potential of the heart is determined by muscle cells, not by nerve cells, nervous elements are present, particularly in the nodal regions (Figs. 8 and 9). These perform a regulatory role, secreting agents which speed up the periodicity and the conduction velocity (e.g., adrenaline) or slow it down (e.g., acetylcholine). Sensory nerves are also present.

Bibliography

Berne, R. M., and Levy, M. N. (1981). "Cardiovascular Physiology." Mosby, St. Louis, Missouri.

Berne, R. M., Sperelakis, N., and Geiger, S. R. (1979). "Handbook of Physiology. Section 2: The Cardiovascular System." Am. Physiol. Soc., Bethesda, Maryland.

Fozzard, H. A., Haber, E., Jennings, R. B., Katz, A. M., and Morgan, H. E. (1986). "The Heart and Cardiovascular System," Sci. Found. Vol. I. Raven, New York.

Nathan, R. D. (1986). "Cardiac Muscle: The Regulation of Excitation and Contraction." Academic Press, Orlando, Florida.

Noble, D., and Powell, T. (1987). "Electrophysiology of Single Cardiac Cells." Academic Press, Orlando, Florida.

TerKeurs, H. E. D. J., and Tyberg, J. (1987). "Mechanics of the Circulation." Nijhoff, Dordrecht, The Netherlands.

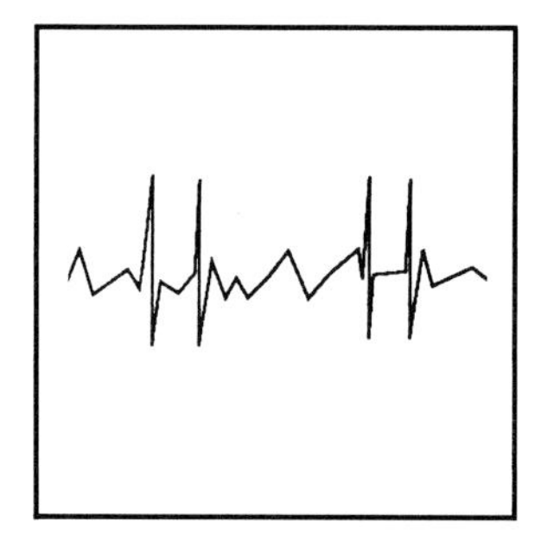

Cardiogenic Reflexes

ROGER HAINSWORTH AND DAVID MARY, *Leeds University*

Glossary

Afferent limb Group of nerve fibers that connect the receptors to the central nervous system. They carry afferent impulses which represent progressive depolarization (i.e., a difference in electrical potential along the surface of the fiber). The number of impulses per unit of time is determined by the intensity of the stimulus to the receptor.

Effector organs Organs that form the target of reflex action via efferent limbs

Efferent limb One or more pathways connecting the central nervous system to organs affected by the reflex. The pathways subserving the connections can involve humoral, neural, or neurohumoral mechanisms.

Myelinated nerves Larger-diameter nerves individually covered by a myelinated sheath. The velocity of conduction in these nerves is more than 2.5 m/sec. Nonmyelinated nerves conduct more slowly

Receptor Small sensory organ with a defined structure and connection via an afferent nerve fiber to the nervous system; in some cases a structure has not been defined, and the receptor is presumed by neural activity in its afferent fiber

Reflex Term that describes reception of a signal by the nervous system and the consequent trigger or modification of organ function; reflexes involve receptors, an afferent limb, central connections, efferent limbs, and effector organs

Threshold Least stimulus that gives rise to activity in the afferent nerve fibers or to a measured response of an effector organ

THE TERM "cardiogenic reflexes" refers to the responses, involving nervous pathways to and from the brain, which result from the stimulation of nerves ending in the heart. Because of difficulties in localization of the stimuli, most of our knowledge of these reflexes has stemmed from experiments involving anesthetized animals. The nervous receptors responsible for these reflexes have been demonstrated in both atria, the ventricles (mainly left), and the coronary arteries. Activation of the cardiac receptors occurs in response to physiological events, including changes in the volumes or pressures within the cardiac chambers. Many of the afferent fibers exhibit tonic activity, being active at normal levels of cardiac filling and pressures. Some only become active in response to chemical stimulation, and this could result from the administration of exogenous pharmacological agents or from chemical changes occurring in some pathological states. The efferent pathway of cardiogenic reflexes controls the rate and force of contraction of the heart itself, the degree of constriction of blood vessels, and the secretion of several hormones. Cardiogenic reflexes thus have an important role in the control of the cardiovascular system.

I. Introduction

In the study of any reflex, certain components must be defined. These are the stimulus (the event that triggers the response), the receptor (the specialized transducer in the body which detects the stimulus and transmits it to a nerve), the afferent nervous pathway (which conveys the information to the brain or the spinal cord), the pathways through the central nervous system, the efferent nervous pathway (which conveys information from the brain or the spinal cord), the effector organ (the heart, blood vessel, etc., which changes as the result of the reflex), and the response. To define these components, complex preparations are required, and these are difficult even in anesthetized animal experiments, and almost impossible in humans. The study of reflexes in the cardiovascular system is particularly difficult, and the study of reflexes arising from the heart in intact humans is nearly impossible. For this reason almost all of our knowledge of cardiogenic reflexes emanates from studies of anesthetized animals, and therefore most of what is contained in this contribution is based on results obtained from animal experiments.

II. Afferent Nerves from the Heart

A. Nerve Endings

The heart, like most organs, is innervated by the autonomic nervous system, and afferent impulses run in the vagal and the sympathetic nerves. Although much is known concerning the afferent nerves from the heart, much less is known concerning the receptor organs attached to these nerves. Two particular structures have been described in the heart: complex unencapsulated endings and a fine network of nerve fibers (Fig. 1).

Complex unencapsulated endings are the discrete branching ends of nerves. They are attached to larger nerve fibers which have a myelin sheath and run in the vagus nerves, and they are located in the cardiac atria on the endocardial (i.e., inner) surface of the heart. The greatest concentrations of these receptors are at the junctional regions between the venae cavae and the pulmonary veins, with the atria. These endings are the only ones which have definitely been shown to be receptors.

A fine network of branching beaded nerve fibers extends over the entire endocardial surface of both the atria and the ventricles. The function of this network is unclear. It might be partially formed from branches of the complex unencapsulated endings. Another possibility is that it is attached to the many fine nonmyelinated nerves known to innervate the heart. It has also been suggested that the nerve net might be concerned with the motor innervation of the heart.

B. Afferent Discharges in Nerves from the Atria

Myelinated vagal afferent nerves have been shown to be attached to the atrial receptors (complex unencapsulated endings). The activity in these nerves has been studied by electrical recording of the discharges traveling toward the central nervous system. The activity in the fibers has been classified as type A, which shows a burst of activity coinciding with contraction of the atria, identified by the *a* wave of the atrial pressure. Type B activity coincides mainly with atrial filling, the *v* wave of pressure. Also present is an intermediate discharge pattern which has the features of both types A and B. There has been much debate concerning the significance of the various discharge patterns. However, it is likely that there is no fundamental difference in the atrial receptors, and the different discharge patterns result from different locations of the receptors. The types of discharge of a single receptor can be changed following hemorrhage, infusions, and changes in heart rate (Fig. 2).

In general, the activity of atrial receptors is related to atrial volume and pressure. They signal the degree of atrial filling, which is dependent on, among other things, blood volume, and hence these receptors have sometimes been referred to as volume receptors.

There are many more small nonmyelinated nerve fibers (C fibers) ending in the atria, compared with the myelinated nerves. The activity of these fibers is also stimulated by increases in atrial pressure and volume. However, in general, they have a higher threshold and require higher pressures for their excitation. Many are silent at normal pressures and only become active at abnormally high levels of atrial pressure.

In addition to vagal nerve endings, there are nerve fibers that run in the sympathetic rami. These afferent sympathetic nerves seem to be stimulated by events similar to those affecting the vagal affer-

FIGURE 1 Example of two types of nervous structures in the dog's atrium. The two frames are whole-thickness preparations of endocardium obtained from the pulmonary venoatrial junction and supravitally stained with methylene blue. (Top) A complex unencapsulated ending believed to be an atrial receptor attached to an afferent nerve. (Bottom) Nerve network.

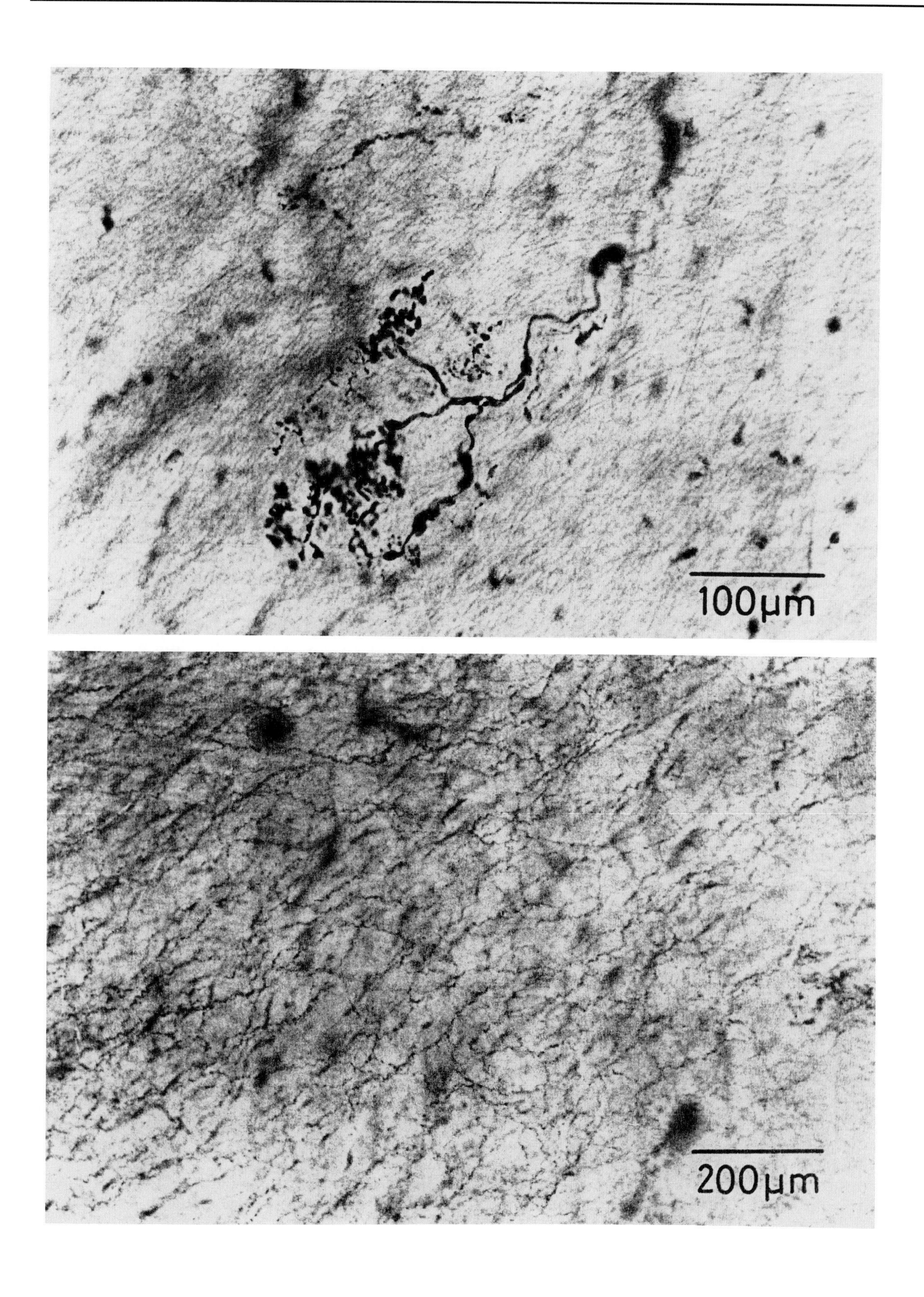
100μm
200μm

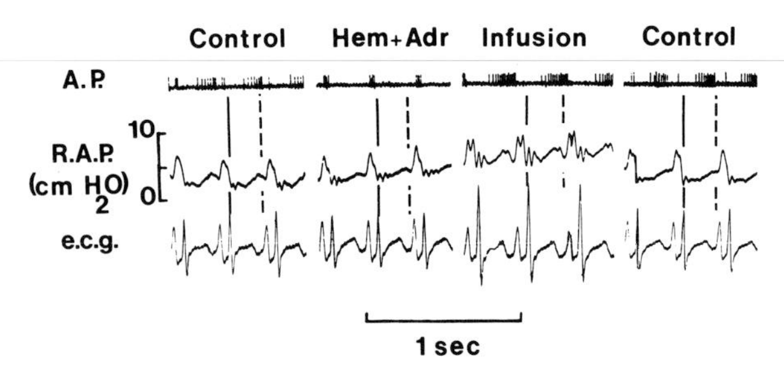

FIGURE 2 Parts of experimental recordings showing changes in the pattern of discharge of an atrial receptor in a cat. Shown are action potentials (A.P.) in a single afferent vagal nerve fiber, right atrial pressure pulse (R.A.P.), and the electrocardiogram (e.c.g.). The vertical lines are drawn to relate temporal events in atria and the afferent nerve: The solid line indicates the end of the *a* wave of atrial systole, and the dashed line indicates the peak of the *v* wave of atrial filling. Control (left), The pattern of activity is of the intermediate type; Hem+Adr, the activity has changed to type A pattern after bleeding and administration of adrenaline; Infusion, the activity has changed to type B pattern after infusion of dextran; Control (right), the receptor has regained its intermediate pattern of discharge.

ents. One interesting feature of these nerves is that frequently they are attached to more than one terminal. Sometimes a single fiber can be stimulated by events in both an atrium and a ventricle.

Some sympathetic afferent nerves can be stimulated by mechanical events, whereas others are excited only by the application of chemicals. Some of the excitant chemicals might be released during injury of the heart (e.g., when blood supply is inadequate in coronary artery disease), and it has been suggested that they might serve as pain receptors.

C. Activity in Nerves from the Ventricles

In contrast to the atria, the ventricles are supplied almost exclusively by small nonmyelinated vagal afferents. Some of these nerves clearly respond to changes in ventricular pressure, although whether they respond mainly to ventricular contraction or to ventricular filling is unclear. Some have a phasic discharge related to ventricular pressure, whereas in others the discharge is apparently random.

Some ventricular nerve fibers are not readily activated by mechanical events, but they are strongly stimulated by toxic and irritant chemical agents applied directly to the receptor site or injected into the coronary circulation. As discussed in Section IV, chemosensitive ventricular afferents can cause powerful reflex responses, but what constitutes their normal stimulus or what is their physiological role is not known.

The ventricles also receive an afferent sympathetic innervation, sometimes coterminal with atrial fibers. These might also be stimulated by mechanical and chemical events.

III. Reflexes Arising in the Atria

A. Methods for Stimulation of Atrial Receptors

Electrophysiological studies outlined above have shown that the discharge from atrial receptors is increased by procedures that increase atrial filling. These include fluid infusion, immersion in water, and changing body position from upright to supine or head down. The disadvantage of these procedures is that they do not provide a localized stimulus to atrial receptors, and the resulting responses would be due to changes in the activity from many cardiovascular reflexogenic areas.

The function of atrial receptors has been studied in experimental animal preparations in which attempts have been made to localize the stimulus to the reflexogenic areas of the atria. Experiments have been performed in which the blood flow into the atria was increased or the outflow from the atria was obstructed. However, the disadvantage of these techniques is that the changes in pressure might affect other regions of the heart.

Since most of the myelinated atrial afferents are situated at the great venoatrial junctions, localized distension of these regions by balloon has provided a useful method for discretely stimulating atrial receptors and studying their function.

B. Responses of the Heart

Discrete stimulation of either right or left atrial receptors results in a reflex increase in heart rate (Fig. 3). The afferent pathway for this reflex has been investigated by graded cooling of the vagal nerves. The response of heart rate to stimulation of atrial receptors is slightly reduced at 16°C, and at 12°C it is nearly abolished. This is the range of temperatures that blocks the increase in the frequency of nerve impulses in the myelinated vagal afferents, but not in nonmyelinated nerves. The afferent pathway of the reflex, therefore, lies in myelinated vagal afferent nerves.

The reflex increase in heart rate can also be abolished by cutting the sympathetic nerves to the

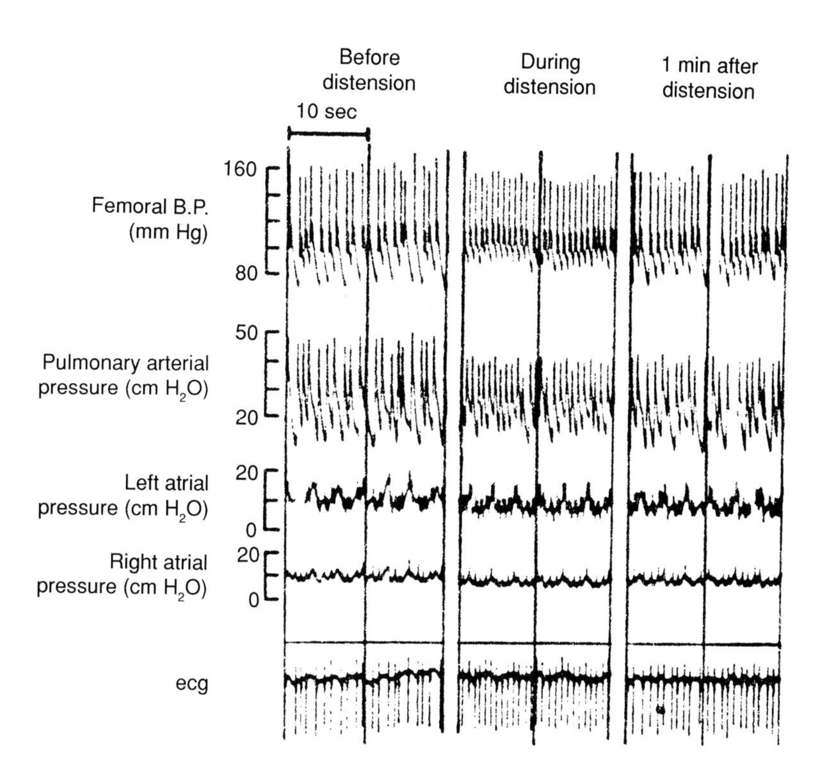

FIGURE 3 Example of the reflex response of an increase in heart rate to discrete stimulation of left atrial receptors in a dog. Shown are femoral arterial pressure (B.P.), pulmonary arterial pressure, left atrial pressure, right atrial pressure, datum line, and the electrocardiogram (e.c.g.). The middle column shows recordings during stimulation of atrial receptors, by distending small balloons at the pulmonary venoatrial junctions, and is to be compared with the control recording (left column) and recovery after the release of balloon distension (right column). During stimulation of atrial receptors, the heart rate increased without significant changes in cardiovascular pressures, indicating the discrete nature of the stimulus. [From J. R. Ledsome and R. J. Linden, eds. (1964). A reflex increase in heart rate from distension of the pulmonary-vein–atrial junctions. *J. Physiol.* **170,** 456–473.]

heart, by blocking the release of norepinephrine from the nerve terminals, or by giving β-adrenoceptor-blocking drugs to prevent their action. The efferent pathway of the reflex, therefore, lies in the cardiac sympathetic nerves.

The efferent limb of the heart rate reflex has several peculiarities. The nerves engaged in the atrial reflex seem to affect only the sinoatrial node, because there is no effect on the force of contraction of the heart which would occur if sympathetic nerves to the myocardium were excited. Also, it has been shown that the particular nerve fibers involved in the atrial receptor reflex are not affected by several other reflexogenic stimuli, including the stimulation of baroreceptors, chemoreceptors, visceral afferents, and cutaneous nerves. Other cardiac sympathetic nerves are affected by all of the other stimuli, but not by atrial receptors. Atrial receptors thus seem to have their own discrete pathway.

There have been several reports that stimulation of atrial receptors results in reflex bradycardia. However, most of the studies on which these reports are based can be criticized due to inadequate localization of the stimulus. Nevertheless, the possibility remains that greater degrees of atrial distension, particularly when the atrial receptors attached to myelinated nerves are for some reason ineffective, might result in reflexes mediated by the nonmyelinated nerves. Thus, bradycardia might occur in some circumstances due to excitation of nonmyelinated afferent nerves.

C. Responses of Blood Vessels

The reflex effects of discrete stimulation of atrial receptors have been studied in several vascular beds. Electrophysiological studies of the efferent activity in sympathetic nerves supplying a number of areas have yielded interesting results. These showed that, whereas sympathetic efferent activity increased in cardiac fibers, it decreased in nerves supplying the kidney (Fig. 4). There was no change in activity in

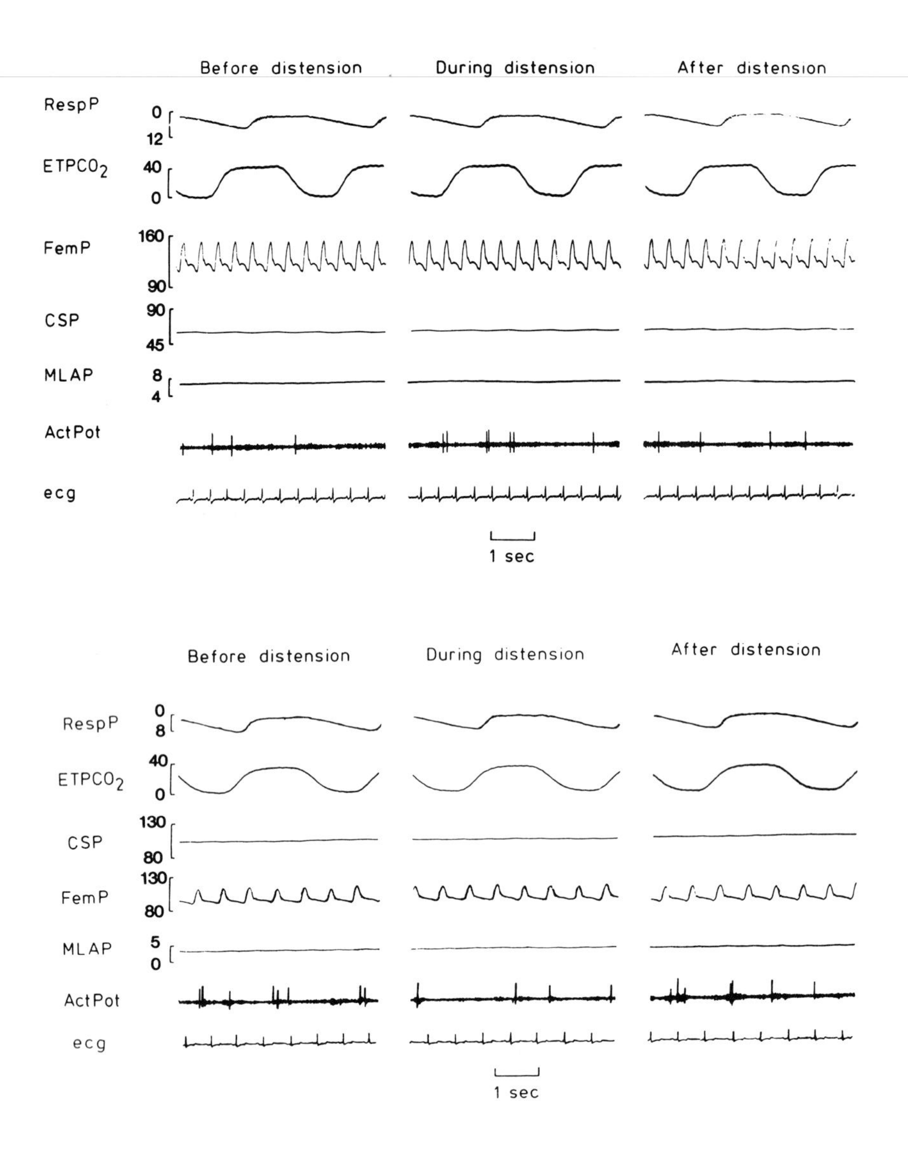

the lumbar and splenic nerves. The effects on the resistance to blood flow reflect this pattern of activity.

Stimulation of atrial receptors has been shown to have no significant effect, in the steady state, on resistance to blood flow in the limb circulation. A transient vasodilatation immediately following distension has sometimes been reported, which is probably due to excitation of nonmyelinated afferent nerves.

As would be expected from the electrophysiological studies of activity in renal nerves, atrial receptor stimulation results in dilatation of renal resistance vessels, which normally would be expected to lead to an increase in renal blood flow. Unlike the variable transient response in the limb, the renal vasodilatation tends to persist for as long as the stimulus is maintained.

FIGURE 4 Examples of a reflex increase in activity of an efferent cardiac sympathetic nerve fiber (top) and a decrease in activity in an efferent renal sympathetic nerve fiber (bottom). In both frames are shown endotracheal or respiratory pressure (RespP), end-tidal carbon dioxide partial pressure ($ETPCO_2$), femoral arterial blood pressure (FemP), carotid sinus perfusion pressure (CSP), mean left atrial pressure (MLAP), action potentials (ActPot), and the electrocardiogram (ecg). The units of pressure are millimeters of mercury. As in Fig. 3, the middle column contains the recording taken during discrete stimulation of atrial receptors and is sandwiched by two columns of initial control and recovery period recordings.

The only other vascular bed shown to be affected by the stimulation of atrial receptors is the coronary circulation (i.e., the blood vessels supplying the heart muscle). Recent studies have shown that stimulation of atrial receptors results in the constriction of coronary vessels, leading to a decrease in their blood flow. It should be appreciated in this context that coronary blood flow is largely dependent on the work of the heart and that factors which increase heart work, including increases in heart rate, greatly increase coronary blood flow. Therefore, the reduction in flow due to atrial receptor stimulation can only be seen if heart rate and heart force are controlled.

D. Effects on Urine Flow

Stimulation of atrial receptors results in an increase in urine flow (Fig. 5). Experiments to demonstrate this have involved discrete distension of the venoatrial junctions, obstruction of the mitral valve by balloons, increases in atrial pressure by volume infusions, or increases in venous return to the heart by procedures such as water immersion.

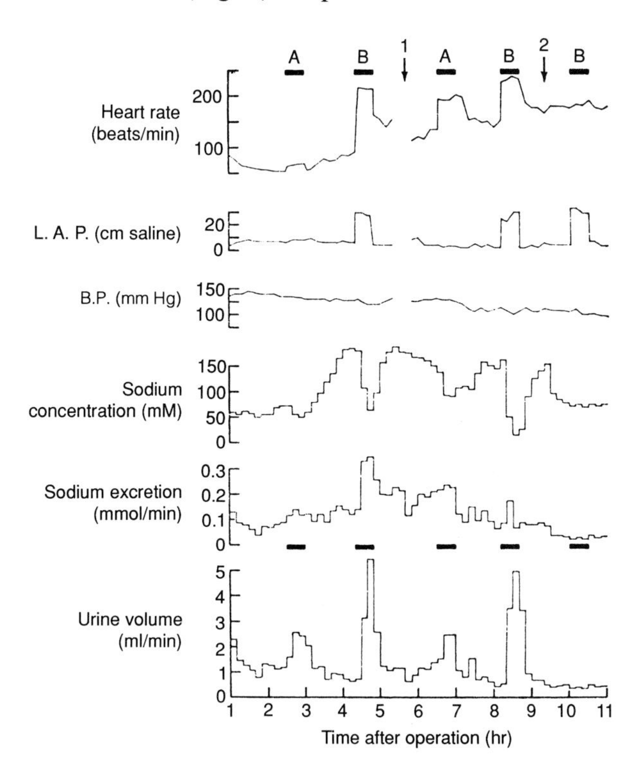

FIGURE 5 Example of the reflex increase in urine flow and the urinary rate of sodium excretion in a dog in response to atrial distension. (A) Small balloons distended in the pulmonary venoatrial junctions to stimulate atrial receptors with no obstruction to blood flow in the heart. (B) Balloon distended in the left atrium to cause mitral obstruction. There is a break in the record at 1, when the atrial balloon was replaced because it leaked. (2) The right vagus nerve was cut in the neck, and the left vagus nerve was cut at the level of the upper border of the aorta. [From J. R. Ledsome and R. J. Linden, eds. (1968). The role of left atrial receptors in the diuretic response to left atrial distension. *J. Physiol.* **198,** 487–503.]

The mechanism responsible for the diuresis is uncertain. It has been shown to occur in animals in which the hypothalamus has been destroyed; therefore, it is not dependent on changes in antidiuretic hormone (ADH), although there is a decrease in the level of ADH, and this could contribute to the response. Renal nerves might also contribute, although an increase in urine flow occurs even in denervated kidneys perfused with blood from a donor animal in which the atria are distended.

The increase in urine flow is sometimes accompanied by an increase in sodium excretion. Whereas the diuretic response seems to be mainly humorally mediated, the natriuresis is dependent on the concomitant changes in heart rate and changes in renal nerve activity.

A humoral agent which has received much attention in recent years is atrial natriuretic peptide. This peptide is released from atrial myocytes in response to stretch. It might be involved in the response to increases in atrial volume, in addition to the responses to stimulation of afferent nerves. However, it is not essential for the diuretic response to atrial distension. It is also possible that an undiscovered hormone(s) might contribute to the response.

In summary, therefore, discrete stimulation of atrial receptors results in increases in urine flow and small and inconsistent increases in sodium excretion. The diuresis is mediated largely by a humoral agent(s). The natriuresis appears to be dependent on changes in renal nerve activity and concomitant hemodynamic changes.

E. Effect on Hormones

Vasopressin or ADH is released by the posterior part of the pituitary gland in response to an increase in the osmolality of blood. Blood osmolality is mainly dependent on the concentration of sodium. There are believed to be osmoreceptors which control the release of this hormone. As mentioned above, there is evidence that the release of ADH into the blood is partly under the control of atrial receptors. An increase in plasma volume results in an increased stimulation of atrial receptors, and this

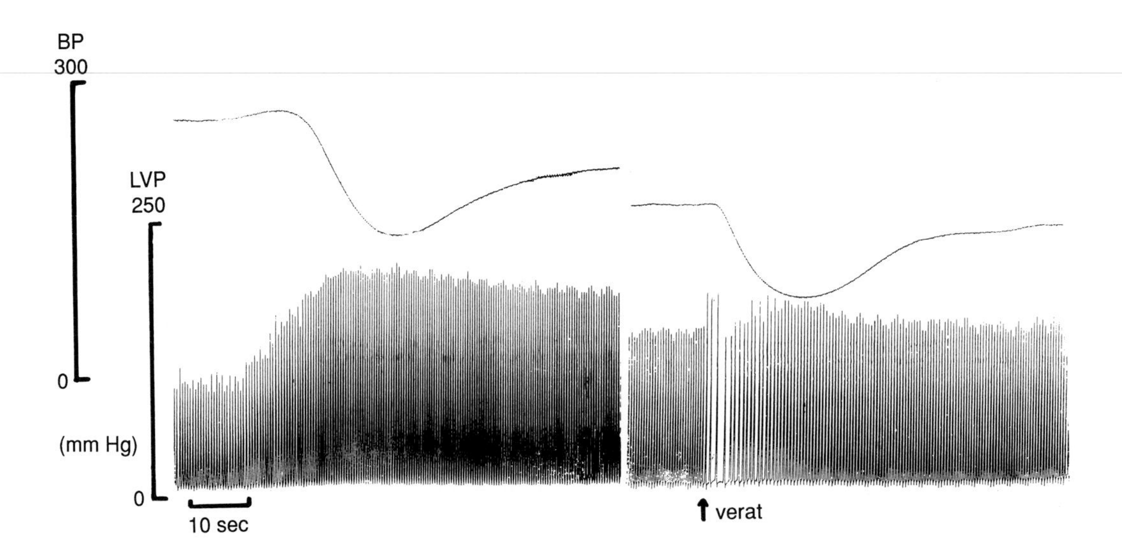

FIGURE 6 Responses to chemical and mechanical stimulation of ventricular receptors. Mechanosensitive ventricular nerves were excited by increasing left ventricular systolic pressure (LVP) and chemosensitive nerves by injection into the coronary circulation of the stimulant chemical veratridine (verat). Note that both methods of stimulation resulted in decreases in blood pressure (BP), whereas only chemical stimulation resulted in a decrease in heart rate.

leads to a reduction in the release of ADH and a consequent water diuresis.

Renin is a hormone which is released by the kidney in response to either a decrease in blood pressure to the kidney or activation of the renal nerves. Renin has no direct cardiovascular effects, but it is responsible for the formation of angiotensin II, a powerful vasoconstrictor agent. The stimulation of atrial receptors results in a decrease in the plasma level of renin. The mechanism of this response has been shown to involve renal nerves. The precise mechanism by which the activity in renal nerves controls renin secretion is not fully understood. It was generally believed to involve solely a β-adrenergic mechanism. However, β-adrenoceptor-blocking drugs do not completely prevent this response. The control of renin release involves an additional pathway which is neither adrenergic nor cholinergic.

Plasma cortisol is controlled by another pituitary hormone, adrenocorticotropic hormone. The stimulation of atrial receptors results in a decrease in this hormone and a consequent decrease in the plasma level of cortisol.

IV. Reflexes from Ventricular and Coronary Receptors

Ventricular afferent nerves can be excited both by increases in ventricular volume and pressure and by the administration of several exogenous chemicals. Unlike arterial chemoreceptors, chemosensitive ventricular nerves are not excited by changes in the tensions of oxygen and carbon dioxide, but they can be strongly stimulated by chemicals such as veratrum alkaloids, nicotine, and bradykinin. Injections of these chemicals result in powerful reflex bradycardia and hypotension. Increases in ventricular systolic pressure also result in vasodilatation. However, unlike chemical stimulation, there is little effect on the heart rate (Fig. 6).

Recent evidence has shown that the control of peripheral vascular resistance by ventricular receptors is likely to be of importance in normal cardiovascular homeostasis, and it seems likely that, in this respect, ventricular receptor function is similar to that of arterial baroreceptors.

The importance of chemosensitive ventricular afferents is much less clear. It is apparent that chemical stimulation of ventricular afferents can result in powerful responses, an intense bradycardia (the heart can stop for several seconds), and an intense vasodilatation. What is not clear is the physiological significance of this response. Bradycardia and hypotension can occur following damage to heart muscle due to thrombosis of the coronary vessels. However, the relevance of this to normal control of the cardiovascular system remains uncertain.

V. Reflexes Mediated by Sympathetic Afferent Nerves

Afferent nerves running in the sympathetic nervous system, like vagal afferents, are activated by chemical and mechanical stimuli. Stimulation of these afferent nerves has been shown to result in spinal reflexes (not involving the brain) which lead mainly to increases in activity in efferent sympathetic nerves. This causes increases in the heart rate and the constriction of blood vessels, and therefore increases in blood pressure. Supraspinal mechanisms are also involved, as seen, for example, by inhibition of efferent vagal activity to the heart.

The reflex effects of stimulation of cardiac sympathetic afferent nerves are thus opposite those from stimulation of vagal afferent nerves. Whereas the vagal reflexes function as a ''negative feedback system'' and consequently inhibit the cardiovascular system, the sympathetic afferents appear to excite it. The role of this mechanism is far from clear, and, at least in anesthetized animals, the vagal inhibitory mechanisms seem to be dominant. However, it is possible that sympathetic reflexes have a modulating role in cardiovascular control.

VI. Physiological Role of Cardiogenic Reflexes

A. Atrial Reflexes

A change in the stimulus to atrial receptors can occur with almost any event, including changes in posture, as well as in less common activities (e.g., water immersion, aviation, and space flight). Whereas arterial baroreceptors appear to be important in the control of blood pressure, atrial receptors have been suggested to be important in the control of heart rate and, consequently, heart size. In this context it has been suggested that atrial receptors might serve to regulate the size of the heart to an optimal level. Thus, if the heart becomes more distended, the rate increase and, consequently, the size become smaller. The effects of atrial receptors on urine flow complement the effects on the heart rate and could contribute to the control of blood volume.

B. Ventricular Reflexes

Ventricular mechanosensitive afferents appear to function mainly as arterial baroreceptors, in that their stimulation results in reflex vasodilatation. There are, however, some differences between the effects of stimulation of ventricular mechanoreceptors and peripheral arterial baroreceptors. Stimulation of baroreceptors results in reflex vasodilatation, bradycardia, and little, if any, effect on respiration. Stimulation of ventricular receptors also results in vasodilatation, but there is little or no effect on heart rate and there is a reduction in respiratory activity.

The responses to chemical stimulation of ventricular afferent nerves are different again. There is also vasodilatation, but in addition there is a powerful slowing of the heart, but little effect on respiration. Although it is likely the mechanosensitive ventricular nerves have an important role in the control of the circulation, the physiological significance of the chemosensitive nerves is unknown.

VII. Possible Role of Cardiogenic Reflexes in Disease States

A. Atrial Receptors

There is evidence that atrial receptors might be destroyed and lose their function in congestive heart failure, thus possibly contributing to the changes which occur in this condition. This functional ''deactivation'' of the atrial receptors might contribute to the water and salt retention, and eventually to the clinical condition of edema, in which extra fluid is obvious in dependent parts of the body. Deactivation of atrial receptors causes decreases in urine flow and in the rate of urinary sodium excretion, increases in the activity of the renin–angiotensin system and vasopressin, and vasoconstriction of the renal vessels.

It is known that shock resulting from extensive bleeding is associated with a higher incidence of renal failure than that associated with acute heart failure. This difference has been suggested to be due in part to deactivation of atrial receptors during bleeding, with a resulting renal vasoconstriction which is greater than that occurring in heart failure. Following bleeding the deactivation of atrial receptors could lead to an increase in the plasma cortisol level, which would protect body tissues and enhance the vasoconstrictive effects of sympathetic nerves on blood vessels to maintain normal arterial blood pressure levels. The increase in the activity of the renin–angiotensin system and in the blood level of vasopressin and the decrease in urine flow would result in fluid retention, to compensate for lost blood, and vasoconstriction to maintain normal arterial blood pressure.

Paroxysmal tachycardia is a condition in which the heart has bouts in which it beats abnormally fast. This condition is sometimes associated with an increase in urine flow. During the rapid heart rate the atria cannot empty normally into the ventricles, and they become distended. This increases the stimulus to atrial receptors, leading to an increase in urine flow.

It is possible to suggest a role for atrial receptor reflexes in the pathophysiological control of the coronary circulation. The stimulation of atrial receptors results in a reflex increase in heart rate, and this, in turn, increases coronary flow. However, the direct reflex effect of the stimulation of atrial receptors is to reduce coronary flow by sympathetic vasoconstriction. This effect might help to maintain uniform perfusion to all layers of the ventricular wall, which is important since the limitation to blood flow to the inner layers is believed to constitute a mechanism for myocardial ischemia during increases in heart rate, particularly in the presence of coronary narrowing. The clinical condition relating to these events is labeled "angina" and is caused by pathological narrowing of the coronary arteries. In such patients the atrial receptor reflex might have a protective role.

B. Ventricular Receptors

Following hemorrhage the stimulus to arterial baroreceptors is reduced, and this leads to increased rate and force of ventricular contraction. This has been suggested to be an effective mechanism for the stimulation of ventricular receptors and might contribute to the so-called "vasovagal" attack, during which the heart rate abruptly slows, blood vessels dilate, and the patient loses consciousness. It must be stressed that this is only one of the suggested mechanisms, and the reason that a subject with low circulatory blood volume should suddenly change from having a fast heart rate and constricted blood vessels to a slow heart rate and dilated vessels is not known.

It has also been suggested that a similar response might occur at the end of a severe bout of physical exercise. The postulated mechanism is that the heart is still being stimulated to contract powerfully, but is no longer being filled so much with blood. Consequently, it contracts abnormally to a small size, and this is what might strongly excite the reflex.

Other inappropriate ventricular reflexes might occur in patients who have narrowing of the aortic valve. This leads to abnormally high ventricular pressures, excitation of ventricular receptors, and a consequent decrease in arterial blood pressure.

Another situation in which ventricular receptors might become abnormally excited is when chemicals are injected into the coronary arteries. This can occur during coronary arteriography, when radioopaque chemicals are injected directly into one of the coronary arteries. This can lead to a profound bradycardia and a decrease in blood pressure. A similar response can occur following a coronary thrombosis, particularly when this affects the coronary arteries supplying the inferior surface of the ventricle.

Bibliography

Hainsworth, R. (1990). Atrial receptors. *In* "Reflex Control of the Circulation" (I. H. Zucker and J. P. Gilmore, eds.). Telford, Caldwell, New Jersey.

Hainsworth, R. (1991). Reflexes from the heart. *Physiol. Rev.*, in press.

Hainsworth, R., McWilliam, P. N., and Mary, D. A. S. G. (eds.) (1987). "Cardiogenic Reflexes." Oxford Univ. Press, Oxford, England.

Linden, R. J., and Kappagoda, C. T. (eds.) (1982). Atrial Receptors. *Monogr. Physiol. Soc.* **39.**

Malliani, A. (1982). Cardiovascular sympathetic afferent fibres. *Rev. Physiol. Biochem. Pharmacol.* **95,** 11–74.

Mary, D. A. S. G. (1989). Cardiovascular receptors and the coronary circulation. *Clin. Sci.* **77,** 1–5.

Cardiovascular Responses to Weightlessness

DWAIN L. ECKBERG, *McGuire Veterans Hospital and Medical College of Virginia*

Glossary

Autonomic nervous system Extensive array of nerves that modulates internal organ function automatically; norepinephrine and acetylcholine are the principal cardiovascular autonomic neurotransmitters

Baroreflex Neural pressure-regulating reflex that opposes changes of blood pressure; during standing, heart rate increases and resistance vessels constrict in response to changes of traffic carried over vagus and sympathetic autonomic nerves

Cardiovascular deconditioning Syndrome that occurs in astronauts after space travel, comprising reduced exercise capacity and, with standing, unusually reduced mean blood pressure, increased heart rate, and, in extreme cases, loss of consciousness

Orthostatic hypotension Unusually large reduction of blood pressure in the standing position

Syncope Loss of consciousness; this extreme manifestation of postspaceflight cardiovascular abnormalities may occur with standing, because of reduction of blood flow to the brain; presyncope signifies lightheadedness and feelings of impending loss of consciousness

Vasovagal reaction Simple faints that occur under a variety of circumstances, including standing; they are due to reflex reductions of sympathetic nerve traffic to vessels and increased vagus nerve traffic to the heart, which lead to abruptly reduced blood pressure and blood flow to the brain

HUMAN CARDIOVASCULAR RESPONSES to microgravity are a source of fascination, in part because they cannot be studied on Earth. Cardiovascular function is normal in space (unless untoward responses are provoked experimentally), and the abnormalities that occur because of space travel are seen only after astronauts (or cosmonauts) return to the Earth's gravitational field. Virtually all astronauts experience unusually large reductions of arterial pressure and increases of heart rate when they stand after sojourns in space. Symptoms with standing range from slight lightheadedness to frank loss of consciousness. These postflight abnormalities are thought to result, in as yet unexplained ways, from a cascade of cardiovascular changes that begins in the first minutes and hours of weightlessness. According to this construct, loss of normal gravitational forces shifts body fluids from the lower to the upper body; translocation of blood stretches cardiac chambers and sequentially provokes release of endogenous diuretic hormones, diuresis, and blood volume reduction. Postflight cardiovascular changes disappear spontaneously within days to weeks and leave no residual abnormalities.

I. History

The earliest animals in space—dogs, rats, mice, and flies—were launched amid predictions of massive organ failure, including particularly cardiovascular collapse. The first human in space, 27-yr-old Yuri Gagarin, placed in Earth orbit by the Soviet Union on April 12, 1961, returned as proof that exposure to weightlessness does not lead to cardiovascular collapse; however, the ninth man in space, the American Wally Schirra, returned from a 9-hr 13-min orbital mission in 1962 and became the first astronaut to report cardiovascular symptoms after a space mission. The symptoms he described have been labeled "cardiovascular deconditioning" and comprise orthostatic hypotension, narrowing of pulse pressure (the difference between systolic and diastolic arterial pressures), lightheadedness or frank syncope, tachycardia, and reduced exercise tolerance. Some of these changes are depicted in Fig. 1. Orthostatic hypotension occurs in almost all astronauts and seems to be independent of mission duration. Symptoms are highly variable. In some instances, symptoms are of trivial importance such as transient lightheadedness after prolonged standing. The Soviet astronaut Levchenko flew an airplane immediately after he returned to Earth from an 8-day mission. In other instances, symptoms may be disabling—two Soviet astronauts could not stand after return to Earth and were carried from their spacecraft on stretchers. Their recovery took at least 11 days.

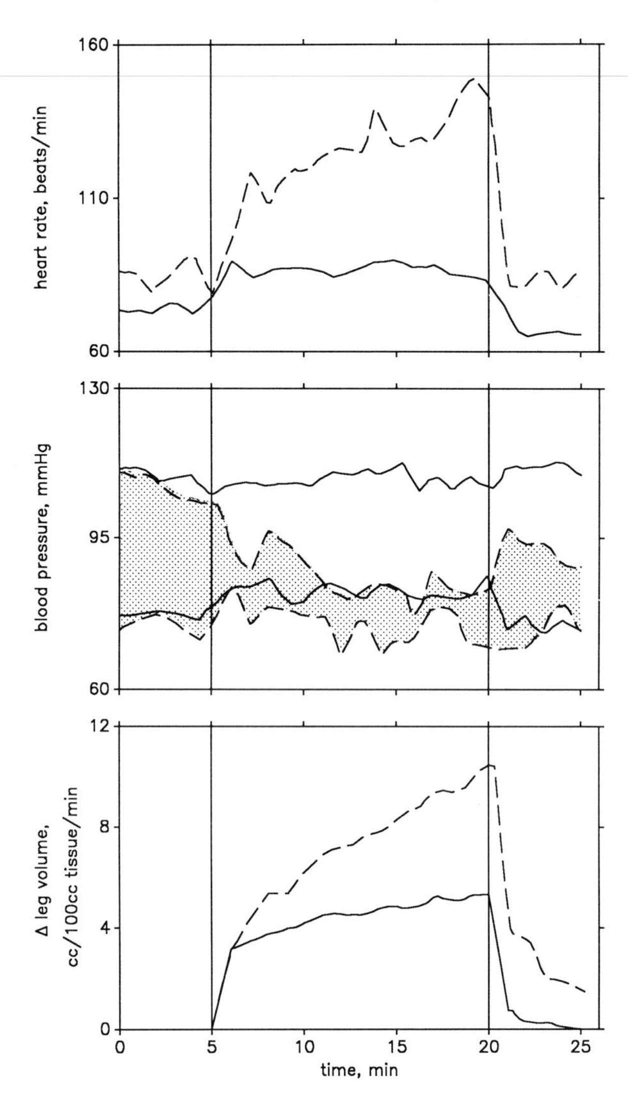

FIGURE 1 Responses of one astronaut to passive 70° tilt before and after a Gemini mission in 1965. Postflight responses are indicated by dashed lines.

II. The Data Base

Humans have been in space for over a quarter century, and a wealth of scientific information has been obtained during human space missions. However, the process of scientific knowing is different for physiologic research done on humans in space than on Earth, and what understanding has been derived from space research must be interpreted in light of the considerable constraints imposed by this unusual species of research. This data base differs from usual scientific data bases in important ways.

First, an abiding problem for human space research has been small sample sizes. Typically, four or fewer astronauts are studied during a single mission, and data obtained are published in terms of responses of individual astronauts. These numbers may be too small to permit statistical analyses. Moreover, the same experiment cannot usually be repeated in space with the same astronaut as the subject. Therefore, the simple requirement that the reproducibility of scientific data be documented may be impossible to fulfill in space. It even is difficult to do the same experiment on *different* subjects during serial space missions; mission durations and the conditions of missions vary.

Second, the types of experiments that can be done with astronauts as subjects are much more limited in space than on Earth. Above all, the experiment must be safe. Many types of measurements that are routine in laboratories on Earth, such as direct measurement of pressure in the right heart or an artery, may be very difficult or impossible in space. As on Earth, it is important that the experiments not alter the physiology being studied. Although knowing how plasma hormone and neurotransmitter concentrations change in space may be of great interest, withdrawal of too much blood for several tests may distort the physiologic measurements.

Related to this issue is the probability that flight surgeons may 'intervene' in the interest of crew safety. One example of this is the treatment of space adaptation syndrome; studies of autonomic function after administration of scopolamine, amphetamine, or metaclopromide may be of uncertain value. Another intervention is the practice of giving crew oral salt and water immediately before they reenter the Earth's atmosphere. This practice is logical, because the combination of blood volume deficit (see below) and increased gravitational forces along the long axis of the body is undesirable. However, this practice may compromise studies of body fluid volumes and autonomic function in the hours immediately following return to Earth.

Third, one of the most scarce commodities in space is astronauts' time. Studies tend to be constrained by an inflexible "time-line," and experiments most likely to succeed are those that can be done quickly. Therefore, the best research is that which can be done by the astronaut on himself or herself. If an experiment must be "tended" by another astronaut, the astronaut time required is doubled.

Fourth, equipment that may be readily available in most laboratories may not be available for use in space, and this constraint may limit the research. For example, during the first quarter century of space travel, freezing plasma samples in space for subsequent analyses of hormones and neurotransmitters was not possible.

Fifth, information on what *is* known from physiologic research in space may be difficult or impossible to gain access to. What measurements were made and what the results were may be difficult to determine. Also, large amounts of in-flight data are buried in reports that may not be readily available. Most of the data regarding human physiology during long-term space missions is published by Soviet scientists in Russian. Even if translations are available, Soviet nomenclature may not be familiar to Western readers.

Sixth, although increasing numbers of women are serving as astronauts, published data does not exist on how responses of women differ from those of men.

Notwithstanding the manifold impediments to development of a sound, statistically robust data base on physiologic changes in humans during weightlessness, published data may be sufficient in many cases to support reasoned conclusions. Moreover, the probability is that many critical missing measurements in the data base will be obtained soon. In the discussion that follows, an effort is made to indicate how solid the scientific data are.

III. Headward Fluid Shift

Some of the earliest observations made during space missions indicated that astronauts in space look and feel different than they do on Earth. They have puffy, ruddy, edematous faces, speak with adenoidal twangs, have difficulty breathing through their noses, and have distended neck veins and small legs ("bird legs," according to actual measurements of leg volumes). Soviet infrared photographs of astronauts in space document dilation of face and neck veins. These changes persist for months in space. All can be explained by headward migration of body fluids during weightlessness. This concept is central to current understanding of cardiovascular changes occurring in space.

Postflight cardiovascular changes are thought to result from a chain of events that begins in the earliest minutes of spaceflight. Weightlessness causes blood and fluid that normally would be pulled toward the feet during most of the day to be shifted toward the head. This shift of blood is believed to trigger a cascade of events that results ultimately in postflight cardiovascular deconditioning. A corollary of this notion, and one that probably is no longer tenable, is that such headward shifts are *sustained* during spaceflight. Evidence supporting (or refuting) these notions is reviewed below; data are arranged in approximate descending order of relevance or scientific credibility.

The simplest way to test the fluid shift hypothesis is to measure volume in heart chambers. Limited data from Soviet, French, and American experiments suggest that left ventricular end-diastolic volume is increased during the first mission day, is reduced to subnormal levels after about the second day, and remains subnormal thereafter, for the duration of the mission. Figure 2 depicts average heart rate and echocardiogram measurements from four American crew members obtained during a Space Shuttle mission in 1985. (These limited data also show that during the first day, stroke volume is elevated. Because heart rate also is elevated, cardiac output is transiently supranormal.) Similar conclusions have been drawn by Soviet and German scientists from measurements of electrical thoracic impedance (called rheography).

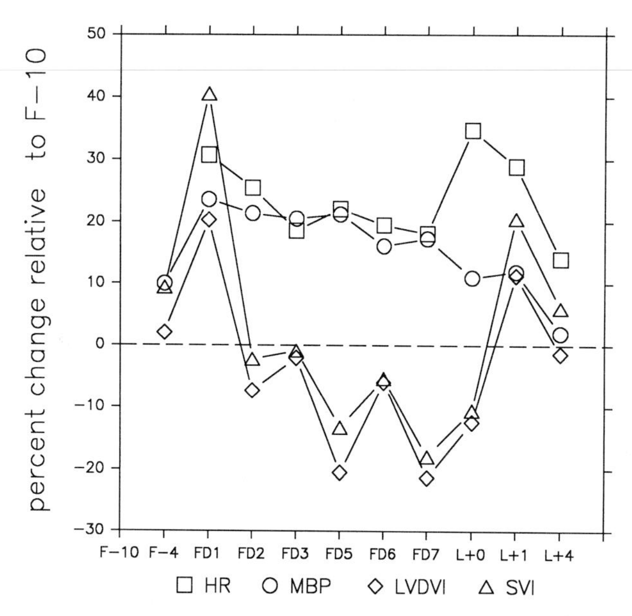

FIGURE 2 Average data obtained from four astronauts during a 1985 Space Shuttle mission. HR, heart rate; MBP, mean blood pressure; LVDVI, left ventricular end-diastolic volume index (volume divided by body surface area); SVI, stroke volume index. [Adapted, with permission, from M. W. Bungo, 1989, Echocardiographic changes related to space travel, *in* "Two-Dimensional Echocardiography and Cardiac Doppler," 2nd ed. (J. N. Schapira and J. G. Harold, eds.), Williams & Wilkins Co., Baltimore.]

A less satisfactory alternative to actual volume measurement is measurement of pressure in intrathoracic veins. Use of pressure measurements to obtain indexes of volumes is based on the assumption that the relation between pressure and volume—compliance—is constant in space. Venous compliance is determined by a complex interplay among viscoelastic and muscular properties, blood volume, and neurohumoral venoconstrictor and venodilator influences. Good evidence indicates that leg vein compliance increases in space; no measurements of compliance of intrathoracic veins exist.

Central venous pressure has been measured directly in one study of weightlessness conducted during parabolic airplane flights. Although central venous pressure has not been measured directly in space, an array of ingenious substitutes has been used. Data from a variety of sources suggest that human central venous pressure does rise, but that increased venous pressure is supplanted very rapidly (within minutes to hours) by reduced venous pressure, relative to measurements obtained preflight with astronauts in the supine position.

At the Institute of Aerospace Medicine in Copenhagen, intrathoracic venous pressure was measured directly in 14 healthy young men during parabolic maneuvers in a Royal Danish Airforce jet. At the top of the parabola, when subjects were weightless, average central venous pressure increased to 6.8 mmHg from its level before the maneuver (recorded in the supine position) of 5.0 mmHg. This small change was highly significant. However, the period of weightlessness at the top of each parabola is very brief (lasting <30 sec) and is bracketed between exposures to increased gravitational forces during ascent and descent. The influence of increased gravitational forces before and after measurements obtained during weightlessness is unknown.

Indirect measurements of central venous pressure have been obtained during actual spaceflight during the first German Space Shuttle mission. Noninvasive measurements of intraocular pressure were used as indexes of central venous pressure in the astronauts. The earliest measurements were obtained in one astronaut 53 min after launch, 44 min after the shuttle entered an Earth orbit. In this astronaut, intraocular pressure was about 18 mmHg during the first measurement and was about 12 mmHg 12 hr later. Average early ocular pressure measurements were reported to be about 20–25% higher than late measurements in the three astronauts studied.

Direct pressure measurements were obtained during two Space Shuttle missions, but in arm rather than intrathoracic veins. The rationale for this experiment was that although there are valves between arm and central veins, without the gravitational pull that normally keeps venous valves closed, pressures should equilibrate throughout the venous system. Vein pressure was found to be reduced (not increased) as early as 45 min after launch. Arm vein pressure rose soon after return to Earth. The rapid increase of pressure was probably due to recruitment of edema fluid from extravascular spaces.

In another indirect approach for estimating astronaut's central venous pressure during spaceflight, astronauts briefly and mildly increased their intrathoracic pressures by straining (a "Valsalva maneuver"). Simultaneously, phasic venous blood flow at the thoracic inlet was measured with a noninvasive Doppler probe, and the intensity of their straining was adjusted until the Doppler signal disappeared. At this point, straining pressure was considered to be equivalent to central venous pressure. Measurements obtained early during missions with this approach suggest that venous pressure is reduced during weightlessness.

Central venous pressures measured in five healthy young men during weightlessness *simulated* by 5° head-down bedrest was found to rise. The pressure elevation, however, occurs very early (at about 30 min), is gone by about 6 hr, and is supplanted by pressure reduction at 24 hr. The average pressure change was from a control level of 5.6 to a maximum of 8.5 mmHg, a highly significant change. These data support the limited findings obtained during weightlessness. This study also underscores one major problem inherent in attempts to study central venous pressure in American astronauts on the Space Shuttle: For about 4 hr prior to launch, astronauts are restrained on their backs in contour chairs.

Finally, the Soviets report two types of venous pressure measurements from space. (1) They made direct pressure recordings during weightlessness in one instrumented monkey (''Bonnie''); these data are somewhat discordant with those referred to above and indicate that central venous pressure is about twice the control levels for the first 5 days of weightlessness. (2) They estimated central venous pressure with a piezo-ceramic sensor affixed to the neck. This device yields complex qualitative waveforms (venous-arterial pulsograms), which include, among others, signals that are caused by changes of jugular venous pressure. The amplitude of these signals falls during prolonged weightlessness.

In summary, the available evidence provides qualified support for the notion that weightlessness leads to headward migration of body fluids. These data point to a brief increase of central venous volume, followed by an early reduction to *below* prelaunch supine levels. Existing data do not explain why facial edema persists despite reduction of central venous pressure to below preflight levels. None of the data explain how this evanescent shift of fluid volume can cause persistent blood volume reduction or postflight orthostatic hypotension. Apparently, however, this temporary increase of the degree of heart chamber stretch leads to an early diuresis. A brief elevation of plasma atrial natriuretic factor was observed early in one Space Shuttle mission. Presumably, this hormone sets in motion changes of fluid balance that lead to the reductions of blood volume, which are discussed next.

IV. Blood Volume Changes

A central component of the cardiovascular deconditioning of weightlessness is a reduction of blood volume that occurs in all missions, including those as brief as 8 hr. Plasma volume falls early in missions and stabilizes at about 12% less than that of preflight levels after 30–60 days of weightlessness. This blood volume reduction probably contributes to postflight orthostatic hypotension; there is a loose, but significant, linear relation between reductions of blood volume and changes of the heart rate responses to standing after weightlessness.

However, it seems unlikely that cardiovascular deconditioning is due simply to reduction of blood volume (which amounts to about 500 ml). Although a minority (about 5%) of healthy people have orthostatic hypotension and faint after they donate blood, most do not. Astronauts uniformly have orthostatic hypotension after exposure to weightlessness. Indeed, many astronauts have orthostatic hypotension, even though they take salt and water prior to reentry. In people studied after head-down bedrest, no correlation seems to exist between the amount of blood volume reduction and the fall of standing blood pressure. Also, if such people are given intravenous saline infusions to replace lost blood volume, or a mineralocorticoid to prevent blood volume reduction, they still have orthostatic hypotension.

A simple question that has not yet been answered is where does the blood volume go? It seems certain that fluid leaves the lower body; leg volumes are consistently less during than before flight. (Also, limited data suggest that arm volumes are unchanged.) Presumably, although it has not been measured, urine production increases very early in space missions. Because astronauts are launched in the Space Shuttle on their backs, increased renal fluid excretion may begin *before* American missions.

Several types of evidence suggest that a diuresis (increased urine production) occurs. The first urine samples obtained during missions are dilute, and this may reflect an excess excretion of water. Also, concentration of blood occurs; this also is consistent with early plasma volume reduction. During space missions, crew members routinely lose weight (on average, about 3 kg). Most weight loss occurs during the first 2 days of missions. The picture regarding fluid balance early in missions is clouded by another factor: Space adaptation syndrome occurs in as many as half of astronauts and may contribute to fluid loss through vomiting and reduction of fluid intake.

V. Cardiac Function

The name cardiovascular deconditioning may be unfortunate because it implies that weightlessness impairs cardiac performance. Evidence for impairment of cardiac function has been sought exhaustively and not found. This is not to say that cardiac function is unchanged; changes occur, but they probably can be fully explained by changes of blood volume and cardiac loading conditions.

Solid evidence indicates that exposure to weightlessness reduces heart size. Postflight chest X-rays document reductions of heart size from preflight levels. Postflight echocardiograms show that left ventricular end-diastolic volume declines by an average of about 25% and stroke volume declines by about 19%. No correlation exists between reductions of cardiac dimensions and flight duration. This provides indirect evidence that blood volume reductions occur during the earliest days of missions and stabilize at some lower level.

Strong evidence also indicates that cardiac function in space is normal. The absolute amounts of blood in the left ventricle before and after contraction are reduced (probably because of reduced blood volume), and the amount of blood ejected with each stroke (stroke volume) is correspondingly reduced. The percentage of the left ventricular volume ejected with each beat (the ejection fraction) is normal in space; this suggests that myocardial contractile properties are normal. At rest, cardiac output (the amount of blood pumped by the heart per minute) changes little if at all, in part because heart rate is faster.

Pre- and postflight echocardiograms indicate that left ventricular wall thickness does not change during weightlessness. Because heart size diminishes, this indicates that total left ventricular mass diminishes. These measurements provide some evidence that the edema of the head and neck observed in all astronauts does not extend to the thorax. Calculated left ventricular mass returns to preflight values very rapidly and, therefore, may be due simply to reductions of interstitial or intracellular fluid.

Many people voiced early concerns that cardiac electrical activity, or rhythm, might be abnormal in weightlessness. Solid evidence from long-term (Holter) electrocardiogram recordings shows that the incidence of cardiac dysrhythmias *is* greater in space than on Earth, particularly during (or, as on Earth, after) exercise. The cause of cardiac dysrhythmias in space is unknown. It is not known if circumstances leading to dysrhythmias in space are reproduced exactly during Holter monitoring on Earth; conceivably, astronauts have more dysrhythmias in space because the work they perform, or the circumstances under which they perform it, is more stressful than on Earth. In two American astronauts who had frequent ventricular premature beats during activities on the moon, cardiac rhythm may have been abnormal because they trained for their mission in the heat and humidity of a Florida summer and probably were launched with body potassium deficiencies.

VI. Autonomic Outflow

Solid evidence shows that heart rate and blood pressure increase in space; on average, the increases are about 10–20%. Because cardiac output tends to be normal (reflecting lower stroke volumes but higher heart rates), increases of blood pressure reflect increases of peripheral resistance. These hemodynamic changes probably result from changes of traffic carried over autonomic nerves to the heart and blood vessels; however, the stimulus for such autonomic changes is unknown. If levels of nerve traffic change in response to reductions of blood volume (or more likely, blood pressure), nerve traffic should be restored to normal levels as soon as blood volume or pressure returns to normal. Also, reflex neural changes should restore blood pressure toward normal but not to levels above normal.

The nature of autonomic neural changes is unknown, and whether or not they occur at all is uncertain. It is not known if the blood pressure elevation and heart rate speeding are due to increased sympathetic activity, reduced vagal activity, or a combination of these two. The evidence that autonomic nerve traffic changes in space is indirect.

During experimentally induced changes of blood pressure on Earth, norepinephrine varies as a linear function of sympathetic traffic carried to the important skeletal muscle sympathetic vascular bed. Therefore, if in space, blood pressure reductions increase sympathetic nerve activity, this should be reflected in increased levels of the principal sympathetic neurotransmitter norepinephrine. There are very limited data regarding sympathetic neurotransmitters in space. Urinary concentrations of norepinephrine and its metabolites were measured during the American Skylab program in the 1960s. There was no major change in excretion (and, therefore, neural release) of norepinephrine. More recently, blood was withdrawn from Soviet astronauts and frozen at −30°C until it could be analyzed on Earth. Blood samples drawn from three astronauts on days 217–219 of a 237-day mission showed that in-flight plasma norepinephrine and epinephrine levels increased in space, but to levels that did not exceed the rather high upper limit of normal. These results do not explain blood pressure elevation that occurs in space; they provide some evidence (from a very small sample size) that sympathetic nervous system activity is normal in space.

Concentrations of both plasma and urinary sympathetic neurotransmitters are supranormal after return to Earth. This suggests that both sympathetic neural activity and adrenal medullary release of epinephrine are increased. These changes can be explained on the bases of hypotension during most of the day (when astronauts are sitting or standing) and, perhaps, stress associated with return to Earth.

The relative high pulse rate (and possibly some of the blood pressure elevation) that occurs in space may be explained also on the basis of reduced vagus nerve restraint of heart rate. Vagal-cardiac nerve traffic has not been measured directly in humans; therefore, it must be quantified with indirect measures. Of these, some measure of the variance of heart rate or, more properly, its reciprocal, the interval between heart beats (R-R interval), and mean R-R interval are used. Faster heart rates during and after weightlessness provide some evidence that shows less vagal-cardiac nervous outflow to the heart. Limited data suggest that variance of R-R interval is less in space and after return to Earth.

Thus, sympathetic outflow appears to be normal, and vagal outflow appears to be reduced during and after exposure to weightlessness. Existing data do not identify the autonomic neural sensors that signal the need for persistent changes of autonomic neural outflow. Some evidence indicates that weightlessness impairs cardiovascular reflex function. This is discussed next.

Astronauts cannot truly stand in space, for lack of gravity. They can and do assume head-up and feet-down positions in space, but such standing differs substantially from standing on Earth. However, there is an experimental analogue that is used to simulate standing in space. Astronauts enter an airtight cylindrical chamber that encloses the lower half of their bodies, below the iliac crest. Pressure in this chamber is lowered stepwise by a vacuum device. The lower body negative pressure draws blood away from the thorax into the pelvis and legs. The amount of blood drawn into the lower body by similar levels of suction is greater in space than on Earth; therefore, the capacity of the venous system is increased. Increased venous compliance may be secondary to reduced neurohumoral constriction of veins or to reduced support of leg veins by atrophic antigravity muscles. For the same amount of suction, heart rate and vascular resistance increase more, and left ventricular stroke volume and cardiac output fall more in space than on Earth.

Not surprisingly, astronauts are more likely to experience hypotension and presyncope or syncope during lower body suction in space than on Earth. Such hemodynamic changes and symptoms are known as "vasovagal" reactions (*vaso-* because sympathetic neural traffic to resistance blood vessels falls off and vessels dilate; *-vagal* because heart rate may slow by virtue of increased traffic carried over the vagus nerves to the heart). Astronauts who have the greatest reduction of blood pressure during lower body suction are likely to have the greatest reduction of blood pressure with standing postflight.

As indicated, vasovagal reactions are due to reflex changes of autonomic nervous activity; however, vasovagal reactions, including syncope or presyncope during lower body suction, occur on Earth as well as in space. On Earth, lower body suction usually is not performed in subjects from whom blood has been withdrawn; when it is, such subjects experience syncope at lower intensities of suction. Therefore, the apparent increased susceptibility of astronauts to presyncope or syncope during lower body suction may result simply from the fact that in space, their blood volumes are lower than they are on Earth.

One hypothesis to explain the orthostatic hypotension that occurs in astronauts after spaceflight is that their baroreflexes do not respond appropriately to standing. This possibility is importantly underscored by research in dogs in whom the nerves that sense arterial pressure changes (the baroreceptors) have been cut and in whom the brain no longer receives accurate information regarding blood pressure. Such dogs experience huge swings of blood

FIGURE 3 Neck chamber and analysis system developed for baroreceptor testing in space by Dwain L. Eckberg at the Medical College of Virginia in Richmond, Virginia and manufactured by Engineering Development Laboratory, Inc., in Newport News, Virginia.

pressure during their daily experiences and large reductions of pressure when they stand.

Baroreceptor reflex function has not been measured in space; however, several lines of evidence suggest that spaceflight may indeed impair reflex circulatory control. Baroreceptor function has been studied in astronauts before and after Space Shuttle missions and in healthy people on Earth subjected to prolonged (in one experiment, 30 days) 6° head-down bedrest. This research employed pressure changes delivered to a neck chamber worn by subjects. Human arterial baroreceptors are located primarily at the bifurcation of the common carotid artery in the neck and the aorta. Suction applied to the neck lowers tissue pressure outside the carotid artery. When this happens, the pressure inside the artery is suddenly pushing against a vacuum, and the artery expands. Carotid artery expansion increases firing rates of baroreceptor nerves in the wall of the carotid artery, increases vagus nerve traffic to the heart and slows the heart rate, and reduces sympathetic nervous traffic and relaxes blood vessels. Positive pressure applied to the neck exerts opposite effects. A neck chamber developed to study baroreflexes in space is shown in Fig. 3.

Results of experiments employing neck chambers show that both spaceflight and simulated weightlessness with head-down bedrest lead to impairment of baroreflex function. A bedrest study showed that volunteers who developed the greatest impairment of baroreflex function were most likely to faint when they stood erect at the end of the bedrest period. The baroreflex function of astronauts was found to be impaired after brief Space Shuttle missions.

VII. Exercise

During comparable levels of exercise in space (if such can be achieved without forces of gravity), heart rate and blood pressure are higher and stroke volume is lower than on Earth. Although stroke volume declines, cardiac output (stroke volume × heart rate) may remain normal, by virtue of more rapid heart rates. Contraction of the left ventricle, as judged by echocardiography, actually may be increased from preflight levels. All of this suggests that changes measured in space are due probably to blood volume changes, and that cardiac muscle function is normal.

In space, oxygen uptake is slightly less during, and slightly more after exercise than before flight. Increases of oxygen uptake after exercise suggest that in space astronauts rely more on anaerobic mechanisms than they do when on Earth. Oxygen uptake during maximal exercise has been measured in only three astronauts during a Skylab mission. Exercise was considered to be maximal because their heart rates exceeded targeted rates (>183 beats/min), which were slightly faster than preflight exercise heart rates. Maximal oxygen consumption *increased* in all three astronauts during weightlessness. This stands in contrast with postflight reductions of maximal oxygen consumption.

Work performed outside the spacecraft (extravehicular activity) constitutes a special case. This kind of work involves primarily arm and upper body muscle groups and may be exceedingly stressful for astronauts; they may have higher heart and respiratory rates and sweat more, and their body heat production may exceed the capacity of space suit cooling systems to dissipate heat. Improvements in space suits have reduced the amount of effort required for extravehicular activity.

Exercise capacity and oxygen consumption clearly are reduced after exposure to weightlessness. Maximal oxygen consumption has not been measured after spaceflight. Reduced aerobic capacity is likely to be multifactorial. (1) Antigravity muscles atrophy from disuse; this was documented directly by microscopic examination of muscles of three Soviet astronauts who died tragically because of spacecraft decompression during reentry. (2) Muscle efficiency is reduced. (3) Blood volume and red blood cell mass and, consequently, oxygen-carrying capacity are somewhat smaller after spaceflight than before. (4) Left ventricular stroke volume during exercise is reduced by about 20%; however, because heart rate during exercise is greater, cardiac output is reduced by only about 10%. Diastolic blood pressure is normal, but systolic pressure is elevated. The combination of increased heart rate and systolic pressure indicates that left ventricular myocardial oxygen consumption is increased during exercise after spaceflight. [*See* EXERCISE AND CARDIOVASCULAR FUNCTION.]

VIII. Countermeasures

For many years, countermeasures have been used in space to try to prevent or ameliorate postweightlessness cardiovascular problems. Many American and Soviet astronauts take oral salt tablets and drink water before reentry. This countermeasure significantly reduces the incidence of postflight orthostatic hypotension and syncope. Many crew members also perform bicycle or other aerobic exercise during missions. Although some evidence suggests that in-flight exercise prevents postflight cardiovascular abnormalities, this has not been proven. Soviet scientists simulate blood volume redistributions that occur with standing with a special suit (''Chiba'') that applies lower body suction during usual activities.

IX. Summary

Weightlessness of even brief durations leads to postflight cardiovascular abnormalities, including reduced exercise tolerance, and inordinate heart rate speeding and blood pressure reduction with standing. In most astronauts, these changes constitute a nuisance—lightheadedness forces them to sit down. In a few astronauts, cardiovascular changes are disabling—they cannot stand without losing consciousness. All of these changes disappear spontaneously within days to weeks after return to Earth, and they appear to leave no residual abnormalities.

Data on human cardiovascular function during and after weightlessness must be pieced together from disparate sources. In most instances, observations are too few to permit proper determination of statistical significance. Nevertheless, extant data suggest that the following changes occur when humans are exposed to a weightless environment: Body fluids shift toward the head, stretch cardiac chambers, provoke release of diuretic hormones, and cause diuresis and blood volume contraction. Upon return to earth, a blood volume deficit and possibly impaired baroreceptor reflex function lead to the cardiovascular changes with standing described above.

Certain critical questions have not been answered: If stretch of cardiac chambers secondary to headward fluid shifts leads ultimately to blood volume contraction, what maintains blood volume reduction after cardiac chamber dimensions return to or below preflight levels? If reduced blood volume and pressure lead to reflex heart rate speeding and increases of blood pressure, what maintains heart rate and blood pressure at high levels after blood pressure has returned to normal? Why do heart rate and blood pressure rise above preflight levels? Ambitious American and Soviet plans for future human physiologic research during space missions are likely to provide answers to these challenging questions and, probably, to questions that have not yet been asked.

Bibliography

Berry, C. A. (1974). Medical legacy of Apollo. *Aerospace Med.* **45,** 1046.

Berry, C. A., and Catterson, A. D. (1967). Pre-Gemini medical predictions versus Gemini flight results. *In* "Gemini Conference Summary." National Aeronautics and Space Administration, Washington, D.C.

Blomqvist, C. G., and Stone, H. L. (1983). Cardiovascular adjustments to gravitational stress. *In* "Handbook of Physiology. The Cardiovascular System. Peripheral Circulation and Organ Blood Flow, Section 2, Part 2" (J. T. Shepherd and F. M. Abboud, eds.). American Physiological Society, Bethesda, Maryland.

Bungo, M. W. (1989). The cardiopulmonary system. *In* "Space Physiology and Medicine" (A. E. Nicogossian, C. L. Huntoon, and S. L. Pool, eds.). Lea and Febiger, Philadelphia.

Bungo, M. W. (1990). Echocardiographic changes related to space travel. *In* "Two-Dimensional Echocardiography and Cardiac Doppler" (J. N. Schapira and J. G. Harold, eds.). Williams and Wilkins, Baltimore.

Catterson, A. D., McCutcheon, E. P., Minners, H. A., and Pollard, R. A. (1963). Aeromedical observations. *In* "Mercury Project Summary Including Results of the Fourth Manned Orbital Flight May 15 and 16, 1963." National Aeronautics and Space Administration, Washington, D.C.

Convertino, V. A. (1990). Physiological adaptations to weightlessness: Effects on exercise and work performance. *Exercise Sport Sci. Rev.* **18,** 119.

Convertino, V. A., Doerr, D. G., Eckberg, D. L., Fritsch, J. M., and Vernikos-Danellis, J. (1990). Simulated microgravity impairs human baroreflex responses and provokes orthostatic hypotension. *J. Appl. Physiol.* **68,** 1458.

Hoffler, G. W., Wolthuis, R. A., and Johnson, R. L. (1974). Apollo space crew cardiovascular evaluations. *Aerospace Med.* **45,** 807.

Johnson, P. C., Driscoll, T. B., and LeBlanc, A. D. (1977). Blood volume changes. *In* "Biomedical Results from Skylab" (R. S. Johnston and L. F. Dietlein, eds.). National Aeronautics and Space Administration, Washington, D.C.

Levy, M. N., and Talbot, J. M. (1983). Cardiovascular deconditioning of space flight. *Physiologist* **26,** 297.

Sandler, H. (1980). Effects of bedrest and weightlessness on the heart. *In* "Hearts and Heart-like Organs," Vol. 2 (G. H. Bourne, ed.). Academic Press, New York.

Cardiovascular System, Anatomy

ANTHONY J. GAUDIN, *California State University, Northridge*

Glossary

Endothelium A thin membrane of flattened cells that lines the cavity of an organ or blood vessel
Pulmonary Pertaining to the lungs and external respiration
Semilunar Shaped like a half-moon
Serous A thin cellular membrane that lacks glands and secretes a watery product similar to tissue fluid
Systemic Pertaining to the entire body

THE CARDIOVASCULAR SYSTEM consists of the heart and blood vessels. The heart is a muscular pump that moves blood through the vessels, which carry the blood to the tissues and return it to the heart.

I. Anatomy of the Heart

The heart lies within the thoracic cavity, just below the sternum. It is cone shaped, with the apex pointing down and to the left edge of the diaphragm (see Color Plate 2). Large blood vessels join the heart at the base of the cone, which is just behind the sternum. The adult heart is usually about the size of a clenched fist. Its dimensions average 13 cm long, 9 cm at the base, and about 6 cm at its thickest anterior–posterior dimension. It weighs about 350 g. These figures vary from person to person, but in general the heart is slightly larger in men than in women. Disease, a prolonged program of physical exercise, or pregnancy can cause an increase in the size of the heart.

A. Pericardium

The heart is surrounded by a double-layered membrane, the pericardium, or pericardial sac. The inner layer of the sac, the visceral pericardium, or epicardium, is a thin serous (i.e., secretory) membrane closely attached to the heart's surface. (It is also considered to be the outermost of the three tissue layers of the heart, described in Section I,B.) The outer layer, the parietal pericardium, consists of two subdivisions—a thin serous pericardium and a thick fibrous pericardium—that provides strength. The pericardial layers are separated by the pericardial cavity. This thin space contains a lubricant called pericardial fluid that is secreted by the serous membranes. Thus, the two membranes slide over the heart during heartbeats, with a minimum of friction. Sometimes these membranes become infected or irritated and inflamed, resulting in pericarditis.

B. Muscular Walls of the Heart

The wall of the heart has three layers of tissue: an internal endocardium, a middle myocardium, and the external epicardium, which is the same as the inner pericardial layer just described (Fig. 1). [*See* CARDIAC MUSCLE.]

The endocardium is a thin layer that lines the interior of the heart, covers all internal structures (such as valves), and extends to the blood vessels that attach to the heart. It consists of two layers, including a thin endothelial layer that forms the free surface of the heart's interior and an underlying fibrous layer that supports the surface layer. The fibrous layer contains blood vessels, elastic and strengthening fibers, and specialized muscle cells (Purkinje fibers) that transmit nerve impulses to coordinate contractions of the heart.

The myocardium, the thickest of the layers, is composed entirely of muscle tissue. The myocardial layer of the ventricles (Fig. 1) is much thicker than that of the atria; its contraction has sufficient force to propel blood through the lungs and the body. In the ventricles, the muscle fibers are arranged in spiraling bands rather than extending straight from the base to the apex, as in the atria. This arrangement produces contractions that squeeze the blood from the ventricles. The myocardial contractions pump the blood from the ventricles through the circulatory system.

The external surface of the myocardium is relatively smooth, but the internal surface in the ventricles is irregular. In these areas are many folds, columns, and ridges, called trabeculae carneae (i.e., fleshy sticks), and conical projections, called papillary muscles. The trabeculae carneae provide additional strength to the myocardium, while adding a minimum of weight. Papillary muscles prevent the backward flow of blood through heart valves. (Both are discussed in detail in Section I,C,2.)

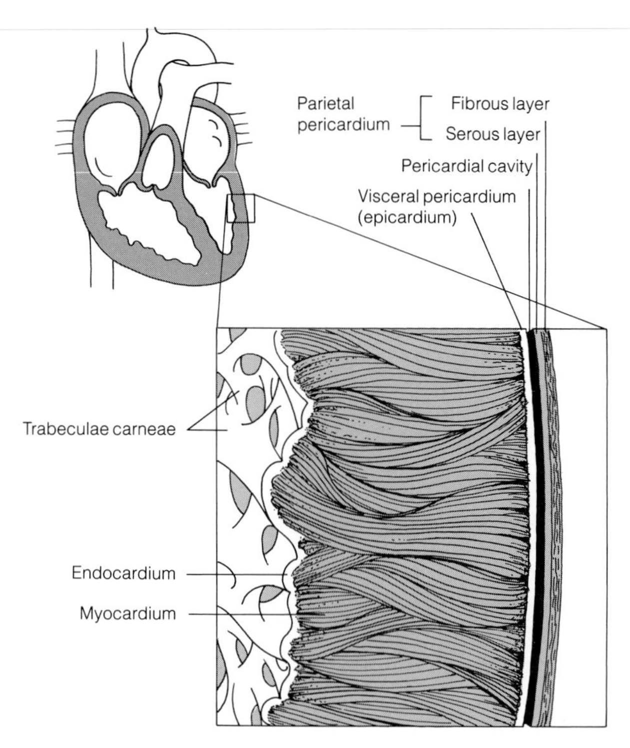

FIGURE 1 The pericardium. [Source: Gaudin, A. J., and Jones, K. C. (1989). "Human Anatomy and Physiology." Harcourt Brace Jovanovich, San Diego, p. 561. Reproduced with permission.]

C. Chambers and Valves of the Heart

The heart has four chambers: right and left atria specialized to receive blood from veins, and right and left ventricles specialized to pump blood into the arteries (see Color Plate 3). The atria have thinner walls than the ventricles and lie at the base of the heart, while the ventricles make up the apex and bulk of the heart. The atria and ventricles are separated by muscular walls referred to as septa. The interatrial septum lies between the left and right atria, while the interventricular septum separates the two ventricles. In addition to muscle, the septa also contain the fibrous skeleton of the heart, an internal network of fibrous connective tissue (Color Plate 4). This skeleton has two functions: First, it provides a place of attachment for cardiac muscle; second, it helps support the valves that separate the atria from the ventricles. These valves are the atrioventricular valves and are attached to a portion of the fibrous skeleton known as the coronary trigone. The fibrous tissue of this structure provides support for the valves as they operate during normal cardiac movements while pumping blood.

The chambers in the two sides of the heart form a natural division of the twofold function of the circulatory system. Blood entering the right side of the heart has traveled through the body, where its oxygen supply has been lowered and its carbon dioxide level has been raised. This blood is pumped through the heart's pulmonary circulation to the lungs. Here, the blood exchanges carbon dioxide for oxygen. Pulmonary circulation restores a high concentration of blood oxygen and lowers the concentration of carbon dioxide. This blood moves from the lungs and enters the left atrium of the heart. The left side of the heart pumps blood through the systemic circulation, the vast system of vessels that carries blood to all tissues of the body. [*See* CARDIOVASCULAR SYSTEM, PHYSIOLOGY AND BIOCHEMISTRY.]

1. Right Atrium

Blood from the body first enters the heart in the right atrium from three major veins: the superior vena cava, inferior vena cava, and coronary sinus (Color Plate 3). The superior vena cava receives blood from the upper part of the body, while the inferior vena cava drains all tissues below the diaphragm. The coronary sinus receives blood from the muscles of the heart itself. The opening of the coronary sinus is an enlarged vein on the posterior surface of the atrium that collects blood coming from the veins of the cardiac walls.

The auricles are ear-shaped extensions of both atria which increase their volume under certain circumstances. During resting conditions, an atrium does not open to its full capacity, expanding only enough to hold the moderate amount of blood entering. During physical activity and exercise, however, the amount of blood entering the heart may double or triple, and at these times the auricles function as a volume reservoir, expanding to their full extent. The auricles thus increase the volume of both atria and the amount of blood they can accommodate. The interior wall of the right atrium is relatively smooth, whereas the auricle contains numerous ridges of cardiac muscle, called the musculi pectinati (i.e., comblike muscles). Musculi pectinati function in much the same way as trabeculae carneae, providing the heart with additional strength with a minimum of weight.

The opening between the superior vena cava and the right atrium is unobstructed, but a flaplike extension from the auricle wall at the base of the inferior vena cava appears to be the remnant of an important functional structure in the fetus. This flaplike structure directs blood through an opening in the fetal heart called the foramen ovale (i.e., the oval hole). This structure has no apparent adult function. In the adult heart, the fossa ovalis, an oval depression, is visible in the interatrial septum. This is a remnant of the former position of the fetal foramen ovale.

2. Right Ventricle

Blood then passes from the right atrium into the right ventricle. The right ventricle is a triangular chamber occupying the right apex of the heart and resting on the diaphragm. The right atrioventricular (tricuspid) valve is a specialized structure that separates the right atrium and ventricle (see Color Plate 5) and consists of two portions: cusps and cords. The three triangular cone-shaped cusps extend across the opening between the chambers. These flexible cusps have their peripheral edges attached to the fibrous skeleton, and the central edges project into the cavity of the right ventricle. The interior surface of the ventricle is not smooth and flat. Instead, it consists of trabeculae carneae and several fingerlike papillary muscles that project into the cavity of the ventricle (Fig. 1 and Color Plate 5). The papillary muscles are attached to the free edges of the cusps of the atrioventricular valve by elongate tendinous cords called chordae tendineae.

As the right atrium fills with blood, it contracts, and the blood enters the right ventricle. The force of blood passing through the tricuspid valve forces the cusps down into the right ventricle, providing a clear pathway into the chamber. After the right ventricle fills with blood, it contracts, squeezing the blood out of the chamber toward the lungs. This action also puts pressure on the undersides of the three atrioventricular cusps and causes them to close by pushing them upward. The simultaneous contraction of the papillary muscles pulls on the chordae tendineae, which prevents the cusps from prolapsing into the right atrium. As a result, the opening is sealed and blood does not leak back into the right atrium.

Blood leaves the right ventricle through the pulmonary trunk. The base of this large artery lies in the upper medial front region of the right ventricle. Trabeculae carneae are absent here. The entrance to this artery contains a three-part pulmonary semilunar valve composed of pocketlike folds (cusps) of endothelium (Color Plates 5 and 7). During ventricular contraction, blood forces the cusps open, but they close rapidly when the ventricle relaxes. Closure of the semilunar valve prevents backward blood flow.

3. Left Atrium

The left atrium resembles the right atrium, but its myocardial layer is slightly thicker. Its auricle projects anteriorly and superiorly, and partially covers the pulmonary trunk. Blood from four pulmonary veins enters the right atrium. Their openings are unobstructed with valves, and relaxation of the atrial walls allows the chamber to fill with oxygenated blood. The musculi pectinati and the left side of the fossa ovalis are also in the left atrium. Contraction of the left atrium sends blood through the atrioventricular opening into the left ventricle (Color Plate 6).

4. Left Ventricle

The left ventricle occupies the left apical position of the heart. Its myocardium is approximately three times as thick as that of the right ventricle. This chamber propels blood throughout the entire body when it contracts. Its lining resembles that of the right ventricle, with numerous trabeculae carneae and several papillary muscles.

The separation between the left atrium and ventricle is maintained by the left atrioventricular valve, also called the mitral, or bicuspid, valve. It has two, instead of three, cusps, connected through numerous chordae tendineae to papillary muscles that project from the wall of the ventricle. The bicuspid valve functions in precisely the same way as the tricuspid valve: Both prevent backflow of blood.

Blood leaves the left ventricle through the aorta, the largest blood vessel in the body. The base of the aorta lies at the superior medial edge of the left ventricle. Its orifice is closed by an aortic semilunar valve (Color Plate 7), similar in function and structure to the pulmonary semilunar valve. It prevents the backward flow of blood from the aorta.

D. Cardiac Intrinsic Blood Supply

An adult heart contracts approximately 75 times each minute, every day of our lives. This effort results in an extensive amount of work by cardiac muscle cells, and they require a constant supply of oxygen and nutrients. They also produce large amounts of waste materials that must be removed. The myocardium has "first call" on the blood leaving the left ventricle. The right and left coronary arteries originate just above the cusps of the aortic semilunar valve (Color Plates 3 and 7). They form the coronary circulation, an intrinsic supply of freshly oxygenated blood to the heart. The left and right vessels mainly supply corresponding sides of the heart, but extensive branching, along with numerous interconnections, called anastomoses, between their branches, results in a complex circulatory pattern as the arteries extend from the base of the ventricles to the cardiac apex. One advantage of extensive branching and anastomoses is that blood arriving at any spot in the myocardium may originate in any one of several large branches. Alternative sources of blood for a specific area are called collateral circulation, and if blockage occurs in one branch leading to a specific area, increased flow through collateral circulation may be sufficient to satisfy the needs of the myocardial cells in that area.

Venous blood in the myocardium is collected by several cardiac veins that anastomose extensively with each other. Almost all of the blood that passes through the myocardium empties into the coronary sinus, an enlarged vein that drains into the right atrium. Only a few small veins empty into the right atrium or into the ventricles directly.

Coronary circulation is essential to the function of cells in the heart. Blood inside the chambers does not supply heart muscle directly; this function is provided instead by the coronary circulation. A partial blockage of one coronary artery can result in a decrease in the critical blood supply to the myocardium distal to the blockage, and, when severe enough, may cause severe chest pain, known as angina pectoris. Cessation of blood flow consequent to complete blockage may result in a myocardial infarction, commonly called a heart attack. An infarction causes the death of the part of the heart cut off from nutrients and oxygen, where the dead muscle is then replaced with fibrous connective tissue. Because this scar tissue lacks the ability to contract, the heart loses some of its function. Massive infarctions affecting large areas of cardiac tissue can result in death.

II. Blood Vessels

The groups of blood vessels are arteries, veins, and capillaries (Color Plate 8). Arteries conduct large quantities of blood from the heart. Distal to the heart, they branch into smaller arteries, called arterioles. These branch into still smaller vessels, which eventually become microscopic capillaries. Capillaries connect arteries and veins. They distribute blood to the tissues and bring it into proximity with the cells. Capillaries form dense networks in practically all body tissues. At the efferent end of a network, the capillaries fuse into slightly larger vessels, called venules, which in turn fuse into veins. Veins collect blood from the tissues and return it to the heart. For the most part, blood is confined within the blood vessels, making the human cardiovascular system a "closed" circulatory system. Some of the blood's fluid component does escape from the capillaries and bathes the body tissues in their immediate vicinity. Most of this fluid reenters capillaries in the same network and is returned to the veins. The rest is collected by the lymphatic system, which in turn returns it to veins.

A. Arteries

Arteries are blood vessels with thick walls (Color Plate 8), ranging from about 2.5 cm to approximately 0.5 mm in diameter. The wall of an artery has layers, or tunicae. The innermost layer, the tunica intima, lines the cavity or lumen of the artery. Also called the tunica interna, it is composed of a thin layer of cells (i.e., the endothelium), whose free border is in contact with the blood. The tunica intima is continuous with the cardiac endocardium and extends throughout the cardiovascular system as a lining only one cell layer thick. The tunica media forms a cylinder of smooth muscle and elastic connective tissue external to the tunica intima. The tunica externa is nearly as thick as the tunica media, but consists of fibrous connective tissue, with only small numbers of smooth muscle fibers.

An arteriole is arbitrarily defined as any artery less than 0.5 mm in diameter.

B. Capillaries

The capillaries are the smallest of blood vessels (Color Plate 8). They are microscopic in size, approximately 0.01 mm in diameter, nearly the diameter of a red blood cell. Some capillaries are smaller than the red blood cells they carry, forcing the cells to bend into a "C" shape as they squeeze through the vessel. This increases the surface of the blood cell in contact with the wall of the capillary. The wall of a capillary consists of a single endothelial layer of cells resembling the pieces of a jigsaw puzzle. Capillaries are usually between 0.5 and 1 mm long. In active tissues, such as skeletal muscles, liver, lungs, nervous system, and kidneys, capillaries are so numerous that few tissue cells are more than one or two cells away from a capillary. In relatively inactive tissues, such as tendons and ligaments, capillaries are not nearly so numerous. Capillaries are absent from certain tissues, including cartilage, the cornea of the eye, the epidermis of the skin, and other epithelial (i.e., covering and lining) tissues.

The capillaries provide the only place where materials can enter and leave an unruptured vessel. Some materials move through the wall of the capillary by means of diffusion or active transport. Serum also can leave the capillaries under hydrostatic pressure, through junctions between the endothelial cells. Certain types of white blood cells also possess the ability to squeeze between the lining cells, leave the capillary, and wander through body tissues. This ability is called diapedesis and usually belongs to white blood cells that engulf and destroy dead body cells and microbial invaders.

C. Veins

The smallest veins, called venules, are formed when several capillaries joint together. They are similar to capillaries, but slightly larger in diameter. Farther from the capillary bed, a thin tunica media, consisting of only a few muscle fibers and some fibrous connective tissue, appears in the venule wall. At the point at which the tunica externa appears, the venules become veins.

Veins conduct blood from the venules back to the heart. The smallest veins have a thin wall consisting of the three tunicae—interna, media, and externa—found in arteries (Color Plate 8). The major structural difference between arteries and veins, however, lies in the relative thickness of the tunicae. In a vein, the tunica media and tunica externa are much thinner than in an artery, and, consequently, the walls of veins tend to bulge more easily under pressure than those of arteries.

Another distinctive feature of veins is the presence of valves, particularly in the veins of the arms and legs (Fig. 2). These valves are folds of the tunica intima, resembling the semilunar valves of the aorta and the pulmonary artery. As shown in Fig. 2, the valves allow the blood to flow in one direction only: toward the heart. A decrease in pressure in the vein on the side of the valve distal to the heart causes the valve to fill and close. On a warm summer day, when the veins just beneath the skin are dilated, their valves may stand out as small lumps on the legs.

III. Anatomy of the Circulatory Pathways

The routes that supply and drain the lungs form the pulmonary circulation, and those that supply and drain other tissues are the systemic circulation.

A. Pulmonary Circulation

Blood in the right ventricle of the heart has come from the tissues. It has surrendered to them much of its oxygen and taken from them carbon dioxide produced by the metabolic activities of the cells. Contraction of the right ventricle sends this blood past the pulmonary semilunar valve into the pulmonary trunk (Color Plate 9). The short pulmonary trunk branches into right and left pulmonary arteries that lead toward the lungs. Shortly before entering the lungs, each pulmonary artery branches into lobar arteries leading to lobes of the lungs, two on the left and three on the right. The lobar arteries undergo additional branching into smaller arteries and arterioles, until they finally branch into capillaries closely associated with the alveoli of the lungs.

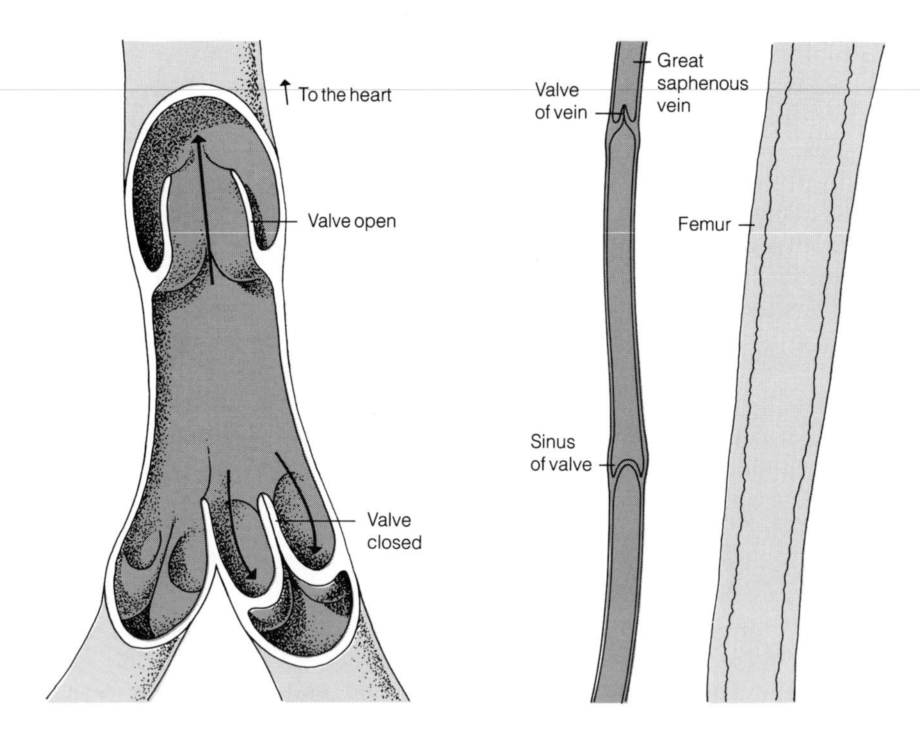

FIGURE 2 Valves in a vein. [Source: Gaudin, A. J., and Jones, K. C. (1989). "Human Anatomy and Physiology." Harcourt Brace Jovanovich, San Diego, p. 587. Reproduced with permission.]

Their close proximity to the lumina of the alveoli facilitates diffusion of carbon dioxide from the blood and into the alveoli and oxygen from the lungs into the capillaries. [*See* RESPIRATORY SYSTEM, ANATOMY.]

After passing through the capillary beds of the lungs, the blood collects in venules, which fuse to form numerous veins. These continue to fuse, eventually forming four large pulmonary veins, two from each lung, which lead to the left atrium. Note that the pulmonary arteries carry deoxygenated blood, while the pulmonary veins carry oxygenated blood. This situation is the reverse of that present in all the other arteries and veins in the adult body.

The pulmonary circuit is responsible for transporting blood from the heart for the purpose of replenishing its oxygen supply and removing its carbon dioxide. This blood is not used to nourish the tissues of the lungs or respiratory passages, nor is it used as a source of oxygen by these tissues. For these purposes, the lungs have their own set of systemic arteries.

B. Systemic Circulation

Systemic circulation is that part of the vascular system that carries blood from the left ventricle to the tissues and returns it to the right atrium. This blood supplies oxygen and nutrients to the body, while simultaneously removing waste materials and, in some cases, important cellular products, such as hormones. It is formed by the systemic arteries and systemic veins.

1. Systemic Arteries

The aorta is the major artery of the systemic circuit. It begins at the superior medial border of the left ventricle, proceeds upward for a short distance, then executes a sharp U-turn (Color Plate 9) and extends down the length of the thorax and abdomen until it reaches the pelvic girdle, where it splits into two branches (Fig. 3).

Table I lists the major branches of the aorta illustrated in Fig. 3, along with the organs they supply.

2. Systemic Veins

The systemic veins collect blood from the tissues and return it to the right atrium (Fig. 4). With few exceptions, capillary blood is collected in venules that aggregate into larger veins. In some places, venous blood collects in sinuses, as in the brain (dural sinuses), in the heart (coronary sinus), and in the liver, spleen, and adrenal glands (where they are called sinusoids). Sinuses and sinusoids are lined with endothelial cells that are continuous with those of capillaries and veins, but lack the muscular

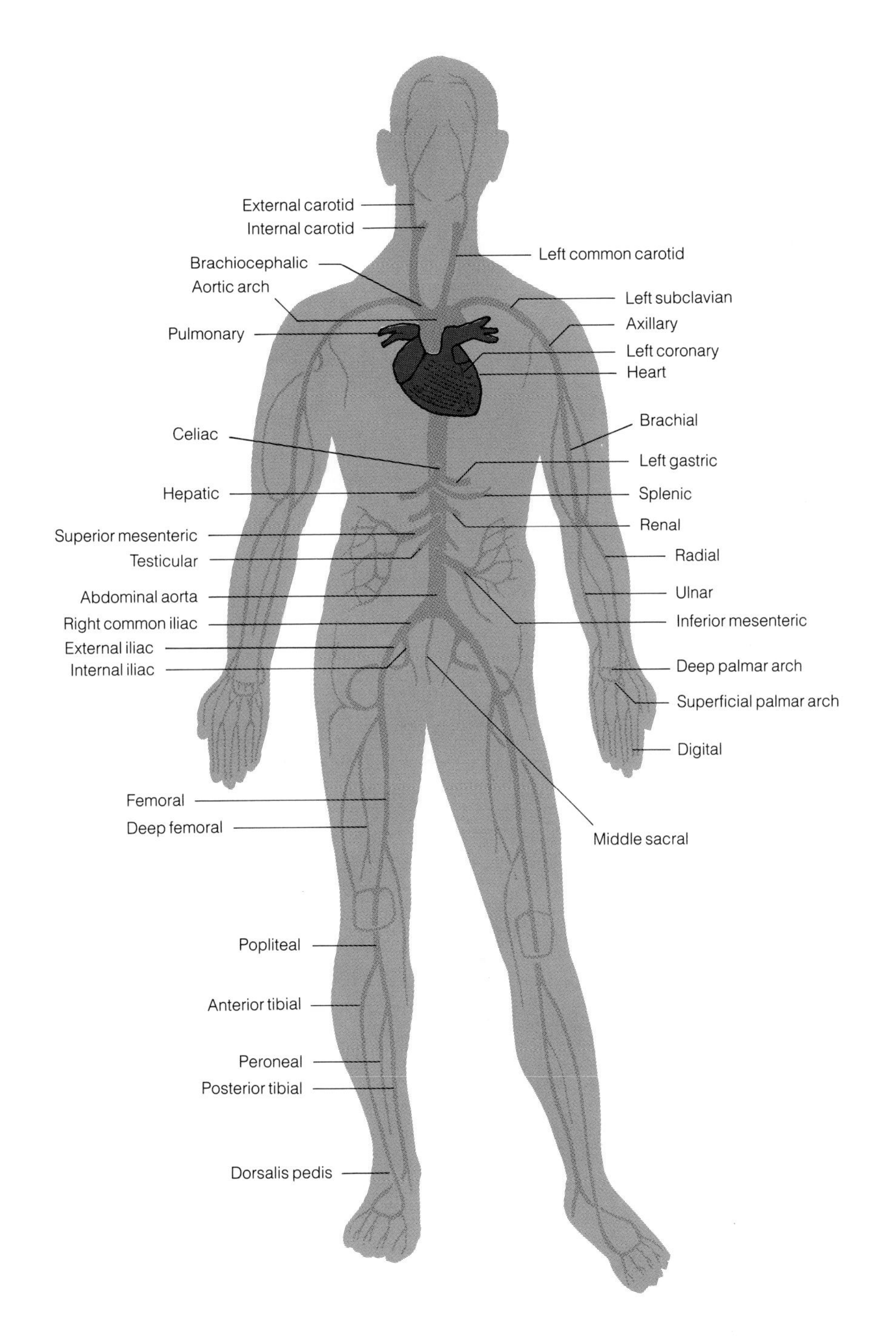

FIGURE 3 Systemic arteries. [Source: Gaudin, A. J., and Jones, K. C. (1989). "Human Anatomy and Physiology." Harcourt Brace Jovanovich, San Diego, p. 590. Reproduced with permission.]

tunica media of veins. Instead, their walls are chiefly fibrous connective tissue. Eventually, all veins drain into the superior vena cava, inferior vena cava, or coronary sinus, which in turn drain into the right atrium.

Superficial (i.e., cutaneous) veins lie just beneath the skin in superficial fascia. They receive blood from superficial tissues, yet freely communicate with deeper veins. Superficial veins are usually visible on the surface of the body, especially in the arms and legs. Deep veins tend to parallel arteries and usually bear the same names.

Table II lists the major veins illustrated in Fig. 4, along with the organs they drain.

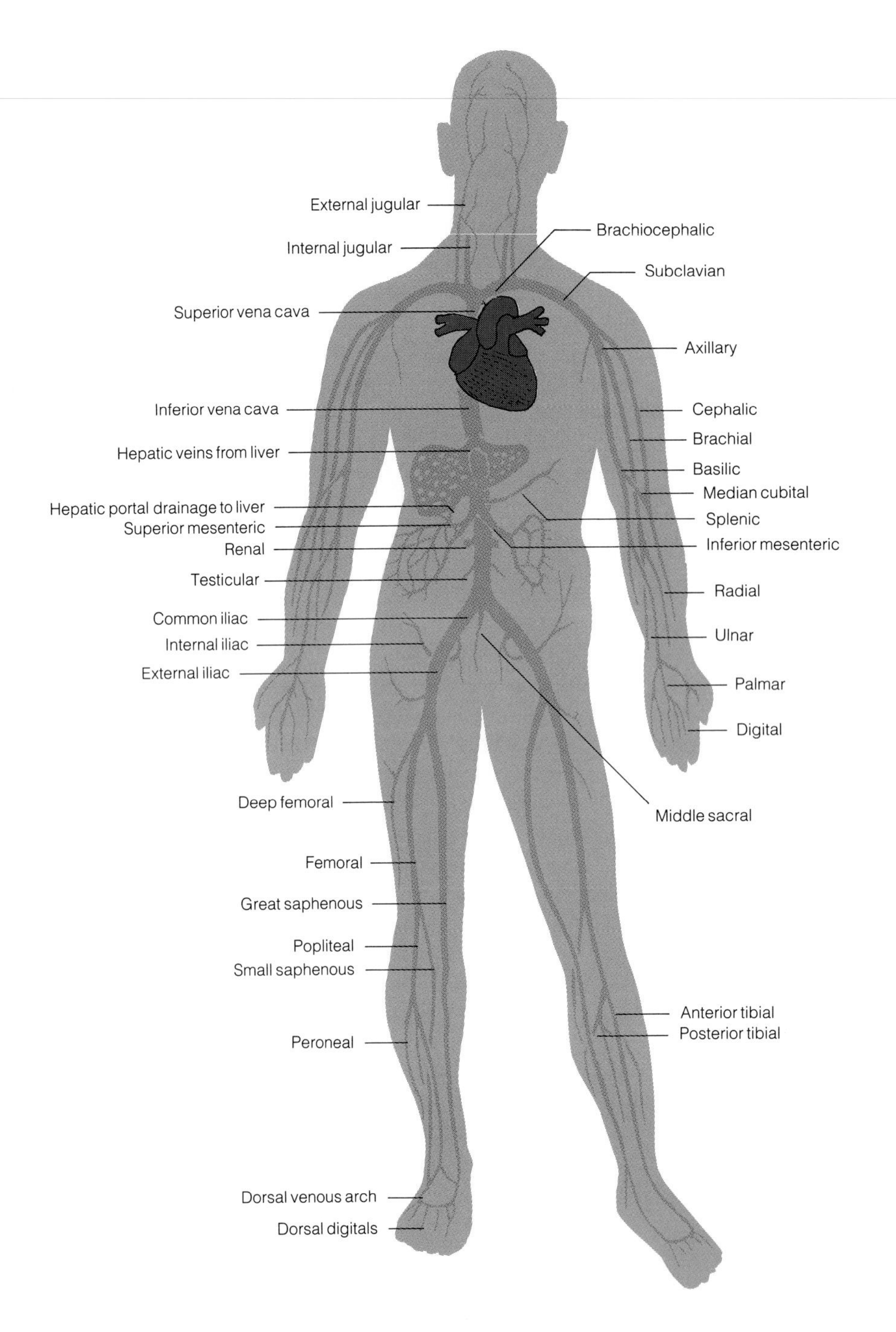

FIGURE 4 Systemic veins. [Source: Gaudin, A. J., and Jones, K. C. (1989). "Human Anatomy and Physiology." Harcourt Brace Jovanovich, San Diego, p. 598. Reproduced with permission.]

C. The Hepatic Portal System

"Portal" systems are circulatory routes that begin and end in capillaries. In the hepatic portal system (Color Plate 10), blood collected from the capillaries of the intestine, spleen, and pancreas flows into veins that reach the liver, where they branch into the sinusoid system of that organ. Two major veins contribute to the hepatic system, namely, the superior mesenteric vein, which collects blood from the small intestine, and the splenic vein, which drains the spleen and pancreas and, through the inferior mesenteric vein, the large intestine. These two veins merge into the large hepatic portal vein (not to be confused with the hepatic vein), which leads into the liver. The portal vein, after collecting several smaller veins that service other digestive organs,

TABLE I Major Systemic Arteries

Artery	Body areas supplied
Axillary	Shoulder and axilla
Brachial	Upper arm
Brachiocephalic	Head, neck, and arm
Celiac	Divides into left gastric, splenic, and hepatic arteries
Common carotid	Neck
Common iliac	Divides into external and internal iliac arteries
Coronary	Heart
Deep femoral	Thigh
Digital	Fingers
Dorsalis pedis	Foot
External carotid	Neck and external head regions
External iliac	Femoral artery
Femoral	Thigh
Gastric	Stomach
Hepatic	Liver, gallbladder, pancreas, and duodenum
Inferior mesenteric	Descending colon, rectum, and pelvic wall
Internal carotid	Neck and internal head regions
Internal iliac	Rectum, urinary bladder, external genitalia, buttocks muscles, uterus, and vagina
Left gastric	Esophagus and stomach
Middle sacral	Sacrum
Ovarian	Ovaries
Palmar arch	Hand
Peroneal	Calf
Popliteal	Knee
Posterior tibial	Calf
Pulmonary	Lungs
Radial	Forearm
Renal	Kidney
Splenic	Stomach, pancreas, and spleen
Subclavian	Shoulder
Superior mesenteric	Pancreas, small intestine, ascending and transverse colon
Testicular	Testes
Ulnar	Forearm

TABLE II Major Systemic Veins

Vein	Body areas drained
Anterior tibial	Shin
Axillary	Shoulder and axilla
Basilic	Superficial regions of upper arm
Brachial	Deep regions of upper arm
Brachiocephalic	Head, neck, and arm
Cephalic	Superficial regions of upper arm
Common iliac	External and internal iliac veins
Deep femoral	Thigh
External iliac	Leg
External jugular	Superficial regions of face, scalp, and neck
Femoral	Thigh and leg
Great saphenous	Thigh and leg
Hepatic	Liver
Inferior vena cava	Tissues and organs below diaphragm
Internal iliac	Urinary bladder and reproductive organs
Internal jugular	Brain and deep regions of face
Internal thoracic	Thoracic wall
Median cubital	Forearm
Ovarian	Ovary
Peroneal	Leg
Plantar arches	Feet
Popliteal	Knee and calf
Posterior tibial	Calf
Radial	Forearm
Renal	Kidney
Small saphenous	Thigh and leg
Subclavian	Shoulder and arm
Superior mesenteric	Pancreas, small intestine, ascending and transverse colon
Superior vena cava	Tissues and organs above diaphragm
Testicular	Testes
Ulnar	Forearm

enters the liver and ramifies into smaller veins and venules, eventually terminating in sinusoids. Blood from the sinusoids collects in the hepatic veins and eventually empties into the inferior vena cava.

BIBLIOGRAPHY

Birnholz, J. C., and Farrell, E. E. (1984). Ultrasound images of human fetal development. *Am. Sci.* **72,** 608–613.

DeVries, W. C., and Joyce, L. D. (1983). The artificial heart. *Clin. Symp.* **35,** no. 2.

Klocke, F. J., and Ellis, A. K. (1980). Control of coronary blood flow. *Annu. Rev. Med.* **31,** 489–508.

McMinn, R. M. H., and Hutchings, R. T. (1988). "Color Atlas of Human Anatomy," 2nd ed. New York Med. Publ., Chicago.

Melloni, J. L., *et al.* (1988). "Melloni's Illustrated Review of Human Anatomy." Lippincott, Philadelphia.

Rigotti, N. A., Thomas, G. S., and Leaf, A. (1983). Exercise and coronary heart disease. *Annu. Rev. Med.* **34,** 391–412.

Rushmer, R. F. (1976). "Structure and Function of the Cardiovascular System," 2nd ed. Saunders, Philadelphia.

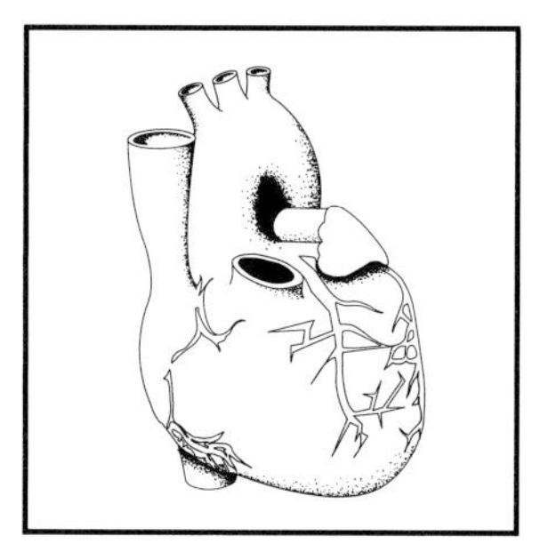

Cardiovascular System, Pharmacology

M. GABRIEL KHAN, *University of Ottawa*

Glossary

Angina pectoris Chest pain caused by severe but temporary lack of blood and oxygen to a part of the heart muscle (see Fig. 1)

Atheroma/atherosclerosis Hardened plaque in the wall of an artery: The plaque is filled with cholesterol, calcium, and other substances. The plaque of atheroma hardens the artery, hence the term "atherosclerosis" (sclerosis-hardening) (see Fig. 2)

Coronary heart disease (CHD) See coronary thrombosis and ischemia; Fig. 1

Coronary thrombosis Blood clot, thrombosis, in a coronary artery blocking flow to a part of the heart muscle; also called a heart attack or myocardial infarction

Hypertrophy Enlargement caused by an increase in size of cells; the muscle cells of the left ventricle enlarge, hypertrophy, under the influence of extra work imposed by prolonged hypertension

Hypercholesterolemia High blood cholesterol

Ischemia Temporary lack of blood and oxygen to an area of cells (e.g., heart muscle) usually caused by severe obstruction of the artery supplying blood to this area of cells. Thus, the term "ischemic heart disease" is synonymous with coronary artery disease or CHD

Myocardial infarction (infarct) Death of an area of myocardium caused by blockage of a coronary artery by blood clot and atheroma; medical term for a heart attack (see Fig. 1)

Platelets Small disc-like particles that circulate in the blood and initiate the formation of blood clots. Platelets clump and form little plugs, which arrest bleeding

CORONARY HEART DISEASE (CHD) is prevalent worldwide. The disease manifests itself as recurrent chest pain, heart attacks, sudden death, and heart failure (Fig. 1). Appropriate treatment of CHD entails the correction or modification of the underlying pathophysiology (Fig. 2). Hypertension is high blood pressure that affects more than 80 mil North Americans causing stroke, kidney failure, left ventricular hypertrophy, heart failure, and CHD. These complications can be prevented by treatment with antihypertensive agents.

Pharmacologic agents used in the management of heart and vascular disease will be discussed with emphasis on their mode of action and pharmacokinetics, as well as their ability to prevent the disease process, ameliorate symptoms, and prevent death.

I. Cholesterol-Lowering Drugs

Cholesterol reaches the blood from food eaten and from biosynthesis in the liver. Cholesterol is insoluble in blood and must attach to a carrier protein that transports it as a lipoprotein molecule. Most of the cholesterol is carried by a particle that has a low density and is thus termed low-density lipoprotein (LDL) cholesterol. LDL cholesterol is the main culprit in the formation of atheromatous plaques in arteries (Fig. 2). In addition, a small amount of cholesterol is carried by a high-density protein called HDL cholesterol. The latter is termed "good" cho-

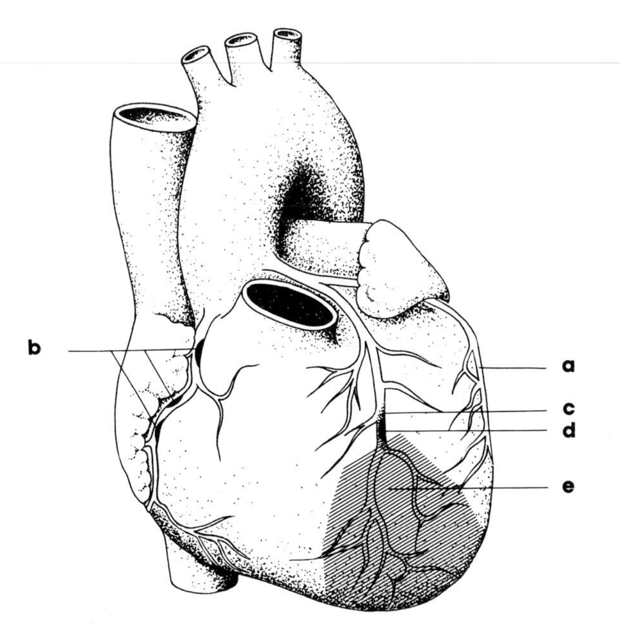

FIGURE 1 Coronary heart (artery) disease. (a) Normal coronary artery; (b) obstruction of a coronary artery by atherosclerosis, causing less blood to reach the heart muscle, producing chest pain, called *angina*; (c) blood clot; (d) complete obstruction of a coronary artery by atherosclerosis and blood clot (coronary thrombosis), causing (e); (e) damage and death of heart muscle cells [i.e., a heart attack (myocardial infarction)]. [From Khan, M. Gabriel (1990). "Heart Attacks, Hypertension and Heart Drugs," 2nd ed. Seal Books, McClelland: Bantam, Inc., Toronto.]

lesterol as a high blood level is associated with a lowered incidence of CHD. An effective drug should cause at least a 20–40% decrease in blood LDL cholesterol. Importantly, drugs that are capable of increasing HDL cholesterol 20–40% are potentially useful. [*See* CHOLESTEROL.]

A. Bile Acid Sequestrants

These drugs are resins that are insoluble in water; they bind bile acids to form insoluble complexes that are not absorbed in the intestine, resulting in increased excretion of bile acids in the feces. Because LDL cholesterol is degraded in the liver to bile acids, the increased excretion of bile stimulates LDL-cholesterol breakdown to bile acids, thus lowering LDL cholesterol in the blood. This therapy also increases the activity of LDL receptors on the membrane of liver cells; the LDL bind to the receptors and are removed from the blood. A severe reduction in LDL-receptor density is present in patients with genetic, familial hypercholesterolemia. In these patients LDL cholesterol cannot be incorporated into cells and remain at high levels in the blood. Bile acid sequestrants are capable of causing a 20–30% reduction in LDL cholesterol, but in practice, only 6–15% reduction is obtained because the unpalatable taste and gastric side effects cause poor patient compliance. A 0–10% increase in HDL cholesterol may occur. Commonly used bile acid binding resins are cholestyramine, colestipol, and divistyramine. Bile acid sequestrants have been shown to decrease obstruction of arteries in animals due to atheromas, and this finding has been confirmed by angiographic studies in humans. Cholestyramine has been shown to reduce slightly the incidence of fatal and nonfatal heart attacks, but a decrease in total mortality rate has not been demonstrated. [*See* BILE ACIDS.]

B. HMG-CoA Reductase Inhibitors

These agents are inhibitors of 3-hydroxy-3-methylglutaryl-coenzyme A (HMG-CoA) reductase, the key enzyme that operates early in the biosynthetic pathway that leads to the formation of cholesterol. This enzyme performs the rate-limiting step in the pathway by catalyzing the conversion of HMG-CoA to mevalonic acid. When the enzymatic step is blocked by a drug, HMG-CoA does not accumulate, because it is water-soluble and is readily broken down to innocuous metabolites. These drugs are very effective: modest oral doses cause a 20–25% reduction in LDL; high doses can reduce levels by about 35% in susceptible individuals. The drugs used are the fungal product lovastatin and simvastatin, which is a semisynthetic compound derived from lovastatin. Pravastatin is produced by microbial transformation of mevastatin. Cataracts have been noted in dogs given high doses of HMG-CoA reductase inhibitors.

C. Fibrates

Fibric acid derivatives activate plasma lipoprotein lipase, which breaks down the lipid component of *very low-density lipoproteins* (VLDL), termed *triglycerides*. Their effect is a decrease in hepatic production of VLDL and an inhibition of the excretion of VLDL, resulting in a reduction in serum triglycerides. However, high blood triglycerides are not a key factor in the production of atheroma and CHD. Fibrates decrease triglycerides by 30–40%, LDL cholesterol by 5–12%, and cause a 5–15% elevation of HDL cholesterol. Available drugs include gemfi-

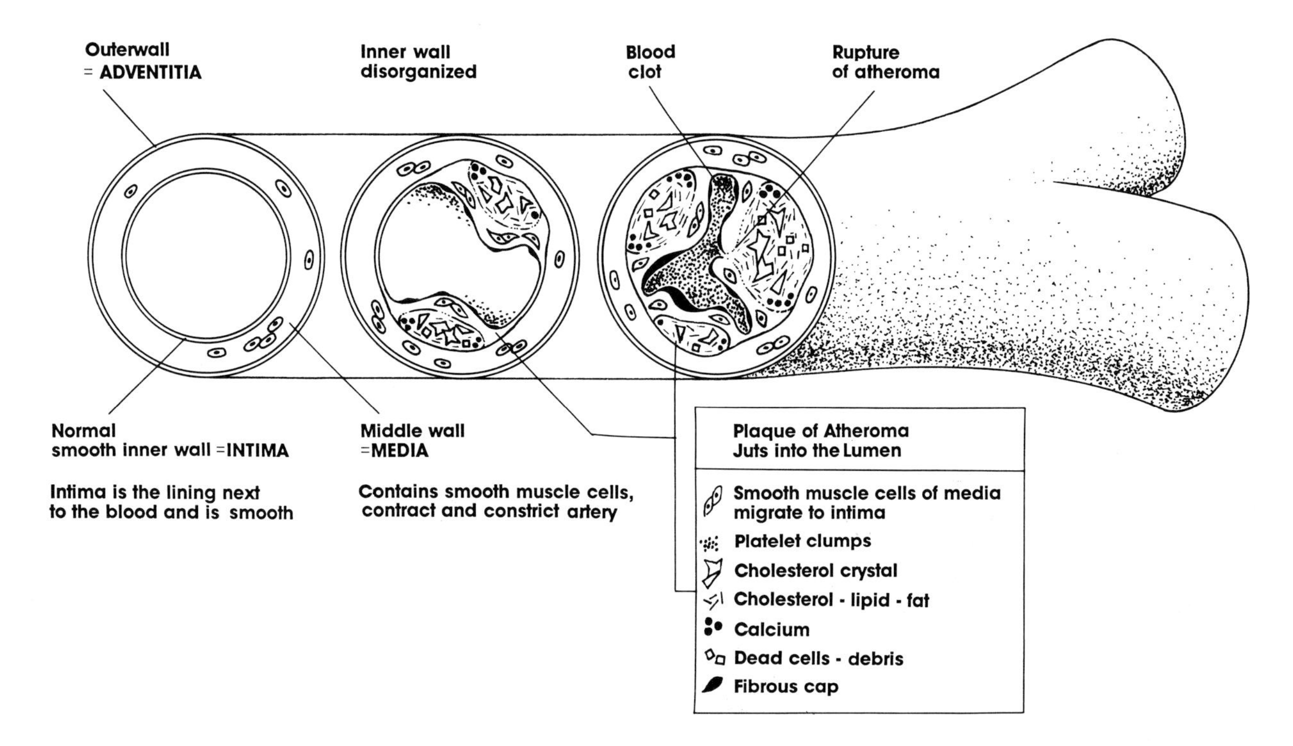

FIGURE 2 Atherosclerosis of artery. [From Khan, M. Gabriel (1990). "Heart Attacks, Hypertension and Heart Drugs," 2nd ed. Seal Books, McClelland: Bantam, Inc., Toronto.]

brozil, bezafibrate, fenofibrate, and etofibrate. Large clinical trials using gemfibrozil for a 5-year study period showed a small decrease in the incidence of nonfatal heart attacks, but no significant reduction in fatal heart attacks or death rate.

D. Nicotinic Acid (Niacin)

The drug inhibits secretion of lipoproteins from the liver, causing a 20–40% reduction in blood triglycerides and about a 12% fall in LDL cholesterol. The drug causes a 5–20% increase in HDL cholesterol. In a large study with patients followed for 15 years, nicotinic acid produced approximately 11% less total deaths and 12% less cardiac deaths than placebo. The salutary effect is so small that in practice the drug is hardly justifiable in view of the large doses required and the bothersome adverse effects it produces.

II. Antiplatelet Agents

When circulating blood platelets come in contact with damaged surfaces, they release a powerful platelet-clumping substance, thromboxane A_2, that causes platelets at the site to clump and stimulate other factors in the blood to form a clot Fig. 2. Antiplatelet agents block the action of thromboxane A_2, thus preventing the formation of platelet clots. Antiplatelet agents are useful in the prevention of clots in arteries but not in veins.

A. Aspirin

Platelets contain arachidonic acid that is converted to thromboxane A_2 by an enzyme, cyclooxygenase. Acetylsalicyclic acid (ASA, Aspirin) irreversibly acetylates cyclooxygenase and thus prevents the formation of an endoperoxide intermediate, which is the precursor of prostaglandins and thromboxanes. A dose of 80–160 mg Aspirin completely destroys cyclooxygenase activity for the 4–7 day life of the exposed platelets. Thus, the action of aspirin on platelets and on bleeding time lasts several days. Aspirin at the dose of 160 mg given to patients at the onset of a heart attack or to patients with severe angina reduces the incidence of myocardial infarction and death.

B. Dipyridamole

Dipyridamole is a weak inhibitor of platelet aggregation but is more effective than ASA in preventing platelet adhesion to foreign material such as implanted heart valves. The drug effect is increased when combined with ASA. Dipyridamole used alone does not reduce the incidence of heart attacks and does not prolong life in any subgroup of patients with heart disease.

C. Omega-3 Fatty Acids

Omega-3 fatty acids, eicosapentaenoic (EPA) and docosahexaenoic (DHA) acids, increase the production of a vasodilator prostaglandin, prostacyclin, in the endothelial lining of blood vessels. The vasodilator effect of prostacyclin counterbalances the constrictor action of thromboxane A_2. EPA and DHA inhibit the formation of thromboxane A_2 by reducing levels of its precursor, arachidonic acid, as the substrate for the enzyme cyclooxygenase, which generates thromboxane A_2. This twofold action in preventing platelet aggregation and maintaining arterial dilatation prevents blood clotting and inhibits atheroma formation. These fatty acids also increase levels of activators [e.g., tissue plasminogen activator (tPA)] that dissolve clots. EPA and DHA concentrate produce about 10–40% lowering of blood triglycerides but do not significantly affect levels of LDL and HDL cholesterol. Omega-3 fatty acids are plentiful in ocean fish, especially mackerel, herring, cod, salmon, and tuna.

A high consumption of EPA and DHA by native Greenlanders confers on this group of people a low incidence of CHD. EPA and DHA given to patients at the time of balloon dilation of obstructive artery lesions (i.e., coronary angioplasty) and continued for a period of 6 months can produce as much as 55% reduction in reocclusion rates.

III. Thrombolytic Agents

Normal arteries produce minute amounts of a clot-dissolving thrombolytic agent, tPA, that dissolves some blood clots. tPA has been produced commercially for use as a thrombolytic agent and is intravenously given to patients with heart attacks. All thrombolytic agents produce their salutary effects by actively converting a circulating blood protein, plasminogen, to plasmin, a powerful fibrin-dissolving, lytic substance. Available thrombolytic agents include streptokinase and tPA.

A. Streptokinase

When introduced in the body by intravenous injection, streptokinase forms an activator complex with circulating plasminogen that converts free plasminogen to plasmin, which then causes dissolution of clots. Streptokinase has a short half-life of about 80 min because it degrades to smaller fragments, resulting in loss of activity. Streptokinase is an effective thrombolytic agent, especially if given within 4 hr of the onset of chest pain that heralds a heart attack. Streptokinase, given by intravenous infusion within 6 hr of onset of a heart attack causes a significant reduction in cardiac death rate. The combination of streptokinase and aspirin has been shown to be superior to the use of either drug alone.

B. Tissue Plasminogen Activator

tPA binds specifically to the fibrin component of a blood clot. tPA interacts with plasminogen through a cyclic fibrin bridge, resulting in conversion of plasminogen to plasmin, which then causes lysis of fibrin. tPA has a short half-life, and about 80% of the drug is removed from the blood within 10 min. The drug is cleared mainly by the liver. tPA given by intravenous infusion to heart attack patients within the first 4 hr of onset has effects similar to those observed with streptokinase.

C. Anisoylated Plasminogen Streptokinase Activator Complex

Anisoylated plasminogen streptokinase activator complex (APSAC) is a 1-to-1 molecular combination of streptokinase and plasminogen, with a catalytic center protected by a chemical group. In the blood, the chemical group is removed, and APSAC becomes active. APSAC has a delayed onset of action because it takes time to remove the protecting group. One advantage of APSAC is its effectiveness when given intravenously as a quick bolus injection.

IV. Anticoagulants

Anticoagulants prevent thrombus (clot) formation. Clot formation is initiated when several clotting factors enter a cascade of enzymatic reactions that finally converts prothrombin to thrombin. Thrombin converts the circulating protein fibrinogen to strands of fibrin, which polymarizes to form a firm clot. The two commonly used anticoagulants are heparin, used intravenously, and warfarin, used orally.

A. Heparin

Heparin is inactivated by gastric acidity and must be given intravenously or, for some situations, subcutaneously. In the circulation, heparin is bound to a plasma protein, antithrombin III, and the complex inhibits the action of thrombin. Thus, heparin neutralizes the action of thrombin and other clotting factors, preventing conversion of fibrinogen to a fibrin clot. The drug takes effect within minutes, with an elimination half-life of 60–90 min.

B. Oral Anticoagulants (Coumarins, e.g., Warfarin)

Prothrombin is a Vitamin K–dependent clotting factor manufactured in the liver. Warfarin competitively inhibits the effects of Vitamin K, thus reducing the availability of prothrombin for conversion to thrombin in the final stages of the clotting cascade. In the absence of thrombin, fibrinogen cannot be converted to a fibrin clot. These agents are well-absorbed orally and have a half-life of 24–36 hr and are given as tablets once daily.

Oral anticoagulants do not prevent heart attacks. They are more effective in preventing the formation and propagation of fibrin thrombi that occur in veins. In the case of venous thrombosis in the legs, oral anticoagulants are given for a few months after a course of intravenous heparin therapy.

V. Antianginal Drugs

The major determinant of oxygen supply to the heart muscle is coronary blood flow. In individuals with atheroma, blood oxygen supply is fixed because of the narrowing of the coronary arteries, and anginal pain occurs when there is an increase in one or more of the three factors that determine myocardial oxygen demand: the heart rate; the velocity of contraction of the heart muscle, and the product of heart rate and blood pressure. Angina, therefore, occurs when there is an imbalance between oxygen supply and demand. The pharmacological agents used in the management of angina are directed at correcting or improving this imbalance. Oxygen supply can be improved only by dilatation of the narrowed artery at the site of obstruction. Pharmacological agents taken orally are not able to dilate continuously segments of diseased and narrowed arteries for days or weeks. Thus, the emphasis is placed on decreasing myocardial oxygen demand. Of the groups of antianginal drugs to be discussed, the beta-blockers are superior in this regard. [*See* ATHEROSCLEROSIS.]

A. Beta-Blockers

The catecholamines, epinephrine and norepinephrine, are released from sympathetic nerve endings and as hormones from the adrenal glands. Catecholamines effect their major action at receptor sites called *beta receptors* that are present on the surface of cells in various organs and tissues, in particular the heart and arteries. Catecholamines are a group of stimulants that cause an increase in the force and velocity of contraction of the heart and an increase in heart rate and blood pressure. Beta-blockers resemble the catecholamines, but block their effects at the beta-receptor sites. They decrease both heart rate and blood pressure, and consequently, the oxygen requirement. By causing a decrease in the force and velocity of contraction of the heart muscle they further decrease the oxygen requirement. An important effect derives from the fact that the coronary arteries are perfused with blood during the short moment of diastole when the heart is not contracting. Beta-blockers, by slowing the heart rate, increase the diastolic period and thus increase blood flow through the coronary arteries. [*See* ADRENAL GLAND; CATECHOLAMINES AND BEHAVIOR.]

Beta-blockers are well-absorbed from the gut; some are metabolized by the liver and others are eliminated by the kidneys. Available preparations include atenolol, acebutolol, metoprolol, nadolol, propranolol, sotalol, and timolol.

Beta-blockers decrease the incidence of heart attacks in patients with ischemic heart disease. They are also effective antianginal agents; about 80% of patients treated are expected to get from 50–75% relief of their anginal pain. Also, bothersome palpitations, abnormal heart rhythms, are abolished. Beta-blockers prolong life in patients who are given the drug immediately after an acute heart attack, continuing for a period of 2 years.

B. Nitrates

Nitroglycerin tablets placed under the tongue quickly relieve the pain of angina. This drug, as well as orally administered nitrate preparations, dilate veins throughout the body that return blood to the heart. This rapid pooling of blood in the peripheral circulation reduces the work of the heart, thus less oxygen is required and chest pain is relieved. Nitrates bind to "nitrate receptors" in the vascular smooth muscle wall of arteries, causing its relaxation and dilatation of the veins. Orally administered nitrates are rapidly metabolized in the liver, limiting their availability to vascular receptors. Sublingual, transdermal, and intravenous preparations partially overcome this problem. The effectiveness of nitrate tolerance commonly decreases after several days of continuous use because of a relative depletion of sulfhydryl groups in vascular smooth muscle cells. A daily 10-hr nitrate-free interval is necessary to allow generation of an adequate supply of sulfhydryl groups and to restore vascular responsiveness. Available preparations include sublingual nitroglycerin, glyceryl trinitrate, oral nitroglycerin tablets, isosorbide dinitrate, and cutaneous preparations; an intravenous preparation is used in severe cases of angina. The antianginal effect of oral nitrates is about half that observed with beta-blockers. Oral or cutaneous nitrate preparations have not been shown to prevent death.

C. Calcium Antagonists

Calcium is transported from the exterior to the interior of cells via a system of tubules called *slow calcium channels*. Calcium reaches the interior of the muscle cell, binds to a regulatory protein, troponin, that removes the inhibitory action of another protein, tropomyosin, and using energy supplied by adenosine triphosphate (ATP), allows the interaction between the muscle filaments, containing myosin and actin, respectively, with consequent contraction of the muscle cell. The slow calcium channels are selectively blocked by a class of agents known as calcium channel blockers or calcium antagonists. Their action prevents calcium influx into the cell, thereby causing relaxation of muscle in arterial walls and consequently dilation of arteries. This effect causes a reduction in blood pressure, and less work is imposed on the heart. These drugs cause modest dilatation of the coronary arteries in some patients, but this may be minimal at the site of atheromatous obstruction.

One group of calcium antagonists, the dihydropyridines of which isradipine, nifedipine and nitrendipine are examples, has their actions mainly on arteries and has no effect on the electrical conduction system of the heart. In contrast calcium antagonists of the phenylalkylamine structure, verapamil, or the benzothiazepine, diltiazem, have actions on the arteries as well as on the heart. Consequently, verapamil and diltiazem cause slowing of the heart rate and a variable decrease in the force of contraction of the heart muscle. Calcium antagonists are well-absorbed when taken by mouth and are mainly metabolized in the liver. The agents reduce chest pain of angina to about the same degree as beta-blockers but do not prevent heart attacks or prolong life. [*See* CALCIUM ANTAGONISTS.]

VI. Antiarrhythmic Drugs

Disturbances of the heart beat, arrhythmias, may arise from the ventricles (i.e., ventricular arrhythmias) or in the atria above the ventricles (i.e., "supra" ventricular arrhythmias). Cardiac cells have an inherent pacemaker potential that is normally suppressed by the powerful, natural generator or pacemaker that resides in the heart's sinus node and produces the heart beat in each living human. The normal impulses (i.e., action potential) generated in the pacemaker reach the ventricles through an intricate network of fibers in which they are conducted at different velocities. Arrhythmias or abnormal heart rhythms are produced by either the generation of abnormal impulses or enhanced automaticity of impulses and/or disturbances of impulse conduction. Antiarrhythmic drugs suppress these mechanisms and stop the abnormal rhythm. Of particular importance are the abnormal rhythms that occur during the onset of myocardial infarction (MI). Such abnormal rhythms termed *ventricular ectopics* or *ventricular premature beats* may occur in runs and suppress the normal heart beat. Multiple ectopic beats occurring consecutively are designated ventricular tachycardia and can worsen, producing ventricular fibrillation (VF), a condition in which the ventricle quivers instead of contracting. During VF there is cardiac arrest, no blood is expelled from the heart, the brain is deprived of blood, and the individual loses consciousness.

The electrocardiogram (ECG) picks up the heart's electrical impulses that are transmitted through the skin of the chest. Figure 3 illustrates

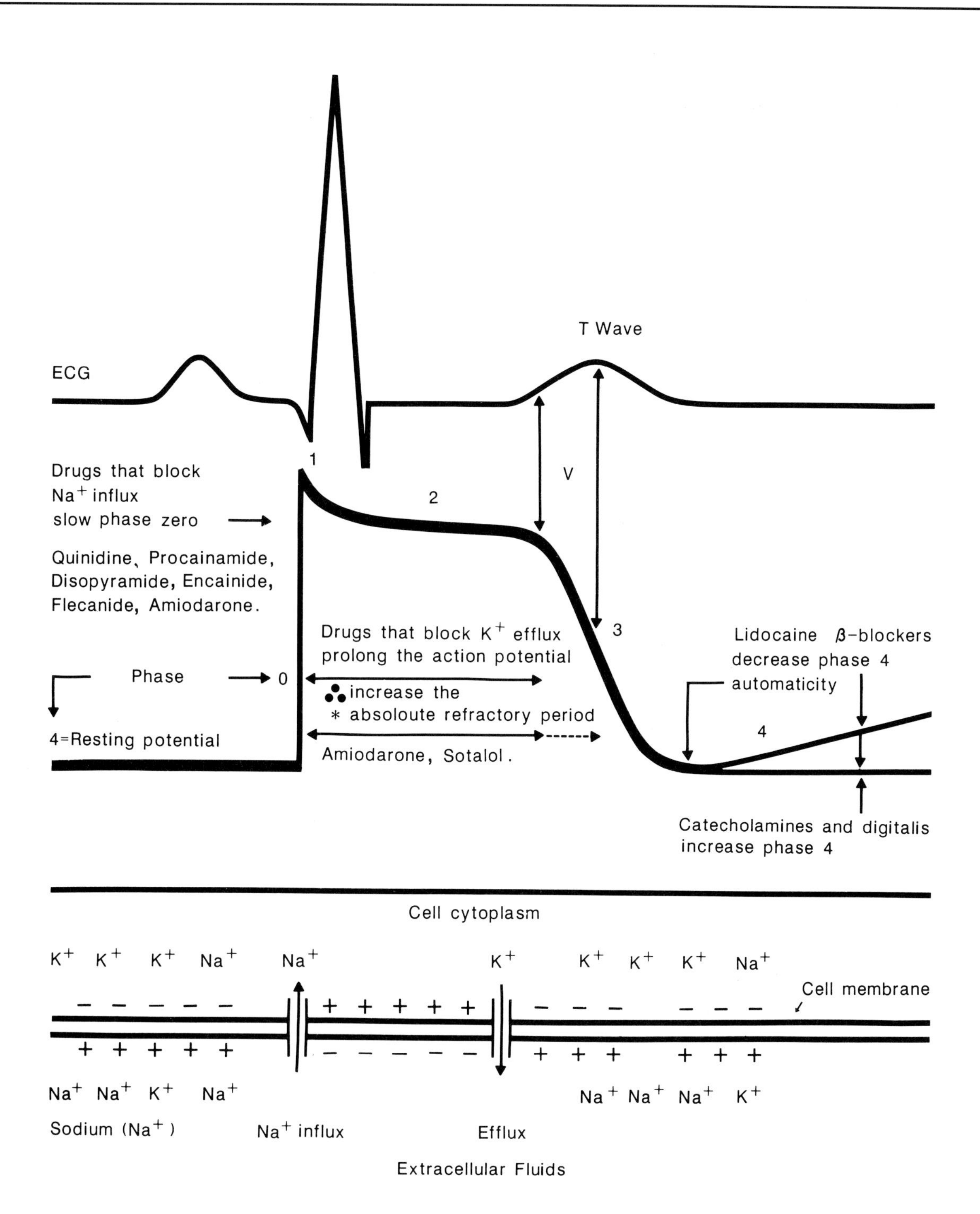

FIGURE 3 Antiarrhythmic drug action. *, Absolute refractory period; during phases 1 and 2 a stimulus evokes no response: an arrhythmia cannot be triggered. V, vulnerable period.

how myocardial cells generate an action potential in phase 0 through a fast influx of sodium ions (Na^+) into cells, which increases the resting potential (voltage) of the cell (depolarization). Later (phase 3), the cell returns to its resting potential with an efflux of potassium ions (K^+). Most antiarrhythmic agents produce their effects by decreasing the rate at which Na^+ enters the myocardial cell in phase 0 (Fig. 3). Thus, generation of the action potential of an abnormal impulse is dampened and does not reach sufficient magnitude to produce abnormal beats. Drugs in this category include quinidine, disopyramide, procainamide, flecainide, encainide, and propafenone. Beta-blockers, lidocaine, and the majority of antiarrhythmic agents decrease the rate of automaticity of abnormal rhythms by depressing phase 4 of the action potential as indicated by the arrow in Fig. 3. Lidocaine is often given intravenously during the first few hours of MI to abolish and prevent serious arrhythmias. The action of lidocaine is immediate, but of short half life (about 10 min) because the drug is quickly metabolized in the liver. Oral derivatives of lidocaine have not proven to be a major success: they do not abolish serious arrhythmias or prolong life and have adverse effects. Other agents (i.e., amiodarone and a unique beta-blocker, sotalol) cause a prolongation of the action potential (phase 2) and thus retard the generation of an abnormal impulse. Also, as indicated in Fig. 3, an increase in the absolute refractory period (phases 1 and 2) protects from dangerous impulse stimuli. During a 20–30-msec vulnerable period in phase 3, in Fig. 3, a strong electrical stimulus or ventricular ectopic beat can readily trigger ventricular tachycardia and VF.

Several of the mentioned antiarrhythmic agents prevent ventricular tachycardia and have the potential to save life. Beta-blockers suppress some serious arrhythmias that occur during an MI; when given to patients after a heart attack and continued for as long as 2 years, beta-blockers have been shown to save lives. Importantly, they reduce the incidence of sudden deaths. Other antiarrhythmics suppress serious arrhythmias but have not been shown to prolong life. Amiodarone appears to increase survival in patients who have survived a cardiac arrest. However, because of side effects, the drug is used as a last resort. In a significant number of patients with concomitant heart failure and a sick heart, some antiarrhythmic agents may actually increase the incidence of dangerous arrhythmias and/or worsen heart failure.

The atrio ventricular (AV) node slightly decreases the rate of conduction of impulses from atria to ventricles. This represents a protective effect. In some individuals, abnormal ectopic beats at the frequency of 150–200/min arise in the atria. They are conducted through the AV node to the ventricles, which then beat at 150–200. This condition is called *paroxysmal atrial tachycardia* (PAT) and is not uncommon in healthy young adults with normal hearts. The fast heart beats can be reduced by drugs such as digitalis, beta-blockers, or a calcium antagonist, verapamil, that slow impulse traffic through the AV node. In a similar condition, atrial fibrillation, instead of regular beats, the atria quivers or fibrillates. Impulses are generated at 400–500/min and bombard the "tollgate" AV node, reaching the ventricles irregularly. The heart beat becomes irregular. Agents (e.g., digitalis) that reduce impulse traffic through the AV node slow the ventricular rate to a normal level of 70–90 beats/min. Beta-blockers and calcium antagonists also slow the heart rate, but digitalis is for its effective and general safety.

VII. Heart Failure

Heart failure is usually the result of a diseased heart. The commonest cause is a weak heart muscle as a result of one or more heart attacks, severe valve disease, hypertension and rare diseases that affect the heart muscle. The weakened heart muscle fails to pump sufficient blood from the left ventricle into the arteries to supply organs and tissues adequately. The proportion of blood that cannot be ejected from the left ventricle backs up into the lungs, causing congestion of veins. Fluid containing sodium and water leaks into the sponge work and air sacs of the lungs, causing severe shortness of breath. [*See* CARDIAC MUSCLE.]

Figure 4 shows the mechanisms that operate during heart failure. The cardiac output falls, less blood reaches the kidney, and sensors initiate the secretion of renin in the kidney. Renin is an enzyme that converts circulating angiotensinogen to angiotensin I. A "converting enzyme" converts angiotensin I to angiotensin II. The latter compound constricts arteries, thereby increasing systemic vascular resistance (SVR), normally a compensatory mechanism when blood volume is reduced, as in blood loss. However, constriction of arteries increases arterial impedance (afterload), thereby increasing the work of the failing heart. This and other compensatory mechanisms therefore become counterproductive. A major medical maneuver in the treatment of patients with heart failure is to block angiotensin "converting enzyme" by drugs called *angiotensin converting enzyme (ACE) inhibitors*, thereby reducing the work of the heart. Another normal compensatory mechanism is reabsorption of sodium and water by the kidney under the effect of the renin-angiotensin aldosterone system and other systems. Sodium and water are returned to the circulating blood, imposing further strain on the failing heart. This is seen as increase in preload in Fig. 4. Congestion in the lungs occurs; sodium and water leak out

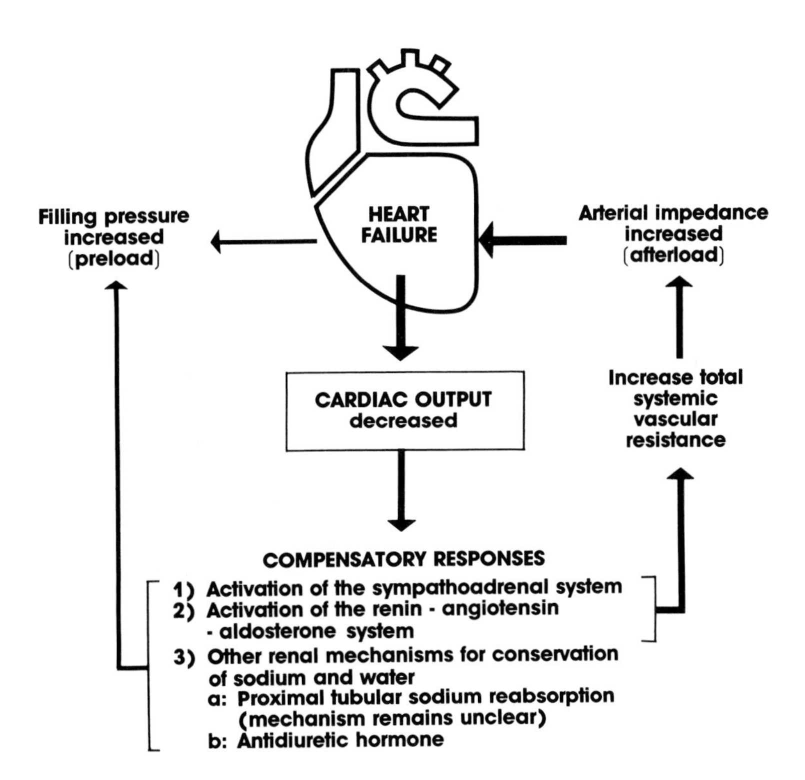

FIGURE 4 Pathophysiology of heart failure. [From Khan, M. Gabriel (1988). "Manual of Cardiac Drug Therapy," 2nd ed. Baillière Tindall/W. B. Saunders, London.]

into the alveoli and into other areas of the body such as the legs, forming edema. In edema, the legs are brine-logged, not just water-logged.

The three main treatments that assist the failing heart are diuretics, which reduce preload; inotropic drugs, which increase cardiac output; and vasodilators, in particular ACE inhibitors, which reduce the arterial impedance, or afterload.

A. Diuretic Drugs

Diuretics cause the kidneys to excrete the several liters of sodium and water that was retained in the body. Available diuretics include furosemide and thiazides. Diuretics do not repair the damaged heart muscle and have no effect on the disease process. However, they are effective in relieving severe shortness of breath and edema. [*See* DIURETICS.]

B. Inotropic Agents

Digitalis, digoxin, increases the force and velocity of myocardial contraction. These compounds inhibit the function of the sodium pump located in the membrane of heart muscle cells. This pump, using ATP energy, pumps Na^+ out of the cell and K^+ into the cell and is mainly responsible for maintaining the resting potential of the cardiac muscle cells. The action of the drugs results in an increase in intracellular sodium accompanied by an increase in intracellular calcium. Calcium within the muscle cell stimulates the force of myocardial contraction. Consequently the cardiac output increases, and some of the compensatory responses that are counterproductive to the body are halted. Occasionally, HF is due to a fast irregular heart beat, atrial fibrillation, a condition in which the top chamber of the heart beats at about 300 beats/min and the ventricles beat at about 150–180 beats/min. Digoxin reduces the heart rate to normal and improves heart failure.

C. Vasodilators

ACE inhibitors dilate arteries and cause some dilatation of veins so that afterload and preload to the heart are reduced (Fig. 4). Available ACE inhibitors include captopril, enalapril, and lisinopril. They prevent further hypertrophy of the heart, are effective in reducing symptoms such as shortness of breath, and improve the exercise capacity of individuals.

D. Nitrates

The actions of nitrates are discussed in the section on antianginal drugs. They dilate veins, pooling blood in the periphery. Because less blood is returned to the overworked heart, they reduce the heart's preload. Available preparations include isosorbide dinitrate and nitroglycerin.

Diuretics, digoxin, and ACE inhibitors individually do not improve survival, but their combination has been shown to prolong life in patients with heart failure.

VIII. Antihypertensive Agents

In most cases, hypertension (i.e., elevated blood pressure) is due to an increase in the state of contraction (tone) of arterial muscles, which produces an increase in systemic vascular resistance (SVR). With the exception of beta-blockers, the final action of all antihypertensive agents is to produce mild to moderate dilatation of arteries, thus lowering the SVR against which the heart must pump. The various groups of drugs achieve this goal by different mechanisms. [*See* HYPERTENSION.]

A. Diuretics

The major action is on the renal tubules, causing excretion of sodium and water and thus a decrease in blood volume and a fall in blood pressure. However, after several months of use, the body compensates for the slight reduction in blood volume, and the major action of chronic diuretic therapy is limited to the loss of sodium, which causes mild dilatation of arteries and a reduction in SVR. Diuretics also cause potassium excretion in the urine. Potassium loss must be prevented because it contributes to abnormal heart rhythms and to a slight increase in cardiovascular deaths. Available preparations include thiazides (e.g., hydrochlorothiazide and bendrofluazide). Diuretics do not prevent heart enlargement or hypertrophy, which carries a risk of sudden death. Also heart attacks that can occur in hypertensives are not prevented. However, they reduce the incidence of stroke to the same extent as other agents.

B. Beta-blockers

The exact mechanism by which beta-blockers reduce blood pressure is not completely understood. Their action on the heart is given in the section on angina. Reduction in heart rate and in the force of cardiac contraction result in a mild decrease in cardiac output, which lowers blood pressure. These agents decrease also renin production by the kidneys and decrease the effects of the sympathetic nervous system on the heart and arteries. They decrease the speed and velocity at which blood is ejected from the heart into the arteries. Thus, there is less hydraulic stress on the walls of arteries, especially at branching points, resulting in less wear and tear. The effect is apparent in the management of aneurysms of the aorta, a condition in which the weakened artery wall dilates and finally ruptures (dissection). In such patients, although blood pressure is usually low, beta-blockers are given intravenously to reduce the ejection velocity of blood and to prevent further dissection and weakening of the arterial wall. Beta-blockers prevent left ventricular hypertrophy and reduce the incidence of fatal and nonfatal heart attacks. Although the effect is modest, they do prolong life.

C. ACE Inhibitors

The action of ACE inhibitors is given in Section VII. They can prevent the production of angiotensin II, resulting in dilatation of arteries, decrease in SVR, and a fall in blood pressure. ACE inhibitors are useful agents both for the management of heart failure and high blood pressure.

ACE inhibitors do not cause an increase in heart rate as seen with other vasodilators. The blocking of angiotensin II causes a decrease in the secretion of a hormone, aldosterone, by the adrenal glands, which results in a favorable excretion of sodium and retention or conservation of potassium. Available ACE inhibitors include captopril, enalapril, and lisinopril. ACE inhibitors prevent left ventricular hypertrophy, but there is no indication that they are superior to diuretics or beta-blockers in prolonging life in hypertensive patients.

D. Calcium Antagonists

The actions of calcium antagonists are given in Section V. They cause peripheral dilatation of arteries, resulting in a decrease in SVR and a fall in blood pressure. Available preparations include nifedipine, nitrendipine, verapamil, and diltiazem.

E. Vasodilators

Calcium antagonists and ACE inhibitors are special groups of vasodilators. Other vasodilators include hydralazine, which has a direct dilating effect on arteries. Prazosin, indoramin, and trimazosin are $alpha_1$-adrenergic receptor blockers that act directly on nerve endings in the walls of arteries. These drugs cause dilatation of arteries, a fall in blood pressure, with a compensatory increase in heart rate as well as an increase in the ejection velocity of the heart. Consequently, they do not prevent left ventricular hypertrophy and are contraindicated in patients with aneurysms. Sodium and water is retained by the kidney, so that the antihypertensive effect usually decreases with prolonged use. The addition of a diuretic is often needed to excrete retained sodium and water, thus potentiating the antihypertensive effect.

F. Centrally Acting Drugs

These drugs modulate or inhibit sympathetic outflow from the central nervous system, which increases tone and constriction of arteries. Available drugs are methyldopa, clonidine, guanfacine, guanabenz, and reserpine. The final action of these drugs results in dilatation of arteries, with consequent retention of sodium and water by the kidney. Often they require the addition of a diuretic to complement the blood pressure lowering effect by removing sodium and water. Centrally acting drugs do not consistently prevent left ventricular hypertrophy.

G. Conclusion

The majority of patients with hypertension can be controlled with a beta-blocker, a diuretic, and/or both. Beta-blockers are as effective as other groups of drugs in lowering blood pressure. All antihypertensive agents, by reducing blood pressure, prevent the occurrence of fatal and nonfatal strokes and heart failure. However, only beta-blocking agents decrease the incidence of fatal and nonfatal heart attacks. Importantly, cigarette smoke increases hepatic degradation of propranolol and other hepatic metabolized beta-blockers and can prevent beneficial effects on mortality. Timolol, a partially metabolized drug, effectively decreases the death rate in smokers and nonsmokers with CHD. Thus, mortality reduction in patients with CHD or hypertension may be improved with the choice of nonhepatic metabolized beta-blockers: timolol, atenolol, or sotalol. Pharmacologic agents should be evaluated not only in terms of their effects on amelioration of symptoms, but also on their ability to prolong life.

Bibliography

ISIS-2 (Second International Study of Infarct Survival) Collaborative Group. Randomised trial of intravenous streptokinase, oral aspirin, both, or neither among 17,187 cases of suspected acute myocardial infarction: ISIS-2. (1988). *Lancet* **2,** 349–360.

Khan, M. Gabriel (1990). "Heart Attacks, Hypertension and Heart Drugs," 2nd ed. Seal Books, McClelland: Bantam, Inc., Toronto.

Khan, M. Gabriel (1988). "Manual of Cardiac Drug Therapy," 2nd ed. Baillière Tindall/W. B. Saunders, London.

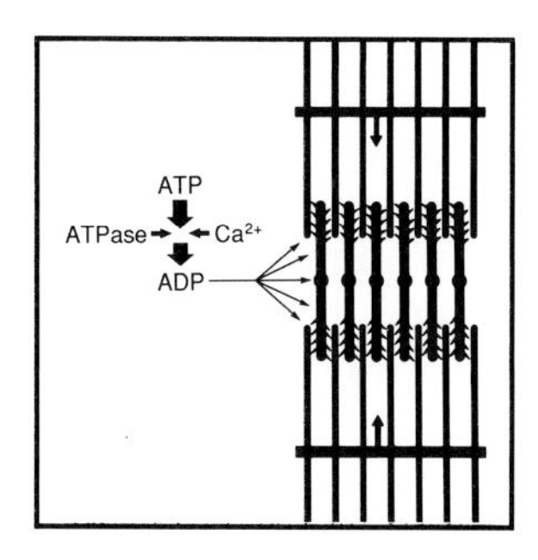

Cardiovascular System, Physiology and Biochemistry

MARC VERSTRAETE, *University of Leuven*

Glossary

Afterload Impedance to ventricular emptying during systole or the sum of all of the loads against which the myocardial fibers must shorten during ventricular contraction

Baroreceptors Receptors or sensors for stretch in the arterial wall

Cardiac index Cardiac output per minute per square meter of body surface

Cardiac output Product of the volume of blood ejected during each heart beat (left ventricular stroke volume) multiplied by the heart rate

Chemoreceptors Sensors in the arterial wall for chemical changes, such as oxygen or carbon dioxide

Inotropic state Vigor of contraction of heart muscle, reflected in the speed and capacity of shortening of the myocardial fibers at a given load

Preload Passive load that establishes the initial length of the myocardial fibers prior to contraction, reflected by the end-diastolic volume of the ventricle (Starling's law of the heart)

Ventricular ejection fraction Left ventricular stroke volume divided by the left ventricular end-diastolic volume

IN SINGLE-CELL organisms there is no need for a respiratory or cardiovascular system, and oxygen can diffuse readily through the cell wall. In small multicelled animals this vital gas can still diffuse through the intermediate tissues in sufficient quantity. When vertebrates struggled on land and became too big to rely on small air-carrying tubes which ramify throughout the body, they required a better method of distributing oxygen. In all higher vertebrates the advanced system consists basically of one air pump, the lungs, and two liquid pumps: the two ventricles of the heart. The system in humans is only a sophisticated extension of the simple diffusion mechanism acceptable to the smallest creatures. The diffusion process is similar, but the distance from the external air into the final destination is greater; blood is the transport agency which compensates for the distance. Blood has to be brought near to the surface (the lungs can be considered an interiorized external surface), so that the oxygen of the inspired air can diffuse into the red blood cells, which are then moved through the pump function of the heart near to all tissues, so that oxygen can then diffuse into them.

I. Heart

The heart lies in the center of the chest, protected by the sternum and flanked on each side by the lungs, and weighs less than one pound in the adult. Starting its pumping system 8 months before birth and continuing to beat thereafter 4000 million times over the course of a lifetime, it is probably the only organ of the human body which never comes to even a temporary rest.

The right and left ventricle beat simultaneously and have the same output, albeit under remarkably different pressure regimens. The blood returns from the body via two large veins (the venae cavae) to the right atrium. Another vein, the coronary sinus, drains the blood present in the heart muscle itself, also to the right atrium. From this compartment the blood flows through an opening and closing device called a valve, made of three leaflets (i.e., the tri-

Encyclopedia of Human Biology, Volume 2.

cuspid valve), and flows to the right ventricle. During contraction of the ventricle, called systole (a Greek word for "contraction"), the blood is pumped at a pressure of approximately 20–30 mm Hg into a short pulmonary circuit ending in the lungs.

There, it flows through capillaries surrounding the lung alveoli, which unite to eventually form four pulmonary veins (two for each lung), which bring the blood back to the heart and enter the left atrium. This blood, oxygenated during its passage through the pulmonary circuit, is passed into the left ventricle through the mitral valve connecting the two compartments of the left heart. The muscularly well-developed left ventricle pumps the blood in the aorta, from which branches distribute into the whole body, forming the systemic circulatory circuit (Fig. 1). The output of both ventricles is about 5 liters per minute, which increases sixfold during severe exercise; however, the left ventricle develops a pressure of about 140 mm Hg, approximately five times higher than the peak pressure generated by the right ventricle, which supplies the much shorter pulmonary circuit. [*See* CARDIOVASCULAR SYSTEM, ANATOMY.]

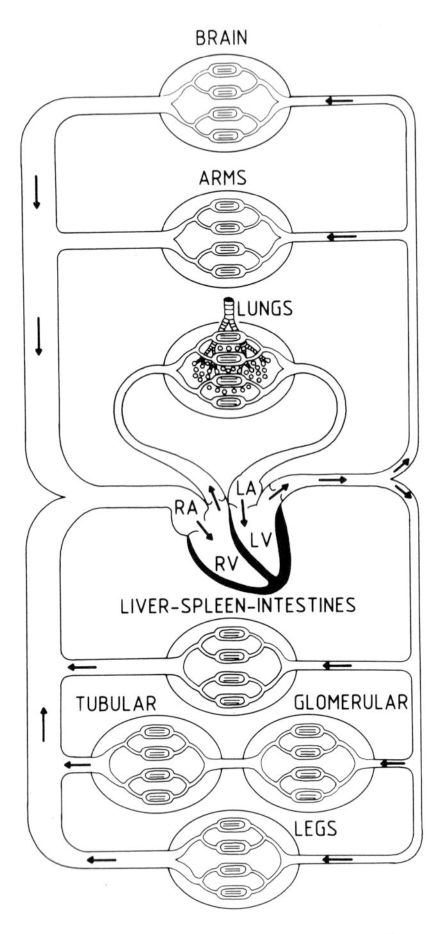

FIGURE 1 Arrangements of the parallel routes by which the circulation passes from the aorta to the venae cavae. For the systemic arterial circulation the arterial pressure is determined by the product of cardiac output and the resistance to flow offered by the various vascular beds arranged in parallel.

A. Cardiac Muscle and the Contractile Process

The heart is essentially a muscular organ. Considering its untiring function, the cardiac muscle, termed the myocardium, is a special kind of muscle. In both its structure and function it is intermediate between striated muscle and smooth muscle. Like the latter, it is able to contract and relax rapidly; like skeletal muscle, it shows cross-striations. The necessity of continuous functioning of the heart demands that the myocardium also resembles the smooth muscle in its ability to carry out long-lasting activities. Microscopically, there is evidence that cardiac muscle is made up of cells with a single nucleus, as is true for smooth muscle. The demand for sustained activity is greater in the case of the heart than for any other organ; however, its activity is not strictly continuous. Although the heart beats without ceasing for over a half-century, there is a very short pause between each contraction of the ventricles, called diastole during which these compartments refill with blood. [*See* CARDIAC MUSCLE.]

The innermost layer of cardiac muscles in the two ventricles is arranged in a circular manner. The adult left ventricle is roughly cylindrical and is made of a thick muscular layer (8–10 mm); the right ventricle is triangular in the frontal projection and its wall is much thinner (3–4 mm). Outside there are two oblique or spiral layers, arranged at almost right angles to one another and tending on contraction to pull the ventricular chamber toward the valve rings or vice versa.

When a ventricle contracts, it does so concentrically, shortening its diameter, and also longitudinally, reducing its length by pulling on its attachment to the valve rings. The total silhouette of the heart appears surprisingly immobile while it is contracting, despite the volume of blood being ejected when the two ventricles contract. This is because of the reciprocal volume changes in the atria and the ventricles; when the latter contract, the right and left atria fill with blood coming from the systemic or pulmonary circulation, respectively. The left ventricle has an ellipsoid shape, shortens more in its short axis than in its long axis, and normally empties about two-thirds of its content during ejection.

The cardiac muscle consists of columns of large cylindrical muscle fibers which branch to form a network. Each fiber consists of a membrane, the sarcolemma, surrounding bundles of small fibers, the myofibrils. Within the myofibrils the contractile proteins are arranged longitudinally into repeating units, the sarcomeres. These are composed of myofilaments, which are macromolecular complexes of contractile proteins. One myofilament is thick and is composed of the protein myosin; the other is thin and is composed of the protein actin. Both slide past each other to produce shortening of the myofibrils. Filaments made up of actin are drawn further and further between the thick myosin filaments, so that the sarcomere shortens. The thin and thick filaments remain constant in length during the process, but interact by virtue of cross-bridges formed between them (i.e., the actomyosin interaction) (Fig. 2). The resulting contraction is rapid, as each muscle fiber shortens at a speed equal to several times its length in 1 second.

B. Energy and Metabolism of the Heart Muscle

As noted above, the "sliding filament" model relates structural changes to functional events during contraction of the myocardium. The movement of the thin filaments (actin) can be accounted for by the cross-bridges which are the projecting parts of the myosin filaments to specific sites on the actin filaments. As force develops between the myofilaments, the actin is pulled toward the center of the sarcomere. To shorten further, cross-bridges must detach and reattach to new binding sites. The velocity of muscle-shortening depends on the speed at which the cross-bridge can attach, pull, detach, and resume the original position. The force development by the muscle depends on the number of cross-bridges attached at a given time.

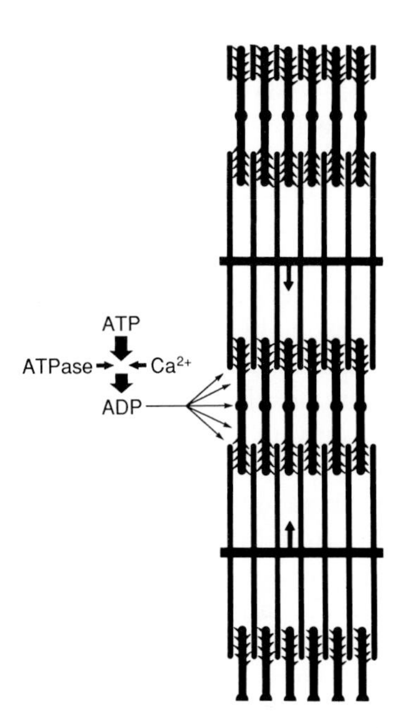

FIGURE 2 The "sliding filament" model of myocardial contraction.

Cardiac muscle can govern its contractility. The force and velocity of myocardial contraction are regulated to a large extent by the amount of free calcium ions released from the sarcoplasmic reticulum. In the relaxed state the actin molecule is not accessible for the cross-bridges because it is embraced by other protein complexes, troponin and tropomyosin, which are located periodically along the actin filaments. In the absence of free calcium ions, troponin works through tropomyosin, which courses along the actin filament to prevent actin from interacting with myosin (Fig. 3). With the arrival of the action potential at the myocyte, calcium is released and binds to the troponin–tropomysin complex, the actin molecules become accessible for the cross-bridges, and the actomyosin interaction starts.

The sequence of events leading to muscle contraction can be schematized as follows: The electrical impulse traveling along the membrane of a muscle fiber reaches the sarcoplasmic reticulum and releases calcium ions. These combine with the troponin component of the thin filament. Troponin acts as a latch, blocking the formation of an active complex between myosin and actin. Calcium binding unlocks the blocked state, allowing ATP-charged myosin to interact with actin. When this occurs, ATP is hydrolyzed by the ATPase contained in myosin (actomyosin ATPase) for the cycling of the cross-bridges and thus for the propulsive force: swiveling of myosin and pulling of thin filaments toward the center of the sarcomere. Successive cycles (binding of ATP to myosin, detachment of myosin from actin, and reattachment in a new position) result in the continued sliding of the thin element (Fig. 2). Relaxation of the actin–myosin myofibrils occurs as the result of an inhibition produced by the troponin–tropomyosin complex in the presence of a low intracellular calcium concentration. Many

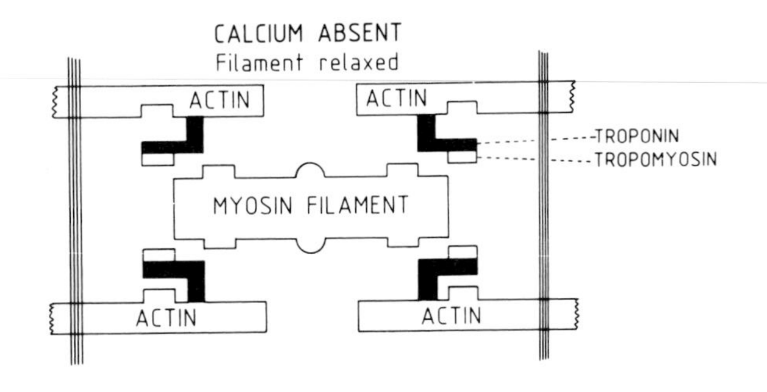

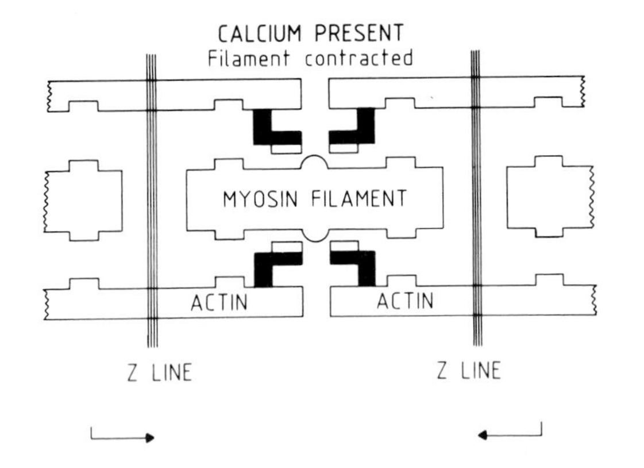

FIGURE 3 In the absence of calcium, the interaction between actin and myosin is prevented by the troponin–tropomyosin complex. In the presence of calcium, the troponin–tropomyosin complex no longer blocks the site at which interaction between actin and myosin takes place, and the contractile event can occur.

drugs, (including digitalis, sympathomimetic amines, calcium channel antagonists, and phosphodiesterase inhibitors) have an influence on myocardial contractility through their effects on available intracellular calcium ions.

To perform its essential function, the myocardium, like all other tissues of the body, must be lavishly supplied with oxygen and Na, K, Ca, glucose and other nutrients. Exchange from the blood flowing through the heart cavities contributes to the nourishment of the inner heart layer, the endocardium, but is inadequate to ensure nutrition to the large mass of the myocardium, due to the rapid transit of the blood. Indeed, although many liters of blood circulate through the heart, this is of no avail to the deeper layers of the cardiac wall. To this end, a separate and complex circulatory system subserves the myocardium and the pericardium, the latter being the outer structure which encloses the heart. This system is called the coronary circuit (from the Latin word "*corona*," which means "crown," as the coronary vessels encircle the heart like a crown).

The principal arteries of this system, the right and left coronary arteries, branch off the aorta near its base, lie on the surface of the heart underneath the epicardium, and supply all of the blood to the myocardium and the epicardium. In keeping with its almost continuing activity, the myocardium is richly supplied with arterioles and capillaries. An intercommunicating network of veins drains 85% of the myocardial blood supply to a collecting venous sinus, the coronary sinus, which empties in the right atrium, just above the tricuspid valve; smaller veins collect the remaining blood and also enter the right atrium.

The energy of the contractile process of the heart is supplied by ATP and creatine phosphate produced by aerobic metabolism. The total energy expenditure of the normal heart can be equated by its oxygen consumption. In the normal heart anaerobic metabolism (not utilizing oxygen) might account for about 5% of the energy utilized. This proportion might increase to 20% or more during exercise in patients with heart failure. Thus, to maintain contractile activity, the cells must regenerate ATP. The muscles contain phosphocreatine, which, in the presence of the enzyme creatine phosphotransferase, can rapidly regenerate ATP according to the equation phosphocreatine + ADP $\rightarrow$ creatine + ATP. Since the amount of ATP is less than that of phosphocreatine and since creatine phosphotransferase is abundant, phosphocreatine plays a major role in the instantaneous regeneration of ATP. Although phosphocreatine provides an immediate reserve of energy, it is not sufficient to sustain prolonged activity. [*See* ADENOSINE TRIPHOSPHATE (ATP).]

Ultimately, regeneration of energy-rich phosphate compounds involves the breakdown of metabolic substrates such as fatty acids, glucose, and lactate. This can occur aerobically or anaerobically. The former results not only in the production of more ATP per molecule of substrate but also in the formation of carbon dioxide and water as end products, which diffuse easily from the contracting cell. By contrast, the anaerobic consumption of substrates terminates with the production of fixed acids. When the concentration of the latter rises, intracellular acidosis becomes unavoidable, since these acids are not freely diffusible. The various end products of metabolism exert a positive feedback on the intracellular enzymatic processes as well as on the oxygen delivery to the working cells, which continuously adjust energy production to its utilization. When the supply of blood to the heart is cut off or reduced, metabolism (but not contraction) can continue at a lower level by anaerobic pathways.

The heart is a lipid-consuming organ, and glucose is used relatively less than in skeletal muscle. Free fatty acids are the chief source of energy. They are derived from plasma-free fatty acids and triglyceride and, in exercise, from myocardial stores of triglyceride. Keto acids and lactate are additional important substrates which supply energy for ATP formation.

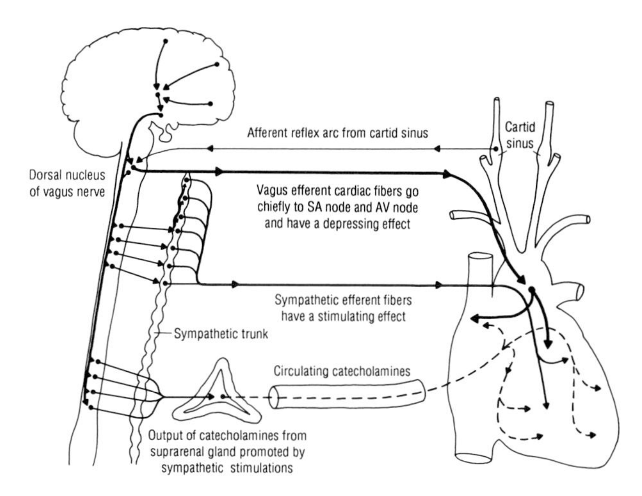

FIGURE 4 Neural and humoral regulation of cardiac function.

C. Functional Characteristics

The heart is inherently rhythmic; this explains why a heart of a mammal will continue to beat for some time when cut off completely from its nerve supply. The initiation of the heart beat is to be found in a specialized neuromuscular tissue, called the sinoatrial (SA) node, which is a small strip of tissue, about 1 × 2 cm in length, and 3 mm at its widest point. This SA node is located in the posterior wall of the right atrium near its junction with the superior vena cava. The SA node sets the heart rate, because its cells have the greatest ability to spontaneously initiate an action potential. This property of being a pacemaker is due to an inherent instability of the cell membrane which, presumably because of a decreased permeability for potassium ions, succeeds in maintaining the resting potential.

As an impulse is generated, it immediately spreads in all directions through the atrial muscle in a ripple pattern, similar to that of waves generated when a stone is thrown in water. The impulses reach a specialized tissue called the atrioventricular node, situated at the base of the interatrial system. To permit sufficient time for complete atrial contraction, the conduction of the impulse in the atrioventricular node is slightly delayed, allowing atrial depolarization to precede ventricular depolarization by approximately 140 msec. From the atrioventricular junction the electrical impulse activates atrioventricular bundles of specialized cells (bundle of His). These nerve bundles pass into the interventricular septum, and its right and left bundle branches project downward and around the tip of the ventricles; their end branches form the network of Purkinje, which terminates in the ventricular muscle. The ventricles contract slightly out of phase. The left ventricle begins to contract about 50 msec after the right one.

To adjust, the heart is connected to the sympathetic and parasympathetic autonomic nervous systems. These two systems provide a reciprocal neural control of the heart. Stimulation of the sympathetic system increases the release of norepinephrine by the nerve endings that are richly distributed throughout the atrium and the ventricles, allowing reflex regulation of the contractility of the myocardium. The sympathetic nerves also heavily innervate the sinoatrial node and the atrioventricular junction, where increases in sympathetic tone accelerate the heart rate and improve conduction velocity. In contrast, stimulation of the parasympathetic vagus nerve is associated with the release of acetylcholine by the numerous nerve endings in the two atria, the sinoatrial node, and the atrioventricular junction, but not beyond the latter or in the ventricles. Activation of the parasympathetic system slows the heart rate (Fig. 4). [*See* AUTONOMIC NERVOUS SYSTEM.]

II. Arterial and Capillary System

A. Structure

The inner layer of the heart, arteries, and veins is lined by endothelial cells. This endothelial lining, the endothelium, provides a smooth low-friction inner surface in contact with the blood. Arteries form a tubing system, taking the blood from the left ventricle and distributing it throughout the body. The much shorter pulmonary artery system similarly distributes the blood from the right ventricle to the lungs. As the left ventricle pumps the blood under high pressure in a rhythmic manner, the receiving large arteries must be elastic and able to recoil during systole, which results in pulsating waves; the recoil imparts energy to the blood, maintaining its flow throughout the diastole. This is possible because the arteries possess strong elastic walls. Elastic tissue provides for the storage of energy by stretching and relaxing; fibrous elements serve to limit the amount of stretch and to prevent overdilatation of the wall or even rupture under a surge of pressure.

As arteries become smaller, the number of elastic fibers decreases and they are gradually replaced by layers of circular or spiral smooth muscle fibers. By reducing in size, the arterioles offer more resistance to blood flow. However, due to their well-developed muscular layers, they can actively vary their wall tension and diameter, and thus the resistance they offer. The constriction or dilatation of arterioles in specific areas of the body (e.g., skeletal muscle, the skin, and the abdominal region) not only regulates blood pressure, but also to vary the distribution of blood in various parts of the body.

Metarteries are still smaller arterial structures connecting the arterioles and the capillaries; they are "thoroughfare channels," because major capillary networks arise from them. The capillary bed is more like a fishnet of interconnecting vessels than a simple parallel arrangement. The flow of blood from the metarteries into the small but numerous capillaries is regulated by precapillary sphincters (ring-shaped muscles surrounding the vessels). Opening and closing of these sphincters alternatively irrigate different capillary networks. It is in the capillaries that the ultimate goal of the cardiovascular system is fulfilled. In these tiny vessels oxygen and other essential substances can leave the bloodstream mainly by diffusion to the tissues, from which metabolites can drain away in the reverse direction. The flow in capillaries is not steady or even uniform in direction. Instead, there is vasomotion, in which individual capillaries are seen to close completely and then reopen. As a result the flow is intermittent, and it often proceeds in the reverse direction when the vessel reopens. This to-and-fro intermittent capillary flow serves as a periodic "rinsing" of the blood. The contents of a capillary are immobilized for a few seconds, while diffusion and exchange of gases and materials can proceed.

B. Function

The low pressure in the pulmonary artery (about one-fifth of the pressure in the systemic circulation) is, in part, accounted for by the large number of open capillary beds in the respiratory surfaces of the lungs. In addition, pulmonary arterioles have thinner muscular coats and larger diameters than do arterioles in the systemic circulation.

There is vigorous pulsatile outflow from the left ventricle at each beat into the aorta. The rise in pressure at each systole causes an expansion of the aorta and the large arteries, because of their well-developed elastic layer and because blood enters the arterial tree faster than it leaves through the small-bore arterioles. The large elastic arteries store blood under pressure during systole and release it again during diastole, so that the run-off of blood in the peripheral circulation is continuous. The effect is to conserve energy in accelerating the blood column and to reduce the pressure which the smallest vessels must withstand. The energy loss from friction as the blood passes through numerous arteries with decreasing diameter is considerable, and a pulsatile square-wave pattern of flow becomes gradually damped and converted in a continuous flow.

The general organization of the systemic circulation is such that the arterial bed serves as a pressure reservoir from which the circulations to the various organs operate in parallel. Each organ takes the blood supply which it requires by regulating its local vascular resistance primarily on the basis of the metabolic needs, whereas the total peripheral vascular resistance is primarily controlled by cardiovascular reflexes and maintains the pressure in the arterial system. For example, during exercise the vascular resistance in exercising skeletal muscles decreases markedly, and, despite the increase in cardiac output, the blood pressure might decrease were it not for a reflex increase in vascular resistance in nonexercising muscles and certain other areas.

III. Venous Network

A. Structure

The small vessels collecting blood from the capillaries are venules. They are simple endothelial tubes supported by some collagenous tissue and, in a larger venule, by a few smooth muscle fibers as well. As venules continue to increase in size, their walls begin to show some characteristics of the arterial wall structure, but are considerably thinner in proportion to the diameter of the vessel.

Smaller veins begin to possess a helical–circular coat of smooth muscle cells located in the middle coat, which contains less muscle and elastic tissue than is found in the arterial wall. Owing to the structure of their walls, veins are much more distensible, can be compressed more easily, and are less well adapted to holding high pressure within them than arteries are. The larger veins eventually collect in one of the two venae cavae, which drain in the right atrium. The venous blood flow is slow due to the greater cross-sectional area, but more continuous than in arteries. Veins are low-resistance channels and only a small pressure gradient exists along the major ones. At intervals in the long veins, particularly in the limbs, the endothelial lining is pulled out to form cup-shaped valve cusps, similar in design to the semilunar cup-shaped valves of the heart. These valves permit unidirectional flow only (i.e., toward the heart).

The systemic veins are not simply a series of passive tubes for the transport of blood back to the heart. They are a reservoir of variable capacity whose compliance is about 30 times greater than the arterial tree. The capacity of the peripheral veins can be modulated by contraction or relaxation of the smooth muscle cells in their walls.

B. Function of Veins

The greater part (i.e., 60%) of the blood volume is contained in the low-pressure side of the circulation, which includes the postcapillary systemic veins, the right heart, and the pulmonary circulation. The total blood volume (one-twelfth of the body weight) is maintained relatively constant by appropriate reflex adjustment of renal hemodynamics and hormonal effects on the renal tubules that preserve or excrete sodium and water. As a consequence, mainly of hydrostatic forces, shifts of fluid do occur between the different parts of the cardiovascular system, especially the cardiopulmonary vessels and the systemic veins. To compensate for these shifts, the veins are capable of active and passive changes in capacity that serve to modulate the filling pressure of the heart by adjusting the central blood volume. The active expulsion of blood is due to contraction of the venous smooth muscle cells, which are under autonomic sympathetic nervous control; the passive change is due to a decrease in venous distending pressure, resulting from constriction of the precapillary resistance vessels. These changes are brought about mainly by alterations in sympathetic nerve signals.

The effective venous distending pressure (i.e., transmural pressure) is determined by the pressure of the blood within the vein minus the counterpressure exerted by the surrounding tissues. The venous blood pressure depends on the arterial pressure and the diameter of the arteriole; when the latter dilate, the venous pressure is augmented, leading to an increase in venous capacity. The venous blood pressure also depends on the hydrostatic load, which increases by the gravitational forces on standing, but, to the same extent as the arterial pressure, nullifying changes in the driving force of flow in the venous and arterial systems. The increased venous pressure on standing, however, increases capillary pressure and thus filtration and will also augment the pooling of blood at the venous side. Peripheral venous pooling can be counteracted by the massage function of the leg muscles during their contraction. Due to the presence of valves, the direction of flow is unilateral, and the resulting translocation of blood from the legs to the large-capacitance veins improves the filling of the heart and augments the stroke volume. Also due to the increase of intraabdominal pressure during inspiration, blood tends to be drawn into the thorax by virtue of the increased pressure gradient; conversely, on expiration the intrathoracic pressure rises and blood flow stops. Forced expiration (e.g., to raise the intrathoracic pressure) might considerably arrest the return of blood to the thorax temporarily. The venous return to the right heart is further enhanced, since the venous pressure inside the thorax decreases below atmospheric pressure during inspiration.

IV. Hemodynamics

A harmonious circulatory system has to allow, under a relatively stable perfusion pressure, for a wide variation in flow, in both the cardiac output and its distribution to different organs and tissues in response to varying demands. To this end, three main features of the circulation have to be kept under control.

A. Regulation of Arterial Blood Pressure

Arterial blood pressure is principally controlled by the activity of the vasomotor and cardiac centers, located in the medulla.

The vasomotor center functions largely as a regulator of the tone of smooth muscles of arterioles by acting through the sympathetic division of the autonomic nervous system. An increase or decrease in tone is brought about by an increase or decrease in the release of norepinephrine at sympathetic nerve endings. Spontaneous activity of the vasomotor center maintains arteriolar tone, which can be increased or decreased by input to the center from peripheral receptors, other brain centers, or substances in the bloodstream. By stimulating specific areas of the vasomotor center, an excitatory (pressor) or inhibitory (dilator) portion can be identified.

The cardiac center governs the heart rate via sympathetic and parasympathetic (vagus nerve) innervation of the sinoatrial node. Increased sympathetic stimulation (which increases the release of epinephrine) and decreased parasympathetic stimulation (which decreases the release of acetylcholine) accelerates the heart rate; the reverse actions slow it. An increase or decrease in the force of ventricular contraction is brought about largely by increasing or decreasing the sympathetic stimulation of cardiac muscle. The best-known and most important regulatory mechanism mediated by these medullary centers is the baroreceptor reflex, adaptive changes in blood pressure initiated by arterial pressor receptors stimulated by tension in the wall of the vessel and thus by pressure within the vessel.

B. Distribution among the Various Local Circulations

The amount of blood flowing to an individual organ is determined by the difference between the arterial and venous pressures in the vessels supplying the organ and by the vascular resistance of the organ. Although the arterial and venous pressures change in situations such as exercise, emotional stress, and eating, most of the alterations in the distribution of blood flow are the result of changes in vascular resistances in the organ (Fig. 1). There are two resistances: the arteriolar and precapillary. The former is controlled by the contraction of smooth muscle in the walls of the small arterioles and is mainly under neural control.

Alterations in the diameter of arterioles affect not only the flow, but also the pressure within capillaries and veins. The precapillary resistance is due to smooth muscle fibers controlling the mouths of the capillary channels and, by governing the number of open capillaries, they control the capillary exchange area. If many sphincters are closed, the blood flows through the main capillary channels, few in number and relatively large. If the sphincters relax, blood is diverted through a much denser network of capillaries and runs slowly in close contact with the tissues. Precapillary sphincters are influenced mainly by local vasodilator metabolites—produced, for example, during exercise—which readily diffuse through the tissues and cause the sphincters to relax.

C. Cardiac Output

At each level of activity, in health or in disease, a complicated interplay automatically adjusts the extent of shortening of myocardial fibers and, consequently, the stroke volume and the cardiac output. There are four principal determinants of the stroke volume: preload (end-diastolic volume or passive load), which establishes the initial muscle length of the cardiac fibers prior to contraction; afterload (impedance to ventricular emptying during systole or the sum of all the loads against which the myocardial fibers must shorten during systole, including the aortic impedance, the peripheral vascular resistance and the viscous mass of the blood); the contractility or inotropic state of the heart (reflected in the speed and capacity of shortening of the myocardium); and the coordinated pattern of contraction. A fifth determinant, the heart rate, sets the cardiac output, which is the product of stroke volume and heart rate.

All of these determinants interrelate, but their relationships are not fixed; they play greater or lesser roles, depending on the functional state of the heart. Thus, when the inherent contractility (i.e., inotropic state) is impaired, stroke volume and cardiac output might be maintained by ventricular dilatation (i.e., the Frank–Starling mechanism). Indeed, according to the Frank–Starling law, there is a direct relationship between end-diastolic fiber length and ventricular work, as the energy of contraction is a function of the length of the muscle fibers.

V. Lymphatic System

A. Structure

In humans the exchange of fluid across the capillaries throughout the body into the interstitial fluid amounts to about 20 liters per day. This moves in both directions, and the great bulk is returned to the systemic circulation through the walls of the capillaries and the venules. Approximately 10–20%, however, is returned to the systemic circulation by a circuitous route, the lymphatic system. The lymph capillaries start with microscopic blind ends and consist of a single layer of overlapping endothelial cells attached by anchoring filaments to the surrounding tissue. With muscular contraction these fine strands might distort the lymphatic vessel to open spaces between the endothelial cells and permit the entrance of protein, large particles, and cells present in the interstitial fluid. The small lymphatic vessels form a network in almost all tissues and have the same relationship to tissue spaces as have blood capillaries. Small lymphatic vessels converge to become larger and unite in lymphatic trunks, which have layered, but still thin, walls. These larger vessels contain valves and join finally to form the two large lymphatic ducts which empty into the left and right subclavian veins. The content of the lymph (Latin for "water") is similar to plasma, except that the lymph only has one-half as much protein. Due to the thinness of their basement membranes, the lymphatic capillaries are much more permeable to macromolecules than are the blood capillaries, so that protein molecules which escape from the latter are taken up into the lymph channels.

There is no pump to move the lymph; its movements are principally due to the squeezing action of the adjacent muscular tissues. Lymphatic circulation is consequently slow and uncertain.

B. Function

Phylogenetically, the need for a lymphatic system arose as soon as a closed cardiovascular system developed. In mammals the necessity for providing a high-pressure system to insure an adequate supply of oxygen to the tissues created a situation favoring the transudation of fluid from capillaries. The role of this mechanism was minimized by the increase in plasma proteins, which exerted a considerable oncotic pressure. However, albumin that leaves the vascular compartment cannot be returned to the blood, since back-diffusion into the capillaries cannot occur against the larger albumin concentration gradient. Were the protein not removed by the lymph vessels, it would accumulate in the interstitial fluid and act as an oncotic force to draw more fluid from the blood capillaries to produce increasingly severe edema. In addition, there still remains the problem of clearing the tissue spaces of substances which had leaked out or which were not absorbed by the blood. As 50% or more of the total circulating protein escapes from the blood vessels, this extravascular protein cannot all be resorbed into the blood capillaries and hence must return by way of the lymphatic vessels, which act as an overflow drainage system. In 24 hours a volume of fluid equivalent to the total plasma volume (approximately 4 liters) and over 50% of the total circulating fluid passes in the thoracic duct alone, emphasizing the importance of the lymphatic system.

Bibliography

Abramson, D. I. (ed.) (1962). "Blood Vessels and Lymphatics." Academic Press, New York.

Berne, R. M., and Levy, M. (eds.) (1972). "Cardiovascular Physiology." Mosby, St. Louis, Missouri.

Braunwald, E. (ed.) (1988). "Heart Disease. A Textbook of Cardiovascular Disease." Saunders, Philadelphia, Pennsylvania.

Burton, A. C. (ed.) (1965). "Physiology and Biophysics of the Circulation. An Introductory Text." Yearbook, Chicago, Illinois.

Forrester, J. M. (ed.) (1985). "A Companion to Medical Students. Anatomy, Biochemistry and Physiology." Blackwell, Oxford, England.

Hardin, G. (ed.) (1961). "Biology. Its Principles and Implications." Freeman, San Francisco, California.

Hurst, J. W., Schlant, R. C., Rackley, C. E., Sonnenblick, E. H., and Wenger, K. (eds.) (1990). "The Heart." McGraw-Hill, New York.

Jacob, S. W., Francone, C. A., and Lassow, W. J. (eds.) (1978). "Structure and Function in Man." Saunders, Philadelphia, Pennsylvania.

Shepherd, J. T., and Vanhoutte, P. (eds.) (1975). "Veins and Their Control." Saunders, Philadelphia, Pennsylvania.

Shepherd, J. T., and Vanhoutte, P. (eds.) (1979). "The Human Cardiovascular System. Facts and Concepts." Raven, New York.

Smith, A. (ed.) (1968). "The Body." Allen & Unwin, London.

Wyngaarden, J. B., and Smith, L. H. (eds.) (1988). "Cecil's Textbook of Medicine." Saunders, Philadelphia, Pennsylvania.

Carnivory

JOHN D. SPETH, *University of Michigan*

Glossary

Australopithecines Extinct hominids of the Pliocene and Lower Pleistocene (ca. 6 to 1 million years ago) in Africa that walked erect (bipedally) and had humanlike hands and teeth, but much smaller cranial capacities and more apelike jaws and skulls. Several species have been recognized (e.g., *Australopithecus afarensis*, *A. africanus*, *A. robustus*, and *A. boisei*).

Hominid Primates belonging to the family Hominidae, which include modern humans (*Homo sapiens sapiens*) and all premodern members of the genus *Homo*, as well as closely related ancestral forms known as Australopithecines

Homo erectus Extinct human species found throughout Africa and southern Eurasia during the Lower and Middle Pleistocene, dating between about 1.6 million and 300,000 years ago; had a cranial capacity intermediate between the Australopithecines and anatomically fully modern humans, and was probably the first hominid to use fire.

Neanderthal Last extinct human form (*Homo sapiens neanderthalensis*) prior to the appearance of anatomically fully modern humans (*H. sapiens sapiens*), dating to the Upper Pleistocene between about 130,000 and 35,000–40,000 years ago

Pleistocene Geological period, known popularly as the Ice Age, that lasted from about 1.75 million to 10,000 years ago; generally divided into three periods: Lower (ca. 1.75 million to 700,000 years ago), Middle (ca. 700,000 to 130,000 years ago), and Upper (ca. 130,000 to 10,000 years ago)

Pliocene Geological period immediately preceding the Pleistocene that lasted from about 6 million to 1.75 million years ago

Primate Mammalian order that includes prosimians (e.g., lemur and loris), New and Old World monkeys (e.g., howler, spider, macaque, and baboon), apes (e.g., gibbon, orangutan, chimpanzee, and gorilla), and hominids (e.g., Australopithecines, *Homo erectus*, Neanderthal, and modern human)

Taphonomy Field which studies the processes (e.g., natural decay, trampling, transport by flowing water, and transport and destruction by carnivores) that affect an assemblage of bones from the time an animal dies until its bones have become incorporated into the fossil record

THE FREQUENT consumption of meat, often in prodigious quantities, by modern humans has attracted the attention of scholars since Darwin. Because humans are believed to have evolved from a primate ancestor, and since most living primates—including chimpanzees, our closest living relatives—subsist largely on vegetal foods (e.g., fruits, leaves, and shoots), the development of the human proclivity to kill and consume the flesh of animals, particularly large and dangerous ones, has come to play a central role in most theories of hominid origins and evolution. These theories, often referred to collectively as the hunting hypothesis, are now being challenged by new insights from living hunters and gatherers, human nutrition, chimpanzee behavior in the wild, archaeology, and taphonomy.

I. Hunting Hypothesis

That humans are descended from a primate ancestor and that we very likely share a common ancestor with the chimpanzee, a great ape found in the tropical forests and woodlands of sub-Saharan Africa, are widely accepted today by most scholars concerned with hominid origins and evolution. Based on genetic, paleontological, geological, and other evidence, there is also a growing consensus that the divergence between the human and ape lines probably occurred around 6 million years ago, plus or minus about 1 million years. Every year, as new hominid fossils are found and previous finds are reanalyzed and more securely dated, as detailed paleoenvironmental and taphonomic studies reveal the context in which these early hominids lived, died, and became incorporated into the fossil record, and as our understanding of the fundamental adaptive and evolutionary processes that give rise to behavioral and genetic differentiation are refined, the when, where, and how of hominid origins and evolution are becoming increasingly clear.

Anthropologists and biologists are also approaching this fascinating problem from the other end of the time scale. Some are studying the behavioral ecology and diet of chimpanzees and other living primates, while others are focusing on the adaptations and subsistence practices of contemporary hunter–gatherers or foragers (living peoples whose life styles are thought to most closely approximate the way our ancestors lived for several million years prior to the emergence of farming and animal husbandry, a series of dramatic economic and social changes that occurred only within the last 10,000 years). These behavioral and comparative studies of living populations help pinpoint important similarities and differences between humans and our closest living nonhuman relatives, and in the process shed light on the very processes that gave rise to the human line. [*See* PRIMATE BEHAVIORAL ECOLOGY.]

One of the most striking contrasts between contemporary hunter–gatherers and most living primates, a difference that was clearly recognized over a century ago by Darwin, is that humans are highly successful predators, even of large and dangerous prey. And while humans are basically omnivores, they are capable of eating (and often do eat) prodigious amounts of meat. Nowhere is this more strikingly apparent than among traditional Eskimos living in the Arctic; these hunters subsist almost entirely on a diet of meat obtained by killing large and often dangerous sea mammals, including seals, walruses, and even whales, as well as caribou, polar bears, and a variety of other terrestrial mammals. Most primates, on the other hand, are largely vegetarian, deriving the bulk of their diet from fruits, leaves, shoots, nuts, resin, and other plant parts.

Darwin, in fact, believed this contrast was so fundamental that he used it as the basis for a theory of human origins that, somewhat modified, remains persuasive today. In its modern form the theory—often known as the hunting hypothesis—holds that during the Miocene in sub-Saharan Africa—roughly 25 to 6 million years ago—climatic conditions gradually deteriorated, becoming drier and more seasonal. These changes favored the expansion of open woodlands and savanna grasslands at the expense of dense tropical rainforests. Many species of great ape were unable to cope with these dramatic habitat changes and became extinct, but some populations of one of these species, thought by most to have been the common ancestor of both chimpanzees and humans, adapted to them by turning to the one obvious new resource that these expanding grasslands offered: large herds of antelopes and other herbivores. But to do this effectively, these "protohominids," known as Australopithecines, would have exposed themselves to new dangers, especially from the many other predators that also preyed on the herbivores. These included lions, leopards, several species of hyena, and many other large dangerous carnivores. To protect themselves and to more effectively kill large herbivores and cut through their thick tough hides, the hominids began to rely on hand-held sharp-edged tools, probably first of wood, bone, or perhaps even shell, then of flaked stone.

Gradually, a positive-feedback relationship emerged in which increasing tool use by the hominids favored greater reliance on a bipedal stance in order to free their hands, as well as greater intelligence (and hence larger brain size) as they came to rely more and more on tools and other forms of learned or culturally mediated behavior. These changes were also accompanied by a gradual reduction in the size of their initially large and intimidating canines as the function of these teeth for both defense and gaining access to tough or inaccessible vegetable foods was supplanted by tools.

More recent variants of this basic hunting hypothesis attribute other characteristically human traits to human carnivory as well. For example, some anthropologists have argued that the bonanza of meat provided by a single large animal carcass favored the emergence of food-sharing, not just between mother and offspring or among siblings, as is common in primates, but among a much broader network of less closely related kin (and perhaps even among nonkin). The hunting of larger and more dangerous prey might also have contributed to the development of the division of labor that typifies most contemporary foraging societies, in which males do the hunting and then transport meat back to a central place to provision females with dependent young and to share with other members of the group. A few anthropologists have even argued that the skills and the knowledge needed to successfully locate and kill large dangerous game favored the development of new and more effective means of communication, a process that culminated in the emergence of true language. These factors are also thought to have favored increased levels of cooperation among larger numbers of hunters, enhancing the importance of group integration beyond the biological family and ultimately giving rise to new and more complex forms of social organization. Hunting and meat-eating, therefore, have often been viewed as the locus of the pivotal selective forces that gave rise to protohominids and gradually transformed these remote ancestors into what we are today. [*See* EVOLVING HOMINID STRATEGIES.]

Despite its elegance and compelling simplicity, various lines of evidence are now beginning to raise serious doubts about the validity of the hunting hypothesis. While no equally comprehensive and compelling new theory has emerged to replace it, our views about the role of carnivory in the evolutionary history of the human line have begun to shift, in some respects dramatically, and they likely will continue to shift. The new evidence and insights are coming from many different sources, including ethnographic studies of modern hunter–gatherers, field studies of wild free-ranging primates (especially chimpanzees), human nutrition, archaeology, taphonomy, and a host of others. In subsequent sections the more important of these new insights are considered and the role that human carnivory might have played in hominid origins and in the subsequent evolution of our species is reevaluated.

II. Protein in Hunter–Gatherer Diet

Hunter–gatherers (or foragers) are peoples who live entirely or largely without the benefits of agriculture or animal husbandry. Prior to the end of the Pleistocene, which occurred about 10,000 years ago, humans everywhere subsisted exclusively by hunting (including fishing) and gathering, and even after the appearance of farming communities, many populations continued to live as foragers, some up to the present. Even as recently as the period of European colonial expansion in the 15th and 16th centuries, there were still hundreds of hunting–gathering groups throughout the world. For example, in the Old World these included Chukchi in Siberia; Ainu in Japan; Agta and Batak in the Philippines; Australian Aborigines; Andaman Islanders; Vedda in Sri Lanka; and Pygmies, Hadza, and Bushmen (San) in sub-Saharan Africa. In the New World there were also many foraging groups, including Eskimos, Dogrib, and Mistassini Cree in the Arctic and Subarctic; Shoshoni, Paiute, Washo, Chumash, and many others in the Great Basin and California in the western United States; Seri in northern Mexico; and Ache, Hiwi, Ona, Yaghan, and numerous others in South America.

All of these historically or ethnographically known foragers do (or did) a great deal of hunting and attribute a lot of social importance to it, although the amount of time they devote to hunting and the actual contribution of animal foods to their diet vary greatly from group to group. Despite this variability, there is a broad positive correlation between geographic latitude and the proportion of meat in the diet. Not unexpectedly, in the Arctic, plant foods contribute only minimally to the diet, while hunting of both land and sea mammals, as well as fishing, constitutes the bulk of the foods consumed, both by weight and in terms of total calories. Outside the Arctic, however (especially as one approaches the equator), the contribution of plant foods to the larder increases sharply, attaining levels in excess of 60–80% by weight of the total food intake in many (although not all) tropical groups.

This observation—that, outside the Arctic, plant foods form a substantial, and often the dominant, component of the foragers' diet—is not new. It was already clearly recognized over a century ago by Friedrich Engels, but for some reason was brushed aside by most anthropologists until the 1960s, when intensive fieldwork among the Kalahari San (Bushmen) in southern Africa and similar work among Australian Aborigines made it impossible for anthropologists to continue to ignore the importance of plant foods in the diet of most hunter–gatherers. In fact, the change in perspective was so dramatic that some anthropologists now insist that we refer to foraging populations as gatherer–hunters, rather than hunter–gatherers.

This change in perspective has also led anthropologists to question some of the basic assumptions of the classic hunting hypothesis. If plant foods constitute the bulk of the forager diet in many areas of the tropics today, perhaps this was also true of the early hominid diet in the distant past, since our earliest ancestors, the Australopithecines, evolved in the tropics and archaeological evidence indicates that hominids probably remained there for several million years, not expanding into the more temperate environments of Eurasia until about 1 million years ago or even more recently, long after the emergence of a more advanced form, *Homo erectus*. This observation, of course, in no way implies that carnivory played *no* role in the origins or evolution of early hominids. It does suggest, however, that the hunting hypothesis might be overly simplistic in downplaying or ignoring the role of plant foods (and, of course, nonsubsistence factors) in this process.

Contemporary hunter–gatherers have also provided other insights that are gradually transforming our views about the nature and evolutionary significance of human carnivory. For example, for many years anthropologists, like their colleagues in the medical and nutritional fields, felt that meat was important first and foremost as a source of high-quality protein. After all, meat is one of the richest natural sources of protein, and the protein is nutritionally ideal, because it possesses essential amino acids which the human body is incapable of synthesizing on its own and because the amino acids are present in proportions that are optimal for use by the body. In comparison, many plant foods are impoverished in protein, the protein might be difficult to digest, and one or more critical amino acids might be underrepresented or missing. [*See* PROTEINS (NUTRITION).]

From this perspective hunting was seen primarily as an efficient strategy for acquiring high-quality protein and protein-derived calories, and much of human evolution was thought to have been driven by technological and social developments that enhanced the success with which foragers were able to capture this critical and presumably limiting macronutrient. Fat—which generally accompanies meat as marbling within the muscle tissue; as subcutaneous layers along the back, neck, chest, and belly; as deposits around the internal organs; and as deposits within the marrow cavities of the limb bones—while an obvious source of calories, was seldom given adequate attention by scholars of human evolution, and (as discussed in the next section) has often received extremely "bad press" from nutritionists because of its probable links to the so-called diseases of civilization: obesity, atherosclerosis, high blood pressure, adult-onset diabetes, and heart disease.

There is no question that protein is an extremely important macronutrient, both as a source of amino acids necessary for normal growth and maintenance and as a source of calories. However, recent studies of living foragers, combined with new insights from medicine and nutrition, are beginning to suggest that we might have placed too much emphasis on the acquisition of high-quality protein while underestimating the importance of fat (and carbohydrate) to peoples whose sustenance came entirely from wild resources. As a consequence it has also become necessary to reexamine some of the basic assumptions of the traditional hunting hypothesis.

The tremendous importance of fat to foragers is immediately evident in their hunting and butchering decisions. For example, they frequently seek out and kill only the fattest animals, ignoring the lean ones, and they often then select just those cuts of meat and marrow bones that have the highest fat content, abandoning the remaining parts or feeding them to their dogs. Sometimes this behavior merely reflects an excess of meat from a particularly successful hunt that permits the participants the luxury of choice. However, foragers also abandon entire animals, if, upon butchering, they prove to be fat depleted, even when the hunters themselves are short of food, and despite the fact that they might already have invested considerable effort to locate and kill the animals. In other words, foragers often seem to be hunting for calories, not protein, and specifically for calories provided by a nonprotein source (i.e., fat). This behavior is evident year-round, but paradoxically it becomes most pronounced not when food is abundant, but during periods of food shortage, when the animals are leanest, the hunters' supply of vegetable foods has dwindled, and the hunters themselves are losing weight. Why don't the hunters eat all of the meat made available by a kill, regardless of its fat content, using the excess protein for energy? The nutritional reasons that underlie this seemingly counterintuitive behavior are providing valuable insights into the nature and evolution of human carnivory.

III. Protein and Human Nutrition

Travel accounts and ethnographies worldwide are filled with graphic descriptions of the almost gluttonous attitude of foragers when it comes to fat; they often devour the fattiest organs and marrow immediately at the kill, including hunks of fat sliced from the back fat, and whenever possible they consume rendered animal fat or fatty marrow mixed in with other foods at almost every meal. Among many foragers animal fat (including fish oil) is so highly prized that it is widely traded, and it is frequently used in rituals.

The intense interest in fat shown by hunter–gatherers seems to stand in stark contrast to the dire warnings of nutritionists and medical researchers, who see a close connection between high-fat diets and many of the diseases of civilization. Their views about the role of fat in the human diet are beginning to change, however, and hunter–gatherers have played an important part in this change. For example, traditional Eskimos consume prodigious amounts of fat, amounts far in excess of what most people in modern industrial nations eat, and yet they display virtually no signs of these diseases. One reason, of course, is that Eskimos are normally far more active than we are and burn off the calories that less active people would store as fat. They also burn off more calories just to keep warm.

But there is another reason that Eskimos and other foragers can consume so much fat without apparent health problems. Nutritionists have found that the fatty acid composition of fat from domestic animals is different from that of fat from wild animals, a reflection of the "unnatural" diet fed to the domestic forms, and it is the "domestic" fat, not fat per se, that appears to be most closely linked to the diseases of civilization. The concern we now show about the composition of fat in our diet is reflected, for example, by the fact that most of us eat margarine, which is high in polyunsaturated fats, rather than butter, which contains more saturated fats. The growing importance of fish oils rich in so-called ω-3 fatty acids reflects a similar concern. [*See* FATS AND OILS (NUTRITION).]

Nutritionists are also concerned, though, about the quantity of fat in the diet, not just its composition. How does one account then for the fact that all hunter–gatherers, not just Eskimos, display such an avid interest in fat? One reason, of course, is that fat makes foods taste good, since many flavor-enhancing substances are soluble only in fat, and it also conveys a feeling of satiety when one has eaten. Fat is also a critical source of both fat-soluble vitamins and essential fatty acids. While foragers are almost certainly unaware of the existence of either nutrient, they are well aware that the inclusion of adequate fat in their diet contributes to an overall feeling of well-being and good health. Moreover, fat is a highly concentrated source of energy, supplying more than twice the calories per gram than either protein or carbohydrate, and fat is more efficiently metabolized than protein.

Unlike the situation in modern industrialized nations, however, where fatty foods are abundant and readily available year-round at the neighborhood supermarket, in the world of foragers living outside of the Arctic fat is actually quite scarce. Most wild animals, even in prime condition, are generally much leaner than domestic ones, and during the late winter and spring (or late dry season and early rainy season in tropical latitudes), as forage abundance and quality decline, wild animals can become very lean. This is especially true for pregnant or nursing female animals supporting a developing fetus or newborn calf during periods when their own food intake is restricted. It is also true for calves, which often have difficulty getting enough forage during stressful seasons, and for older animals, whose teeth have begun to wear out, reducing their ability to masticate food.

Thus, fat is usually abundant only during seasons when animals are in prime condition, and at such times it is consumed by foragers with great zeal. During ensuing seasons of shortage, however, both animals and foragers become lean again. As a consequence obesity and other diseases of civilization that are linked to excessive fat intake are not a problem to foragers as long as they are living under traditional conditions in which food intake follows an annual "boom–bust," or feast–famine, cycle. Only when foragers become settled and acquire the "benefits" of a Western diet do they begin to show the same diseases of civilization that we do.

There is another important reason that animal fat is critical to hunter–gatherers, particularly during times of stress, when plant foods become scarce or unavailable. There appears to be an upper limit to the total amount of protein that one can safely consume on a regular basis. This limit—best expressed as the total number of grams of protein per unit of lean body mass that the body can safely handle—is about 300 g, or roughly 50% of one's total calories under normal nonstressful conditions. Protein intakes above this threshold, especially if they fluctuate sharply from day to day, can exceed the rate at which the liver can metabolize amino acids and the body can synthesize and excrete urea, leading ultimately to hypertrophy and functional overload of the liver and the kidneys, elevated (even toxic) levels of ammonia in the blood, dehydration and electrolyte imbalances, severe calcium loss, micronutrient deficiencies, and lean tissue loss.

Perhaps of greater significance for the success and long-term viability of foraging groups, however, is the recent suggestion by nutritionists that the safe upper limit to total protein intake for pregnant women might actually be considerably lower than 300 g. Several studies have shown that supplementation of maternal diets with protein in excess of about 20% of total calories usually leads to declines, not gains, in infant birth weight, and often to increases in perinatal mortality as well. Infants who are born prematurely appear to be most severely affected by such high maternal protein supplements. Declines in infant birth weight might be seen in situations in which maternal weight increases in response to the dietary supplementation, and the decline becomes more pronounced the greater the mother's protein intake above the 20% threshold. Birth weight also declines when the mother's total calorie intake is restricted, but is most extreme when the diet is both low in energy and very high in protein.

Thus, adult foragers must normally get more than one-half of their total daily calories from nonprotein sources in order to remain healthy, and pregnant women must get considerably more. When plant foods are abundant, this poses little problem (as long as the plant foods themselves are not excessively rich in protein), but during seasons or interannual periods of food shortage, when vegetable resources dwindle and wild animals become very lean, finding adequate nonprotein energy sources becomes increasingly difficult.

One obvious way that modern foragers cope with this problem is to become extremely selective in the animals they hunt and in the body parts they consume, discarding those parts that are too lean. They also invest considerable effort to extract the grease that is dispersed within the spongy bones, by smashing up joints and vertebrae and boiling out the precious lipids. Despite these efforts, by the end of the dry season foragers like the Kalahari San might be forced to consume almost 2 kg of lean meat per person per day (roughly 420 g of protein), and yet their body weight declines. This astounding meat intake, not uncommon among contemporary foragers, means that protein at times can contribute up to 70% or more of their daily caloric intake, a level far in excess of the 300 g suggested earlier as the safe upper limit, and therefore almost certainly a sign of stress, not affluence, as many anthropologists once thought. And for pregnant women, such high protein intakes could have devastating reproductive consequences.

Modern hunter–gatherers have the benefit of stone or steel axes to smash up bones and clay or iron pots to boil them in to extract the grease. These seemingly mundane items of material culture, absolutely critical to modern foragers during periods of resource stress, are actually comparatively recent additions to their technological inventory. Stone axes and ceramic vessels did not make their appearance until the very end of the Pleistocene at the earliest, and of course metal items are very recent introductions. The use of stone-boiling to extract the grease, a much more labor-intensive process in which stones are heated in a fire and then transferred to a watertight hide, gut, or basketry container, probably also dates to the latter part of the Pleistocene or later. Thus, prior to the Upper Pleistocene the grease in the spongy bones might have been largely, if not entirely, inaccessible to foragers, and as a consequence their reliance on hunting and meat-eating might have been quite different from the pattern we see among modern hunter–gatherers.

IV. Chimpanzee Carnivory

If the chimpanzee is our closest living relative, as most scholars now believe, then patterns of behavior shared by both chimpanzees and modern foragers should provide invaluable clues about the behavior of protohominids. Until the early 1960s, however, when Jane Goodall began her pioneering work in the Gombe National Park—a strip of rugged forested terrain along the eastern shore of Lake Tanganyika in Tanzania—little was actually known about free-ranging chimpanzees. Over the three decades since her studies began, our understanding of chimpanzees and other primates under natural conditions has improved immeasurably, and, not surprisingly, our views about the origins and evolutionary significance of human carnivory have been altered in the process.

In Darwin's time, and in fact right up to the 1960s, chimpanzees were thought to be vegetarians, actually frugivores, animals whose diet includes large quantities of fruits. As field studies proceeded, however, it became clear that chimpanzees also ate insects, especially termites, which they collected from nests using simple "fishing" tools made from grass stems. Perhaps the most dramatic change in our views of the chimpanzee diet came when researchers first began to observe these supposedly peaceful vegetarians killing and eating mammals, especially the young of several species of monkeys and small antelopes. Chimpanzees were clearly omnivores, not vegetarians.

Until the chimpanzees became habituated to the presence of their human observers, however, researchers had great difficulty following them closely enough to get reliable information on the frequency of meat-eating or the conditions which triggered the behavior. Thus, in the early stages of the Gombe research, the chimpanzees were provisioned with bananas at a central feeding station so that they could be observed more easily. Unfortunately, the concentrated food resource provided by the bananas also attracted baboons, a situation which often led to intense competition and conflict between the two primate species. Many of the cases of meat-eating seen in the early stages of research at Gombe occurred when chimpanzees killed and consumed immature baboons near the provisioning site. Most researchers therefore tended to attribute chimpanzee meat-eating to the competitive environment artificially created by food provisioning.

Over the years, however, more and more incidences of killing and meat-eating were observed, and they continued to be reported long after provisioning had ceased. Moreover, cases of killing and meat-eating began to be reported from other chimpanzee study sites as well, in both Central and West Africa. As field studies progressed it became evident that chimpanzee predation was neither just an occasional event nor an artifact of provisioning, but occurred repeatedly and with design. It is now clear that chimpanzees actually hunt deliberately, often cooperatively, and share meat with other group members. In fact, recent work in the Ivory Coast has shown that West African chimpanzees hunt, on average, almost once every 3 days, far more frequently than do their counterparts in Gombe. They also hunt in larger groups, focus more on adult prey, share the kill more actively and widely, and in at least one case consumed an estimated 1.4 kg of meat and bone per individual. And while chimpanzees do not use tools to kill their prey, the Ivory Coast chimpanzees occasionally use sticks to extract marrow from the limb bones.

These ongoing studies of chimpanzees in the wild clearly demonstrate that our closest living relatives hunt and share meat, behaviors that until quite recently anthropologists had reserved solely for humans. In fact, it now seems likely that the common ancestor of both chimpanzees and humans already hunted, making it improbable that protohominids were the first primates to "discover" carnivory, a discovery which supposedly then transformed them from apes into hominids. The insights from chimpanzee studies have in no way made the hunting hypothesis obsolete, however. Instead, many anthropologists now feel that the explanation for hominid origins lies not in the discovery of hunting per se, but in the quantity of meat that early hominids were able to procure; in other words, in their success at locating and killing large game.

V. Hunting and Scavenging in Prehistory

Since the pioneering work by Mary and Louis Leakey in the 1950s and 1960s at Olduvai Gorge in Tanzania, our understanding of the archaeological record of early hominids has improved dramatically. Many well-preserved sites have been painstakingly excavated, mapped, and analyzed, providing tantalizing "snapshots" of the behavior of our earliest ancestors as much as 2 million years ago. In case after case the sites revealed the close juxtaposition of sharp-edged stone flakes and heavy-duty chopping or pounding tools with the bones of many different species of animal, ranging in size from rodents, birds, and turtles to huge dangerous ones such as buffaloes, hippopotamuses, and relatives of the modern elephant. The repeated association of tools and bones seemed to be incontrovertible proof that early hominids were avid and highly successful hunters and provided one of the principal cornerstones of the traditional hunting hypothesis.

While archaeology was uncovering the hard evidence of past human activities, another field—taphonomy—was beginning to explore other issues that would ultimately force archaeologists to reexamine many of their basic assumptions. Taphonomy is the study of the processes that can alter and distort an assemblage of bones from the time an animal dies until its bones become incorporated into the fossil record. Thus, for example, grassland-dwelling animals (e.g., wildebeest) might drown while crossing a stream, becoming fossilized in fluvial deposits alongside the bones of crocodiles, hippos, and other water-loving species with which wildebeest, in life, are obviously unassociated; softer less dense animal bones, such as the proximal ends (i.e., those closest to the body) of femurs, tibias, and humeri, might decay more readily than more durable ones, such as distal humeri or teeth; smaller bones (e.g., those of rodents) are more likely than larger ones (e.g., from antelopes or elephants) to be trampled into the substrate and preserved; lighter less dense bones are more likely than compact ones to be winnowed away by a flowing stream; hyenas might selectively destroy some of the marrow bones and remove others from a hominid campsite after the hominids have abandoned the locality; and so forth.

Taphonomic studies began to make it clear that the mere juxtaposition of stone tools and bones did not demonstrate their functional association. Flowing water could have brought the two together in a channel deposit, or hyenas and humans could each have taken advantage of the same shade tree, but at slightly different times, producing a fortuitous association of bones and tools. Because of these basic taphonomic questions, archaeologists found themselves almost having to start from scratch, demonstrating through painstaking analyses what they had previously merely assumed. While tedious, this process of reevaluation, which is still going on, has proven to be extremely fruitful.

For example, we now know that some, but by no means all, of the early sites where stone tools and bones occur together do, in fact, represent places where hominids butchered animal carcasses. The best evidence is provided by unambiguous cut marks on many of the bones, produced by sharp-edged flakes, and by use–wear studies of some of the stone tools themselves, which reveal distinct polishes on their edges, shown by experimental work to be the product of meat-cutting. Other, somewhat less direct and hence more controversial, evidence includes the fact that bones of many species occur together in a single place, a pattern different from that normally found at hyena or lion kills. The proportions of different skeletal elements in some of these sites make it unlikely that the bones were transported there by hyenas or other carnivores.

The taphonomic studies of these sites have raised a different issue, however—one that has far-reaching implications for the hunting hypothesis. Many of the bones at these sites have gnaw marks, punctures, and other clear evidence of carnivore damage. Moreover, even in sites where it can be shown convincingly that flowing water and differential decay played no significant role in altering the composition of the bone assemblages, less dense limb elements with lots of spongy tissue (e.g., the proximal ends of femurs, tibias, and humeri) are, nevertheless, conspicuously underrepresented. Detailed studies of the feeding behavior of many different carnivores suggest that these bones have most likely been destroyed or carried off by hyenas.

These observations have raised an even more fundamental question, one that had not been seriously considered until quite recently. Did humans kill the animals, and hyenas scavenge the remains that littered the campsite once its human occupants had left? This scenario, of course, is entirely compatible with the traditional archaeological view. Or did lions or hyenas kill the animals, and humans merely scavenge the carcasses for edible scraps of meat and marrow after the carnivores had finished feeding or were driven off by the humans? Many scholars now believe the latter is more likely.

If early hominids were basically scavengers, not hunters, what role did scavenging play in hominid origins and evolution? The answer to this question remains far from resolved, but certain facets of the problem are beginning to become clear. One facet concerns the kinds of scavenging opportunities that would have been available to early hominids as they began to exploit the expanding woodlands and open grasslands of sub-Saharan Africa. Ongoing studies of hyena scavenging in the wild are providing some interesting insights.

Hyenas are voracious carnivores, hunting as well as scavenging from carcasses killed by other predators, and in a matter of minutes they can consume the entire carcass, bones and all, of a small-sized antelope, such as a gazelle. Not unexpectedly, however, the larger the carcass, the more of it the hyenas are likely to leave behind, other things being equal (e.g., the level of competition among hyenas, or between hyenas and other predators, at the site of the carcass). Thus, hyenas might consume most of the carcass of medium-sized animals, such as wildebeest and zebras (live weights between about 100 kg and 350 kg), but abandon the vertebrae, skull, and lower limbs partially or largely intact. From these elements early hominids would have been able to scavenge scraps of edible muscle tissue, marrow, and, of course, brain. Sharp flakes would have made it possible for them to open the skin around the limb bones in order to get at the marrow within the shaft; a stone or limb bone could have been used as a hammer to break open the shaft or to open the skull to get at the brain. Without appropriate boiling technology, however, grease in the spongy tissue of the limb bones and vertebrae probably would have been largely inaccessible.

Hyenas often abandon the largest carcasses—those weighing in excess of about 350 kg (e.g., the buffalo, rhinoceros, giraffe, and elephant)—more or less intact. These would have provided early hominids with a bonanza of meat and marrow, but kills or natural deaths of these megafauna are infrequent events.

These observations suggest that early hominid scavengers would have been most likely to encounter the partial remains of medium-sized animals, from which they could have gleaned scraps of meat and marrow, particularly from the lower limbs, and perhaps brain. If they transported the edible parts back to a central place to process them in comparative safety (and perhaps to share with other members of the group), these are the skeletal elements that we can expect to find in greatest abundance in Pliocene–Pleistocene archaeological sites.

An important issue that remains unresolved, however, is whether early hominids were passive or active scavengers, that is, whether they had to wait until hyenas and other predators had finished feeding on a carcass or instead were able to drive them away and thereby get earlier access to it. If they were active scavengers, they would have been able to transport back to their home base many more of the less marginal meaty parts, such as the upper limbs and the rump.

The difficulty in resolving whether early hominids were primarily active or passive scavengers is largely methodological at this point. Understandably, faunal analysts generally focus on those bone fragments that can be identified reliably to species. Unfortunately, usually only the joints (i.e., epiphyses) of the limbs are suitable; shaft fragments are virtually impossible to identify to species. The proximal epiphysis of the humerus and the tibia and both epiphyses of the femur, however, contain lots of spongy tissue that is relished by carnivores and easily destroyed. This means that even if early hominids regularly transported these meaty elements back to their camp, hyenas would almost certainly carry them off as soon as the camp was abandoned. The result is that, to the faunal analyst, the archaeological bone assemblage would appear to be devoid of meaty upper limbs.

To get around this problem, archaeologists are now beginning the painstaking task of identifying the thousands of shaft fragments, obviously not to species, but to approximate carcass size. The initial results have been intriguing but controversial. They suggest that, at least on some early hominid sites, upper limbs originally might have been fairly well represented, but only shaft fragments now remain. This indicates that early hominids might have been active scavengers capable of driving away predators from carcasses before the meatiest portions had been totally devoured. On the basis of these shaft fragment studies, some archaeologists have even suggested that early hominids might have hunted these medium-sized animals, a position that, if correct, brings us full circle to where we began when taphonomic insights first challenged the hunting hypothesis.

Given the considerable evidence now available for successful and regular hunting by chimpanzees, most notably by those in the Ivory Coast, it seems extremely likely that our Australopithecine ancestors were, in fact, successful hunters, at least of the small and medium-sized classes, although it remains a matter of conjecture whether these small hominids were capable of intimidating hyenas and lions and driving them away from freshly killed carcasses.

Scavenging studies have revealed another interesting facet of hyena behavior. During the rainy season many herbivores move away from permanent water sources into the open grasslands, where they rely on temporary pools of water. In these open habitats hyenas are extremely aggressive scavengers (and hunters) and often leave little remaining of a carcass that could then be scavenged by hominids. Groups of hyenas might even drive lions away from a carcass before they have finished feeding. In contrast, during the dry season, when many herbivores stay much closer to permanent water sources, hyenas often leave lion kills untouched or only partially devoured. The reason for this seasonal difference in hyena feeding behavior appears to be related to their fear of lions, which are much more likely to ambush them in the dense thickets and woodlands near water courses than out in the open. This suggests that the opportunities for hominid scavenging are likely to have been greater during the dry season than during the rainy season and close to water courses (assuming, of course, that early hominids were less intimidated in vegetated areas by lions than modern hyenas seem to be). In addition, the dry season is also the time when very young and very old or sickly animals are most likely to die of hunger, disease, or other natural causes, providing additional potentially scavengable carcasses.

The dry season is also the time when both hominids and herbivores are most likely to be suffering from resource shortages and hence are relatively lean. This would also be the time, therefore, when hominids would have to be most selective in their scavenging, targeting especially the few parts remaining on a carcass that retain fat—the brain, the mandible, and the marrow bones of the lower limbs, the last elements in the legs of a severely stressed animal to become fat depleted.

Thus, answers to these critical and interesting questions—whether early hominids were hunters or scavengers and, if the latter, whether they were active or passive scavengers; also, whether they procured meat year-round or focused primarily on scraps of fatty tissue gleaned from carcasses during periods of food stress—hinge ultimately on the ability of archaeologists to find ways to determine, among other things, whether upper limbs were actually present in substantial numbers at early hominid sites and on their success in establishing the seasonality of these ancient occupations. Both are seemingly straightforward methodological tasks, but both so far have proven to be exceedingly difficult.

The many profound changes in our ideas about early hominid carnivory are also affecting our views about hunting and meat-eating in later stages of human evolution. Until quite recently, the traditional evolutionary scenario went more or less like this: Once early hominids had learned to hunt animals, much of subsequent human evolution revolved around improvements in the effectiveness with which they were able to kill more and larger game, a positive-feedback relationship that involved increasing intelligence, better forms of communication (and ultimately language), more effective intragroup cooperation, and, of course, better weaponry. Already by 500,000 years ago, or even earlier, hominids—by this time *Homo erectus*—were believed to have been highly successful big-game hunters, killing animals as large and dangerous as buffaloes, rhinosceroses, and even elephants. One site of this time period in Spain, known as Torralba, was even thought to have been the location of a communal elephant drive, where many of these huge beasts were supposedly stampeded into a bog and killed there by hunters.

In the past few years, however, many of the sites on which this evolutionary scenario was based have been reexamined from a taphonomic perspective, and, perhaps not surprisingly, the evidence for big-game hunting is being called into question. The consensus that has emerged seems to be that even *Homo erectus* was primarily a scavenger rather than a hunter, at least of larger animals, although this conclusion remains far from proven.

Debate now centers more on the hunting prowess of Neanderthals (*Homo sapiens neanderthalensis*), the last occupants of Europe and the Near East prior to the appearance of anatomically fully modern humans, some 35,000–40,000 years ago. Sites dating to the period of the Neanderthals—referred to by archaeologists as the Middle Paleolithic (or middle Old Stone Age)—often contain the remains of large mammals, including wild cattle, horses, bison, mammoths, and giant cave bears. Some of these sites, in fact, contain dozens or even hundreds of individuals of a single species, suggesting that Neanderthals not only could kill these large animals, but perhaps did so in large communal drives.

At the moment, however, opinion is divided concerning the proper interpretation of the faunal remains. Some archaeologists feel that many of the large mammal remains, particularly those found in cave sites, have little or nothing to do with humans, but were instead dragged into the cave by hyenas that made their dens there after the human occupants had left these sites. These archaeologists argue that Neanderthals probably hunted only the smaller more docile species (e.g., deer), scavenging from the carcasses of the larger ones. True big-game hunting, in this view, emerged late in human evolution, probably not until the appearance of anatomically fully modern humans, ca. 35,000 years ago. Others feel that the transition to full-fledged hunting economies occurred somewhat earlier, perhaps midway through the Middle Paleolithic, roughly 70,000–90,000 years ago.

At stake in this debate is not just the issue of whether anatomically premodern humans, perhaps as recently as the Upper Pleistocene, were scavengers rather than hunters, but whether they actually possessed the necessary cognitive (and, of course, technological) sophistication to successfully plan, coordinate, and carry out a communal hunt of extremely large dangerous prey. Answers to questions such as these will bring us much closer to understanding how a small-brained quadrupedal ape was transformed into the unique creature that we are today.

Bibliography

Binford, L. R. (1981). "Bones: Ancient Men and Modern Myths." Academic Pess, New York.

Blumenschine, R. J. (1987). Characteristics of an early hominid scavenging niche. *Curr. Anthropol.* **28,** 383.

Boesch, C., and Boesch, H. (1989). Hunting behavior of

Bunn, H. T. (1986). Patterns of skeletal representation and hominid subsistence activities at Olduvai Gorge, Tanzania, and Koobi Fora, Kenya. *J. Hum. Evol.* **15,** 673.

Eaton, S. B., Shostak, M., and Konner, M. (1988). "The Paleolithic Prescription." Harper & Row, New York.

Goodall, J. (1986). "The Chimpanzees of Gombe." Belknap, Cambridge, Massachusetts.

Gordon, K. D. (1987). Evolutionary perspectives on human diet. *In* "Nutritional Anthropology" (F. E. Johnston, ed.), pp 3–39. Liss, New York.

Lee, R. B., and DeVore, I. (eds.) (1968). "Man the Hunter." Aldine, Chicago, Illinois.

O'Connell, J. F., Hawkes, K., and Jones, N. B. (1988). Hadza scavenging: Implications for Plio/Pleistocene hominid subsistence. *Curr. Anthropol.* **29,** 356.

Speth, J. D. (1987). Early hominid subsistence strategies in seasonal habitats. *J. Archaeol. Sci.* **14,** 13.

Speth, J. D. (1989). Early hominid hunting and scavenging: The role of meat as an energy source. *J. Hum. Evol.* **18,** 329.

Cartilage

JOSEPH A. BUCKWALTER, *University of Iowa*

Glossary

Appositional growth Enlargement of cartilage by the addition of new cells and matrix to the surface of the tissue

Collagens A family of 12 or more proteins that consist, at least in part, of helical amino acid chains. Some assume the form of fibrils and give cartilages their form and tensile strength; others do not form fibrils and might help organize and stabilize the meshwork of fibrillar collagens.

Elastic cartilage Type of cartilage distinguished by a high concentration of elastin in the matrix; forms parts of the external ear, epiglottis, and laryngeal and bronchiolar cartilages

Elastin Protein that forms fibrils and sheets that can be deformed, without rupturing or tearing, and then return to their original shape and size

Enchondral ossification Replacement of cartilage by bone through cartilage mineralization, cartilage resorption, and bone formation

Fibrous cartilage Type of cartilage distinguished by a high concentration of type I collagen fibrils in the extracellular matrix; forms parts of the intervertebral disk, pubic symphysis, tendon and ligament insertions, and intraarticular menisci

Glycoproteins and noncollagenous proteins Molecules consisting of protein and a small number of monosaccharides or oligosaccharides; some might help organize and maintain the macromolecular framework of the cartilage matrix and help chondrocytes adhere to the matrix framework

Glycosaminoglycans Polysaccharide chains formed from repeating dissacharide units containing a derivative of either glucosamine or galactosamine and at least one negatively charged carboxylate or sulfate group; those found in cartilage include hyaluronic acid, chondroitin 4-sulfate, chondroitin 6-sulfate, dermatan sulfate, and keratan sulfate

Hyaline cartilage Type of cartilage distinguished by high concentrations of type II collagen fibrils, large aggregating proteoglycans, and water. In the fetus it forms most of the skeleton; during skeletal growth it forms the cartilaginous growth plates, or physes; in adults it persists as the nasal, laryngeal, bronchial, articular, and costal cartilages; and at any age it participates in healing some types of bone and cartilage injuries.

Interstitial growth Enlargement of cartilage by the addition of new cells and matrix within the substance of the tissue

Link proteins Small noncollagenous proteins that stabilize proteoglycan aggregates, increase the degree of aggregation, increase the size of aggregates, and influence the spacing of proteoglycan monomers in aggregates

Perichondrium Layer of cells and matrix that covers some cartilage surfaces

Permeability Ease with which water can flow through the cartilage matrix; measured by the force required to cause water to flow through cartilage

Proteoglycan aggregates Molecules formed by the noncovalent association of multiple proteoglycan monomers, multiple link proteins, and a hyaluronic acid filament

Proteoglycan monomers Molecules consisting of glycosaminoglycans covalently bound to protein. Cartilage proteoglycan monomers exist in the form of large aggregating monomers, large nonaggregating monomers, and small nonaggregating monomers

Viscoelasticity State of having both viscous and elastic properties. Viscous fluids tend to flow slowly because of friction between component molecules as they slide past one another. Elastic solids tend to regain their original shape after being compressed. The response of a viscoelastic material to loading can be modeled as the combined responses of a viscous fluid and an elastic solid. When subjected to constant load or constant deformation, the response of the material varies with time.

BEGINNING WITH fetal life, cartilage, a connective tissue consisting of an abundant extracellular matrix and specialized cartilage cells (chondrocytes) demonstrates remarkable capacities for interstitial and appositional growth, changing shape as it grows. In some locations the development of cartilage, followed by cartilage mineralization and resorption, makes possible the formation of bone. Throughout life the unique material properties of cartilage—including durability, stiffness, resiliency, and viscoelasticity—make possible the normal function of the musculoskeletal, auditory, and respiratory systems.

Cartilage or precartilaginous tissues first appear in the fifth week of human prenatal life, form the initial wholly cartilaginous skeleton, grow rapidly, and remodel as they grow. At about the eighth week of prenatal life, the cartilage begins to be replaced by bone in many areas. During fetal life, infancy, and adolescence, growth plate or physeal cartilage produces longitudinal bone growth and is replaced by bone. When skeletal growth ceases, the growth plates disappear, but other kinds of cartilage remain as essential structural components of the external ear, the respiratory system (including the nasal, laryngeal, tracheal, and bronchial cartilages), the skeletal system (including intervertebral disks, menisci, costal cartilages, parts of the insertions of tendon and ligaments into bone, and the remarkably durable, almost frictionless, surfaces of synovial joints). At any age cartilage participates in the healing of bone and cartilage injuries.

Like other dense organized connective tissues, including tendon, ligament, and bone, cartilage consists of a sparse population of mesenchymal cells embedded within an abundant extracellular matrix. In most mature cartilage the cells contribute about 5% to the total tissue volume, and the matrix contributes about 95%. The roughly spherical shape of the cartilage cells, or chondrocytes, and the unique composition of the matrices they synthesize and assemble distinguish cartilage from the other connective tissues. Although cartilage lacks nerves, lymph vessels, and blood vessels, chondrocytes are metabolically active and respond to changes in hormonal balance, availability of nutrients, oxygen tension, and mechanical loads. With the exception of the exposed surfaces found in synovial joints and the junctions of cartilage and bone, a thin layer of fibrous tissue, or perichondrium, covers cartilage and separates it from other tissues.

The matrix component molecules—including collagens, elastin, proteoglycans, and noncollagenous proteins—and the organization of these matrix macromolecules give the tissue its material properties. Differences in matrix composition and organization, with resulting differences in appearance and material properties, distinguish three types of human cartilage: hyaline, fibrous, and elastic. Hyaline cartilage, the most abundant type, has been more extensively studied than the others, so most concepts of cartilage are based on this type. As a result many authors refer to hyaline cartilage simply as "cartilage," obscuring the differences among the three types of cartilage.

I. Formation, Growth, and Maturation

Cartilage forms from groups of mesenchymal cells that cluster together. Each cell then assumes a spherical shape and begins to synthesize and secrete macromolecules that form a cartilaginous matrix. Subsequent promotion and maintenance of chondrogenesis depend on interactions among the cells, and between the cells and the molecules of extracellular matrix. Matrix accumulation separates the cells, and a perichondrium usually forms, marking the boundary between the cartilage and the surrounding tissues. The outer layer of perichondrium consists of fibroblastlike cells and a fibrous matrix; the inner layer contains more spherical cells, with the appearance of immature chondrocytes (chondroblasts). [*See* EXTRACELLULAR MATRIX.]

Cartilage volume can increase either by appositional growth (i.e., the addition of new cells and matrix at the tissue surface from the perichondrium) or by interstitial growth (i.e., growth within the tissue). Interstitial growth occurs by cellular proliferation, synthesis of new matrix, cell enlargement due to synthesis of new cell organelles and cytoplasm, and cell, and possibly matrix, swelling. The plasticity of the cartilage matrix makes interstitial growth possible. Tissues with rigid matrices (e.g., bone or mineralized cartilage) can only increase their volume by appositional growth and therefore usually cannot grow as rapidly or change shape as easily.

During cartilage formation and growth, the cell density is high, as the chondrocytes proliferate rapidly and synthesize large volumes of matrix. With maturation matrix synthesis slows, cell density declines, and, once growth ceases, chondrocytes rarely divide. In mature tissue the cells decrease their synthetic activity to the level required to maintain the matrix. With aging cell death further decreases cell density and the composition and structure of at least some matrix macromolecules change. [*See* CELL DEATH IN HUMAN DEVELOPMENT.]

II. Composition

Cartilage consists of three components: chondrocytes, extracellular water, and macromolecules that form the framework of the extracellular matrix.

A. Chondrocytes

Unlike tissues such as liver, muscle, or kidney, the primary functions of cartilage depend on the matrix, not the cells, and the cells form only a small part of the tissue. However, the cells form, maintain, and modify the matrix. Like most mesenchymal cells, chondrocytes surround themselves with the matrix they synthesize and assemble, and they do not form contacts with other cells. They contain the organelles responsible for matrix synthesis, including endoplasmic reticulum and Golgi membranes, and also frequently contain intracytoplasmic filaments and glycogen. At least some chondrocytes have a cilium that extends into the matrix and might sense changes in the matrix.

A close association exists between chondrocyte shape and cell function. With the exception of the flattened ellipsoid cells found immediately deep to the perichondrium, in the superficial zone of articular cartilage and in the proliferative zone of growth plate, chondrocytes have a spherical shape. During chondrogenesis, when undifferentiated mesenchymal cells assume a spherical shape, they generally begin to synthesize the cartilage-specific collagens and proteoglycans. Experimentally, when they lose their spherical shape, they stop making cartilage-specific molecules.

The relationship between chondrocytes and their matrix does not end with the synthesis and assembly of the matrix molecules. Maintenance of normal composition, structure, and function of the tissue depends on continual complex interactions between the chondrocytes and their matrix. Normal degradation of matrix macromolecules, especially proteoglycans, forces the cells to synthesize new matrix components to preserve the tissue. Chondrocytes sense the loss of matrix molecules caused by disease or injury and attempt to replace them. If they fail, the matrix deteriorates and the chondrocytes could die. The matrix limits the types and concentrations of nutrients, hormones, or drugs that can reach the chondrocytes and protects the cells from injury due to normal mechanical loading.

In addition, the matrix acts as a mechanical signal transducer for the chondrocytes. Deformation of the matrix generates signals that are transmitted through the matrix and cause cells to alter their synthetic activity. The signals might be purely mechanical (e.g., pressure or tension on the cell membrane), but deformation of the matrix could generate other types of signals, including changes in electrical potential or ion fluxes. Through these mechanisms, or possibly others, the synthetic function of chondrocytes and, thus, maintenance of the normal matrix composition depend on at least some loading of the tissue.

B. Tissue Fluid

The tissue fluid forms the largest component of cartilage; depending on the type of cartilage and its age, water can contribute up to 80% of the wet weight of cartilage. It contains dissolved gases, small proteins, and metabolites. Because the cartilage water is not contained by cell membranes, its volume, concentration, organization, and behavior depend on its interaction with the structural macromolecules. The nutrients and metabolites essential for chondrocyte function must pass through matrix fluid, and, if the macromolecular framework did not maintain and organize the tissue water and impede its flow through the matrix, cartilage would lose its resiliency and ability to resist compression.

C. Structural Macromolecules

The structural macromolecules that contribute 20–40% of the wet weight of cartilage include fibrillar molecules (e.g., collagens and elastin) and nonfibrillar, or ground substance, molecules (e.g., pro-

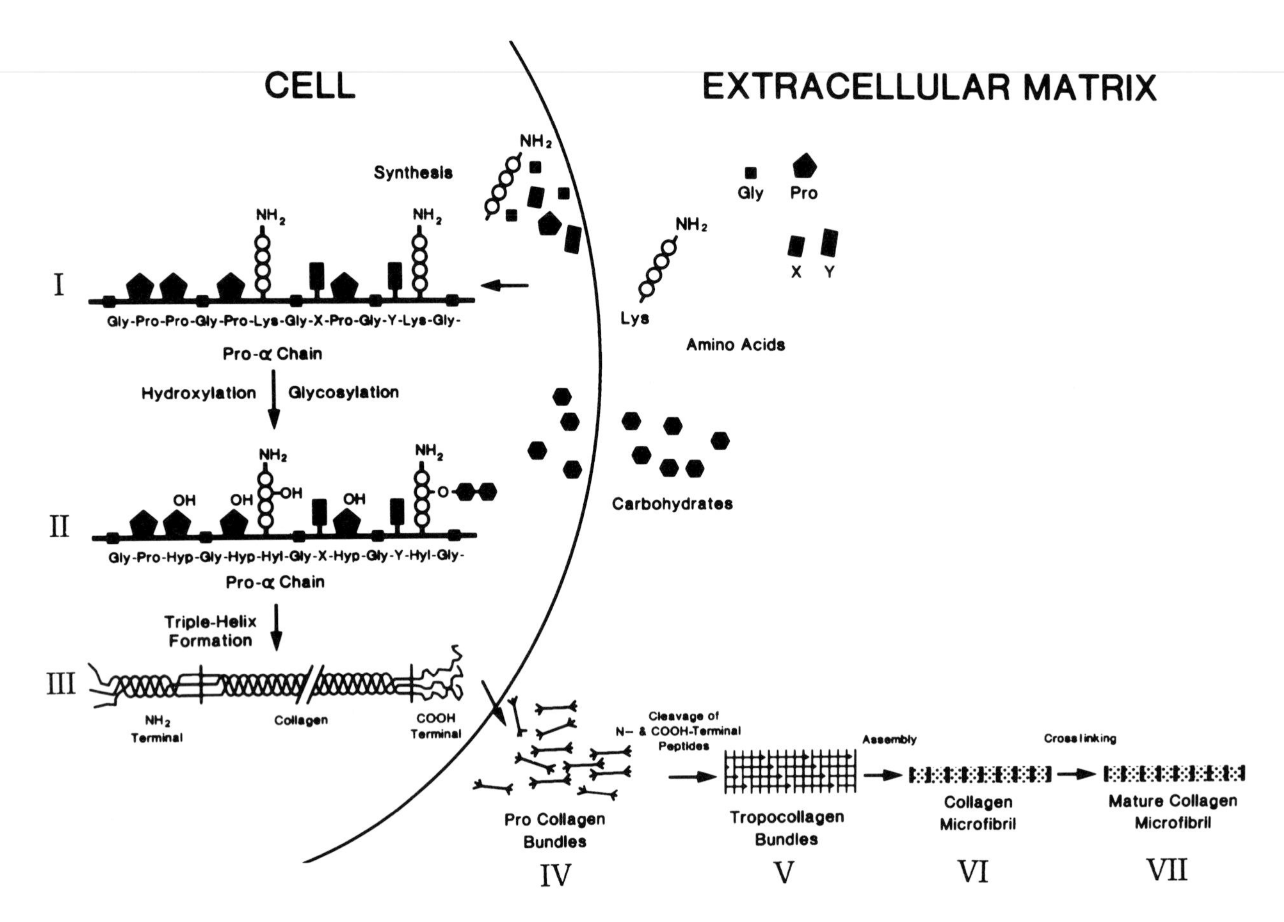

FIGURE 1 The synthesis, secretion, and matrix assembly of fibrillar collagen. (I) The cell links amino acids into Pro-α amino acid chains and then (II) adds carbohydrates. (III) Three Pro-α chains wind around each other to form the collagen triple helix, and (IV) the cell secretes the procollagen into the matrix. Here (V), procollagen peptidases cleave the amino (N)- and carboxy (COOH)-terminal peptides to form the collagen or tropocollagen molecule. (VI) Tropocollagen molecules then align themselves in a staggered arrangement to create collagen microfibrils. (VII) Crosslinking within and among collagen molecules strengthens the microfibril. Microfibrils aggregate to form the collagen fibrils seen by electron microscopy. [Reproduced from J. A. Buckwalter and R. R. Cooper, The cells and matrices of skeletal connective tissues. *In* "The Scientific Basics of Orthopaedics" (J. A. Albright and R. A. Brand, eds.), p. 19. Appleton & Lange, East Norwalk, Connecticut, 1987.]

teoglycans and noncollagenous proteins). The cells synthesize these macromolecules from amino acids and sugars; available evidence suggests that all chondrocytes can synthesize each type of molecule. [*See* COLLAGEN, STRUCTURE AND FUNCTION; ELASTIN.]

1. Collagens

Collagens, a family of 12 (types I–XII) or more distinct protein molecules, consist, in part, of helical amino acid chains (Fig. 1). In addition to classification by differences in amino acid composition, collagens can be classified into fibrillar interstitial collagens and nonfibrillar collagens by their form and location. In most tissues, including cartilage, fibrillar collagens form a much larger proportion of the matrix than do nonfibrillar collagens, and for this reason the nonfibrillar collagens have been referred to as "quantitatively minor collagens."

Fibrillar collagens appear as crossbanded ropelike fibrils when examined by electron microscopy (Figs. 1 and 2). They give cartilage its tensile strength and form and create an extracellular meshwork for cell attachment and binding or entanglement of the other matrix macromolecules. The recognized fibrillar collagens include types I, II, III, V, and XI. Other collagens, including types IX and XII, associate with the cross-striated fibrils formed by fibrillar collagens. The two most common fibrillar collagens—types I and II—have different distributions within the tissues. Type I collagen forms the fibrillar matrix meshwork of skin, tendon, ligament, bone, meniscus, fibrous cartilage, and annu-

FIGURE 2 Electron micrograph of a human infant annulus fibrosus showing the ropelike crossbanded collagen fibrils paralleling an immature elastic fiber consisting of amorphous elastin and glycoprotein microfibrils. Notice that patches of amorphous elastin (*) have accumulated within the more darkly stained microfibrils. [Reproduced from J. A. Buckwalter and R. R. Cooper, The cells and matrices of skeletal connective tissues. *In* "The Scientific Basics of Orthopaedics" (J. A. Albright and R. A. Brand, eds.), p. 21. Appleton & Lange, East Norwalk, Connecticut, 1987.]

lus fibrosis. Type II collagen fibril forms the collagen meshwork of hyaline cartilage, nucleus pulposus of the intervertebral disk, and vitreous body of the eye and has a higher hydroxylysine content. In general, tissues such as hyaline cartilage, containing primarily type II collagen, have a relatively high concentration of water and ground substance molecules, which gives them a glassy, translucent, or clear appearance. Tissues such as fibrous cartilage, containing primarily type I collagen, tend to have a lower concentration of water and ground substance molecules, which makes them more opaque.

The nonfibrillar collagens might lie close to cells, form part of basement membranes, or associate with other matrix macromolecules, including fibrillar collagens. When examined by electron microscopy, they can appear as fine threads or filaments that can form networks or matlike structures. Some types of nonfibrillar collagens can be found directly adjacent to or in the region of the chondrocyte cell membranes, where they form part of the pericellular matrix. Others help organize and stabilize the extracellular matrix, including the fibrillar collagen meshwork.

2. Elastin

Although elastin, like collagen, can form protein fibrils (Figs. 2 and 3), the fibrils lack the crossbanding pattern and differ in amino acid composition, conformation of the amino acid chains, and material properties. Elastin can also form sheetlike structures. In any shape, elastin can undergo some deformation, without rupturing or tearing, and when unloaded it can return to its original shape and size.

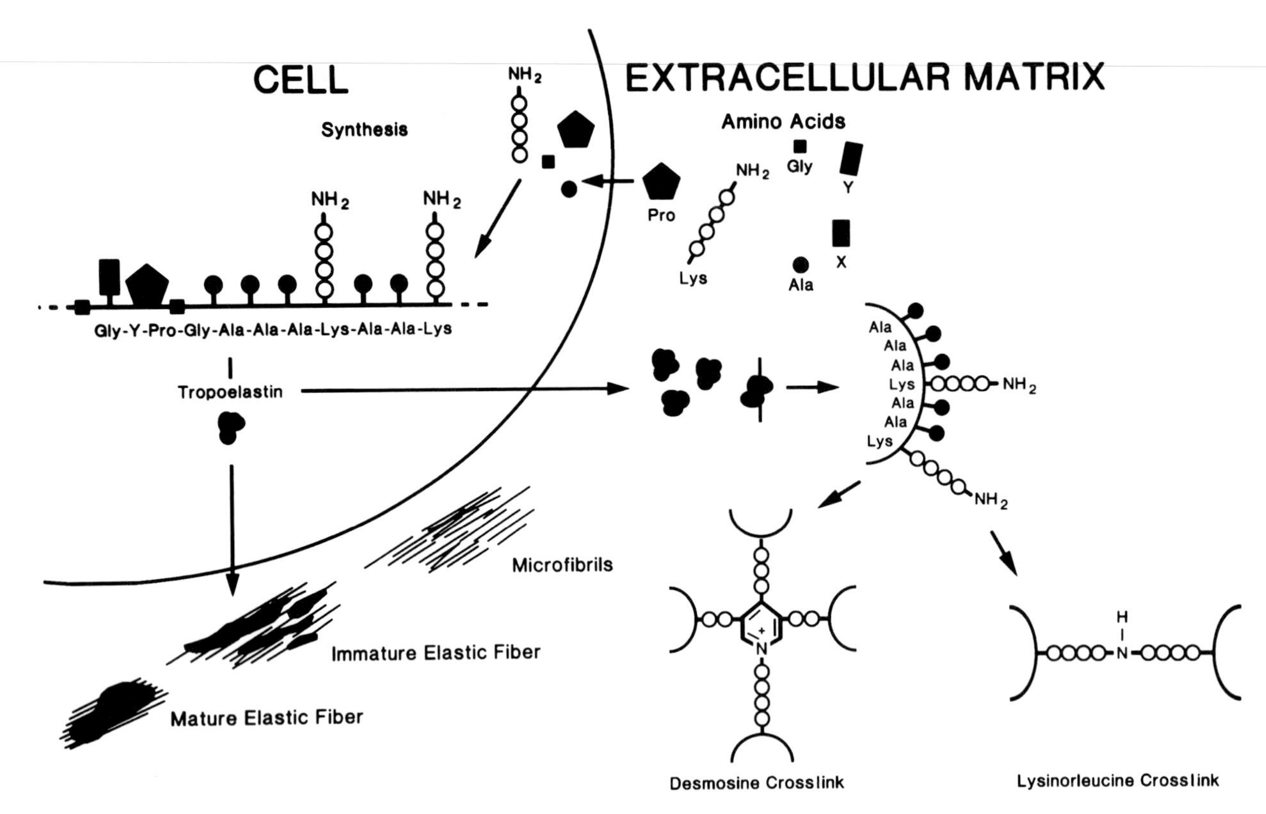

FIGURE 3 The synthesis, secretion, and matrix assembly of elastic fibers. Immature elastic fibers consist almost entirely of glycoprotein microfibrils. Amorphous elastin progressively accumulates in the region of the microfibrils until the mature elastic fiber consists almost entirely of elastin, with a few peripheral microfibrils. The crosslinking of the amino acid chains help give elastin its special material properties. [Reproduced from J. A. Buckwalter and R. R. Cooper, The cells and matrices of skeletal connective tissues. *In* "The Scientific Basics of Orthopaedics" (J. A. Albright and R. A. Brand, eds.), p. 21. Appleton & Lange, East Norwalk, Connecticut, 1987.]

Like collagens, elastin consists primarily of a protein with a small carbohydrate component. However, elastin contains little hydroxyproline, no hydroxylysine, and two unique amino acids: desmosine and isodesmosine. Unlike the precise ordered arrangement of the amino acid chains in helical collagen molecules, the elastin amino acid chains assume a variety of crosslinked random coil conformations that allow the molecule to stretch and recoil without rupturing. Elastin usually first appears within aggregates of glycoprotein microfibrils (Fig. 3). Elastin then accumulates until the enlarging elastic fiber appears to consist of a central region containing only elastin surrounded by a thin layer of microfibrils. Only in elastic cartilage, large blood vessels, and some ligaments (e.g., the nuchal ligament and ligamentum flavum of the spine) does elastin significantly contribute to the structure and material properties of human tissue.

3. Proteoglycans

Proteoglycan monomers, molecules consisting of polysaccharide chains bound to protein, are the major macromolecule of the cartilage ground substance. They contain little protein (about 5%) and consist primarily of special polysaccharide chains called glycosaminoglycans. These, in turn, consist of repeating dissacharide units containing a derivative of either glucosamine or galactosamine. Each dissacharide unit also contains at least one negatively charged carboxylate or sulfate group, so that the glycosaminoglycans form long strings of negative charges. Cartilage glycosaminoglycans include hyaluronic acid, chondroitin 4-sulfate, chondroitin 6-sulfate, dermatan sulfate, and keratan sulfate. Proteoglycan monomers have multiple forms (Fig. 4), including large nonaggregating proteoglycans,

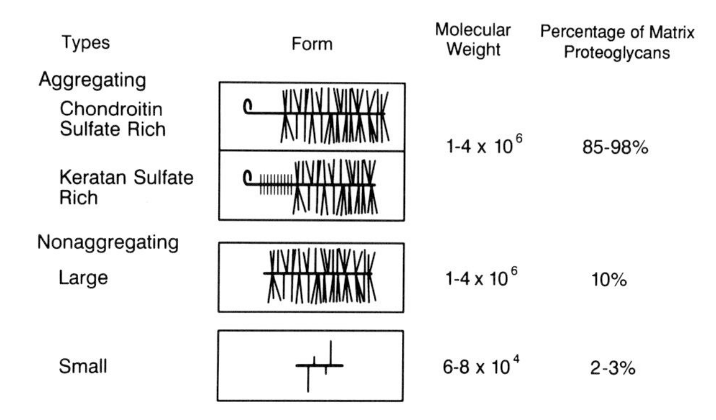

FIGURE 4 The known types of cartilage proteoglycan monomers. The percentage of matrix proteoglycans given in the last column refers to hyaline cartilage. Each monomer type consists of a protein core (the central line of each drawing) and covalently bound glycosaminoglycan chains (the projecting side arms). The longer side arms of the aggregating monomers represent chondroitin sulfate; the shorter side arms, keratan sulfate. In hyaline cartilage, aggregating monomers rich in chondroitin sulfate and aggregating monomers rich in keratan sulfate have been identified. The aggregating monomers are shown with a folded region of their protein core to represent the region that binds to hyaluronic acid.

small nonaggregating proteoglycans, and large aggregating proteoglycan monomers. The latter molecules can exist as individual monomers or as aggregates consisting of multiple monomers, hyaluronic acid, and small proteins called link proteins.

The source and function of the large nonaggregating proteoglycans (Fig. 4) remain uncertain. They might derive from the breakdown of aggregating proteoglycans or they might represent a distinct population of proteoglycans that have a function similar to that of the aggregating proteoglycans. The small nonaggregating proteoglycans (Fig. 4) might contain chondroitin sulfate, and at least some small proteoglycans contain another glycosaminoglycan, dermatan sulfate. The dermatan sulfate proteoglycans might form specific associations with collagen fibrils.

Aggregating monomers consist of protein core filaments with multiple covalently bound oligosaccharides and longer chondroitin and keratin sulfate chains (Figs. 4 and 5). Because each of the glycosaminoglycan chains creates a string of negative charges that bind water and cations, proteoglycans in solution fill a large volume. Because water molecules are bipolar, with a negative charge at the central oxygen atom and positive charges at the hydrogen atoms, the ordered array of negative charges interacts with large volumes of water. Since adjacent negatively charged glycosaminoglycan chains repel each other, they tend to maintain the molecule in an extended form and draw water into their molecular domain, creating swelling pressure within the matrix. An intact collagen fibril meshwork limits the swelling of the proteoglycans, but loss or degradation of the collagen fibril meshwork allows the tissue to swell, increasing the water concentration and decreasing the proteoglycan concentration.

In the matrix most of the aggregating proteoglycan monomers form a noncovalent association with hyaluronic acid and link proteins. The hyaluronic acid filaments form the backbone of aggregates that can reach a length of more than 10,000 nm, with more than 300 monomers (Fig. 6). The link proteins stabilize the association between monomers and hyaluronic acid and might have a role in directing the assembly of aggregates in the matrix. Aggregates might help anchor monomers within the matrix, preventing their displacement, and organize and stabilize the macromolecular framework of the matrix. They also might help control the flow of water through the matrix and help maintain water within it. Because of their ability to interact with water, proteoglycans help give cartilage both its stiffness to compression and its resilience (Fig. 7) and might contribute to its durability.

4. Noncollagenous Proteins

Less is known about the noncollagenous proteins and glycoproteins than about collagens, elastin, or proteoglycans. Only a few noncollagenous proteins of cartilage have been identified, and although their functions are not been defined, they form a significant part of the macromolecular framework of cartilages. These proteins, which might have a small number of attached monosaccharides or oligosaccharides, might help organize and maintain the macromolecular structure of the matrix and the relationship between chondrocytes and the matrix.

Link proteins help organize and stabilize the matrix through their effects on proteoglycan aggregation. Other noncollagenous proteins, including chondronectin, and fibronectin, anchorin CII, might influence the behavior of the cartilage cells. Chondronectin is thought to mediate the adhesion of chondrocytes to the matrix and help stabilize the phenotype of hyaline cartilage chondrocytes, while fibronectin, under experimental conditions, can cause chondrocytelike cells to assume the form and function of fibroblastlike cells. Anchorin CII might have functions similar to those of chondronectin.

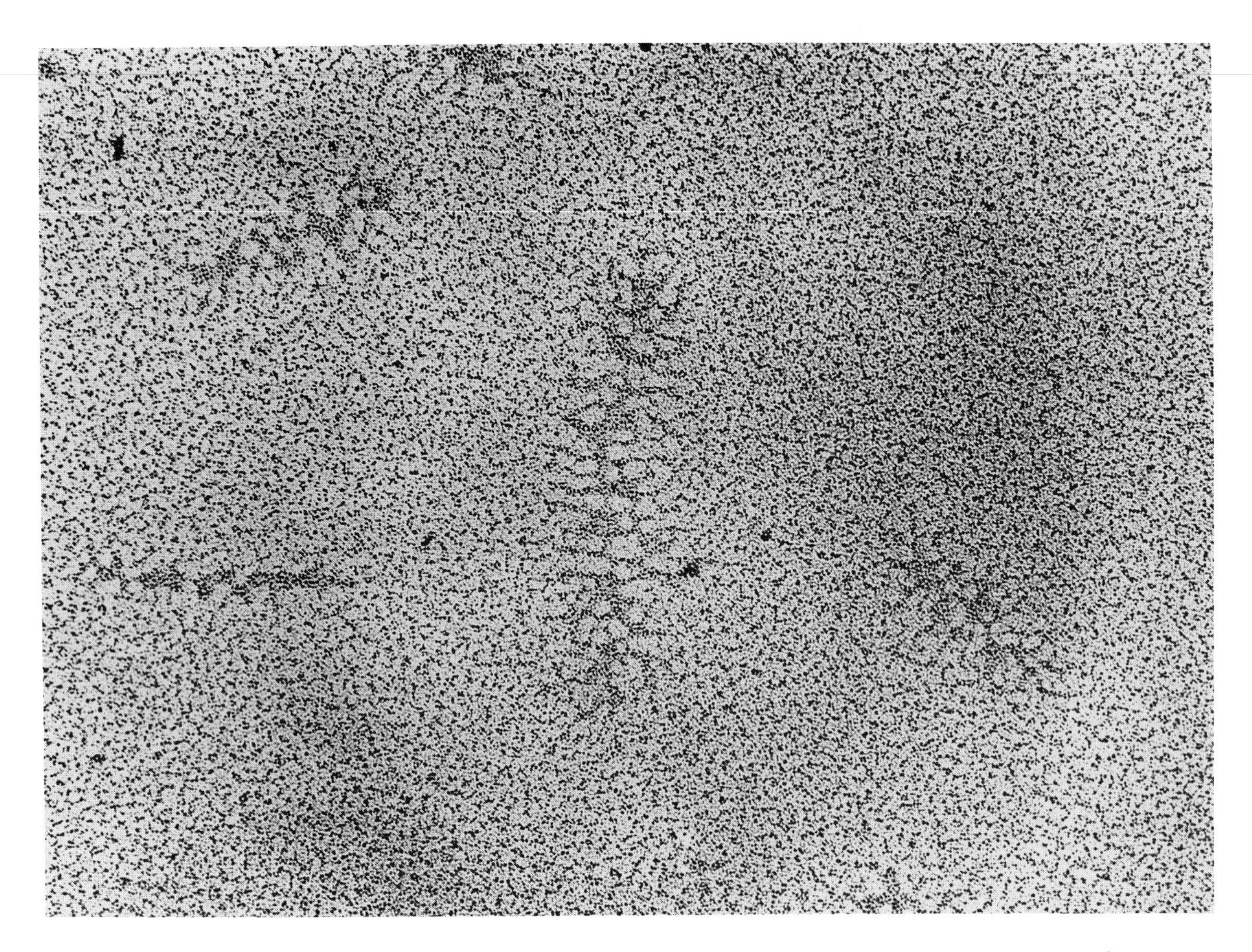

FIGURE 5 Electron micrograph of a large proteoglycan monomer. It is not possible to determine from this micrograph whether this monomer can aggregate. The densely packed glycosaminoglycan chains attached to the central protein filament obscure the filament in many areas. [Reproduced from J. A. Buckwalter, Articular cartilage. *Am. Acad. Orthop. Surg., Instr. Course Lect.* **32,** 355 (1983).]

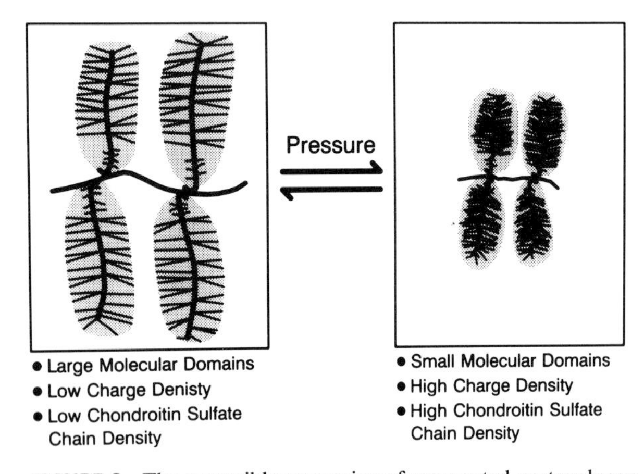

FIGURE 6 The reversible expansion of aggregated proteoglycan monomers in solution. Forcing the negatively charged glycosaminoglycan chains closer together drives water from the molecular domain and increases the resistance to further compression. Release of the compression allows water to return to the molecular domain.

III. Material Properties

Articular cartilage provides load bearing, a low friction surface, and resilience and distributes loads across synovial joints. Although only a few millimeters thick, it usually performs these functions for 80 years or more without significant deterioration. Intervertebral disks withstand the large loads caused by bending and lifting, usually without rupturing. When loaded, or when the spine twists, flexes, or extends, they deform and then regain their original size and shape. Tendon and ligament insertions have great tensile strength, yet they remain pliable and rarely tear from repetitive bending and loading.

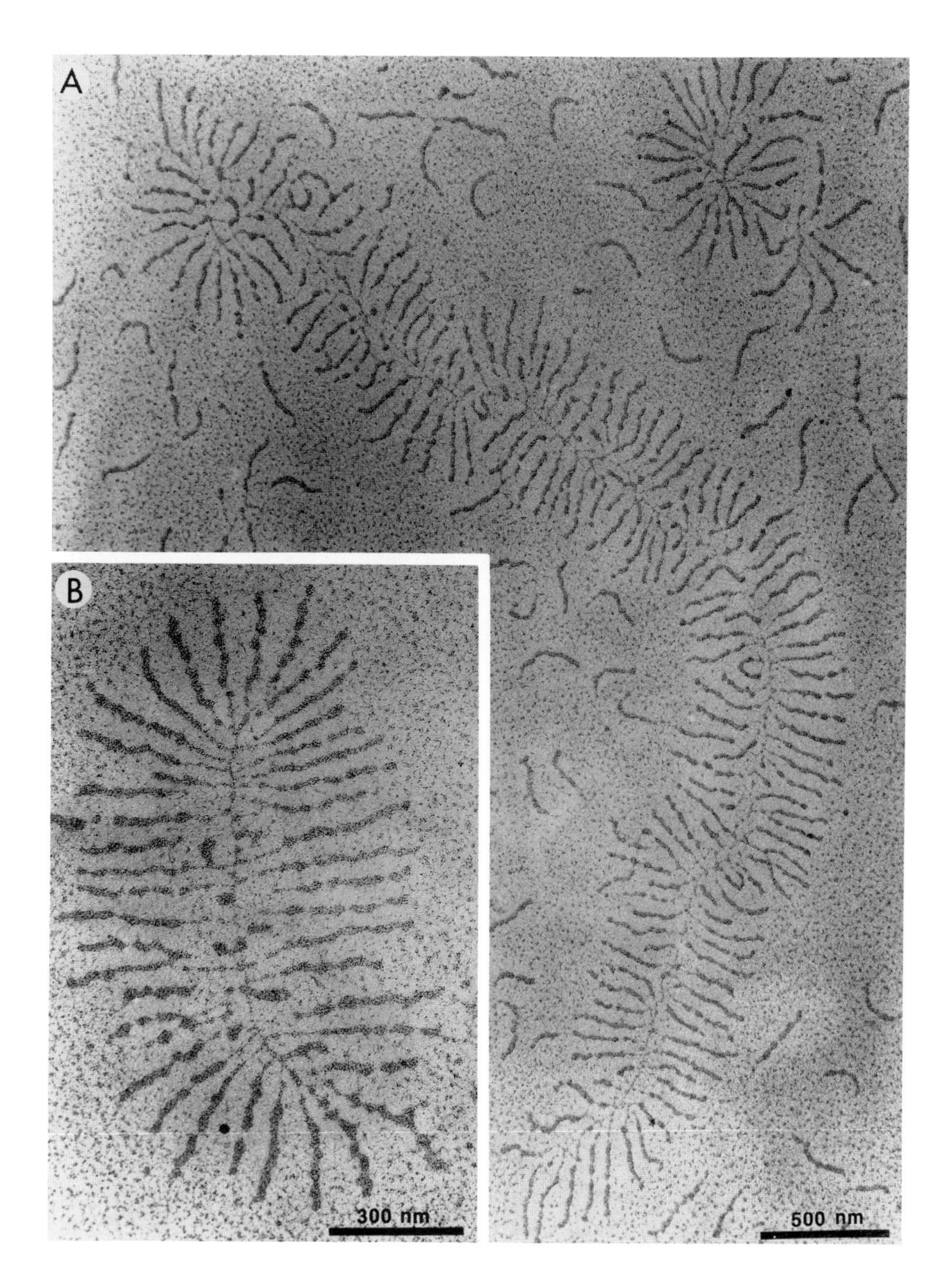

FIGURE 7 Electron micrographs of proteoglycan aggregates. The central filaments are hyaluronic acid. The projecting side chains are proteoglycan monomers. In this preparation the glycosaminoglycan chains are collapsed around the monomer protein cores. (A) A large proteoglycan aggregate, (B) A moderate size proteoglycan aggregate. [Reprinted, by permission VCH Publishers, Inc., 220 East 23rd St., New York, 10010 From: *Collagen and Related Research,* Vol. e489–504, 1983; Figure 1, p. 492.]

The cartilage of the nose, respiratory passages, ears, larynx, and ribs preserves the shape of these structures, resists deformation, yet remains flexible.

A. Matrix Composition, Structure, and Material Properties

The behavior of cartilage when loaded depends on the composition and organization of the matrix, that is, the concentration, properties, and organization of the matrix macromolecules, the water content, and the physical and electrical interactions between water and the macromolecular framework. Since cartilage matrix consists of a macromolecular framework filled with water, the tissue can be considered a fluid-filled porous solid, and porosity (i.e., the ratio of the pore volume within the tissue to the total volume of the tissue) can be estimated by the water content. Since some cartilages have a water content of 60–80%, their porosity is relatively high. Interactions between the water and the pore walls of the macromolecular framework of the matrix generate resistance to the flow of water and determine the permeability of cartilage (i.e., the ease with which water can flow through the matrix).

Because cartilage matrix consists of a solid macromolecular framework filled with water it behaves as a viscoelastic material; that is, its response to loading combines viscosity, a property characteristic of fluids, with elasticity, a property characteristic of solids. The response of viscoelastic materials to constant load or a constant deformation varies with time. When subjected to a constant load, a viscoelastic material responds with an initial deformation, followed by slow progressive deformation, until it reaches an equilibrium state; this behavior is called creep. When subjected to constant deformation, a viscoelastic material responds with high initial stress, followed by a slow progressive decrease in the stress required to maintain the constant deformation; this behavior is called stress relaxation. Creep and stress relaxation might be caused by fluid flow through the matrix or by deformation or movement of the matrix macromolecules. To a point cartilage can restore its original form after deformation by reversing fluid flow and restoration of the macromolecular framework (Fig. 6).

B. Effect of Differences in Matrix Composition

Differences in matrix composition cause differences in the material properties of cartilage. Collagen fibrils provide tensile strength, but little resistance to compression; elastin provides less resistance to tensile loads, but can be deformed without damage and then regain its size and shape; and the interaction of proteoglycans and water provides resistance to compression, swelling pressure, and resilience, but little tensile strength. Therefore, cartilage with a high concentration of collagen fibrils oriented parallel to an applied tensile load has great tensile strength. A high concentration of elastin creates a flexible tissue, such as the cartilage of the external ear, which can be repetitively stretched, bent, or twisted without permanent damage and almost instantly regains its former shape and size after deformation. A tissue with a high concentration of large aggregated proteoglycans can have great compressive stiffness, swelling pressure, and resilience. In general a higher proteoglycan concentration in hyaline cartilage is associated with a lower water concentration, decreased permeability and porosity, and increased compressive stiffness, while a lower proteoglycan concentration is associated with a higher water concentration, increased permeability and porosity, and decreased compressive stiffness.

C. Effect of Differences in Matrix Organization

The organization of matrix macromolecules is not necessarily uniform throughout the tissue, and therefore, even if there are no differences in composition, differences in matrix organization can cause differences in material properties. For example, the material properties of the same region of articular cartilage differ with the orientation of that region relative to the axis of joint motion. Presumably, differences in collagen fibril organization and orientation cause the differences in tensile strength and stiffness at different orientations. The material properties of articular cartilage also differ with depth from the articular surface. These differences might be related to changes in the collagen fibril orientation, the relationship between collagen fibrils and proteoglycans, matrix composition, or other factors. It is likely that more subtle alterations in matrix organization—such as the degree of proteoglycan aggregation, the size of proteoglycan aggregates, the length of chondroitin sulfate chains, the size of proteoglycan monomers, and the strength and stability of the interactions among collagens, proteoglycans, and other matrix molecules—also affect the material properties of the matrix.

IV. Nutrition

At all ages chondrocytes have significant nutritional requirements. Their proliferative and synthetic activities in growing cartilage require a steady supply of nutrients; but even chondrocytes in mature cartilage are metabolically active. Although they rarely divide, they must synthesize molecules to compensate for degradation of those in the matrix, particularly proteoglycans. Vascular canals have been reported within cartilage, but most of them probably allow vessels to pass through the cartilage, rather than supply the chondrocytes. Therefore, for chondrocytes to survive and maintain the matrix, nutrients and metabolites must efficiently and rapidly pass through the matrix for significant distances. This might occur by simple diffusion, convection, or a combination of the two. Transport by convection is based on interstitial fluid flow caused by cartilage deformation as a result of cartilage loading; repetitive loading might therefore help maintain chondrocyte nutrition.

V. Cartilage Types

Adult human cartilages—hyaline, fibrous, and elastic—differ in distribution within the body, matrix composition (Fig. 8A and B), material properties, gross and microscopic appearances, and function. Within each type cartilages vary considerably, and intermediate forms exist, one of which is the repair cartilage that forms during healing of cartilage and bone.

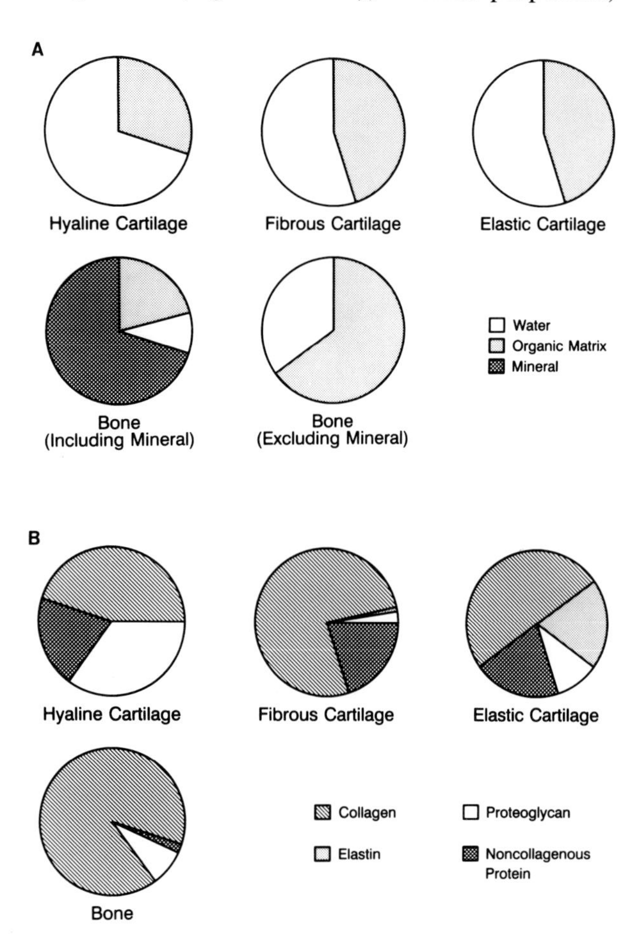

FIGURE 8 (A) Composition of the three cartilage types and bone. Notice that hyaline cartilage has the highest concentration of water. Excluding the mineral component, bone has the lowest concentration of water. (B) The organic matrices of the three cartilage types and bone. Notice that hyaline cartilage has a much greater concentration of proteoglycans and that a high concentration of elastin distinguishes elastic cartilage. The organic matrices of bone and fibrous cartilage have similar compositions: a high concentration of fibrillar type I collagen and a relatively low concentration of proteoglycans.

A. Hyaline

Hyaline cartilage, the most widespread and extensively studied human cartilage, received its name because of the clear appearance of the matrix (from the Greek *"hylos"* meaning "glass"), as seen by light microscopy. It is smooth, slick, and firm to the touch. In the fetus hyaline cartilage forms most of the skeleton before it is resorbed and replaced by bone through the process of enchondral ossification. It forms the physeal cartilages that produce longitudinal bone growth, until growth ceases and they are resorbed and replaced by bone. In adults hyaline cartilage persists as the nasal, laryngeal, bronchial, articular, and costal cartilages.

The highly specialized macromolecular framework of hyaline cartilage (Fig. 8B) consists of fibrillar type II collagen, large aggregating proteoglycans, large nonaggregating proteoglycans, small nonaggregating proteoglycans (Fig. 4), and specific noncollagenous proteins, including chondronectin and anchorin CII. Hyaline cartilage matrix also contains at least two quantitatively minor collagens—types IX and XI—and might also contain types V and VI. Type IX collagen associates with the surface of type II-containing collagen fibrils and could be involved in the organization and stabilization of the type II collagen meshwork and the interaction of the collagen meshwork with other matrix molecules. Type XI collagen might form part of the type II fibrils and help determine their diameter. In addition, hyaline cartilage has a higher water content than the other cartilage types or bone (Fig. 8A), and it does not contain elastin (Fig. 8B). This specialized matrix gives hyaline cartilage optimal material properties for serving as the articular surface of synovial joints, including stiffness, viscoelasticity, resiliency, and durability.

B. Fibrous

Fibrous cartilage forms important structural parts of the intervertebral disk, pubic symphysis, and tendon and ligament insertions into bone. A specialized form of fibrous cartilage makes up the intraarticular menisci. Unlike hyaline cartilage, it is opaque and densely collagenous; its cut surface is often rough, and its matrix appears fibrillar when examined by light microscopy. Also, unlike hyaline cartilage, the fibrillar collagen of fibrous cartilage is type I, the concentration of fibrillar collagen is higher, and the concentrations of water and proteoglycans are lower (Fig. 8A and B). Furthermore, few fibrous cartilages have a high concentration of large aggregating proteoglycans, and at least some fibrous cartilages contain small amounts of elastin. Fibrous cartilage generally has great tensile strength, but does not form smooth, durable, low-friction surfaces, as does hyaline cartilage. Its composition and appearance make it difficult to distinguish it from other dense fibrous tissues, including some regions of tendons and ligaments. For example, in tendon and ligament insertions the transition between the substance of the tendon or ligament and fibrous cartilage is almost imperceptible.

At other locations fibrous cartilage is distinguished by the presence of spherical chondrocytes, rather than the flattened fibroblasts or fibrocytes of dense fibrous tissue. It is likely that the composition, matrix organization, and material properties of fibrous cartilage differ from other dense fibrous tissues, but these possible differences have not been extensively examined.

C. Elastic

A high concentration of elastin (Fig. 8B) gives elastic cartilage, the rarest form of human cartilage, a yellowish hue that distinguishes it from the bluish-white, often translucent, appearance of hyaline cartilage or the off-white color of fibrous cartilage. Elastic cartilage forms the auricle of the external ear, a major portion of the epiglottis, and some of the laryngeal and bronchiolar cartilages. Bending or twisting the external ear shows that elastic cartilage lacks the stiffness of fibrous or hyaline cartilage, but that it has remarkable flexibility and the ability to be deformed and then regain its original shape.

D. Repair

Bone or cartilage injuries frequently result in the formation of repair cartilage. This tissue usually has a matrix composition and a light-microscopic appearance intermediate between hyaline and fibrous cartilages. It could contain both types I and II collagen and regions with high concentrations of hyaline cartilage like proteoglycans. In healing bone fractures new bone replaces the repair cartilage. If a bone fracture fails to heal, cartilage might persist at the nonunion site, form a fibrous or cartilaginous union of the bone fragments, or cover the ends of the bone fragments with hyalinelike cartilage to form a false synovial joint, called pseudoarthrosis. The formation of repair cartilage from undifferentiated mesenchymal cells demonstrates that, even in adult humans, these cells have the potential to differentiate into chondrocytes or chondrocytelike cells and synthesize the components of the cartilage matrix.

VI. Cartilage Mineralization and Enchondral Ossification

Mineralization of cartilage (i.e., deposition of relatively insoluble calcium phosphate in the matrix, particularly in hyaline cartilage) is part of skeletal formation and growth and the healing of some bone fractures. Mineralization, which makes the pliable cartilage matrix rigid, precedes the replacement of cartilage by bone during enchondral ossification. Mineralization might also occur with aging and in some diseases, such as pseudogout. By adversely affecting the material properties of the matrix, it might accelerate deterioration of the tissue. The presence of calcium and phosphate ions alone does not cause cartilage to mineralize. In addition, the matrix must be prepared for mineralization. Probably, matrix components that inhibit cartilage mineralization must be altered or removed, and, possibly, matrix components that promote mineralization must be added.

The process that converts cartilage into bone, called enchondral or intracartilaginous ossification, includes cartilage mineralization as part of a complex sequence of cell and matrix changes. As the chondrocytes enlarge, the surrounding matrix mineralizes. Capillary sprouts then invade the matrix, some cartilage is removed, and osteoblasts lay seams of osteoid (the organic matrix that mineralizes to form bone) over the mineralized cartilage. Shortly thereafter, the osteoid mineralizes to form primary bone, and osteoblasts add further layers of osteoid. Eventually, chondroclasts and osteoclasts remove the calcified cartilage and primary bone, and osteoblasts form mature lamellar bone in their place.

VII. Specialized Forms of Hyaline Cartilage

Although all human cartilages have special features, two forms of hyaline cartilage have particularly high degrees of organization and complex functions.

A. Articular

The function of synovial joints (e.g., the knee or the shoulder) depends on articular cartilage, which forms their bearing surfaces. It distributes loads, minimizing peak stresses on subchondral bone; it can be deformed and regain its original shape; it has remarkable durability; and it provides a load-bearing surface with unequalled low friction. In articular cartilage the chondrocytes organize the matrix components of hyaline cartilage in a unique way to allow it to perform these essential functions. The most apparent organization of the matrix is the changing structure and composition between the joint surface and the subchondral bone. Although the changes are not abrupt, this organization can be described by dividing articular cartilage into four successive zones, beginning at the joint surface: the superficial, or gliding, zone; the middle, or transitional, zone; the deep, or radial, zone; and the zone of calcified cartilage (Fig. 9). Within the zones distinct matrix regions or compartments can generally be identified: the pericellular matrix, the territorial matrix, and the interterritorial matrix (Fig. 9). [*See* ARTICULAR CARTILAGE AND THE INTERVERTEBRAL DISC; ARTICULATIONS, JOINTS BETWEEN BONES.]

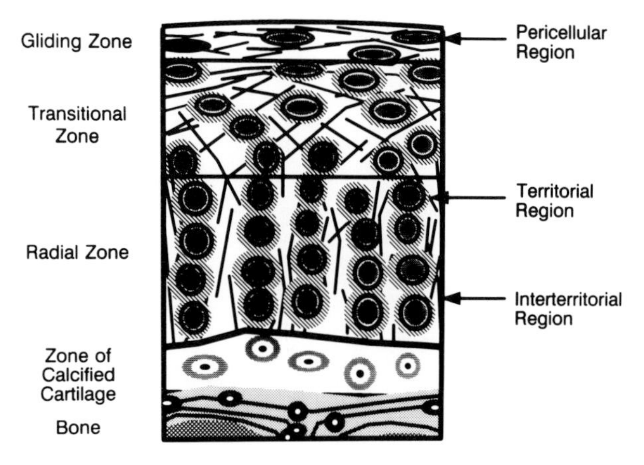

FIGURE 9 Articular cartilage. The tissue can be divided into four zones: the superficial or gliding, zone; the transitional, or middle, zone; the radial, or deep, zone; and the calcified cartilage zone. In most areas three matrix regions can be identified: pericellular, territorial, and interterritorial. Interterritorial matrix collagen fibrils, represented by lines, tend to lie parallel to the joint surface in the superficial zone and perpendicular to the joint surface in the deep zone.

1. Cartilage Zones

The thinnest articular cartilage zone—the gliding, or superficial, zone—forms the surface of the joint (Fig. 9). A thin cell-free layer of matrix, consisting primarily of fine fibrils and relatively little polysaccharide, lies directly adjacent to the synovial cavity. Immediately deep to this layer, elongated flattened chondrocytes are arranged with their major axis parallel to the articular surface. These relatively inactive cells contain small volumes of endoplasmic reticulum, Golgi membranes, and mitochondria. Little or no hyaluronic acid is present in this region, and the proteoglycans associated with the collagen resist extraction to a greater degree than those from other zones, suggesting a particularly strong association between collagen and proteoglycan.

The transitional, or middle, zone has several times the volume of the superficial zone. Its more spherical cells contain a greater volume of endoplasmic reticulum, Golgi membranes, mitochondria, and glycogen. They also contain occasional cytoplasmic filaments. The larger interterritorial matrix collagen fibrils are more randomly oriented than are those of the gliding zone.

In the deep, or radial, zone the cells resemble the spherical cells of the transitional zone, but tend to arrange themselves in a columnar pattern perpendicular to the joint surface. This zone has the largest collagen fibrils, the highest proteoglycan content, and the lowest water content, so that, from the superficial zone to the deepest portion of the deep zone, water content decreases and proteoglycan content as well as the diameter of the fibrils of interterritorial matrix collagen increase.

The zone of calcified cartilage separates the softer hyaline cartilage from the stiffer subchondral bone. Collagen fibrils penetrate from the deep zone directly into the calcified cartilage, anchoring the articular cartilage to the bone. The cells are smaller than are those found in other regions and appear relatively inactive.

It seems reasonable to speculate that the differences in matrix composition and organization among articular cartilage zones reflect differences in mechanical function. The superficial zone forms a thin tough layer that might primarily resist shear. The transitional zone might allow a change in the orientation of the collagen fibrils from the superficial zone to the deep zone, and the deep zone might primarily resist compression and to distribute compressive loads. The calcified cartilage zone might provide a transition in material properties between hyaline cartilage and bone, as well as anchor the articular surface to the bone.

2. Matrix Regions

The matrix regions (Fig. 9) differ in their proximity to chondrocytes, their collagen content, their collagen fibril diameter, the orientation of their collagen fibrils, and their proteoglycan and noncollagenous protein content and organization.

Chondrocyte cell membranes appear to attach to a thin layer of pericellular matrix that surrounds the cell. This smallest matrix region appears to contain little or no fibrillar collagen, although it could contain some of the quantitatively minor nonfibrillar collagens. The predominant macromolecules of the pericellular matrix appear to be proteoglycans, noncollagenous proteins, and glycoproteins such as anchorin CII and chondronectin. Chondrocyte cell membranes seem to adhere to the pericellular matrix, thereby indirectly anchoring the cells to the macromolecules of the other matrix compartments. This arrangement might have a role in transmitting mechanical signals from the matrix to the cell.

An envelope of territorial matrix surrounds the pericellular matrix of each chondrocyte and, in some cases, pairs or clusters of chondrocytes and their pericellular matrices. For example, in the radial zone a territorial matrix surrounds each chondrocyte column. The thin collagen fibrils of the territorial matrix nearest the cells appear to bind to the pericellular matrix, but at a distance from the cell they spread and intersect at various angles, forming a fibrillar "basket" around the cells that could provide mechanical protection for the chondrocytes when cartilage is deformed.

The largest matrix compartment of articular cartilage, the interterritorial matrix, is distinguished from the territorial matrix by an increase in collagen fibril diameter and a transition from the interlacing basketlike orientation of fibrils to a more parallel arrangement. The organization and orientation of the collagen fibrils as they pass from the articular surface to the deeper zones are important features of articular cartilage organization. In the superficial zone the fibrils are oriented primarily parallel to the joint surface, in the transition zone they have a more random orientation, and in the deep zone they line up perpendicular to the joint surface (Fig. 9).

B. Physeal

Human bones elongate not by growth of bone tissue, but by growth of the cartilage forming the physes, or growth plates. The complex organization of these structures makes it possible for them to increase their volume so that they cause longitudinal bone growth and then to convert the tissue they produce into bone. Like articular cartilage, they have a layered or zonal organization and a regional organization. The zones differ from those of articular cartilage (Fig. 10), but the matrix regions, like those of articular cartilage, can be identified as pericellular, territorial, and interterritorial.

The layers, or zones, are the reserve, proliferative, and hypertrophic (Fig. 10). Reserve zone cells show relatively little evidence of metabolic activity. Their functions have not been clearly established, but some of them might serve as stem cells for the proliferative zone. In the proliferative zone cells divide, synthesize extracellular matrix, and assume a highly oriented, flattened, disklike shape. In rapidly growing bones they create long columns of highly ordered cells that resemble stacks of plates. At the bottom of the proliferative zone, the chondrocytes begin to enlarge, rapidly increasing their volume five- to 10-fold, and assume a more spherical or polygonal shape. In the lowermost portion of the hypertrophic zone, or the zone of provisional calcification, the longitudinal septae of interterritorial matrix that lie between chondrocyte columns begin to mineralize, the enlarged chondrocytes condense, and metaphyseal capillary sprouts penetrate the unmineralized territorial matrix, invading the cell la-

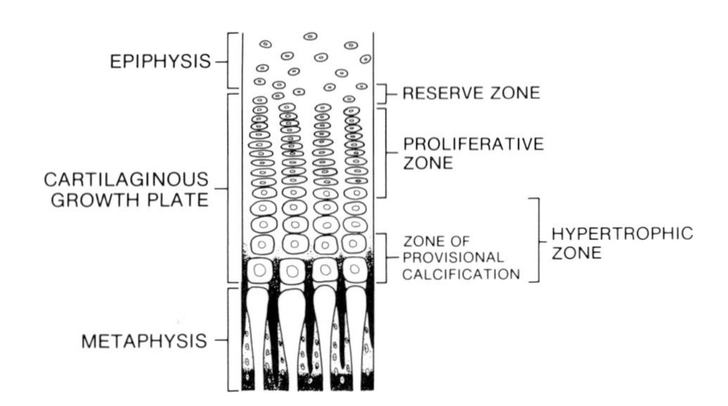

FIGURE 10 Growth plate cartilage. The cartilaginous physis lies between the epiphysis and the metaphysis of the bone and consists of three zones: reserve, proliferative, and hypertrophic. Between the upper proliferative zone and the lower hypertrophic zone cell volume increases five– to 10-fold. In the lowermost region of the hypertrophic zone, the zone of provisional calcification, the interterritorial matrix begins to mineralize.

cunae. In the metaphysis the mineralized cartilage bars are covered with new bone, resorbed, and eventually replaced by mature bone, duplicating the process of enchondral ossification that converts much of the embryonic cartilaginous skeleton into bone and helps heal some bone fractures.

As in articular cartilage, the orientation of the interterritorial matrix collagen fibrils changes among zones. In the reserve and upper portions of the proliferative zone, the collagen fibrils have little apparent orientation, but in the middle and lower regions of the proliferative zone and throughout the hypertrophic zone, the interterritorial matrix collagen fibrils lie parallel to the long axis of the bone.

Growth of long bones requires a directed increase in the volume of the physeal cartilage. The chondrocytes accomplish this by synthesizing new matrix and increasing their volume by the synthesis of cytoplasm and organelles and the accumulation of water. In addition, territorial matrix swelling could contribute to the growth of bone. The swelling of the cells and territorial matrix might be directed to produce longitudinal growth by the physeal perichondrial tissues and the internal organization of the physes, including the transphyseal collagen fibrils of the interterritorial matrix. Through mechanisms that remain poorly understood the growth plate chondrocytes coordinate these activities to produce symmetrical skeletal growth and prepare the matrix they produce for mineralization.

Bibliography

Arnoczky, S., Adams, M., DeHaven, K., Erye, D., and Mow, V. (1988). Meniscus. *In* "Injury and Repair of the Musculoskeletal Soft Tissues" (S.-L. Woo and J. A. Buckwalter, eds.), pp. 487–537. American Academy of Orthopaedic Surgeons, Park Ridge, Illinois.

Buckwalter, J. A., and Rosenberg, L. C. (1988). Electron microscopic studies of cartilage proteoglycans. *Electron Microsc. Rev.* **1,** 87–112.

Buckwalter, J. A., Mower, D., Ungar, R., Schaeffer, J., and Ginsberg, B. (1986). Morphometric analysis of chondrocyte hypertrophy. *J. Bone Joint Surg.* **68A,** 243–255.

Buckwalter, J. A., Hunziker, E., Rosenberg, L., Coutts, R., Adams, M., and Eryre, D. (1988). Articular cartilage: Structure and composition. *In* "Injury and Repair of the Musculoskeletal Soft Tissues" (S. L. Woo and J. A. Buckwalter, eds.), pp. 405–425. American Academy of Orthopaedic Surgeons, Park Ridge, Illinois.

Caplan, A. I. (1984). Cartilage. *Sci. Am.* **251,** 84–94.

Fawcett, D. W. (1986). Cartilage. *In* "A Textbook of Histology," 11th ed., pp. 188–198. Saunders, Philadelphia.

Gosline, J. M., and Rosenbloom, J. (1984). Elastin. *In* "Extracellular Matrix Biochemistry" (K. A. Piez and A. H. Reddi, eds.), Ch. 6, pp. 191–227. Elsevier, New York.

Kosher, R. A. (1983). The chondroblast and chondrocyte. *In* "Cartilage" (B. K. Hall, ed.), Vol. 1, pp. 59–85. Academic Press, New York.

Mayne, R., and Irwin, M. H. (1986). Collagen types in cartilage. *In* "Articular Cartilage Biochemistry" (K. E. Kuettner, R. Schlegerbach, and U. C. Hascall, eds.), pp. 23–28. Raven, New York.

Moss, M. L., and Moss-Salentijn, L. (1983). Vertebrate cartilages. *In* "Cartilage" (B. K. Hall, ed.), Vol. 1, pp. 1–30. Academic Press, New York.

Mow, V., and Rossenwasser, M. (1988). Articular cartilage: Biomechanics. *In* "Injury and Repair of the Musculoskeletal Soft Tissues" (S. L. Woo and J. A. Buckwalter, eds.), pp. 427–463. American Academy of Orthopaedic Surgeons, Park Ridge, Illinois.

TYROSINE

Catecholamines and Behavior

ARNOLD J. FRIEDHOFF, *New York University School of Medicine*

Glossary

β-Adrenergic receptors One type of receptor specifically responsive to norepinephrine and norepinephrinelike substances

Antipsychotic drugs Drugs such as haloperidol or chlorpromazine used to treat patients with psychotic symptoms of diverse origin. These drugs often produce blunting or attenuation of emotional responses as a side effect

Catecholamines Customarily refers to the three endogenous compounds—dopamine, norepinephrine, and epinephrine—which serve as neurotransmitters in postnatal life and probably as neurohormones in embryonic development

Depolarization inactivation Loss of the ability of presynaptic neurons to fire and thereby release transmitter. This phenomenon occurs after prolonged treatment with antipsychotic medication and might be involved in the mechanism of action of these drugs

Gilles de la Tourette's syndrome Disorder beginning in early childhood manifested by multiple tics of the face, trunk, or extremities and by vocalizations, frequently of an obscene or scatological nature

Limbic system Group of brain structures important in the regulation of emotion and the association of emotion with behavioral and mental function

Locus coeruleus Pigmented area in the floor of the brain which is a major center of noradrenergic neurons

Neurotransmitter (transmitter) Chemical substance released from the axon terminal of one neuron onto a specific receptor of a following neuron. This is the means by which chemically mediated information transfer takes place in the brain

Nigrostriatal system Major system mediating the action of dopaminergic neurons located in the substantia nigra of the brain. These neurons project to the corpus striatum. Both nuclei are important centers of the extrapyramidal nervous system, which coordinates involuntary aspects of voluntary muscle movement

Reticular activating system Ascending system in the brain stem and the spinal cord believed to be involved in alertness and attention

Stereotypy Persistent repetition of movements or behaviors

Substantia nigra Pigmented subcortical structure which is a major center of the dopaminergic neurons that project to the corpus striatum

Supersensitivity Increased responsivity of a neuron to a given concentration of transmitter, usually occurring through an increase in the density of specific receptors

THE CATECHOLAMINES—dopamine, norepinephrine, and epinephrine—are powerful chemicals that begin to shape the function of the brain before birth, during fetal development. These compounds continue to play a key role after birth in both mediating and modulating functions as diverse as mood regulation, thinking, unconscious control of muscle movements, postural reflexes, and emotional responsivity. In this article the focus is on the role of these substances in the brain, although they are also found in the adrenal gland, in blood vessels, in skin, and in the ganglia and terminals of the sympathetic nervous system that regulates internal organs. In the brain these substances help to set emotional tone, alter the level of arousal and attention, and regulate the flow of associations, thus enhancing or diminishing the impact of life events.

Dopamine alone among the three catecholamines has a central role in regulating motor behavior, in addition to its involvement in mental and emotional functions. Disturbance in its function leads to tremors and loss of fine control of movement, a condition known as parkinsonism. Dopamine plays a similar role in keeping associative thinking on track, a function that is disturbed in schizophrenia and other psychotic illnesses. Dopamine and norepinephrine, as well as another amine, serotonin, not belonging to the catecholamine family, appear to mediate systems that are responsible for maintaining mental stability in the face of persistent nonremediable stress or adversity. These dopaminergic circuits stabilize the brain under conditions of serious duress.

Thus, the catecholamines have powerful influences on the most vital life functions—on movement, on thinking, and on mood and emotions. As discussed here, much of what we know about the action of catecholaminergic systems in the brain has been learned from studying various psychotropic drugs that act on one of these chemicals.

I. Characteristics of Catecholaminergic Systems in the Brain

The catecholamines all originate from a dietary amino acid, tyrosine. They are formed through the following reaction (Fig. 1).

All of these compounds are simple chemicals of relatively low molecular weight. Dopamine and norepinephrine are synthesized and stored in neurons and released as needed in response to electrical signals. The cellular localization and role of epinephrine in the brain are not as clear, although it is the major catecholamine found in the adrenal gland. Dopamine not only serves its own function, but it is also the precursor in the synthesis of norepinephrine. The latter compound plays an important role in the brain and, in turn, also serves as a precursor in the biosynthesis of epinephrine.

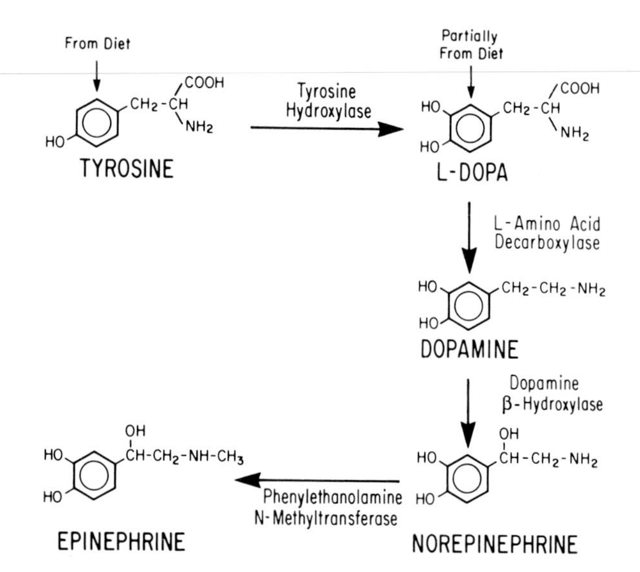

FIGURE 1 Pathway for biosynthesis of the major catecholamines.

II. Neuronal Pathways and Role in the Neuron

In the human brain the dopaminergic system is organized into several systems, including the nigrostriatal and limbic dopaminergic systems. The nigrostriatal system is the system mainly disturbed in parkinsonism, while the limbic system, which involves the prefrontal cortex and other higher centers, might be more influential in modulating mental activity. In addition to these two major systems, dopaminergic neurons project to other areas, including the pituitary gland. Additionally, it appears that the prefrontal dopaminergic system could have a downward influence on lower limbic centers and on the nigrostriatal system. Most noradrenergic pathways originate in the locus coeruleus and project to all parts of the brain, although another major center lies in the lateral tegmental nucleus. Adrenergic cells containing epinephrine are colocalized with at least two branches of the noradrenergic system. There are many excellent reviews of the details of the centers and pathways of the catecholaminergic system. In relation to their mediation and modulation of behavioral and mental functions, their diffuse projections are compatible with their important influence.

In the neuron either dopamine or norepinephrine, depending on the type, is stored in various-sized vesicles found in proximity to the synaptic membrane. In response to an electrical signal moving down the axon, the substance is released, diffusing rapidly across the small distance of the synaptic junction and transmitting a signal to the next neuron in the brain. These neurotransmitters propagate an impulse in accordance with the specific function of the neurons involved. Important for this specificity are the receptors onto which the transmitter impacts, which are designed to respond only to the specific amine or to substances mimicking critical aspects of specific amines. Some of these latter substances have been made into psychotropic drugs.

When transmitter impacts on a specific receptor on the receiving neuron, it alters one or more ion channels or activates a system that makes a second chemical messenger which, in turn, influences the postsynaptic neuron in specific ways. It is through this alternation of electrical signals and chemical messengers that information is propagated through the brain, ultimately controlling all of our behavior and most other bodily functions.

In the described systems an electrical signal is passed to the next neuron via a chemical messenger. In order for this system to be effective, the action of the transmitter on the receptor must be terminated rapidly. This is done through a reuptake system which senses the amine in the synapse and takes it back into the neuron from which it was released. Reuptake serves the dual function of terminating the signal and conserving the transmitter. The transmitter remaining in the synapse is altered by several enzymes into substances which are relatively inert biologically and which are ultimately excreted.

III. Functions of Catecholamines in Regulation of Behavior

Primarily through the use of psychotropic drugs that modify the action of dopamine or norepinephrine, it has been possible to infer a great deal about the role of these neurotransmitters in mediating and regulating behavioral function. Drugs that alter dopaminergic function are used in the treatment of parkinsonism and various psychotic disorders, while psychotropic drugs that modify the action of norepinephrine are used to treat depression. No psychotropic drug acting by affecting the function of epinephrine has been developed, so that the behavioral functions of that substance are more obscure. [*See* NEUROPHARMACOLOGY.]

IV. Dopamine

The compound dopamine is best known because of its association with parkinsonism, which is a dopamine deficiency disease; however, our understanding of this neurotransmitter evolved mainly through another route. Reserpine, a drug derived from *Rawoulfia serpentina* and used originally in the treatment of hypertension, was found to have antipsychotic properties. In the course of treating patients with schizophrenia and other psychotic illnesses, some developed muscle rigidity and tremors, a syndrome that bore a startling similarity to naturally occurring parkinsonism. Ultimately, the mechanism for this effect was found to involve the depletion of dopamine from the brain. Reserpine had the ability to deplete all of the catecholamines, but the Parkinsonian syndrome resulted specifically from the depletion of dopamine. These findings led to the study of patients with parkinsonism and of postmortem brain tissues from patients with parkinsonism, and it was discovered that the illness was associated with a loss of dopamine-producing neurons in the substantia nigra. [*See* PARKINSON'S DISEASE.]

Not long after the introduction of reserpine as an antipsychotic drug, it was found that an unrelated class of drugs, the phenothiazines, were also useful in the treatment of psychosis. These drugs also decreased dopaminergic activity, not by depleting it, but by blocking dopamine receptors. From observations of the action of this and other dopamine antagonists, it has been possible to infer a great deal about the behavioral functions of dopamine itself.

One observation made early in the use of antipsychotic drugs was that they produce a flattening or blunting of emotional responsivity (i.e., affect). Patients taking these medications lose their normal expressivity. This is interesting because many patients with schizophrenia, particularly those who do not respond to antipsychotic drug treatment, have flat affect as one of their symptoms. The affect-blunting effect of reserpine and of the newer antipsychotic drugs, all of which reduce dopaminergic activity, coupled with the fact that patients with parkinsonism, a dopamine deficiency disease, have flat emotional responses, points to the important role of dopamine in regulating emotions.

Antipsychotic drugs also normalize associative processing, which is disturbed in psychotic individuals. Psychotic patients tend to have loose associations and other disturbances which interfere with the orderly progression of normal thinking. Antipsychotic drugs tighten associative processing and help keep thinking on track. It might be coincidence that antipsychotic drugs order associative processes and also produce constriction of emotional responses; however, it might also be that some of the functions of emotions are to lend color to thinking, to permit freer reign of thought processes, and to contribute to the creative process. This could be why patients under treatment with antipsychotic medication often complain that they have lost their motivation and their creative capacity, perhaps the price that must be paid for stabilizing abnormal thinking processes. Thus, enhanced emotionality, while it might favor creativity, might also induce mental instability. In fact, this could account for the association—in myth, if not in science—between insanity and genius, there allegedly being a thin line between the two. Emotional expression, under important dopaminergic control, certainly makes people more vulnerable to psychological trauma. The investment of emotional energy in another person gives that person the power to hurt in a way not generally resulting from negative interactions with a more distant individual.

Patients with schizophrenia also have attentional problems, finding it difficult to maintain focus. Reducing dopaminergic activity with antipsychotic medications improves the ability to attend. Thus, dopamine also seems to be involved in the regulation of attentional processes. Apart from its function in mentational and emotional processes, dopamine plays a key role in the extrapyramidal nervous system, which maintains unconscious control over important aspects of voluntary movement. For example, rising to a standing position requires simultaneous adjustment of many muscle groups. It is the extrapyramidal system that coordinates this activity and enables us to maintain posture upright without thinking about which muscles to contract and which to relax. [*See* SCHIZOPHRENIC DISORDERS.]

The functions described above are all regulated primarily by a class of dopamine receptors known as D_2. There is also a class of receptors called D_1 which are positively coupled to the enzyme adenylate cyclase, increasing the activity of this enzyme and the production of cAMP when the receptors are activated. This D_1 system produces a behavioral syndrome known as rapid jaw movements in rats, and also appears to facilitate D_2-mediated behaviors, but its role in humans is poorly understood. Dopamine, therefore, exercises important regulatory control over three key areas of human function: emotions, associative thinking processes, and movement and posture. Clearly, it is one of the most important chemicals in the brain.

V. Norepinephrine

Norepinephrine, like dopamine, has had its function defined through its association with particular brain structures and its involvement in the action of drugs. Because of its presence in areas of the brain constituting the reticular activating system, it is believed to be involved in activation or arousal mechanisms. Drugs such as amphetamines increase the release of norepinephrine, increase alertness, and interfere with sleep. Amphetamines also trigger the release of dopamine, but their effect on norepinephrine is believed to be responsible for their activating effects. On the other hand, their ability to reduce appetite might involve dopamine, or norepinephrine together with dopamine, inasmuch as dopamine is known to be involved in appetite regulation.

A big boost in our understanding of the role of norepinephrine in behavioral function came about through the discovery of antidepressant drugs. Major classes of these drugs act by blocking the reuptake of norepinephrine after its release into the synapse. Because of the blockade, the transmitter cannot be inactivated by the fast-acting reuptake system. This leaves its inactivation to the enzymes associated with synaptic function, which act more slowly than the reuptake system. Slower inactivation means, in turn, a higher concentration of norepinephrine in the synapse. This increase, however, occurs shortly after administration of these drugs. In contrast, mood elevation produced by antidepressant drugs typically does not occur for 1 week or more, and frequently as long as 3 weeks after effective blood levels are achieved. [*See* ANTIDEPRESSANTS.]

By the time mood elevation occurs, the chronic increase in norepinephrine concentration in the synapse has produced a secondary decrease in the number of β-adrenergic receptors, one of the key types of receptors responsive to norepinephrine. At present it is not known with certainty what is the net effect of the increase in norepinephrine and the decrease of its receptors. What appears clear is that norepinephrine is involved in the regulation of mood, but whether mood elevation involves increased or decreased noradrenergic activity is not entirely clear.

Perhaps limiting attention to norepinephrine alone is itself too simple. Many antidepressant drugs block the reuptake of serotonin as well as norepinephrine, and some block the reuptake of serotonin only. In any event the brain must be seen as a highly integrated network of many types of transmitter systems. Each impacts on many of the others. Therefore, when we speak of the specific action of a transmitter such as norepinephrine, we really mean that it can impact on the ongoing activity of one or more other systems to modify them and to produce a specific effect.

VI. Epinephrine

Neurons containing epinephrine have been identified in the brain and, as mentioned earlier, are colocalized in several areas, with noradrenergic neurons. Epinephrine might function, in the brain, in blood pressure and neurohormonal regulation; however, in the absence of psychoactive drugs specifically mediated by systems dependent on epinephrine, it has been difficult to assign behavioral or mental functions to this substance.

VII. Catecholamines as Mediators of Brain Adaptive Systems

The brain must be capable of maintaining stable function in the face of continuing or repeated stress or other adverse conditions. At least in part, stability is maintained through adaptive changes in the relationship of brain systems and subsystems. These adaptive changes in the brain can be mediated in many ways (e.g., by a shift of function from damaged to intact parts of the brain after local trauma); however, adaptation is necessary not only in response to trauma, but also in order to cope with chronic adverse psychological situations. These latter adaptations can involve fight, flight (coping by fleeing from adversity), shifts of attention, and use of so-called psychological defense mechanisms. Although little studied, changes at the physiological level could include changes in the sensitivity of sensory processing mechanisms, changes in the level of arousal, and longer-term changes in transmitter synthesis or receptor number or affinity. Many of these effects alter the level of ongoing endogenous activity. Both dopamine and norepinephrine appear to play an important role in maintaining or restoring stable mental function.

Typically, endogenous brain neuronal activity is modulated through downward inhibition and release of inhibition, the cortex inhibiting systems below it and so on down the line. Conscious inhibition of some functions is presumably mediated by the cortex, the region of conscious control. That is why, in a person who has had a stroke, cortical inhibition of lower motor centers is often lost. Such stroke victims experience disinhibition of the motor functions controlled by the area of motor cortex damaged by the stroke, which causes spastic paralysis of the disinhibited muscles. [*See* CORTEX; STROKE.]

Some awareness of these inhibitory systems has existed for millenia, having been known even to the ancient Egyptians, who were able to relate specific traumatic brain injuries to disturbances in motor function (see the Edwin Smith papyrus original in the library of the New York Academy of Medicine, New York City.) It was through these observations of traumatic injuries that the Egyptians deduced that the motor function of the extremities was controlled by the cortex on the opposite side.

As mentioned in Section IV, dopamine is prominently involved in the maintenance of ambulatory control of the extrapyramidal motor system. This control is illustrated by the function of the human finger, which is always in a state of oscillation, or tremor. When the finger must move in a given direction, the appropriate muscles are disinhibited at a time when the finger, owing to its tremor, is already moving in that direction. Thus, the finger gets a "running start," making for shorter reaction times and increased dexterity. This tremor is under dopaminergic inhibitory control, exercised through the action of brain dopaminergic systems on the spinal neurons that maintain the endogenous tremor.

Equivalent inhibitory control of intellectual functions such as thinking or speech has not been established; however, it is likely that such control exists.

An example of a disturbance in inhibitory function can be provided by the disorder Gilles de la Tourette's syndrome. This syndrome occurs early in childhood and is characterized by multiple motor tics and by a strange disturbance in which the affected children often shout obscene words or phrases. The afflicted children, who are not psychotic, are quite humiliated by these outbursts, which can occur at socially inappropriate times. The motor tics appear to occur through transient loss of inhibitory control of smooth motor movements, and the vocalizations can also occur through loss of inhibitory control of associative thinking.

It is common experience that we think many things that are inappropriate to say in a given social context, and most of these thoughts are automatically suppressed and not translated into speech. High on the list of thoughts that frequently remain unspoken are obscenities. Through weakening of automatic inhibitory control in Tourette's syndrome, taboo thoughts are vocalized at inappropriate times.

Some direct evidence that the brain exercises inhibitory control of mental and vocal, as well as motor, processes comes from the fact that the drug haloperidol, which is an antagonist of the neurotransmitter dopamine, generally suppresses both the motor and vocal tics in patients with Tourette's syndrome, whereas dopaminergic agonists exacerbate these symptoms. From this it appears that both might be regulated by the dopaminergic system.

Catecholamines might also be important in the action of psychotropic drugs. In an experiment which has now become classic, Arvid Carlsson, in Sweden, found that antipsychotic drugs act by blocking dopamine receptors. Inasmuch as these drugs produce improvement in the symptoms of schizophrenia, it was felt that overactivity of the dopaminergic system might be the cause of schizophrenia; however, over the years it has been recognized that the role of dopamine in this illness is more complex. On the other hand, understanding the mechanism of action of antipsychotic drugs has enabled us to greatly expand our knowledge of the role of the dopaminergic system in normal mental function.

The initial response to administration of antipsychotic drugs (i.e., dopaminergic receptor blockers) is a massive outpouring of dopamine from axon terminals, as a compensatory response to the blockade, so that the net effect of receptor blockade coupled with increased release has been difficult to determine. Moreover, receptor blockade is maximum within hours after the administration of antipsychotic drugs, whereas optimal therapeutic effects do not occur for days or even weeks.

More recently, it has been discovered that after prolonged blockade the compensatory increase in dopamine release ceases and a phenomenon called depolarization blockade, or depolarization inactivation, occurs, in which transmitter release slows. It could be that it is the combined effect of the reduction in transmitter release, through depolarization inactivation or some similar process, plus the receptor blockade, which results in a net reduction in dopaminergic activity. Inasmuch as depolarization inactivation does not occur for days or weeks, it coincides better, temporally, with the onset of therapeutic effects of antipsychotic drugs, effects that are characteristically delayed.

The observation that a significant action of antipsychotic drugs involves not only rapid receptor blockade, but also a delayed reduction in transmitter release, adds additional significance to recent studies showing unusually high correlations between the therapeutic effects of antipsychotic drugs and plasma levels of the principal dopamine metabolite, homovanillic acid (HVA).

The relationship of dopamine metabolites in the plasma to those of interest in the brain has been the subject of study and speculation for a number of years. The principal difficulty in assessing brain activity from plasma metabolites is that dopamine and its metabolites are also released into the plasma from peripheral neurons. Thus, plasma metabolites are a mix of those originating in the brain and those originating from peripheral sources. Release from the brain, but not the periphery, is increased by the administration of antipsychotic drugs. Thus, the increase in plasma HVA in response to antipsychotic drugs might more nearly reflect the release of dopamine from the brain, presuming that metabolism and clearance are not also significantly affected by these drugs.

These studies of plasma HVA response to antipsychotic drugs have proved to be informative for our understanding of the normal brain. Responders to antipsychotic drugs have a bigger initial increase of plasma HVA, after treatment is begun, than do nonresponders. Despite the problems in extrapolating from plasma HVA to the status of the dopaminergic system in the brain, uniquely high correlations have been obtained between measures of neuroleptic-induced plasma HVA changes and the therapeutic effect of subsequent treatment.

The brain, being an exquisitely regulated organ as

well as a regulator of virtually all bodily functions, must have systems to prevent its own destabilization. The dopaminergic system appears to stabilize brain function in the face of a variety of insults. In the context of this hypothesis, the plasma HVA response of neuroleptic-treated patients becomes understandable. We know from many treatment studies that the administration of dopamine antagonists often leads to suppression of psychotic symptoms. Looking at it another way, inasmuch as the cause of the psychosis has not been removed, lowering dopaminergic activity raises the threshold for expression of these symptoms.

The dopaminergic system can increase or decrease its own activity in the absence of drugs; It is possible that it functions physiologically to suppress emerging psychotic symptoms by reducing dopaminergic activity. Reduced dopaminergic activity, as indicated earlier, is associated with flat affect. Thus, effective maintenance of mental stability in the face of persistent destabilizing factors would be expected to be associated with flat affect. This, in fact, corresponds with psychological studies. Subjects under chronic stress react with emotional withdrawal and loss of affective warmth.

Norepinephrine might also mediate an adaptive system in the brain. The fact that antidepressant drugs, which block the reuptake of norepinephrine, do not produce therapeutic effects for several weeks probably reflects the fact that these effects result from adaptive changes in the brain. In contrast to psychosis, which affects only a small percentage of the population, depression is known to all of us. It is a normal reaction to tragic events or significant losses. Grief or depression, therefore, does not become pathological unless it fails to lift in a reasonable period of time or unless it occurs without known causal event. Thus, a system in the brain meant to prevent serious disruption from depression would help lift depressive symptoms rather than prevent them. Norepinephrine, as well as serotonin and probably other transmitters, would be involved in such a system. [*See* DEPRESSION: NEUROTRANSMITTERS AND RECEPTORS.]

VIII. Prenatal Programming of the Dopaminergic and Noradrenergic Systems

Reducing dopamine at the receptor site in the rat, prenatally, by either blocking the receptor with antipsychotic drug or blocking dopamine synthesis, produced a significant decrease in the number of dopamine receptors postnatally. This effect occurred only during a sensitive period between gestational days, (GD) 15.5–17.5 of a 21-day gestation period. During this sensitive period dopamine receptors develop the kinetic properties of mature receptors, so that after gestational day 17.5 they have all of the properties of their postnatal counterpart. Interestingly, after this time antipsychotic drugs produce a response similar to that found in the postnatal animal. The changes produced by intervention during the sensitive prenatal period persist for the life of the animal, so that although the total number of dopamine receptors increases during early postnatal life, receptor number in animals subjected to neuroleptic treatment during fetal life always remains somewhat behind that in their untreated counterparts.

Prenatal treatment with neuroleptics during the sensitive period also produces a decrease in dopamine-dependent behaviors, such as apomorphine-induced stereotypy. Animals exposed to antipsychotic drugs prenatally also have a marked decrease in the postnatal ability to develop supersensitivity. Normally, receptor number increases about 25% in response to 1–4 weeks of postnatal treatment with an antipsychotic dopamine antagonist. This capacity is lost if animals are exposed prenatally.

One final functional implication of the prenatally induced decrement has to do with the relationship of the D_1 and D_2 receptor subtypes of the dopaminergic system. D_1 receptors are positively coupled to adenylate cyclase, and stimulation of these receptor subtypes activates this enzyme. D_2 receptors, on the other hand, inhibit cyclase activity. Although dopamine receptors have limited selectivity during the sensitive prenatal period, prenatal treatment with a D_2-selective drug such as haloperidol produces a greater postnatal decrement in D_2 receptors than in D_1. Thus, postnatally, there is an increase in the D_1/D_2 ratio. This imbalance in D_1/D_2 activity results in spontaneous rapid jaw movements. In rats exposed prenatally the movements occur early in postnatal life, in contrast to young age-matched controls, which show little of this behavior. Inasmuch as D_1 receptors are positively coupled to the enzyme adenylate cyclase, cAMP levels are affected. cAMP plays a key regulatory role in cells.

The prenatal environment, therefore, can have a profound influence on the postnatal function of the dopaminergic system and on the role it plays in regulating mental and behavioral functions.

The noradrenergic system is also modified prenatally, depending on the prenatal environment. Early treatment (i.e., between gestational days 8 and 11) in the rat produce persistent postnatal changes in a key noradrenergic receptor known as the β-adrenergic receptor, which is especially sensitive to the maternal environment. Thus, maternal stress during the sensitive period produces a significant decrement in the number of β-receptors. Inasmuch as this is the effect of many antidepressant drugs given postnatally, it could be that these rats are born with a greater resistance to the stressors that provoke a depressive response in humans.

IX. A Look to the Future

The catecholamines are key substances in regulating mood, emotional tone, associative processing, and motor activity. This regulation occurs by on-line or adaptive changes in the level of activity of these systems. Our ability to alter their level of activity and to influence the functions they regulate is currently limited to the use of selective antagonists or agonists. Increasingly, however, we are developing knowledge about regulation at the genomic level. Genes for receptors in both the dopaminergic and noradrenergic systems have been cloned and sequenced, as has the gene for the enzyme tyrosine hydroxylase, which is the rate-limiting enzyme in the biosynthesis of catecholamines. These developments have made possible the study of regulatory elements controlling the activity of these genes. In the not-too-distant future peptide regulators of these genes will be identified, and after development of appropriate delivery systems it might be possible to increase or decrease the activity of the system. Thus, for better or worse, it is likely that in the foreseeable future we will be able to adjust mood, emotions, and looseness of associations at will.

Bibliography

Abelson, P. H., Butz, E., and Synder, S. H. (1985). "Neuroscience." Am. Assoc. Adv. Sci., Washington, D.C.

Cooper, R. J., Bloom, F. E., and Roth, R. H. (eds.) (1986). "The Biochemical Basis of Neuropharmacology," 5th ed. Oxford Univ. Press, New York.

Friedhoff, A. J. (1986). A dopamine dependent restitutive system for maintenance of mental normalcy. *Ann. N.Y. Acad. Sci.* **463,** 47.

Meltzer, H. Y. (1987). "Psychopharmacology: The Third Generation of Progress." Raven, New York.

Miller, J. C., and Friedhoff, A. J. (1988). Prenatal neurotransmitter programming of postnatal receptor function. *Prog. Brain Res.* **73,** 509.

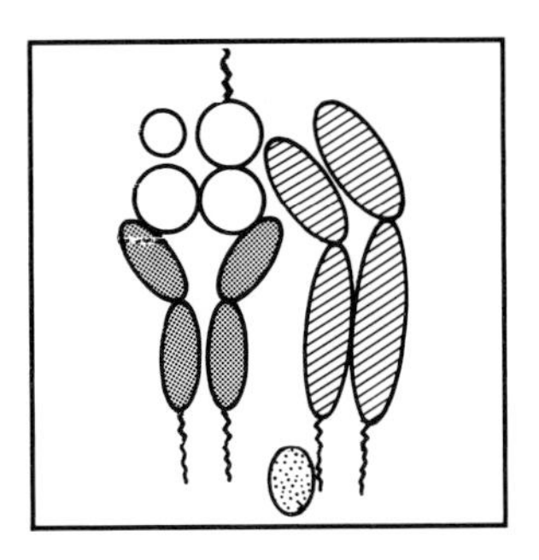

CD8 and CD4: Structure, Function, and Molecular Biology

JANE R. PARNES, *Stanford University*

Glossary

Antigen Foreign substances that are specifically recognized by the immune system

Cytotoxic T lymphocyte T lymphocyte that specifically kills cells expressing the combination of foreign antigen bound to a major histocompatibility complex protein for which its T-cell receptor is specific

Helper or inducer T lymphocytes T lymphocyte that provides signals to induce the further differentiation of either B lymphocytes (to secrete antibody) or other T lymphocytes on recognition of the foreign antigen bound to a major histocompatibility complex protein for which its T-cell receptor is specific

Major histocompatibility complex (MHC) Large genetic region on human chromosome 6 containing genes mediating a variety of immune functions, including the highly polymorphic (i.e., variable from individual to individual) class I (classical transplantation antigens), class II (immune response gene products), and class III (certain complement components) proteins

T-cell receptor (TCR) Heterodimeric T-lymphocyte surface protein that varies from cell to cell and mediates the specific recognition patterns of T cells

Thymocytes T-lymphocyte precursors in the thymus gland

CD8 AND CD4 are cell surface glycoproteins expressed primarily on mature T lymphocytes and on their developmental precursors in the thymus. T lymphocytes are responsible for cellular immune responses and are important in protecting the individual against infections by parasites, fungi, and intracellular viruses, as well as against cancer cells and foreign tissues. Although the T-cell receptor (TCR) for antigen is responsible for specific immune recognition, CD8 and CD4 also play important roles in T-cell responses to foreign antigens. These proteins divide mature T lymphocytes into two distinct functional subsets on the basis of their recognition patterns. CD8 and CD4 have been referred to as "accessory molecules" or "co-receptors" because they enhance T-cell responses and as "differentiation antigens" because of their changing pattern of expression during T-cell development.

I. Introduction

As T cells differentiate in the thymus, the earliest precursors (about 5% of thymocytes) express neither the CD8 nor the CD4 cell surface protein. Such thymocytes are referred to as "double-negatives." Based on studies in the mouse, it is believed that thymocytes next pass through a transient phase of expression of only CD8 and not CD4. These cells, which consist of only a small percentage of thymocytes, can be differentiated from the more mature stage of CD8 "single-positive" cells by the absence of surface expression of the TCR and CD3, a complex of at least five proteins associated with the TCR. The next major developmental stage contains the majority of thymocytes (approximately 80%) and is characterized by simultaneous expression of both CD8 and CD4. It is during this stage that cells begin to express surface TCR and CD3. Recent studies have provided a large body of evidence to suggest that it is at this "double-positive" stage that one form of self-tolerance occurs, namely, elimination of thymocytes that are strongly reactive to self-antigens ("negative selection"). It is also thought that a positive selection occurs on double-positive cells to allow continued differentiation of only those thymocytes that express TCR molecules that will later recognize foreign peptides bound to self major histocompatibility complex (MHC) proteins, (i.e., those present in cells of the same individual). It should be recalled that because MHC proteins are highly polymorphic, different individuals in most cases have two different forms of each MHC protein. It is not at all clear how positive selection occurs, because the foreign antigens that may later be encountered by mature T cells are certainly not present. One suggestion is that thymocytes are positively selected by binding weakly to peptides generated in their own cells and bound to self-MHC proteins [but not strongly enough to be eliminated by negative-selection, which operates to remove cells that might later be autoreactive (i.e., reactive against cells of their own body)]. What is clear is that most thymocytes die at the double-positive stage probably because they fail these tests. Those that do differentiate further lose expression of either CD8 or CD4 and become mature single-positive CD4 or CD8 cells, with an approximate 2 : 1 ratio of CD4 : CD8. These cells are released into the periphery, and hence mature T cells in peripheral blood consist of approximately 65% CD4 cells and 35% CD8 cells. [*See* LYMPHOCYTES; T-CELL RECEPTORS.]

Soon after the identification of CD8 and CD4 through the use of specific antibodies, it was recognized that expression of these proteins on the surface of mature T cells correlates reasonably well with the type of functional activity exhibited by the T cell. In general, CD8 cells are cytotoxic (or possibly suppressor) T cells, whereas CD4 cells are helper or inducer T cells. However, there are some notable exceptions to this generalization, and it is now accepted that the best correlation with CD8 versus CD4 expression is with the recognition properties of the specific TCR expressed by a given T cell. In contrast to antibodies, which recognize external features of foreign proteins, TCRs, which mediate the specific recognition properties of T cells, recognize either foreign peptides bound to self-proteins of the MHC or foreign MHC proteins. This requirement for foreign peptide antigen to be bound to self-MHC proteins for T-cell recognition is referred to as MHC restriction of T cells. The MHC proteins involved in this restriction are highly polymorphic from one individual to another and fall into two major categories: class I proteins (HLA-A, -B, and -C), which are the classical transplantation antigens and are responsible for rejection of foreign tissues, and class II proteins (HLA-DP, -DQ, and -DR), which are the products of "immune-response" genes. T cells with receptors that recognize foreign peptides bound to self-class I MHC proteins (or to foreign class I MHC proteins in the absence of foreign peptide antigen) express CD8. Such cells are usually, although not always, cytotoxic (or suppressor) in function. In contrast, T cells with receptors that recognize foreign peptides bound to self-class II MHC proteins (or foreign class II proteins in the absence of foreign peptide antigen) express CD4 and are generally helper or inducer in function. Importantly, certain cytotoxic T cells express CD4 and not CD8, and are specific for class II MHC proteins.

Early studies with monoclonal antibodies (mAbs) specific for CD8 and CD4 indicated that these proteins are important in the function of the T cells that bear them, because T-cell function could be blocked by incubation of T cells with such mAbs. These findings, coupled with the knowledge of the correlation between accessory molecule expression and class of MHC protein recognized suggested the hypothesis that CD8 and CD4 might themselves be receptors for class I and class II MHC proteins, respectively (Fig. 1). However, in contrast to the TCR, which recognizes polymorphic regions on MHC proteins, CD8 and CD4 were hypothesized to recognize invariant regions on these proteins, or at least regions that are almost identical in all individuals, because CD8 and CD4 themselves did not vary in cells of a given individual or from one individual to another. Strong evidence in favor of this hypothesis comes from studies of cells, which, as a result of genetic engineering, have large amounts of either CD8 or CD4 proteins on their surface. These cells, as well as artificial membrane vesicles displaying high levels of these proteins, can bind to other cells or membrane preparations bearing surface class I or class II MHC proteins, respectively, but not to cells or membranes lacking the appropriate class of MHC protein. These interactions can be inhibited by mAbs specific for either the accessory molecule or the MHC protein to which it binds. The

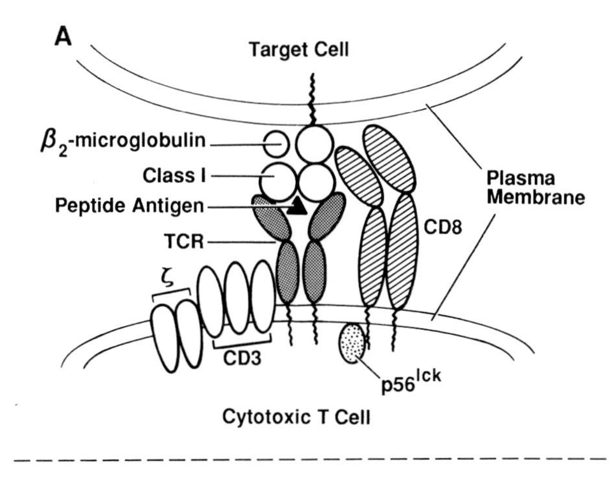

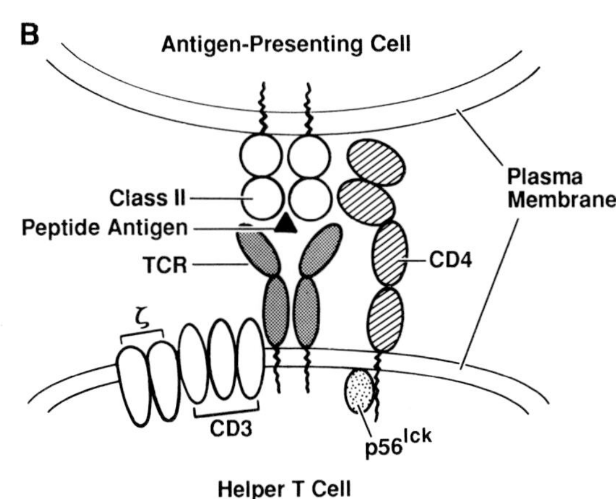

FIGURE 1 Model for interaction between CD4 or CD8 and major histocompatibility complex (MHC) proteins. (A) CD8 on a cytotoxic T cell is shown binding to a nonpolymorphic region of a class I MHC protein on a target cell and interacting with a T-cell receptor (TCR) molecule that is binding both to the same class I MHC protein and to a foreign peptide. CD8 is also shown to be associated with the T cell–specific tyrosine kinase p56^{lck}. (B) CD4 on a helper T cell is shown binding to a nonpolymorphic region of a class II MHC protein on an antigen-presenting cell and interacting with a TCR molecule that is binding both to the same class II MHC protein and to a foreign peptide. As in the case of CD8, CD4 is also shown to be associated with the T cell–specific tyrosine kinase p56^{lck}.

ability of CD8 and CD4 to bind to MHC proteins is thought to aid T-cell responses in two ways: either by increasing the strength of the interaction between the T cell and its target or antigen-presenting cell (i.e., an adhesive role) and/or by facilitating signal transduction in the interacting T cell to elicit its specific response. It has also been suggested that these molecules may under certain circumstances transmit negative signals to T cells, blocking their effects. The functional properties of CD8 and CD4 are described in greater detail in Section IV.

IV. CD8 Protein and Gene Structure and Expression

A. Polypeptide Structure and Subunit Composition

Early biochemical studies of human CD8 led to the conclusion that this molecule consists of disulfide-linked dimers and higher multimers of a single 34-kDa polypeptide chain on peripheral blood T cells, but, as discussed below, it is now known that it consists of dimers of two different polypeptides. In the thymus it was noted that the higher multimers contained an additional, larger polypeptide of 46 kDa. The latter has been identified as CD1, a protein of unknown function expressed on some human immature thymocytes and on some tumor cell lines. Although CD1 is related to class I MHC proteins, it is encoded on a different chromosome. The significance of the association of a portion of thymic CD1 with CD8 is not clear. However, CD1 is not associated with CD8 in human mature peripheral T cells (which do not express CD1) or in mouse thymocytes. Recent studies have suggested that a portion of CD8 on human peripheral T cells is disulfide-linked to class I MHC proteins. Again, the significance of this finding is not known and the geometry of such an interaction would have to be distinct from that of the noncovalent interaction between CD8 on a T cell and class I MHC proteins of a target cell or antigen-presenting cell.

The initial conclusion that CD8 on human T cells is primarily a homodimer was surprising when one considers that both mouse and rat CD8 had been identified as heterodimers of two distinct polypeptide chains, CD8α and CD8β. The one human CD8 polypeptide chain that had been identified and studied with mAbs was shown to be the homologue of the mouse CD8α chain. Molecular genetic studies have since demonstrated that the physiological form of human CD8, like its rodent counterparts, is a disulfide-linked heterodimer. The vast majority of CD8$^+$ human peripheral blood T cells express both chains of CD8 (α and β). Those few CD8$^+$ cells that are CD8β^- (1–10%) may be a different cell type, perhaps natural killer (NK) cells or another subset of T cells. Introduction of isolated genes into CD8$^-$ cells have demonstrated that CD8α can be expressed on the cell surface as a homodimer, whereas CD8β can only be expressed together with CD8α as a heterodimer. [*See* NATURAL KILLER AND OTHER EFFECTOR CELLS.]

B. CD8α Complementary DNA and Gene Structure

The isolation of complementary DNA (cDNA) clones encoding human CD8α has permitted the prediction of the amino acid sequence of human CD8α from the DNA sequence. The most striking conclusion to be drawn from the protein sequence was that CD8α is a member of what has been called the "immunoglobulin (Ig) gene superfamily," a large family of genes that are evolutionary related to immunoglobulins. This conclusion is based on the presence in CD8α of an amino-terminal domain that is homologous to Ig variable (V) regions, especially those of the Ig light chain. This domain of CD8α is of the same size as Ig V and constant (C) regions (~100 amino acids) and contains many of the conserved residues of members of the Ig gene superfamily, including the centrally placed disulfide loop that is so characteristic of Ig "homology units." Computer analyses of the structural characteristics of this V-like domain suggest that this region can fold in a similar manner to Ig domains. CD8α was the first example of a T-cell differentiation antigen that was found to be a member of the Ig gene superfamily, a finding that has now been extended to many other T-cell surface markers.

The predicted CD8α protein has a signal peptide of 21 amino acids needed for insertion of the protein into the membrane and subsequently cleaved off, followed by 214 amino acids, which constitute the protein in its final form. The protein crosses the cell membrane. Its external portion consists of the 96-amino-acid V-like domain and, close to the membrane, a 65-amino-acid hinge-like region or connecting peptide. This is followed by a 24-amino-acid hydrophobic segment embedded in the membrane and a highly basic 29-amino-acid cytoplasmic tail. The sequence suggests the existence of a site for N-glycosylation (Asn-X-Ser or Asn-X-Thr), which, however, is not used owing to the presence of proline as the variable residue (Asn-Pro-Thr) in the CD8α sequence. The protein does contain O-linked glycosylation, and this is thought to be primarily in the hinge region. CD8α contains 9 cysteine residues: three in the V-like region (two of which form the Ig-like disulfide loop), and two each in the hinge, transmembrane region, and cytoplasmic tail. Biochemical studies of the human protein have demonstrated that the cysteines in the V-like region are not involved in interchain disulfide bridges, and studies in the mouse system indicate that dimerization relies on external cysteine residues. It is therefore likely that either or both of the cysteine residues in the hinge region contribute to dimerization with CD8β or with another CD8α chain in the absence of CD8β.

The complete human CD8α gene spans approximately 6.5 kb of DNA and consists of six coding regions (exons) separated by five introns (noncoding regions). The gene structure supports the evolutionary relation between CD8α and Ig genes. During the splicing for removing introns a different form of CD8α mRNA is generated in human T cells and cell lines, which lacks the sequence encoding the transmembrane region. This mRNA is predicted to encode a form of protein that would be secreted from the cell. Indeed such a secreted form of CD8α has been identified, but its function, if any, remains unknown.

The human CD8α gene resides on the short arm of chromosome 2 (2p12) at a location closely linked to the human Ig κ light-chain locus. A close linkage to κ is also observed in mice, supporting the hypothesis that CD8α and κ derived from a common ancestral gene. Despite their similarities to Ig genes, the mouse and human CD8α differ from them in two important respects: they are present in a single copy and do not require rearrangement for expression.

C. CD8β cDNA and Gene Structure

Human thymocyte cDNA clones corresponding to CD8β encode a mature protein of 189 amino acids with 143 residues external to the cell, and transmembrane and cytoplasmic domains of 27 and 19 residues, respectively. The predicted protein is similar to mouse and rat CD8β, with approximately 56% identical amino acids. Computer analyses of the CD8β sequence have shown that it contains not only an amino-terminal Ig V-like domain, but also a 12-amino-acid sequence with marked similarity to an Ig joining segment (J) immediately thereafter.

On normal human $CD8^+$ T cells, CD8β is present as a heterodimer with CD8α. That this is the most common form was shown using a CD8-specific monoclonal antibody, which binds to the surface of cells expressing both CD8α and CD8β cDNA at their surface, but not cells expressing either alone. This mAb was shown to bind to the surface of almost all human peripheral blood T cells expressing CD8.

From mRNA studies there seems to be a potentially secreted form also of CD8β protein, but it is not known whether the protein is actually made and if so, whether it is secreted.

The human CD8β gene is located on the same chromosome as the CD8α gene (i.e., chromosome 2). Similarly, the two genes are on the same chromosome in mice, where they are very close. The human CD8β gene spans approximately 20.5 kb and contains six exons. As in the case of CD8α, the gene organization contains many of the features characteristic of members of the Ig gene superfamily. Like CD8α the CD8β gene does not rearrange.

D. Regulation of CD8 Expression

CD8 has a rather restricted tissue distribution of expression. To date it has been described only on thymocytes and mature T cells, and in some instances on NK cells. The major level of control of CD8 expression is transcriptional, but definitive studies have not been done.

There are no data regarding the mechanisms by which the CD8α and CD8β genes are turned on early in thymocyte development or how their expression is later either maintained (in the case of cells bearing TCRs recognizing class I MHC proteins) or turned off (in the case of cells bearing TCRs recognizing class II MHC proteins). Although the expression of the CD8α and CD8β genes is coordinate in most instances, there are some examples of cells that express only CD8α and not CD8β. The two genes must therefore also be subject to certain independent regulatory mechanisms, which are not yet defined.

III. CD4 Protein and Gene Structure and Expression

A. Polypeptide and cDNA Structure

CD4 is a glycoprotein of approximately 55 kDa, which, in contrast to CD8, immunoprecipitates as a single polypeptide chain. The isolation of human cDNA clones encoding CD4 allowed the prediction that the mature human CD4 polypeptide chain has 435 amino acids, preceded by a 23-amino-acid signal peptide. The protein has two potential N-linked glycosylation sites. There are six cysteine residues external to the cell, and these are connected by disulfide bonds forming three loops. CD4 has a hydrophobic transmembrane segment of 26 amino acids followed by a basically charged cytoplasmic tail of 38 amino acids.

CD4, like CD8α and CD8β, it is a member of the Ig gene superfamily, with an amino-terminal portion (approximately 100 amino acids) homologous to Ig V regions and two cysteine residues characteristic of Ig homology units. Computer predictions of secondary structure potential suggest that this region may fold in a similar fashion to Ig. This V-like region is followed by a short sequence that is similar to Ig J segments, although it is not nearly as closely related as that in CD8β. In contrast to CD8, which is shorter, CD4 has a second severely truncated V-like domain (V′) just downstream of the J-like segment. This truncated V′ domain may have originated as a duplication of part of the main V-like domain.

B. CD4 Gene Structure

CD4 is encoded by a single gene on human chromosome 12. As is the case for CD8α and CD8β, this gene does not rearrange before expression. The gene spans approximately 35 kb and is believed to consist of nine exons separated by eight introns. The most unusual feature of the gene is the presence of a large intron of about 12 kb dividing the sequence encoding the amino-terminal V-like domain into two exons (exons 2 and 3) approximately halfway through the predicted protein domain. This is an unusual feature for members of the Ig gene superfamily in that each Ig homology unit is usually encoded within a single exon. However, the same feature is present in another Ig-homologous gene, the gene for the neural cell adhesion molecule (N-CAM). The remainder of the intron/exon structure of the CD4 gene correlates well with the predicted protein domains, a feature characteristic of members of the Ig gene superfamily.

C. CD4 Expression

Although CD4 protein has only been identified on thymocytes and mature T lymphocytes in mice, in both rats and humans it has additionally been identified on the surface of macrophage/monocyte cells and the related Langerhans cells of the skin. Full-length human CD4 mRNA and protein has also been identified in brain tissue. Macrophages are likely to be the predominant class of CD4^{+} cells in brain, but the protein also appears to be expressed on at least some neuronal and glial cells. Normal human B cells do not express CD4, but some Epstein-Barr virus–transformed human B-cell lines do. Full-length CD4 transcripts have also been demonstrated in human granulocytes, although these are not known to express cell surface CD4. The physiological role (if any) of CD4 on non-T cells is unknown. The regulatory mechanisms governing CD4 gene expression have not yet been elucidated.

IV. Function of CD8 and CD4

A. Role of CD8 and CD4 in T-Cell Responses

The initial evidence that CD8 and CD4 play a role in T-cell function came from studies using antisera and subsequently monoclonal antibodies specific for these proteins. These antibodies block all antigen-driven functions (e.g., cytotoxicity, proliferation, lymphokine release) in T cells that express the corresponding molecule and are specific for the appropriate class of MHC protein, although there is clear heterogeneity in the ability of various T-cell clones to be blocked. In the case of cytotoxicity, anti-CD8 mAbs block the formation of conjugates (relatively stable cell–cell contacts) between class I–specific cytotoxic T lymphocytes (CTL) and target cells. Anti-CD4 mAbs block cytotoxicity by class II–specific $CD4^+$ CTL and induce dissociation of preformed conjugates between such CTL and target cells. These results suggested a role for CD8 and CD4 in the recognition step as opposed to the lytic machinery and led to the hypothesis that the function of CD8 and CD4 is to increase the strength of the interaction between T cells and antigen-presenting or target cells by binding to class I or class II MHC molecules, respectively. As would be predicted by this model, T cells bearing TCR molecules with apparently low affinity for antigen/MHC were found to be more dependent on CD4 or CD8 interactions (i.e., more easily blocked), whereas T cells with higher-affinity TCRs were less dependent on CD4 or CD8. Finally, as discussed above, binding studies have indeed shown that CD8 and CD4 can act as receptors (although with apparently low affinity) for class I and class II MHC proteins, respectively. In the case of CD8, binding studies using cells expressing extremely large amounts of surface CD8 (because of genetic manipulation) have demonstrated that amino acids in the third external domain (i.e., the domain of class I MHC proteins closest to the membrane) are important for binding of these cells to B-cell lines expressing human class I MHC proteins. Two important points emerge from the localization of the binding site CD8-MHC-I. First, the site is in the third external domain of class I MHC proteins, which is the most highly conserved domain, a feature that would be expected for binding of the nonpolymorphic CD8 protein. Second, binding of CD8 to this region would theoretically allow a TCR molecule and a CD8 molecule to bind simultaneously to the same class I protein because current evidence indicates that the TCR engages the more polymorphic external two domains. The residues on class II MHC proteins important for binding of CD4 have not yet been identified; however, amino acids in both of the first two external domains of CD4 appear to be important for this interaction.

Recent studies have suggested that CD8 and CD4 may not be simply cellular adhesion molecules and that they may additionally be involved in pathways of signal transduction. CD8 has been found to play some role in the initial, antigen-independent formation of conjugates between T cells and other cells, although other molecules (CD2 and LFA-1) may play a more important role in this process. The CD8 role is more important at a later step because mAbs specific for CD8 block cytotoxicity not only during the phase of conjugate formation, but also during the cytolytic phase. Similarly, mAbs specific for CD4 inhibit cytotoxicity by class II–specific $CD4^+$ CTL clones at a postbinding step and have only a small effect on the initial formation of conjugates. Anti-CD4 mAb can also induce the dissociation of preformed conjugates, although this dissociation is much slower and requires higher temperatures than that induced by anti-LFA-1 mAb.

A variety of studies have shown that mAbs specific for CD8 or CD4 can inhibit T-cell activation induced by lectins or by mAbs specific for CD3 or the TCR, although these activating reagents do not interact with CD8 or CD4. These findings led to the hypothesis that CD8 and CD4 might function to deliver a negative signal to the T cell, thereby inhibiting activation. In support of this hypothesis, anti-CD4 mAbs have been shown to block the lectin- or antigen-induced rise in cytoplasmic free Ca^{2+} in a $CD4^+$ T-cell clone, and the anti-CD3 mediated mobilization of cytoplasmic free Ca^{2+} in human peripheral blood T cells. However, there is still no direct evidence for transmission of a negative signal via CD8 or CD4, and gene transfer studies support the concept that the primary physiological function of these molecules is to enhance rather than to block T-cell function. The studies supporting the negative signal model involve mAb blocking, which might interfere sterically with the function of neighboring molecules.

The possibility that mAbs binding to CD8 or CD4 could interfere with T-cell activation sterically is supported by the growing body of evidence that there may be a physical association between these proteins and the TCR/CD3 complex (Fig. 1). For example, a variety of studies have demonstrated that CD8 and CD4 are internalized into the cell together with the TCR/CD3 complex on T-cell stimulation with either antigen or phorbol esters, or with mAbs specific for the TCR/CD3 complex. CD4 has also been shown to cluster together with the TCR to the area of intercellular contact between T cells and antigen-presenting cells during antigen-specific recognition. Finally, a recent study has demonstrated that both CD4 and CD8 can be immunoprecipitated together with the TCR.

An association between CD4 and/or CD8 and the TCR/CD3 complex raises the possibility that CD4 and CD8 might be involved directly or indirectly in the transmission of positive signals to activate T cells. CD4 and CD8 have both been shown to be rapidly phosphorylated at serine and then dephosphorylated on T-cell activation, but the role of this phosphorylation (if any) is not known. Cross-linking of CD4 or CD8 to the TCR/CD3 complex with mAbs can activate resting T cells and induce proliferation under conditions in which anti-CD3 alone does not, and can markedly increase other measures of T-cell activation (e.g., mobilization of intracytoplasmic free Ca^{2+} or synthesis of inositol phosphates). Perhaps most convincing, CD8 and CD4 have both seen shown to be associated with a T cell–specific tyrosine kinase, $p56^{lck}$, which is localized on the inner surface of the plasma membrane (Fig. 1). Cross-linking of CD4 has been shown to activate this tyrosine kinase and to be associated with phosphorylation of the ζ chain of the TCR/CD3 complex, an event that occurs during T-cell activation. Although such results do not imply direct transmission of a signal via CD8 or CD4, they do suggest that these molecules may enhance signal transduction by their association with the TCR/CD3 complex and that this is accomplished at least in part by means of the association of CD8 and CD4 with $p56^{lck}$.

The most direct demonstration of a physiological role for CD8 and CD4 in T-cell activation has come from recent gene transfer studies. The CD8 or CD4 genes have been introduced into functional T cells, and the effects of cell surface expression of CD8 or CD4 on antigen responses have been examined. These studies have shown that CD8 or CD4 can markedly increase antigen responses as long as the appropriate ligand (i.e., class I or class II MHC proteins, respectively) is present on the antigen-presenting or target cell. The requirement for and effects of CD8 or CD4 expression vary depending on the particular T cell and its TCR. In some instances, no response to antigen is present in the absence of one of these "co-receptor" molecules. In such cases, the expression of CD8 or CD4 is absolutely required for a measurable antigen response. In other cases, a basal response is present even in the absence of CD8 or CD4 but is increased by its presence. Interestingly, all the functional studies with gene transfer of CD8 have involved only CD8α and not CD8β. Clearly, homodimers of CD8α are capable of enhancing T-cell antigen responses in the absence of the CD8β chain. It is not known whether the physiological heterodimer (CD8$\alpha\beta$) would function better or differently. The question arises whether the major role played by the CD8β chain is only evident during T-cell differentiation in the thymus. If such a role involves expression of surface CD8β in isolation from CD8α, the hypothesis cannot be readily examined, because CD8β requires heterodimer formation with CD8α for surface expression.

The ability to assay CD8 and CD4 function by gene transfer into functional T cells has provided a mechanism for establishing the role of different portions of these molecules in enhancing T-cell responses. Thus, it has been demonstrated that these proteins function optimally when their cytoplasmic tails are present, although in some instances they can enhance T-cell responses in their absence. This is likely to be at least partially related to the importance of the cytoplasmic tail for the association of CD8 and CD4 with $p56^{lck}$. The cytoplasmic tail of CD4 has also been shown to be important for internalization of CD4 into the cell. The external portions of these molecules are clearly required for appropriate ligand binding.

Gene transfer experiments have provided strong support for the notion that CD8 and CD4 can enhance responses by either of two mechanisms: increased adhesion and signal transduction. They have also demonstrated that CD4 and CD8 can have different effects depending on whether they can bind to the same MHC molecule as the TCR and/or only at a different site. In the mouse system, studies with blocking by monoclonal antibodies had suggested that CD8 and CD4 could enhance responses in certain but not all instances by binding to MHC proteins to which the specific TCR could not bind. This has been demonstrated in a more direct fashion by gene transfer. For example, gene transfer of human CD8 into a mouse T cell specific for a human class II molecule can stimulate the response to class II as long as class I MHC proteins (the CD8 ligand) are expressed on the antigen-presenting cell. Similarly, a CD4–class II MHC interaction can stimulate the antigen response of a T-cell hybridoma specific for a class I MHC protein. However, this is not always the case, and there are clear instances when only the appropriate co-receptor–MHC protein interaction can stimulate a response. These appear to be cases in which the affinity of the TCR for its ligand is low, and hence the cell is more dependent on a co-receptor. It appears likely (although definitive proof is lacking) that enhanced signal transduction mediated by CD8 or CD4 will only occur when one of these molecules and the TCR bind to the same MHC protein and/or functionally associated with the TCR and that responses are stimulated only by an adhesion component when binding of CD4 or CD8 to its ligand is independent of the TCR. A strictly adhesion function of CD4 or CD8 may not be sufficient to allow detectable responses in the case of low-affinity TCRs but may provide clear enhancement of responses mediated by TCRs with somewhat higher affinity.

Regardless of whether there is a clear division of function based on binding to the same or different MHC molecules, the bulk of evidence indicates that depending on the specific TCR involved (particularly its affinity for a given antigen/MHC) and the antigen/MHC density (or concentration), some responses appear to require binding of CD4 or CD8 to the same MHC molecule as the TCR, some only need binding at sites apart from the TCR/CD3 complex, and some are totally independent of CD4 or CD8 interactions. What remains unclear is the mechanism(s) by which these molecules enhance function. The stimulation from their adhesion function appears to be a result of increasing the strength of the interaction between the T cell and target or antigen-presenting cells. The mechanisms involved in enhancing responses when the TCR and CD4 or CD8 bind to the same MHC protein molecule are still not entirely clear. It appears likely that the association of CD4 and CD8 with the tyrosine kinase $p56^{lck}$ allows increased signal transduction when CD4 and CD8 are able to associate with the TCR/CD3 complex. It is also possible that an association between CD4 or CD8 and the TCR/CD3 complex alters the conformation of the latter either to increase its affinity for antigen/MHC or to facilitate signal transduction (perhaps in ways other than the proximity of $p56^{lck}$). It has not been excluded that CD4 or CD8 might directly transmit positive signals to the T cell (i.e., in a manner independent of the TCR/CD3 complex). Another suggestion has been that internalization of CD4 or CD8 into the cell in response to activating stimuli might lower the threshold for T-cell triggering. However, it has been shown that internalization of neither CD4 nor the TCR is an absolute requirement for T-cell activation, at least when activation is induced by mAbs.

B. Role of CD4 as the Receptor for HIV-1

In addition to its critical role in T-cell function, CD4 has been shown to serve another important, albeit nonphysiological function as the cell surface receptor for human immunodeficiency virus-1 (HIV-1), the retrovirus that is responsible for causing acquired human immunodeficiency syndrome (AIDS). *In vitro* infection of $CD4^+$ cells by HIV-1 can be inhibited by some mAbs directed against CD4 but not by mAbs specific for other cell surface molecules. Expression of CD4 introduced by gene transfer confers susceptibility to HIV-1 infection on human cells that otherwise lack CD4 expression and hence are resistant to HIV-1. Cell surface expression of CD4 is required not only for viral infection, but also for cell fusion (syncytium formation) mediated by HIV-1. CD4 binds to gp120, the exterior envelope glycoprotein of HIV-1. Notably, mouse CD4 does not bind HIV-1 despite its homology to human CD4, and mouse cells expressing human CD4 bind HIV-1 but are not susceptible to infection. Numerous studies have shown that genetically engineered soluble forms of human CD4 block infection by HIV-1, leading to the testing of such forms as a therapeutic agent for treatment of AIDS. A variety of studies have indicated that the binding site for gp120 is located within the amino-terminal Ig V-like domain, and amino acids between residues 40 and 55 appear to be of major importance. It remains possible that other regions within the amino-terminal half of the protein play an additional role. Although the gp120 binding site also seems to be important for CD4 interaction with class II MHC proteins, mutations in residues that do not affect gp120 binding can inhibit interaction of CD4 with class II MHC proteins, implying that the two sites are different. On gp120 the binding site(s) for CD4 is located within the carboxyl-terminal half of gp120. Amino acids 410–421 have been shown to be critical for binding, but other sequences and tertiary structure may also be important. Binding of HIV-1 to cell surface CD4 has been found to induce a rapid increase in the phosphorylation of CD4, similar to that seen with antigen activation. However, binding of isolated gp120 is not sufficient to induce CD4 phosphorylation, and elimination of the phosphorylation sites in the CD4 cytoplasmic tail (or even most of the cytoplasmic tail) does not block viral entry.

HIV-1 infection has also been shown to result in a loss or decrease of cell surface expression of CD4. Loss of surface CD4 correlates with the presence of intracellular complexes between the envelope glycoproteins (gp120 and gp160) and CD4, suggesting that the reduction of surface CD4 may be a consequence of altered processing and localization of this protein in cells infected with HIV-1.

HIV-1 infection may disrupt the function of $CD4^+$ cells in a variety of ways. The absolute number of $CD4^+$ cells decreases during HIV-1 infection. This may be a result of several mechanisms: cell death after syncytium formation mediated by CD4–gp120 interaction, direct cell death from infection, antibody-dependent cellular cytotoxicity (ADCC) of cells bearing surface gp120, and disappearance of CD4 from the cell surface of infected cells as described in this section. Immune function would be compromised in all cases because of the central role of $CD4^+$ cells in immune responses and the functional importance of CD4 on those cells. However, additional mechanisms are most likely involved, because only a small proportion of $CD4^+$ cells are infected by the virus, and immune responsiveness of $CD4^+$ cells decreases before cell numbers decrease. Soluble gp120 has been shown to inhibit phytohemagglutinin-induced proliferation of $CD4^+$ T cells and to block the enhanced responsiveness of T cells induced by the expression of human CD4 after gene transfer. Thus gp120 mimics the inhibitory activity of many anti-CD4 mAbs. These findings suggest that gp120 released in the bloodstream of infected patients could bind to cell surface CD4 and decrease the immune responsiveness directly, as well as make noninfected $CD4^+$ cells targets for destruction by ADCC.

V. Summary

The cloning of the cDNA and genes encoding CD8 and CD4 has provided the first detailed knowledge of their protein sequences and predicted structures. Both chains of CD8 and the single chain of CD4 were found to be members of the Ig gene superfamily, and the amino-terminal domain of each appears to be capable of folding like an Ig V region. There is strong evidence that these molecules do indeed bind to class I and class II MHC molecules, respectively. It is likely that it is the V-like portions of these molecules that are responsible for the binding. The site on class I MHC proteins that has been identified as being important for CD8 binding is within the third external (membrane-proximal) domain. The CD4 binding site on class II molecules is currently under investigation. Human CD4 has also been demonstrated to have another ligand, namely, the gp120 protein of HIV-1, and the availability of the cloned CD4 cDNA has led to a proliferation of studies that have increased our understanding of the interaction between HIV-1 and $CD4^+$ cells.

For both CD4 and CD8, the isolation of molecular clones has added a new dimension to studies of the normal function of these proteins, providing an important complement to the mAb blocking studies. It is evident from the results of both types of analyses that these molecules play more than a single role in the complicated processes of T-cell recognition and activation. At a minimum they are involved both in adhesion and in T-cell triggering. It is likely that the latter function involves an association between CD8 or CD4 and the TCR/CD3 complex. It is also clear that with some but not all T cells, CD8 and CD4 can enhance T-cell responses whether they bind to the same or different MHC molecules as the TCR. However, the role played by these molecules may be different in the two cases [i.e., signal transduction (plus adhesion) or only adhesion, respectively.]

Despite the strong progress on the understanding of CD4 and CD8 function, there are still many open questions regarding how these molecules work. We know little about the mechanisms of regulation of expression of CD8 and CD4 during T-cell development or the means by which T cells that are restricted by class I or class II MHC proteins become CD8 or CD4 single-positives, respectively. The latter most likely involves a selection process related to the ability of CD4 or CD8 to bind the same MHC molecule as the expressed TCR, but how this happens mechanistically and developmentally remains obscure. It is likely that the combination of molecular genetics, biochemistry, and cell biology will eventually fill in the gaps in our current knowledge.

Bibliography

Biddision, W. E., and Shaw, S. (1989). CD4 expression and function in HLA class II–specific T cells. *Immunol. Rev.* **109,** 5–15.

Bierer, B. E., Sleckman, B. P., Ratnofsky, S. E., and Burakoff, S. J. (1989). The biological roles of CD2, CD4, and CD8 in T-cell activation. *Annu. Rev. Immunol.* **7,** 579–599.

Doyle, C., Shin, J., Dunbrack, R. L., Jr., and Strominger, J. L. (1989). Mutational analysis of the structure and function of the CD4 protein. *Immunol. Rev.* **109,** 17–37.

Lifson, J. D., and Engleman, E. G. (1989). Role of CD4 in normal immunity and HIV infection. *Immunol. Rev.* **109,** 93–117.

Parnes, J. R. (1989). Molecular biology and function of CD4 and CD8. *Adv. Immunol.* **44,** 265–311.

de Vries, J. E., Yssel, H., and Spits, H. (1989). Interplay between the TCR/CD3 complex and CD4 or CD8 in the activation of cytotoxic T lymphocytes. *Immunol. Rev.* **109,** 119–141.

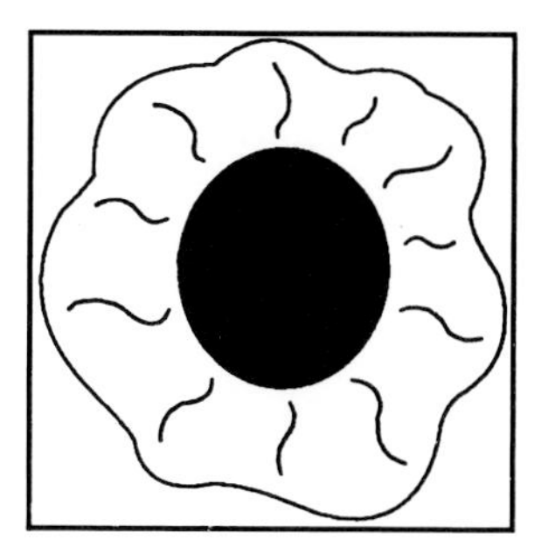

Cell

HENRY TEDESCHI, *State University of New York at Albany*

Glossary

Axon Elongated process of nerve cells; nerve fibers of most vertebrates contain many axons forming a bundle

Cell culture Growth of cells in a tissue-culture dish or flask in a partially defined medium

Cytoskeleton Complement of fibers and tubules forming a network in the cytoplasm

DNA Substance carrying the genetic information of cells

Energy transduction Transformation of energy from one form to another (e.g., from a chemical reaction (adenosine triphosphate hydrolysis) to a mechanical event (muscle contraction)

Gamete Sperm or ovum

Messenger RNA Ribose nucleic acid carrying the transcript of the genetic information

Ribosomes RNA-protein particles involved in protein synthesis

Polysomes Complex of messenger RNA and several ribosomes in which each ribosome is attached to a nascent peptide chain

Transcription Synthesis of RNA complementary to the DNA strand, which serves as its template

Translation Synthesis of polypeptide using the instructions contained in the nucleotide sequence of messenger RNA

Transfer RNA RNA serving to translate the triplet nucleotide code into an amino acid sequence

A CELL IS A COMPARTMENT enclosed by a membrane and capable, at least under circumscribed conditions, to exist independently. All living organisms are composed of cells—either one or many. Unicellular organisms are those in which a single cell manages the metabolic processes involved in nutrition, growth, maintenance, and vegetative reproduction. In multicellular organisms, different cell types carry out specific physiological functions, which amplify some of their abilities and restrict others; i.e., they are differentiated. Because their genetic makeup is usually identical (the notable exception being B lymphocytes of the immunological system, where hypermutation is a mechanism allowing for the production of antibody diversity), cells of different groups are said to differ in gene expression.

Maintenance of the living cell requires the precise orchestration of many complicated chemical reaction sequences. The level of regulation and specificity of these reactions can only be achieved in a carefully designed environment and on intricately ordered catalytic structures. These two requirements can only be fulfilled by cells that are, as often stated, the basic units of life. The cell or plasma membrane enables cells to maintain an internal environment suitable for those reactions that they need to build, repair, and replicate themselves. Without exception, this environment is different from the outside environment. Nevertheless, cells are not isolated from their environment, but rather they continuously communicate with it, exchanging materials and energy with it. Furthermore, in multicellular organisms, communication is continuous among the different cells, just as there is communication among the various intracellular compartments.

I. Cell Systems

Organisms are classified as prokaryotes or eukaryotes on the basis of their cellular organization. In contrast to eukaryotes, prokaryotes, which include bacteria and cyanobacteria, are most frequently unicellular and have no organelles or well-defined nucleus. Prokaryotic cells lack a cytoskeleton, and their DNA, present in a circular structure, is attached to the plasma membrane. Some prokaryotes can function metabolically either aerobically or anaerobically, and others only in one of these two modes. A variety of bacteria can carry out photosynthesis. Some specialized bacteria are chemolithographs; i.e., they function metabolically by oxidizing inorganic compounds.

Eukaryotes include protists, plants, and animals. Their cells, most frequently elongated, are generally larger than in prokaryotes, ranging in length from a few micrometers to several centimeters, or even longer in the case of special processes such as the axons of neurons. The reactions and structures of cells are present in separate compartments or organelles surrounded by selective membranes. The DNA is present in the nucleus in very long linear molecules generally bound to histones. During cell division, the DNA condenses to form chromosomes, which are pulled apart by the spindle apparatus (mainly made up of microtubules), to form two daughter cells of identical genetic complement (see below).

Most of our knowledge about human cells has been derived from studies on eukaryotic cells, frequently mammalian cells in culture; however, prokaryotic cells have been easier to deal with in research and have permitted rapid advances. Their usefulness results in part from the short generation time (as little as 60 min in some bacteria), in part from their simplicity and the possibility of growing them in chemically defined media. These advantages have allowed rapid genetic and molecular studies, which would have been intractable in more complex systems.

II. Eukaryotic Cells

Eukaryotic cells are compartmentalized, where each compartment or organelle is enclosed by one or more membranes. The membranes form a sheet constituted by phospholipid bilayers with protein spanning them, the integral membrane proteins. The membranes offer a selective barrier to the diffusion of solutes. The nucleus, mitochondria, and, in plant cells, chloroplasts are surrounded by double membranes. Many other organelles and vacuoles have a single membrane. A variety of processes take place in these compartments in isolation from the rest of the cytoplasm.

III. The Membranes

A. The Plasma Membrane

Some of the plasma membrane lipids and proteins are glycosylated (i.e., conjugated to carbohydrate components, which project into the extracellular space). The position of integral membrane proteins across the phospholipid bilayers of the membranes allow them to play a fundamental role in linking the surface of the cell to the cytoskeletal network of the cytoplasm, thus integrating the entire cytoplasmic volume. Furthermore, they function in communication among cells, between cells and the extracellular environment (either the extracellular fluid or extracellular matrix), and in response to chemical signals.

Channels made up of integral membrane proteins permit the passage of specific solutes or water through the membrane. In specialized junctions between cells—the gap junctions—channels actually connect the cells by traversing the two plasma membranes, allowing intimate communication between the cytoplasms. Transport proteins (or transporters), also known as carriers, are integral proteins that transport solutes from one side of the membrane to the other. They have a major role in creating a distinct cytoplasmic environment, which differs from the extracellular medium. Many of these transport proteins act as ion pumps, notably the Na^+,K^+-adenosine triphosphatase (ATPase) of the plasma membrane and the Ca^{2+}-ATPase of the sarcoplasmic reticulum of muscle cells. The transport ATPase couples the uphill transfer of ions against an electrochemical gradient (both a concentration and an electrical potential) to ATP hydrolysis. In the test tube under special unphysiological conditions, all these pumps can be run on reverse (i.e., in the direction of the electrochemical gradient) to synthesize ATP from adenosine diphosphate (ADP) and inorganic phosphate (P_i). Some other

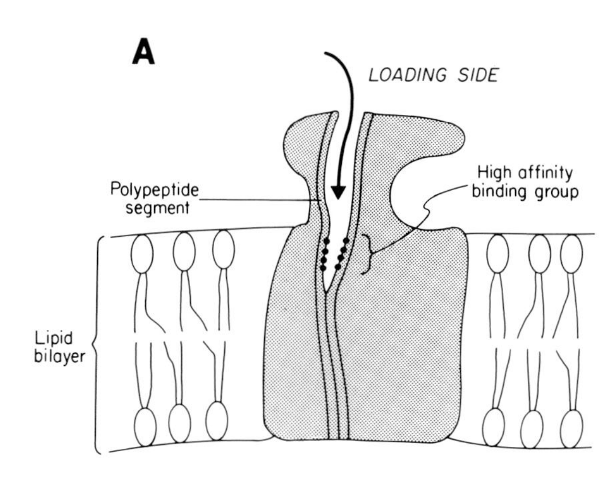

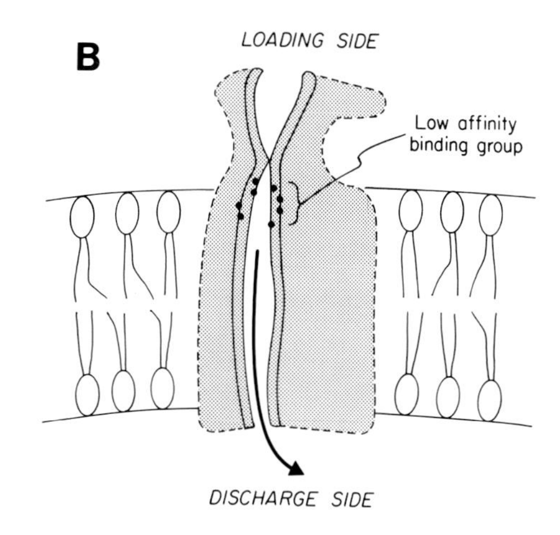

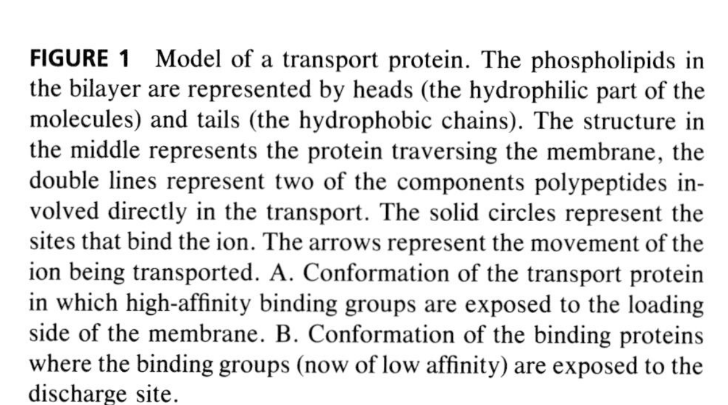

FIGURE 1 Model of a transport protein. The phospholipids in the bilayer are represented by heads (the hydrophilic part of the molecules) and tails (the hydrophobic chains). The structure in the middle represents the protein traversing the membrane, the double lines represent two of the components polypeptides involved directly in the transport. The solid circles represent the sites that bind the ion. The arrows represent the movement of the ion being transported. A. Conformation of the transport protein in which high-affinity binding groups are exposed to the loading side of the membrane. B. Conformation of the binding proteins where the binding groups (now of low affinity) are exposed to the discharge site.

pumps do not depend on the hydrolysis of ATP to provide the necessary energy; the passage of a solute is coupled to the flux of another solute (frequently the downhill influx of Na^+) downhill in the direction of its electrochemical gradient (i.e., the concentration gradient and the gradient imposed by the membrane potential [negative inside]). [*See* ADENOSINE TRIPHOSPHATE (ATP); CELL JUNCTIONS; CELL MEMBRANE TRANSPORT.]

The uphill pumping of ions (an active transport) coupled to ATP hydrolysis is not completely understood but apparently involves (1) the binding of the ion to high-affinity sites of the transporter molecule on the loading side of the membrane; (2) a phosphorylation of the transporter; (3) a conformational change in the transporter molecule, so that the binding sites acquire a lower affinity for the ion and they become accessible to the discharge site; and (4) a hydrolysis of the phosphorylated transporter. A model for transport is shown in Fig. 1. The structure in the middle of the phospholipid bilayer corresponds to the transporter molecule. The solid circles in the inside correspond to the groups that bind the transported molecule. Fig. 1A shows the availability of the binding groups (high affinity) on the loading side of the membrane. A conformational change of the transporter is represented in Fig. 1B; the shape of the transporter molecule has changed, and the binding groups (now low affinity because they are no longer clustered together) are now available to the discharge side of the membrane. The diagram does not show the phosphorylation of the transporter nor the hydrolysis of the phosphorylated protein responsible for the conformational change and required for an uphill transport.

The Na^+,K^+-ATPase is responsible for the pumping of Na^+ out of cells and K^+ into cells, so that the intracellular concentration inside cells is high for K^+ and low for Na^+. The tendency of K^+ to diffuse out of cells through K^+-channels is responsible for the resting potential (inside negative) of cells. In excitable cells, the influx of Na^+ through the Na^+-channels depolarizes the cell and reverses the potential so that the inside is now positive, a phenomenon known as the action potential. Recovery to the original resting potential occurs when the K^+ efflux compensates for the Na^+ entry. In elongated excitable cells such as nerve cells, or neurons, a depolarization wave corresponding to the action potential travels the length of the axon in the transmission of the electrical signal. The Ca^{2+}-ATPase of the sarcoplasmic reticulum (S.R.) of striated muscle pumps Ca^{2+} out of the cytoplasm into the S.R. to maintain a low cytoplasmic concentration of the ion (<1 μM). When released, Ca^{2+} triggers muscle contraction and other metabolic alterations. [*See* ION PUMPS.]

Many growth factors and hormones are peptides or chemicals unable to pass through the membrane. They act by binding to membrane receptors. These are integral membrane proteins that generally activate enzymes such as adenylate kinase or phospholipase C to produce second messengers at the cytoplasmic face of the cell membrane. The activation requires the mediation by another kind of protein, the G-proteins. Second messengers are intracellular chemicals whose signals produce a cascade of enzymatic events, which trigger and magnify a response. For example, the formation of cyclic adenosine monophosphate (cAMP) catalyzed by adenylate cyclase in response to the binding of epinephrine to its membrane receptor induces, through a series of enzyme-catalyzed reactions, the phosphorylation of phosphorylase *b* to produce the active form of the enzyme, phosphorylase *a*, which in turn catalyzes the breakdown of glycogen. Receptors and bound ligands are taken up by endocytosis (see Section IV.A). The physiological function of the receptor-ligand complex is generally independent of this uptake, which is thought to have a regulatory role. [*See* G PROTEINS.]

B. The Nuclear Envelope

The nucleus is enclosed by a specialized double membrane—the nuclear envelope. The nuclear envelope allows exchanges with the cytoplasm primarily through the nucleopore complexes, specialized protein complexes forming intricate passageways that link the cytoplasm with the interior of the nucleus. [*See* CELL NUCLEUS.]

Most proteins are synthesized in the cytoplasm by a process that follows a genetic blueprint provided by messenger RNA (mRNA). mRNA and the RNA of the protein-synthesizing machinery of the cytoplasm, such as ribosomes or transfer RNA, are transcribed from the DNA of the nucleus; therefore, a continuous traffic takes place between the nucleus and the cytoplasm through the nucleopore complexes. In addition, proteins have been shown to shuttle in and out of the nucleus by a similar mechanism. These proteins are thought to play a regulatory role during development. [*See* NUCLEAR PORE, STRUCTURE AND FUNCTION.]

C. Energy-Transducing Membranes

Specialized portions of the plasma membrane of bacteria, the inner mitochondrial membrane, or the *thylakoid vesicles* inside chloroplasts, house protein complexes (made up mostly of integral proteins), which have the unique function of transducing chemical oxidation-reduction reactions to the synthesis of ATP from adenosine diphosphate (ADP) and P_i. In photosynthetic bacteria and chloroplasts, the absorption of light powers the movement of electrons through protein complexes in the membranes—the electron transport chain (see Fig. 2)—which in turn translocates H^+ (i.e., protons) uphill across the membranes to form a proton electrochemical gradient. Similarly, in other bacteria and mitochondria, the oxidation of substrates powers the movement of electrons and the translocation of protons. The passage of protons downhill in the opposite direction, the direction of their electrochemical gradient, through the ATP-synthase complex (F_0F_1) produces ATP from ADP and P_i. In contrast to the transport ATPases, the ATP-synthases synthesize ATP under physiological conditions. The F_0 portion of the ATP-synthases is embedded in the membrane, whereas the F_1 portion is present in the water phase. The components of the F_1 complex of most living cells have been highly conserved, indicating a common evolutionary origin.

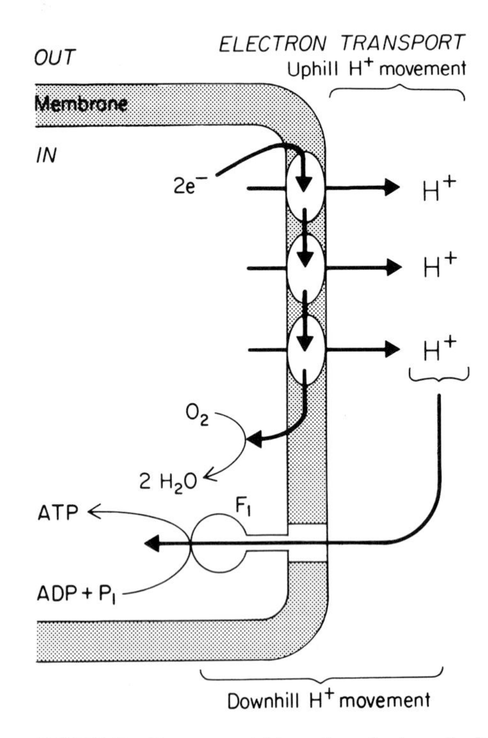

FIGURE 2 Processes taking place in transducing membranes. The process represented in this diagram corresponds to the events taking place in mitochondria. The arrows in the membrane represent the passage of electrons in the electron transport chain. Their passage is coupled to the uphill efflux of protons (powered by the electron transport). The downhill return of protons to the mitochondrial interior through ATP-synthase is coupled to the synthesis of ATP from ADP and P_i.

IV. The Intracellular Transport System

The cytoplasmic system of membranes is very complex and in a dynamic state. Material is taken up in endocytosis by invagination of the plasma membrane. In contrast, a variety of proteins synthesized in the endoplasmic reticulum (ER) are transferred to different sites in the cell or secreted. Secretion occurs by a process called exocytosis, in which the membrane of the vesicles becomes continuous with the plasma membrane and discharge their contents to the outside. Endocytosis, exocytosis, and intracellular transport are illustrated in Fig. 3.

A. Endocytosis

During endocytosis, a portion of the cell membrane first invaginates and then pinches off to form vesicles, referred to as the endosomes. Receptor-mediated endocytosis has been intensely studied. In this form of endocytosis, a ligand, such as a hormone or growth factor, is first bound to a specific receptor. This binding in some way triggers the formation of coated vesicles. On the surface of the plasma membrane, the receptors are either present in specialized structures, the coated pits, or they move to the coated pits after binding the ligand. The coated pits are small, specialized indentations at the surface. The coated vesicles are surrounded by clathrin arranged in a characteristic polyhedral basket structure. The coated pits invaginate to form vesicles and fuse to endosomes, in which the contents are processed to recycle either the receptor or the ligand, or both depending on the system. Alternatively, either ligand, receptor, or both are digested together with material taken up nonspecifically in endocytosis. Digestion occurs when the endosomes fuse with lysosomes, organelles containing hydrolytic enzymes.

Phagocytosis is a form of endocytosis in which larger particles (e.g., bacteria) are engulfed by cells (e.g., macrophages) to form large vacuoles.

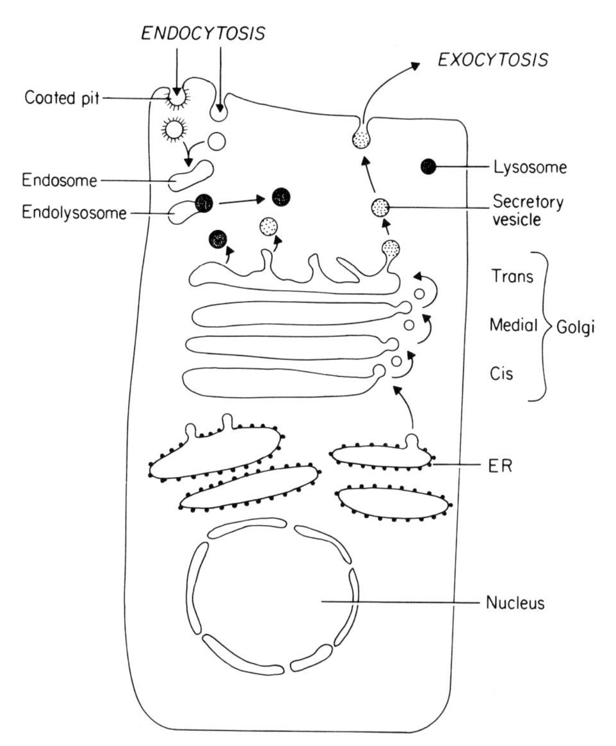

FIGURE 3 Diagram of the intracellular transport system of the cell. Materials continuously enter the cell by endocytosis (shown on left). Materials synthesized by the endoplasmic reticulum (ER) are either present in the lumen or in the ER membrane. They are exported to the Golgi system, and they are moved and processed biochemically from the *cis* to the *trans* end. The various components are segregated into vesicles such as lysosomes or secretory vesicles. The secretory vesicles are discharged by exocytosis (shown on right).

B. Transport of Newly Synthesized Proteins

Polypeptides destined for intravesicular or plasma membrane sites are inserted through the membrane of the ER while being translated by polysomes. They remain in the membrane when they are to become integral proteins. Alternatively, they are delivered to the interior of the ER vesicle. In either case, the polypeptide is guided through the membrane by a special amino acid sequence of approximately 20 amino acids at the amino end, the signal or leader sequence, which is cleaved almost immediately to produce the mature form of the protein. Oligosaccharides are attached by *N*-glycosylation (i.e., to an asparagine residue of the polypeptide) in the ER.

Mitochondrial or chloroplast polypeptides are mostly synthesized in the cytoplasm and subsequently are transferred into the organelles. Some are guided by special signal sequences, but others are thought to be already in their mature form and to be guided by targeting sequences in the interior of the molecule.

The Golgi complex (also called Golgi) (see Fig. 3) consists of a stack of cisternae (as many as six) distributed apically in cells. The *cis* face of the Golgi complex is closely associated with elements of the ER, whereas the *trans* side is generally toward the periphery of the apical cells. The oligosaccharide portion of the peptides are processed in the Golgi complex by the enzymatic removal of some of the carbohydrate in a process referred to as trimming. Other oligosaccharides are added in the Golgi apparatus to serine or threonine residues by O-linked glycosylation (e.g., in the case of proteoglycans). Processing starts at the *cis* portion of the Golgi system and progresses to the medial portion, and the processed material leaves at the *trans*-cisternae. From here it moves to a reticular network that segregates them specifically into separate vesicles for delivery to their final destination. The processing through the Golgi complex corresponds to highly localized, sequential enzymatic steps characteristic of each lamellar element. The transfer through the Golgi (ER to *cis* Golgi, to medial Golgi, to *trans* Golgi etc.) to the final destination (i.e., secretory granules, plasma membrane, lysosomes, etc.) occurs exclusively in vesicular compartments. The process of marking the various materials for delivery to a specific fate is referred to as targeting. [*See* GOLGI APPARATUS.]

The nature of targeting is not always clear. Lysosomal enzymes are selected to be packaged into lysosomes by their mannose-6-phosphate (M6P), which have been added to the N-linked oligosaccharides in the *cis* portion of the Golgi system. The selection is carried out in a process in which the M6Ps attached to the enzymes are bound to M6P receptors clustered on the inner surface of the membrane of vesicle budding-off from the network of tubular vesicles connected to the Golgi apparatus. Proteins destined to the plasma membranes or continuously secreted (in a process known as constitutive secretion) apparently do not need a special signal. In exocytosis, which discharges the contents of the vesicles to the outside, their membrane is incorporated into the plasma membrane so that the internal portion of the vesicle becomes external in relation to the surface of the plasma membrane. For example, the carbohydrate portion of integral glycoproteins, which originally faced the interior of the ER, becomes external. Some secretory proteins are stored to be released following an appropriate physiological message, such as the presence of a hormone (in a process known as regulated secretion). In regulated secretion, some kind of sorting signal is needed to segregate the proteins into the appropriate secretory vesicles. The nature of this signal, presumed to be in the amino acid sequence of the protein, is still unknown. It is unlikely to correspond to an amino acid sequence at the amino end of the protein, as shown by substitution experiments. Furthermore, it must be common to several proteins because several secretory products can be stored in the same vesicle.

V. Metabolism of Mammalian Cells

Virtually all chemical reactions in cells are catalyzed by enzymes. The energy generated by metabolism is primarily coupled to the formation of ATP from ADP and P_i. Conversely, most reactions requiring energy are powered by ATP hydrolysis.

A. Catabolic Reactions

The oxidative breakdown of glucose, fatty acids, and amino acids provides the energy needed by cells. The energy reserves are primarily in the form of glycogen (a storage polymer of glucose) and neutral fat deposits. Glycogen serves only as a temporary source of energy, whereas fats can provide energy for longer periods provided some carbohydrate is available. Glycogen is present in the liver and skeletal muscle; fats primarily in adipose tissue. The breakdown of glucose (glycolysis) in the absence of the respiratory reactions provides some of the energy, two ATPs per glucose molecules. In contrast, 36–38 ATPs are generated by the oxidation of glucose to CO_2. The intermediate common to all reactions involving oxidative metabolism, other than glycolysis, is acetyl-CoA. Its oxidation to CO_2 proceeds via the tricarboxylic acid (TCA) cycle, which also generates nicotinamide adenine dinucleotide, reduced, and flavin adenine dinucleotide, reduced, which is oxidized by the electron transport chain. The TCA cycle corresponds to a cyclic set of enzyme-catalyzed reactions that take place in the mitochondrial interior.

The degradative and synthetic pathways are generally catalyzed by different enzymes. This difference allows for separate regulation.

B. Regulation of Metabolism

Regulation can take place through changes in the specific activity of enzymes (i.e., the activity per mole of enzyme). Changes in specific activity can result from allosteric interactions where the binding of certain metabolites to the enzyme changes its activity by modifying its conformation. Feedback inhibition is an allosteric regulation of a metabolic pathway where a metabolic end product decreases the activity of the initial reaction of the pathway.

Enzymes can also be covalently modified so that they are converted from an inactive to an active form. In many cases, this takes place in a mechanism involving phosphorylation of the regulated enzyme. Phosphorylation can be the end result of a complicated cascade of events. For example, as discussed above, an epinephrine can trigger the production of a second messenger such as cAMP, which in turn facilitates a number of subsequent enzymatic events culminating in the activation of phosphorylase *b* by phosphorylation (see above). The active form of the enzyme (phosphorylase *a*) catalyzes the breakdown of glycogen to produce glucose.

In other mechanisms, the level of an enzyme can be regulated in response to hormonal or dietary factors either by changing the rate of degradation or synthesis of an enzyme. The most frequent mechanism is alteration of the transcription rate of the corresponding gene; this in turn increases or decreases the production of the enzyme.

C. Metabolic Specialization of Cells

Cells of various organs and tissues are metabolically specialized. In addition, liver and adipose tissue serve as processors of substrates used as fuel by other cells. Because most of the circulation from the intestine goes to the liver first, the liver receives the substances taken up by intestinal absorption and processes them. The liver can take up large amounts of glucose and convert it into glycogen. Adipose cells contain the fat reserves of the body in the form of triacylglycerides. When mobilized, the triacylglycerides are hydrolyzed by a lipase to form glycerol and fatty acids, and the fatty acids are bound by albumin in the bloodstream. Fatty acids can be metabolized directly by several tissues. The fatty acids and glycerol can also be processed by the liver, and fatty acids released by the liver are contained in low-density lipoprotein. When fatty acids are used in metabolism they form acetyl-CoA, which can be oxidized via the TCA cycle. When acetyl-CoA is present in excess in the liver, acetoacetate and 3-hydroxybutyrate (the ketone bodies) are produced. Ketone bodies are metabolized primarily in the heart and in the renal cortex, which prefer it to glucose. Liver cells metabolize mostly keto acids produced from the breakdown of proteins. [*see* FATTY ACID UPTAKE BY CELLS.]

Brain cells use exclusively glucose, although ketone bodies can partially replace glucose during starvation. In contrast, muscle can use glucose, fatty acids, and ketone bodies. During bursts of activity, muscle functions primarily by metabolizing the glucose-6-phosphate generated from glycogen. In contrast, fatty acids are the substrate of preference in resting muscle. Heart muscle generally favors ketone bodies over glucose.

VI. Cell Movement

The movements of cells follow two basic prototypes: the actin–myosin (usually called actomyosin) system and the tubulin–dynein system. The actomyosin system is epitomized by striated muscle. The system works by cyclic formation and release of cross-bridges between fibers of actin and myosin. Tilting of the myosin heads while attached to the actin fibers are thought to make the actin move in each cycle. The repetition of this process summated for many fibers would correspond to the muscle contraction. The process is illustrated in Fig. 4. The fiber made up of globular subunits is the actin. A myosin bundle is shown with a projection containing a double-headed myosin head (Fig. 4). A myosin head attaches to actin forming a cross-bridge. The movement of the myosin head (the tilt in Fig. 4) produces the movement of the actin. The myosin head (no longer tilted) detaches and then binds to the next cross-bridging site, restarting the cycle. Each cycle hydrolyzes one ATP. When the actin is attached to a structure (in muscle the Z-line), the movement can be converted into work. The head portion of the myosin is responsible for the movement, the cross-bridges, and the energy expenditure, whereas the actin molecule plays essentially a passive role. Molecules such as myosin, which are directly responsible for transduction of chemical energy into mechanical work, have been referred to as motors.

Similar to the actin–myosin system, in the tubulin–dynein system, the tubules formed by tubulin have a passive role, whereas dynein, which in cilia or flagella is attached to a tubule, acts as a motor. When the microtubules are attached (in cilia and flagella in the basal bodies), the dynein–microtubule complexes slide in relation to each other, so that the

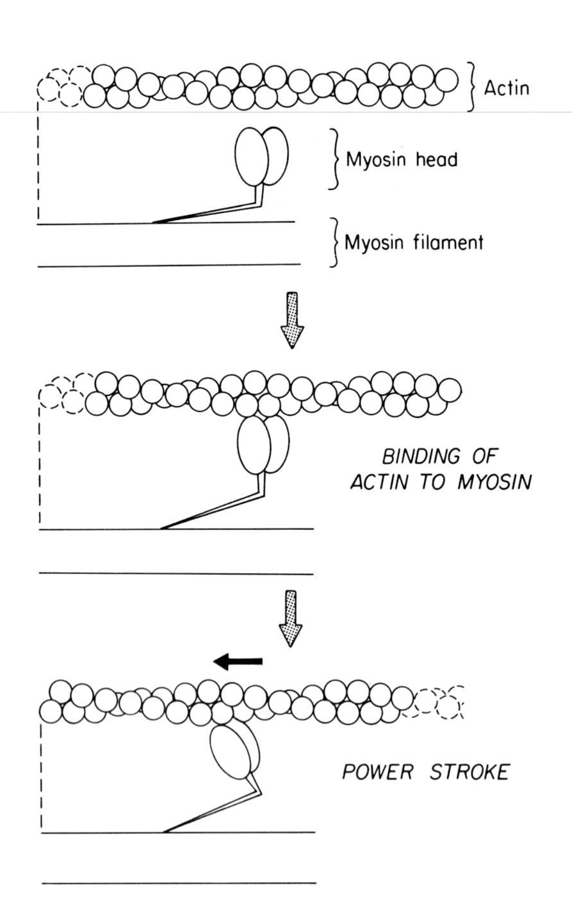

FIGURE 4 The actin–myosin system: interaction of myosin with actin in skeletal muscle. In the first panel, the myosin heads are not attached to the actin. In the second panel, the myosin head has attached to the actin filament. In the third panel, the myosin head tilts, producing a motion in the actin filament (indicated by black arrow). The cycle starts again when the myosin head reattaches to the actin filament in the next cross-bridge site.

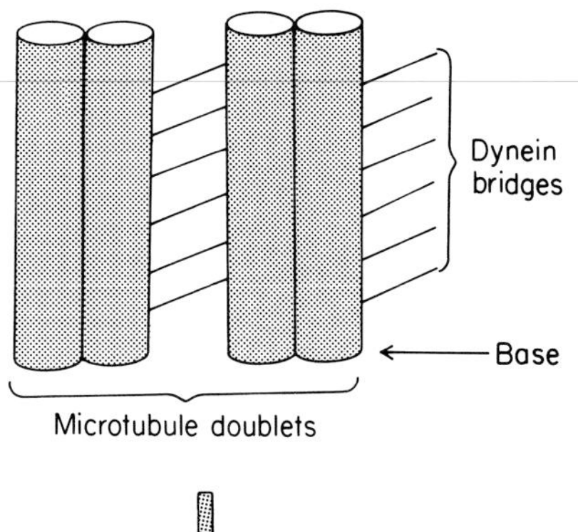

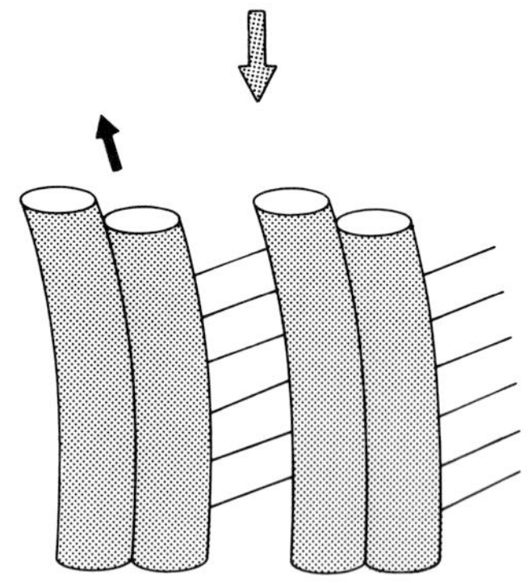

FIGURE 5 The tubulin–dynein system: bending of the microtubules forming part of the cilium or flagellum. The double filament on the left is sliding in relation to that on the right, forcing the whole unit to bend. The sliding mechanism is thought to resemble, in principle, that shown in Figure 4 for actomyosin.

cilia or flagella, which are made up of highly organized tubule bundles, are forced to bend (Fig. 5).

Analogous systems are present in the cytoplasm where microtubules (Fig. 6) or actin fibers (Fig. 7) serve as conduits for the movement of the motors, either the myosins or dyneinlike molecules such as kinesin. Vesicles or other structures attached to the motors are moved along either actin or microtubular linear elements. In microtubular movement, different motors are responsible for movement in different directions along a microtubule.

In muscle, myosin is two-headed and tailed (now called myosin II). Myosin II is also involved in cytokinesis (the actual separation of two daughter cells by contractile events during cell division) and is found in the tail region of ameboid cells, suggesting a role in retraction during ameboid movement.

Other myosins are involved in some forms of cell movement. Myosin I, a single-headed, tailless variety, has a role in interactions with the cell surface. In ameboid cells, it has been found in the leading edge of moving cells and at sites of phagocytosis. In vertebrate cells, it has been found associated with intestinal microvilli. A general role of this myosin I or myosin I-like molecules is also suggested by their presence in the microvilli of the eye of the fruitfly *Drosophila*. Various possible roles of actin and myosin in movement are illustrated in Fig. 7.

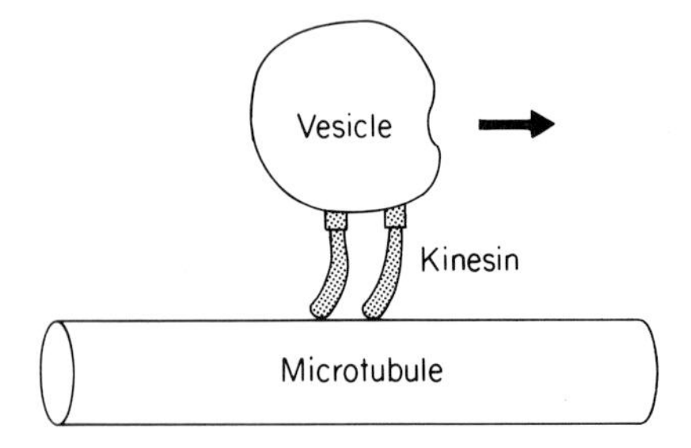

FIGURE 6 Movement of a vesicle along a microtubule by means of a dyneinlike motor.

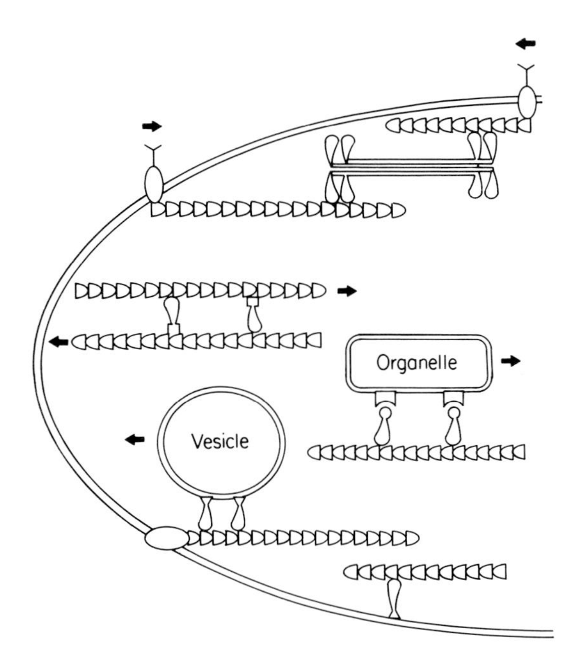

FIGURE 7 Speculative schematic diagram of various actin-based motors. The actin filaments are represented as polar structures. Conventional myosin (myosin II) forms bipolar filaments, which can pull two points in the cell together. When binding two actin filaments, myosin II motors can cause shear (shown in middle). Alternatively, myosin I can attach to organelles or vesicles indirectly or directly or to actin molecules allowing movement. [Reproduced, with permission, from J. A. Spudich, 1989, In pursuit of myosin function—Minireview. *Cell Reg.* The American Society of Cell Biology, **1,** 1–11.]

VII. The Cell Cycle

In somatic cell division (i.e., mitosis), a tetraploid cell, produced by a duplication of the chromosomes, divides to produce two diploid daughter cells. The nuclear envelope is disrupted and the chromosomes condense. They are then pulled apart by the mitotic spindle. The daughter cells are separated out by a contractile event, cytokinesis, where actin and myosin in the contractile ring, pull the two sides of the membrane together to separate out the two daughter cells. These events have been labelled the M-phase (M for mitosis). The period between M-phases is the interphase.

In cells that are periodically dividing, the interphase nucleus is far from inactive. Although there is a pause, the G_1-phase (G for gap), DNA is synthesized during the S-phase, followed by another pause, G_2, which precedes the M-phase. These events ($M \rightarrow G_1 \rightarrow S \rightarrow G_2$) are referred to as the cell cycle. Although it varies with cell type and physiological conditions, typically the cycle lasts from 16 to 24 hr; the M-phase represents 1–2 hr, whereas the S-phase may last as long as 8 hr.

Cells that stop dividing are said to be in G_0, and they can stay in that phase indefinitely. Several signals, which are not completely understood, regulate the cycle. A cell leaves G_0 by the action of external growth factors, which turn on a series of internal activations such as S-phase activator. Another activator (the M-phase factor [MPF]) drives the nucleus into the M-phase. The production of MPF is thought to be triggered by another protein, cyclin.

Normal cells undergo a limited number of divisions. For example, fibroblasts from a human fetus will undergo approximately 50 divisions. Cells of older individuals are limited to fewer divisions. Transformed cells (e.g., cells that have been made tumorlike by infection with some viruses) can divide indefinitely. This release from inhibition apparently results by activation or malfunction of oncogenes. Oncogenes are genes present in all cells, which are thought to play a role in the regulation of normal growth. When the gene is altered, growth becomes uncontrolled. [*See* ONCOGENE AMPLIFICATION IN HUMAN CANCER.]

Mitosis is generally considered to proceed in five M-stages and the final step cytokinesis (Fig. 8). The M-stages are prophase, prometaphase, metaphase, anaphase, and telophase. Interphase corresponds to the quiescent period between M-phases and, therefore, corresponds to $G_1 + S + G_2$. In interphase, the centrosome, or cell center, located on one side of the nucleus, contains a pair of self-replicating centrioles, structures that are arranged at right angles from each other. The centrosome material serves as a microtubule-organizing center and is attached to microtubules. Plants and mouse oocytes lack centrioles. In interphase, arrays of microtubules radiate from the centrosome. The centriole pair replicates during the S-phase. In prophase, the centrosome splits, and the two resulting centrosomes move to opposite sides of the cell, where they serve as focal points for polymerization of microtubules to form an aster. At prometaphase, the nuclear membrane breaks down, and the microtubules attach to the chromosomes to form a spindle where the centrosomes have become the spindle poles. Chromatids, the two identical components formed by the replication of each chromosome, remain closely attached. The spindles of higher animals and plants contain several thousand microtu-

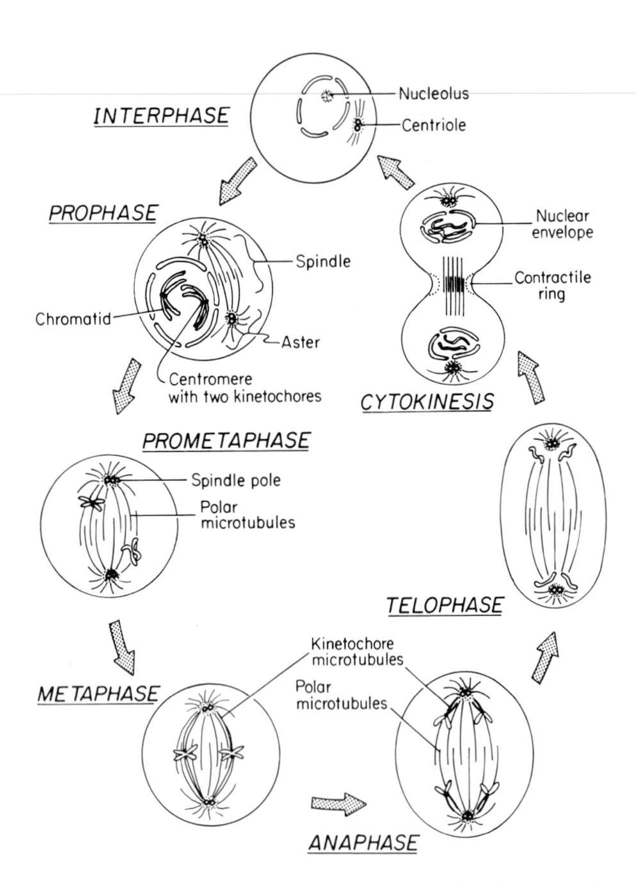

FIGURE 8 Diagrammatic representation of mitosis and cytokinesis. For simplicity, only two chromosomes are shown at the beginning of mitosis. There are 46 chromosomes in human cells, i.e., 23 pairs.

bules, whereas in some fungi there may be as few as 50. [*See* MITOSIS.]

The two chromatids are attached to each other near their centromere. The centromere corresponds to specific DNA sequences needed for chromosome separation. A specialized structure, the kinetochore, develops at each centromere. By prometaphase, some of the microtubules (typically 20–40), the kinetochore microtubules, have connected to the kinetochores so that the two sister chromatids are attached to opposite spindle poles. The spindle microtubules, which are not attached to the kinetochores, are known as the polar microtubules. Those that are not in the spindle are called astral microtubules. The chromosomes appear to be pushed to the equator, which they reach at metaphase. At anaphase, the sister chromatids are separated, each with one kinetochore. Two simultaneous processes seem to take place. At anaphase A, the chromatids are moved toward their respective pole in a process in which the kinetochore microtubules are shortened. In anaphase B, the poles are separated further as the spindle elongates by an assembly of microtubules at their distal ends. This polymerization accompanied by sliding of the astral microtubules is thought to push the poles apart. At telophase, the chromosomes arrive at the poles, the kinetochore microtubules disappear, the daughter cells separate, the condensed chromatin expands, and the nuclear envelope reforms in the two cells.

VIII. Meiosis

With the exception of the sex-determining chromosomes, cells are generally diploid (i.e., they have two versions of each chromosome—homologous pairs—each derived from one parent). In contrast, the gametes are haploid, and they contain only one chromosome of each homologous pair. Meiosis is the special cell division that produces the gametes. Meiosis differs significantly from mitosis. First, each homologous pair replicates to produce sister chromatids just as in mitosis. However, the four chromatids are physically attached before lining up in the spindle. The first division (division I of meiosis) separates out the homologous pairs to produce two diploid daughter cells. A second division (division II of meiosis) occurs without DNA replication and separates out the sister chromatids to form a haploid gamete. At prophase, the close association between homologous pairs preceding division I allows recombination. In this process, equivalent parts of the homologous chromosome pair are exchanged in an event known as crossing-over. [*See* MEIOSIS.]

Bibliography

Alberts, B., Bray, D., Lewis, J., Raff, M., Roberts, K., and Watson, J. D. (1989). "Molecular Biology of the Cell," 2nd ed., Chapters 1, 6–9, 11. Garland Publishing, Inc., New York and London.

Darnell, J., Lodish, H., and Baltimore, D. (1990). "Molecular Cell Biology." W. H. Freeman, New York.

Gennis, R. B. (1989). "Biomembranes, Molecular Structure and Function," Chapters 1, 6, 8, 9. Springer-Verlag, New York and Berlin.

Lackie, J. M. (1986). "Cell Movement and Cell Behaviour," Chapters 2, 3. Allen and Uniwin, London and Boston.

Tedeschi, H. (1991). "Cell Physiology." Hemisphere, Washington, D.C.

Cell Death in Human Development

RALPH BUTTYAN, *Columbia University*

Glossary

Apoptosis Physical changes in a cell undergoing "programmed death." Nuclear chromatin condenses, cells lose water and shrink in size

Apoptotic body Histologic appearance of the condensed remains of cells that have undergone "programmed death"

Atresia Process by which a subset of ova degenerate and die during follicular development

Necrotic cell death Passive form of cell death, resulting from direct physical trauma to a cell, especially at the outer membrane. Typically associated with intense hyperthermia, caustic chemicals, ice crystal formation, hypoxic conditions, or membrane-damaging agents

Programmed cell death Active form of cell death, resulting from a cell's genetic response to specific developmental, hormonal, or environmental stimuli. Typically involved in embryonic restruction or tissue remodeling processes

Regression Tissue degeneration (benign or malignant) often related to a change in hormonal status or to sensitivity to a chemotoxic agent

THE CONCEPT OF CELL DEATH might seem to be best associated with studies of human disease states or the progress of human aging. However, as paradoxic as it may seem, regulated cell death is critically important for many aspects of normal human development. The requirement for controlled cell death pervades all stages of human life. The human embryo undergoes periods of massive restructuring during which entire populations of embryonic cells die on cue of poorly understood developmental signals. This restructuring enables the proper morphogenesis of the human fetus. Later, during postnatal life, cell killing directed by a diverse network of cells of the immune system allows for the functional maintenance of the human body. Finally, all through life, reproductive and immunologic tissues remain sensitive to the hormone levels of the body. Natural fluctuations in the levels of these hormones can drastically alter the cellular content and functional ability of these organs. Therefore, regardless of the role that abnormal cell deaths play in human diseases and aging, the mechanisms by which the body normally controls the onset of cell death should provide a significant focus for biomedical research.

Aside from the function of immune surveillance, the role of cell death in human development has been recognized only during the past 40 years. Much more recently we have accepted the existence of a genetic program for death, as one of the inbuilt regulatory systems of every cell. In essence, scientific evidence indicates that cells of the human body have the ability to "commit suicide" on command of specific hormonal or developmental signals and that many types of cells are normally required to follow this command. Studies of simple organisms, such as the small nematode *caenorhabditis elegans,* to far more complex mammalian organisms are just now helping us unravel the genetic mechanisms involved in the active process of cell death. Ultimately, we can expect that our understanding of this process will open new avenues toward treating birth defects or provide novel cancer therapies or help protect against modern environmental stresses such as toxic chemical or radiation exposure. More important, the acceptance of this concept emphatically underscores the significance of balance in all biological systems. Even in our own body, a natural rate of cell growth must be balanced by a corresponding rate of cell death.

I. Cell Death During Embryogenesis

As with other higher organisms, regulated cell death acts as a sculpting force in early human life. In a manner similar to removing pieces of clay or rock from a sculpture in progress, cell death allows for the acquisition of adult features from differentiating masses of embryonic cells. This force probably affects every developing tissue to some extent. Studies of rodent and avian embryogenesis models have already shown that the formation of the functional heart and eye and the development of active neural networks all depend on morphogenetic changes involving embryonic cell death. More prominently noted examples of this phenomenon include the development of functional fetal limbs and hands (or feet) from embryonic limb bud tissues (Fig. 1). This morphogenesis proceeds with the death of several distinct cell types at different stages of human embryonic limb bud development. As the forelimb buds elongate from the embryonic thorax, defined regions of this rapidly growing tissue begin to degenerate. Within 5 weeks of conception, dying cells

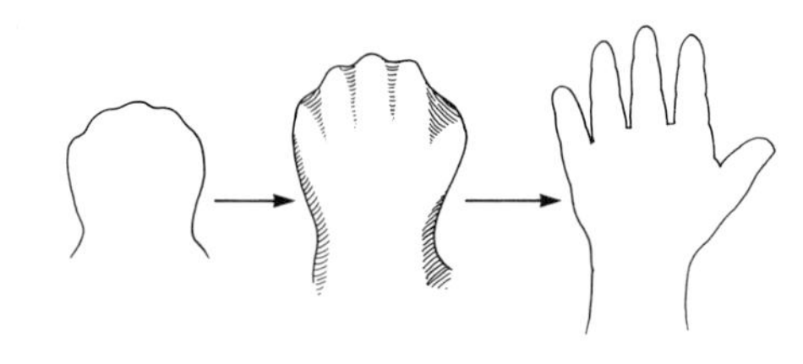

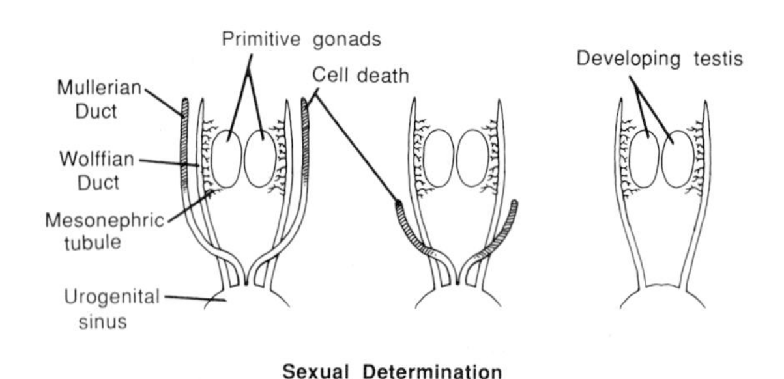

FIGURE 1 Cell death in embryogenesis.

in the regions of the wrist, elbow, and under the forearm give the growing limb bud definite structural characteristics of an arm. At this time, the tip of the limb bud flattens to form the paddle-shaped hand plate. During the sixth week of development, the hand plates acquire four notches and the appearance of "rays," regions of mesenchymal cells that will ultimately define the space between the digits of the hand. The rays demarcate active regions of interdigital cell death so that by the eighth week, the forelimb of the fetus has acquired all the superficial features that define the functional adult forelimb, including distinct and separate digits, narrowed wrists, the elbow region, and increased separation from the thorax. Development of the hind limbs (legs and feet) reiterates this pattern with a delay of about 1 week. To date, the biochemical signals that regulate the onset of cell death in developing limb bud tissues remain poorly understood. It is clear, though, that these signals arise from the embryonic limb bud mesenchymal-organizing regions because physical interference with these regions can cause profound changes in the subsequent pattern of cell death. [*See* EMBRYO, BODY PATTERN FORMATION.]

Another noted region of embryonic cell death occurs during the development of the human reproductive tract. In the early forming embryo of both sexes, primordial tissues with potential to develop into the male sexual accessory organs (the mesonephric or Wolffian ducts) are formed within 30 days of conception, whereas tissues capable of developing into the female sexual accessory organs (the paramesonephric or Mullerian ducts) can be recognized within 45 days postconception. For the genotypically male embryo, sexual development proceeds with the growth and differentiation of the Wolffian ducts into the epididymis, vas deferens, prostate gland, seminal vesicles, and ejaculatory ducts under the influence of the male hormone testosterone secreted by the Leydig cells of the primitive testis. In addition to testosterone, the fetal testis also produce a polypeptide known as the Mullerian inhibiting substance (MIS). This factor, synthesized and secreted by Sertoli cells, is highly toxic to the cells in the Mullerian duct. Therefore at the time of its synthesis (~6–8 weeks postconception) it initiates the rapid degeneration of the embryonic Mullerian duct. The effects of MIS are apparently exerted in a highly local manner because removal of a single fetal testis before the testicular cord develops will allow the formation of a unilateral oviduct on the side of the excised testis. Although the mechanism of action of MIS on the cells of the Mullerian duct remains unclear, its synthesis ensures that the male fetus retains a distinct male identity during the rest of its development. Based on its toxicity to cells of Mullerian duct in origin, MIS is currently under study as a potentially useful agent in cancers of the female genital tract.

II. Cell Death in Adult Tissues

The stability of all adult tissues is related to two specific factors: the endogenous rate of cell growth and division, and the rate of cell death. Homeostatic maintenance in adult tissue implies that these two factors must equal each other (Table I). This means that tissues with a rapid rate of cell growth must be balanced by an equivalently rapid rate of cell death. High radiation sensitivity often defines tissues with rapid cell division rates. The intestines, skin, bone marrow, and testis are organs that are most adversely affected by radiation exposure. As might be expected, these are also the tissues with the highest rate of cell death under normal conditions. The function of each of these tissues is to produce a highly differentiated cell of extremely specialized activity. For the testis, it is the sperm; for the skin, it is the keratinocyte; for the bone marrow, it is the leucocyte and erythrocyte. In each of these cases, the terminally differentiated end product cell no longer has the capacity to undergo cell division. Because the end cells of these organs have an extremely short life span, the process of terminal differentiation becomes an ultimate commitment for cell death. It should be pointed out that not all such highly specialized, terminally differentiated cells have such brief life spans. Individual neuronal cells, for example, will survive and function through an entire human lifetime. [*See* Radiosensitivity of the Integumentary System; Radiosensitivity of the Small and Large Intestines.]

Other tissues of the adult body, however, can experience rapid and transient changes in the rate of cell death, and in general, these changes are linked to fluctuations in body hormone levels. Hormonal regulation of tissues is most often considered in terms of their effects on cell growth rates or secretory activity in response to hormonal stimulation. But for many tissues, hormones also control the life span of the resident cells. In certain cases, as was discussed for MIS action on embryonic Mullerian duct tissue, a hormone may act as if it were toxic to a particular cell. This same type of reaction is seen in both the fetus and adult, where thymocytes (precursors of T cells) are sensitive to the concentrations of circulating steroid hormones, especially the glucocorticoids. High doses of cortisone will rapidly stimulate their death. This sensitivity helps explain the immunosuppressive side effects of glucocorticoid therapy and why these types of drugs are effective in treating conditions involving hyperactive immune responses or autoimmune conditions. Under more natural conditions, this explanation has been used to account for the decrease in immune function seen in chronic stress, a condition associated with increased production of corticosteroids. [*See* Steroids.]

Unlike the lethal effects of glucocorticoids on immature thymocytes, for many other tissues, cell life span is dependent on the availability of a hormone. In the absence or depletion of a required hormone, certain tissues will undergo massive restructuring. This is frequently noted in the sexual accessory organs of the male and female reproductive tract, where low levels of circulating sex steroids will result in atrophy of these tissues subsequent to the death of hormone-dependent cells. Large natural fluctuations in hormone levels are most associated with the female reproductive cycle. The regular fluctuations in hormone levels of the postpubescent female are controlled by feedback loop interactions of steroid hormones (estradiol and progesterone) produced by the ovary and polypeptide hormones [follicular-stimulating hormone (FSH) and luteinizing hormone] produced by the central nervous system. These hormone cycles are necessary for the successive maturation and release of small numbers of mature ovum at regular intervals during the reproductive years of the female. One interesting aspect of human female reproduction requiring the regulated onset of cell death regards the spontaneous degeneration of subsets of ova during follicular maturation. This phenomenon, known as atresia, is related to the requirement of FSH for ovum maturation and survival. During each menstrual cycle, multiple follicles will begin the pathway toward

TABLE I Changes in the Rate of Cell Death Can Have Drastic Effects on Tissue Dynamics

Proliferation rate > death rate	=	growth
Proliferation rate < death rate	=	regression
Proliferation rate = death rate	=	homeostasis

ovum maturation and release. At the start of the cycle, peak estradiol concentration within an early developing follicle will promote the local retention of FSH by increasing FSH binding. As the cycle progresses, systemic FSH concentration begins to decrease. Near the end of the cycle, ova within the follicles that have retained the highest amount of FSH will survive the period of FSH decline, continue to mature, and be released. However, the ova in smaller, less differentiated follicles with low levels of FSH retention will instead atrophy and die. It is of interest that this developmental system consistently allows the survival of only one mature ovum per ovary during the menstrual cycle of humans. [*See* POLYPEPTIDE HORMONES.]

Apart from the maturation and release of the ovum, the condition of the female sexual accessory tissues is also tied into these hormone cycles. Epithelial cells of the uterine endometrium, for example, are highly dependent on the levels of circulating estradiol and progesterone (Fig. 2). At the times during the menstrual cycle in which the levels of these steroids decline, the mature endometrial cells begin to die. Their loss reaches a maximum during menstruation with a commensurate decrease in the thickness of the uterine wall. Elevated prolactin synthesis during late pregnancy, in conjunction with estrogens, stimulates the activity and further growth of the mammary glandular system. On the termination of lactation, prolactin production drops, allowing for the degeneration of the most active glands. The breast tissue developed to nourish the offspring then atrophies to the state before pregnancy. Although males do not show such exaggerated hormone cycles, it is known that the male sexual accessory organs are, in general, highly dependent on circulating testosterone levels. Surgical or chemical castration will result in a similar atrophy of the male internal organ system.

The hormone sensitivities shown by the normal tissues of the body are often retained when cancers arise in these same tissues. For this reason, hormonal therapies have proven to be the most useful treatment for many types of cancers. As examples, lymphatic cancers are treated with high doses of glucocorticoids, whereas estrogen-blocking therapy is used in breast cancer and androgen-withdrawal therapy in advanced prostate cancer. All these therapies have the advantage of being relatively nontoxic to normal tissues of the body. However, because the ability to induce death by hormones is a genetically programmed response, all these therapies suffer in that tumor cells can frequently develop a genetic resistance to the hormone signals, which induce cell death. More effective cancer therapies may come when we learn to bypass this genetic block to hormone sensitivity.

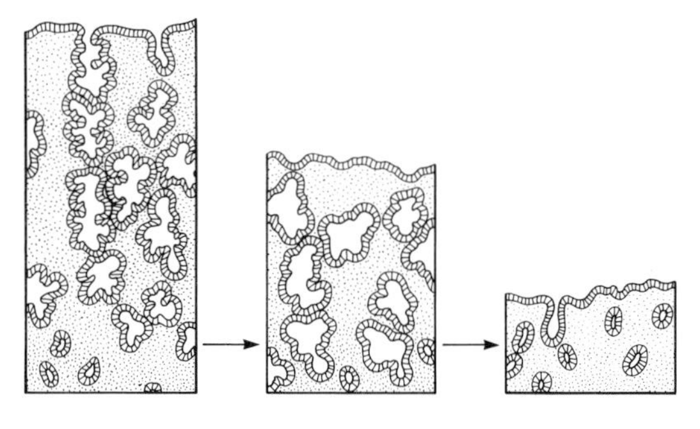

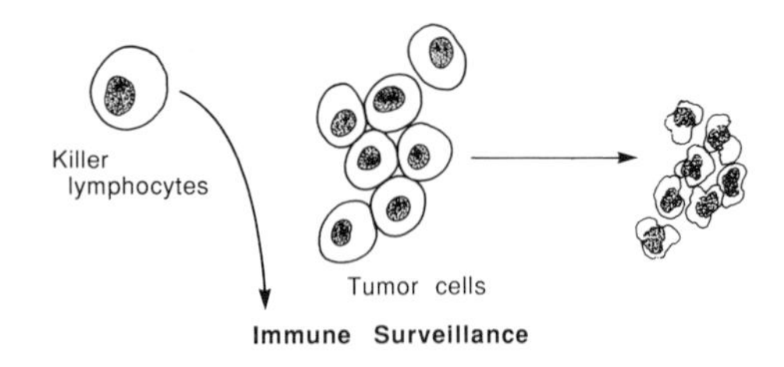

FIGURE 2 Cell death in adulthood.

III. Immunological Control of Cell Death

Protection against invading organisms requires an immune system capable of killing these organisms. Protection against more insidious invaders (e.g., viruses and cancer cells) requires an immune system capable of recognizing and killing cells of the body showing any sort of abnormality. Thus, as might be suspected from the numerous occasions at which the immune system of the human body must stimulate the onset of cell death, a variety of strategies have evolved to accomplish this process. Complement-mediated lysis, although complex in mechanism, is one of the most simple in concept. The binding of antibodies to the surface of a cell initiates a cascade reaction by which a variety of soluble serum complement proteins become sequentially activated by proteolysis, bind to the surface of the target cell, and cooperatively force the formation of pores through the cell membrane. Although this system depends on the production of specific antibodies by the B cells and complement proteins by the endothelial cells of the blood vessels or by the liver, no direct contact with lymphocytes is required to kill the target cell. Complement-mediated lysis is most effectively used in protecting the body against invading microorganisms and virus-infected cells.

In many conditions of immune surveillance, more direct lymphocyte participation is required to stimulate cell death. In some cases, migrating lymphocytes attracted to the target area secrete toxic factors in the vicinity of the offending cells. These factors include a variety of simple compounds such as peroxides or free radical ions and more complex substances including toxic peptides such as interferons and tumor necrosis factor. This nonspecific method has the drawback in that it can injure adjacent normal cells. Alternatively, more direct killing by cytotoxic T lymphocytes and natural killer cells may be accomplished by a unique mechanism involving the active insertion of proteins (perforins) into the membrane of the target cell. In the presence of calcium ion, the perforin proteins are believed to polymerize to become large transmembrane pores, thus directly destroying the integrity of the target cell membrane. [*See* IMMUNE SURVEILLANCE; LYMPHOCYTES.]

IV. Mechanisms of Cell Death

Cell survival depends on the ability of a cell to segregate its complex biochemical systems from the entropic conditions of the outside environment. This segregation is achieved through a network of membranes that effectively limit and regulate the communication between the external milieu and the internal environment. Therefore, one of the simplest means of causing cell death is to breech mechanically the integrity of the external membrane. As was described, several of the methods used by lymphocytes in the immune system act at this level. In addition, a cell depends on the ability of its energy-generating systems to provide a constant level of high-energy products for the building processes needed to maintain its integrity. Disruption of the energy-generating systems can be as harmful as any direct physical trauma. Hypoxic conditions, energy poisons, caustic chemicals, lytic viral infections, hypoosmotic conditions, intense heat, and membrane disruption related to detergent action or ice crystal formation can cause widespread cell death in affected areas. Pathologists refer to this form of cell death as "necrotic cell death" and describe it in terms of specific histologic characteristics. In general, the cell increases in size before death, experiences massive disruption of mitochondrial and lysosomal structure, and ultimately appears to "explode" from within. The intracellular contents are spilled out into the extracellular space.

The cell deaths associated with human development, however, differ significantly with regard to the histologic appearance of the cell in death and in the manifestation of certain biochemical markers during the death process. For one thing, rather than enlarging to lysis, these cells tend to shrink drastically in size. Also, there are profound physical changes in the structure of the nucleus that harbinge the onset of this developmental death process. Nuclear chromatin condensation usually precedes cell shrinkage and the dense nucleus of a dying cell becomes highly irregular in shape, often showing fragmentation. Biochemical analysis of cells stimulated to die with hormones or toxic polypeptides shows that the nuclear DNA is rapidly degraded to regular nucleosome-sized fragments. This degradation can apparently occur even before the cell shows morphologic signs of changes related to death. Pathologists refer to this developmental type of cell death process as "apoptosis," and the dense, shrunken, dying cell is called an "apoptotic body." Under normal conditions, the apoptotic body will be phagocytized, either by a neighboring cell or by macrophages.

Because this form of cell death is characteristic both of cells "programmed" to die during development or cells stimulated to die because of changes in hormone concentrations or because of specific environmental insults, it is also referred to as a "genetic cell death." The concept of genetic cell death implies that certain gene products of the cell are required for the cells to die. Because genetic cell death is usually initiated either by a known hormone or at least a hormone-like stimulus, one part of the internal genetic apparatus regulating the onset of programmed death must be the cellular receptor for the hormone or stimulus. As best understood for certain human tumor models, genetic unresponsiveness to the signals for cell death is often related to a lack of the signal receptor molecule. As an important example, modern clinical decisions regarding the usefulness of estrogen-blocking therapy for breast cancer patients are based on the concentration of estrogen and progesterone receptor proteins in the cells of the primary and metastatic tumors. Patients whose tumors demonstrate low or undetectable levels of these receptors are not likely to experience tumor regression in response to hormonal therapies and are therefore not likely candidates for these treatments. [*See* CELL RECEPTORS.]

Because a large number of biochemically distinct stimuli can induce cells to undergo apoptosis, each of these stimuli are likely to initiate the program of cell death through different receptor molecules. Although there is a great diversity in the mechanisms by which intracellular signals are transmitted by various hormone receptors, the end result of effective hormonal stimulation, regardless of the receptor, is to alter the pattern of gene activity in the responding cell. The changing configuration of gene products synthesized by the stimulated cell dictate the pathway of cellular response. Based on this concept, when the response to a stimulus is genetic cell death, in a real sense, the cell is considered to be "committing suicide," presumably through the action of the cadre of gene products synthesized in response to the death stimulation. Experiments in animal models of genetic cell death confirm the participation of the cell in its own death. Drugs that block the synthesis of either new RNA or new protein will delay or prevent cells from undergoing genetic death in response to specific hormonal and developmental stimuli. These experiments imply that at least one of the gene products synthesized after death stimulation is a potential lethal factor.

Analysis of acute changes in gene activity during programmed death in several different experimental systems has allowed for the current identification of at least two specific gene products whose expression is highly associated with the general onset and progression of the genetic death response in mammalian cells: the proto-oncogene c-*fos*, and a unique gene currently referred to as testosterone-repressed prostate message-2 (TRPM-2). In a similar manner to its rapid and brief expression during other cell response pathways (proliferation, differentiation, or hypertrophy), the induction of c-*fos* expression is an early and specific marker of cells stimulated to die. The protein encoded by the c-*fos* proto-oncogene is thought to act as a transcriptional cofactor. By binding to any of a variety of other nuclear proteins, the c-*fos*–nuclear protein complex becomes a functional transcription factor, enabling the regulated synthesis of specified messenger RNAs. Thus, even though the activity of the c-*fos* gene during all cellular response programs might imply that it does not play a distinctively important role in the process of cell death, its early synthesis is likely to be required for a cell to respond to the signals initiating the cell death reaction. The synthesis of TRPM-2 gene products, however, is much more specific for dying cells. This gene encodes a highly glycosylated protein of approximately 62,000 kDa molecular weight, which is proteolytically processed and secreted by the dying cells. Structural analysis of the protein product of this gene shows a strong resemblance to a family of complement-like proteins. Although the ultimate function of this protein in the pathway toward programmed death is not yet defined, it is synthesized in great abundance by mammalian cells in the later stages of programmed death, regardless of the stimuli used to induce the death program. Continuing research into the gene products involved in the process of programmed death can be expected to provide more molecular markers of this cellular event. [*See* *FOS* GENE, HUMAN.]

V. Abnormalities in Cell Death—Birth Defects, Diseases, and Aging

Considering the importance of cell death to the development of the human organism, it would be expected that any abnormalities in this process are severely detrimental. Furthermore, because regulated cell death is involved in so many different aspects of human development, singular abnormalities in this process can be manifest in any of a variety of conditions. Birth defects, immune deficiencies, endocrine disorders, and spontaneous atrophic diseases each represent a different class of problems with a potential basis of a disorder in regulated cell death (see Table II). Many of these conditions are genetic in nature, and it may be years before we even understand the basis of the genetic defects. However, certain conditions resulting from acute traumatic injury or environmental insult also fit well into a category related to abnormal onset of cell death. Transient periods of anoxia (lack of oxygen in tissues) caused by acute vascular ischemia (myocardial or renal infarction) have an interesting side effect in that many cells of the heart or kidney become committed to undergo apoptosis, especially if the ischemia is relieved after a brief period. Chronic pressure insults, such as occur in a kidney

TABLE II Human Disease States Associated with Abnormal Rates of Cell Death

Birth Defects
Cleft palate
Syncactylism
Pseudohermaphroditism
Spontaneous Degenerative Disorders
Huntington's chorea
Muscular dystrophy
Acute Traumatic Injury
Ischemia/reperfusion injury
Ureteral obstruction
Radiation exposure
Toxic chemical exposure
Conditions of aging
Alzheimer's disease
Parkinson's disease
Benign prostatic hyperplasia

when a ureter becomes blocked by stone disease, have this same effect. Even ionizing radiation and toxic chemicals (as best understood for chemotherapeutic agents) demonstrate the capacity to induce cells to undergo a programmed death, often at doses lower than would be necessary to cause substantial cellular damage directly. Thus much of the cellular loss that occurs in these types of injuries can be more related to the abnormal induction of cell death processes rather than to the direct action of the trauma or poison.

At first thought, the typical dysfunctions associated with human aging might seem to be caused by cumulative or accelerated cell deaths in various human tissues. However, the modern concept of aging at the cellular level is probably more closely related to a decrease in the activity or responsiveness of living cells in the body (senescence) rather than with any acceleration in the rate of cell death. Nonetheless, several specific diseases associated with aging appear to be directly related to abnormalities in the cellular death rate of specified tissues. The diverse nature of these diseases delineate the complex role that cell deaths play in maintaining the human body. At one end of the spectrum lies such conditions as Parkinson's disease or Alzheimer's disease, in which the normal rate of death in specific areas of the brain is abnormally accelerated. Alzheimer's disease is a global brain degenerative disorder affecting many of the cognitive functions of the brain, whereas for Parkinson's disease patients, the specific loss of dopamine-producing cells in the substantia nigra results in severe imbalances in local and systemic neurotransmitter levels. [*See* ALZHEIMER'S DISEASE; PARKINSON'S DISEASE.]

At the other end of the spectrum, we can consider the condition of benign prostatic hypertrophy (BPH), a disease whose affliction is only apparent because an age-associated increase in the mass of the male prostate gland restricts the urethral passage below the bladder. This restriction results in chronic symptoms related to urine retention. Although BPH is somewhat heterogeneous in character, cytological studies of hypertrophied prostate glands have not shown enhanced rates of cellular proliferation compared with normal prostate glands. Therefore, as indicated in Table I, if the proliferation rate of the cells in the hypertrophied gland is unchanged, yet the gland is gaining in cell number, we might suspect that the normal rate of cell death has slowed significantly. As a simplistic explanation of this phenomenon, it has been suggested that during aging, the normal progression of prostate cells from the undifferentiated stem cell to the highly differentiated prostatic secretory cell becomes slowed or interrupted. Because terminal differentiation represents a potential commitment to cell death, any interruption in this progression could change the glandular balance by decreasing the rate of cell death. This change would result in a net accumulation of prostate cells and, ultimately, the increased size of the hypertrophied prostate gland.

In summary, we now know that cell death is as active a process as cell growth. Both of these processes contribute significantly to human development and maintenance. Therefore we can also understand that imbalances in the equation between cell growth and cell death can explain a variety of human disease states.

Bibliography

Bowen, I. D., and Lockshin, R. A., eds. (1981). "Cell Death in Biology and Pathology." Chapman and Hill, New York.

Buttyan, R., Olsson, C. A., Pintar, J., Chang, C., Bandyk, M., Ng, P.-Y., and Sawczuk, I. S. (1989). Induction of the TRPM-2 gene in cells undergoing programmed death. *Mol. Cell. Biol.* **9,** 3473–3481.

Coffey, D. S., Berry, S. J., and Ewing, L. L. (1987). An overview of current concepts in the study of benign prostatic hypertrophy. *In* "Benign Prostatic Hypertrophy," vol. III. (C. H. Rodgers, D. S. Coffey, G. Cunha, J. T. Grayhack, S. Hinman, and J. R. Horton, eds.). U.S. Department of Health and Human Services, N.I.H. publication #87-2881, pp. 1–13.

Saunders, J. W. (1966). Death in embryonic systems. *Science* **154,** 604–609.

Umansky, S. R. (1982). The genetic program of cell death. Hypothesis and source applications: Transformation, carcinogenesis, ageing. *J. Theor. Biol.* **97,** 591.

Wyllie, A. H., Kerr, J. F. R., and Currie, A. R. (1986). Cell death: The significance of apoptosis. *Int. Rev. Cytol.* **68,** 251–306.

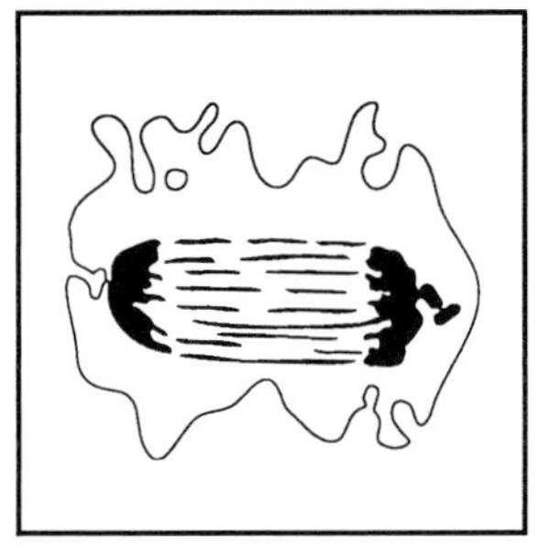

Cell Division, Molecular Biology

RENATO BASERGA, *Temple University Medical School*

Glossary

cDNA libraries Libraries of plasmids or phage containing DNA inserts derived from mRNA extracted from cells (cDNA)
Cell cycle Interval between two successive mitoses
DNA synthesizing machinery Complex of proteins necessary for the replication of DNA
Growth-regulated Gene or gene products whose expression is modified by growth stimuli in the environment
Proto-oncogenes Genes that are the cellular equivalent of retroviral transforming genes

CELL DIVISION (mitosis in eukaryotes) is the process by which a progenitor cell divides into two sibling cells. It is found in bacteria (which divide by fission) throughout the evolutionary tree, up to higher animals, and it is the mechanism by which populations of cells expand. Cell division is characterized by morphological changes and is accompanied and preceded by a series of biochemical events that are a prerequisite for division. The genes and the gene products that are necessary and/or specific for cell division constitute the subject of the molecular biology of cell division.

I. Background

A. Growth in Size and Growth in Number

The tissues and organs of animals are made up of populations of cells bathed in fluid and often surrounded by various amounts of intercellular substance. The intercellular substance of a tissue is usually a secreted product of the cell (e.g., collagen) and can be considered an extracellular extension of the cytoplasm. If we consider the intercellular substance as part of the size of a cell, we can say that in multicellular organisms the growth of tissues and organs may occur by (1) an increase in the number of cells, or (2) an increase in size of the cells, or (3) both. Indeed, both processes occur in nature. For instance, bamboo shoots grow rapidly (several centimeters overnight), largely by increasing the size of their cells. Conversely, the deer's antlers in the spring grow rapidly because of a continuous proliferation of progenitor cells at the base of the antlers. In general, in animals, growth in number predominates over growth in size, although some growth in size also occurs during development. At any rate it does not take a profound observer to realize that in the case of humans, growth in cell number is by far the most important component in development. Each individual originates from one single cell (i.e., the fertilized egg) and grows to an adult organism who contains, on the average, 5×10^{13} cells. During development there is also a modest increase in the size of the cells, modest, at least, in respect to the increase in cell number. For instance, myocardial fibers in humans double in diameter from birth to adulthood. A two- to threefold increase in the size of cells from newborn to adult humans also occurs in the skeletal muscle and in the kidneys. In the liver, the cell mass increases from birth to adulthood three- or fourfold, but the cell number during the same period increases by a factor of 12. Once humans reach adulthood, the number of cells remains essentially constant. Even in adult humans, though, cell division continues at a high rate. Each day, about 10^{12} cells die and have to be replaced. Most of the cells that die come from certain tissues and organs (e.g., the gastrointestinal tract, the skin, and the bone marrow). These organs are endowed with a large capacity for cell proliferation to replace the cells that continuously are eliminated. In this case, cell proliferation is said to be in a steady state (i.e., the number of cells that is produced equals the number of cells that die). This simple equation can break down in either direction. If the number of cells that are produced exceeds the number of cells that die in a given period of time, there is growth. Tumors are an example of an excess of production of cells, localized to a specific tissue or compartment. If fewer cells are produced than die, the organism shrinks, and we have what the pathologist calls *atrophy*. This occurs often in old age, especially in certain organs such as the brain. In humans the brain loses, with age, an average of 150 g of tissue. This is roughly the equivalent of 10^{10} cells. We should not worry too much though about the loss of these cells. [*See* CELL; CELL DEATH IN HUMAN DEVELOPMENT.]

Another important point to remember is that the amount of DNA per cell is constant. In humans the amount of DNA per somatic cell is 6.7×10^{-12} g and is called the *diploid amount*. Germ cells, of course, have half that amount and are called *haploid*. In some tissues and organs, the amount of DNA per cell is twice the normal amount. This occurs especially in certain organs such as heart and liver. However, when the amount of DNA per cell increases in normal cells, it is always a multiple of the diploid amount. The constancy of the DNA amount per cell is an important point because it means that when a cell divides, it has to double its amount of DNA so that each daughter cell can have the same normal amount. Indeed, the cell before division has to double all its components to generate cells of its original size. If it were not so, obviously at each generation, cells would get smaller and smaller and eventually vanish.

Because growth depends largely on cell division, if we wish to understand how organs and tissue grow, it is first necessary to understand the mechanism(s) of cell division.

B. Populations of Cells

In terms of cell proliferation, in every population of cells, whether *in vivo* or in tissue cultures, normal or abnormal, we can always distinguish three cell populations (Fig. 1):

1. Cells that continuously proliferate, going from one mitosis to the next one, which we call *cycling cells*.
2. Terminally differentiated cells that leave the growth cycle and are destined to die without dividing again.
3. A third population of nondividing cells, which we call *G_0 cells,* which can reenter the cell cycle and then leave it again.

We have already met the first two populations of cells, those that are continuously dividing and those that leave the growth cycle and die without dividing again. These are the cells in the bone marrow, the skin, or the gastrointestinal tract that continue to proliferate to replace the fraction of cells that die daily.

We have now to define the G_0 cells, which are outside the cell cycle and yet are capable of further proliferation. These G_0 cells are an interesting subpopulation of cells and are present in the living animal as well as in tissue cultures. For instance, in the adult liver most of the cells are in G_0, (i.e., cell division is a rare occurrence in hepatocytes, which, however, are alive and carry out a great number of metabolic processes). If two-thirds of the liver are surgically removed, the remaining liver cells quickly reenter the cell cycle, proliferate, and more or less restore the liver to its original size. This is observed in humans when a surgeon must remove a part of the liver because of a neoplastic growth.

There are several other types of G_0 cells in the body. One that is of particular interest to us is the stem cell of the bone marrow. These stem cells are capable of producing all the different lineages of hemopoietic cells from lymphocytes to erythrocytes to megakaryocytes. Most of the bone marrow stem cells are in G_0, and this is fortunate because when we try chemotherapeutic agents on individuals who have disseminated forms of leukemia or lymphoma, or metastatic cancer, we have to use agents that kill proliferating cancer cells. Unfortu-

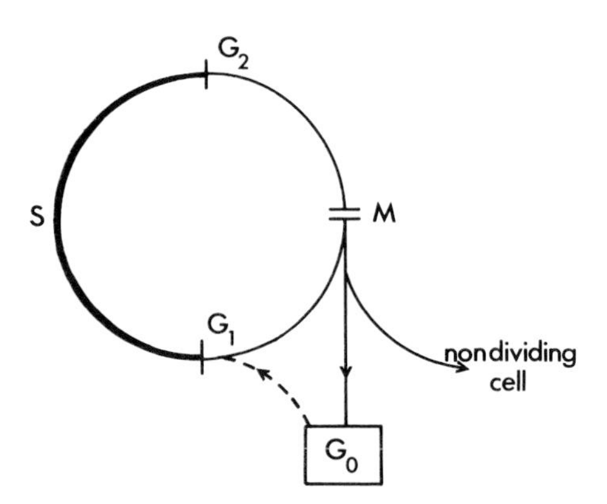

FIGURE 1 From the point of view of cellular proliferation, all populations of cells in multicellular organisms are made up of three subsets of cells: (1) continuously dividing cells, that go from one mitosis (M) to the next one, passing through the classic phases of the cell cycle, G_1, S, and G_2; (2) nondividing cells (i.e., terminally differentiated cells that leave the cell cycle and are destined to die without dividing again); and (3) G_0 cells, which are noncycling, nondividing cells but are capable to reenter the cell cycle if an appropriate stimulus is applied.

nately, these agents also kill the normal proliferating cells, and the patient would be left without bone marrow cells if it were not for the fact that 75% of the stem cells are in G_0. Because cells in G_0 are not proliferating, they can escape the effect of chemotherapeutic agents. The bone marrow depletion stimulates stem cells to enter the cell cycle so that they can eventually repopulate the bone marrow by their proliferation. Thus the presence of G_0 cells in cell populations is not only of scientific interest, but is useful at a practical level in optimizing our therapeutic approaches. [*See* HEMOPOIETIC SYSTEM.]

A state of G_0 can also be induced in cell cultures simply by decreasing the amount of growth factors. If properly done, cells in cultures that are partially deprived of growth factors are in a perfect physiological state (i.e., like the liver cells, they vigorously metabolize and, although not dividing, can do so if an appropriate stimulus, in this case the re-addition of growth factors, is applied).

C. The Cell Cycle

Figure 1 needs a little more explanation. It shows that cycling cells go through four different phases that are defined as G_1, S phase, G_2, and mitosis. DNA synthesis in dividing cells occurs in a discrete phase of the interphase (i.e., the period between one mitosis and the next one) (Fig. 2). This is a fundamental property because it not only localizes

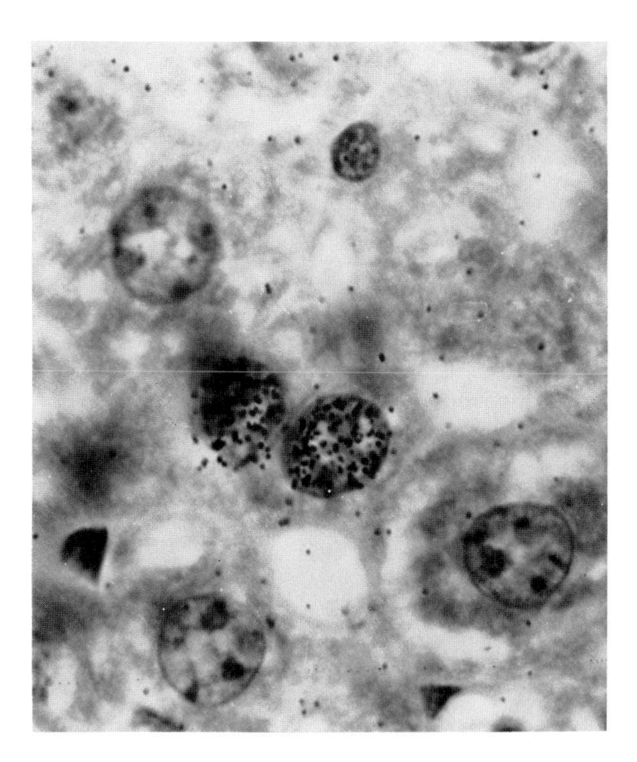

FIGURE 2 DNA synthesis is detected in single cells by autoradiography after exposure to [^{3}H]thymidine. Thymidine is exclusively incorporated into DNA and, if radioactive (tritium-^{3}H is the radioisotope of choice), the cells in the process of synthesizing DNA become radioactive. Radioactivity in DNA can be detected by autoradiography, which consists in overlaying the cells with a photographic emulsion, which is subsequently developed and fixed like a photographic film. In this picture, the nuclei of two liver cells are radioactive, as shown by the presence of developed silver grains in the emulsion above them.

the DNA synthetic period in the interphase, but it also clearly shows that DNA replication occurs in a discrete period of the interphase leaving intervals between mitosis and the initiation of DNA synthesis and between the completion of DNA replication and mitosis. These two intervals are called G_1 and G_2, where the G stands for gap, a gap in our knowledge. As knowledge accumulated in subsequent years, it became clear that a number of important biochemical and molecular events occur during G_1 and G_2. It is now clear that G_1 is the period during which the cell prepares for DNA synthesis and that G_2 is the period during which the cell prepares for mitosis. G_1 and G_2, however, are not necessary. There are some types of cells in animals, including humans, that do not have a G_1 or a G_2. These cells prepare all their components for DNA synthesis or mitosis in advance so that the cell cycle is reduced to its three fundamental components: DNA replication, doubling of cellular components, and cell division. This reduces the critical steps in the control of cell proliferation to three, namely, (1) the transition from a quiescent G_0 state into a cycling state, (2) the onset of DNA replication, and (3) mitosis. It should be remembered that the onset of DNA synthesis in each cell is sudden and occurs simultaneously in all chromosomes except the inactivated X chromosome that, for some reason, always replicates last.

D. Parameters of Growth

A glance at Fig. 1 reveals how a population of cells can grow in number (Table I). Any one of the three mechanisms outlined in Table I can increase the number of cells in a population. As it turns out, all three mechanisms are operative in living animals as well as in tissue cultures. With tumors, for instance, sometimes (but not always) there is a shortening of the cell cycle in respect to normal cells, or there is an increase in the growth fraction (i.e., in the fraction of proliferating cells) or occasionally a decrease in the rate of cell loss. In most cases all three parameters are operative and determine the aggressiveness of the tumor. Thus, when all three mechanisms are involved a tumor can grow quickly. For instance, there is a form of childhood lymphoma called *Burkitt lymphoma* that can double its size in as little as 38 hr. Conversely, there are other tumors that grow much more slowly and double their size only in weeks or months. However, although the length of the cell cycle can be shortened and the rate of cell death can also be modified, it is the general consensus that the most important parameter in regulating the extent of cell proliferation in a cell population is the fraction of proliferating cells.

The question now becomes: What are the factors that determine the extent of cell proliferation in a cell population? From the point of view of the molecular biologist, the question becomes: What are the genes and the gene products that determine the extent of cell proliferation? The problem can be stated in simple terms. Cells in tissues or in cultures cease to proliferate when the environmental conditions become nonconducive to cell proliferation. Addition of growth factors or other manipulations induce nonproliferating cells to reenter the cell cycle (i.e., to double their size and divide). With the problem reduced to these terms, we can ask two questions: (1) What are the environmental signals, stimulatory and inhibitory, in the environment that regulate cell proliferation by decreasing the fraction of G_0 cells and increasing the fraction of cycling cells? and (2) What are the genes and the gene products of the cell that interact with and respond to these growth factors? These two questions constitute the fundamental basis of the molecular biology of cell division.

TABLE I How a Population of Cells Can Grow in Number

Action	Result
By shortening the length of the cell	More cells produced
By decreasing the rate of cell death	Fewer cells die
By moving G_0 cells into the cell cycle	More cells produced

II. Genes and Gene Products for Cell Division

Table II summarizes the four major approaches to the identification of genes that regulate the proliferation of animal cells. Proto-oncogenes such as c-*fos*, c-*myc*, and c-*myb* are the usual equivalents of the cancer-inducing genes carried by certain viruses (oncogenes). Viral oncogenes are proto-oncogenes modified by a variety of mutations. They are also overexpressed or, at any rate, permanently expressed, because they are usually under the control of strong viral promoters. What is clear is that when proto-oncogenes are modified or overexpressed or in any other way altered, the growth regulation of cells is also altered. It therefore makes sense to believe that cellular proto-oncogenes also play a role in the control of cell proliferation. Indeed this has been formally demonstrated at least for some of the proto-oncogenes. [*See* ONCOGENE AMPLIFICATION IN HUMAN CANCER.]

The second approach is centered on growth factors. Factors present in the environment, whether of a stimulatory or inhibitory nature, control the extent of cell proliferation. Although this is easier to demonstrate in cells in culture than in the living animal, evidence is mounting that at least some of these growth factors also operate in the living animal. To operate, growth factors need to bind to the appropriate receptors so that the growth stimulus can be transmitted to the gene in the cell nucleus. Clearly, without cellular receptors, growth factors would have no effect on cells. This is, indeed, what happens when the absence of appropriate receptors makes a cell insensitive to the stimulation by a specific growth factor. For instance, lymphocytes do

TABLE II Four Approaches to the Identification of Genes and Gene Products for Cell Division

Source	Examples
Proto-oncogenes (retroviral transforming genes)	C-*fos*, c-*myc*
Growth factors and their receptors	Insulin and its receptor
Temperature-sensitive mutants of the cell cycle	Mutants for DNA ligase or thymidylate synthase
Growth-regulated genes	Proliferation cell nuclear antigen (PCNA)

not respond to epidermal growth factor (EGF) because they lack receptors for EGF. However, if the gene for the EGF receptor is artificially introduced into lymphocytes, they become sensitive to the effect of EGF.

The third approach is the use of temperature-sensitive mutants of the cell cycle. These are mutant cells in which a point mutation has occurred in a protein necessary for cell cycle progression. This point mutation has made the protein defective in a way that when the temperature of the cell culture is raised, the protein loses its function, although it is perfectly normal at a reduced temperature. In practice, we use two temperatures: a permissive one, which for animal cells is 34°, and a nonpermissive one, which is 39.6°. These temperature-sensitive mutants grow well at 34°, but when they are shifted to the restrictive temperature, they are blocked at a specific point in the cell cycle. Obviously, the gene coding for the mutated protein is relevant to the control of cell proliferation because when this protein is nonfunctional (at the restrictive temperature), the cell cannot progress through the cell cycle. The molecular cloning of such genes therefore would give us a good insight on the genes and gene products that are necessary for cell proliferation. There are a great number of temperature-sensitive mutants of the cell cycle, both from yeast and from animal cells. Temperature-sensitive mutants of the cell cycle have been described for every phase of the cell cycle (i.e., there are mutants that stop at the restrictive temperature in G_1, others in G_2, others again in S phase, and even mutants that block in mitosis at the restrictive temperature). Each of these temperature-sensitive mutants constitutes a source of knowledge for the molecular biology of cell cycle.

The fourth approach is to identify genes that are differentially expressed in a specific phase of the cell cycle. This is usually done by differential screening of cDNA libraries, and with this method we can identify genes that are expressed in one phase of the cell cycle, and not in others. In practice, we make a comparison between two different phases of the cycle, for instance, cells that are in G_0 and cells that have been stimulated to proliferate by growth factors (i.e., cells in G_1 or in the S phase). In this way we identify genes that are preferentially expressed in G_1 or in the S phase and not in G_0, or vice versa; we identify genes that are preferentially expressed in G_0 and are no longer expressed when the cells are stimulated to proliferate. One important thing to keep in mind, though, is that this approach does not give us growth regulatory genes. It gives us only growth-regulated genes (i.e., genes whose expression is regulated by growth factors). They may or may not be growth regulatory. One of the growth-regulated genes that has been most extensively studied is the thymidine kinase gene, which is a gene coding for an enzyme of the DNA synthesizing machinery. This gene is growth-regulated (i.e., it is induced by growth factors in quiescent cells), but its role as a growth regulatory gene is extremely unlikely. There are many cells that grow well without expressing the thymidine kinase gene, and in fact, there are animals such as squirrels and woodchucks that simply do not express the thymidine kinase gene and yet grow into adult animals. This distinction is important. In the case of the first three approaches, we actually identify genes and gene products that are certainly growth regulatory, whereas in the fourth approach, we identify genes that are growth-regulated and only hope that among them are also growth regulatory genes. The rationale for the last approach is based on the fact that some proto-oncogenes are growth-regulated.

Growth-regulated proto-oncogenes are

c-*fos*
c-*myc*
c-*myb*
c-*ras*
c-*fgr*
c-*jun*
ets-1 and *ets*-2

All these proto-oncogenes (except the *ets* oncogenes) are induced early after exposure to growth stimuli (see Section IV,A). The reasoning is that if some proto-oncogenes are growth-regulated, it may be possible that other growth-regulated genes may be potential oncogenes. Because we recognize that proto-oncogenes certainly play a role in the regulation of cell proliferation, we can conclude that perhaps among other growth-regulated genes that have not been picked up by retroviruses, some genes may play a major role in cell cycle progression and cell division.

One more question has to be answered: Is gene expression necessary for the control of animal cell proliferation? Actually the expression of many genes keeps going on, even in quiescent cells, as the hepatocytes clearly demonstrate. The question is whether the expression of certain genes specific for cellular proliferation is necessary for cell cycle progression and cell division. In animal cells, it seems clear that gene expression is necessary for the transition of cells from a quiescent to a cycling state. This has been formally shown by experiments indicating that a nonfunctional RNA polymerase II blocks cells in G_1. RNA polymerase II is the enzyme that transcribes genes that produce mRNAs. This indicates that the synthesis of appropriate mRNAs and therefore gene expression are necessary for the transition of cells from a quiescent to a growing stage. Once established that gene expression is a necessary step, it becomes legitimate to ask which genes are activated in this transition, in the hope, of course, to find the genes that are rate-limiting in the process.

We shall now return to the original statement that cellular proliferation is controlled by the environmental signals and by the genes and gene products that respond to these environmental signals, and we will start looking at the environmental signals that control the extent of cellular proliferation.

III. Growth Factors

The extent of cell proliferation in a population of cells is regulated by signals in the environment that are of two general types: spatial restrictions, and chemical signals. Spatial restrictions occur when cells come in contact with each other. The contact of cells between each other results in a decreased capacity for cellular proliferation as if the cells were sending each other negative signals for growth. These signals are poorly understood, but the other environmental signals (i.e., the chemical signals in the environment) are better known. The chemical signals that regulate cell proliferation can be more easily studied in the controlled environment of cell cultures. However, we will also discuss the importance of growth factors in the control of cellular proliferation in the living animal.

Growth factors can be divided into two large groups: factors that stimulate the proliferation of cells, and factors that inhibit their proliferation. Both of them are, strictly speaking, growth factors. However, it is the general consensus to refer only to stimulatory factors as growth factors and to use the term *inhibitory factors* for the second category. I am excluding here, from the environmental signals, the nutrients, inorganic ions, trace elements, etc., that are necessary for the life of a cell. The distinction between growth factors and the nutritional requirements of cells in culture is not always clear. Obviously, cells need amino acids, inorganic ions, and other compounds that are necessary for the well-being of a cell. Certain cells need high concentrations of calcium for proliferation; other types of cells, on the contrary, will grow only in low concentrations of calcium or in the absence of it. It is necessary, though, to separate these nutritional requirements from the need for growth factors, among which, though, we shall also include hormones.

A. Growth Factors for Cells in Culture

Table III gives examples of some types of cells and the growth factors that are required for optimal growth *in vitro*. Platelet-derived growth factor (PDGF) is one of the active principles in serum for the growth of fibroblasts. It is a highly cationic, heat stable, small protein present in relatively small quantities in platelets but is sufficiently potent, so that small quantities of it are sufficient to stimulate a large number of cells to synthesize DNA and divide. There are two forms of PDGF, A and B, which have sequence similarities and also share sequence similarities with the proto-oncogene c-*sis*. The EGF was originally purified from the mouse maxillary gland as a low molecular weight protein that when injected into newborn mice caused precocious opening of the eyelids and early eruption of the incisors. EGF was rapidly shown to be an important growth factor that had mitogenic action *in vitro* on cultured cells. Its molecular weight is 6,045 daltons, and it is a heat stable, nondialyzable polypeptide containing only 53 amino acid residues. EGF is present in large amounts in human milk. It is present in lower levels in the serum, and although it

TABLE III Different Growth Factors are Required by Different Types of Cells

Cell type	Growth factors
Fibroblasts	Platelet-derived growth factor, insulin-like growth factor I
Epithelial cells	Epidermal growth factor, Insulin
T lymphocytes	Phytohemagglutinin, Interleukin 2
B lymphocytes	Antigen, B-cell growth factor
Bone marrow cells	Interleukin 3 and other hemopoietic growth factors

is a good stimulant for fibroblasts, it is a growth factor that is particularly required by epithelial cells. Indeed, for some epithelial cells EGF is an absolute requirement for growth. Growth hormone and insulin-like growth factors (IGF) are also important regulators of cell growth. Growth hormone is one of the principal regulators of balanced post-natal growth *in vivo,* and closely related to it are prolactin and the placental lactogen. These three hormones have a high degree of amino acid homology and overlapping biological activities. IGFs, especially IGF-1, constitute a group of peptide factors that are important for the growth of cells *in vivo* and *in vitro*. The major characteristics that distinguish them from other growth factors are (1) their serum concentrations are growth hormone–dependent; (2) they are mitogenic for a variety of cell types, and (3) they produce insulin-like actions. IGF-1 is probably one of the most important growth factors in the serum. Insulin is also a growth factor, but in general it is a poor mitogen, and although its role in cell division is far from being negligible, in the great majority of cases, insulin acts *in vitro* in high concentrations by binding to the receptor for IGF-1. In other words, most of the effects that insulin has on the growth of cells *in vitro* can be replaced by IGF-1. However, it is possible that insulin potentiates the effect of other growth factors, although it has little effect by itself. [*See* INSULIN AND GLUCAGON; INSULIN-LIKE GROWTH FACTORS AND FETAL GROWTH.]

After we leave fibroblasts and epithelial cells, we enter into a totally different category of growth factors (i.e., the growth factors for hemopoietic cells). There is a great variety of these because each of these growth factors specifically stimulates the proliferation and promotes the differentiation of different hemopoietic cells.

We can say that for the stimulation of totipotential stem cells, we need Interleukin 3 or similar factors such as Interleukin 1, Interleukin 6, and the GM colony-stimulating factor (CSF). The totipotential stem cells promptly separate into two large lineages: the lymphocytes, and the myeloid stem cell. The growth factors needed by lymphocytes for proliferation include the B-cell growth factor and Interleukin 2, as shown in Table III, whereas for the growth of other lineages, the myeloid and erythroid stem cells require other growth factors besides GMCSF, which include Interleukin 5 and erythropoietin. A combination of these growth factors determines the extent of cell proliferation and simultaneous differentiation of the cells of the various hemopoietic lineages. We should also remember the existence of transforming growth factors (TGF), which, despite their name, are sometimes inhibitory growth factors (e.g., TGF-β). This lead us into the field of inhibitory growth factors, which are becoming more and more popular as they are isolated and purified. Some of these inhibitory growth factors are active on fibroblasts; others are active on hemopoietic cells. In defining a factor as an inhibitor growth factor, we should be able to exclude toxic factors, and in general, an inhibitory growth factor is considered a molecule that arrests the cell in a physiological state, possibly in G_0 or in G_1, from which the cells can reenter the cell cycle if an appropriate stimulus is applied. [*See* INTERLEUKIN-2 AND THE IL-2 RECEPTOR; LYMPHOCYTES; TRANSFORMING GROWTH FACTOR-α; TRANSFORMING GROWTH FACTOR-β.]

One important rule that should be kept in mind is that as shown in Table III, most cell lines, if not all, require more than one growth factor for growth. Occasionally a cell can be stimulated to proliferate by a single growth factor, but it is not difficult to show that large doses of the single growth factor simply stimulate the cells to secrete a second growth factor or to increase the number of receptors for a growth factor. For instance, high concentrations of PDGF stimulate fibroblasts to secrete IGF-1, thus reducing or even abrogating the requirement for added IGF-1 that most fibroblasts in culture display. Some growth factors are important for the transition of cells from G_0 to G_1; others are important for the onset of cellular DNA synthesis and mitosis. Their role in these three different critical steps can be better understood when we study the expression of genes in cells that are stimulated to proliferate.

B. Growth Factors *In Vivo*

The question that is often raised is whether the growth factors that are active on cells in cultures are also important in the control of cellular proliferation in the living animal. Despite the enormous advantages that are offered by tissue cultures in the analysis of life processes, we should not forget that the environment of cells in culture is highly artificial and no more resembles the environment of cells *in vivo* than a zoological garden may resemble an African jungle. There is no question, however, that hemopoietic growth factors are active *in vivo* as well as *in vitro*. Indeed, there is some evidence that they may act even better *in vivo* than they act *in vitro*. These growth factors are responsible for both the proliferation and differentiation of hemopoietic cells, and they are so effective that they have even been used in humans in the therapy of disease. Their use in human medicine has, of course, been facilitated by recombinant DNA technology that has allowed the preparation of the individual growth factors in sufficiently large amounts. Their action on cell proliferation and differentiation is undisputed.

Epidermal growth factor is active *in vivo* as well as *in vitro,* and in fact, it has already found some practical uses. For instance, in New Zealand, injections of EGF are used as a way of fleecing sheep, as it is so effective in stimulating the proliferation of follicular epithelial cells. It has also found its way in the treatment of burns and wounds in humans, incorporated into ointments. Growth hormone, the absence of which is the cause of dwarfism, is actually more active *in vivo* than *in vitro*. Growth hormone acts largely indirectly (i.e., by stimulating the liver to produce IGF-1). In fact, in patients with acromegaly or hypopituitarism, clinical assessments of disease activity correlate better with blood levels of IGF-1 than of growth hormone. IGF-1 can therefore be considered as a growth factor in the living animal and, indeed, as an important growth factor that participates in the regulation of the extent of cell proliferation. PDGF may be involved in wound healing, but its function here is not as yet clear-cut as with some of the other growth factors mentioned above.

C. Receptors for Growth Factors

Each growth factor binds to its receptor on the cell surface. There are extensive molecular structure similarities among the receptors for various growth factors, and sometimes, the receptors can be interchangeable, as for instance, insulin, which at high concentrations binds to the IGF-1 receptors and thus mimics its action. In general, receptors for polypeptide growth factors have three domains: an extracellular domain for the binding of the specific growth factor, a transmembrane domain, and an intracellular domain, which transmits the growth factor signal to the interior of the cell. Alterations of these domains can transform the receptor into an oncogene. These receptors can be internalized, but at least in most cases they do not reach the nucleus. By contrast, there are other receptors (e.g., steroid receptors) that are not anchored to membranes, are distributed in the cytosol, and when they bind the ligand, move to the nucleus, where they induce changes in gene expression. Changes in gene expression, of course, are also induced by polypeptide growth factors, but we do not know how the signal moves from the membrane to the nucleus, whereas with steroid receptors, the messages to the nucleus is carried by the receptor itself. [*See* RECEPTORS, BIOCHEMISTRY.]

IV. Molecular Biology

A. G_0 to G_1 Transition

We have mentioned that the G_0 to G_1 transition is one of the critical steps that control the extent of proliferation in a population of cells. With the exception of cells that are destined to die without dividing again, the ratio of G_0 cells to cycling cells is what largely determines the extent of cell proliferation. The recruiting of cells from the G_0 pool is therefore a crucial step, which is signaled by a variety of stimuli (growth factors and other mitogenic stimuli), and results in the activation of the expression of several genes.

A great number of genes are activated when G_0 cells are stimulated to proliferate by mitogenic stimuli. We can divide them into three large categories: the ones in which the expression is increased promptly after stimulation, the ones whose expression increases only at the G_1-S boundary, and the ones in between (Fig. 3). First we will discuss the genes that are induced early after stimulation of G_0 cells to proliferate. These genes go under the collective name of early growth regulated genes (Table IV). These genes have generally been identified by differential screening of cDNA libraries (i.e., by

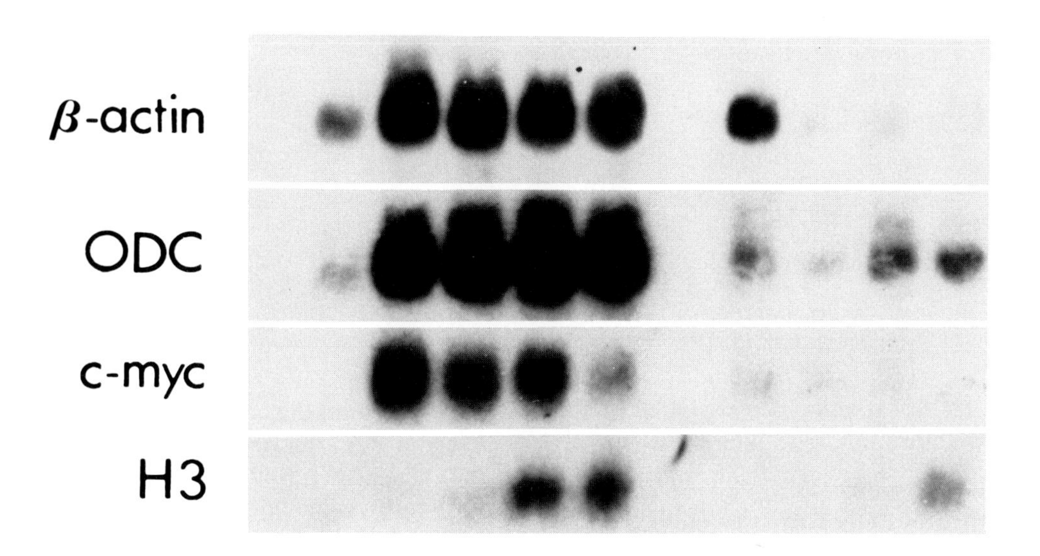

FIGURE 3 Gene activation in cells stimulated to proliferate. Northern blot of mRNAs obtained from mouse fibroblasts at various hours after stimulation with growth factors. (A) cells stimulated with serum, which stimulates cell proliferation. (B) cells incubated in platelet-poor plasma. The mRNAs recognized by the hybridizing probes are those for β-actin, ornithine, decarboxylase (ODC), c-*myc*, and histone H3. The first three genes are activated early by serum, whereas histone H3 mRNA is not detectable until the cell reaches S phase (16–24 hr after stimulation). Barely any gene activation occurs with platelet-poor plasma.

techniques that allow the investigator to identify genes that are preferentially expressed in early G_1 and are not expressed in resting cells). In the beginning, only a few such genes were identified by a handful of laboratories, but as more and more laboratories got involved in the identification, the list has been rapidly increasing. A glance at Table IV and the previous list of early growth-regulated proto-oncogenes will reveal that these early growth-regulated genes constitute a rather heterogeneous group of considerable interest. They include, besides proto-oncogenes such as c-*myc*, c-*fos*, and c-*jun*, other interesting genes, some of which are related to cytokines, transcription factors, DNA binding proteins, cytoskeletal proteins, and proteins that are involved in the general metabolism of the cell (e.g., the ADP/ATP translocase). Early growth-regulated genes share certain characteristics, namely, (1) the expression of these genes (mRNA levels) increases early, essentially in the first 2–3 hr after G_0 cells are stimulated to proliferate. (2) The expression increases even in the presence of high concentrations of cycloheximide, which is a potent inhibitor of protein synthesis. In fact, cycloheximide itself sometimes induces the expression of some of these genes. This indicates that the increased expression of these early growth-regulated genes is not dependent on *de novo* protein synthesis. (3) These genes are also expressed in G_1 specific, temperature-sensitive mutants of the cell cycle at the restrictive temperature (see Section II). These mutant cells have a thermosensitive defect that blocks them in the G_1 phase of the cell cycle when shifted to 39.6°. Despite the block, the mRNA's of early growth-regulated genes are normally expressed when these cells are stimulated

TABLE IV Growth-Regulated Genes

G_0 G_1	G_1	G_1/S
c-*myc*	Calcyclin	Core histones
c-*fos*	Dihydrofolate reductase	DNA topoisomerase I
p53	β-Actin	Thymidine kinase
Ornithine decarboxylase	Mitogen-regulated protein	Thymidylate synthase
jun-Related proteins	Major excreted protein	Ribonucleotide reductase
fos-Related proteins	VL-30 sequences	PCNA/cyclin
ADP/ATP translocase	Calmodulin	Transferrin receptor
IL-2 receptor	Protease inhibitors (2)	DNA polymerase-α
Vimentin	Transin	
Glycolytic enzymes (4)	Collagen and collagenase	
Egr-1	Fibronectin	
JE (cytokine ?)	Proliferin	
Egr-2	Progressin	
KC	Osteopontin	
Nur 77 (NGFI-B)		
Krox − 20		
3.7 kb gr		
thombospondin		
bcl-2		
Bravo's 82 clones		

with growth factors at either permissive or nonpermissive temperature.

B. Significance of Early Growth-Regulated Genes

An obvious question at this point is: Are these genes necessary and sufficient for cellular proliferation? The answer is simple. Some of them, at least, are necessary; none of them is sufficient. That some of them are necessary is demonstrated by the fact that monoclonal antibodies to their protein products injected into cells, or anti-sense RNA, or anti-sense oligodeoxynucleotides to their transcripts inhibit cell cycle progression and cell proliferation in general. When the mRNA or protein are neutralized by appropriate and specific inhibitors, cell cycle progression is also inhibited. For instance, an antibody, or an anti-sense RNA to c-*myc* will inhibit cellular proliferation. Several other examples are available for some of these genes that clearly play a necessary role in cell division. None of them has yet to be found to be sufficient (i.e., by themselves, none of these gene products can stimulate the proliferation of cells, although they can effect the G_0-G_1 transition). The cells then stop in G_1 and will not progress further toward the S phase. This has been illustrated in a dramatic way by the microinjection technique. For instance, if quiescent mouse 3T3 cells are microinjected with roughly 800,000 copies of the c-*myc* protein, the cells do not enter S phase. However, if, after microinjection of the c-*myc* protein, the cells are exposed to a sufficient concentration of IGF-1, the cells enter S phase. These experiments demonstrate that c-*myc* can replace the first growth factors, in the case of fibroblasts, PDGF, but cannot replace the effect of the second growth factor, which, in this case, is IGF-1. A similar situation occurs in other types of cells, as described in Table III. For instance, in the case of B lymphocytes, the first mitogen, the antigen, induces the appearance of c-*myc* and ATP/ADP translocase, and the cells progress to G_1 but do not progress to S phase. A second mitogen is necessary for entry into S phase and cell division. Thus we can conclude that some of these early growth-regulated genes are necessary prerequisites but are not sufficient for cellular proliferation. Of course, only a handful of them have thus far been demonstrated to perform a necessary function. Some of these growth-regulated genes may not have a regulatory function but may simply accompany stimulation of transcription that follows the addition of mitogenic stimuli. In other words, we should again keep in mind that growth-regulated is not the same thing as growth regulatory.

C. G_1-S Boundary

This is the second critical step for cellular proliferation (i.e., the activation of the genes for the DNA synthesizing machinery). In reality there are several other genes whose expression increases somewhere in mid-G_1 (i.e., before the G_1-S boundary). These genes, most of which are sensitive to inhibition by protein synthesis, are also listed in Table IV together with the genes whose expression increases sharply at the G_1-S boundary. Most of the latter genes are genes of the DNA synthesizing machinery or, like the genes for histones, are coding for proteins that are part of the genetic material.

Are these genes necessary? There is no question that the G_1-S boundary genes are often necessary because most of them deal with the DNA synthesizing machinery, and if any one of them is inhibited, DNA synthesis does not occur. There are, of course, exceptions (e.g., thymidine kinase, which we have already mentioned), but others are definitely crucially involved in DNA replication and they are not dispensable. For instance, if one adds to the culture medium of exponentially growing cells, anti-sense oligodeoxynucleotides to the proliferating cell nuclear antigen [PCNA], which is a cofactor of DNA polymerase δ), DNA synthesis and cell cycle progression completely stop.

The picture that emerges is that a great number of genes are activated when cells are stimulated to proliferate and that many of these genes play an important role in the regulation of cellular proliferation (i.e., they are necessary for cellular proliferation). A question that should be raised at this point is: Are they specific? The present evidence is that most of the early growth-regulated genes may not be specific for cell proliferation, although they are necessary. For instance, a number of events occur in stimulation of cellular proliferation that also occur in a number of situations in which there is no stimulation of cellular proliferation. Other events (e.g., the activation of tyrosine kinase) also occur in situations in which cellular proliferation is actually inhibited. The most spectacular example, of course, is the induction of c-*fos* by differentiating agents (i.e., by agents that actually inhibit the proliferation of cells). Many other examples are available, clearly indicating that the expression of certain proto-oncogenes and the early events of cellular proliferation also occur in situations that are not conducive to cell division. In other words, we can picture a situation in which G_0 cells are aroused by more or less aspecific stimuli in the environment. The induction of c-*fos* or similar genes prepares the cells for a second set of growth factors that are more proliferation-specific. This second set of growth factors includes those we have mentioned above that can lead the cell into S phase. It is important, though, to realize that proliferation-specific events in mid-G_1 or late G_1 cannot take place unless the cells are first primed by early growth-regulated genes. The cells expressing early growth-regulated genes may still have the option of differentiating or otherwise responding in ways that are not related to cellular proliferation. The previous history of the cells, the presence of receptors for other growth factors, or the presence in the environment of a second set of signals are the parameters that finally determine whether a cell, which has been previously aroused, may actually enter S phase (i.e., begin DNA synthesis) or not.

D. Onset of S Phase

The onset of DNA synthesis in each cell is sudden and occurs simultaneously in all chromosomes except the inactivated X chromosome. Despite a few contrary reports, there is no gradual build-up. Instead, DNA synthesis in each cell starts suddenly at nearly maximal rates and continues throughout the S phase with little variation. DNA synthesis itself requires the cooperation of several proteins. In eukaryotic organisms, each DNA fiber is divided into many replicating units called *replicons*. Each replicon has a center, the origin, from which growing forks extend outward in both directions. Apart from the various enzymes that are necessary for the synthesis of the building blocks for DNA synthesis, the best model for the study of cellular DNA synthesis is actually the replication of SV40 DNA. Replication of SV40 DNA can be done in a cell-free system that is capable of replicating the plasmid DNA containing the SV40 origin of replication. For replication of the SV40 DNA, one of its own products is necessary, the T antigen, which is coded by the SV40 T-antigen coding gene of the SV40. It is a multifunctional protein that plays a central role in replication of SV40 DNA. This is specific for SV40 and is not necessary for the replication of cellular DNA, but it serves as a model in which one can study the other proteins that are necessary for cellular DNA synthesis. Among the proteins that are necessary for cellular DNA synthesis, the following have been implicated: DNA polymerase α-primase, topoisomerases I and II, single-stranded binding proteins, and the PCNA that we have already discussed. However, at least seven factors are required, in addition to T antigen, for efficient DNA replication *in vitro*. Four of the proteins have been highly purified: the replication protein A, PCNA, the DNA polymerase α-primase complex, and topoisomerase I. The remaining three fractions thus far have only been partially purified. Some questions on DNA replication, unfortunately, remain unsolved. For instance, why does DNA synthesis occur with such a sudden onset? Why is it multifocal and takes several hours, often 7–8 hr, in a single cell for complete replication of DNA? In the developing, fertilized egg, the S phase can last a little as 5 min. Why such a marked difference? Another important puzzle is that the time at which each gene is replicated is constant, (i.e., there are certain genes that are constantly replicated in the early part of S phase, whereas others are replicated in the late part of the S phase). There is a general tendency for active DNA to be replicated early, but this is not an absolute rule. [*See* DNA SYNTHESIS.]

E. G_2 and Mitosis

The molecular biology of G_2 and mitosis has been little known until recently. For a long time it was known that protein synthesis in G_2 was necessary for the entry of cells into mitosis, but little was known about the proteins or gene products in general that are necessary for this transition. A series of experiments had shown that histone H1 phosphorylation played some role in the transition of cells from S phase to mitosis, and some laboratories were able also to isolate monoclonal antibodies that were specific for G_2 cells.

β-Tubulin is, of course, necessary for mitosis, and this is an interesting protein because it autoregulates its own production directly on the translation of its messenger RNA. Some insights on the molecular biology of mitosis have emerged recently, stemming from investigations on the so-called maturation-promoting factor that was originally isolated from meiotic cells. A number of studies from several laboratories using mitotic mutants, as well as cells from frogs and yeast, have come to the conclusion that there is a protein of molecular weight 34,000 daltons, called *p34,* which is inactive in interphase. It then interacts with another protein called *cyclin,* which is also called *p50–60,* and the interaction of cyclin with the p34 protein activates it during mitosis. The p34 protein is a protein kinase that has as a substrate histone H1. Another protein that is involved in the control of mitosis is a potential inhibitor, the p13 protein. What is most interesting in this partial elucidation of the molecular biology of mitosis is that the p13 protein turns out to be the equivalent of the cdc 13 gene of yeast and the p34 protein turns out to be the equivalent of cdc 2 gene of yeast, which is the "start" gene of yeast, the gene that seems to regulate the extent of cellular proliferation in yeast cells. In animal cells the equivalent of the cdc 2 gene, the p34 protein, seems to be involved only in the regulation of G_2 and mitosis, rather than the transition of cells from G_0 to G_1. [*See* MITOSIS.]

V. The Transformed Cell

The definition of a transformed cell is relatively easy if we limit ourselves to cells in culture transformed by certain viruses. It becomes more complicated if cells in culture are transformed by chemical or physical agents, and quite difficult if we wish to make transformed cells the equivalent of neoplastic cells. The criteria of transformation for fibroblasts or fibroblastlike cells are shown in Table V. The first thing we should notice is that transformed cells are characterized by irregular growth, unrestrained growth, and infinite growth. Irregular growth is due to lack of contact inhibition of movement, so that cells can move over each other to form randomly arranged multilayers or, as it is often stated, "they can pile up." Normal cells in culture form monolayers and do not pile up. Unrestrained growth transformation is the loss of contact inhibition of cell proliferation, so that the transformed cells reach a higher saturation density than the normal cell population. In general, when we compare a cell line with its virally transformed counterpart, we can say that transformed cells grow at a higher saturation density, especially at low concentrations of serum. Indeed, we can say that as a rule the serum requirements are lower in transformed cells, so that transformed cells grow in concentrations of serum or growth factors that do not allow the growth of normal cells. Infinite growth transformation is the loss of aging properties permitting cells to undergo unlimited division in culture. That is, transformed cell lines have an indefinite growth potential and are said to be "immortalized." We should probably define these cells as simply having an indefinite growth potential. The question that has to be answered, though, is whether a transformed cell is the equivalent of a cancer cell. The ultimate test of transformation is the ability to induce tumors in ap-

TABLE V The Transformed Cell

Minimal transformation	Intermediate transformation	Full transformation
In vitro		
High saturation density Indefinite growth potential Growth in low serum	Clonal growth in agar (anchorage-independence)	Ability to produce tumors in animals
In animals		
Benign tumors		Malignant tumors

propriate host animals. Syngenic animals or nude mice are often used to test the tumor-inducing ability of tumor cells. As shown in Table V, the ability to produce tumors in animals is really what makes a transformed cell a cancer cell. We should view transformed cells as progressing slowly from a minimal transformation phenotype to intermediate transformation and, finally, to a clearly open cancer cell. In between the minimal transformation and the full transformation, is the ability to grow unattached to a supporting glass or plastic surface. This ability to grow in suspension is certainly a further step in the direction of the acquisition of the malignant phenotype.

Oncogenes certainly play a considerable role in the transformation of cells. As we already mentioned, the modification of proto-oncogenes results in alteration of the regulation of cell growth. This modification can take place in a variety of ways, and the following mechanisms have been invoked to explain the transforming activation of normal proto-oncogenes (Table VI): (1) insertion mutagenesis, in which a proto-oncogene is activated by the integration near it of viral DNA, especially a viral promoter. (2) Amplification of the proto-oncogene. A typical example is the c-Ki-*ras* gene, which is amplified in certain tumors. C-*myc* is also amplified in human promyelocytic leukemia cell lines, and c-*abl* in human, chronic, myelogenous leukemia cell line. (3) Translocation of the proto-oncogene. Translocation of c-*myc* is important in Burkitt lymphoma, and translocation of other genes is important in other forms of human leukemias (e.g., c-*abl* in chronic myeloid leukemia). (4) In the three instances mentioned above, an increased level of expression of the proto-oncogene is either demonstrated or postulated, but a fourth mechanism of oncogene activation depends not on the amount of protein, but on the amino acid sequence of the oncogene protein. Mutations can transform a normal proto-oncogene into an oncogene that confers to the cell the transformed phenotype. The most famous example is the point mutation in the Ha-*ras* proto-oncogene in which a G $\rightarrow$ T mutation causes glycine, normally present at the 12th residue of the encoded 21,000-dalton protein to be replaced by valine. The *ras* proto-oncogene becomes activated and causes transformation of cells. Similar mutations have been demonstrated in a great variety of other oncogenes that eventually result in the transformed phenotype.

TABLE VI Mechanisms of Activation of Cellular Protooncogenes

Mechanisms	Example
Insertion mutagenesis	c-*myc*
Amplification of proto-oncogenes DNA	c-Ki-*ras*, c-*myc*
Translocation	c-*myc*, c-*abl*, *bcl*-1
Mutations in the coding sequence	c-*ras*

More recently, the concept of anti-oncogenes is emerging. In certain forms of human tumors (e.g., retinoblastoma), chromosomal deletions occur, and these chromosomal deletions are clearly related to the tumor phenotype. In the case of retinoblastoma, a deletion in chromosome 13 is regularly found. A gene has been identified in this deleted area of chromosome 13, which has been called the *retinoblastoma gene,* or briefly, the *RB gene*. Although it is not absolutely certain that this RB gene is the one responsible for the genesis of retinoblastoma, some interesting findings have already emerged. What we know, for instance, is that the product of the RB gene binds to the products of certain DNA oncogenes, specifically, to the E1A protein of adenovirus and the T antigen of SV40. This binding between the products of oncogenes of DNA oncogenic viruses and the product of the RB gene is rather specific. Of course, we are tempted to speculate that transformation by DNA oncogenic viruses may occur simply because the oncogene neutralizes the anti-oncogene that, by itself, controls the proliferation of cells. This is an attractive proposition, and we, of course, have to keep in mind that these anti-oncogenes could have an important function in the regulation of cell proliferation. For instance, it has been reported that PCNA is necessary for the synthesis of the replication of DNA because it neutralizes an inhibitor of DNA synthesis (i.e., PCNA could therefore work as an inhibitor of an anti-oncogene). In our understanding of the molecular biology of cell division, we should therefore keep in mind not only that we need effector genes (i.e., new genes that are activated and whose products can stimulate the proliferation of cells) but also that these gene products may simply act by neutralizing inhibitors that otherwise control the proliferation of cells. It is likely that in the future it will become more and more apparent that the control of cell proliferation, as well as the transformed phenotype, rests on a delicate balance between effector genes that cause cellular proliferation and inhibitor genes that inhibit cell proliferation. [*See* RETINOBLASTOMA, MOLECULAR GENETICS; TUMOR SUPPRESSOR GENES.]

Bibliography

Baserga, R. (1985). "The Biology of Cell Reproduction." Harvard University Press, Cambridge, Massachusetts.

Baserga, R., and Surmacz, E. (1987). Oncogenes, cell cycle genes and the control of cell proliferation. *Biol Technology* **5,** 355–358.

Crabtree, G. R. (1989). Contingent genetic regulatory events in T lymphocyte activation. *Science* **243,** 355–361.

Dunphy, W. G., and Newport, J. W. (1988). Unraveling of mitotic control mechanisms. *Cell* **55,** 925–928.

Huberman, J. A. (1987). Eukaryotic DNA replication: A complex picture partially clarified. *Cell* **48,** 7–8.

Lohka, M. J. (1989). Mitotic control by metaphase-promoting factor and cdc proteins. *J. Cell Sci.* **92,** 131–135.

Pegoraro, L., Avanzi, G., and Lista P. (1988). Growth and differentiation factors in human hematopoiesis. *Haematologica* **73,** 525–543.

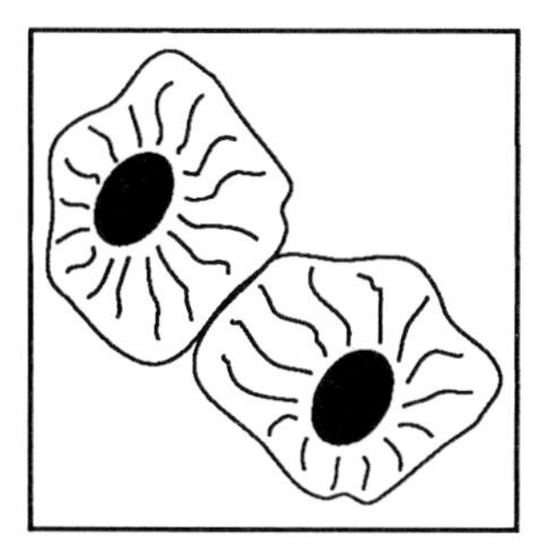

Cell Junctions

JUAN C. SÁEZ, DAVID C. SPRAY, *Albert Einstein College of Medicine*

Glossary

Desmosomes and adhesive belts Vertebrate adhesive junctions with dense membrane underlining
Cytoplasmic bridges Areas of cytoplasmic continuity resulting from incomplete mitosis
Gap junctions Communicating junctions of animal cells generally occurring as plaques of particles
Plasmodesmata Intercellular communicating junctions of plant cells, extending through the cell walls
Septate junctions Invertebrate adhesive membrane contacts characterized by parallel rows of intramembrane particles
Tight junctions Occluding junctions that separate apical and basal faces and occur as interlaced webs of membrane particles

HIGHER ORGANISMS ARE MADE up of differentiated tissues, which consist of cells with various phenotypes. Evolution of multicellular organisms required cells to establish polarity, adhesion, and pathways for intercellular signaling. Polarity, in which properties of the apical surfaces are different from those of the basal surface, achieved the separation of the external environment from the intra-aggregate milieu and allowed survival in media with various salinities and availability of nutrients. Adhesion maintained the multicellular aggregate and permitted the establishment of intracellular signaling between specific cellular subsets. Intercellular signaling provided mechanisms by which nutrients could be shared, toxins could be buffered, and cells could signal one another to achieve coordinated behavior of similar cells or to recruit cells with differentiated functions into arrays of responsive elements.

Each of these requirements for establishment, maintenance, and function of cellular aggregates has been accomplished by the expression of various types of cellular contacts or junctions. Extracellular matrix components and a variety of cell adhesion molecules also contribute to the interplay between cells that comprise tissues and organs. Although these elements provide a lattice in which tissue function is maintained and can certainly influence the modes of cell contact, these components are generally not elaborated into specific structural elements between cells and are not considered further here.

Cellular junctions of vertebrates are categorized into four types: desmosomes, adherens junctions, tight junctions, and gap junctions. They are usually divided according to their functions of occlusion (tight junctions), adhesion (adherens, desmosomes, and septate junctions of invertebrates), or communication (gap junctions, cytoplasmic bridges and plasmodesmata in plants). All junctions were discovered using electron microscope techniques, and, although experimental evidence has strengthened the association of function with some of the structures, the role that each junctional type plays has most often been inferred from location on cell surfaces and by which cell type expresses each type of junctional contact. Recently, biochemical and immunological approaches have been brought to bear on the identification of molecular components for some junctional types. As more is learned about the conditions that control expression of each junctional type, the dynamics of assembly and cell and regional specificity will become clearer.

I. Adhesive or Anchoring Junctions

A. Septate Junctions

Septate junctions were first discovered in a coelenterate (Hydra) and have subsequently been found throughout invertebrate phyla. In thin-section electron micrographs, septate junctions are quite variable in appearance but are generally recognized as alternating regions of close contact or septum bridging the wide space between the membranes of adjacent cells (Fig. 1A). Freeze fracturing reveals details of the interior of the membrane, as well as its surface topography and commonly occurs transversely along the hydrophobic central plane of a cellular membrane (i.e., between the two layers of its bilayer). Freeze-fracture replicas of septate junctions reveal quite parallel rows of usually P-face particles and corresponding rows of E-face depressions or pits (Fig. 1B) (P indicates the fracture face of the protoplasmic leaflet and E indicates that of the axoplasmic leaflet). Although these junctions have been isolated, their biochemical nature is unknown.

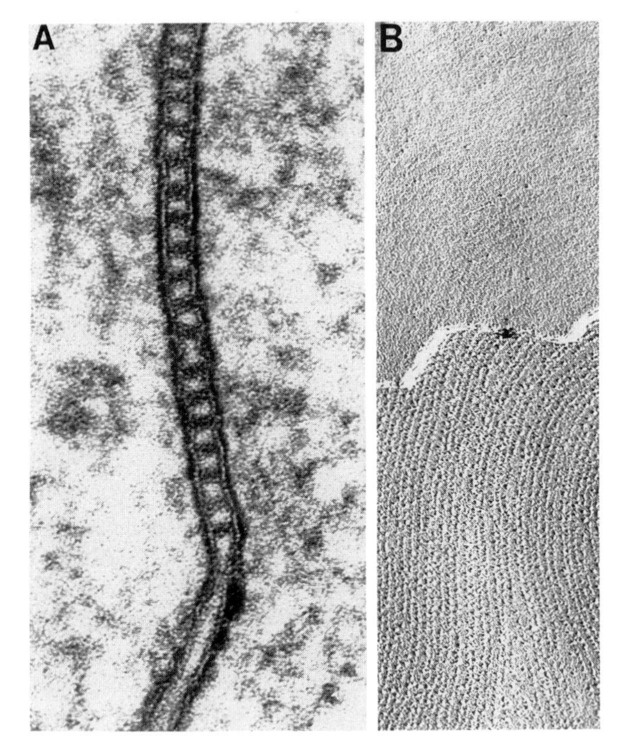

FIGURE 1 Structure of septate junctions. A. Thin-section appearance of septate junctional arrangement between two lateral epithelial cells of clam gill. ×210,000. B. Freeze-fracture replica of septate junction between two epithelial cells of clam gill. ×69,000. [Micrographs courtesy of Dr. Peter Satir.]

Septate junctions presumably serve primarily as regions of intercellular adhesion, but a role in establishing transepithelial permeability barrier (as the invertebrate equivalent to tight junctions) is not rigorously excluded. A role in intercellular communication was assigned early to septate junctions, but subsequent studies revealed that gap junctions were located nearby; this latter junctional type is now accepted as the major type of intercellular communicating pathway in animal cells (see below). Terminology often confuses the issue, because junctions between crayfish axon segments are termed septate, although their ultrastructure is that of the gap junction.

B. Desmosomes and Adherens Junctions

Desmosomes and adherens junctions are intercellular junctions found prominently in epithelial, cardiac, and arachnoid tissues. Two types of symmetric intercellular junctions are distinguishable morphologically: macula adherens, or spot desmosomes, and fascia adherens, or intermediate junctions. In addition, there are asymmetric junctional types called hemidesmosomes and focal adhesions, which are adhesive junctions between epithelial cells and the extracellular matrix. All types of adhesive junctions can be distinguished by their location, elaboration, and relationship with cytoskeletal elements.

1. Structural Features

Spot desmosomes are symmetric membrane specializations between adjacent cells. They are generally oval, 0.2–0.5 μm in length, with 25–35 nm intercellular spacing. These structures contain a central dense stratum lying parallel to the cell membranes and dense fibrillar plaques located in the subjacent cytoplasm. The intercellular junctional space is continuous with the nonjunctional space but can be distinguished from the latter by the electron-dense stratum, which appears to be connected to the junctional membranes by arrays of filaments (Fig. 2A). The cytoplasmic plaques are separated from the cytoplasmic membranes by a slightly less electron-dense region and are sites of contact for tono- or intermediate-size filaments (10 nm diameter; Fig. 2B).

In freeze-fracture replicas, spot desmosomes are seen as aggregates of intramembrane particles that

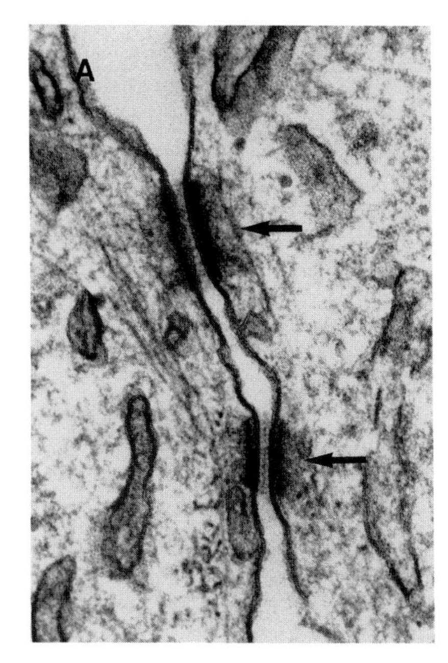

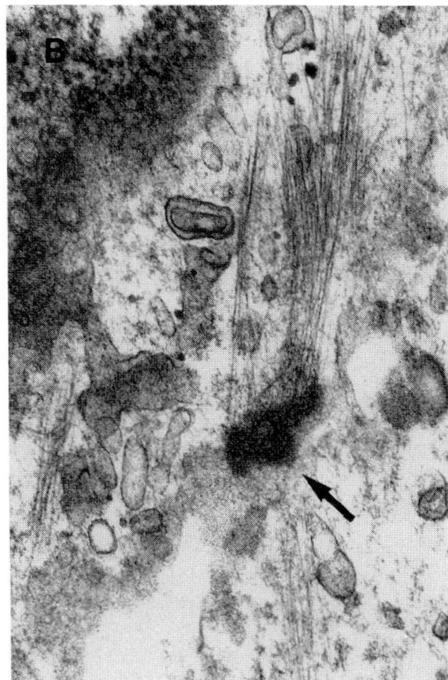

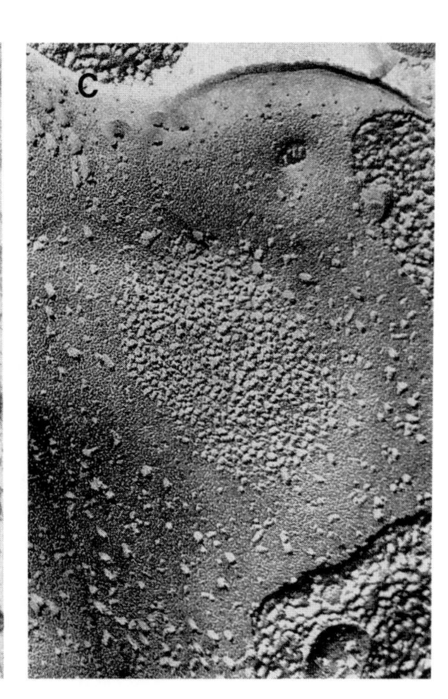

FIGURE 2 Structure of spot desmosomes. A. A thin-section electron micrograph showing two desmosomes (arrows) between cells of a hepatoma cell line (HepG2 cells). B. A thin-section micrograph of the cytoplasmic aspect of a desmosome (arrow) showing the interaction of keratin filaments with the cytoplasmic plaque. ×59,700. [Micrographs A and B courtesy of Dr. Phyllis Novikoff.] C. Freeze-fracture replica of a desmosome between cytoplasmic membrane cells showing particles in the E-face. ×50,000. [Contributed by Dr. David Hall.]

usually, but not always, cleave with the E-face (Fig. 2C). Variability in particle distribution can be found in freeze-fracture replicas obtained at different stages of development and is introduced by fixatives. The particles have irregular forms with diameters of about 12–30 nm. The intercellular stratum is seen as an etch-resistant line 5–7 nm wide, which in some cases appears to be made up of numerous small particles.

Studies on the molecular components of desmosomes suggest that these sites of cell contact are multimolecular complexes. The protein components of spot desmosomes can be distinguished as transmembrane proteins, which are connected to the cytoplasmic plaques and hold the cytoplasmic membrane of the adjacent cell through their extracellular region, and a complex of intracellular proteins that form the plaques and interact with tonofilaments. Several polypeptides are localized morphologically with desmosomes and are referred to using various nomenclatures. The nonglycosylated desmosomal proteins (DP; localized in the cytoplasmic plaques), in order of decreasing molecular mass, are termed DP-I (desmoplakin I; M_r 250 kDa), DP-II (desmoplakin II; M_r 215 kDa), DP-III (plakoglobin; M_r 83 kDa), and DP-IV (M_r 75 kDa). Desmosomal glycoproteins, or desmogleins (DG; localized in the junctional cytoplasmic membrane), are called DG-I (triplet, M_r 165–175 kDa) and DG-II, (doublet, M_r 110–115 kDa). Moreover, another component copurifies with desmosomes isolated from some cell types and has been termed desmocalmin (M_r 240 kDa) to indicate its calmodulin binding. Research on the biochemistry of desmosomes is still progressing. Recently, a high molecular mass protein (desmoyokin; 680 kDa) of spot desmosomes has been characterized and localized in the cytoplasmic plaques of stratified epithelium.

Although desmosomes of different sources are morphologically similar, identity of their molecular components cannot be assumed. Crossreactivity between antibodies that recognize desmosomal components in different species suggests that at least some antigenic epitopes are conserved. The biochemical and immunological analysis of preparations enriched in desmosomes from different tissues and cell lines show variation in the spectrum of desmosomal proteins and the apparent molecular mass of desmogleins, confirming the existence of a degree of diversity between the molecular components of desmosomes from different cell types. In spite of the molecular diversity, the mechanism of adhesiveness of desmosomes is rather conserved because they can be formed between cells of different tissues and species.

Cell–cell adherens junctions are seen on internal surfaces of the plasma membrane. In epithelial tissues, they are beltlike structures (zonula adherens or belt desmosomes), and in several nonepithelial tissues they have an interrupted topography (fascia adherens, abundant in the intercalated disc of cardiac myocytes). The belt desmosomes found in epithelial cells are localized immediately below the tight junctions. In thin sections of cell–cell adherens junctions, the intermembrane separation is 15–20 nm, with no electron-dense stratum between the membranes. Three major domains have been distinguished for zonula adherens: (1) membrane receptors, (2) cytoplasmic plaques that contain vinculin and, in some cases, talin, and (3) cytoskeletal structures containing actin filaments and actin-associated proteins (α-actinin). Because intramembrane elaboration is only poorly observed in freeze-fracture replicas, it has been suggested that belt desmosomes should be designated regions of intercellular contact rather than intercellular junctions, despite the clear cytoplasmic and intercellular differentiation.

The use of immunological methods localized some unique proteins in the undercoat of adherens junctions. Vinculin, α-actinin, and $pp60^{src}$ are found in both cell–cell and cell–substrate adherens junctions, but plakoglobin is only in cell–cell junctions. Very recently, adherens junctions were successfully isolated from rat liver. From these isolated junctions, a 82-kDa protein has been purified and shown to inhibit actin filament assembly. Moreover, a 135-kDa protein has been identified in fascia adherens isolated from chicken hearts. It is localized in cell–cell adherens junctions but not in cell–matrix junctions. Cardiac intercalated discs and areas of intercellular contact of cultured myocytes and lens cells are particularly enriched in this protein.

Hemidesmosomes are asymmetric structures adhering the membranes of epithelial cells to the basal lamina. Hemidesmosomes display an extracellular electron-dense line parallel to the membrane and the basal lamina. The cytoplasmic plaques are sites of contact or end attachment of tonofilaments. Although hemidesmosomes are morphologically similar to one half of spot desmosomes, they differ in molecular composition (i.e., DP-IV is found in spot desmosomes but not in hemidesmosomes).

Focal adhesions, also called cell–substrate adherens junctions, attach cells and their actin filaments to the extracellular matrix. They were first identified in electron microscope studies of cultured fibroblasts. Focal contacts are confined, typically 2–10 μm long, and 0.25–0.5 μm wide. In cultured cells, they are seen as regions of the ventral cell surface that are much closer to the substrate than others. These regions exhibit an increased electron density, and actin microfilaments are associated with local, dense plaques on the cytoplasmic face of the membrane.

A variety of techniques have been developed for isolating focal adhesions based on the fact that these regions are the most adherent parts of cultured cells. In most of these methods, a stream of buffer is used to detach the bulk of the cell leaving behind the focal adhesions attached to the substrate. Some protein components were studied in more detail in focal adhesions isolated from chicken gizzard. Transmembrane proteins (i.e., antigen 30B6 (M_r, 130 kDa) and integrin) as well as cytoplasmic components, (i.e., α-actinin, fibrin, vinculin, and talin) were identified. Moreover, in cells infected with Herpes simplex virus, the viral glycoprotein D accumulates in focal adhesions. Talin is localized only in cell–substrate junctions, and the 130-kDa protein is present predominantly in cell–substrate junctions in close association with α-actinin localized at the ends of stress fibers.

2. Regulation of Cell-Anchoring Structures

An agent that influences cell adhesion most dramatically is Ca^{2+}. In various cell types, the formation of spot desmosomes and zonula adherens does not occur in low extracellular calcium (1–5 μM), and formation can be induced when the extracellular calcium is raised to 1–2 mM. During the process of formation, the first local change that has been visualized occurs in the cytoplasmic surface of the junctional cytoplasmic membrane and corresponds to the assembly of the plaques. Calcium ions are also necessary for stabilization of spot desmosomes and zonula adherens because a decrease in its concentration effectively separates epithelial cells by splitting the junctional complex symmetrically. Initially, "half desmosomes" structures are seen on the surface of the dissociated cells followed by internalization of half desmosomes. During internalization, the cytoplasmic plaques associated to filaments are retrieved from the membrane as single units; then they contract and are translocated to a perinuclear region where they are gradually disintegrated. Although the mechanism of action of Ca^{2+} in the biogenesis of several components of the junctional complex is unknown, it is believed that Ca^{2+} favors the interaction between molecules of cell adhesion (uvomorulin-like molecules). Moreover, at least in the formation of hemidesmosomes, the effect of Ca^{2+} seems to be mediated by calmodulin, because calmodulin blockers reversibly inhibit the formation of these structures. Although it is difficult to prove whether or not Ca^{2+} ions regulate desmosomes under conditions *in vivo*, variations in the availability of local extracellular Ca^{2+}, competitive binding, or modulation of Ca^{2+} uptake could possibly provide sensitive regulatory systems affecting epithelial interaction.

Studies on the assembly kinetics of some of the protein components of desmosome plaques indicate that DP-I and DP-II exist in two molecular forms, a soluble and an insoluble pool. Upon cell–cell contact, the capacity of the insoluble pool rapidly increases, while that of the soluble pool declines. Then the insoluble pool is stabilized, while the soluble pool continues to be degraded rapidly. The stabilization of DP-I and DP-II in their respective insoluble forms seems to require the participation of other proteins because, in the presence of cycloheximide, a blocker of protein synthesis, both proteins are rapidly degraded.

Formation of focal adhesions can be affected by second-messenger pathways, suggesting that hormonal regulation may occur. The elevation of 3′,5′-cyclic adenosine monophosphate levels could have opposite effects depending on the cell type. After elevation of 3′,5′-cyclic adenosine monophosphate levels, several cell types adhere more tightly to the substratum but in Balb/c3T3 cells this is followed by disappearance of vinculin. The activation of protein kinase C is generally associated with reduced number of focal adhesions. Although this effect could be related to phosphorylation of vinculin or talin (both of which are substrates for protein kinase C) this has not been shown yet.

3. Function of Adhesive Junctions

Desmosomes and adherens junctions provide cell adhesion. Apparently, this is accomplished by the interaction of fibers that project from the cell surface to the extracellular matrix or to similar fibers that project from the cell surface of an adjacent cell. Moreover, these structures serve as anchoring sites for intermediate and actin filaments, which provide a structural framework for the cytoplasm and provide tensile strength. The actin filaments that interact with adherent junctions allow the transmission of tension through the cell. The noncontractile filaments seem to form a framework that possibly participates in the passive distribution of forces within a tissue. Presumably due to these properties, the adhesive structures are more abundant in tissues that are subject to mechanical stress such as skin, cervix, and urinary bladder; moreover, they are expressed during developmental stages when stable and strong intercellular adhesion is required.

The importance of desmosomes in cell adhesiveness is illustrated by the potentially fatal skin disease pemphigus, in which antidesmosomal glycoproteins are generated. These antibodies bind to and disrupt desmosomes between epithelial cells, causing severe blistering as a result of the leakage of body fluid to the skin. The effect is somewhat specific for skin epithelium, suggesting that desmosomes of other tissues may be antigenically different.

II. Occluding (Tight) Junctions

Epithelia form the boundary between biologically distinct compartments and maintain a selective permeability barrier to small molecules across this boundary. To establish a semipermeable barrier, structures that occlude the intercellular space are required. Those structures are specializations of the cytoplasmic membrane called tight junctions, or zonula occludens.

A. Structure of Tight Junctions

In epithelial cells, tight junctions are localized just below the apical region. In thin sections, they appear as short, fused regions of the external leaflets of the cytoplasmic membranes of adjacent cells (Fig. 3A). The structural components of tight junctions are seen in replicas of freeze-fractured membranes as P-face membrane particles that tend to be fused into short ridges forming a network of branching and anastomosing strands; complementary groves are seen in the E-face of the membrane (Fig. 3B). Whether these particles and ridges are proteins or lipid micelles remains unresolved.

The molecular components of tight junctions are not yet well known. Tight junctions are resistant to detergents that dissolve the lipid environment of the membrane (Triton X-100 and sodium deoxycholate), suggesting that the molecular components responsible for the sealing effect are not just lipids but also contain protein components. Two tight junction-associated proteins have been isolated: ZO-1 has been purified from isolated tight junctions, and cingulin was found during preparation of antibodies against brush-border myosin. Biochemical studies indicate that cingulin is not a transmembrane protein because it is extracted with solutions of low ionic strength without detergents and is localized nearby but not within the junctional membrane. The ZO-1 protein is localized in the junctional membrane, and an antibody against it recognizes a cytoplasmic region. Moreover, ZO-1 is partially extracted with solutions of high ionic strength at alkaline pH, suggesting that this is a peripheral and not a transmembrane protein.

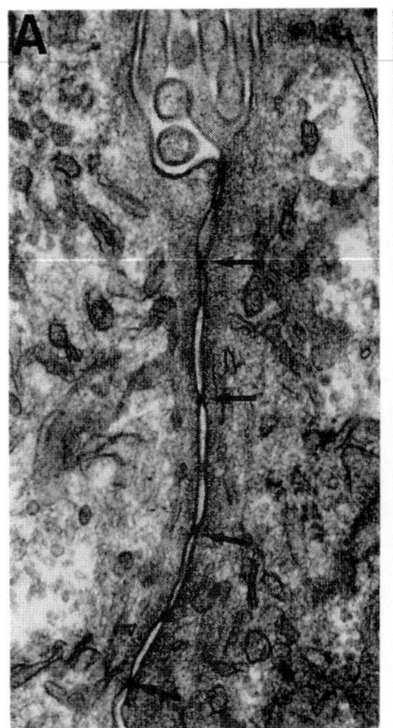

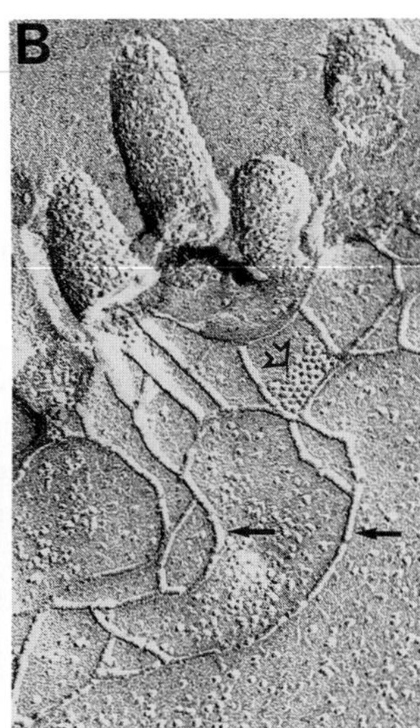

FIGURE 3 Structure of tight junctions (zonula occludens) between rat hepatocytes. A. Electron micrograph of a thin section showing tight junctions that are seen as a series of focal connections between the outer leaflets of the two interacting plasma membranes (arrows). ×43,000. [Contributed by Dr. Phyllis Novikoff.] B. Freeze-fracture replica of the plasma membranes of two adjacent hepatocytes; tight junctions surround the bile canaliculi and are seen as ridges of intramembrane particles on the cytoplasmic fracture face of the membrane (the P-face; arrows). The arrowhead is pointing to a small gap junction plaque. [Contributed by Ms. Chris Roy.]

B. Regulation of Tight Junctions

Formation of tight junctions depends on protein synthesis. Monoclonal antibodies that recognize uvomorulin-like proteins (E-cadherin or L-CAM) block formation of the junctional complex, including tight junctions present in epithelial cells. Because these proteins are localized in the lateral cell membranes, they have been suggested to participate in the biogenesis of tight junctions.

Tight junctions can be disrupted by the proteolytic action of trypin or when the extracellular Ca^{2+} concentration is decreased to 0.1 μM. Intracellular second messengers also affect tight junctions. The transepithelial resistance of amphibian gallbladder epithelium increases substantially after exposure to 3′,5′-cyclic adenosine monophosphate analogues, and the resistance changes are paralleled by tight junction elaboration. Exposure to Ca^{2+} ionophore produces a substantial and sustained rise in transepithelial resistance that is paralleled with an increase in number of tight junction strands detected in freeze-fracture replicas. Tight junctions are made more permeable by exposure to tumor-promoting phorbol esters, which activate protein kinase C. Several lines of evidence indicate that the effect of second messengers might be mediated by cytoskeletal changes. Cytochalasins D or B, compounds that depolymerize microtubules, increase transepithelial ion flux, which correlates with alterations of tight junction structure. Changes in the extracellular medium including osmolarity, pH, ionic strength, and temperature can induce major changes in the electrical resistance, selectivity to electrical charges, and structure of tight junctions. It has been suggested that brief changes in extracellular osmolarity trigger intercellular signaling that leads to cytoskeletal alterations as well as structural and functional changes in tight junctions. Activation of glucose or amino acid transcellular transport may likewise perturb the structure of tight junctions to increase paracellular permeability, possibly triggered by condensation of perijunctional actinomyosin ring. Deposition of actively transported glucose or amino acids below the tight junctions results in osmotically driven water flow across the tight junctions, which can pull glucose and amino acids from lumen to apical compartments.

Function of Tight Junctions

Although tight junctions may play a structural role by helping to maintain cell shape, their main function is of intercellular adhesiveness and sealing the paracellular space forming a selective permeability barrier across epithelial cell sheets, thereby separating the fluids of each side that has different chemical composition. Physiological studies indicate that different epithelia can be classified as impermeable (e.g., epithelium of the urinary bladder) or semipermeable (e.g., epithelium of kidney proximal tubule). The degree of transepithelial occlusion, measured as the transepithelial electrical resistance, is directly related to the abundance of tight junctions. The transcellular pathway depends on membrane-bound carriers that actively pump selected ions and molecules into the epithelial cells from the apical compartment, and on carriers on the basolateral compartment that allow the same molecules to leave the cells by facilitated diffusion into the extracellular fluid. Although tight junctions present in polarized epithelia are not responsible for membrane polarization, they segregate the cytoplasmic membrane into apical and basolateral domains. Lectins, plant proteins that bind to the carbohydrate region of membrane glycoproteins, and fluorescently labeled lipids can diffuse from one membrane domain to the other when tight junctions are disrupted, e.g., by removing the extracellular Ca^{2+} required for tight junction integrity.

III. Intercellular Communication

A. Chemical Signaling

Coordinated cell function depends on long and short range signaling via substances secreted by individual cells. Such signaling can be humoral, in which case the substances are secreted at some distance from their binding sites, or transmitter, in which case the substances are released close to their targets. Release of humoral factors and neurotransmitters is considered in detail elsewhere in this volume. Chemical signaling is accomplished by specializations of the membrane termed receptors, which bind the chemicals secreted from the other cell. Binding can trigger internalization of the ligand to act intracellularly, activate second messenger systems, or open ion channels. Direct intercellular exchange of nutrients, ions, and signaling molecules is another mode of chemical signaling and is accomplished through cytoplasmic bridges, plasmodesmata, and gap junctions. [*See* CELLULAR SIGNALING.]

B. Cytoplasmic Bridges

Incomplete cell division, in which daughter cells remain attached to one another by bridges of cytoplasm, is found in the earliest cleavage stages of many vertebrate and invertebrate embryos and can persist until late developmental stages in some species, such as squid. Cytoplasmic bridges are also found between Leydig cells in the mammalian testis and in certain cell lines (such as the Kc line of *Drosophila*). In many cases, cytoplasmic bridges (also termed midbodies) are apparently filled with cytoskeletal elements, including microtubules, which may function to occlude intercellular transport of organelles and macromolecules. Thus, although the bridges may be $>1\ \mu m$ in diameter, tortuosity of the pathway may limit the apparent size of permeant molecules to those crossing plasmodesmata or gap junctions. The apparent function of cytoplasmic bridges is to coordinate subsequent cell division. Thus, the intermitotic intervals of daughter Kc cells, which are connected via cytoplasmic bridges, is much shorter than for cousins, where the bridges are lost.

C. Plasmodesmata

The private pathway present between plant cells equivalent to the gap junction in animal cells is the plasmodesmatum. Although many morphologic variations have been described between plasmodesmata of different plants or different anatomical parts of a same plant, they all share functional features with gap junctions. Plasmodesmata communicate the cytoplasm of adjacent cells, allowing the diffusion of ions and small molecules ($M_r < 1.0$ kDa) and excluding organelles and macromolecules. They are present in inferior and superior plants, and they are more abundant in cells that participate actively in translocation of water, ions, and molecules. Cells located far from the source of nutrients can be fed by the movement of molecules (i.e., carbohydrates and amino acids) through a permeable pathway formed by plasmodesmata. This pathway also serves for diffusions of second messengers and electrical signaling which would allow integration of syncytial activity stimulated by hormones or other environmental factors.

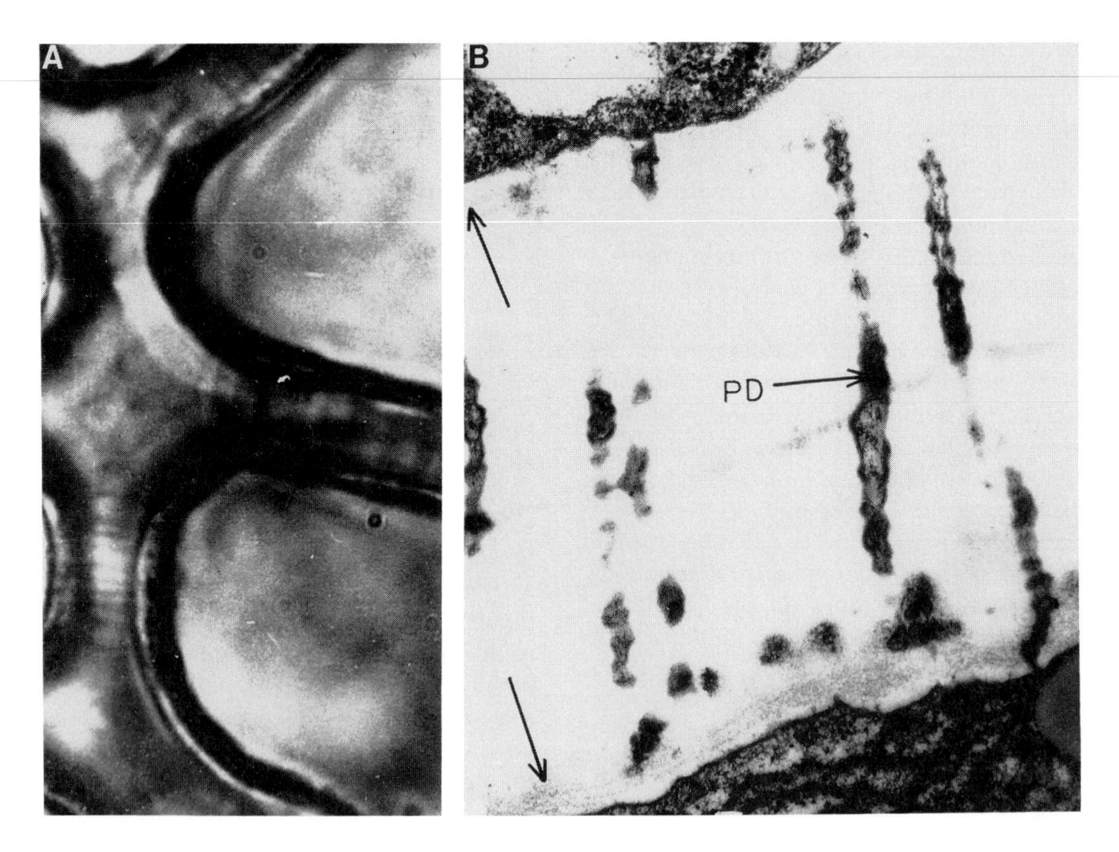

FIGURE 4 Structure of plasmodesmata. A. Light micrograph of freshly cut aleurone tissue stained by the IKI-crystal violet procedure. Threadlike strands interconnect the resistant envelopes and the cell wall matrix. ×1,500. B. Electron micrograph of sections of aleurone cell showing plasmodesmata (PD). The region of the cell wall closest to the plasmodesmata is distinguished on the basis of its staining properties (arrow). ×35,000. [Contributed by Dr. Russell L. Jones.]

Plasmodesmata are the only intercellular junctions visible under the light microscope (Fig. 4). This feature is due in part to the fact that most plant cells are surrounded by an intercellular matrix, called the cell wall, which can be as thick as 1 μm. This type of communicating junction was the first discovered, but the slow development of efficient methods to isolate plasmodesmata has delayed more insight into the molecular architecture. Intriguing recent reports suggest that plasmodesmata components may share antigenic determinants with gap junctions, but interpretation of these findings awaits more efficient plasmodesmata isolation procedures.

D. Gap Junctions

These intercellular communicating junctions have been referred to as the nexus, macula communicans, or gap junctions. The latter term, although contrasting with the function of this type of junction, emphasizes a morphological feature—the gap between the junctional membranes seen in thin section.

The first experimental evidence suggesting the existence of direct cytoplasmic communication between animal cells came from studies on strips of cardiac muscle, where the space constant of propagation of electrical current was larger than predicted for the length of a single cell. Morphological evidence of membrane specializations corresponding to intercellular communicating junctions were obtained in septate axons of crayfish, and later these structures were correlated to functional studies, identifying the electrical synapse or nexus. Later, gap junctions were found between neurons of fish central nervous system, in smooth and cardiac muscle, and in neurons of mammalian central nervous system. Although gap junctions were first discovered in electrically excitable cells, they were later found between nonexcitable cells such as epithelial, glandular, and glial cells, and it is now believed that only a few cell types do not form gap junctions, e.g., red blood cells, sperm, and skeletal muscle cells.

The finding of gap junctions and determination of their permeability led to modification of the concept of cell biology that considered the cell as the functional unit of a tissue with respect to ions and small molecules. The physiologically relevant functional unit is the syncytium formed by cells coupled by gap junctions.

1. Structure of Gap Junctions

In 1963, Robertson working on the ultrastructure of electrical synapses, identified the existence of membrane subunits, which are know recognized as the structural unit of gap junctions. In thin sections, gap junctions are seen as regions of the cytoplasmic membrane in close apposition separated by an extracellular gap of 2–3 nm (Fig. 5A). In stained thin sections, gap junctions are recognized as septilaminar linear membrane appositions. The seven lamina are the clear extracellular gap between each cell's membrane and consist of two stained (electron opaque) lipid head groups on each side of the electron-lucent membrane interior. In freeze-fracture replicas, gap junctions are recognizable as arrays or plaques of approximately 8–9 nm intramembrane particles localized in the E-face in arthropod cells and in the P-face in vertebrates (Fig. 5B). The plaques are of variable shape and size in different cell types.

Because gap junctions were the only cell junction found between some types of coupled cells, they have been considered to be responsible for the direct communication between the cytoplasm of adjacent cells. En face views of isolated gap junctions negatively stained with lanthanum show that gap junctions are aggregates of hexagonally arranged structures with 9–10 nm center-to-center spacing. An electron opaque region (2 nm in diameter) is discerned at the center of each unit, which is believed to be hydrophilic stain that penetrated the lumen of the gap junction channel; the clear annulus surrounding the central opacity is believed to be the channel wall.

X-ray diffraction studies of isolated liver gap junctions indicate that each hemichannel, contributed by one cell, is formed by six subunit polypeptides (termed connexins) arranged in a hexagonal configuration perpendicular to the membrane plane. Each hemichannel or connexon is in register with a

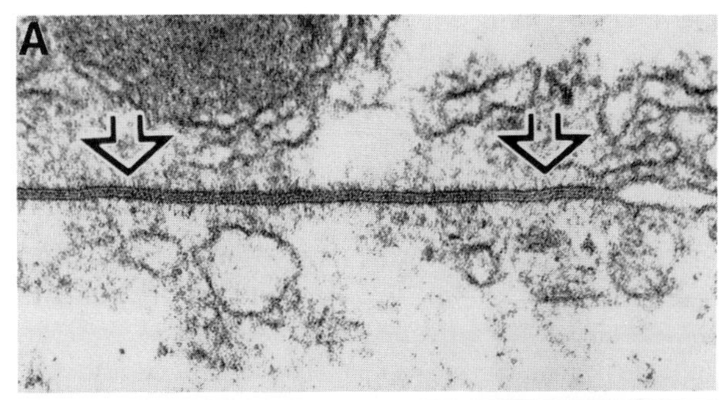

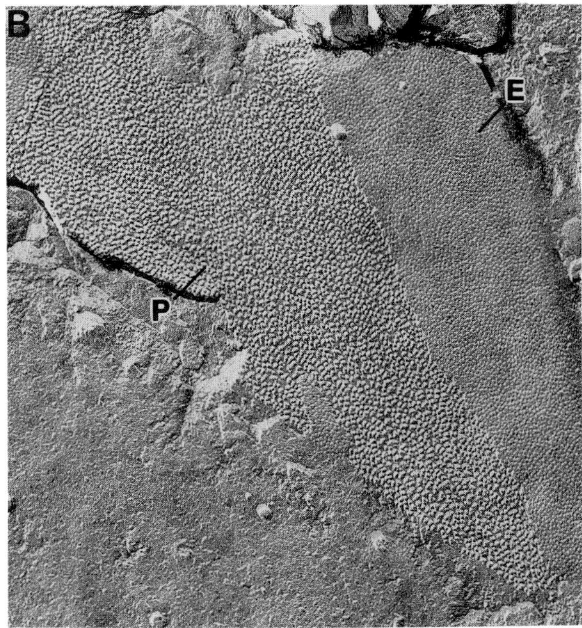

FIGURE 5 Structural features of gap junctions. Thin-section (lateral view) (A) and freeze-fracture (face view) (B) electron micrographs of gap junctions between rat hepatocytes. Arrowheads in A indicate the parallel nature of the gap junction with its septilaminar appearance. P-face particles and E-face pits are indicated in B. ×50,000. [Courtesy of C. Roy.]

connexon located in the cytoplasmic membrane of the adjacent cell, forming a gap junction channel. Ultrastructural analyses have revealed similarity between gap junctions in different tissues and species, although an electron-dense cytoplasmic fuzz is observed on both sides of gap junctions between cardiac myocytes and on one or both sides of gap junctions between some neurons.

The gap junction between rat liver cells was a prototype to study the biochemical components of gap junctions. In this tissue, gap junctions occupy a large fraction of the total membrane surface. Their abundance and resistance to alkali and detergent have permitted isolation of membrane fractions highly enriched in gap junctions. Biochemical analysis of these fractions show a main protein component with an apparent molecular weight of 27 kDa (connexin 32) and a dimer of 47 kDa, and antibodies that recognize the main gap junction protein have been developed. A common feature of liver gap junction preparation is the presence of a protein with an apparent molecular weight of 21 kDa (connexin 26), which is in low amount in rat liver gap junctions, but it can represent as much as 50% of the total protein in mouse liver gap junctions. Immunocytological evidence indicates that both connexins 32 and 26 coexist within the same gap junction plaque. Biochemical analyses of these connexins show that both connexins are acetylated by myristate and palmitate, they are not glycosylated, and that connexin 32 is a phosphoprotein but connexin 26 is not.

It is now accepted that connexins are the structural units of connexons, and it has become increasingly clear that connexins are a closely related group of membrane proteins, encoded by a gene family. They have >50% amino acid sequence homology and similar motif. Full-length complementary DNAs have been obtained for several connexins, and analysis of the predicted amino acid sequences for these proteins suggests the presence of four transmembrane segments, three of which consist predominantly of hydrophobic amino acids. The third transmembrane segment fulfills expectations of an amphipathic helix with a hydrophilic and hydrophobic face, the hydrophilic face of which would line the intercellular pore of the gap junction channel. Several aspects of this model have been proven experimentally indicating that the amino and -COOH termini are located at the cytoplasmic surface of the gap junction and that a hydrophilic domain connects the second and third transmembrane segments. Differences among these proteins are more pronounced in the cytoplasmic segments of the proteins, the -COOH terminus, and a cytoplasmic loop connecting the second and third transmembrane segments. These differences are likely responsible for different mechanisms by which channel activity may be regulated (see below).

2. Regulation of Gap Junctions

The degree of electrical or metabolic coupling between cells depends on the expression and regulation of gap junctions. Junctional conductance depends, in part, on the total number of channels, the fraction of channels that are open, and the permeability of the open channels.

The number of available channels depends on the half-lives of the molecular components of each type of channel. The half-life of connexin 32 studied *in vivo* and *in vitro* is around 3 hr. Agents that elevate the intracellular levels of 3′,5′-cyclic adenosine monophosphate, and activate protein kinase A, stimulate the synthesis of connexin 32, which is partially due to an increase in stability of the messenger RNA encoding connexin 32. Agents that stimulate protein kinase C decrease the number of gap junctions in several cell types, but the locus of action is unknown. The metabolic control of some gap junction channels also occurs at posttranslational level because some connexins (32 and 43, but not connexin 26) are phosphoproteins, and their phosphorylation state can be modified by the activation of protein kinases. Moreover, the hormonal effect on gap junction expression is dramatically affected by components of the extracellular matrix in cultures of rat hepatocytes. Different connexins are expressed by parenchymal cells of different tissues, probably to provide an appropriate channel structure for the function and regulation required. The activation of the same transduction system can lead to opposite effects in different cell types. These differences may reside in the type of connexins expressed or in the stoichiometry of combination in heteromeric channels.

The functional state of gap junctions is affected by changes in intracellular concentration of H^+ or Ca^{2+} ions. Both ions lead to closure of the channels when their concentrations are raised above physiological levels, and their effect is usually reversed when concentrations are reduced to normal levels. Gap junction channels are also reversibly blocked by anesthetics such as octanol, heptanol, and halothane and in several cell types by halomethanes and arachidonic acid. Moreover, gap junctions in several tissues are voltage-dependent, junctional conductance in some systems being reduced by voltage imposed across the junctional membrane and, in other systems, by voltage from inside to outside the cells. Transjunctional voltage dependence is most strongly displayed in the crayfish rectifying synapse and in early stages of invertebrate and vertebrate embryos. In crayfish, this effect is unidirectional, allowing orthodromic, while prohibiting antidromic conduction. Inside–outside voltage-dependence exists in cells that include insect salivary gland cells, in salamander rod photoreceptors, and after pharmacological treatments in squid blastomeres. In these systems, depolarization of either cell, or both

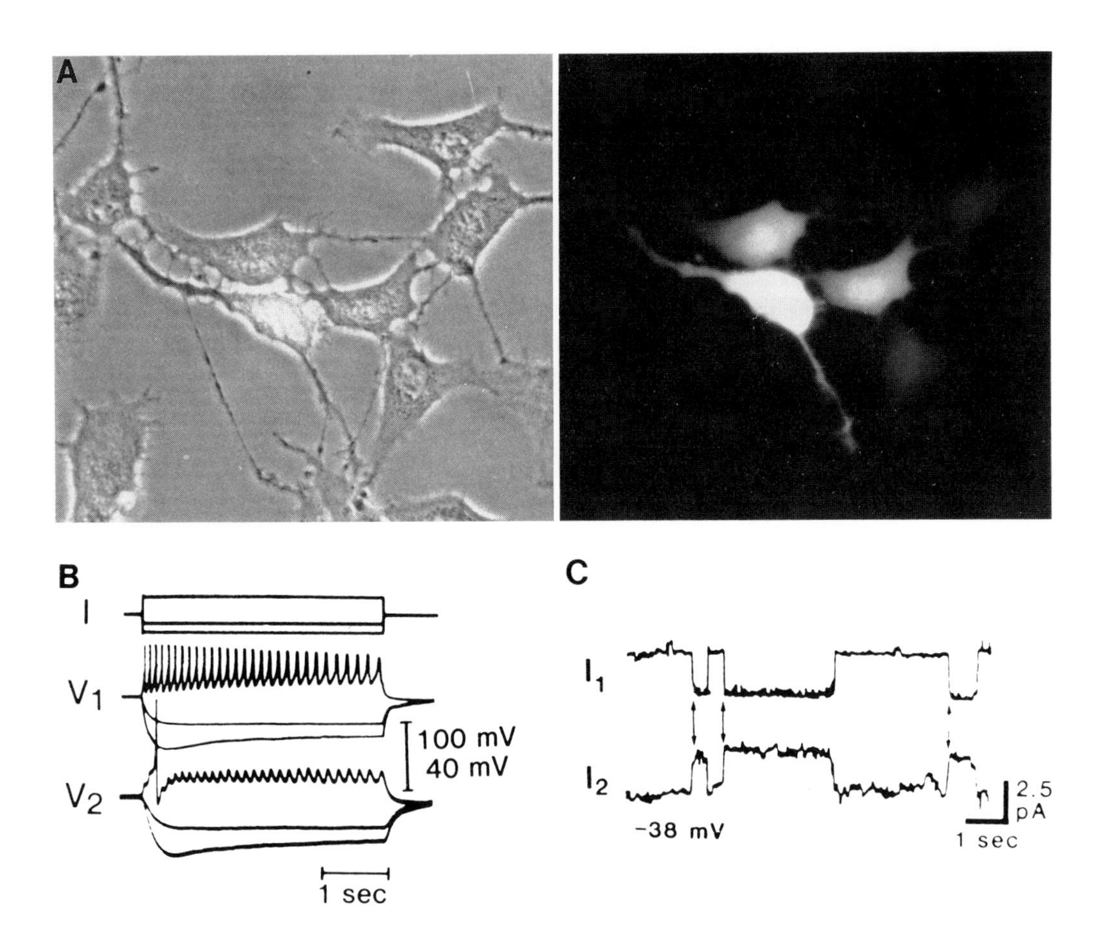

FIGURE 6 Intercellular communication via gap junction channels. A. Dye-coupling among cultured Ito cells from liver. Lucifer Yellow was injected into the brightest cell in the fluorescence micrograph and 30 sec later had diffused to adjacent cells and their neighbors. B. Electrical coupling between neurons in the mollusc, *Aplysia*. Current (I) injected into one cell causes that cell (middle traces) to fire an action potential that leads to firing of the lower trace cell (V_2). Hyperpolarizing pulses spread. C. Single gap junction channels seen between neonatal heart cells. One cell was held at −38 mV and the other was depolarized to 0 mV. Openings of single channels are indicated by arrows.

cells together, reduces junctional conductance, and hyperpolarization increases it. Voltage-dependence exists to a variable extent in other tissues, although strong sensitivity is absent in most adult mammalian tissues.

Function of Gap Junctions

Most gap junction functions can be inferred from their permeability and regulatory features. Gap junctions are permeable to ions and small molecules with molecular mass around 1,000 daltons (Fig. 6A). Gap junctions are believed to be critical in the embryonic development of an organism, allowing the diffusion of morphogens that determine cell differentiation, and to be fundamental to the normal function of different organs in adulthood.

Gap junctions are the structural substrate for propagation of electrical signals (Fig. 6B), which in general are mediated by K^+ ions. This property is particularly important in normal function of excitable tissues such as heart or of secretory organs such as the pancreatic islet. Gap junctions allow the synchronic contraction of cardiac myocytes, and perturbation of their functional state may be one cause of cardiac arrhythmias. In smooth muscle of myometrium, regulation of expression of gap junctions leads to synchronous electrical activity necessary for efficient and coordinated contraction of the uterus during delivery. Although many types of neurons are coupled to each other, the specific functional meaning may be different in different neuronal types and is not yet well understood. In metabolically active tissues such as liver, gap junctions allow the propagation of second messenger signals to cells that contain the target metabolic pathways but are deficient in hormone receptors that trigger the synthesis of the second messenger. Gap junctions can serve a buffering function by uniting the cytoplasms of coupled cells. For example, glial cells form a syncytium, which may provide a sink for extracellular ions and neurotransmitters accumulated during intense neuronal activity. In liver, gap junctions may also play a role in detoxification by allowing dilution of intake from cells exposed to higher concentrations of exogenous or locally generated toxins.

All functions mentioned above imply that increased coupling is beneficial, but decreased coupling could also play an important role during conditions such as cell death and tumorigenesis. Cell uncoupling during cell injury would prevent loss of necessary metabolites from surviving cells to dying ones and prevent the spread of toxic substances in the opposite direction. It is believed that gap junctions play a significant role in the regulation of proliferation and differentiation, and this role is supported by inhibition of gap junctional communication by tumor promoters in some systems and the correlation of oncogene expression with cell uncoupling.

In summary, gap junctions are gated ionic channels (see Fig. 6C) that open and close in response to physiological and pathological stimuli, thereby allowing control of the extent of intercellular communication. Furthermore, expression of gap junctions can be regulated, allowing possible changes in development and during such pathological conditions as tumorigenesis.

Bibliography

Bennett, M. V. L., and Spray, D. C. (eds.) (1985). "Gap Junctions." Cold Spring Harbor, New York.

Boek, G., and Clark, S. (1987). "Junctional Complex in Epithelial Cells." Ciba Foundation Symposium 125. John Wiley and Sons Ltd., Buffins Lane, Chichester, Sussex.

Burridge, K., and Connell, L. (1983). A new protein of adhesion plaques and ruffling membranes. *J. Cell Biol.* **97,** 359–367.

Burridge, K., Fath, K., Kelly, T., Nuckolls, G., and Turner, C. (1988). Focal adhesions: transmembrane junctions between the extracellular matrix and cytoskeleton. *Ann. Rev. Cell Biol.* **4,** 487–525.

Cereijido, M., Ponce, A., and Gonzalez-Mariscal, L. (1989). Tight junctions and apical/basolateral polarity. *J. Membr. Biol.* **110,** 1–9.

Cowin, P., Kapprell, H.-P., and Franke, W. W. (1985). The complement of desmosomal plaque proteins in different cell types. *J. Cell Biol.* **101,** 1443–1454.

Edelman, G. M. (1986). Cell adhesion molecules in the regulation of animal form and tissue pattern. *Ann. Rev. Cell Biol.* **2,** 81–116.

Farquhar, M. G., and Palade, G. E. (1963). Junctional complex in various epithelia. *J. Cell Biol.* **17,** 375–412.

Gilula, N. B. (1974). Junctions between cells. *In* "Cell Communication" (R. P. Cox, ed.), pp. 1–29. Wiley Publishing Co., New York.

Hertzberg, E. L., and Johnson, R. G. (1988). "Gap Junctions." Alan R. Liss, Inc., New York.

Lane, N. J., and Skaer, H. leB. (1980). Intercellular junctions in insect tissue. *Adv. Insect. Physiol.* **15,** 35–213.

McNutt, N. S., and Weinstein, R. S. (1973). Membrane ultrastructure at mammalian junctions. *Prog. Biophys. Mol. Biol.* **26,** 45–101.

Robards, A. W. (1975). Plasmodesmata. *Ann. Rev. Plant Physiol.* **26,** 13–29.

Robards, A. W., Lucas, W., Pitts, J. D., Jongsma, H., and Spray, D. C. (1990). "Parallels in Cell to Cell Junctions in Plants and Animals." Springer-Verlag, Berlin.

Sperelakis, N., and, Cole, W. C. (1989). Cell Interactions and Gap Junctions. *CRC Press, Inc.*, Florida.

Spray, D. C., and Bennett, M. V. L. (1985). Physiology and pharmacology of gap junctions. *Ann. Rev. Physiol.* **47,** 281–303.

Staehelin, L. A. (1974). Structure and function of intercellular junctions. *Int. Rev. Cytol.* **39,** 191–283.

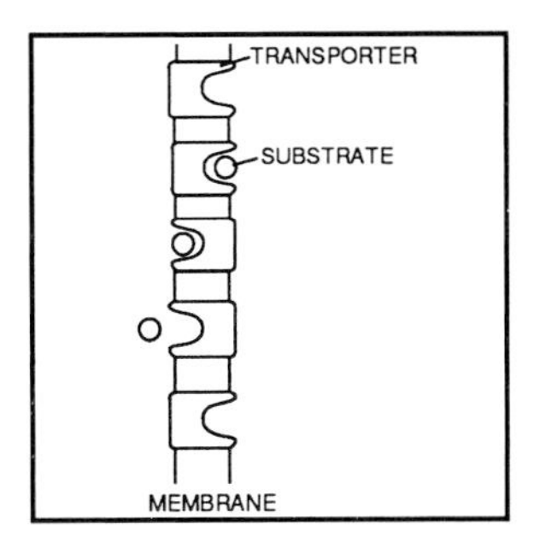

Cell Membrane Transport

ADRIAN R. WALMSLEY, *The University of Leicester*

Glossary

Active transport Utilization, by a transporter, of the energy of hydrolysis of adenosine triphosphate to drive the transmembrane movement of an ion against its concentration gradient

Amphipathic helix Helix composed of both hydrophobic and hydrophilic amino acid residues

Cotransporters Transporters involved in the process of secondary active transport

Facilitated diffusion Transporter-mediated transmembrane movement of a molecule or ion down its concentration gradient

Ion channel A type of transporter that acts as an ion specific membrane pore, opening in response to either the binding of an effector ligand or a change in membrane potential. The movements of these ions is driven by the electrochemical gradients created across the membrane by active transport

Ion pumps Transporters that mediate the active transport of an ion

Passive transporters Transporters involved in the process of facilitated diffusion

Secondary active transport Utilization, by a transporter, of the concentration gradient of an ion, to drive the transmembrane movement of a molecule or second ion against its concentration gradient; this is achieved by coupling the movement of the transported species to the movement of the ion, down its concentration gradient

Transporter Membrane-spanning protein that facilitates the movement of specific molecules and/or ions across the membrane

THE SOLUTE COMPOSITION of the cell is controlled by the plasma membrane, which acts as a semipermeable hydrophobic barrier to the movement of water-soluble molecules and ions. The translocation of selected species across the membrane is facilitated by a network of membrane-spanning proteins. Depending on their function these "transporters" can be divided into four major classes.

(1) Transporters, which facilitate the passive movement of molecules and ions down their concentration gradient, such as the erythrocyte glucose and anion transporters.

(2) Ion pumps, which couple the hydrolysis of adenosine triphosphate (ATP) to the movement of ions against their concentration gradient. The ubiquitous Na^+–K^+ pump and the Ca^{2+} pump of the sarcoplasmic reticulum are typical examples.

(3) Cotransporters, which couple the movement of an ion down its concentration gradient to that of another molecule against its concentration gradient. Typically, the Na^+ gradient produced by the Na^+–K^+ pump is utilized in pumping, for example, glucose into the epithelial cells of the intestine via the Na^+-glucose symporter.

(4) Ion channels, which open in response to either (a) the binding of a ligand or (b) a change in membrane potential, to allow the free movement of selected ions down their electrochemical gradient. For example, in nerve and muscle cells the binding of acetylcholine to its receptor opens a channel, allowing Na^+ and K^+ ions to permeate the mem-

brane, causing it to depolarize. Depolarization of nerve and muscle cell membranes in turn activates voltage-dependent Na^+ and Ca^{2+} channels, mediating a nervous impulse and muscular contraction, respectively.

I. The Structure and Function of Transporters

Although these transport systems may have evolved to perform a diversity of roles in eukaryotic cells, it is now clear that they share certain common structural and mechanistic features.

Most, if not all, mammalian transporters that have been subjected to biophysical methods, such as circular dichroism and infrared spectroscopy, have been shown to have a membrane-spanning domain, which is largely α-helical in structure. These helices are thought to be packed into a cylindrical arrangement to form an aqueous channel through the membrane. If this channel is not to allow the leakage of nonspecific ions and small solutes through the membrane then it must be blocked, possibly by a constriction of the helices. To allow the passage of specific solutes and ions through the channel, this blockage must be transiently removed from the pathway during translocation. In this manner, the translocation of a solute or ion through the transporter is said to be "gated." This gating mechanism may merely involve the movement of a few specific aminoacids, which restrict translocation, away from the transported species allowing its passage.

With the advent of cDNA cloning techniques, it has been possible to sequence the genes for many transport proteins and to deduce the number of putative helices. Basically, the procedure involves scanning the amino acid sequence for stretches of 21 amino acids, which are sufficiently hydrophobic to interact with the membrane lipids as membrane-spanning α-helices. A stretch of 21 hydrophobic amino acids is generally sought, because this represents the minimum number of amino acids capable of spanning the membrane as an α-helix. In this way, models for the secondary structure of various transporters have been obtained. Although speculative these models can be tested by a combination of vectorial proteolytic digestion, covalent labeling, and sequence-specific antibody-binding studies.

For most transporters, several α-helices have been predicted (12 and 25 for the glucose transporter and acetylcholine receptor, respectively). Because the dimensions of the channel produced by having all the helices arranged into a single ring would be excessively large for most transporters, the helices are envisaged as being arranged into an inner and outer ring. The most hydrophobic helices form the outer ring and act as a membrane anchor, whereas the more amphipathic helices form the inner core of the cylinder and are responsible for forming the substrate-binding site and channel. Most models also predict a relatively large cytoplasmic globular domain. This may be involved in holding the helices together in a tight cylindrical arrangement and/or acting as the site for the binding of nontransported ligands, which modulate transport activity.

This view of transporter structure has to some extent been substantiated by studies of the bacterial photosynthetic reaction center, the only membrane-bound protein, to date, to have been crystalized and its three-dimensional structure determined to atomic resolution. The protein is composed of three subunits (denoted L, M, and H). The L and M subunits, each consisting of five-helices, are arranged into a cylindrical structure of eliptical cross section. The outer ring of the cylinder is composed of six helices, three donated from each subunit. The inner ring is assembled from the remaining helices. The H subunit consists of a single helix and a large globular domain, which is thought to play a role in stabilizing the L-M complex.

In view of the similarities in structure of the different transporters, it is perhaps not surprising to find that they also share common features in their mechanism of action. All are believed to operate by alternating between two principal conformations. For both passive and active transporters, the substrate translocation event involves a conformational change in the transporter such that the binding site is alternately presented at the two faces of the membrane. Reorientation of the transporter is invariably the rate-limiting step in the transport cycle. The situation is somewhat different for ion channels, because the transporter cycles between open and closed conformations. While in the open conformation, ions are allowed to pass through the ion channel, as opposed to being bound and translocated to the opposite membrane face.

II. Passive Transporters

Certain small molecules and ions, such as glucose, nucleosides, amino acids, and anions, are passively taken up by cells in a nonenergy-requiring process.

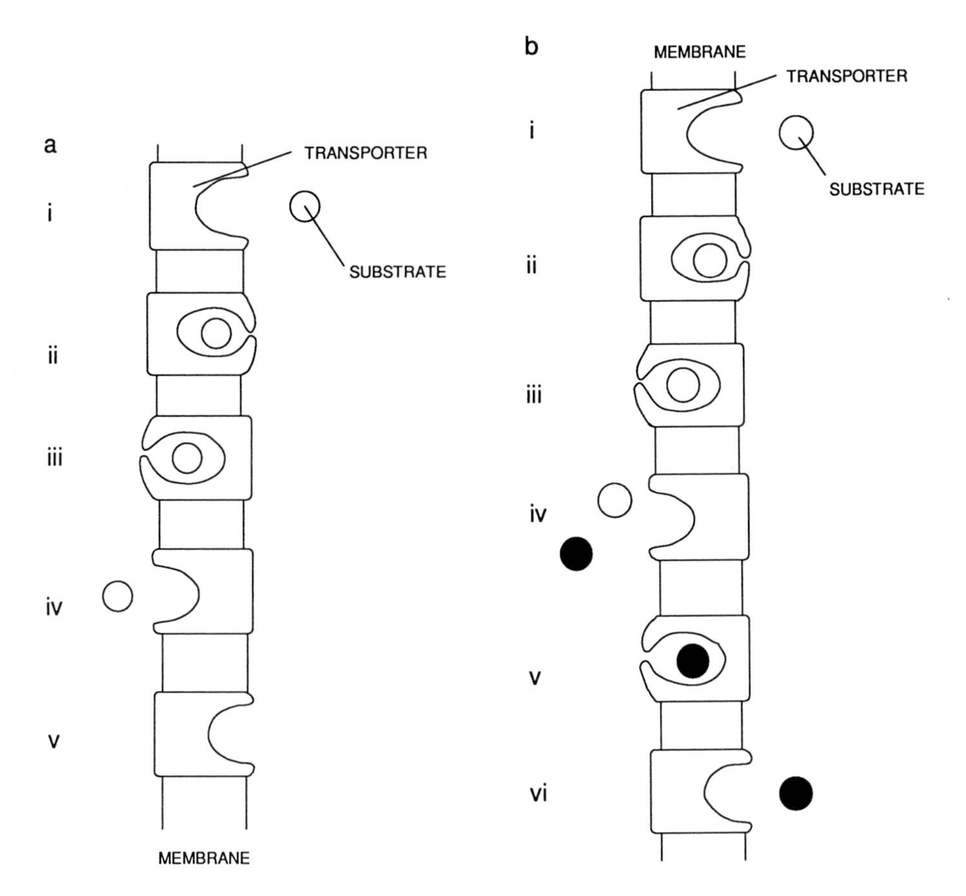

FIGURE 1 A model for net and exchange transport by passive transporters. (a) Net transport: (i) the transporter binding site is exposed at one face of the membrane; (ii) substrate binds and is occluded from the bathing medium; (iii) the transporter reorientates; (iv) the substrate dissociates at the opposite face of the membrane; and (v) the unloaded transporter reverts to its original conformation. There is a net movement of one substrate molecule across the membrane. (b) Exchange transport: substrate traverses the membrane according to the translocation cycle described by (i) to (iv) of (a); (v) the transporter binds a second substrate molecule from the opposite side of the membrane and then (vi) reverts to its original conformation. There is a one-for-one exchange of substrate across the membrane.

Translocation of these hydrophilic species across an essentially hydrophobic membrane barrier is facilitated by specific transporters. These transporters serve to accelerate the rate of translocation of specific species across the membrane, which would otherwise be a negligibly slow process. The driving force for the movement of these species is due simply to a concentration gradient of the relevant species across the membrane. These transporters essentially operate by cycling between conformations in which the binding site is inward- or outward-facing, a process that is accelerated by bound substrate. During net transport, the translocation of a substrate molecule is followed by reorientation of the unloaded transporter. In contrast, during exchange transport, the transporter returns to its original conformation after binding a second substrate molecule so that a one-for-one exchange of substrates occurs (Fig. 1). This type of transport is commonly termed facilitated diffusion. Two classic examples of this type of transport are exhibited by the glucose and anion transporters.

A. The Glucose Transporter

Most, if not all, cells possess a transport system for the facilitated diffusion of glucose across the plasma membrane. The transport system serves simply to equilibrate glucose across the cell membrane, providing the substrate for glycolysis and subsequent ATP production. One of the most extensively characterized glucose transporter is that from human erythrocytes.

The glucose transporter has been identified as a 492-amino acid glycoprotein (55 kDa), which has been purified and reconstituted into artificial membranes and vesicles, with retention of transport activity. The transporter, which consists of a single polypeptide, has been cloned and sequenced and its membrane disposition deduced (Fig. 2). In essence, the transporter is thought to be arranged into three major domains (Fig. 2): (1) 12 α-helices, which span the membrane; (2) a large, highly charged, internal domain; and (3) a smaller external domain bearing the carbohydrate moiety. The membrane domain can be further subdivided: Helices 7, 8, and 11 are believed to form the binding site and aqueous channel, because they contain many serine and glutamine residues capable of hydrogen-bonding to glucose; helices 9 and 10 are thought to form a hydrophobic cleft, which envelopes the glucose molecule during binding.

Mechanistically, the transporter is thought to operate in the following manner. Consider a glucose molecule approaching the transporter from the extracellular medium. The transporter opens up around the C-1 end of the sugar, allowing hydrogen-bond formation between the C-1, C-3, C-4, and C-6 sugar hydroxyls and the binding site. Subsequently, the hydrophobic cleft closes around the C-6 end of the glucose molecule, transiently occluding it from the bulk aqueous environment. This binding process may involve the dissociation of water from both glucose and the binding site, as glucose effectively exchanges hydrogen bonds with water for a similar number of hydrogen bonds with the transporter. Translocation occurs as a result of the transporter opening up in front of the bound glucose, allowing it to diffuse away from the transporter into the cytosol. Dissociation of glucose from the transporter is likely to occur, because water will enter the binding site to rehydrate the glucose molecule. Such a simple mechanism is consistent with the slight conformational change detected by infrared spectroscopy during transporter reorientation and also with the rapid rate of transport. The transporter then returns to its original conformation.

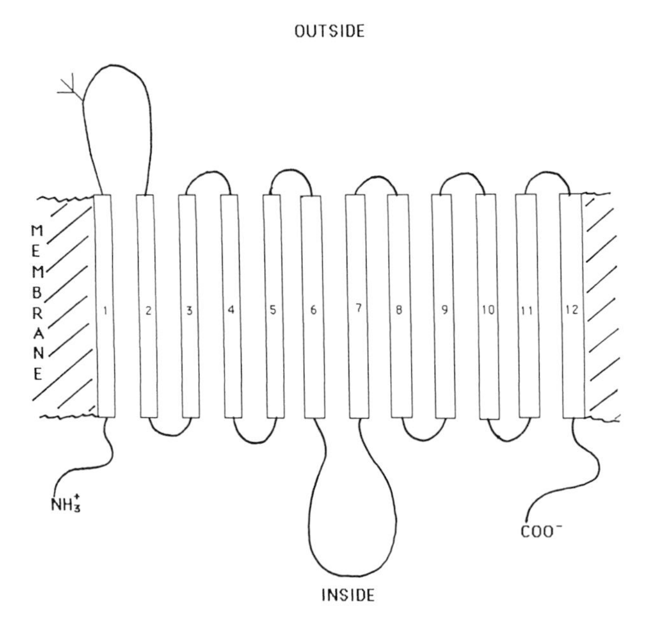

FIGURE 2 A schematic diagram of the predicted topography of the glucose transporter. The 12 membrane-spanning helices are depicted as large numbered rectangles, the internal and external domains as large loops, and the carbohydrate moiety as a "tree."

B. The Anion Transporter

The anion transporter catalyzes the one-for-one electroneutral exchange of anions, principally chloride for bicarbonate, across the plasma membrane. This role is of particular physiological importance in erythrocytes, where it allows the blood to act as an efficient medium for the carriage of carbon dioxide from the tissue to the lungs. Carbon dioxide released from respiring cells has only a poor solubility in aqueous plasma, but upon entering the erythrocyte is rapidly converted by carbonic anhydrase to the highly soluble bicarbonate anion. The anion transporter then allows this bicarbonate ion to exit the cell in exchange for a chloride ion, so that the plasma also becomes available for bicarbonate carriage.

The anion transporter is a 90–100-kDa glycoprotein, consisting of 929 amino acid residues. This transporter can be divided into two distinct structural domains: a highly negatively charged amino-terminal domain of about 400 residues, facing the cytoplasm, and a membrane-spanning domain of about 500 residues (Fig. 3). The cytoplasmic domain is a highly extended structure possessing high-affinity binding sites for seveal glycolytic enzymes, hemoglobin, and the membrane cytoskeletal protein ankyrin. Its function remains unclear but could involve modulation of anion transport activity, which resides entirely with the transmembrane domain of the transporter. This membrane domain is thought to consist of 10 α-helices, eight of which span the membrane, with the remaining two helices forming a loop within the membrane. The ion channel is probably formed by helices 4–7, because they possess several charged and polar residues. In addition, almost all the charged or polar side chains possessed by these particular helices lie along the same

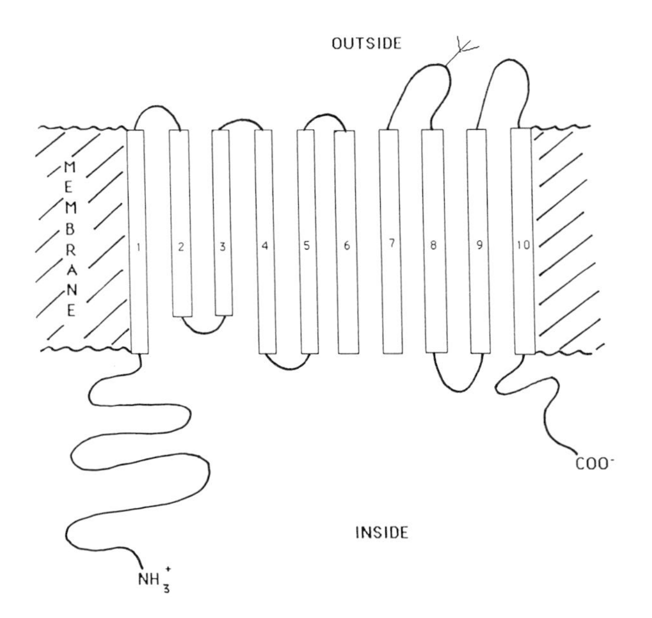

FIGURE 3 A schematic diagram of the predicted topography of the anion transporter. The 10 membrane-bound helices are depicted as large numbered rectangles, the internal and external domains as large loops, and the carbohydrate moiety as a "tree." Helices 2 and 3 are only partially membrane-spanning, forming a loop within the membrane.

face of the helix. Consequently, they could be envisaged as forming a central hydrophilic cavity, lined by the charged and polar residues of these helices. With the helices arranged in this manner, the transporter would appear as a positively charged funnel that narrows down to two or three negatively charged residues when viewed from either side of the membrane. The positively charged residues may be involved in anion selection, while the negatively charged residues could block anion translocation and be responsible for gating the transporter. Furthermore, the gating mechanism may involve the formation of two alternative salt bridges between a carboxylate and two positively charged residues at the anion-binding site. The binding of an anion breaks the salt bridge by competing for the carboxylate group, which forms a second salt bridge with a nearby positively charged residue. This conformational change in the transporter probably has two net results. First, it closes the transporter behind the anion, precluding the further binding of anions. Second, it shifts the negatively charged residues from their position, constricting the movement of the anion. The anion is then free to diffuse away from the transporter at the opposite face of the membrane. For the transporter to return to its original conformation, it must bind a second anion from the opposite side of the membrane, to break the newly formed salt bridge and reverse the conformational change. Transport is said to occur by a "ping-pong" mechanism. Again, as with the glucose transporter, such a simple mechanism is consistent with the observed slight conformational changes associated with reorientation of the transporter and the rapidity of the transport.

III. Ion Pumps

A ubiquitous property of eukaryotic cells is the maintenance of concentration gradients for certain ions, such as Na^+, K^+, Ca^{2+}, and H^+, across the plasma and intracellular membranes. This function is largely performed by ion pumps, which utilize the energy of hydrolysis of ATP to drive the movement of these ions against their respective electrochemical gradients. These ion pumps essentially operate in a manner similar to passive transporters, in that ion translocation involves the pump protein alternating between inward- and outward-facing conformations. However, they differ from passive transporters in that this conformational change is driven by phosphorylation and dephosphorylation of the pump protein. Two extensively characterized examples of this type of transport are the sodium–potassium pump and the calcium pump. [*See* ION PUMPS.]

A. The Sodium–Potassium Pump

The sodium-potassium pump ($Na^+ - K^+$ pump) is responsible for maintaining the high internal K^+ and low internal Na^+ concentration characteristic of most eukaryotic cells. This is generally accomplished by the pump exchanging three Na^+ ions from the cytosol for two K^+ ions from the extracellular medium. Maintenance of this gradient is fundamental to several diverse cellular functions. These include (1) regulation of cellular volume; (2) generation of the transmembrane Na^+ gradient necessary for driving the uphill transport of sugars, amino acids, and ions; (3) maintenance of the ion gradients essential for the membrane potential and excitability of the membrane; and (4) the transport of Na^+ across the epithelia.

The Na^+–K^+ pump consists of two noncovalently linked polypeptides: a 1,016-amino acid residue α-subunit, with a molecular mass of about 110, and a glycosylated β-subunit, consisting of

302-amino acid residues, with a molecular mass of about 55. Most of the functions of the Na^+–K^+ pump have been localized to the α-subunit. It contains the binding sites for Na^+, K^+, and ATP and also for ouabain, a potent inhibitor of the pump. Ion translocation is coupled to phosphorylation of this subunit. The exact function of the β-subunit remains unknown.

The α-subunit is essentially thought to be arranged into two major domains (Fig. 4): a membrane-spanning domain consisting of eight α-helices and a large cytoplasmic domain. The latter domain is composed of three cytoplasmic segments of the pump protein, a large segment connecting helices 4 and 5, and two smaller segments connecting helices 2 and 3, and helices 7 and 8. Specific functions of the subunit have been tentatively related to distinct epitopes. The Na^+-binding site is associated with the amino-terminus and the cytoplasmic end of helix 1, while the K^+-binding site and the ouabain-binding site may be located toward the ectoplasmic end of helices 2 and 3. ATP binds to the cytoplasmic domain, phosphorylating an aspartate residue within this region.

The β-subunit consists of a single membrane-spanning helix, which links a short cytoplasmic amino-terminal domain, with a large extracellular globular domain containing glycosylation sites (Fig. 4). The amino-terminal domain has been shown to

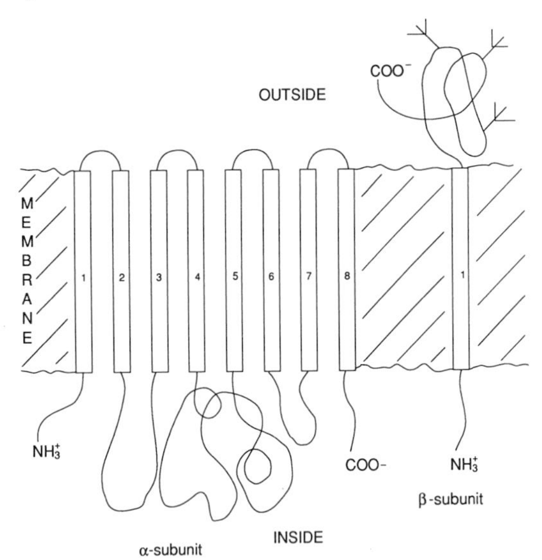

FIGURE 4 A schematic diagram of the predicted topography of the α- and β-subunits of the sodium–potassium pump. The eight membrane-spanning helices of the α-subunit are depicted as large numbered rectangles and the internal and external domains as large loops. The β-subunit is depicted as a single membrane-spanning helix with the carbohydrate moieties as "trees."

possess one or more disulfide bonds. Reduction of these bonds results in a loss of ATPase activity of the pump, showing that the β-subunit is essential for the functioning of the pump, even if its exact role remains a mystery at present.

The pump is thought to operate according to the following sequence of events (Fig. 5). Cytoplasmic Na^+ binds to a high-affinity site on the pump catalyzing phosphorylation of the protein. Two distinct events then follow. First, a minimal conformational change in the protein separates the ions from the cytoplasmic medium. The formation of this occluded state may occur by a similar process to that envisaged for the binding of glucose to its transporter, in that water associated with the cations is replaced by interactions with carbonyl groups on the protein. These groups are probably located within a cavity on the cytoplasmic surface of the transporter and dehydrate the Na^+ ion by their movement around the ions as the cavity closes to the cytosol. A more substantial rearrangement of the protein then spontaneously occurs. This is termed the E1-E2 conformational transition and is directly involved in cation translocation. It is a process thought to involve the movement of the binding groups between cytoplasmic and extracellular ion-binding cavities. The gating mechanism may simply involve an outward displacement of these groups as they move between cavities. Their movement effects an opening of the extracellular cavity while reducing the affinity of these residues for Na^+ ions. The sodium ions are then rehydrated and diffuse away from the pump into the extracellular medium. In this new position, the binding groups may be suitably placed for binding K^+ ions. The binding of potassium ions from the extracellular medium catalyzes the dephosphorylation of the protein. This dephosphorylation induces a conformational change of the pump in which the K^+ ions are separated from the extracellular medium. Subsequently, there is a reversal of the E1-E2 conformational transition, and the K^+ ions are exposed to the cytoplasm. In this manner, there is an exchange of Na^+ ions for K^+ ions across the membrane.

B. The Calcium Pump

The function of the Ca^{2+} pump is to maintain a low cytosolic Ca^{2+} concentration, a role vital to the proper functioning of eukaryotic cells. The regulation of intracellular levels of calcium may be achieved either by pumping Ca^{2+} ions out of the cell

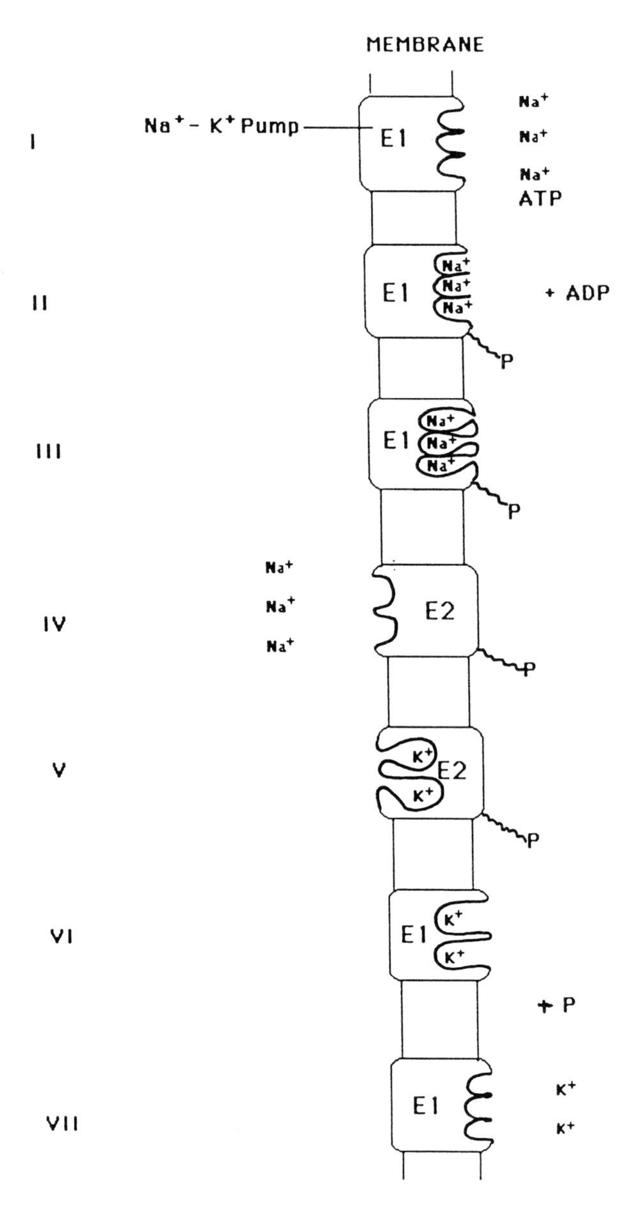

FIGURE 5 A model for the mechanism of action of the sodium–potassium pump. El and E2 are conformational states of the pump. (i) Three Na^+ ions and ATP bind to sites on the cytoplasmic surface of the pump, which are exposed when the pump is in the El conformation. (ii) The pump is phosphorylated and adenosine diphosphate released. (iii) The ions become occluded from the cytosolic medium. (iv) The protein then undergoes a conformational change from El to E2. The result is that the Na^+ ions are pumped out of the cell and two K^+-binding sites are exposed on the external surface of the pump. (v) Two K^+ ions bind and are occluded. (vi) The pump then dephosphorylates, triggering an E2-E1 conformational change. (vii) The K^+ ions are pumped into the cytosol and the cytoplasmic Na^+ binding sites are re-created.

or by internalizing them within the vesicular endoplasmic reticulum. A transient increase in the cytosolic CA^{2+} concentration elicited by some external stimuli, such as the binding of various hormones to their receptors, is involved in triggering a diversity of cellular functions. For example, the binding of acetylcholine to its receptor in muscle cells leads to the rapid release of Ca^{2+} ions from the sarcoplasmic reticulum, via Ca^{2+} channels, ultimately leading to muscular contraction.

The Ca^{2+} pump responsible for maintaining the Ca^{2+} ion gradient across the membrane of the sarcoplasmic reticulum in the muscle has been studied extensively. It has been shown to consist of a single polypeptide of 997 amino acid residues, with a molecular mass of about 110 kDa. This polypeptide is thought to be arranged into several distinct domains: three globular cytoplasmic domains, which form the headpiece of the pump; a membrane bound domain, composed of 10 helices; and a pentahelical stalk, which connects the head and membrane domains (Fig. 6).

Specific functions have been attributed to distinct parts of the Ca^{2+} pump. The first cytoplasmic globular domain is involved in binding ATP, which is used to phosphorylate the second cytoplasmic domain. In fact, the predicted secondary structure of these cytoplasmic domains resembles the known structure of several kinases, enzymes which are involved in protein phosphorylation. The stalk, which contains a preponderance of charged amino acid

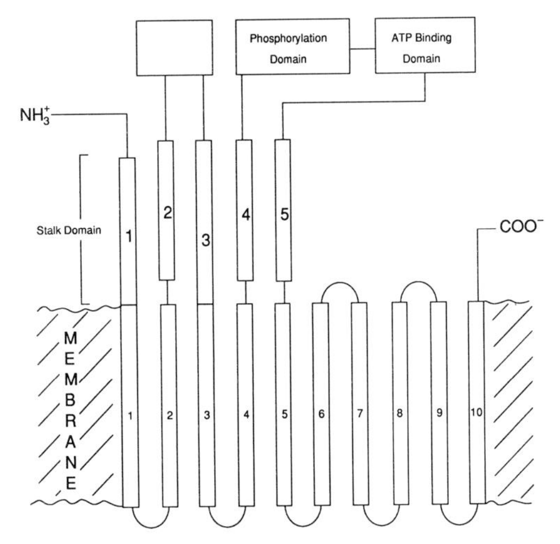

FIGURE 6 A schematic diagram of the predicted topography of the calcium pump. The 10 membrane-spanning helices of the transmembrane domain and the 5 helices of the stalk are depicted as numbered rectangles and the three cytoplasmic globular domains as large boxes, with the ATP-binding and phosphorylation sites indicated.

residues, is probably involved in funneling Ca^{2+} ions to the binding site formed by the transmembrane helices. In this role, the stalk would be performing a function similar to that of the putative positively charged funnel formed by the transmembrane helices of the anion transporter.

Mechanistically, the Ca^{2+} pump resembles the Na^+–K^+ pump. There is a sequential binding of two Ca^{2+} ions from the cytoplasm, which catalyzes phosphorylation of the pump. These ions then become dehydrated by occlusion from the cytoplasm and subsequently there is conformational change in the pump (a so-called E1-E2 transition), which exposes them to the luminal side of the sarcoplasmic reticulum. Dissociation of the Ca^{2+} ions triggers dephosphorylation and a reverse conformational change.

IV. Cotransporters

Certain eukaryotic cells can accumulate solutes and deplete the cytosol of specific ions by secondary active transport. This process involves coupling the translocation of solutes or ions to the simultaneous translocation of another ion, usually Na^+. The movement of solutes and ions against their concentration gradients is driven by the movement of Na^+ down an electrochemical gradient, created by the Na^+–K^+ pump. Cotransport, in which the Na^+ ion and the transported species move in the same direction is termed symport. This type of cotransport is involved in the cytosolic accumulation of sugars, amino acids, carboxylic acids, and neurotransmitters. Cotransport in which the Na^+ ion and the transported species move in opposite directions is termed antiport. Typical examples of this type of cotransport system are the Na^+–Ca^{2+} exchanger of cardiac muscle, which regulates the cytosolic Ca^{2+} level and, therefore, muscular contraction of the heart, and the Na^+–H^+ exchanger, involved in cytoplasmic pH regulation.

A. The Na^+–Glucose Cotransporter

The Na^+-glucose transporter is a typical, well-characterized, cotransport system. The transporter is responsible for the symport of Na^+ and glucose into the epithelial cells of the intestine and kidney tubules, where its role is to mediate the cytosolic accumulation of glucose from the small intestine and urine, respectively.

The transporter is a 662-amino acid glycoprotein, with a molecular mass of about 75. Secondary structure predictions indicate that the transporter is basically composed of four major domains (Fig. 7). There are two extracellular globular domains; one of these domains bears the carbohydrate moiety and the other is highly charged. In addition, there is a transmembrane domain that consists of 11 α-helices and a highly charged cytoplasmic globular domain. Five of these helices are amphipathic (helices 2, 5, 6, 7, and 9) and may be involved in forming the channel or channels. The two polar extramembraneous domains are both located toward the carboxyl-terminus. One links helices 10 and 11 on the cytoplasmic face of the membrane, and the other links helices 7 and 8 on the extracellular face. These domains may be involved in forming the glucose- and Na^+-binding sites. Biophysical studies have shown that these sites are about 30–40 Å apart. With these binding sites so far apart, it seems unlikely that glucose and Na^+ are conducted through the same channel.

At present, little is known of the mechanism of operation of the cotransporter. It has been shown that the binding of Na^+ to the transporter produces a long-range conformational change in which the affinity of the binding site for glucose increases.

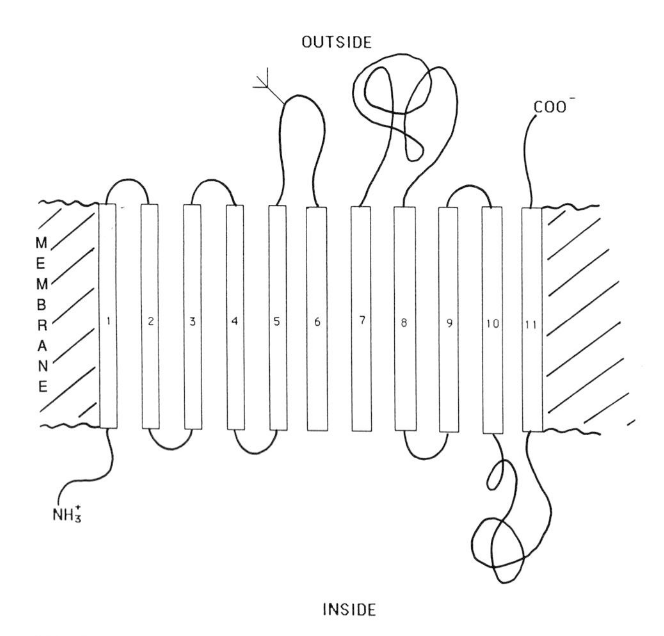

FIGURE 7 A schematic diagram of the predicted topography of the Na^+–glucose cotransporter. The 11 membrane-spanning helices are depicted as large numbered rectangles, the internal and external domains as large loops, and the carbohydrate moiety as a "tree."

This may occur by an opening up of the glucose site, since studies have shown that the increase in affinity is concomitant with an increase in exposure of the site to the aqueous medium. The transporter probably then follows a sequence of events similar to those suggested for the glucose transporter. Bound glucose and Na^+ become occluded from the aqueous medium, and subsequently, there is a conformational change in the transporter, which exposes these species to the cytosolic medium; they then dissociate and the transporter reverts to its original conformation.

V. Ion Channels

Ion channels play a crucial role in the process of intracellular communication. In general, they function by allowing the passive movement of ions across the membrane. The driving force for translocation is powered by the electrochemical gradient of the ion generated by the corresponding active transport system. Ion channels operate in a manner similar to passive and active transporters in that they can alternate between open and closed conformations. On the other hand, they differ in that the ions move freely through a water-filled pore rather than being selectively bound, occluded, and translocated. This conformational change is driven either by the binding of an effector ligand or by a change in membrane potential. These two types of ion channel can be distinguished by their differing structure: ligand-dependent ion channels are characteristically made up of several polypeptides, whereas voltage-dependent ion channels principally consist of a single polypeptide. However, the single polypeptide of voltage-dependent channels usually contains several strongly homologous repeat sequences, so that the two types of ion channel have essentially parallel structures.

The role of ion channels in cell–cell signaling is typified by the sequence of events leading up to the contraction of a vertebrate skeletal muscle. Stimulation of a sensory neuron triggers a wave of depolarization down the cell, mediated by voltage-dependent sodium channels. These channels open in response to a localized change in the membrane potential maintained by the Na^+–K^+ pump. The change in potential deriving from the flux of Na^+ ions through a particular channel causes neighboring channels to open, thus propagating a wave of potential change (depolarization). Arrival of the depolarization wave at the neuromuscular synapse activates voltage-dependent Ca^{2+} channels within the neuron. The resulting dissipation of the Ca^{2+} gradient, usually maintained by the Ca^{2+} pump, triggers release of the transmitter acetylcholine into the synaptic cleft. Free transmitter diffuses across the cleft and binds to specific receptors on the muscle cell membrane. These receptors then act as ligand-dependent cation channels, permeable to both Na^+ and K^+ ions. The flux of these ions causes depolarization of the muscle cell membrane, activating voltage-dependent Ca^{2+} channels largely located within the sarcoplasmic reticulum. The resulting increase in cytosolic Ca^{2+} triggers muscular contraction as the Ca^{2+} ions bind to the troponin component of the muscle filament.

A. Acetylcholine Receptors

The acetylcholine receptor is a pentameric complex of five similar glycoprotein subunits. The individual subunits consist of around 500 amino acids and show a high degree of sequence homology. They are distinguished from one another by a series of insertions and deletions, which causes their molecular masses to vary from around 55–66 kDa. Each subunit has five membrane-spanning α-helices, an extracellular globular N-terminal domain, and a globular cytoplasmic domain connecting helices 3 and 4 (Fig. 8). The principal difference between the subunits is in the size of the globular domains. Interestingly helix 4 is amphipathic, with charged and hydrophobic faces. It is thought that the charged face of this helix is contributed by each subunit in forming the ion channel.

The three-dimensional structure of the receptor has been determined at low resolution by electron diffraction. The receptor is clearly partitioned into three domains (Fig. 9). There is a long cylindrical domain, 25 Å wide with an internal diameter of about 10–13 Å, which extends about 60 Å into the synaptic cleft. The receptor subunits form the walls of this structure. The diameter of the receptor rapidly decreases as it traverses the membrane, until it reaches a value of less than 10 Å toward the center of the lipid bilayer. This structure probably represents the central ion channel, which has been shown to have a minimal diameter of about 7 Å by studying the conductance of ions of known size. In fact, it has been shown that a tight packing of the five amphipathic helices with a 20° tilt would produce a structure of the correct dimensions. The ion

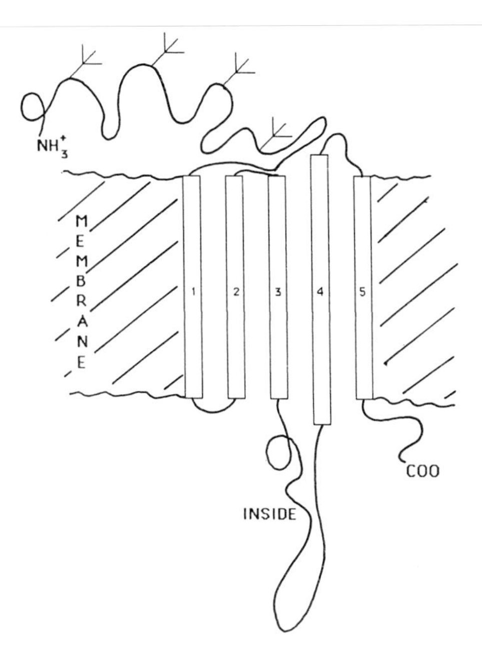

FIGURE 8 A schematic digram of the predicted topography of a subunit of the acetylcholine receptor. The five membrane-spanning helices are depicted as large numbered rectangles, the amino and internal domains as large loops, and the carbohydrate moiety as a "tree." The principal difference between subunits is in the size of the amino and cytoplasmic domains.

gate is probably located toward the center of the membrane, since terbium ions, which can be located within the receptor by low resolution X-ray analysis, have been shown to bind to the closed channel in this region.

The receptor is thought to operate as an ion-conducting channel in the following manner: the co-

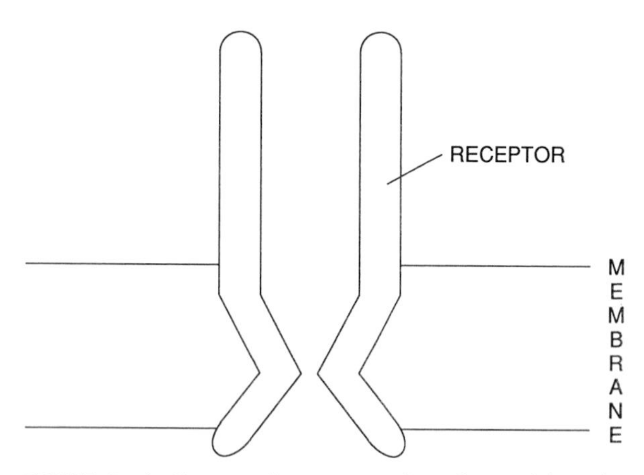

FIGURE 9 A digrammatic representation of an axial section through the acetylcholine receptor.

operative binding of two acetylcholine molecules triggers an extensive conformational change, involving all the subunits, opening the channel largely to conduct Na^+ ions but also K^+ ions. This event may involve a rotational or tilting of the helices relative to one another. In adopting the open conformation, positively charged amino acid residues blocking the channel are apparently removed from the ion-conducting pathway, allowing the free movement of ions. In contrast to passive and active transporters, ion channels do not posses high-affinity binding sites for specific ions but select them merely on the basis of their size and charge. Selection may be carried out by charged residues lining the channel, a role analagous to that of the charged funnel of the anion transporter and the stalk of the Ca^{2+} pump. The channel remains open for between 0.1 and 10 msec, before closing spontaneously, during which time it conducts about 10,000 ions per millisecond. [*See* CELL RECEPTORS; RECEPTORS, BIOCHEMISTRY.]

B. The Sodium Channel

The functional voltage-dependent Na^+ channel consists of a glycoprotein of 1,820 amino acids, which has a molecular mass of about 260 kDa. In mammalian cells, this glycoprotein is noncovalently associated with one or two smaller glycoproteins, each with a molecular mass of about 30–40, located at the extracellular surface of the membrane. The three glycoproteins have been termed the α, β_1-, and β_2-subunits, respectively.

Analysis of the sequence of the α-subunit has shown that there are four homologous multiple membrane-spanning domains, interconnected by three large, negatively charged, globular cytoplasmic domains (Fig. 10). Each transmembrane domain can be divided into distinct segments. Three of the transmembrane domains have been predicted to consist of six fully and two partially membrane-spanning helices. A small globular extracellular segment connects helix 5 with the first partial transmembrane helix (helix 6). The fourth domain is differentiated from the other three in that the partial transmembrane helices are thought to be replaced by a β-strand, because this segment contains several prolines, which would make a helical conformation unlikely. Helix 4 is amphipathic, with positively charged residues, predominantly arginine, at every third position. Consequently, one face of the helix is polar and the other apolar. Helices 2 and 7

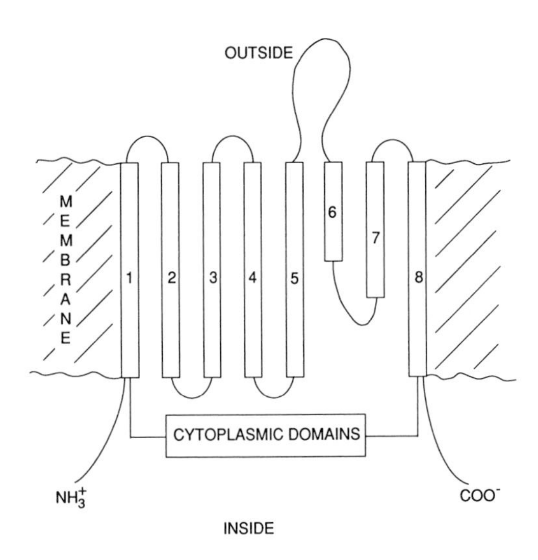

FIGURE 10 A schematic diagram of the predicted topography of the sodium channel. One of the four homologous sections of the channel is shown. These sections are linked by three large cytoplasmic domains, represented as boxed cytoplasmic domains. The eight membrane-bound helices of any one homologous section are depicted as large numbered rectangles and the extracellular domain as a large loop. Helices 6 and 7 are only partially membrane-spanning and in the fourth domain they are thought to be replaced by a β strand.

have several negatively charged residues, while the remaining helices are relatively apolar. These features of the Na^+ channel have also been shown to be relatively well conserved in the Ca^{2+} channel. It has been proposed that helix 4 and possibly helix 7 are involved in detecting changes in the membrane potential, while helix 2 may be contributed by each domain in forming the ion channel.

The following mechanism has been proposed for the mode of operation of the Na^+ channel: The positive charges of helix 4 are paired with negative charges on helix 7, to give a metastable closed conformation for the channel at the resting potential of the membrane. Depolarization of the membrane exerts a force on these charges, inducing a conformational change in the protein as it attempts to attain a more stable configuration. Channel activation may involve a movement of helix 4 toward the extracellular surface and helix 7 toward the cytoplasm, by a screwlike movement of the helices. This movement of helices 4 and 7 is conveyed to the ion channel, which then adopts an open conformation. It remains open for about 1 msec, during which time about 6,000 Na^+ ions pass through the channel. Subsequently, the channel adopts an inactive closed conformation, in which it is unable to open again until the membrane is repolarized. These features of the channel imply that the open conformation is merely a transient state en route to the closed inactive conformation. A similar situation must also prevail for the acetylcholine receptor.

Bibliography

Catterall, W. A. (1986). Molecular properties of voltage-sensitive sodium channels. *Annu. Rev. Biochem.* **55,** 953–985.

Hedigier, M. A., Coady, M. J., Ikeda, T. S., and Wright, E. M. (1987). Expression cloning & cDNA sequencing of the Na^+/glucose co-transporter. *Nature* **330,** 379–381.

Jay, D., and Cantley, L. (1986). Structural aspects of the red cell anion exchange protein. *Annu. Rev. Biochem.* **55,** 511–538.

Jennings, M. L. (1989). Topography of membrane proteins. *Annu. Rev. Biochem.* **58,** 999–1029.

Jorgenson, P. L., and Anderson, J. P. (1988). Structural basis for E1-E2 conformational transitions in Na/K-pump and Ca-pump proteins. *J. Membr. Biol.* **103,** 95–120.

Stroud, R. M., and Finer-Moore, J. (1985). Acetylcholine receptor structure, function and evolution. *Annu. Rev. Cell Biol.* **1,** 317–351.

Walmsley, A. R. (1988). The dynamics of the glucose transporter. *Trends Biochem. Sci.* **13,** 226–231.

Wright, J. K., Seckler, R., and Overath, P. (1986). Molecular aspects of sugar : ion co-transport. *Annu. Rev. Biochem.* **55,** 225–248.

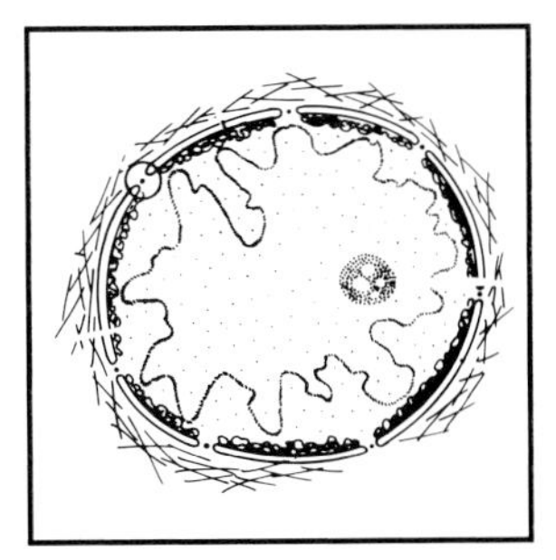

Cell Nucleus

HENRY TEDESCHI, *State University of New York at Albany*

Glossary

Diploid Containing two homologous copies of each chromosome
Gametogenesis Formation of gametes, i.e., sperm or oocytes by meiosis
Interphase Period between cell divisions
Mitosis Cell division in which two diploid daughter cells are produced
rRNA RNA component of the ribosome
Spindle Collection of microtubules forming part of the apparatus needed for cell division
Transcription Synthesis of RNA complementary to the DNA strand serving as a template
Translation Synthesis of polypeptides using the instruction contained in nucleotide sequence of mRNA

THE NUCLEUS IS THE cellular compartment of eukaryotes that contains nearly all the DNA of the cell. Eukaryotic cells have a distinct nuclear envelope, which separates two distinct compartments, the cytoplasm and the nucleoplasm. Because the transcription and maturation of mRNA occurs in the nucleus and the translation of the mRNA (i.e., the synthesis of polypeptides) occurs in the cytoplasm, the nuclear envelope is important in regulating the transport of macromolecules. The structural organization of the nucleus is complex and comprises a number of structures, many of them associated with the envelope.

I. Nuclear Organization

The envelope of the nucleus is in contact with a network of filaments (see Fig. 1) on both the cytoplasmic and nucleoplasmic sides. The filaments on the nucleoplasmic side line the inner face of the membrane and form a compact lining, the nuclear lamina.

Prominent in the nucleus is the nucleolus, which corresponds to DNA loops formed by several chromosomes (10 in human cells). The nucleolus contains a cluster of rRNA genes known as the nucleolar organizing region.

The centrosome is a cytoplasmic organelle on one side of the nucleus. The centrosome contains a pair of centrioles at right angles to each other. It is a major microtubule organizing center with a special role in cell division and in the separation of the chromosomes during mitosis. Each centriole begins dividing during the S-phase of the cell cycle, and eventually the two daughter centrioles migrate to opposite sides of the dividing cell. They provide organizing centers for the microtubules, which form the mitotic (or meiotic) spindle. [*See* MITOSIS.]

Most of the density inside the nucleus when viewed with the electron microscope corresponds to chromatin, the DNA-protein complex of eukaryotic chromosomes. The denser areas represent heterochromatin, an unusually compact form of chromatin that is transcriptionally inactive.

Encyclopedia of Human Biology, Volume 2.

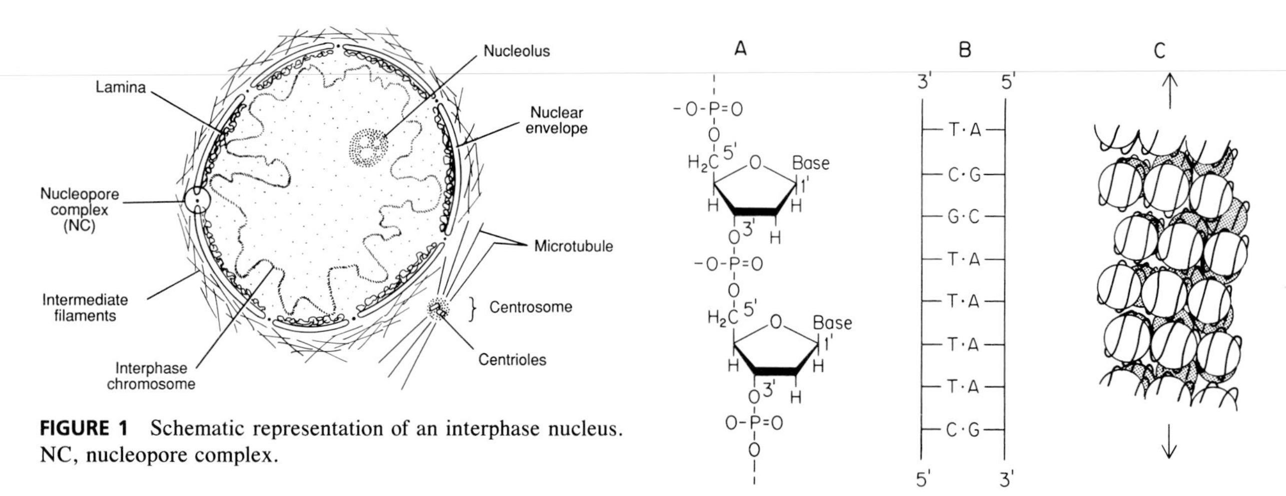

FIGURE 1 Schematic representation of an interphase nucleus. NC, nucleopore complex.

FIGURE 2 Diagram of DNA organization. A. The backbone of DNA strands and the numbering system used to identify atoms in the deoxyribose component. The numbers also indicate the direction of the DNA strand. B. Antiparallel arrangement of DNA, which is shown here in an extended form. C. Possible model of the arrangement of DNA (the thread) in relation to the histones (the balls) in chromatin. This unit would then be folded further in condensed chromatin. Each ball with the associated thread corresponds to one nucleosome.

II. The Chromosomes

A. Arrangement of the Chromosomal Components

Each human chromosome contains a single linear molecule of DNA estimated to be several centimeters in length. DNA has been shown to occur in three different helical structures, all formed from antiparallel strands, i.e., one strand has a 5′ end facing the 3′ end of the other strand. The form present in cells is probably mostly the B-form (B-DNA). The B-DNA is a right-hand helix with the base pairs normal to the helix axis. B-DNA corresponds to the original DNA model originally proposed by Watson and Crick. The antiparallel arrangement is illustrated in Fig. 2. Part A of the figure illustrates the numbering system used to distinguish the atoms present in the deoxyribose of the deoxyribonucleotides joined to form a DNA molecule; part B shows the antiparallel arrangement (shown in a straight model rather than a helix). The helices can bend, kink, and unravel. The bends can be rather sharp at sites where special nucleotide sequences are present. [*See* DNA AND GENE TRANSCRIPTION.]

The most abundant proteins of chromatin are the histones. The remaining proteins of chromatin are called nonhistone proteins. Histones are tightly attached to the DNA. They are relatively small proteins, generally containing between 100 and 200 amino acids with a high proportion of the positively charged amino acids lysine and arginine. Histone plays a major role in the orderly folding of DNA so that molecules, which in a fully extended form would be several centimeters long, are packed in structures a few micrometers in length. The DNA-histone complexes have a beaded appearance when extended and viewed with the electron microscope. Each bead is a nucleosome. Each nucleosome is formed by a histone octamer core on which the DNA helix is wound twice (see the individual beads in Fig. 2C). It has been estimated that an average eukaryotic gene contains as many as 50 nucleosomes. Nucleosomes are thought to be absent from areas from which they have been competitively displaced by sequence-specific DNA-binding proteins. [*See* HISTONES AND HISTONE GENES.]

Chromatin is highly condensed so that the nucleosomes are packed on top of each other producing a thread 30–40 nm (see Fig. 2C) in diameter, which is then folded further. The most condensed states are present during cell division, when the chromosomes can be clearly seen with the light microscope. [*See* CHROMATIN FOLDING.]

During interphase, the chromosomes of most cells are not visible with the light microscope. Two specialized kinds of chromosomes, which are visible with light microscopy, have been very useful in studying interphase chromosomes. The lamp brush chromosomes of growing oocytes form distinguishable chromatin loops. These loops apparently correspond to very active transcription sites. The polytene chromosomes of the secretory cells of fly larvae undergo multiple replication cycles without cell division, and the various chromosomal copies, several thousand more than the normal amount, remain side by side. After staining, the chromosomes appear banded and the arrangement shows many landmarks, which have permitted extensive cytogenetic studies with the light microscope. The heavily stained areas correspond to highly condensed chromatin.

B. Replication and Repair

One of the major functions of the nucleus is to maintain chromosomal DNA as intact and unaltered molecules. DNA repair mechanisms mostly depend on repairing a strand from the information contained in its complementary strand. When both are damaged in the same location, the genetic information is lost. The process of DNA repair is represented diagrammatically in Fig. 3. The mechanisms of repair involve recognition and removal of the damaged nucleotides by repair nucleases, which remove the damaged segment (Fig. 3B). The DNA polymerase binds to the 3′ OH end of the cut DNA and add one nucleotide at a time using the complementary strand as a template (Fig. 3C). DNA ligase in a final step attaches the restored segment to the damaged strand (Fig. 3D). [*See* REPAIR OF DAMAGED DNA.]

In addition to repair, the nucleus must provide mechanisms for the replication of the chromosomes and for the appropriate distribution to the daughter cells during mitosis or meiosis so that the homologous chromosomes are present in both daughter cells. The latter functions are associated with particular sequences in the DNA.

The huge eukaryotic DNA molecule is efficiently duplicated in segments—the replicons. The number of separate replicons in mammalian DNA is as high as 20,000–30,000. To replicate, a DNA molecule needs a replication origin, where an initiator protein will bind to catalyze a replication fork after which a multienzyme system proceeds to replicate the DNA. In DNA replication, each strand serves as a template for the replication of the new strand. Because the template strand remains as part of the new double strand, it is said to be conserved. The replication if semiconservative because each daughter cell inherits chromosomes containing an old and a new DNA strand. Replication involves separation of the two complementary strands. As new strands are being formed, the resulting structure is Y-shaped and is called a DNA replication fork. The

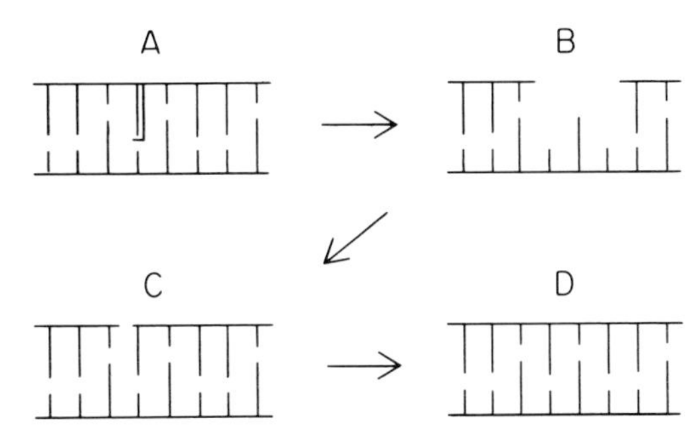

FIGURE 3 Diagram illustrating the DNA-repair mechanism. The backbone of the DNA is represented by horizontal lines, whereas the bases are represented by vertical lines. A. An altered nucleotide is indicated by the open rectangle. B. The damaged piece is removed by repair nucleases. C. The missing part is resynthesized by DNA polymerase using the complementary strand as a template. D. The resynthesized piece is reattached to the strand by DNA ligase.

replication proceeds asymmetrically at each replicon. Both strands are synthesized from the 5′ to the 3′ direction (see Fig. 4). Only one of the two has a 3′ end at the initiation site (the strands are antiparallel) so that only this one (the leading strand) can serve as a template for a continuous 5′ to 3′ synthesis. The other strand (the lagging strand) is replicated in pieces 1,000–2,000 nucleotides long in the 5′ to 3′ direction, the so-called Okasaki fragments. The fragments are then joined by DNA ligase, the same enzyme that functions in repair. The replication is said to be semidiscontinuous.

The DNA polymerase can extend a previously started chain at the 3′ OH end; however, it cannot initiate the replication without a primer. In this case, the primer is an RNA strand about 10 bases long, paired to one strand of DNA to provide the free 3′ OH end. Four DNA polymerases have been recognized in eukaryotes and only two are present in the nucleus: polymerase α, which is involved in DNA replication and priming, and polymerase β, which is involved in repair.

The centromeres play an important role in mitosis. They are specific DNA sequences needed for the attachment of the chromatids to the mitotic spindle during cell division.

C. Chromosomes and Transcription

The total DNA in a mammalian nucleus has enough nucleotides to code for as many as 3 million proteins; however, it has been estimated that only a tiny fraction, perhaps as few as 50,000–100,000, are actually produced. This is because only a small por-

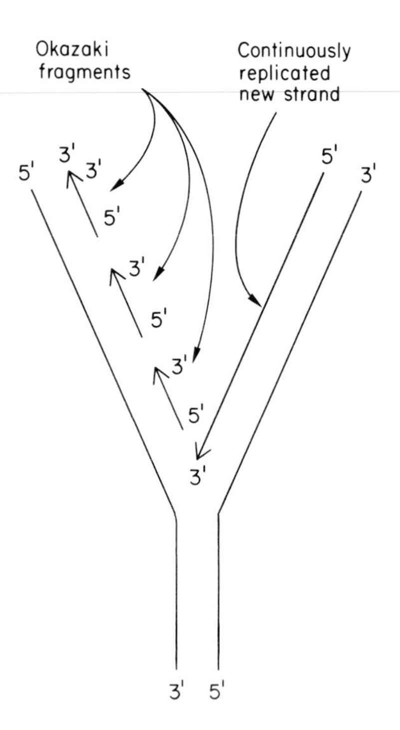

FIGURE 4 DNA replicating fork. The arrows indicate the direction of the duplication. The large arrow indicates the continuous duplication of one of the strands from its 5′ end. The smaller arrows show the replication of the other strand discontinuously in small fragments, also from the 5′ ends of the segments.

tion of the DNA sequences are transcribed and, furthermore, during RNA processing, the nucleus eliminates much of the transcribed RNA.

RNA synthesis is the transcription of DNA. One of the two DNA strands acts as a base-pairing template for the nascent RNA strand. Transcription converts genetic information into a form that can be used by the cytoplasmic machinery to produce proteins of the amino acid sequence predetermined by the genes. Transcription occurs very rapidly, much faster than DNA replication.

Transcription is catalyzed by RNA polymerases. RNA polymerase II transcribes protein coding genes, whereas polymerase I synthesizes the large rRNA molecules, and polymerase III a variety of small and stable RNAs such as transfer RNA (tRNA) or small nuclear RNA (snRNA) (see below). Transcription factors (TF) are sequence-specific DNA-binding proteins needed for the initiation of RNA synthesis. The TFs bind specific segments of the DNA and selectively attract RNA polymerase molecules so that the enzyme repeatedly transcribes the appropriate portion of the DNA.

Portions of the DNA of the chromosome are specialized to form regulatory DNA segments capable of binding regulatory proteins, which either enhance or inhibit transcriptional activity. Many of these segments are upstream in relation to the transcribed segment (i.e., on the 5′ side of the DNA), but other locations are also possible.

Initiation involves the binding of RNA-polymerase to the double-stranded DNA. This requires unwinding the DNA and recognition of a specific DNA sequence, the promoter. The RNA chain grows as the RNA-polymerase moves along the DNA, a process known as elongation, forming a DNA–RNA hybrid. Termination involves recognition of a terminator sequence. At this point, transcription ceases, and the DNA–RNA hybrid is disrupted.

A diagrammatic representation of the transcription of DNA and the processes needed to produce mature mRNA is shown in Fig. 5. Polymerase II produces a large transcript, heterogeneous nuclear RNA (hnRNA), which eventually forms mRNA. Newly synthesized hnRNA is immediately complexed to proteins, forming particles larger than nucleosomes. The 5′ end of the nascent RNA strand is capped by the addition of methylated guanylate. The cap plays a role in protein synthesis and protects the RNA from degradation. The 3′ end, the last to be synthesized, is also modified by the addition of a poly-A tail, a segment of 100–200 residues of adenylate. The role of the poly-A tail is unknown, but it may prevent degradation. Most of the RNA produced in this way is degraded; only approximately 5% of the RNA originally in the hnRNA reaches the cytoplasm. The sequences retained are those responsible for coding the amino acid sequence of proteins (corresponding to exons), whereas those degraded are the noncoding segments (corresponding to introns). The DNA corresponding to a gene is completely transcribed. This primary transcript is then processed so that the intron sequences are removed, a process known as splicing. First, the RNA has to be cleaved, a process carried out by nucleases. Then the ends of the exons have to be joined, a process known as ligation, and catalyzed by RNA-ligase. In addition, splicing requires special snRNA molecules, which recognize the appropriate sequence where splicing is to take place. The various components, snRNAs, and enzymes appear to be present in a large complex (the spliceosome) analogous to ribosomes.

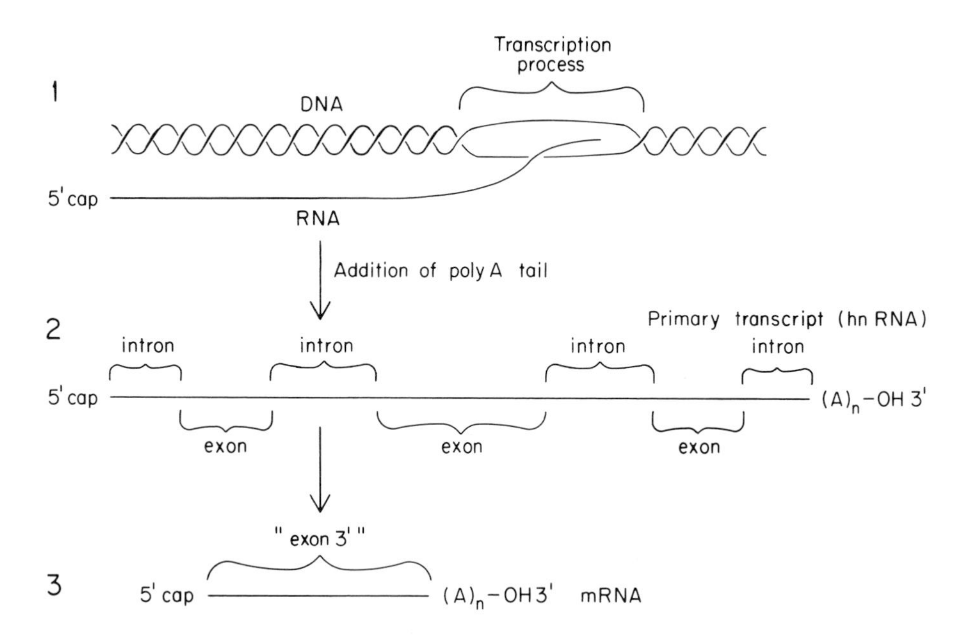

FIGURE 5 Diagrammatic representation of transcription and RNA maturation. 1. The RNA is transcribed from one of the DNA strands and immediately capped. 2. A poly-A tail is attached. 3. The introns are removed from the primary transcript.

III. The Nuclear Envelope

The nuclear envelope is double, containing an inner and an outer membrane (see Fig. 1). The envelope is discontinuous at the nucleopore complex where the two sets of membranes are joined. The outer membrane has been frequently found to bind ribosomes, and the envelope resembles the endoplasmic reticulum with which it is sometimes continuous. The nuclear envelope is disassembled into vesicles during cell division.

Chromatin is associated with the nuclear envelope. Much of the nuclear heterochromatin, the much condensed form of chromatin present at interphase, is attached to the inner membrane. Although this DNA is not involved in transcription or DNA replication, it may have a role in the overall organization of the chromatin. The arrangement of the interphase chromosomes in relation to the membrane are easiest to follow in polytene chromosomes—for example, in the salivary glands of *Drosophila*. At least in this case, each chromosome occupies a separate spatial domain, and specific loci in the chromosomes are attached to the nuclear envelope. Different kinds of polytene cells from the same organism seem to have a different kind of chromosomal arrangement.

The inner membrane is lined with a fibrous protein meshwork, the nuclear lamina, which has been characterized in some eukaryotic cells. The fibers of the lamina have strong similarities to the cytoplasmic intermediate filaments (e.g., similar amino acid sequences) and serve as a point of attachment of the chromosomes (Fig 2). Furthermore, the lamina is attached to the nucleopore complex (NC) of the nuclear envelope.

A. The Lamina

The lamina is a ubiquitous component associated with the nuclear envelope of eukaryotes. It has been found in mammals, the African frog *Xenopus*, and even invertebrates such as the clam or the fruitfly *Drosophila*. The lamina are 10–50 nm in thickness. Three protein components, named *lamins*, have been identified and found to be of 60–80 kDa. At least five different lamin species have been recognized in *Xenopus* oocytes. The fibers of the lamina are 10 nm in thickness and have an axial repeat of 25 nm.

In mammalian cells, lamin A and C are identical in their first 566 amino acids. Lamin B has not been studied as thoroughly. A and C both have a 350-amino acid segment, which is very similar in sequence to a segment of the intermediate filaments present in the cytoplasm. These two lamins form dimers from side-to-side association of monomers.

The resulting structure is rod-shaped with two globular heads at one end.

Different lamins seem to predominate depending on the developmental stage. In mammals, a lamin (either lamin B or a very similar protein) is present before implantation, whereas A and C make an appearance during organogenesis. The distribution of the lamins is also affected by gametogenesis.

At least in mammals, the lamina is associated with the nuclear envelope through lamin B. Lamin B also remains attached to vesicles formed when the nuclear envelope dissociates during cell division.

B. The Nucleopore Complex

The nuclear envelope is traversed by specialized channel structures, the NCs. In vertebrates there are 10–20 NCs per μm^2 or 2,000–4,000 per nucleus. Lamellae resembling the nuclear envelope, the annulate lamellae, are often found in the cytoplasm. Their relation to the nuclear envelope is not clear.

The NC corresponds to a circular opening in the nuclear envelope about 80 nm in diameter. In this region, the inner and outer membranes are joined. The NC has a nuclear and a cytoplasmic ring (Fig. 6), both about 120 nm in diameter and composed of eight subunits each. Eight spokes restrict the nuclear opening to about 80 nm. A central structure (central granule or plug, labeled T in the diagram) in the pore opening is thought to be involved in mediated transport of macromolecules. The nucleoplasmic side of the NC is attached to the lamina.

A number of pore complex proteins, mostly glycoproteins, have been identified. In contrast to other glycoproteins, where the carbohydrate portion is only on one side of the membrane (e.g., the outer surface of the plasma membrane and the luminal side of the endoplasmic reticulum), some of the carbohydrate portion of these proteins are exposed to the cytoplasm and some to the nucleoplasm. [*See* NUCLEAR PORE, STRUCTURE AND FUNCTION.]

The NC is the site of transport through the nuclear envelope. Transport takes place by two distinct mechanisms. Diffusion of a variety of molecules occurs through the NC passively. In this process, the NC functions as a sieve, allowing small molecules through and blocking the passage of molecules above a critical size. In addition, a specialized active transport is mediated by NC components.

In the passive passage through the NC, macro-

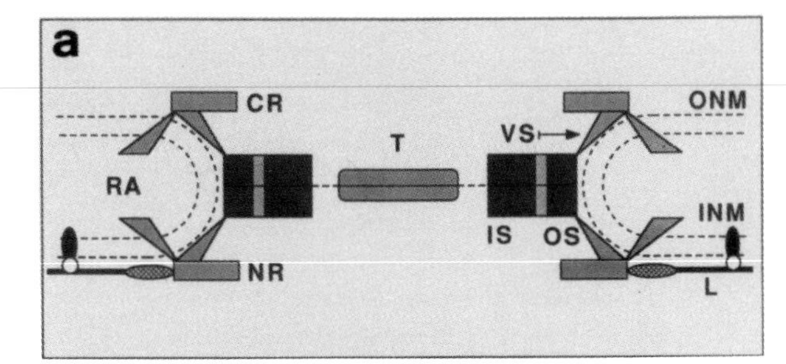

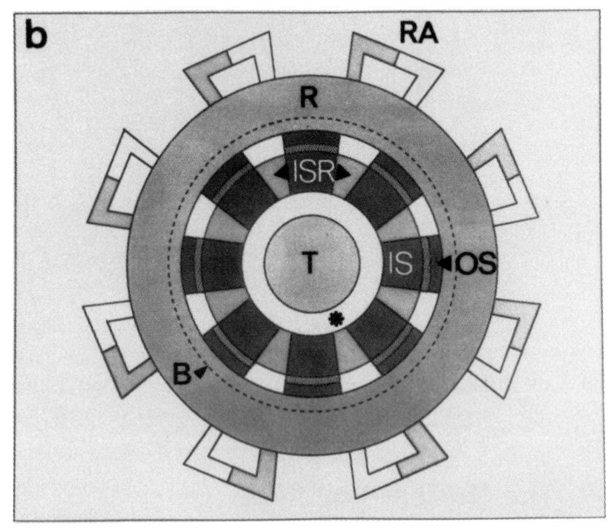

FIGURE 6 Model of the NC. a. Cross section. The major structural domains are shown, including inner spokes (IS), outer spokes (OS), vertical supports (VS), cytoplasmic and nucleoplasmic coaxial rings (CR and NR, respectively), and radial arms (RA). INM and ONM, inner and outer nuclear membrane, respectively; L, lamina; T, transporter. b. En face. B, maximum radius of the membrane border. *, positions where passive exchange of small molecules may occur. [Reproduced, by copyright permission of the Rockefeller University Press, from C. W. Akey, 1989, *J. Cell Biol.* **109,** 955–970.]

molecules used as probes diffuse at a rate inversely related to their size. Molecules (e.g., non-nuclear proteins) of approximately 20–40 kDa or above have been found not to enter significantly. The evidence suggests a functional diameter for the NC of about 10 nm. In contrast, some proteins (e.g., some proteins extracted from the nucleus) enter rapidly from the cytoplasmic side and can be accumulated by the nucleus even when very large. In fact, large particles such as colloidal gold can be transferred when coated with these proteins (see Fig. 7).

The transport requires the presence of a special sequence, a nuclear localization sequence. In contrast to the signal sequence of nascent polypeptides transferred into the endoplasmic reticulum's lumen, the nuclear localization sequence is present in the

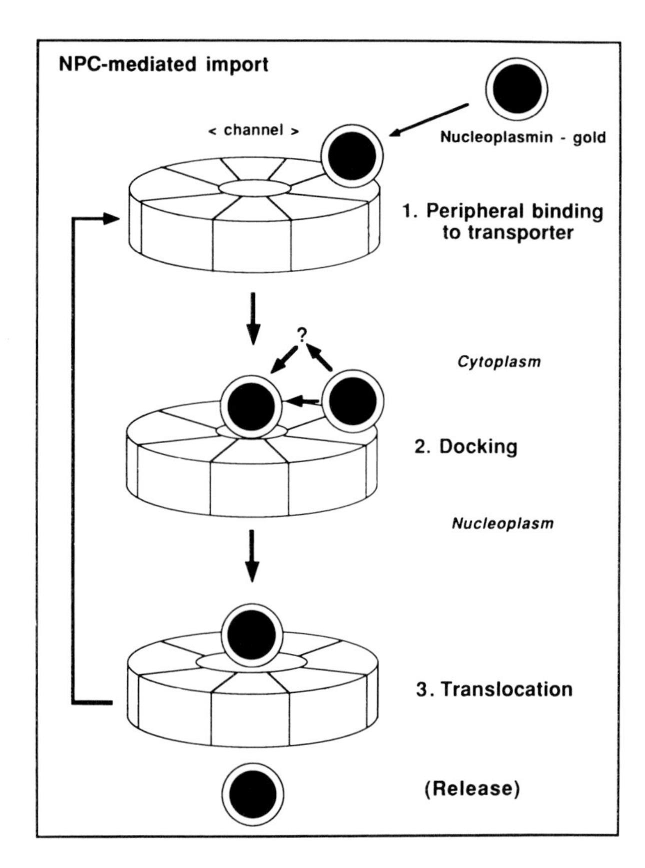

FIGURE 7 Possible three-step mechanism of nuclear import. [Reproduced, with permission, from C. W. Akey and D. S. Goldfarb, 1989, *J. Cell Biol.* **109,** 971–982.]

mature protein and is, therefore, not split off after or during the transfer. In some cases, the needed sequence has been found to contain as few as seven amino acids. Synthetic nuclear localization peptides covalently attached to non-nuclear proteins allow them to be accumulated in the nucleoplasm even if they are very large (e.g., in the case of ferritin, 465 kDa). The details of this transport have been recently examined with isolated preparations. Apparently, the process requires the hydrolysis of adenosine triphosphate as an energy source.

Because the movement of proteins in and out of the nucleus seems to reflect developmental stages, the transport system is thought to have a role in the regulation of gene expression. Thus, the transfer of the steroid hormone–receptor complex, which is necessary for the hormone action, occurs through this mechanism.

The RNA transcribed in the nucleus undergoes many processing steps. Following these the RNA is transferred to the cytoplasm in a very selective manner. In some cells, only a few mRNA transcripts are transferred by this same active mechanism. The remainder remains in the nucleus where it is degraded. These results suggest a role of the NC-transport system in posttranscriptional regulation of gene expression.

The active transfer or either proteins, RNA, or RNA-protein complexes is vectorial, i.e., in a single direction. A variety of evidence indicates that they all involve the same pathway. Figure 7 represents diagrammatically the findings based on experiments in which colloidal gold coated with nucleoplasmin (a nuclear protein) serves as a probe of mediated transport. The colloidal gold is visible with electron microscopy. The particles first attach peripherally in the nucleopore complex. They are then moved to the transporter, which opens to allow the passage of the particle.

Bibliography

Alberts, B., Bray, D., Lewis, J., Raff, M., Roberts, K., and Watson, J. D. (1989). "The Molecular Biology of the Cell," 2nd ed., Chapters 3, 5, 9. Garland Publishing, Inc., New York and London.

Darnell, J., Lodish, H., and Baltimore, D. (1990). "Molecular Cell Biology," 2nd ed., Part II. W. H. Freeman and Co., New York.

Gerace, L., and Burke, B. (1988). Functional organization of the nuclear envelope. *Ann. Rev. Cell Biol.* **4,** 336–374.

Lewin, B. (1990). "Genes IV," Chapters 17–19, 29–31. Oxford University Press, Oxford and New York.

Tedeschi, H. (1991). "Cell Physiology," Chapter 2. Hemisphere Publishing Corp., Washington, D.C.

Cell Receptors

P. MICHAEL CONN, University of Iowa College of Medicine

Glossary

Analogue Chemical structure that is similar to another

Autoimmune disease Disorder in which an animal (or human) produces antibodies directed against its own molecules

Binding site Locus at which a biologically significant molecule under consideration attaches

Coated pit Specialized part of the plasma membrane at which endocytosis is believed to begin. These appear as invaginations in the membrane coated with a material that appears furry in the electron microscope

Effector Molecule, group of molecules, or structure that transduces a signal from a receptor to produce an alteration in cellular function

Endocytosis Process by which particles and solutes enter eukaryotic cells. Frequently, this process is subdivided into pinocytosis (cell drinking) and phagocytosis (cell eating)

Ligand Molecule that occupies a receptor or binding site

Receptor Binding site that produces a physiologic response when it is occupied by an agonistic ligand

TO PRODUCE the appropriate responses to changes in the environment around them, cells rely on mechanisms that sense chemical signals such as drugs, hormones, and nutrients. Receptors are responsible for the specific recognition of such signals. Pancreatic cells that release insulin in response to elevated blood sugar, photoreceptors that produce electrical and chemical responses to light, and taste cells that recognize "sweet" tastes all rely on receptors. In the same highly specific fashion that individual locks respond to unique shapes of keys, so do receptors recognize and bind their specific ligands. Frequently receptors may be found in the membrane delimiting cells with the ligand recognition site facing outward, monitoring the extracellular environment. They may also be found inside the cells (or even inside subcellular organelles), where they bind chemicals that enter the cell or chemicals that are synthesized or released within the cell. Receptors are coupled to effector systems that alter cellular metabolism, shape, or development when their receptor-specific ligands are bound.

Interactions of Receptors with Ligands

The simplest receptor–ligand interaction can be described as

$$R + L \leftrightarrow RL$$

In more complicated situations a receptor may bind more than one (identical or nonidentical) ligand and the binding of one ligand may influence (positively or negatively) the binding of another. Chemical analogue of ligands are frequently divided into agonists and antagonists. The former provoke cellular responses as a result of occupancy of the receptor, whereas the later do not. Antagonists are useful to block the receptor from occupancy by endogenous agonists when it is desired to stop a target cell from responding. When this is done, the antagonist is

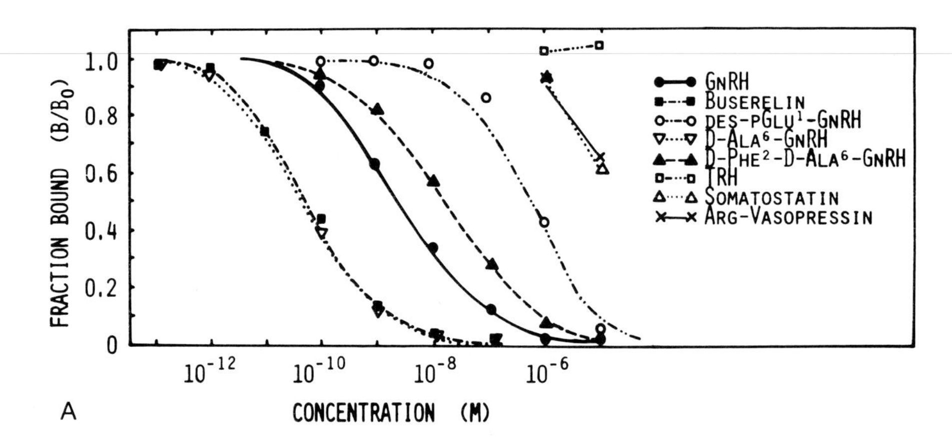

viewed as being a competitive inhibitor of agonist binding.

FIGURE 1 Specificity of ligand binding to reception. Pituitary cell membranes, which contain receptors for the gonadotropin-releasing hormone, GnRH, were incubated with a constant amount of a radioactively labeled analogue of GnRH called Buserelin. (Hoescht-Roussel Pharmaceuticals, Inc.), together with various concentrations (given on the horizontal scale) of other competitors. Some of these (e.g., unlabeled Buserelin, GnRH itself, and other analogues), have varying abilities to displace binding of the labeled analogue, whereas others (e.g., TRH, somatostatin, and Arg-vasopressin) are scarcely able to do so. The results show that the receptor recognizes only peptides closely related to GnRH. Data (vertical scale) are expressed as ratios of radioactivity that remains bound at the concentration of competition given in the horizontal scale to total amount of radioactivity that can be maximally bound to receptors. The more displaced to the left the curve for a given competitor is, the more effective the competition. [Reprinted from Marian J., et al. (1981). *Mol. Pharmacol.* **19**, 399, with permission.]

II. Biological Criteria Used to Identify Receptor

It is generally agreed that criteria must be fulfilled to identify a receptor.

A. Ligand Specificity

The hallmark of a receptor is the ability to recognize and bind its ligand with high specificity. A receptor for any given hormone will not recognize other chemically unrelated hormones (see Fig. 1). This characteristic is important so that receptors will not be "fooled" into becoming activated by an inappropriate stimulus.

B. Tissue Specificity

We expect to find receptors for particular hormones or drugs only in tissues that produce a response to these agents. A receptor that is located in all cells would not provide selective responses in individual tissues.

C. Finite Saturability

The number of receptors on a particular cell should be finite and, therefore, saturable by hormone. Saturability allows a distribution between specific receptors and nonspecific binding sites (e.g., a hormone simply dissolving in a cell membrane rather than binding to a specific receptor). Typically, nonspecific binding occurs in such a way that a constant percentage of the added ligand binds to the sites, irrespective of the amount added. In contrast, receptors show finite and saturable binding, (i.e., beyond certain value of the amount added, there is no further binding). A cell has a certain number of receptors for a specific ligand; all of them are occupied by ligand, no additional specific binding is possible.

D. Ligand Affinity

Receptors bind their ligands with high affinity, (i.e., the binding takes place readily) and dissociation of the bound ligand is rare. Typically the affinity of hormone receptors and drug receptors for their ligands is higher than that of enzymes for their substrates.

E. Coupling to Biologic Response

The characteristics described in the above items, when fulfilled, constitute the definition of a "binding site." A receptor requires the fulfillment of an additional characteristic (i.e., coupling to a biologic response). When ligands bind to receptors, there is a measurable alteration in some activity in the cell (e.g., secretion, metabolism, or contraction).

F. Metabolic Conversion of Receptors

An additional criteria has been sometimes ascribed to receptors to distinguish them from enzymes, which also bind specific substances to their substrates. The enzyme-bound substrate is altered: It is metabolized, whereas it is frequently held that the ligand bound to a receptor is not metabolized. More and more evidence, however, has been associated with ligand metabolism as a result of receptor binding internalization of the receptor–ligand complex and routing to lysosomes. Typically, this is related to degradation of the ligand, but occasionally the ligand is converted to a more active form. Accordingly, fulfillment of this criterion is generally not required of a receptor. On entering the cell, for example, testosterone is enzymatically converted to dihydrotestosterone, which is believed to be the active principle.

III. Receptor-Mediated Diseases

It is clear that anything that interferes with the ability of a ligand to bind to a receptor (or, alternatively, mimics the binding of a ligand in its absence) is capable of interfering with the recognition process necessary for the biological functioning of the receptor. In the disease myasthenia gravis, the affected individual produces antibodies to the acetylcholine receptor, through which muscles receive contraction signals from nerves. The antibodies eventually bind to these acetylcholine receptors, interfering with the binding of acetylcholine itself. As a result, there are severe deficiencies in the ability of muscles and others of the neuronal system to function. This disease can be produced experimentally by injecting acetylcholine receptors into rabbits. The receptors are obtained from the electric organ of the eel, which contains them in large quantities. The injected rabbits produce antibodies to the receptors and develop symptoms similar to those of the human myasthenia gravis. [*See* MYASTHENIA GRAVIS.]

Other diseases have a similar origin [i.e., antibodies to individual's hormone receptors (autoimmune diseases)]. Some individuals produce antibodies to the thyroid-stimulating hormone (TSH) receptor. This receptor, on binding TSH, usually stimulates the gland to produce thyroid hormone. The antibodies bind to the receptor in the thyroid gland, and in some instances, activate it, mimicking the action of TSH and causing the gland to synthesize and release thyroid hormones in excess, causing disease. In other instances, the antibodies have the opposite effect: They bind to the receptor without themselves activating it, but preventing TSH from binding to it. The second type of antibodies cause deficiency in thyroid hormone production, resulting in a different disease. Antibodies may also interact with the insulin receptor, usually by blocking its action, occasionally activating it. [*See* AUTOIMMUNE DISEASE.]

Another type of disease is caused when imperfect receptors are formed. Individuals producing defective androgen (testosterone) receptors suffer from androgen insensitivity (also called *testicular feminization*). Such individuals are genetically male; however, the defective receptors are unable to bind and recognize the masculinizing hormone, testosterone. As a result these "males" never masculinize and grow up with the (phenotypic) appearance of a female. They frequently do not realize until puberty (when they seek medical attention because menstruation does not begin) that they are, in fact, genetically male.

In other disorders such as pseudohypoparathyroidism, there is a defect in the coupling of the receptor to the intracellular signaling system, which mediates the biological effect of the hormone. Without this coupling mechanism, receptors are unable to mediate the action of the hormone, altering cellular functions even though the receptor itself recognizes its ligand with appropriate fidelity and affinity.

IV. Receptor Assay

Historically, receptors were identified in terms of ability of a tissue to respond to particular ligands. More recently it has been possible to characterize receptors by using radioactivity labeled ligands. Based on the amount of material bound, the number and binding affinity of the receptor can be identified. Radioactive ligands are also useful for autoradiography in which case receptors on cells are allowed to interact with radioactive ligand, and then the cells are overlayed with a thin photographic emulsion.

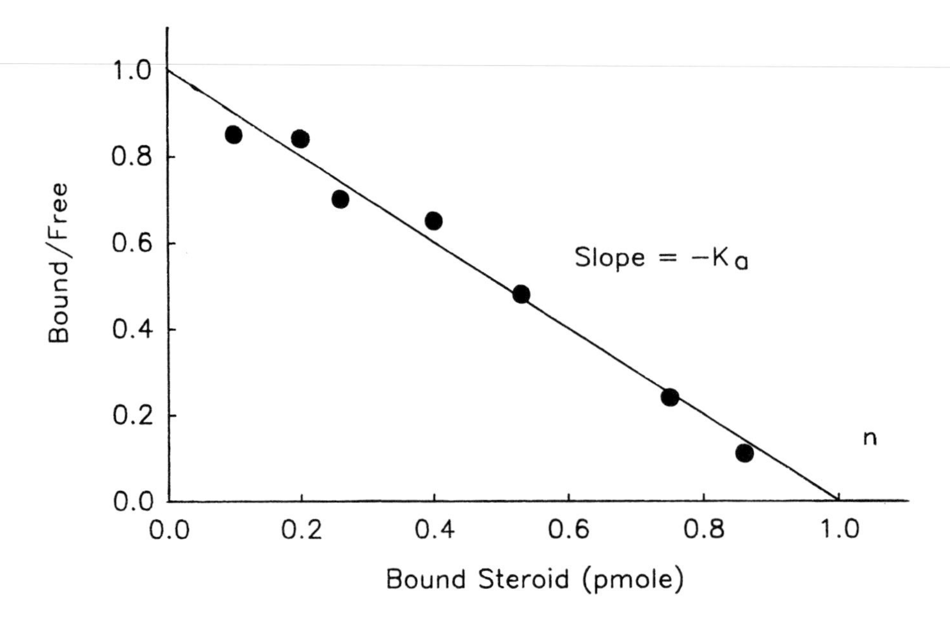

FIGURE 2 Example of a Scatchard plot. To construct a Scatchard plot, a constant quantity of receptor (present as purified material, a cellular fraction, or intact cells) is incubated with varying amounts of ligand (here a steroid hormone). The amount of bound ligand is plotted (horizontal axis) against the ratio of bound to free ligand (vertical axis). The affinity constant, K_a, is determined from the slope of the curve. The number of receptors is calculated from the intercept of the line drawn through the points with the horizontal axis. Here it corresponds to a picomole (10^{-12} moles) of receptors.

After adequate exposure to the radiation emitted by the ligand, the emulsion is developed photographically. Now ligands can be suitably labeled to be detected by their fluorescence or can be coupled to colloidal gold or ferritin, which makes the ligand recognizable by electron microscopy after they are bound to their receptors.

With the use of radioligands, the numbers of receptors on a cell and their affinity for the ligand can be measured. The most common method of analysis is the so-called *Scatchard Plot* Fig. 2 shows an example of this analysis.

V. Synthesis and Disposal of Receptors

Many receptors that have been identified to date are proteins. Like other proteins they are synthesized by ribosomes on the rough endoplasmic reticulum, and the precursor, polypeptide, usually by virtue of a specific targeting sequence, is routed specifically to the plasma membrane or to other cellular location. After interaction with their ligands, some receptors located in the plasma membrane internalize by *endocytosis,* frequently through a coated pit, and may be degraded inside the cell in proteolytic lysosomes (Fig. 3) [*See* PROTEINS.]

VI. Signal Transmission from Receptors

Cellular receptors are coupled to effector mechanisms, which result in the production of *intracellular messengers* (e.g., certain lipids, cyclic AMP, cyclic GMP); calcium released from storage or entering the cell can also be a messenger. Such compounds then regulate other intracellular enzymatic systems such as *kinases* (which phosphorylate proteins) or are recognized by other intracellular receptor proteins such as *calmodulin* (a calcium binding protein) or other metabolic regulating proteins (e.g., G proteins) (Fig. 4). There is a great deal of interest on the mechanisms by which receptors can be coupled (and uncoupled) from their effector systems, because they have important clinical implication for human disease.

VII. Chemical Modifications of Receptors

Receptor proteins may have virtually all chemical modifications found in other proteins (e.g., glycosylation and phosphorylation). These modifications may be important to activate or inactivate receptors. By analogy to other cellular proteins modifications, they may determine the longevity of a recep-

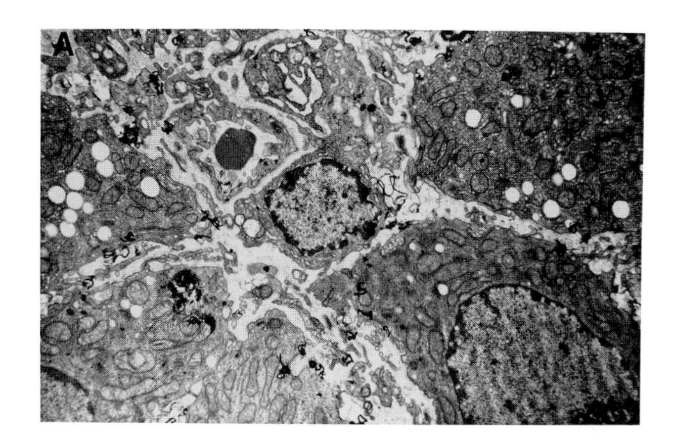

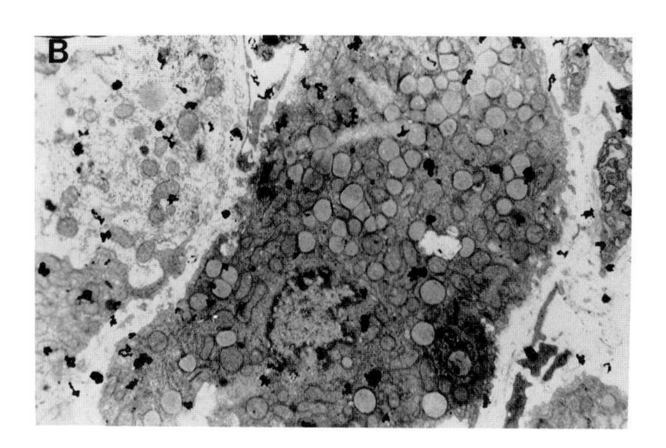

FIGURE 3 Demonstration of cellular uptake of a ligand through a surface receptor. A radioactive hormone (^{125}I-hCG) was injected into animals, and the ovary (containing receptors) was removed at 30 min (A) or 4 hr (B). Tissue was thinly sectioned and placed under a photographic emulsion, which was developed to show the location of radioactivity (*dark curved lines,* distinct from the otherwise clear emulsion). At earlier times the radioactivity is predominantly at the receptors in the cell membrane. At later times, the radioactivity is internalized. It can be shown that the internalized radioactivity represents hormone still bound to its receptor.

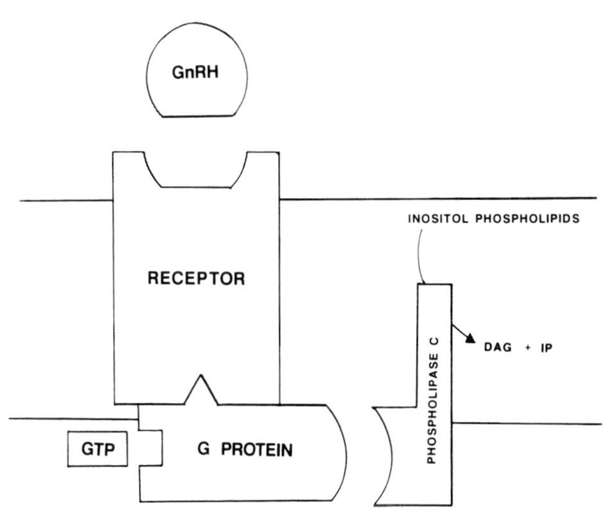

FIGURE 4 *Model of GnRH-receptor association with a G protein.* Schematic representation of GnRH- and GTP-mediated activation of phospholipase C. Receptor is visualized as spanning the plasma membrane and having segments that have functions for binding hormone and for lining with the guanine nucleotide regulatory protein (G protein) that binds GTP. The G protein forms a bridge between the receptor and the catalytic component of phospholipase C, thereby activating the whole enzyme. The result is the splitting of inositol phospholipids present in the cell membrane, with production of diacylglycerol (DAG) and inositol phosphates (IP). These two substances, in turn, after the state of regulation of the cell. From William V. Andrews and P. Michael Conn, University of Iowa College of Medicine.

tor or modify its actions (ligand binding, effector activation).

VIII. Receptor Structure

From studies on purified proteins and sequences and structures deduced from cDNA cloning, it is clear that there are classes of receptors with certain structural similarity. For example, most receptors for steroid hormones have three domains. One, at the N-terminus, is variable in terms of amino acid content; the central one is both cystine-rich and well-conserved; the third one, the C-terminus portion, is also well-conserved. The N-terminal does not bind the hormone; the central domain probably binds to DNA, using zinc atoms connected to the Cys residues. In general, specific functions (i.e., hormone binding, nuclear translocation, DNA binding, dimerization, effect transcription) appear to be the properties of specific regions of the receptor. The same is probably true for sites for binding of adenylate cyclase, guanyl nucleotide binding proteins, and other effectors.

Acknowledgment

The work reported from the author's laboratory was supported by NIH HD19899.

Bibliography

Blecher, M., and Bar, R. (1981). "Receptors and Human Diseases" Williams and Wilkins, Baltimore, Maryland.
Conn, P. M. (1984–1986). "The Receptors" vol. I–IV. Academic Press, Orlando
Limbird, L. (1987). "Cell Surface Receptors: A Short Course on Theory and Methods." Martinnus Nijhoff, Boston, Massachusetts.

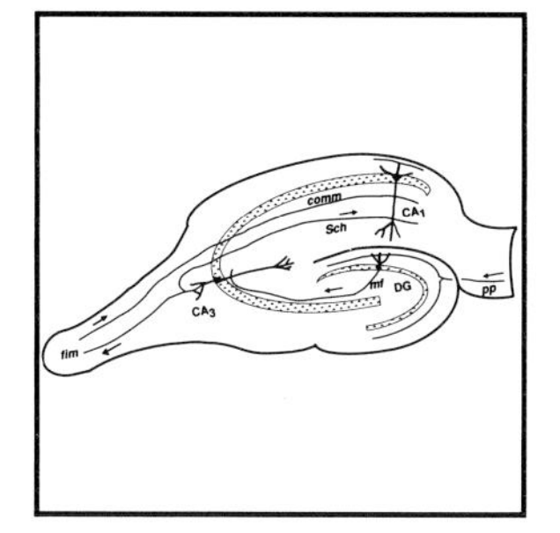

Cellular Memory

PATRICK L. HUDDIE, DANIEL L. ALKON, *National Institutes of Health*

Glossary

Agonist Drug which produces an effect at a receptor site

Antagonist Drug which blocks the effect of an agonist

Associative learning When an organism learns relationships among environmental events; there are two types, classical conditioning and instrumental conditioning

Classical conditioning (Pavlovian conditioning) Two stimuli are presented to an animal in such a way that a response normally evoked by one of the stimuli eventually is elicited by the other stimulus as well. Thus the unconditioned stimulus would normally evoke the unconditioned response, but after associative learning the previously neutral conditioned stimulus will also produce a similar conditioned response. The classic example is the salivation of Pavlov's dogs. The unconditioned stimulus was food presented to the dog. Pavlov limited his observation to the unconditioned response of salivation. An auditory conditioned stimulus (metronome click) was presented in advance of the unconditioned stimulus and after several trials, the dog learned to associate the conditioned stimulus with forthcoming food, and salivated before the food was presented (conditioned response).

Conductance Conductance of an ion channel represents the quantity of ions which can pass through the channel in unit time; conductance of a cell membrane represents the sum of all the ion channels in that membrane

Delayed rectifier Type of potassium current which turns on slowly upon depolarization and rectifies because current flows more readily out of the cell than into the cell; hence it is also known as an outward rectifier

Depolarization and hyperpolarization Describe changes in membrane potential; depolarization is a shift to less negative membrane potentials, hyperpolarization is a shift to more negative membrane potentials

Habituation Occurs when a stimulus elicits a diminished response on successive trials; sensory receptor adaptation or fatigue are not regarded as habituation. An oft-quoted example is the behavior of the polychaete worm *Nereis pelagica,* that contracts back into its burrow when stimulated mechanically or by a shadow. Repeated presentation of one type of stimulus evokes smaller and smaller contractions, but fatigue can be excluded because a novel stimulus produces full contraction after habituation to the original stimulus. Habituation may be a mechanism that limits fatigue; the habituation gradually disappears after a period without the habituating stimulus: this is known as dishabituation

Heterosynaptic Those neuronal interactions in which a third neuron modulates synaptic transmission between two others, by modifying the functions of the presynaptic terminal or the postsynaptic membrane

Ion channels Gated protein pores which penetrate lipid bilayers and select which ions may pass through on the basis of size and valence. Some ion channels are opened by changes in membrane potential; others are opened selectively by specific neurotransmitters and second messengers

Iontophoresis Movement of charged drug molecules from a drug filled electrode to intracellular or extracellular spaces by application of an electric field
Instrumental conditioning Experimenter that aims to modify the frequency or intensity (or both) of an innate behavior using reinforcement chosen by the experimenter; reinforcement strategies include: rewarding a specific behavior (reward conditioning), punishing a specific behavior (aversive conditioning), and avoidance learning, where the aversive stimulus is delayed if the behavior is displayed
Phosphorylation Attachment of inorganic phosphate to a protein; this modifies the conformation of the protein and frequently changes the function of the protein
Phototaxis Movement of an organism towards light
Post synaptic potentials Caused by a neurotransmitter released from the presynaptic terminal that opens ion channels in the postsynaptic neuronal membrane and causes ionic currents to flow; the currents produce transient depolarizations or hyperpolarizations known as postsynaptic potentials
Quantal release The hypothesis that neurotransmitters are released from the presynaptic terminal in small packets, or quanta, believed to involve the fusion of a single transmitter-containing vesicle with the terminal membrane
Sensitization Gradual increase in the response to a repeated stimulus. Generally, the sensitizing stimulus is noxious or particularly strong
Vestibular organs Detect acceleration; in *Hermissenda* the organ comprises 13 ciliated cells (hence hair cells) surrounding a central fluid-filled space which contains calcium carbonate stones, or statoconia. When accelerations shift the statoconia the hair cell cilia detect the movement
Voltage clamp Technique used to resolve the difficulty of knowing which membrane currents are involved in electrical activity of excitable cells; many classes of membrane ion channels are voltage-dependent, and as the membrane potential changes, the ionic current that passes also changes, and this in turn affects the membrane potential; thus, it is desirable to hold (clamp) the membrane potential at known voltages and measure the resulting currents in isolation; this is done with electronics that rapidly detect very small voltage changes and respond by passing enough current to oppose the change; this current is equal and opposite to the membrane current involved

CELLULAR MEMORY can be defined as the change in neuronal properties consequent upon the association of environmental events, and causal, necessary and sufficient for the expression of the learned response. This is a narrow view, but it is crucial for the study of learning that the chain of events between behavioral modification and molecular changes be made explicit and be causally related.

I. Cellular Memory and Behavior

This chapter's primary example of cellular memory is that displayed by the marine mollusc *Hermissenda crassicornis;* however, our discussion is informed by the extensive studies of the cellular basis of behavior performed on other animal models, with the restriction that model systems where no discrete cellular component has been implicated in learning-dependent changes have not been considered. We have also considered evidence emerging from the study of other types of cellular transformation, including development and oncogenesis. In addition, we note that many principles of cellular memory are applicable across phyla.

II. A Model System—Associative Learning in *Hermissenda*

The nudibranch gastropod *Hermissenda* normally exhibits phototaxis. In turbulent ocean conditions, the phototaxis is inhibited: the animal slows or ceases movement, contracts its foot, and adheres more strongly to the substrate. This simple organism, with simple behavior, can be associatively conditioned in a classical conditioning paradigm. The animal can be trained to respond to light (the conditioned stimulus) with inhibition of movement and foot contraction (the conditioned response). [*See* CONDITIONING.]

The sensory systems involved in these responses are the visual and vestibular organs and are located in the head of *Hermissenda*. The visual system consists of two caudal eyes, each of which comprises five photoreceptors, screening pigments, and a simple lens. The eyes are located above the circumoesophageal nervous system and receive inhibitory input from the vestibular organs, the statocysts, located similarly juxtaposed to the circumoesophageal nervous system, and in close proximity to the

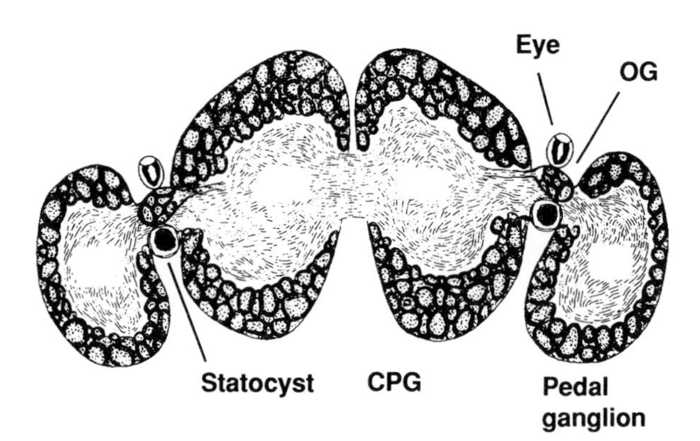

FIGURE 1 *Hermissenda* nervous system. The circumoesophageal nervous system has four major ganglia, connected by commissural nerve tracts. Between the cerebropleural ganglion (CPG) and the pedal ganglion is a triad of structures. The eye is caudal to the small optic ganglion and the statocyst. Axons from the eye pass through the optic ganglion (OG) and are contacted by axons from the statocyst within the CPG.

optic ganglion (Fig. 1). The photoreceptors and statocysts interact through direct synaptic connections and via additional synaptic connections through interneurons with the optic ganglion and cerebropleural ganglion.

During training wild *Hermissenda* are maintained in the laboratory under constant temperature and dim illumination before being placed in the training chamber. A number of animals can be trained in the chamber, which consists of eight seawater-filled tracks, each wide enough to accommodate one animal (Fig. 2). The chamber is attached to a mechanical agitator that moves the plate in the horizontal plane. Above the chamber is a light source with an intervening shutter. The agitator and the shutter are controlled by a computer, so that precise periods of light and turbulent agitation can be provided repeatedly.

The animals are trained by multiple training trials; after 100 pairings of light and rotation, *Hermissenda* respond to test flashes of light with reduced locomotion and foot shortening. Animals trained with explicitly unpaired stimuli, or naive animals, respond to light with phototaxis and foot lengthening. Thus, the unconditioned response to agitation (foot contraction, etc.) has been transferred to the conditioned response (light) and the animal anticipates the agitation on the basis of exposure to light. Thus, the *Hermissenda* have been conditioned to associate light with forthcoming agitation.

The cellular basis of this behavioral change has been investigated effectively in *Hermissenda* because of the relative simplicity of the sensory and neuronal processing. The elements of this simple nervous system show less specialization than a complex nervous system like our own; that is each cell (in a simple system) can subserve several functions. This elegant parsimony gives way to neuronal specialization in vertebrate brains. Thus *Hermissenda* photoreceptors transduce light into electrical signals and perform significant processing via their synaptic inputs; with only two Type A photoreceptors and three Type B photoreceptors available in each eye, much of the integration of sensory input occurs at the periphery in the "simple" nervous system.

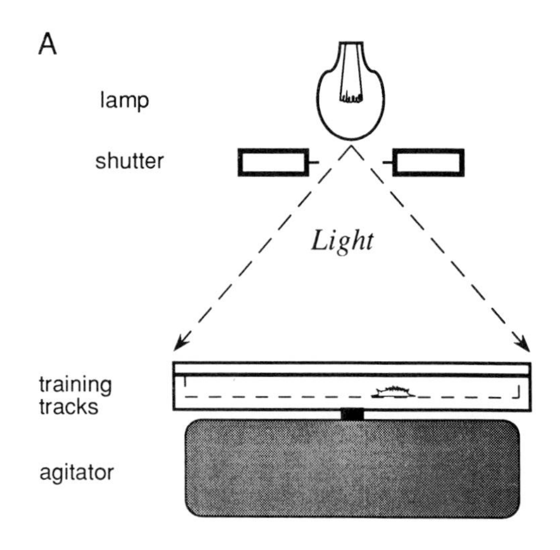

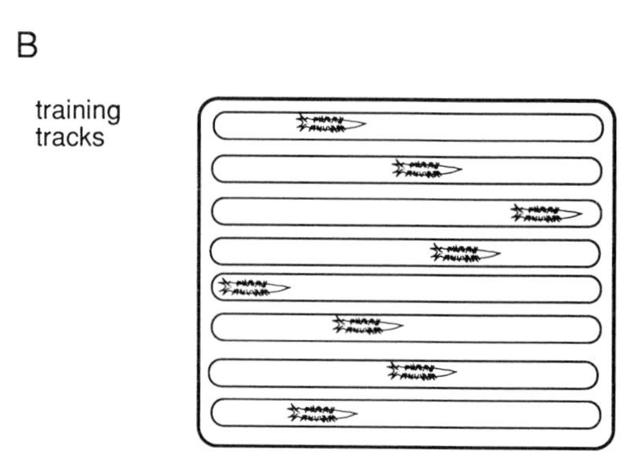

FIGURE 2 Training apparatus for *Hermissenda*. A. *Hermissenda* held inside the training tracks are illuminated from above when the computer-controlled shutter opens. The training tracks are attached to a mechanical agitator that moves the plate in the horizontal plane; the agitator is also controlled by the computer. B. Plan view of training chamber. A number of animals can be trained at one time in the chamber, which consists of several seawater-filled tracks, each wide enough to accommodate one animal.

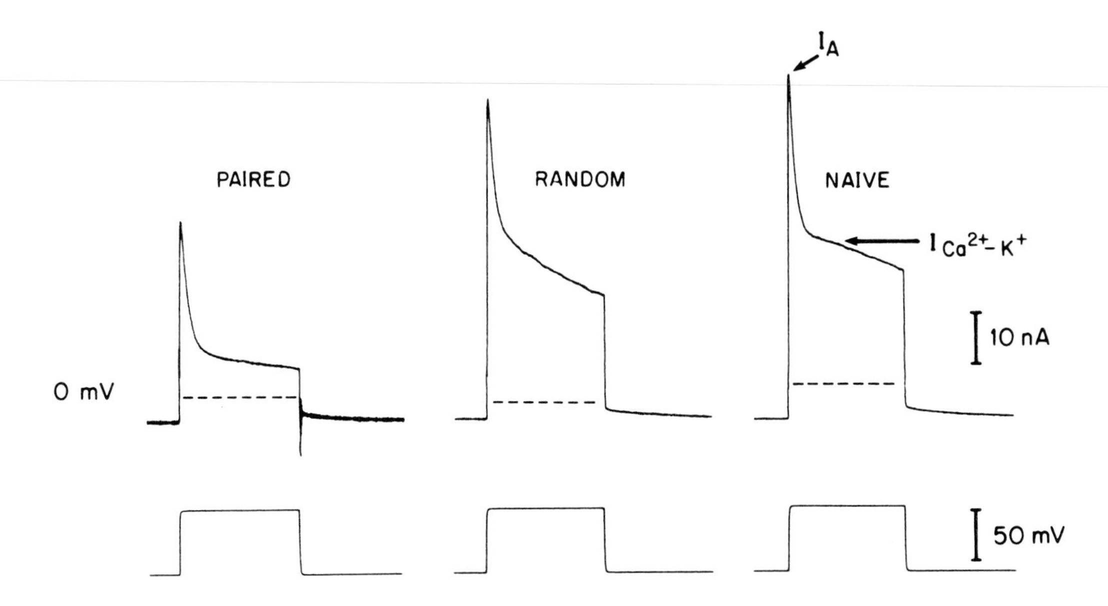

FIGURE 3 Conditioning-induced K^+ current charges in type B photoreceptors. Voltage-clamp records from type B photoreceptors of *Hermissenda*. The lower trace shows that the photoreceptor membrane potential was held at −60 mV, then stepped to 0 mV for 1 sec. The upper trace shows the resulting currents; the initial spike is the transient voltage-dependent K^+ current (I_A). The slowly declining late component of each current trace is the slower calcium-dependent K^+ current ($I_{Ca^{2+}K}$). The magnitude of both currents is markedly reduced in animals that received paired light and agitation (record on the left), when compared with naive animals or animals that received explicitly unpaired light and agitation. (The dashed lines indicate the amount of current that was not due to the K^+ current, also called the leak current).

After training, the membrane resistance of the Type B photoreceptors is increased, which renders the cells more excitable for a constant stimulating current input. This change in the excitability of the B photoreceptor following rotation is a consequence of the properties of the local neuronal network. During rotation, the hair cells increase their output and release more inhibitory transmitter and the B cell is hyperpolarized. After rotation hair cells undergo an intrinsic poststimulus hyperpolarization and therefore release less inhibitory transmitter onto the B cell and onto an interneuron in the optic ganglion. Both these actions result in net depolarization of the B cell, since less inhibition comes from the hair cell, and the excitatory input to the B cell from the optic ganglion cell is disinhibited.

After the pairing of light and rotation, the B-cell membrane is depolarized, and calcium-selective voltage-dependent ion channels open. These channels do not inactivate as long as the depolarization is sustained, and while they are open calcium moves into the cell. The total influx of calcium into the B cell is dependent on both the duration of each post-trial depolarization and on the number of training trials.

This influx of calcium must be related to the separate effects of light on the B photoreceptor. Light alone causes the release of diacylglycerol from the cleavage of membrane lipid via photosensitive proteins, and phospholipase C. The diacylglycerol is effective in stimulating the membrane-associated activity of a calcium-dependent protein kinase (protein kinase C, [PKC]). PKC is stimulated by both diacylglycerol and the increased level of intracellular Ca^{2+}, and goes on to cause the increase in membrane resistance. A better understanding of the action of PKC has been gained from studies of voltage-clamped Type B photoreceptors.

The Type B photoreceptor is penetrated with two microelectrodes, one electrode to measure membrane potential and the other to pass current into the cell. The cell is voltage-clamped and the ionic currents that flow during activity are measured.

The dominant outward currents in Type B photoreceptors are potassium currents; K^+ ions are more abundant in the cell than in the extracellular medium, and tend to flow down the electrochemical gradient, out of the cell, when K-selective ion channels in the membrane are opened, by changes in membrane voltage or intracellular calcium concentration. Striking reduction of the transient voltage dependent K^+ current (I_A), and the slower calcium-dependent K^+ current ($I_{K\text{-}Ca}$) can be seen after conditioning (Fig. 3). This change in outward currents is the major cause of the increase in input resistance consequent upon classical conditioning of *Hermissenda*. Further voltage-clamp experiments suggest a role for PKC in the reduction of K^+ currents following conditioning. PKC is activated *in vitro* by a class of lipid-soluble chemicals known as phorbol esters. Application of phorbol ester to the Type B cell together with elevation of Ca^{2+} produces reduction of I_A and $I_{K\text{-}Ca}$ analogous to the effect of conditioning.

The enzymatic activity of PKC in the cell is phosphorylation of appropriate substrates. Whether PKC acts directly to phosphorylate K^+ channels or acts on an intermediary is an open question. Recent evidence suggests that an important intermediary may be cp20. a G-protein of 20-kDa molecular weight, which is phosphorylated after phorbol ester treatment and which inhibits potassium channels when injected into naive Type B cells. [*See* PROTEIN PHOSPHORYLATION.]

To understand the consequences of the increased excitability of the Type B cell we must refer to the known network (Fig. 4). The B photoreceptor inhibits the medial Type A photoreceptor. The medial Type A photoreceptor excites an interneuron (I), that excites a motoneuron that causes locomotion. In addition, conditioning enhances the Type A photoreceptor K^+ currents, and further inhibits the Type A response to light; therefore after conditioning, the Type B cell inhibits the Type A cell and locomotion is inhibited.

An important question is the causal relationship of these cellular changes to the behavioral changes. We know that for *Hermissenda,* at least, the induction of the cellular change is partially responsible for the predicted behavioral change. Intact *Hermissenda* can be restrained and the B cell impaled with a microelectrode. Pairing of light and depolarizing current causes a significantly slower phototactic response, when compared with control unpaired and sham-operated animals in subsequent behavioral tests. Therefore, training-induced changes at the convergence of the visual and vestibular sensory systems are probably causal for the behavioral change.

Having outlined the salient features of a promising animal model of cellular memory, we would like to consider in more detail the various cellular elements that have been implicated in cellular memory in our own studies and in other systems.

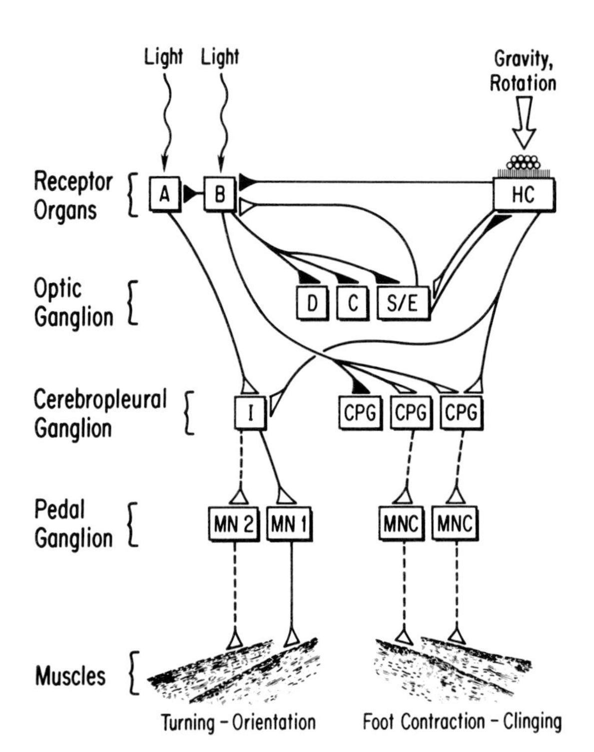

FIGURE 4 Schematic of the *Hermissenda* neural system. This simplified circuit diagram shows the interaction between sensory and motor systems, which underlies associative conditioning. Excitatory synapses are represented by open triangles, and inhibitory synapses by filled triangles. Light stimulates both type A and type B photoreceptors. The Type A cell excites interneuron I and, thus, stimulates motoneurons (MN1, MN2), which produce muscle contraction, and movement towards light. The type B photoreceptor inhibits the type A cell. Training makes the B cell more excitable and increases inhibition of the A cell, thus limiting phototaxis.

The type B photoreceptor also sends inhibitory input to neurons D, C, and S/E of the optic ganglion, and both inhibitory and excitatory inputs to neurons of the cerebropleural ganglion (CPG). Excitatory inputs to some of the CPG neurons cause activation of motoneurons (MNC) and contraction of the *Hermissenda's* foot.

The hair cells of the statocysts detect acceleration due to gravity or motion and interact with the visual system at several sites. The hair cells directly inhibit the type B cells and the S/E cell of the optic ganglion. The S/E cell reciprocally inhibits the hair cell but excites the type B photoreceptor. The hair cells also excite interneurons I and CPG of the cerebropleural ganglion.

After hair cell stimulation, the intrinsic postburst hyperpolarization of the hair cell and the slow, depolarizing component of the synaptic potentials at the hair cell to type B cell synapse combine, in a pairing-specific, temporally constrained fashion, with increased S/E cell activity and the intracellular sequelae, which follow illumination in the type B cell to produce an enhancement of type B photoreceptor excitability specific to conditioning. After conditioning, the type B cells are more excitable and tend to activate the foot contraction side of the network and suppress the phototactic side.

III. Membrane Processes

The cell membrane is the interface between the cell and the world. Input reaches the cell via the membrane, and output is expressed by membrane processes.

A. Ion Channels

1. Ca^{2+} Channels

The slower inward Ca^{2+} current component of the action potential is present to a greater or lesser extent in various types of neurons. The influx of Ca^{2+} is an important signal, and the role of Ca^{2+} as an intracellular messenger is discussed below. Here we consider the changes in Ca^{2+} channels occurring during learning. In *Hermissenda* after excessive training (e.g., 300 trials), voltage-dependent Ca^{2+} currents are reduced.

Ca^{2+} currents may also be reduced in another model system, reflex habituation of *Aplysia* sensory neurons. *Aplysia* is a gastropod that has been widely used to study the cellular basis of nonassociative learning behaviors such as habituation and sensitization.

In the intact animal, gentle touch of the mantle shelf or siphon causes withdrawal of the gill and the siphon. This gill and siphon withdrawal (GSW) reflex can be habituated, so that the response decrements with repeated presentation of a constant stimulus, and shows recovery and dishabituation. The GSW reflex is controlled by both the peripheral and central nervous systems in *Aplysia;* the relative contribution of each is unclear. However, changes in identified neurons of the abdominal ganglion of the central nervous system are correlated with the central component of the GSW reflex. In the abdominal ganglion, sensory neurons from the skin synapse upon identified gill and siphon motor neurons. Sensory neurons also synapse upon abdominal ganglion interneurons, that make synaptic contact with the motor neurons. Facilitatory neurons, a fourth type of neuron, synapse upon the presynaptic terminals of the sensory neurons (Fig. 5).

In an *in vitro* preparation that retains the sensory and effector organs and the abdominal ganglion, habituating responses can still be induced, and this preparation is suitable for electrophysiological recording from the motor neurons. The biophysical correlate of habituation is a progressive reduction in the frequency of action potentials evoked by sensory neuron stimulation, and when action potentials

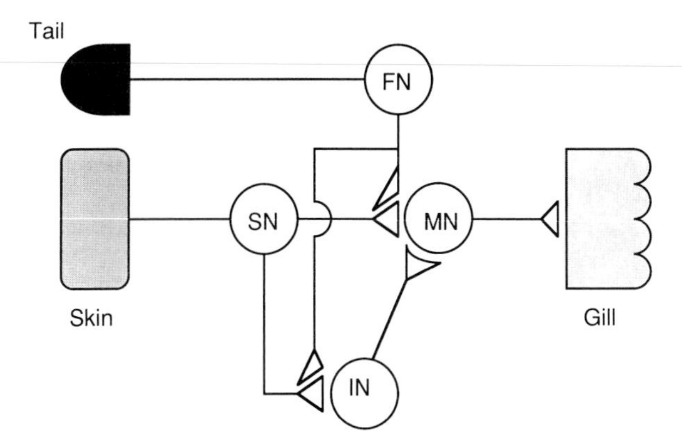

FIGURE 5 Schematic of the abdominal ganglion and sensory organs of *Aplysia*. A simplified circuit diagram of the major inputs, outputs, and neuronal connections of the *Aplysia* abdominal ganglion involved in the gill-siphon withdrawal (GSW) reflex. Sensory information from the skin is transmitted by sensory neurons (SN) to motoneurons (MN) and interneurons (IN). The interneurons can be either excitatory or inhibitory. Gill withdrawal follows activation of the motoneuron. Sensory input from tail stimulation modifies the action of the sensory neurons via facilitatory neurons (FN) that synapse upon the presynaptic terminals of the sensory neurons.

are inhibited by hyperpolarizing the motor neurons, the depolarizing excitatory postsynaptic potentials (EPSPs) are revealed; *in vitro* habituation is correlated with reduction of the amplitude of these EPSPs. Dishabituation is correlated with recovery of the EPSP amplitude.

The decline of the EPSP may be caused by a reduction in the number of transmitter quanta released rather than changes in postsynaptic membrane sensitivity; however, because quantal analysis requires many comparable EPSPs, and because the EPSP amplitude varies with activity so much in this preparation, it is difficult to know if the number of quanta is the only changing factor in synaptic depression. In habituated synapses, fewer vesicles are seen; this supports the hypothesis of reduced transmitter release following habituation. Given these considerations, a role for Ca^{2+} ions in the habituation process is possible.

Pharmacological isolation of the inward Ca^{2+} currents of voltage-clamped sensory and motor neurons, with the sodium channel blocker tetrodotoxin and the potassium channel blocker tetraethylammonium, reveals that the Ca^{2+} current of the sensory neuron soma is inactivated more during synaptic depression; the relationship of this phenomenon to habituation is not established. The mechanism by which the inactivation occurs also is unclear; a possible mechanism is Ca^{2+} inactivation of Ca^{2+} channels, which has been described for *Hermissenda* neurons but not specifically in learning. One caveat is that the cellular habituation of *Aplysia* sensory neurons appears to occur at the presynaptic terminal, whereas the voltage-clamp analysis is only descriptive of the cell body of the neuron. Thus, the investigation of the biophysical correlates of the behavioral phenomenon of habituation in *Aplysia* is still at an early stage.

2. K^+ Channels

Neuronal excitability is primarily regulated by K^+ channels. The K^+ channels determine the duration of the refractory period that follows an action potential. During an action potential, voltage-dependent K^+ channels open and K^+ ions flow out of the neuron down the electrochemical gradient. This outward current is responsible for repolarizing the cell and returning the membrane potential to the resting value after the depolarizing Na^+ and Ca^{2+} currents have terminated. However, immediately following the action potential, the cell is hyperpolarized beyond the normal resting potential, because the dominant membrane conductance is to K^+ ions, and the membrane potential approaches the equilibrium potential for potassium. The postaction potential after hyperpolarization is crucial in determining when the neuron will be able to respond to a new excitatory input or, in the case of spontaneously active neurons, the frequency of action potentials.

Potassium channels are many and diverse. A useful generalization is to divide the K^+ channels into 3 classes: A currents (I_A), delayed rectifiers, and calcium-dependent K^+ channels. I_A is a fast transient outward currents activated by depolarization, that inactivates rapidly even if the depolarization is sustained. The delayed rectifier is a K^+ channel that opens slowly when a threshold depolarization is attained, and remains open while the membrane is depolarized. The Ca-dependent K^+ channels ($I_{K\text{-}Ca}$) are weakly activated by voltage in some but not all neurons, and strongly activated by increased calcium concentration at the intracellular face of the channel. Such an increase in intracellular Ca^{2+} concentration occurs after Ca^{2+} channels open during membrane depolarization, or when intracellular stores of Ca^{2+} are released; thus, the Ca-dependent K^+ channel can be thought to be gated by a second messenger.

Experiments on *Hermissenda* show that both I_A and $I_{K\text{-}Ca}$ are important for the acquisition and expression of cellular memory, and the changes in these currents persist for days following training.

Similar profound changes in K^+ currents are implicated in classical conditioning of rabbits. Nictitating membrane response conditioning of the rabbit is a robust quantifiable behavior modification in which the conditioned stimulus (a tone or light flash) is paired with the unconditioned stimulus (an air puff to the cornea, or a para-orbital shock). The unconditioned stimulus elicits eyeball retraction, and the rabbit's third eyelid (the nictitating membrane) closes as a consequence. The animal rapidly learns to blink upon the conditioned stimulus: Nictitating membrane closure is used as a convenient measure of the conditioned response.

Analysis of the cellular mechanisms underlying nictitating membrane response conditioning is at an early stage. The hippocampus is not essential but, when present, permits more complex nictitating membrane response conditioning, and under less than optimal conditions. The firing of action potentials by neurons in the CA1 region of the hippocampus (see Fig. 7) is increased during training, and the increase in firing precedes the nictitating membrane response by 30–40 msec. The pattern of firing is correlated with the profile of the nictitating membrane response. A cellular mechanism for the increase in excitability of these CA1 neurons may be inhibition of $I_{K\text{-}Ca}$. In hippocampal slices from conditioned animals, the after hyperpolarization that followed action potentials elicited by depolarizing current injection was reduced when compared with control animals; resting potential and input resistance were unaffected. The after hyperpolarization is due to $I_{K\text{-}Ca}$ and is apparently reduced in conditioning. Single electrode voltage clamp studies of CA1 neurons from conditioned and control animals show directly that $I_{K\text{-}Ca}$ is reduced by conditioning. Thus the membrane potential returns more rapidly to the resting potential in conditioned animals, and excitability is enhanced. Additionally, synaptic potentials are enhanced in conditioned animals, also consistent with a reduction in $I_{K\text{-}Ca}$. Thus, the changes seen in *Hermissenda* may be generalized to mammalian brains. [*See* HIPPOCAMPAL FORMATION.]

A third system in which K^+-channel modification in learning is suggested is instrumental motor learning in locusts. Both reinforcement and punishment protocols have been used to train locusts to hold the metathoracic leg in particular positions. A simpli-

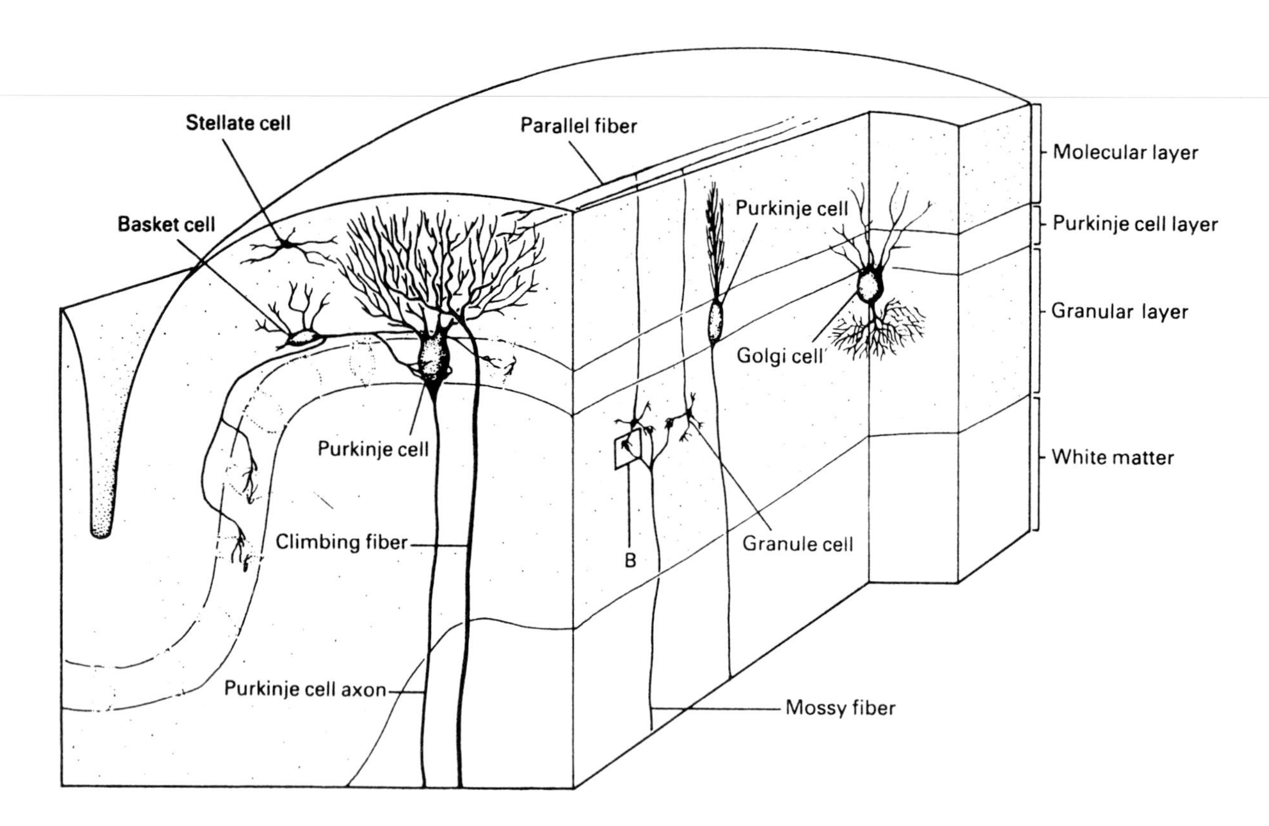

FIGURE 6 Neuronal architecture of the cerebellum. The cerebellar cortex has a regular structure, which aids interpretation of the information flow. Vestibular information comes in via mossy fibers. These synapse upon granule neurons (region marked B). The granule neuron axons bifurcate to become the parallel fibers that synapse on the distal (or upper) dendrites of the Purkinje neurons in the "molecular layer" of the cerebellum. The climbing fibers carry visual information via the inferior olive, to the proximal (or lower) dendrites of Purkinje neurons. The inhibitory output axon of the Purkinje cells goes to the vestibular-ocular relay nuclei. [Reprinted, by permission of the publisher, from E. R. Kandel and J. H. Schwartz. The cerebellum. *In* "Principles of Neural Science," 2nd ed. (C. Ghez and S. Fahn, eds.) p. 507, Elsevier Science Publishing Co, Inc., New York]

fied preparation was also used, where firing frequency of the anterior adductor coxa motoneuron (AAdC) was considered as the response. The AAdC can be "trained" to increase or decrease its output to the anterior adductor muscle.

In trained AAdC neurons firing frequency was increased when the preparation was trained to elevate the leg (up-learning) and decreased when the preparation was trained to lower the leg (down-learning). The evidence suggests that up-learning is correlated with a decrease in K^+ conductance of the AAdC membrane and down-learning with increased K^+ conductance. However, the evidence is indirect, and voltage-clamp studies are required to clarify this interesting finding. Moreover, the data also support changes, correlated with learning, in the synaptic inputs to the AAdC neuron. Despite these considerations, the findings suggest that both increases and decreases in K^+ conductance may occur in different types of learning in the same nervous system.

B. Neurotransmitter Receptors

1. Glutamate

Modification of glutamatergic neurotransmission is implicated in two systems used in the study of cellular modifiability: the vestibulo-ocular reflex (VOR), and the *in vitro* short-term synaptic plasticity model, long-term potentiation (LTP).

We will focus on the possible cellular mechanisms underlying the VOR. Briefly, however, the VOR keeps eye position constant as the head moves and depends on head position information generated by the vestibular organs of the inner ear. In the cerebellum, parallel fibers carry vestibular information, and climbing fibers carry visual information via the inferior olive to the dendrites of Purkinje neurons in the flocculus (Fig. 6). The output of the Purkinje cells is inhibitory and projects to the vestibular-ocular relay nuclei. The modifiability of this system is exhibited when reversing prisms are worn. Initially, eye movement is inappropriate

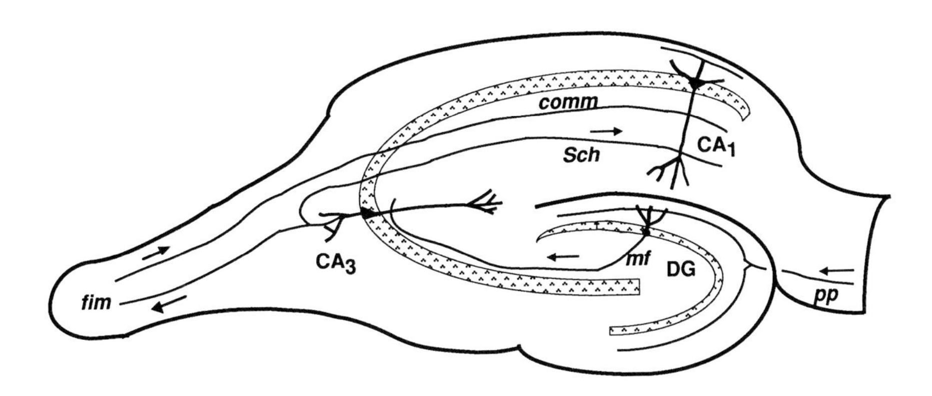

FIGURE 7 Neural pathways and regions of the hippocampus. The hippocampus also has a well-defined architecture. When sliced in a particular way, the connections between hippocampal neurons and the input pathways are preserved. Here the excitatory pathways are delineated; the inhibitory pathways are susceptible to damage when the slice is prepared and have not been well characterized yet. The cell bodies (somata) of the CA1 and CA3 neurons are found in the larger speckled area; the somata of the dentate gyrus neurons (DG) are found in the smaller speckled region (which is the dentate gyrus). The excitatory input of the perforant path (pp) synapses upon the dendritic field of DG neurons that project to the dendritic field of CA3 neurons. The output of the CA3 neurons goes out of the hippocampus via the fimbria (fim) and to the CA1 neurons via the Schaeffer collaterals (Sch). The commissural nerve tract (comm) also synapses on CA1 neurons.

for head rotation; the VOR rapidly reverses and again the animal successfully tracks the visual target during rotation of the head. The learned response is retained for times between hours and days. [*See* EYE MOVEMENTS.]

A cellular locus for this modifiable behavior may be the convergence of the glutamatergic parallel–mossy fibers and climbing fibers on the Purkinje neurons of the flocculus. When glutamate iontophoresis onto Purkinje cells is paired with olivary stimulation (which excites the climbing fibers), the response to glutamate is depressed compared to unpaired controls. Synaptic responses elicited by parallel–mossy fiber stimulation are also depressed by iontophoretic glutamate paired with climbing fiber stimulation. This long-lasting depression is comparable with the long-term depression (LTD) of parallel fiber–Purkinje cell transmission that is caused by pairing parallel fiber and climbing fiber stimulation.

There are three major glutamate receptor subtypes in neuronal tissue, distinguished by agonist specificity: kainate, quisqualate, and N-methyl-D-aspartate (NMDA). Normal excitatory transmission in the hippocampus is mediated by quisqualate and kainate type receptors.

The glutamate receptors involved in long-term depression appear to be of the quisqualate subtype, because quisqualate, but not kainate nor aspartate can mimic the effects of glutamate. The quisqualate receptor antagonist kynurenic acid also blocks the induction of LTD. Therefore, LTD is thought to be a consequence of quisqualate receptor responses. A role for Ca^{2+} ions in LTD is also suggested by experiments in which the Ca^{2+} chelator EGTA was injected into Purkinje cell dendrites and LTD was abolished. The precise interactions of glutamate, Ca^{2+}, and the climbing fiber input have not yet been elucidated. Ca^{2+} may directly promote glutamate receptor desensitization, by analogy with Ca^{2+}-linked desensitization of nicotinic acetylcholine (ACh) receptors. Ca^{2+} influx into Purkinje cells may be promoted by climbing fiber activity, and then the Ca^{2+} could elevate intracellular cyclic guanosine monophosphate (cGMP) levels by activating guanyl cyclase (a known action of Ca^{2+}). However, cGMP is itself excitatory on Purkinje neurons, and the hypothesis is difficult to test. An unexplored idea is that Ca^{2+} may activate PKC thus changing the sensitivity of postsynaptic glutamate receptors.

A glutamatergic system widely employed for studying modifiability is LTP in the hippocampal slice *in vitro*. The hippocampus is linked to learning because of clinical and experimental studies of anterograde amnesia. Hippocampal neuroanatomy is highly structured, with laminar organization of neuronal cell bodies and dendritic fields (Fig. 7). This has facilitated electrophysiological investigation. However, the *in vitro* studies of the physiology of hippocampal synaptic modifiability suffer from lack of a clearly defined behavioral frame of reference. For example, in most cases, the incoming stimuli are electrical shocks to nerve tracts, defined by the experimenter, rather than sensory stimuli appropriate to those nerve tracts. Thus, the wider applicability of these interesting changes in synaptic function remain uncertain for the most part.

LTP was first demonstrated by stimulating dentate gyrus neurons via the perforant path, and the activity of a population of these neurons was recorded with extracellular electrodes; it was found that high frequency trains of stimuli (up to 100 Hz for 4 sec) would enhance the subsequent population response (the population EPSP and population action potential). The potentiation persisted for 30 min to several hours.

Subsequently, LTP has been investigated *in vivo* and *in vitro* and a wealth of information is available. Here we focus on the involvement of glutamate receptor-linked ion channels in LTP.

LTP is induced in the *in vitro* hippocampal slice preparation by brief high-frequency stimulation of an excitatory monosynaptic pathway to both pyramidal and granule neurons in the slice and can be observed for up to 24 hr. LTP requires depolarization of the postsynaptic neuron. LTP of the pyramidal neurons of the CA1 is blocked by the selective NMDA receptor antagonist aminophosphovalerate. In the CA1 and dentate gyrus regions, the NMDA receptor is present and presumably binds glutamate released from presynaptic terminals, but at resting membrane potentials most NMDA receptor-linked ion channels are blocked by Mg^{2+} ions. Depolarization of the neuron expels the Mg^{2+} ion from the NMDA receptor channel and permits the passage of Na^+, K^+, and Ca^{2+} ions. Under physiological conditions, the fraction of the current carried by Ca^{2+} ions is small. Although the calcium permeability of the NMDA receptor channels of spinal cord neurons exceeds the sodium permeability by about fivefold, the relative physiological extracellular concentrations of Ca^{2+} and Na^+ mean that the net sodium current exceeds the calcium current. In addition, the voltage-activated calcium channels present in these neurons are activated by depolarization to −30 mV. However, the NMDA receptor-linked channel could provide a significant calcium signal to the intracellular compartment under conditions of LTP, and the NMDA receptor can be considered an important neuromodulatory system for hippocampal LTP in the CA1 and dentate gyrus and in the commissural–associative CA3 pathway. However, LTP is not exclusively linked with NMDA. Kainate appears to serve a similar role in LTP of the mossy fiber to CA3 neuron pathway. Kainate receptor-linked channels have negligible Ca^{2+} permeability, and their role in LTP may be to overcome endogenous inhibition so that postsynaptic depolarization, and consequent Ca^{2+} influx through voltage-dependent Ca^{2+} channels, can occur. Thus, a general role for NMDA receptors in LTP cannot be assumed. This is particularly the case since because NMDA and quisqualate are equally effective in elevating Ca^{2+} in cultured hippocampal neurons, but both are less effective than glutamate or aspartate. In fact, the NMDA antagonist 2-aminophosphovalerate is as effective as NMDA itself in elevating intracellular Ca^{2+}.

2. ACh

Conditioning of the cat eye-blink response has been linked with the action of ACh. The response is a short latency eye blink and can be conditioned by pairing an audible "click" with a glabella tap. The sensorimotor cortical neurons that reach the facial nuclei show long-term changes (weeks) in unit activity; their threshold for excitation is lower, and it is suggested that their membrane conductances are altered after conditioning. Pairing depolarization of these neurons with application of ACh increases membrane resistance for approximately 10 min; intracellular injection of cGMP or calcium-calmodulin protein kinase can substitute for ACh. Whether or not the short-term effects of ACh are involved in the conditioning process remains an open question.

3. Dopamine

Long-lasting dopaminergic modulation of neurotransmission has been found in the relatively accessible mammalian sympathetic ganglion preparation. The slow muscarinic EPSP is mimicked by cGMP application and is potentiated heterosynaptically for hours by dopamine acting at dopamine receptors of the D_1 class. This long term enhancement is mimicked by D_1 agonists and blocked by D_1 antagonists specifically, and appears to involve elevations of the concentration of the intracellular second messenger cyclic adenosine monophosphate (cAMP) in the postsynaptic cell. Thus, dopamine released by adrenergic interneurons within the ganglion facilitates the action of ACh released from cholinergic terminals.

4. Neuropeptides

The gill-siphon withdrawal reflex (GSW) of *Aplysia* is modified by a tetrapeptide transmitter, FMRF-amide (Phe-Met-Arg-Phe-NH_2), released by an interneuron that contacts the presynaptic terminal of the sensory neuron, which innervates the motoneuron responsible for gill withdrawal. Tail shock stimulates the interneuron, hyperpolarizes the sensory neuron, reduces the width of the action potential, and depresses transmitter release. FMRF-amide is found in this interneuron and exogenous FMRF-amide mimics the effects caused by stimulating the interneuron. The behavioral response (inhibition of GSW) is only apparent with weak tail shock; increasing the tail shock causes sensitization. The behavioral paradigm uses a weak tactile stimulation of the siphon as the test stimulus, and tail shock strength is varied. If the frequency of tactile stimulation is increased, the response to weak shock habituates, masking the inhibition. Thus, the inhibition by the FMRF-amide pathway can be masked by habituation. This suggests that FMRF-amide could contribute to the habituation process described above. FMRF-amide action involves the second messenger arachidonic acid, which has been shown to activate a potassium channel in the sensory neurons (the S channel).

5. Serotonin

Studies of sensitization of the GSW reflex and the related tail withdrawal in *Aplysia* have led to the suggestion that serotonin, also known as 5-hydroxytryptamine (5HT), is crucial for this behavioral change.

The behavioral paradigm involves two stimuli. The GSW reflex elicited by gentle touch of the mantle or siphon is enhanced by mechanical or electrical shock stimulation of the head or tail of the animal. In the isolated abdominal ganglion, costimulation of the sensory neuron and the connective bundle from the head facilitates the EPSP observed in the motoneuron during sensory stimulation (Figure 5). The facilitation persists for several minutes, and quantal analysis of the EPSP suggests that the number of quanta per EPSP is increased. Because of intrinsic variability of EPSP amplitude, low EPSP frequency, and polysynaptic and polyneuronal innervation, quantal analysis in this system is not as straightforward as at the neuromuscular junction, for example; therefore, to assert that a reduction in the number of quanta is the only change occurring in this *in vitro* preparation during synaptic facilitation is difficult. Technical difficulties preclude recording from the presynaptic terminals of the sensory neuron, so the electrophysiological analysis of the sensory neuron and the changes induced by connective stimulation has focused on the properties of the neuronal soma, with the assumption that changes at the soma will reflect presynaptic terminal behavior faithfully.

Intracellular recording from the soma of the sensory neuron showed that facilitatory stimulation broadened the action potential (the opposite to the action of the FMRF-amide input to these neurons) by 10–30% under physiological conditions. A probable consequence of spike broadening is increased influx of Ca^{2+}, known to be essential for transmitter release. Under voltage-clamp, short-term facilitation (10–15 min) was found to reduce the steady state K^+ conductance, and also to reduce a transient K^+ current evoked by depolarization. The consequence of reducing membrane conductance and voltage-dependent K^+ current is to render the neuron more excitable and to slow repolarization following the action potential, which would increase Ca^{2+} influx through voltage-gated Ca^{2+} channels. Exogenous 5HT mimics the effect of connective stimulation, as do agents that block cAMP degradation. A role for elevated intracellular cAMP levels is also supposed. Recently, serotonergic neurons with cell bodies in the cerebral ganglia, which project to the abdominal ganglion, have been identified; stimulation of these neurons produces facilitation of the siphon sensory neurons.

Single channel recordings from sensory neurons show that 5HT can suppress the activity of the S channel, and that cAMP also suppresses this channel, as does the catalytic subunit of the cAMP-dependent protein kinase (protein kinase A [PKA]). However, this is not the only effect of 5HT in these sensory cells. Exogenous 5HT applied to voltage clamped sensory neurons elevates intracellular Ca^{2+} by a mechanism that is independent of the 5HT effects on the S channel. Despite the attractive simplicity of 5HT-linked facilitation, other studies support a much more complex cellular basis for the sensitization process.

For example, other sets of interneurons are capable of producing the presynaptic facilitation phenomenon. The L28 and L29 group of neurons also have cell bodies located in the abdominal ganglion. The L29 neurons are activated by tail shock and depressed by head stimulation. The L28 group of neurons are excited by head or tail stimulation. 5HT is not present in these neurons, but stimulation of these neurons in the absence of the cerebral ganglia produces presynaptic facilitation of the sensory neurons. The transmitter at these facilitatory synapses has not been identified to date.

Another group of facilitatory neurons, that synapse upon the sensory neurons use small cardioactive peptide transmitters, SCP_A and SCP_B; the cell bodies are not localized yet. Therefore, it is probably premature to consider 5HT as the major facilitator in this system.

6. Gamma-Aminobutyric Acid

Gamma-aminobutyric acid (GABA) has recently been implicated in learning in *Hermissenda crassicornis*. As discussed above, contiguous stimulation of the Type B photoreceptor by light and the statocyst by agitation cause reduction of photoreceptor K^+ conductance and of phototaxis. Stimulation of the hair cells causes inhibition of action potentials in the Type B cell, and the inhibitory post synaptic potential recorded in the B cell is blocked by bicuculline, a specific $GABA_A$-receptor antagonist; the $GABA_A$-receptor is commonly linked to chloride ion channels. The hair cells contain GABA, and exogenous GABA also inhibits action potentials in the B cell; however, the situation is more complex. A slowly developing increase in input resistance follows the application of GABA; when a chloride channel blocker is present, the GABA induced increase in input resistance persists, and a slow depolarization is revealed. The depolarizing response can be blocked by H7, an antagonist of the intracellular enzyme, PKC. An as yet untested hypothesis is that the joint effect of light and GABAergic hair cell activity is to increase intracellular calcium and increase the inhibition of K^+ channels by PKC, and that the contiguity of the two stimuli is required to achieve the necessary threshold for a persistent increase in excitability of the B cell. [*See* NEUROTRANSMITTER AND NEUROPEPTIDE RECEPTORS IN THE BRAIN.]

V. Intracellular Messengers and Enzyme Systems

A. Calcium

The role of calcium as a message critical for synaptic transmission is implicit in the discussion of the involvement of Ca^{2+} channels above. However, Ca^{2+} has a plethora of intracellular effects, and the regulation of intracellular Ca^{2+} concentration ($[Ca^{2+}]$) is an obvious target for mechanisms that affect cellular modifiability. Agents that increase $[Ca^{2+}]$ and are implicated in neuronal modifiability include excitatory amino acids on hippocampal neurons, light and depolarization of *Hermissenda* photoreceptors and sensitization of *Aplysia* sensory neurons. Habituation of *Aplysia* neurons is proposed to reduce intracellular calcium. [*See* CALCIUM, BIOCHEMISTRY.]

Calcium is present in the extracellular medium at concentrations between 1 m*M* (mammals) and 10 m*M* (marine gastropods). However, the internal free Ca^{2+} is regulated to submicromolar concentrations. Because of this concentration difference, a relatively small change in inward transmembrane Ca^{2+} flux serves as a significant signal. Additionally, cells sequester Ca^{2+} into intracellular compartments, notably the endoplasmic reticulum and mitochondria. Here we consider the evidence for changes in $[Ca^{2+}]$ levels in learning. Later on we examine the cellular targets of Ca^{2+}, that effect changes in neuronal function, including I_{K-Ca}, multifunctional Ca-calmodulin kinase II (CaM-kinase II), calcium–phospholipid-dependent kinase (PKC), calpain, and neurotransmitter release.

Increasing the Ca^{2+} concentration without depolarizing the neuron can be achieved by balanced iontophoresis of Ca^{2+} from intracellular microelectrodes. This has been done for *Hermissenda* and produces a persistent reduction of K^+ currents I_A and I_{K-Ca}. This is in contrast to the early effect of raising intracellular $[Ca^{2+}]$ (i.e., increase in $I_{K\text{-}Ca}$,) and suggests that the long-term effects of Ca^{2+} on K^+ channels are mediated by a third messenger.

It was noted above that excitatory amino acids elevate $[Ca^{2+}]$ in hippocampal neurons. The dendritic Ca^{2+} level remains elevated for minutes after repeated applications of glutamate; PKC blockers inhibit the sustained increase in Ca^{2+}, but not the initial transient elevation caused by glutamate. GABA acting at the $GABA_A$ receptor inhibits the rise in $[Ca^{2+}]$; GABA may have other covert effects via the $GABA_B$ receptor. These findings are suggestive in terms of LTP; Paired application of glutamate is more effective than continuous single application. PKC activity sustains the Ca^{2+} level, which reciprocally stimulates the activity of PKC; lastly, GABA apparently is able to reset the Ca^{2+} level to normal, perhaps by closing the Ca^{2+} channels held open by PKC. Ca^{2+} is required for LTP, because removing or increasing extracellular Ca^{2+} can suppress or enhance the development of LTP, and the Ca^{2+} chelator EGTA will block LTP when injected into pyramidal neurons. LTP also increases the uptake of radioactive Ca^{2+} into slices.

In *Aplysia* sensory neurons, as noted earlier, intracellular Ca^{2+} levels are increased in a PKC-dependent fashion by exogenous 5HT, independently of effects on the S channel.

B. Cyclic Nucleotides

Good evidence indicates that cAMP is important in facilitation of *Aplysia* sensory neurons and the GSW reflex. Injection of the catalytic subunit of cAMP-dependent protein kinase (PKA) into sensory neurons simulates facilitation, and blockade of adenylate cyclase or PKA blocks facilitation. A cellular action of cAMP and PKA is to close the S channel, thus increasing excitability. In this model, 5HT and the other transmitters implicated in facilitation are postulated to elevate intracellular cAMP levels, causing PKA activation and the closure of S channels.

Learning-deficient *Drosophila* mutants exhibit deficiencies of cAMP metabolism. The short-term memory capacity of the flies was measured in associative conditioning paradigms using olfactory stimuli and electrical shock. The mutant *rut* has low adenylate cyclase levels, and CaM-kinase can no longer stimulate adenylate cyclase. The mutant *dnc* has reduced or absent cAMP phosphodiesterase activity; this enzyme degrades cAMP, and its absence would increase cAMP levels. The fact that both of these mutations show learning deficiencies, but have opposite absolute cAMP levels, shows that the role of cAMP in learning involves a complex interplay between synthesis, the target molecules, and degradation. However, it is not clear from the *Drosophila* studies whether cAMP metabolism is critical for learning or disrupting cAMP metabolism has general effects that also impinge on learning capacity.

C. Protein Kinases

These enzymes, which phosphorylate their targets, and whose activity is modulated by a range of second messengers, are linked to memory in several systems. In *Hermissenda,* PKC causes the I_A and I_{K-Ca} reduction that is causal for the behavioral change. Injection of purified PKC into Type B photoreceptors elevates I_A and I_{K-Ca}; however, pretreatment of the photoreceptors with the PKC activator, phorbol ester, reveals an opposite effect of PKC injection, in that K^+ currents are reduced. The phorbol esters cause translocation of cytoplasmic PKC to the cell membrane, and it seems that PKC must be in the membrane to cause inhibition of K^+ channels, whereas soluble cytoplasmic PKC is a K^+ channel activator. The translocation step also increases PKC sensitivity to Ca^{2+}.

After nictitating membrane response conditioning, rabbit hippocampal CA1 neurons show reductions in K^+ currents, and phorbol ester treatment mimics the effect of conditioning. Conditioning also causes translocation of PKC to the membrane. Using tritiated phorbol ester as a ligand for PKC, it has been shown that soon after training PKC levels are elevated in the cell body, but after 3 days the label is localized to the apical dendrites of the CA1 neurons, implying compartmentalization as consolidation progresses.

K^+ channels may be only one target for PKC; the growth associated protein GAP-43 (variously known as F1 or B50) is also phosphorylated by PKC. Phorbol esters are known to sustain synaptic potentiation in the hippocampus and PKC is translocated during LTP; therefore it is interesting that phosphorylation of GAP-43 is increased in LTP. GAP-43 is found in presynaptic membranes and growth cones and appears to be important for neurite extension. However, because GAP-43 is not present in dendrites, it may not be directly involved in synaptic changes in dendrites.

Calcium–calmodulin-dependent kinase II (CaM-kinase II) is probably involved in producing LTP, because LTP is blocked by calmidazolium, which is a calmodulin antagonist, and the kinase inhibitor H7 (that also inhibits PKC). One hypothesis is that autophosphorylation of CaM-kinase II is permissive for the proteolytic activity of calpain, a Ca^{2+}-dependent protease, and that calpain proteolyzes cytoskeletal and membrane skeleton proteins such as spectrin, leading to changes in dendritic morphology. Both the number and shape of dendrites may be changed during LTP, but a correlation with calpain action has not been shown.

The cAMP-dependent protein kinase, PKA, has been mentioned above in connection with facilitation in *Aplysia* sensory neurons. Behavioral sensitization is followed by a decrease in the number of regulatory subunits of PKA in sections of the whole abdominal ganglion, which would presumably lead to unregulated protein phosphorylation by the catalytic subunit of PKA. This has not yet been shown to occur, although phosphorylation of the S-channel is a critical element of the molecular model of *Aplysia* sensitization.

In the *Drosophila* learning mutant *tur,* the cAMP-binding affinity of PKA is reduced. As discussed in relation to cAMP, it is difficult to know if this deficiency is specific to mnemonic mechanisms.

Both PKA and CaM-kinase II have been injected into *Hermissenda* Type B photoreceptors and cause reductions of the potassium currents I_A and I_{K-Ca}. However, the profile of the changes induced by PKA does not mimic the changes produced by conditioning, and are probably less specific in this system than those of PKC and CaM-kinase II. Interestingly, these experiments illustrate how essential a clear understanding of the biophysical effects caused by training is in isolating the molecular mechanisms underlying behavior; all these kinases are potent reagents and will almost invariably have nonspecific effects, that must be quantitatively distinguished from memory-specific effects.

D. G-Proteins

Recently, evidence for involvement of guanosine triphosphate (GTP) binding proteins (G-proteins) in memory processes has emerged. In trained *Hermissenda,* the photoreceptors contain increased levels of a phosphorylated 20-kDa protein, known as cp20. This protein binds GTP and also has GTPase activity, features characteristics of G-proteins. Injection of cp20 into the Type B photoreceptors of naive animals produces reductions in I_A and I_{K-Ca}, that mimic the effect of conditioning. It has been suggested that PKC phosphorylates cp20, which then modulates the K^+ channels. [*See* G PROTEINS.]

E. Oncogenes and Protooncogenes

Both learning and carcinogenesis involve long-term modifications of cellular function, and this has stimulated researchers to seek common mechanisms. Phorbol esters for example, stimulate PKC, are tumor promoters, and elevate expression of proto-oncogene proteins, which are involved in cell growth and division. Many proto-oncogenes are now known, and several have been suggested to play a part in memory.

Ras proto-oncogenes are G-proteins with 21-kDa molecular weight. Injection of *ras* into Type B photoreceptors modulates K^+ currents and also inhibits axonal transport; this is comparable to the actions of cp20. Whether *ras* is acting at sites that would normally be targets for cp20, or whether *ras*-like proteins exist in *Hermissenda,* is not known.

The proto-oncogene c-*fos* can be induced by LTP, as well as many other agents, including heat-shock and high-potassium depolarization. The *fos* protein and the related oncogene protein *jun* form a dimer, that binds to the DNA regulatory site AP1 and induces transcription of various proteins; however, a specific protein product consequent upon LTP is not known yet. Another protein elevated after hippocampal LTP is egr-1, which is one of a family of proteins with zinc-finger DNA binding and regulatory activity. The role of changes in transcription and translocation associated with behavior modification and learning is discussed below. [*See* FOS GENE, HUMAN.]

IV. Gene Expression

Modification of gene expression is an attractive model for sustaining the lifetime of memories. Many early experiments showed that long-term memory was especially impaired by systemic administration of protein synthesis inhibitors; however, the physiological insult provided by inhibiting protein synthesis makes it difficult to infer much from these studies.

More precise information has been gained from molluscan nervous systems. In *Hermissenda,* for example, at least 21 species of mRNA are elevated in the eyes of conditioned animals, and conditioning also increases ^{32}P incorporation into mRNA immediately after training and for up to 4 days after the end of training.

Polyribosomes located in dendritic spines thus can translate mRNA at the site of synaptic activity. Selective transport of mRNA in dendrites in culture has also been observed; thus there could be site-specific mRNA translation at synapses. Such selective site-specific mRNA translation has not yet been observed in the context of learning.

Much attention has been given to the nuclear early genes (for example c-*myc* and c-*fos*) that are rapidly activated following stimulation and could go on to regulate transcription of other "late" genes that could be responsible for maintenance of memories. Again, this is a tempting idea, but has not been shown to occur in a behaviorally relevant context. Another potential mechanism could be through protein kinase C, which has been shown to phosphorylate and activate the nuclear enzyme topoisomerase II; this enzyme is implicated in the regulation of DNA and the transcription of specific genes.

Possible changes in gene expression include

novel proteins from previously unexpressed genes, novel proteins formed by alternative splicing of mRNA sourced from distinct DNA exons, and changes in the proportion of isoforms of extant proteins, in addition to simply changing the level of expression of an extant protein. Alternative splicing has been shown to modify potassium channel function in *Drosophila* nervous systems. Isoforms of regulatory proteins, that show different levels of expression in different tissues, include PKC and PKA; however, isoform differences between particular neurons remain to be demonstrated. One area in which changes in protein levels could be expressed markedly is in changes in cellular morphology, and in fact several proteins are linked to morphological changes relevant to behavioral modifiability. The extracellular glycoproteins ependymins are synthesized more after training, in both goldfish and rodents; the ependymins polymerize when the Ca^{2+} concentration falls. It is possible that ependymins would undergo polymerization at active synapses as the flux of Ca^{2+} ions into the cell transiently depleted the local Ca^{2+} concentration. Another protein class linked to modifiability are the S-100 polypeptides, which increase neurite extension in culture. The last example of an extracellular protein implicated in modifiability is the proteoglycan produced by the *per* gene in *Drosophila,* mutations of which impair courtship behavior. Such extracellular proteins could play a role in the changes in dendritic spine shape observed in LTP.

In summary, there is evidence that gene expression changes during learning. Membrane processes do alter the activity of intracellular messengers such as PKC, which have nuclear targets, and changes in the transcription of mRNA and translation of proteins have been seen in learning. Major questions remain; how is the mRNA or its protein product specifically directed to the active region of the neuron? Are common mechanisms available, or do different types of gene expression regulation occur in different neurons within the same animal? And is translation regulated independently of transcription?

V. Integrating Experience and Output

The adaptive advantage of memory is presumably that an organism learns from previous events to respond appropriately and successfully to new environmental challenges. The types of learning most amenable to cellular analysis are relatively simple when compared with human learning, yet may be different only in degree, rather than in kind, at the cellular level.

Generating a model for cellular memory processes assumes that the same events occur in all neurons capable of modification. From the preceding discussions this is patently not the case. However, several general principles do emerge. First, the importance of temporal association of stimuli for achieving a critical threshold of excitation mirrors the behavioral constraint of contiguity.

Second, the spatial dimension is particularly relevant for neuronal function. PKC activity in the plasma membrane is different from PKC activity in the cytosol; however, which synapse out of thousands of synapses acquires translocated PKC is critical for the specificity of the function of that neuron.

Finally, many varieties of incoming signals are transduced to evolutionary ancient intracellular signals, that are shared by neurons from different phyla; this observation argues for common targets for the intracellular messengers and supports the use of the experimental models to explore the basis of our own biology. The second messengers of interest include Ca^{2+}, cAMP, phospholipids, and perhaps arachidonic acid. Additionally, third messengers are prominent in molecular mnemonics, the protein kinases, and the proto-oncogene proteins in particular and may be especially important in regulating gene expression and thus the memories that persist for our lifetimes.

Bibliography

Alkon, D. L. (1987). "Memory Traces in the Brain." Cambridge University Press, Cambridge.

Alkon, D. L. (1989). Memory storage and neural systems. Sci. *Am.* **260(7),** 42–50.

Bank, B., LoTurco, J. J., and Alkon, D. L. (1989). Learning induced activation of protein kinase C. *Mol. Neurobiol.* **3,** 55–70.

Dudai, Y. (1989). "The Neurobiology of Memory." Oxford University Press, Oxford.

Farley, J., and Alkon, D. L. (1985). Cellular mechanisms of learning, memory, and information storage. *Ann. Rev. Psychol.* **36,** 419–494.

Glickstein, M., Yeo, C., and Stein, J. (1987). "Cerebellum and Neuronal Plasticity." Plenum, New York.

Gormezano, I., Prokasy, W. F., and Thompson, R. F.,

(eds.) (1987). "Classical Conditioning." Lawrence Erlbaum, Hillsdale.

Ito, M. (1989). Long-term depression. *Ann. Rev. Neurosci.* **12,** 85–102.

Kandel, E. R. (1976). "Cellular Basis of Behavior: An Introduction to Behavioral Neurobiology." Freeman, San Francisco.

Lynch, G. (1986). "Synapses, Circuits, and the Beginnings of Memory." MIT Press, Cambridge.

Nelson, T. J., and Alkon, D. L. (1989). Specific protein changes during memory acquisition and storage. *BioEssays* **10,** 75–79.

Rudy, B. (1988). Diversity and ubiquity of K channels. *Neuroscience* **25,** 729–749.

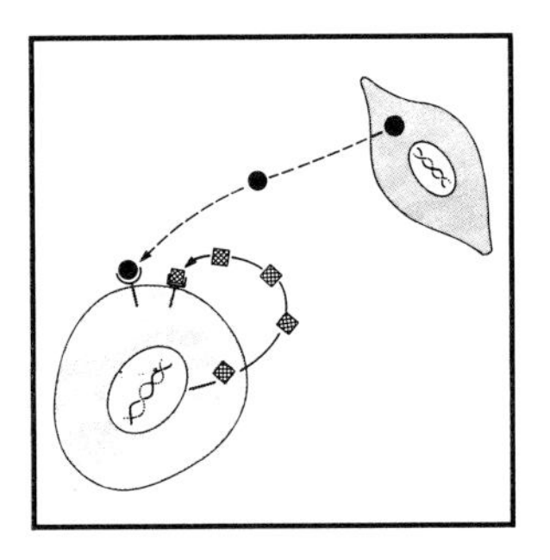

Cellular Signaling

ALEXANDER LEVITZKI, *Institute of Life Sciences, The Hebrew University of Jerusalem*

Glossary

Acetylcholine Neurotransmitter of the nervous system

Agonist Compound that activates a receptor; this compound can be a hormone, a neurotransmitter, or a synthetic drug

Autocrine Form of self-regulation by a cell, which secretes a hormone or a growth factor that affects the same cell through a receptor

Cerevisiae Species of yeast; baker's yeast—*Saccharomyces cerevisiae*

Chemotaxis Movement of a cell or a microorganism up a gradient of a chemical attractant or down the gradient of a repellent

Growth factor Molecule that stimulates cell growth through a specific receptor

Kinase Enzyme that utilizes adenosine triphosphate or another nucleoside triphosphate to phosphorylate other molecules including proteins

Ligand Molecule that binds to a binding site on a receptor or on an enzyme

Muscarinic Class of acetylcholine receptors to which molecules such as muscarine bind specifically

Paracrine Form of regulation of a cell by a neighboring cell that secretes a hormone or a growth factor

Saccharomyces *Sacchoramyces cerevisia;* see *cerevisiae*

Second messenger Molecule generated by a biochemical effecter system coupled to a receptor; when the receptor binds the agonist, it activates a biochemical system that generates a second messenger, which transmits the message by activating other biochemical systems

Serotonin 5-Hydroxytryptamine, a neurotransmitter

Steroids Class of hormones whose structure is derived from cholesterol

Triiodothyronine Hormone also known as T_3, secreted by the thyroid gland

Tyrosine *p*-hydroxyphenylalanine, an aromatic amino acid

SINGLE CELLS AS WELL AS whole tissues respond to environmental signals such as nutrients, hormones, growth factors, chemoattractants, and toxic materials. In almost all cases, the first event involves interaction of the signaling molecule with a cell-surface receptor. This interaction is followed by a transmembrane signaling event triggered by the activated receptor. A biochemical signal is thus transmitted through the cell membrane into the cytoplasm awakening the cell to produce a characteristic biochemical response. Many signals, such as those produced by growth factors, antigens, or steroids, are further propagated to the nucleus. Signal transduction to the nucleus induces expression of specific sets of genes inducing the cell to divide or differentiate. More is known about transmembrane signaling events from cell-surface receptors into the cytoplasm than about signaling from the cytoplasm into the nuclear DNA. The molecular mechanisms that underlie cellular signaling have become an important focus in biological research.

I. Cell–Cell Communication

At the turn of the twentieth century, surgical removal of an organ (i.e., pancreas) from an animal resulted in a severe physiological deficiency. This deficiency could be remedied by implanting the appropriate glandular tissue of the pancreas or by injecting pancreatic extract into the animal from which the pancreas was removed. The extracts were then purified by biochemical procedures to yield pure hormone. These findings led to the understanding that separate (Grk. *krinen*) organs within (Grk. endos) the body control the action of other organs by secreting an active chemical substance into the blood vessels. This, in turn, led to the definition of a new branch of science known as endocrinology, the scientific discipline that explores the mode of action of hormones. The term hormone (from Grk., meaning arousing substance) was first introduced to describe the action of a specific chemical substance extracted from intestinal cells, which stimulates the pancreas to secrete its enzymatic content; this specific hormone was named secretin. Secretin is carried by the blood from the intestinal cells to the pancreas, where it stimulates the latter to secrete its enzymes into the intestine. Hormone refers, in general, to a chemical substance secreted by a gland remote from the organ that it affects. The hormone travels through the bloodstream from the secreting organ to the target tissue (Fig. 1). [*See* ENDOCRINE SYSTEM.]

The discovery of acetylcholine as the chemical transmitter of nerve stimuli was a historical landmark and accelerated the acceptance of the notion that cellular signaling is mediated by chemical transmission. Until the discovery of acetylcholine, many believed that nerve stimulation was a purely electrical phenomenon with no involvement of transmitter molecules. Once both hormones and neurotransmitters were recognized, the stage was set for the discovery of many other chemical signals. Indeed, growth factors and local mediators (which are very similar to hormones but can also travel through the tissue fluids, for short distances) were identified. All four classes of chemical signals—neurotransmitters, hormones, growth factors, and local mediators—coordinate the activities of all organs in a multicellular organism. [*See* NEUROENDOCRINOLOGY; TISSUE REPAIR AND GROWTH FACTOR.]

Some signals are mediated through contact signaling: One cell touches the next cell so that a membrane-bound signaling molecule interacts directly with a receptor on the neighboring cell. Direct cell–cell signaling can also be mediated through a gap junction. This molecular device is like a tube within the cell membrane, through which signaling molecules can flow from one cell to another. [*See* CELL JUNCTIONS.]

Not only mammalian cells respond to chemical signaling; bacteria, fungi, yeast, and plant cells also communicate with each other through chemical signals. The signaling cell itself can be a target for another signaling cell (Fig. 2). Thus, a complex network of signals allows the cell, the tissue, and the

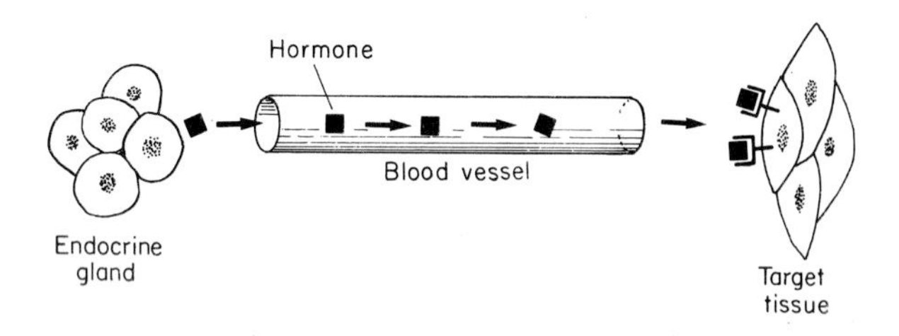

FIGURE 1 Hormones and their action. Hormones are synthesized by glands remote from their target organs and reach their receptors through the bloodstream. The hormone affects the target tissue by interacting with specific receptors. Binding of the hormone to its specific receptor triggers the cell to respond.

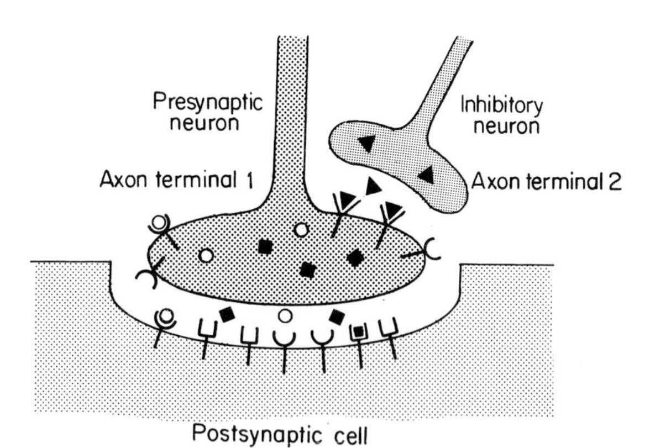

FIGURE 2 The action of neurotransmitters. Axon terminal 1 can release two different neurotransmitters, schematically represented by filled squares (■) and empty circles (○). These diffuse across the synaptic cleft to bind to their respective receptors at the postsynaptic cell. The neurotransmitter (○), for example, can bind to specific receptors that reside on the nerve terminal itself and, therefore, are termed autoreceptors. The nerve terminal can be innervated by yet another nerve (e.g., an inhibitory neuron). The axon terminal of the inhibitory neuron, axon terminal 2, releases an inhibitory transmitter (▲) that binds to a specific receptor on axon terminal 1. The binding of the inhibitory neurotransmitter attenuates the action of the affected neuron by inhibiting the release of its neurotransmitters.

whole organism to function efficiently in its environment. The signaling molecules are named according to the functions they convey, as detailed below.

II. Signaling Molecules

All cellular signals are transmitted by chemically defined molecules, which interact with their target cells through specific receptor molecules. In this section, we discuss in some detail the three principal classes of signaling molecules: neurotransmitters, hormones, and growth factors, with a brief discussion of other chemical signals, such as chemotactic molecules and pheromones.

A. Neurotransmitters

Neurotransmitters are molecules produced by nerve cells and released at the nerve terminal by specific mechanisms, whose molecular details are far from being fully understood. The released neurotransmitters interact with specific receptors on the innervated target tissue (Fig. 2) and also with receptors located at the nerve terminal itself (Fig. 2). The neurotransmitters travel by diffusion no more than 500–1,000 A to their target and can thus signal the target cells within milliseconds. Indeed, the nervous system is known for its ability to transmit signals rapidly and precisely.

The role of the receptors at the nerve terminal is to regulate and modulate release of neurotransmitters; therefore, they are referred to as autoreceptors (or presynaptic receptors). One nerve terminal can synthesize and release two chemically different neurotransmitters, as represented in Fig. 2. Neurotransmitters are either small-molecular weight (MW 100–300) molecules, such as noradrenaline (norepinephrine), serotonine, and glycine, or medium-size neuropeptides, such as enkephaline (MW 550), substance P (MW 1,200), and β-endorphin (MW 3,500). Neurotransmitters travel by diffusion over a very short distance across the synaptic cleft to interact with their receptor on a nearby cell (Fig. 2). The nerve terminal is often innervated by yet another nerve cell. In this case (Fig. 2), the cell possesses another set of receptors, which respond to a different neurotransmitter (e.g., an inhibitory neuron). Thus, when nerve cell 1 receives a signal from nerve cell 2 through the inhibitory neurotransmitter, the output of its own signaling neurotransmitters is attenuated. Therefore, the postsynaptic response will be lower when cell 2 is active than when it is quiescent. Clearly, both nerve cell 1 and nerve cell 2 can be innervated by still other nerve cells, generating the network of signaling. Indeed, understanding the signaling network in the brain is essential for understanding how different behavioral patterns emerge. Elucidation of the mechanism of action of each signal is insufficient; one needs to investigate the network of cellular signals to understand the complete signaling pattern of a whole brain area. To some extent, this observation is also true for individual cells, because each cell possesses an array of receptors that respond to different chemical stimuli. Integration of the various responses to the different stimuli, acting through specific receptors on the cell surface, results in a characteristic pattern of response. [*See* NEUROTRANSMITTER AND NEUROPEPTIDE RECEPTORS IN THE BRAIN; RECEPTORS, BIOCHEMISTRY.]

B. Hormones

Hormones are molecules that are released from an endocrine gland into the bloodstream and elicit a response at any cell that expresses a specific receptor to the hormone (Fig. 1). Tissues, because they are very different from each other, can respond simultaneously to the same hormone. In contrast to neurotransmitters, hormones act at a distance and reach their target cells through the bloodstream. This process is much slower than signaling by neurotransmitters released at nerve terminals. Indeed, the response time to hormones is minutes, or even hours, subsequent to their release into the bloodstream. The chemical variety of hormones is much larger compared with that of neurotransmitters. Hormones range from small hydrophilic molecules, such as (−)noradrenaline (norepinephrine), to very hydrophobic steroid hormones to medium-size polypeptide hormones, such as insulin (MW 6,000) and the multisubunit human chorionic gonadotrophin (MW of subunit 39,000). Many cells possess more than one receptor for different hormones; therefore, the net response of the cell depends on the spectrum of hormones interacting with the cell at any particular time. Unicellular yeasts such as *Saccaharomyces cerevisiae* also respond to specific peptide pheromones, which induce mating between two mating (sex) types of the yeast. The response through receptors is a regulatory element in unicellular organisms as well as in cells of multicellular

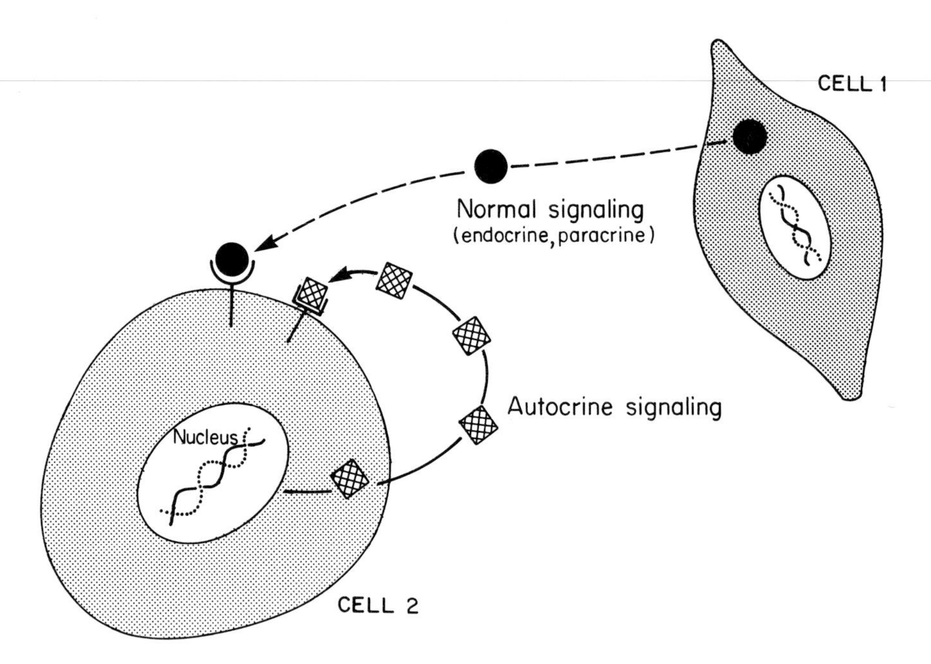

FIGURE 3 Autocrine signaling is compared with endocrine and paracrine signalings, which are more widespread.

organisms. The molecular mechanisms that underlie signal transduction in unicellular organisms are similar to those found in multicellular organisms. [*See* POLYPEPTIDE HORMONES; STEROIDS.]

C. Growth Factors

Growth factors, such as epidermal growth factor (EGF), nerve growth factors (NGF), and platelet-derived growth factor (PDGF), are peptides of molecular weight 6,000–30,000. Different cell types require different growth factors for growth, especially at early stages of development. For their growth or differentiation, different cells are now thought to require unique combinations of growth factors, each of which interacts with a specific receptor on the cell surface. Growth factors are often secreted in a tissue close to the target tissue but may also travel like a hormone via the bloodstream to a distant organ. In contrast to hormones and neurotransmitters, growth factors induce cells to proliferate (divide) and also induce characteristic biochemical transformations within the cell (differentiation). Some cells respond to the growth factor that they secrete. This type of regulation is known as autocrine signaling (Fig. 3). There are also factors that inhibit cell growth and, therefore, counteract the action of growth factors. These are frequently called antigrowth factors, or growth inhibitors, which probably play a role in growth arrest. [*See* NERVE GROWTH FACTOR.]

D. Other Chemical Signals

Although neurotransmitters, hormones, and growth factors are the major signaling molecules in multicellular organisms, other types of signaling molecules exist for unicellular organisms. For example, simple nutrients such as glucose can be signal molecules in addition to being major metabolites. When the unicellular yeast *S. cerevisiae* is exposed to glucose, it is induced to grow and divide. Thus, aside from being a substrate for energy-producing reactions, glucose signals the yeast to "wake up" by stimulating proliferative biochemical pathways. Also, bacteria respond to chemical signals such as nutrients and repellents. For example, the amino acid aspartic acid interacts with specific receptors on the surface of the bacterial membrane and induce the bacteria to swim toward the higher concentrations of the nutrient. Thus, nutrients can act as chemotactic signals for the bacteria.

E. Signaling in Plants

Knowledge of the molecular basis of cellular signaling in plants is still in its infancy, but the emerging pattern is similar to that found in the animal kingdom. Plant hormones induce cellular growth and differentiation of specialized tissues. The molecular

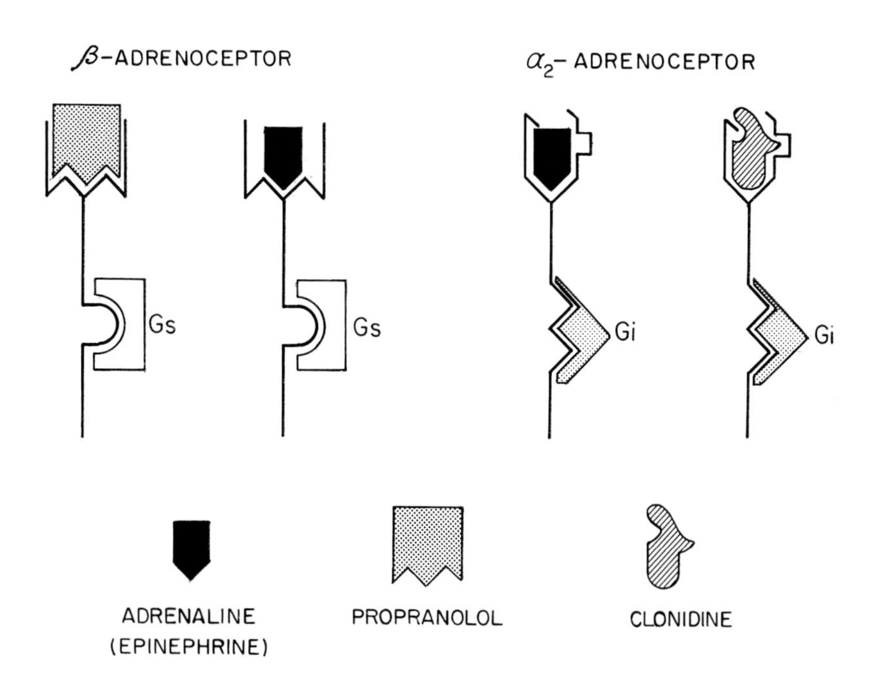

FIGURE 4 Selectivity in the design of receptors. Both the β-adrenoceptor and the α_2-adrenoceptor bind adrenaline. Propanolol, however, is a selective drug for β-adrenoceptor, whereas clonidine is selective to the α_2-adrenoceptor. The schematic representation shows that it is possible to design selective drugs that exploit the difference in the binding domain of the two receptors. The common structural denominator of the β-adrenoceptor and the α_2-adrenoceptor allows adrenaline (and noradrenaline) to bind to both. G_s, stimulatory GTP binding protein which mediates the activation of adenylyl cyclase by stimulatory receptors like the β-adrenergic receptors; G, inhibitory GTP binding protein, which mediates the inhibition of adenylyl cyclase by inhibitory receptors like the α_2-adrenergic receptor.

mechanisms of plant cellular signaling are currently under intensive investigation, although our knowledge of this mode of action is currently at a very rudimentary level.

III. Receptors—The Selectors of Signals

A signal can trigger a cell to respond only if the cell possesses a receptor that can selectively bind the signaling molecule. For each chemical signaling molecule there is at least one type of protein receptor molecule. With the improvement of biochemical techniques and the advent of gene cloning, it is now possible to purify and crystallize receptor proteins, clone and express receptor genes in various cell systems, and explore in detail their properties and mechanisms of action. Receptors are no longer defined only by pharmacological assays or by their ability to bind radioactively labelled drugs. At present, some of the receptors can be expressed even in large amounts and eventually be crystallized, such that their structure can be analyzed using X-ray methods. For most signaling molecules one finds a family of receptors, where each member is expressed in different cell types. These different species of receptors, which respond to the same molecule, can be either very different from each other or closely related. Thus, two types of cells can respond to the same signaling molecule, but the signaling molecule triggers different responses in the two different cell types, which express the two different types of receptors. A well-studied example (Fig. 4) is the β-adrenoceptor and the α_2-adrenoceptor, both of which bind and respond to noradrenaline (norepinephrine) and adrenaline (epinephrine). The β-adrenoceptor *activates* adenylyl cyclase, inducing elevation of intracellular cyclic adenosine monophosphate (cAMP), whereas α_2-adrenoceptor *inhibits* adenylyl cyclase, causing a decrease of cAMP production. The two receptors are closely related in their gross structural features but differ in the details of the structure.

The β-adrenoceptor activates the adenylyl cyclase system by activating guanosine triphosphate-binding protein (Gs), whereas the α_2-adrenoceptor inhibits adenylyl cyclase through its inhibitory G-protein. The two receptors also differ in their hormone-binding protein domain, although both bind the same adrenaline or noradrenaline molecules.

The structural differences in the binding sites allows the design of selective drugs for each type of receptor. For example, l-(−)-propranolol binds to β-adrenoceptors but not to α_2-adrenoceptors, whereas clonidine and yohimbine bind to α_2-adrenoceptors and not to β-adrenoceptors.

Another example is the existence of two closely related receptors with subtle structural differences: the two types of β-adrenoceptors. β_1- and β_2-adrenoceptors differ in their selective affinity toward adrenaline and noradrenaline, but both activate adenylyl cyclase. It is interesting and of physiological relevance that the heart greatly expresses β_1-adrenoceptors, whereas the lung expresses β_2-adrenoceptors. The difference between the two subtypes of β-adrenoceptors in the heart and lung allows the design of β_1 and β_2 selective drugs, which act selectively on either the heart or lung.

IV. Mechanisms of Transmembrane Signaling

Most receptors reside on the cell membrane. These receptors transmit their signal across the membrane. Some receptors, however, are cytoplasmic or nuclear, and the signaling molecule must penetrate the cell membrane to reach its target receptor. The two classes of receptors function differently (Fig. 5). Membrane receptors are physically coupled to an intracellular biochemical apparatus, which is activated once an agonist binds to the receptor. Cytoplasmic receptors also activate an intracellular biochemical machinery they bind or agonist.

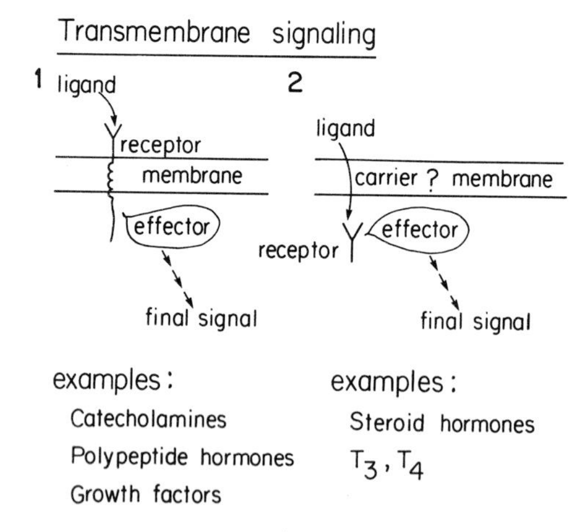

FIGURE 5 Transmembrane signaling, in principle, can occur by two mechanisms. 1. The ligand binds to a membrane-bound receptor, and signal transduction is a transmembrane molecular event. 2. The ligand crosses the membrane to reach an intracellular receptor, which resides in the cytoplasm or the nucleus.

A. Intracellular Receptors

Steroids and triiodothyronine (T_3) are hydrophobic hormones that bind to their intracellular receptors after crossing the cell membrane. Whether these hormones penetrate the cell membrane passively or through a specific membrane protein is still unclear. The intracellular steroid or T_3 receptor changes its conformation upon binding of the hormone and is activated. Following this process, the activated steroid or T_3 receptor binds to specific regulatory DNA regions, leading to the expression of specific genes. Different genes are activated in different cell types because the steroid or T_3-responsive DNA elements are linked to different genes. A more detailed discussion on the mode of action of steroid hormones is given elsewhere.

B. Mechanisms of Signal Transduction by Membrane Receptors

Until the early 1970s, signaling events were defined by characteristic pharmacological responses or biochemical events distal to the initial binding event. For example, the secretion of enzymes from the parotid gland is triggered by acetylcholine, which interacts with a particular class of muscarinic acetylcholine receptors. Twitching of guinea pig ileum induced by serotonin defines a class of serotonin receptors. These signals are distal to the initial binding event. In recent years, more has been learned about the early biochemical events triggered by the binding of signaling molecules and the subsequent events that follow, which bring about the distal, final response.

The biochemical system that is physically activated by the ligand-bound receptor is usually called the effector system. From a biochemical perspective, one can identify a limited number of effector systems. The number of receptors that can be defined pharmacologically, namely according to their ligand specificity, exceeds by far the number of biochemical effectors. This observation immediately suggests that receptors for different ligands may have common structural features, which allow them to interface and activate identical effector systems. Indeed, distinct receptors that activate adenylyl cyclase have similar effector domains but very different ligand-binding domains. Distinct receptors that inhibit adenylyl cyclase also seem to share similar effector domains (Fig. 6). Similarly, different receptors that all activate phospholipase C seem to have similar effector domains. The design principle extends to all families of receptors. Table I summarizes the types of biochemical effector systems coupled to receptors; a few examples of pharmacologically very different receptors that couple to the same effector system are listed. Table II summarizes the classes of chemicals that activate different effector systems. The table demonstrates that the same effector systems can be activated by different ligands.

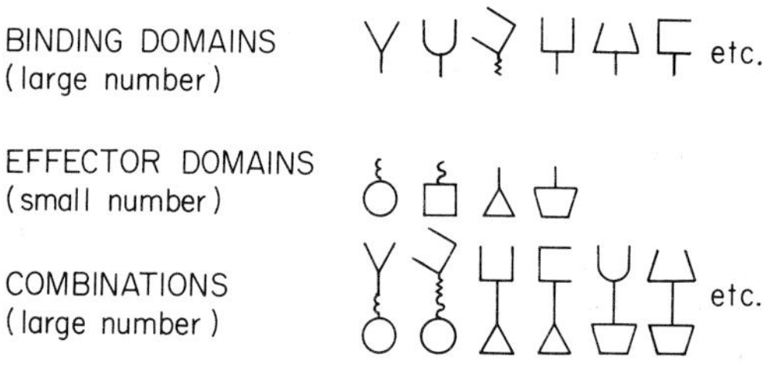

BINDING DOMAINS (large number) etc.
EFFECTOR DOMAINS (small number)
COMBINATIONS (large number) etc.

FIGURE 6 A limited number of effectors are coupled to a large number of receptors. Many receptors that differ markedly in their ligand-binding properties are functionally linked to similar or identical effectors. Therefore, a limited biochemical repertoire may be used to generate the diversity of response found. Different combinations can be expressed in a variety of cells and tissues, thus generating the response characteristic to that cell or tissue.

C. Basic Features of Cellular Signaling

For all types of cellular signaling, there are a number of common underlying principles.

1. The signaling molecule binds to a specific receptor, which is always a protein molecule.

2. Binding of the signaling molecule to the receptor triggers a structural change at the receptor, which in turn triggers a cascade of biochemical reactions, culminating in the final cellular response. In many instances, the first biochemical reactions lead to the formation of a small molecule such as cAMP, cyclic guanosine monophosphate or diacyl glycerol, or IP3. The ligand that triggers receptor activation is commonly known as the first messenger, whereas the small molecules produced or released inside the cell that transmit the message further are known as second messengers.

3. The activated receptor triggers a response, which eventually triggers a mechanism that terminates the primary response. This biphasic nature of the response lends its transient nature (Fig. 7), which is the essence of a signaling event.

TABLE I Types of Chemical Signalings

No.	Primary signaling event	Examples of final response	Types of receptors involved	Mechanisms of signal termination
1	Activation of adenylyl cyclase (elevation of cAMP)	enzyme secretion, steroid biosynthesis, glycogen breakdown, activation of ion channels, etc.	β-adrenoceptors, glucagon, ACTH, prostaglandins, serotonin, adenosine (A2), etc.	inactivation of the receptors by phosphorylation and/ or removal from the cell membrane
2	Inhibition of adenylyl cyclase (inhibition of cAMP production)	attenuation of effects triggered by cAMP	α_2-adrenoceptors, adenosine (A1), somatostatin	as in No. 1
3	Activation of phospholipase C (elevation of inositol-trisphosphate and diacylphycerol)	neurotransmitter release, platelet aggregation	muscarinic acetylcholine receptors, α_1-adrenoceptors, glucagon	probably similar to No. 1
4	Activation of Cl^- channels	inhibitory nervous signals, sedation	glycine, GABA	unknown
5	Opening of Na^+/K^+ channel	prepheral muscle contraction	nicotinic (acetylcholine)	conformational change
6	Opening of Na^+ channels	excitation	glutamate	unknown
7	Opening of Ca^2 channels	inotropic effect in the heart	adrenaline	as in No. 1
8	Activation of protein tyrosine kinases	proliferation of cells, glucose uptake into cells and its utilization	EGF, insulin	internalization of the ligand-bound receptor
9	Activation of protein tyrosine phosphatases	growth arrest(?), entry into mitosis, T-cell response	T-cell receptor Protein tyrosine phosphatase receptors	unknown

ACTH, adrenocorticotropin; GABA, gamma-aminobutyric acid.

TABLE II Types of Signaling Molecules and Their Signals

Types of signaling molecules	Classification	Types of biochemical signal triggered (examples)
Amino acids	neurotransmitters, hormones	opening of channels (glycine, GABA, glutamate)
Amino acid derivatives, catecholamines (adrenaline, noradrenalin, dopamine, octopamine), serotonin, histomine	neurotransmitters, hormones	activation and inhibition of adenylyl cyclase, activation of phospholipase C
Small peptides (MW $\leq$1,000)	neurotransmitters, hormones, growth factors	activation and inhibition of adenylyl cyclase, activation of the pheromone response in *S. cerevisiae,* chemotaxis of neutrophils, activation of phospholipase C
Large peptides and proteins	hormones, growth factors	activation and inhibition of adenylyl cyclase, activation of phospholipase C, activation of protein tyrosine kinases
cAMP	second messenger in high eurcaryotes, primary messenger in *Dictiostelium discoideum*	aggregation of the slime mold *D. discoidum*
Prostaglandins	local (paracrine) mediators	activation of adenylyl cyclase, other unknown effects
Steroids, vitamin D_3	hormones	DNA binding and transcription of certain genes
Retinoic acid, T_3	hormones	DNA binding and transcription of certain genes

D. Amplification, Adaptation, and Sensitivity

Triggering of a response by the ligand-bound receptor results from a structural change within the receptor molecule. The conformational change in the receptor places the receptor to its active form. The active form interacts with the effector system, activating it to produce the primary biochemical signal (Table I). Often, a small number of receptors activate a large number of enzyme molecules as an intermediate step in the signaling event, or as an end result of the primary response, or both. This phenomenon is known as amplification. For example, the β-adrenoceptor system has one receptor that activates a dozen adenylyl cyclase molecules, each of which produces about 100–250 molecules of cAMP before the activity of the enzyme decays. Each cAMP molecule subsequently activates the enzyme cAMP-dependent protein kinase, which, once activated, phosphorylates at least hundreds of protein molecules. Some of these molecules are themselves enzymes, whose activity is either activated or shut off. The overall amplification of the signal in this system is about 10^6. Amplification is found in almost every system in which a receptor on a cell surface transmits a signal through the membrane into the cell. In ion-gated receptors, many ions pass through a single channel opened upon receptor activation, also producing an amplified signal. The very significant amplification produced by receptors allows nature to economize on the number of receptors and enables the cell membrane to accommodate many kinds of receptors, in addition to other proteins.

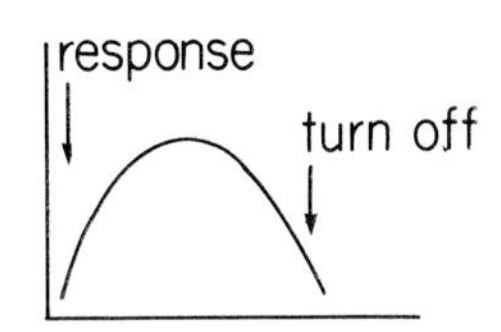

FIGURE 7 The transient nature of signaling. Binding of the ligand to its receptor elicits a response that is turned off after a time delay. The time scale is different for different receptor systems. For some receptors, the response is maximal within milliseconds, and the whole signal is over within well below a second. For other systems, the time scale is in minutes or hours. The signal must have a transient nature.

Sensitivity of a receptor system is not only controlled by amplification but also by the number of receptors on the cell. Thus, cells with a high number of receptors display enhanced sensitivity to the stimulating ligand, whereas cells with a low number of receptors are less sensitive. Another possible mechanism to increase the sensitivity of a receptor system to its stimulating ligand is cooperativity. In this case, the receptor is composed of a number of protein subunits, with two or more ligand-binding sites. The response is triggered only when all the receptor-binding sites are bound with the stimulating ligand; therefore, the response of the receptor system becomes cooperative. For example, the nicotinic acetylcholine receptor is activated only when two molecules of acetylcholine bind to the receptor. This property allows the nicotinic receptor system to respond fully within a narrow range of acetylcholine concentrations and in a cooperative fashion. This is similar to the cooperative binding of oxygen to hemoglobin, which allows hemoglobin to undergo full oxygenation of deoxygenation over a narrow range of oxygen concentration.

Adaptation is another hallmark of many receptor systems. Following a time delay, the activated receptor triggers, following a time delay, a secondary cascade of biochemical events, which lead to the termination of the signal, thus lending it its transient nature (Fig. 7). For example, the depolarization response of the nicotinic receptor is triggered within 1 msec, whereas the desensitization of the response sets in within 20–100 msec. In this case, the acetylcholine-occupied receptor is slowly changing its conformation, so that its Na^+/K^+ channel (Table I) is closed. In the β-adrenoreceptor system activation of adenylyl cyclase occurs within 40 sec, whereas desensitization occurs within 2 min. In this case, the receptor becomes phosphorylated at specific sites, a modification that uncouples the receptor from the adenylyl cyclase systems. Tyrosine kinase receptors (Table I) activate intracellular events within 1 min, following which the ligand-bound receptor becomes internalized and removed from its topographical location, thus preventing it from interacting with target molecules.

V. Aberrant Cellular Signaling

Abnormal cellular signaling can lead to pathological conditions. It can result from a few factors: abnormal secretion of the signaling molecule, its complete absence, mutational alterations in the receptor molecule, a change in the number of receptor molecules, or a combination of the phenomena cited. For example, diabetes can result from the absence of insulin secreting cells, blockade of proinsulin processing to insulin. Severe forms of diabetes may result from a mutated insulin receptor or the presence of antiinsulin receptor antibodies where, in both cases, the receptor cannot respond to the hormone. The consequences of aberrant signaling are varied and result in diseases of various kinds. The diminished secretion of the hormone insulin or its complete absence leads to the development of diabetes. Amplification of the gene coding for EGF receptor is associated with a number of cancers.

Psoriasis represents a good example of aberrant autocrine regulation, namely the phenomenon in which a cell that secretes a growth factor also responds to it. The psoriatic skin cells overproduce the transforming growth factor α (TGFα), which interacts in an autocrine fashion with EGF receptor of the cell secreting it. This feature causes these cells to hyperproliferate and induces the psoriatic state of the skin. [*See* TRANSFORMING GROWTH FACTOR-α.]

Autocrine stimulation stems from partially relaxed, underregulated expression of the genes that code for growth factors and results in cells becoming independent of external growth factors. This phenomenon is very characteristic of malignant cells and is most probably partially responsible for their escape from the regulated growth characteristic of normal cells.

Aberrant signaling can result not only from the constant exposure of the receptor to its activating ligand but also from mutations which result in an altered, constitutively active receptor. In the normal cell, the growth factor receptor is inactive as long as the growth factor is not bound to it. The abnormal receptor is in its active form all the time; therefore, its growth promoting activity is constitutive or "on" all the time.

VI. Cross Talk Between Signaling Pathways

Signaling events are, in essence, unidirectional although divergent signaling may occur (see below). Basically, triggering a receptor leads to a linear cascade of biochemical events, culminating in the final response. It has, however, been found that signaling systems can cross talk. Steroids acting through stimulation of gene expression by the steroid-bound receptor induce the biosynthesis of β-adrenoceptors, thus leading to enhanced response to catecholamines. Growth factor receptors have been demonstrated to activate a certain form of phospholipase C, thus elevating intracellular Ca^{2+}, a typical effector system for other receptors (Table I). Activation of protein kinase C through certain receptor systems can cause phosphorylation of growth factor receptors and attenuate their response to the growth factor.

Because each cell possesses a large number of receptors, they are usually challenged by combinations of chemical signals, which activate interacting biochemical systems. For example, adrenaline activates both β-adrenoceptors and α_1-adrenoceptors on the surface of liver cells. Therefore, both cAMP and Ca^{2+} levels are elevated within the cell, activating glycogen breakdown through phosphorylase kinase, which is activated by both Ca^{2+} and cAMP-dependent protein kinase. For this response, the effect of the two intracellular messengers is synergistic. Actions of two signals may also be antagonistic. For example, if a cell harbors a receptor that activates adenylyl cyclase and also a receptor that inhibits it, and if the two receptors are activated simultaneously, the level and pattern of cAMP production depend on the net actions of the two opposing receptors. The examples above illustrate the complexity of cellular signaling even at the level of a single cell. Clearly then, understanding cellular signaling within a tissue or organ is a formidable task. Investigation of both response pattern of the cell, as well as the molecular details of each signaling event, is essential for understanding how a single cell responds to signaling molecules. Because cells are usually in contact with each other and communicate with each other through gap junctions, the response of the whole tissue to its environment is a complex phenomenon. [*See* CELL.]

VII. Direct Cell–Cell Signaling

In multicellular organisms, cells of the same tissue or of different tissues physically interact with each other. Cells can allow passage of nutrients and metabolites as well as of signaling molecules such as cAMP, through gap junctions. For example, allowing cAMP to pass between cells enables cells not in direct contact with an adenylyl cyclase-activating hormone to respond to it, if the neighboring cell possesses a receptor system coupled positively with adenylyl cyclase. The oligomeric protein device, which constitutes gap junction, allows cooperation between the cells and coordination of their biochemical behavior. Since permeability through gap junctions is regulated, these devices become dynamic coordinating elements, which contribute to the response of the tissue as a whole. Another mechanism of direct cell–cell signaling is via cell-adhesion molecules (CAMs). These are transmembrane proteins expressed on one cell that interact with neighboring cells through specific receptors. Certain CAMs provide the signal to both cells to stop dividing and, therefore, are responsible for the well-known phenomenon of contact inhibition. Indeed, it has been recently suggested that the biochemical activity of some of these CAMs may be a protein tyrosine phosphatase, whose biochemical activity opposes the tyrosine phosphorylation associated with growth factor receptors.

VIII. Modifying Chemical Signaling by Drugs

Understanding cellular signaling allows pharmaceutical chemists and clinicians to intervene and modulate the natural patterns of cellular signaling. Understanding the pathophysiology of signaling systems can lead to the design of improved drugs. For example, recognizing that myasthenia gravis is caused by a deficiency of acetylcholine receptors at the neuromuscular junction has enabled a more sophisticated treatment to relieve and even abolish the clinical symptoms. Recognizing the difference between β_1- (cardiovascular type) and β_2- (lung type) adrenoceptors enables researchers and clinicians to constantly improve drugs for heart and asthma conditions. Similarly, increasing knowledge on Ca^{2+} channels and their mode of action enables researchers to produce improved Ca^{2+} blockers, known for their beneficial effects on heart diseases and other conditions. Even ulcers can be cured by novel drugs rather than the surgeon's scalpel, due to the understanding of the signaling events that lead to the overproduction of acid within the stomach. Refined understanding of how dopamine and gamma-aminobutyric acid act in the brain and of their involvement in depression and moods has caused the development of novel tranquilizers. Active investigation is also being carried out on growth factor receptors and their modes of action are currently being deciphered. With the unraveling of their biochemical pathways, new ways to combat proliferative diseases such as atherosclerosis, psoriasis, and cancers may soon be discovered. [*see* MYASTHENIA GRAVIS.]

Bibliography

Berridge, M. J. (1985). The molecular basis of communication within the cell. *Sci. Am.* **253**(4), 142–152.

Evans, R. M. (1988). The steroid and thyroid hormone receptor superfamily. *Science* **240,** 889–895.

Levitzki, A. (1984). "Receptors: A Quantitative Approach." Benjamin Cummings. 140 pp. Menlo Park, California.

Levitzki, A. (1987). Regulation of hormone-sensitive adenylate cyclase. *Trends Pharmacol. Sci.* **8,** 299–303.

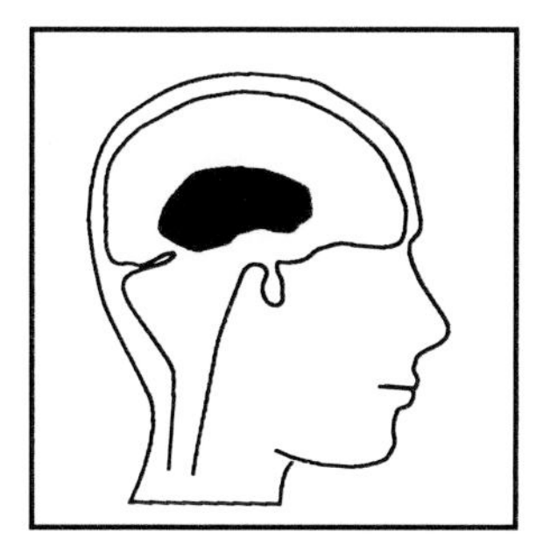

Central Gray Area, Brain

ALVIN J. BEITZ, *University of Minnesota*

Glossary

Afferent Fibers or impulses leading to a predetermined site in the nervous system
Cytoarchitecture Arrangement of nerve cell bodies in the central nervous system
Efferent Fibers or impulses leading away from a predetermined site in the nervous system
Neuromodulator Substance released from neurons that modifies presynaptic or postsynaptic function
Neurotransmitter Substance synthesized in neurons and usually released from an axon terminal, which acts on a receptor site to produce either excitation or inhibition of the target cell
Nociception Appreciation or transmission of a sensation produced by damage to body tissue
Nucleus Well-defined cluster of nerve cell bodies
Receptor Protein entity existing within the cell cytoplasm or in conjunction with the cell membrane that specifically binds neurotransmitters and other mediators of cell signaling to ultimately produce a biological effect

THE CENTRAL GRAY AREA, also known as the periaqueductal gray, is a midline region which encircles the mesencephalic aqueduct of the human midbrain. It comprises several types of neurons (i.e., nerve cells) and neuroglial cells and can be separated into four anatomical subdivisions based on its cytoarchitecture, histochemical staining, and functions.

The ventrolateral subdivision contains the largest neurons and plays a major role in the involvement of the central gray in pain modulation and analgesia. The dorsolateral subdivision contains the greatest number of neurons per unit area and appears to be involved in vocalization and behavioral reactions. The medial subdivision comprises elongated neurons that form a rim around the mesencephalic aqueduct. These neurons are oriented predominantly parallel to the edge of the aqueduct and are purported to participate in aversive behavior (i.e., escape responses are induced by stimulation of this region in animals). The dorsal subdivision is anatomically the most distinct subregion. It has the lowest number of neurons per unit area, the smallest neurons, and the largest glia (i.e., nonneuron)–neuron ratio.

The neurons of the human central gray possess a plethora of neurochemicals (i.e., neurotransmitters and neuromodulators) and receptors. The relationship of these components to the numerous functions of the central gray is only beginning to be defined.

I. Anatomical Organization

A. Location and Subdivisions

The human central gray consists of a mass of cells and fibers that surround the mesencephalic aqueduct of Sylvius and traverse the entire longitudinal extent of the midbrain. It is bordered anteriorly by the posterior commissure, a structure that constitutes a fibrous arch at the transition of the midbrain into the diencephalon, and posteriorly it extends to a point immediately anterior to the decussation of the trochlear nerve. As indicated in Fig. 1, the central gray is bounded dorsally and dorsolaterally by the superior and inferior colliculi, ventrally by the oculomotor and trochlear nuclei, and laterally by the mesencephalic trigeminal nucleus and several reticular nuclei.

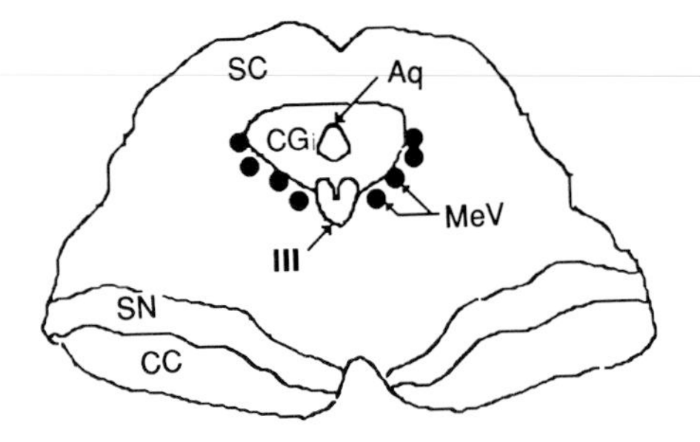

FIGURE 1 Transverse (coronal) section through the caudal midbrain, illustrating the location of the central gray (CG). SC, Superior colliculus; III, oculomotor nucleus; MeV, mesencephalic trigeminal nucleus; CC, crus cerebri; SN, substantia nigra; Aq, mesencephalic aqueduct.

Although the question of anatomical subdivisions within the central gray has been the source of considerable controversy, recent studies indicate that the human central gray is divisible into four distinct regions. The dorsal subdivision, illustrated in Fig. 2, is the most distinct subdivision observed in stained sections. As indicated in Table I, this subdivision contains the smallest neurons (mean neuronal area, 128.17 μm^2), has the lowest neuronal packing density, and exhibits the highest glia–neuron ratio. Thus, in contrast to other subdivisions, the dorsal subdivision is characterized by a marked accumulation of glial nuclei, among which occasional small neurons are found. It should be noted that the glia density is highest in the medial part (toward the midline) of the dorsal subdivision and decreases progressively in the lateral part.

The medial subdivision forms a ring around the mesencephalic aqueduct and, as noted in Table I, contains a larger number of neurons per cubic millimeter than the dorsal subdivision. The hallmark of

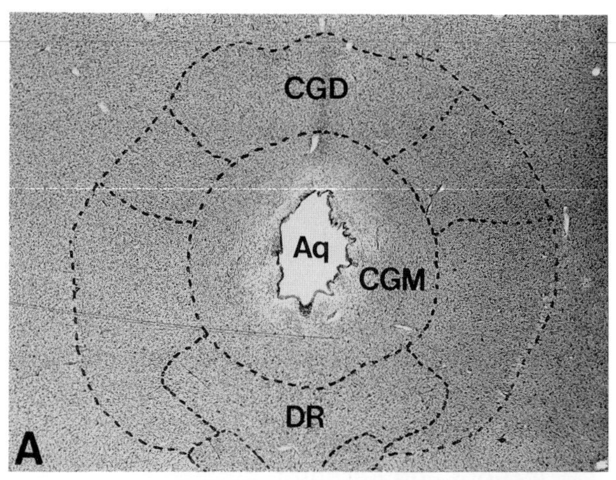

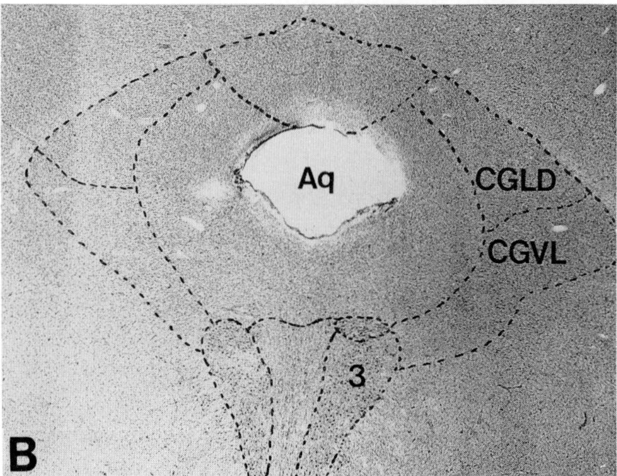

FIGURE 2 Coronal sections through the human midbrain central gray at the levels of the trochlear (A) and oculomotor (B) nuclei. The dorsal (CGD), medial (CGM), dorsolateral (CGLD), and ventrolateral (CGVL) subdivisions are illustrated. DR, Dorsal raphe nucleus; 3, oculomotor nucleus; Aq, mesencephalic aqueduct. [Reproduced with permission from A. J. Beitz, Central gray. *In* "The Human Nervous System" (G. Paxinos, ed.). pp. 307–320. Academic Press, San Diego, California, 1990.]

TABLE I Neuronal Data for the Four Human Central Gray Subdivisions[a]

Central gray subdivision	Mean neuronal packing density (cells/mm³)	Mean neuronal area (μm^2)	Mean neuronal diameter (μm)	Axial ratio[b]	Orientation preference[c]	Glia–neuron ratio
Medial	6801.1 (307.6)	147.41 (1.21)	9.31 (0.05)	1.95 (0.03)	70–130°	6 : 1
Dorsal	4079.8 (234.2)	127.17 (0.96)	9.78 (0.07)	1.76 (0.04)	50–110°	19 : 1
Dorsolateral	8894.1 (393.3)	142.28 (0.51)	9.90 (0.02)	1.71 (0.01)	10–50° and 150–170°	6 : 1
Ventrolateral	7123.5 (181.6)	166.54 (0.87)	10.81 (0.04)	1.73 (0.01)	130–170°	8 : 1

[a] Data are based on the measurement of over 20,000 neurons. [Reproduced with permission from A. J. Beitz, Central gray. *In* "The Human Nervous System" (G. Paxinos, ed.). pp. 307–320. Academic Press, San Diego, California, 1990.]
[b] Axial ratio is derived from the ratio of cell body length to width and provides an indication of shape, since round cells have a ratio of approximately 1, whereas more elongate neurons have a ratio greater than 1.
[c] Orientation is relative to a vertical axis through the center of the mesencephalic aqueduct.

the medial subdivision is the distinctive orientation of its neurons. The majority of the neurons in this subdivision exhibit an orientation preference between 70° and 130°; that is, their long axis is arranged parallel to the edge of the aqueduct. By contrast, the majority of the neurons in the ventrolateral and dorsolateral subdivisions are oriented at obtuse or acute angles to the edge of the aqueduct (Table I). The medial subdivision is also characterized by a prevalence of elongated neurons with a mean axial ratio (i.e., the ratio of length to width of the cell body) of 1.95, indicating the elliptical shape of these nerve cells.

The ventrolateral and dorsolateral subdivisions comprise the lateral regions of the central gray, as illustrated in Fig. 2. These two divisions have the greatest neuronal density and contain the largest neurons of the four subregions (Table I). Although these two subdivisions are difficult to distinguish in routine stained midbrain sections, several histochemical and immunocytochemical staining procedures, as well as certain *in vitro* receptor binding assays, provide additional evidence to support their existence in the central gray.

B. Neuronal Types and Cytoarchitecture

The neurons of the human central gray can be classified into four broad categories: fusiform, stellate, multipolar, and pyramidal. The fusiform cells have an elliptical cell body, with at least one process emerging from each pole. Although they predominate in the medial subdivision, they are present throughout the central gray and exhibit rather distinctive orientations in the four subdivisions. Fusiform cells have a vertical orientation in the dorsal subdivision, are arranged parallel to the edge of the aqueduct in the medial subdivision, and are arranged either vertically or horizontally in the two lateral subdivisions.

Both stellate and multipolar cells are distributed throughout the central gray. Stellate cells are characterized by an oval soma which gives rise to four to six randomly oriented primary dendrites. The multipolar neurons are distinguished by an extensive dendrite arbor, spreading preferentially in the transverse plane.

The fourth neuronal type, the pyramidal cell, is most numerous in the lateral subdivisions of the central gray. It has a pyramidal cell body and is characterized by an extensive dendritic arborization, which often penetrates well into the overlying superior colliculus or into the adjacent cuneiform nucleus.

Although neurons in the central gray can be classified into these four broad neuronal categories, it should be emphasized that the neurons of this region constitute a heterogenous population, and some neurons are not easily classified.

C. Connections with Other Brain Regions

1. Afferent Connections

The central gray receives neural inputs from a large number of central nervous system regions, which allow this midbrain area to be influenced by motor, sensory, autonomic, and limbic system structures. The major inputs to the central gray are summarized in Fig. 3 and are based on the results of studies of the monkey central gray.

The hypothalamus provides the greatest descending input to the primate central gray. Major projections arise from the anterior, posterior, and lateral hypothalamus and probably allow the central gray to play a key integrative role in descending limbic (i.e., those portions of the brain involved in learning, emotions, and behavior) and autonomic (i.e., those portions of the brain and the spinal cord controlling heart rate, respiration rate, and other visceral functions) multisynaptic pathways. [*See* HYPOTHALAMUS.]

In addition to the hypothalamus, the importance of other descending limbic system input to the primate central gray should be emphasized. Anatomical studies have shown direct projections from the central and basolateral amygdala, bed nucleus of the stria terminalis, and cingulate gyrus (a component of the limbic system) to this midbrain region, while electrophysiological studies have demonstrated that over 50% of the neurons in the central gray are influenced by stimulation of the hippocampus and the amygdala. These diverse limbic system inputs to the central gray underscore the involvement of this region in certain aspects of human behavior (see Section III,E). [*See* HIPPOCAMPAL FORMATION.]

The central gray also receives a substantial projection from areas 6, 8, 9, and 10 of the cerebral cortex, as well as from the insular and prefrontal cortices. The prefrontal cortex has a rather selective projection, sending fibers predominantly to the dorsolateral subdivision of the central gray. It is likely that this projection allows neocortical regulation of our perception of noxious stimuli. [*See* CORTEX.]

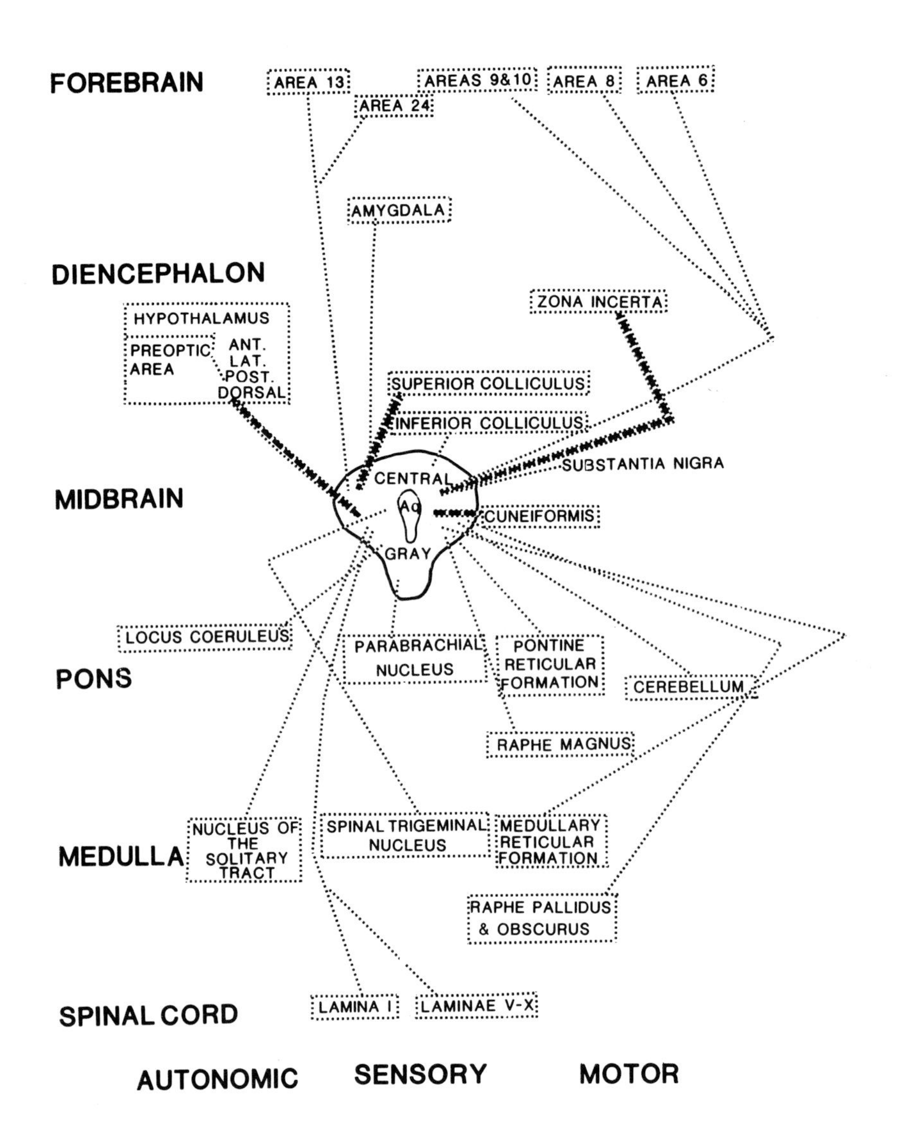

FIGURE 3 Origin of the major afferent projections to the central gray from different levels of the central nervous system. The largest inputs are indicated by heavy dashed lines and include the hypothalamus, zona incerta, and cuneiform nucleus. The various regions are also broadly classified under motor, limbic, autonomic, and sensory systems. Obviously, some regions (e.g., the reticular formation and the superior colliculus) can be grouped under more than one functional system. [Reproduced with permission from A. J. Beitz, Central gray. *In* "The Human Nervous System" (G. Paxinos, ed.). pp. 307–320. Academic Press, San Diego, California, 1990.]

The central gray receives an important descending projection from the zona incerta, as well as inputs from the mesencephalic reticular formation and the deep layers of the superior colliculus. In addition, the primate central gray receives a large number of ascending projections, including afferent connections from the locus coeruleus, parabrachial nuclei, raphe magnus, raphe pallidus, and a variety of brain stem reticular nuclei. Neurons in the spinal trigeminal nucleus and the spinal cord also project to this midbrain region.

It is likely that these ascending projections convey autonomic, nociceptive, and other somatosensory information to the central gray. This is supported by electrophysiological data demonstrating the presence of neurons in the primate central gray which respond to genital, rectal, innocuous somatosensory, or various forms of noxious stimulation.

2. Efferents

A review of the literature indicates that the central gray is reciprocally linked with most brain regions. Thus, the primate central gray projects to the anterior, dorsal, lateral, and posterior hypothalamic nuclei. All of these nuclei have been demonstrated to provide reciprocal projections back to the central gray region, as indicated above. The central gray also sends efferent fibers to the zona incerta; mesencephalic, pontine, and medullary reticular formation; the superior colliculus; the lateral parabrachial nucleus; the raphae magnus and pallidus; the spinal trigeminal nucleus; and the spinal cord. Each of these regions sends reciprocal projections back to the central gray. This two-way communication system between the central gray and numerous other brain structures underlies its unique role in integrating numerous functional systems (see Section III).

II. Neurotransmitters and Neuromodulators

The central gray contains a myriad of neurotransmitters and neuromodulators. Unfortunately, our knowledge of the presence and distribution of these neurochemicals within the central gray far exceeds our understanding of their relationship to the neural circuitry of this region or their functional roles within this midbrain area. The four following sections summarize the existing data concerning the localization of the neurochemicals in the human central gray.

A. Monoamines

The catecholamines dopamine, norepinephrine, and epinephrine and the indoleamine serotonin are neurotransmitters in the central nervous system. Immunocytochemical studies using antibodies against tyrosine hydroxylase or phenylethanolamine *N*-methyltransferase, catecholamine-synthesizing enzymes, have demonstrated catecholamine-containing neurons and fibers in the ventrolateral subdivision of the central gray in both fetal and adult human midbrains. Biochemical studies have shown that epinephrine and norepinephrine are the key neurotransmitters of the catecholamine-containing cell bodies and fibers in this area.

This is further supported by *in vitro* receptor binding data demonstrating the presence of relatively high densities of receptors for these neurotransmitters (called adrenergic receptors) in the human central gray, while both D_1 and D_2 dopamine receptors are present in low densities. Although catecholamines could play a role both in behavioral responses related to central gray stimulation and in the production of analgesia in lower mammals, their role in the human central gray remains to be elucidated. [*See* ADRENERGIC AND RELATED G PROTEIN-COUPLED RECEPTORS; CATECHOLAMINES AND BEHAVIOR.]

Serotonin-containing neuronal cell bodies are confined to the ventrolateral central gray, while serotoninergic fibers are most numerous in the medial subdivision of this midbrain structure. Serotoninergic neurons are round or fusiform, with their longitudinal axis oriented dorsoventrally. Both serotonin-1 and serotonin-2 receptors have been demonstrated in the human central gray, and there is some evidence suggesting that the serotonin system in this midbrain region could play a role in human behavior.

B. Acetylcholine

In addition to catecholamines and indoleamines, acetylcholine probably also plays an important role in the normal function of the human central gray. The human brain contains all of the elements of the cholinergic synapse (which releases the transmitter acetylcholine): the biosynthesizing enzyme choline acetyltransferase, the catabolizing enzyme acetylcholinesterase, uptake and storage mechanisms, and both postsynaptic and presynaptic receptors.

The human central gray contains a high density of one type of acetylcholine receptor, the muscarinic receptor, especially in the dorsolateral and ventrolateral subdivisions. Acetylcholinesterase is also present in the central gray, where it appears to be more prominent in the dorsolateral and ventrolateral subdivisions. This is consistent with the receptor binding data and suggests that acetylcholine might play an important role in these two subdivisions. In fact, injection of cholinergic agonists (i.e., drugs that stimulate acetylcholine action) into the ventrolateral central gray of animals produces a dose-dependent bradycardia (i.e., slowing of the heart beat), implicating this transmitter in autonomic regulation by the central gray.

C. Amino Acid Transmitters

Glutamate and possibly aspartate represent two of the most important excitatory neurotransmitters in the central nervous system, while γ-aminobutyric acid (GABA) and glycine represent important inhibitory transmitters in the brain. Based on their widespread distribution throughout the central nervous system, it is not surprising that these four simple amino acid transmitters are present in the human central gray.

Using immunocytochemical procedures, glutamate-immunoreactive cell bodies are found predominantly in the ventrolateral and dorsal subdivisions. Aspartate-immunoreactive cell bodies are typically fusiform or triangular and are localized to the dorsolateral and ventrolateral subdivisions. The presence of glutamate- and aspartate-immunoreactive neurons in the human central gray could be functionally significant in light of the results of animal studies which demonstrate that the central gray projection to the raphe magnus utilizes excitatory amino acids as transmitters. This projection plays an important role in the descending pain modulation system discussed below.

GABA-immunoreactive neurons are scattered throughout the rostrocaudal extent of the central gray and are concentrated in the ventrolateral and dorsal subdivisions. These neurons are predominantly fusiform and display distinct orientations dependent on their location. Although GABA receptors have not been studied in the human central gray, benzodiazepine binding sites which are related to GABA receptors are found in relatively high amounts in this midbrain region.

Analysis of glycine immunoreactivity in the central gray reveals predominantly fiber and terminal staining. The dorsal and dorsolateral subdivisions contain the highest density of both fibers and glycine receptors.

D. Neuropeptides

The large class of neuropeptides is structurally quite diverse and plays important neurotransmitter or neuromodulator functions in the mammalian central nervous system. Although a majority of these neuropeptides have now been detected in the human brain, only recently have immunohistochemical techniques been used to localize peptides within the human brain stem. Opioid peptides were among the first detected in the human midbrain. Both β-endorphin and Met-enkephalin are present in the central gray, as well as several peptides derived from proenkephalin A. The μ, δ, and κ opiate receptor binding sites have also been identified in this midbrain area, reinforcing the hypothesis that opioid peptides play an important role in the human central gray, especially with regard to the activation of descending pain modulatory systems (see Section III,A).

Substance P- and somatostatin-immunoreactive cells and fibers are found throughout the central gray. Substance P fibers are concentrated in the dorsal half of the central gray, while somatostatin fibers are distributed homogeneously. Cholecystokinin- and vasoactive intestinal polypeptide-immunoreactive fibers have also been described throughout the human central gray. The functional roles of these four peptides in the human central gray remain to be elucidated.

III. Functions and Receptors

A. Pain Modulation and Opiate Receptors

The initial demonstration by Reynolds in 1969, that electrical stimulation of the central gray induces a profound analgesia (i.e., loss of pain sensation), coupled with the discovery of an opioid system which is dependent on the central gray for inducing pain relief, resulted in an explosion of interest in the role played by this region in both nociception and analgesia. It is now well established that an endogenous pain modulation system exists in the central nervous system and that the central gray, together with the nucleus raphe magnus and several other brain stem nuclei, is a key component of this system. The original studies by Reynolds and others demonstrating behavioral antinociception (i.e., inhibition of pain sensation) in animals following central gray stimulation laid the foundation for determining the therapeutic effect of such stimulation in humans. [*See* PAIN.]

Clinical evidence has now established that stimulation of this region in humans is effective in the control of pain syndromes that are responsive to opiates (e.g., morphine). Initial studies indicated that stimulation-produced analgesia was reversed by the opiate antagonist naloxone, which blocks the interaction of opiates with their receptors. It is not clear, however, whether pain relief from electrical stimulation of the human central gray involves an opioid mechanism or whether the response to opiates is predictive of the effectiveness of central gray stimulation in a given patient.

Studies in animals have indicated that the μ opiate receptor is responsible for descending pain inhibition originating in the central gray. Since β-endorphin, an opioid peptide, is present in the human central gray and is one of the most potent endogenous ligands of the μ receptor, it could represent one of the most important opioid peptides for central gray-initiated antinociception. β-Endorphin fibers in the central gray all arise from cell bodies located outside of this region, implying that if opioids are involved in stimulation-produced analgesia, the critical feature of such stimulation within the central gray is the stimulation of β-endorphinergic axons. However, this does not exclude the possibility that central gray stimulation might also operate via activation of nonendorphinergic systems. Consistent with this possibility are studies in human patients indicating that pain relief obtained by stimulation of the dorsal central gray is not reversed by naloxone and is not accompanied by alterations in the endorphin level in ventricular cerebral spinal fluid.

These data imply a possible heterogeneity of the analgesic system existing at the level of the central gray in humans. Although further work is necessary to clarify the role of opioid and nonopioid systems in analgesia induced by electrical stimulation in the central gray, stimulation of this region provides a viable alternative means of long-term pain control.

B. Vocalization and Amino Acid Receptors

Evidence over the past two decades has indicated that excitatory amino acids are the most prevalent excitatory transmitters in the central nervous system. Glutamate- and aspartate-immunoreactive cell bodies, fibers, and terminals, as well as the three major classes of excitatory amino acid receptors (i.e., *N*-methyl-D-aspartate, quisqualate, and kainate), have been localized to the central gray. Based on their extensive distribution in this region, it is not surprising that excitatory amino acids play a role in several proposed functions of the central gray, including its involvement in vocalization.

It has been known since the late 1930s that stimulation of the primate central gray yields species-specific calls in rhesus monkeys. Vocalization, the nonverbal production of sound, can be elicited in many vertebrates by stimulation in several regions of the brain, but most easily in the central gray. Furthermore, recent data indicate that the posteriolateral central gray, but not the anterior part, is involved in vocal motor control. Vocalization can be induced by microinjections of excitatory amino acid agonists into the dorsolateral primate central gray, implicating excitatory amino acid receptors in the vocalization process. Anatomical and physiological data indicate that the central gray projects to the nucleus retroambiguus, which, in turn, projects to vocalization motor neurons, and that this projection system forms the final common pathway in vocalization.

C. Eye Movements

The central gray has also been implicated in eye movements, and the dorsolateral subdivision appears to be especially important in regulating presaccadic activity (i.e., neural activity occurring immediately before the quick jump of the eyes from one fixation point to another). Bilateral lesions of the human dorsolateral central gray have been reported to cause selective paralysis of the downward gaze, while lesions of the dorsal subdivision have been reported to cause upward gaze paralysis. However, because the central gray is located close to several nuclear groups and fiber bundles known to be directly involved in eye movement, such reports must be interpreted with some caution.

D. Autonomic Function and Opiate Receptors

Several autonomic reactions can be elicited by stimulation or lesions of the central gray, including temperature regulation and changes in stomach and bladder tone, respiratory pattern, pressor responses, and pupillary dilation. Recent work demonstrating that morphine microinjected into the ventral central gray produces hyperthermia (i.e., increased body temperature) provides further evidence for a role in temperature regulation. Microinjection of morphine into the central gray has also been reported to inhibit intestinal propulsion. In addition to their role in analgesia, it appears that the opioids could play an important role in the involvement of the central gray in autonomic function. [*See* AUTONOMIC NERVOUS SYSTEM; BRAIN REGULATION OF GASTROINTESTINAL FUNCTION.]

With regard to effects on respiration and blood pressure, stimulation of the central gray causes increases in respiratory rate and blood pressure. Hyperventilation and apneic periods (i.e., without breathing) have also been reported in humans following stimulation of this region. Finally, it should be noted that excitatory amino acids microinjected into the central gray of animals cause changes in both blood pressure and heart rate, while injections of cholinergic agonists produce a dose-dependent bradycardia, as indicated in Section III,B. This suggests that excitatory amino acid receptors and cholinergic receptors might also be involved in autonomic functions associated with the central gray.

E. Role in Behavior

A possible role for the central gray in human behavior can be predicted from the rich interconnections that exist between this midbrain area and limbic structures. Thus, it is not surprising that the central gray has been implicated in sexual behavior, rage and fear reactions, and memory storage. Interestingly, stimulation of the human central gray produces feelings of intense fear. Similarly, stimulation of the dorsal central gray induces aversive behavior in animals, and it has been suggested that aversive stimulation of this region might serve as a model for human anxiety. Indeed, benzodiazepines (which are commonly used to treat human anxiety) have been shown to diminish the aversive behavior produced by stimulation of the central gray in animals. With regard to anxiety, it is possible that the autonomic responses produced by central gray stimulation, as described in the previous section, reflect an "anxiety" component common to both fear and pain.

The central gray transmitters and receptors that mediate the involvement of this region in behavior are only beginning to be elucidated, but it appears likely that a large number of transmitters and their receptors are involved. Microinjection of GABA antagonists into the dorsal central gray produces escapelike behavior in animals, and this behavioral effect is modulated (i.e., modified) by preinjection of morphine into the same central gray site. Similarly, affective defensive behavior produced by electrical stimulation of the dorsal central gray is suppressed by microinjection of the enkephalin analog D-Ala^2-Met^2-enkephalinamide into the site of stimulation.

Excitatory amino acids and serotonin have also been implicated in the modulation of defensive behavior. Thus, GABA benzodiazepine, opiate, excitatory amino acid, and serotonin receptors appear to be involved in the mediation of certain types of behavior by the central gray, indicating the chemical complexity of this particular aspect of central gray function.

Bibliography

Beitz, A. J. (1990). Central gray. *In* "The Human Nervous System" (G. Paxinos, ed.). pp. 307–320. Academic Press, San Diego, California.

Depaulis, A., Penchnick, R. N., and Liebeskind, J. C. (1988). Relationship between analgesia and cardiovascular changes induced by electrical stimulation of the mesencephalic periaqueductal gray matter in the rat. *Brain Res.* **451,** 326.

Fields, H. L., and Besson, J. M. (eds.). (1988). Pain Modulation. *Prog. Brain Res.* **77.**

Jurgens, U., and Richter, K. (1986). Glutamate-induced vocalization in the squirrel monkey. *Brain Res.* **373,** 349.

Mayer, D. J., and Frenk, H. (1988). The role of neuropeptides in pain. *In* "Neuropeptides in Psychiatric and Neurological Disorders" (C. B. Nemeroff, ed.). pp. 199–280. Johns Hopkins Univ. Press, Baltimore, Maryland.

Reynolds, D. V. (1969). Surgery in the rat during electrical analgesia induced by focal brain stimulation. *Science* **164,** 444.

Young, R. F., and Chambi, V. I. (1987). Pain relief by electrical stimulation of the periaqueductal and periventricular gray matter. *J. Neurosurg.* **66,** 364.

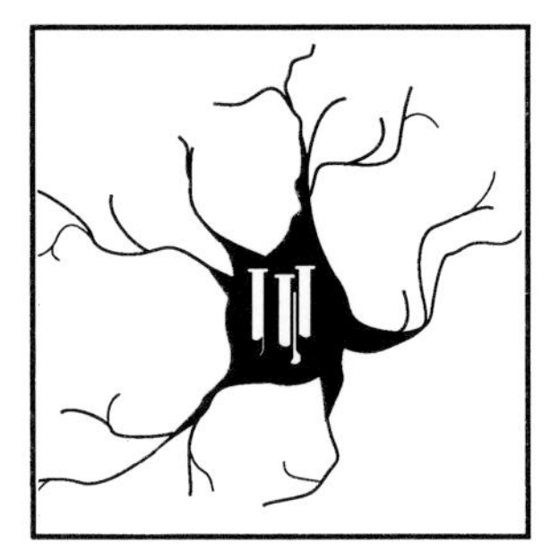

Central Nervous System Toxicology

STATA NORTON, *University of Kansas Medical Center*

Glossary

Astrocyte The most common nonneuronal cell of the brain, often called a "nurse cell" for its role in sustaining the metabolic integrity of the neuron

Axon Extension of the neuron through which impulses, in the form of sodium and potassium ion shifts, are transmitted from one neuron to another

Dendrites Processes of the neuron which receive information from other neurons

DNA Deoxyribonucleic acid; carries the genetic information of the cell

Encephalopathy Clinical condition in which the functions of the brain are disorganized; signs vary from mild (drowsiness or confusion) to severe (coma or seizures)

Endoplasmic reticulum Cell organelle involved in the synthesis of some cellular proteins

Mitochondria Cell organelles that contain part of the synthetic mechanism for energy production in the cell using oxygen

Mitosis Process by which a cell doubles and through which cell numbers increase during growth of the organism

Myelin Lipid component of the cellular sheath surrounding some axons; produced by oligodendrocytes and insulates axons from each other

Neuron Cell in the nervous system specialized for receiving, storing, and transmitting information

Neuropathy Clinical condition in which the function of the nervous system is abnormal; sensory neuropathies involve increased or decreased sensations of vibration, temperature, pressure, or pain; motor neuropathies include rigidities, spasms, and weakness

Neuropil Process of cells, including neurons, astrocytes, and oligodendrocytes, that fills the tissue of the brain between the cell bodies

Oligodendrocyte Nonneuronal cell in the brain which produces processes that surround the axons of many neurons

Synapse The junction of an axon with another neuron. Specialized chemicals (neurotransmitters) are stored in vesicles on the presynaptic (transmitting) side and are liberated into the synaptic cleft by a nerve impulse. The postsynaptic (receiving) side contains receptors that combine with the transmitter, causing a response in the postsynaptic cell.

THROUGHOUT LIFE, the human body is exposed to toxic substances from various sources. A toxic substance is a chemical that, in sufficient concentration, causes damage to bodily structure or function. Useful, or even essential, chemicals may be toxic at concentrations greater than those required by the body. The term "toxic agents" includes not only chemicals, but physical agents, such as X-rays, which can be harmful at high exposure levels. These definitions contain two important concepts in the study of toxic agents: Dose response is the concept that the intensity of the response of living organisms to a toxic agent is related to the dose; the second concept, duration response, is that the intensity of the response is related to the duration of exposure. Our awareness of the effects of toxic agents on the body is through the nervous system. It has been noted that the two most common adverse effects of drugs are headache and nausea, both of which originate in the central nervous system, regardless of the organ damaged by the toxic agent. Furthermore, the nervous system is the di-

Encyclopedia of Human Biology, Volume 2.

rect target for some types of toxic agents, and damage to the nervous system can result in a wide range of effects on an individual. The cells of the central nervous system as targets for toxic agents and the functional consequences are discussed here.

I. Historical Background

The 16th century physician and alchemist, Paracelsus, has been credited with first proposing the concept that all chemicals are poisons at some dose. His interest in chemistry and that of his contemporaries marked the beginning of the discovery of thousands of chemicals to benefit humans. The science of toxicology is then an outgrowth of the science of chemistry and the growing understanding of the chemicals, both natural and man-made, which can, on one hand, save lives and, on the other, damage cells. In the development of the biological sciences, the central nervous system has been called the last frontier, because the principles governing the relationships between its complex structures and their functions are still being elucidated. The study of toxicology as it relates to the central nervous system depends on the knowledge of the biology of the central nervous system and also has added to the knowledge of the way in which the nervous system functions.

II. Exposure to Toxic Agents

A. Unintentional

The environment contains many man-made or naturally occurring chemicals to which living organisms are sometimes exposed at levels that result in toxicity. The types of exposure are highly varied, from brushing the skin against a poison ivy leaf to inhaling tetraethyl lead in vaporized gasoline. The most serious exposures in air include lead, carbon monoxide, ozone, and oxides of nitrogen and sulfur. These chemicals are present in the atmosphere of cities primarily as a result of combustion for the production of energy in various forms. Exposure to toxic chemicals also may occur from ingestion of food or water contaminated from agricultural chemicals, industrial effluent, or the disposal of hazardous substances. Although concern is often expressed for the increased incidence of types of cancer as a result of environmental exposure, the central nervous system is a target in some of these exposures.

The possibilities of exposure in the workplace are numerous. For chemicals used in the production of many consumer goods, standards for acceptable exposure of workers have been set, which are revised as new information is available on the effects of different levels and durations. Good sources for current information are the publications of the American Conference of Governmental Industrial Hygienists, which include threshold limit values and biological indices of exposure for many chemical and physical agents in the workplace.

B. Intentional

The intent in the therapeutic use of drugs and in the addition of chemicals to food is human benefit. Much study has been devoted to insure that chemicals used as drugs or food additives lack serious toxicity to the nervous system. In the case of drugs, the benefit is sometimes balanced against an acceptable risk. For example, in determining the optimal dose for the treatment of acute leukemia by vincristine and similar drugs, the possibility of development of peripheral neuropathy, which results in weakness in leg or arm movement, must be balanced against the therapeutic effect on the cancer. In recent years, a most notable example of concern over the effects of food additives on the central nervous system has been the proposed link between certain food dyes and hyperactive behavior in children. After considerable controversy, substantial evidence against such a link was obtained by controlled studies of hyperactive children. Currently, there are no major studies implicating food additives with disorders of the central nervous system. However, the controversy did focus on an important consideration, that is, the greater susceptibility of segments of the population to some toxic chemicals. Not only children, in whom the nervous system is still developing, but also aged individuals may be uniquely sensitive.

Intentional exposure to toxic substances also takes place when individuals ingest, inject, or inhale chemicals with effects on the central nervous system, such as ethyl alcohol, cocaine, heroin, toluene, and benzene. Each of these chemicals has the ability to produce acute toxic changes. High doses may cause death through depression of the brain or other effects. Single doses of low levels have reversible effects. Prolonged use may cause irreversible changes.

III. Responses of the Nervous System

Effects of toxic agents on the nervous system can be evaluated in two basic ways: examination of structure and examination of function. Either structure or function or both may be altered by exposure, and the effects may be reversible or irreversible, depending on the agent, dose, and duration of exposure.

A. Structure of the Nervous System

The tissue of the central nervous system of all mammals, including humans, is composed of three major cell types (neurons, astrocytes, and oligodendrocytes), and between these cell bodies, the processes of the cells constitute the neuropil. Blood capillaries and some other specialized cells, such as microglia, complete the tissue of the brain. Each of the three major cell types has been recognized as a selective target for some toxic substances.

Neurons are the cells responsible for transmission, interpretation, and storage of information in the central nervous system. Their characteristic appearance, when the cell body and all of its processes are visualized, shows processes that extend from a central core, called the perikaryon, or cell body (Fig. 1). The large nucleus in the cell body is surrounded by cytoplasm richly filled with the cell organelles necessary for producing the energy to sustain very active metabolism. Energy is required by the neuron for the transport of nutrients down the processes, for the synthesis of chemical transmitters of information, and for maintaining ion homeostasis. Not surprisingly, neurons are at risk of damage whenever the levels of oxygen or glucose, supplied by the blood capillaries, are even temporarily reduced. Oxygen and glucose are the basic chemicals required for the extensive aerobic metabolism which neurons perform.

Astrocytes have been termed the "nurse cells" for the neurons, tending to cluster around large neurons (Fig. 2). Astrocytes are more numerous than neurons and, in some parts of the brain, outnumber them 10 : 1. The processes of the astrocyte are found against the cell body of the neuron and also wrapped around the endothelial cells lining the blood capillaries. [*See* ASTROCYTES.]

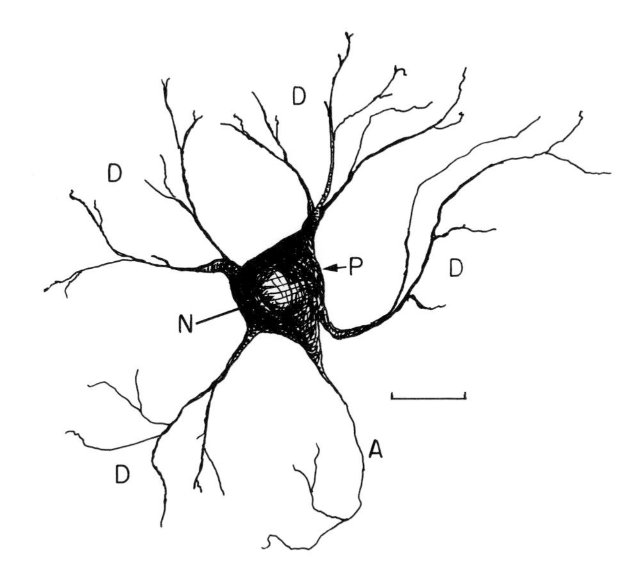

FIGURE 1 A neuron showing the perikaryon (P), nucleus (N), dendrites (D), and axon (A). Bar, 50 μm.

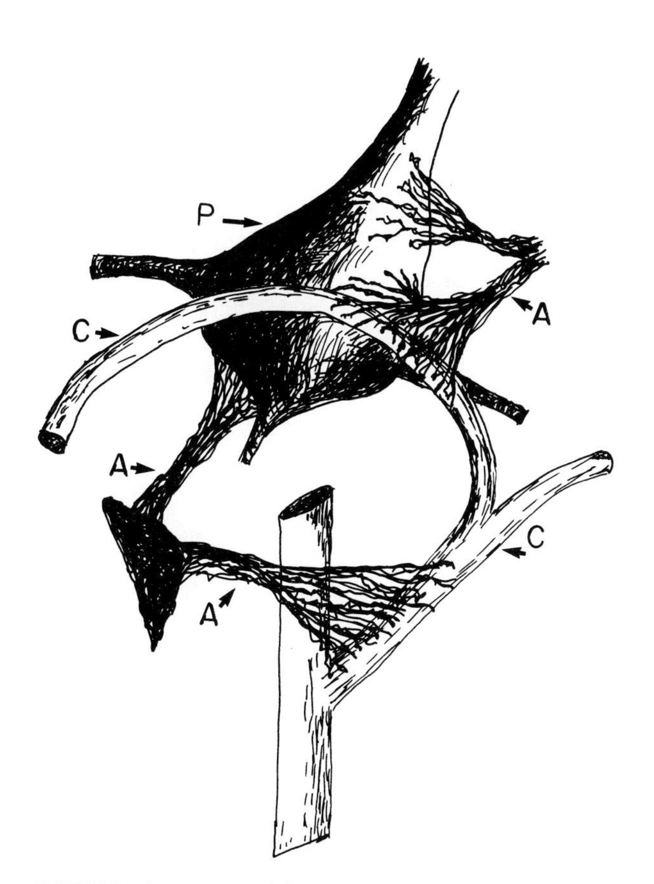

FIGURE 2 A neuron partially enclosed by processes from astrocytes (A). The astrocytes also contact the adjacent capillaries (C). P, Perikaryon.

Oligodendrocytes produce spiral wrappings of myelin which insulate axons, neuronal processes that may extend long distances in the nervous system, often in tracts with other axons insulated from each other by myelin.

B. Morphological Response of Neurons

Examination of a neuron with the light microscope or, at higher magnification, with the electron microscope, reveals characteristic changes in structure in response to high doses of many toxic chemicals. Any serious interference with the neuronal production of energy from glucose will distort the fluxes of calcium ions, which require energy. Moreover, the pH of the cell will decrease, from the accumulation of lactic acid. Both increased intracellular calcium levels and lowered pH activate damaging proteolytic enzymes within the cell. The appearance of the neuron under these conditions is diagrammed in Fig. 3. The nucleus becomes eccentric, the endoplasmic reticulum becomes dispersed, and the cell body and the mitochondria swell, owing to an influx of water. If the process is not rapidly reversed by restoration of an adequate source of energy, the nucleus will become pyknotic (i.e., condensed, with irregular margins), surrounded by a clear area of cytoplasm in the swollen dying cell. Since neurons cannot be replaced by cell division in the adult brain, death of the neurons may result in serious functional disability. However, there appears to be redundancy in carrying out function, so that function is not lost unless large numbers of neurons are affected.

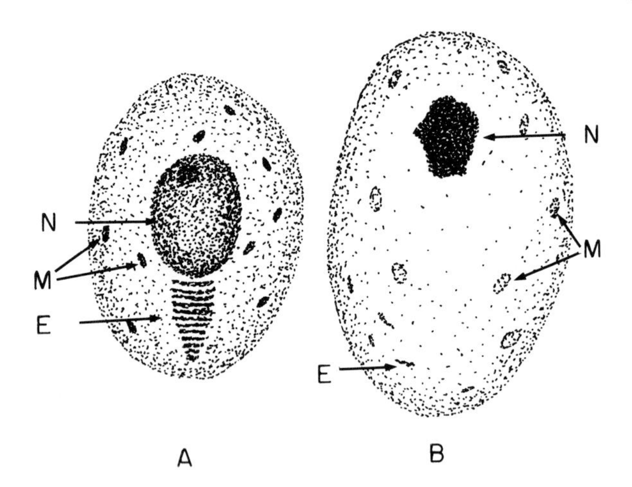

FIGURE 3 The changes in a neuron (A) following exposure to a toxic substance (B) that interferes with energy production. N, nucleus; E, endoplasmic reticulum; M, mitochondria.

In addition to cell damage by alterations in energy production, there are other targets for toxic chemicals in the neuron. Reversible effects are common in the synaptic area, where two neurons communicate. As noted above, the axon, often myelinated, is the process that transmits impulses to neurons in other areas. The axon makes synaptic contacts with the cell bodies, axons, or dendrites of other neurons. At each synapse, packets of special chemicals, called neurotransmitters, are liberated from the terminals of the axon into the synaptic cleft (Fig. 4). These transmitters act on the postsynaptic neuron by combining with specialized proteins called receptors. The synapse is, therefore, a target for chemicals that may alter the synthesis or release of transmitters, interfere with the postsynaptic actions of transmitters, or even mimic the normal transmitter.

C. Functional Response

The types of response to exposure to toxic agents are related to the areas of the nervous system which are affected. Their intensity ranges from headache, which disappears when exposure ceases, to permanent damage, with motor weakness or paralysis. Fatal exposures occur when neuronal function is severely compromised. For chemicals whose exposure occurs in the workplace at levels that result in recognizable pathology, the cause is usually correctly identified and measures are taken to reduce

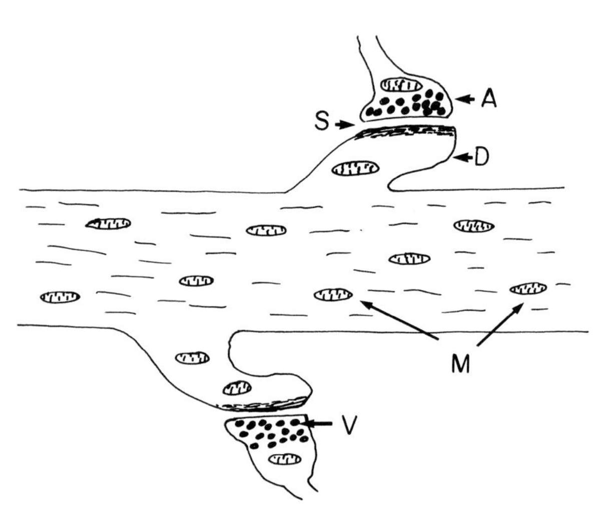

FIGURE 4 Synapse between an axon (A) and a dendrite (D). Vesicles (V) containing neurotransmitter are located in the terminal of the axon. The transmitter is released into the synaptic cleft (S). M, Mitochondria.

further exposure. The Federal government sets accepted occupational exposure levels for these chemicals. Low-level exposures may result in the insidious or slow development of damage, which is difficult to diagnose correctly. There is, however, controversy about "threshold" effects, that is, effect at the level below which no damage occurs following prolonged exposure. For example, although lead exposure from leaded gasoline and lead-containing paint and putty has been studied extensively, the lowest level of lead with a deleterious effect on learning in children is not certain. The complexity of obtaining precise epidemiological data compounds the problem of identifying toxic effects at low levels of exposure.

Table I lists some substances known to cause different functional types of damage, together with the mechanism and site of action in the nervous system. The site of damage varies from the synapse to the axon, to the insulating sheath of myelin, or to the cell body. For some agents, the action is indirect instead of damaging neurons directly (e.g., an agent which alters the oxygen supply to the neuron).

Botulinum toxin is formed by the bacterium *Clostridium botulinum,* which grows anaerobically in improperly canned foods. The toxin reaches the synapse of axons innervating skeletal muscle and blocks the release of acetylcholine, the neurotransmitter for skeletal muscle, resulting in muscular paralysis.

Carbon monoxide is formed during incomplete combustion. When inhaled, the gas combines with hemoglobin in the blood to form carboxyhemoglobin, preventing the formation of oxyhemoglobin, necessary for transporting oxygen to the central nervous system and other tissues. Some neurons die from lack of oxygen at levels which do not kill many other cells.

Methyl bromide is a gas used for fumigation of stored grain. It is fatal to insects and, in high concentrations, damages all organisms. When inhaled, the capillaries in the lungs are first affected. Damage to brain capillaries follows as methyl bromide is transported in the blood, preventing proper oxygen supply to the neurons. Since the damage to the capillaries of the central nervous system is generalized, an acute encephalopathy develops, characterized by a loss of consciousness and repeated seizures.

Methyl mercury is an organic form of mercury. It is used for protection of seed grain from insects and is not retained in the growing plant. Methyl mercury, formed from mercury by bacteria in natural bodies of water, can be picked up by organisms in the food chain and finally by fish, which are consumed by humans. This process resulted in a serious episode of human and animal poisoning by methyl mercury in Minimata Bay in Japan in the 1950's.

Tetrodotoxin is a toxic chemical derived from improperly prepared pufferfish. This fish is considered a delicacy in Japan and is commercially prepared by specially trained workers. Tetrodotoxin is present only in the liver, which must be removed without contaminating the meat. Its effect is to block the sodium channels of the nerve axon, thus preventing transmission of its nerve impulse. This chemical has been of great interest and usefulness for studying the physiological mechanisms surrounding transmission of the nerve impulse.

Triethyltin is one of several organic tin compounds used in industrial processes. Its target is the oligodendrocyte. The rings of myelin around axons

TABLE I Examples of Responses to High-Level Exposures

Agent	Functional response	Site of effect	Mechanism of effect
Botulinum toxin	Paralysis	Peripheral synapse of nerve with muscle	Block of the release of transmitter
Carbon monoxide	Coma	Neurons throughout the central nervous system	Depletion of oxygen through formation of carboxyhemoglobin
Methyl bromide	Encephalopathy	Capillaries throughout the central nervous system	Block of blood flow
Methyl mercury	Sensory loss, especially visual	Small neurons of the brain and the spinal cord	Death of small neurons
Tetrodotoxin	Paralysis	Peripheral synapse of nerve with muscle	Block of the transmission of nerve impulse
Triethyltin	Encephalopathy	Myelin throughout the central nervous system	Damage to oligodendrocytes

are split apart and fluid collects between the rings, resulting in generalized brain edema, which can be serious. However, when the acute phase is passed, the oligodendrocyte can resynthesize the myelin wrapping of the axons.

IV. Toxic Agents

A list of toxic agents causing adverse responses of the nervous system would be lengthy; however, the list of agents causing major problems with irreversible or fatal outcomes is much shorter. Table II encompasses agents important for either their frequency of occurrence or the seriousness of the consequence of exposure. A more complete listing is available in the books given in the bibliography. Each of the agents in Table II illustrates a different type of toxicity, which is discussed briefly here.

Ionizing radiation includes emissions of high-energy subatomic particles from radioactive chemicals, both naturally occurring and man-made, and from cosmic radiation. In the body, high-energy particles give rise to "reactive radicals," reactive chemical species which can damage cells. If the repair mechanisms of the cell are overwhelmed, cell death can occur. The most sensitive cells are those in the process of cell division. Thus, the fetus, with large numbers of dividing cells, is especially at risk.

Lead is a contaminant of the atmosphere from the exhaust of automobiles burning leaded gasoline. It may also be emitted by some industrial processes. In the body, lead, like calcium, can be stored in, or mobilized from, bone. Blood levels of lead are a measure of current exposure, but not of body burden. When high blood levels of lead are reached, the central nervous system becomes seriously involved. The primary target is probably the capillaries. Generalized damage to the vascular system of the brain results in encephalopathy, a condition of brain edema with coma and recurring motor seizures. Children are more likely to develop encephalopathy than adults. Atmospheric lead is insufficient to achieve such high blood levels, which require other sources of exposure, usually the ingestion of lead. A major concern with atmospheric levels is

TABLE II Toxic Agents Affecting the Central Nervous System

Agent	Type of effect
Environmental contaminants	
Ionizing radiation	Kills dividing cells; adult resistant; fetus sensitive
Lead	Atmospheric lead inhaled; chronic exposure may cause slowed learning in children
Methyl mercury	Chronic or acute exposure to high levels causes sensory loss; fetus and child more sensitive than adults
Nitrates	Converted to nitrites in intestine; high levels can cause methemoglobinemia in babies
Occupational exposures	
Acrylamide	Causes peripheral neuropathy; axons of large sensory and motor nerves degenerate from chronic or acute exposure
Carbon disulfide	Chronic inhalation results in peripheral neuropathy and encephalopathy
Manganese	Chronic inhalation causes emotional disturbances, followed by parkinsonian-like muscle rigidities
Organophosphate insecticides	Acute exposure through skin contact or inhalation results in accumulation of the transmitter, acetylcholine
Foods and food additives	
Absinthe	Severe encephalopathy has developed with continued ingestion of wormwood-containing liquor
Lathyrus sativus	Excessive consumption of this pea causes peripheral neuropathy
Methyl alcohol	Contamination of some alcoholic beverages (improperly prepared "moonshine" liquor) has caused blindness or generalized encephalopathy
Triorthocresyl phosphate	Contamination of food has occurred from reuse of containers; peripheral neuropathy results
Drugs and chemicals of abuse	
Alcohol (ethyl alcohol)	Acute intoxication from ingestion; chronic use is addictive; selective brain damage and peripheral neuropathy in adults; fetal alcohol syndrome in offspring of alcoholic mothers
Atropine	Acute overdose reversibly blocks part of the autonomic nervous system and causes hallucinations
Benzene	Chronic inhalation causes neuronal death and affects blood-forming tissues
Vincristine	When used therapeutically as an anticancer drug, may cause neuropathy as a side effect

that the continued exposure of children to levels even below those causing encephalopathy may have deleterious effects on brain function.

Methyl mercury selectively damages small neurons in the central nervous system, especially those of the area of the brain involving vision, the visual cortex, and the sensory cell bodies of the spinal cord in the dorsal root ganglia. The functional consequences are constriction of the visual field, or even blindness, and various abnormalities of sensation, such as numbness or tingling in the arms and legs.

Nitrates occasionally are found in high concentrations in well water, usually as a result of the agricultural use of nitrate-containing fertilizer. Nitrates in drinking water are not toxic to children or adults with normal liver function, but babies, in whom liver function is incompletely developed, may be unable to convert nitrites (which are formed from nitrates by bacteria in the gut) to nontoxic nitrates. Nitrites combine with hemoglobin in the blood to produce methemoglobin, a non-oxygen-carrying form. With severe methemoglobinemia, babies develop a bluish color of the skin and may become comatose from lack of oxygen.

Acrylamide is used industrially in the formation of plastic polymers. Occupational exposure tends to be chronic at low levels. With prolonged exposure, some workers develop a peripheral neuropathy associated with damage to the large axons, and both sensory and motor alterations. There may be some involvement of the autonomic nervous system, since the clinical signs include peripheral vasodilation (reddening of the skin) and sweating in the hands and feet. As with most toxicity from chemical exposure, there is recovery upon withdrawal from exposure, unless damage to the axons is severe.

Carbon disulfide is of historical importance, because the effects of chronic exposure were documented carefully in studies of workers exposed to the chemical in processing rubber. This episode is an example of the development of acceptable standards of exposure to protect workers. It was found that, after exposure for long periods (months or years, depending on levels in the air), workers developed neuropathies involving both sensory and motor fibers, and signs of brain involvement, including fatigue, sleep disturbances, and psychological depression.

Manganese is mined for industrial use in several countries. A central nervous system disorder has been characterized in miners exposed chronically to dust during production. The most specific effect is a type of parkinsonism with muscle stiffness and facial rigidity. Lesions have been found in areas of the brain involved in the control of motor movement. Emotional disturbances may accompany or precede the motor rigidity. [*See* PARKINSON'S DISEASE.]

Organophosphate insecticides, such as parathion, are agricultural chemicals which must be used with care to prevent excessive skin contact or inhalation during application. Exposure by either route causes a block of the enzyme, cholinesterase, that destroys the neurotransmitter, acetylcholine. This is the same as the mechanism of toxicity to insects, which also use acetylcholine as a neurotransmitter. When the cholinesterase is blocked, the synapse continues to be activated by the neurotransmitter, and the postsynaptic tissue continues to respond. In the autonomic nervous system, this results in excessive salivation and bronchial secretions, pupillary constriction, and bradycardia. In skeletal muscle, there are repeated small contractions, called fasciculations.

Absinthe is an anise-flavored liquor which, in the 19th century, was flavored with wormwood, an essential oil from the plant *Artemisia absinthium*. Absinthe oil caused neuronal death in the brain and, with repeated exposure, serious psychological disturbances and madness. The oil is no longer used in flavoring.

Lathyrus sativus is a form of pea which is of toxicological importance to countries where this pea is occasionally consumed in large quantities, for instance, during famine. The condition of lathyrism is a spastic paralysis of the legs caused by damage to the large motor neurons controlling the skeletal muscle. The small amounts present in most diets have no effects.

Exposure to methyl alcohol has occurred when it was a contaminant of some improperly prepared alcoholic beverages, such as "moonshine." It is not present in significant concentrations in commercial beverages. Acute exposure to large amounts may cause death from respiratory failure, or blindness in those who survive.

Triorthocresyl phosphate ingestion has resulted in several outbreaks of paralysis. Exposure has been the result of unintentional contamination of food or drink, when stored in industrial containers previously used for triorthocresyl phosphate and still containing residual amounts. The chemical causes damage to large axons in the arms and legs, with weakness and sensory disturbances. Triorthocresyl phosphate is an example of a chemical which causes delayed neurotoxicity. A few chemicals, such as this one, do not cause effects in a few hours. A single exposure may cause severe paralysis, but only after a delay of about 2 weeks. This delay is not shortened by larger doses. Recovery is slow and may be incomplete when poisoning is severe.

Alcohol (ethyl alcohol) is the most commonly used drug of abuse. The central nervous system consequences of acute intoxication with alcohol are well known. Chronic ingestion results in peripheral neuropathy, which is partially reversed by dietary thiamine (vitamin B_1); damage to the brain results in Korsakoff's syndrome, in which there is a severe loss of memory. The fetal alcohol syndrome can develop in a baby exposed repeatedly *in utero* to alcohol from maternal ingestion. Since there is no placental barrier to alcohol, the fetal blood level is the same as the mother's. Damage to the baby includes mental and growth retardation. The precise blood levels and length of exposure that are deleterious to the fetus are not known. [*See* ALCOHOL TOXICOLOGY.]

Atropine and related chemicals have many therapeutic uses, but have also been abused for their disorienting effects on the brain. Atropine blocks the action of the neurotransmitter, acetylcholine. The effects on the autonomic nervous system are dry mouth, flushing, tachycardia, and pupillary dilation. Hallucinations are common during overdose. Recovery is complete in a few hours.

Benzene and some other solvent gases are abused by inhalation to obtain effects on the central nervous system, which resemble those produced by ethyl alcohol. The effects are reversible, but repeated exposures have been linked with gradual deterioration of personality. Benzene has additional toxic actions on other organs, notably damage to blood-forming tissue. Industrial exposure levels are kept below those causing toxicity.

Vincristine is a drug with valuable therapeutic actions at levels that may cause irreversible toxicity, an example of the need to balance risk and benefit to the individual. Vincristine is effective against some types of cancers at levels apt to cause irreversible peripheral neuropathy. In combination with other anticancer drugs, it can be used at lower doses, which are still effective but have a lower risk of neuropathy.

V. Special Considerations

A. Immature Nervous System

It has been noted that the immature nervous system is more sensitive to toxic agents than the adult nervous system. The dividing cell is at risk because mitosis puts high metabolic demands on a cell and because DNA is more sensitive to damage during replication. The division of neurons ceases shortly after birth in the human, but maturation proceeds after birth for some time. Neuronal processes grow and synaptic contacts are completed postnatally. Astrocytes continue to divide. The formation of myelin by oligodendrocytes is a prolonged process, paralleling the growth of axons. The entire period of formation and maturation of the nervous system is a time of increased risk from exposure to toxic substances.

B. Geriatric Nervous System

Altered susceptibility to toxic substances with increasing age is an important consideration. Since the liver and the kidney are the major organs of destruction and excretion, respectively, of toxic agents, any deterioration of these organs will result in increased levels of the chemicals during exposure, increasing the risk of a toxic response. In addition, there is some loss of neurons with the aging process, although the rate of loss is uncertain. With depletion of neurons, a toxic response is again more likely to result.

C. Research on Central Nervous System Toxicology

The emphasis in the manifestations of toxicity reported here has been on those agents and exposures causing marked functional and cellular damage; they are the best known. It is important, however, to study the consequences of exposures below levels producing dramatic clinical signs and pathology, because the level of exposure to many toxic agents cannot be reduced to zero and must be controlled to minimize risk.

Bibliography

Adams, J. H., Corsellis, J. A. N., and Duchen, L. W., eds. (1984). "Greenfield's Neuropathology," 4th ed. Wiley, New York.

American Conference of Governmental Industrial Hygienists (ACGIH) (1987). ''Threshold Limit Values and Biological Exposure Indices for 1987–1988.'' ACGIH, Cincinnati, Ohio.

Hathcock, J. N., ed. (1982). ''Nutritional Toxicology,'' Vol. 1. Academic Press, New York.

Klaassen, C. D., Amdur, M. O., and Doull, J., eds. (1986). ''Casarett and Doull's Toxicology,'' 3rd ed. Macmillan, New York.

Mitchell, C. L., ed. (1982). ''Nervous System Toxicology.'' Raven, New York.

Narahashi, T., ed. (1984). ''Cellular and Molecular Neurotoxicology.'' Raven, New York.

Spencer, P. S., and Schaumburg, H. H., eds. (1980). ''Experimental and Clinical Neurotoxicology.'' Williams & Wilkins, Baltimore, Maryland.

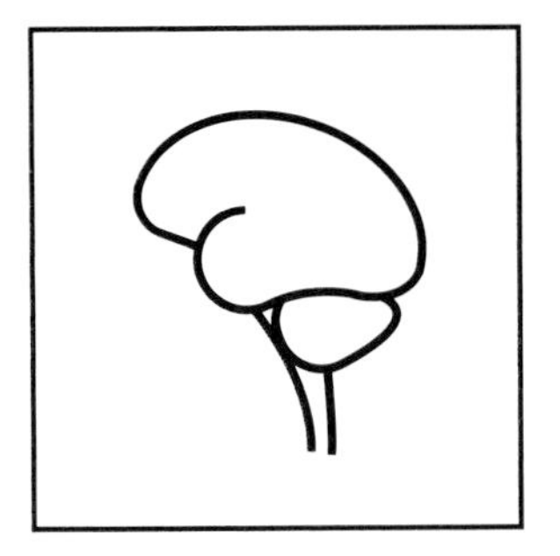

Cerebral Specialization

ROBIN D. MORRIS, WILLIAM D. HOPKINS, AND DUANE M. RUMBAUGH, *Georgia State University*

Glossary

Asymmetry Unequal, biased; one side of the brain is better at a given function than another

Laterality Side of the brain that provides primary control of a given function

Cerebral hemispheres Two main sides of the brain which are typically referred to as the left and right hemispheres. Each neocortical hemisphere is divided into four anatomical regions called lobes (temporal, parietal, occipital, and frontal).

CEREBRAL SPECIALIZATION, also referred to as cerebral dominance, cerebral asymmetry, or functional asymmetry, represents the concept that certain mental abilities or processes are performed more efficiently by one cerebral hemisphere (i.e., one part of the brain) compared to another. Globally, this concept is referred to as cerebral lateralization.

I. Historical Background and Concepts

Over 100 years ago clinical scientists such as Broca, Dax, and Wernicke described patients who, because of damage to their left cerebral hemispheres caused by strokes or injuries, were unable to talk or understand what others said to them, but at the same time were able to perform other non-language-related activities. As further studies of such patients were undertaken, the classic tenet, called cerebral dominance, was formed, which suggested that the left cerebral hemisphere was crucial for language functions and was the dominant hemisphere. These early ideas suggested that language-related skills were the most important mental abilities and that the left hemisphere controlled the right hemisphere, which came to be known as the nondominant, or minor, hemisphere. In addition, a special link between this left hemisphere dominance for language and right-handedness was suggested.

About 50 years later researchers began to find that damage to the right cerebral hemisphere also created specific types of mental deficits—not in the area of language functions, but in the area of visuospatial abilities. Because of these new findings, it became clear that the right hemisphere also contained important abilities. As evidence for other specialized abilities within each hemisphere began to accumulate, it was realized that each hemisphere was "dominant" for unique skills, and the concept of a dominant hemisphere slowly lost meaning. Currently, if one is speaking of cerebral dominance, the specification of cerebral dominance "for what" is required. Because of this history, it is probably more accurate to discuss such relationships as evidence of cerebral specialization.

In addition to the changing view of cerebral dominance and hemispheric specialization over history, a more complex model of lateralization of such functions has also developed. Numerous studies have shown that the extent of hemispheric differences for most mental functions is not as pure as was once suggested. For example, although most studies of right-handed subjects present evidence that the left cerebral hemisphere is primarily involved in language-related functions, it has been shown that the right cerebral hemisphere has the capacity to understand some language and perform some language-related functions, although not with the effectiveness or capacity of the left. Within this framework most mental or cognitive functions can be performed by either hemisphere to some extent, but among individuals one hemisphere is more specialized, dominant, and efficient than the other.

Finally, recent studies have suggested that cerebral specialization is found in species other than humans, including nonhuman primates, birds, and rats. Although these studies are recent and require additional confirmation, they suggest that environmental and genetic factors could influence the development of different patterns of cerebral specialization in different species and that cerebral specialization is not unique to humans.

II. Anatomical Asymmetries

Since evidence for cerebral specialization began to be observed, there has been interest in whether the actual asymmetries in mental functions were related to asymmetries in the underlying brain anatomy. Although such relationships would not prove that the anatomical differences caused the pattern of cerebral specialization, they might lead to further evidence regarding the ontogeny or phylogeny of such patterns. It should be noted, though, that there are competing and interactive theories suggesting that such cerebral specialization is affected more by environmental factors, or an interaction between environmental and genetic factors, than by morphological factors.

A wide range of anatomical brain structures has been examined for left–right asymmetries. These studies have included measurements of the length, width, volume, weight, and angulation of numerous brain structures. Such studies have been performed directly on postmortem brains or via neuroimaging techniques (e.g., computerized tomography, nuclear magnetic resonance imaging, and arteriograms) *in vivo*. [*See* MAGNETIC RESONANCE IMAGING.]

Anatomical areas which have been shown to be larger or longer in the left cerebral hemisphere compared to the right have included the lateral posterior nucleus, frontal operculum, fusiform gyrus, insula, occipital lobe, occipital horn of the lateral ventricle, parietal operculum, planum temporale (i.e., the posterior and superior surface of the temporal lobe), and sylvian fissure. Right hemisphere structures which have been shown to be larger or longer than those in the left have included the frontal lobe, Heschl's gyrus, and the medial geniculate. Overall, the most consistent and asymmetric region is the temporoparietal area, most right-handers showing larger left hemisphere structures in this area. The right hemisphere, on the other hand, has been described as being larger overall and extending more forward in the skull than the left hemisphere. Besides these anatomical asymmetries, there is also evidence for asymmetries in the underlying neurotransmitter systems (e.g., dopamine). These anatomical asymmetries have been observed in infants and children, and although there is evidence that each hemisphere could develop at different rates, the sequence of developmental neuroanatomical changes has not been clearly detailed.

III. Functional Asymmetries

A. Studies of Patients with Neurological Disorders

There have been two main groups of patients with neurological disorders who have provided the mainstay of research on cerebral specialization: patients with lateralized brain lesions or damage and patients who have undergone a commissurotomy (i.e., split-brain surgery) because of uncontrolled seizures. Patients with damage in one cerebral hemisphere due to a stroke, injury, surgery, or some other problem have been studied extensively to identify those mental functions which might be affected following such and injury.

The classic experimental paradigm for this type of research is to show that patients with one damaged hemisphere cannot perform a certain task or function, while patients with the other hemisphere damaged can perform the task. Even more specific localization of the effects of the damage can be made by comparing patients with damage in various areas of the same hemisphere. For example, patients with left frontal lobe damage might be compared to patients with left temporal lobe damage on certain language tasks. By this process the components of the system, and their links, can be identified. Unfortunately, such studies do not directly provide evidence for the exact functions of the damaged area, but provide information regarding how the remaining and intact parts of the brain work without that part. Because of this limitation, studies of neurological patients might not always provide accurate information regarding how a normal (i.e., nonneurologically damaged) brain might be specialized and might function.

Split-brain patients are those who have had their corpus callosum, the main bundle of fibers connecting the two cerebral hemispheres, surgically separated because of spreading seizures which cannot be controlled by other means. This surgery results in each of the two cerebral hemispheres being more independent and being unable or only partially able to communicate with each other. By special procedures researchers are able to present information to one hemisphere at a time and to study its ability to process it without the typical enlistment of the contralateral (i.e., opposite) hemisphere. For example, although the right hemisphere might be able to see a picture presented to it, the patient cannot name the picture; but if given a multiple-choice array of pictures which represent the possible answers, the patient can point to the correct answer using his or her left hand. If the same picture is presented to the left hemisphere, the patient, can name the picture but cannot point to it with his or her right hand. Such a demonstration shows the different capabilities of the two hemispheres when they cannot pass information between them, as is typical.

B. Studies of Normal Subjects

There are many problems with using brain-damaged patients to study normal brain functioning. Therefore, a number of methods have been developed to study the cerebral specialization of people without brain damage. One of these is the visual half-field technique, which uses the neuroanatomical features of the visual system to present visual information to one hemisphere or the other. If a picture is presented very quickly (usually in less than 200 msec) using a special machine called a tachistoscope, in one visual field (not to one eye, but to either the left or right of the visual midline), the information is directed into the contralateral hemisphere. By comparing how well the left hemisphere performs such a task compared to the right hemisphere, one can identify processing asymmetries. In other words, by comparing how fast or how well one hemisphere performs the presented task compared to the other one can document patterns of cerebral specialization.

Another method is the dichotic listening technique, which uses the anatomical features of the auditory system to assess cerebral asymmetries. This is typically done by presenting different words, sounds, music, or related auditory information to each ear at the same time. Therefore, a person might hear the word "dog" in his or her right ear, but "pig" in the left ear. How fast a person responds, which words are heard and recalled, or how well a specific stimulus is identified are measures used to compare the processing of information given to the right and left ears. Given that a majority of the auditory pathways are contralateral (i.e., crossed) in nature, stimuli presented in the right ear go directly to the left hemisphere, while stimuli presented to the left ear go directly to the right hemisphere.

Another method of study (i.e., dihaptic study) involves a similar paradigm using tactually presented stimuli which are presented bilaterally to each hand. There have also been studies using olfactory stimuli. These laterality paradigms have yielded results consistent with those from studies of brain-injured patients.

IV. Differences in Cerebral Specialization

One of the more basic questions addressed within the area of cerebral specialization is whether men and women differ in their patterns of abilities. Besides the typically cited findings that females are better in verbal-related functions while males are better in spatial-related functions, most research suggests that women might exhibit more bilateral (i.e., in both hemispheres) representation of various abilities when compared to men, although all of these differences are small. In other words, men appear to be more lateralized, or show greater asymmetry in specialization, than women.

There has also been a great deal of interest in differences in patterns of cerebral specialization between people with right- and left-handed preferences. A relationship between handedness and the lateralization of language has been identified since the earliest studies in this area. A large majority of the human population (i.e., >90%) is right-handed, and almost all of these right-handers (i.e., >98%) show a left cerebral hemisphere specialization for language-related functions. On the other hand, there is great debate about the pattern of cerebral specialization in left-handers. Most studies suggest that over 60% of left-handers have left cerebral hemisphere specialization for language-related functions, while there is great debate about how the remaining 40% of the left-handers are specialized. Most studies suggest that about 20% of the left-handers have right cerebral hemisphere specialization for language-related functions, while the remaining 20% are more bilateral in their representation of language abilities. The degree of specialization for non-language-related abilities does not appear to be as strong or specific, although most right-handers (i.e., 70%) show right hemisphere specialization for visuospatial abilities, while only about 33% of left-handers show such a relationship.

V. Asymmetries in Other Species

The question of whether animals other than humans manifest lateral asymmetries has been of considerable interest to scientists. Recent evidence suggests that anatomical asymmetries exist in nonhuman primate brains, similar to those found in humans. For example, the length of the left sylvian fissure is longer than that of the right in both monkeys and chimpanzees. Also, patterns indicate a right frontal–left occipital extension in nonhuman primates. Thus, if anatomical asymmetries are correlated with functional asymmetries, then some manifestation of these asymmetries should exist in the behavior of the organisms.

Two issues have been the central focus of studies. The first involves the distribution of lateral asymmetries within a given species. The second involves the homologous or analogous relationship between asymmetries observed in one species relative to another, including humans.

Many individual subjects within a given species show hand, or paw, preferences. However, the overall distribution of hand/paw preference appears to be equal between left- and right-preference individuals. This is different from the pattern observed in humans, which has prompted many researchers to conclude that asymmetries do not exist in species other than *Homo sapiens*.

Recently, some have suggested that species, particularly nonhuman primates, might show population levels of hand preference, but the direction and degree of asymmetry differ from those observed in humans. Thus, the features of hand preferences in nonhuman species at this time do not seem well defined. Moreover, the pattern of asymmetry might differ among different species. Thus, until further research is conducted, the features and parameters which account for hand preferences in nonhuman animals remain inadequately defined.

Although most studies of hemispheric specialization in nonhuman animals have focused on hand/paw preferences, there is a growing body of literature suggesting that certain cognitive asymmetries exist in other species. The principal technique for evaluating these processes has been to teach an animal to perform a particular task and then lesion an area of the brain. If that particular area is involved in the task being studied, then decrements in performance should be observed. For example, songbirds have been shown to have a left hemisphere dominance in the production of songs learned early in life. If the left hypoglossal nerve is severed in these birds, their capacity to produce songs becomes severely disrupted, while if the right nerve is severed, song is not affected. Similar findings have been reported in rodents as well as some primate species in terms of auditory processing of species-specific sounds.

Although one could argue that the processing of species-specific sounds is identical or homologous to that of language processing in humans, this comparison is difficult. The cognitive operations and component analysis involved in language processing might be different in humans relative to other species. Thus, the processing of auditory stimuli presented in a sequential manner might be an underlying asymmetry shared among many species, but the process of extracting meaning, syntax, or intonation from this stream of sounds could involve different aspects of neuropsychological functions not found in other species. Thus, perception of species-specific sounds might be a function analogous to language processing in humans and might have evolved by means of different selection processes.

VI. Evolution of Hemispheric Specialization

In the conceptualization of hemispheric specialization, one must think about its evolutionary significance and what prompted its emergence in animals, including humans. Many theories have been espoused elucidating the evolutionary precursors to hemispheric specialization and the environmental factors which shaped its development. Some have argued that language and cerebral dominance evolved in a parallel manner due to the close link between manual specialization and language processing. For example, some people argue that the first language used by *Homo sapiens* was a manual, or gestural, language. If a left hemisphere dominance for hand preference already existed in early humans, then assuming they would have utilized their dominant hand for a gestural language, language could have evolved in the left hemisphere. Still others argue that early humans did not use a manual or gestural language, but, rather, utilized an auditory communication system. They argue that the left hemisphere had a specialization for processing and producing sequential movements. Since auditory signals (e.g., speech) are sequential, the left hemisphere subsequently took control of processing speech sounds. Although both theories can account for current findings in the literature, neither explain how the left hemisphere initially came to have a hemispheric specialization. In other words, it is unknown why the brain operates asymmetrically at all. [*See* LANGUAGE, EVOLUTION.]

It has been suggested that perhaps the greatest advantage to having asymmetrical processing in the brain is because it nearly doubles the cognitive capacity of an organism. If each half of the brain subsumes separate operations but communicates with the other half, it potentially provides for twice as many capacities. Without complementary specialization, cognitive function is redundant for both halves of the brain. Within this theoretical framework neither language nor other cognitive functions need to be assumed as major selection pressures for the evolution of hemispheric specialization in humans. Instead, at both an individual and a species level, the selection pressure for asymmetrical processing of relevant information within the niche of that species would serve as the basis for differential processing of stimuli within the brain.

Whatever the case, hemispheric specialization has emerged as a most fascinating and controversial issue in the realm of science. Its applications and implications for the fields of education, biology, and psychology are only beginning to emerge. One goal of science is to explain and understand phenomenan. Scientists have only touched the surface with regard to understanding the evolution, phylogeny, and process of hemispheric specialization in animals and humans.

Acknowledgment

The work reported here was supported by National Institutes of Health grants RR0165 and NICHD:06016; and NASA grant NAG 2-438.

Bibliography

Beaton, A. (1986). "Left Side, Right Side: A Review of Laterality Research." Yale Univ. Press, New Haven, Connecticut.

Gazzaniga, M. (1985). "The Social Brain: Discovering the Networks of the Mind." Basic Books, New York.

Geschwind, N., and Galaburda, A. M. (1985). Cerebral lateralization: Biological mechanisms, associations, and pathologies. A hypothesis and a program for research. *Arch. Neurol.* **42,** 428–459.

Harnad, S., Doty, R. W., Goldstein, L., Jaynes, J., and Krauthamer, G. (1977). "Lateralization in the Nervous System." Academic Press, New York.

Kolb, B., and Whishaw, I. Q. (1980). "Fundamentals of Human Neuropsychology." Freeman, San Francisco, California.

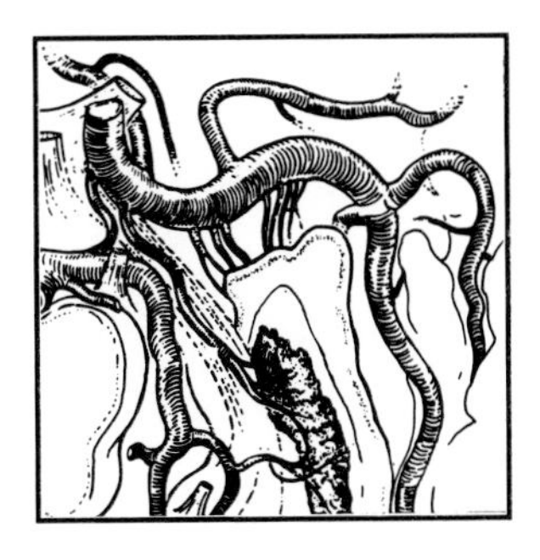

Cerebrovascular System

TOMIO SASAKI, *Japanese Neurosurgical Society*

NEAL F. KASSELL, *American Association of Neurological Surgeons*

Glossary

Anastomosis Natural communication between two blood vessels by collateral channels
Autoregulation Relative constancy of blood flow during alterations in arterial pressure
Innervation Supply of nerves to a body part

CEREBRAL BLOOD VESSELS differ structurally, physiologically, and pathologically from those of extracranial sites. These differences appear to be related to be phenomena responsible for the unique functions of the highly specialized tissues of the central nervous system. From a basic as well as a clinical point of view, currently available information on the features of the cerebrovascular system is provided in this chapter.

I. Development of the Cerebrovascular System

The development of the cerebral circulation system is divided into five stages.

1. The formation of a primordial endothelial vascular plexus from cords of angioblasts.
2. The differentiation into primitive capillaries, arteries, and veins.
3. The stratification of the vasculature into external, dural, and leptomeningeal or pial circulation.
4. The rearrangement of the vascular channels to conform to the marked changes in surrounding head structures.
5. The late histologic differentiation into adult vessels.

When the sequential order of this process fails, clinically significant lesions may occur. Faulty formation of the capillary bed during the second stage results in arteriovenous malformations. The changes during stage 3 would then determine whether the arteriovenous malformations would be situated in the scalp, the dura, or the brain. [*See* BRAIN.]

During the first 8 wk of fetal life, transitory arteries appear and serve as anastomoses between the primitive internal carotid artery and the bilateral longitudinal neural arteries. Bilateral longitudinal neural arteries will later fuse to form the basilar artery. The trigeminal, otic, and primitive hypoglossal arteries are transitory. These transitory arteries become progressively atrophied (first the otic, then the hypoglossal, and finally the trigeminal) and then disappear completely. Occasionally, these vessels persist into adult life.

II. Anatomy

The arterial supply of the brain is derived from two pairs of arterial trunks: the internal carotid arteries and the vertebral arteries.

A. The Carotid System

1. The Internal Carotid Artery

The internal carotid artery may be divided into four segments: cervical, intrapetrosal, intracavernous, and supraclinoid. The fine carotico-tympanic

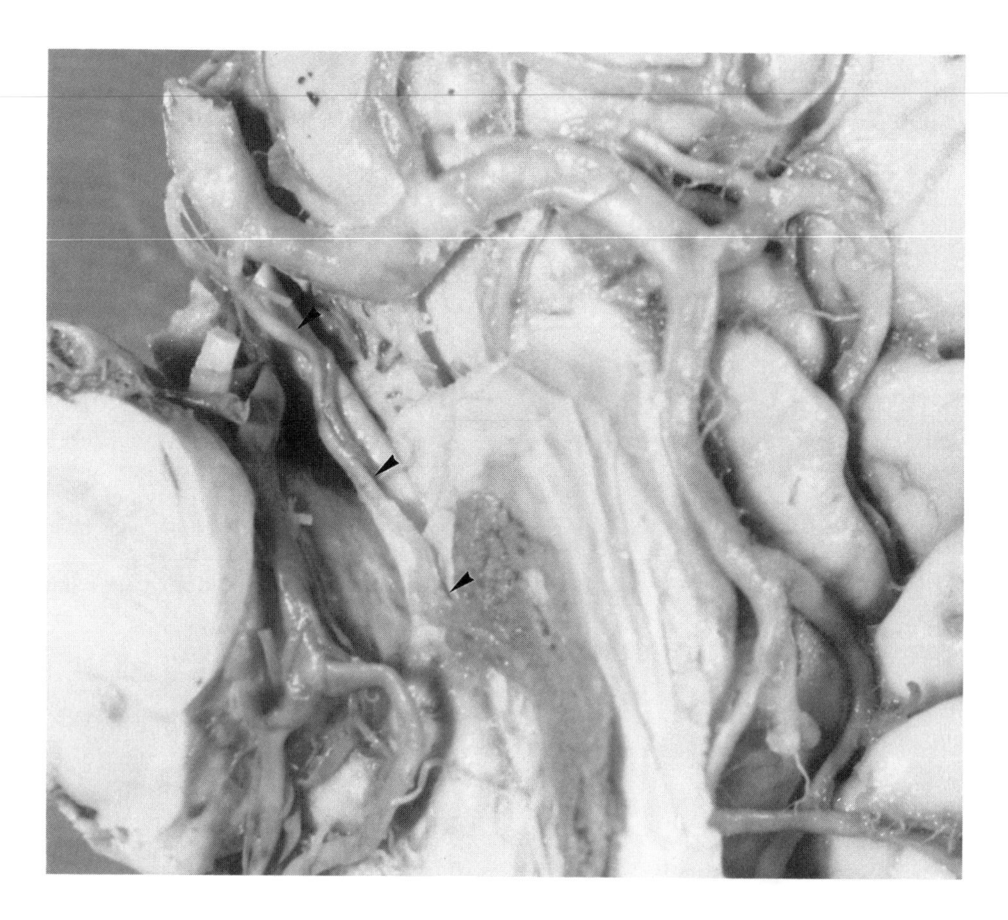

FIGURE 1 Photograph of the base of a brain. The tip of the temporal lobe is removed and the anterior choroidal artery (arrowheads) is exposed. Top is rostral.

artery arises from the intrapetrosal segment. Numerous small branches arise from the intracavernous segment. There are three main trunks: the meningohypophyseal trunk, the inferior cavernous sinus artery, and the capsular arteries. The meningohypophyseal trunk divides into the inferior hypophyseal artery, the dorsal meningeal artery, and the tentorial artery. The inferior cavernous sinus artery supplies the third, fourth, and sixth nerves, the meninges of the middle fossa, and the gasserian ganglion. Three major branches derive from the supraclinoid portion of the internal carotid artery: the ophthalmic artery, the posterior communicating artery, and the anterior choroidal artery.

The ophthalmic artery generally originates from the medial wall of the internal carotid artery. It enters into the orbit through the optic canal, ventrolaterally to the optic nerve.

The posterior communicating artery courses posteriorly and medically to join the posterior cerebral artery. The thalamoperforating arteries arise from the posterior communicating artery.

The anterior choroidal artery arises from the posterolateral aspect of the internal carotid artery approximately 2 mm proximal to the origin of the anterior cerebral artery (Figs. 1 and 2). From its origin, the anterior choroidal artery courses posteromedially in the chiasmatic cistern. It then follows the optic tract running between the cerebral peduncle and the medial surface of the uncus. It continues laterally and passes through the choroidal fissure to enter the temporal horn of the lateral ventricle. There are connections between the anterior choroidal artery and the lateral posterior choroidal artery in the region of the lateral geniculate body and the choroid plexus of the lateral ventricle. The anterior choroidal artery supplies blood to the posterior two-thirds of the optic tract, the external part of the lateral geniculate body, part of the piriform cortex, the uncus of the temporal lobe, the amygdaloid body, the tail of the caudate nucleus, the anterior third of the base of the cerebral peduncle, the substantia nigra, the upper parts of the red nucleus, a lateral portion of the ventral anterior and ventral

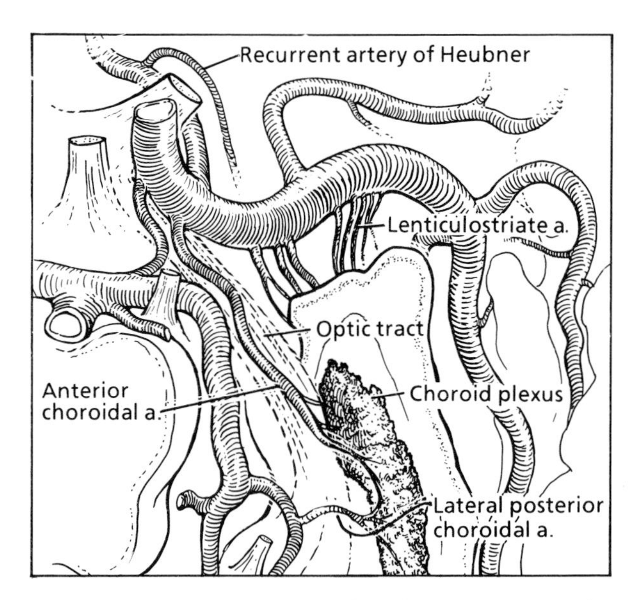

FIGURE 2 Schematic representation of the anterior choroidal artery, the posterior choroidal artery, the recurrent artery of Heubner, and the lenticulostriate arteries (based on Fig. 1).

lateral nuclei of the thalamus, the inferior half of the posterior limb of the internal capsule, and the retrolenticular fibers of the internal capsule.

2. The Anterior Cerebral Artery

From the point of bifurcation, the anterior cerebral artery courses medially and anteriorly above the optic nerve and chiasm. In the interhemispheric fissure, it is joined to the opposite anterior cerebral artery by the anterior communicating artery. Several small branches arise from the anterior communicating artery to supply the infundibulum, optic chiasm, and preoptic area of the hypothalamus. The anterior cerebral artery then ascends in front of the lamina terminalis to the level of the genu of the corpus callosum and passes backward above the corpus callosum in the pericallosal cistern.

Eight cortical arteries derive from either the pericallosal artery or the callosomarginal trunk: the orbitofrontal, frontopolar, anterior internal frontal, middle internal frontal, posterior internal frontal, paracentral, superior internal parietal, and inferior internal parietal arteries (Fig. 3). The anterior cerebral arteries supply the internal part of the orbital surfaces of the frontal lobe, the corpus callosum, the anterior two-thirds of the internal surfaces of the hemispheres, and a parasagittal segment of the external surfaces of the hemispheres.

The recurrent artery of Heubner originates at the precommunical segment of the anterior cerebral artery proximally to the anterior communicating artery (usually immediately before the origin of the anterior communicating artery) and, in a few cases, more distally (Fig. 2). It courses laterally to enter the anterior perforated substance just medial to the lenticulostriate arteries. The recurrent artery of Heubner supplies the head of the caudate nucleus, putamen, and anterior limb of the internal capsule.

3. The Middle Cerebral Artery

The middle cerebral artery passes laterally in the lateral cerebral fissure to the sylvian fissure (Fig. 2). From this horizontal segment derive the lenticulostriate arteries and the anterior temporal arteries. The lenticulostriate arteries enter the anterior perforated substance and supply the substantia innominata, the lateral portion of the anterior commissure, most of the putamen and the lateral segment of the globus pallidus, the superior half of the internal capsule and the adjacent corona radiata, and the body and head of the caudate nucleus (except for the anteroinferior portion).

The anterior temporal branches of the middle cerebral artery supply the temporal pole and the lateral aspect of the temporal lobe. Distal to the horizontal segment, the middle cerebral artery turns upward and posteriorly, and enters the depths of the sylvian fissure to reach the insula. In this region, the middle cerebral artery divides into its major branches (frontal, parietal, and temporal). These branches leave the sylvian fissure, circumnavigate the frontal, parietal, or temporal opercula, and give off 12 cortical branches.

1. Frontal branches: orbitofrontal, prefrontal, precentral, and central arteries.
2. Parietal branches: anterior parietal, posterior parietal, and angular gyrus arteries.
3. Temporal branches: temporo-occipital, posterior temporal, middle temporal, anterior temporal, and temporal polar arteries.

The small perforating branches derive from the insular segment of the middle cerebral artery and supply the claustrum and external capsule.

B. Vertebrobasilar System

1. The Posterior Cerebral Artery

The posterior cerebral arteries derive from the rostral end of the basilar artery and course posteriorly in the perimesencephalic cisterns to encircle the cerebral peduncle. This proximal trunk of the

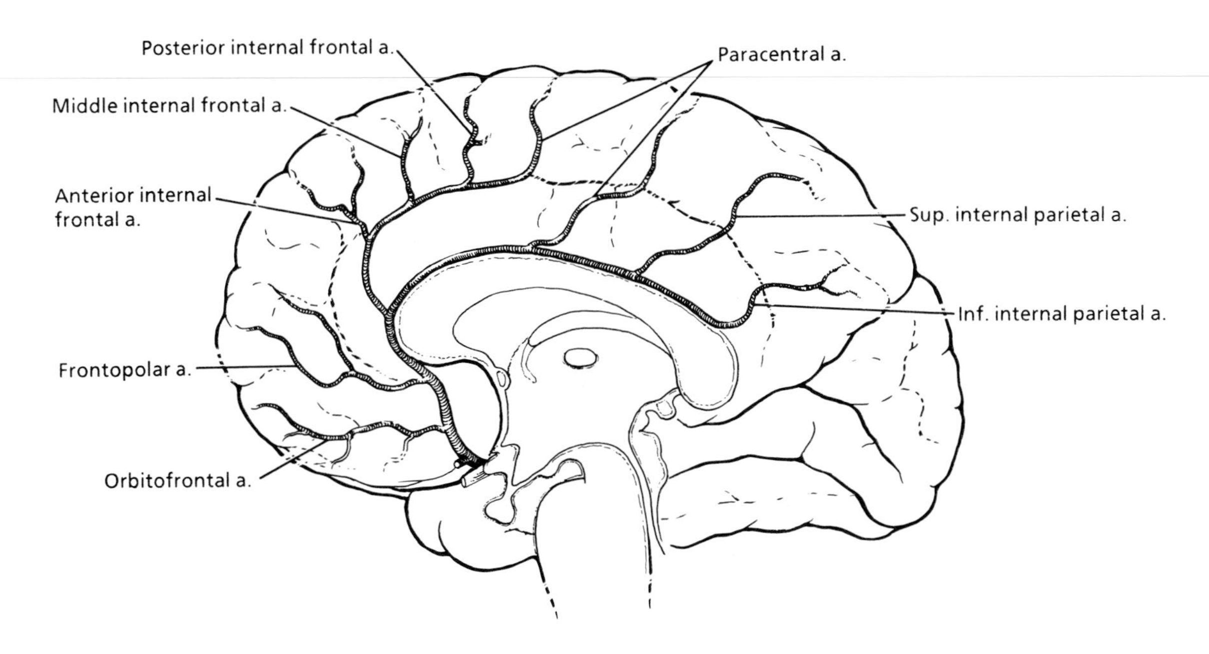

FIGURE 3 Schematic representation of the anterior cerebral artery and its branches.

posterior cerebral artery may be divided into peduncular, ambient, and quadrigeminal segments. The posterior communicating artery arises from the midportion of the peduncular segment. The quadrigeminal segment continues posteriorly beneath the splenium of the corpus callosum to terminate in cortical branches. Four main cortical branches (anterior temporal, posterior temporal, parieto-occipital, and calcarine arteries) arise usually from the ambient segment or from the quadrigeminal segment. These cortical arteries supply the inferior and internal surfaces of the temporal lobe, the internal surface of the occipital lobe, and the posterior part of the precuneus. The posterior cerebral artery also provides the perforating branches to the midbrain and the thalamus.

The mesencephalic branches include the interpeduncular perforating branches, the peduncular branches, and the circumflex mesencephalic branches. These mesencephalic branches arise from the peduncular segment of the posterior cerebral artery and supply the cerebral peduncle, the red nucleus, the substantia nigra, and the tegmentum.

The thalamic branches include the thalamoperforating arteries and the thalamogeniculate arteries. The thalamoperforating arteries arise from the posterior communicating artery and the proximal peduncular segment of the posterior cerebral artery. The thalamoperforating arteries supply the posterior chiasm, the optic tract, the posterior hypothalamus, part of the cerebral peduncle, and the anterior and medial portions of the thalamus.

The thalamogeniculate arteries usually derive from the ambient segment of the posterior cerebral artery and supply the lateral geniculate body and the posterior and lateral portions of the thalamus.

One medial posterior choroidal artery and at least two lateral posterior choroidal arteries belong to the group of the posterior choroidal arteries arising from the posterior cerebral artery (Figs. 2 and 4). The medial posterior choroidal artery usually arises from the proximal segments (peduncular or ambient) of the posterior cerebral artery and curves around the midbrain. It approaches the region lateral to the pineal body and then courses forward to the roof of the third ventricle. This vessel supplies the tectum, the choroid plexus of the third ventricle, and the dorsomedial nucleus of the thalamus.

In most instances, the lateral posterior choroidal arteries originate from the ambient segment of the posterior cerebral artery. These vessels initially course laterally to enter the choroid fissure. The anterior branch extends forward to supply the choroid plexus of the temporal horn. The posterior branch courses posteriorly around the pulvinar, supplying the choroid plexus of the trigone and the body of the lateral ventricle. The posterior branch also supplies the fornix, most of the dorsomedial nucleus of the thalamus, the pulvinar, and part of the lateral geniculate body.

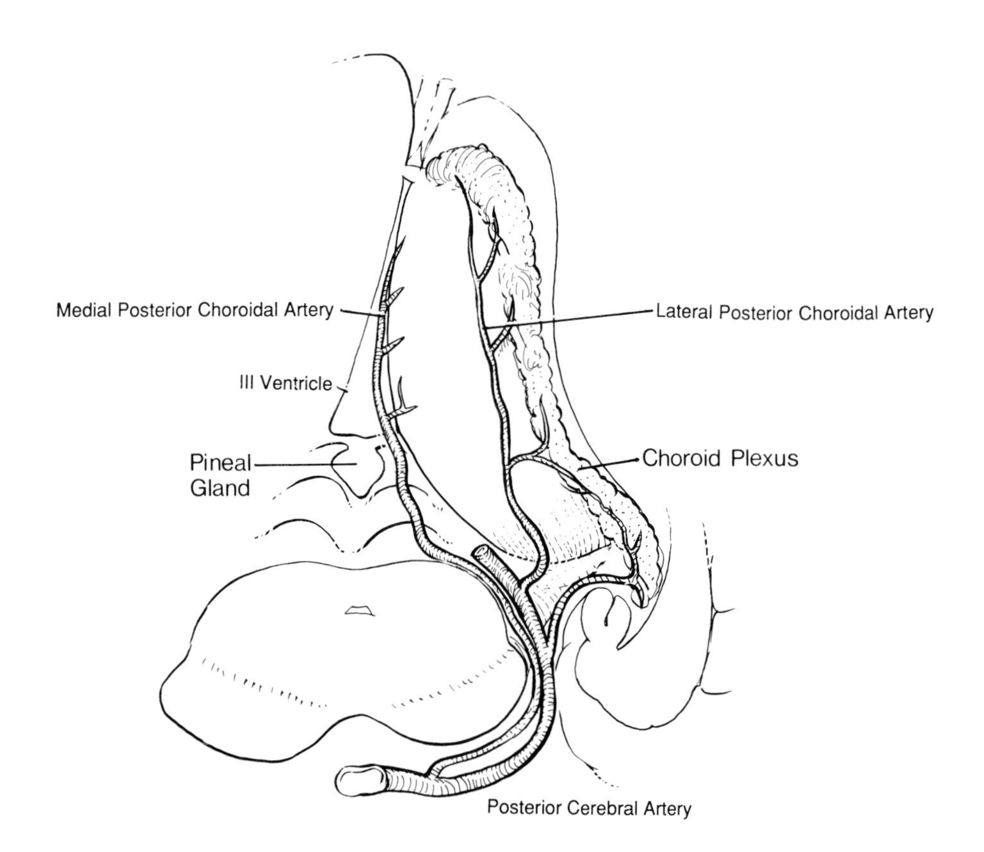

FIGURE 4 Schematic drawing of the medial posterior choroidal artery and the lateral posterior choroidal artery, as viewed from a dorsal and posterior perspective.

2. The Superior Cerebellar Artery

The superior cerebellar artery originates from the rostral part of the basilar artery. It courses posterolaterally in the perimesencephalic cisterns to encircle the upper pons and lower mesencephalon. On reaching the cerebellum, it divides into several cortical branches. These cortical branches supply the lateral and superior surface of the cerebellar hemispheres, and the superior vermis. From these cortical arteries numerous branches extend deeply into the cerebellum and supply the superior medullary velum, the superior and middle cerebellar peduncles, and the intrinsic cerebellar nuclei, including parts of the dentate nucleus.

3. The Anterior Inferior Cerebellar Artery

The anterior inferior cerebellar artery arises from the inferior third of the basilar trunk and courses laterally and downward. It sends small branches into the pons that supply the lateral aspect of the pons from the junction of the upper and middle third of the pons down to the upper part of the medulla. Within the cerebellopontine angle cistern, the anterior inferior cerebellar artery is usually situated ventral to the acoustic-facial nerve bundle. The internal auditory artery usually originates from this proximal portion of the anterior inferior cerebellar artery. After crossing the seventh and eighth nerves, the anterior inferior cerebellar artery courses medially toward the cerebellopontine angle and then extends to the cerebellar hemisphere.

4. The Posterior Inferior Cerebellar Artery

The posterior inferior cerebellar arteries originate from the vertebral arteries at an average distance of 16 mm below the vertebrobasilar junction. The main trunk of the posterior inferior cerebellar artery may be divided into four segments: the anterior medullary, the lateral medullary, the posterior medullary, and the supratonsillar (Fig. 5).

The anterior medullary segment courses posteriorly within the medullary cistern and winds around the lower end of the olive of the medulla oblongata. It continues posteriorly around the lateral aspect of the medulla as the lateral medullary segment. On reaching the posterior margin of the medulla oblongata, the posterior inferior cerebellar artery (posterior medullary segment) ascends to the anterior aspect of the superior pole of the tonsil

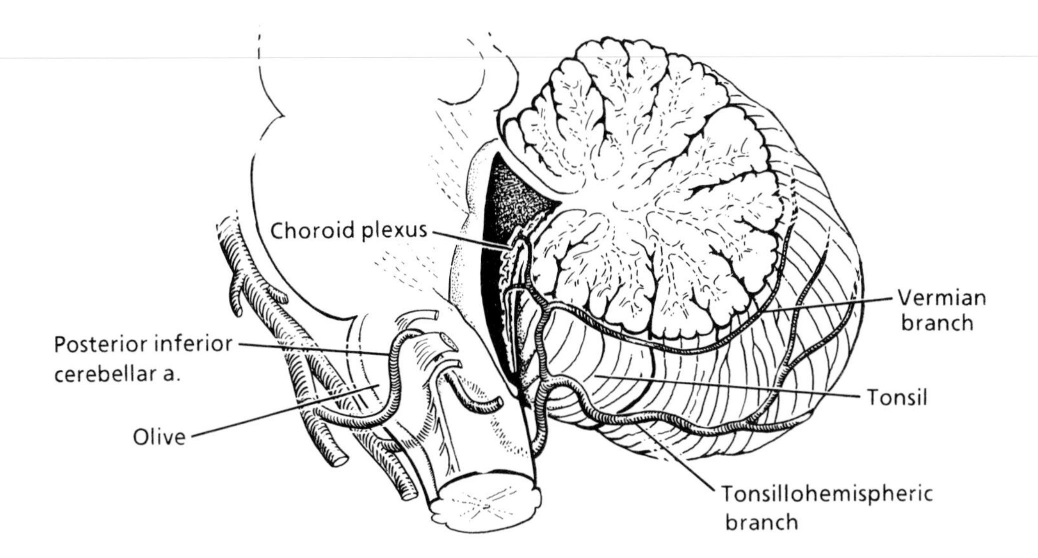

FIGURE 5 Schematic drawing of the posterior inferior cerebellar artery.

behind the posterior medullary velum. The posterior inferior cerebellar artery continues dorsally to the superior pole of the tonsil as the supratonsillar segment. It gives off small branches to the choroid plexus of the fourth ventricle. The supratonsillar segment of the posterior inferior cerebellar artery then bifurcates into two terminal branches: the tonsillohemispheric and vermis branches. The tonsillohemispheric branches descend along the posterior margin of the medial aspect of the tonsil and divide into tonsillar branches and hemispheric branches. In some cases, tonsillohemispheric branches arise adjacent or ventral to the cerebellar tonsil. Vermis branches course below the inferior vermis in the sulcus vallecula between the inferior vermis and the cerebellar hemisphere.

The basilar artery provides three series of perforating branches (paramedian, short circumferential, and long circumferential arteries) to the ventral portion of the pons. The paramedian arteries supply the most medial pontine area, including the pontine nuclei and the corticopontine, corticospinal, and corticubulbar tracts. The short circumferential arteries supply a wedge of tissue along the anterolateral pontine surface. The long circumferential arteries supply most of the tegmentum.

Although the arterial supply to the brain is derived from two carotid and two vertebral arteries, the relative importance of these vessels varies in different species. In primates, the carotid and vertebral circulations participate in approximately equal manner. The carotid system provides a much larger share of the blood supply than the vertebral arteries in horses, and a smaller share in rats. Also, the relative importance of the internal and external carotid arteries varies between species. In dogs, the internal carotid artery is small, and the external carotid artery branches into a complex network of anastomoses, the rete mirabile. Branches of the external carotid artery provide more flow to the brain than the internal carotid artery in that species. In contrast, the internal carotid artery is larger than the external carotid artery and no rete in primates. There are numerous collateral pathways in the brain circulation.

5. The Circle of Willis

At the base of the brain, the anatomical pattern of vessels permits collateral blood flow from the internal carotid arteries and the vertebrobasilar system through the circle of Willis. This arterial circle is formed by anterior and posterior communicating arteries and proximal portions of the anterior and posterior cerebral arteries.

6. Leptomeningeal Anastomoses

Anastomoses other than those of the circle of Willis occur over the surfaces of the cerebral hemispheres and cerebellum. Predominantly, leptomeningeal collateral circulation is found in the anastomoses between the anterior cerebral and middle cerebral, and between the middle cerebral and posterior cerebral arteries. On the cerebellar surface leptomeningeal anastomoses occur between each of the cerebellar arteries and also across the midline to connect with the corresponding arteries of the op-

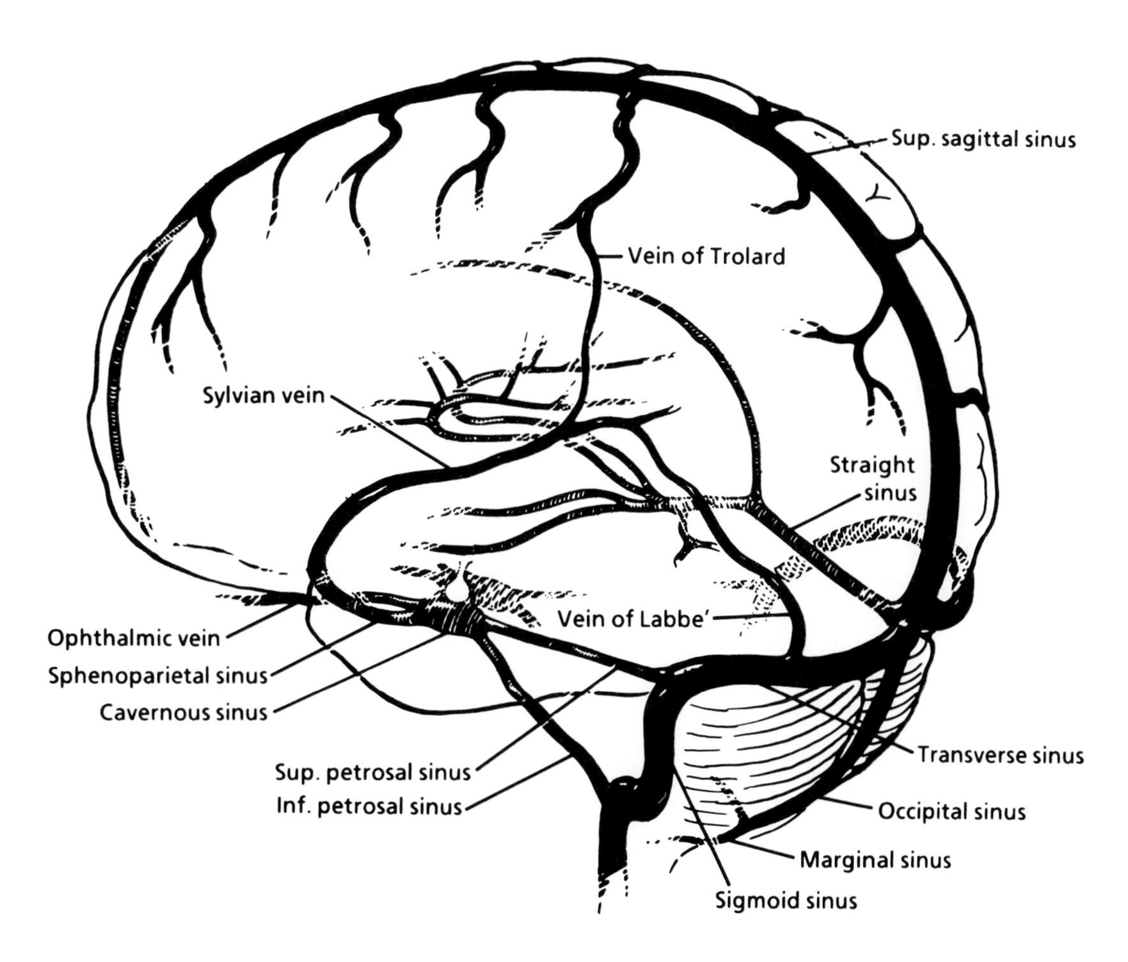

FIGURE 6 Diagram of the intracranial venous sinuses and superficial cerebral veins and deep cerebral veins.

posite side. Communications between adjoining branches of the cerebellar arteries are more frequent than in the case of the cerebral arteries.

C. Dural Venous Sinuses

The superior sagittal sinus begins rostrally at the foramen cecum of the frontal bone and proceeds in a caudal direction toward the internal occipital protuberance (Fig. 6). It receives venous outflow from the lateral lacunae and from the ascending superficial cerebral veins. The inferior sagittal sinus courses above the corpus callosum in the free edge of the falx cerebri. This sinus receives small veins that drain the roof of the corpus callosum, the cingulate gyrus, and the adjacent medial hemisphere. The straight sinus receives blood from the inferior sagittal sinus and from the vein of Galen. As it proceeds caudally, this sinus communicates with the superior sagittal sinus and the bilateral transverse sinuses, forming a torcular Herophilis or confluence of sinuses. When the transverse sinuses are poorly developed, the occipital sinus situated in the dorsal aspect of the falx cerebelli and terminating in the confluence of sinuses may persist as a large sinus.

The transverse sinuses course laterally and anteriorly along the attached margin of the tentorium cerebelli and along the groove of the squamous portion of the temporal bone. Near the petrous temporal bone, the transverse sinus receives blood from the superior petrosal sinus, thus communicating with the cavernous sinus. The transverse sinus is continuous with the sigmoid sinus. After traversing the jugular foramen, the sigmoid sinus becomes continuous with the internal jugular vein. The inferior petrosal sinus may enter the most distal segment of the sigmoid sinus, although it usually drains directly into the internal jugular vein. The sphenoparietal sinus receives blood from the superficial sylvian vein. It courses in the inferoposterior ridge of the lesser wing of the sphenoid bone and ends in the antero-inferior angle of the cavernous sinus.

The cavernous sinus also receives the ophthalmic veins, the inferior and superior petrosal sinuses, and the basilar plexus of veins.

D. Cerebral Veins

The superficial cerebral veins may be divided into three groups: the ascending superficial cerebral veins, the superficial sylvian veins, and the inferior cerebral veins (Fig. 6).

The ascending superficial cerebral veins drain the medial, superolateral, and frontal regions of the cerebral hemisphere. Most of these veins empty into the superior sagittal sinus. The superficial sylvian veins originate near the posterior limb of the sylvian fissure and drain into the sphenoparietal sinus. There are two major anastomotic veins, which connect the superficial sylvian veins with other venous sinuses. The great anastomotic vein of Trolard connects the superficial sylvian veins with the superior sagittal sinus. The lesser anastomotic vein of Labbé connects the superficial sylvian veins with the transverse sinus. The inferior cerebral veins drain the portions of the lateral convexity not drained by the superficial sylvian veins, the ventral surface of the temporal lobe, or the ventral surface of the occipital lobe. These veins drain into the transverse sinus.

The deep cerebral veins drain the blood from the deep white matter, the basal ganglia, and the diencephalon. The medullary veins carry blood in the centripetal direction from the deep structures of the cerebrum toward the lateral ventricles, where they coalesce to form the subependymal veins. The thalamostriate vein is one of the largest subependymal veins. It receives venous return from the caudate nucleus, the lenticular nuclei, the internal capsule, the deep white matter of the posterior frontal lobe, and the deep white matter of the anterior parietal lobe. The subependymal veins empty into the internal cerebral veins at the level of the foramen of Monro. The internal cerebral veins from each side unite under the splenium of the corpus callosum and join the great cerebral vein of Galen. The great cerebral vein of Galen also receives the basal vein of Rosenthal, which originates near the optic chiasm and courses caudally and dorsally around the cerebral peduncle.

The veins of the posterior fossa can be divided into the following three groups: the Galenic draining group, the petrosal draining group, and the tentorial draining group.

The Galenic group of veins receives most of the venous return from the midbrain and from the medial and paravermal regions of the superior cerebellar surface. The major veins of the superficial brainstem include the tectal veins, the lateral mesencephalic vein, the posterior mesencephalic vein, and portions of the longitudinally directed anterior pontomesencephalic vein. The medial and paravermal regions of the superior cerebellar surface are drained by the superior vermian vein and by the precentral cerebellar vein.

The petrosal vein is a major vessel of the petrosal draining group. The veins of the great horizontal fissure, the transverse pontine vein, the brachial vein, the vein of the lateral recess of the fourth ventricle, and the lateral pontine vein converge toward the anterior angle of the cerebellum, where they form a single stem, the petrosal vein. The petrosal vein usually drains into the superior petrosal sinus.

The tentorial draining group includes the inferior vermian vein and the superior and inferior hemispheric veins. The inferior vermian vein is formed by the union of the superior and inferior retrotonsillar tributaries behind the cerebellar tonsil. It proceeds posterosuperiorly through the inferior paravermian sulcus toward the transverse sinus. The superior hemispheric veins run posteriorly and inferiorly on the superior surface of the cerebellar hemisphere and open into the transverse sinus. The inferior hemispheric veins, on the other hand, run superiorly on the inferior aspect of the cerebellar hemisphere and drain into the transverse sinus.

III. Innervation

Cerebral vessel receives dual sympathetic and parasympathetic innervation. The adrenergic innervation emanates primarily from the ipsilateral superior cervical ganglion, but arteries in basal and medial areas receive bilateral innervation. The preganglionic nerves originate in the intermediolateral column of the spinal cord. The innervation of arteries in the internal carotid system is more extensive than the innervation in the vertebrobasilar system. A central noradrenergic pathway, originating primarily in the locus ceruleus, also innervates cerebral vessels, particularly smaller arteries and capillaries.

The parasympathetic (cholinergic) innervation emanates mostly from the sphenopalatine ganglion and possibly from the otic ganglion. The preganglionic nerves originate in the superior salivatory nucleus and project via the greater superficial petrosal nerve, a branch of the seventh cranial nerve, to the sphenopalatine ganglion. The lesser superficial petrosal nerve may be the preganglionic projection to the otic ganglion and originates in the inferior salivatory nucleus. Like the sympathetic innervation, cholinergic innervation is more dense in the anterior circulation than in the posterior circulation.

Recent immunohistochemical studies have shown that neurotransmitters within the sympathetic and parasympathetic innervation to the blood vessels coexist with vasoactive peptides, i.e., with vasoactive intestinal polypeptides (VIPs) in the cholinergic nerves and with neuropeptide Y in the adrenergic nerves. Nerve fibers immunoreactive for calcitonin gene-related peptide (CGRP) have also been identified in the cerebral arteries. CGRP coexists with substance P. CGRP- and substance P-containing fibers originate in the trigeminal ganglion of the same side. CGRP, substance P, and VIP dilate cerebral arteries, the relative potency being CGRP–substance P–VIP. On the other hand, neuropeptide Y constricts cerebral arteries; however, the functional role of neuropeptides in the control of the cerebral circulation is currently a matter of speculation.

IV. Regulation of Cerebral Blood Flow

There are several mechanisms by which cerebral blood flow (CBF) may be regulated: autoregulation, neurogenic regulation, chemical regulation, and metabolic regulation.

A. Autoregulation

Autoregulation is defined as the relative constancy of blood flow during alterations in arterial pressure (Fig. 7). CBF is maintained constant over a wide range of blood pressure than in other organs. Lower and upper limits of autoregulation are approximately 40 and 160 mmHg, respectively. Decreases in arterial pressure cause vasodilation. Maximum dilation of cerebral pial arteries has been observed at blood pressures of around 40 mmHg. On the other hand, increases in arterial pressure cause vasoconstriction within the range of autoregulation. However, a marked increase in arterial pressure (usually >200 mmHg) induces dilation of pial arterioles and is irreversible for several hours. In such situations, CBF is markedly increased, a phenomenon called breakthrough of autoregulation. In hypertensive patients, the pressure–flow curve is shifted to the right with higher limits of autoregulation (Fig. 7).

B. Neurogenic Control

Cerebral vessels are richly innervated, but the role of sympathetic nerves in regulation of CBF is unclear. The α-adrenoceptors of cerebral vessels are less sensitive to norepinephrine than the α-receptors of other vessels. Transmural electrical stimulation produces much less contraction in cerebral arteries than in peripheral arteries, and electrical stimulation of sympathetic nerves has little or no effect on CBF under normal conditions.

Several studies have shown that acetylcholine produces concentration-dependent vasodilation of cerebral vessels. However, the role of cholinergic nerves in the regulation of CBF has been highly controversial.

C. Chemical Control

Hypercapnia induces pial arteriolar dilatation and increases CBF. Topical application of solutions with low pH on the brain surface produces pial arteriolar dilatation. This vasodilatory response can be obtained when changes in pH are induced either by changes in CO_2 at constant HCO_3^- concentration or when changes in HCO_3^- concentration are made at

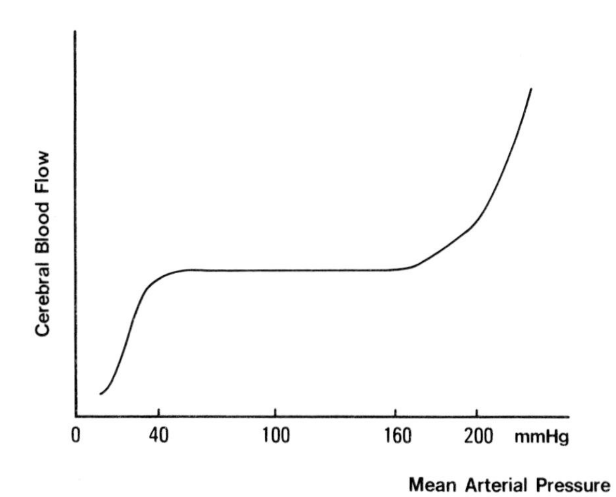

FIGURE 7 Schematic representation of the autoregulation of cerebral blood flow. In normotensive patients, lower and upper limits of autoregulation are approximately 40 and 160 mmHg, respectively.

constant pCO_2. However, pCO_2 or HCO_3^- concentrations have no effect on pial arteriolar caliber unless the pH is allowed to change. The most important mechanism of action of CO_2 seems to be its local effect on vascular smooth muscle mediated via a change in extracellular fluid pH.

D. Metabolic Regulation

It is well established that there is a tight coupling between brain metabolism and CBF. Alterations in the level of activity of the brain are accompanied by parallel changes in both brain metabolism and CBF. For instance, motor activity in the arm is accompanied by parallel increases in both blood flow and O_2 consumption in the contralateral sensorimoter area of the cortex. The mechanisms by which the brain metabolism influences CBF are not clearly understood. However, it has been assumed that the active neurons release vasodilator metabolites, which presumably reach the blood vessels by diffusion. Among several vasodilator metabolites, adenosine and K^+ have been considered to be promising candidates. Adenosine has a vasodilatory effect on pial arterioles at very low concentrations, and the concentration of adenosine in the brain increases during increased functional activity of the brain. K^+ produces vasodilation of pial arterioles between concentrations of 0 and 10 m*M*. Several studies have shown that K^+ is released into the extracellular fluid during increased functional activity of the brain, to an extent adequate to achieve maximum vasodilation of the pial arterioles.

E. Role of Vascular Endothelium

Recent studies have demonstrated that vascular endothelium continuously synthesizes or releases a large number of vasoactive substances such as prostacyclin, endothelium-derived relaxing factor (EDRF), endothelium-contracting factor (EDCF), and others. These discoveries have led to important new concepts in vascular pathophysiology.

Prostacyclin, synthesized in endothelial cells of cerebral (Fig. 8) as well as peripheral vessels, relaxes vascular smooth muscle cells and inhibits platelet aggregation and adhesion. EDRF is released when muscarinic receptors on endothelial cells are stimulated by acetylcholine (Fig. 9). Many other substances release EDRF from endothelial cells. They include arachidonic acid, bradykinin, histamine, substance P, norepinephrine, 5-hydroxy tryptamine, calcium ionophore A23187, adenine nucleotides, thrombin, vasopressin, and changes in shear stress. Vascular smooth muscle is the target for the biological action of EDRF, which inhibits the contractile process by stimulating guanylate cyclase leading to the production of cyclic guanosine monophosphate. EDRF also potentially inhibits platelet aggregation and adhesion. EDRF has a very short half-life (approximately 6 sec), and it may be simply nitric oxide.

EDCF was identified as a vasoconstrictor peptide in the conditioned medium from cultured bovine aortic endothelial cells. It is probably identical to endothelin obtained from the supernatant of cultured endothelia of porcine aorta. Endothelin is probably produced *in vivo* because its mRNA is expressed in the cultured endothelial cells of cerebral microvessels. Endothelin produces potent constriction of cerebral microvessels as well as major cerebral arteries (Fig. 10).

Prostacyclin, EDRF, and EDCF synthesized in cerebral vascular endothelium could contribute to the physiological regulation of cerebral circulation, and the absence or dysfunction of endothelial cells may play a role in pathological vascular events associated with atherosclerosis, thrombogenesis, and cerebral vasospasm.

V. The Blood–Brain Barrier

The unique barrier function has been observed not only in the brain capillaries but also in the major cerebral arteries (Fig. 11). Cerebral capillaries differ from other capillaries owing to the presence of tight junctions between contiguous endothelial cells and a paucity of pinocytotic vesicles. These morphological features constitute the blood–brain barrier. There is also an enzymatic component in the cerebral vessels; a high activity of monoamine oxidase, which presumably results in effective degradation of the catecholamines that pass the endothelium. The adenosine triphosphatase (ATPase) activity in the brain capillaries is localized to the basal lamina and to the membranes of glial end feet, whereas in nonbarrier capillaries ATPase activity is located in the endothelial pinocytotic vesicles. Such features also may affect the permeability properties and the metabolic function of the blood–brain barrier. It is thought that the development of the blood–brain barrier is induced by contact between cerebral

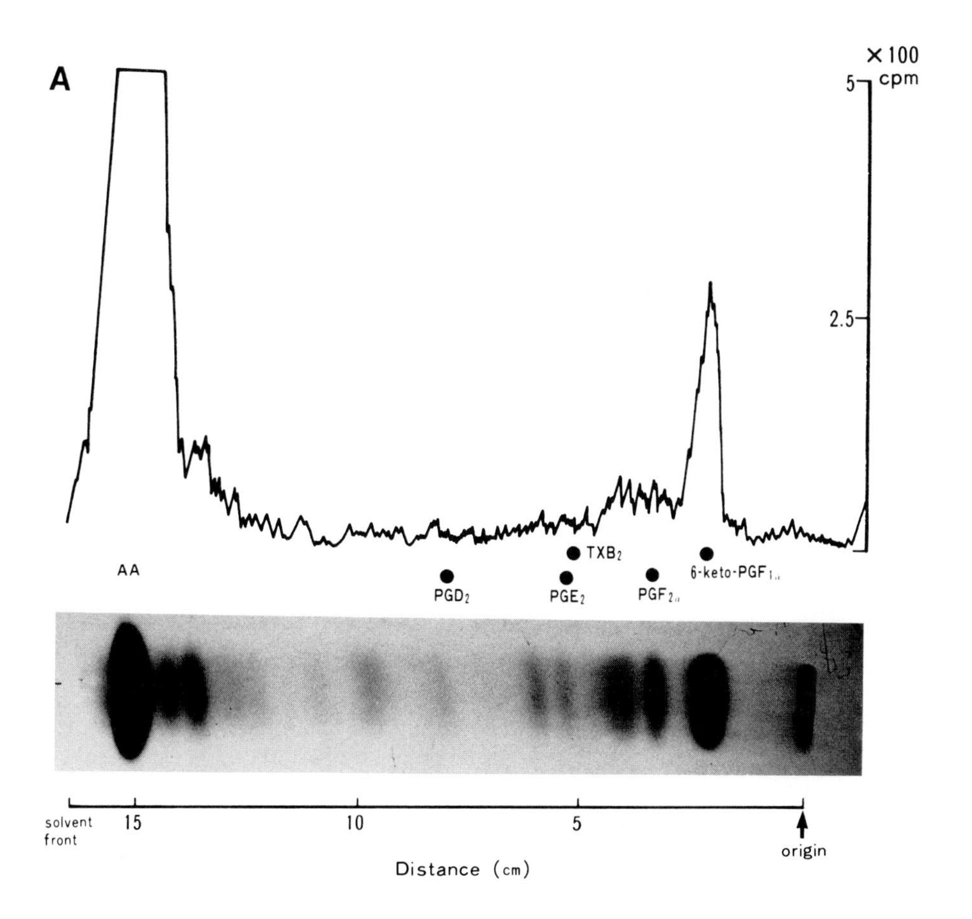

FIGURE 8 Thin-layer radiochromatograms (A) and autoradiograms (B) obtained after incubation of arterial specimens with 0.2 μCi ^{14}C-arachidonic acid (AA) for 30 min at 37°C. ^{14}C-arachidonic acid was mainly transformed to 6-keto-$PGF_{1\alpha}$, the end product of PGI_2 metabolism, and a very small amount of $PGF_{2\alpha}$ was also noted.

endothelium and astrocytes. In fact, γ-glutamyl transpeptidase, which plays an important role in amino acid transport across the blood–brain barrier, often disappears in cultures of cerebral endothelium, but it can be induced by coculturing the endothelium with glial cells. [*See* BLOOD–BRAIN BARRIER.]

VI. The Blood–Arterial Wall Barrier

Endothelial cells in the intradural major cerebral arteries have a barrier function, which is similar to the parenchymal microvessels of the brain. Proximal segments extending 1–4 mm from the origin of the intradural segments of both internal carotid and vertebral arteries have incomplete barrier function. Focal barrier deficiency was also noted at the branching sites of intradural major cerebral arteries and may relate to atherogenesis.

VII. Pharmacological Characteristics of Cerebral Arteries

The responses of cerebral vessels to vasoactive substances differ from those in systemic vessels. Cerebral vessels, although receiving sympathetic innervation equal to that of the systemic vessels, are relatively insensitive to norepinephrine, perhaps because they have fewer α-adrenoceptors. The differences may relate to the different embryonic origin of the two types of vessels, which arise from different primordial cells. The site of fusion of the arteries corresponds to the abrupt change in responsiveness to norepinephrine. The sensitivity of cerebral arteries to norepinephrine differs between species; norepinephrine is a much more potent cerebral vasoconstrictor in the human and monkey than in rabbits, dogs, or cats. It has been recently proposed

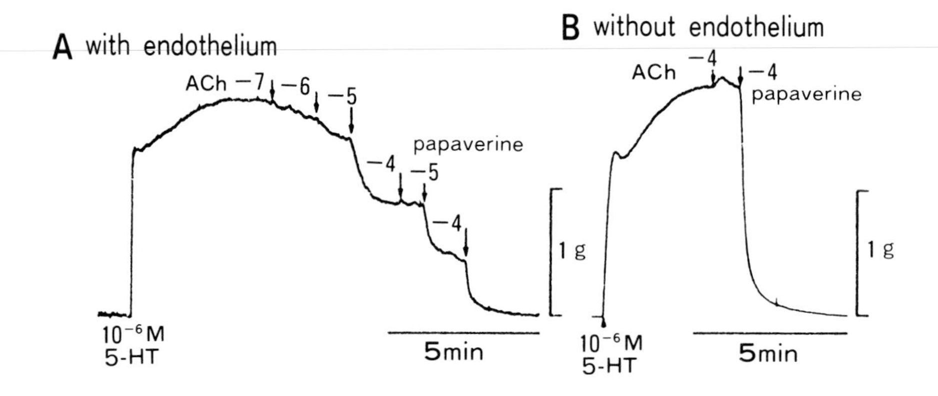

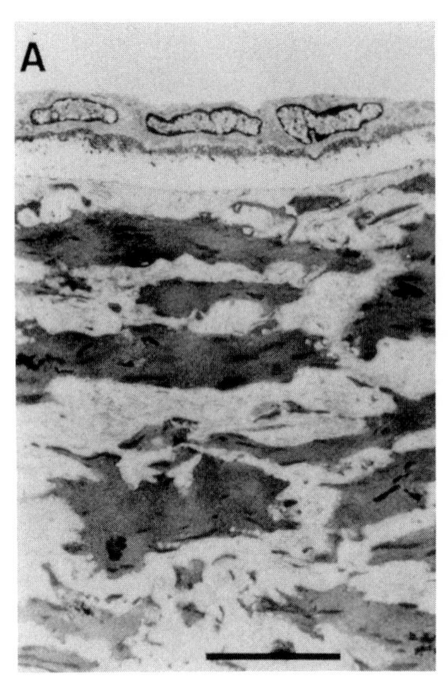

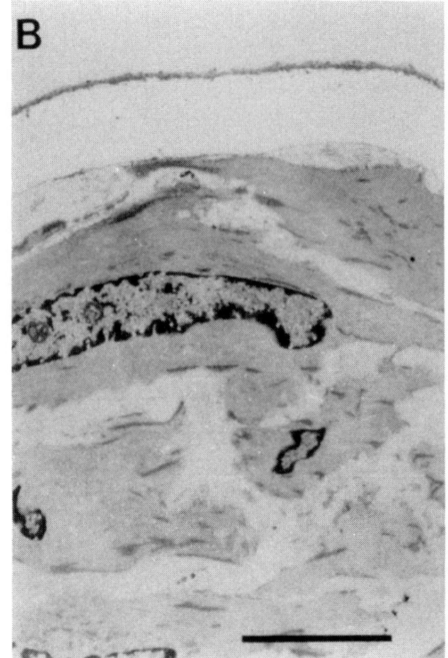

FIGURE 9 Upper: Effect of acetylcholine (ACh) on serotonin (5-HT)-induced contraction of rabbit basilar artery with or without endothelium. A. Typical pattern of dilator response induced by ACh in basilar artery with intact endothelium. B. Typical pattern of effect of ACh on basilar artery after removal of endothelium. Ordinate indicates absolute contractile tension of rabbit basilar artery. The number with the arrow indicates the log molar concentrations of ACh or papaverine in the bath. Lower: Electron microscopic views of the rabbit basilar artery with (A) or without (B) endothelium.

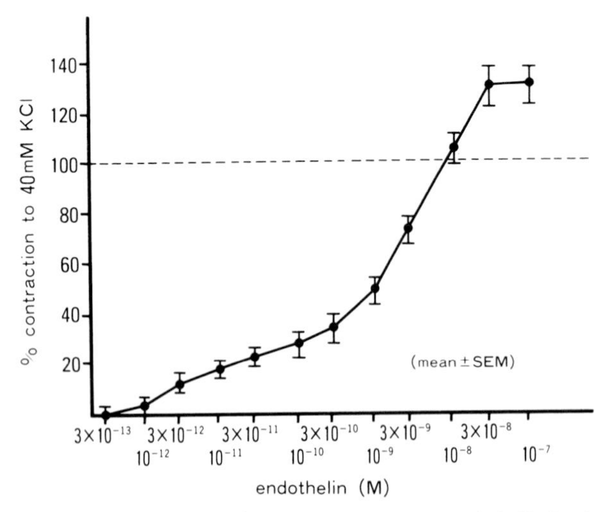

FIGURE 10 Concentration–response curve to endothelin for the canine basilar arteries.

that the different sensitivity of species differences may be due to differences in postjunctional α-adrenoceptors. Systemic arterial smooth muscles contain mainly postjunctional α_1-adrenoceptors, whereas systemic venous smooth muscles contain both α_1- and α_2-adrenoceptors. Postjunctional α-adrenoceptors in canine and feline cerebral arteries resemble the α_2-subtype, whereas in monkey and

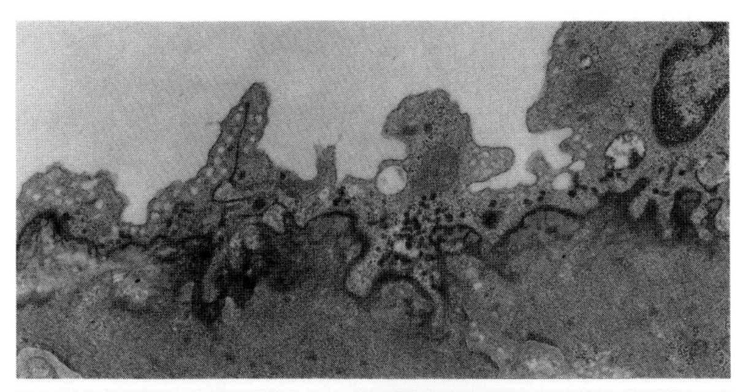

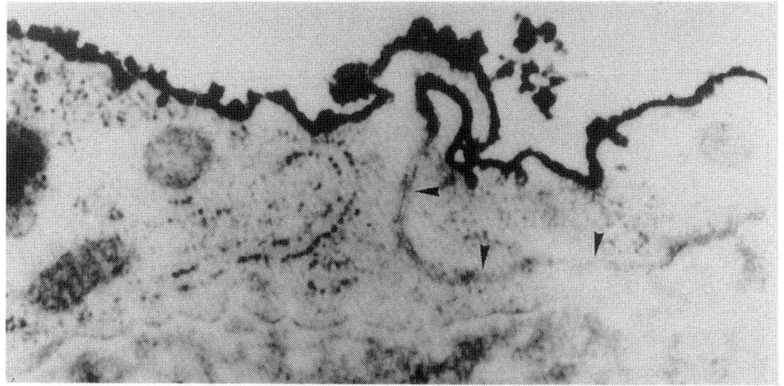

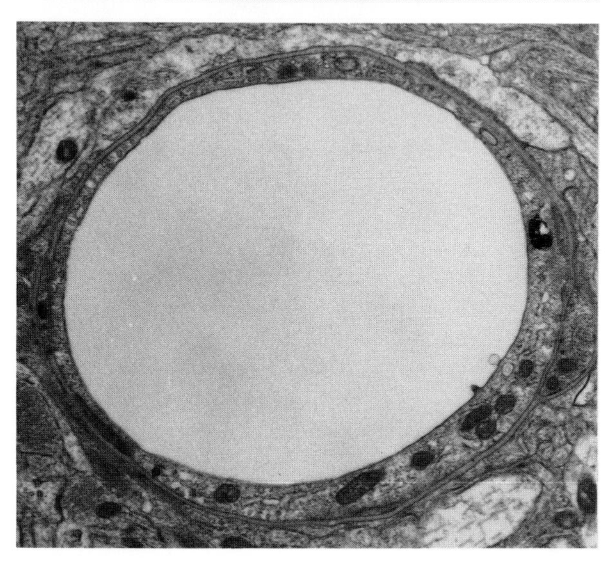

FIGURE 11 Blood–brain barrier and blood–arterial wall barrier. Horseradish peroxidase (HRP) was administered intravenously, and the animal was sacrificed by perfusion-fixation 10 min after the infusion. Upper: The arteriole in the outer layer of small intestine of a rat. HRP permeated into the subendothelial space. Note the HRP reaction products in the interendothelial cleft, plasmalemmal vesicles, and the subendothelial space. Middle: The basilar artery of a rat. Interendothelial cleft (arrowheads) was devoid of HRP reaction products. HRP did not permeate into the subendothelial space. Lower: The brain capillary of a dog. HRP reaction products did not permeate into the basement membrane and into the adjacent extracellular space.

human cerebral arteries, they resemble the α_1-subtype. In pig cerebral arteries norepinephrine is a potential vasodilator. This is probably due to the dominancy of β-adrenoceptors in pig cerebral arteries.

Species differences in cerebral artery responses to other vasoactive substances have also been reported. Histamine is a prominent vasodilator in human cerebral arteries, whereas it is a strong vasoconstrictor in rabbit cerebral arteries. Human and monkey cerebral arteries are more sensitive to thromboxane A2 than canine cerebral arteries.

Recent studies using calcium channel-blocking agents have revealed that contractions of cerebral arteries are more dependent on an influx of extracellular calcium than other systemic arteries. Thus, contractile responses of the basilar artery to serotonin depend almost entirely on the influx of extracellular calcium. In contrast, contraction of the saphenous artery in response to serotonin is produced by release of intracellular calcium as well as by influx of extracellular calcium. These findings suggest that the inhibition of the influx of extracellular calcium into cerebral arterial smooth muscle cells with calcium channel-blocking agents may increase CBF after an ischemic stroke or may prevent the occurrence of cerebral vasospasm following subarachnoid hemorrhage.

Bibliography

Bevan, J. A. (1981). A comparison of the contractile response of the rabbit basilar and pulmonary arteries to sympathomimetic agonists: Further evidence for variation in vascular adrenoceptor characteristics. J. Pharmacol. Exp. Ther. **216,** 83–89.

De Mey, J. G., and Vanhoutte, P. M. (1981). Uneven distribution of postjunctional alpha$_1$- and alpha$_2$-like adrenoceptors in canine arterial and venous smooth muscle. *Circ. Res.* **48,** 875–884.

Edvinsson, L., Ekman, R., Jansen, I., McCulloch, J., and Uddman, R. (1987). Calcitonin gene-related peptide and cerebral blood vessels: Distribution and vasomotor effects. *J. Cereb. Blood Flow Metabol.* **7,** 720–728.

Furchgott, R. F., and Zawadzki, J. V. (1980). The obligatory role of endothelial cells in the relaxation of arterial smooth muscle by acetylcholine. *Nature* **288,** 373–376.

Heistad, D. D., and Kontos, H. A. (1983). Cerebral circulation. *In* "Handbook of Physiology" (J. T. Shepherd, F. M. Abboud, and S. R. Geiger, eds.), pp. 137–182. American Physiological Society, Bethesda, Maryland.

Parmer, R., Ferrige, A. G., and Moncada, S. (1987). Nitric oxide release accounts for the biological activity of endothelium-derived relaxing factor. *Nature* **327,** 524–526.

Sasaki, T., Kassell, N. F., Torner, J. C., Maixner, W., and Turner, D. (1985). Pharmacological comparison of isolated monkey and dog cerebral arteries. *Stroke* **16,** 482–489.

Streeter, G. L. (1918). The developmental alterations in the vascular system of the brain of the human embryo. *Contrib. Embryol.* **8,** 5–38.

Toda, N. (1983). Alpha adrenergic receptor subtype in human, monkey and dog cerebral arteries. *J. Pharmacol. Exp. Ther.* **226,** 861–868.

Yanagisawa, M., Kurihara, H., Kimura, S., Tomobe, Y., Kobayashi, M. Mitsui, Y., Yazaki, Y., Goto, K., and Masaki, T. (1988). A novel potent vasoconstrictor peptide produced by vascular endothelial cells. *Nature* **332,** 411–415.

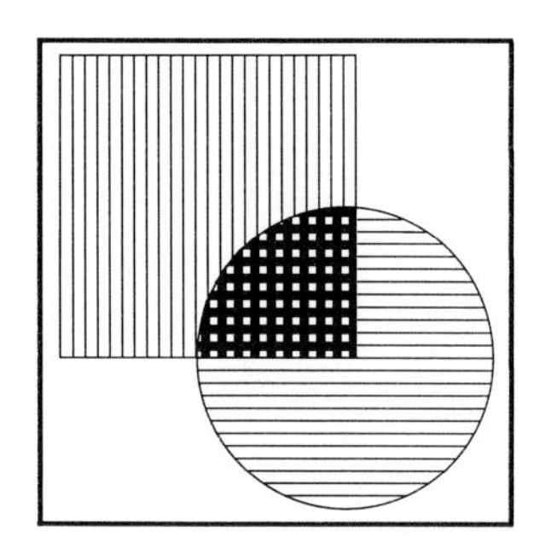

Chemotherapeutic Synergism, Potentiation and Antagonism

TING-CHAO CHOU, *Memorial Sloan-Kettering Cancer Center, Cornell University*

DARRYL RIDEOUT, *Research Institute of Scripps Clinic*

JOSEPH CHOU, *Harvard University*

JOSEPH R. BERTINO, *Memorial Sloan-Kettering Cancer Center, Cornell University*

Glossary

Antagonism Smaller than the expected additive effect

Combination index (CI) Introduced to provide a quantitative measure of the degree of drug interaction for a given end point of effect measurement; CI = 1, < 1, and > 1 indicate additive, synergistic, and antagonistic effects, respectively

Dose-reduction index Introduced in 1988, provides a measure of how much the dose of each drug in a synergistic combination may be reduced at a given effect level compared with the doses of each drug alone; toxicity toward the host may be avoided or reduced when the dose is reduced

Isobologram Graph representing the equipotent combinations of various concentrations (or doses) of two drugs that produce the same effect; it can be used to identify synergism, additivism, or antagonism

Potentiation, inhibition In drug combinations, a potentiator or inhibitor will augment or depress the effect of other drug(s), whereas by itself it has no effect. The presentation of results are straightforward: they are usually presented by a percent or fold of changes incurred as a result of the administration of the potentiator or inhibitor.

Selectivity index, therapeutic index Selectivity index is the ratio of the median-effect concentrations (or doses) for toxicity to the host in relation to that concentration for the target, such as tumor cells or pathogens (e.g., the ratio of the median-effect toxic dose versus the median-effect therapeutic dose: ID_{50}/ED_{50}); the chemotherapeutic index has similar meaning but usually refers to a therapeutic situation and may refer to different effect levels (e.g., 10% toxic dose versus 90% therapeutic dose: TD_{10}/ED_{90}); the higher the values of these indices, the safer the drug

Synergism Greater than the expected additive effect

I. Why Combination Chemotherapy?

When two or more treatment modalities are used for combination therapy (e.g., chemotherapy, radiation therapy, thermotherapy), the effects can be synergistic, additive, antagonistic, or a mixed effect (e.g., antagonistic at low-dose levels, synergistic at high-dose levels). For combination chemotherapy, a multiple target approach with several drugs can be used, or multiple drugs active with different mechanisms against a single target can be used. For example, in the treatment of cancer, combination chemotherapy attempts to deal with the heterogeneous cell populations or to avoid the development of resistance to make the eradication of the malignant cells feasible.

A. Main Goals of Combination Chemotherapy

1. To increase therapeutic efficacy against the target(s) by synergism, thus lowering the dose of one or both drugs for a given effect.
2. To decrease toxicity or untoward side effects against the host by reducing the dose(s) in synergistic combinations.
3. To minimize or delay the development of drug resistance by use of lower doses and/or multiple targets for attack. The criteria for comparing the development of resistance by a single drug and by multiple drugs are based on a given degree of drug effect and a given duration of exposure to the drugs. Increased therapeutic efficacy decreases the length of therapy for a given degree of therapeutic effect; in turn, the shorter duration of therapy may decrease the risk of developing resistance.
4. To achieve selective synergism against the target (pathogenic microorganisms, cancer cells, or parasites) and/or selective antagonism toward the host, hence increasing the therapeutic index.

Chemotherapy is a double-edged sword: It can harm both the target and the host. Efficacy must be balanced against toxicity. Combination chemotherapy offers options to exploit treatment advantages with respect to efficacy, toxicity, and schedules of treatment.

If the mass-action law is strictly followed, the combined effect of two drugs should be additive. If synergism or antagonism occurs, one or more mechanisms may be involved, which may or may not be understood. A study of dose–effect relationships permits the determination of whether or not synergism exists and, if so, the quantification of the degree of synergism. However, such a study does not provide information with regard to how and why synergism occurs. Therefore, two issues must be addressed separately: the determination of synergism–antagonism and the interpretation of synergism–antagonism.

B. Some Selected Examples of Drug Combinations

Numerous examples of synergism and antagonism in biomedical literature and reviews are available. The following are some selected examples of drug combinations.

(1) Drugs that inhibit two enzymes forming part of a single, vital metabolic pathway may exhibit synergism or antagonism due to their combined effects on the synthesis of the final metabolic product or products. Because of branching metabolic pathways, feedback inhibition, and potentiation by metabolites, it is usually impossible to predict whether synergism or antagonism will occur in such antimetabolite combinations without detailed knowledge of the pathways and extensive calculations. One clinically useful example of antimetabolite combinations involves sulfamethoxazole and trimethoprim, both of which inhibit reactions required for bacterial DNA synthesis with an end result of strong synergistic effect.

(2) A combination of the prodrugs leucovorin and 5-fluorouracil exhibits a clinically relevant synergistic effect against gastrointestinal and other carcinomas because of simultaneous binding of methylene tetrahydrofolate and 5-fluorodeoxyuridine monophosphate (the active metabolites) to the enzyme thymidylate synthase. The combination, in contrast to either methylene tetrahydrofolate or FdUMP alone, forms a tighter binding ternary complex with the enzyme, thus preventing *de novo* DNA synthesis.

(3) Almost all drugs are metabolized to some extent in the human body, due mainly to the activity of enzymes found in normal tissues and/or at the site of the disease. For example, penicillin and related antibacterial agents (β-lactams) can be converted to inactive metabolites by the enzyme β-lactamase found in some resistant strains of bacteria. Hence, although β-lactams alone are not effective for the treatment of infections caused by these bacteria, combinations of β-lactams with β-lactamase inhibitors such as sulbactam are effective. Sulbactam potentiates the antibacterial activity of β-lactams by preventing the breakdown of β-lactams by the bacterial β-lactamase, thereby overcoming drug resistance.

(4) Most chemotherapeutic agents must reach targets inside cells (as opposed to surface targets) to exert their effect. Tumor cells and microbes may be resistant to drugs because of slow rates of drug influx and/or rapid rates of drug efflux. The action of such drugs can be enhanced by adding a second agent that affects the outside of the cell. For example, penicillin inhibits the synthesis of cell wall components in bacteria such as *Streptococcus mitis,* thereby enhancing the permeability of streptomycin. When streptomycin and penicillin are used in combination, relatively more streptomycin reaches its intracellular target (the ribosome). As a result, streptomycin and penicillin exhibit therapeutically useful antibacterial synergism. Some tumors possess or develop resistance to vinblastine, doxorubi-

Dose-Effect Curves

AZT

Mix

IFN

Fraction Affected (Fa)

Dose

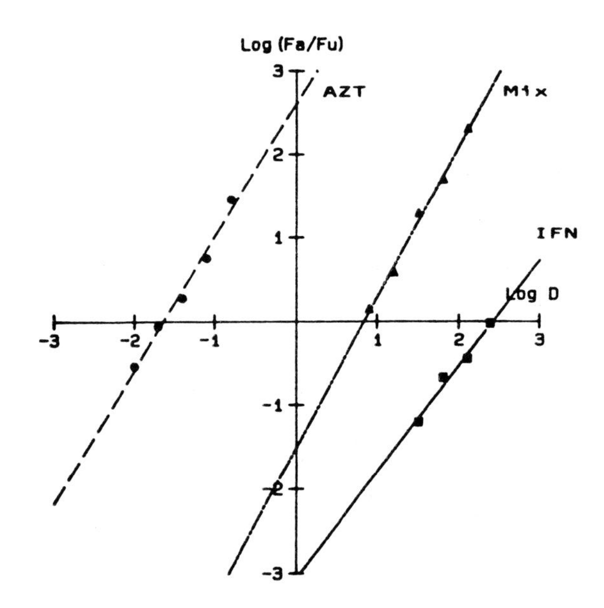

FIGURE 1 Dose–effect curves and the median-effect plot for inhibition of human immunodeficiency virus, type 1, (HIV-1) by recombinant interferon α (IFN) and 3′-azido-3′-deoxythymidine (AZT), and AZT (in μM) and IFN (in U/ml) mixture (1 : 800). Data were obtained using reverse transcriptase assays. The parameters obtained are $(D_m)_{AZT} = 0.0233$ μM, $(m)_{AZT} = 1.593$, $(D_m)_{IFN} = 263.06$ U/ml, $(m)_{IFN} = 1.262$, $(D_m)_{mix} = 6.856$, $(m)_{mix} = 1.803$. M and Dm are obtained from the slope and intercept, respectively.

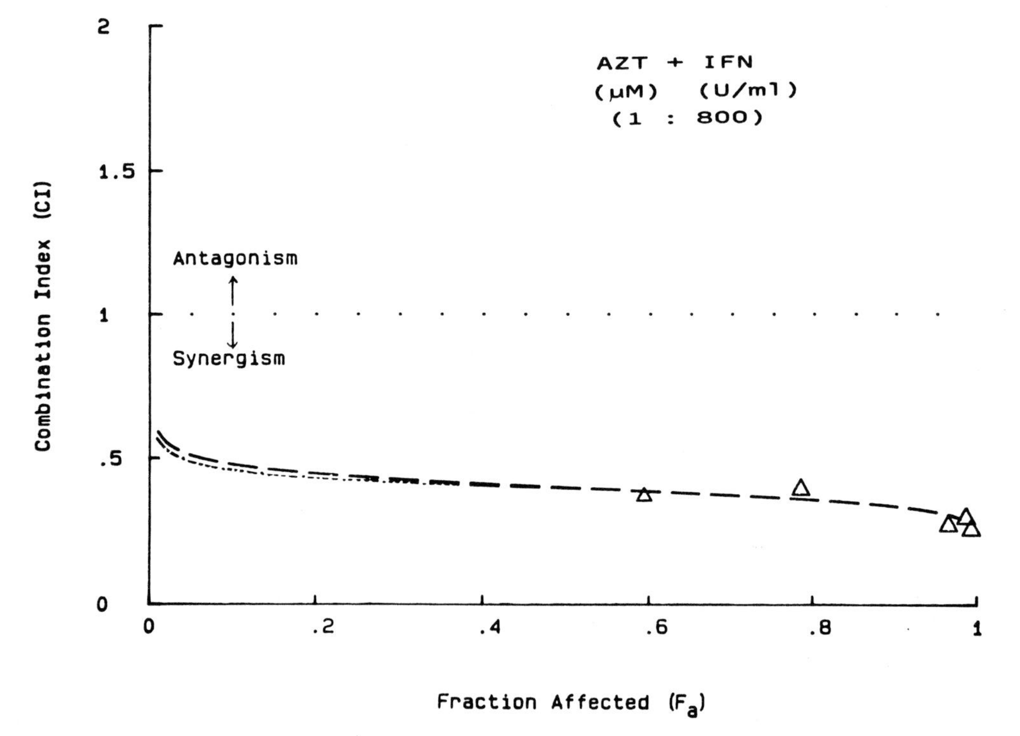

FIGURE 2 The F_a–CI plot showing synergism–antagonism with combination index (CI) as a function of fraction affected (f_a). Parameters shown in Figure 1 were used for the calculations. CI values for each combination data point are shown with triangles. Dashed and dotted lines are computer-simulated results of CIs assuming mutually exclusive and mutually nonexclusive interactions, respectively, between AZT and IFN. The results show strong synergism between AZT and IFN with CI < 1. In this example, both exclusive and nonexclusive assumptions give very similar results.

cin, and other drugs because these molecules are actively pumped out of the cell by an efflux pump known as the p-glycoprotein. This resistance can be reversed by clinical agents such as verapamil and quinidine, which inhibit the activity of the pump.

(5) The simplest possible mechanism of synergism involves direct covalent combination between two agents to form a more bioactive molecule. A number of aldehyde–hydrazine derivative combinations exhibit antineoplastic or antibacterial synergy *in vitro* because cytotoxic hydrazones, formed *in situ*, are more cytotoxic than each drug alone.

(6) Combinations of drugs exhibiting synergism can be superior to single drugs in preventing the development of drug resistance in tumors and microbes. If the drugs act on two different macromolecular targets, the probability of a mutation or mutations leading to resistance to both drugs simultaneously is quite small. Two drugs that exhibit collateral sensitivity can also be very useful in this regard: If the target cell or microbe becomes resistant to one agent, it may simultaneously become hypersensitive to the second agent in the combination. For example, tumors that become resistant to 6-mercaptopurine lose their ability to synthesize purines from hypoxanthine and thus, become highly susceptible to methotrexate, a folate antagonist that blocks the *de novo* biosynthesis of purines.

(7) 3′-Azido-3′-deoxythymidine (AZT) and recombinant interferon (IFN) have been shown to be strongly synergistic against human immunodeficient virus, type 1 (HIV-1), replication. It is well established that HIV-1 is an etiologic pathogen for acquired immunodeficiency syndrome. AZT inhibits reverse transcriptase, which is required for HIV-1 replication, whereas IFN affects the late-stage, viral particle assembly in the HIV-1 life cycle. Examples of the quantitative determination of synergistic interaction between AZT and IFN against HIV-1 are given in Figs. 1–3.

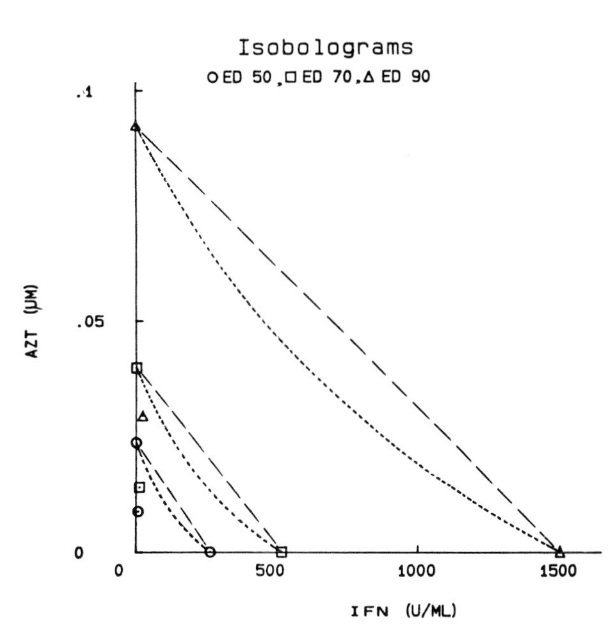

FIGURE 3 The ED_{50}, ED_{70}, and ED_{90} isobolograms. Their construction is based on experimental data, and parameters are shown in Figure 1. Symbols shown are calculated ED_{50} (○), ED_{70} (□), and ED_{90} (△) for each drug alone (IFN on *x*-axis and AZT on *y*-axis) and their equieffective combinations. The points that fall into the lower left indicate synergism, and the points that fall into the upper right antagonism. Dashed and dotted lines depict equieffective simulations for additive effect based on mutually exclusive and mutually nonexclusive assumptions, respectively. The results showed strong synergism between AZT and recombinant interferon α (IFN) against HIV-1 replication.

II. What is Synergism?

The most common objective for combination chemotherapy is to achieve synergistic drug effects, although in special situations antagonism is exploited. An example involves the use of a high dose of methotrexate to produce drastic therapeutic and toxic effects followed by citrovorum rescue (from toxicity) for cancer treatment.

While the pursuit of synergy among therapeutic agents extends as far back as the recorded history of medicine, the concept of synergy was not clearly and vigorously defined in mathematical terms until recently. By definition, synergism is more than an additive effect and antagonism is less than an additive effect. Therefore, the crucial question is: what is an additive effect? Unless additivity of drug effects is unambiguously defined, concepts of synergism and antagonism are meaningless.

To simplify the discussion, the combination of two drugs will be considered first. The terms synergy, synergism and synergistic effect are frequently mentioned in biomedical literature, but synergy is not clearly defined. Many factors need to be considered for drug combination studies: If drug 1 has a hyperbolic dose–effect curve and drug 2 has a sigmoidal dose–effect curve, how do we predict the expected additive effect? If two drugs are synergistic at ED_{50} (the dose for 50% effect), are they necessarily synergistic at ED_{90} (the dose for 90% effect)? Among different methods frequently used in drug combination studies, what are the limitations of the fractional product method of Webb (1963) for calculation of expected additive effect? Is the classical isobologram of Loewe (1953) generally valid?

III. Underlying Theory: The Median-Effect Principle of the Mass-Action Law

Pharmacological receptor theory and enzyme kinetics are based on the mass-action law. Both have evolved into well-developed disciplines. By using enzyme kinetic systems as models and by carrying out systematic derivation of equations for single substrate–single product reactions and multiple substrates–multiple products reactions with various mechanisms, in the absence and in the presence of single or multiple inhibitors (competitive, noncompetitive and uncompetitive) and their permutations, hundreds of equations have been derived and amalgamated into a general median-effect principle (MEP). A few general equations that describe relationships between dose and effect are presented below. These equations are no longer restricted to different reaction mechanisms or different mechanisms of inhibition, because the kinetic constants such as K_m, K_i, V_{max}, etc. are canceled out.

A. The distribution equation

$$\frac{K_i}{I_{50}} = \frac{E_x}{E_t}, \tag{1}$$

where K_i is the inhibitor ligand-enzyme dissociation constant, I_{50} is the concentration required for 50% inhibition, E_x is the amount or concentration of enzyme receptors or species available for ligand binding at the steady state, and E_t is the amount or concentration for the total enzyme receptors or species. The equation shows that K_i will never be greater than I_{50}. The distribution equation provides the basis for deriving numerous equations that eventually lead to the generalized dose–effect equations (discussed below).

B. The Median-Effect Equation

$$f_a/f_u = (D/D_m)^m, \tag{2}$$

where f_a and f_u are the fractions affected and unaffected, respectively, by a dose (D); D_m is the median-effect dose (ED_{50}); and m is a coefficient signifying the shape of the dose–effect curve. The values of $m = 1$, > 1, or < 1 indicate a hyperbolic, sigmoidal, or negative sigmoidal dose–effect curve, respectively.

In the median-effect equation, both sides represent dimensionless quantities that relate dose (right side) and effect (left side) in the simplest possible way. Thus, when two basic parameters, m and D_m, are determined, the entire dose–effect curve is described; i.e., by rearranging the median-effect equation, the dose required for any given effect may be calculated and the effect for any given dose estimated.

The logarithmic form of the median-effect equation yields

$$\log(f_a/f_u) = m \log(D) - m \log(D_m). \tag{3}$$

Therefore, a plot of dose–effect data with $x = \log(D)$ with respect to $y = \log(f_a/f_u)$ determines m (the slope) and D_m (the antilog of x-intercept) values. This plot is referred to as the median-effect plot. These procedures provide a mathematical basis for simulating dose–effect relationships with a computer.

The MEP, as depicted by its equation, can be readily used for deriving basic equations in biochemistry. Thus, the Michaelis–Menten equation in enzyme kinetics, the Hill equation for higher-order allosteric interactions, the Scatchard equation for receptor binding, and the Henderson–Hasselbalch equation for acid-base equilibrium can all be derived from the MEP in one or two steps. In retrospect, it is not surprising that the fractions affected and unaffected (f_a and f_u) are equivalent to the fractions saturated and unsaturated, the fractions occupied and unoccupied, the fractions bound and free, and the fractions ionized and unionized, respectively. The median-effect dose (D_m) for 50% effect is equivalent to half-saturation (K_m), half-occupied (K), half-bound (K_D), and half-ionized (K_a). Thus, four seemingly different equations are based on the same underlying principle.

The Hill equation and the median-effect equation have similar mathematical forms; however, they have elements of basic differences: (1) The Hill equation was derived for a primary ligand such as substrate or agonist, whereas the median-effect equation was derived for a reference ligand, such as inhibitor or antagonist. The equations were obtained from completely different routes. (2) The Hill equation was derived relative to V_{max} as 100%, which determination requires extrapolation and approximation, whereas the median-effect equation relates the observed effect to the uninhibited control value, which can be precisely determined.

C. Multiple Drug–Effect Equation

The median-effect equation described above is for single drugs. A similar approach can be used for multiple drug situations by systematic derivation of equations and conversion into fractional inhibitions. Depending on whether the effects of two drugs are mutually exclusive (i.e., same or similar modes of action) or mutually nonexclusive (i.e., different or independent modes of action), two different equations can be combined as shown by Eq. 4:

$$\left[\frac{(f_a)_{1,2}}{(f_u)_{1,2}}\right]^{1/m} = \left[\frac{(f_a)_1}{(f_u)_1}\right]^{1/m} + \left[\frac{(f_a)_2}{(f_u)_2}\right]^{1/m} + \left[\frac{\alpha(f_a)_1(f_a)_2}{(f_u)_1(f_u)_2}\right]^{1/m}$$
$$= \frac{(D)_1}{(D_m)_1} + \frac{(D)_2}{(D_m)_2} + \frac{\alpha(D)_1(D)_2}{(D_m)_1(D_m)_2}, \quad (4)$$

where subscripts for the parameters f_a, f_u, D, and D_m denote single drugs (1 or 2) or the combination of two drugs (1, 2). When the effects of two drugs are mutually exclusive, the combined effect is the sum of the first two terms (i.e., $\alpha = 0$), and when the effects of two drugs are mutually nonexclusive, the combined effect is the sum of three terms ($\alpha = 1$). When $m = 1$, regardless of whether $\alpha = 0$ or $\alpha = 1$, the equation provides an exact solution for describing dose–effect relationships based on the proposed model. When $m \neq 1$ and $\alpha = 0$ (mutually exclusive effects of two agents), the equation gives a precise solution; however, when $m \neq 1$ and $\alpha = 1$ (mutually nonexclusive effects of two agents), it gives a first-degree approximation in which the most prominent species of ligand–target interaction in the distribution is taken into account.

When $\alpha = 0$ (mutually exclusive), if lines for drug 1 and drug 2 in the median-effect plots are parallel, the median-effect plot for the mixture with serial dilution will give lines parallel with those of the parent compounds. When $\alpha = 1$ (mutually nonexclusive), if the theoretical graphs of the median-effect plots for drug 1 and drug 2 are parallel, the median-effect plot for the mixture of drug 1 and drug 2 with serial dilution will concave upward. In this case, exclusivity of two drugs can be diagnosed. However, the median-effect plots for each drug alone in actual experiments are frequently not parallel, and exclusivity cannot be determined in this way. Therefore, in actual dose–effect analyses with microcomputers, the calculations are carried out both ways in appreciation of the fact that the method does not always lead to an unequivocal diagnosis for exclusivity or nonexclusivity of action.

D. The Isobologram Equation

Although the use of isobolograms has a long history in diagnosis of additivity, synergism, and antagonism, there is no explicit derivation of equations from defined models to justify their use. By using the multiple drug–effect equation described in Eq. 4, the ED_{50} isobologram can be derived simply by setting $(f_a)_{1,2} = (f_u)_{1,2} = 0.5$. Therefore,

$$\left[\frac{(f_a)_1}{(f_u)_1}\right]^{1/m} + \left[\frac{(f_a)_2}{(f_u)_2}\right]^{1/m} + \left[\frac{\alpha(f_a)_1(f_a)_2}{(f_u)_1(f_u)_2}\right]^{1/m} = 1 \quad (5)$$

or

$$\frac{(D)_1}{(D_m)_1} + \frac{(D)_2}{(D_m)_2} + \frac{\alpha(D)_1(D)_2}{(D_m)_1(D_m)_2} = 1. \quad (6)$$

The multiple drug–effect equation (Eq. 4), as well as the isobologram equation (Eq. 5), describes the additive effect of drug combinations. If experimental results deviate from the additive effect, synergism or antagonism is evident. The left side of the isobologram equation (Eq. 5) has been designated as the drug combination index (CI), wherein CI = 1, < 1, and > 1 represent additivism, synergism, and antagonism, respectively. Computer software for automated construction of isobolograms has been developed according to Eqs. 5 and 6.

The ED_{50} isobologram equation has been extended to the isobologram for other effect levels. The generalized equation is given by

$$\frac{(D)_1}{(D_x)_1} + \frac{(D)_2}{(D_x)_2} + \frac{\alpha(D)_1(D)_2}{(D_x)_1(D_x)_2} = 1. \quad (7)$$

When $\alpha = 0$ and $\alpha = 1$, the classical isobologram and the conservative isobologram are obtained, respectively.

From Eq. 7 it becomes evident that the conservative isobologram equation (when $\alpha = 1$) always yields slightly less synergism than the classical isobologram equation (when $\alpha = 0$). Thus, if the effects of two drugs are mutually nonexclusive, then the indiscriminate use of the classical isobologram will result in overestimation of synergism.

The sham test, which assumes that two drugs under study have the same identity, provides an internal check of the validity of the classical isobologram but not of the conservative isobologram. If a drug is divided into two test tubes in a double-blind experiment for drug combination studies, the results will be necessarily additive because they are the same compound. Because they are the same compound, they are necessarily mutually exclusive as adopted by the classical isobol concept. In real experimental situations, however, there is little chance that two drugs will have identical effects.

IV. Methods of Quantitation of Synergism–Antagonism

At least three major methods with sound theoretical bases can be used for determining synergism, additivism, or antagonism. These methods are (A) the fractional product method of Webb, (B) the classical isobologram method of Loewe, and (C) the CI method of Chou and Talalay.

Because the equations for A and B can be derived from C, the usefulness and limitations of A and B can be clearly defined (Table I). Although isobolograms have a long history in drug combination studies, there are still empirically modified proposals in this field. Thus, a pocket isobologram method has been proposed by some investigators. This method includes reading off from dose–effect curves, and by adding and subtracting the distances on the graphs, it estimates the displacement, referred to as the additive effect, as a range rather than a specific value or index. This empirical method is similar to $2 + 2 = 3$ to 5, which has no obvious theoretical basis. Another proposed isobologram method by some investigators uses the following equation: $P_{ab} = P_a + P_b(1 - P_a)$, where P_a and P_b are the fractions of the organisms responding, respectively, to agents A and B, and P_{ab} is the fraction responding to their combination. However, no evidence indicates that this equation actually describes the classical isobologram of Loewe. In fact, it is identical to the fractional product method of Webb, which although widely used has severe limitations, as shown in Table I.

TABLE I Comparison of Applicability of Different Methods

Methods	Dose–effect curve characteristics			
	Mutually exclusive[a] (similar mode of action)		Mutually nonexclusive[b] (independent mode of action)	
	First order[c]	Higher order[d]	First order[c]	Higher order[d]
Webb's fractional product method[e]	no	no	yes	no
Loewe's isobologram method[f]	yes	yes	no	no
Multiple drug effect equation[g]	yes	yes	yes	yes

[a] Mutually exclusive drugs in a mixture give a parallel median-effect plot with respect to the parent compounds.
[b] Mutually nonexclusive drugs in a mixture give an upwardly concave dose–effect curve with respect to the parent compounds.
[c] Hyperbolic dose–effect curve.
[d] Sigmoidal dose–effect curve.
[e] Webb (1963): $i_{1,2} = 1 - [(1 - i_1)(1 - i_2)]$ or $(f_u)_{1,2} = (f_u)_1(f_u)_2$, where i is fractional inhibition and f_u is the fraction unaffected.
[f] Loewe, 1957, *Pharmacol. Rev.* **9,** 237–242.
[g] Chou and Talalay (1984). Also known as the median–effect principle and the combination index method.

V. Computerized Automation of Analysis

Since the median-effect principle of the mass-action law and its equation and the multiple drug–effect equation and its isobologram provide theoretical bases, it has been possible to interrelate the experimental scheme of design to equations, the parameters, the graphics, and computerized automation of data analysis. The algorithms and examples of stepwise procedures for the determination of synergism, additivism, and antagonism have been reviewed, and computer software for IBM-PC and Apple II computers has been developed.

For a typical *in vitro* combination study of two drugs, the concentrations of each drug alone and their mixture are serially diluted in five to six steps (e.g., twofold series) to give a proper dose range and dose-density spacing. Ideally, the dose range is set between two concentrations below IC_{50} (D_m) and four concentrations above IC_{50}, and the combination ratio for drug 1 and drug 2 is initially kept at their IC_{50} ratio (i.e., equipotency ratio). Because the mixture is serially diluted, the combination ratio remains constant. Mixtures at other ratios can also be made, and they may provide an opportunity to estimate the optimal combination ratio for maximal synergy.

The dose and effect numerical data are then entered into a microcomputer (e.g., IBM-PC) using a software (Chou and Chou, 1987), the median-effect plot (Fig. 1) can be readily constructed, in which $x = \log(D)$ is on the abscissa and $y = \log(f_a/f_u)$ is on the ordinate. The plot is then subjected to linear regression analysis for the determination of the two basic parameters m and D_m and related statistics (e.g., the linear correlation coefficient (r) and mean ± SE for the slope and intercepts. By using the m and D_m values for each drug and their combination, the dose (D_x) required for any degree of effect (f_a) can be calculated from the rearrangement of the median-effect equation:

$$D_x = D_m[f_a/(1 - f_a)]^{1/m}. \quad (8)$$

The D_x values are then substituted in the multiple drug effect equation (Eq. 7) for calculating the CI value:

$$\text{CI} = \frac{(D)_1}{(D_x)_1} + \frac{(D)_2}{(D_x)_2} + \frac{\alpha(D)_1(D)_2}{(D_x)_1(D_x)_2}, \quad (9)$$

in which $(D)_1$ and $(D)_2$ (in the numerators) in combination ($D_{1,2}$) give the same x% inhibitory effect (f_a). If $(D)_1$ and $(D)_2$ concentration ratio is P : Q, then $(D)_1 = (D)_{1,2} \times P/(P + Q)$ and $(D)_2 = (D)_{1,2} \times Q/(P + Q)$. $(D)_{1,2}$ for x% inhibition (f_a) can be similarly calculated from $m_{1,2}$ and $(D_m)_{1,2}$. In this way, the CI values for any degree of effect (f_a) can be simulated to construct an F_a–CI table or an F_a–CI plot (Fig. 2). The classic isobolograms and the conservative isobolograms are also automatically constructed (Fig. 3). Furthermore, the CI values for the actual experimental combination data points will also be calculated automatically to indicate the degrees of synergism and/or antagonism.

It should be noted that isobologram and the F_a–CI plot should give identical conclusions about synergism, additivism, or antagonism. The isobologram is dose-oriented, whereas the F_a–CI plot is effect-oriented. Isobolograms show synergism–antagonism at several effect levels, whereas F_a–CI plots show synergism–antagonism at all effect levels simultaneously.

Assuming that two drugs are mutually exclusive, the dose reduction index (DRI) can be calculated by:

$$\text{CI} = \frac{(D_x)_{1,2} \times P/(P + Q)}{(D_x)_1} + \frac{(D_x)_{1,2} \times Q/(P + Q)}{(D_x)_2}$$

$$= \frac{(D)_1}{(D_x)_1} + \frac{(D)_2}{(D_x)_2} = \frac{1}{(\text{DRI})_1} + \frac{1}{(\text{DRI})_2}, \quad (10)$$

where $(D)_1 : (D)_2 = P : Q$; $(D_x) = D_m \; [F_a/(1 - F_a)]^{1/m}$; $(\text{DRI})_1 = (D_x)_1/(D)_1$; and $(\text{DRI})_2 = (D_x)_2/(D)_2$.

DRI depicts by how many folds the dose can be reduced because of synergism for a given degree of effect when compared with a single drug alone. When the therapeutic efficacy is maintained, reduction of the dose will result in lower toxicity toward the host and may also reduce or delay the development of drug resistance.

The method for dose–effect analysis for the combination of three drugs has been similarly developed and has also been subjected to computerized automation by the procedures of Chou and Chou.

Bibliography

Berenbaum, M. C. (1977). Synergy, additivism and antagonism in immunosuppression: A critical review. *Clin. Exp. Immunol.* **28,** 1.

Bertino, J. R., and Mini, E. (1987). Does modulation of 5-fluorouracil by metabolite or metabolites work in the clinics. *In* "New Avenues in Developmental Cancer Chemotherapy," Vol. 8, pp. 163–184. Academic Press, New York.

Chou, T.-C. (1976). Derivation and properties of Michaelis–Menten type and Hill type equations for reference ligands. *J. Theor. Biol.* **59,** 253.

Chou, J., and Chou, T.-C. (1987). "Dose–Effect Analysis with Microcomputers: Quantitation of ED_{50}, Synergism, Antagonism, Low-Dose Risk, Receptor Ligand Binding and Enzyme Kinetics." Manual and Software Disk for Apple II and IBM-PC Series. Elsevier-Biosoft, Cambridge, United Kingdom.

Chou, T.-C., and Rideout, D. (eds.) (1991). "Synergism and Antagonism in Chemotherapy." Academic Press, San Diego.

Chou, T.-C., and Talalay, P. (1981). Generalized equations for the analysis of multiple inhibitions of Michaelis–Menten and higher order kinetic systems with two or more mutually exclusive and nonexclusive inhibitors. *Eur. J. Biochem.* **115,** 207.

Chou, T.-C., and Talalay, P. (1983) Analysis of combined drug effects: A new look at a very old problem. *Trends Pharmacol. Sci.* **4,** 450.

Chou, T.-C., and Talalay, P. (1984). Quantitative analysis of dose–effect relationships: The combined effects of multiple drugs or enzyme inhibitors. *Adv. Enzyme Regul.* **22,** 27.

Chou, T.-C., and Talalay, P. (1987). Applications of the median-effect principle for the assessment of low-dose risk of carcinogens and for the quantitation of synergism and antagonism of chemotherapeutic agents. *In* "New Avenues in Developmental Cancer Chemotherapy" (K. R. Harrap and T. A. Connors, eds.), pp. 37–64. Bristol-Myers Symposium series. Academic Press, New York.

Goldin, A., and Mantel, N. (1957). The employment of combinations of drugs in the chemotherapy of neoplasia: A review. *Cancer Res.* **17,** 635.

Hitchings, G., and Burchall, J. (1965). Inhibition of folate biosynthesis and function as a basis for chemotherapy. *Adv. Enzymol.* **27,** 417–468.

Loewe, S. (1953). The problem of synergism and antagonism of combined drugs. *Arzneimittel-Forsch.* **3,** 285.

Rideout, D., Calogeropoulou, T., Jaworski, J., and McCarthy, M. (1990). Synergism through direct covalent bonding between agents: A strategy for rational design of chemotherapeutic combinations. *Biopolymers* **29,** 247.

Rothenberg, M., and Ling, V. (1989). Multi-drug resistance: Molecular biology and clinical relevance. *J. Natl. Cancer Inst.* **81,** 907–913.

Steel, G. G., and Peckham, M. J. (1979). Exploitable mechanisms in combined radiotherapy-chemotherapy: The concept of additivity. *Int. J. Radiat. Oncol.* **5,** 85.

Webb, J. L. (1963). Effect of more than one inhibitor. *In* "Enzyme and Metabolic Inhibitors," Vol. 1, pp. 66, 488. Academic Press, New York.

Chemotherapy, Antiparasitic Agents

WILLIAM C. CAMPBELL, Merck Institute for Therapeutic Research

Glossary

Chemotherapy Treatment of disease by chemical means (i.e., by drugs)

Cyst Thick-walled quiescent stage of a protozoan parasite; in the case of the intestinal amebae, the stage shed in the feces of the host

Trophozoite Feeding and growing stage of a protozoan parasite

PARASITIC DISEASES are amenable to control by such means as proper sewerage, clean water, and in the case of some parasites, by controlling the intermediate hosts or invertebrate vectors that transmit them. In practice, however, these means are successful only where political and economic resources allow, and that excludes a large portion of the modern world. In the case of a few parasites, immunoprophylaxis may soon become practicable, but at present it is not in routine use for any human parasitic disease. Thus the control of these diseases still devolves largely on the age-old concept of chemotherapy (i.e., the prevention or cure of disease by administration of chemicals known to destroy or inactivate the infectious agent).

Central to the concept of chemotherapy is the concept of "differential toxicity." As Sir William Osler pointed out, it is dosage that determines whether a substance is a medication or a poison. Modern antibacterial medications generally enjoy a wide margin of safety; their toxicity for the microorganism is vastly greater than for humans. The armanentarium of antiparasitic drugs, however, still contains some weapons of marginal safety as well as those that can be deployed with confidence. In the accompanying tables dosages of particular drugs are given as a matter of reference, but the proper use of drugs requires consideration of the prevailing clinical circumstances as well as an awareness that recommendations change in the light of new information. The dosages in the tables are therefore intended to be representative rather than definitive.

Many parasitic infections are diseases of the developing nations of the tropics. The use of chemotherapy in these countries is often constrained by lack of adequate medical, paramedical, and economic resources. Under these conditions, chemotherapy, like sanitation and vector control, becomes a political as well as scientific problem.

For the present purpose, parasitic diseases will be considered under the headings of diseases caused by protozoa, roundworms, flukes, tapeworms, and arthropods. By convention, bacteria, viruses, and fungi, no matter how parasitic their existence, are not included and may be found elsewhere. To provide greater accessibility for the nonspecialist, individual diseases are listed under the name of the disease rather than the parasite that causes it. This is a significant consideration only for the "older" diseases such as malaria and amebiasis, in which the names of the disease and the parasite are dissimilar.

I. Diseases Caused by Protozoa

A. General Comments

The success of chemotherapy in the treatment of protozoal infections remains highly variable. On the one hand, there are several diseases for which a standard and fairly acceptable treatment has been adopted. These include amebiasis, vaginal trichomoniasis, and giardiasis. For a second group of protozoal diseases, effective drugs are available, but their practical application has severe limitations. These may be in the form of unacceptable toxicity, as in the case of the drugs currently in use for the treatment of African trypanosomiasis, South American trypanosomiasis (Chagas' disease), and leishmaniasis in its various forms. Alternatively, the limitations may be in the form of drug resistance, as is notably the case in malaria. At the other end of the spectrum are protozoal diseases for which treatment is either totally unavailable or amounts to no more than a few experimental drugs of varying degrees of clinical promise and investigational legitimacy. Diseases in this unfortunate category include cryptosporidiosis, *Pneumocystis carinii* pneumonia (PCP), and amebic meningoencephalitis.

B. Protozoal Diseases Primarily of the Intestinal Tract

1. Amebiasis

The mainstay of treatment for amebiasis is the nitroimidiazole compound, metronidazole. (Structure shown below.)

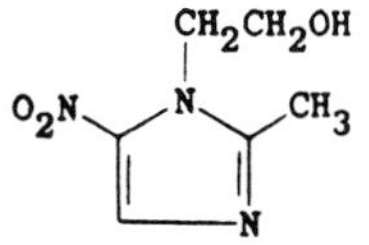

CH_2CH_2OH O_2N N CH_3 N

This drug is active against the so-called luminal amebae in the colon (those thought to be more-or-less free in the canal) and against the amebae invading the wall of the colon and inducing abcesses in other organs, especially the liver. The side effects of metronidazole are fairly common but not usually severe. Like other nitroimidazoles, it has been associated with carcinogenicity and mutagenicity in animal and microbial test systems. In practice, treatment with metronidazole is generally followed by a course of treatment with one of the "luminal" drugs, such as iodoquinol. When asymptomatic "cyst passers" are treated with either kind of amebacidal drug, cysts disappear from the stool of the patient, but this is believed to represent activity against the trophozoite or growing stage of the parasite (and consequent cessation of cyst production) rather than activity against the cysts themselves. [*See* AMEBIASIS, INFECTION WITH *Entamoeba histolytica*.]

2. Cryptosporidiosis

There is no known effective treatment for cryptosporidiosis. Spiramycin (1,000 mg tid/p.o. × 14 days) has been associated with clinical amelioration, but clear-cut evidence of efficacy is not presently on hand.

3. Giardiasis

Giardiasis may be treated with the old antimalarial compound quinacrine hydrochloride shown here.

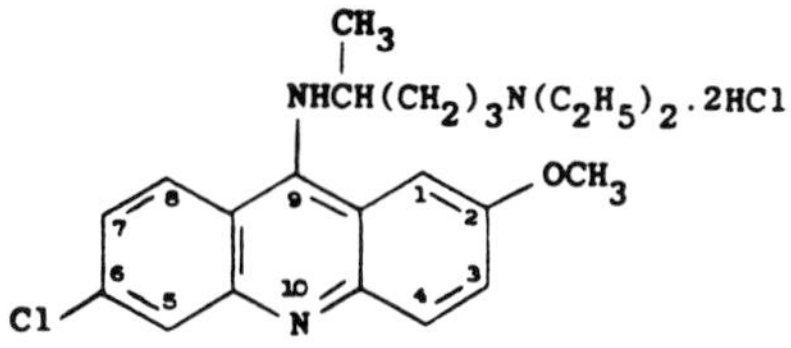

CH_3 $NHCH(CH_2)_3N(C_2H_5)_2 \cdot 2HCl$ OCH_3 Cl N

The nitroimidazole compounds, metronidazole and tinidazole (below), are also effective and are approved for this use in some countries.

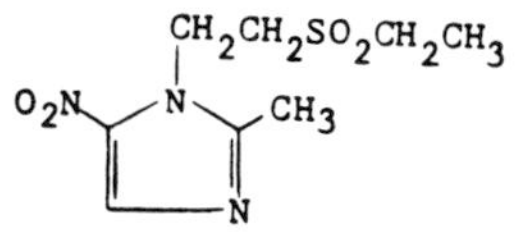

$CH_2CH_2SO_2CH_2CH_3$ O_2N N CH_3 N

C. Protozoan Diseases Primarily of the Urogenital Tract

1. Trichomonal Vaginitis

The mainstay of treatment of trichomoniasis is metronidazole. It is highly effective, but sexual partners of treated patients should also be treated. If this is not done, treatment is likely to be defeated by reinfection.

D. Protozoan Diseases Primarily of the Blood

1. Malaria

Malaria is the chemotherapist's dream—and nightmare. From ancient South American folklore, Western medicine learned that the bark of a certain tree would alleviate the torment of "intermittent

TABLE I Chemotherapy of Diseases Caused by Protozoa[a]

Disease	Representative treatment	Other drugs
Protozoan diseases primarily of the digestive tract		
Amebiasis		
intestinal	metronidazole (750 mg tid, 5–10 days) followed by iodoquinol (650 mg tid, 20 days)	tinidazole, dehydroemetine, paramomycin, diloxanide furoate
hepatic	metronidazole (750 mg tid, 10 days)	chloroquin, dehydroemetine
Cryptosporidiosis	none (see text)	
Giardiasis	quinacrine HC1 (100 mg tid, 5 days)	metronidazole, tinidazole, furazolidone
Protozoan diseases primarily of the urinogenital tract		
Trichomoniasis	metronidazole (250 mg tid, 7 days or 2,000 mg once)	tinidazole
Protozoan diseases primarily of the blood		
Malaria (prevention)	chloroquine phosphate (500 mg, weekly)	pyrimethamine plus sulfadoxine; proguanil, doxycycline, mefloquine
Malaria (cure)	chloroquine phosphate (1,000 mg, followed 6 hr later by 500 mg; then 500 mg daily for 2 days)	quinine dihydrochloride (i.v.); quinine sulfate with pyrimethimine, sulfadiazine, or tetracycline; mefloquine
Trypanosomiasis (African)	eflornithine (see text)	suramin, pentamidine, melarsoprol, tryparsamide
Protozoan diseases primarily of other tissues		
Trypanosomiasis (American)	nifurtimox (8–10 mg/kg/day in 3 or 4 divided doses × 60 days)	benznidazole
Leishmaniasis	stibogluconate-sodium (20 mg/kg/day i.m. or i.v. × 20 days)	pentamidine, amphotericin-B
Toxoplasmosis	sulfadiazine (1 g qid × 3–4 wk) plus pyrimethamine (25 mg daily × 3–4 wk)	trisulfapyrimidines, spiramycin
Pneumocystis carinii pneumonia (PCP)	pentamidine isethionate (4 mg/kg i.v. over 2 hr, daily × 14 days)	trimethoprim plus sulfamethoxazole

[a] Treatment regimens (oral unless otherwise stated) are given for illustrative, not prescriptive, purposes. tid, thrice daily; qid, four times daily; i.v., intravenously, i.m., intramuscularly.

fevers,'' and for more than a century it has been known that a chemical substance named *quinine* was the active ingredient. Subsequent political events led to a search for alternative remedies, and highly effective manmade antimalarial compounds were eventually developed; but the success of the synthetic chemist has been thwarted in part by the success of the malarial parasite in developing ways of eluding the effects of treatment. Drug resistance is not rare in the world of chemotherapy, but because of the enormity of malaria as a disease of humans, and because its treatment has depended so heavily on one or two drugs, the emergence of drug resistance has brought about an enduring crisis. [*See* MALARIA.]

a. Prevention Until recently, the prevention of malaria has rested almost entirely on the 4-aminoquinoline drug chloroquine. (shown here)

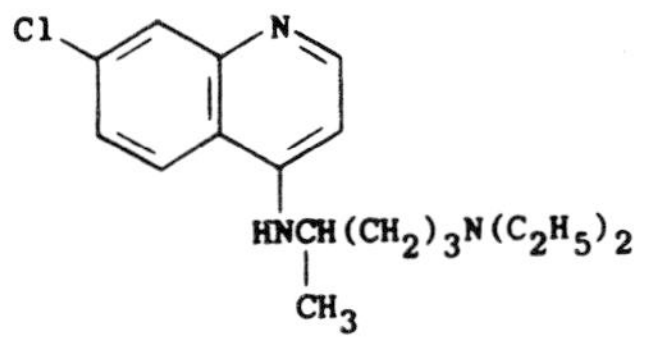

Cl N

$HNCH(CH_2)_3N(C_2H_5)_2$

CH_3

For the most part (that is, when the situation is not complicated by drug resistance), malaria can be prevented by swallowing a tablet of chloroquine once a week. The drug is not a ''causal prophylactic'' (i.e., it will not prevent the sporozoites inoculated by mosquitoes from reaching the liver and developing into asexual multiplicative stages). It is a ''clinical prophylactic,'' because it is active against the stages that multiply in the erythrocytes and cause attacks of malaria. It is, moreover, active against the erythrocytic stages of all four major species of human malaria. If chloroquine medication is continued for 6 weeks after departure from a malarious region, any infection of *Plasmodium falciparum* or *P. malariae* that may have been acquired will "burn itself out" [i.e., all the liver stages will have matured and released the next (asexual) generation of parasites into the blood, where they will take up residence, and chloroquine vulnerability, in the erythrocytes]. In the case of *P. vivax* and *P. ovale*, however, some of the hepatic stages will persist for long periods and may become activated long after the traveler has left the endemic region. To prevent this, the 8-amino-quinoline drug primaquine can be taken to kill the hepatic stages and so prevent their "seeding" the erythrocytes and causing attacks of malaria. Because of the potential toxicity of primaquine, it may be more prudent for the erstwhile traveler to eschew primaquine treatment and simply wait to see if fever ensues. If this happens, the erythrocytic phase can be cured with chloroquine, and radical cure can be achieved with primaquine. In that way, primaquine will be used only when the risk of toxic reaction is offset by a definite need for treatment. In regions where chloroquine resistance is known or suspected, the standard chloroquine prophylactic regimen should be supplemented with, or replaced by, other antimalarial drugs. Because of the constantly changing pattern of drug resistance, chemoprophylactic recommendations are becoming increasingly "tailor-made" to suit a given time and place.

b. Cure Chloroquine (or a related 4-amino-quinoline) is an effective cure for all four species of malaria in humans, but there are qualifications that may be a matter of life and death. If the malaria is due to *P. vivax* or *P. ovale*, chloroquine should be supplemented with primaquine to prevent relapses (the principle involved being the same as that discussed above). If the malaria is due to *P. falciparum*, which may rapidly prove fatal, consideration must be given to the question of whether the particular strain of parasite has become resistant to the drug or is likely, on geographical grounds, to have become so. Where resistance is known or suspected, quinine is the drug of choice. It may be given orally (as the sulfate) or parenterally (as the dihydrochloride). In critically ill patients, parenteral administration is often advisable. In many instances, quinine treatment will not prevent recurrence of malarial attacks (in the 6 weeks or so in which *P. falciparum* remains capable of reinvading the erythrocytes) and so the quinine treatment is often supplemented with a pyrimethamine–sulfadiazine combination or with tetracycline. Where parasite strains are resistant also to quinine, resort should be made to other drugs, including mefloquine (the hydrochloride salt is shown here), tetracycline, or clindamycin.

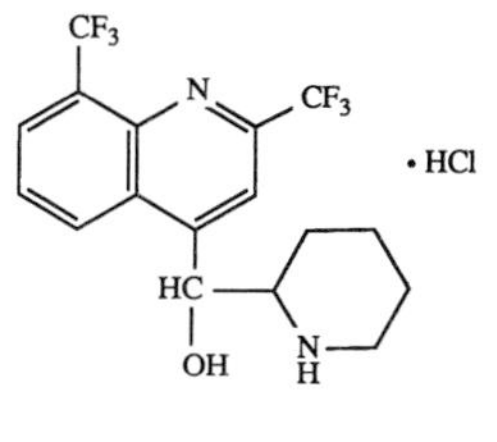

CF_3 N CF_3 · HCl HC OH N H

Because of the serious threat posed by chloroquine resistance, the discovery and development of new antimalarial drugs is of the utmost importance. One such drug, mefloquine, is a quinoline methanol that has been introduced (sometimes in combination with pyrimethamine and sulfadoxine) mainly in southeast Asia. Unfortunately, resistance to mefloquine has already begun to appear. Some antibiotics (e.g., tetracyclines) have antimalarial activity and need to be further evaluated. A drug of exceptional promise is artemisinin, derived from the Chinese herbal extract quinghaosu, which is now undergoing clinical trials.

It has recently been found that certain tricyclic psychotropic drugs (e.g., desipramine) can reverse chloroquine resistance. A combination of the desipramine and chloroquine was curative in monkeys infected with resistant *P. falciparum*, and this discovery holds great promise for the clinical management of life-threatening cases in humans.

2. African Trypanosomiasis (Sleeping Sickness)

African trypanosomiasis, caused by subspecies of the *Trypanosoma brucei* complex, may be prevented by intramuscular injections of the diamidine compound pentamidine. Cure of the clinical disease relies on intravenous injections of the naphthylamine sulfonic acid compound suramin or, especially if the disease has progressed to the point of involvement of the central nervous system, intravenous injections of melarsoprol (below, left) or tryparsamide (below right).

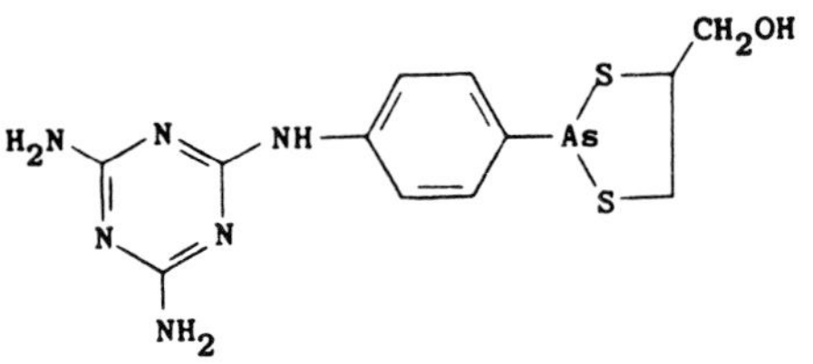

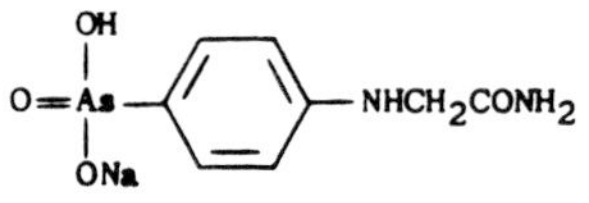

These old drugs are very toxic, especially the last, and much effort has been made to develop safer medications. Among the most promising are eflornithine, allopurinol, bleomycin, and certain nitrofurans. The efficacy of eflornithine is particularly striking; it now appears to be the drug of choice in Gambian Sleeping Sickness, especially in cases in which the central nervous system has become involved and the organisms are resistant to the arsenical compounds. Pending further clinical experience, a representative dosage is 400 mg/kg (p.o. or i.v.) daily for 5 weeks. [*See* TRYPANOSOMIASIS.]

E. Protozoan Diseases Primarily of Other Tissues

1. American Trypanosomiasis (Chagas' Disease)

The drugs used for South American trypanosomiasis are quite different from those used for its African counterpart, but they share the liability of serious toxicity. Indeed the drugs used (i.e., nifurtimox [shown below] and benznidazole) are not only hazardous but also cumbersome to use and of somewhat debatable efficacy, especially in the case of chronic cases.

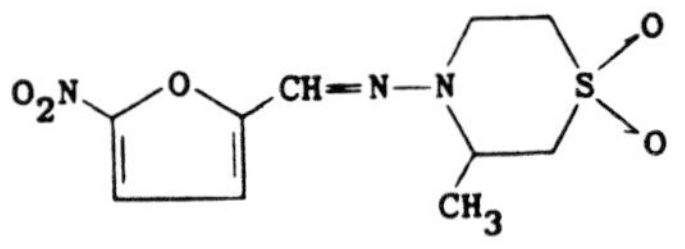

2. Leishmaniasis

The use of pentavalent antimonials has transformed visceral leishmaniasis from a routinely fatal disease to one that is generally curable. This represents a major achievement of 20th Century chemotherapy but must not obscure the fact that these drugs are very toxic and inconvenient to use. Treatment typically consists of parenteral administration of stibogluconate sodium. For certain types of leishmaniasis, pentamidine or amphotericin B may be used. [*See* LEISHMANIASIS.]

3. Toxoplasmosis

The selection of drugs for the treatment of toxoplasmosis is based largely on laboratory experimentation and uncontrolled clinical trials. Nevertheless, the use of sulfa drugs synergized with dihydrofolate reductase inhibitors (e.g., sulfadiazine plus pyrimethamine or sulfamethoxazole plus trimethoprim) has become standard. In pregnant women the macrolide antibiotic spiramycin is sometimes used to obviate the risk of teratogenicity associated with other treatments. Clindamycin might similarly be useful.

4. *Pneumocystis carinii* Pneumonia

Pneumocystis carinii pneumonia is essentially a disease of immunosuppressed patients. It has come to the fore because of the emergence of AIDS as an epidemic, and thus the evaluation of drug efficiency has been of short duration and has been fraught with the difficulties of drug evaluation in severely ill patients and complicated clinical situations. Recent molecular studies suggest that *P. carinii* is a fungus rather than a protozoon, so its inclusion in this section is provisional. The distinction is not trivial because awareness of a phylogenetic relationship is important in guiding the direction taken by chemotherapeutic studies.

Two drugs have been found effective in PCP and are in current clinical use: pentamidine (shown here) and a combination of trimethoprim and sulfamethoxazole.

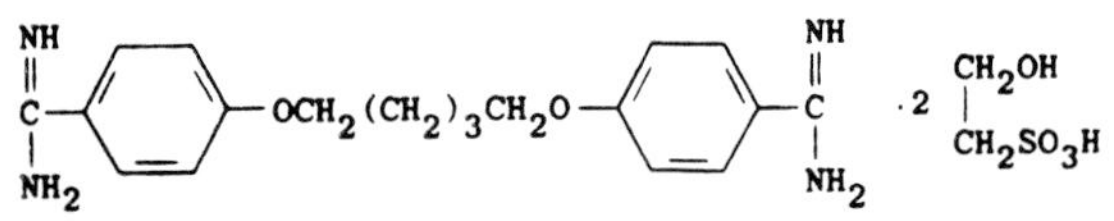

The trimethoprim–sulfamethoxazole combination (given p.o. or i.v.) is generally preferred, but it is not well tolerated in AIDS patients. For reasons that are not understood, some patients do not respond favorably to either drug. In the treatment of various infections, pentamidine is given by intravenous or intramuscular injection; but in cases of PCP, it is now being given by aerosol inhalation to achieve maximum local effect with minimum systemic toxicity. Both forms of treatment (i.e., pentamidine and various combinations of trimethoprim and sulfas or sulfones) may be used to prevent recurrence of pneumonia in treated patients or to prevent the disease in high-risk (immunosuppressed) groups of people. Eflornithine has given promising results in the treatment of PCP in AIDS patients.

II. Diseases Caused by Roundworms

A. General Comments

For roundworm infections of the gastrointestinal tract, current chemotherapy is generally very satisfactory, but for extraintestinal roundworm infections, it is not. The drugs (anthelmintics) used for gastrointestinal helminthiasis have mostly been developed for the routine control of comparable infections in domestic animals and have reached a high degree of chemotherapeutic sophistication. Similar transfer of technology from animal to human is just beginning to occur in the case of the worms living in extraintestinal sites.

B. Roundworm Infections of the Intestinal Tract

1. Common "Soil-Transmitted" Nematode Infections

Narrow-spectrum anthelmintics, directed at one or two particular species, are sometimes used— a prime example being piperazine, which is inexpensive and widely used to eliminate *Ascaris* in children. In recent times, however, broad-spectrum

TABLE II Chemotherapy of Diseases Caused by Helminths[a]

Disease	Representative treatment	Other drugs
Roundworm infections of the intestinal tract		
Common "soil-transmitted" nematode infections	mebendazole (100 mg bid × 3 days)	pyrantel, albendazole
Pinworm infection (enterobiasis)	pyrantel pamoate (1 mg/kg, once; repeat after 2 wk)	mebendazole, albendazole
Roundworm infections of the extraintestinal tissues		
Invasive strongyloidiasis	thiabendazole (25 mg/kg bid × 2 days)	albendazole
Trichinellosis (trichinosis)	mebendazole (300 mg tid × 3 days, then 400 mg tid × 10 days)	albendazole
Onchocerciasis		
adult worms	suramin (200 mg i.v. to test tolerance; then 1 g i.v. weekly × 5 wk)	
microfilariae	ivermectin (0.15 mg/kg once; repeat every 6 or 12 months)	diethylcarbamazine
Filariasis (lymphatic)		
adult worms	diethylcarbamazine (50 mg once, increasing over 3 days to 2 mg/kg tid × 21 days)	
microfilariae	diethylcarbamazine (as for adult worms)	ivermectin
Loiasis		
adult worms	diethylcarbamazine (5 mg/kg daily × 21 days)	—
microfilariae	diethylcarbamazine (5 mg/kg daily × 21 days)	—
Dracunculosis	metronidazole (250 mg tid × 10 days)	thiabendazole, mebendazole
Fluke infections		
Schistosomiasis, clonorchiasis, paragonimiasis	praziquantel (20 mg/kg tid, 1 day)	oxamniquine (*S. mansoni*)
Fasciolopsiasis, metagonimiasis	praziquantel (25 mg/kg tid, 1 day)	—
Fascioliasis	bithionol (40 mg/kg every other day for 12 doses)	dehydroemetine
Tapeworm infections		
Intestinal infections, various species	praziquantel (20 mg/kg once)	niclosamide
Extraintestinal larval infections	(see text)	

[a] Treatment regimens (oral unless otherwise stated) are given for illustrative, not prescriptive, purposes. tid, thrice daily; bid, twice daily; i.v., intravenously.

anthelmintics have become so effective and so safe that they are increasingly used, especially in the tropics where multiple infections are extremely common. The leading broad-spectrum anthelmintics are pyrantel pamoate, mebendazole, and albendazole (structures shown below). All these compounds are effective against the common species of *Ascaris, Ancylostoma, Necator, and Trichostrongylus*. Efficacy against the hookworms (*Ancylostoma* and *Necator*) is seldom perfect, and larger or more frequent doses may have to be used. Albendazole enjoys an advantage in that it is generally effective as a single dose, and is becoming increasingly more widely used. Mebendazole and albendazole are also effective against the whipworm, *Trichuris*. All three are inadequately active against *Strongyloides,* and for this helminth, the older drug, thiabendazole, is still used despite its greater frequency of side effects.

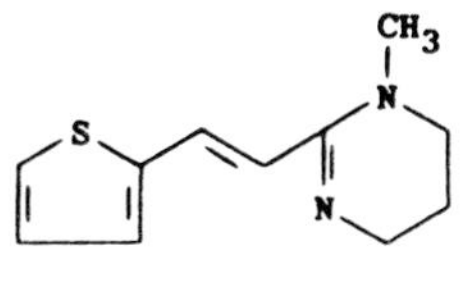

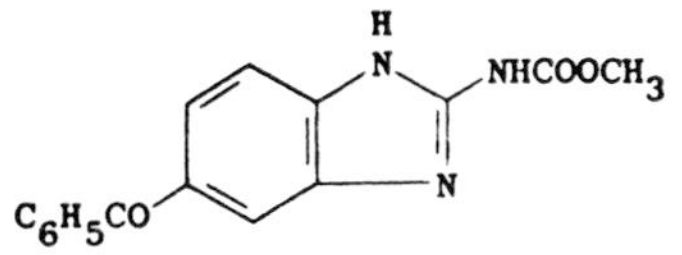

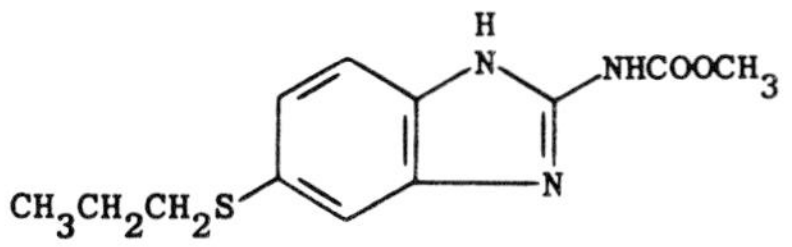

2. Pinworm Infection (Enterobiasis)

The broad-spectrum anthelmintics pyrantel pamoate, mebendazole, and albendazole are all highly active against the pinworm *Enterobius vermicularis* and are much more convenient to use (and usually much safer) than the drugs used previously.

3. Miscellaneous

Infections caused by *Capillaria philippinensis* may be treated with mebendazole, but a prolonged treatment regimen may be needed. No treatment is known for anisakiasis caused by ingestion of uncooked fish harboring larvae of various anisakine nematodes.

C. Roundworms of the Extraintestinal Tissues

1. Invasive Strongyloidiasis

Although *Strongyloides stercoralis* is a parasite of the small intestine, its most dramatic and life-threatening aspect is its propensity to invade other tissues, especially in immunosuppressed patients. Thiabendazole appears to be less effective in such "invasive strongyloidiasis" than in the intestinal disease, but albendazole has shown much promise in early trials. Ivermectin, although active against the intestinal *Strongyloides* in preliminary trials, has not yet been adequately tested against the invasive condition.

2. Trichinellosis (Trichinosis)

The treatment of trichinosis, apart from symptomatic treatment, which is beyond the scope of this article, depends on the use of benzimidazoles. Thiabendazole, the first to be introduced, has been superceded by mebendazole as the drug of choice. On the basis of early clinical trials, it would appear that mebendazole in turn will be replaced by albendazole. The benzimidazoles can be used prophylactically in those rare circumstances in which they can be administered within a few hours or days after ingestion of infected meat. If given within a few hours, the drugs can be expected to prevent maturation of the worms in the intestine. If given within a few days, they will suppress the reproduction of the worms and so prevent the shedding of larvae and subsequent invasion of the musculature. When used therapeutically (i.e., when the progeny of the worms have already settled in the patient's muscles), the benzimidazoles provide clinical improvement, although it is not clear whether this is due to direct destruction of the larvae or to inhibition of their metabolic processes with consequent reduction in the response of the host to metabolic products. In extremely severe cases of trichinosis, treatment must naturally be considered in the context of the patient's overall condition and the possibility, at least theoretical, or host reaction to the protein from dead parasites. [*See* TRICHINOSIS.]

3. Onchocerciasis

Once *Onchocerca volvulus* has matured in the human body and the female worms have released their numerous offspring (microfilariae) to populate the skin, destruction of the adult worms may not prevent or cure the skin lesions or ocular damage associated with the infection (because these are caused by the microfilariae). Killing the adults will, however, limit the number of microfilariae produced, and therefore the severity of the lesions. Moreover, if the microfilariae are killed by drug treatment (see below) destruction of the adult worms will prevent reinvasion of the skin and eyes by newly shed microfilariae. Suramin (shown here) is the only drug used in humans for the destruction of the adult worms (macrofilaricidal effect)

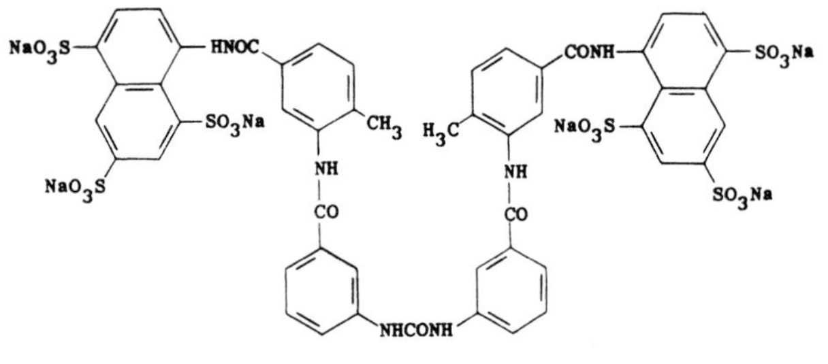

It requires multiple intravenous injections (a serious liability in endemic regions) and is very toxic.

To attack the microfilariae in the dermal and ocular tissues, diethylcarbamazine (DEC) (below, *left*) was used for many years but is now being replaced by ivermectin (below, *right*).

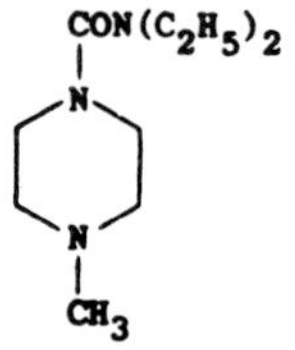

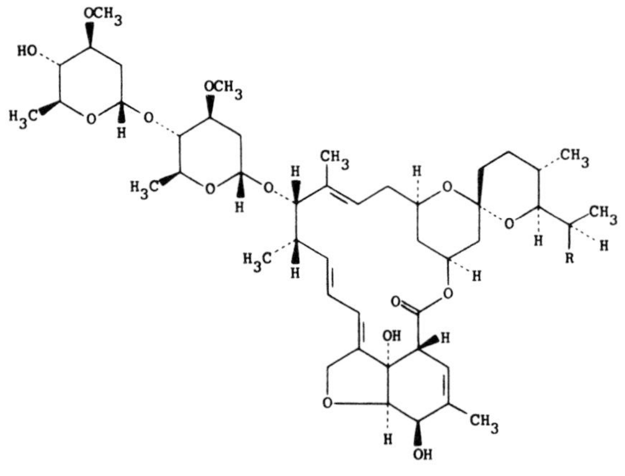

It has long been believed that the adverse reactions associated with DEC therapy (intense itching, etc.) have been caused by host reaction to protein liberated when microfilariae are killed, and thus they would follow the use of any microfilaricidal drug. Nevertheless, reactions after ivermectin therapy, although they do occur and are probably caused by dead microfilariae, are both fewer and less severe than those after DEC. Moreover, ivermectin is given as a single oral dose, whereas DEC requires daily oral doses for a period of about 2 weeks. Reinvasion of the skin by newly shed microfilariae appears to take longer after ivermectin than after DEC. For all these reasons, ivermectin has been used in community-based trials, and if good results continue to be reported, it will become the drug of choice both for individual patients and community control programs.

4. Lymphatic Filariasis

In lymphatic filariasis (caused by *Wuchereria bancrofti* or *Brugia malayi*), the adult worm is the primary pathogen. Multiple doses of DEC give clinical amelioration and probably cause the death of at least some of the adult worms. DEC is also effective against the microfilariae in the bloodstream and so can be used to prevent transmission from human to mosquito. It has been successfully employed in this way to control the disease in geographically isolated regions. As in the case of onchocerciasis, ivermectin is active against the microfilariae when given as a single oral dose and may eventually replace DEC for that use.

5. Other Filariases

Diethylcarbamazine is effective against the microfilariae of *Dipetalonema streptocerca* and *Loa loa* and may induce serious allergic reactions to liberated parasite antigens. However, DEC has little activity, or at least little documented activity, against the microfilariae of *Mansonella perstans* or *M. ozzardi,* and this is in accord with the general lack of hypersensitivity reactions after treatment. Mebendazole or ivermectin may prove to be the microfilaricidal drug of choice for some or all of these four filarial infections. No drug is known to be effective against the adult worms.

6. Dracunculus medinensis (Guinea Worm)

Metronidazole and thiabendazole have been found useful in treating guineaworm infection. The adult worms are more easily extracted from the subcutaneous tissue after treatment, but it is not clear to what extent this is due to their anthelmintic efficacy or to their nonspecific anti-inflammatory effect. Use of either drug requires consideration of potential toxicity.

7. Miscellaneous

Visceral larva migrans, caused by migratory larvae of dog ascarids, is not diagnosed frequently enough to allow systematic evaluation of therapy. There have been reports that thiabendazole treatment has resulted in clinical amelioration of the disease. Cutaneous larva migrans, caused by the migratory larvae of dog or cat hookworms, is readily treated with thiabendzole. Oral treatment is effective, but topical treatment is preferable because it is highly effective and obviates the side effects associated with systemic treatment. No treatment is known for infections due to *Angiostrongylus costaricensis* or *Gnathostoma* sp.

III. Diseases Caused by Flukes (Trematodes)

A. General Comments

The treatment of trematode infections of humans has been revolutionized in recent years by the discovery of praziquantel (shown here).

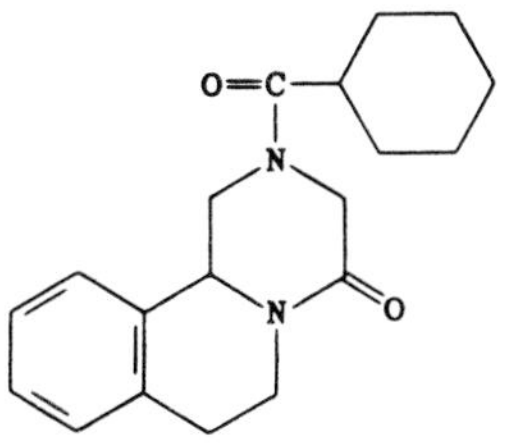

The drugs used previously for such infections were characterized by poor efficacy and awesome toxicity. Some of them had to be administered in difficult and dangerous ways. Praziquantel, however, is given orally, and although adverse reactions are by no means uncommon, they are seldom severe.

B. Intestinal Infections

Praziquantel is effective against *Fasciolopsis buski, Heterophyes heterophyes,* and *Metagonimus yokogawai.*

C. Extraintestinal Infections

1. Clonorchiasis, Opisthorchiasis, Paragonimiasis

Praziquantel is effective against the liver flukes *Clonorchis sinensis* and *Opisthorchis viverrini* and against the lung fluke *Paragonimus westermani.*

2. Fascioliasis

Remarkably, praziquantel does not seem to be effective in the treatment of infections caused by the liver fluke *Fasciola hepatica*. This is especially remarkable because the flukes do take up the drug when exposed to it *in vitro*. Bithional and dehydroemetine are effective, and encouragement may be found in a report that triclabendazole, used for the treatment of fascioliasis in domestic animals, was effective in cases of human fascioliasis.

3. Schistosomiasis

Praziquantel is the drug of choice for infections caused by *Schistosoma haematobium, S. mansoni, S. japonicum,* and *S. mekongi.* For *S. mansoni* infection, oxamiquine (shown here) is an effective and well-tolerated alternative.

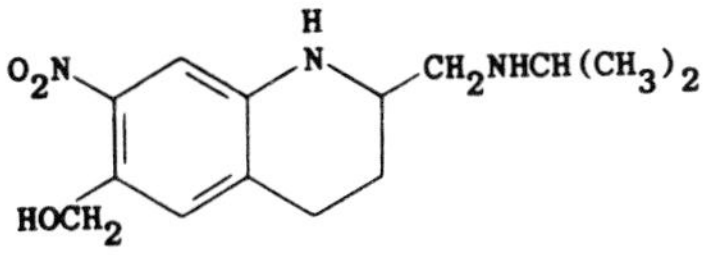

IV. Diseases Caused by Tapeworms (Cestodes)

A. General Comments

The tapeworms share with their platyhelminth relatives, the flukes, a susceptibility to praziquantel, and this has greatly simplified the clinical management of these infections. Unfortunately, however, the most serious tapeworm pathogens are the cystic larvae of certain species, and the cystic infections are extremely difficult to manage in the clinic even with the availability of drugs with some degree of specific activity.

B. Intestinal Infections

Praziquantel is effective against the "adult" or strobilate stage of *Taenia saginata* (beef tapeworm), *T. solium* (pork tapeworm), *Diphyllobothrium latum* (fish tapeworm), and *Dipylidium caninum* (dog tapeworm that occasionally infects humans). The salicylanilide compound niclosamide may be used as an alternative and is well-tolerated. Both drugs cause disruption of the tapeworm "segments" or proglottids, and in the case of *T. solium* infection, this is a cause for some concern. The eggs liberated by the breakdown of gravid proglottids may be swept forward into the stomach as a result of retching and vomiting, and the eggs may thus be activated by gastric fluid. As they pass back into the small intestine, such eggs are capable of hatching and releasing larvae that penetrate the wall of the small intestine and become cysticerci in the extraintestinal tissues, as would normally happen in the pig host. It is not clear whether there is a significant probability of this happening, but cysticercosis, if induced by this or other means, is a disease of such clinical significance that even a low probability cannot be dismissed lightly.

C. Extraintestinal (Cystic) Infections

The clinical management of cystic tapeworm infection is difficult, and both surgical and chemotherapeutic interventions must take into account the size and location of the cysts. Drugs capable of destroying the cysticerci of *T. solium* are known (i.e., praziquantel, flubendazole, albendazole, and metrifonate), but the optimum dosage regimen and guidelines for clinical use have yet to be standardized.

Similarly, the hydatid cysts of *Echinococcus granulosus* are known to be susceptible to the action of the benzimidazole compounds mebendazole, flubendazole, and albendazole, but their use remains experimental; they should be used only when surgical removal is technically impracticable or unduly hazardous. Release of protoscolices from ruptured cysts is always a hazard, but it may be minimized by local application of praziquantel or albendazole. There is evidence from studies in laboratory animals that parenteral administration of benzimidazole compounds may be more effective than oral administration against cystic tapeworms, but this has not been confirmed or extended to human use.

V. Diseases Caused by Arthropods (Ectoparasites)

A. General Comments

The bites of bedbugs, fleas, and flies may require topical treatment to relieve local inflammation, but the control of these insect pests depends on environmental measures rather than human chemotherapy. However, itch mites and lice are more truly parasitic, and human infection may call for specific treatment.

B. Diseases Caused by Mites, Ticks, and Lice

For scabies (infection with *Sarcoptes scabiei*), the standard treatment is topical application of lindane (*gamma*-benzene hexachloride) or the pyrethroid compound permethrin (where registered for such use).

Infections with the follicle mite, *Demodex folliculorum,* rarely need treatment, but some infections have been treated successfully with lindane lotions or sulfur ointments. Severe skin lesions associated with *D. folliculorum* have responded well to treatment of the patient with oral doses of metronidazole (200 mg daily for 6 weeks). This may have been the result of a drug-induced change in the microhabitat (sebum) rather than a direct effect on the mites.

Chigger mites (*Trombicula* sp.) induce local reactions that can be alleviated by local analgesics, and chigger bites can be prevented by application of repellents containing *N, N*-diethyl-*m*-toluamide (deet). Impregnation of clothing with permethrin will protect against chigger bites, not by repellency but by acaricidal action.

Ticks (Ixodidae and Argasidae) are more important as transmitters of disease than as causes of disease. They are commonly removed mechanically, and their bites may be prevented by applying repellents such as deet or dimethylphthalate (CMP) to skin or clothing. Prevention can also be achieved by impregnating clothing with the pyrethroid compound permethrin, which will kill at least some ticks that crawl onto a treated garment. It cannot be assumed that these repellents and acaricides will protect against all stages of all tick species.

For lousiness (infestation with *Pediculus capitis* or *Phthirus pubis*), the standard topical lindane treatment is being replaced in some quarters by topical application of malathion or pyrethroids (synergized with piperonyl butoxide).

Ivermectin, which is effective against itch mites and biting lice in domestic animals, has not yet been evaluated for efficacy against these ectoparasites in humans.

VI. Discovery of Antiparasitic Drugs

The drugs that are in routine clinical use for the treatment of parasitic infections were discovered empirically (i.e., by trial-and-error testing in laboratory systems or by chance observation made in the course of clinical or laboratory investigation). Some of the drugs now entering clinical trial are the products of research into the comparative biochemistry of parasite and host. Both approaches to new drug discovery are thoroughly entrenched, as also, to a lesser extent, is the "semirational" approach in which compounds are empirically screened in an assay designed to detect activity against some particular biochemical function of the parasite.

Regardless of the approach taken to the discovery of compounds with antiparasitic activity, it should be remembered that such discovery is only the first step in a long and difficult process. The probability of finding a new substance with antiparasitic activity is high; but there is only a low probability of carrying it, over innumerable manufacturing, economic, toxicological, and regulatory hurdles, to the point of clinical or commercial success.

VII. Mode of Action of Antiparasitic Drugs

A. General Comments

The biochemical mechanism by which a drug destroys a parasite is known in some instances with reasonable certainty. For other drugs, one or more mechanisms have been proposed with some plausibility, whereas for others the mechanism remains almost wholly mysterious. It is important to remember that the clinical usefulness of a drug, although dependent on its biochemical effect on the parasite and relative lack of effect on the host, is in no way dependent on our knowledge of those effects. Such knowledge, however, is important in many ways (e.g., for understanding in its own right, for the discovery of new compounds with similar mode of action but different properties or effects, for developing assays in which compounds are screened empirically for a given kind of biochemical action, and as an aid in obtaining the approval of regulatory agencies).

Another pitfall in the consideration of mode of action is the tendency to conclude that a particular biochemical effect is the mode of action of a drug just because the effect has been discovered. To take just one of many possible examples, the discovery that benzimidazoles (structure of thiabendazole shown below) inhibit the fumarate reductase of certain helminths led to widespread, although not universal, acceptance of that effect as the mode of action.

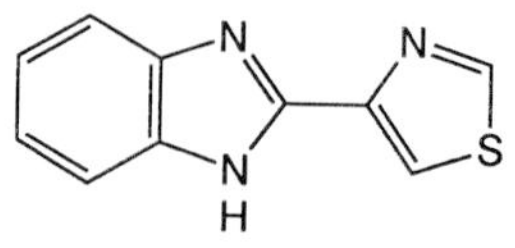

N N S N H

The opinion was bolstered by the observation that the enzyme from a benzimidazole-resistant strain of worm was less easily inhibited than the corresponding enzyme from a sensitive strain. Yet later studies led to the replacement of that hypothesis with an entirely different one (see below). An observed effect that turns out not to be the mode of action (i.e., not the effect that destroys the parasite) may be unconnected with the actual mode of action or may be secondary to it. A drug may, for example, inhibit the uptake of glucose by a worm parasite, but that might be merely a consequence of the drug's primary action and a mere reflection of the "ill health" of the worm. Then, too, an effect observed *in vitro* may not be relevant to the *in vivo* situation. Similarly, an effect observed *in vitro* or *in vivo* under experimental conditions might have little relevance because the concentrations of drug employed in the experiment might vastly exceed those obtained under clinical conditions.

The effect by which a parasite is destroyed may be one that is directly lethal; but for internal parasites, a nonlethal effect may be just as good (that is, just as bad for the parasite) as a lethal effect. As explained by a well-known Australian veterinary parasitologist, a narcotized ascarid worm in the gut of a pig will recover to find that the pig has gone! Intestinal worms will readily be expelled from the host if they are paralyzed and unable to hold their position in the gut; but even extraintestinal worms may die and be resorbed by the host if they are displaced from their normal microhabitat (as schistosomes, for example, are driven from the mesenteric veins to the liver sinusoids after antimonial therapy) or if they are physically or physiologically altered in such a way as to make them susceptible to the host's immune response.

B. Mode of Action of Some Antiprotozoal Drugs

Quinine, chloroquine, and perhaps mefloquine appear to have a common mode of action. In parasitized erythrocytes, a lytic degradation product of hemoglobin, ferriprotoporphyrin IX, is sequestered in a vacuole within the malarial parasite in a nontoxic form (the "malaria pigment"). When chloroquine is taken up by the malaria parasite, the drug binds to the heme compound, forming a complex that apparently cannot be sequestered and that aggregates into clumps that destroy the parasite. The action of mefloquine may be similar, but it is apparently not identical, because mefloquine is effective against many chloroquine-resistant strains of the malaria parasite. The mode of action of the new antimalarial drug artemisinin is not known, but it has been suggested that the mechanism may involve the production of toxic oxygen intermediates or the inhibition of protein synthesis.

The mechanism of chloroquine resistance is not fully understood. Resistant strains of the parasite accumulate less chloroquine than ordinary strains, and this appears to depend not on a decrease in the rate at which the parasite takes up the drug, but rather on an increase in the rate at which it pumps the drug out. The phenomenon can be reversed *in vitro* by the calcium blocker verapamil and can be reversed *in vitro* and *in vivo* by the antidepressant drug desipramine, which is a calcium antagonist and has weak antimalarial activity in its own right. Pyrimethamine (structure shown below) inhibits dihydrofolate reductase and is much more potent against the enzyme of the malarial parasite than against the corresponding mammalian enzyme.

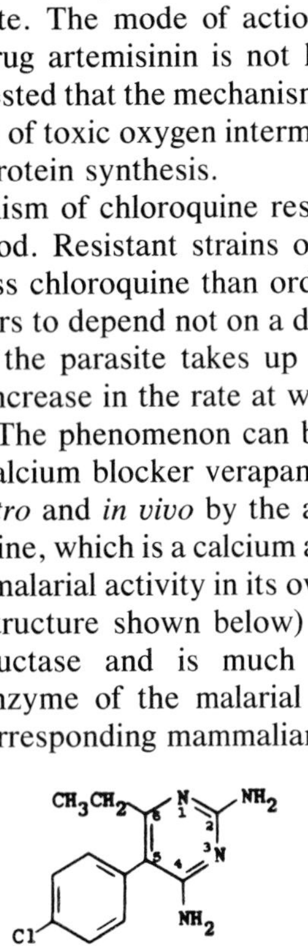

By preventing the reduction of dihydrofolate to tetrahydrofolate (a cofactor essential for the synthesis of thymidylate, purine nucleotides, and certain amino acids), the drug disrupts the multiplication process (shizogony) of the parasite within the host erythrocyte. Resistance to the drug, at least in some strains of *P. falciparum,* is associated with a point mutation in the gene encoding dihydrofolate reductase-thymidylate synthase. The sulfonamides that are used in combination with pyrimethamine are PABA analogues, acting at an earlier stage of folate metabolism.

The mode of action of suramin remains unclear, but it may act by inhibiting the phosphorylation of regulatory proteins in trypanosomes or by inhibiting enzymes involved in glucose metabolism. The drug does not readily penetrate trypanosomes *in vitro,* and its effects occur slowly *in vivo*. The trivalent arsenicals, however, enter trypanosomes quickly, and death (probably caused by the inhibition of enzymes of glucose metabolism) occurs rapidly *in vivo* and *in vitro*. Several compounds have promising activity against trypanosomes but have not yet reached the stage of clinical use. Among these, purine analogues such as allopurinol may kill trypanosomes because those organisms cannot synthesize purines and so must acquire them from exogenous sources.

The investigational drug eflornithine (DFMO) has shown exceptional promise in human trypanosomiasis, and its efficacy is attributed to the inhibition of polyamine synthesis.

C. Mode of Action of Some Antinematodal Drugs

The benzimidazoles, which include several important broad spectrum antinematodal drugs, appear to act by blocking the assembly of the microtubules essential for maintaining cell structure and function. Most of the other antinematodal drugs act by interfering with neurotransmission. Some (e.g., levamisole, pyrantel and morantel) depolarize cell membranes by acting on cholinergic receptors.

The organophosphate anthelmintics (not used to a significant extent in human medicine) inhibit the acetylcholinesterase of nematodes more readily than they inhibit the corresponding enzyme in the host and so induce paralysis of the worm at dosages tolerated by the host. Piperazine, which has been used widely for the treatment of ascaris infection, causes a flaccid paralysis because of blockade of GABA-mediated neurotransmission. The use of ivermectin in humans has so far been limited almost entirely to the treatment of *Onchocerca* infection, and its mechanism of action against this parasite is essentially unknown. It is believed to act against various nematode and arthropod parasites by increasing chloride ion influx and thereby interfering with neuronal function. In some neurobiological test systems, ivermectin is known to cause an increased release of the neurotransmitter GABA and the enhanced binding of GABA to its postsynaptic receptors, with consequent opening of chloride ion channels and decrease in cell function. In other test systems, however, ivermectin affects chloride channels that are independent of GABA. Much remains to be done to clarify the mechanism, or mechanisms, by which ivermectin paralyzes nematode worms. The mechanism of its effect on the reproduction of adult *Onchocerca* is particularly obscure.

D. Mode of Action of Some Drugs Used Against Flukes and Tapeworms

Praziquantel, which is highly active against both flukes and tapeworms, has multiple and only poorly understood effects. In both flukes and tapeworms, the drug causes a rapid paralysis of muscle and a severe vacuolization of the tegument. The biochemical mechanisms involved, however, are apparently not identical. In the case of flukes, or at least schistosome flukes, calcium and magnesium ions are involved in both the muscle and the tegument effects. It has been suggested that the impaired muscle function and damaged tegument make the worms more vulnerable to the host immune response. In the case of tapeworms, the efficacy of praziquantel depends on the availability of endogenous calcium ions rather than on the uptake of exogenous calcium ions as in the case of schistosomes.

The results of *in vitro* and *in vivo* studies suggest that oxamniquine destroys schistosomes by irreversibly inhibiting the synthesis of nucleic acids and proteins. Niclosamide, a salicylanilide still used for the treatment of tapeworm infections, is probably an uncoupler of oxidative phosphorylation in cestode mitochondria.

E. Mode of Action of Some Drugs Used Against Ectoparasites

The modern drugs used to treat ectoparasitic diseases act on the nervous system of arthropods (unlike the older and more toxic compounds that blocked energy metabolism).

Lindane (gamma-benzene hexachloride), like several other insecticides, affects the central nervous system of insects and results in tremors, convulsions, and paralysis. It apparently causes excess release of the neurotransmitter acetylcholine at nerve terminals, and the effects can be blocked by a cholinergic blocking agent. The precise mechanism of action, however, is unclear, and it is likely that calcium ions are involved in the effect on neurotransmitter release. Lindane may, in fact, antagonize the inhibitory neurotransmitter GABA and so lead to increased excitatory action at the synapse, with consequent hyperactivity typical of this and many other insecticides. The mode of action as it applies to human lice and itch mites is conjectural, because the pertinent biochemical studies have been done on other arthropods.

Malathion, like other organophosphates, is thought to inhibit the acetylcholinesterase that is needed to inactivate the neurotransmitter acetylcholine arthropod synapses. Disruption of sodium ion flow across cell membranes is probably involved.

Pyrethroids block neurotransmission by depolarizing the cell membranes, and this in turn is brought about by interfering with the flow of ions across those membranes. The compounds may exert this action directly on the membranes or they may affect the operation of ion pumps. Some pyrethroids may act on the GABA receptor or the ionophore portion of the receptor complex, thus interfering with the operation of chloride ion channels. The action of ivermectin on arthropods also appears to involve chloride ion channels and to be similar to the action seen in nematodes (above), except that nerve transmission is blocked, at least in some insects, at the neuromuscular junction rather than the interneuron–neuron synapse as in the case of nematodes.

Bibliography

Campbell, W. C. (1983). Progress and prospects in the chemotherapy of nematode infections of man and other animals. *J. Nematol.* **15,** 608–615.

Campbell, W. C. (1986). The chemotherapy of parasitic infections. *J. Parasitol.* **72,** 45–61.

Campbell, W. C., and Rew, R. S., eds. (1986). "The Chemotherapy of Parasitic Diseases." Plenum Press, New York.

Denham, D. A., ed. (1985). "Chemotherapy of parasites." Symposia of the British Society for Parasitology, vol. 22. *Parasitology* **90**(4), 613–721.

Gustafsson, L. L., Beerman, B., and Abdi, Y. A. (1987). "Handbook of Drugs for Tropical Parasitic Infections." Taylor and Francis, London.

Gutteridge, W. E. (1985). Existing chemotherapy and its limitations. *Br. Med. J.* **41,** 162–168.

Hooper, M., ed. (1987). "Chemotherapy of Tropical Diseases." John Wiley, Chichester, United Kingdom.

James, D. M., and Gilles, H. M. (1985). "Human Antiparasitic Drugs: Pharmacology and Usage." John Wiley, Chichester, United Kingdom.

Mansfield, J. M. ed. (1984). "Parasitic Diseases: The Chemotherapy," vol. 2. Marcel Dekker, New York.
Peters W., and Richards, W. H. G., eds. (1984). "Antimalarial Drugs," vols. I and II. "Handbook of Experimental Pharmacology," vol. 68. Springer-Verlag, New York.
Sturchler, D. (1982). Chemotherapy of human intestinal helminthiases: A review with particular reference to community treatment. *Adv. Pharmacol. Chemother.* **19,** 129–154.
Vanden Bossche, H., Thienpoint, D., and Janssens, P. G. eds. (1985). "Chemotherapy of Gastrointestinal Helminths." Springer-Verlag, Berlin.

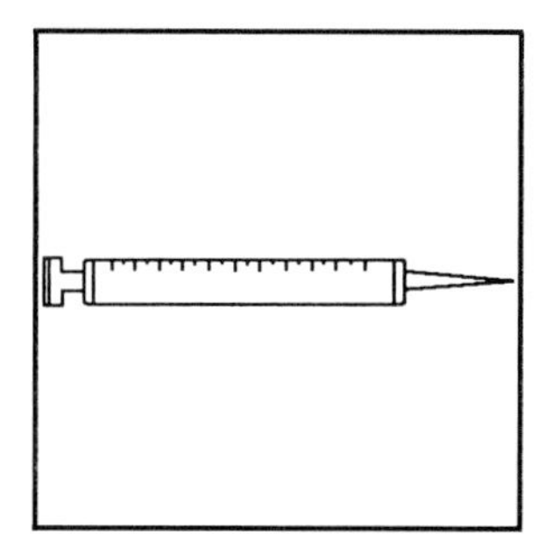

Chemotherapy, Antiviral Agents

THOMAS W. NORTH, *University of Montana*

Glossary

Latency State in which viral genome is present in cell but not expressing genes necessary for virus replication

Nucleoside analogue Compound with a structural modification of the base or sugar of commonly occurring nucleosides

Resistant mutants Mutant virus that can replicate in the presence of a drug that is inhibitory to wild-type virus

Selectivity Degree to which a drug inhibits virus replication more effectively than it inhibits cell growth or function

Target Viral enzyme or component against which the active form of a drug exerts its inhibitory effect

CHEMOTHERAPY has become important in treatment of viral infections that cannot be prevented with vaccines. Antiviral drugs are available for treatment of herpesvirus and influenza A infections in humans. There is now hope for development of antiviral drugs to combat other viruses that afflict humans, including human immunodeficiency virus (HIV), the causative agent of the acquired immune deficiency syndrome (AIDS). Advances in chemotherapy of herpesvirus infections, most notably with nucleoside analogues, have been of particular importance because they point the way to general strategies that may be used against any virus.

I. Antiviral Strategies

The most effective strategy yet employed for viral chemotherapy is the inhibition of a virus-encoded enzyme that is essential for its replication. The successful antiherpes drugs are nucleoside analogues that are metabolically activated to their respective nucleotides, and these selectively inhibit the herpesvirus-encoded DNA polymerase. Each herpesvirus encodes a DNA polymerase that is uniquely required for replication of its DNA and cannot be replaced by a cellular DNA polymerase. Although the herpes and cellular DNA polymerases catalyze similar reactions, they are different in physical properties and, most importantly, in their sensitivities to inhibitors. This has enabled development of selective inhibitors, which block herpesvirus DNA synthesis without affecting DNA synthesis of uninfected cells. Similarly, the most promising agents for treatment of AIDS are nucleoside analogues, including 3′-azidothymidine (AZT, zidovudine), which are metabolized to nucleotides, and these are selective in their action on reverse transcriptase. This enzyme, which is necessary for the replication of retroviruses, has no counterpart in uninfected cells. [*See* DNA SYNTHESIS; HERPESVIRUSES.]

The selectivity of some of the antiherpes nucleosides can be acheived in another way. This strategy takes advantage of the fact that some herpesviruses encode an enzyme (thymidine kinase) that can phosphorylate certain nucleoside analogues. Several of the antiherpes nucleosides derive at least some of their selectivity from the fact that they are phosphorylated by this enzyme but not by any cellular enzymes. Thus, the active forms of these nucleosides, their corresponding nucleotides, are formed in virus-infected cells but not in uninfected cells.

Although the viral enzymes required for DNA synthesis have received much attention, there are

idoxuridine vidarabine (ara A) acyclovir

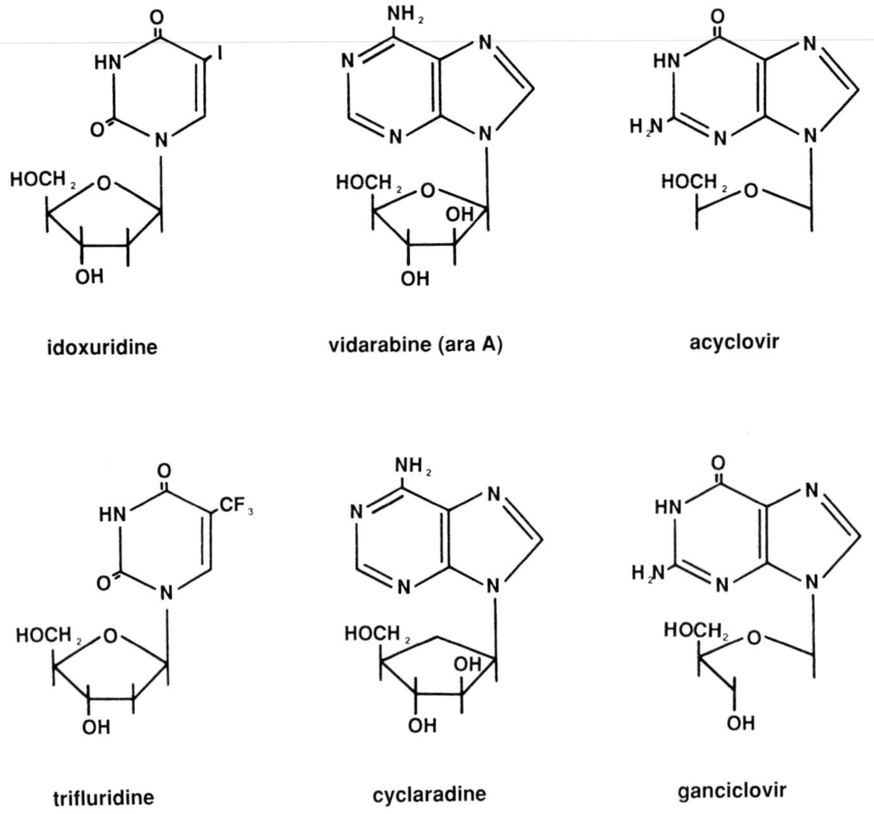

trifluridine cyclaradine ganciclovir

FIGURE 1 Nucleoside analogues with selective anti-herpesvirus activities. All these except cyclaradine have been approved for use in humans.

numerous other viral targets that might be exploited for chemotherapy. Each virus encodes a specific set of proteins that are required for its replication, and each of these might be used as a target for chemotherapy. Many of these proteins are unique and have no counterpart in uninfected cells. These include structural proteins as well as enzymes. Viral nucleic acids might also be considered chemotherapeutic targets for "drugs" that will recognize specific viral sequences and inactivate or destroy them.

Attempts have been made to develop antivirals that inhibit cellular enzymes. A rationale for this approach is that certain pathways or enzymes are more critical for virus replication than for cellular functions. If such events are critical to viruses in general, it may provide a strategy for development of broad spectrum antivirals. Of course, this approach is more likely to result in toxicity to host cells and tissues.

II. Antiherpes Drugs

All the drugs that have been approved or show promise for use in treatment of human herpesvirus infections exert their activity by inhibition of the herpesvirus-encoded DNA polymerase. Most of these are nucleoside analogues that must be phosphorylated to their corresponding triphosphates to exert antiviral activity. This metabolism is critically important because the activity of an antiviral nucleoside is dependent on the amount of the triphosphate formed within cells. Also, selectivity can be achieved at the level of activation if a viral-encoded enzyme is required to perform one of the phosphorylations.

The first two drugs that were approved for treatment of herpetic infections in humans were nucleoside analogues, 5-iododeoxyuridine (IUdR, idoxuridine) and 9-β-D-arabinofuranosyladenine (araA, vidarabine) (Fig. 1). These were approved for treatment of disorders caused by herpes simplex virus (HSV). Both of these nucleosides are metabolized to their corresponding 5′-mono-, di-, and triphosphates by cellular kinases. The 5′-triphosphates (IdUTP and araATP) are the biologically active forms. IdUTP can substitute for dTTP and be incorporated into DNA by DNA polymerases. However, it is not a selective drug because it can be incorporated into DNA by both cellular and herpesvirus DNA polymerases. Accordingly, IUdR has proven too toxic for systemic use but has been used topically for many years in treatment of herpetic keratitis caused by HSV. Another nucleoside analogue, trifluorothymidine (F_3-TdR, trifluridine) (Fig. 1) has more recently been approved for treatment of herpes keratitis. Its mechanism and lack of selectivity are similar to that of IUdR.

In contrast to these, araATP is a selective inhibitor of herpesvirus DNA polymerases. It takes substantially higher concentrations of araATP to inhibit cellular polymerases than are necessary to inhibit herpes DNA polymerases. This selectivity enabled araA to become the first drug approved for systemic use in treatment of human herpesvirus infections. The metabolism of araA is shown in Fig. 2. In addition to its activation to araATP, it is metabolically inactivated by adenosine deaminase to 9-β-D-arabinofuranosylhypoxanthine (araHx). This occurs rapidly in humans because of high levels of adenosine deaminase present in blood and most tissues. This, combined with a low solubility, has limited the therapeutic efficacy of araA. The antiviral activity of araA has been successfully enhanced through combination with inhibitors of adenosine deaminase, such as erythro-9-(2-hydroxy-3-nonyl)adenine or 2′deoxycoformycin, although such combinations have not been approved for clinical use. Nevertheless, this approach has stimulated development of analogues of araA that are not substrates for adenosine deaminase. One of the most promising of these compounds is carbocyclic araA (cyclaradine, Fig. 1). This drug has excellent activity against herpes simplex virus types 1 and 2; the reason it has not been developed for clinical use is unclear.

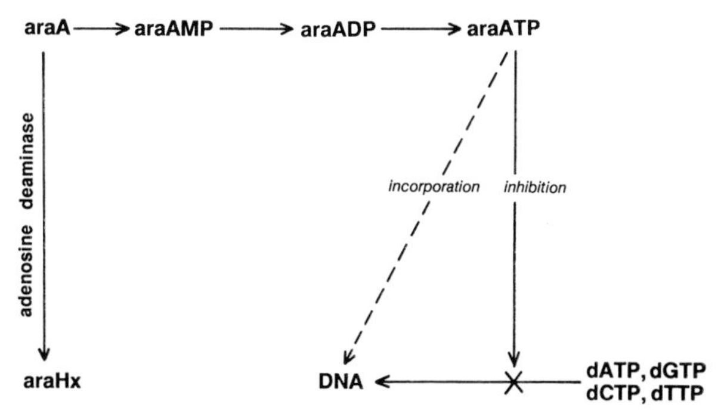

FIGURE 2 Metabolism and actions of araA.

AraATP exerts its selective antiviral activity through effects on the herpesvirus DNA polymerase. It is both an inhibitor of this enzyme and a substrate that is incorporated into DNA. It is not clear which of these two events is more important for the antiviral activity. Unlike some other antivirals that will be discussed below, araA is not a chain terminator. It is incorporated internally into DNA chains.

A major advance in viral chemotherapy occurred with development of an acyclic nucleoside analogue, 9-(2-hydroxyethoxymethyl)guanine (acyclovir) by the Burroughs Wellcome Co. This nucleoside analogue is metabolized to the corresponding triphosphate in a manner similar to the nucleosides described above. But there is one critical difference in the metabolism of this analogue: the first phosphorylation is catalyzed *only* by a virus-encoded enzyme. This viral-specified activation enables a tremendous increase in selectivity over that achieved with analogues that are activated by cellular kinases.

The metabolism of acyclovir is shown in Fig. 3. The first step, formation of acyclovir monophosphate, is carried out by the herpesvirus-encoded thymidine kinase. Three of the human herpesviruses (HSV-1, HSV-2, and herpes zoster) encode this enzyme. They have a broad substrate specificity and will phosphorylate thymidine, 2′-deoxycytidine, and many nucleoside analogues, including acyclovir. Cellular thymidine kinases have a much narrower substrate specificity and do not effectively phosphorylate 2′-deoxycytidine, acyclovir, or many other of the nucleoside analogues that are phosphorylated by the viral thymidine kinases. In fact, acyclovir is not phosphorylated at all by cellular thymidine kinase and is not effectively phosphorylated to a monophosphate by any cellular enzyme. Thus, acyclovir triphosphate is formed only in cells infected with these herpesviruses, and acyclovir is effective in treatment of infections by these three herpesviruses.

Acyclovir monophosphate is further phosphorylated to the corresponding di- and triphosphate by cellular kinases. The 5′-triphosphate is a competitive inhibitor of herpesvirus DNA polymerases. This inhibition is also selective. Acyclovir triphosphate inhibits the HSV DNA polymerase at a concentration 10–50-fold lower than is required to inhibit cellular DNA polymerases. This selectivity, combined with the selective phosphorylation discussed above, gives acyclovir a wide margin of

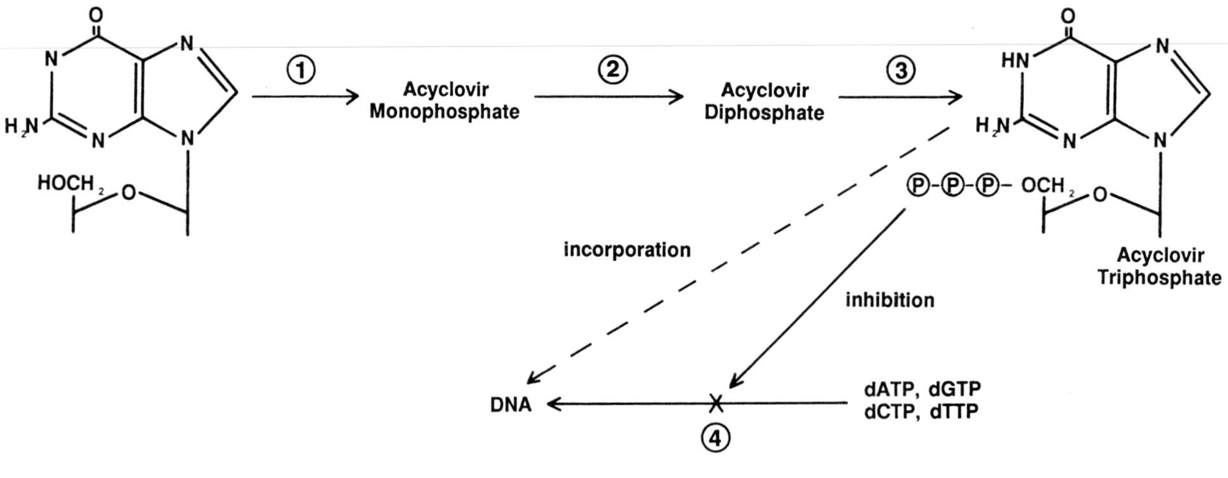

FIGURE 3 Metabolism and action of acyclovir. The enzymes involved are (1) the herpes-encoded thymidine kinase, (2) cellular GMP kinase, (3) cellular nucleoside diphosphokinase, and (4) herpes-encoded DNA polymerase.

safety. Acyclovir is also incorporated into DNA and serves as a chain terminator because the acyclic sugar analogue lacks a suitable acceptor for addition of the next nucleotide. Recent evidence indicates that incorporated acyclovir also irreversibly inactivates the HSV DNA polymerase.

Acyclovir is widely used to treat infections by HSV-1, HSV-2, and herpes zoster. It shows little or no activity against viruses that do not encode a thymidine kinase capable of phosphorylating it. Thus, it is much less active against some other members of the herpesvirus family (e.g., cytomegalovirus and Epstein Barr virus). However, another acyclic nucleoside analogue, 9-(1,3,-dihydroxy-2-propoxymethyl)guanine (ganciclovir, DHPG) (Fig. 1), has recently been approved for treatment of cytomegalovirus infections in humans. This compound must also be phosphorylated to a triphosphate to exert its activity. The enzymes responsible for this activation are not known. Attempts to find a cytomegalovirus-induced thymidine kinase have been unsuccessful. Moreover, uninfected cells cannot effectively phosphorylate ganciclovir. Nevertheless, ganciclovir is phosphorylated in cytomegalovirus-infected cells, and the drug is an effective inhibitor of cytomegalovirus replication. Recently, cytomegalovirus mutants that are resistant to ganciclovir have been reported, and the drug is not phosphorylated in cells infected by these mutants. This exciting discovery should help identify the mechanism by which ganciclovir is activated.

Nucleoside analogues are not the only selective antiherpes drugs that are targeted against the herpes DNA polymerase. Several pyrophosphate analogues, including phosphonoacetate and phosphonoformate, are selective inhibitors of herpesvirus DNA polymerases. Clinical use of these pyrophosphate analogues has been limited because of their accumulation in bone.

It should be noted that all these antiherpes drugs are replication inhibitors. However, herpesviruses are able to establish a state of latency in which the viral genome is established within cells and is not replicating. During latency the virus is not susceptible to replication inhibitors. Nevertheless, antiherpes drugs are effective in treatment of disease, but infections can recur when drug is removed. This will be discussed further in Section VI.

The advances that have occurred in development of antiherpes drugs have been facilitated by the availability of target enzymes to study (DNA polymerase and thymidine kinase) and by the availability of animal models to test drugs and therapeutic strategies. The rabbit eye model was particularly important in early studies for treatment of ocular herpes infections because it enabled one eye to serve as a control for drug therapy of the other. For systemic therapy a mouse model in which HSV causes rapid encephalitis has been important. A guinea pig model for vaginal infections by HSV-2 has been particularly important in development of therapeutic strategies for genital herpes. The coordination of *in vitro* and *in vivo* approaches has been particularly important in development of chemotherapy for herpesvirus infections.

III. Chemotherapy for Other Viruses

There are relatively few other viruses for which there are approved drugs or rational chemotherapeutic strategies. Amantadine has been approved for prophylaxis and symptomatic management of influenza A infections since 1966. It and a closely related compound, rimantadine, block replication of influenza A viruses. The exact mechanism is unclear, although a study of amantadine-resistant mutants suggests that it involves the viral matrix (M) protein. Amantadine is a weak base that accumulates in acidic vacuoles within cells and may prevent proper uncoating of the virus. Attempts to develop antiviral drugs for rhinoviruses and other picornaviruses have also focused on compounds that bind to the virion and prevent uncoating. These attempts have been limited by lack of a suitable animal model for rhinovirus infection. [*See* INFLUENZA VIRUS INFECTION.]

Several thiosemicarbazones have antiviral activity and one of these, 1-methyl-β-isatin-thiosemicarbazone (methisazone) is effective in treatment of smallpox (although eradication of smallpox has eliminated the need for this).

IV. Broad-Spectrum Antivirals

It is improbable that broad-spectrum antivirals can be developed by using viral enzymes as targets because viruses are a diverse group that does not share a common target. Nevertheless, there are compounds that have broad spectrum antiviral activity. These compounds exert their effects through inhibition of cellular enzymes. Presumably, these pathways are more critical to viral replication than to cellular functions. It might be expected that these approaches would not be as selective as targeting of drugs to viral enzymes. Ribavirin (1-β-D-ribofuranosyl-1,2,4-triazole-3-carboxamide), another nucleoside analogue, inhibits a broad spectrum of RNA and DNA viruses. Its exact mechanism is not clear, although it is known to inhibit a cellular enzyme, IMP dehydrogenase, and to cause decreases in levels of GTP. Nucleotides of ribavirin may also inhibit specific viral functions such as 5′-capping of messenger RNA. Several inhibitors of 5-adenosylhomocysteine hydrolase have also been shown to have activity against a broad spectrum of viruses.

Perhaps the most promising approach in development of broad-spectrum antivirals is through inducers of interferons. Interferons are broad-spectrum, antiviral glycoproteins that are synthesized by the body in response to a variety of stimuli, including virus infections. A large number of different types of compounds can induce interferons. Another promising approach is the development of immunomodulators that can enhance the ability of the immune system to protect against viral infections. [*See* INTERFERONS.]

V. Chemotherapy of AIDS

The rapid spread of HIV and problems in vaccine development have triggered massive efforts to develop drugs and chemotherapeutic strategies for treatment or management of AIDS. Hope has been provided by AZT (Fig. 4), a nucleoside analogue that works in a manner analogous to many of the antiherpes drugs. The urgency of the AIDS problem has also stimulated a variety of other approaches and evaluation of all HIV components as potential chemotherapeutic targets. Some of these may provide new general strategies for viral chemotherapy. [*See* ACQUIRED IMMUNODEFICIENCY SYNDROME (VIROLOGY).]

Not surprisingly, the most promising target (and the first one exploited) is the HIV reverse transcriptase. This enzyme, which catalyzes synthesis of a DNA copy of the virion RNA, is absolutely essential for replication or expression of the HIV genome. Reverse transcriptase is the target for active forms (the corresponding 5′-triphosphates) of several nucleoside analogues that are able to block selectively replication of HIV. These include AZT and the dideoxynucleosides.

AZT was the first drug effective enough in inhibition of HIV replication to be approved for treatment of AIDS patients. Although its use has been limited somewhat by toxicity to bone marrow, AZT enables marked improvement and increased life expectancy in some patients.

Like other nucleoside analogues, AZT must be activated by phosphorylation to its corresponding 5′-mono-, di-, and triphosphates. This metabolism is accomplished by cellular kinases. The 5′-triphosphate (AZ-TTP) is a competitive inhibitor of the HIV reverse transcriptase. It is also a substrate for this enzyme and is incorporated into DNA. On in-

zidovudine (AZT) AZdU d4T ddC

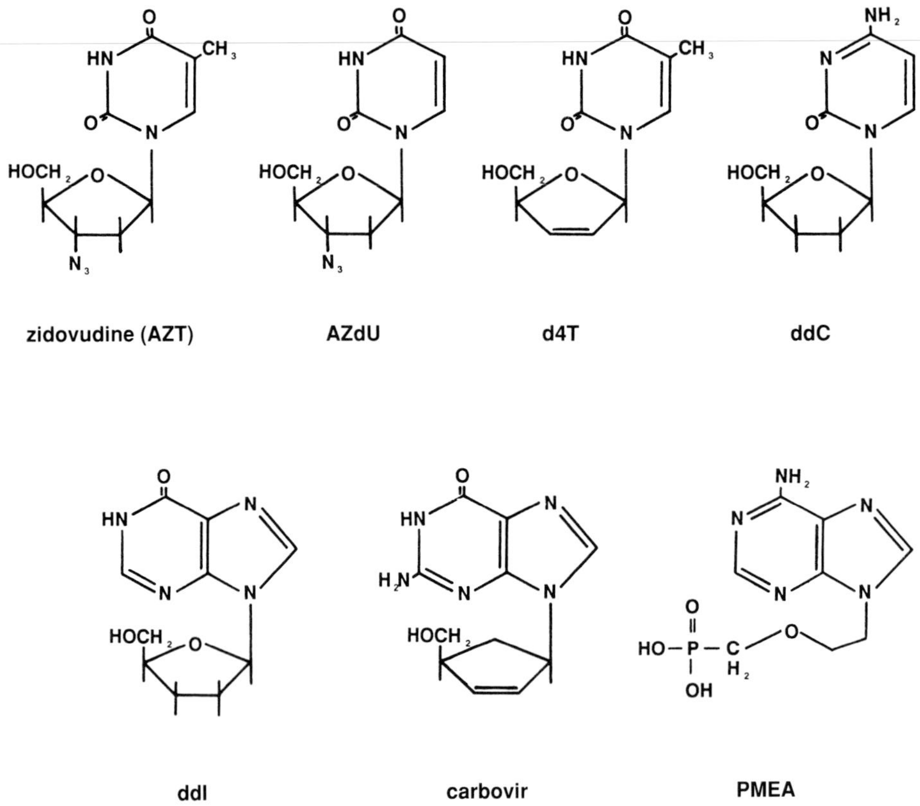

ddI carbovir PMEA

FIGURE 4 Some of the nucleoside analogues that are selective inhibitors of the human immunodeficiency virus.

corporation, AZT is a chain terminator because the 3′-azido group cannot serve as an acceptor for the next incoming nucleotide. The selectivity of AZT is due to the fact that AZ-TTP is not an effective inhibitor or a substrate for cellular DNA polymerases. Thus, treatment with AZT leads to a selective inhibition of viral DNA synthesis. It has not been determined whether it is the inhibition of reverse transcriptase, the chain termination, or both, that is (are) responsible for the antiviral activity.

Several dideoxynucleosides have been shown to be selective inhibitors of HIV replication, and these work in a manner similar to that of AZT. These are metabolized to their corresponding nucleotides and the 5′-triphosphates are competitive inhibitors of the HIV reverse transcriptase. The dideoxynucleotides are also incorporated into DNA and are well-characterized chain terminators in bacterial systems (as evident from their use in sequencing of DNA). The antiviral activity of these compounds is presumably due to their inhibition of reverse transcriptase and/or their incorporation into and termination of viral DNA. Two dideoxynucleosides (Fig. 4) have shown promise in AIDS therapy. These are 2′,3′-dideoxycytidine (ddC), which is metabolized to the corresponding 5′-triphosphate ddCTP, and 2′,3′-dideoxyinosine (ddI), which is metabolized to ddATP. The limiting toxicity of these dideoxynucleosides, peripheral neuritis, is different from that of AZT; neither of them causes suppression of bone marrow.

Many other nucleoside analogues have shown promise as inhibitors of HIV replication, and some of these are shown in Fig. 4. These include 3′-azido-2′-3′-deoxyuridine (AZdU), 2′,3′-dideoxy-2′,3′-didehydrothymidine (d4T), carbocyclic-2′,3′-dideoxy-2′,3′-didehydroguanine (carbovir), and a group of phosphonyl compounds such as 9-(2-phosphonylmethoxyethyl)adenine (PMEA). Interestingly, some of the phosphonyl derivatives have excellent antiherpes activity.

Two other HIV enzymes have been identified and are candidate targets for chemotherapeutic approaches. One of these is RNase H, which is an activity that is present on the same protein as reverse transcriptase. RNase H is required to degrade the RNA strand of the RNA-DNA hybrid that occurs after reverse transcriptase has made a DNA copy of viral RNA. Removal of the RNA is necessary to enable synthesis of double-stranded DNA. Although the HIV RNase H activity has been characterized, selective inhibitors have not been identified.

HIV also encodes a protease, which has recently received much attention as a target for chemotherapy. The HIV protease is essential for replication of HIV. It cleaves large polyproteins into the individual viral enzymes and structural proteins and may also be necessary for assembly and/or maturation of the virion. The HIV enzyme is an aspartyl protease; it has been purified and its crystal structure determined. Much effort is directed toward development of selective inhibitors of this protease.

Several other targets have been proposed as having potential for chemotherapy of AIDS. One approach is to block attachment of HIV to cells. This attachment is to a specific cellular protein, the CD4 receptor, that is found on the surface of susceptible lymphocytes and other cells that can be infected with HIV. Attachment of HIV to this receptor can be blocked by providing an excess of soluble CD4 protein, and attempts to develop this into a viable chemotherapeutic strategy are underway. A problem of this approach is that soluble CD4 will not inhibit virus replication once the virus enters the cell, and so the block must be complete (or nearly so) to be effective. HIV also encodes several regulatory genes, which control expression of viral genes. These have been suggested as chemotherapeutic targets, but a complication is that some proteins from host cells or from other viruses may be able to substitute for these regulatory elements. Yet another approach being initiated uses a viral nucleic acid as a target. This involves development of reagents that will react with and inactivate or destroy specific sequences of viral RNA or DNA. This approach, although potentially selective, is only in early stages of development. [*See* CD8 AND CD4: STRUCTURE, FUNCTION, AND MOLECULAR BIOLOGY.]

A serious limitation in the development of strategies for chemotherapy of AIDS has been the lack of a suitable animal model. Much work is in progress to develop such a model, and some of the approaches look promising. A simian immunodeficiency virus has been identified, and its reverse transcriptase is similar to that of HIV. This will be useful for some studies, but nonhuman primates are not available in sufficient numbers for large-scale chemotherapeutic studies. A number of rodent models are in development, and some of these should provide useful information. However, the murine retroviruses used in these studies are different from HIV, as are their reverse transcriptases and the clinical course of their infections. One promising development is transplantation of the human immune system into immunodeficient mice to produce SCID/hu mice. These mice produce human lymphocytes and are susceptible to infection by HIV. It is not clear whether the mice develop an AIDS-like disease; nevertheless, these mice may be useful for chemotherapeutic studies if they can be generated in sufficient numbers and if the properties of the transplanted immune system are reproducible. Yet another model with which the author is working is feline immunodeficiency virus (FIV). This virus, like HIV, is a lentivirus, and it causes an immune deficiency in cats that is similar to AIDS in humans. The FIV reverse transcriptase is similar to the HIV reverse transcriptase in physical properties and sensitivities to the active forms of several antivirals, including AZ-TTP. This model promises to be useful for coordination *in vitro* and *in vivo* studies.

If successful chemotherapeutic strategies for AIDS can be devised, it is likely they will point the way for chemotherapy of other human retroviral infections. Successful strategies should also provide insight for general approaches to combat other classes of viruses.

VI. Latency, Resistance, and Other Challenges

All the effective antivirals currently available are replication inhibitors. However, many viruses (including retroviruses and herpesviruses) are able to establish latent states where the viral genome is present in infected cells, but it is not actively replicating. Latent virus is not susceptible to these replication inhibitors; after the drug is removed, the virus can reactivate and undergo replication. It is evident from the situation with herpesviruses that the existence of latency does not preclude success with replication inhibitors. The situation with AIDS awaits further evidence. However, if the goal is to eliminate virus totally, ways must be devised to eliminate latent virus. One possibility is to devise drugs that will reactivate the virus; it might then be eliminated with a replication inhibitor. For this to be successful, the activation from latency will have to be complete. It is not known whether this will be possible. Another approach is to kill all virus-infected cells, including those harboring latent virus. If no viral genes are expressed during latency, this approach might require agents directed at the viral genome. If latency cannot be alleviated, another alternative might be continuous or periodic use of a replication inhibitor.

Another major problem in viral chemotherapy is the emergence of drug-resistant mutants. HSV mutants resistant to each of the antiherpes drugs have been isolated *in vitro*. Resistance to acyclovir (or other analogues that require activation *via* the viral thymidine kinase) can occur from mutations in either the thymidine kinase gene or the DNA polymerase gene. Resistance to araA or phosphonoformate is due only to mutations in the DNA polymerase gene. Resistance to acyclovir occurs clinically, and some of these mutants are cross-resistant to other antivirals. Resistance to phosphonoformate also occurs readily. However, clinical resistance of HSV to araA has not been reported, and most of the araA-resistant HSV were isolated from their resistance to other drugs. Although araA is much less selective than acyclovir, this slower development of resistant virus might be attractive in some instances.

AZT-resistant mutants of HIV have already emerged from therapy of AIDS patients. These mutants have not been fully characterized so it is not known how much of a problem this will pose. Because HIV and other retroviruses have high mutation frequencies, it is expected that drug resistance will be a serious problem.

Drug resistance is encountered in other areas of chemotherapy (e.g., cancer and tuberculosis). The most successful approach to combat this is combination chemotherapy. The likelihood of multiple resistance occurring during combination chemotherapy is a product of the probabilities of resistance to each drug; combinations of two or three drugs directed at different targets are usually sufficient to eliminate problems of resistance. It is generally recognized that long-term therapy of AIDS will require such combinations of drugs directed at two or more targets.

Despite these problems, there is a great deal of promise for the future of viral chemotherapy. In a relatively short period, this field has emerged from an era of random screening of compounds and a feeling by many that selective antivirals might be impossible because of the reliance of viruses on host machinery for much of their metabolism. Now there are selective antivirals for certain viruses and strategies that should enable development of drugs for other classes of viruses. Nucleoside analogues have been developed that lead to selective inhibition of the DNA polymerase of DNA viruses or the reverse transcriptase of retrovirus. A similar approach might be directed for development of selective inhibitors of the RNA-dependent RNA polymerases of many RNA viruses. The large scale efforts toward AIDS chemotherapy should provide other general strategies. Many disciplines are contributing to the development of chemotherapeutic strategies and identification of viral targets. It is likely that many further developments will be facilitated by collaborative, multidisciplinary approaches.

Bibliography

Elion, G. B. (1984). Acyclovir. In "Antiviral Drugs and Interferon: The Molecular Basis for Their Activity" (Y. Becker, ed.), pp. 71–88. Martinus Nijhoff Publishing, Boston.

Hirsch, M. S., and Kaplan, J. C. (1987). Treatment of human immunodeficiency virus infections. *Antimicrob. Agents Chemother.* **31,** 839–843.

Hirsch, M. S., and Kaplan, J. C. (1987). Antiviral therapy. *Sci. Am.***256,** 76–85.

North, T. W., and Cohen, S. S. (1984). Aranucleosides and aranucleotides in viral chemotherapy. In: "International Encyclopedia of Pharmacology and Therapeutics, Section III. Viral Chemotherapy," vol. 1 (D. Shugar, ed.), pp. 303–340. Pergamon Press, Oxford.

Öberg, B. (1989). Antiviral Therapy. *J. Acquired Immune Deficiency Syndrome* **1,** 257–266.

Yarchoan, R., and Broder, S. (1987). Development of antiretrovirus therapy for the acquired immunodeficiency syndrome and related disorders. *N. Engl. J. Med.* **316,** 557–564.

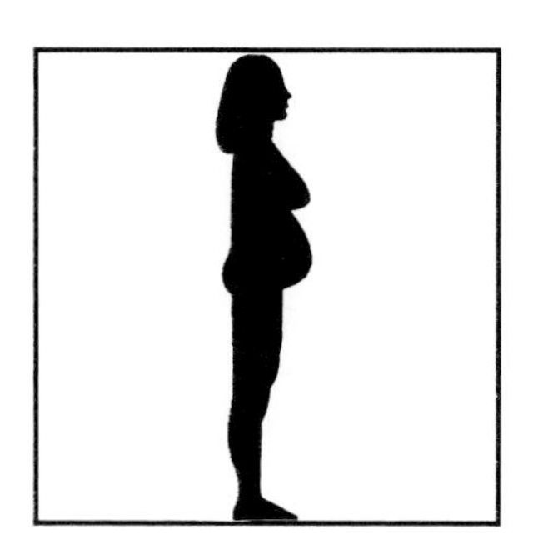

Childbirth

VANDA R. LOPS, *UCSD Nurse-Midwifery/Navy Resource Sharing Program*

Glossary

Dilatation Expansion of an orifice—in labor, the cervix
Fontanel Soft spot found between the cranial bones of the skull of an infant
Fundus Top of the uterus
Gravida Pregnant woman
Multigravida Woman who has had two or more pregnancies
Primagravida Woman pregnant for the first time

CHILDBIRTH, AS DEFINED in the most general sense, is the bringing forth of young. Variously referred to as accouchement, confinement, and/or parturition, it is the culmination of a gestational period, which can last from 19 days in the mouse to over 1 yr in the elephant. The purpose of this article is to present a description of the childbirth process as it occurs in the human gravida after a gestational period of 10 lunar mo, 9 calendar mo, or 280 days.

I. Labor

In most species, childbirth is the end result of labor, a process by which the products of conception are expelled to the outside world. Labor, for descriptive purposes, can be divided into three stages. Stage I of labor theoretically begins with the first uterine contraction, which causes the cervix to dilate and efface, and continues on to full dilatation (10 cm or 5 finger breadths in diameter) and complete effacement (the cervix shortens and thins, becoming continuous with the lower uterine segment).

Stage II begins with full dilatation and ends with the delivery of the infant. The period of time from the delivery of the infant to the delivery of the placenta is referred to as Stage III. Many authorities include a fourth stage during which the mother's physical conditions (vital signs, bleeding, etc.) stabilizes after the demanding process of birth.

There are numerous theories as to the initiation of labor; the most currently accepted involves the interaction of the hormones estrogen, oxytocin, and prostaglandins (PGE), a group of compounds derived from unsaturated fatty acids. Oxytocin receptors, under the influence of estrogen, a female hormone, appear to increase with gestational age, peaking at term or just before the initiation of labor. Additionally, the uterine lining begins to synthesize prostaglandins, which further appear to enhance the uterine threshold to oxytocin, a uterotonic hormone, produced by the pituitary gland, by sensitizing the uterine muscle (myometrium) to the developing oxytocin receptors and increasing myometrial gap junctions. This process makes the uterus sensitive to the very low levels of circulating oxytocin, and thus contractions are initiated (Fig. 1).

During pregnancy, uterine contractions are isometric and irregular in duration, frequency, and intensity. In labor, these contractions achieve a marvelous sense of coordination and regularity characterized by a contraction and relaxation phase with the uterus always maintaining a certain degree of resting muscle tone. The fibers of the uterine musculature interlace (Fig. 2), and the contraction wave is propagated from the top (fundus) across and down the body (corpus) of the uterus. This characteristic is referred to as the decreasing gradient of the contractions. Additionally, the myome-

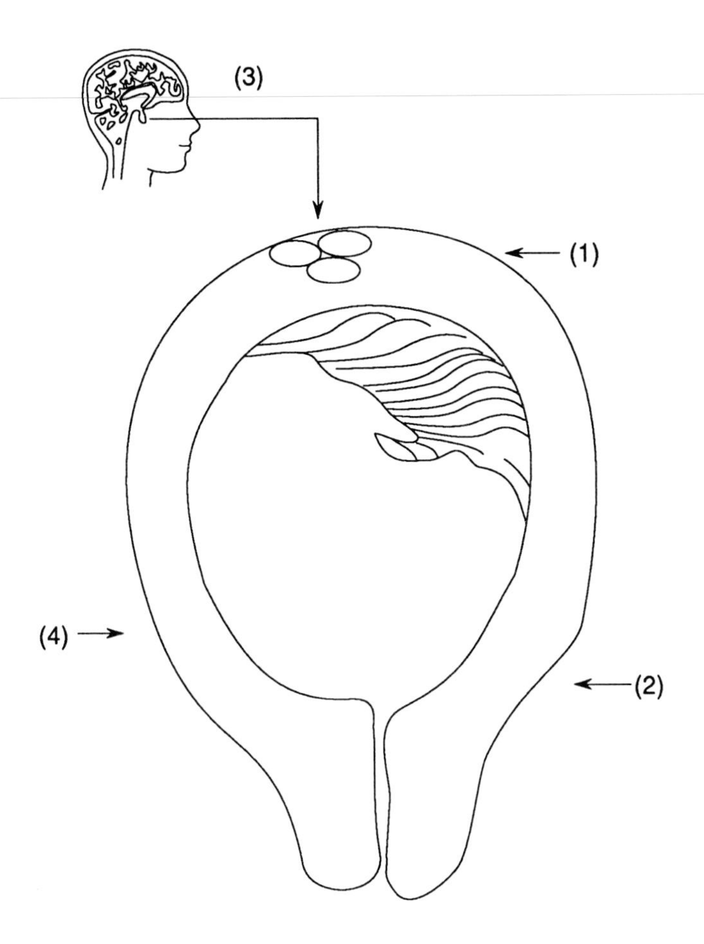

FIGURE 1 Initiation of labor. (1) Increased estrogen levels cause increased oxytocin receptors; (2) decidua produces PGE_4, which sensitizes myometrium to increasing oxytocin receptors; (3) oxytocin is produced by posterior pituitary; (4) uterine contractions starts.

trial cells of the uterine smooth muscle contain filaments, which have the ability to shorten with each contraction, causing the uterus to thicken its upper working segment and efface and dilate its lower, passive segment (Fig. 3). Subsequently, the fetus is propelled down and through the bony pelvis and eventually is born.

II. Stage I

The first stage has been divided into three phases: latent, active, and deceleration (Fig. 4). The latent phase, which extends from the beginning of labor to approximately 3-cm dilatation, is characterized by cervical effacement rather than aggressive dilatation. In this phase, contractions can be 30–60 min apart, lasting only 30 sec. These contractions follow a regular pattern and can be accompanied by abdominal cramps, backache, increased vaginal discharge, and/or ruptured membranes. During pregnancy and labor the fetus floats in amniotic fluid within the fetal membranes, or bag of waters. These membranes can spontaneously rupture at any time during labor or, in some cases, be artificially ruptured to stimulate labor forces. This closed fetal environment serves not only as a protection against infection but as a cushion against the stress of uterine contractions during labor. In the primagravida, this phase can last anywhere from 6–8 hr, whereas in the multipara the average length of the latent phase is approximately 5 hr. Time limits establishing normalcy of this phase have been set at 20 and 13 hours, respectively.

Most dilatation occurs during the active phase, which lasts from 3 cm to approximately 9–10 cm. In this phase, the uterine contractions increase in intensity and frequency and last a longer period of time so that they can be 2–3 min apart, lasting 40–45 sec. Subjectively, they may be experienced as

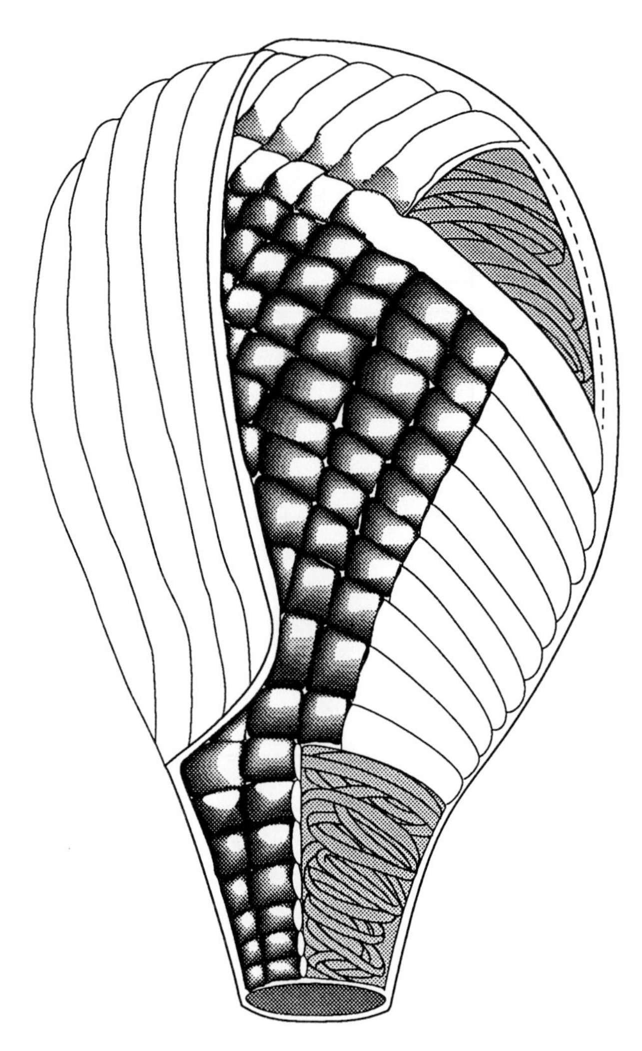

FIGURE 2 Interlacing of the uterine muscle fibers. [Adapted, with permission, from J. S. Malinowski, C. G. Pedigo, and C. R. Phillips, 1989, "Nursing Care during the Labor Process," 3rd ed., F. A. Davis Company, Philadelphia.]

painful and felt in the abdomen, above the pubic bone and in the lower back.

A short deceleration phase following the active phase, and ending with full dilatation, is characterized by progress in the descent of the presenting part, i.e., that part of the fetus coming down through the bony pelvis first. Many believe this phase is integrated into and a part of Stage II. It must be remembered that although cervical dilatation and effacement is viewed by many as progress in labor, true progress must also include descent of the presenting part through the planes of the bony pelvis.

During this phase, the laboring patient will commonly experience the most discomfort. If the membranes have not ruptured previously, they may do so at this point. The contractions may be accompanied by severe low backache, leg cramps, due to the pressure of the descending fetus, and a stretching sensation deep in the pelvis with the beginning urge to bear down.

III. Stage II

Stage II extends from full dilatation to the delivery of the infant. The most usual presentation of the fetus to the birth canal is the vertex presentation in which the head (vertex) or, more specifically, the occipital region of the head is entering the bony pelvis first. The fetal head will maneuver through the pelvic diameters by executing the cardinal movements or mechanisms of labor. These mechanisms are (1) engagement, (2) further descent with flexion, (3) internal rotation, (4) extension, (5) external rotation, (6) restitution, and (7) expulsion.

To fully understand the character and exquisite logistics of these mechanisms, a brief description of the bony pelvis and fetal head is in order. The bony pelvis consists of the inlet, or entrance, much like a door leading into a room; the cavity, which can be compared to the spatial diameters of the room; and an outlet, or exit, much like the door leading from the room. Each of these components have diameters through which the fetal head must fit. Figure 5 demonstrates the actual numerical value in centimeters of these diameters.

The fetal head must also be described in terms of diameters. It must be remembered that the bones of the fetal head are not fused as in the adult. Figure 6 lists the nomenclature for the various bony processes as well as the dividing lines (sutures) and areas (fontanels) between them. The fontanels, commonly referred to as the "soft spots" are two in number and differ in shape. The diamond-shaped anterior fontanel and the triangular-shaped posterior fontanel allow the practitioner, via vaginal examination, to determine the actual position or placement of the vertex in relation to the quadrants of the bony pelvis during labor. In the most usual position, the fetal head is well flexed, chin to chest. This presenting diameter (suboccipital bregmatic) from the occiput to the anterior fontanel is the smallest diameter of the fetal head and, thus, the most favorable to fit through the various pelvic di-

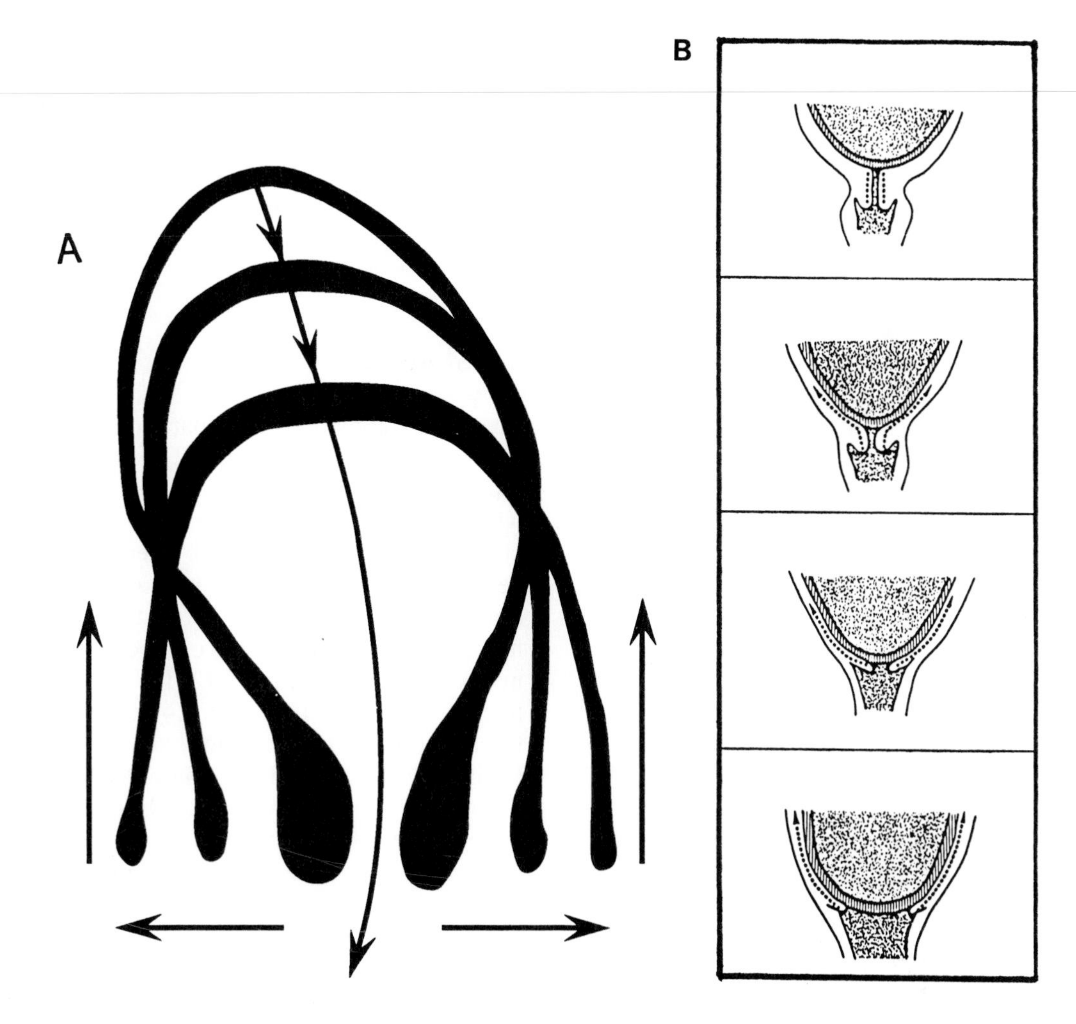

FIGURE 3 Sagittal (A) and anterior-posterior (B) views of dilatation and effacement. [Adapted, with permission, from J. A. Ingalls and C. N. Salerno, 1983, "Maternal and Child Health Nursing," C. V. Mosby Company, St. Louis.]

ameters. As the head deflexes, the presenting diameter, which must maneuver through the bony pelvis, changes. Figure 5 allows comparison of the possible fetal vertex diameters with the bony pelvic diameters. It also clearly demonstrates that certain positions of the fetal head will result in presenting diameters greater than the pelvic diameters, thus impeding the ability of the head to descend through the bony pelvis. The *raison d'étre* of the mechanisms of labor is that the smallest fetal diameters will attempt to conform to the largest pelvic diameters, so that the passage through the birth canal is accomplished as easily and efficaciously as possible.

Engagement, the first mechanism, is said to occur when the lowermost bony portion of the fetal vertex has reached the midpoint of the pelvic cavity as demarcated by the ischial spines, bilateral bony structures protruding into the cavity of the pelvis. As the vertex continues to descend, flexion of the head to the chest, if not complete, will also continue. As these two mechanisms are occurring, the vertex is in a transverse position (OT) with the fetal face looking out across the pelvis. Upon reaching the ischial spines and muscles of the pelvic floor, which offer resistance to further descent, the vertex begins to pivot or rotate away from these obstacles, seeking the path of least resistance. It thus rotates so that the vertex is no longer in the transverse but now in the anterior-posterior position, most commonly with the occiput anterior and the fetal face toward the mother's spine. Because of this internal

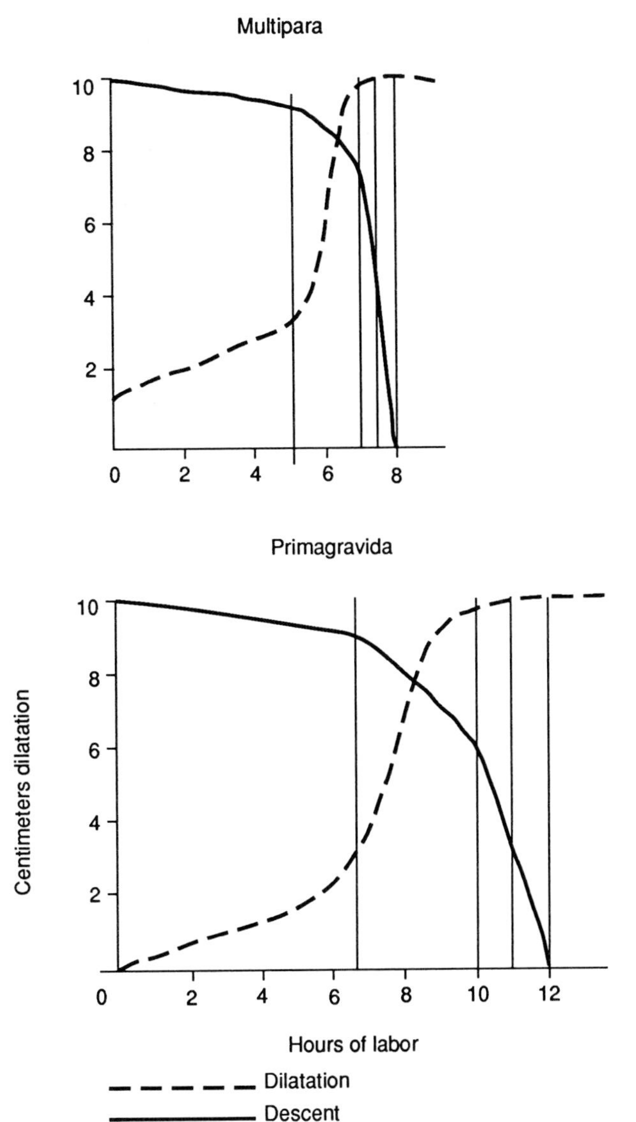

FIGURE 4 Mean labor curves. [Adapted, from E. A. Friedman, 1978, "Labor: Clinical Evaluation and Management," 2nd ed., Appleton-Century Crofts, New York.]

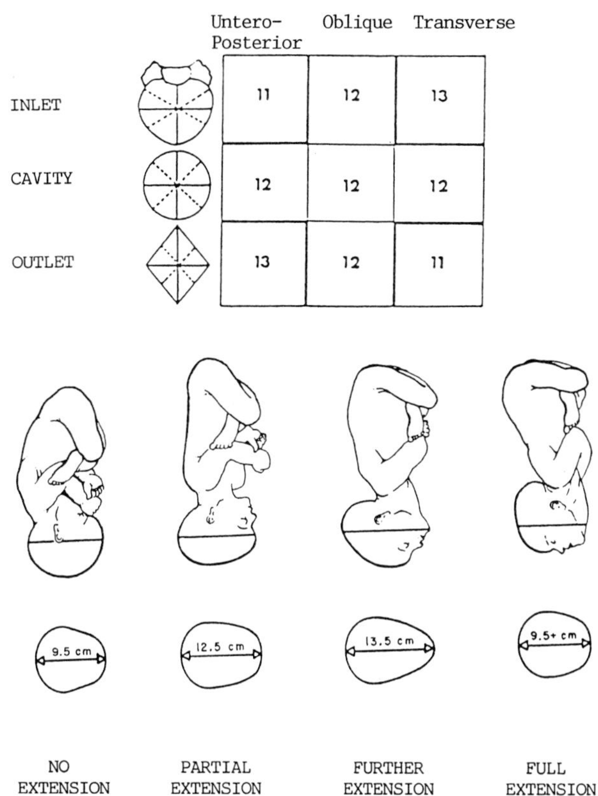

FIGURE 5 Comparison of pelvic diameters and fetal head diameters with degrees of extension. [Adapted, from H. Oxorn and W. R. Foote, 1986, "Human Labor and Birth," Appleton-Century Crofts, New York.]

rotation of the vertex, the shoulders are now in the transverse so that their diameter as they enter the pelvis is passing through the largest diameter of the pelvic inlet.

The fetal head continues to descend through the pelvis under the forces of the uterine contractions and the next mechanism, extension, occurs when the vertex is at the maternal perineum. At this point, the birth canal slopes upward so that with each contraction and maternal pushing force, the chin slowly extends away from the chest as if attempting to avoid the resistance offered by the muscles and soft tissue of the maternal pelvic floor. Figure 7 demonstrates how increasingly more of the fetal head becomes visible at the introitus as extension occurs.

Once extension is complete, the head is delivered and will begin an external rotation maneuver duplicating internal rotation, which will ultimately place it in the same position (OT) in which it first engaged. In completing this combination of external rotation and restitution, the shoulders, which had previously engaged in a transverse position, will now, because of this external rotation of the head, internally rotate and assume an anterior-posterior position to pass easily between the ischial spines. They, and the rest of the infant, then deliver spontaneously via the mechanism of expulsion (see Color Plate 11). Stage II can last anywhere from 30 min to 2 hr in the primagravida and 10 min to 1 hr in the multigravida.

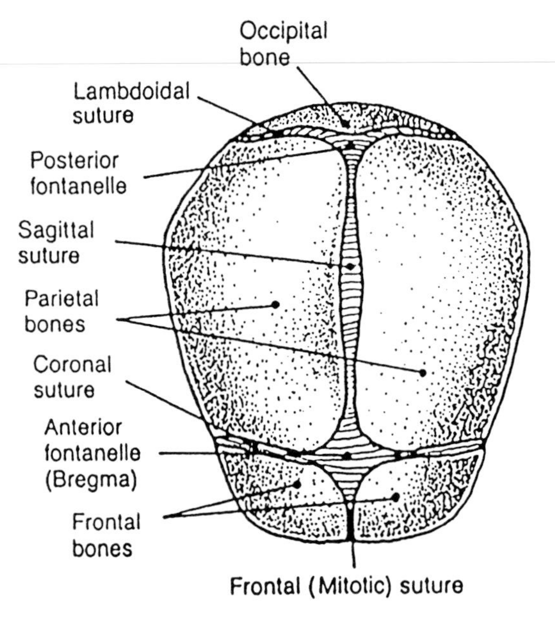

FIGURE 6 Superior view of the fetal skull. [Reproduced, with permission, from J. A. Ingalls and C. M. Salerno, 1983, "Maternal and Child Health Nursing," C. V. Mosby Company, St. Louis.]

IV. Stage III

After delivery of the infant, uterine contractions temporarily cease. When they resume, they are usually painless and directed toward causing the placenta to separate from the uterine wall and be delivered. As the uterus contracts, there is a sudden and dramatic decrease in the size of the placental site. Because the placenta is semi-rigid, it cannot shrink to compensate for this and, thus, begins to detach from the uterine wall.

Once completely separated, the placenta takes the same route as the fetus down through the birth canal to the outside world. This emptying of the uterus allows contractions of interlacing fibers of the myometrium, in turn causing constriction of the large maternal blood vessels, which had provided circulation to the placental sites. This results in maternal hemostasis. Stage III usually lasts approximately 5–10 min; however, if the placenta has not delivered by 30 min, appropriate medical intervention to assist in the process is warranted.

The end of stage III marks the completion of the labor and childbirth process. It is the phenomenon by which each generation duplicates itself, beguiling in its simplicity and rationality and overwhelming in its perfection. For many, it is the most precarious and exquisitely orchestrated journey they will ever undertake.

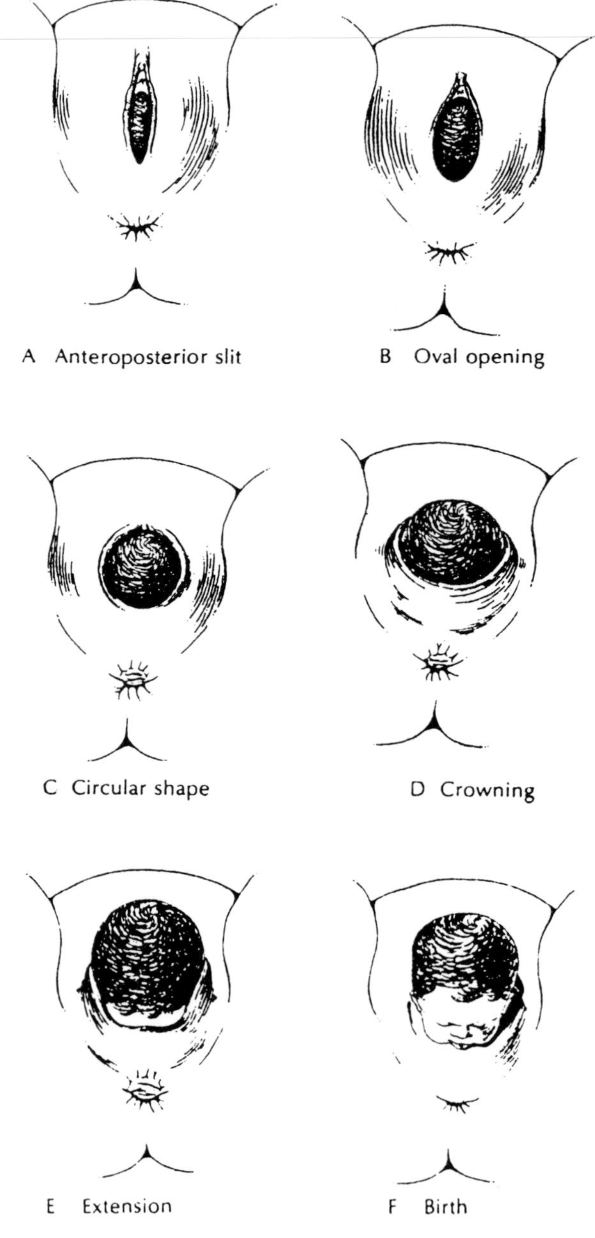

FIGURE 7 Dilation of the introitus due to extension of the head. [Reprinted, from H. Oxorn and W. R. Foote, 1986, "Human Labor and Birth," Appleton-Century Crofts, New York.]

Bibliography

Danforth, D. N., Scott, J. R. (eds.) (1986). "Obstetrics and Gynecology," 5th ed. J. B. Lippincott Company, Philadelphia.

Friedman, E. (1970). An objective method of evaluating labor. *Hosp. Pract.* **July,** 82–82 (in syllabus).
Friedman, E. (1971). The functional division of labor. *Am. J. Obstet. Gynecol.* **109,** 274–280.
Friedman, E. (1978). "Labor: Clinical Evaluation and Management," 2nd ed. Appleton-Century-Crofts, New York.
Gabbe, S. G., Niebyl, J. R., Simpson, J. L. (1986). "Obstetrics, Normal and Problem Pregnancies." Churchill Livingston, New York.
Gay, J. (1978). Theories regarding endocrine contributions to the onset of labor. *JOGN Nursing* **September/October,** 42–46.
Hendricks, C., and Brenner, W. (1970). Normal cervical dilatation patterns in late pregnancy and labor. *Am. J. Obstet. Gynecol.* **106,** 1105.
Ingalls, J. A., and Salerno, C. N. (1983). "Maternal and Child Health Nursing." C. V. Mosby Company, St. Louis.
Ledger, W. (1969). Monitoring of labor by graphs. *Obstet. Gynecol.* **34,** 174–181.
Liggins, G. (1983). Initiation of spontaneous labor. *Clin. Obstet. Gynecol.* **26(1),** 47–55.
Miller, F. (1983). Uterine motility in spontaneous labor. *Clin. Obstet. Gynecol.* **26(1),** 78–86.
Myles, M. (1985). "Textbook for Midwives." Churchill Livingston, Edinburgh.
Oxorn, H., and Foote, W. (1986). "Human Labor and Birth," 5th ed. Appleton-Century Crofts, New York.
Peisner, D. R., and Rosen, M. G. (1986). Transition from latent to active labor. *Obstet. Gynecol.* **68(4),** 448–451.
Varney, H. (1987). "Nurse-Midwifery," 2nd ed. Blackwell Scientific Publications, Boston.

Cholesterol

N. B. MYANT, *Royal Postgraduate Medical School, Hammersmith Hospital, London*

I. Structure and properties
II. Free and esterified cholesterol
III. Distribution
IV. Cholesterol as a component of membranes
V. Conversion into bile acids and steroid hormones
VI. Synthesis
VII. Uptake of cholesterol by cells
VIII. Reverse cholesterol transport
IX. Absorption from the intestine

Glossary

Apolipoprotein B Protein component of low-density lipoprotein (LDL) responsible for the recognition and binding of an LDL particle by an LDL receptor

Chylomicrons Triglyceride-rich particles secreted by the small intestine into the blood circulation during the absorption of fat

Foam cells Cells derived from macrophages and containing cytoplasmic droplets filled with esterified cholesterol

High-density lipoprotein Small, dense lipoprotein particle secreted into the plasma in nascent form by the liver and intestine

HMG-CoA reductase Membrane-bound enzyme that catalyzes the reduction of HMG-CoA to mevalonic acid, the rate-limiting step in the synthesis of cholesterol from acetyl-CoA

3-Hydroxy-3-methylglutaryl-CoA (HMG-CoA) Substrate for HMG-CoA reductase

Low-density lipoprotein Lipoprotein that carries most of the cholesterol in human plasma

Low-density lipoprotein receptor Glycoprotein attached to the plasma membrane of most cells and possessing a specific region that binds lipoproteins containing apolipoprotein B or apolipoprotein E in a suitable conformation

Mevalonic acid Branched C_6 compound, formed by the reduction of HMG-CoA, that serves as intermediate in the synthesis of cholesterol and of several essential nonsterol metabolites

Remnant particle Cholesterol-rich particle produced by the partial intravascular degradation of chylomicrons and very low-density lipoproteins

Very low-density lipoprotein Triglyceride-rich lipoprotein secreted into the plasma by the liver

CHOLESTEROL IS AN ESSENTIAL component of the membranes of all animal cells, of the myelin sheaths of nerves, and of the outer shell of plasma lipoprotein particles. In certain specialized cells, cholesterol acts as precursor of bile acids and steroid hormones. Most cells satisfy the bulk of their cholesterol requirements for maintenance, growth, and specialized functions by two regulated processes: (1) the intracellular synthesis of cholesterol from small molecules and (2) the receptor-mediated uptake, from the external medium, of cholesterol-rich particles called low-density lipoprotein(s) (LDL). LDL is the terminal product of the intravascular metabolism of very low-density lipoprotein (VLDL), a lipoprotein secreted by the liver.

The body as a whole acquires exogenous cholesterol by absorption of dietary cholesterol from the lumen of the small intestine. Cholesterol absorbed from the intestine is secreted into the intestinal lymphatics and, thence, into the bloodstream, as a component of large, fat-enriched particles called chylomicrons. As soon as they enter the circulation, chylomicrons are degraded by lipoprotein lipase (an enzyme bound to the luminal surfaces of the blood capillaries) to produce cholesterol-rich "remnant" particles, from which most of the fat has been removed. Chylomicron remnants are taken up by the liver and degraded to release free cholesterol within liver cells. Some of this cholesterol is re-excreted into the intestine in the bile and is then reabsorbed after mixing with dietary cholesterol. Thus, biliary and dietary cholesterol participate together in an enterohepatic circulation. Dietary cholesterol, taken up by the liver as chylomicron remnants, inhibits hepatic synthesis of cholesterol. As a result of this regulatory mechanism, and of the inhibition of cholesterol synthesis in nonhepatic tissues by the uptake of LDL, the amount of cholesterol in the whole body of a fully grown individual remains roughly constant from day to day.

I. Structure and Properties

Cholesterol is a member of a class of naturally occurring compounds called sterols. Figure 1 shows the structural formula of cholesterol. The cholesterol molecule has a nucleus of four fused rings common to all sterols, methyl groups at C-10 and C-13, a branched side-chain, an OH group at C-3, and a double bond at C-5. The presence of the polar group at C-3 enables cholesterol to interact with other lipids such as fatty acids and phospholipids. Pure cholesterol is a solid at body temperature (melting point 150°C) and is essentially insoluble in water. However, the cholesterol in normal cells and in plasma is prevented from crystallizing by its association with phospholipids, whereas cholesterol in normal bile is held in micellar solution by the detergent action of bile salts and lecithin (see Section IV). The presence of the nuclear double bond makes the ring system flatter than it would be otherwise. This flattening effect facilitates the insertion of cholesterol between the molecules of a biological membrane and may have played a part in the natural selection of cholesterol as the predominant sterol of vertebrates. [*See* LIPIDS.]

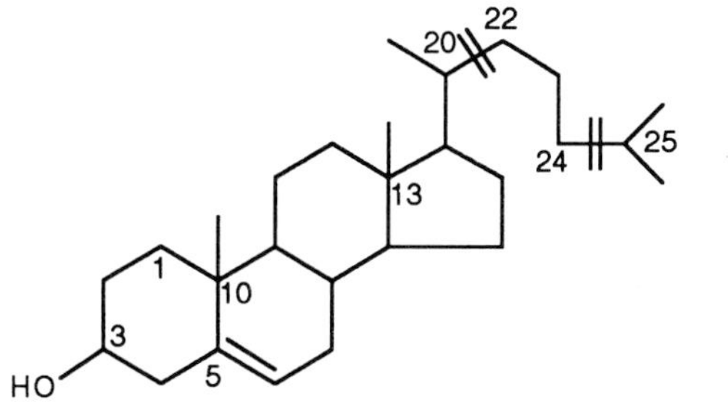

Cholesterol

FIGURE 1 Structural formula of cholesterol ($C_{27}H_{45}O$) drawn with the side-chain at upper right in folded conformation. Carbon atoms mentioned in the text are numbered. The H atoms attached to carbons are omitted; stroked at C-10, -13, -20, and -25 are methyl groups. Double bars show the positions of cleavage in the formation of bile acids and steroid hormones.

II. Free and Esterified Cholesterol

Most of the cholesterol in the body as a whole is present in free, or unesterified, form; however, in certain tissues and in plasma, a significant proportion of the total is esterified at the 3 position with long-chain fatty acids (those with 12 or more carbon atoms). Cholesterol esterified with fatty acids is less polar than free cholesterol because the 3-OH group is masked by the fatty acyl chain. Esterification of cholesterol within cells is catalyzed by the enzyme acyl-CoA–cholesterol acyltransferase (ACAT), so called because it transfers the fatty-acid residue of acyl-CoA to cholesterol. The CoA ester of oleic acid is the preferred acyl substrate for ACAT. Cholesteryl esters are also formed in plasma by the transfer of a fatty-acid residue from lecithin (a phospholipid) to cholesterol. The enzyme catalyzing this reaction is called lecithin–cholesterol acyltransferase (LCAT). The preferred fatty-acid residue for human LCAT is linoleic acid.

III. Distribution

A. In the Whole Body

Cholesterol is present in all tissues, but the concentration varies widely among different tissues, from <100 mg/100 g fresh weight in adipose tissue and muscle to >3 g/100 g in brain, nerve, and adrenal glands. An adult human has about 1 g of cholesterol per kg body weight, corresponding to some 60 g in the whole body. About 1 g of cholesterol is lost from the body each day by conversion into bile acids and steroid hormones, and by fecal excretion of unabsorbed biliary cholesterol. In the steady state, this loss is exactly balanced by endogenous synthesis and absorption of dietary cholesterol. Nearly one-fifth of all the cholesterol in the body is present in the myelin sheaths of nerve fibers. Despite its importance in health and disease, the cholesterol in plasma accounts for $<10\%$ of the total in the body. More than 80% of the total cholesterol in the body is unesterified. This reflects the predominant role of cholesterol as a membrane component (see Section IV). In plasma and adrenal glands, about 70% of the total cholesterol is esterified with fatty acids.

B. Within Cells

In most cells, cholesterol is present largely as the free sterol in the plasma membrane and, to a much smaller extent, in the endoplasmic reticulum and the mitochondrial and lysosomal membranes. In the cells of most tissues, small amounts of esterified cholesterol are dispersed throughout the cytoplasm. In contrast, the hormone-producing cells of the adrenal cortex contain large quantities of esterified cholesterol confined within cytoplasmic droplets. When these cells are called upon to secrete steroid hormone, cholesteryl esters in the storage droplets are rapidly hydrolyzed to release free cholesterol (see Section V.B). Cytoplasmic droplets filled with esterified cholesterol also occur in cells of the macrophage class, giving them a foamy appearance. Focal accumulations of foam cells are responsible for the presence of raised fatty streaks in the arterial wall. Fatty streaks may be the precursors of atherosclerotic lesions.

C. Developmental Aspects

The total amount of cholesterol in the human body increases throughout fetal and postnatal development. This increase is due mainly to the laying down of new cell membranes required to keep pace with the multiplication and growth of cells in all tissues. In addition, there is a relative increase in the cholesterol content of certain tissues at particular stages of development, as in the developing brain. Myelination in the white matter of the human brain is most rapid from the 7 month of fetal life to the end of the first few months of postnatal life. During this period, the concentration of cholesterol in whole brain increases twofold and the total amount of brain cholesterol increases severalfold.

IV. Cholesterol as a Component of Membranes

The basic component of all biological membranes is a double layer of phospholipid molecules arranged with their fatty-acid chains parallel to one another and facing inward into the hydrophobic interior of the bilayer (Fig. 2).

The physical properties of an artificial phospholipid bilayer containing no sterol are such that it could not function efficiently as a biological membrane. At temperatures below a critical phase-transition temperature (T_c) the membrane would exist in a "crystalline" state in which the fatty-acid chains of the phospholipids are immobilized. In this state, permeability to solutes is limited and the viscosity of the bilayer is too high to permit lateral diffusion of proteins and other macromolecules. At temperatures above the T_c, the bilayer exists in an unstable "liquid-crystalline" state in which the fatty-acid chains are freely mobile, the viscosity of the bilayer is greatly reduced, and the phospholipids are held together only by noncovalent interaction between their polar head groups.

When free cholesterol is added to a phospholipid bilayer, the cholesterol molecules enter the spaces between the phospholipid molecules, interacting noncovalently with their fatty-acid chains. The effect of this is to stabilize the bilayer, over a wide range of temperature, in an intermediate-gel state. In this state, the first 8–10 carbon atoms of the fatty-acid chains are immobilized and the remaining

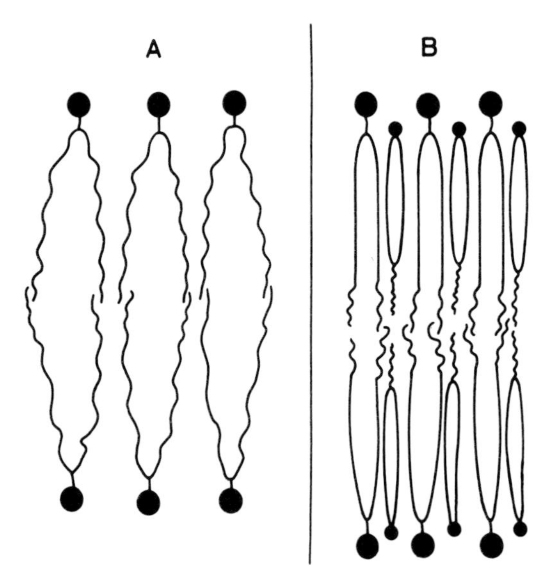

FIGURE 2 Effect of cholesterol on the physical state of a phospholipid bilayer at a temperature above the phase-transition temperature. In the absence of cholesterol (A), the chains are fluid. In the presence of cholesterol (B) (50 moles %), the first 8–10 carbon atoms of the chains are rigid and the bilayer is condensed. The cholesterol molecules are shown with their hydroxyl groups (●) close to the head groups of the phospholipids (⬤).

carbons are freely mobile (Fig. 2). The cholesterol–phospholipid ratio required for optimal functioning of cell membranes under physiological conditions varies from one type of membrane to another. In humans, there is about one molecule of cholesterol for every phospholipid molecule in the plasma membrane of the red cell, whereas in mitochondrial membranes the cholesterol–phospholipid molar ratio is <0.1.

In addition to influencing the physical state of the bulk phospholipids of a cell membrane, free cholesterol may also affect the fluidity of a limited zone of membrane in the immediate vicinity of a membrane-bound enzyme. This may alter the conformation of the enzyme and, thus, affect its state of activation. [*See* MEMBRANES, BIOLOGICAL.]

V. Conversion into Bile Acids and Steroid Hormones

A. Bile Acids

1. Formation and Secretion of Bile Salts

Bile acids are formed from unesterified cholesterol in the liver by a series of enzymic reactions culminating in the oxidative cleavage of the side-chain between C-24 and C-25 (see Fig. 1). The C_{24} bile acids resulting from these reactions are converted into bile salts by conjugation with glycine or taurine. The bile salts, together with free cholesterol and lecithin, are secreted in the bile. In normal bile, all the free cholesterol is dissolved by the combined detergent actions of bile salts and lecithin. The bile is stored in the gallbladder in the fasting state and is discharged into the duodenum and, thence, into the jejunum, in response to a fatty meal. [*See* BILE ACIDS.]

2. Enterohepatic Circulation

After participating in the absorption of fat and cholesterol from the jejunum, about 95% of the bile salts secreted by the liver are reabsorbed from the ileum into the blood circulation and are taken up by the liver, to be resecreted with newly synthesized bile salts. The small quantity that is not reabsorbed is excreted in the feces. Bile salts taken up by the liver down-regulate the synthesis of bile acids from cholesterol by feedback inhibition.

To maintain a constant amount of bile salt in the biliary system and small intestine, the liver makes just enough bile acid to replace the daily loss in the feces. In the presence of a normal enterohepatic circulation, this is equal to about 400 mg/day. This rate of synthesis is equivalent to nearly half the total daily turnover of cholesterol (about 1 g/day in humans). If the reabsorption of bile salts is diminished, the conversion of cholesterol into bile acids is partially released from feedback inhibition, so that bile-acid synthesis increases to keep pace with increased fecal loss. In the extreme case, when all the bile is diverted from the small intestine by an external fistula, the rate of conversion of hepatic cholesterol into bile acid increases 10-fold; consequently, the liver is partially depleted of cholesterol. This results in an increase in hepatic LDL-receptor activity.

B. Steroid Hormones

Steroid hormones are formed from unesterified cholesterol in the gonads, placenta, and adrenal cortex. The initial step is the cleavage of the cholesterol side-chain between C-20 and C-22 (see Fig. 1) to give pregnenolone, the precursor of all steroid hormones. The free cholesterol used as substrate for the cleavage reaction is derived from the intracellular stores of esterified cholesterol (referred to in Section III.B). Release of free cholesterol is mediated by the hydrolytic action of an enzyme (cholesteryl-ester hydrolase), whose activity is regulated by adrenocorticotrophic hormone in the adrenal cortex and by gonadotrophic hormones in the gonads. The amount of cholesterol converted into steroid hormones is much less than the amount converted into bile acids. In a normal man, <50 mg of cholesterol is converted into steroid hormones each day. During the luteal phase of the sexual cycle in women, 50–100 mg of cholesterol per day may be used for hormone production. In the adrenal cortex and gonads, the stores of cholesterol used for hormone production are replenished partly by intracellular synthesis of cholesterol, but mainly by receptor-mediated uptake of LDL from the external medium. [*See* STEROIDS.]

VI. Synthesis

A. HMG-CoA and the Rate-Limiting Step

Cholesterol is synthesized in all animal cells by a pathway in which the primary precursor is acetyl-

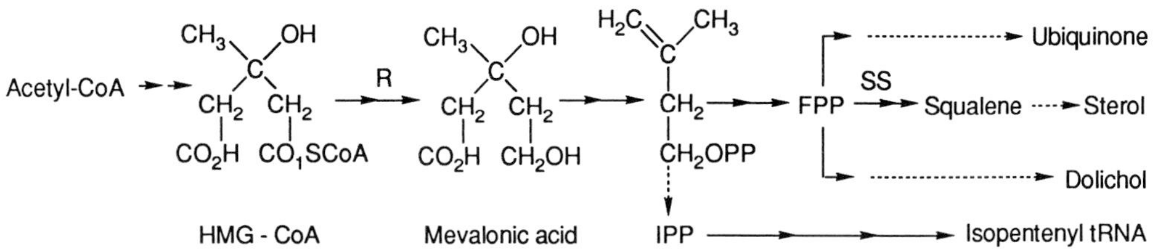

FIGURE 3 Formation of mevalonic acid and the products of its metabolism in animal cells. FPP, farnesyl pyrophosphate; IPP, isopentenyl pyrophosphate; R, reductase; SS, squalene synthetase.

CoA, the methyl and carboxyl carbons of the acetyl residue supplying all 27 carbon atoms of the cholesterol skeleton. The initial steps in this pathway result in the formation of HMG-CoA, the CoA thioester of 3-hydroxy-3-methylglutaric acid (a branched C_6 acid). HMG-CoA is reduced by the enzyme HMG-CoA reductase to mevalonic acid. This compound is then decarboxylated to give a branched C_5 isoprenoid unit (Fig. 3). The isoprenoid unit is the building block used in the formation of cholesterol and of various nonsterols, including ubiquinone, dolichols, and tRNAs containing an isopentenyl residue.

The rate at which cholesterol is synthesized in the cell is determined by the rate at which mevalonic acid is generated from HMG-CoA. This, in turn, is determined by the activity of HMG-CoA reductase, the enzyme with the lowest capacity in the biosynthetic chain leading to cholesterol.

B. HMG-CoA Reductase and the Regulation of Cholesterol Synthesis

HMG-CoA reductase is present in all animal and plant tissues. This is not surprising in view of its role in the generation of a precursor required for the formation of sterols and of several nonsterols essential for the growth and multiplication of most eukaryotic cells. The mammalian enzyme is a glycoprotein with its N-terminal portion anchored to the smooth endoplasmic reticulum and its C-terminal portion, which contains the catalytic site, projecting into the cytoplasm.

HMG-CoA reductase is a highly regulated enzyme, its activity varying over a several hundred-fold range under different conditions in the intact cell. Regulation of reductase activity is achieved by modulation of the rates of enzyme synthesis and breakdown and by changes in the state of activation of existing enzyme molecules. Reductase activity undergoes reciprocal changes in response to alterations in the amount of cholesterol present in an intracellular regulatory pool of the unesterified sterol. For example, reductase activity declines when the cholesterol content of the regulatory pool is increased by receptor-mediated uptake of LDL. The inhibitory effect of cholesterol on HMG-CoA reductase is thought to be mediated partly by oxysterols formed within the cell by the enzymic oxidation of cholesterol. Indirect evidence suggests that oxygenated sterols such as 25-hydroxycholesterol activate a specific protein that, in its activated form, suppresses transcription of the reductase gene by binding to a short segment in the promoter region. Mitogens, growth factors, and hormones such as insulin and thyroxine also influence the rate of synthesis of cholesterol by modulating reductase activity.

The rates of synthesis of cholesterol and nonsterol metabolites of mevalonic acid are regulated independently by multivalent feedback inhibition, a process analogous to the independent regulation, in some bacteria, of the synthesis of several end products derived from a common intermediate.

VII. Uptake of Cholesterol by Cells

A. Composition, Origin, and Metabolism of LDL

Most cells in the human body use LDL as their major source of extracellular cholesterol. An LDL particle consists essentially of a hydrophobic core of triglyceride and esterified cholesterol surrounded by a polar shell of phospholipid, free cholesterol, and protein. The phospholipids and free cholesterol are arranged as a monolayer with the phospholipid head groups and the OH groups of cholesterol in contact with the aqueous medium. Each LDL particle has one molecule of a protein called apolipoprotein B (apoB) associated noncovalently with the polar shell. Each molecule of apoB has a single region that binds specifically to an LDL receptor on the surface of a cell (see Section VII.B).

All the LDL in the circulation is derived from remnant particles resulting from the incomplete degradation of VLDL by lipoprotein lipase, an enzyme attached to the luminal surfaces of the blood capillaries. VLDL is secreted by the liver as a triglyceride-rich lipoprotein. During the intravascular conversion of VLDL into LDL, cholesteryl esters, produced by the action of LCAT in the plasma and with linoleate as their predominant fatty acid, are transferred from high-density lipoprotein (HDL) to LDL. [*See* LIPOPROTEINS.]

Between one half and two thirds of the LDL in plasma is degraded by the LDL-receptor pathway, mainly in the liver (see Section VII.B). Most of the remainder is degraded by other routes not involving specific receptors. However, a small but significant proportion of the plasma LDL is taken up and degraded by cells of the macrophage class after the LDL has been modified by oxidation (see Section VII.C).

B. The LDL-Receptor Pathway

1. Structure and Orientation of the Receptor

The intracellular degradation of LDL by the LDL-receptor pathway begins with the binding of LDL particles to receptors present on the surfaces of most cells in the body. The LDL receptor is a transmembrane glycoprotein with the N-terminus outside the cell, a single membrane-spanning segment, and a C-terminal segment projecting into the cytoplasm. The N-terminal segment contains a sequence of seven imperfect repeats that together constitute the binding domain, which binds the apoB molecule in an LDL particle. The LDL receptor also has high-binding affinity for another apolipoprotein called apoE. The orientation of the receptor in relation to the plasma membrane, and the probable arrangement of the seven repeats, are shown in Fig. 4. Most of the LDL receptors on the surface of a cell are clustered in specialized regions called coated pits. The coated pit is a shallow depression on the cell surface with a cytoplasmic coat consisting mainly of a protein (clathrin) arranged in the form of a lattice. LDL receptors are anchored in coated pits by an interaction between the protein coat and the cytoplasmic extension of the receptor.

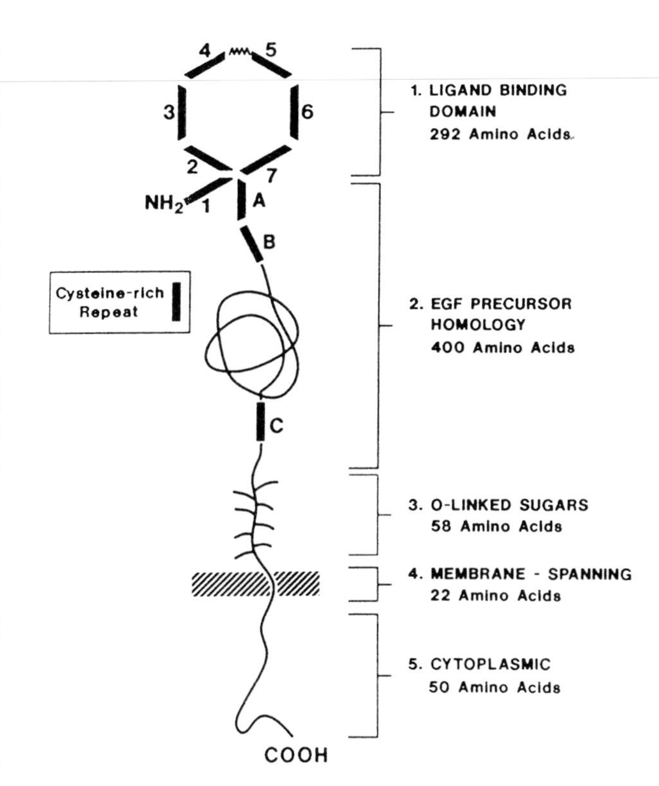

FIGURE 4 Model showing the possible arrangement of elements of the ligand-binding domain of the human LDL receptor and the orientation of the C-terminal and N-terminal portions in relation to the plasma membrane. The essential feature of the binding domain is a hexagonal structure composed of repeats 2, 3, and 4 joined to repeats 5, 6, and 7 by a linker sequence of eight amino acids. EGF, epidermal growth factor. [Reproduced, with permission from Esser *et al.*, 1988, *J. Biol. Chem.* **263,** 13282–13290.]

2. Binding, Internalization, and Degradation of LDL

LDL particles bound to receptors are carried into the interior of the cell by invagination of coated pits, which pinch off from the plasma membrane to form vesicles enclosing bound LDL (Fig. 5). The vesicles fuse with lysosomes, exposing the internalized LDL to digestion by lysosomal enzymes. Before fusion takes place, the receptors in the vesicles dissociate from their bound LDL and return to the plasma membrane, where they diffuse laterally until they become anchored in coated pits, to begin a new cycle of endocytosis. Lysosomal digestion of LDL results in the complete hydrolysis of its protein component and the hydrolysis of esterified cholesterol to yield free cholesterol. The free cholesterol leaves the lysosomes and becomes available to the cell for membrane formation; conversion into bile

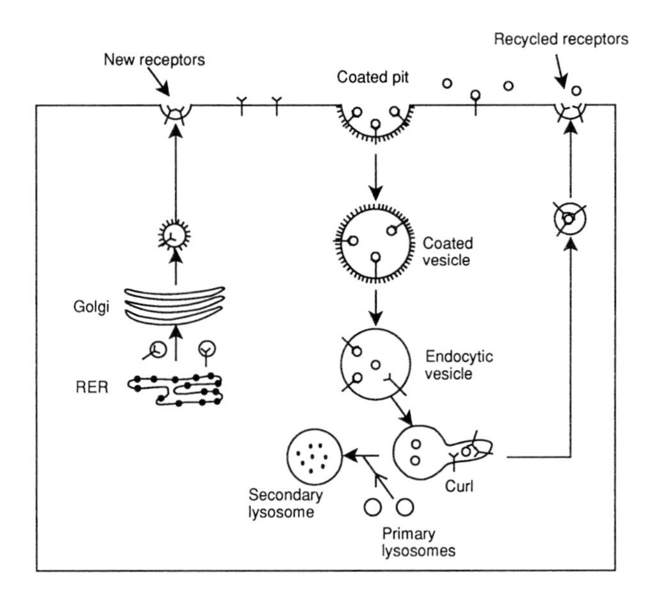

FIGURE 5 Diagram showing the probable sequence of events in the LDL-receptor pathway. The right-hand side shows the endocytosis of LDL (○) bound to LDL receptors (Y) in a coated pit, the recycling of receptors, and the delivery of LDL to a lysosome in which the particles are digested. The left-hand side shows the synthesis of new LDL receptors in the rough endoplasmic reticulum (RER), their transport to Golgi cisternae for processing, the transport of mature new receptors to the plasma membrane, and their lateral diffusion into a coated pit. Endocytosed receptors are thought to return to the plasma membrane in specialized vesicles called compartments of uncoupling of receptor and ligand (CURL). A few internalized LDL particles fail to dissociate from their receptors and are returned to the cell surface by retroendocytosis. [Reproduced, with permission, from N. B. Myant, 1990, "Cholesterol Metabolism, LDL, and the LDL Receptor," Academic Press, San Diego.]

acids, steroid hormones, or lipoproteins; or esterification with fatty acids by ACAT in the cytoplasm.

When the amount of cholesterol brought into the cell by the receptor pathway exceeds that required to satisfy its metabolic needs, HMG-CoA reductase activity is suppressed and the synthesis of new LDL receptors declines. Thus, the LDL-receptor pathway fulfills two biological functions. It is the major route for the irreversible removal of LDL from the plasma, and it provides cells with a regulated supply of cholesterol. In the inherited absence of LDL receptors, LDL accumulates in the plasma at levels that give rise to premature heart disease, but the tissues do not become cholesterol-deficient because their requirement for cholesterol is met by intracellular synthesis.

The quantitative importance of the LDL-receptor pathway varies widely from one tissue to another. In some tissues, such as skeletal muscle, the cell's requirement for cholesterol is satisfied almost entirely by synthesis *in situ*. In others, such as the adrenal cortex in the resting state, >90% of the cholesterol required is supplied by receptor-mediated uptake of LDL. As noted above, the liver is responsible for most of the receptor-mediated uptake and catabolism of LDL in the body as a whole. For this reason, liver transplantation has been used successfully in the treatment of patients with an inherited absence of LDL receptors.

C. Receptor-Mediated Uptake of Other Lipoproteins

Some cells obtain extracellular cholesterol by receptor-mediated uptake of lipoproteins other than LDL. Macrophages and other cells of the reticuloendothelial system express receptors (acetyl-LDL receptors) with high affinity for LDL that has been modified by acetylation or peroxidation. Acetyl-LDL is not produced in the body. However, certain cells, including those of the vascular endothelium, generate free radicals capable of oxidizing LDL *in vivo*. Oxidized LDL produced locally may be taken up via acetyl-LDL receptors on macrophages in the arterial wall and may, therefore, be responsible for the formation of foam cells (see Section III.B).

LDL receptors on liver cells, in addition to binding LDL itself, are also responsible for the hepatic uptake of some of the remnants produced by the action of lipoprotein lipase on VLDL. Uptake of VLDL remnants by hepatic LDL receptors is mediated by the presence of apoE in the particles. Decreased activity of hepatic LDL receptors leads to a fall in the rate of removal of VLDL remnants from the circulation, with a consequent increase in the rate of production of LDL (the end product of the intravascular metabolism of VLDL).

The cholesterol-rich remnant particles produced by the partial hydrolysis of chylomicrons (see Section IX) are taken up by the liver by receptors (called apoE receptors) that recognize lipoproteins containing apoE but do not recognize LDL. ApoE receptors mediate the rapid removal of chylomicron remnants that appear in the circulation after a fatty meal.

VIII. Reverse Cholesterol Transport

The major route for the removal of cholesterol from the body is by secretion in the bile as bile salts or free cholesterol, followed by excretion in the feces after partial reabsorption from the small intestine. The only tissues outside the liver that are capable of degrading cholesterol are the adrenal cortex and the gonads. Hence, net removal of cholesterol from extrahepatic tissues must largely depend on the transport of cholesterol through the plasma from the periphery to the liver, i.e., in a direction opposite to that of the flow of cholesterol into cells by uptake of LDL. This process, known as reverse cholesterol transport, is mediated by the transfer of free cholesterol from cells to small, dense HDL particles present in the interstitial fluid. The free cholesterol is esterified by LCAT, giving rise to cholesteryl ester-rich HDL particles that acquire apoE. These HDL particles deliver their cholesterol to the liver, probably by a combination of receptor-mediated uptake of whole particles (involving LDL and apoE receptors) and direct transfer of cholesterol to hepatocytes without internalization of HDL.

IX. Absorption from the Intestine

Cholesterol is absorbed from the jejunum in unesterified form. Dietary cholesteryl esters are hydrolyzed by pancreatic cholesteryl ester hydrolase in the lumen of the jejunum. The free cholesterol so produced, together with that already present in the diet and the bile, is incorporated into mixed micelles containing bile salts, phospholipids, and other lipids. Free cholesterol is then transferred from the micelles to the mucosal cells lining the jejunum. Within the cells, cholesterol is esterified with fatty acids by ACAT, and the esterified cholesterol is incorporated into the triglyceride core of nascent chylomicrons. The chylomicrons are secreted into the intestinal lymphatics and, thence, into the blood circulation. On reaching the circulation, chylomicrons are rapidly converted into remnant particles by the action of lipoprotein lipase (see Section VII.A). Chylomicron remnants have lost nearly all the triglyceride present in the parent particles but have retained most of the esterified cholesterol. These cholesterol-rich particles deliver their load of cholesterol to the liver by uptake mediated by apoE receptors. Thus, the immediate destination of absorbed cholesterol is the liver.

Absorption of cholesterol from the human intestine is incomplete, even when the intake of dietary cholesterol is very low. Under experimental conditions, less than half the cholesterol taken in the diet is absorbed when the daily intake is varied from about 50 mg to >2 g. Because cholesterol is not completely absorbed, the bile provides the body with an exit for cholesterol.

Bibliography

Brown, M. S., and Goldstein, J. L. (1986). A receptor-mediated pathway for cholesterol homeostasis. *Science* **232,** 34–47.

Gibbons, G. F., Mitropoulos, K. A., and Myant, N. B. (1982). "Biochemistry of Cholesterol." Elsevier, Amsterdam.

Myant, N. B. (1981). "The Biology of Cholesterol and Related Steroids." Heinemann, London.

Myant, N. B. (1990). "Cholesterol Metabolism, LDL, and the LDL Receptor." Academic Press, New York.

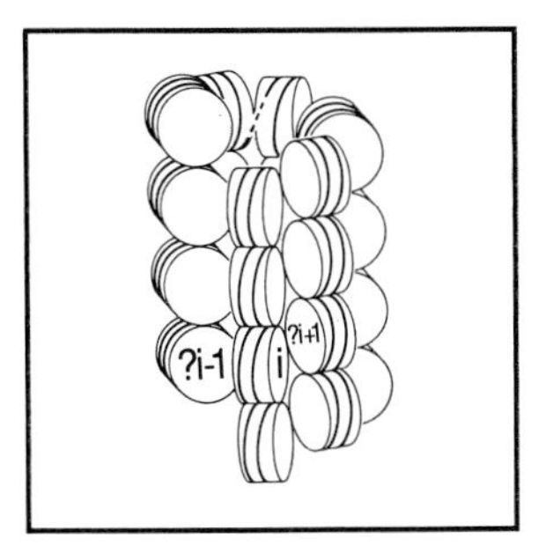

Chromatin Folding

JONATHAN WIDOM, *University of Illinois at Urbana–Champaign*

Glossary

Chromatin Complex of DNA and histone proteins; the material of which chromosomes are constituted

Core histones Collective name given to four proteins (H2A, H2B, H3, and H4), which make up the core of the nucleosome

Dyad symmetry Object possessing dyad symmetry (2-fold rotational symmetry) is invariant upon rotation by 180 about the dyad axis

Histones Family of small, basic proteins that are highly conserved throughout evolution and are responsible for packaging DNA

Histone octamer Structured complex containing two each of the four core histones, forming the core of the nucleosome on which DNA is wrapped

Synaptonemal complex Specialized chromosomal structure formed during meiosis. A pair of homologous chromosomes, in an intermediate stage of compaction, are brought into alignment along their length. The structure is believed to facilitate meiotic recombination

HUMAN CELLS have two homologs, each of 23 distinct chromosomes. Each chromosome in a cell contains one molecule of double-stranded DNA that is ~20 Å wide but extremely long; a typical human chromosomal DNA molecule is a few centimeters in length. Such lengths are orders of magnitude greater than the diameter of a cell nucleus. *In vivo,* these DNA molecules are closely associated with a number of proteins that fold the DNA in a hierarchical series of stages and eventually produce a 10,000-fold linear compaction of the DNA preparatory to cell division. Thus, the substrates for important genetic processes such as transcription, replication, recombination, and chromosome division are not bare DNA, but rather are these protein–DNA complexes (chromosomes) in one or another stage of compaction. This chapter summarizes what is known about the structures and mechanisms of chromosome folding.

I. Subunit Structure

The lowest level of chromosome structure is based on a repeated motif called a nucleosome. In each nucleosome, a short stretch of DNA, 165–245 bp in length, is locally folded and compacted in a complex with nine proteins: two molecules each of histones H2A, H2B, H3, and H4 and one molecule of histone H1. The length of DNA in a single nucleosome is only a small fraction of the overall length of a typical DNA molecule, so this motif is repeated hundreds of thousands of times along the entire DNA length. Thus, nucleosomes can be considered chromosome subunits, except that, owing to the continuity of the DNA, they are connected together in a chain. [*See* DNA IN THE NUCLEOSOME.]

Stretches at each end of the DNA associated with a single nucleosome in chromatin are accessible to double-stranded DNA endonucleases such as micrococcal nuclease. The earliest stage of digestion of chromatin by such a nuclease produces a size distribution of soluble chromatin fragments (oligonucleosomes). These fragments are the source of chromatin used in the structural studies discussed in subsequent sections of this article. More extensive digestion releases individual nucleosomes, and further digestion degrades the nucleosomes into derivative particles: first into chromatosomes, and ultimately into nucleosome core particles.

A. The Nucleosome Core Particle

The nucleosome core particle is the best understood of these various particles, because it has been crystallized and its structure has been determined by X-ray crystallography at 7 Å resolution. The core particle contains 145 bp of DNA and an octamer of the core histones: two each of the histone proteins H2A, H2B, H3, and H4. [*See* HISTONES AND HISTONE GENES.]

At 7 Å resolution, one cannot discern the locations of individual atoms; therefore, little information is available at this time concerning the structure of the protein components. On the other hand, the alternating pattern of major and minor grooves in DNA makes the DNA plainly visible at this resolution. Objects in the structure that are not DNA are presumed to correspond to one or another of the histone proteins. The particle resembles a disk, with a diameter of 110 Å and a thickness of 57 Å. It is believed to possess dyad symmetry, although this is distorted in the crystals owing to packing considerations. The DNA is in the B-form, and is wrapped around the surface of the histone octamer in ~1.8 turns of an approximate flat, left-hand, irregular superhelix. The pitch of this superhelical wrapping varies along the trajectory, with an average of ~28 Å. The location of each of the histones within the octamer core has been deduced from protein–DNA cross-linking data. Each stretch of DNA in the core particle appears to interact predominantly with one histone. The histones appear, from either DNA end, in the sequence H2A, H2B, H4, H3, H3, H4, H2B, and H2A. The central turn of DNA (~80 bp) interacts predominantly with a histone tetramer ($H3_2$–$H4_2$). The remaining ~20 bp of DNA at each end interacts predominantly with one H2A-H2B heterodimer each.

B. The Chromatosome

The chromatosome, from which the core particle is derived, consists of a nucleosome core particle with an additional 10 bp of DNA at each end (giving 165 bp total) and one molecule of histone H1. It is not known where on a nucleosome core particle these additional components are located. At this time, all attempts at crystallization of chromatosomes or larger chromatin fragments have failed, and direct imaging by electron microscopy has not yet achieved adequate resolution; consequently, no direct structural evidence exists concerning the locations of these extra components, and one must instead rely on indirect biochemical or biophysical studies.

In the absence of other information, two of the most likely possible trajectories for the extra 10 bp of DNA at each end include the following: (1) the DNA continues its superhelical wrapping, giving two complete superhelical turns with the 165 bp; or (2) the DNA exits tangentially from the core particle immediately after its 1.8 superhelical turns. Other trajectories are also possible. As discussed below, even these two rather similar trajectories lead to significantly distinct predictions for the three-dimensional structure of the nucleosome filament.

Histone H1 is known to sit on the surface of the chromatosome, and several lines of evidence indicate that it is located on a chromatosome in the vicinity of the region where DNA enters and exits the particle. Moreover, H1 has a three-domain structure, consisting of a central, folded 80-amino acid region referred to as the globular domain ("GH1"), flanked by extended, highly basic, N- and C-terminal tails, and it is the globular domain that is apparently responsible for determining the ability of H1 to recognize and bind to its site in a chromatosome. The amino acid sequence of the globular domain of H1 is well conserved throughout evolution, and its structure has been determined by two-dimensional nuclear magnetic resonance (NMR) methods.

The two possible DNA trajectories discussed above each lead to a distinct model for the nature of the binding site for GH1. Trajectory 1 leads to a model in which GH1 binds to the surface of the chromatosome in the region where the DNA completes its putative two complete superhelical turns; in this case, the GH1 binding site consists of three coplanar DNA segments. Trajectory 2 creates a pocket on the chromatosome surface in the vicinity of the DNA entry–exit region that is delineated by three DNA segments: one at the back, from the central turn of DNA on the core particle, and one at each side, from the two DNA segments entering and leaving the chromatosome. The expected dimensions for this pocket match the actual diameter of GH1 as measured by small-angle neutron scattering and as seen from the two-dimensional NMR structure.

C. The Nucleosome

A nucleosome consists of a chromatosome together with part of each of the two DNA segments, called linker DNA, that connect that chromatosome to its two immediate neighbors on the chromatin fiber. Little structural information is available at this level *per se*. Linker DNA varies in length from one nucleosome to the next about an average value that is characteristic for cells of a given type; this average value itself varies from one cell type to another, within the range ~0–80 bp. The trajectory taken by the two partial linker regions in an isolated nucleosome is not known. The N- and C-terminal domains of histone H1 adopt extended, largely α-helical, structures in solution, and are believed to interact chiefly with linker DNA.

Nucleosomes are heterogeneous in several respects, in addition to the variability of the linker DNA. The sequence of DNA in a nucleosome is of course variable (because essentially all of the genome is packaged in nucleosomes), and this variability seems certain to affect the structure at high resolution. The histones are present in many variants, coded by different genes; moreover, each histone variant is subject to a large number of post-translational modifications, such as acetylation, methylation, phosphorylation, and the attachment of polyadenosine diphosphate (ADP) ribose and of ubiquitin (a complete small protein). Any of these variations could affect structure.

II. Structure of the Nucleosome Filament

The lowest level of organization of a complete chromosome is the nucleosome filament. The available data suggest a one-dimensional picture of the chromatin filament in which chromatosomes, having a generally conserved composition and structure but differing in detail, are spaced along the DNA at irregular intervals that vary statistically about a well-defined average. At the level of one-dimensional structure, the chromatin filament is ordered but is not regular in the geometric sense.

Chromatin suitable for physical studies is isolated from purified cell nuclei in two steps. First, the extremely long chromosomal DNA molecules are randomly digested *in situ* with an enzyme such as micrococcal nuclease that preferentially attacks the linker DNA. Then, the nuclei are lysed and chromatin fragments (nucleosome oligomers) are released into solution and separated from residual nuclear debris. The chromatin fragments tend to adopt more highly folded states, which are discussed in the following sections of this chapter. However, in the absence of any multivalent cations and provided that the concentration of monovalent cations is sufficiently low, these more highly folded states unfold and allow the lowest level of organization to be studied.

A. Nucleosome Filaments versus 100-Å Filaments

Electron microscopic studies led initially to considerable uncertainty regarding the structure of the nucleosome filament. Depending on details of specimen preparation, the chromatin could appear either as a filament of distinct nucleosomes (i.e., with chromatosomes separated by visibly extended linker DNA), now referred to as the nucleosome filament, or as a continuous filament having a width of ~100 Å (presumably, actually 110 Å), referred to as the nucleofilament or the 100-Å filament. It was simply not possible to determine from these studies, which, if either, of these two classes of images were representative of the real structure in solution. One could in principle distinguish between two such models by quantitative analysis of hydrodynamic properties—sedimentation coefficients or translational diffusion coefficients—and many such studies were carried out. In practice, these experiments suffered from too many uncertainties (such as the unknown locations and hydrodynamic properties of linker DNA and extended domains of histones) to be definitive.

The question of which of the two models—nucleosome filament or 100-Å filament—more accurately represents the species present in solution has now been settled by small-angle X-ray and neutron-scattering experiments. These techniques allow one to measure key model-independent properties of extended structures: the radius of gyration of the cross section (R_c), which is related to the actual diameter or width, and the mass per unit length (m/l), which can be expressed as the number of nucleosomes per unit displacement along the filament. The two different models lead to very different predictions for R_c and m/l, and the experiments are in agreement with the predictions of the nucleosome filament model. It now seems that it was the presence of uranyl acetate, used to enhance contrast, that caused nucleosome filaments to appear in the form of 100-Å filaments. Uranyl acetate solutions contain significant concentrations of mono- and divalent cationic species and, as discussed below, these may alter the folded state of chromatin.

B. The Nucleosome Filament

The nucleosome filament is the conformation that might be expected on purely physical grounds. Linker DNA segments are, in general, very short compared with the persistence length of naked DNA. (The persistence length is a lengthscale of polymer stiffness; for DNA it is ~500 Å, or ~150 bp). If linker DNA had the properties of naked DNA, then it would be essentially straight and inflexible and, therefore, might be expected to extend straight from one chromatosome to the next.

Once it was learned that the nucleosome filament was the appropriate structural model, it was possible to give a physical interpretation to a peak observed in the X-ray and neutron-scattering patterns: the diffraction peak arises from (three-dimensional) distance correlations between consecutive nucleosomes along the filament and allows that distance to be measured approximately. The average nucleosome center–center distance measured in this way is found to be in agreement with those expected if the linker DNA is extended (given the known average linker length), and it changes as expected when the average linker DNA length is varied by using chromatin from differing cell types.

Three models for the structure of the nucleosome filament that are consistent with available data are shown in Fig. 1; the structures are shown flattened onto a surface, as occurs when samples are prepared for electron microscopy. The three related structures correspond to three different models for possible trajectories taken by linker DNA (see above). Model C corresponds to trajectory 1, in which the DNA on a chromatosome is organized in two full turns and then leaves the chromatosome surface tangentially. Model B corresponds to trajectory 2, in which DNA leaves the chromatosome surface tangentially after only 1.8 turns. Model A corresponds to a trajectory in which the DNA leaves the chromatosome surface tangentially after only 1.5 turns. No experimental evidence supports

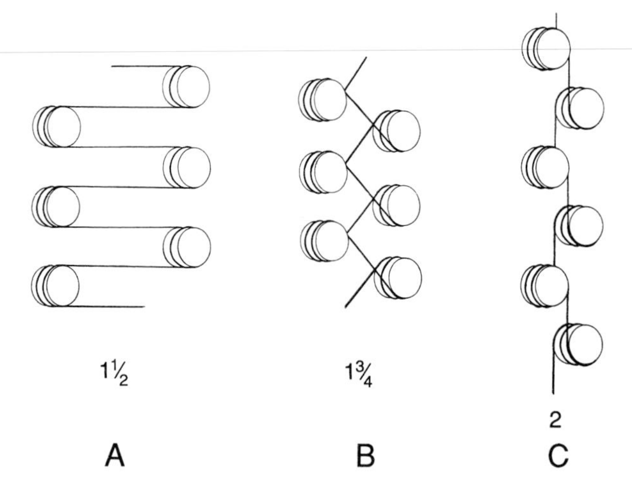

FIGURE 1 Three possible models for the structure of the nucleosome filament. The structures are shown flattened into two dimensions, as happens during electron microscopy. The number underneath each model is the number of superhelical turns made by the DNA as it wraps on the histone surface. The different models lead to different locations in three-dimensional space for the nearest-neighbor nucleosomes and to different numbers and spatial arrangements of DNA segments at the proposed binding site for histone H1. [Drawing by S. Holz.]

model A; it is included here only because such a model for the nucleosome filament is implied in certain models for higher levels of folding, discussed below.

In solution, the natural twist of DNA causes the nucleosome filament to rotate to a variable degree about the axis of each linker, so that the flat faces of the chromatosomes need not be coplanar as illustrated. Such rotations give model B the appearance of a three-dimensional zig-zag. Optical studies of oriented nucleosome filaments in solution suggest that the nucleosomes have their flat faces parallel (±20°) to the axis of the nucleosome filament. Models B and C have important consequences for chromatin types having a very short average linker DNA length. Because of the resistance to torsion and bending of DNA, certain very short linker lengths must be disallowed because they would cause consecutive nucleosomes to overlap in space. This restriction can be alleviated by adding or subtracting ~5 bp of DNA (about half of a helical turn). Many subtly different conformational isomers will exist, even along the same nucleosome filament, depending on the exact length or torsional stress of each linker DNA.

Histone H1 is an important element in the structure of the nucleosome filament. Both electron microscopy and X-ray scattering studies show that when H1 is removed, the distance between consecutive nucleosomes is slightly increased, and the filament is disordered. The characteristic zig-zag appearance in electron micrographs is lost, and the X-ray scattering peak, which arises from distance correlations between consecutive nucleosomes is reduced in intensity. These results indicate that H1 imposes a locally ordered structure on a nucleosome and its immediate neighbors along the chain.

III. Folding of the Nucleosome Filament

The next higher level of chromosome structure, referred to as the 30-nm filament, is achieved by intramolecular folding and compaction of the nucleosome filament to produce a shorter, wider fiber having a diameter of approximately 30 nm. The 30-nm filament is a particularly important level of chromosome structure because this is the folded state in which most chromatin is maintained throughout most of the cell cycle. It must be unfolded to allow transcription and replication, and it is further folded to allow meiosis and mitosis. Before considering the structure of the 30-nm filament, it is helpful to consider what is known about the folding process itself.

In vitro, folding of chromatin in the nucleosome filament state into 30-nm filaments is induced by increasing the concentration of cations. The 30-nm filaments remain soluble and can be studied by numerous different physical methods.

An important question is whether or not the 30-nm filaments produced in this way are similar in structure (and not just in appearance to the 30-nm filaments *in vivo*. Two lines of evidence suggest that they are. First, low-angle X-ray diffraction patterns from 30-nm filaments *in vitro* show the same set of bands characteristic of chromatin in nuclei and in whole living cells (certain cell types have very little cytoplasm, and X-ray diffraction patterns from solutions of these cells turn out to be dominated by diffraction from the chromatin). This is good evidence that the 30-nm filaments produced *in vitro* are similar in internal structure to those produced *in vivo*. Second, chromatin can be isolated from cells without ever exposing it to conditions in which 30-nm filaments would unfold into the nucleosome filament state. Electron microscopic and hydrodynamic studies of such 30-nm filaments have led to the conclusion that there are no significant differences between the 30-nm filaments extracted in that manner and 30-nm filaments formed by refolding from the nucleosome filament state.

One important conclusion from all of these studies is that, at sufficiently high cation concentrations, 30-nm chromatin filaments are no longer soluble and precipitate out of solution. Further addition of multivalent cations causes the aggregated (insoluble) 30-nm filaments to pack very closely together, so that electron density contrast between them is lost.

A. Role of Cations

Essentially, any cation can apparently suffice to cause nucleosome filaments to fold into 30-nm filaments, as judged by electron microscopy, hydrodynamic, optical, and diffraction methods. Inorganic monovalent cations such as Na^+ and K^+, divalent cations such as Ca^{2+}, Mg^{2+}, and Mn^{2+}, trivalent cations such as $Co(NH_3)_6^{3+}$, and organic multivalent cations such as spermidine^{3+} and spermine^{4+} all lead to closely similar 30-nm filament states of chromatin. There remains some debate whether or not any differences do in fact exist.

It is found that the valence of the cation is of primary importance in determining the concentration of cation necessary to induce chromatin folding. Monovalent cations are required at concentrations of 60–80 mM, divalent cations at 100 μM–1 mM, trivalent cations at 10–100 μM, and tetravalent cations at 1–10 μM. The relatively wide concentration ranges given are due in part to experiments with multivalent cations being done in the presence of a range of monovalent cation concentrations (e.g., buffer cations such as $Tris^+$ or buffer counterions such as Na^+ or K^+) because of cation competition (see below). When the monovalent cation concentration is kept constant (and $\leqslant 60$ mM), cations of the same valence are typically effective within a twofold concentration range. Exceptions to this rule may arise when a cation is particularly bulky or when the cation can bind nonionically to chromatin (e.g., Cu^{2+}).

These data are sufficient to define the role of cations in the stabilization of 30-nm filaments. *A priori,* one can list three possible mechanisms by which cations might act: (1) by binding to a particular site that has chemical selectivity, such as the metal binding site in metalloenzymes, (2) by screening repulsions between small numbers of charges, such as amino acid side chains, or (3) by acting as general DNA counterions, reducing the effective charge per

DNA phosphate and screening repulsions between adjacent DNA helices. Model 1 is ruled out by the observed lack of sensitivity to the size or chemical nature of the cation. If model 2 were correct, chromatin folding would be governed by the ionic strength in accord with Debye-Huckel theory. However, the ionic strength necessary to stabilize 30-nm filaments is found to decrease by three orders of magnitude as the valence of the cation is increased from +1 to +4.

The theory governing model 3 is called counterion condensation theory and applies to linear systems having a charge density that exceeds some critical value. In chromatin, roughly one-half of the DNA phosphates (which each carry a formal charge of −1) are neutralized by the positively charged histones; nevertheless, the net linear charge density of the nucleosome filament still exceeds the critical threshold for counterion condensation, and this charge density will increase when nucleosome filaments compact into 30-nm filaments. Counterion condensation theory predicts that the effectiveness of cations as DNA counterions primarily depends on the valence of the cation, with an effect much greater than for the Debye-Huckel theory. This behavior is in qualitative accord with the results obtained for chromatin folding. The theory further predicts that the concentration of a multivalent cation required to induce folding will depend on the monovalent cation concentration because of cation competition. As discussed below, this surprising prediction is verified experimentally. Thus, one concludes that cations induce chromatin folding by reducing repulsions between DNA segments within 30-nm filaments.

B. Cation Competition

An important prediction of counterion condensation theory is that monovalent and multivalent cations compete with each other for binding to chromatin and for stabilizing the 30-nm filament state. This surprising prediction is verified experimentally and leads to complicated behavior. Na^+ (and presumably any other monovalent cation) has dual effects. Sufficiently high concentrations of Na^+ (~60–80 m*M*) cause chromatin to fold into 30-nm filaments even in the absence of Mg^{2+} or other multivalent cations; however, at low concentrations of Na^+ (less than ~45 m*M*) and when Mg^{2+} (or some other multivalent cation) is present and stabilizing the 30-nm filament state, Na^+ has the opposite effect: it competes with the higher valence cation for binding to chromatin and destabilizes the 30-nm filament state. Cation competition appears to underlie chromatin folding and also the aggregation of 30-nm filaments.

It may seem paradoxical that Na^+ (or any other monovalent cation) can both stabilize and destabilize 30-nm chromatin filaments; however, several points are worth noting. Such behavior does not violate any rules of thermodynamics and, indeed, is expected theoretically whenever any cations interact with nucleic acid. Furthermore, identical behavior is observed for two other analogous but completely different systems: Na^+ can both stabilize and destabilize the Z-form of DNA relative to the B-form, in the presence of $Co(NH_3)_6^{3+}$; and Na^+ can both stabilize and destabilize double-stranded DNA relative to single-stranded (melted), in the presence of Mg^{2+}. In the latter case, there exists a semiquantitative theoretical analysis.

C. Titrations without End Points

Various physical experiments used to monitor the folding of nucleosome filaments into 30-nm filaments have given two very different classes of results. Electron microscopy leads to the conclusion that cation titrations of nucleosome filaments reach a definite end point, in which all of the chromatin is folded into 30-nm filaments that do not change further in appearance on further addition of cations. Neutron-scattering studies show that R_c and m/l reach plateau values at similar cation concentrations, as do optical parameters of oriented chromatin filaments and the rotational relaxation times. A very different view is reached from measurements of sedimentation coefficients, translational diffusion coefficients, light-scattering intensities, and from the sharpness of peaks in X-ray solution-scattering patterns. These studies suggest that 30-nm filaments continue to change in structure even after the end point detected by other methods, suggesting that there may be no single 30-nm filament state, and that the only end point of a cation titration of chromatin is the point at which no detectable chromatin remains soluble.

These results collectively have two implications. First, chromatin can apparently continue to change in structure even past the end points detected in certain experiments; however, such changes must be quite subtle, because some measures do give definite end points, such as the overall appearance in electron micrographs. Perhaps the 30-nm filament should be regarded as a continuum of closely related states. Second, ionic conditions that stabilize 30-nm filaments also stabilize their aggregation. This undesirable property makes it difficult to carry out properly many physical studies, because one cannot follow the folding transition reversibly between two definite end points.

D. Continuous versus Two-State Folding

If a sufficient concentration of some cation is added to chromatin in the nucleosome filament state, the chromatin refolds into 30-nm filaments. The folding process (as monitored by stopped-flow X-ray scattering) is very fast. In a wide range of physical experiments, it is found that addition of less of the folding cation yields physical parameters (e.g., R_c, m/l, dichroism, sedimentation coefficient) that are intermediate in magnitude between those of nucleosome filaments and those of 30-nm filaments. The question arises: is the folding of each chromatin filament a gradual (i.e., continuous) process, or is each molecule in one of two limiting states (the nucleosome filament or the 30-nm filament), with the intermediate values of various parameters reflecting cation concentration-dependent fractions of the population of chromatin molecules in each state?

With many physical experiments, it is not possible to settle this question—particularly because a definite 30-nm filament end point may be experimentally inaccessible. However, electron microscopy allows measurements to be made on individual molecules and, therefore, should allow one to distinguish between continuous and two-state folding. Early electron microscopic studies were qualitative, but showed that, at intermediate points in titrations, each chromatin filament appeared to be intermediate in width. This has been confirmed by a more recent quantitative study (using scanning transmission electron microscopy) of m/l during a Na^+ titration. At intermediate points in the titration, the chromatin was plainly in intermediate states of folding; the two limit states were largely unpopulated. Chromatin folding is continuous, not two-state.

E. Structural Consequences

Electron microscopy shows that, in the early stages of a cation titration of nucleosome filaments, the nucleosomes move visibly closer together. Typical results are illustrated schematically in Fig. 2. The open zig-zag appearance (obtained in 1 mM Na^+) collapses to a closed zig-zag in which the nucleosomes often appear to be in two closely paired parallel rows (in ~20 mM Na^+). As the cation concentration is further increased, the chromatin takes on a range of diverse appearances, implying diverse models for the structure of the 30-nm filaments that will be formed. When the concentration of cations is sufficiently high (~60 mM Na^+), the chromatin reaches the 30-nm filament state and becomes more uniform in appearance. Hydrodynamic studies show that the sedimentation coefficient, and (equivalently) the translational diffusion coefficient continues to increase throughout the folding transition, providing evidence that chromatin folding in solution is accompanied by the packing together of nucleosomes. Similarly, an X-ray scattering peak, which monitors the average internucleosomal distance, is found to move to wider angles, indicating a shortened average internucleosomal distance and, ultimately, to disappear, indicating that most of the nucleosomes are in physical contact with other nucleosomes.

Unfortunately, none of these experiments reveal *which* nucleosomes are packing together in space

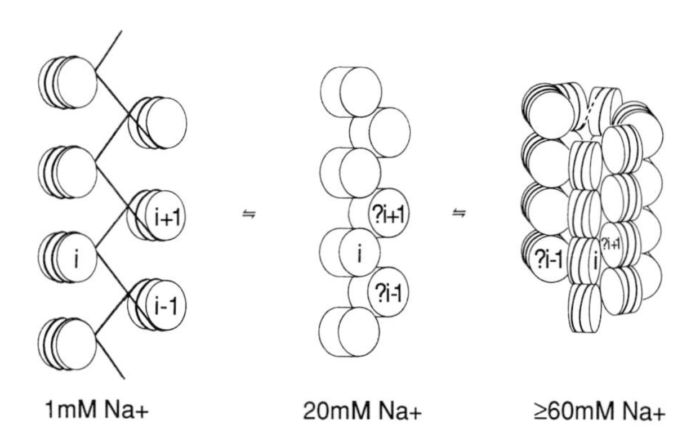

FIGURE 2 Schematic illustration of the chromatin-folding process as seen by electron microscopy. Typical structures are shown for three different concentrations of Na^+ (as indicated). Increasing concentrations of Na^+ cause the chromatin to fold up. The most compact structure (right side, 60 mM Na^+) is approximately 30 nm in diameter, and is called the 30-nm filament. The particular structure shown in the solenoid model; other models are closely similar in appearance (see text). The starting nucleosome filament model (1 mM Na^+) is the one thought most likely to be correct, but this is not known. Nucleosome i refers to an arbitrary but particular nucleosome in the chain; $i-1$ and $i+1$ are its immediate neighbors. The question marks (?) shown in the other structures indicate that those nucleosomes might or might *not* be the one-dimensional neighbors of nucleosome i. [Drawing by S. Holz.]

during chromatin folding. For example, Fig. 2 illustrates that the nearest neighbors (in three dimensions) of nucleosome i may or may not be $i - 1$ and $i + 1$. It is simply not possible to answer this question with currently available data. It will be seen in the following section that there are currently several distinct models for the structure of 30-nm filaments that differ chiefly in their connectivity—i.e., in their assumptions regarding which nucleosomes along the chain are brought into proximity in the 30-nm filament.

F. Role of Histone H1

Chromatin that is biochemically stripped of histone H1 no longer undergoes cation-dependent folding into 30-nm filaments. Electron micrographs of H1-stripped chromatin in solutions containing sufficient Na^+ or Mg^{2+} for folding of native chromatin into 30-nm filaments show regions of random nucleosome strings interspersed with compact disordered clumps of nucleosomes. The sedimentation coefficient of stripped chromatin does increase during Na^+ or Mg^{2+} titrations, but much less than that of native chromatin. These observations lead to the conclusion that histone H1 thermodynamically stabilizes the 30-nm filament state relative to less compact states. Moreover, H1 appears to confer structural specificity on the compact chromatin, since the nucleosome clumps found with stripped chromatin appear to lack even local order.

Histone H1 has the ability to bind cooperatively to DNA. It now appears that the globular domain of H1 ("GH1") on its own has this same capability. At this time, it is not known with certainty whether or not H1 binds cooperatively to chromatin. H1 exhibits a preference for longer chromatin oligomers, in accord with a cooperative binding mechanism, but simple experiments in which H1-stripped chromatin is titrated with increasing concentrations of H1 fail to reveal any hints of cooperativity. This is a very important issue, because cooperative binding of H1 potentially provides a mechanism for the cooperative folding or unfolding of *regions* of chromosomes, possibly containing several genes. Because RNA and DNA polymerases are large in size compared with nucleosomes, one presumes that 30-nm filaments must be unfolded prior to transcription or replication. Thus H1 could potentially regulate the ability of groups of genes to be transcribed or replicated.

Studies of the binding of GH1 to DNA suggest that cooperativity in binding is mediated by direct contacts between adjacent molecules of GH1. An important question currently under investigation is whether or not such contacts between GH1s exist in chromatin. This is closely related to the question of whether or not 30-nm filaments are cooperatively stabilized; moreover, as discussed in a following section, certain models of 30-nm filament structure can be distinguished experimentally through study of how many GH1s can be simultaneously in contact.

IV. Structure of the 30-nm Filament

While a great deal of progress has been made toward elucidation of the structure of the 30-nm filament, several key features of the structure remain unknown. Before considering what is known, it is useful to consider why the problem has been so difficult to study.

There are two methods that are potentially capable of providing direct, high-resolution, structural information for objects on this size scale: X-ray crystallography and electron microscopy. There have been no reports of successful crystallization of chromosome fragments in the 30-nm filament state; this effectively precludes the application of X-ray crystallographic methods for structure determination. The reasons that 30-nm filaments have not yet been crystallized and that electron microscopy has not yet solved the structure are the same. First, because nucleosome filaments are heterogeneous in length and in local composition along the filament, 30-nm filaments are too. This greatly reduces the chances for successful crystallization. In electron microscopic studies, structural heterogeneity along the 30-nm filament may sufficiently distort the internal symmetry of the filament as to preclude the application of symmetry-based image reconstruction methods.

The second, and most important, problem comes from the phase diagram for the cation-induced folding of chromatin. The ionic conditions that one wishes to use, to best stabilize the 30-nm filament state, cause the 30-nm filaments to aggregate. Electron microscopic studies of 30-nm filaments under conditions where they are properly soluble show that the forces between nucleosomes in 30-nm filaments are not large compared with various distorting forces (e.g., surface effects) that are unavoidably encountered during sample preparation. These studies have produced a wide range of mutually exclusive structures, with the likely possibility that none of them are correct. The aggregation that accompanies the further addition of cations is noncrystallographic; crystallization is not possible in such conditions. When internal symmetry may not exist, structures can still be solved by electron microscopy, using tomographic methods. However, tomographic analysis of the structure of 30-nm filaments in aggregates is extraordinarily difficult, if not impossible. When the concentration of multivalent cations is sufficiently high (giving presumably the most stably folded 30-nm filaments), electron density contrast between 30-nm filaments in the aggregates is lost altogether. It is not known if this problem can ever be solved or circumvented. The aggregation of 30-nm filaments is believed to be a built-in property, necessary for subsequent stages of folding.

A. Nucleosome Packing

The problems associated with methods for directly determining the structure of 30-nm filaments have led to an alternative approach. Here, one attempts only to specify the packing of nucleosomes within the 30-nm filament; one can then impose on this packing model the detailed structure of the nucleosome core particle, as determined from crystallographic studies. The location of histone H1 and the path of the linker DNA (i.e., the connectivity) need to be determined separately, because these components are not present in nucleosome core particles.

The packing of nucleosomes within 30-nm filaments has been addressed through a wide range of experiments. Electron microscopy suggested a number of distinct, mutually exclusive possibilities. It has been possible to distinguish between these through X-ray diffraction studies of partially oriented samples of 30-nm filaments. Since the structure of individual nucleosome core particles was known (from crystallographic studies, as discussed above), the diffraction patterns from the 30-nm filaments could be deciphered. Other important data were provided by measurements of optical properties such as dichroism or birefringence of 30-nm filaments that were oriented by electric or fluid-flow fields. Hydrodynamic studies and small angle-scattering measurements of R_c and m/1 provide additional constraints on possible models.

There is now general agreement that, in 30-nm filaments, nucleosomes are packed edge-to-edge in the direction of the 30-nm filament axis, and radially around it, as illustrated in Fig. 2. The nucleosomes are oriented with their flat faces parallel $\pm 20°$ to the filament axis. The packing is believed to be helical, but key helical parameters such as the handedness and the number of "starts," remain matters of debate. In the best studied case, there are roughly six nucleosomes per 11 nm (110 Å) translation along the filament axis. The number need not be an integer and, indeed, may vary along the filament: evidence indicates that it depends on the length of linker DNA, which varies from one nucleosome to the next.

A fundamental problem is that most of the mass of chromatin is contained in the nucleosome core particles (i.e., H1 and the remaining DNA are relatively small in mass). Core particles have the shape of disks, and analysis of the core particle structure shows that key properties such as the appearance in electron micrographs, the X-ray diffraction, and the optical dichroism are quite insensitive to rotation of a core particle about its disk axis. Therefore, while some aspects of nucleosome packing are known, three key unknown features remain: the rotational setting of each nucleosome in the 30-nm filament about its disk axis, the linker DNA trajectory, and the location within the 30-nm filament of the globular domain (and other domains) of histone H1. These three unknown structural features are all closely related. Geometric considerations imply that, for the case of chromatin having a very short (~0 bp) average linker length, specifying any one of these determines the others implicitly.

Our current inability to specify the rotational setting of nucleosomes has an important consequence. We wish to know which groups on nucleosomes in 30-nm filaments interact to stabilize the structure. Presumably this is a subset of the groups that neighbor in space. As discussed earlier, we know the location within a nucleosome of each of the different proteins. Unfortunately, since the rotational settings are not known, we cannot specify which of these groups are neighbors in 30-nm filaments.

Different possibilities for these three related unknown properties lead to differing models for 30-nm filament structure. As discussed in a following section, experiments can be devised to distinguish between the different possibilities and, in that way, to test and distinguish the current models of 30-nm filament structure.

B. Current Models of 30-nm Filament Structure

Three different models of 30-nm filament structure have received the greatest attention in the recent literature and are the most likely to be correct: the solenoid model, the twisted ribbon model, and the crossed linker model. Each of these models is derived by compacting a nucleosome filament in a different but apparently natural manner. All of these models look quite similar, because they are based on the same data for nucleosome packing. In each case, their appearance is similar to that shown in Fig. 2. They differ, however, in their connectivity, their implied location of the globular domain of H1, and in their implied rotational settings of nucleosomes. Each model is consistent with much, but not all, of the available data.

In the solenoid model, the nucleosome filament is compacted by first bending the linker DNA and thus bringing consecutive nucleosomes together in space, producing a closely opposed nucleosome filament. This filament is then compacted progressively through a range of one-start helical structures having a constant pitch of 110 Å (nucleosomes packed edge-to-edge with their faces roughly parallel to the direction of the filament axis) and an increasing number of nucleosomes per turn, until the limiting 30-nm filament structure is achieved. The helix could be right- or left-handed. Linker DNA connects laterally neighboring nucleosomes around the filament axis. The lateral neighbors of nucleosome i are $i+1$ and $i-1$. This model is the one illustrated in Fig. 2.

In the revised twisted ribbon model, the zig-zag form of the nucleosome filament is considered as a ribbon, forming the 30-nm filament by wrapping helically about an axis, which then becomes the 30-nm filament axis. The fundamental unit of this model is two nucleosomes in length (a zig-zag pair), so the filament that results is two-start. The connectivity is vertical: linker DNA, which must be bent, connects between nucleosomes that are in contact edge-to-edge in the direction of the 30-nm filament axis. The lateral neighbors of nucleosome i are $i+2$ and $i-2$.

The crossed linker model allows the linker DNA to remain straight. A zig-zag nucleosome filament is simultaneously twisted about its axis and compressed along the axis to produce the 30-nm filament. The starting nucleosome filament axis and the final 30-nm filament axis are coincident. Depending on how the twisting and compression are coordinated, both one- and two-start helices can result. The handedness is specified. The lateral neighbors of nucleosome i are again $i+2$ and $i-2$.

These three different models each make different predictions regarding the locations of GH1, linker DNA, and the rotational settings of nucleosomes. The solenoid model and the crossed linker model imply the same location for GH1 and, therefore, the same rotational setting of nucleosomes: nucleosomes are oriented about their disk axis with the proposed GH1-binding site—the DNA entry–exit region—facing inward toward the center of the 30-nm filament. By contrast, the revised twisted ribbon model implies that nucleosomes are rotated about their disk axis such that the GH1-binding site points up or down. Consecutive nucleosomes are packed vertically; the GH1-binding site of the lower nucleosome faces up and that of the higher nucleosome faces down, so that the two GH1s are in contact.

For a detailed critique of the experimental evidence that bears on these models, the interested reader should consult the recent review articles listed in the bibliography. A main conclusion is that the only compelling evidence supporting any one of these models and apparently contradicting the others comes from qualitative or quantitative image analysis of selected electron micrographs. Depending on which images are selected (within the constraints imposed by the X-ray diffraction data discussed above), one or another of these models is implied.

C. Tests of Current Models

Despite the close similarity of each of the current models for 30-nm filament structure, it has become apparent that the models can be distinguished through a number of different physical experiments. Many such experiments are currently in progress; a few examples are described here.

The solenoid model and the revised twisted ribbon model require linker DNA to bend during chromatin folding. If the relevant physical properties of linker DNA segments are those of naked DNA, such bending would be unexpected. By contrast, the crossed linker models allow linker DNA to remain straight. Can linker DNA bend as required by the solenoid and revised twisted ribbon models? This question has recently been addressed in the author's laboratory by studying the physical properties of chromatin fragments containing just two nucleosomes, connected by one linker ("dinucleosomes''). It is found that during cation titrations of the dinucleosomes, the two nucleosomes go from an average center–center separation of ~150 Å, as expected for fully extended linker DNA, to an average separation of 0 Å: linker DNA can bend to bring *consecutive* nucleosomes in chromatin into contact.

The twisted ribbon model can be tested by determining the extent to which one can make chemical cross-links between the globular domains of histone H1. Simple geometric considerations for this model lead to the conclusion that globular domains of H1 can be in contact only in pairs (on two vertically connected nucleosomes). Therefore, the longest polymer of globular domains that could be produced given this model is of length two. By contrast, the solenoid model and crossed linker models allow for contacts between GH1s along the entire chromatin filament length. (Indeed, such contacts are suspected to be of key importance in stabilizing these structures.) An experiment can be devised, using chemical cross-linking followed by selective proteolysis, to determine whether the maximum number of globular domains that can be crosslinked is two or greater than two.

The crossed linker model can be tested by determining the diameter of ''30-nm'' filaments produced by chromatin having ~0 bp of linker DNA. The maximum possible diameter in this case is 250 Å, and even this wide a fiber can only be produced by unwrapping DNA off the surface of a nucleosome so that there are only $1\frac{1}{2}$ turns on each nucleosome. The extra linker DNA created in this way can then be used to span an increased filament width. This implies the structure for the nucleosome filament shown in Figure 1, model A, for which there is no experimental support. If the DNA on each nucleosome is wrapped in $1\frac{3}{4}$ turns or 2 turns (Figure 1, models B and C), as believed, the maximum filament diameter for the crossed linker model is reduced toward ~220 Å. By contrast, the solenoid and revised twisted ribbon models do not place limits on possible filament widths. Chromatin sources (having ~0 bp of linker DNA) that can be used for this test include the yeast *S. cerevisiae* and neuronal cells from mammalian brain.

V. Higher Levels of Structure

During meiosis and mitosis, chromosomes are folded into highly compact and specialized structures. It is useful to characterize these by a ''packing ratio,'' which is defined as the ratio of the contour length of a chromosomal DNA molecule relative to the length of that same DNA molecule packaged in chromatin, in one or another stage of folding. For meiotic chromosomes in the form of synaptonemal complexes, the packing ratio is typically in the range of 300–1,000-fold. For mitotic chromosomes, the packing ratio is ~10,000-fold. For comparison, the packing ratio for the 30-nm filament level of structure is ~40-fold. [*See* MEIOSIS; MITOSIS.]

It is believed that both meiotic and mitotic chromosomes are constructed by further folding of chromatin in the 30-nm filament state. For meiotic chromosomes, the evidence is limited and comes from qualitative electron microscopic studies of the ultrastructure of synaptonemal complexes. These show 25–30-nm-diameter fibers, which are interpreted as chromatin in the 30-nm filament state, that loop out and then rejoin the body of the synaptonemal complex.

Much more information is available regarding the structure of mitotic chromosomes. Electron microscopic and X-ray diffraction studies have shown convincingly that mitotic chromosomes are produced by further folding of 30-nm filaments; however, light and electron microscopy have led to two completely different classes of model for the nature of this folding.

Light and electron microscopic studies of mitotic chromosomes that have been treated with standard methanol–acetic acid mixtures or with NaCl–polyglutamic acid have provided compelling evidence that the final level of compaction consists of helical coiling of a fiber that probably has a diameter of roughly 200 nm. The biochemical effects of the methanol–acetic acid treatment are not known; the chief effect of the NaCl–polyglutamic acid treatment is probably to remove some or most of the histone H1 from the chromosome. The manner in which the 30-nm filament is compacted into the 200-nm filament is not clearly established, but many investigators believe that the 30-nm filament is first gathered into loops, at intervals, which are subsequently coiled into a 200-nm filament. Also, good evidence suggests that, in two sister chromatids attached at their centromeres, these helical gyres may have opposite handedness, giving the two chromatids mirror symmetry. This model is illustrated in Fig. 3A.

A very different view is provided by quantitative three-dimensional tomographic studies of mitotic chromosomes *in situ* or after biochemical isolation.

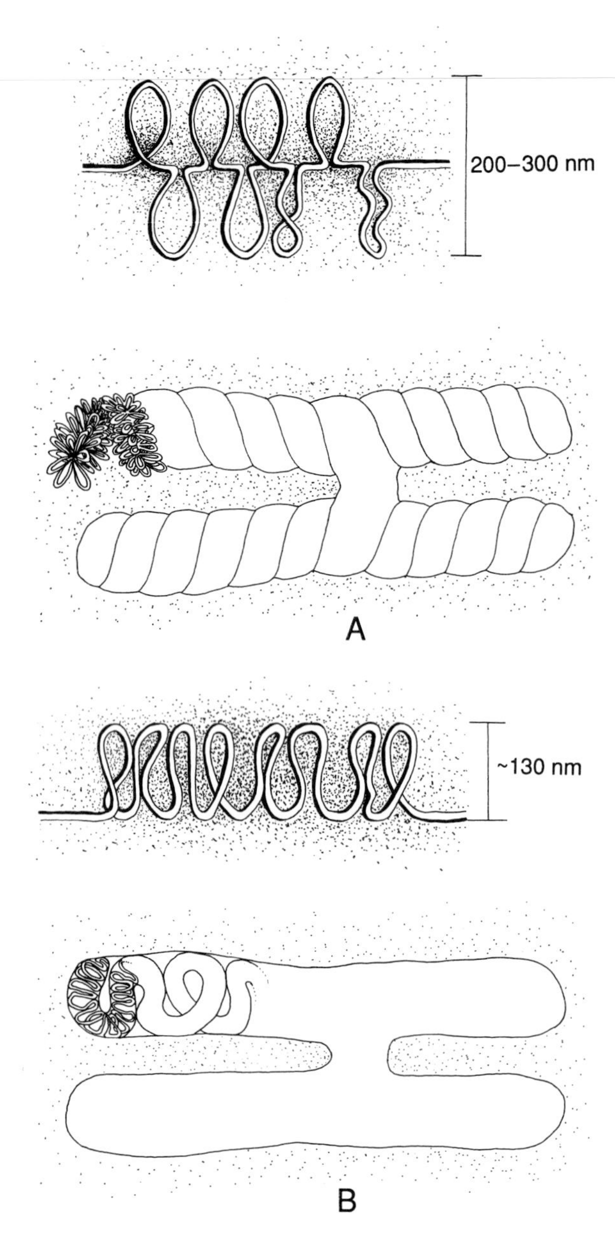

FIGURE 3 Models of higher levels of folding. A is adapted from the work of J. B. Rattner. *Top:* the 30-nm filament (shown as a featureless thread) is locally organized into loops, giving a shorter fiber having a width or diameter of ~200–300 nm. The loops of the 30-nm filament may twist about themselves. *Bottom:* The 200–300-nm-wide fiber is coiled helically into a mitotic chromatid. The figure shows a pair of chromatids that are connected at the centromeres. These two chromatids will separate to the two daughter cells at cell division. The two chromatids may have opposite handedness for this final stage of helical coiling, as shown. B is adapted from the work of A. S. Belmont. *Top:* the 30-nm filament is locally folded (not necessarily organized in loops) into a shorter fiber having a width of ~130 nm. *Bottom:* the 130-nm fiber is locally folded into the mitotic chromatid. This could be accompanied by twisting of the 130-nm fiber about its long axis (not illustrated). Again, two chromatids are shown connected at their centromeres. [Drawings by A. Ellingwood.]

These studies fail to reveal any evidence for a final stage of helical coiling. Evidence of substructure on a 100–130-nm scale suggests the existence of filaments having that diameter. There is also some indication that these 100–130-nm fibers are produced by repeated folding of the 30-nm filament. This view of mitotic chromosome folding is illustrated in Fig. 3B.

At this time it is not possible to reconcile these two apparently mutually exclusive views of mitotic chromosome structure. It is possible that the final stage of compaction really is via helical coiling and that this level of structure is simply obscured by the general close packing of 30-nm filaments within mitotic chromosomes that are maintained in a native-like state. It is equally possible that the helical coiling is not an aspect of native mitotic chromosome structure, but that it arises artifactually after removal of some of the histones (e.g., H1) and a subsequent unfolding of loops or folds of 30-nm filament. This question may be resolved as the tomographic reconstructions are taken to higher resolution. It could presently become possible to trace long stretches of 30-nm filament through three-dimensional space in a chromatid, despite the generally high density of material. Alternatively, it should be possible to find ionic conditions in which 30-nm filaments within mitotic chromosomes are stable, but where subsequent levels of compaction are made artificially less dense (i.e., the chromatids swollen slightly) by reducing the strength of interactions between 30-nm filaments. Promising conditions can be identified from the phase diagram for chromatin folding and aggregation.

Bibliography

Adolph, K. W. (ed.) (1988). "Chromosomes and Chromatin," Vols. 1–3. CRC Press, Boca Raton, Florida.

Adolph, K. W. (ed.) (1989). "Chromosomes: Eukaryotic, Prokaryotic, and Viral," Vols. 1–3. CRC Press, Boca Raton, Florida.

Alberts, B., Bray, D., Lewis, J., Raff, M., Roberts, K., and Watson, J. D. (1989). "Molecular Biology of the Cell," 2nd ed. Garland Publishing, New York.

van Holde, K. E. (1989). "Chromatin." Springer-Verlag, New York.

Wassarman, P. M., and Kornberg, R. D. (eds.) (1989). "Nucleosomes." *In Methods in Enzymology* **170,** Academic Press, San Diego.

Widom, J. (1989). Toward a unified model of chromatin folding. *Annu. Rev. Biophys. Biophys. Chem.* **18,** 365–395.

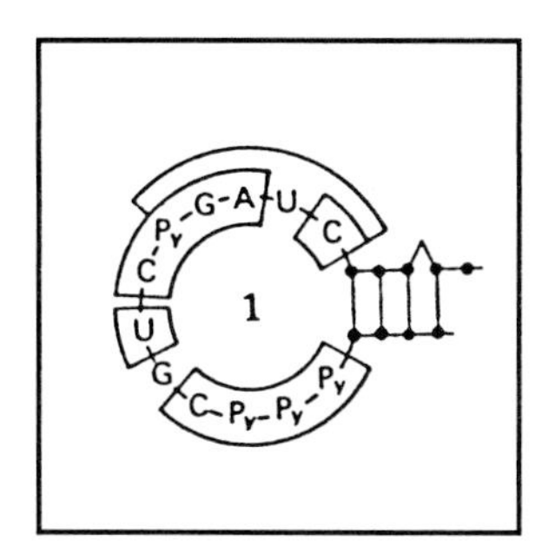

Chromatin Structure and Gene Expression in the Mammalian Brain

IAN R. BROWN and TINA R. IVANOV, *University of Toronto*

Glossary

Chromatin conformation Folding of nucleosomes into a solenoid chromatin fiber
DNA-binding protein Class of nonhistone nuclear proteins that bind to DNA regulatory sequences, thereby activating gene transcription
DNA regulatory sequence DNA sequence involved in the control of gene transcription
DNase I Enzyme that preferentially digests an open or decondensed chromatin conformation
Northern blot Method for determining the size and tissue distribution of specific RNA species
Nucleosome Organization of DNA and histones into repeating nucleoprotein units

IN THE MAMMALIAN brain up to one-third of all single-copy genes (i.e., protein-coding sequences) can be transcribed as nuclear RNA. The proteins encoded by these genes are responsible for the complexity of brain structure and function. In cortical neurons high levels of gene transcription may be facilitated by modifications in chromatin organization which render genes more accessible to the transcriptional machinery. Gene transcription also requires the interaction of nuclear proteins with DNA regulatory sequences. Approaches to investigating this aspect of gene regulation in the brain include the mapping of DNase I-hypersensitive sites, the introduction of cloned genes into cells grown in tissue culture, gel retardation assays, and DNase I protection "footprinting."

I. Transcription

A. Analysis of Brain RNA by Nucleic Acid Hybridization

The complexity of brain structure and function dictates that many different proteins be synthesized. An estimate of the number of genes transcribed in the mammalian brain can be calculated by first determining the sequence complexity of brain RNA. This value can be obtained by the use of nucleic acid hybridization assays which measure the extent of RNA hybridization either to single-copy DNA (i.e., primarily protein-coding sequences) or to a cDNA population (i.e., reverse transcripts of brain mRNAs). One study, which investigated the complexity of adult rat brain RNA, reported that approximately 20% of single-copy DNA is expressed as cytoplasmic RNA. If one assumes a mammalian genome size of 1.8×10^9 base pairs (bp), the RNA sequence complexity in the brain is 3.6×10^8 bp, which could encode approximately 240,000 different proteins of molecular mass 50,000 daltons. A fraction of these DNA sequences, however, do not encode RNAs. [*See* DNA AND GENE TRANSCRIPTION.]

Studies in which a cDNA population derived from brain poly(A)$^+$ RNA is hybridized to RNA from the liver or the kidney indicate that 65% of these transcripts are brain specific. When similar experiments are performed using nonneural tissue, significantly lower estimates of sequence complexity are obtained. For example, in the liver and the kidney approximately 57,000 and 38,000 genes, respectively, are transcribed and expressed in a total cytoplasmic RNA fraction. These experiments indicate that the level of gene transcription is at least fourfold higher in the brain than in nonneural tissues and that the majority of the neural transcripts are unique to the brain.

B. Clonal Analysis of Brain RNA

An estimate of RNA complexity in the adult rat brain can also be obtained through the analysis of clones from a cDNA library derived from brain mRNA. One hundred ninety-one clones were randomly selected, the only restriction being that the cDNA be longer than 500 bp. The cDNA clones were classified based on their tissue distribution by Northern blot analysis. Of these clones 154 detected mRNA species expressed in rat tissues (the remaining clones could represent cloning artifacts).

Class I clones (29 of the 154) hybridize to RNAs which are present in essentially equivalent amounts in the brain, liver, and kidney. These clones could correspond to genes coding for "housekeeping" proteins, which perform common functions in most cell types. Class II clones (41 of the 154) correspond to RNAs which are expressed in the three tissue types, but are differentially regulated. Class III and IV clones (43 and 41 of the 154, respectively), approximately one-half of all clones, correspond to mRNA species which are uniquely expressed in the brain. Those of class IV are of particular interest, in that they are thought to correspond to rare brain-specific mRNAs present at low levels in only a small population of brain cells.

To calculate RNA sequence complexity and obtain an estimate of the number of genes transcribed in the brain, all four classes of clones must be considered. Calculations for class I–III clones indicate that approximately 2640 different mRNAs are present in the brain at an abundance of 0.01% or greater and have a sequence complexity of 8×10^6 nucleotides. Accurate estimates of abundance and mRNA size are difficult to obtain for class IV clones, since the majority of these clones do not detect mRNA species by Northern blot analysis. If a value of 1.4×10^8 is taken for total brain mRNA sequence complexity, class IV clones should account for a complexity of 1.3×10^8. The rarest detectable class III clones detected mRNAs with an average length of 3900 nucleotides. Given the trend that rare brain mRNAs tend to be longer than more abundant species, one can estimate that 30,000 class IV mRNAs ($1.8 \times 10^8/3900$) exist in the brain.

C. Structure of Brain-Specific mRNAs

By analyzing 39 of the brain-specific clones, particularly a subset of 15 clones corresponding to rare mRNA, several unique characteristics of brain-specific mRNAs have been observed. Rare brain-specific transcripts which account for most of the mRNA complexity and mass are, on average, 5000 nucleotides in length, compared to 1600 nucleotides in nonneural transcripts. Structurally, the brain transcripts contain both longer coding and noncoding regions. The longer coding regions could indicate that some brain-specific proteins are synthesized as complex polyproteins (e.g., the proopiomelanocortin gene, which gives rise to many hormones and neuropeptides). This arrangement could allow for the coordinate expression of sets of proteins in the brain. Such a mechanism could be suitable for a highly specialized tissue which has recently undergone extreme expansion in function and has not had sufficient evolutionary time to scatter coding regions throughout the genome and set up coordinate regulation. The presence of longer noncoding regions in brain mRNA species could provide signals for molecular regulatory events which are unique to the brain.

II. Organization of Chromatin in Cortical Neurons

A. Introduction

Hybridization analysis of neuronal and glial nuclear RNA to single-copy DNA suggests that the high sequence complexity of brain RNA is primarily due to neuronal nuclear RNA. To accommodate this complex transcriptional activity, modifications to neuronal chromatin structure might be required. In cortical neurons in the mammalian brain, a temporal correlation has been observed between a rearrangement of chromatin structure and an increase in the RNA template activity of isolated nuclei.

B. Developmental Shortening of the Nucleosomal DNA Repeat Length

Eukaryotic DNA is organized into repeating nucleoprotein units, termed nucleosomes. Each nucleosome is composed of a core particle consisting of two copies of each of the four histones H2A, H2B, H3, and H4, about which are wrapped $1\frac{3}{4}$ turns of double-stranded DNA. Histone H1 is associated with a variable length of linker DNA, which bridges two adjacent core particles. The total length of DNA associated with a nucleosome (i.e., the core particle plus the linker region) is determined by digesting purified nuclei with the enzyme micrococcal nuclease. This enzyme cuts chromatin within the linker region. By electrophoresing the digestion products through a low-percentage polyacrylamide or agarose gel, the nucleosomal DNA repeat length can be determined. For a wide variety of cell types this value is approximately 200 bp. Cortical neurons in the adult mammalian brain demonstrate an atypically short nucleosomal repeat length of 165–170 bp, which reflects a shorter length of the linker DNA. [*See* DNA IN THE NUCLEOSOME; HISTONES AND HISTONE GENES.]

In mouse, rat, and rabbit, the short nucleosomal repeat length in cortical neurons is not present at birth, but appears during the first week of postnatal development, when cortical neurons are postmitotic, postmigratory, and about to enter a differentiation phase. Observations in guinea pig cortical neurons and rat hypothalamic neurons support this hypothesis. For these two neuronal populations the shift to the short repeat length occurs *in utero*. In guinea pigs the gestation period is three times as long as in rat or mouse, such that the guinea pig brain is more advanced in differentiation at birth. A similar situation occurs for rat hypothalamic neurons, in that the shortening of the repeat length occurs by day 19 of gestation, which developmentally correlates to the differentiation of the hypothalamus.

A shortening of the nucleosomal repeat length from 200 bp to 167 bp has also been demonstrated for cultured neurons from 16-day fetal rat in response to the thyroid hormone triiodothyronine (T3). The shortening of the nucleosomal repeat length follows a lag period of 15 days. This rearrangement in chromatin structure could result from a series of molecular events mediated by the binding of the T3 hormone–receptor complex to specific chromatin regions. It is of interest that the number of chromatin-associated T3 receptors has been reported to significantly increase at birth.

C. Analysis of Histones in Neurons of the Cerebral Cortex

The shortening of the nucleosomal repeat length in cortical neurons has been investigated in relation to possible changes in histones of the nucleosome core and/or linker region. By polyacrylamide gel electrophoresis no significant qualitative differences in histones have been detected between cortical neurons, which possess the short repeat length, and glial or kidney nuclei, which possess the typical 200-bp repeat length. The major difference in histone composition is quantitative. Cortical neuronal nuclei have a low H1 content, with approximately 0.5 H1 molecule per nucleosome. This 50% reduction in the amount of linker histone in cortical neurons could affect the second order of chromatin packing, which involves the folding of nucleosomes into a solenoid structure (see Section II,E).

Fluctuations in the relative proportions of H1 subtypes have also been reported during postnatal development. These changes, however, are not unique to cortical neurons and are detected in nondividing cell cultures. A histone H1 variant, $H1^0$, accumulates in cortical neurons during postnatal development following the shift in repeat length (8–18 days postnatally). In nonneural systems $H1^0$ has been implicated in the maintenance of the differentiated cell type.

D. DNase I Sensitivity and Transcriptional Activity of Cortical Neurons

The DNase I digestability and the RNA template activity of cortical neurons before and after the shift to the atypical short repeat length have been investigated. Pancreatic DNase I recognizes and preferentially digests open or decondensed chromatin. Experiments were performed to compare the digestion kinetics of ^{3}H-thymidine labeled neuronal nuclei isolated from rat cortex at 1 and 8 days postnatally. Similar digestion kinetics are observed for cortical neuronal nuclei and kidney nuclei isolated 1 day after birth. At 8 days of postnatal development, however, the DNase I digestibility of the neuronal nuclei, but not that of the kidney nuclei, is increased. This suggests that neuronal chromatin is in a less condensed state after the shift in nucleosomal repeat length. Support for this concept comes from studies on yeast which demonstrate a short DNA repeat length and decondensed chromatin. In addition, electron-microscopic studies indicate that adult neuronal nuclei contain little, if any, discernible heterochromatin (i.e., condensed chromatin).

An analysis of the *in vitro* transcriptional activity of isolated rat neuronal nuclei relative to the appearance of the short nucleosomal repeat length has also been performed. These experiments show that neuronal nuclei isolated after the shift (i.e., at 11 days) incorporate higher levels of [^{3}H]UTP into RNA, relative to 1.5-day-old neuronal nuclei. Assays for RNA transcription of isolated kidney nuclei, which do not undergo postnatal shortening of the nucleosomal repeat length, revealed no developmental increases in template activity. These experiments indicate that the postnatal appearance of the short repeat length in cortical neurons is accompanied by a decondensation of chromatin and an elevation of transcriptional activity.

E. Chromatin Conformation of Genes in Cortical Neurons

The DNase I digestion experiment described in the previous section analyzes total neuronal chromatin and does not address whether differences in chromatin conformation exist between genes which are transcribed and genes which are transcriptionally silent in cortical neurons. To determine whether differences exist in chromatin conformation (i.e., the folding of nucleosomes into a solenoid) between these two gene classes, Southern blot analysis of DNase I digests has been performed. In this analysis DNA is purified from DNase I-digested nuclei and cut with a restriction enzyme, and the products are analyzed by agarose gel electrophoresis. The DNA is then transferred to a nylon membrane and hybridized with labeled DNA probes which correspond to genes which are (e.g., the neuron-specific enolase gene) or are not (e.g., the albumin gene) transcribed in cortical neurons.

Neuron-specific genes demonstrate an enhanced sensitivity to DNase I in cortical neurons compared to liver nuclei. This result is in agreement with experiments which show a correlation between gene transcription and increased DNase I sensitivity. The albumin gene, however, is also relatively sensitive to DNase I in cortical neurons. This suggests that both transcribed and nontranscribed genes are relatively sensitive to nuclease digestion in cortical neurons. The short repeat length chromatin in cortical neurons therefore could result in a chromatin solenoid fiber of reduced stability which renders all DNA sequences sensitive to DNase I. This reduced stability might be caused by conformational restraints imposed by the short nucleosomal repeat length and/or the 50% reduction in linker histone H1. A role for H1 in higher-order chromatin packing is suggested by the observation that an H1 polymer lies close to the fiber axis and stabilizes the helical folding of the nucleosomes into the solenoid structure. A partial removal of H1 might be sufficient to disrupt the H1 polymer scaffold and partially unfold the chromatin solenoid.

The functional significance of the relatively unfolded chromatin conformation in cortical neurons is not known. One possibility is that this chromatin conformation allows cortical neurons to regulate gene transcription more efficiently by bypassing an initial step in gene activation—namely, the unfolding of chromatin regions, which renders genes accessible to the transcriptional machinery. This state of transcriptional readiness in cortical neurons could relate to the complexity of gene transcription in the postnatal mammalian brain, the plasticity of postnatal brain development, and the ability of the brain to respond quickly to environmental stimuli.

III. Nuclear Regulatory Proteins

A. Introduction

Histones are a relatively simple class of nuclear proteins involved in the coiling of DNA into nucleosomes and the folding of nucleosomes into a solenoid. They are considered to have an inhibitory effect on gene transcription. The positioning of a nucleosome at transcription regulatory sequences can prevent the formation of the transcription initiation complex. Also, considering their involvement in the higher-order folding of nucleosomes into a solenoid, histones might inhibit transcription by packaging a gene within the chromatin solenoid such that it is not accessible to the transcriptional machinery. Nonhistone nuclear proteins, by comparison, are a complex class of proteins involved in a variety of nuclear events, including gene activation.

B. Gel Analysis of Brain Nonhistone Nuclear Proteins

Analyses of brain nonhistone nuclear proteins by one- or two-dimensional gel electrophoresis have demonstrated differences, primarily quantitative, between brain regions and cell types and during brain development and aging. Brain nuclei contain more high-molecular-weight proteins than do other tissue types. Large differences also exist in the amount of nuclear protein within the different classes of brain nuclei. Cortical neuronal nuclei contain almost twice as much nonhistone nuclear protein compared to nonastrocytic glial nuclei. These quantitative differences could reflect differences in template activity among cell types, given the observation that chromatin more actively engaged in RNA synthesis has a higher nonhistone protein content.

Developmentally, differences in nonhistone nuclear proteins have been detected in cerebellar and cortical neurons when neurons have stopped dividing and have begun to differentiate. The significance of these changes is difficult to determine, because the functions of the majority of these proteins are not known. These developmental differences could represent fluctuations in the level of a variety of proteins with diverse enzymatic or regulatory functions. In cortical neurons, for example, fluctuations likely occur in the level of the DNA replicative enzyme, given the cessation of cell division at birth. Also, during early postnatal development, when terminal cortical neuron differentiation initiates, changes in gene expression occur (i.e., an increase in brain RNA sequence complexity). Fluctuations therefore could occur in the synthesis of the enzyme DNA-dependent RNA polymerase II or transcription regulatory proteins.

C. Methods for Identifying Regulatory Sequences Associated with Brain-Specific Genes

Analyzing brain nonhistone nuclear proteins by gel electrophoresis, while providing evidence for cell type and developmental differences, does not allow for determination of their functional significance. This experimental approach is also limited in that only the most abundant proteins are studied. One important group of nuclear proteins, which, due to their low abundance, would be difficult to detect by gel analysis, are those involved in the transcriptional regulation of brain-specific genes. Due to their low abundance, these nuclear proteins are frequently characterized by their target DNA binding sites.

One method of identifying DNA regulatory sequences is the mapping of DNase I-hypersensitive sites. These sites encompass short stretches (i.e., 100–200 bp) of chromatin which are very sensitive to low levels of DNase I and are situated primarily, but not exclusively, at 5′ flanking sequences of active and potentially active genes. Structurally, these sites correspond to a discontinuity in the repeating arrangement of nucleosomes along the DNA strand. In nonneural systems DNase I-hypersensitive sites are located near gene regulatory sequences, where the binding of nuclear proteins could be responsible for the generation of these sites.

Another method for identifying regulatory DNA sequences involves (1) the introduction of a cloned gene (including its flanking regulatory sequences) into cells grown in tissue culture and (2) subsequently assaying for transcriptional activity. Such transfection experiments utilize cultured cells that can transcribe that gene. Modifications must be made to the cloned gene such that its transcription can be distinguished from the cells' endogenous mRNA. Usually, putative regulatory sequences are physically linked to a "reporter" gene, whose activity can be easily monitored (e.g., bacterial genes coding for either chloramphenicol acetyltransferase or β-galactosidase activity). To analyze the role of a short DNA sequence as a transcriptional regulatory element, deletions are performed. The absence of reporter gene activity indicates the involvement of the deleted sequence in transcriptional control.

These gene constructions can also be introduced into animals by injection into fertilized eggs (i.e., transgenic mice). This method, although technically more difficult, is superior to the transfection of cell cultures, since the introduced cloned gene is exposed to a complete set of cell-specific transcriptional signals during embryonic development.

D. Analysis of DNA-Binding Proteins

After identifying putative DNA regulatory regions, experiments can be performed to determine whether they contain protein-binding sequences. The most direct method of investigation is gel retardation assays. These experiments involve the incubation of a protein containing nuclear extract with a short labeled DNA fragment containing the putative regulatory sequence. Following electrophoresis through a low-ionic-strength polyacrylamide gel, DNA complexed to protein is detected as bands with reduced mobility (relative to the free unbound DNA). To identify the protein binding site at the nucleotide level, DNase I protection footprinting experiments are performed. DNA sequences which bind nuclear proteins are protected from digestion by DNase I and, as such, generate a footprint, or gap, within the nucleotide ladder following electrophoresis through a denaturing polyacrylamide gel. Once the specific binding sequence has been identified, the next step is to purify and characterize the regulatory nuclear protein. This can be accomplished by passing a nuclear extract through a chromatography column bearing a matrix to which is bound the target DNA with the binding sequence for that protein. Application of the techniques just described should allow for the rapid accumulation of data on brain regulatory DNA sequences and their corresponding *trans*-acting proteins.

Bibliography

Brown, I. R. (1983). The organization of DNA in brain cells. *In* "Handbook of Neurochemistry" (A. Lajtha, ed.), Vol. 5. Plenum, New York.

Brown, I. R., and Greenwood, P. D. (1982). Chromosomal components in brain cells. *In* "Molecular Approaches to Neurobiology" (I. R. Brown, ed.). Academic Press, New York.

Cestelli, A., Di Liegro, I., Castiglia, D., Gristina R., Ferraro, D., Salemi, G., and Savettieri, G. (1987). Triiodothyronine-induced shortening of chromatin repeat length in neurons cultured in a chemically defined medium. *J. Neurochem.* **48,** 1053.

Chaudhari, N., and Hahn, W. E. (1983). Genetic expression in the developing brain. *Science* **220,** 924.

Chikaraishi, D. M. (1979). Complexity of cytoplasmic polyadenylated and nonpolyadenylated rat brain ribonucleic acids. *Biochemistry* **18,** 3249.

Greenwood, P. D., and Brown, I. R. (1982). Developmental changes in DNase I digestability and RNA template activity of neuronal nuclei relative to the postnatal appearance of a sort DNA repeat length. *Neurochem. Res.* **7,** 965.

Greenwood, P. D., Silver, J. D., and Brown, I. R. (1981). Analysis of histones associated with neuronal and glial nuclei exhibiting divergent DNA repeat lengths. *J. Neurochem.* **37,** 498.

Heizmann, C. W., Arnold, E. M., and Kuenzle, C. C. (1980). Fluctuations of non-histone chromosomal proteins in differentiating brain cortex and cerebellar neurons. *J. Biol. Chem.* **25,** 255, 11504.

Ivanov, T. R., and Brown, I. R. (1983). Developmental changes in the synthesis of nonhistone nuclear proteins relative to the appearance of a short nucleosomal DNA repeat length in cerebral hemisphere neurons. *Neurochem. Res.* **9,** 1323.

Ivanov, T. R., and Brown, I. R. (1989). Genes expressed in cortical neurons—Chromatin conformation and DNase I hypersensitive sites. *Neurochem. Res.* **14,** 129.

Milner, R. J., and Sutcliffe, J. G. (1983). Gene expression in rat brain. *Nucleic Acids Res.* **11,** 5497.

Sutcliffe, J. G. (1988). mRNA in the mammalian central nervous system. *Annu. Rev. Neurosci.* **11,** 157.

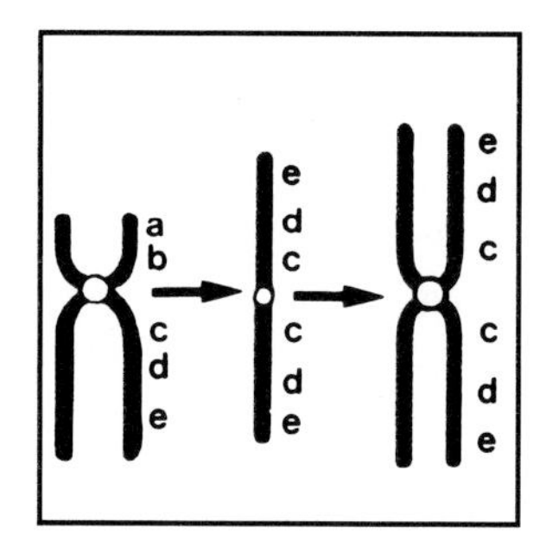

Chromosome Anomalies

JEANNE M. MECK, *Georgetown University Medical Center*

ROBERT C. BAUMILLER, *Georgetown University Medical Center*

Glossary

Aneuploidy In human genetics, refers to loss or gain of whole chromosome(s)
Genome Haploid set of chromosomes
Haploid 23 chromosomes (i.e., 1–22 plus X or Y)
Isochromosome Chromosome whose p and q arms are identical genetically
Karyotype Arrangement of the chromosomes of a single cell in pairs in ordered array
p Arm Short arm
q Arm Long arm
-Somy With prefix, indicates the number of copies of a specific chromosome in a cell line (e.g., trisomy 21 or monosomy X)

CHROMOSOME ANOMALIES occur spontaneously in all organisms at a low rate. Radiation, certain chemicals, and other environmental stresses can increase the frequency of such changes. The human genome is remarkably free of any major structural chromosomal polymorphisms even in populations that have been breeding isolates for many hundreds of years. This uniformity argues for a unique origin of *Homo sapiens*. This argument is strengthened by the fact that the higher apes have a genome remarkably similar to humans in its content but differ by a number of structural alterations, each of which can occur spontaneously but is almost invariably lost from a population.

Chromosome alteration resulting in loss or gain of autosomal material is generally lethal. Alterations that are balanced but produce altered quantities of genetic material in some fraction of the gametes subsequent to normal recombination or to normal meiotic division are rapidly lost from a population, usually persisting for three generations or less.

I. Polyploidy

The human is a diploid species with a relatively brief haploid stage in the male sperm cell and a transient haploid stage in the female, which occurs post–sperm penetration and presyngamy.

Triploidy (three genomes, 69,XXX; 69,XXY; 69,XYY) usually occurs because two sperm enter the egg and two male pronuclei join with the haploid female pronucleus. Triploidy may also result if the female product after meiosis I fails to undergo meiosis II successfully and the diploid female nucleus joins with the male pronucleus. Rarely, a fetus can survive to birth, and cases have been reported to have survived for several weeks.

Tetraploidy (92,XXYY; 92,XXXX) occurs because of chromosome doubling without nuclear division after syngamy. The diploid XY or XX zygote becomes tetraploid and almost always fails to reach delivery. Other mechanisms leading to tetraploidy can be postulated, but these would necessitate the congeries of two or more exceedingly rare phenomena. Polyploidy is an evolutionary mechanism in plants but is not tolerated in higher organisms.

II. Autosomes

A. Changes in Chromosome Number

There are 23 pairs of chromosomes in the normal human karyotype. The autosomes, or nonsex chromosomes, have been numbered 1 through 22, approximating decreasing size order. The 23rd pair is the sex chromosomes. In the normal male, one X and one Y chromosome are contained in each cell; in the normal female, there are two X chromosomes per cell and no Y chromosomes. [*See* CHROMOSOMES.]

Approximately one of every 200 liveborns has a numerical chromosome abnormality. The most common autosomal trisomy is trisomy 21 (i.e., three copies of the #21 chromosome in each cell). This chromosome abnormality is responsible for Down syndrome and affects approximately one of every 800 liveborns. Other less common autosomal abnormalities include trisomy 13 (Patau syndrome), trisomy 18 (Edward syndrome), and, rarely, trisomy 8, trisomy 9, and trisomy 22. No other autosomal trisomies in every cell in the liveborn have been definitively documented. [*See* DOWN'S SYNDROME, MOLECULAR GENETICS.]

By contrast, trisomies for every autosome have been documented in spontaneous abortions. Half of all miscarried fetuses have a chromosome abnormality, and the most common of these is the autosomal trisomy. Trisomy 16 is the most common and is found in 32% of chromosomally abnormal spontaneously aborted pregnancies. Autosomal monosomies (i.e., only one copy of a particular chromosome) are believed to occur in early first trimester spontaneous abortions. Monosomy 21 and 22 are the only exceptions and have been reported in rare instances in liveborns. [*See* ABORTION, SPONTANEOUS.]

Mitotic and meiotic numerical errors occur by the mechanisms of nondisjunction and anaphase lag. Mitosis is the cell division process that occurs in somatic cells. Before mitosis, replication of the chromosomal material occurs so that each chromosome consists of two strands or chromatids connected at the centromere. During the process of mitosis, the chromatids of each chromosome are separated and pulled to opposite poles. A new nuclear and cytoplasmic membrane then surrounds each set of chromosomes. The resulting daughter cells will have identical genetic material. When an error in the separation of the two strands of the chromosome occurs (nondisjunction), this can lead to the existence of two karyotypically different cell lines. This condition is known as mosaicism. For example, a chromosomally normal embryo may undergo a single nondisjunctional mitotic event during embryogenesis in which a trisomy 21 cell and a monosomy 21 cell are formed amidst the normal cells. The monosomic cell line cannot reproduce itself competitively; however, the trisomy 21 cell line can. Many individuals who have a mixture of normal and abnormal cells are believed to be the result of a trisomic fertilized egg that has lost the extra chromosome in one cell, which then proceeds to form a lineage of normal cells among the abnormal cells. An individual with such a mixture of normal cells and trisomy 21 cells is mosaic for Down syndrome. Such individuals may have a higher mental function and less severe physical abnormalities than those who have no normal cells. However, this is not always the case because the distribution of normal:abnormal cells as observed in a lymphocyte chromosome analysis may not be the same as that found in other tissues, which may have a greater effect on phenotype such as the central nervous system. [*See* MITOSIS.]

When all cells of an individual are karyotypically abnormal, it is usually the result of a meiotic error. Meiosis is a process of cell division in germ cells, which leads to the formation of the haploid egg or sperm. Meiosis consists of two divisions: MI and MII. Before MI, the oogonia or spermatogonia undergo DNA replication so that there are 46 chromosomes, each consisting of two chromatids. During MI, the homologous chromosomes are paired. Then, the homologues migrate to opposite poles. Nuclear division takes place and results in two daughter nuclei. Each of the two daughter nuclei should contain only 23 chromosomes, each with two chromatids. MII is a mitotic-like division in which there is separation of the chromatids in all 23 chromosomes. The chromatids of each chromosome pair go to opposite poles and different daughter nuclei. [*See* MEIOSIS.]

A numerical chromosome abnormality occurs as a result of a mistake in segregation in MI or MII in the male or female. However, the products of maternal meiotic errors are found more frequently than paternal meiotic errors (81% versus 19%), and furthermore, MI errors are more common than MII errors (72% versus 28%). During nondisjunction, there is failure of the homologues to migrate to opposite poles in MI or the chromatids to do so in MII. This results in one daughter cell gamete becoming disomic and the other becoming nullisomic for a particular chromosome. Nullisomy, or lack of a particular chromosome in a gamete, leads to a monosomic embryo. A disomic gamete, however, when joined with a normal monosomic gamete of the other partner, leads to trisomy.

An unusual phenomenon resulting from abnormal cell division known as uniparental disomy has been shown to be a causal mechanism in genetic disease. In uniparental disomy, both homologues of a chro-

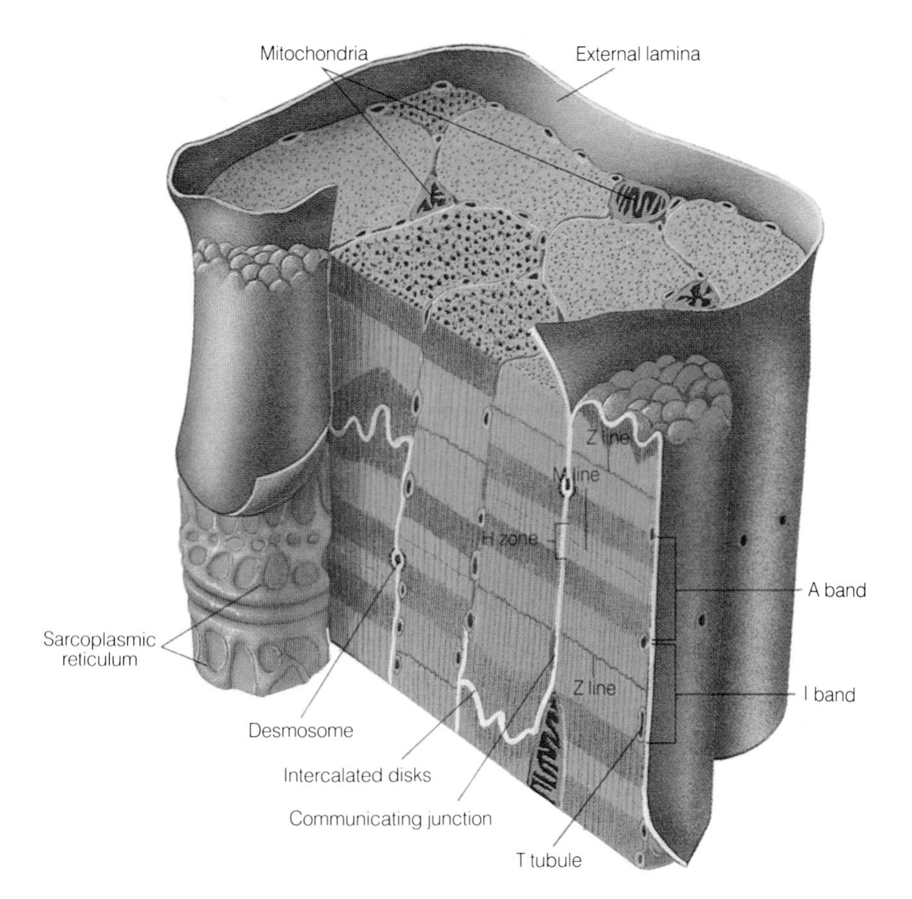

COLOR PLATE 1 Perspective diagram of the region containing the contractile elements in a ventricular myocardial cell. An intercellular abutment cuts across the fibrillar continuity, where intercalated disks are formed. The orange structure corresponds with the transverse (T) system. [Source: Gaudin, A. J., and Jones, K. C. (1989). "Human Anatomy and Physiology," Harcourt Brace Jovanovich, San Diego, p. 198. Reproduced with permission.]

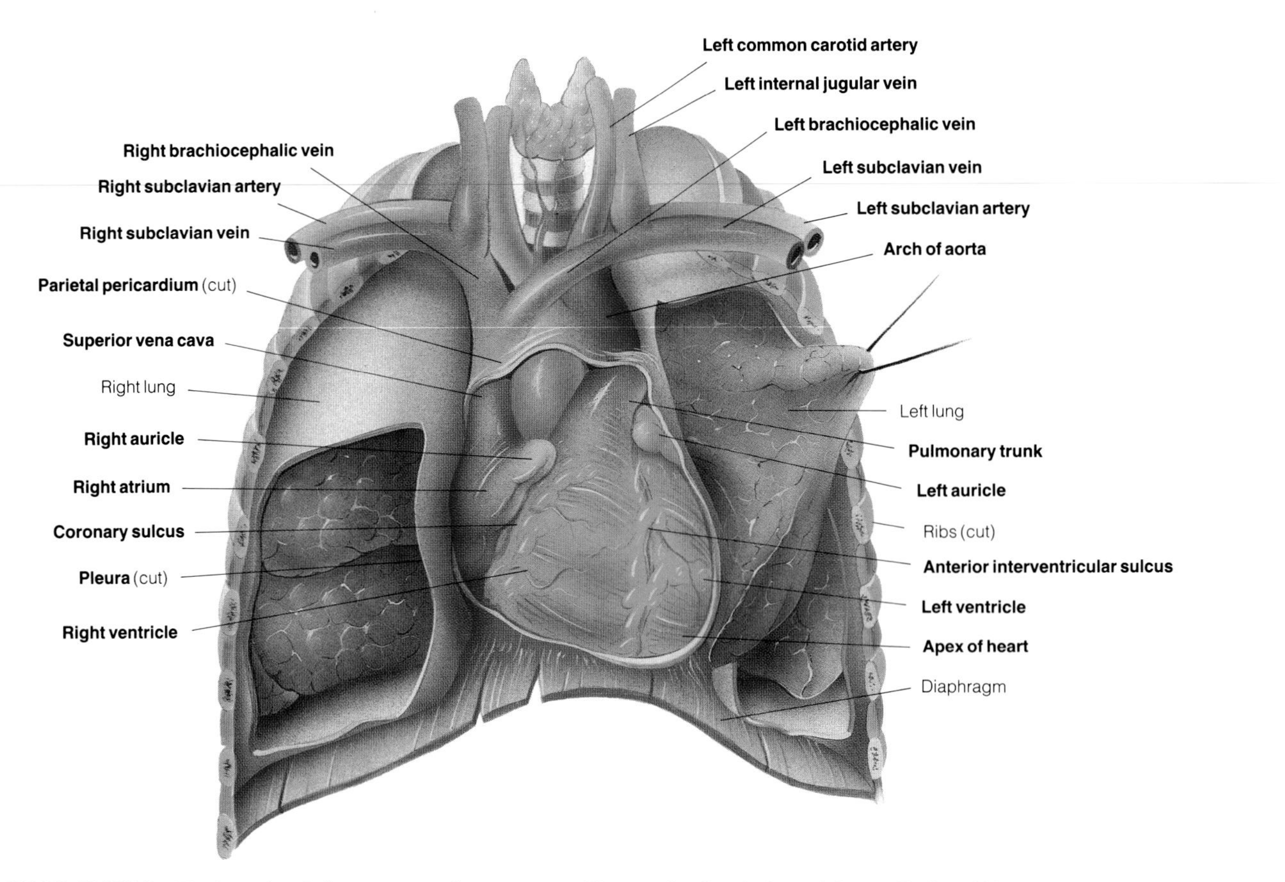

COLOR PLATE 2 The heart in relation to surrounding structures. [Source: Gaudin, A. J., and Jones, K. C. (1989). "Human Anatomy and Physiology." Harcourt Brace Jovanovich, San Diego, p. 559. Reproduced with permission.]

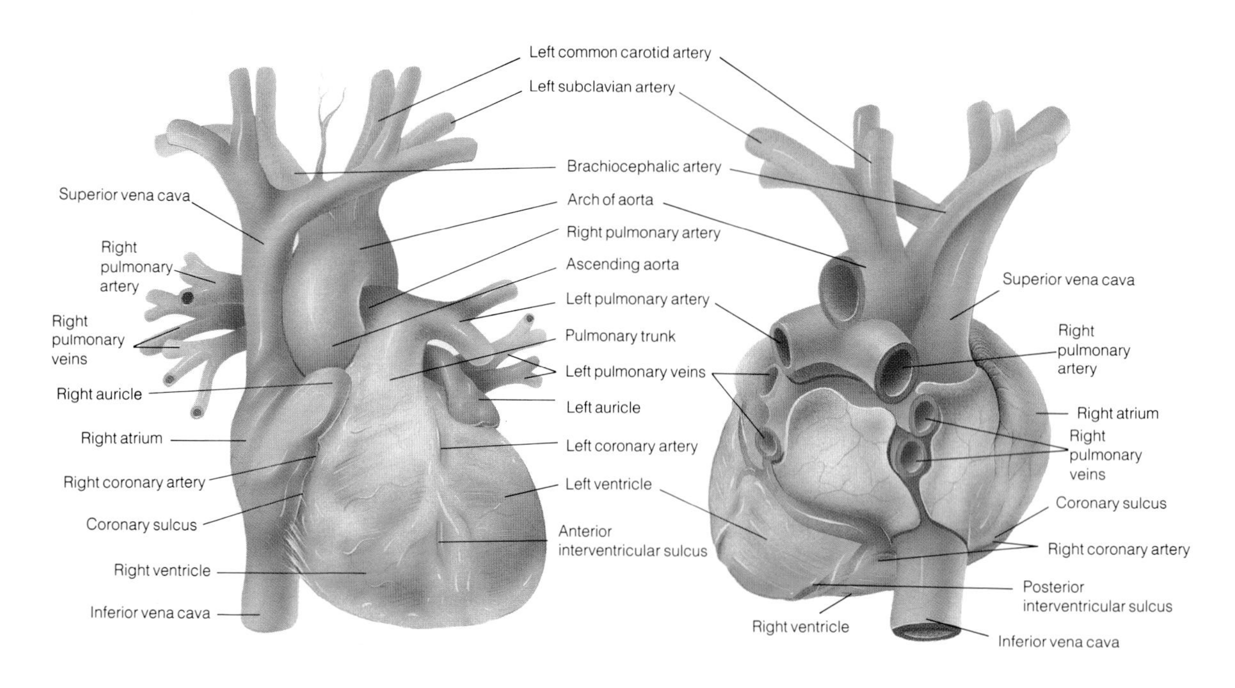

COLOR PLATE 3 The heart in detail. [Source: Gaudin, A. J., and Jones, K. C. (1989). "Human Anatomy and Physiology." Harcourt Brace Jovanovich, San Diego, p. 562. Reproduced with permission.]

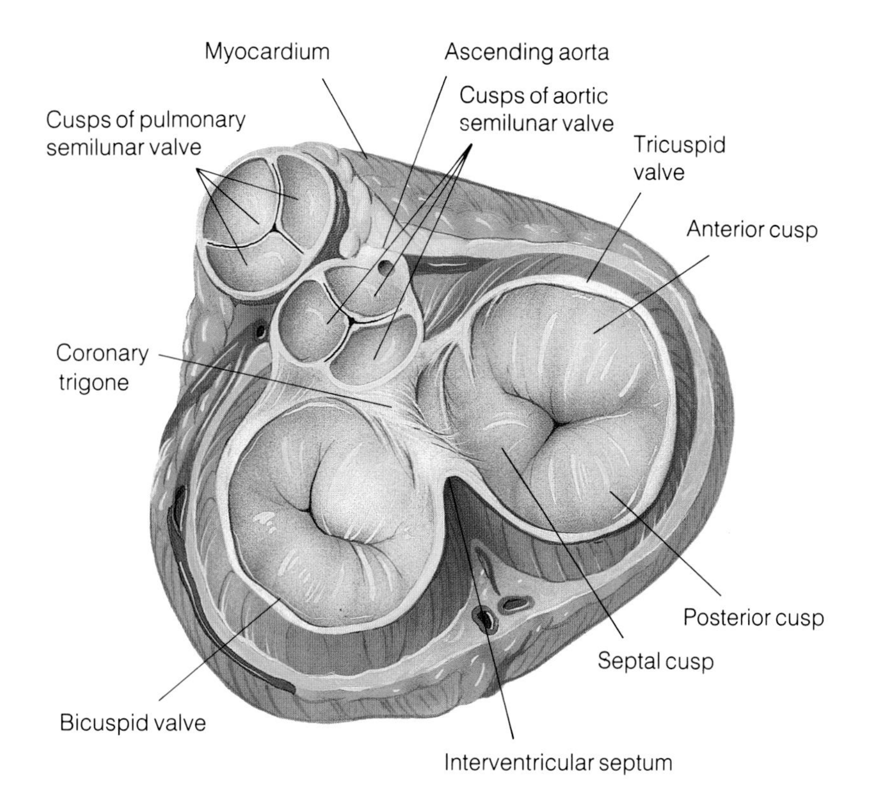

COLOR PLATE 4 The fibrous skeleton of the heart. [Source: Gaudin, A. J., and Jones, K. C. (1989). "Human Anatomy and Physiology." Harcourt Brace Jovanovich, San Diego, p. 563. Reproduced with permission.]

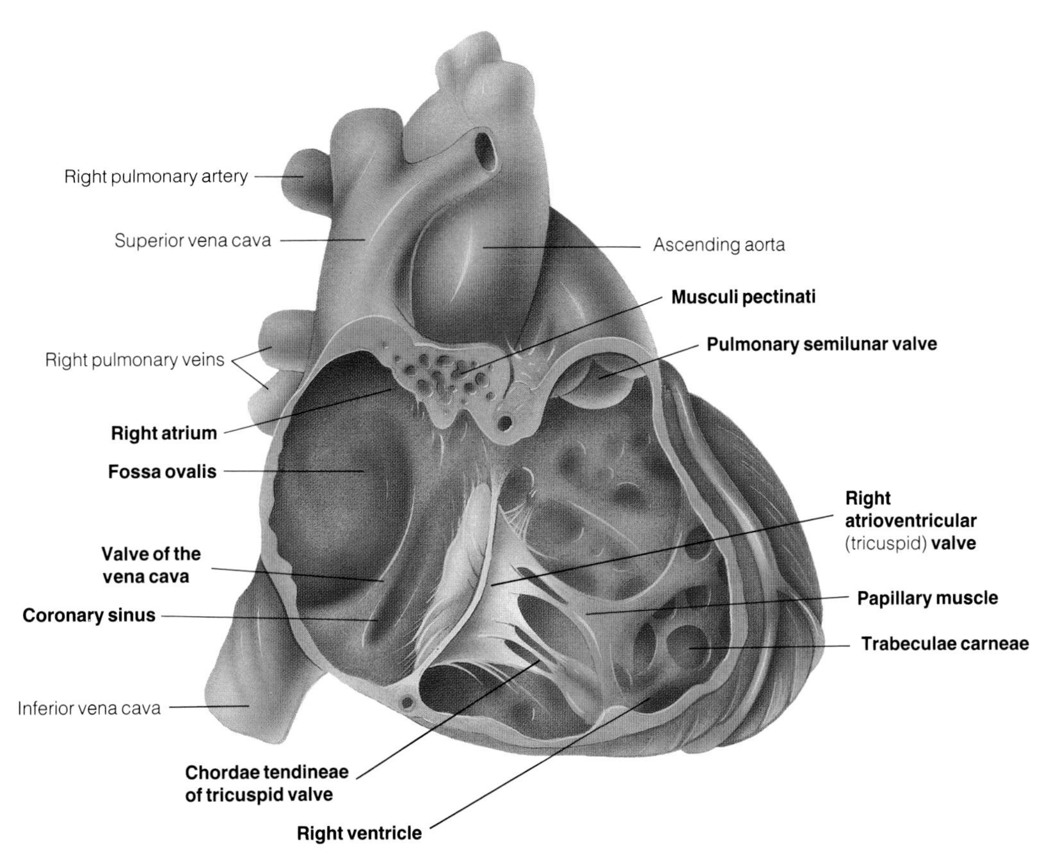

COLOR PLATE 5 A frontal view of the interior of the right atrium and right ventricle. [Source: Gaudin, A. J., and Jones, K. C. (1989). "Human Anatomy and Physiology." Harcourt Brace Jovanovich, San Diego, p. 565. Reproduced with permission.]

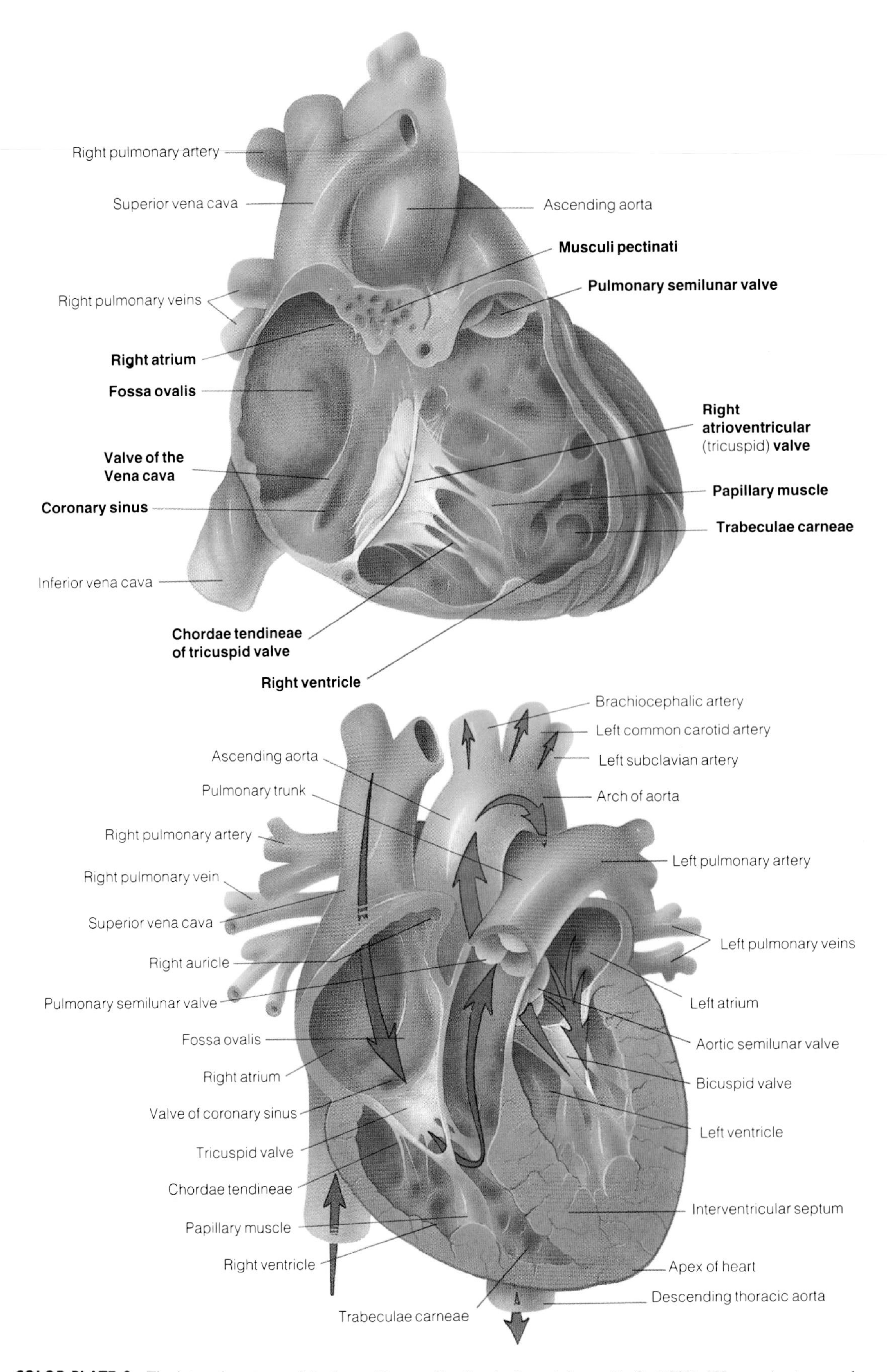

COLOR PLATE 6 The internal anatomy of the heart. [Source: Gaudin, A. J., and Jones, K. C. (1989). "Human Anatomy and Physiology." Harcourt Brace Jovanovich, San Diego, p. 565. Reproduced with permission.]

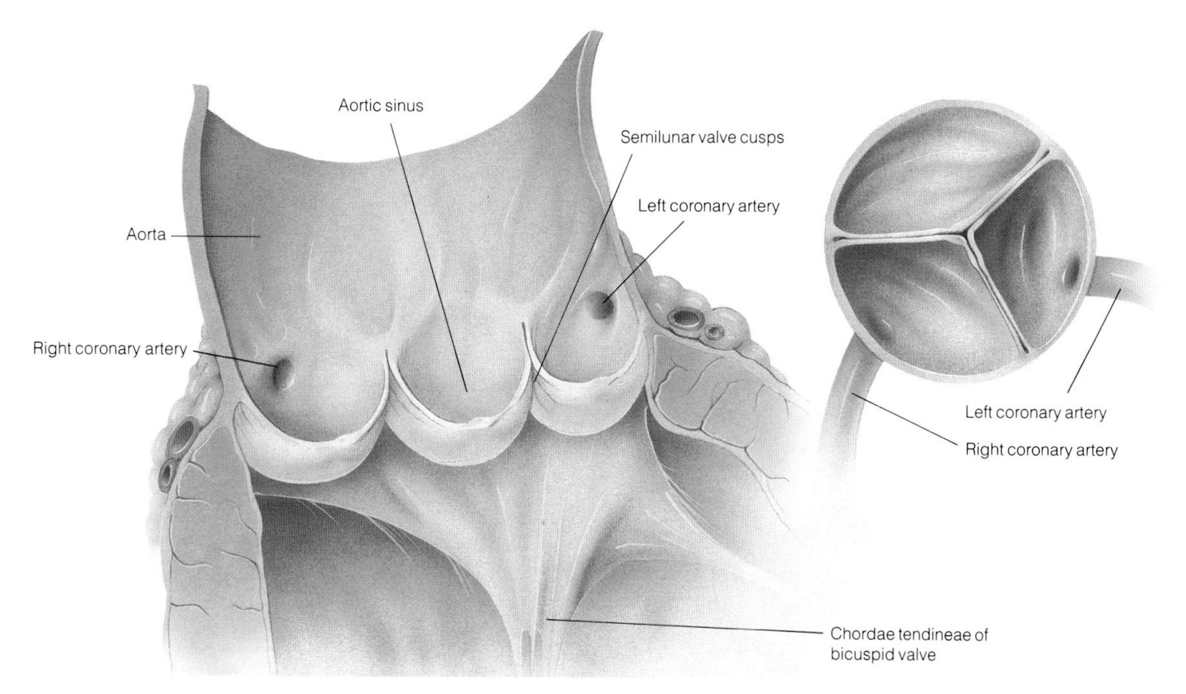

COLOR PLATE 7 The aortic semilunar valve. [Source: Gaudin, A. J., and Jones, K. C. (1989). "Human Anatomy and Physiology." Harcourt Brace Jovanovich, San Diego, p. 566. Reproduced with permission.]

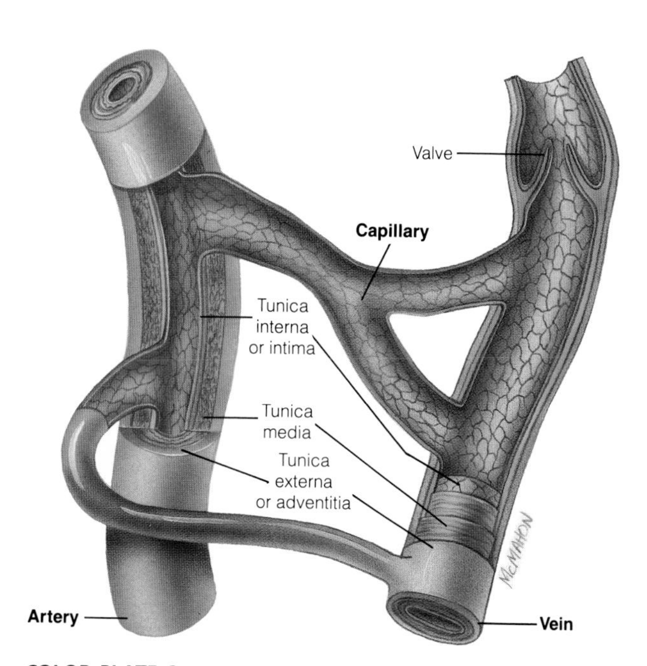

COLOR PLATE 8 Structure of blood vessels. [Source: Gaudin, A. J., and Jones, K. C. (1989). "Human Anatomy and Physiology." Harcourt Brace Jovanovich, San Diego, p. 585. Reproduced with permission.]

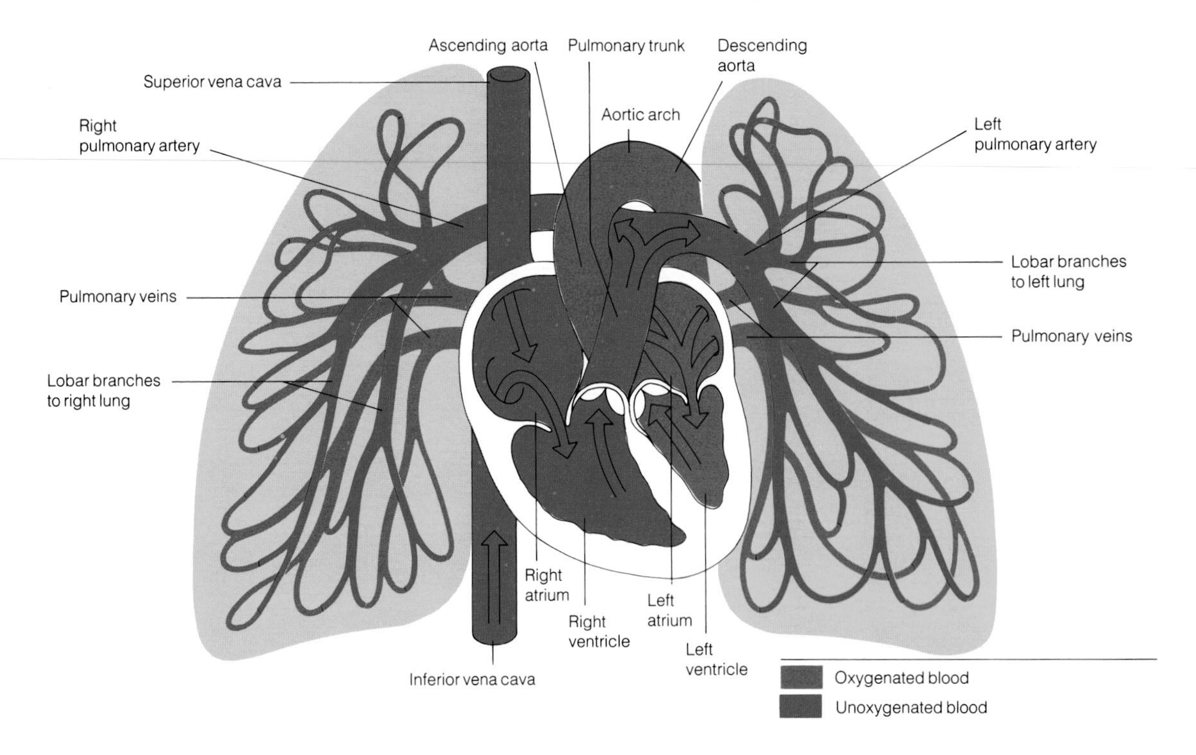

COLOR PLATE 9 Pulmonary circulation. [Source: Gaudin, A. J., and Jones, K. C. (1989). "Human Anatomy and Physiology." Harcourt Brace Jovanovich, San Diego, p. 589. Reproduced with permission.]

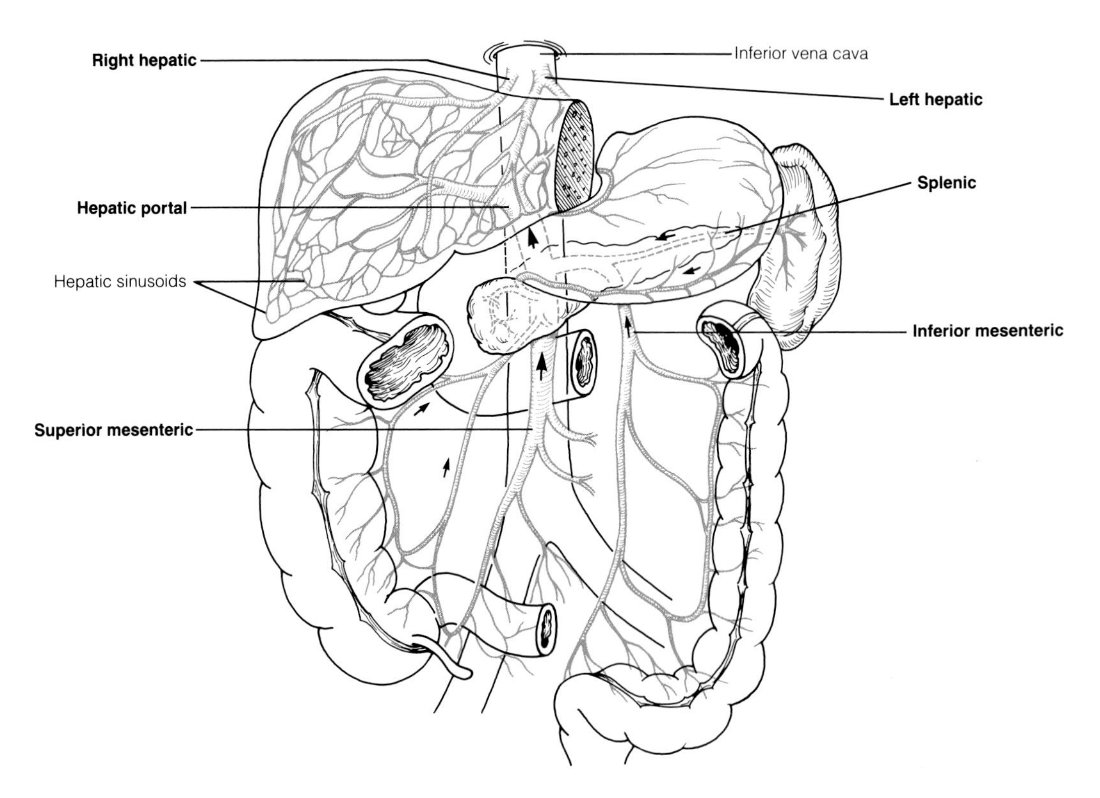

COLOR PLATE 10 Veins of the hepatic portal system. [Source: Gaudin, A. J., and Jones, K. C. (1989). "Human Anatomy and Physiology." Harcourt Brace Jovanovich, San Diego, p. 602. Reproduced with permission.]

1
Engagement, Descent, Flexion

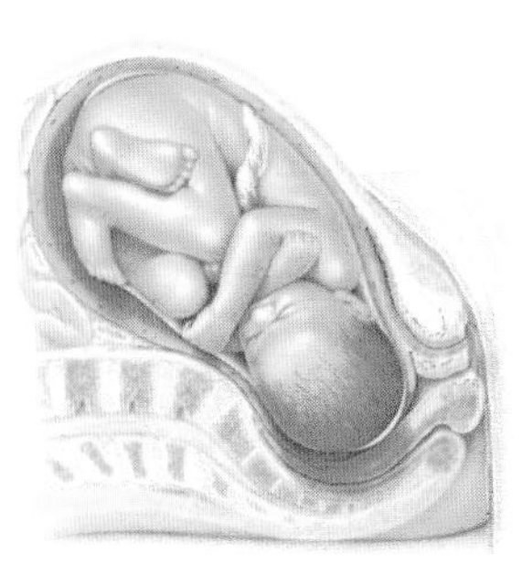

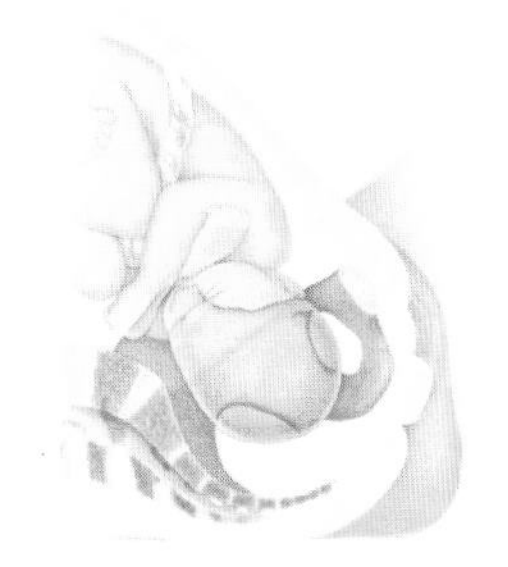

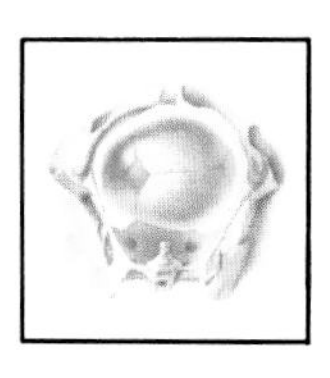

2

Internal Rotation

3

Extension Beginning (Rotation Complete)

4

Extension Complete

5

External Rotation (Restitution)

6

External Rotation (Shoulder Rotation)

7

Expulsion

COLOR PLATE 11 Mechanisms of labor. [Reprinted, with permission from Ross Laboratories, Columbus, Ohio 43216, from Clinical Educ. Aid, #13 © 1964 Ross Laboratories.]

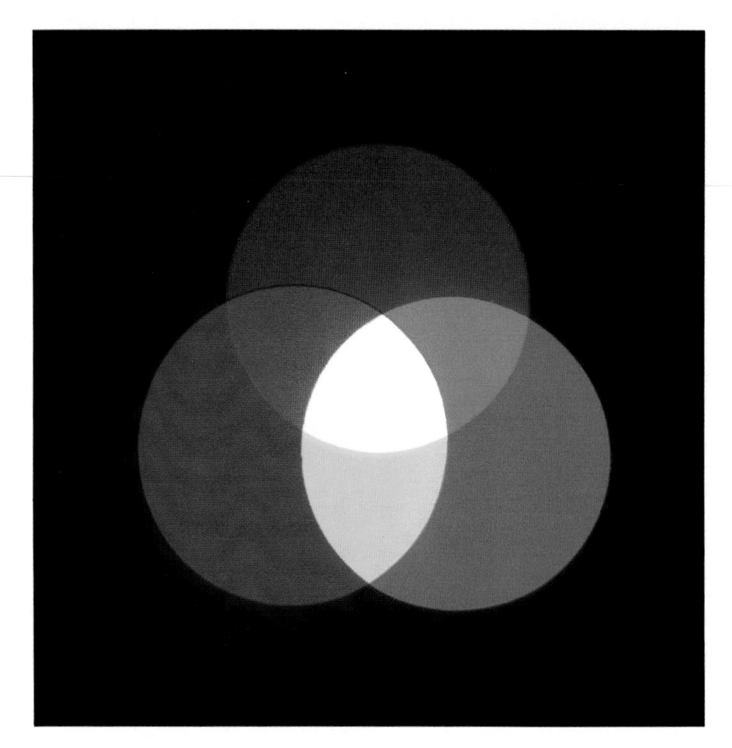

COLOR PLATE 12 Depiction of a dramatic demonstration of color mixture and the law of trichromacy. Circles of light of three wavelengths (e.g., 460, 530, and 650 nm) are partially overlapped on the screen. The three outer crescents show the perceived colors of each of the three original wavelengths (blue, green, and red). The three small triangles show pair-wise mixtures of the three wavelengths, and the central triangle shows the mixture of all three. By varying the intensities of the three light beams, the central triangle can be set to look white. By intuition, further variations of intensities can yield any desired color in the central triangle (e.g., gradually turning down the intensity of the 460 nm light will create colors graduating from white to yellow). The demonstration can be set up with three slide projectors (preferably with variable voltage inputs) and three narrow-wavelength-band filters from a photography store. [Reproduced, by permission, from Anthony W. Young.]

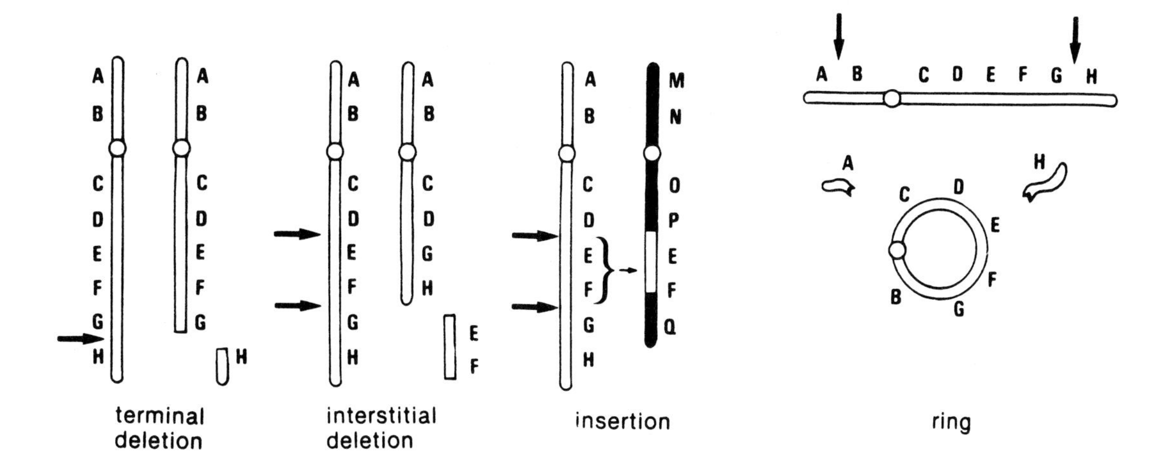

FIGURE 1 Examples of alterations in chromosome structure: terminal deletion, interstitial deletion, insertion, and ring chromosome. (From Thompson, M. (1986). Genetics in Medicine." W. B. Saunders, Philadelphia.)

mosome pair in a diploid cell originate from the same parent, whereas the other parent makes no contribution for that chromosome pair. One mechanism for uniparental disomy is the union of a nullisomic and a disomic gamete. If the chromosomes in question are identical to each other, perhaps resulting from an error in the second meiotic division, this is known as 'isodisomy.'' However, if the chromosomes from the same parent are different, perhaps resulting from a first meiotic division error, this is termed ''heterodisomy.'' Uniparental disomy cannot usually be detected by means of routine cytogenetic analysis. In general, molecular techniques must be used to track the parental source of the genes for a particular pair of homologues. In a landmark case study, isodisomy was shown to be the cause for the autosomal recessive disease, cystic fibrosis, in a child who had only one parent that carried the cystic fibrosis gene. Both of the child's #7 chromosomes that carried the cystic fibrosis gene were shown to be maternal in origin.

Meiotic chromosome errors have been demonstrated to be more frequent in children of women of advanced maternal age. For example, a 26-year-old woman has an age-related risk of 1/1,176 to have a child with trisomy 21 (Down syndrome), whereas a 40-year-old woman has a 1/100 risk. There is no such correlation known with errors leading to mosaicism.

There appears to be no influence of socioeconomic status, race, or parity in this phenomenon. There does not appear to be a substantially increased risk for chromosomally abnormal children associated with advanced paternal age. The reason for the increase in chromosomally abnormal offspring in older women is not known. There are, however, several hypotheses.

B. Changes in Autosome Structure

Alteration in the structure of the chromosomes is not associated with advanced maternal age. Instead, it is more likely the result of exposure to radiation, chemicals, or other mutagens.

1. Structural Rearrangements Involving a Single Chromosome

a. Deletion (Fig. 1) If a chromosome is broken in one place and the piece not containing the centromere is lost, this is known as a terminal deletion. Cri du chat (5p-) and Wolf-Hirshhorn syndrome (4p-) are examples of terminal deletions.

If a chromosome is broken in two places and the middle section is lost, this is known as an interstitial deletion. Prader-Willi syndrome and Angelman syndrome involve a small interstitial deletion of 15q; aniridia-Wilms tumor complex involves an interstitial deletion of 11p.

A ring chromosome is a specialized type of deletion in which the telomeres of the p and q arms are lost. The result is that the ''sticky ends'' remaining on either end of the chromosome join together to form a ring. Ring chromosomes are rare because they are usually unstable in mitosis and are easily lost.

Individuals who have deletions have partial monosomy for the portion of chromosome material that is missing.

"Contiguous gene syndrome" is a term that refers to the microdeletion of a portion of a chromosome that affects more than one gene and thus combines the phenotypes of the disorders involved. Some of these deletions are large enough to be detectable cytogenetically; others require the use of molecular techniques for detection. Examples of contiguous gene syndromes on the autosomes include the aniridia-Wilms tumor gene complex (11p13) and Langer-Giedion and the trichorhinophalangeal syndromes (8q24).

b. Duplication Segments of chromosomes are sometimes duplicated by errors in DNA replication or exchange and arranged in a tandem fashion. The duplicated portion can be oriented in the same direction as the original piece of chromosome (direct duplication) or it may be inverted (inverted duplication).

Individuals who have duplications have partial trisomy for a portion of the chromosome.

c. Inversion If there are two breaks in a chromosome and the middle segment is flipped around, this structural change is a chromosomal inversion. There is no extra or missing material, simply a rearrangement. Most inversion carriers are phenotypically normal. Inversions are of two types: pericentric and paracentric. In pericentric inversion, one of the breaks is in the short arm, and the other break is in the long arm. The inverted segment, therefore, contains the centromere. Individuals who have pericentric inversions are at an increased risk of having chromosomally unbalanced offspring, which have duplication of some of the chromosome and deficiency of another portion of the chromosome. The most common pericentric inversion involves the #9 chromosome. It is found in a frequency of 0.13% in the Caucasian population and 1.07% in the Black population. In paracentric inversion, both breaks are in the same arm of the chromosome (i.e., the broken portion is only in the p arm or only in the q arm). The centromere is not contained in the inverted segment.

Individuals who have paracentric inversions are also at increased risk for having chromosomally unbalanced offspring. However, the risk is lower than for pericentric inversion carriers because the resulting abnormalities in the case of a paracentric inversion are dicentric chromosomes (ones with two centromeres) and acentric fragments (no centromeres). Both of these situations are unstable and lead to early embryonic death because of monosomy.

d. Isochromosome An isochromosome is one in which the two arms of the chromosome are identical (Fig. 2). For example, an isochromosome of the long arm of chromosome 21 consists of two copies of 21q joined by a centromere. It is believed that isochromosome formation can arise by misdivision of the centromere. Instead of dividing longitudinally so that the one chromatid goes to each daughter cell, the chromatids divide transversely to that the p and q arms go to different daughter cells.

2. Structural Rearrangements Involving Two or More Chromosomes

a. Translocation A translocation involves breakage and rearrangement of two or more chromosomes. An individual who has a translocation and maintains the correct amount of chromosome material is known as a balanced carrier. Such individuals are usually phenotypically normal. However, they are at an increased risk of having chromosomally unbalanced (abnormal) offspring. There are two basic types of translocations: (1) robertsonian and (2) reciprocal.

A robertsonian translocation involves breakage at or near the centromeres of two acrocentric chromosomes with fusion of their long arms and loss of their short arms. (An acrocentric chromosome is one in which the centromere is near the end of the chromosome. The p arm of an acrocentric chromosome is not composed of essential structural and

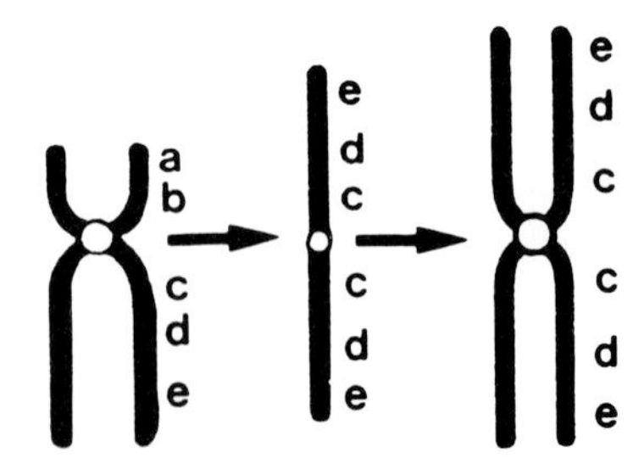

FIGURE 2 Formation of an isochromosome by the mechanism of misdivision of the centromere leading to an isochromosome or duplication of the long arm and complete absence of the short arm. After DNA replication the isochromosome consists of two chromatids joined together. (From Thompson, M. (1986). "Genetics in Medicine." W. B. Saunders, Philadelphia.)

regulatory genes.) The acrocentric chromosomes are chromosomes 13, 14, 15, 21, and 22. Any two acrocentrics may join together to form a robertsonian translocation. A balanced carrier of a robertsonian translocation has only 45 separate chromosomes per cell, but one is really two chromosomes joined together. For example, if an individual is a balanced carrier of a robertsonian translocation involving chromosomes 13 and 14, that person will have in addition to one normal chromosome 13 and one normal 14, the translocation chromosome involving the long arms of 13 and 14 joined together. Such individuals can have normal children—some with normal karyotypes and others with balanced carrier karyotypes. They also may conceive fetuses with trisomy 13 or 14 or monosomy 13 or 14. Most of these will end in early spontaneous abortions with only the trisomy 13 conceptuses having a chance to survive gestation.

A reciprocal translocation involves the exchange of chromosome material between any two chromosomes at any point on the chromosome. In this situation, the balanced carrier has 46 chromosomes; however, the two chromosomes involved in the translocation are structurally rearranged (Fig. 3). Like the robertsonian translocation carriers, reciprocal translocations carriers can have normal children—some with normal karyotypes and some with balanced carrier karyotypes. However, they also have an increased risk of having conceptions in which there is partial monosomy of some chromosome material and partial trisomy for other chromosome material. The risk of having chromosomally abnormal offspring varies from one translocation to another and occasionally varies depending on the sex of the carrier parent. In general, the larger the pieces of chromosome material involved in the translocation, the smaller the chance of having a liveborn with a chromosome abnormality. There are also translocations in which three or more chromosomes may be involved. These are known as complex translocations and are rare.

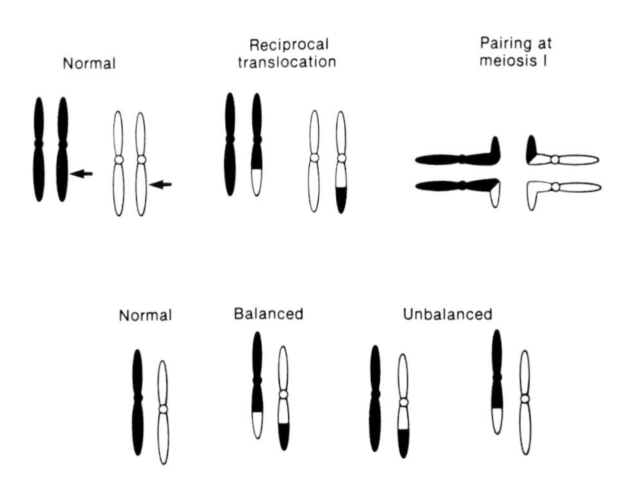

FIGURE 3 Breakage of two nonhomologous chromosomes leading to a reciprocal translocation. Pairing during the first meiotic division results in a cross-shaped figure. Four different types of gametes that may result include: (1) a normal chromosome complement; (2) a balanced chromosome complement; (3) and (4) two different types of chromosome imbalances, each resulting in partial monosomy and partial trisomy. (From Thompson, M. (1986). "Genetics in Medicine." W. B. Saunders, Philadelphia.)

b. Insertions A piece of chromosome material that is derived from an interstitial deletion may insert itself into another chromosome that has suffered a single break or into another part of the same chromosome from which it came. For example, there may be two breaks in the long arm of a #7 chromosome and one break in the short arm of the #12 chromosome. The portion of 7q can insert itself into the 12p. The terminus of the 7q may join with the remainder of 7, which includes the centromere (Fig. 1).

3. Acquired Versus Constitutional Changes

Chromosome abnormalities associated with birth defects (e.g., Down syndrome) are constitutional abnormalities (i.e., a karyotypic anomaly that occurred before or during embryogenesis). There are also other chromosome abnormalities that occur later in life and are associated with malignancies. For example, most people who develop chronic myelogenous leukemia (CML) also develop a change in chromosomes 9 and 22 of some of the nucleated cells of their hematopoietic system. This change is a balanced reciprocal translocation between the long arms of chromosomes 9 and 22 and involves a change in the position of the "abl" oncogene normally located on the number 9 chromosome. The derivative 22 chromosome formed from this translocation is known as the "Philadelphia chromosome." The reorientation of the abl oncogene in its new location on chromosome 22 is believed to be directly associated with the development of the malignancy. As the disease progresses, other karyotype anomalies are also noted in individuals affected with CML. These anomalies include: a second Philadelphia chromosome, trisomy 8, trisomy 19, and isochromosome 17q, all in the malignant line only. Some of the acquired chromosome anomalies associated with malignancies are quite specific. In addition to hematologic malignancies, solid tumors have also been shown to undergo karyotypic changes. In contrast to constitutional abnormalities, which usually affect only one or two chromosomes, acquired abnormalities are often quite complex and involve many chromosomes. Frequently, the modal number of chromosomes per cell is different from 46 and may show extreme hypo- or hyperdiploidy. [*See* CHROMOSOME PATTERNS IN HUMAN CANCER AND LEUKEMIA.]

The human organism tolerates little autosomal aneuploidy. Addition or loss of even a small amount of genetic material is usually not compatible with development. A conceptus with an imbalance is often lost before or in the process of implantation. Even trisomy 21, which occurs in about one in 800 live births (when prenatal diagnosis and termination is not practiced), represents at best only 12% of those who reach 8–12 weeks gestation.

III. Sex Chromosomes

A. Changes in Number

The sex chromosomes (X and Y) allow major deletion (monosomy X) and multiple major duplication (XXX, XXY, XYY up to at least XXXXX). The reasons that such gross cytogenetic anomalies are found are twofold: (1) inactivation of all X chromosomes except one, and (2) the almost agenic constitution of the Y.

The human Y chromosome has a small portion at the tip of the short arm, which pairs with the tip of the short arm of the X (Fig. 4). The pairing function and subsequent exchange is considered to be necessary to ensure proper disjunction of X and Y at meiosis. That portion of the X that pairs with the Y remains active on every X chromosome. Also on the short arm of the Y is a locus that is necessary for male development initiation, presently termed the "sex determining region of the Y chromosome." Closely linked to the centromere is the H-Y antigen regulator. This gene was named from the observation that male to female grafts in isogenic mice did not succeed because of this Y-linked regulator gene(s). Although male specific, neither the product of the H-Y regulatory gene nor its effect on differentiation is known as of this writing. The long arm of the Y has been suspected of containing genes that control spermatogenesis and a gene or genes relating to stature based on height differences between

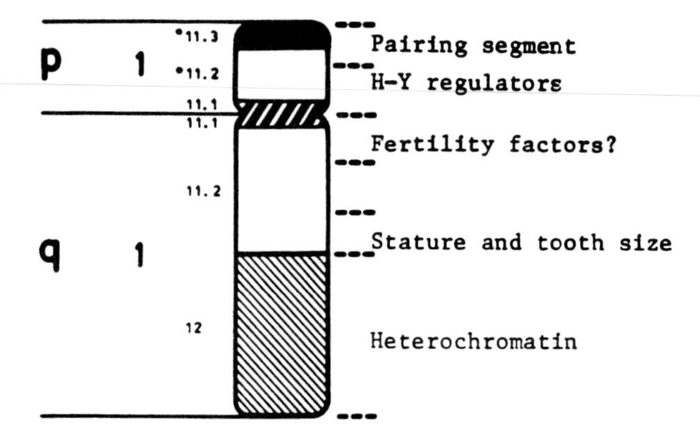

FIGURE 4 Cytogenetic diagram of the human Y chromosome. (From Therman, E. (1986). "Human Chromosomes." Springer-Verlag, New York.)

X, XY, and XYY individuals. There are two pseudogenes on the long arm of the Y and a variable length heterochromatic portion.

The variable length of the heterochromatic region makes the Y polymorphic and thus able to be traced in families and populations at the chromosomal level and on the DNA level. A single Y causes male development regardless of the number of X chromosomes present, but XXY, XXXY, etc., males are sterile. XYY males are not sterile and are taller than their XY sibs on the average. The children of XYY and XY males have a normal sex-determining chromosome complement at similar frequencies (i.e., XX or XY), not XYY or XXY as might be expected. The greater the genetic imbalance (i.e., XYY, XXY, XXYY, XXXXY, etc.), the greater the physical and mental impairment. The latter two are severely affected whereas the XYY individual is only slightly lower in IQ than would have been expected. Socialization of XYY individuals can be normal but is more complicated for XXY, XXYY, and XXXXY individuals because of their sterility, skewed IQ, and sexual ambiguity.

The X chromosome, unlike the Y, is genic, having a gene density similar to the autosomes. Genes on the X were mapped before the autosomes because of their phenotypic/genotypic exposure in the male and the ability of knowing about the linkage state in the grandfather and therefore in the mother. Mammalian males and females are amazingly similar. This similarity is ensured by an early gene dosage compensating mechanism, which randomly inactivates the maternal or paternal X chromosome. This activation occurs at the 1,000–2,000 cell stage in the blastocyst. X inactivation is termed "lyoniza-

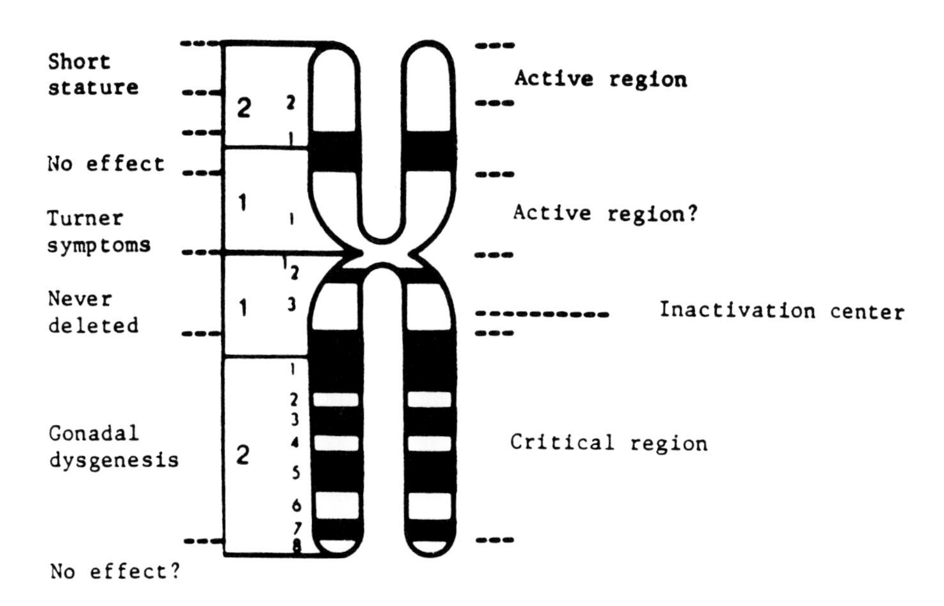

FIGURE 5 Cytogenetic diagram of the human X chromosome showing the presumably always-active regions, the inactivation center, and the critical region. At the *left* are the main phenotypic effects of various deletions. (From Therman, E. (1986). "Human Chromosomes." Springer-Verlag, New York.)

tion" after Mary Lyon who is credited with first expounding the hypothesis in 1961.

Monosomy X, clinically called "Turner syndrome," is the only monosomy compatible with viable development. X monosomy females are as frequent as one in 2,500 females by some reports but rarer by others (one in 10,000), suggesting polymorphic population factors. Data indicate that only 2.5% of monosomy X embryos conceived reach delivery. Thus the relatively agenic Y has a genic constitution that is nonetheless capable of rescuing the X-containing egg. It is thought that the pairing segment of the Y and the always active area of the X are the essential portions.

The Turner female has short stature and gonadal dysgenesis and a wide array of dysmorphic features, each of which occurs with relatively high frequency. The monosomy X female is almost always sterile. The few Turner females who are fertile have a high risk of having aneuploid offspring. The condition is usually diagnosed at birth with confirmation by chromosome analysis. The trisomy X individual (XXX) is not recognized at birth and often escapes detection throughout life. Menopause is often early. The individual is fertile, and children do not have XXY or XXX chromosome constitution as would be predicted by chance. A lowering of expected IQ occurs in the XXX female as in the XYY male but nevertheless is usually in the normal range. The XXXX and XXXXX females have increasing mental retardation and increasing physical malformations, both in number of systems affected and degree of effect.

A part of the aging process in humans and perhaps contributing to it is the loss of the inactivated X in the female and of the Y in the male. No correlation between any aspect of aging and this loss has been reported.

B. Sex Chromosome Structural Changes

The X chromosome diagrammed identifies general regions of the X and their most easily identified phenotypic functions (Fig. 5). Of peculiar importance for the X and proper balancing in the genic compensation is the presence of the inactivation center of Xq13. No X chromosome with a deletion of this area has been found. Thus presence of two active Xs or portions of two active Xs is lethal. (The exception being the small, always active region on the short arm of the X, which is capable of pairing with the Y.) Therefore any chromosome abnormality resulting in the loss of this area is lethal.

In cases of a balanced X chromosome translocation, most cells have the normal X chromosome inactivated, thus leaving one full X active and not influencing the inactivation of part of the autosome attached to an inactivated X. This contrasts with an unbalanced X autosome translocation carrier in whom the normal X usually remains active and the rearranged X becomes inactive. It is believed that the inactivation is random at the time of occurrence but that selection for proper balance is intense and by birth almost all dividing cells show the appropriate X inactivated.

Various regions of the X are associated with symptoms associated with Turner syndrome. Loss of the Xp2 region results in short stature, whereas loss of the entire p arm results in the full Turner syndrome. Chromosome translocations that are balanced usually have little phenotypic effect except when a breakpoint is in the Xq2 region. This position of the breakpoint leads to gonadal dysgenesis. Such a position effect has been recorded even in the case of a paracentric inversion within the Xq2 region.

Contiguous gene syndromes have also been detected on the X chromosome. Males with these deletions express the abnormal phenotype; females may or may not be affected depending on the percentage of cells in which the X bearing the deletion is active. Whereas structurally abnormal unbalanced X chromosomes are generally selectively inactivated, there is no such selection in females who carry contiguous gene syndromes. Deletions in the region of Xp21 have been associated with a syndrome that includes the phenotypes of Duchenne's muscular dystrophy, glycerol kinase deficiency, adrenal hypoplasia, and mental retardation.

Dicentric chromosomes typically have one active centromere, whereas the other one is inactive or suppressed. In cases in which both centromeres remain active, dicentrics are usually lost in mitosis unless the two centromeres are in close opposition and necessarily travel to the same pole at anaphase. An idic(X)(p22) observed by the author had almost two complete Xs joined together, but only one active centromere. The isodicentric chromosome was inactivated leaving the normal X as functional.

Because only a small portion of the Y is necessary for normal development, many grossly deleted functional Y chromosomes have been seen in families. Pericentric inversions have been described as well as ring Ys. Translocation of a small portion of the Y to the X or autosome can lead to males with an XX chromosome karyotype. All XX males are thought to be caused by the presence of the testis-determining factor being exchanged to the X by uneven crossing over or translocational insertion into the X or an autosome.

Y isochromosomes have rarely been described, and only as mosaics. An isochromosome for Y short arms i(Yp) would be expected to have a phenotype of XYY and thus usually escape study. Chromosome techniques in the past were inadequate to distinguish such a chromosome from a partial q arm deletion. Methods for identification of i(Yp) are now available. Yq duplication and X monosomy phenotype may not be compatible with successful development.

Bibliography

Borganonkor, D. (1989). "Chromosome Variation in Man," 6th ed. Alan R. Liss, New York.

Gardner, R., and Sutherland, G. (1989). "Chromosome Abnormalities and Genetic Counseling." Oxford University Press, New York.

Klein, J., Stein, Z., and Sussen, M. (1989). "Conception to Birth." Oxford University Press, New York.

Therman, E. (1986). "Human Chromosomes," 2nd ed. Springer-Verlag, New York.

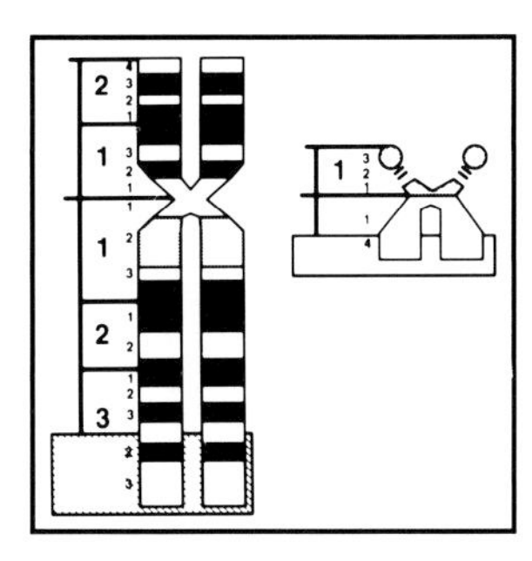

Chromosome Patterns in Human Cancer and Leukemia

SVERRE HEIM AND FELIX MITELMAN, *Department of Clinical Genetics, University of Lund*

Glossary

Band A chromosomal area distinguishable from adjacent segments by appearing lighter or darker by one or more banding techniques. Bands are numbered consecutively from the centromere outward along each chromosome arm. In designating any particular band, four items of information are therefore required: the chromosome number, the arm symbol, the region number, and the band number within that region. These items are given in consecutive order without spacing or punctuation. For example, 12q13 means chromosome 12, the long arm, region 1, band 3.

Centromere Constricted portion of the chromosome, separating it into the long (q) and short (p) arms

Chromosomes Structures in the nucleus, classified according to size, the location of the centromere, and the banding pattern along each arm (Fig. 1). The autosomes are numbered from 1 to 22 in descending order of length; the sex chromosomes are referred to as X and Y. Both the long and short chromosome arms consist of one or more regions.

Clone A cell population derived from a single progenitor. It is common practice in tumor cytogenetics to infer a clonal origin when a number of cells are found that have the same karyotypic characteristics. Since subclones may evolve during the development of a neoplasm, clones are not necessarily completely homogeneous.

Deletion Abbreviated "del." Loss of a chromosomal segment. del(1)(q23) means loss of all material distal to band q23 on chromosome 1, while del(5)(q13q33) means loss of the interstitial segment between bands 13 and 33 on the long arm of chromosome 5.

Inversion Abbreviated "inv." A 180° rotation of a chromosomal segment, so inv(16)(p13q22) indicates that the centromeric portion of chromosome 16, between bands p13 and q22, is inverted (Fig. 5)

Karyotype The chromosome complement of cells; the first item to be recorded in describing a karyotype is the total number of chromosomes. The sex chromosome constitution follows next. A normal male karyotype is thus written 46,XY; the normal female complement is 46,XX.

Karyotypic alterations Changes that can be either structural, implying that the banding pattern or size of a chromosome is altered, or numerical, which means additional or missing whole chromosomes. A plus or minus sign, when placed before a chromosome number, indicates the gain or loss of that particular chromosome. When placed after a symbol, the sign indicates an increase or decrease in the length of a chromosomal arm. Thus, 47,XY,+21 means a male karyotype with an extra (trisomy) 21, whereas 46,XX,5q− means a female complement that is normal except for loss of chromosomal material from the long arm of one chromosome 5.

Landmarks Consistent distinct morphological features of importance in chromosome identification. Landmarks include the ends of chromosome arms (i.e., the telomeres), the centromere, and certain characteristic bands

Region Area consisting of one or more bands and lying between two adjacent landmarks. Regions are numbered consecutively from the centromere outward along each chromosome arm (Fig. 2). Thus,

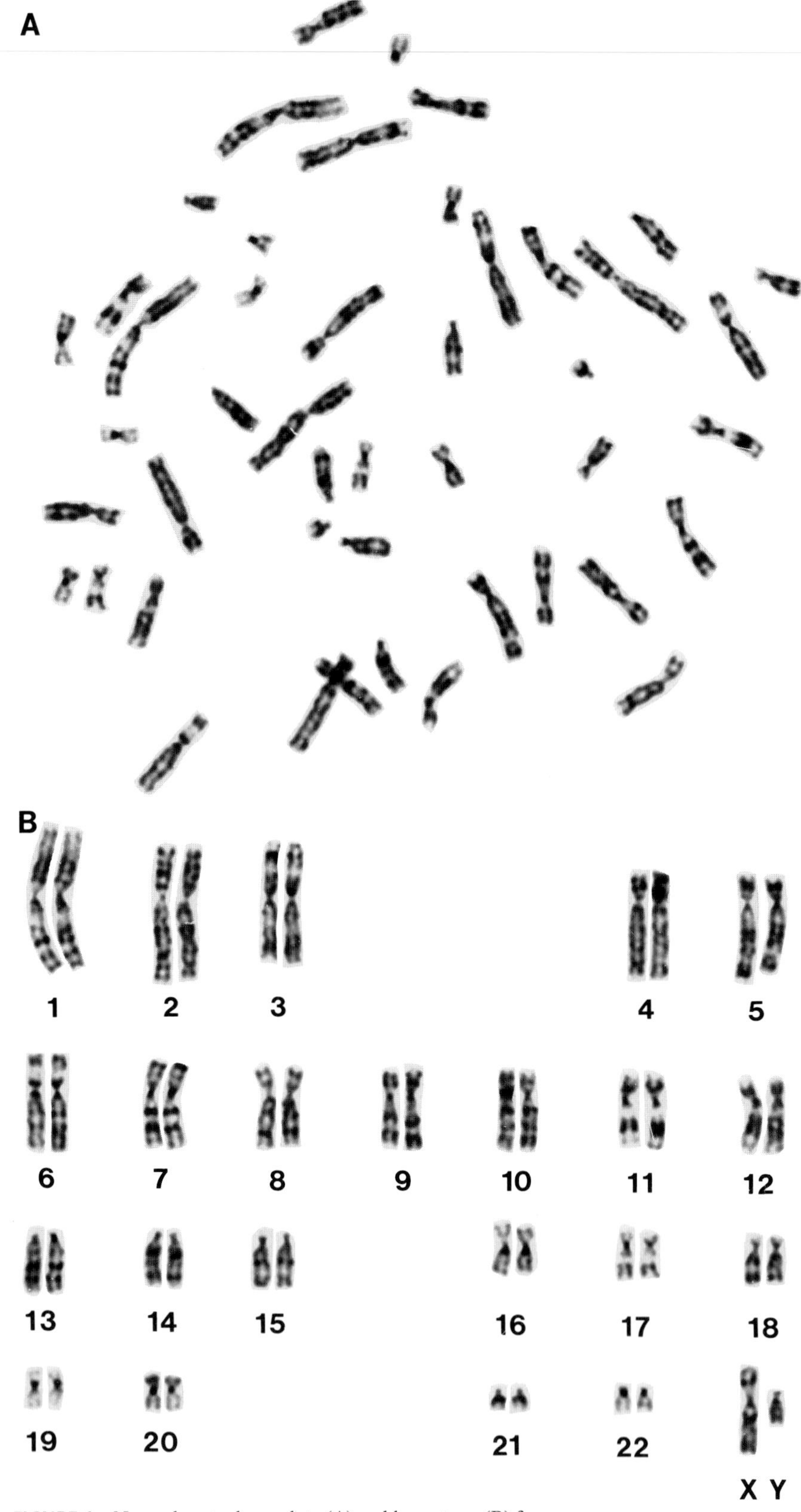

FIGURE 1 Normal metaphase plate (A) and karyotype (B) from a man (46,XY).

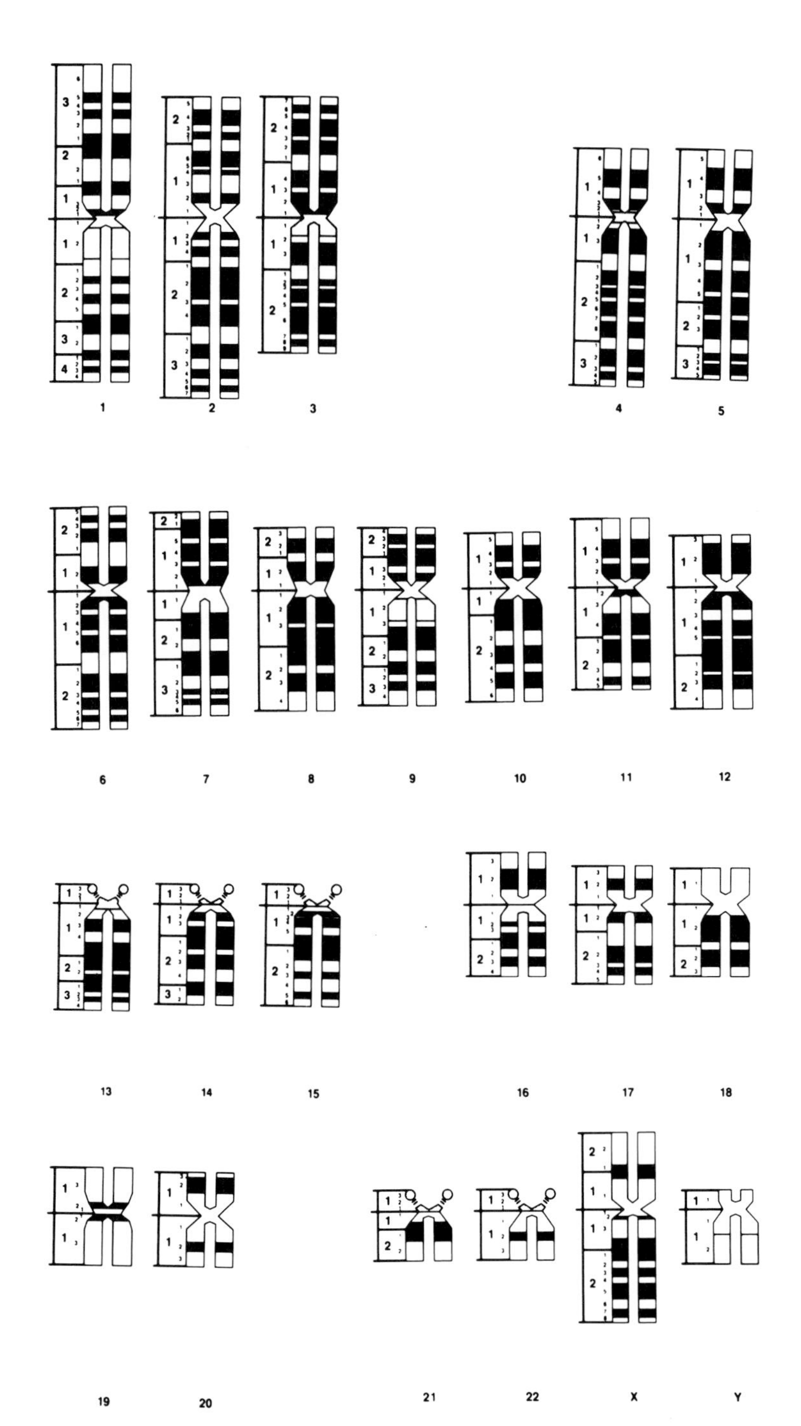

FIGURE 2 Idiotype of the human chromosome complement, illustrating the description of chromosomes as consisting of arms, regions, and bands.

the two regions adjacent to the centromere are number 1 in each arm; the next, more distal regions are number 2, and so on.

Structural aberrations Deviations such as translocations, deletions, and inversions. The chromosome number(s) is specified immediately following the symbol, indicating the type of rearrangement. If two or more chromosomes are altered, a semicolon is used to separate their designations. The breakpoints, given within parentheses, are specified in the same order as the chromosomes involved, and a semicolon is again used to separate the breakpoints.

Translocation Abbreviated "t." Movement of a chromosomal segment from one chromosome to an-

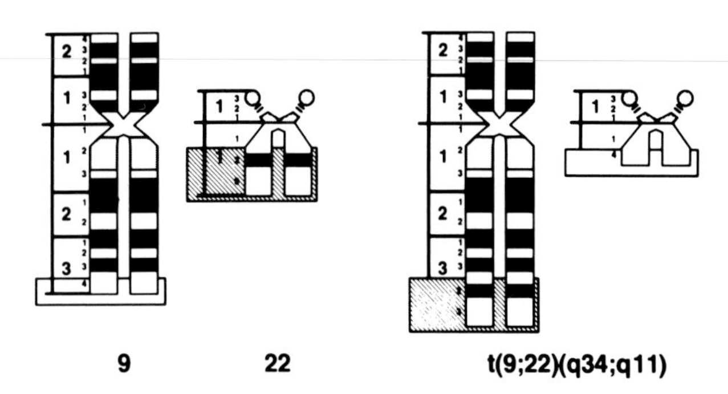

FIGURE 3 t(9;22)(q34;q11) is a characteristic anomaly of chronic myeloid leukemia, but it also occurs in acute leukemia.

other. 46,XX,t(9;22)(q34;q11) thus describes an otherwise normal female karyotype containing a translocation between chromosomes 9 and 22 (Fig. 3), with the breakpoint in chromosome 9 in the long arm, region 3, band 4, and the breakpoint in 22 in the long arm, region 1, band 1.

CANCER AND LEUKEMIA are genetic diseases not at the organism level, although occasionally this is also the case, but at the level of individual cells. The crucial difference between a neoplastic cell (whether benign or malignant) and a normal one is that the former has undergone DNA change(s) that enable it to escape the organism's normal proliferation control mechanisms. These changes are stable mutations, reproduced with every cell division, and hence confer upon daughter cells the same relative growth advantage that originally established the tumor.

One essential organizational level of the human genome is the chromosome. Acquired cytogenetic changes (chromosomal gains, losses, and relocations) characterize all tumor types that have been extensively studied. The aberrations may be primary, meaning that they occur early in the tumor's life and presumably are essential in its establishment, or secondary, meaning that they occur during later tumor evolution. Their overall distribution is markedly nonrandom, and some tumor or leukemia subtypes are characterized by specific changes. The study of neoplasia-associated chromosomal anomalies teaches us which are the essential genetic changes in different phases of tumorigenesis and in different neoplasms, and also provides some insight into how the pathogenetic effect is achieved. The aberrations also serve as disease markers that help reach a more precise diagnosis, including a better assessment of the prognosis in individual cases.

I. Neoplasia as a Genetic Disorder

The essential defect in tumorigenesis is an imbalance between the tendency of neoplastic cells to divide and spread throughout the body and the organism's attempts, through a wide range of control mechanisms, to restrain such growth. Modern medical thinking holds that the reason for the disrupted equilibrium must be sought mainly in the neoplastic cells themselves, not in their regulatory environment. Cancer (and the corresponding malignancies of the hemato- and lymphopoietic organs) is primarily a cellular disease; inherent derangements of the tumor cells, rather than systemic failure to provide appropriate proliferation control, are of the essence.

When tumor cells divide, the tumorigenic quality is faithfully passed on from mother cells to their progeny. The obvious way to achieve this would be via the cells' hereditary material. The first evidence that this is indeed so dates back to the end of the 19th century, when pathologists studying histopathological preparations described a multitude of nuclear abnormalities—in fact, chromatin abnormalities—in cancer cells. Numerous methodological improvements during the last four decades—in particular, (1) the introduction of improved tissue and cell culture techniques, (2) the discovery that the drug colchicine causes mitotic arrest in metaphase, (3) the use of hypotonic solutions to separate chromosomes and thereby facilitate their analysis, and (4) the introduction of banding techniques that enable identification of individual chromosomes and chromosomal rearrangements—have revolutionized tumor cytogenetics and confirmed the early suspicions: Tumor cells are indeed characterized by chromosomal rearrangements. Furthermore, the regularity of the aberration patterns encountered in the different types of neoplasia indicates that the abnormalities are pathogenetically important, not epiphenomena occurring as by-products during tumorigenesis.

The chromosome is obviously an important organizational level of the human genome, but so is the gene, a much smaller structure. Given that the haploid complement contains about 3×10^9 base pairs of DNA and that, in chromosome preparations of tumor tissue, approximately 300 bands can be discerned, it follows that each band, which roughly corresponds to the unit below which changes will not be cytogenetically detectable, contains, on average, 10^7 base pairs. Since the number of genes in the human genome is probably between 50,000 and

100,000, each band will then harbor something on the order of 200 genes. Extensive DNA-level alterations, not only point mutations but deletions or relocations that may encompass several genes, are thus possible without being cytogenetically visible.

Even if mutations were ubiquitous in neoplasia, cases would therefore probably exist in which the aberrations would not be of a sufficiently large-scale nature to show up as microscopically recognizable chromosomal lesions. As it turns out, perhaps two-thirds of all investigated neoplasms (but with large variations between tumor types) are found to contain acquired chromosome aberrations. This must be interpreted as a minimum estimate of the true frequency; it is highly likely that, in a minority of cases, chromosomal changes really do exist, but the available techniques fail to unravel them.

Granted neoplasia is a cellular disorder, even a genetic one; but does it develop from one or many mother cells? Immunological data and studies of protein and DNA polymorphisms in tumors indicate that, in most instances, the cells derive from a single clone. The cytogenetic evidence bearing on this question is not unequivocal. In hematological and lymphatic malignancies, the chromosomal rearrangements almost always exhibit striking similarities, or may even be identical, in all cells, so the overwhelmingly likely conclusion is that the whole neoplasm resulted from the expansion of a single cellular progenitor. The solid tumor data point in the same direction, with the possible exception of epithelial tumors with squamous cell differentiation, in which several clones with cytogenetically unrelated abnormalities have repeatedly been detected. The most straightforward explanation for the latter findings would be to accept a polyclonal origin. If so, a systematic pathogenetic difference seems to exist between squamous cell tumors and the majority of other human neoplasms, in which the cytogenetic data strongly favor monoclonal tumorigenesis.

II. Primary and Secondary Chromosomal Aberrations

Current thinking about carcinogenesis visualizes the process as consisting of three main stages: tumor initiation, promotion, and progression. Although mutational changes are presumably essential only at the initiation stage, genetic alterations are thought to play a role in the other phases of multistep tumorigenesis as well. This theoretical framework is in agreement with observed cytogenetic facts: More often than not, tumor cells contain multiple, not single, genomic aberrations. The important questions then become: Which of the observed alterations are late additions? Which represent the fundamental changes that caused the transition from a normal to a neoplastic cell, or from benign to malignant neoplastic proliferation?

In principle, the acquired aberrations of neoplastic cells may belong to three different categories. Primary aberrations are essential in establishing the neoplasm. They presumably represent rate-limiting steps in tumorigenesis, may occur as solitary cytogenetic changes, and are, as a rule, strongly correlated with tumor type. Secondary aberrations are incurred during later stages of tumor development. They are thus important after the tumor has been established, in tumor progression, and reflect the clonal evolution (see below) during this disease phase. Much evidence indicates that tumor cells are genetically less stable than their normal counterparts, a circumstance that facilitates the emergence of secondary changes. Cytogenetic noise may be a useful term for the massive and bizarre karyotypic rearrangements found in some tumors. In these genomically highly unstable cells, rearrangements are produced at a high rate, and, though some or perhaps most are without selective advantage or indeed may be detrimental to the cells that harbor them, their sheer number may obscure the pathogenetically important changes and completely dominate the karyotype.

It is well known that during its lifetime a tumor undergoes a variety of phenotypic changes, usually leading to increasingly aggressive behavior, with infiltration of surrounding tissues and the establishment of distant metastases. The diversification of phenotypic traits is frequently accompanied by genotypic alterations, often involving an overall increase in the genetic complexity of the tumor cells. The modern view on tumor progression holds that additional mutations regularly occur, whereupon the selection pressure confronting the new variant cells determines which will have an evolutionary edge and outgrow their neighbors. Thus, Darwinian selection determines the relative prominence of the various subpopulations of slightly different cells within a tumor: The most fit increase in number; the less fit die away.

The direction of this clonal evolution is determined by the balance between the inherent genomic instability of the tumor cells, which tends to diversify the tumor's genetic constitution, and the selection pressure confronting the cells, which tends to reduce the genetic heterogeneity in favor of maximum adaptation to existing growth conditions. These conflicting tendencies will be operative irrespective of whether the tumor starts out as a monoclonal proliferation or represents a confluence of several simultaneously transformed cells. If genetic instability predominates, then the clonal evolution will lead to increased heterogeneity within the tumor cell population (i.e., genetic divergence), a situation seen in several neoplasms that have been cytogenetically monitored throughout their course. On the other hand, radical alterations in the selection pressure (e.g., the introduction of a new cytotoxic therapy) may well lead to diminished genetic complexity (i.e., genetic convergence) of the tumor. It is important to bear in mind that every cytogenetic investigation of a neoplasm only represents a "snapshot" of the tumor in question, and inferences about its preceding history, let alone about its future characteristics, are necessarily uncertain. This cautionary note also applies to conclusions about the tumor's clonal nature.

III. Specific Cytogenetic Abnormalities in Human Neoplasms

Close to 150 acquired chromosomal aberrations have been consistently associated with human neoplastic disorders. The majority are known in the leukemias and lymphomas, a fact that may be surprising, considering that these diseases—from the points of view of both mortality and morbidity—are much less important than solid tumors. The skewed cytogenetic knowledge reflects the technical difficulties encountered: Whereas leukemic cells have been easy to sample and are readily brought to divide *in vitro,* solid tumors have been a more problematic material to work with. Only in the last few years have data on the more common human cancers been reported in any number, promising that a clearer picture of solid tumor cytogenetics may soon emerge.

A systemic description of the various specific chromosome aberrations encountered in different tumors would be outside the scope of an encyclopedia of human biology, in particular, because rather extensive use of specialized medical terminology would be unavoidable. We restrict ourselves to a few examples of the many cytogenetic–medical associations presently known; for detailed information of this nature, the reader may consult the textbooks and articles in the bibliography.

One of the principal lessons learned from the study of chromosomal aberrations in hematological malignancies is that different types of preleukemia, leukemia, and lymphoma are characterized by different anomalies (Table I). Sometimes the specificity of the association is quite high [e.g., between t(15;17) and promyelocytic differentiation of the leukemic cells]; in other instances, it is broader (5q− may be found in a variety of myeloid neoplasms). The normal differentiation pattern of the cell type may be reflected in the acquired aberration: Lymphocytic cells of the T lineage have a tendency to have rearrangements of 14q11 (where the T cell α-chain receptor locus is found), while B-lineage lymphocytes regularly have changes affecting the heavy-chain immunoglobulin (*Ig*) locus in 14q32.

In addition to the basic research interest inherent in the mapping of the various aberrations, chromosomal aberrations also convey information of direct clinical importance. First, the finding of an acquired chromosome abnormality proves that a bone marrow or lymph node disorder is neoplastic, which may not have been certain before sampling. Second, the existence of specific or typical anomalies associated with many of the hematological malignancies often enables a more precise diagnosis than might otherwise have been possible. Finally, the presence or absence of certain karyotypic anomalies affects the prognosis of a patient, either adversely or by indicating a more favorable outlook than is generally associated with that particular disorder. Even today, therefore, purely by functioning as disease markers and without taking into account

TABLE I Examples of Consistent Karyotypic Anomalies in Hematological Neoplasms

Cytogenetic abnormality	Disease
t(1;19)(q23;p13)	Acute lymphatic leukemia
del(5q)	Myelodysplasia
t(8;14)(q24;q32)	Burkitt's lymphoma
t(8;21)(q22;q22)	Acute myeloid leukemia
t(9;22)(q34;q11)	Chronic myeloid leukemia
+12	Chronic B cell lymphatic leukemia
inv(14)(q11q32)	Chronic T cell lymphatic leukemia
t(14;18)(q32;q21)	Malignant lymphoma
t(15;17)(q22;q11)	Acute promyelocytic leukemia
inv(16)(p13q22)	Acute myelomonocytic leukemia

TABLE II Examples of Consistent Karyotypic Anomalies in Solid Tumors

Cytogenetic abnormality	Disease
t(X;18)(p11;q11)	Synoviosarcoma
del(1p)	Neuroblastoma
t(2;13)(q35–37;q14)	Rhabdomyosarcoma
t(3;8)(p21;q12)	Mixed tumor of the salivary gland
t(3;12)(q27–28;q13–14)	Lipoma
del(3p)	Carcinomas of kidney and lung
del(11p)	Nephroblastoma (Wilms' tumor)
t(11;22)(q24;q12)	Ewing's sarcoma
t(12;14)(q14–15;q23–24)	Uterine leiomyoma
t(12;16)(q13;p11)	Myxoid liposarcoma
del(13q)	Retinoblastoma
−22	Meningioma

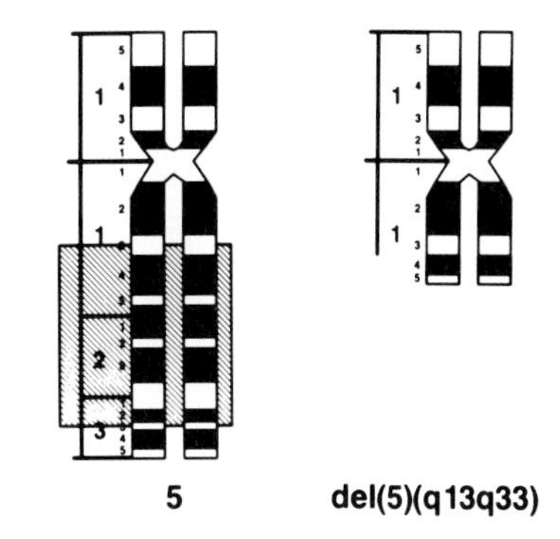

FIGURE 4 del(5)(q13q33), an interstitial deletion of the long arm of chromosome 5, is a common anomaly in myelodysplasia, but it is also seen in acute myeloid leukemia.

what they reveal about the pathogenetic mechanisms of various cancers and leukemias, the karyotypic aberration patterns may, in given instances, decide the choice of treatment and ultimately the outcome in individual cases.

On average, solid tumors (Table II) have more complex karyotypes, with more massive aberrations than have hematological neoplastic disorders. Probably, this only reflects the later sampling of solid tumors, although a more profound biological difference cannot be ruled out. The general lesson from solid-tumor cytogenetics resembles that from the leukemias and lymphomas: Different tumor types are characterized by different cytogenetic aberrations. Here, too, the anomalies function as disease markers that make the diagnoses more precise and the assessment of prognosis more reliable; hence, they also indirectly influence the choice of therapy.

IV. Pathogenetic Consequences of Chromosome Anomalies

The genomic unit currently thought to be the most important is the gene. Everything we know about the functional consequences of genomic rearrangements we explain in terms of gene-level changes; any alterations of larger genomic structures—such as the addition of whole chromosomes or polyploidies—have effects that are presently incomprehensible. Thus, the selective advantage provided by most of the changes we see in cancer cells remains conjectural; we guess that they have somehow made the tumor cells fitter, or we would not have seen them. Correct though this reasoning may be, it offers little new insight into the mechanisms that may be operative.

There are now two groups of genes known to play a direct role in neoplastic proliferation: the oncogenes and the tumor suppressor genes, or antioncogenes. Research on these gene classes is covered elsewhere in this encyclopedia; here, we concentrate only on the cytogenetic aspects of their tumorigenic importance.

The oncogenes may malignantly transform suitable target cells in a dominant fashion (i.e., a single activated oncogene allele is sufficient). Knowledge about them owes much to the study of tumor-producing RNA viruses (i.e., retroviruses). Later research has shown that the tumor genes—the oncogenes—of the viruses also exist in mammalian, including human, cells, but in a nontumorigenic form, as protooncogenes. The normal function of these genes is presumably in the regulation of cellular proliferation and differentiation, and only when they are inadvertently activated (see below) do they contribute to tumor growth. The proteins encoded by the oncogenes, of which currently about 50 are

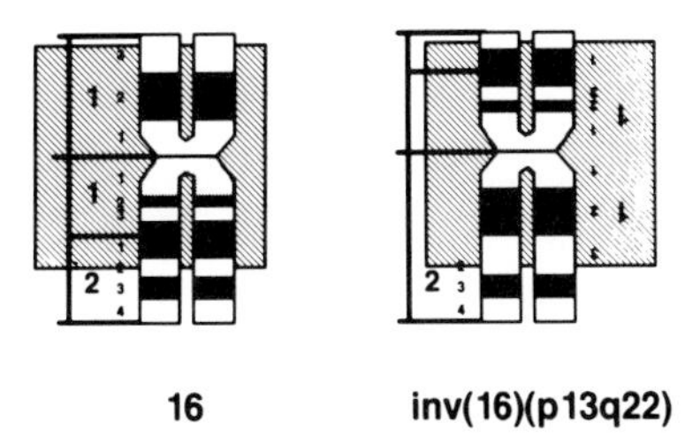

FIGURE 5 inv(16)(p13q22) characterizes a specific myelomonocytic subtype of acute myeloid leukemia.

known, either are growth factors, have similarities with the cytoplasmic membrane growth factor receptors, interact with the system of secondary messengers that transduce information from the cell's surface to the nucleus, or bind to DNA, usually in complexes with other proteins, and so presumably directly affect transcription. Interference with any of these levels of proliferation control could set off unchecked cell division. [*See* ONCOGENE AMPLIFICATION IN HUMAN CANCER; TUMOR SUPPRESSOR GENES.]

It is now known that the loci of protooncogenes within the genome to a remarkably high degree coincide with the breakpoints of cancer-associated chromosome rearrangements. This observation was pivotal in bringing about the understanding that one of the major carcinogenic mechanisms for the activation of protooncogenes is chromosomal change. Thus, although some oncogenes are activated by submicroscopic changes, even point mutations, others normally require chromosome-level alterations to turn them on. In principle, either oncogene activation can involve a qualitative change, meaning that the coding segments of the gene are rearranged, causing the production of an altered protein product, or it may be quantitative, in which case the protein is only insignificantly altered if at all, but is instead produced at a too high rate or is eliminated too slowly. Overproduction can be achieved either by increasing the number of gene copies (i.e., amplification, which may sometimes be cytogenetically surmised when double minute chromosomes or homogeneously staining regions are detected) or by heightened activity on a normal number of loci. Here, we exemplify both qualitative and quantitative oncogene activation by means of cytogenetic rearrangements.

Practically all patients with Burkitt's lymphoma or Burkitt-type acute lymphatic leukemia have in their tumor cells a t(8;14)(q24;q32) or one of the two variant translocations t(2;8)(p12;q24) and t(8;22)(q24;q11) (Fig. 6). The common breakpoint 8q24 harbors the protooncogene *MYC*, which appears to hold the pathogenetic key to the three Burkitt-associated rearrangements. In t(8;14), *MYC* is relocated to the breakpoint region of 14q32, which is within the *Ig* heavy-chain locus. The juxtaposition of *MYC* with constitutively active promoting or enhancing sequences of the *Ig* locus leads to inappropriate *MYC* transcription. In the two variant translocations, the 8q24 breakpoint is slightly more distal (closer to the telomere), so that *MYC* remains on the derivative chromosome 8. The functional

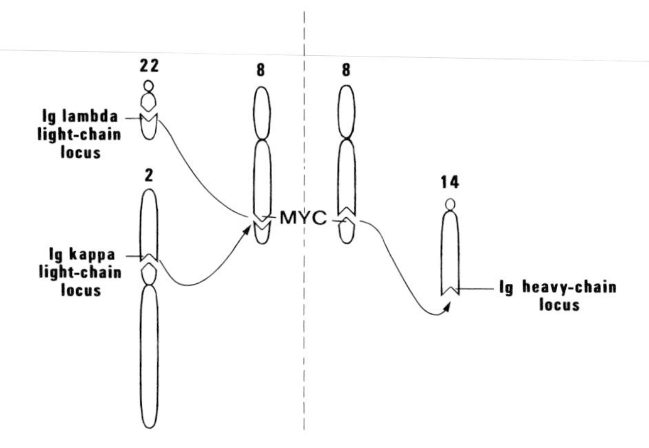

FIGURE 6 t(8;14)(q24;q32) or the variant translocations t(2;8)(p12;q24) and t(8;22)(q24;q11) are characteristic of Burkitt's lymphoma and Burkitt-type acute lymphatic leukemia. The *MYC* protooncogene, which is located in 8q24, is through the translocations brought under the control of regulatory elements whose normal function is to ensure the constitutive activity of the *Ig* genes.

results seem to be indistinguishable from the outcome of the more common 8;14 translocation, however. To the immediate vicinity of *MYC* is translocated the *Ig* κ light-chain, in the 2;8 translocation, or, in t(8;22), the *Ig* λ light-chain locus. Again, the protooncogene comes under the influence of controlling sequences whose normal function is to maintain a high level of *Ig* locus activity, and this results in untimely or increased production of the *MYC*-encoded protein.

The function of this protein, which binds to the nucleus, is unknown. Presumably, it takes part in the regulation of transcriptional activity, and, through interference with this function, *MYC* activation plays a role in neoplastic transformation. It should be emphasized that, in all three translocations, the coding sequences of the *MYC* protooncogene remain essentially unchanged. The activation therefore exemplifies the quantitative route, that genetic alterations of regulatory sequences may lead to increased or sustained production of an otherwise normal protein and, through this means, exert a major tumorigenic influence.

t(9;22)(q34;q11) (Fig. 3) was the first rearrangement to be specifically associated with a human neoplasm, when the Philadelphia chromosome (named after the city where it was discovered and referring to the derivative chromosome 22 resulting from the 9;22 translocation) was described in patients with chronic myeloid leukemia. Later research has revealed that the breakpoint in 9q34 oc-

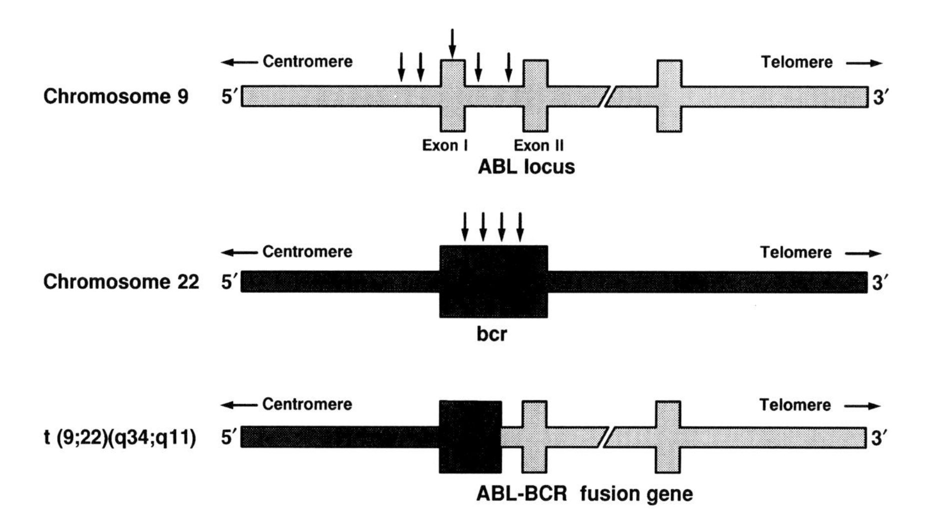

FIGURE 7 The essential molecular consequence of the t(9;22)(q34;q11) in chronic myeloid leukemia is the creation of an *ABL–BCR* fusion gene that, instead of the normal ABL product, encodes a novel 210-kDa protein with strong tyrosine kinase activity.

curs within an approximately 200-kb region at the 5′ end of the *ABL* protooncogene, and always so that the most 3′ exons (exons 2–11) of the oncogene are translocated to chromosome 22. In 22q11, the breakpoints are restricted to a breakpoint cluster region (bcr) of 5.8 kb, which is part of a much larger *BCR* gene. Through the 9;22 translocation, a novel *ABL–BCR* fusion gene is created (Fig. 7) in which the most 5′ end of *ABL* now consists of the most 5′ *BCR* sequences. Instead of the normal 145-kDa *ABL*-encoded protein, the new fusion gene encodes a 210-kDa protein that is substantially altered in its amino-terminal end and has pronounced tyrosine kinase activity (i.e., it attaches phosphate residues to tyrosine). The cellular target for this new chimeric protein is unknown. The protein was created by the activation of the *ABL* protooncogene by a qualitative change and may represent the first truly malignancy-specific protein known in humans.

The molecular specificities of the Philadelphia chromosome-positive leukemias may be greater than revealed by the cytogenetic findings. A substantial subset of patients with acute lymphatic leukemia and a few patients with acute myeloid leukemia also have t(9;22)(q34;q11). It appears that, in many of these patients, the breakpoints are slightly different from those seen in chronic myeloid leukemia, so that a 190-kDa fusion protein is formed instead of the 210-kDa *ABL* product associated with this disease. How the phenotypic effects of the novel proteins are achieved, especially how they vary in acute and chronic leukemia, is still unknown.

Knowledge about the other major group of cancer-relevant genes—the tumor suppressor genes, or antioncogenes—was originally derived from two sources: cell fusion experiments and the study of hereditary childhood cancers. When cancer cells were fused with normal cells, they were often found to lose the malignant phenotype, indicating that the normal genome contained genes capable of suppressing the genetic changes responsible for malignancy. By sequential loss of chromosomes from such hybrid cells, the malignant behavior often returns, implicating those chromosomes that harbor tumor-suppressing capacity, by inference, tumor suppressor genes. The studies of childhood cancers relevant in this context lead to the finding of small deletions in patients with retinoblastoma or with nephroblastoma (Wilms' tumor) and additional developmental anomalies, consistent with the view that loss of chromosomal material (or, at the submicroscopic level, of genes) may unleash tumor growth. [*See* RETINOBLASTOMA, MOLECULAR GENETICS.]

Later examinations of both autosomal dominant childhood cancers and the common sporadic malignancies have confirmed the early hypotheses about the existence of tumor suppressor genes. Whereas the oncogenes seem to bring about neoplastic transformation in a dominant manner (i.e., only one of the protooncogenes of a homolog pair needs to be activated), loss of tumor suppressor genes is a recessive trait at the cellular level. Both loci must be inactivated in order to elicit tumor growth; if one of the alleles is still in the normal wild-type configuration, the phenotype remains nonneoplastic. Inactivation of one of the two suppressor genes in the germ line (thus, all somatic cells have only one functioning allele) generates a dominant predisposition to cancer; tumor development starts only after a chance event inactivates the second allele. The pathogenetic importance of tumor suppressor gene inactivation has been proven for some relatively rare tumor types, such as retinoblastoma (the suppressor locus is in 13q14) and nephroblastoma (a nephroblastoma locus is found in 11p13) and seems likely for several of the more common cancers (e.g., loss of 3p loci in lung and kidney cancers or loss of a 5q locus in large-bowel cancer). The further generality of these and similar findings remains uncertain.

The molecular nature and the precise function of the tumor suppressor genes are unknown. It is also unknown whether they are functionally coupled to the oncogene system of proliferation control, for example, by exerting some kind of regulatory influence on them. The extent to which tumorigenesis through simultaneous oncogene activation and antioncogene inactivation exists, or whether it may indeed be essential in some tumor types, likewise remains wholly conjectural.

In elucidating both classes of cancer genes, the combination of chromosome analyses and a host of other scientific techniques has been necessary. Prominent among the latter have been molecular genetic investigations, which have added a novel dimension to the cytogenetic analysis of cancer-associated rearrangements. It should be emphasized that the development of the new recombinant DNA techniques by no means obviates the need for continued cytogenetic studies of neoplastic cells, for two principal reasons. First, the cytogenetic and molecular genetic techniques assess two widely different organizational levels of the genome. Functional aspects may exist on both levels that are difficult or even impossible to grasp with too crude or too detailed methods. Today, we know next to nothing about the functional role of higher-order structural units of the genome. The point stressed here is similar to the need, obvious to anyone experienced in histopathology, to use both low and high magnification when tissues or cells are examined under the microscope. Second, only by chromosome analyses can the changes that have occurred in individual cells be examined. This makes cytogenetics uniquely suited to investigation of the clonal nature of tumor cell populations and to analysis of the evolutionary changes they undergo with the passage of time.

Bibliography

Green, A. R. (1988). Recessive mechanisms of malignancy. *Br. J. Cancer* **58,** 115–121.

Heim, S., and Mitelman, F. (1987). "Cancer Cytogenetics," 309 pp. Liss, New York.

Heim, S., and Mitelman, F. (1989). Primary chromosome abnormalities in human neoplasia. *Adv. Cancer Res.* **52,** 1–43.

Klein, G. (1987). The approaching era of the tumor suppressor genes. *Science* **238,** 1539–1545.

Klein, G., and Klein, E. (1985). Evolution of tumours and the impact of molecular oncology. *Nature (London)* **315,** 190–195.

Knudson, A. G., Jr. (1985). Hereditary cancer, oncogenes, and antioncogenes. *Cancer Res.* **45,** 1437–1443.

Le Beau, M. M., and Rowley, J. D. (1986). Chromosomal abnormalities in leukemia and lymphoma: Clinical and biological significance. *Adv. Hum. Genet.* **15,** 1–54.

Nowell, P. C. (1986). Mechanisms of tumor progression. *Cancer Res.* **46,** 2203–2207.

Ponder, B. (1988). Gene losses in human tumours. *Nature (London)* **335,** 400–402.

Weinberg, R. (1985). The action of oncogenes in the cytoplasm and nucleus. *Science* **230,** 770–776.

Yunis, J. J. (1983). The chromosomal basis of human neoplasia. *Science* **221,** 227–236.

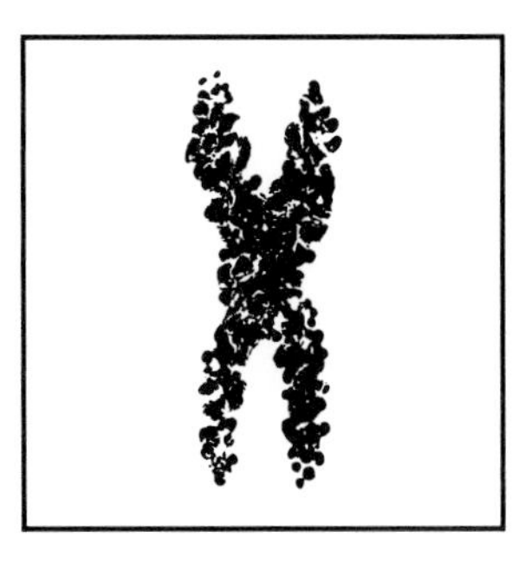

Chromosome-Specific Human Gene Libraries

LARRY L. DEAVEN *Los Alamos National Laboratory*

Glossary

Cloning Isolation of a single cell and propagating it into a population of cells that are identical to the ancestral cell; in recombinant DNA technology, the procedures used to produce multiple copies of a fragment of DNA

DNA library Set of cloned DNA fragments that represent all or a portion of a genome or the DNA transcribed in a particular tissue

Flow cytometry Method for measuring properties such as DNA content or size of cell or subcellular particles and sorting pure fractions from these cells or particles according to those measured properties

Genome One copy of all the DNA in a cell

Vector DNA molecule modified to carry a segment of foreign DNA into a host cell where it is replicated in large quantities

GENE LIBRARIES ARE SETS of cloned DNA fragments that represent all or a portion of the genetic information in an organism. The term library is somewhat misleading because it implies order, but a gene library is simply a random collection of DNA fragments. The fragments may or may not contain a complete gene. They are maintained and propagated in a biological cloning vector, usually a bacterial virus or plasmid. If the DNA used to construct a gene library comes from a single human chromosome, the resulting library will be chromosome-enriched or chromosome-specific. Such libraries represent a subset of the genetic information in human cells, and they provide a rich source of DNA fragments from each chromosome for mapping and sequencing genes and for studying genome organization. The most successful approach to making libraries from the DNA in individual human chromosomes has been through the use of chromosomes purified by flow sorting. This chapter describes the construction process and the application of these libraries to human genetics research, especially the elucidation of molecular detail in the human genome. [*See* DNA AND GENE TRANSCRIPTION; GENES.]

I. Background

Two major technological developments have made the construction of chromosome-specific gene libraries possible. The first of these, flow sorting, enables the purification of millions of copies of each human chromosome. These sorted chromosomes are required as sources of DNA for library construction. (The subject of flow sorting will be discussed in detail in Section II.C.)

The second development necessary for library construction was recombinant DNA technology. Although a library was made from DNA isolated from the cells of fruit flies (*Drosophila melanogaster*) in 1974, application of these methods to larger and more complex genomes had to await the development of more efficient cloning techniques and methods for selecting specific DNA sequences from large numbers of cloned fragments. A series of improvements in these areas was reported over the next 3 years, and in 1978, the first library of human DNA fragments was made. Individual clones containing human γ- and β-globin genes were isolated from this library, clearly establishing a new approach to mapping the human genome and to studies of the structure and function of human genes.

[*See* GENETIC MAPS; GENOME, HUMAN; RECOMBINANT DNA TECHNOLOGY IN DISEASE DIAGNOSIS.]

Since 1978, numerous improvements in recombinant DNA technology have been reported. It is now possible to make a variety of different types of libraries and, therefore, to design the construction of a library to be advantageous for a specific application. For example, by selecting the proper cloning vector, a library can be made that contains small (1–9 kilobases [kb]), medium (15–40 kb), or large (200–1,000 kb) pieces of inserted DNA. Some vectors have distinct advantages over others with regard to the quantity of starting material or target DNA required to make a library. Some have features that facilitate the recovery of the inserted DNA fragments, and some are better than others at retaining the inserted DNA without deleting or rearranging it. (A more complete discussion of the use of different types of libraries is in Section IV.) A major constraint in constructing libraries from sorted chromosomes is in the relatively small amount of available target DNA. This limitation requires the use of vectors with high cloning efficiency (see Section II.D).

While it is possible to make many different types of DNA libraries, with some types more useful than others for a given application, subdividing the DNA of an entire genome before library construction is almost always advantageous. The resulting set of libraries covers all of the genomic DNA, but each library is less complex than a single library containing all of the cellular DNA. A convenient way to make subsets of human DNA is on a chromosome-by-chromosome basis. To include all of the nuclear DNA in human cells, 24 different libraries are necessary (22 autosomes plus the X and Y chromosomes) (Table I). The libraries will vary in size with the largest (for chromosome 1) being five times as large as the smallest (for chromosome 21).

TABLE I DNA Content of Human Chromosomes

Chromosome	Mass DNA per chromosome ($\times 10^{-13}$ grams)
1	5.14
2	5.03
3	4.19
4	3.97
5	3.81
6	3.56
7	3.31
X	3.17
8	3.02
9	2.86
10	2.82
11	2.81
12	2.77
13	2.26
14	2.14
15	2.08
16	1.93
17	1.76
18	1.68
20	1.38
19	1.31
Y	1.16
22	1.08
21	0.98

II. Library Construction

A. Sources of Chromosomes

Human chromosomes may be isolated from three different types of cultured cells for sorting on flow cytometers. Two of these, human fibroblast cells and human lymphoblastoid lines, contain only human chromosomes. The third, rodent–human hybrid cells, contain a background of mouse or hamster chromosomes with one or a few human chromosomes in each cell. Each of these cell types has advantages and disadvantages as sources of chromosomes for sorting. These characteristics are briefly described in the following paragraphs. [*See* CHROMOSOMES.]

Human diploid fibroblasts are cells cultured directly from a small piece of human tissue. The tissue is either cut into fragments or digested into single cell units and placed in a culture flask in growth medium at 37°C. The fibroblast cells begin to divide under these conditions, and the growing cells are propagated by serial transfer into new flasks. The chromosomes in fibroblasts are normal and stable, with little or no tendency to undergo structural rearrangements, provided that the donor of the tissue had a normal karyotype and the tissue was normal. These cells have a finite life span. They grow rapidly for approximately 15 generations, then progressively slower, until they stop dividing entirely at about 50 generations. This is a major disadvantage for isolating large quantities of chromosomes for sorting purposes. Fibroblast cells are useful only during the rapid growth phase when many cells are dividing.

Human lymphoblastoid cell cultures are initiated by transforming human lymphocytes with Epstein–Barr virus. They do not undergo cell senescence and continue to divide indefinitely, thus having a distinct advantage over fibroblast cells. The chromosomes in these cell lines are not as stable as those in fibroblasts; however, a number of available lymphoblastoid lines show little or no karyotypic change over long periods of time in culture. They grow as suspension cultures, which is an easy and inexpensive way to propagate cells, but it is difficult to separate mitotic from interphase cells, and isolated chromosomes from these lines contain high concentrations of interphase nuclei. Another potential disadvantage to the use of chromosomes from these lines for library production is the incorporated Epstein–Barr virus. Although the chromosomes appear to maintain stable morphology, the viral incorporation may induce subtle changes in the DNA of the host cells. [*See* EPSTEIN–BARR VIRUS; LYMPHOCYTES.]

Hybrid cell lines are constructed by inducing two different types of cells to fuse together to become one cell. Hybrids between human cells and mouse cells or Chinese hamster cells are useful for sorting human chromosomes. Initially, the hybrid cell contains two complete sets of chromosomes, one from the human cell and one from the rodent cell. However, when a hybrid cell begins to divide, it loses some of its human chromosomes. Some of the hybrid cells lose all of their human chromosomes, while others retain one or more of them for various lengths of time. Single cells can be cloned from a population of hybrid cells, grown as a cloned culture, and analyzed for the specific human chromosomes they contain. Alternative methods for selecting a specific human chromosome involve growth in special culture medium that selectively kills cells without the desired chromosome. With the use of these techniques, a series of cell lines are available that contain only one to three human chromosomes in a rodent background. These lines are necessary to sort those human chromosomes that have very similar DNA content and that flow sorters cannot distinguish from each other (chromosomes 9–12), and they are advantageous for sorting many other chromosomes. A distinct advantage is that any contaminating DNA present in the sorted chromosomes will be hamster or mouse, which can easily be screened out of a library. On the other hand, these cell lines tend to be unstable, and the desired human chromosome can be lost or rearranged at any time in culture. This undesirable feature requires frequent cytogenetic analysis to make sure the line contains the human chromosome to be sorted in an intact, unrearranged state.

Metaphase chromosomes for sorting are obtained from any of the cell types discussed above by blocking exponentially growing cultures with Colcemid. Colcemid holds the dividing cells in metaphase for up to 12 hr, the longest block of time used for this work. Hybrids and fibroblasts are grown as flat monolayers of cells attached to plastic culture flasks. When the cells enter metaphase, they become round and are attached to the surface less securely than the other cells. They can be dislodged from the monolayer by gentle treatment with trypsin or by shaking the flask before removing growth medium, leaving the interphase cells still attached to the flask. The portion of mitotic cells (mitotic index) among the dislodged cells is usually >90%, providing a rich source of chromosomes for isolation and sorting. As mentioned earlier, lymphoblastoid cells grow in suspension and do not attach to culture flasks. They are also blocked with Colcemid, but at the end of the Colcemid treatment the entire population is collected for chromosome isolation. In this case, the percentage of mitotic cells is generally 20–60%, depending on the growth rate of the cells. After chromosomes are isolated from these cells, interphase nuclei can be removed by gentle centrifugation. This step is important because all sorters have some error rates. If even a few interphase nuclei are sorted into the collection tube, they have serious effects on the purity of sorted DNA. By reducing the concentration of nuclei with differential centrifugation, the chances of sorting errors involving nuclei become negligible.

B. Isolation and Staining of Chromosomes

The mitotic cells collected from culture flasks are pelleted by centrifugation and the growth medium is removed. The first step in the chromosome isolation process is to resuspend the cells in a hypotonic solution (usually 75 m*M* KCl). This swells the cells, helps to dissolve the mitotic apparatus, and weakens the cell membrane. After the swelling step, the cells are gently pelleted in a centrifuge, the hypotonic solution is removed, and the cells are resuspended in a chromosome isolation buffer. This suspension is swirled in a vortex apparatus to break open the cell membranes and disperse the chromosomes.

The chromosome isolation buffer is probably the most critical element in this process. It must stabilize the chromosomes so that they remain intact during sorting, it must protect the chromosomal DNA from nucleases released when the cells are disrupted, and it must be compatible with the fluorescent stains used for flow cytometric analysis. Several chromosome isolation techniques have been developed involving the use of different kinds of buffers for chromosome stabilization; however, only one of them yields DNA of sufficient molecular weight to construct medium and large insert libraries. This buffer stabilizes chromosomes from many different cell lines for up to 6 mo after isolation. It contains 15 m*M* Tris–HCl, 2 m*M* EDTA, 0.5 m*M* EGTA, 80 m*M* KCl, 20 m*M* NaCl, 14 m*M* beta-mercaptoethanol, 0.2 m*M* spermine, 0.5 m*M* spermidine, and 0.1% digitonin, pH 7.2. Although the general procedure outlined above has wide applicability, subtle differences among individual cell lines require optimization of each step to obtain ideal results. Variables that can be adjusted include: Colcemid concentration, blocking time, cell concentration during isolation, hypotonic swelling buffer, swelling time, and time of vortex agitation of cells in isolation buffer.

Isolated chromosomes can be stained with a variety of fluorescent dyes for analysis on a flow cytometer. If the chromosomes are stained with a single dye that is specific for DNA, the resolution of single chromosomes is inadequate for sorting at the highest levels of purity. For example, if diploid human chromosomes are stained with propidium iodide, a fluorescent dye that intercalates with DNA, the 24 human chromosomal types will resolve into 17 or 18 peaks based on the total DNA content of each chromosome. However, most of the peaks overlap one another, so an attempt to sort one chromosomal type from a single peak would result in as much as 50% contamination from flanking peaks. This level of purity would be unacceptable for most library applications. [*See* Flow Cytometry.]

Improvements in resolution can be obtained by staining the chromosomes with two stains, one with an affinity for AT-rich DNA (Hoechst 33258) and one with an affinity for GC-rich DNA (chromomycin A_3). When chromosomes from diploid human cells are stained with these dyes and analyzed in a dual laser flow cytometer, the 24 human chromosomal types are resolved into 21 peaks (chromosomes 9–12 overlap in one peak) (Fig. 1). These stains also differentiate many human chromosomes from the hamster chromosomes in hybrid cells (Fig. 1). This staining combination differentiates chromosomes on the basis of total DNA content and also on the basis of the relative AT- or GC-

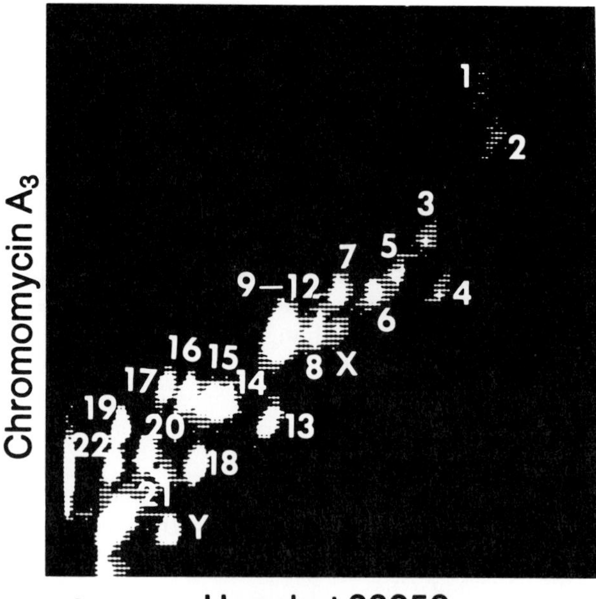

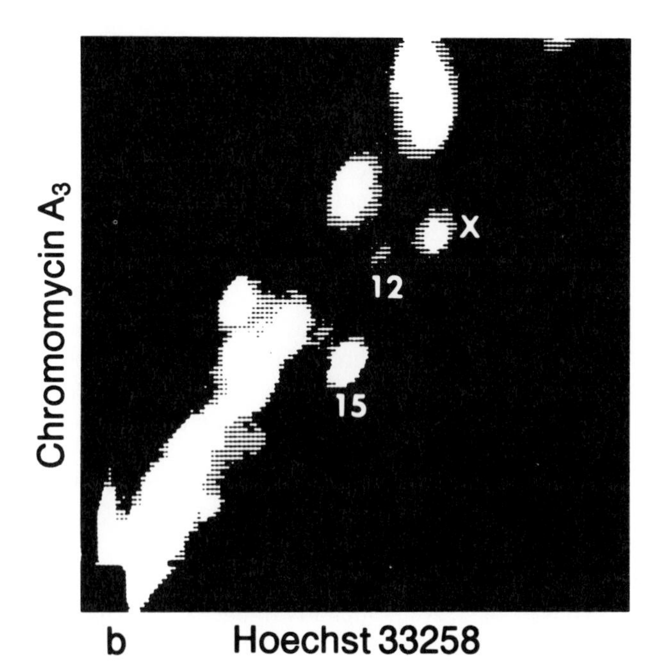

FIGURE 1 (a) A flow histogram of chromosomes isolated from a diploid human fibroblast cell strain (HSF-7). The histogram is composed of the number of fluorescence events versus the fluorescence intensities of the stains chromomycin A_3 and Hoechst

richness of each chromosome. Staining is accomplished by adding chromomycin A_3 (final concentration 62.5 μM) to the isolated chromosomes for 3 hr and then adding Hoechst 33258 (final concentration 3.7 μM). The stains are usually added to the chromosome suspension 1 day before sorting.

C. Sorting

Flow cytometers were developed at the Los Alamos National Laboratory during the late 1960s, primarily to measure the DNA content of single cells. Since that time, many improvements have been made, and these instruments can now be used in other applications. The DNA content of organisms as small as bacteria can be detected, and cells or subcellular components can be measured for a number of properties and sorted on the basis of variations in these properties. The basic operating principles of a flow cytometer are illustrated in Figure 2. The objects to be analyzed and sorted (chromosomes) are carried through the instrument in a fluid stream. The chromosomes enter a flow chamber where the size and speed of the stream is controlled to permit only one chromosome at a time to cross beams of light from two lasers. The fluorescent dyes, Hoechst 33258 and chromomycin A_3, fluoresce as they cross the ultraviolet beam from one laser and the 458-nm beam of the second laser. The resulting fluorescence signals are converted from optical signals to electrical signals by photomultipliers. The signal intensities are proportional to the AT and GC base-pair content of each chromosome. The electrical signals are stored in histograms (Fig. 1) obtained on a pulse-height analyzer; they are also passed to the sorting circuitry of the instrument. After passing through the flow chamber, the fluid stream containing chromosomes is broken up into droplets by a piezoelectric transducer driven by a high-frequency oscillator. If the two fluorescence signals from a chromosome fall within a range defined by a preset sorting window, a charging circuit is activated, which applies a short voltage pulse to the portion of the stream containing the chromosome. This happens just before the stream is broken into droplets, so that the droplet containing a desired chromosome is electrically charged. Actually, three droplets are charged to allow for small timing uncertainties. When these droplets pass through a pair of deflection plates, they are deflected into a

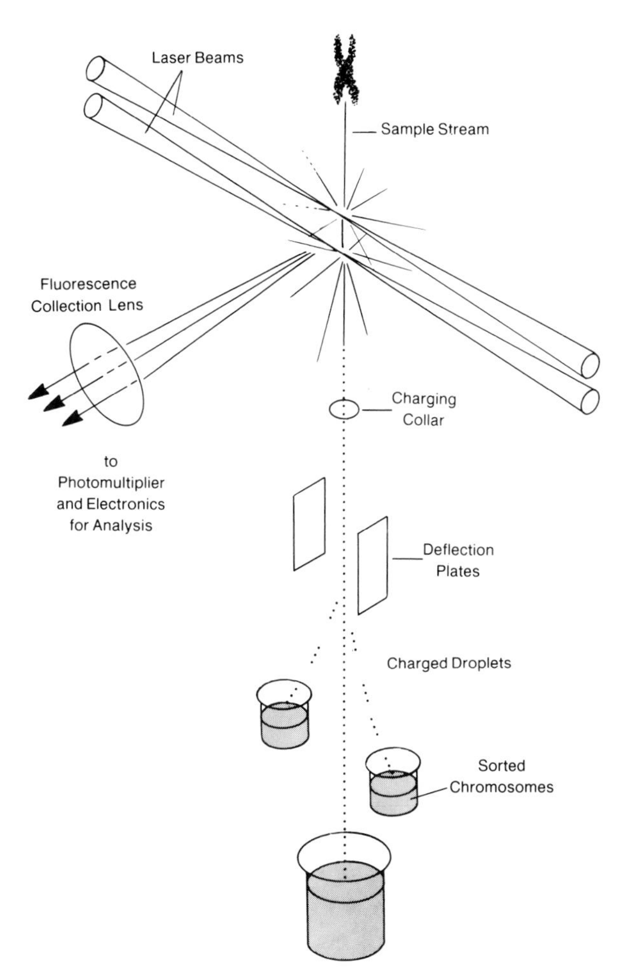

FIGURE 2 A diagram illustrating the operating principles of a dual-laser flow cytometer. A narrow stream of a suspension of stained chromosomes flows across the focused region of two laser beams, where fluorescence properties of each chromosome are measured and stored as electronic signals. If the fluorescence intensities define a desired chromosome, an electric charge is put on the stream, which is then broken into droplets. When the droplets containing the desired chromosome pass through charged deflection plates, they are deflected according to the charge imparted to them. Two chromosomes may be selected for sorting from each suspension. [Reproduced, with permission, from Los Alamos Science, 1985, Spring/Summer, No. 12.]

33258. Chromosomes 9–12 are so similar in total DNA content and in base composition that they form a single, overlapping peak. To sort these chromosomes, hamster–human hybrid cells, containing a single (or a few) human chromosome, must be used as chromosome sources. (b) A flow histogram of the chromosomes in a hybrid retaining human chromosomes X, 12, and 15. The hamster chromosomes fall on a separate axis from the human chromosomes due to differences in base composition between the chromosomes of the two species. [Reproduced, with permission, from Los Alamos Science, 1985, Spring/Summer, No. 12.]

separate collection vessel, while uncharged droplets containing the unwanted chromosomes go to a waste collection vessel.

For library construction, there are two critical aspects to the sorting process. Chromosome sorting must be relatively fast, and for most libraries it must be accomplished with a high level of purity. Two different types of sorters have been used to purify chromosomes for library construction. They sort at either the conventional rates typical of commercial instruments or at speeds that are approximately an order of magnitude higher. These latter instruments have been developed at the Lawrence Livermore and Los Alamos National Laboratories specifically to aid research projects requiring high numbers of sorted objects. They are called high-speed sorters. Conventional sorters are capable of analyzing 1,000–2,000 chromosomes/sec and of sorting up to 50 chromosomes of a particular type each second. Optimal chromosome preparations and instrumental performance will yield 1×10^6 copies of a chromosome in 8 hr of operating time. High-speed sorters can analyze 20,000 chromosomes/sec and sort up to 200 chromosomes/sec. Under optimal operating conditions, they will yield 5×10^6 chromosomes in 8 hr of operating time. Thus, high-speed sorters have a distinct advantage over conventional sorters, especially if relatively large amounts of DNA (microgram quantities) are required as starting material for library construction.

The purity of sorted chromosomes is critical to the construction of chromosome-specific DNA libraries, and a large amount of effort has gone into purity determinations of sorter outputs. The capability of sorters to differentiate and sort specific objects from a mixed population with high levels of accuracy and precision can be determined with fluorescent microspheres. In test runs, sorters used in this project can sort a single population of microspheres from mixed populations with 95–100% purity at operating speeds. Purity determinations of sorted chromosomes are more difficult and require direct examinations of sorted chromosomes when practical, as well as indirect measurements such as cytological examination of cells used for chromosome sources and library purity determinations.

Indirect examinations that have been employed include:

1. Karyotype analysis (G-banding, Q-banding, G-11 staining, 4′,6-diamidino-2-phenylindole [DAPI] Netropsin staining) of metaphase cells from cell strains and lines used as chromosome sources: This analysis confirms that a chromosome of interest is present in the cell culture, the frequency of its presence, and whether it is normal or rearranged.
2. Isozyme analysis of hybrid lines to support karyotype analysis.
3. Measurements of chromosome peak locations and peak volumes in flow histograms and comparisons of that information with cytological karyotype analysis: This type of information on chromosome frequency (peak volume) and chromosome DNA content (peak location) is especially useful when hybrid cells are being used and the human chromosome content is unstable.
4. Examinations of libraries for the presence of Chinese hamster sequences or for extraneous human DNA sequences.

Direct measurements of the purity of sorted chromosomes include:

1. Banding analysis of chromosomes sorted directly onto microscope slides: This method is useful; however, isolated chromosomes are often too condensed to provide high resolution G-band patterns (series of dark and light bands along the length of a chromosome, produced by a staining agent called Giemsa dye).
2. DAPI-Netropsin staining of sorted chromosomes: This method stains only the human chromosomes with large blocks of centromeric heterochromatin (chromosomes 1, 9, 16, and Y), but it is useful if isolation methods are used that do not permit G-band analysis.
3. Molecular hybridization of human and hamster genomic DNA to sorted chromosomes: This procedure can be done *in situ* on sorted chromosomes and used to identify hamster chromosomes or human–hamster translocations.
4. Hybridization of human chromosome-specific DNA probes to sorted chromosomes: The major limitations of this approach are the small number of probes available (there are no suitable probes for approximately half of the chromosomes in the human karyotype), and the chromosomal region covered by these probes is usually limited to the centromere or telomere areas. In chromosomes sorted from hamster–human hybrids, it is possible to miss the detection of hamster DNA in a translocation that does not involve the centromere or telomere area of a human chromosome.

D. Cloning

The final step in the construction of libraries specific for the DNA of a single human chromosome is to insert fragments of the sorted chromosomal DNA into a self-replicating biological vector. Although many different types of libraries can be produced, there are four basic steps in the cloning process that are common to all of them: preparation of DNA fragments from DNA isolated from sorted chromosomes and from the vector system; formation of recombinant DNA molecules involving both chromosomal and vector fragments; assembly of self-replicating units containing the recombinant DNA molecules; and the amplification of these units into large numbers of copies, perhaps a million times the original starting material. The major challenge in cloning DNA from sorted chromosomes is to maximize the efficiency of each of these steps to obtain large libraries from small amounts of target DNA. For conventional cloning, total cellular DNA is easy to obtain and easy to replace if a cloning procedure fails. On the other hand, it may take several weeks to accumulate 1 μg of sorted DNA, so it is essential to utilize it with minimal loss.

The steps involved in constructing a small insert library are illustrated in Fig. 3. In this example, the biological vector is a modified λ bacteriophage or bacterial virus called Charon 21A. It can accept fragments of human DNA up to 9,100 base pairs in length. The first step is to extract DNA molecules from the sorted chromosomes and from the cloning vector, and prepare them for cloning. To prepare the large chromosomal DNA molecules, they are fragmented to provide pieces of suitable size, according to the vector used. This is accomplished by digesting them with a restriction enzyme called *Eco*R1. Restriction enzymes recognize specific sequences in the DNA molecules and cut the DNA at each of these sequences. The enzyme *Eco*R1 recognizes a sequence of 6 bases (-G-A-A-T-T-C-) and cuts the phosphodiester bond between the -G-A- bases in this sequence. The cuts on the two strands do not coincide, separated by the A-A-T-T bases; therefore, the two cut ends overlap. In addition to the enzymes that recognize 6 bases, there are also enzymes that recognize 4 bases or 8 bases. Assuming a random distribution of the four bases A, T, G, and C in DNA, a 4-base recognition sequence would occur on average every 256 bases, a 6-base sequence every 4 kb and a 8-base sequence every 65 kb. By selecting the proper restriction enzyme, it is possible to cut large DNA molecules into fragments of different average lengths. In our example, in a digest run to completion, *Eco*R1 will reduce the DNA to an average length of 4 kb; however, approximately 33% of the DNA will remain in fragments longer than 9 kb, the upper size in the acceptance range for Charon 21A. This will result in a

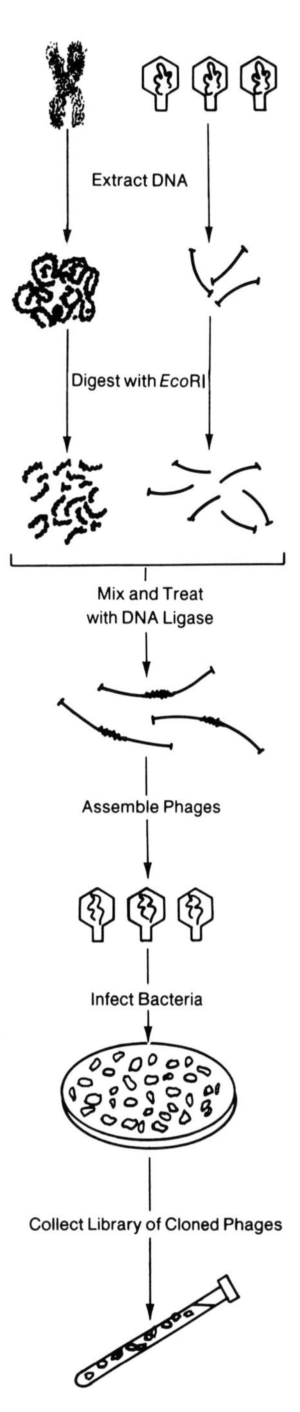

FIGURE 3 A diagram illustrating the major steps involved in the construction of a chromosome-specific DNA library in a λ-phage vector. The DNA in the sorted chromosomes and in the vector is isolated and digested with a restriction enzyme called *Eco*R1. The fragments of DNA are ligated together and packaged into infectious phage particles. These recombinant phage are then amplified by propagating them on a lawn of host bacterial cells. Aliquots of these amplified phage represent the DNA of the sorted chromosomes. [Reproduced, with permission, from Los Alamos Science, 1985, Spring/Summer, No. 12.]

library that contains only 67% of the chromosomal DNA. To increase the coverage of the chromosomal DNA, a second library would have to be constructed using a different restriction enzyme that recognizes a different 6-base sequence in the chromosomal DNA as well as in the vector.

The linear DNA molecule of Charon 21A is about 42 kb in length, and it contains a single sequence recognized by *Eco*R1. These molecules are digested with *Eco*R1, which cuts them into two "arms." These phage arms are then treated with the enzyme alkaline phosphatase to remove a phosphate group. This prevents most of the arms from recombining into intact vector molecules without an insert.

Next, the two DNA preparations are mixed, and because they were both cut with *Eco*R1, they both have AATT overlapping (cohesive) ends that become linked by hydrogen bonds. The result is the formation of recombinant DNA molecules consisting of two vector arms linked by a fragment of human DNA. When these recombinant molecules are treated with the enzyme DNA ligase, they are permanently rejoined by the formation of new phosphodiester bonds. To maximize the yield of recombinant molecules, an excess of phage arms is used in the mixing reaction. Because the arms were treated with alkaline phosphatase, they cannot be rejoined to each other with DNA ligase, and this prevents the formation of large numbers of nonrecombinant contaminates of the library.

The recombinant DNA molecules can now be "packaged" into intact Charon 21A phages by providing the required phage proteins. The resulting infectious phage particles are then seeded onto a monolayer, or "lawn," of a suitable host strain of bacteria. Here they inject their DNA into a bacterium, causing the metabolic and replicative apparatus of the host cell to produce more phage protein and DNA. When the cell has produced between 100 and 200 new phage particles, it ruptures and the infection process continues with surrounding cells. The final result is a clear area or plaque on the bacterial lawn. Each plaque contains from 1 to 10 million infective phage particles, and each phage contains a copy of the original recombinant molecule with its human insert. Together they form a library for the chromosomal DNA used as starting material.

For the construction of libraries with larger inserts, the basic steps described above are repeated, but with different vectors and modified methods of preparing the chromosomal target DNA. Libraries with medium-sized inserts of up to 45 kb are cloned into cosmid vectors. Cosmids are cloning vehicles that incorporate functional properties of λ-phage and plasmids, hence their name. The "cos" part comes from a property of phage λ. During the λ-phage replication cycle, hundreds of copies of λ DNA form a long chain or concatamer, in which the copies of the λ genome are joined by a short sequence, known as the cohesive end or cos site. In the phage particles, they constitute the two ends of the genome. The λ-packaging enzymes recognize two cos sites 35–45 kb apart in the concatamer, cleave this unit, and package it into a phage head. The cos sites are the recognition system for reducing concatamers into λ genome units, as part of the packaging process. Plasmids are tiny circular DNA molecules that often contain genes for antibiotic resistance. They reside in bacteria where they confer antibiotic resistance to the host cells. They are useful as vectors because when a plasmid with an antibiotic resistance gene is used as a vector in plasmid-free host cells that are antibiotic-sensitive, the presence or absence of the plasmid vector can be readily detected by treating the host cells with antibiotic. By incorporating the cos sites of λ-phage to plasmids, a new class of cloning vectors (cosmids) was created.

For cosmid cloning, the sorted chromosomal DNA is partially cleaved with a restriction enzyme to yield relatively large pieces of DNA as compared with the complete digestion used for small-insert phage cloning. The cosmid vector is cleaved at the same restriction site to produce cloning arms, each with a cos site at one end. The cosmid arms are then ligated to the partially digested human target DNA fragments. Recombinant molecules consisting of a cloning arm ligated to each end of a fragment of human DNA, where the cos sites are approximately 48 kb apart, can be packaged to produce infectious phage particles. These phage can then be used to introduce their recombinant molecules into a suitable bacterial host. Once inside the bacteria, the recombinant cosmid DNA circularizes and is maintained within the bacteria as a plasmid.

The first cosmid vectors were not as efficient as phage vectors, and the requirements of large amounts of target DNA prevented the construction of cosmid libraries from sorted chromosomes. A number of improvements, especially the addition of a duplicated cos sequence, now make it possible to construct large libraries from submicrogram quantities of DNA. When these new vectors were combined with improved methods for isolating chromosomes with high molecular weight DNA, it became possible to construct chromosome-specific libraries in cosmids.

The largest-capacity cloning vectors currently available have an insert acceptance size over 10 times the limit of cosmids. These vectors are called yeast artificial chromosomes (YAC). They contain a cloning site within a gene to facilitate detection of recombinant clones, a yeast centromere, and two sequences that give rise to functional yeast telomeres. All of these sequences are carried in a single plasmid that can replicate in *Escherichia coli* cells. The basic approach to YAC cloning is similar to phage and cosmid cloning. The vector is cleaved at the cloning site and between the telomere sequences. This produces two vector arms that are treated with alkaline phosphatase to prevent religation among themselves and then ligated with large molecules of target DNA. The yeast artificial chromosome is then transformed into yeast cells where it becomes a stable chromosome. As in the case of cosmid vectors, the initial results suggested that this cloning method would not be practical to combine with chromosomal DNA purified by flow sorting (25 μg of insert DNA was used). However, a partial YAC library from sorted human chromosome 21 DNA has now been constructed. The results from this work suggest that a complete library could be made with a few micrograms of sorted DNA. A set of YAC libraries for each human chromosome will likely be available in the next few years.

III. Uses of Libraries

The application of human chromosome-specific DNA libraries to problems in human genetics research has been dependent on two major components: (1) the capability to construct different types of libraries from flow-sorted chromosomes and (2) the parallel development of new techniques in molecular genetics that resulted in novel applications for libraries and bold new directions in human genetics research.

When it became feasible to construct the first chromosome-specific libraries (1983), the most urgent need for them was for human gene mapping and genetic disease diagnosis. A basic approach to locating genes involved with disease was to screen a genomic DNA library for fragments of DNA or probes that were in close proximity to the actual disease-producing gene. These probes are used in family studies to detect changes in the location of restriction sites in the DNA of closely related individuals who do or do not have the inherited disease. These changes are called restriction fragment-length polymorphism, and they are inherited as simple genetic traits that obey the Mendelian laws of inheritance. The chromosome-specific libraries help to overcome the "needle-in-the-haystack" dilemma by providing rich sources of probes from only one chromosome as opposed to all chromosomes. Human DNA contains families of repetitive sequences that are dispersed throughout the genome. These repeat sequences are not useful as probes because they occur as many as 300,000 times per genome. They must be screened out of a library to isolate unique sequence or single-copy probes that map to only one site. Because the repeat sequences are dispersed throughout the genome, the larger the insert size in a library, the more likely it is to contain a repeat, hence the decision to make the first libraries from complete digest DNA cloned into a vector that accepts DNA fragments up to 9 kb. The small inserts in these libraries have a relatively high probability of containing only unique sequence DNA. [*See* DNA MARKERS AS DIAGNOSTIC TOOLS.]

A limitation of these small insert libraries is that any fragment >9 kb will not be represented in the library. This deficiency was partially overcome by constructing two libraries for each chromosome. One set of libraries was made using *Eco*R1 as the restriction enzyme and a second using a different 6-base cutter, *Hin*dIII. A fragment of DNA excluded from one library has a high probability of being included in the other.

These small insert libraries are stored in liquid nitrogen at the American Type Culture Collection in Rockville, Maryland. Aliquots of the original amplification are available to research groups throughout the world. They have been used extensively in the search for unmapped genes, especially those genes responsible for genetic disease. For example, several hundred probes have been isolated and mapped from the chromosome 4 and 7 libraries in the search for the defects responsible for Huntington's disease and cystic fibrosis. Although improved methods now permit the construction of larger insert libraries, these complete digest libraries continue to be useful. Over 3,500 of them have been sent to research laboratories.

A major disadvantage of small insert libraries is

that individual fragments may not contain complete genes. Although many genes are 5–10 kb in length, the gene for factor VIII, which encodes the blood-clotting factor deficient in humans with hemophilia A, spans at least 190 kb, and the defective gene for Duchenne's muscular dystrophy covers over 1 million base pairs. In addition to the advantages of having whole genes or even groups of genes in a single-cloned insert for molecular studies of gene structure and expression, new research initiatives in human genetics made chromosome-specific gene libraries with larger inserts highly desirable. The ultimate aim of this latter work is to determine the sequence of the 3 billion base pairs in the human genome. One way to accomplish this is to organize the fragments of DNA in a library into the linear sequence that exists in an intact chromosome. This ordered library could then be sequenced one insert at a time until the known sequence extended to the entire length of a chromosome. Library ordering is a formidable task because an average-sized human chromosome will require approximately 3,500 cosmid inserts to cover its length. Nevertheless, work on several cosmid libraries is underway, and an ordered cosmid library for chromosome 16 is now about 40% complete.

Yeast artificial chromosome libraries will be easier to order than cosmid libraries because each yeast insert covers 5 or 6 cosmid inserts; however, the YAC inserts are too large to sequence without first subcloning them into cosmids. These libraries will also permit functional studies of large tracts of DNA not capable of being cloned into other systems. The large size of these inserts will facilitate reliability checks of physical maps derived from other approaches and will aid in establishing continuity of map data from different laboratories.

It should be emphasized that none of the cloning systems are perfect and each type of library will probably turn out to be incomplete with regard to coverage of all of the DNA in a chromosome. However, by constructing libraries in a variety of vectors the chances of complete coverage are considerably enhanced. Taken together these sets of libraries provide powerful tools for analyzing and defining the functional properties of the human genome.

Acknowledgment

This work has been supported by the U.S. Department of Energy under contract number W-7405-ENG-36 to the Los Alamos National Laboratory. I would like to thank the following individuals for scientific contributions or for help in preparing this manuscript: R. Archuleta, N. C. Brown, E. W. Campbell, M. L. Campbell, L. S. Cram, J. J. Fawcett, C. E. Hildebrand, J. L. Longmire, L. J. Meincke, P. L. Schor, and M. A. Van Dilla.

Bibliography

Deaven, L. L., Hildebrand, C. E., Fuscoe, J. C., and Van Dilla, M. A. (1986). Construction of human chromosome specific DNA libraries: The National Laboratory Gene Library Project. *Genet. Eng.* **8,** 317.

Deaven, L. L., Van Dilla, M. A., Bartholdi, M. F., Carrano, A. V., Cram, L. S. Fuscoe, J. C., Gray, J. W., Hildebrand, C. E., Moyzis, R. K., and Perlman, J. (1986). Construction of human chromosome-specific DNA libraries from flow sorted chromosomes. *Cold Spring Harbor Symp. Quant. Biol.* **51,** 159.

Van Dilla, M. A. and Deaven, L. L. (1990). Construction of gene libraries for each human chromosome. *Cytometry* **11,** 208–218.

Van Dilla, M. A., Deaven, L. L., Albright, K. L., Allen, A. A., Aubuchon, M. R., Bartholdi, M. F., Brown, N. C., Campbell, E. W., Carrano, A. V., Clark, L. M., Cram, L. S., Fuscoe, J. C., Gray, J. W., Hildebrand, C. E., Jackson, P. J., Jett, J. H., Longmire, J. L., Lozes, C. R., Luedemann, M. L., Martin, J. C., Meyne, J., McNinch, J. S., Meincke, L. J., Mendelsohn, M. L., Moyzis, R. K., Munk, A. C., Perlman, J., Peters, D. C., Silva, A. J., and Trask, B. J. (1986). Chromosome-specific DNA libraries: Construction and availability. *Biotechnology* **4,** 537.

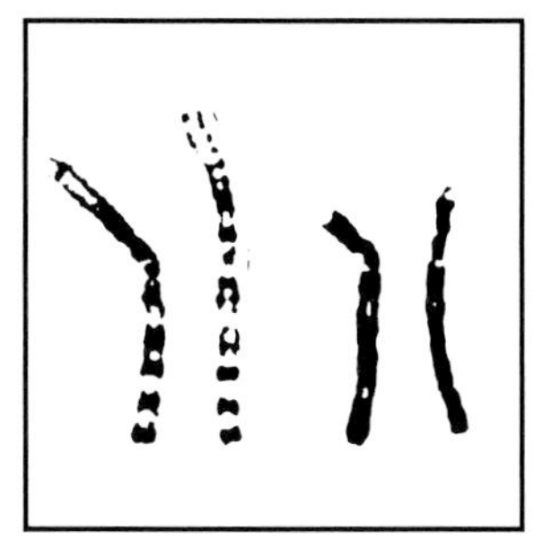

Chromosomes

BARBARA A. HAMKALO, *Department of Molecular Biology and Biochemistry, University of California–Irvine*

Glossary

Centromere Region of a chromosome involved in segregation; the site of attachment of the spindle apparatus

Chromatin DNA–protein complex which makes up eukaryotic chromosomes

Eukaryotic Possessing a double-membrane nuclear envelope separating the genome from the cytoplasm

Heterochromatin Condensed chromatin

Histones Small, basic, evolutionarily conserved chromosomal proteins

In situ hybridization Technique used to locate specific DNA sequences in cytological preparations

Karyotype Display of mitotic chromosomes in a diploid cell

Kinetochore Specialized, trilamellar, platelike structures organized on the outer surface of each chromatid that are the sites of attachment of microtubules between the chromatids and the poles of the mitotic apparatus

Nucleosomes Fundamental nucleoprotein subunit of eukaryotic chromosomes

Polymorphism Property of existing in several forms; in molecular genetics, single base changes in the same sequence that are detected via base sequence-specific restriction enzymes

Supercoiling/helicity Coiled coil of double-stranded DNA

Telomere Molecular end of a chromosome

THE GENETIC INFORMATION in eukaryotic organisms is partitioned into units (i.e., chromosomes) which vary in size and number among species. Each chromosome consists of a linear piece of double-stranded DNA complexed with a large number of proteins, some of which are structural, others of which function in DNA duplication, RNA synthesis, or chromosome maintenance. This article deals with the molecular and macromolecular structures of eukaryotic chromosomes.

I. General Features of Chromosome Structure—Chromosome Identification

Chromosomes are visible as distinct entities only transiently, as a consequence of extensive condensation at the time cells are preparing to undergo division. During the remainder of the cell cycle, these structures are sufficiently decondensed to be invisible as defined units, even by high-resolution microscopy. As a result, individual chromosomes can be identified only during mitosis, when cells are dividing and chromosome condensation is maximal. Gross morphological differences, which provide a crude way to distinguish among chromosomes, include their total length and the ratio of the length of the short arm to that of the long arm; convention names the short arm "p" and the long arm "q," relative to the centromere. The only apparent distinction at this level is that centromeres are relatively more condensed (i.e., heterochromatic) than are chromosome arms. This scheme of identification has low discrimination, because many chromosomes look virtually identical by these criteria. However, with a variety of experimental protocols, it is possible to reveal subtle differences in chromosome composition or folding which are reflected

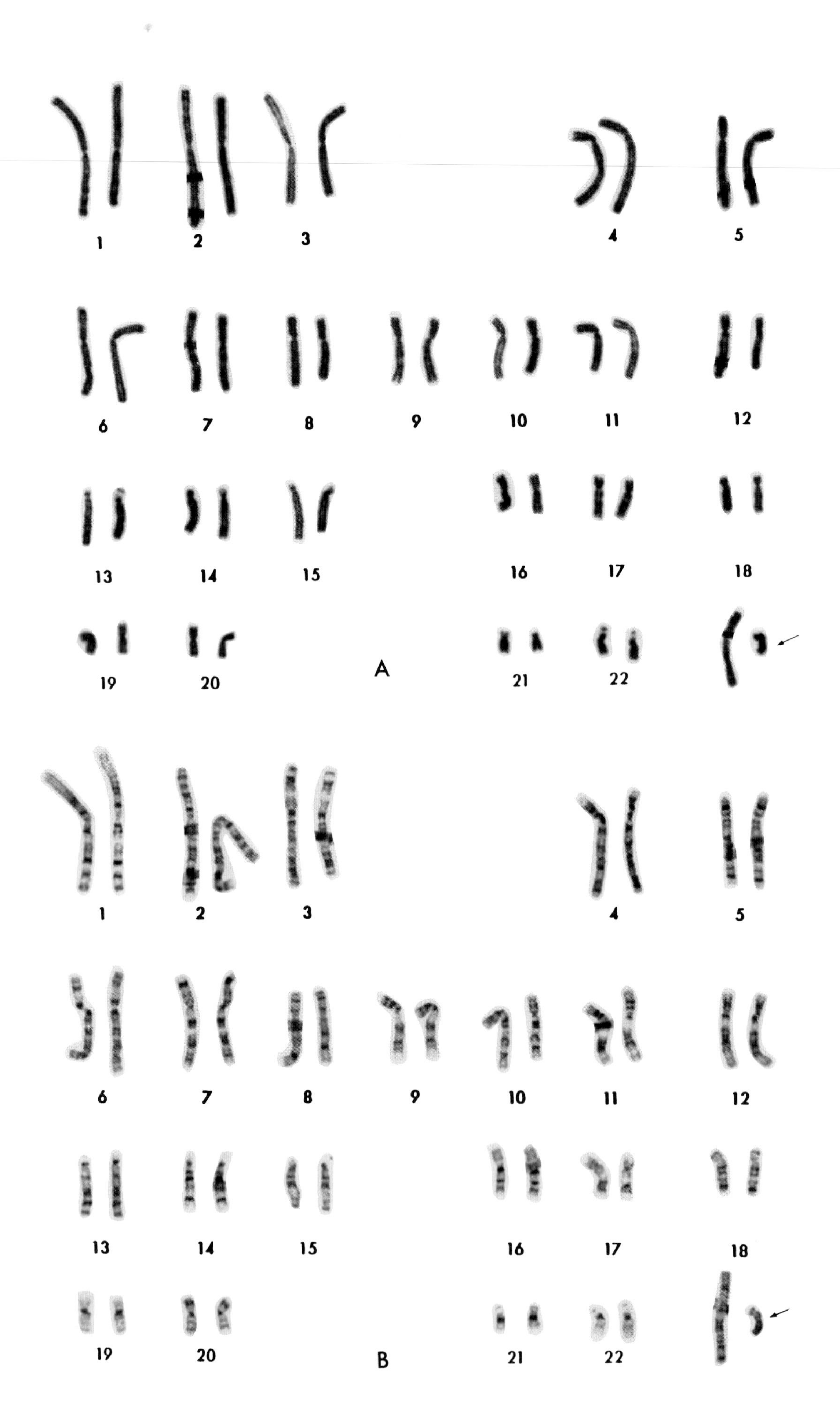
1
2
3
4
5
6
7
8
9
10
11
12
13
14
15
16
17
18
19
20
A
21
22
1
2
3
4
5
6
7
8
9
10
11
12
13
14
15
16
17
18
19
20
B
21
22

in chromosome-specific patterns of alternating densely and lightly stained bands.

Chromosome banding represents a major advance in the identification of individual chromosomes, because the banding pattern of a given chromosome is essentially a "fingerprint" of that chromosome, which will be identical for any number of individuals, provided that the chromosome is not aberrant. In fact, one significant application of chromosome identification by banding is to reveal chromosomal abnormalities, such as reciprocal translocations and small deletions, which could not be identified by size–arm ratio comparisons. Fig. 1A represents a typical human karyotype stained for DNA, and an equivalent karyotype after a banding protocol is shown in Fig. 1B. It is clear that chromosomes that do not look visibly distinguishable in Fig. 1A can be identified by their characteristic and unique banding patterns.

II. Gene Mapping

It is possible to combine chromosome identification, as described above, with the use of cloned DNA probes to map the position of a probe along the length of a chromosome and, therefore, to correlate specific sequences with specific bands. Such cytological sequence mapping is carried out by *in situ* hybridization. This technique is based on the fact that an appropriately labeled DNA or RNA probe can form a specific hybrid with complementary sequences in a chromosome, provided the chromosomal DNA is denatured. The reaction is formally analogous to hybridization of a probe in solution to purified nucleic acid immobilized on a solid support. Radioactively labeled probes can be mapped relative to chromosome positions after hybridization, emulsion autoradiography, and banding. [*See* GENETIC MAPS.]

Although *in situ* hybridization with radioactive probes has been invaluable for much early gene mapping, it is slow, because the shorter the target, the longer the exposure time required for a detectable reliable signal. In addition, sites of hybridization using low-abundance probes may be detected in considerably less than half of the relevant chromosomes, making tedious statistical analysis essential.

A major breakthrough was made with the synthesis of nucleotides covalently coupled to biotin. These modified nucleotides are substituted for their radioactive counterparts in standard labeling protocols, and the probes generated are detected rapidly in one of several ways. Detection is possible by bright-field microscopy using either avidin or antibodies against biotin coupled to an enzyme (e.g., horseradish peroxidase or alkaline phosphatase) that generates a colored precipitate as reaction product. Alternatively, if the same ligands are coupled to a fluorescent indicator, hybridized probes can be localized by fluorescence microscopy. Fig. 2A is an example of fluorescence detection of a human single-copy DNA probe cloned from chromosome 11. Finally, it is possible to substitute colloidal gold particles for either an enzyme or a fluorochrome to map sequences at high resolution in the electron microscope, as shown in Fig. 2B.

Subsequent to the introduction of biotin-substituted nucleotides, a number of other nonradioactive labeling schemes have been developed which can be used in conjunction with biotin to simultaneously localize several probes. These innovations should be particularly useful in ordering probes as they are generated in conjunction with human genome mapping, provided their locations are separated sufficiently.

It is possible to identify the chromosome to which a genetic marker maps even without a molecular probe by exploiting somatic cell genetics. The fusion of two somatic cells from different species generates interspecific hybrid cells which can be grown in culture. During this growth, chromosomes of one of the parents are preferentially lost. After losing most of the chromosomes of one species, hybrids retaining a single chromosome of that species can be selected, if the retained segment contains a selectable gene.

The chromosome containing the selectable gene can be identified even though each hybrid may retain more than one chromosome, because that would be the only chromosome in common among several independent isolates. The region of interest can be further delineated by combining X-rays to cause chromosome breaks in conjunction with con-

FIGURE 1 Representative male human karyotypes (A) shows a Giemsa-stained karyotype in which chromosomes of similar dimensions are indistinguishable. (B) is a karyotype treated to reveal GTG bands (G-bands by trypsin using Giemsa stain). Each chromosome displays a unique and characteristic series of light and dark-staining bands. Arrows denote Y chromosomes. [Photos courtesy of Lauren Jenkins (University of California, San Francisco).]

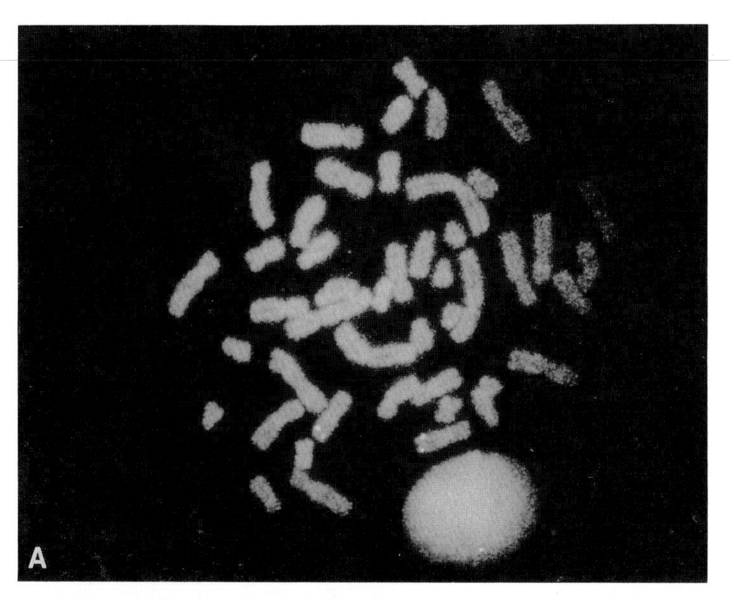

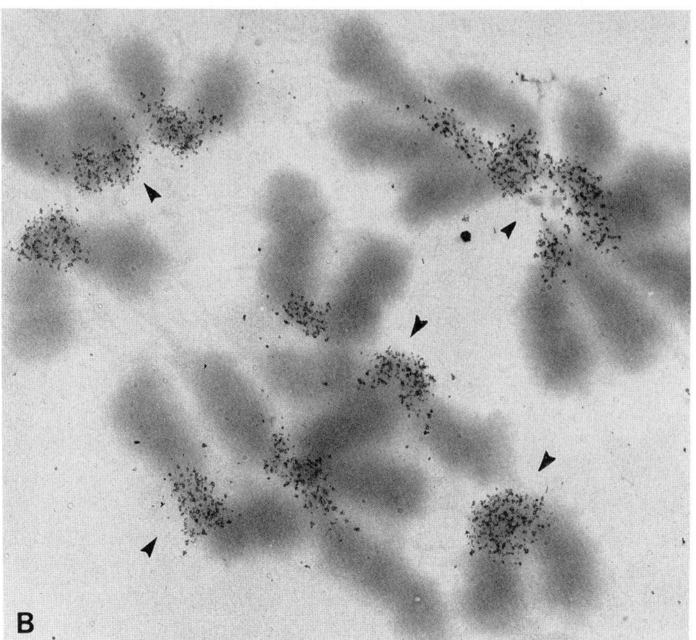

tinued selection, thus obtaining hybrids with small chromosome fragments containing the selected genes. This approach has been successful in many cases, but the production of chromosomal rearrangements between and within chromosomes resulting from X-rays can cause ambiguities.

Another somatic cell genetic approach for chromosome mapping involves treatment of cells to induce the formation of micronuclei, each containing one or a small number of chromosomes. Fusion of small pieces of membrane-bound cytoplasm, each containing a micronucleus (i.e., cytoplasts), with

FIGURE 2 (A). A complete human metaphase chromosome spread after chromosomal *in situ* suppression (CISS) hybridization with a biotin-labeled cosmid DNA probe. The probe is detected via fluorescein labeled avidin (yellowish-green fluorescence) whereas the chromosomes are counterstained with propidium iodide (red fluorescence). The cosmid DNA maps to the long arm of chromosome 11 as seen by the highly specific signals on both chromatids of both chromosome 11 homologs. [Photo courtesy of Peter Lichter, Thomas Cremer, Laura Manuelidis, and David Ward (Yale University) in collaboration with Katherine Call, David Housman (MIT), and Glen Evans (Salk Institute).] (B) Electron micrograph of a group of mouse metaphase chromosomes after *in situ* hybridization with a biotin-labeled probe for centromere-specific DNA sequences and de-

cells from a different species, results in the rapid generation of hybrids containing a small number of chromosomes for use in mapping, as described above.

Once a gene is regionally mapped on a chromosome, the obvious next step is to isolate it in cloned form. Many approaches have been used for this purpose, but a particularly useful one is the generation of chromosome-specific recombinant DNA libraries. The most powerful method is to produce a library made from isolated chromosomes. Single-chromosome isolation is now possible by taking advantage of fluorescent dyes that recognize subtle differences in DNA content and base composition to separate the chromosomes by fluorescence-activated cell sorting. [*See* CHROMOSOME-SPECIFIC HUMAN GENE LIBRARIES.]

A more direct, but demanding, approach to cloning a region of a chromosome involves microdissection of the region of interest. This technique has been used with considerable success for cloning segments of polytene chromosomes from *Drosophila* (fly) salivary glands, which are amplified as a result of endoreduplication without chromatid separation. Although its application to diploid chromosomes is only beginning, it shows great promise.

III. Molecular Structure of Chromosomes

The overall molecular composition of chromosomes is rather simple: DNA represents about one-half to one-third of the total mass, the remainder being protein. The amount of DNA per equivalent somatic cell is typically invariant, but the total amount of protein varies, depending in large part on whether the cells are proliferating. This apparent simplicity masks a tremendous complexity when each molecular component is analyzed in detail. The power of molecular biology has resulted in great progress in defining this complexity and in analyzing certain subcomponents in exquisite detail. Nevertheless, many questions remain with respect to the relationships between particular molecular species and their respective functions.

tection via antibiotin antibodies and secondary antibody-coated colloidal gold particles. The centromere regions are labeled with a large number of gold particles (arrowheads). [Photo courtesy of Sandya Narayanswami and Barbara Hamkalo (University of California, Irvine).]

A. DNA

Each chromosome contains one long, linear, double-helical DNA molecule; the longest, if extended, would measure about 7.3 cm. Despite the apparent monotony of the molecule, composed of only four bases, it is possible to molecularly characterize three distinct classes of DNA sequences which coexist in a typical chromosome of higher eukaryotes. They are referred to as unique (i.e., single-copy), moderately repetitive, and highly repetitive sequences. The proportions of the genome represented by each class vary widely among different organisms, and, to a certain extent, in the same species. In general, in the haploid genome, a unique sequence exists in one or a few copies, moderately repetitive sequences occur in from a few hundred to a few thousand copies, and highly repetitive sequences appear in tens of thousands to millions of copies.

Additional complexity is apparent within a class, because each is composed of a collection of distinct DNA sequences which are considered together simply for the similarity of their numbers of copies. This heterogeneity shows that these three molecular classes do not correspond to three functional components of the chromosome.

Despite the fact that it is not yet possible to define functions (if they exist) for all types of sequences molecularly identified in the genome, some generalities have emerged. For example, the unique sequence class includes the protein coding sequences, which are distributed throughout most of the length of a given chromosome, interspersed with other DNA classes. This class includes both functional genes and so-called pseudogenes, which are copies of functional genes rendered nonfunctional by mutations. An issue that has yet to be resolved is the distribution of coding sequences relative to the physical map of a chromosome. That is, are genes relatively uniformly spaced on average, or is the arrangement dictated by unidentified rules? The nature and the possible function of the bulk of single-copy DNA also remain enigmas. It certainly contains many unknown genes, but it may also include other types of sequences with different functions.

The middle repetitive sequence class contains both coding and noncoding components. Genes that code for ribosomal RNAs are middle repetitive, as are those coding for histone proteins or immunoglobulin variable regions. To date, however, the majority of moderately repetitive sequences have not been assigned functions. These sequences typically are interspersed in the genome, having varying average periodicities and repeat units of different lengths. Functions suggested for interspersed repeats include *cis*-acting regulatory sequences involved in the coordinate regulation of unlinked genes, and/or origins of replication. In fact, many *cis*-acting regulatory sequences have now been identified and, based on an ever-expanding data base, it is clear that many genes do share common *cis*-acting regulatory elements which are, by definition, members of the interspersed repeat class.

The highly repetitive component of the higher eukaryotic genome consists of numerous distinct members of varying lengths and sequences, usually in long tandem arrays interspersed in the genome. The most extensively studied of such sequences tend to be localized at chromosomal centromeres and telomeres. In different organisms, these sequences vary enormously in length and in their number of copies. Since these sequences are not transcribed, it has proved difficult to determine their function, beyond inferences based on their chromosomal positions. For example, although centromeric tandem repeats have been characterized in large numbers of organisms, there is no evidence to support the contention that these sequences actually function to effect chromosome segregation directly; they may play a role in organizing the centromeric chromatin so that it can assemble structures necessary for spindle attachment. [*See* TELOMERES, HUMAN.]

Many other functions have been proposed for these sequences, including a role in speciation or in chromosome pairing. The latter possibility is attractive because chromosome-specific versions of at least one centromeric repeat family have been identified for several human chromosomes. The extreme alternate view that these sequences have no function has also been put forward. The sequences located at the ends of chromosomes undoubtedly function to stabilize those sites, as discussed below.

Two human interspersed repeats which appear to be indicators of chromosome organization based on their distribution are the Alu and L1 repeats. The former are short repeats (approximately 300 bp in length) which are relatively (G+C) rich; the others are several kilobases in length and relatively (A+T) rich. Recent *in situ* hybridization data show that both Alu and L1 are nonuniformly distributed along chromosomes with Alu sequences concentrated in regions defined as R bands and L1 enriched in G bands. The bands result from treatments such as those mentioned earlier. This differential distribution in relation to independent chromosomal markers may tell us something about the organization of chromosomes.

Finally, a distinct class of human highly repetitive sequences, with a repeat length between 11 and 60 bp arranged in tandem arrays, has been identified. These repeats exhibit exceedingly high variability in the number of repeats in an array from individual to individual. Their high degree of polymorphism is exploited in human molecular genetics, to map disease loci, and in forensic science, to define a genetic fingerprint unique to each person.

B. Chromosomal Proteins and Chromosome Organization

1. Basic Chromatin Subunits

The proteins associated with chromosomal DNA are divided into two major groups: the histones and the nonhistone chromosomal protein (NHCPs). The former class is defined biochemically as a set of five low-molecular-weight highly evolutionarily conserved basic proteins, which exist in amounts strictly proportional to DNA content. The primary role of the histones is structural, although there is some evidence that they may also affect gene expression. As implied by its name, the second group is composed of all other proteins associated with the DNA and is, therefore, a large, heterogeneous, and relatively poorly defined collection of diverse proteins with diverse function. Among the NHCPs are enzymes involved in DNA replication, RNA synthesis, protein modification, and hydrolysis; factors required for the positive and negative regulation of gene expression; structural proteins other than the histones; and proteins that function in the maintenance of stable chromosomes (e.g., telomere-binding proteins) and in their faithful segregation (e.g., kinetochore components). With the exception of the proteins with enzymatic activity, only a small proportion of total NHCPs have been isolated and analyzed in detail. In the past few years, however, procedures for isolating DNA-binding proteins have advanced dramatically. These techniques, coupled with biochemical and molecular analyses, allow rapid progress in determining the structure and function of many NHCPs involved in gene expression and chromosome organization.

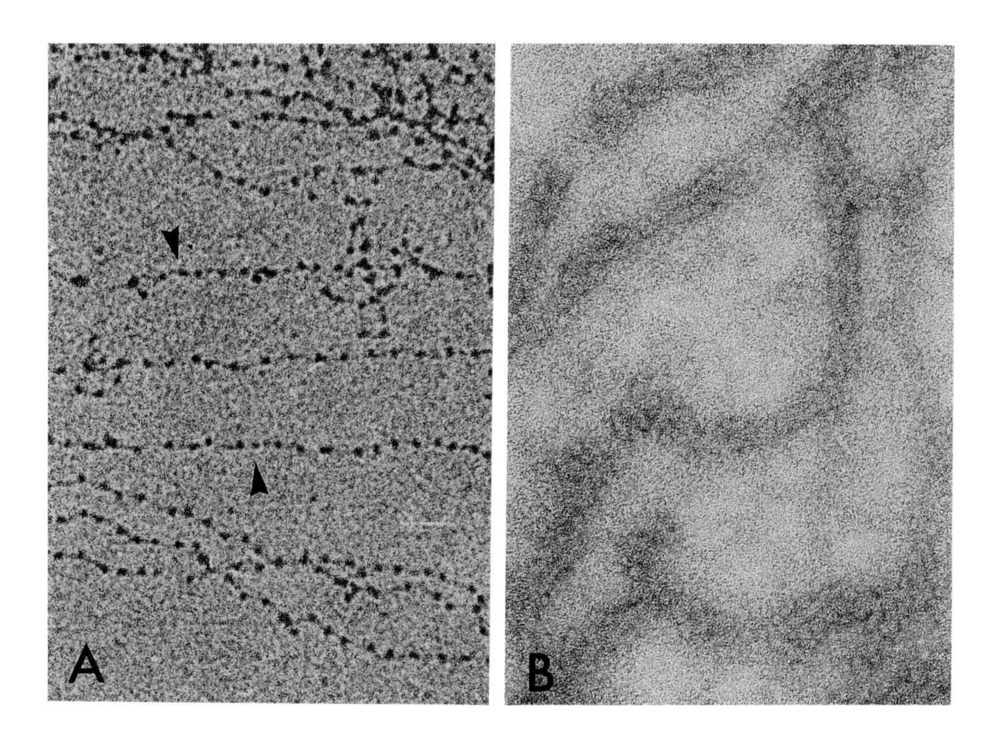

FIGURE 3 Electron micrographs of chromatin fibers released from lysed cultured mammalian cells. A , the fundamental "beads-on-a-string" (arrowheads) fiber; B, a specimen prepared under conditions which stabilize 20–30 nm higher order fibers. [Photos courtesy of J. B. Rattner (University of Calgary) and Barbara Hamkalo (University of California, Irvine).]

Histones are the fundamental structural proteins of a chromosome. The core histones (i.e., H2A, H2B, H3, and H4) are found universally in equal stoichiometry, from lower eukaryotes, such as yeast, to mammals. These proteins are remarkably conserved throughout evolution, particularly with respect to their structure in solution and, to a somewhat lesser extent, their primary amino acid sequences. They are rich in basic amino acids, such as lysine and arginine, and can interact with one another in solution with precise stoichiometry to form an octamer containing two molecules of each core histone. The octamer is a compact structure which, in turn, interacts with 146 bp of DNA to form the fundamental structural subunit of all eukaryotic chromosomes, referred to as the nucleosome core particle.

The DNA is wrapped around the outside of the octamer to form a nucleoprotein complex whose structure, based on X-ray diffraction of crystals, approximates a flattened disk. Core particles are arranged along the DNA to give a fiber with the appearance of beads on a string; adjacent beads are connected by thin fibers referred to as linkers. Fig. 3A illustrates this configuration as seen in the electron microscope. Adjacent core particles are closely packed *in vivo* and the fifth histone, H1, has been implicated in this and the next higher order of packing.

Linker regions are relatively susceptible to digestion by certain DNases providing a convenient way to isolate the chromatin subunits. Analysis of digestion products with time reveals a relatively DNase-resistant nucleoprotein structure defined as a monomer nucleosome. Although core particles of virtually identical composition represent the universal chromatin subunit, monomer nucleosomes contain from as little as 165 to over 200 bp of DNA. Differences of DNA length in monomers have been reported between organisms and in different cells of the same organism. The variability in DNA lengths observed in nucleosome preparations is attributed to differences in the linker lengths. Since the fifth histone, H1, is associated with linker DNA, it may have a role in determining the repeat length. Although the significance of variable-length linkers is not yet clear, this variability could be involved in determining how the fundamental nucleosome-containing fiber is organized at the next level of packing. The formation of nucleosomes effects an approximately sevenfold compaction of the DNA relative to its linear length, results in DNA supercoiling, and is the first step in achieving the estimated 10,000-fold packing required to create a visible mitotic chromosome.

Despite the apparent lack of complexity of the histones (i.e., only five proteins), they are, in fact, quite complex, because there are multiple primary amino acid variants of each, and histones can undergo a large number of postsynthetic modifications (e.g., phosphorylation, methylation, acetylation, and ADP ribosylation). If one considers these variations, it becomes clear how complex and mosaic the basic chromatin fiber can be. Although much progress has been made in defining the fundamental chromatin structure, analysis of chromatin fiber mosaicism and its possible relationship to function is a problem for the future.

2. Higher-Order Structure

Investigations of the higher-order organization of chromosomes have not yet generated a unitary model for the way the fundamental fiber is folded. A number of distinct models have been proposed, based on a variety of types of data, but conflicts among the models have yet to be resolved. One major stumbling block in this area is the inability to isolate higher-order structures under conditions known not to perturb the subtle features of folding. One generalization that is agreed on is that the dimensions of the folded basic fiber is between 20 and 30 nm in width, regardless of the mode of the folding. Fig. 3B shows an example of a folded fiber in which individual nucleosomes can be resolved. It has been estimated that formation of the 20- to 30-nm fiber results in a net condensation of the DNA of about 40-fold, which is not sufficient to create a mitotic chromosome. Thus, several additional levels of folding must occur and, again, there is no general agreement as to exactly how they are effected, although coils and/or loops are frequently proposed as intermediate structures. Some data argue that chromatin folding at these higher levels is mediated by a small number of nonhistone chromosomal proteins, including the enzyme topoisomerase II, which catalyzes reactions that change the superhelicity of DNA.

The organization of chromosomes in interphase nuclei is another unresolved question. The results of molecular and biochemical approaches suggest that the genome is organized or partitioned into domains which can be defined operationally, because of their resistance to attack by specific DNases, presumably as a result of specific protein associations. In some cases, the nuclease-resistant sequences identified in interphase nuclei and in mitotic chromosomes are the same, implying a commonality in the basic organization of the chromosome, regardless of its state of condensation. The issue of nuclear organization is being studied intensely, so that general rules of organization, if they exist, should emerge in the next few years. Equally important questions related to higher-order chromosome structure, presently under study, are the structural basis for chromosome banding and whether there are functional correlates to this organization.

IV. Essential Functional Elements of a Chromosome

The integrity and stability of eukaryotic chromosomes are dependent on three noncoding genetic elements: sites for the initiation of DNA replication ("origins"), sequences that mediate chromosomal attachment to spindle microtubules (centromeres), and sequences that delineate the molecular ends of chromosomes (telomeres). Progress in identifying these elements has been dramatic in lower eukaryotes, such as yeast. The problem of developing assays for the functions of untranscribed sequences in higher eukaryotes has been a major stumbling block in this regard. It is compounded by the large size of higher eukaryotic genomes, since the elegant experimental approaches that were so successful in lower systems are not readily adapted to larger, more complex, genomes.

A variety of different experimental approaches support the statement that each eukaryotic chromosome possesses a large number of sites at which DNA replication can be initiated. These sites may not be uniformly distributed along the DNA, and the precision with which the bacterial replication machinery recognizes a well-defined origin sequence may not be a feature of eukaryotic initiation. An interesting aspect of eukaryotic origin structure and function is highlighted by the organism's need to regulate DNA replication during development. This point is best illustrated in *Drosophilia,* in which, at an early stage in development, cells replicate their DNA in a very short time compared to the time it takes in somatic cells. In these embryonic nuclei, active replication origins are very close together, whereas, in somatic cells,

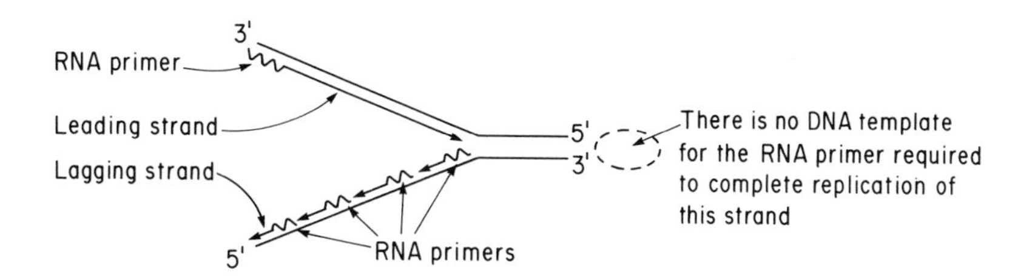

FIGURE 4 Schematic representation of the dilemma faced when replicating the very end of a linear DNA molecule.

many potential origins are not utilized, resulting in distantly spaced replication sites. Cloning and characterization of origin sequences are being attempted by many groups, and successful progress will help define what an origin is, assess the diversity of origin sequences, and define developmental versus somatic origins, if they are distinguishable.

The functional centromere of a higher eukaryotic chromosome has not yet been defined in molecular terms. A centromere is cytologically defined as the primary constriction, a heterochromatic region where sister chromatids remain in close apposition until anaphase, when they begin to move to opposite poles of the mitotic spindle. An essential component of a functional centromere is the kinetochore, a proteinaceous structure that mediates attachment of a sister chromatid to spindle microtubules and orients the direction of chromatid movement. A longstanding hypothesis for the way kinetochores participate in faithful chromosome segregation suggests that they interact specifically, in a currently ill-defined way, with a subset of the DNA sequences found in centromere heterochromatin. Molecular characterization of DNA sequences from yeast, which confer segregation to linked DNA and which, in turn, are associated with distinct chromosomal proteins, supports the general features of this model.

Proof of this model for higher eukaryotic centromeres requires the identification of DNA sequences essential for segregation and a detailed biochemical characterization of the kinetochore, neither of which has been accomplished. The presence of large amounts of tandemly repeated DNA in the centromeric heterochromatin of most eukaryotes led some investigators to propose that these sequences confer the segregation function, but, at the present time, there is no evidence to support this contention. The sera of patients with certain autoimmune diseases contain antibodies that react with various chromosomal regions, including centromeres. Several groups have exploited these antibodies to attempt to define kinetochore polypeptides, but none of the molecules identified has been unambiguously defined as an integral kinetochore component. Thus, the molecular identification of origins of replication and centromeres of higher eukaryotic chromosomes represents fertile areas for future research.

Better understood are the nature and the role of telomeres, or chromosome ends. There are two reasons to suppose that telomeres have unique structural properties. First, it has been known for many years that a broken chromosome end behaves differently than a telomere: Telomeres are inert, whereas broken ends are usually "sticky," resulting in the formation of rearranged, often dicentric, unstable chromosomes. Second, if one considers the replication of the end of a linear DNA molecule, it is obvious that DNA polymerase could not replicate it to the very end, for lack of a primer on the discontinuously synthesized strand (Fig. 4). Therefore, telomeres must possess special structural features to allow the complete replication of chromosomes; otherwise, they would be shortened at each replication.

Telomere-functional sequences have now been cloned from a large number of organisms, from yeast to humans. A comparison of the telomere sequences from such evolutionarily divergent organisms reveals striking similarities and remarkable evolutionary conservation, which is explained by their crucial function. They are tandemly repeated (G+C)-rich short sequences, with the 5′ to 3′ polarity of the strand rich in guanine residues, always in the same direction relative to the chromosome end. Chromosomes have different numbers of the simple repeat, and, in some organisms, there is evidence for changes in the number of repeats with time.

The telomere repeat unit (TTAGGG) identified in human DNA is identical to that found in trypanosomes, an acellular slime mold, and all other vertebrates analyzed arguing for conservation of an essential function. The identification of human telomeres will be invaluable for the cloning of genes located close to the ends of chromosomes (e.g., Huntington's Disease on chromosome 4).

Bibliography

Bernardi, G., and Bernardi, G. (1986). *Cold Spring Harbor Symp. Quant. Biol.* **51,** 479.

Blackburn, E. H. (1984). *Annu. Rev. Biochem.* **53,** 163.

Hamkalo, B. A. (1985). *Trends Genet.* **1,** 255.

McKusick, V. A. (1986). *Cold Spring Harbor Symp. Quant. Biol.* **57,** 423.

Van Holde, K. E. (1989). "Chromatin." Springer-Verlag, New York.

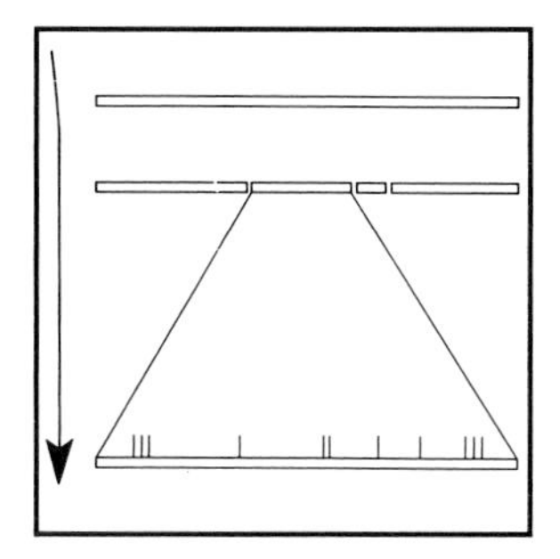

Chromosomes, Molecular Studies

RAFAEL OLIVA, *Lawrence Berkeley Laboratory*

SIMON K. LAWRENCE, *Scripps Clinic, California*

YUE WU, *Lawrence Berkeley Laboratory*

CASSANDRA L. SMITH, *University of California, Berkeley and Lawrence Berkeley Laboratory*

Glossary

Allele One of several alternative DNA sequences of a gene

Autosomal chromosome Chromosome other than a sex chromosome

Consensus sequence DNA sequence containing the most frequently occurring bases as from comparing DNA sequences derived from different sources

C, T, G, A DNA bases: cytosine, thymine, guanine, and adenine respectively

DNA hybridization Annealing of an oligonucleotide or polynucleotide to its complementary sequence to form a stable double-stranded structure

Genetic imprinting Inherited DNA methylation patterns imposed by parental germline cells

Gradient gel electrophores A technique that separates DNA on the basis of composition, thus, allowing single base differences to be detected

Inverse polymerase chain reaction Method that allows the region adjacent to a known sequence to be amplified

Incomplete penetrance Lack of expected phenotype as predicted from a genotype

Karyotype Chromosomal constituents of a genome

Linkage Disequilibrium High-frequency cosegregation of linked genes due to suppressed meiotic recombination between the genes

Multiplexing Any methodological approach where different independent members are treated *en masse* to save labor

Open reading frames DNA sequence that could code for a protein because of the absence of translational stop codon

Phenotype Observable properties of an organism, produced by the genotype in conjunction with the environment

Polymerase chain reaction *In vitro* exponential amplification of DNA molecules by multiple cycles of DNA denaturation, primer annealing, and synthesis

Pulsed field gel electrophoresis Electrophoresis technique that fractionates megabase DNA by exposing them to alternating electrical fields

Restriction enzyme site DNA sequence recognized and usually cleaved by its corresponding restriction enzyme

Somatic cell Cell of the eukaryotic body other than those destined to become sex cells

THE DEVELOPMENT OF PULSED field gel electrophoresis (PFG) and related large DNA techniques expanded 100-fold the size range of DNA molecules readily accessible to analysis and manipulation. The ability to study DNA molecules as large as 10 megabases (Mb-million bases) encouraged the development of a variety of molecular and genetic approaches for analyzing the human genome. This chapter will review the structure and constituents of human chromosomes and the molecular techniques and approaches currently being used to characterize the human genome.

I. Composition of the Human Genome

A. Chromosomes

The human genome consists of 23 pairs of nuclear chromosomes and a circular cytoplasmic chromosome located in mitochondria. The nuclear genome consists of 22 autosomal pairs and 1 pair of sex chromosomes designated X and Y. The autosomal chromosomes are numbered on the basis of size: the largest chromosome is number 1; the smallest chromosome is number 21, rather than 22, because chromosome 22 was initially thought to be the smallest.

Each chromosome contains a single linear double-stranded DNA molecule with a 40% GC content. The low GC content is due to a deficiency in the dinucleotide sequence 5′ CpG 3′. The rarity of this sequence appears to be due to two factors. First, *in vivo*, the C residue in this sequence is frequently converted to 5-methyl cytosine (5-MeC). Second, deamination, the most frequent naturally occurring DNA damage, converts C to U and 5-MeC to T. A DNA repair system removes U from DNA, restoring the original C-G base pair. However, repair of a G-T mismatched base pair does not always lead to restoration of the original base pair. Thus, the human genome like other cytosine methylated genomes, is low in CpG sequences and high in TpG sequences.

The state of methylation of a particular region has been correlated with gene expression, carcinogenic changes, and genetic imprinting. Methylation may influence gene expression by promoting the formation of an unusual DNA structure and/or differential protein binding. Carcinogenic changes and genetic imprinting effects are very likely due to changes in gene expression. Maternal and paternal methylation patterns imposed by genetic imprinting in germ cells are not entirely equivalent and in some cases lead to differential gene expression in offspring. Genetic imprinting may explain the need for progeny to inherit particular chromosomal regions from a particular parent.

Individual chromosomal DNA molecules are estimated to range in size from 50 to 250 Mb. The entire human haploid genome contains approximately 3,000 Mb. This means that a single chromosomal DNA molecule may be as long as 70 mm and that the entire human genome would be nearly 2 m in length if all the chromosomal DNA molecules were laid end-to-end. Compaction of nuclear DNA by proteins allows the entire DNA complement to fit within nuclei that are only 5 μm (5×10^{-6} m) in diameter.

Chromosomes consist of DNA complexed with histone and nonhistone proteins into a structure called chromatin. Histones constitute a group of five low-molecular weight basic proteins each encoded by multiple genes. Histones H2A, H2B, H3, and H4 package DNA into a structure called the core particle. A second level of packaging is provided by the histone H1, which organizes a 10-nm nucleosomal fiber into a 30-nm fiber with a packaging ratio of 40 : 1. Most DNA is complexed with the same five histone proteins, although some histone genes clearly code for unique subtypes. Some tissue-specific subtypes have been identified, but, in general, the role and deposition of histone subtypes is not clearly understood.

The remaining protein components of chromatin, globally called nonhistone proteins, are heterogeneous. Some are regulatory proteins that control gene expression; others are involved in macromolecular processes such as DNA replication and transcription. Some nonhistone proteins probably contribute to the nuclear scaffold (or nuclear matrix) responsible for higher order organization of the interphase nuclear chromosomes.

During cell division, chromosomes condense further into highly ordered structures that may be observed by light microscopy (Fig. 1). A constricted portion, called the centromere, serves to correctly partition replicated chromosomes during cell division. The centromere divides the chromosome into small and large arms designated p (for petite) and q (q follows p in the alphabet), respectively. Giemsa

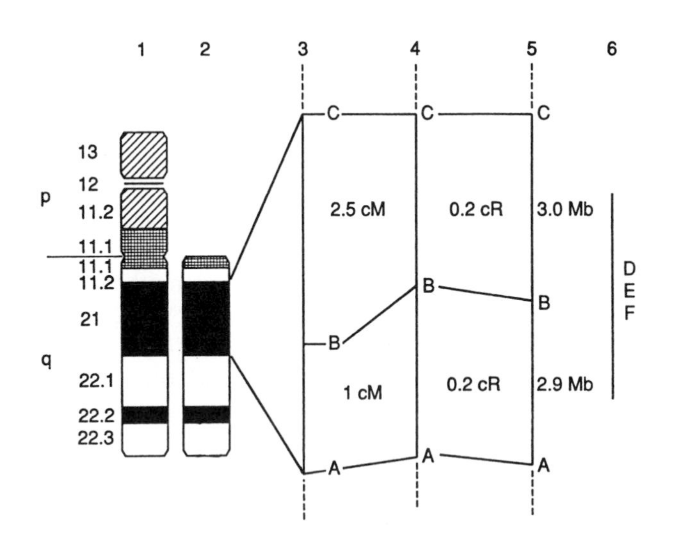

FIGURE 1 Types of chromosomal landmarks, which need to be integrated into a genomic data base. Lane 1: Distinct cytogenetic bands of chromosome 21 observed by light microscopy (note the distinct banding pattern of the rDNA repeat 21p12). Lane 2: Fragmented chromosome 21. Lane 3: Genetic linkage map. Lane 4: X-irradiated chromosome map. Lane 5: Physical maps (including restriction maps, STARs maps, clones from overlapping libraries, and DNA sequence). Lane 6: Localization of a marker to a chromosomal region.

and other staining techniques are used to subdivide the arms into an increasingly finer characteristic set of bands. These bands are numbered on each arm from the centromere to the telomere. For example, the major histocompatibility complex (MHC) is located at 6p21.2 (i.e., the short arm of chromosome 6, band p21.2). Although bands are useful for identifying and subdividing chromosomes and a number of molecular characteristics have been attributed to them, the precise physical basis for banding is not well understood. Giemsa light bands are believed to contain higher amounts of GC base pairs, of genes, of earlier replicating genes, and of *Alu* repetitive elements (see Section I, B).

Chromosomes contain special structures at their ends called telomeres. These structures are formed from a simple repeat element 5′ TTAGGG 3′ (Table I). Multiple copies of this repeat are added to the 5′ ends of DNA in the absence of DNA template by an enzyme called telomerase. The added repeat serves as a template for synthesis of the second strand by DNA polymerase. The addition of this repeat and associated histone and nonhistone proteins function to preserve the ends of the chromosome during DNA replication and to prevent their degradation within the nucelus. Telomeres may also help to position chromosomes in resting nuclei.

B. Repetitive DNA

In prokaryotes, genome size accurately reflects gene number. This is not the case in most eukaryotes where repeated DNA elements and other noncoding sequences account for substantial portions of the genome. Thus, even though the human genome is large enough to contain a million genes encoding proteins that average 300 amino acids in length, the actual number of genes is probably between 50,000 and 100,000.

A number of human repetitive DNA sequences have already been identified and characterized to one extent or another. However, at this time, there is no coherent picture of the organization and function of DNA repeats. In general, human repetitive sequences may be divided into two types, those that are tandemly repeated and those that are interspersed (Table I). Tandemly repeated simple sequences, called satellite DNA, are located near centromeres. In humans there are four families of satellite DNA. Each type consists of thousands of tandem and inverted repeats of short (5–10 bp) sequences. As much as 5% of the human genome may consist of satellite sequences. Although not yet demonstrated, these sequences probably play a role in determining chromosome structure and in partitioning chromosomes during mitosis and meiosis. Tandemly repeated centromeric alphoid sequences may be divided into groups, some of which are specific for sets of chromosomes.

Hypervariable minisatellites sequences (also called variable number of tandem repeats [VNTRs]) are dispersed highly polymorphic sequences composed of tandemly repeated sequences about 10–15 bp in length, VNTRs are useful for DNA fingerprinting and for genetic mapping experiments (see below). The polymorphisms within these, and other tandem repeated sequences, are due to insertions and deletions that arise from recombination promoted by the close proximity of multiple copies of the same sequence.

TABLE I Repetitive DNA Elements in the Human Genome

DNA element	% of genome	Number in the genome	Repeating unit	Type of element	
Alu	5.0	5×10^5	300[a]	SINE	Interspersed
Kpn	3.3	2×10^4	5,000[b]	LINE	
(CA)n	0.02	5×10^4	~50[b]		
stDNA I	0.5	3×10^6		Satellite I	Tandemly repeated
stDNA II	2.0	1×10^7		Satellite II	
stDNA III	1.5	9×10^6	~5[c]	Satellite III	
stDNA IV	2.0			Satellite IV	
Alphoid	0.75	6×10^{-4}	340[d]	centromeric	
(TTAGGG)n	0.01	7×10^4	6[e]	telomeric	

[a] Occurs once every 5 kb.
[b] Occurs once every 50 kb.
[c] A single block of repeat is usually 2 kb in size.
[d] A single block may range from 0.5 to 5.0 Mb.
[e] A single block may be from 5 to 15 kb.

The most common interspersed repeat is called *Alu* because it contains a recognition site for the restriction enzyme *Alu* I. Different *Alu* consensus sequences divide the *Alu* repeats into different families. The 300-bp *Alu* repeat is a member of a group of repeats called short interspersed nucleotide repeats (SINEs). *Alu* sequence elements account for as much as 5% of the human genome. Long interspersed nucleotide repeats (LINEs) are characterized by the 5,000-bp *Kpn* element, which occurs at one-tenth the frequency of *Alu* elements.

Repeat sequences like *Alu*, are dispersed through the genome by retrotransposition. Retrotransposition occurs through the formation of an RNA intermediate as a rare transcription event at particular integration sites. Here, the enzyme reverse transcriptase, forms a complementary DNA, which is then inserted at another location in the genome. Transposition into a gene would cause its inactivation and would be very likely of negative adaptive value. Nonetheless, *Alu* sequences have been tolerated during human evolution, suggesting that their presence in the human genome has not had a negative adaptive value. Some have called retrotransposing sequences parasitic DNA.

Some medium repetitive DNA sequences have unexpected characteristics. For instance, one repeat was found to be inserted in chromosomes and also as an extrachromosomal circular form. The latter form is probably generated by recombination events between adjacent chromosomal elements. Recently, rearrangement of one repeat was associated with differentiation.

Because little or no coding information appears to be contained within repetitive DNA, some researchers have called repetitive DNA sequences "junk" DNA. Repetitive DNA, as well as other noncoding "junk" DNA (see below) has the potential to reveal totally unexpected aspects of the human genome because few (if any) interesting roles have been attributed to them.

C. Genes and Gene Families

The remainder of the genome consists of single and low copy DNA sequences. The development of gene cloning and sequencing techniques has provided detailed pictures of the structure of eukaryotic genes. Yet genes are still difficult to recognize from DNA sequence data alone. The presence of a gene may be inferred from the identification of sequences shared with previously identified control elements (e.g., promoters, enhancers, polyadenylation sites, and splice sites) and long coding elements (open reading frames [ORF]). In addition, putative gene functions sometimes can be predicted from the identification of particular protein motifs with a functional correlate (e.g., adenosine triphosphate binding motif, protease motif, immunoglobulinlike domains, the zinc-finger motif characteristic of DNA-binding proteins).

Eukaryotes have three types of genes defined by the type of RNA polymerase that transcribes them. Each type has distinct structural and regulatory characteristics. RNA polymerase I transcribes the genes for 5.8S, 18S, and 28S ribosomal RNAs (rRNA) which are required for protein synthesis. Genes transcribed by RNA polymerase II include all known protein coding genes as well as some small nuclear RNA genes (snRNA), which are involved in RNA processing (see below). RNA polymerase III genes synthesize 5S ribosomal RNAs as well as the transfer RNAs (tRNA), which are required for translation of RNA polymerase transcripts into proteins.

The cell contains about 100,000 copies of the RNA polymerase II enzyme. This large amount of RNA polymerase II is needed to support generalized transcription. Large amounts of translation components are also required to support a reasonable protein synthetic rate. In fact, the ribosomal RNA and ribosomal protein genes are the most highly expressed genes and the major cytoplasmic components.

The 5.8S, 18S, and 28S rRNA genes are contained within a 45-kb monomer sequence that is tandemly repeated an unknown number of times on chromosomes 13, 14, 15, 21, and 22. Intensive transcription of the 45-kb monomer units containing 5.8S, 18S, and 28S rDNA at specific nuclear locations, called nucleoli, permits their visualization by light microscopy. There are an estimated 100–200 5S rRNA genes. Most 5S rRNA genes are located in the chromosome 1q telomere region; although, other copies might be dispersed over the genome. Genes that code for ribosomal proteins and tRNA are found at multiple locations in the genome. Thus, the genes encoding the various components involved in protein synthesis are not only located on different chromosomes but are also transcribed by different RNA polymerases. It is not clear how synthesis of the various ribosomal components is coordinately regulated.

Genes consist of promoter sequences, regulatory

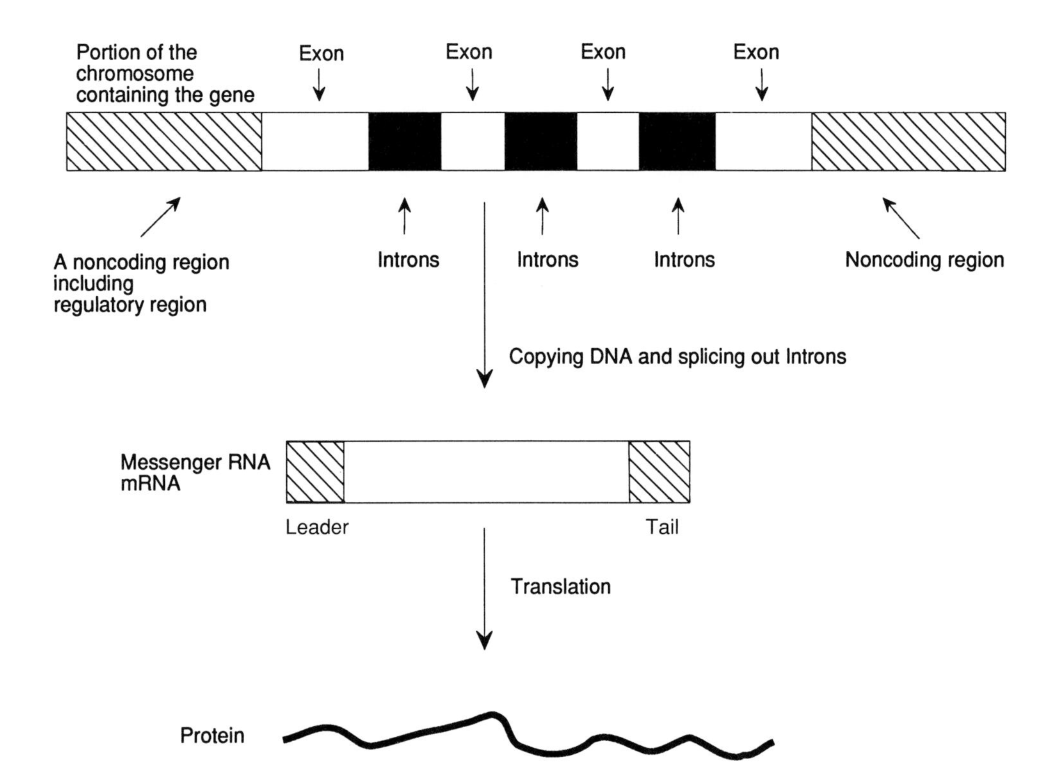

FIGURE 2 Eukaryotic genes are composed of coding and noncoding sequences (introns). Introns are removed from heterogeneous nuclear RNA before translation.

sequences, expressed or coding sequences (exons), and intervening sequences (introns). Promoters are defined by sequences near transcriptional start sites which serve as RNA polymerase recognition sites. Generally, promoters are located at the 5′ end of genes transcribed by RNA polymerase I and II and within genes transcribed by RNA polymerase III. Cis-acting regulatory sequences are bound by trans-acting factors which regulate the amount, the timing, and the tissue specificy of gene expression.

Regulatory sequences are separated from each other and from promoter sequences: they may be located upstream, downstream, or within a gene. Enhancer sequences promote nearby gene expression, whereas silencer elements decrease gene expression. Enhancer and silencer sequences do not appear to be gene-specific and may be located at large distances, 5′ or 3′, from a gene.

DNase I hypersensitivity sites near genes are thought to mark open chromatin configurations at regulatory elements, necessary for RNA polymerase accessibility. For example, a 20-kb region of DNA located more than 40-kb upstream from the human β-globin gene cluster contains multiple DNase I hypersensitivity sites and is required for both tissue specificity and developmentally regulated expression of the beta-globin gene cluster. Termed a dominant control region it may include multiple enhancers. The region may also contain sequences necessary for the attachment of the globin gene complex to the nuclear matrix.

A eukaryotic gene may consist of as little as 10% exons and 90% introns (Fig. 2). Introns are removed from high molecular weight nuclear RNA, called hnRNA (heterogeneous nuclear RNA), before translation by the splicing machinery. This means that the average gene coding for a protein of average size (about 30,000 mass molecular weight) is considerably larger than the size needed to actually code for the polypeptide product (e.g., 900 bp). Most genes that code for proteins range in size between 10 and 40 kb; however, some genes are very large. The dystrophin gene, whose mutant forms account for Duchenne and Becker Muscular Dystrophy, is over 2 Mb long, codes for a 400,000-molecular weight protein, and has about 100 introns. In contrast, the protamine gene has a single intron and is only 0.5 kb in size.

The function and evolutionary origin of introns is an unresolved issue. Some believe introns are also genomic "junk." However, introns may provide several functional and evolutionary advantages to the cell. For instance, alternative forms of the same protein can be generated by splicing together different exons, e.g. alternative use of an exon encoding a transmembrane domain leads to the formation of either a membrane bound or secreted forms of immunoglobulin molecules. Another possibility is that novel genes may evolve by "exon shuffling," in which an exon from one gene links to an exon of another gene, generating proteins with new and adaptive properties. If the exon shuffling occurs through a gene duplication the original exon functions will be preserved. Finally, in some cases ORF within introns may encode small proteins.

Although DNA–RNA renaturation experiments indicate that most genes coding for proteins are present in single copies, many genes share sequence and structural characteristics. For example, the CD8, Thy-1, T-cell receptor, and histocompatibility genes are all members of the immunoglobulin supergene family. Genes within supergene families are believed to have evolved through duplication of a primordial gene which then diverged in structure and function. Members of supergene families are dispersed throughout the genome often as multigene families. In most cases, close linkage is maintained between multigene family members possibly reflecting a requirement for coordinate regulation, as in the case of the developmentally regulated hemoglobin gene cluster. In the case of the immunoglobulin and T-cell receptor gene families, linkage must be retained to facilitate the somatic rearrangements necessary for the generation of functional gene products. The significance of linkage is less clear in such cases as the HLA gene complex.

Pseudogenes are inactive copies of functional genes that have accumulated one or more mutations (e.g., splice site defects, deletions, premature termination codons) that preclude their expression. Pseudogenes frequently occur in multi-gene families. Like parasitic DNA, processed pseudogenes are dispersed in the genome and most likely result from integration of a DNA copy of a processed RNA transcript containing a poly (A) tail. Pseudogenes provide a reservoir of genetic material for evolution.

D. Mitochondrial Genome

Mitochondria are cytoplasmic organelles that are the major energy producers of the cell. This organelle contains approximately 10 copies of a small, 16.5-kb, circular chromosome. Unlike nuclear chromosomes, the mitochondrial DNA is only inherited from the mother. The complete sequence of the mitochondrial genome has been determined. The mitochondrial genome encodes its own rRNAs and tRNAs, enzymes of the electron transport system, and a number of as yet unidentified gene products predicted from long ORFs. Mitochondrial genes do not contain introns and contain only small tracts of spacer sequences. Surprisingly, the mitochondrial genetic code is slightly different from the universal genetic code. Several neuromuscular diseases have been associated with mutations found in the mitochondrial genome perhaps because of the high energy requirements of neurons and muscle cells, which cannot be met when the underlying energy producing machinery is not functional.

II. Genetic Mapping of Human Chromosomes

Genetic and physical maps of human chromosomes consist of the linear representation of genetic loci. A genetic locus is a location on a chromosome. This location may be defined by a gene, a restriction enzyme site, a phenotype, a band, a rearrangement, a DNA sequence, or a molecular clone. The various types of loci are identified by distinct, but sometimes overlapping or complementary technologies, including linkage analyses, somatic cell techniques, molecular cloning in bacteriophage, *E. coli* or yeast, DNA sequencing, chromosome restriction mapping, chromosome walking, chromosome linking, and chromosome jumping. The new field of genomics strives to integrate information obtained from all of these studies into one coherent picture (see Fig. 1).

A. Linkage Analysis

The classical form of a genetic map is derived from studies measuring the frequency of meiotic recombination, which occurs during the formation of germ cells, between chromosomal loci. Two loci that are close together on a chromosome display linkage by cosegregating during meiosis more frequently than loci that are further apart. In the simplest case, linkage of the two phenotypes, sex and the disease hemophilia, mapped the hemophilia gene to the X chromosome. In the 1930s, the cosegregation frequency of a chromosome X color blindness gene with hemophilia was measured. This allowed an estimate of the distance between the genes and represented the first map of a human chromosome. In practice, phenotypic markers useful for genetic mapping are infrequent in the human genome. Furthermore, many phenotypes are due to multiple genes and some genetic alleles exhibit incomplete penetrance.

About 0.1 to 1% of DNA sequences varies between two individuals. The identification of sequence polymorphisms and the study of their inheritance provides useful genetic markers which are independent of phenotypes. DNA polymorphisms may be identified using restriction enzymes, which cleave DNA at specific DNA sequences and generate DNA fragments of defined sizes. Most restriction fragment-length polymorphisms (RFLPs) result when two individuals differ in a sequence defining a restriction enzyme site (i.e., restriction site polymorphism). Some RFLPs arise when two individuals differ in the size of DNA between two restriction sites e.g., VNTRs (discussed above). Hybridization experiments are used to assess the status of VNTRs and RFLPs in human pedigrees. Polymorphisms are ordered and organized into genetic linkage groups corresponding to each of the chromosomes. Recently, two groups have established RFLP maps encompassing the entire human genome. Currently, approximately 1,500 RFLPs are cloned and mapped in the human genome. Although this number is impressive, it represents, on average, only one RFLP every 2 Mb. This means that genetic mapping can usually only narrow rather than precisely locate a gene.

A recent advance in genetic mapping is the application of the polymerase chain reaction (PCR) to analyzing meiotic recombination in sperm. This approach allows highly accurate mapping because each sperm represents a unique meiotic event and large numbers of samples are analyzed. This approach eliminates the need to obtain DNA samples from large numbers of human pedigrees. This approach only provides recombination frequencies for male meiosis. Recombination frequencies are different in females. Obviously, the possibility of obtaining the equivalent samples from females (e.g., ovum polar bodies) is quite limited. This method cannot be used to map disease phenotypes.

B. Somatic Cell Genetics

A second category of genetic mapping employs somatic cells rather than examine inheritance patterns. In its simplest application, loci can be identified in somatic cells by observing abnormal karyotypes in mitotic chromosomes. A familiar example is the correlation of Down's syndrome with trisomy (three copies instead of two) chromosome 21. In some cases, somewhat finer mapping is possible when a phenotype involves a naturally occurring chromosomal rearrangement. For instance, the rare occurrence of X-linked muscular dystrophy in a female was traced to a translocation between X and the ribosomal repeat DNA on chromosome 21. This information was ultimately instrumental in the cloning of the muscular dystrophy gene by chromosomal walking (discussed below) from the rDNA.

Genetic loci can also be visualized by *in situ* hybridization of labeled DNA clones to metaphase chromosomes. Recently, highly sensitive methods employing fluorescence-labeled DNA probes have enabled not only the detection of 5-kb sequences but also the ordering of sequences less than 500 kb apart. DNA sequences may also be localized by electron microscopy using immuno-gold labeling of hybridized probes. Although single copy sequences have not yet been detected by this technique electron microscopy has considerable potential for revealing new details of chromosome structure.

A number of different somatic cell variants have been generated for use in genetic mapping experiments. Interspecies somatic cell hybrids or heterokaryons generated by Sendai virus or polyethylene glycol mediated fusion contain a complete complement of rodent chromosomes plus one or more human chromosomes. The chromosomes may be distinguished by their banding patterns and/or hybridization experiments with chromosome specific probes. Hybrid cells can be assayed for the presence of a human gene or its expression, thus, mapping a gene to the human chromosome present in the somatic cell hybrid. For example, human glucose-6-phosphate dehydrogenase can only be detected in rodent–human hybrid cells containing a copy of human chromosome X. Rodent cell lines harboring mutations in genes such as those involved in DNA repair and purine metabolism have been useful for selecting and identifying chromosomes containing the homologous human gene. Some hybrid cell lines contain fragments of human chromosomes. A panel of cell lines containing different chromosomal subregions may be used to regionally assign genes. For example, a map of the MHC was constructed by correlating karyotypic losses following gamma irradiation with losses in expression of cell surface proteins encoded by this complex.

An alternative mapping method involves the creation of a panel of clonal cell lines derived from the fusion of the gamma irradiated hybrid cells with unirradiated rodent cells. The panel is tested for the presence of markers of interest. Genetic linkage in this map is established in a manner similar to mapping by recombination by assuming the closer two genes are, the more likely they will stay together during X-ray-irradiation induced chromosomal fragmentation. The map distance is in centirays (cR) and should closely correlate with physical distance since chromosome breaks due to gamma irradiation are random.

III. Physical Mapping

The ultimate physical map of a chromosome is its complete nucleotide sequence. At the moment this goal remains elusive. Meanwhile a variety of approaches are being used to construct chromosomal restriction maps. These approaches are the same that were used to construct complete genomic restriction maps for the 5-Mb genome of the bacterium *Escherichia coli*.

Genomic mapping strategies can be broadly classified into "top down" and "bottom up" approaches (Fig. 3). Top down approaches are directed approaches that produce complete maps but are difficult to automate and do not provide directly cloned DNA sequences. Bottom up approaches, are easy to automate and provide ordered genomic libraries. Bottom up random approaches rarely produce complete maps unless more directed "endgame" strategies are used such as conventional chromosome walking (see Section III, A).

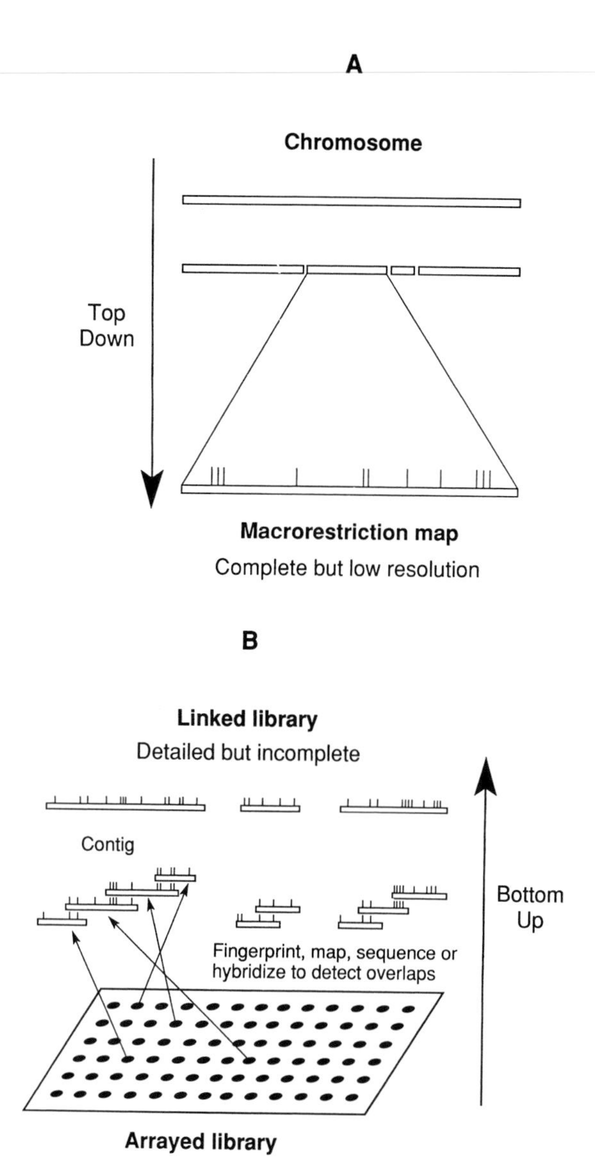

FIGURE 3 Two basic genomic mapping strategies. A. Top down approaches produce complete low-resolution restriction maps. B. Bottom up approaches produce high-resolution restriction maps and overlapping libraries.

A. High-Resolution Physical Maps and Overlapping Libraries

Traditional bottom up approaches build a map by overlapping cloned sequences. A complete bottom up map will consist of an ordered set of overlapping DNA clones representing all the DNA in a chromosome or a chromosome subregion. Libraries of DNA fragments from human chromosomal DNA can be isolated in a variety of simple organisms by recombinant DNA technology. These libraries are distinguished by the size of the genomic fragments, vector and host cell. For instance, libraries constructed in *E. coli* plasmid vectors usually contain short (~4 kb) inserts. Lambda-phage libraries contain inserts ranging from 5 to 15 kb in size, cosmid libraries have inserts ranging up to 40 kb in size, and yeast artificial chromosome (YAC) libraries have inserts of up to 500 kb (Fig. 4). In general, transformation efficiencies decline as insert sizes increase. YAC transformation frequencies are extremely low (i.e., only several hundred clones/μg DNA) but fewer clones are required for complete representation of the genome. Any library is potentially incomplete because of unclonable sequences. For instance, somes sequences may be unstable as recombinant molecules or may specify toxic products.

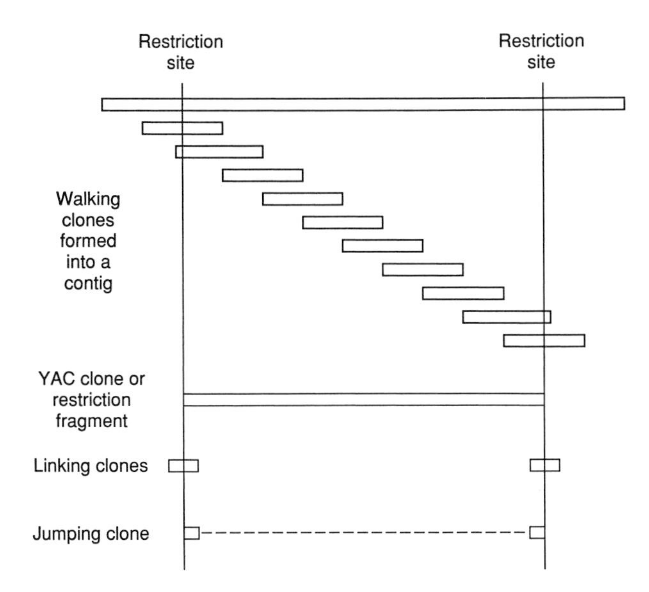

FIGURE 4 Physical characterization and cloning of chromosomes. Originally overlapping libraries formed by contigs of small DNA pieces were the only methods available for expanding large regions of the genome. Recent advances in PFG and YAC technology greatly facilitate the characterization and isolation of large regions of the genome. Large regions may also be spanned by small linking and jumping clones.

Chromosome-specific or region-specific libraries can be constructed using human DNA isolated from flow-sorted chromosomes or from physically dissected chromosome regions. Usually genomic DNA is partially digested with a restriction enzyme or physically sheared at random to the appropriate size for cloning to ensure that the library contains clones that overlap (Fig. 4). Randomly chosen clones from such a library can then be fingerprinted by one of a variety of methods in order to identify overlaps between clones. Fingerprinting methods include the identification of (1) unordered, shared restriction fragments, (2) ordered restriction sites, (3) complete DNA sequencing, (4) sequencing around rare cutting sites and the detection of shared sequences by hybridization with (5) specific oligonucleotides or repeats, or (6) with the clones themselves. The overlapping clones can then be ordered into contiguous overlapping clone sets called contigs (Fig. 3B). The final contig spans the entire region of DNA from which the library was constructed.

Thus far, only three random methods (1, 2, and 6) have been successful in generating large genomic restriction maps. However, map closure with the first two approaches required chromosome walking. Here, probes from the ends of contigs were used to identify clones to specifically span gaps (Fig. 4). The number of chromosome walking (discussed below) steps is minimized if single clones span gaps. Hence, large YAC clones are especially useful.

Chromosome walking experiments (method 6) have been used to produce a complete overlapping library of a small 1-Mb prokaryotic genome. The straightforward application of chromosome walking requires repeated cycles of hybridization and probe isolation; consequently, it is extremely laborious and has sometimes been called chromosome crawling. This approach has been eased by the construction of cloning vectors which allow easy end probe generation. The amount of work involved in chromosome walking may be further reduced by multiplexing. Nonetheless, whether or not this laborious approach alone could ever efficiently yield a complete map of the human genome is not clear.

B. Low-Resolution Restriction Mapping

The first step in a top down mapping strategy is the naturally occurring division of the genome into chromosomes. Chromosomal DNA can be further divided by cutting with a restriction enzyme that generates large DNA fragments (Table II). The dinucleotide CpG is contained with the recognition sequence of these restriction enzymes. This dinucleotide is underrepresented in the human genome and is usually methylated (discussed above) . Since unmethylated CpG-rich islands of DNA have been identified near the 5′ end of many genes, and most enzymes with CpG in their recognition sequence are inhibited by cytosine methylation, most of the restriction enzymes that generate large DNA fragments will preferentially cut at the 5′ end of genes. This means that restriction maps with such enzymes will locate genes (discussed below).

The restriction enzymes *Not* I and *Mlu* I are unique in that they produce the largest fragments from the human genome. Some evidence indicates that the recognition sequences for these enzymes tend to occur in the same CpG islands; thus, they may detect a particular subset of genes. Although both *Not* I and *Sfi* I have 8 bp recognition sequences, *Sfi* I produces much smaller restriction fragments, probably because its recognition sequence lacks a CpG sequence.

Large restriction fragments can be size-fractionated with PFG. After fractionation, the fragments can be ordered and a map constructed by hybridiza-

TABLE II Restriction Enzymes that Produce Megabase Fragments

Enzyme	Size of recognition site (bp)	Recognition sequence	Average fragment size (kb)	
			In a random sequence	Detected by PFG
Not I	8	GC/GGCCGC	64	1,000
Sfi I	8	GGCCNNNN/NGGCC	64	250
Mlu I	6	A/CGCGT	4	1,000
Sat I	6	G/TCGAC	4	500
Nru I	6	TCG/CGA	4	300
Pvu I	6	CGAT/CG	4	200
Xho I	6	C/TCGAG	4	200
Nar I	6	GG/CGCC	4	100
Apa I	6	GGGCC/C	4	100
BssH II	6	G/CGCGC	4	100
Sac II	6	CCGC/GG	4	100

PFG, pulsed field gel electrophoresis.

tion experiments using cloned genomic DNA sequences as probes. These probes will hybridize to an unknown location on a large restriction fragment. Hybridization probes may be of several types. For instance, single-copy, genetically mapped, probes can serve as anchor points between genetic and physical maps. The more precisely a probe has been genetically mapped, the more precisely it will map a fragment. In the best case, a genetically mapped probe will locate a restriction fragment to within 5–20 Mb on the chromosome. Obviously, the sole use of this approach requires many hybridization experiments with closely spaced probes to ensure that fragments, especially small ones, are not missed.

Telomer-specific clones are especially important because they define the ends of the chromosomes and therefore the ends of the physical maps. In addition, telomeric clones used as hybridization probes to PFG fractionated DNA partially digested with a restriction enzyme will identify a series of bands that extends in one direction from the end of the chromosome (Fig. 5). Thus, this type of partial restriction enzyme data is very easy to interpret and allows an extensive map construction in a single experiment. On smaller molecules this type of experiment has been called a Smith–Birnstiel or an indirect-end labeling experiment.

Human telomeres were cloned in yeast by assuming that they would functionally complement half of a yeast artificial chromosome (YAC). Chimeric YAC-human telomere clones were detected in YAC libraries using an oligonucleotide probe specific for the simple sequence, 5′TTAGGG3′, which is present at the ends of all higher eukaryotic chromosomes (see Table I). Nearby the simple telomeric repeat is a series of subtelomeric human-specific repeats. The human-specific subtelomeric repeats have been used in hybrid cell lines to identify specific human telomeric restriction fragments.

Linking clones allow complete *Not* I restriction maps to be constructed with a relatively small number of hybridization experiments (see Fig. 4). These small clones contain recognition sequences for a particular restriction enzyme. A complete *Not* I linking library of the entire human genome would consist of as few as 3,000 clones, whereas a com-

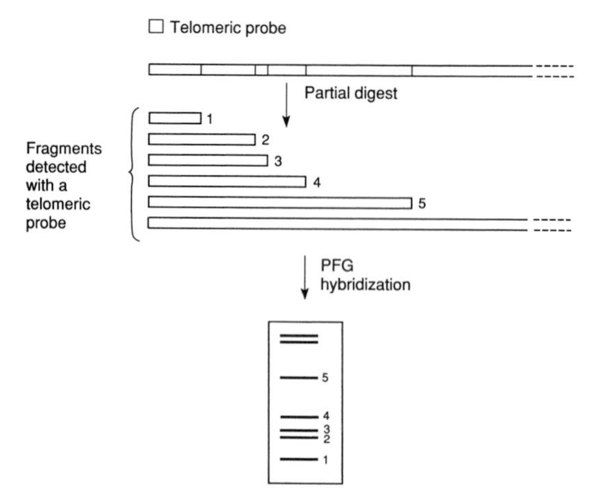

FIGURE 5 Indirect-end labeling experiment with telomeric sequences reveals the restriction sites close to the end of the chromosome. Partial digestion experiments are easy to interpret with telomeric probes because fragments extend in only one direction.

plete *Not* I library for chromosome 21 would consist of only about 50 clones. A *Not* I linking clone used as a hybridization probe identifies two adjacent *Not* I restriction fragments. Linking clones identifying adjacent *Not* I sites will share a common restriction fragment, thereby establishing continuity in the map. The 50 *Not* I restriction fragments expected from chromosome 21 are not likely to overlap in size as long as PFG conditions are adjusted to maximize resolution. Thus, the detection of the same sized restriction fragment by two chromosome 21 probes usually indicates commonality. However, a PFG fractionation of the total human genome cannot resolve all genomic fragments. Each PFG size class will probably contain several genomic fragments. Therefore, common fragments need to be linked by additional fingerprinting methods.

Megabase restriction fragment length polymorphism (MRFLP) detected in a series of different cell lines fingerprints a chromosomal region. The fingerprint is then used to prove continuity between two loci. This method has been termed polymorphism link-up (see Fig. 6). It may be used with any type of clone. MRFLPS may be due to either genetic differences or to methylation differences between the cell

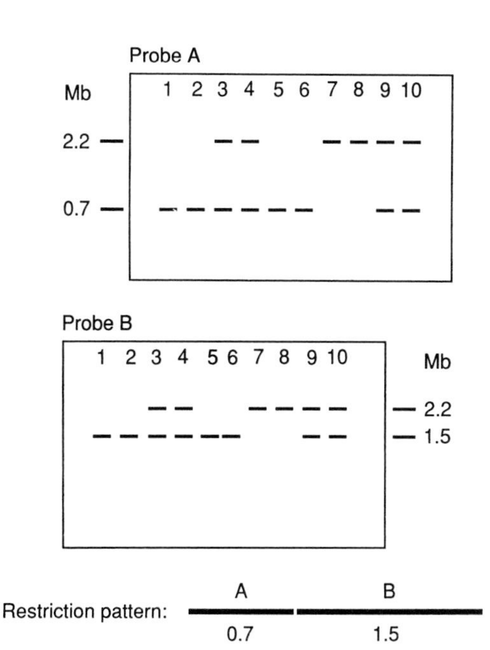

FIGURE 6 Physical map construction using polymorphism link-up. Hybridization of DNA from different cell lines (lanes 1–10) with two putatively linked probes (A and B) leads to the detection of different and common fragment sizes but identical polymorphism patterns.

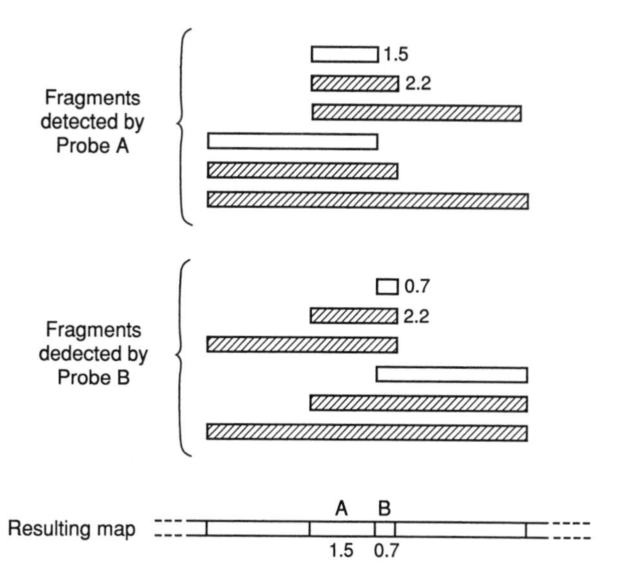

FIGURE 7 Use of partial restriction enzyme digestion to construct physical maps. Single probes used in hybridization experiments to partially digested DNA detect bidirectional partial digestion products. Probes on adjacent fragments detect common (shaded) and unique (unshaded) bands.

lines. This approach provides an overview of the diversity of the human genome, because physical maps are made for multiple genomes. An even more powerful approach involves characterizing restriction enzyme cutting sites by determining the DNA sequence surrounding them (see Section III, C).

Partial digestion strategies can be used to identify neighboring *Not* I fragments and fingerprint chromosome regions (Figs. 4 and 7). Linking clones are particularly useful for partial mapping strategies. Here, each half of a linking clone is used independently as a hybridization probe to partially digested genomic DNA. A series of partial digestion fragments are identified that expand bidirectionally from the smallest fragment containing the probe. While more than one restriction map will be compatible with the hybridization data obtained with each half linking clone, the true restriction map must be consistent with the data obtained with both halves of the linking clone (Fig. 7).

Jumping clones are another mapping tool useful for characterizing large genomic regions (Fig. 4). Jumping clones contain two short noncontiguous genomic sequences. They are made by deleting the DNA between two distant genomic sequences. Jumping clones are used for rapid chromosome walking over long distances. A *Not* I jumping clone contains DNA sequences from the two ends of a *Not* I fragment (Fig. 4).

C. DNA Sequencing

Current technical limitations preclude large-scale genomic sequencing; however, short runs of unique DNA sequences, called sequence tagged sites (STS) can be used as clone identifiers and serve multiple purposes. The use of STS eliminates the need to store of large number of clones because the STS of a clone allows the design of PCR primers to reisolate the sequence from any genomic sample. This means that both small and large laboratories can contribute and use an STS data base.

A collection of STS markers around restriction sites (Sequence TAgged Restriction sites [STARs]) helps construct maps (Fig. 8). For instance, sequencing a complete *Not* I linking library for the human genome would require sequencing about 2 Mb of DNA. This is a large, but not unreasonable, amount of sequencing. Sequencing of a complete jumping library of the human genome would require the same amount of work. Simple sequence comparisons between the two libraries will order all the *Not* I sites but not give the distances between them unless the jumping clones were generated from DNAs of known size. A *Not* I YAC library whose ends are sequenced could be anchored to this map or be used to help create it. Alternatively, end sequences of size-fractionated *Not* I fragments could also provide linkage between two *Not* I sites (Fig. 8). DNA sequence from the ends of YAC clones or DNA fragments can be obtained by several PCR strategies. The identification and sequencing of *Not* I sites in contigs will link these sites to particular genomic *Not* I STARs.

In practice, *Not* I sites may be good starting points for collecting STARs sequences, but other more frequently occurring STARs must also be collected. The restriction enzyme *Eag* I recognizes the inner 6 bp of the *Not* I site. *Eag* I sites occur three times more frequently than *Not* I sites. Collecting *Eag* I STARs eliminates problems associated with manipulating very large *Not* I fragments, i.e., the inability to routinely make YAC clones 400 kb and jumping clones greater than sequences 500 kb apart.

This approach can also be used for finer restriction mapping. For instance, *Not* I–*Eag* I jumping fragments, will position sequences adjacent to a *Not* I site next to a nearby *Eag* I sites (Fig. 9). Then

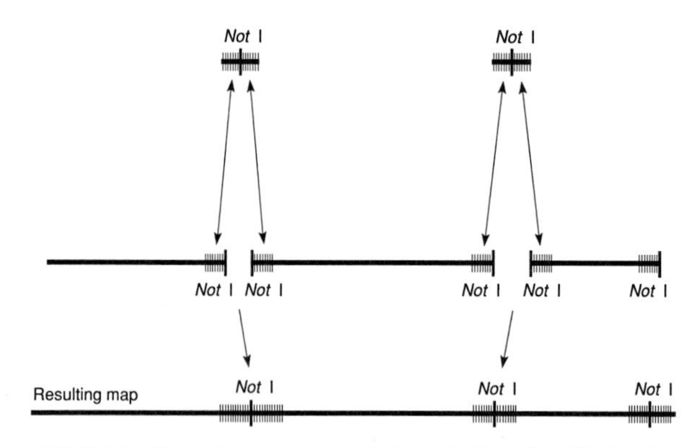

FIGURE 8 Use of sequence tagged restriction sites (STARs) to construct physical maps. DNA sequences from a linking clone library are compared (using a computer) with the end sequences of large DNA fragments, YAC clones, or jumping clones. The different DNAs are then linked by sequence similarity to construct a STARs genomic map.

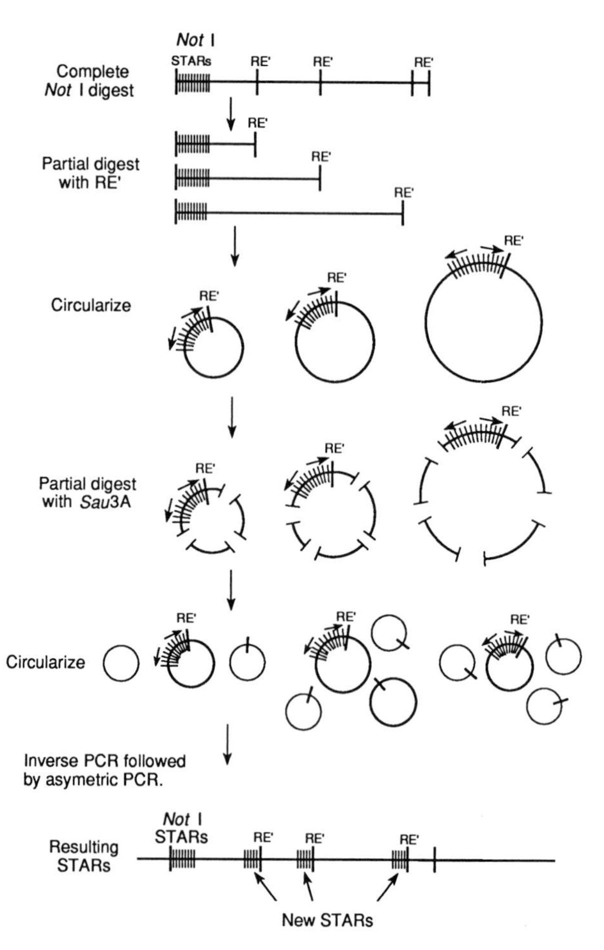

FIGURE 9 Use of existing STARs to define other nearby STARs. Chromosome jumping is used to move sequences around a restriction site to a nearby *Not* I site. Inverse PCR then allows the sequence of the DNA moved close to the *Not* I site to be amplified and subsequently sequenced.

a type of PCR, called inverse PCR, can be used to obtain the sequence around the *Eag* I site using primers from sequences around the *Not* I site. This approach can be extended by using partial digests as well as more frequently cutting restriction enzymes until, eventually, the entire sequence of a genome region is obtained. This approach, unlike many other approaches, provides data that will be continually added to as more sequence data accumulates. There are many more advantages; for instance, the same PCR primers may be used many times. The strategy can be implemented on genomic DNA and, thus, requires little or no cloning and can be done with raw sequence data with high mistake levels. Lastly, this strategy can be implemented either in a random or directed way or by some combination of approaches.

IV. Reverse Genetics

One major interest in studying the human genome involves the identification and isolation of disease genes. Previous studies on disease genes have usually involved the isolation of the protein product. The subsequent purification and protein sequencing of the gene product allowed both antibody and oligonucleotides probes to be produced, which were then used for gene searches in genomic libraries. Now, RFLP mapping allows gene searches to be initiated at the genetic level. This approach has been called reverse genetics. This approach begins by identifying a particular RFLP, which segregates with the disease gene. Subsequently, a variety of methods are employed to identify the gene of interest. For instance, candidate genes may be identified by mapping and cloning of CpG islands in a genetically defined region. The isolation of sequences for these loci allows finer genetic mapping, which may narrow the region if polymorphic loci are cloned. Denaturing gradient gel electrophoresis, allows single base differences between any DNAs to be detected and potentially should allow all markers to be made polymorphic enough for genetic studies. In some cases, such as cystic fibrosis, a more precise genetic location was defined by linkage disequilibrium studies since most patients had the same mutation. However, this is not always the case. Disease phenotypes may originate from one or more mutation at a single locus or from multiple mutations in several loci.

Molecular studies of cloned or PCR amplified sequences, within candidate gene regions, allow the determination of the sequences that are expressed, conserved through phylogeny (i.e., the assumption is that important and expressed sequences are conserved), and reveal whether or not any physical chromosomal abnormalities are associated with a particular disease phenotype. The identification of molecular abnormalities with disease phenotypes has been instrumental in identifying a number of disease genes. Thus, methods for physically characterizing human chromosomes provide tools for mapping genes.

Acknowledgment

The authors thank Sylvia Spengler for contribution and Jesus Sainz for discussion. This work was supported by DOE grant DG-FG-02-87ER60582.

Bibliography

Bickmore, W. A., and Sumner, A. T. (1989). Mammalian chromosome banding—An expression of genome organization. *Trends Genet.* **5,** 144–148.

Bird, A. P. (1987). CpG islands as gene markers in the vertebrate nucleus. *Trends Genet.* **3,** 342–347.

Cantor, C. R., Pevny, L. and Smith, C. L. (1988). Organization and evolution of genomes as seen from a megabase perspective. *In* "Proceedings of the NATO/FEBS Workshop, Spetsai, Greece, August 29–31, 1988." Plenum Press, New York (in press).

Cantor, C. R., and Smith, C. L. (1989). Large DNA technology and possible applications in gene transfer. *In* "Gene Transfer in Animals: UCLA Symposium on Molecular and Cellular Biology," Vol. 87, new series (I. Verma, R. Mulligan, and A. Beauset, eds.), pp. 269–281. Alan R. Liss, Inc., New York.

Cantor, C. R., and Smith, C. L. (1989). Perspectives on the human genome project. *In* "Biotechnology and Human Genetic Predisposition to Disease: UCLA Symposia on Molecular and Cellular Biology," new series (C. R. Cantor, T. Caskey, L. Hood, D. Kamely, and G. Omenn, eds). Alan R. Liss, Inc., New York (in press).

Cantor, C. R., Smith, C. L., and Mathew, M. (1988). Pulsed field gel electrophoresis of very large DNA molecules. *Annu. Rev. Biophys. Biophys. Chem.* **17,** 287–304.

Condemine, G., and Smith, C. L. (1990). New approaches for physical mapping in small genomes. *J. Bact.* (in press).

Gasser, S. M., and Laemmli, U. K. (1987). A glimpse at chromosomal order. *TIGS* **3,** 16–21.

Orkin, S. H. (1986). Reverse genetics and human disease. *Cell* **47,** 845–850.

Reik, W. (1989). Genomic imprinting and genetic disorders in man. *Trends Genet.* **5,** 331–336.

Ruddle, F. H. (1981). A new era in mammalian gene mapping: Somatic cell genetics and recombinant DNA methodologies. *Nature* **294,** 115–119.

Oliva, R., Fong, H., and Smith, C. L. (1990). Construction of genomic restriction maps. *In* "Mapping the Human Genome" (S. Naylor, ed.). Academic Press, San Diego (in press).

Smith, C. L., and Cantor, C. R. (1987). Purification, specific fragmentation and separation of large DNA molecules. *In* "Recombinant DNA" (R. Wu, ed.). *Methods in Enzymology*, Academic Press, New York, **155,** 449–467.

Smith, C. L., Lawrance, S. K., Gillespie, G. A., Cantor, C. R., Weissman, S. M., and Collins, F. S. (1987). Strategies for mapping and cloning macro regions of mammalian genomes. *In* "Molecular Genetics of Mammalian Cells (M. Gottesman, ed.). *Methods in Enzymology*, Academic Press, New York, **151,** 461–489.

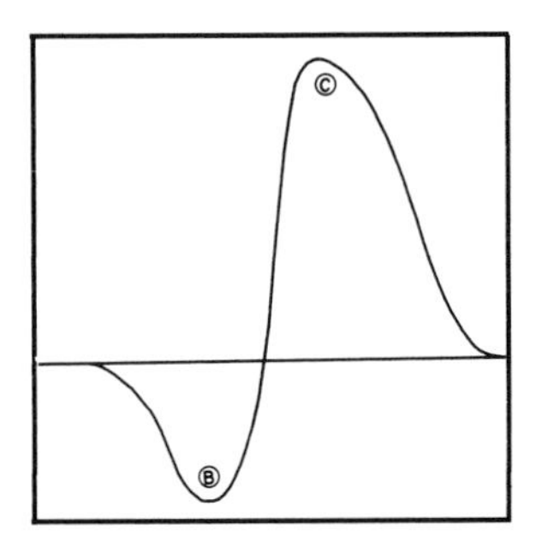

Circadian Rhythms and Periodic Processes

JANET E. JOY, *National Institute of Mental Health*

Glossary

Entrainment Steady state in which an endogenous rhythm runs synchronously to another rhythm (i.e., *Zeitgeber*) with a constant phase relationship, or phase angle difference

Free-running period Fundamental period of a biological clock when it is not entrained to some forcing oscillation (*Zeitgeber*); often abbreviated as τ

Pacemaker Localizable entity capable of self-sustaining oscillations and of synchronizing other oscillations

Period Time required for completion of one cycle

Phase Any instantaneous state of a cycle (e.g., peak or trough)

Phase angle Relative term measuring the relationship between a particular phase in a cycle and some arbitrarily chosen reference point or phase of another cycle. For example, the phase angle between the rhythm of body temperature and the sleep–wake cycle could be described as the difference in hours between the body temperature minimum and the onset of sleep. If the temperature minimum occurs at 10:30 P.M. and sleep onset at 11:30 P.M., then the phase angle differences between these two rhythms is 1 hour

Phase–response curve Waveform plot indicating how the amount and direction of a phase shift induced by a single stimulus depends on the time, or phase, at which it is applied

Phase shift Single displacement of an oscillation along the time axis; the means by which a biological clock is reset

Zeitgeber Time giver; any external rhythm that will entrain, or synchronice, an endogenous biological rhythm

BIOLOGICAL RHYTHMS are a fundamental element of human physiology. Although we are usually most aware of rhythms associated with eating and sleeping, our lives are also governed by rhythms in hormonal fluctuations, cellular metabolism, and neural activity. Many of these rhythms are driven by internal biological "clocks." The study of biological rhythms can be divided into two distinct areas. The first area includes studies of rhythmic processes themselves and addresses questions about the implications of specific patterns of temporal organization. The second area is primarily concerned with the nature of biological clocks and how rhythms are generated. For example, how is the clock itself regulated, where is it located, and what are the elements that make up a biological clock?

Introduction

We live in a periodic universe. The passage of our days, months, and years is dictated by the periodic motion of the planets. Life on Earth evolved in this periodic environment and biological rhythmicity emerged as a basic element of physiological organization. The most prominent rhythms are, of course, the daily cycles that arise from the earth's rotation about its axis, and by far the most widely reported biological rhythms are those that show daily cycles. Daily cycles have been reported for manual dexterity, pain sensitivity, drug metabolism, and cell division—to name only a few of the many rhythms that rule our lives. Biological rhythms are associated with other geophysical cycles as well. Marine animals living in tidal zones have rhythms that follow the lunar cycle, and most animals living away from equatorial regions, where seasonal changes in the physical environment are particularly pronounced, undergo yearly cycles. Biological rhythms also include frequencies that are not associated with known planetary motions. For example, pacemaker cells of the mammalian heart have an intrinsic rhythm of electrical activity, with cycles lasting several 100 msec, and certain hormones are secreted episodically every few hours.

Human physiology is governed by rhythms and periodic processes that influence every level of organization, from the biochemical to the behavioral. This article focuses on the most prominent type of rhythms, namely those associated with daily fluctuations. The role of biological rhythms in human biology is outlined, although many of the general principles are illustrated by using examples based on animal research, since most of the advances in this field necessarily come from such studies. Section II discusses the endogenous nature of biological rhythms and how they are regulated. Section III covers the neural basis of biological rhythms. Later sections explore the development of biological clocks within the individual and the role of biological clocks in selected processes, such as sleep and reproduction. Section IX discusses rhythm disorders in human physiology.

II. Endogenous Nature of Biological Rhythms

A remarkable feature of many biological rhythms is that, although they evolved in response to a periodic environment, they are driven by mechanisms that are independent of periodic environmental input. That is, most biological rhythms are self-sustaining and persist even when all known environmental cycles are eliminated. Such rhythms are said to be endogenous, meaning that they are intrinsic to the organism and are not derived from environmental fluctuations. The pattern of wheel-running activity in the golden hamster is a classic example of an endogenous rhythm. A hamster housed in constant darkness and constant temperature will show a precise rhythm of running activity, with a period (τ) that is likely to be about 24.2 hours (Fig. 1). Such a rhythm is said to free-run, because it continues to run free from input from external signals.

The term "circadian" is derived from the Latin words for "about" (*circa*) "a day" (*dies*). The observation that the period of the rhythm is close to, but not exactly equal to, 24 hours is typical of circadian rhythms. For example, humans living in temporal isolation with no time cues nonetheless show regularly alternating patterns of sleeping and waking—generally with a period of about 25 hours (Fig. 2). The terms "daily" and "circadian" are not synonymous, however. A circadian rhythm is seen as a daily rhythm under natural daylight cycles; it does not follow that daily rhythms necessarily show a circadian cycle under constant lighting conditions.

Free-running periods are genetically influenced and show characteristic differences between spe-

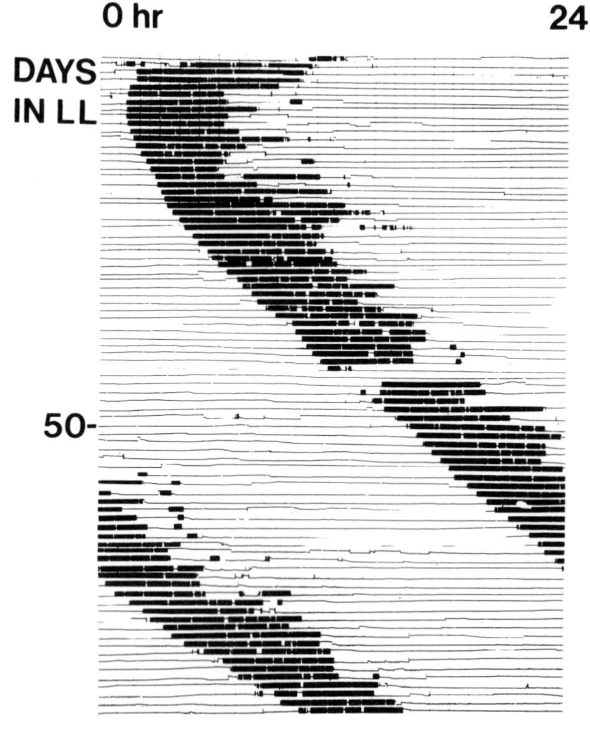

FIGURE 1 Free-running rhythm of wheel-running activity in a golden hamster living in constant light (LL). This record is made by dividing a continuous activity record into 24-hour periods and plotting successive days beneath each other. Activity bouts are indicated by vertical tick marks, which blend together to form a solid band during periods of intense activity. Note the precision of the daily onset of activity. The free-running period (τ) of this animal is 24.4 hours.

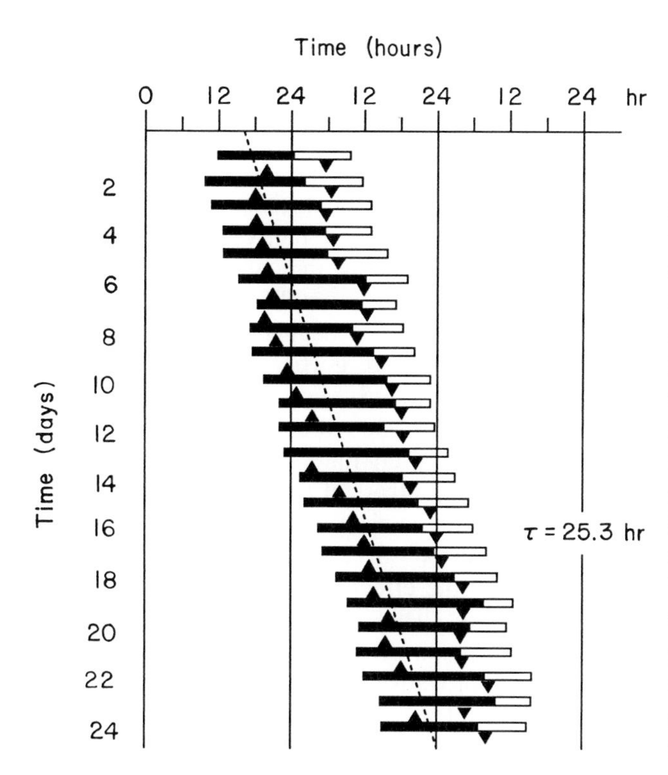

FIGURE 2 Circadian rhythms of wakefulness and sleep (solid and open bars, respectively) and rectal temperature (triangles above bars for maxima, below bars for minima) recorded in a subject living alone in an isolation unit under constant conditions. The free-running period (τ) is 25.3 hours. [From J. Aschoft, *Ergonomics* **39,** 739–754 (1978).]

cies as well as between individuals of the same species. The deviations from a strict 24-hour periodicity provide compelling evidence that these cycles are truly endogenous, since there are no known environmental cycles with periodicities that could hypothetically drive them.

Many, if not all, daily rhythms are driven by endogenous circadian oscillators. To confirm that a daily rhythm is under the control of an endogenous oscillator (as opposed to being passively driven by daily environmental fluctuations), the rhythm in question must be shown to free-run under constant conditions. Circadian oscillators in animals share certain basic properties: They free-run under constant conditions, they are encoded in the genome, they are generated from neural structures, and they can be entrained by light.

Circadian rhythms are not unique to higher vertebrates. They are found in plants as well as lower animals, including unicellular organisms. In fact, they have been found in every eukaryotic organism thus far studied. Processes as diverse as spore formation in fungi and color changes in fiddler crabs are known to be regulated by circadian rhythms.

Endogenous rhythms also occur in noncircadian ranges, such as circannual cycles (about 1 year) or ultradian cycles (less than 1 day). Hibernation in ground squirrels and migration patterns of garden warblers show circannual cycles that generally have a period of 10–11 months. At the other extreme are the 60-second ultradian rhythms in the courtship songs of fruit flies. A prominent ultradian rhythm in human physiology is the 90- to 100-minute alternation between the rapid eye movement (REM) and non-REM stages of sleep. (REM is the stage of sleep usually associated with dreaming.) Ultradian rhythms occur in a wide range of frequencies, and individual rhythms show considerable cycle-to-cycle variation in length. They are often described as "pseudoperiodic," whereas circadian rhythms, which typically show less than a 5% variation in period length, are strictly periodic. As a rule, ultradian rhythms seem to be independently generated as consequences of the particular physiological systems of which they are a part, and they are not under the control of an endogenous ultradian oscillator. In contrast, circadian rhythms appear to be coupled to an integrated system of circadian oscillators.

III. Entrainment of Biological Clocks

The circadian oscillator acts as a biological clock. However, since the endogenous period of this clock is typically close to, but not equal to, 24 hours, it must be reset each day. It must be synchronized, or entrained, to the 24-hour day. The most powerful entraining agent, or *Zeitgeber* (German for "time giver"), is the light–dark cycle. Circadian rhythms are exquisitely sensitive to small amounts of light. As little as 1 second of light per day is sufficient to entrain the circadian period of locomotor activity in the golden hamster. A single light pulse can reset, or phase shift, the rhythm to an earlier phase (phase advance) or a later phase (phase delay) or cause no change in the rhythm at all (Fig. 3, upper panel). The critical determinant in the ability of a light pulse to reset the rhythm is not the duration of light, but rather the phase of the circadian cycle at which it is applied. The phase–response curve, in which the size and direction of the phase shift are plotted against the phase at which the pulse is delivered, illustrates the circadian rhythm of sensitivity to

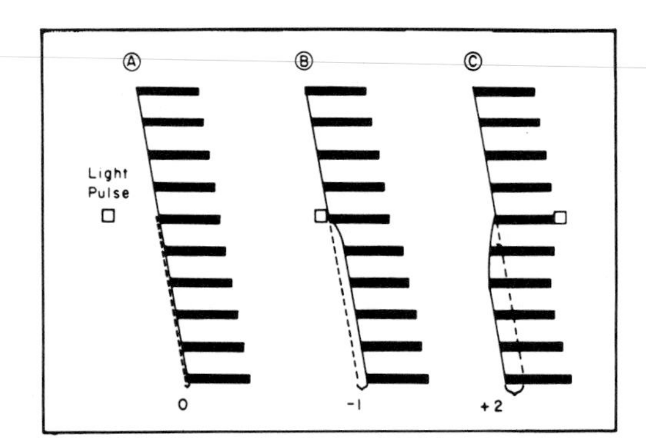

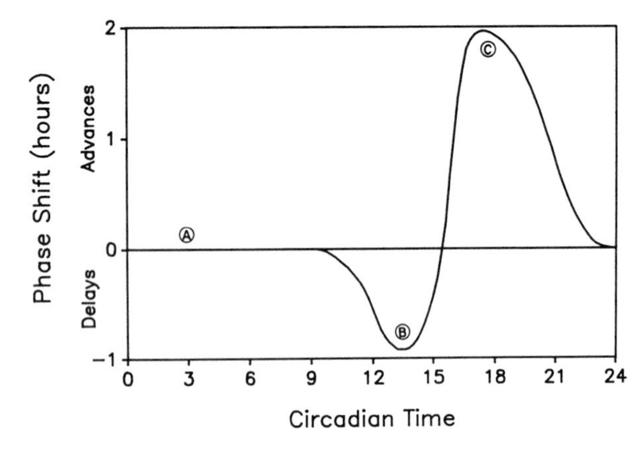

FIGURE 3 Derivation of a phase–response curve. (Top panel) Results from an experiment in which a golden hamster is exposed to a 1-hour light pulse at different circadian times. At all other times the animal is maintained in constant darkness. Data are plotted as in Figs. 1 and 2. In nocturnal animals (e.g, the hamster) circadian time 12 is arbitrarily defined as the daily onset of activity. A light pulse given early in the subjective day (A) has no effect on the clock, but when given early in the subjective night (B), the light pulses causes a 1-hour phase delay, and when given later in the subjective night (C), it causes a 2-hour phase advance. (Lower panel) Phase–response curve that summarizes the effect of individual light pulses given at different times spanning a 24-hour period.

light (Fig. 3, lower panel). To measure this rhythm, the phase–response curve must be generated from animals in free-running conditions. The convention of circadian time is used to accommodate individual differences seen in free-running periods. One circadian hour is defined as one-twenty-fourth of an individual's free-running period (1 circadian hour = τ/ 24).

The phase–response curve illustrates the behavior of the underlying circadian oscillator, which is often described as the pacemaker of the circadian rhythms it drives. It is important to make the distinction between an oscillator and its output (i.e., the rhythm it drives). Circadian rhythms represent the hands of the clock, as opposed to being part of the timing mechanism itself. A rhythm that is coupled to a circadian pacemaker provides a marker of the pacemaker's state. Only two properties of a rhythm are reliable indicators of the state of the clock: period length and phase. Amplitude of a rhythm is not a reliable indicator. If a particular rhythm is suppressed or amplified, it does not necessarily follow that the oscillator itself is altered. For example, although cortisol secretion from the human adrenal gland shows a clear circadian rhythm, the suppression of cortisol by the drug dexamethasone is entirely independent of the circadian pacemaker. In contrast, if the pacemaker is phase-shifted, then the cortisol rhythm is also shifted.

Phase–response curves provide important information about the effect of different stimuli on biological clocks. They are useful in determining whether particular drugs or neurotransmitters (i.e., chemicals that transmit signals in the nervous system) are capable of phase-shifting biological clocks. Phase–response curves can also be used to indicate whether a particular stimulus is capable of entraining a biological clock and what the pattern of entrainment will be. Finally, the phase–response curve can be used to predict the phase angle of entrainment. The term "phase angle" is used to describe the temporal relationship between two separate rhythms; it is defined as the difference between selected phase reference points of any two rhythms (e.g., the entraining rhythm and a biological rhythm). For example, the difference in hours between the time of lights on and the time a hamster beings to run in its activity wheel is a conventional way to measure the phase angle of entrainment in a nocturnal animal. This phase angle varies, depending on the light–dark cycle to which the animal is entrained.

Phase–response curves are typically based on phase shifts that occur after a single pulse of light (or any other stimulus that can cause a phase shift). This type of phase–response curve to light for humans has not been established, although a phase–response curve based on successive light pulses (i.e., 5-hour light pulses given at the same time of day for 3 days in a row) has been demonstrated. It is not clear how this compares to the phase–response curves established for nonhuman species. It does appear, however, that a relatively high-intensity light pulse (approx 5000 lux) is needed to cause phase shifts in humans, whereas maximal phase shifts with 1-hour light pulses of 300 lux can be obtained in rodents. Comparisons of absolute threshold between nocturnal rodents and humans must, of course, be viewed with caution. In studies on humans, background illumination is generally about 100–400 lux (i.e., the intensity of ordinary indoor room light), whereas in rodent studies light pulses are given against a background of complete darkness.

IV. Nonphotic Effects on Biological Rhythms

A variety of nonphotic (i.e., unrelated to light) agents, including social factors, feeding schedules, drugs, and numerous neurotransmitter-related substances, have been shown to alter either the phase or the period of circadian rhythms.

The importance of social factors was first recognized in human studies, in which subjects who were unable to entrain to experimental light–dark schedules were later found to entrain to the same schedules when they were periodically signaled by the sound of a gong to perform routine tasks for the experimenter. The addition of the gong, which was interpreted by the subjects as a social contact, enabled them to entrain to an otherwise weak entraining stimulus. Further support for the hypothesis that social factors can influence human circadian rhythms is the observation that when people are tested in groups, their free-running periods tend to become synchronized with each other.

The role of social cues is particularly, but not uniquely, important in humans. Differences between species most likely reflect differences in social biology. Thus, social interactions do not seem to entrain the relatively asocial squirrel monkey, but do entrain bats that live in large social colonies. The effect of social cues on circadian rhythms may be quite subtle and probably depends on the nature of the social contact.

Periodic feeding schedules have been shown to entrain circadian rhythms of food activity in rats, but this effect appears to be restricted to the entrainment of a rhythm of "food-anticipatory" activity that is distinct from other components of activity. Feeding schedules have been reported to entrain circadian rhythms in the squirrel monkey. Although the possible role of meal schedules as a *Zeitgeber* for humans is unknown, such an effect might be expected at least partially because of the traditional association of mealtimes with social contacts.

Many drugs and endogenous substances have been found to affect the mammalian circadian system. Some of these agents are capable of causing phase shifts, whereas others cause changes in period, but not phase. Chronic treatments with lithium and clorgyline, drugs that are prescribed in affective disorders, as well as ethanol have been shown to cause slight increases in the length of the free-running period of rodents; whereas the sex hormones estradiol and testosterone have been shown to decrease period length. None of these substances has been shown to cause phase shifts in mammals. The benzodiazepines, triazolam (Halcion, Upjohn, Kalamazoo, Michigan) and diazepam (Valium, Roche Laboratories, Nutley, New Jersey) cause either phase advances or phase delays, depending on the time of treatment. Numerous neurotransmitter-related substances have also been shown to reset rodent clocks, including muscimol [which simulates the action of the neurotransmitter γ-aminobutyric acid (GABA)], carbachol (which simulates acetylcholine), and the neurotransmitter neuropeptide Y.

It is interesting to note that several of these drugs are used to treat sleep and/or mood disorders. Benzodiazepines are widely prescribed as anxiolytics and for insomnia, whereas lithium and clorgyline (a type I monoamine oxidase inhibitor) are prescribed for the treatment of depressive disorders. The abilities of muscimol and carbachol suggest that neurons containing GABA and acetylcholine, respectively, are important in the regulation of mammalian circadian oscillators. This is discussed further in the following section.

V. Neural Basis of the Biological Clock

A group of nerve cells, called the suprachiasmatic nucleus (SCN), located at the base of the mammalian brain, is thought to be the master pacemaker that sets the pace for circadian rhythms in mammals. The SCN is a bilaterally paired structure in the anterior hypothalamus. The two halves of the human SCN lie on the lateral walls of the third ventricle (one of a group of small cavities within the brain) and just above the optic chiasm. The SCN is a heterogenous and complex structure containing extensive connections between nerve cells. Over 25 different types of neurotransmitter-related substances have been found within the SCN regions of rats and hamsters. The most abundant neurotransmitter in this area is GABA. Other prevalent neuroactive substances in this region, which probably also function as neurotransmitters, are vasopressin, neuropeptide Y, somatostatin, and serotonin.

The SCN is necessary for normal circadian functions and fulfills the basic requirements for an endogenous pacemaker. In rodents destruction of the SCN abolishes circadian rhythmicity in locomotor activity, sleep–wake cycles, feeding, sexual activity, many hormone rhythms (including pineal melatonin, pituitary prolactin, and gonadotropins), and numerous other functions. The mechanisms regulating these functions appear to remain intact after SCN lesions; they simply become temporally disorganized. Thus, after SCN lesions animals continue to use running wheels, but instead of restricting most of their activity to a single bout lasting approximately 8 hours, they run for short bursts at irregular intervals. Pathological evidence in humans suggests a similar pacemaker role for the SCN. Tumors that affect the SCN can cause sleep disorders wherein patients repeatedly fall asleep at any time of day, much like SCN-lesioned animals. Sleep in such patients is otherwise normal and is unlike the comalike sleep that is seen in later stages of such illness.

Three other lines of evidence based on studies in rats and hamsters provide compelling evidence that the SCN contains a self-sustained circadian oscillator: First, the SCN will continue to show circadian rhythms in neuronal activity even after surgical isolation from surrounding brain tissue. Such isolation eliminates rhythmic activity in other brain areas, demonstrating their dependence on neural signals from the SCN. Second, SCN tissue, isolated from the brain and kept *in vitro* under constant lighting conditions, shows circadian rhythms in vasopressin release and neuronal firing rates. Finally, circadian rhythmicity in arrhythmic SCN-lesioned animals can be restored by neural grafts of SCN tissue from donor animals.

A. Input and Output Pathways

Light entrainment of circadian rhythms in mammals occurs only through the eyes. (Entrainment that occurs independently of the eyes is seen in birds and reptiles, but not in mammals.). The primary light entrainment pathway is via a bundle of nerve fibers called the retinohypothalamic tract (RHT), which transmits signals from the retina to the SCN. The RHT pathway has been established in all major mammalian orders, but has not been confirmed in humans, because the neuroanatomical tracing methods, requiring the injection of radioactive substances in the retina shortly before death, cannot be applied.

Entrainment is also mediated through a secondary visual projection, which leads indirectly from the retina to the SCN through intermediary pathways. It involves the geniculohypothalamic tract (GHT), which arises from the intergeniculate leaflet, which is surrounded by the lateral geniculate nucleus, and terminates in the SCN, overlapping extensively with the RHT terminals (at least in rats and hamsters). Lesions of the intergeniculate leaflet or ablation of the GHT does not eliminate entrainment, but does alter the phase–response curve to light and, under certain conditions, also alters the length of the free-running period. The GHT is thought to be important in mediating the effects of light intensity on the circadian system.

The neuronal pathways that relay signals to the SCN (afferent pathways) and the pathways that relay signals from the SCN to the rest of the brain (efferent pathways) have been characterized in hamsters and rats. As mentioned above, the RHT and the GHT are the major afferent pathways from which the SCN receives light input. Both the RHT and the GHT send signals from the eye on the same side of the body (ipsilateral) as well as from the other side (contralateral) to each side of the SCN. About two-thirds of the fibers go to the contralateral SCN. The SCN also receives afferent connections from many areas of the brain: from the midbrain (e.g., the raphe nuclei and the tegmentum), the hypothalamus (e.g., the anterior and ventromedial nuclei, arcuate nucleus, and preoptic area), the thalamus (e.g., the paraventricular nucleus), the septal area, and the hippocampus.

In addition, the two halves of the SCN are connected to each other via an extensive system of nerve fibers that traverse the midline in both directions. The efferent pathways go primarily to other areas within the hypothalamus (e.g., the anterior and ventromedial nuclei, retrochiasmatic area, periventricular area, ventral tuberal area, arcuate nucleus, median eminence, and paraventricular nucleus), as well as other brain centers (e.g., the lateral habenula, midbrain central gray, and septal areas). These widespread connections provide the basis for extensive SCN communication with the rest of the brain and are consistent with the role of the SCN as the temporal coordinator of many diverse rhythms: the master pacemaker.

B. Circadian Neural Oscillators outside the Suprachiasmatic Nucleus

Evidence pointing to the existence of non-SCN oscillators in mammals comes from experiments with rats which suggest the presence of a "food-entrainable" oscillator lying outside the SCN region. This oscillator is thought to be coupled to a "light-entrained" oscillator, presumably within the SCN. Exposure of rats to daily schedules of food availability will entrain a rhythm of food-anticipatory activity which is distinct from other locomotor activity rhythms and does not appear to be entrained by the light-sensitive oscillator. This rhythm is entrainable by feeding schedules even in SCN-lesioned animals.

The pineal gland (situated just beneath the skull and between the two brain hemispheres) of certain birds and reptiles contains an endogenous self-sustained circadian oscillator. The pineal gland of many birds and reptiles contains both photoreceptive and endocrine cells, which secrete the hormone melatonin. (Although the role of melatonin is not fully understood, it is known to be involved in regulation of the timing of reproduction in many animals, particularly seasonal breeders.) Interestingly, the pineal glands of many birds and reptiles show circadian rhythms of melatonin release in culture, which can be phase-shifted by light pulses. In contrast, the mammalian pineal gland contains primarily melatonin-secreting endocrine cells and no photoreceptive cells. The mammalian pineal gland does not show circadian rhythms in culture and appears to be under direct control of the SCN. The patterns of variation seen in the pineal glands of different species suggest that there has been an evolutionary trend away from the role of this gland in lower vertebrates as an endogenous circadian oscillator toward a role in higher vertebrates as an organ whose rhythmic output is directly under SCN control.

Circadian oscillators have been found in the eyes of many animals, from mollusks to mammals. In the rabbit a circadian rhythm of electroretinogram (a measurement of electrical current generated by retinal nerve cells) of the isolated eye free-runs under constant conditions. In the human eye, electroretinogram measurements, intraocular pressure (which influences diagnostic tests for glaucoma), and the turnover rate of photoreceptive membranes all show diurnal rhythms (i.e., how rapidly they are renewed). It is likely that, like other vertebrate retinal rhythms, these rhythms are governed by a circadian oscillator within the eye itself.

VI. Development of Circadian Rhythms within the Individual

The development of the mammalian circadian system does not depend on exposure to environmental fluctuations. Rat pups born and reared under constant lighting conditions nonetheless show circadian rhythms as adults. Circadian rhythmicity develops even in pups whose mothers are housed in constant conditions throughout gestation or whose mothers' own circadian rhythmicity has been abolished by an SCN lesion.

The biological clock in the SCN begins to oscillate during fetal life. Circadian rhythms can be detected in the fetal rat by the 19th gestational day (3 days before birth), when the SCN is not yet innervated by the RHT and when it is almost completely lacking in functional synapses (i.e., connections at which signals are transmitted between nerve cells). The vast majority of SCN afferent and efferent pathways are formed postnatally and are apparently not necessary for the generation of circadian rhythmicity.

Fetal circadian rhythms are entrained by the mother through unknown mechanisms. Maternal entrainment of postnatal circadian rhythms is less well established. Rat pups are born blind, and maternal rhythmicity appears to act as a weak entraining agent for about the first 6 days of life. After this the pups are able to open their eyes and can be independently entrained by light.

Studies in human infants demonstrate that circadian rhythms in sleeping patterns develop before they are entrained to a 24-hour cycle (Fig. 4). The development of regular 24-hour sleeping patterns in infants is reportedly enhanced by mother–infant interactions, suggesting that maternal factors are also important for the entrainment of circadian rhythms in the human infant.

Daily rhythms appear at different developmental stages, presumably reflecting the maturation of output pathways downstream from the circadian pacemaker. The maturation of a rhythm is typically accompanied by an increase in its amplitude. In the rat most behavioral and hormonal rhythms are not

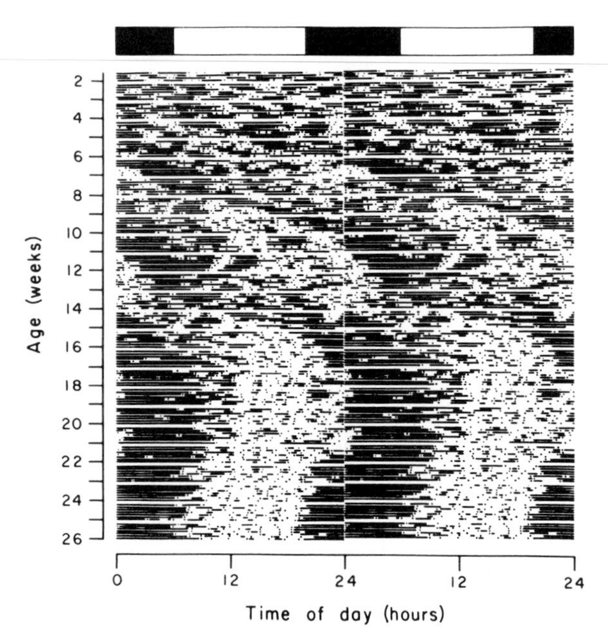

FIGURE 4 The development of sleep–wake rhythmicity in a human infant on a self-demand feeding schedule and a light–dark cycle as indicated at the top of the figure. The black bars in the record indicate time asleep, and the dots represent feedings. The record is repeated and shifted up 1 day on the right to aid visual appreciation of the free-running rhythm. Weekly sets of data are separated by white gaps of no data. [From N. Kleitman and T. G. Eaglemann (1953) *J. Appl. Physiol.* **7,** 269. Reproduced with permission.]

expressed until 2–3 weeks after birth, when daily rhythms in SCN neuronal firing rates are first detectable. The majority of human circadian rhythms do not become evident until well into the postnatal period. Daily sleep–wake rhythms appear during the second postnatal month, whereas daily rhythms in plasma adrenal hormone (i.e., corticosteroid) levels are not established until about 2 years of age.

VII. Sleep and the Biological Clock

The sleep–wake cycle dominates our lives more than any other biological rhythm. Sleep cycles are influenced by ultradian cycles superimposed on circadian rhythms. There are two major components of sleep: patterns of sleep, which refers to the temporal relationship between sleep onset and waking, and sleep structure, which refers to the distribution of different sleep stages during the time spent in sleep. Sleep is divided into five stages, generally related to the "depth" of sleep. Stages 1–4 describe increasingly deeper sleep. A fifth stage is typified by REMs and is called REM sleep.

Patterns of sleep are timed by the circadian system, although external events (e.g., daily activities or ambient temperature) can alter either the timing or the structure of sleep. Circadian cycles of fatigue are apparent even during sleep deprivation, explaining why one is likely to feel less tired during the morning after a lost night's sleep than during the night—even though the accumulated hours of lost sleep are greater in the morning.

Three primary variables characterize sleep patterns: (1) sleep latency, or the duration of wakefulness preceding sleep onset; (2) sleep length; and (3) timing of sleep onset and termination. Sleep latency is inversely correlated with, and is primarily determined by, the duration of prior wakefulness. Sleep length is only slightly modified by extended periods of wakefulness. Recovery sleep after deprivation rarely matches the amount of sleep lost. For example, a 17-year-old boy chose to undergo 11 days without sleep. When he finally went to bed, he slept for only 14 hours and then resumed his regular 8-hour nighttime sleeping patterns. Further, human subjects living in temporal isolation show no consistent relationship between sleep length and the duration of previous wakefulness. The time of waking is the most regular aspect of sleeping and is though to be most strongly under circadian control. Sleep onset and termination show consistent relationships with circadian rhythms of body temperature. The most frequently chosen bedtime among experimental subjects is just after the lowest point in the body temperature cycle, and wakeup times typically occur during the rising phase in the temperature cycle. Subjects who went to bed at the low point in the temperature cycle slept for short episodes, whereas subjects who retired close to the peak of the temperature cycle slept much longer.

Ultradian rhythms in sleep structure occur in the 90–100 cycles of alternation between REM and non-REM sleep, with the first REM episode generally occurring 70–90 minutes after sleep onset. In young adults REM sleep accounts for 20–25% of total sleep. Unlike circadian rhythms in sleep patterns, the period of the ultradian REM–non-REM cycles does not change when allowed to free-run in constant conditions. In addition, these ultradian rhythms show considerable cycle-to-cycle variation in period length and do not appear to be produced by an endogenous oscillator, but appear to result from interactions between neurons in various areas of the brain stem. In contrast, the proportion of REM sleep within each REM–non-REM cycle increases as the night progresses and appears to be under circadian control.

The secretion of various hormones is regulated by both the timing of sleep and the circadian clock. The relative degree of control by sleep and the circadian clock varies with different hormones. The secretion of growth hormone, which shows a clear diurnal rhythm, appears to be primarily dependent on sleep. During sleep deprivation growth hormone secretion is suppressed, whereas the onset of sleep elicits a large secretory burst of growth hormone, whether sleep is delayed, advanced, interrupted, or fragmented.

In contrast, secretion of the pituitary hormone prolactin appears to be controlled jointly by sleep and by the circadian oscillator. Plasma prolactin levels are usually highest during the last 2 hours of a normal sleep period, fall rapidly after sleep ends, and remain relatively low during the day. Increases in prolactin levels are, however, also associated with daytime naps, which has been interpreted as indicating that diurnal rhythms in prolactin are directly controlled by sleep patterns. However, it has since been demonstrated that prolactin secretion is also under circadian control.

The secretory rhythm of adrenocorticotropic hormone (ACTH), the pituitary hormone that triggers cortisol release, exemplifies a hormonal rhythm that is under strong circadian control. ACTH and cortisol are secreted in episodes about once ever 1–2 hours, and the circadian profile appears to be produced by modulating the size of successive secretory pulses. Studies in rodents suggest that, in addition to being primarily dependent on the circadian rhythm of ACTH release, the adrenal cortisol secretion rhythm is amplified by daily variations in the responsiveness of the adrenal gland to ACTH. Both rhythms persist during sleep deprivation.

VIII. Rhythms in Reproduction

Biochemical, neural, hormonal, and behavioral rhythms play a central role in many different aspects of reproduction. Rhythms in reproduction cover a wide spectrum of frequencies: from the 60-second courtship songs of fruit flies to circannual rhythms of reproductive status in ground squirrels.

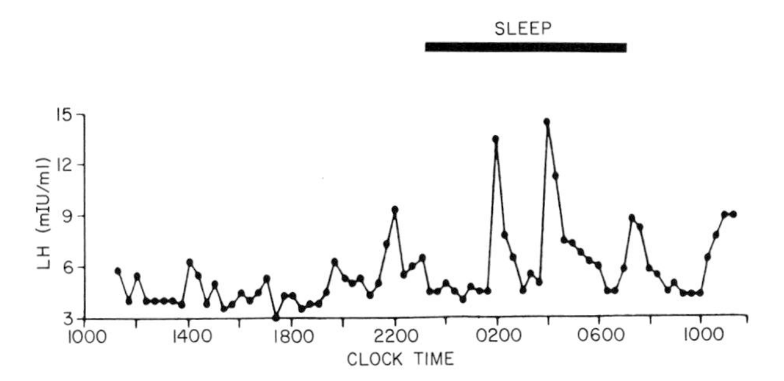

FIGURE 5 Twenty-four plasma luteinizing hormone (LH) profiles of a 14-year-old boy during a normal sleep–wake cycle. LH release occurs in pulsatile bursts. The nighttime augmentation of LH release is influenced by sleep, as well as by circadian signals that occur independently of sleep. [From S. Kapen, R. M. Boyer, J. W. Finkelstein, L. Hellman, and E. W. Weitzman. (1974). *J. Clin. Endocrinol. Metab.* **39,** 293.]

All of the major reproductive hormones are known to fluctuate throughout the day. Hormonal release is predominantly pulsatile, or episodic, with secretory bursts occuring at 1- to 4-hour intervals (Fig. 5). Patterns of hormonal release can be influenced by diurnal modulations in pulse frequency and/or pulse amplitude.

Reproductive processes are regulated by the hormones of the hypothalmic–pituitary–gonadal axis. Many aspects of the reproductive process are regulated through precisely timed changes in pulse frequency and/or amplitude. Pulsatile hormone release is, however, often undetected or underestimated, because blood samples are typically taken too infrequently to map rapid fluctuations in serum hormone levels. Hormonal pulsatility is influenced by many factors in addition to circadian signals—particularly developmental changes and feedback effects from other hormones. For example, testosterone (the male sex hormone) has a major effect on the pulsatile pattern of the release of the pituitary luteinizing hormone (LH), which stimulates testosterone release. Castration of both rats and monkeys results in dramatic increases in the frequency and amplitude of LH pulses. In women, the pulse of gonadotropin-releasing hormone (GnRH) is reduced during the luteal phase of the menstrual cycle (the phase after ovulation), and this reduction in frequency is thought to be due to the neural actions of progesterone secreted by the corpus luteum, which is formed after ovulation.

The onset of puberty is characterized by profound changes in diurnal, as well as pulsatile, variations in the pituitary gonadotropic hormones LH and follicle-stimulating hormone. In prepubertal children of both sexes, gonadotropins are secreted in low-amplitude pulses during the night. This nocturnal rise is thought to be partially influenced by the circadian system, but is primarily sleep dependent. As a child approaches puberty, the amplitude of the nighttime pulses increases, thus amplifying the overall circadian variation. In adult men LH and FSH continue to be released in pulses every few hours, but the diurnal variation is damped and, in most cases, undetectable. Serum testosterone levels fluctuate, with about 17–18 pulses per 24-hour period in young adult men, and are modulated by diurnal rhythms, with lowest levels generally at night and highest levels in the morning.

Since there is little, if any, diurnal variation in serum LH levels (which provide the primary stimulus for testosterone secretion), the diurnal fluctuations in testosterone levels must be due to factors that modify the effectiveness of the LH signal on a daily basis. In women the patterns of pituitary LH release are modulated by the menstrual cycle. Ovulation is triggered by a surge in LH release. During the early follicular phase (the phase before ovulation) LH pulses are more frequent during the night. As ovulation approaches this diurnal modulation of pulsatility disappears and remains absent during the subsequent luteal phase. The timing of the preovulatory LH surge is determined by an interaction between hormonal changes of the menstrual cycle and the circadian system. In normal women the onset of the LH surge usually occurs in late sleep or early morning. The extent of circadian control has not, however, been precisely defined in humans.

In rodents, which show 4- to 5-day estrous (i.e., ovarian) cycles, estrous-related events (e.g., timing of the surge in LH release, ovulation, postovulatory increase in progesterone, and onset of sexual receptivity) are tightly linked to the circadian system and all occur at specific times of day on specific days during the cycle. Interestingly, the neural signal that induces the LH surge occurs late in the afternoon every day, but only results in a surge every fourth day. The occurrence of the LH surge requires high levels of estrogen, which occur only on the day before ovulation.

In rodents the timing of birth is under circadian control. Human births have not been studied in temporal isolation, but are undoubtedly influenced by circadian rhythms. The onset of labor shows a clear diurnal rhythm, with labor beginning most frequently about 1:30–2:30 A.M. Natural births are more variable in timing than the onset of labor, but also show a diurnal rhythm, with births occuring most frequently between 3:00 A.M. and 6:00 A.M.

Although the 28-day human menstrual cycle closely matches the 29.5-day lunar cycle, it does not appear to be influenced by the moon's gravitational pull. The correlation is merely coincidental and is not seen in other mammals. Duration of estrous or menstrual cycles in mammals varies widely in length, from the 4-day cycle of the golden hamsters to the 126-day cycle of the Indian elephant.

A number of reproductive disorders are associated with abnormal circadian or ultradian hormonal profiles. Pulsatile patterns of gonadotropin releasing hormone (GnRH) release from the brain (hypothalamus) are critical for the maintenance of normal profiles of gonadotropin release from the pituitary. In monkeys whose endogenous supply of GnRH has been eliminated, normal patterns of gonadotropin release can be restored with exogenous GnRH, but only under certain treatment schedules. If replacement GnRH is given continuously or at too high a frequency (e.g., once every half-hour), gonadotropin levels are suppressed. However, the same overall dose of GnRH, when given in hourly pulses, is sufficient to restore normal gonadotropin profiles.

Anovulatory cycles in women are sometimes traceable to inadequate GnRH pulse frequency, and pulsatile GnRH has been shown to be an effective treatment in hypothalamic amenorrhea. Pulsatile GnRH therapy has also been successfully used for the treatment of idiopathic hypogonadotropic hypogonadism in men. Abnormal patterns in pulsatile release of LH are often associated with abnormal reproductive function in both men and women. In amenorrhea associated with anorexia nervosa, the secretory patterns of LH regress to the pubertal pattern, with low daytime pulsatility and increased secretion at night, or to infantile patterns.

Humans are capable of reproducing throughout the year. Most other mammals, however, breed only during restricted seasons. Seasonal cycles in reproduction are generated from both endogenous and exogenous mechanisms. As mentioned above, ground squirrels show circannual cycles. They are only capable of breeding for one period of several weeks during each circannual cycle. When housed at a constant temperature and day length, or photoperiod (i.e., with no seasonal changes in day length), their seasonal breeding cycles will free-run, with a period of 10–11 months. A different type of seasonal cycle is seen in the golden hamster, whose reproductive status is determined by photoperiod. In hamsters exposure to short photoperiods (i.e., less than 12 hours of light per 24-hour period) induces gonadal regression. This effect is reversible by transfer to a long photoperiod (i.e., more than 12 hours of light per 24-hour period).

Photoperiodic regulation of reproduction relies on a physiological system for the measurement of seasonal time that involves integration of changing environmental information with the biological clock. In addition to regulating seasonal cycles, photoperiod also regulates the timing of the onset of puberty in many mammals. The photoperiodic response depends on the rhythmic pattern of melatonin release, the primary pineal hormone. Among vertebrates melatonin release is highest during the night and low during the day. In addition, melatonin release is directly suppressed by exposure to light. Although the role of melatonin rhythms in nonphotoperiodic animals remains unclear, there are hints that melatonin could play some role in human reproduction. Puberty in children with pineal tumors is usually disrupted and could be either precocious or delayed.

There are conflicting reports on seasonal fluctuations in human reproductive physiology. The onset of puberty occurs more often during months of longer day lengths. Annual fluctuations in serum testosterone levels have been reported in men, although the basis of such fluctuation is unclear. In human studies it is difficult to independently evaluate separate influences from the physical and social environments. Fluctuations in hormonal levels, as well as sexual activity and birth rate, are likely to be influenced by social and behavior factors—particularly by the timing of work schedules, annual holidays, and vacations.

IX. Biological Rhythms and Human Health

A. Jet Lag

The most abrupt dislocation of the biological clock we are likely to experience is after traveling across several time zones by plane. Depending on the direction of travel, the biological clock must be either phase-advanced or phase-delayed to synchronize with the new time zone. Several days to several weeks might be necessary to become fully resynchronized to the new schedule. During this period one is likely to experience symptoms of jet lag, which include sleep disruption, gastrointestinal disturbances, decreased alertness and attention span, and general feelings of malaise. The process of reentrainment is characterized by internal temporal disorder. Phase relationships between different rhythms are disrupted, or internally desynchronized, because some rhythms might take longer than others to become fully reentrained. After a 6-hour phase shift, activity rhythms in human subjects are usually reentrained within 3 days, whereas it might be 1 week or more before the rhythms of body temperature or sodium excretion rates are reentrained.

Rates of reentrainment are influenced by the number of time zones crossed, the direction of travel, individual differences, and environmental circumstances. Most people adjust more rapidly to westward travel, which involves delaying the biological clock, than they do to eastward travel. Because the average free-running period of human circadian rhythms is about 25 hours, most people find it easier to sleep one hour later in the morning than to rise 1 hour earlier. Differences between individuals in the number of days needed to adjust to new time zones are correlated with the amplitude of their body temperature rhythms. The strength or abundance of entraining agents also influences rates of reentrainment. People who arrive in a new time zone and spend their time indoors do not readjust as quickly as people who go out during the day—thereby exposing themselves to more intense light, social contacts, meal cues, and other reinforcing environmental factors. Despite popular claims, there is no clear evidence that particular diets minimize jet lag.

B. Shift Work

While jet travel induces the most abrupt dislocation of the biological clock, the disruptive schedules of shift work affect far more people: 15–25% of the working population in industrialized countries are engaged in shift work. Most shift workers are subjected to rapidly changing schedules that do not allow them to become synchronized. They generally remain at least partially entrained to a normal day–night cycle. Individuals differ greatly in their tolerance for shift work: Some report no complaints in adjusting to abnormal shifts, while others suffer sleep and digestive disorders. Differences in circadian physiology may account for at least some of these individual differences.

Human performance of many different tasks shows pronounced daily rhythms, particularly in tasks involving repetition and vigilance. It is not surprising, then, that the incidence of work errors is highest between 3:00 A.M. and 5:00 A.M., the low points in daily performance rhythms. This is a problem for worker safety, as well as public safety. We rely on the vigilance of thousands of shift workers who, for example, operate nuclear power plants, pilot airplanes, and monitor life support machines in hospitals, whose shifts cover hours when their performance rhythms are lowest.

A less common type of shift work subjects workers to abnormal period lengths. Nuclear submarine operators in the U.S. Navy work in 6-hour shifts, alternating with 12-hour rest periods. These 18-hour cycles are beyond the limits of entrainment for human circadian rhythms and these workers endure long-term disruption of their biological clocks. Experimental subjects exposed to similar cycles have problems with insomnia, emotional disturbances, and impaired coordination. Animal studies have associated such abnormal cycles or frequent phase shifts (e.g., weekly shifts of 6 hours or more) with decreased longevity, although it remains unclear whether this also applies to human physiology.

C. Rhythm Pathologies

Most known disorders of biological rhythms are those that interfere with the sleep–wake cycle. The most common complaint is difficulty in falling asleep at night. People with insomnia often have no problem once they fall asleep; the problem lies in the timing of sleep. However, many of these people have no problems sleeping when they are able to follow their own schedules, in contrast to the schedule imposed by a typical work day. Certain patients with this "delayed sleep phase" type of disorder can be successfully treated by phase-delaying their sleep cycles until they are in synchrony with "normal" schedules and therefore maintaining a strict 24-hour sleep–wake cycle.

Early morning waking is a classic symptom of clinical depression. Depressed psychiatric patients also show an earlier onset of REM sleep relative to nondepressed patients. Many rhythms are advanced in depressed patients. These include the daily profiles of prolactin, cortisol, growth hormone, melatonin, and body temperature. These atypical profiles are possibly due to circadian abnormalities (e.g., unusually short cycle lengths) or, alternatively, altered phase relationships between the pacemaker and the observed rhythms. Interestingly, a 6-hour advance of the sleep–wake cycle or sleep deprivation for 1 night can induce a remission lasting several days to 2 weeks.

Seasonal affective disorder (SAD) is characterized by seasonally recurring episodes of major depression. In the typical pattern patients become depressed during the fall and winter. The opposite case, in which depression occurs during the summer, has also been reported, but is much more rare. Bright light therapy (i.e., over 5000 lux, with indoor lighting ranging from about 100-1500 lux) is often effective in alleviating depression in SAD patients within several days, although relapse usually follows within 1 or 2 weeks.

The biological mechanisms underlying the therapeutic effects of bright light are unknown. Proposals that light therapy works by altering melatonin secretion or, alternatively, by causing a phase shift in the circadian oscillator have been shown to be insufficient to fully explain the effects of light in SAD patients. The causes of SAD are also unclear. Symptoms of SAD might simply represent the extreme end of a continuum of normal seasonal fluctuations. Studies of healthy individuals show clear seasonal variation in mood, with lows occurring during the fall and winter. The timing of seasonal mood cycles for SAD patients and healthy individuals is similar, but the cycles of SAD sufferers show a much greater amplitude. The peaks of both groups are similar, but the lows of the SAD cycles are much lower.

D. Medical Implications

Understanding the role of biological rhythms in human physiology has important implications for both the diagnosis and treatment of disease. As mentioned in Sections VII and VIII, disruptions in timing are associated with sleep and reproductive disorders. In some cases the key to a correct diagnosis lies in the temporal aspects of a particular variable. The diagnosis of adrenal dysfunction provides a clear illustration. In Cushing's disease the adrenal gland is hyperactive, whereas Addison's disease is caused by adrenal failure. In both cases serum cortisol levels are within the normal daily range, but show a loss in circadian rhythmicity. The critical difference is that in Cushing's disease cortisol levels remain near the normal daily maximum, while in Addison's disease cortisol levels remain near the daily minimum. Thus, information about the normal range of cortisol variation is insufficient to understand the nature of adrenal function in these cases. It is also essential to understand the temporal structure of cortisol secretion.

Daily rhythms have been shown in the therapeutic effectiveness of many drugs, including anesthetics, toxins, histamines and antihistamines, and drugs used in chemotherapy. Circadian rhythms could influence drug effectiveness in a number of ways. (1) Uptake: The rate of absorption from the intestine into the circulation could show circadian variation. This might also be true for intramuscularly administered drugs. (2) Metabolism: The rate of drug inactivation may show circadian variation. (3) Delivery to the target tissue: Blood volume and extracellular fluid volume show circadian variations. (4) Target tissue response: The target tissue may show circadian variation in the number or affinity of receptor binding sites for the drug.

The influence of circadian rhythms in human physiology is pervasive, and temporal organization is a fundamental element of human physiology. Our well-being relies on the biological clocks that maintain this temporal order.

Bibliography

Aschoff, J. (ed.) (1981). "Handbook of Behavioral Biology. Vol. 4: Biological Rhythms." Plenum, New York.

Moore-Ede, M. C., Sulzman, F. M., and Fuller, C. A. (1982). "The Clocks That Time Us." Harvard Univ. Press, Cambridge, Massachusetts.

Rosenwasser, A. M., and Adler, N. T. (1986). Structure and function in circadian timing systems: Evidence for multiple coupled circadian oscillators. *Neurosci. Biobehav. Rev.* **10,** 431–448.

Turek, F. W., and Van Cauter, E. (1988). Rhythms in reproduction. *In* The Physiology of Reproduction" (E. Knobil, J. Neill, L. L. Ewing, G. S. Greenwald, C. L. Market, and D. W. Pfaff, eds.). Raven, New York.

Van Cauter, E., and Aschoff, J. (1988). Endocrine and other biological rhythms. *In* "Endocrinology" (L. J. DeGroot, ed.), pp. 2658–2705. Saunders, Philadelphia, Pennsylvania.

Wehr, T. A., and Goodwin, F. K. (1983). "Circadian Rhythms in Psychiatry." Boxwood, Pacific Grove, California.

$$\begin{array}{l} H_2C—COO^- \\ \quad | \\ HO—C—COO^- \\ \quad | \\ H_2C—COO^- \end{array}$$

Citrate Cycle

DANIEL E. ATKINSON, *University of California, Los Angeles*

Glossary

Adenylate energy charge Effective mole fraction of ATP in the adenine nucleotide pool: ([ATP] + 0.5[ADP])/([ATP] + [ADP] + [AMP]), this function is a linear measure of the energy status of the adenylate nucleotides, the main energy-transducing system in the cell

Anabolism Synthesis or building up of cellular constituents

Catabolism Breaking down or degradation of molecules from the diet or from the cell's stores or structure

Flux Rate of flow of material through a metabolic sequence; when the sequence is at steady state, the rate of each reaction is equal to the flux through the sequence as a whole

Modifier Metabolite that modulates the catalytic properties of an enzyme, usually affecting the affinity with which the enzyme binds substrate; a positive modifier causes an increase in affinity for substrate and a negative modifier causes a decrease

THE CITRATE CYCLE (also known as the tricarboxylic acid or TCA cycle and as the Krebs cycle) is the central metabolic sequence in most aerobic heterotrophic organisms, including humans and other mammals. Although often thought of as merely the final sequence in the oxidation of glycogen, glucose, and other carbohydrates, the citrate cycle is the most important metabolic crossroads in aerobic metabolism and is involved in both the catabolism (breaking down) and the anabolism (synthesis, or building up) of most of the major classes of metabolites. The cycle or some of its constituent enzymes participate in the oxidation of carbohydrates, proteins, and fats, in the conversion of carbohydrate to amino acids or fats and of proteins and amino acids to carbohydrates or fats, and in the generation of starting materials for many biosynthetic sequences.

I. Reactions and Stoichiometry of the Cycle

In the citrate cycle, oxaloacetate condenses with the acetyl group of acetyl coenzyme A (AcSCoA), forming citrate and liberating free coenzyme A (HSCoA). Citrate is then oxidized in seven steps, with loss of two molecules of CO_2, to oxaloacetate. Thus, in each turn of the cycle, from oxaloacetate back to oxaloacetate, one acetyl group is fully oxidized to carbon dioxide. The reactions of the cycle are diagrammed in Fig. 1. The equations for the reactions and the structures of the intermediates are shown in the Fig. 1 legend.

II. Oxidations

A. Oxidation of Acetate

In the simplest case, the citrate cycle is the route for oxidation of acetate units that are derived from the oxidation of carbohydrates, fats, and some of the amino acids that are liberated in the hydrolysis of proteins. To enter the cycle, acetate units must be activated by formation of a thiol ester with coenzyme A. This activation is built into the degradation

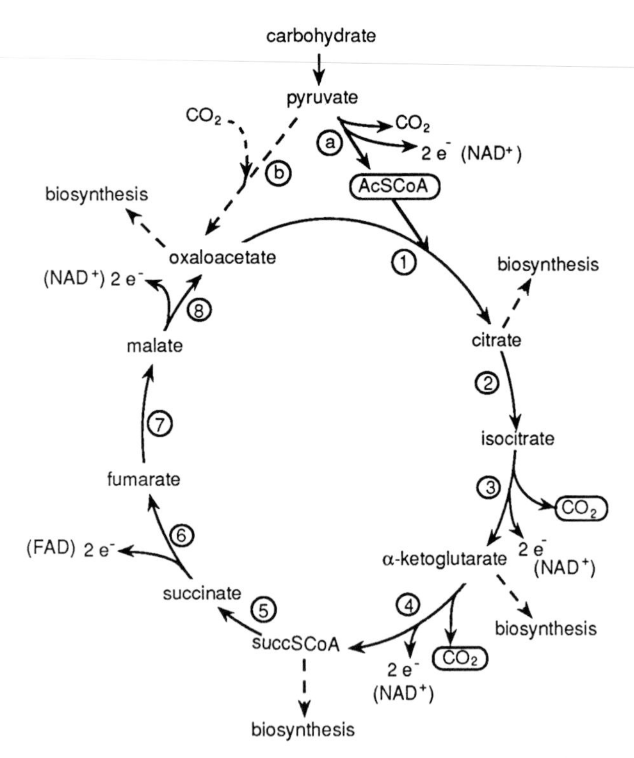

FIGURE 1 The citrate cycle. The input, acetyl coenzyme A, and the output, carbon dioxide, are indicated by circling. The electron transfer agents that accept electrons (e^-) at the oxidative reactions of the cycle are indicated in parentheses. Broken arrows represent reactions that contribute to the generation of biosynthetic intermediates.

The reactions, keyed to the identifying numbers in the figure, and the enzymes that catalyze them:

① citrate synthase:
$AcSCoA + OAA^{2-} + H_2O \rightarrow citrate^{3-} + H^+ + HSCoA$

② aconitase:
$citrate^{3-} \leftrightarrows isocitrate^{3-}$

③ isocitrate dehydrogenase:
$isocitrate^{3-} + NAD^+ \rightarrow \alpha\text{-ketoglutarate}^{2-} + CO_2 + NADH$

④ α-ketoglutarate dehydrogenase:
$\alpha\text{-ketoglutarate}^{2-} + HSCoA + NAD^+ \rightarrow succSCoA^- + NADH + CO_2$

⑤ succinate thiokinase:
$succSCoA^- + GDP^{3-} + P_i^{2-} + H_2O \rightarrow succinate^{2-} + GTP^{4-} + HSCoA$

⑥ succinate dehydrogenase:
$succinate^{2-} + FAD \rightarrow fumarate^{2-} + FADH_2$

⑦ fumarase:
$fumarate^{2-} + H_2O \rightarrow malate^{2-}$

⑧ malate dehydrogenase:
$malate^{2-} + NAD^+ \rightarrow OAA^{2-} + NADH + H^+$

Stoichiometry of the cycle (sum of reactions 1–8):

$$AcSCoA + 3NAD^+ + FAD + 3H_2O + GDP + P_i \rightarrow 2CO_2 + HSCoA + 3NADH + FADH_2 + GTP + 3H^+ \quad (9)$$

Although three protons (H^+) are shown to be generated for each molecule of acetyl coenzyme A oxidized, the citrate cycle does not cause acidification in the living cell. The protons are consumed in the course of the oxidation of NADH:

$$3NADH + 3H^+ + 1.5\ O_2 \rightarrow 3NAD^+ + 3H_2O$$

Subsidiary reactions shown in Figure 1 and discussed below are:

ⓐ pyruvate dehydrogenase:
$pyruvate^- + NAD^+ + HSCoA \rightarrow AcSCoA + NADH + CO_2$

ⓑ pyruvate carboxylase:
$pyruvate^- + CO_2 + ATP^{4-} + H_2O \rightarrow oxaloacetate^{2-} + ADP^{3-} + P_i^{2-} + 2\ H^+$

Structures of the intermediates of the cycle:

citrate^{3-}:
$H_2C—COO^-$
|
$HO—C—COO^-$
|
$H_2C—COO^-$

isocitrate^{3-}:
$H_2C—COO^-$
|
$HC—COO^-$
|
$HO—CH—COO^-$

α-ketoglutarate^{2-}:
$H_2C—COO^-$
|
CH_2
|
$O═C—COO^-$

succSCoA$^-$:
$CH_2—CO—S—CH_2—CH_2$ . . .
[coenzyme A]
|
$CH_2—COO^-$

succinate^{2-}:
$CH_2—COO^-$
|
$CH_2—COO^-$

fumarate^{2-}:
$CH—COO^-$
‖
$CH—COO^-$

malate^{2-}:
$HO—CH—COO^-$
|
$H_2C—COO^-$

oxaloacetate^{2-}:
$O═C—COO^-$
|
$H_2C—COO^-$

pyruvate$^-$:
$CH_3—CO—COO^-$

sequences; both in glycolysis and in fatty-acid oxidation, the two-carbon unit is generated directly as acetyl coenzyme A. [*See* COENZYMES, BIOCHEMISTRY.]

Carbohydrates are converted to pyruvate by the reactions of glycolysis. In the process, two molecules of pyruvate are produced from each molecule of glucose, with the transfer of four electrons to the metabolic electron carrier oxidized nicotine adenine dinucleotide (NAD^+). Pyruvate is oxidized to acetyl coenzyme A and CO_2 in a reaction catalyzed by the three-enzyme pyruvate dehydrogenase complex. In this reaction, two electrons are transferred to NAD^+ for each molecule of pyruvate, or four for each starting molecule of glucose; thus, eight electrons are transferred to NAD^+ in the course of conversion of one molecule of glucose, or one glucosyl unit of glycogen, to two molecules of acetyl coenzyme A and two molecules of CO_2. Sixteen additional electrons (eight for each molecule) are lost in the further oxidation of this acetyl coenzyme A to CO_2. This oxidation is carried out through the citrate cycle, which thus is responsible for the removal of two-thirds of the electrons that are lost in the total oxidation of glycogen or hexoses to CO_2. [*See* GLYCOLYSIS.]

Storage fats are hydrolyzed to fatty acids and glycerol. The fatty acids are oxidized to acetyl coenzyme A. Because four electrons are lost per two-carbon unit in the conversion of fatty acids to acetyl coenzyme A and eight remain to be removed in the cycle, it can be seen that also in fat degradation the reactions of the citrate cycle are responsible for most of the oxidation. [*See* FATTY ACID UPTAKE BY CELLS.]

In the digestion of dietary proteins or the breakdown of proteins of the cell, the first step is hydrolysis. The resulting amino acids are degraded by specific individual catabolic sequences. Among the products of some of these sequences is acetyl coenzyme A, which is oxidized through the citrate cycle.

In the oxidation of each molecule of acetyl coenzyme A, three pairs of electrons are transferred to NAD^+ and one pair to oxidized flavin adenine dinucleotide (FAD). The benefit that the cell derives from oxidation of carbohydrates, fats, and proteins is the provision of these electrons to the oxidative phosphorylation system, which regenerates adenosine triphosphate (ATP) from adenosine diphosphate (ADP) and inorganic phosphate at the expense of the free energy drop that is involved in the transfer of electrons from reduced nicotinamide adenine dinucleotide (NADH) or reduced flavin adenine dinucleotide ($FADH_2$) to molecular oxygen. The reduced carriers NADH and $FADH_2$ that are derived from the oxidative reactions of the citrate cycle are the major source of electrons to drive ATP synthesis in the cells of humans and most other aerobic organisms. [*See* ADENOSINE TRIPHOSPHATE (ATP).]

B. Oxidation of Intermediates of the Cycle

Proteins contain 20 kinds of amino acid residues. Each amino acid is catabolized, or degraded, by a specific pathway. Thus, unlike the catabolism of carbohydrates and fats, which in the main follow single pathways and lead to acetyl coenzyme A as the metabolite requiring further oxidation, the degradation of proteins proceeds by many pathways and leads to a number of metabolites. Among these, in addition to acetyl coenzyme A, are α-ketoglutarate, oxaloacetate, and succinyl coenzyme A (succSCoA). Fruits, vegetables, and green parts of plants contain considerable amounts of salts of citrate, isocitrate, succinate, and malate. Thus, a means of oxidation of cycle intermediates is needed.

Reactions 1–8 in Fig. 1 can accomplish only the oxidation of acetyl coenzyme A to CO_2. If any intermediate is supplied from outside the cycle, it will, as far as the cycle proper is concerned, be oxidized only to oxaloacetate. Because every molecule of oxaloacetate that is used in reaction 1 leads to its own replacement by reaction 8, the figure shows no way that a cell could metabolize oxaloacetate itself.

The auxiliary reactions that are called into play to deal with a metabolic or dietary supply of cycle intermediates are diagrammed in Fig. 2. Oxaloacetate is decarboxylated and phosphorylated at the expense of guanosine triphosphate (GTP) (ATP in some species) to produce phosphoenolpyruvate (PEP). Phosphoenolpyruvate is the immediate precursor of pyruvate in glycolysis, and it is converted to pyruvate with generation of a molecule of ATP. Thus, in the conversion of oxaloacetate to pyruvate (equation 10) the net conversion is only the loss of CO_2. One GTP or ATP is used and one ATP is produced, so there is no net effect on the nucleoside triphosphate energy supply.

III. Generation of Synthetic Intermediates

A. Use of Citrate Cycle Intermediates as Biosynthetic Starting Materials

Humans and other mammals are heterotrophs; i.e., they cannot produce carbohydrates and other cell

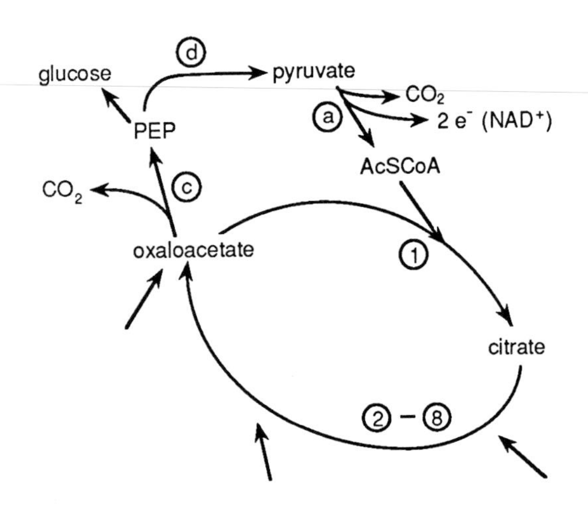

FIGURE 2 Reactions that allow the oxidation of intermediates of the citrate cycle. Two decarboxylations and loss of two electrons (e^-) convert oxaloacetate to acetyl coenzyme A, which enters the cycle and is oxidized in the normal way. Reactions 1–8 correspond to those in Fig. 1. The arrows pointing to the cycle represent entry of cycle intermediates derived from the diet or from catabolism of cell constituents.

Additional reactions:

ⓒ PEP carboxykinase:
oxaloacetate^{2-} + GTP^{4-} + H$^+$ → phosphoenolpyruvate^{2-} + GDP^{3-} + CO_2

ⓓ pyruvate kinase:
phosphoenolpyruvate^2 + ADP^{3-} → pyruvate$^-$ + ATP4^{4-}

Sum of reactions c and d:

(10) oxaloacetate^{2-} + H$^+$ → pyruvate$^-$ + CO_2

Structure of phosphoenolpyruvate:

```
 O⁻              COO⁻
   \            /
HO—P—O—C
   /            \\
 O⁻              CH₂
```

With the addition of reactions c and d, the citrate cycle is able to participate in the oxidative degradation of any compound that can be converted to acetyl coenzyme A or to any intermediate of the cycle.

constituents from carbon dioxide, but must depend on autotrophic organisms, mainly green plants, for their nourishment. Therefore, in animal metabolism, the starting materials for all syntheses must be derived from the breakdown of constituents of the diet. All of the many synthetic sequences start from intermediates of the main catabolic pathways: glycolysis, the pentose phosphate pathway, and the citrate cycle. Of the intermediates of the citrate cycle, citrate, α-ketoglutarate, succinyl coenzyme A, and oxaloacetate serve as starting materials for biosynthetic sequences.

Citrate leaves the mitochondrion (see Section IV) and, by way of the reaction catalyzed by citrate lyase (equation 11), serves as the source of cytosolic acetyl coenzyme A.

$$(11)\quad \text{citrate}^{3-} + \text{ATP}^{4-} + \text{HSCoA} \rightarrow \text{AcSCoA} + \text{OAA}^{2-} + \text{ADP}^{3-} + \text{P}_i^{2-}$$

Because acetyl coenzyme A supplies all of the carbon atoms of storage fats, of cholesterol, which is an essential component of membranes, and of the steroid hormones, supplying citrate to support cytosolic production of acetyl coenzyme A is an important function of the citrate cycle. [*See* CHOLESTEROL; STEROIDS.]

It will be noted that the reaction catalyzed by citrate synthetase (reaction 1 of Fig. 1), the movement of citrate through the mitochondrial membrane, and the reaction catalyzed by citrate lyase, acting together, have the consequence that acetyl coenzyme A and oxaloacetate are moved from the inside of a mitochondrion to the cytosol at the expense of the hydrolysis of a molecule of ATP. As noted, the beneficial consequence is the provision of acetyl coenzyme A for biosynthetic sequences in the cytosol. The oxaloacetate is generally not needed there, and most of it is reduced to malate and taken back into the mitochondria. This sequence of reactions illustrates the generalization that reactions can be made to go in the physiologically useful direction, and materials can be moved to other compartments, by use of ATP. Although it is citrate that actually crosses the mitochondrial membrane, the reactions discussed in this and the preceding paragraph are in effect a pump by which acetyl coenzyme is moved from the mitochondrion to the cytosol, with hydrolysis of ATP supplying the energy.

α-Ketoglutarate is converted by transamination to glutamic acid, an amino acid that is required in the synthesis of proteins. Glutamic acid is also the starting material for synthesis of two other amino acids, glutamine and proline. In addition to its use in protein synthesis, glutamine supplies amino groups in the synthesis of several other types of compounds, including the purines required in nucleic acid synthesis, and is also the primary form in which nitrogen is transported in the blood between tissues and organs in the animal body. [*See* DNA SYNTHESIS.]

Succinyl coenzyme A is required in the synthesis of heme, which is a component of hemoglobin and of cytochromes [*See* HEMOGLOBIN.]

Oxaloacetate is converted by transamination to aspartic acid, an amino acid required in protein synthesis. Aspartic acid is also the precursor to asparagine, another amino acid constituent of proteins, and of lysine, threonine, methionine, and isoleucine in organisms that can synthesize those amino acids.

B. Replenishment of Cycle Intermediates

In the basic citrate cycle, each molecule of oxaloacetate that is consumed in the synthesis of citrate is replaced by a molecule that is generated when the citrate is oxidized, as shown in Fig. 1. There is no net use of oxaloacetate in that case, as shown by equation 9, but each molecule of citrate, α-ketoglutarate, succinyl coenzyme A, or oxaloacetate that is removed from the cycle for use in biosynthesis prevents the regeneration of one molecule of oxaloacetate. There is thus a loss of oxaloacetate equimolar with the sum of all cycle intermediates used in synthesis. If that loss is not balanced by production of oxaloacetate, the cycle will soon come to a stop because of lack of oxaloacetate for reaction 1.

In mammals, the most important route by which oxaloacetate is replenished is the carboxylation of pyruvate (reaction b in Fig. 1). Thus, pyruvate has two routes of entry into the cycle. When the cycle is operating almost exclusively as an oxidative sequence supplying electrons for regeneration of ATP, the usual situation in muscle, only the entry by way of acetyl coenzyme A (reaction a in Fig. 1) is of importance. However, when a significant amount of synthesis is occurring, as is usually true in liver, kidney, and various other organs, pyruvate becomes an important partition point (see Section VI). Enough pyruvate must be carboxylated to exactly balance the amount of intermediates that are lost from the cycle.

Many kinds of organisms, including plants and many bacteria, replenish oxaloacetate by a different but equivalent reaction. Rather than pyruvate, they carboxylate its immediate precursor in the glycolytic pathway, phosphoenolpyruvate. Because enol phosphates are thermodynamically unstable, this carboxylation is driven by the loss of phosphate, and no ATP is required. Whether pyruvate carboxylase or PEP carboxylase is used, the end result in both cases is that oxaloacetate is generated from a three-carbon metabolite derived from carbohydrate by the glycolytic pathway.

IV. Intracellular Localization of the Enzymes of the Cycle

All of the enzymes of the citrate cycle are found on the inner mitochondrial membrane or in the matrix space inside the membrane. The enzymes of glycolysis are in the cytosol, so pyruvate is produced in the cytosol and must enter the mitochondrion before it can be further oxidized. The entire oxidative phosphorylation system, which converts ADP to ATP at the expense of the energy drop as electrons are transferred from NADH and $FADH_2$ to molecular oxygen, is located on the mitochondrial inner membrane. This location in the interior of the same organelle of both the oxidative phosphorylation system, which is powered by electrons, and the citrate cycle, which supplies most of the electrons (in the form of NADH and $FADH_2$), presumably enhances the efficiency of coupling between the two systems. In addition, the enzymes that convert fatty acids to acetyl coenzyme A are also located in the mitochondrion. Long-chain fatty acids enter the mitochondrion as esters of carnitine, a metabolite that is apparently specialized to participate in the transfer of fatty acids across the inner mitochondrial membrane. Thus, both electrons released in the conversion of long-chain fatty acids to acetyl coenzyme A and those released in the oxidation of acetyl coenzyme A to CO_2 are transferred to NAD^+ or FAD in the immediate vicinity of the system that uses the reduced carriers in the generation of ATP.

V. The Glyoxylate Cycle

Mammals convert carbohydrates to fat or protein (however, they can make only about half of the amino acids that are needed) and protein to fat or carbohydrate. As we have seen, the reactions of the citrate cycle play important roles in all of these interconversions. Mammals cannot convert fats to carbohydrates or protein, however, because the reactions of the cycle provide no route from acetyl coenzyme A, the product of the first stage of fatty-acid degradation, to anything except carbon dioxide.

The glyoxylate cycle (Fig. 3) is a modification of the citrate cycle that allows acetyl coenzyme A to supply carbon atoms for the production of all of the intermediates of the cycle and, hence, for all of the biosynthetic activities of a cell. Although it does not

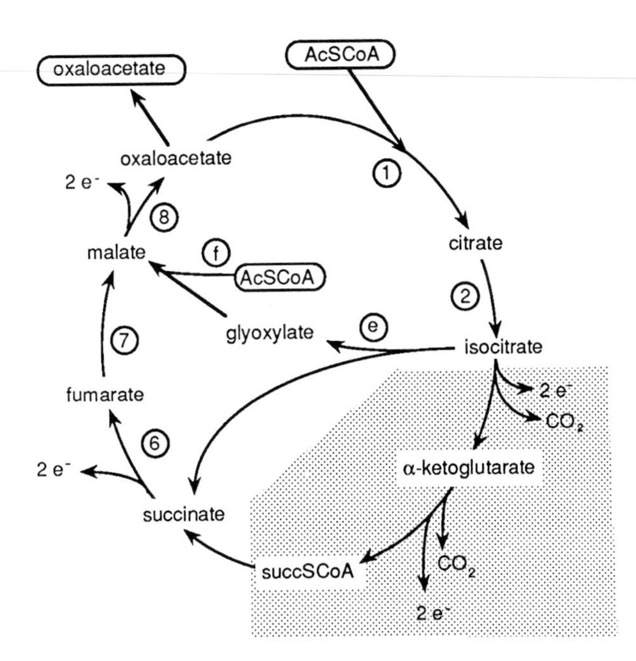

FIGURE 3 The glyoxylate cycle. The decarboxylation reactions of the citrate cycle (stippled area) are bypassed, and a second molecule of acetyl coenzyme A enters at reaction f. The resulting cycle converts two molecules of acetyl coenzyme A to one molecule of oxaloacetate.

exist in humans and other mammals, the glyoxylate cycle deserves mention in this treatment because of its great importance in the biosphere generally and its interest as an example of the fact that small changes can allow the same enzymes to participate in sequences that have different functions.

The two enzymes specific to the glyoxylate cycle are isocitrate lyase and malate synthase (Fig. 3).

ⓔ isocitrate lyase
isocitrate^{3-} → glyoxylate^{-} + succinate^{2-}

ⓕ malate synthase
glyoxylate^{-} + AcSCoA + H_2O → malate^{2-} + H^+ + HSCoA

The structure of glyoxylate is O═CH—COO$^-$.

When reactions e and f are combined with reactions 1, 2, and 6–8 of Figure 1, the overall conversion represented by Figure 3 is given by reaction 12.

(12) 2AcSCoA + FAD + 2NAD$^+$ + 3 H_2O →
oxaloacetate^{2-} + $FADH_2$ + 2NADH +
2HSCoA + 4H$^+$

The citrate cycle oxidizes acetyl coenzyme A to CO_2 and thus, serves as the final stage in the oxidation of many metabolites and cell constituents. It seems strange that five of the eight enzymes of the citrate cycle can participate in a metabolic sequence with a chemical direction and physiological function opposite those of the citrate cycle: catalyzing the first stage in the use of acetyl coenzyme A as the starting material for synthesis of all of the constituents of the cell. Figure 3 shows that this reversal of function is achieved by the substitution of reactions e and f for reactions 3–5. Both of the reactions in which CO_2 is released are bypassed (the stippled area of Fig. 3), and a second molecule of acetyl coenzyme A enters at reaction f. Each turn of the glyoxylate cycle starts with one oxaloacetate and generates two, resulting in a net gain. Reaction of the product oxaloacetate with a third molecule of acetyl coenzyme A allows synthesis of citrate and thus of all other cycle intermediates from acetyl coenzyme A. Reactions 3 and 4 are still available, so that a cell that is using acetyl coenzyme A as its sole source of carbon and energy can employ part of its isocitrate in making the α-ketoglutarate and succinyl coenzyme A that it requires for its biosynthetic activities.

Phosphoenolpyruvate, which can be made from oxaloacetate by the action of PEP carboxykinase (Fig. 2), can be converted to glucose and hence to storage glycogen and all of the structural carbohydrates that are required in cell walls and other cell components.

VI. Regulation

Because the reactions of the citrate cycle have oppositely directed functions of participating in all oxidative degradations and of providing materials for most biosynthetic sequences, regulation of the cycle and its component and associated reactions is of central importance to the economy of the cell. The cycle is related to or affects nearly all of the activities of an organism, so it is not surprising that it should be regulated by many factors that do not impinge on it directly. A complete description of regulatory influences on the cycle would include most of the regulatory and correlative interactions of the cell or organism. We are far from having the information necessary for such a complete description, and only a sampling of what is known can be included here.

Acetyl coenzyme A is the primary fuel of the citrate cycle. Therefore, the flux of material through the cycle must be determined in part by factors that control the rate of generation of acetyl coenzyme A.

The metabolism of both carbohydrates and storage fats is regulated by complex systems that respond to many inputs, among them the energy status of the cell, which can be expressed as the adenylate energy charge or the ATP/ADP ratio. As a result of those interactions, the rate at which acetyl coenzyme A is made available by catabolic sequences increases when the energy status of the cell decreases slightly. Thus, the supply of fuel to the cycle increases when an increased rate of ATP production is desirable. Those sequences also react appropriately to organism-wide needs by responding to hormonal signals.

At the level of the citrate cycle itself, citrate synthase (reaction 1 of Fig. 1) is regulated by the adenylate system (ATP, ADP, and AMP) in some or all organisms, and by the degree of reduction of the NAD^+/NADH electron-carrier system in some species. The direction of the responses is again such that a greater rate of entry of acetyl coenzyme A into the cycle is favored when the energy status of the cell is slightly below its normal value.

Isocitrate dehydrogenase (reaction 3 of Fig. 1) is sensitively regulated by the adenylate energy-transducing system (ATP, ADP, and AMP). When the energy status is low, this enzyme has high affinity for its substrate isocitrate. Thus, nearly all of the isocitrate that is produced by reactions 1 and 2 will proceed through reaction 3, maximizing the flow through the remainder of the cycle and the rate of donation of electrons to the oxidative phosphorylation system to support regeneration of ATP. When the energy charge is high, the affinity of isocitrate dehydrogenase for its substrate is much reduced. Thus, isocitrate must accumulate to a higher concentration before it can be converted to α-ketoglutarate by action of the dehydrogenase. When the concentration of isocitrate rises, so will that of citrate, with which it is in near-equilibrium. The resulting high level of citrate favors its movement from the mitochondrion to the cytosol, where it is cleaved to supply the acetyl coenzyme A that is required for synthesis of storage fats, steroids, and other cell constituents. Citrate is also a strong negative modifier or inhibitor of phosphofructokinase, the most important regulatory enzyme in glycolysis.

To summarize, it appears that if the energy status is poor, nearly all of the isocitrate is channeled along the citrate cycle. The rate of ATP regeneration is maximized. Because of the low concentration of citrate, little of it moves into the cytosol. The rates of biosyntheses that use acetyl coenzyme A are limited by lack of substrate (there are also direct effects of the energy charge on the enzymes of these sequences). This decrease in the rates of syntheses conserves energy (ATP). Because of the low concentration of citrate in the cytosol, the activity of phosphofructokinase is not diminished by citrate binding, and a rapid rate of glycolysis, which supplies acetyl coenzyme A to the citrate cycle, is favored. The result is an increase in the rate of regeneration of ATP.

In contrast, when the energy charge is high, the accumulation of isocitrate and citrate lead to movement of citrate into the cytosol. Under those conditions, energy is available for biosyntheses, and the increased supply of cytosolic acetyl coenzyme A appropriately favors increased rates of synthesis. If the rate of supply of citrate to the cytosol exceeds the level required to meet the cell's synthetic needs, citrate tends to accumulate and inhibit phosphofructokinase, thus decreasing the rate at which glycolysis supplies acetyl coenzyme A to the citrate cycle. Thus, the rate of regeneration of ATP and of citrate itself decreases. The interaction of these and many other related control systems maintains the energy charge, the ATP–ADP ratio, and the concentrations of cell constituents within fairly narrow physiological ranges.

It can be seen from Fig. 1 that pyruvate is an important metabolic branchpoint. If all of the pyruvate that is supplied by glycolysis is oxidatively decarboxylated to form acetyl coenzyme A, the cycle will furnish the maximal number of electrons to the ATP-regenerating machinery, but it will not be able to supply any starting materials for biosynthesis. That condition is approximated closely in a working muscle.

However, under most conditions the cycle must also supply some of its intermediates to biosynthetic sequences, and it is necessary that enough pyruvate be carboxylated to oxaloacetate to exactly match those losses. The responses of pyruvate dehydrogenase and pyruvate carboxylase (reactions a and b in Fig. 1) to variations in the concentration of acetyl coenzyme A are major factors in the partitioning of pyruvate between the two reactions. When intermediates are being withdrawn from the cycle for use in biosynthesis, the concentration of oxaloacetate will decrease. This will limit the rate of reaction 1 (the formation of citrate). Because it is being used more slowly, the concentration of acetyl coenzyme A will rise. Acetyl coenzyme A is a strong negative modifier for pyruvate dehydrogenase (reaction a) and a strong positive modifier for pyruvate carboxylase (reaction b); therefore, when the concentration of acetyl coenzyme A rises slightly, the partitioning of pyruvate changes: More is directed toward oxaloacetate and less toward acetyl coenzyme A. This response will automatically hold the concentration of oxaloacetate at a level that is adequate for citrate synthase, which catalyzes reaction 1. By supplying oxaloacetate at the rate needed to stabilize its concentration, the system is also automatically adjusting the rate of input of oxaloacetate to exactly match the sum of the rates of removal of cycle intermediates.

Most metabolic correlation and control is exerted by adjustment of partitioning ratios at branchpoints. The pyruvate branchpoint involves only partitioning between two alternate routes of entry into the citrate cycle, but it is nevertheless the most centrally important of the metabolic control points in metabolism, because it deals with the fundamental distinction between degradation and biosynthesis. Use of cycle intermediates for biosyntheses, including syntheses of storage carbohydrates and fat, is possible only to the extent that pyruvate is directed toward oxaloacetate, while energy for synthesis and other cell activities results in very large part from oxidation of the pyruvate that is directed toward acetyl coenzyme A. Constant adjustment of the balance between the rates of these two reactions that compete for pyruvate allows the citrate cycle to meet the shifting momentary demands of the cell for energy and for biosynthetic starting materials, and to support production of storage glycogen and fat when resources are available in excess of present demand.

Bibliography

Atkinson, D. E. (1988). The tricarboxylic acid cycle. *In* "Biochemistry," 2nd ed. (G. Zubay, ed.). Macmillan, New York. pp. 481–511.

Kornberg, H. L. (1987). Tricarboxylic acid cycles. *BioEssays*. **7,** 236–328.

Mathews, C. K., and van Holde, K. E. (1990). *In* "Biochemistry." Benjamin/Cummings, Redwood City, California. pp. 467–503.

Rawn, J. D. (1989). *In* "Biochemistry." Neil Patterson Publishers, Burlington, North Carolina. pp. 329–358.

Voet, D., and Voet, J. G. (1990). *In* "Biochemistry." Wiley, New York. pp. 506–527.

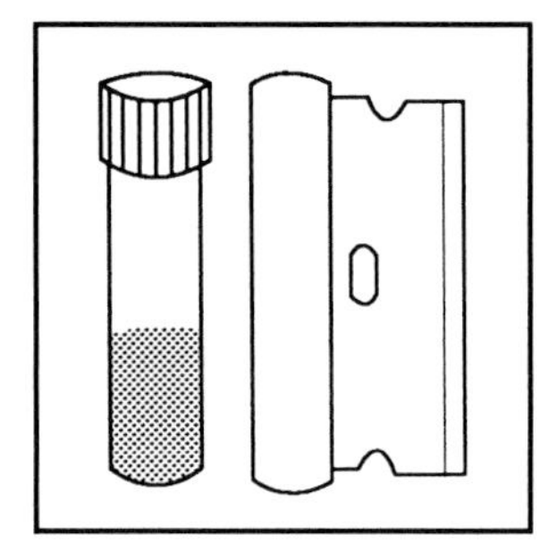

Cocaine and Stimulant Addiction

TONG H. LEE, EVERETT H. ELLINWOOD, *Duke University Medical Center*

Glossary

Anhedonia Loss of ability to obtain pleasure from hobbies and activities

Binge Pattern of stimulant abuse characterized by repetitious administration of high doses so as to sustain stimulant euphoria

Cocainisinus Turn-of-the-century term for the cocaine abuse–addiction syndrome leading to subsequent excessive abnormal behavior

Dysphoria Extreme, uncomfortable, disagreeable feeling

Freebase Free form of cocaine that is produced from cocaine hydrochloride by chemically removing the hydrochloride ions. Unlike the hydrochloride form, the freebase is heat stable and can be smoked

Hypersomnia Intense and prolonged sleep characterized by its excessive duration often with multiple intense dreaming episodes

Psychomotor Pertaining to combined physical and mental activity, or to movement that is psychically determined

Reinforcement Term used in the learning theories to designate a stimulus, drug, or situation that brings on an increase in incidence of a behavior (e.g., milk that is used to train a hungry rat to bar-press)

Tachyphylaxis Rapid appearance of decreased response to the drug in the body as a result of prior exposure to the drug

STIMULANTS, IN THEIR BROADEST definition, are agents that excite or stimulate various systems in the body. Thus, central stimulants are classified as such for their common ability to stimulate functions mediated by the central nervous system. According to this broad definition, central stimulants may include convulsants, which produce convulsions by excessive stimulation of the brain; however, the term central stimulants is commonly used in a more restrictive sense for those agents that produce alertness, elevated mood and interest, decreased appetite, and increased motor and speech activity. Defined in this fashion, central stimulants include, among others, cocaine, a spectrum of amphetaminelike drugs, caffeine (as well as a group of xanthine drugs sharing a similar chemical structure with caffeine), and some nonamphetamine-type diet pills. Central stimulants have legitimate medical uses; on the other hand, the current cocaine "epidemic" and consequent declaration of a "war on drugs" in the United States have focused on abuse problems associated with certain central stimulants presently in the limelight. The scope of this chapter is primarily to review selective central stimulants, namely cocaine and amphetamine, which have specific medical indications but also have been repeatedly abused in history. Because of the limited scope of this chapter, the term stimulant will denote those two drugs; when the term is used in a more general sense, the usage will be clear from the context.

Encyclopedia of Human Biology, Volume 2.

I. History

Uses of plants for their central nervous system-stimulating effects date back to prehistorical time. The earliest written record of psychomotor stimulant use is ascribed to the Chinese emperor Shen Nung (c. 3100 B.C.), who described the medicinal use of the herb Ma Huang, a plant containing the central stimulant ephedrine.

Cocaine also has historical precedents: the native coca plant (*Erythroxylon coca*) has been cultivated in South America since prehistoric time. Archaeological evidence suggests that its leaves were chewed for mental and physical energy, religious or sacramental reasons, and even nutritive sustenance throughout western South America, especially in the high Andes. Bags containing coca leaves and flowers have often been unearthed in burial sites even dating back to c. 2500 B.C. In the Incan Empire, the plant was considered a gift from the gods and served important social and religious functions. The Temples of the Sun were adorned by solid gold models of coca sprigs, and their altars could be approached only by the elite with coca in their mouths. In addition, use of coca was restricted to the aristocracy and other personages designated by royalty. Among those privileged were priests, doctors and the fabled long-distance runners who relayed messages along the well-developed Incan road system.

Following the Spanish conquest of the Incan Empire, use of coca was initially prohibited by the conquistadores as satanic idolatry. However, soon recognizing that the stimulant effect of coca enabled the enslaved natives to endure forced labor, the Spanish reinstated its use; workers were both paid and taxed in coca.

Despite favorable reports from South America about the effects of coca leaves, they were not widely used in Europe until the nineteenth century. Modern investigators have speculated that this lack of popularity was primarily due to the unavailability of fresh coca leaves. Delays during long voyages from South America led to decay of coca leaves, rendering them useless; in addition, the European climate was not conducive to local production of leaves. As expected, isolation of a stable, active compound (cocaine) from coca leaves by Niemann (in 1855) was followed by widespread use of the stimulant.

Perhaps the scientist who did more than any other to elucidate behavioral effects of cocaine was Sigmund Freud. He published several reports in which, from experimentation on himself, he correctly identified cocaine as both a central nervous system stimulant and a euphoriant. At the same time, he recommended its use as an antifatigue formula or aphrodisiac, and for treatment of a wide range of disorders, including asthma, digestive disorders, and alcohol and morphine addictions.

The late nineteenth century can be considered the heyday for cocaine. Positive opinions within the scientific community abounded, and cocaine or cocaine-containing formulations (many of which were patented medicines) were indiscriminately prescribed; sporadic reports of adverse reaction to these products were generally ignored as atypical reaction. Freud himself considered the toxic reactions as a manifestation of character defect. In the public, the high demand for cocaine was captured by entrepreneurs in the form of concoctions. Both in Europe and the United States, a tonic mixture of coca extract and wine was used and indeed endorsed by the elite, including United States President William McKinley, Thomas Edison, August Rodin, and Pope Leo XVIII. In 1896, the need for a nonalcoholic substitute for the above preparation (alcohol prohibitionist sentiment was high during this period) was exploited by Atlanta entrepreneur John Pemberton, who introduced such a preparation containing extract of coca leaves and caffeine-rich African kola nuts; Coca Cola® was advertised as the "intellectual beverage and temperance drink."

The early twentieth century marked a period in which society finally began to take note of the ever-increasing number of reports on cocaine toxicity. In response to a rise in public concern, the United States federal government began to regulate the manufacture and sale of cocaine-containing preparations. The Pure Food and Drug Act of 1906 required listing cocaine on the labels of all cocaine-containing patent medicines. The Harrison Narcotic Act in 1914 put a further restriction on coca products by forbidding the use of cocaine in proprietary medicines and requiring registration of those involved in coca product trade. With the advent of these restrictive measures, use of cocaine "went underground." Except for a brief resurgence in the 1920s, its use was limited to Bohemian, art, and music subcultures until the early 1970s, when the current epidemic began.

Amphetamine was first synthesized in 1887, and its pharmacology was first studied in 1910. However, because the experiments were performed in anesthetized animals, its central stimulant properties were not discovered until it was independently resynthesized in 1927 as part of a search for synthetic substitutes for ephedrine in the treatment of asthma. In the 1930s, the drug began to be used in nasal inhalers and for treatment of narcolepsy in tablet form; the central stimulant (e.g., euphorogenic) effects of amphetamine became apparent in patients being treated with the above preparations. Not surprisingly, the latter discovery was followed by increasing nonmedical uses by the public; it was as if the public had substituted the cheaper, more widely available amphetamine for cocaine, which was by now expensive and tightly regulated. The general public as well as the medical community had forgotten about the previous cocaine epidemic, and major amphetamine epidemics occurred in Japan (1950–1956), Sweden (1964–1968), and the United States (1965–1969) during the following 30-year period.

II. Pharmacology

A. Catecholamine Effects

Many communications between cells (e.g., nerve cells) are mediated by chemicals, which, following release from a group(s) of cells, produce specific effects on sensitive cells. This chemical mode of transmission carries out much of the information transfer in the nervous system. The actions of the compounds or neurotransmitters are terminated by, among others, their degradation or reuptake back into the originating cells for reutilization. Amphetamine and cocaine share the property of potentiating transmission mediated by a particular class of chemicals, namely catecholamines. Catecholamines, so named because they are amines containing the catechol moiety in their structures, act as mediators of a wide range of neural functions. The two stimulants potentiate responses mediated by catecholamines by increasing their extracellular concentrations; however, the exact mechanisms of action are different. Amphetamine directly causes increased release of catecholamines from nerve terminals, whereas cocaine acts by blocking reuptake of these neurotransmitters. Because reuptake plays a major role in terminating the actions of catecholamines, its blockade leads to results similar to those produced by a direct releaser.

Catecholamines have a wide variety of critical functions in the body. One set of catecholamines, norepinephrine and epinephrine, mediate actions of the sympathetic nervous system, which plays a major role in preparing the body for coping with stress and emergency. Activation of sympathetic activity leads to increased heart rate, dilation of the airways, and diversion of blood from the digestive tract to the muscles. The sympathomimetic effects of the stimulants (i.e., mimicking the action of the sympathetic pathway) is thought to be responsible for many of the complications produced by stimulants.

On the other hand, another catecholamine, dopamine, is involved with mediating incentive reinforcement (see below) and modulating a variety of motor and cognitive functions by the brain. Dopamine mechanisms are most often associated with reinforcing effects in incentive-type behaviors. Innate behaviors in lower animals are represented by exploratory stalking, hunting, and foraging behaviors, which appear very early in the young animal's development; these innate behaviors are energized–motivated during this developmental phase without any consumatory reinforcement (e.g., food, specific objects). After interactions with the environment over time, these behaviors become integrated into cascades of object-related behavior. Increasingly, they are integrated during learning into automatic sequences of behavior with attendant emotional responses. High-dose stimulant intoxication in lower animals provokes these exploratory intrinsic behaviors into long trains of stereotyped repetitions of a behavior without object relatedness, which can be triggered by cues associated with injection (e.g., saline injection). If the experimental animal is bar-pressing for small injections of stimulant into a vein or the brain, the drug-administering behavior usually becomes compulsively stereotyped. A drug-associated cue can trigger compulsive drug administration (e.g., monkeys self-administering cocaine when given free access will compulsively dose themselves to death). [*See* CATECHOLAMINES AND BEHAVIOR.]

Cocaine and amphetamine stimulants at low doses induce pleasurable reinforcement, and at high doses intense orgiastic reinforcement. These reinforcement effects have been pinpointed to be secondary to the release of dopamine from terminal areas of very specific dopamine tracts. A variety of studies including (1) electrical stimulation of these tracts, (2) infusion of dopamine agonists into the terminal area, and (3) blockade of dopamine agonist reinforcement with antagonists have validated this basis of stimulant-induced reinforcement.

In the early stages of human stimulant use, many everyday behaviors become pleasantly reinforced within an environmental condition. Later, the environmental repertoire becomes increasingly constricted to the stereotyped incentive patterns often focused on the procurement (foraging) and use of the stimulant. Other human stimulant-induced innate patterns of behavior successively evolve from initial exploratory behaviors associated with curiosity to intense stereotyped suspicious behaviors, which then evolve into paranoid stimulant psychoses. The means by which stimulants act on dopamine incentive behavioral systems is currently under intensive scientific study.

B. Local Anesthetic Effect

Local anesthetics block nerve conduction when applied in one area, and although they affect all types of nerves, their clinical use is to block nerves carrying information from pain receptors (e.g., their use by dentists to numb pain). Unlike other stimulants, cocaine also acts as a local anesthetic; in fact, it is the first of its kind to be discovered. Clinical application of cocaine as a local anesthetic was initiated in the late nineteenth century by a Viennese physician who discovered its utility in eye surgeries. However, this practice in ophthalmological surgeries has been abandoned because of its toxic effect on the eye. Current application is limited to topical application in the upper respiratory tract (e.g., nasal mucous membrane [see below]).

C. Behavioral Stimulant Effects in Human

Despite their structural and other differences, cocaine and amphetamine are similar in many of their behavioral effects in humans; some experienced abusers may have difficulty distinguishing between the two drugs. Stimulants are among the most powerful euphorogenic drugs known. At even moderate doses, they produce alertness and a sense of well-being and heighten self-esteem as well as increased pleasure derived from various activities. At high doses taken by a rapid route of administration (e.g., "crack" smoking), the euphoria produced by stimulants is often described as consisting of two stages: an intense, "orgiasticlike," short-lasting euphoria (e.g., cocaine "rush"), followed by a more sustained sensation of well-being ("high").

The main difference between cocaine and amphetamine appears to be the duration of their effects. This length of time depends on various factors including the specific drug's half-life and development of tachyphylaxis and tolerance (see below). Practically defined, half-life is the time for drug concentration in the blood to decline to one-half of its initial value. Everything else being equal, a stimulant with a longer plasma half-life lasts longer in the body and, consequently, produces longer-lasting euphoria. Thus, based on cocaine's rapid metabolism and plasma half-life of 90 min or less, it can reasonably be predicted that the drug's effects will dissipate much faster than those of amphetamine with an average half-life of 6–8 hr.

Decreased sensitivity to a drug acquired as a result of prior exposure to the drug is termed drug tolerance; usually, it develops following multiple doses of the drug over a period of time (e.g., days). When it develops more rapidly (e.g., sometimes with the second dosing), it is called tachyphylaxis. Another term, acute tolerance, was also used to describe a similar phenomenon of rapid decrease in sensitivity. Classically, acute tolerance refers to a development of decreased sensitivity while the drug is still present in the body (i.e., the effect of a drug subsides faster than would be predicted by the blood level of the drug). Tachyphylaxis, on the other hand, describes a decreased sensitivity to a *subsequent* dose of a drug (i.e., the same amount of the drug as on the previous occasion produces a lesser effect). Over the years, the distinction between the two terms has become blurred, probably because it is not clear whether or not the two represent *mechanistically* different phenomena.

Much evidence, both clinical and basic, suggests that the central stimulant effects of cocaine and amphetamine are prone to a development of tolerance and tachyphylaxis (acute tolerance). Thus, decreasing sensitivity to cocaine is usually manifested by the user taking multiple, increasing doses over a short period of time (to maintain the cocaine "high"). A similar phenomenon is also observed in amphetamine users (i.e., tachyphylaxis in its classical definition). In addition, because of the longer plasma half-life of amphetamine, a faster decrease in the amphetamine can be readily observed while its blood level falls over the next few hours (acute tolerance). Further reduction in stimulant-induced euphoria over days or weeks (tolerance) is also noted in both legitimate (e.g., treatment of narcolepsy) and illicit uses.

D. Medical Application

In medical practice, amphetamine (along with related compounds methamphetamine and methylphenidate) are used for their alertness-producing and appetite-suppressant effects. Clinical practice of utilizing stimulants for sympathomimetic effects (e.g., for airway dilation in asthmatic patients) has been abandoned because of the high-abuse potential and the availability of safer drugs. Stimulants are, at present, prescribed for narcolepsy, which is characterized by a sudden uncontrollable disposition to sleep, and for short-term treatment of obesity, which has been unresponsive to alternative forms of therapy (e.g., repeated dieting and/or other drugs). Interestingly, another major use of the above agents is to calm down children with attention-deficit disorder (also called hyperactive child syndrome). The disorder is characterized by impulsive behaviors, low frustration tolerance, distractiblity, hyperactivity, and memory difficulties—a set of problems that may be expected to be exacerbated by stimulants at higher doses. It is emphasized that the potential for significant side effects or abuse is high even when stimulants are used under medical supervision; consequently, physicians are warned against imprudent use of stimulants.

Cocaine, which shares the common properties of stimulants, is not used in the above fashion because of, among various factors, its short half-life (see above). The requirement for frequent dosage for the desired therapeutic effects greatly enhances toxic and abuse liability. Currently, cocaine is used as a topical application as a local anesthetic, mostly in the upper respiratory tract. For example, emergency physicians can directly examine a patient with nasal trauma by applying cocaine to the mucous membrane in the nose. The drug reduces pain as well as bleeding (cocaine's vasoconstrictive property constricts capillaries, allowing less blood to flow through the affected area), thus greatly enhancing visualization and treatment of the traumatized area.

III. Epidemiology of Stimulant Abuse

The United States is currently experiencing a cocaine epidemic. In the 1980s, millions of people tried cocaine for the first time, as the positive effects of its use (e.g., alertness, euphoria, increased mental and physical energy) were espoused, indeed proselytized. In 1986, the National Institute on Drug Abuse estimated that more than 3 million people in the United States abused cocaine regularly, more than five times the number of heroin addicts. The spread of cocaine use was associated with a precipitous 15-fold increase in cocaine-related emergency room visits over the preceding 10-yr period. In addition to a general upswing in cocaine use, increasing utilization of intravenous injection or freebase ("crack" or "rock") smoking has become a major health policy issue because of the higher propensity of these forms to induce addiction as well as more severe toxicity. The advent of the highly purified form of cocaine freebase known as crack is especially alarming because (1) the form is inexpensive and widely available, (2) it is simple to use, and (3) its rapid absorption and subsequent high blood level is associated with an increased propensity toward compulsive use. Report of a dramatic rise in first-time hospitalization for cocaine abuse (224 in 1984 compared with just 1 in 1980) shortly following the introduction of crack in the Bahamas illustrates the increased risk.

Past experience indicates that several sequential events occur during the establishment of a stimulant epidemic. These factors for amphetamine epidemics could well apply to the 1980s cocaine epidemic.

1. Introduction of the stimulant to the population for recreational purposes, for medical purposes, or for its antifatigue properties.
2. Widespread dissemination of knowledge and, at times, proselytizing of the intensely euphoric stimulant experience.
3. Development of a sufficient core of buying and selling abusers who establish a reliable illegal market for the stimulant.
4. The illegal supply routes providing for immense profits.
5. The proliferation of multiple illegal drug sources compensating for government curbs.
6. Increasing use of rapid onset routes of administration associated with an intensified stimulant effect (e.g., intravenous injection).

One sociological factor aiding in the successful establishment of a stimulant epidemic appears to be the prevailing misguided positive opinions among the general public and scientific community about the particular drug(s). These opinions are often enhanced by positive portrayals from the media, folk heroes, entertainers, etc., undaunted by lessons of historical experience. For example, as discussed above, the cocaine epidemic of the late nineteenth century occurred in the context of highly positive opinions about the drug, both in the general and scientific communities. In addition to being "unequaled as a tonic-stimulant for fatigued or overworked body and brain" (taken from an advertisement for a nostrum containing cocaine), it was espoused as a cure for a variety of ailments including opium addiction, alcoholism, asthma, tuberculosis, impotence, and digestive disturbances. During this time, reports of adverse reactions to the drug were largely ignored. The subsequent stimulant epidemics of the 1920s, 1950s, and 1960s were again characterized by initial enthusiasm for the virtues of the stimulants untempered by historical perspective.

With respect to the most recent cocaine epidemic, well-documented medical literature from the early twentieth century on adverse effects of cocaine abuse (i.e., cocainisinus) was later dismissed in the 1970s and 1980s as Victorian moralizing. Even recent medical literature was dismissed by the media, which heralded cocaine as the champagne of drugs used by cultural heroes and the elite, and described it as quite safe. For example:

> "Cocaine is less harmful than any legal and illegal drugs popular in America. Most of the evidence is that there aren't adverse effects to normal cocaine use. It looks to be much safer than barbiturates and amphetamine and there is no evidence that it has the body effects of cigarettes or alcohol. If I were going out to sell a drug to the public it would probably be cocaine. We might be better off using it as a recreational drug than marihuana." (San Francisco Chronicle, Oct. 21, 1976, p. 4)

According to historical experience, the current cocaine epidemic may well abate within the next few years. Society will gradually become wary of the negative aspects of the drug (e.g., its toxicity, the increasingly criminal and violent drug subculture) and will increasingly exert specific legal and medical countermeasures. The restrictive Harrison Narcotic Act in 1914 in the United States was passed largely in response to increasing newspaper publicity about the problems associated with cocaine abuse in the late nineteenth and early twentieth centuries (e.g., aggressive, sex-crazed cocaine "dope fiends"). With respect to the current epidemic, current survey studies indicate that in 1989 the incidence of first-time cocaine and casual use has indeed begun to decline. Eventually, negative opinions about cocaine will take a strong hold on the public, and a hiatus will ensue. Unfortunately, history also warns us that in the future people may again forget about the "cocaine epidemic of the 1980s." Will another major stimulant epidemic with one of the known stimulants or a new "safe one" subsequently ensue? Is there a way to prevent this periodic reenactment of this history?

IV. Clinical Characteristics of Abuse

A. Initial, Low-Dose Use

Typically, individuals are initially exposed to single low to moderate doses of stimulants for therapeutic, recreational, and other purposes. The initial recreational form of cocaine use usually takes the form of snorting fine cocaine crystals into the nasal passages, where it is absorbed into the bloodstream by the mucous membrane lining. Taken through this route, cocaine retards the rate of its own absorption because of its powerful vasoconstrictive effect (narrowing of blood vessels). Swallowing pills (e.g., the usual route of administering amphetamine diet pills) is also a relatively slow absorption process. Consequences of delayed absorption are a relatively slow onset of behavioral responses and decreased peak drug concentrations in the blood. The latter results because the body's ability to break down the drug can better keep up with the slow absorption. During the initial low-dose stimulant use, the main factor maintaining the drug-taking behavior appears to be positive responses from others to the user's increased energy and productivity, rather than the stimulant-induced (pharmacological) euphoria *per se,* which is reduced in its intensity anyway due to the lower-peak concentrations.

Most stimulant users do not become compulsive stimulant addicts, whose main daily concern is attainment and maintenance of a stimulant "high." The National Institute on Drug Abuse has estimated that 80% of 30 million Americans who have tried cocaine intranasally do not progress to regular users and, thus, either stop or remain periodic recreational users. Predisposing factors to continuing use and then to compulsive stimulant abuse in the remaining 20% are not well known, although desires to re-experience the stimulant high and to avoid the subsequent dysphoria and stimulant craving appear to play a role.

B. Compulsive Abuse

Some low-dose users discover that certain routes of administration intensify euphoria. Thus, they discover that intravenous use or crack smoking can produce extreme pleasure. The individual's daily activities soon become devoted to a search for pharmacological euphoria. Reaction from others, which plays a major role in maintaining the low-dose use, no longer matters, and the individual becomes more socially isolated.

In its most severe form, compulsive stimulant abuse is characterized by binges, in which high doses are repetitively administered in an attempt to "chase" the stimulant high. Binging episodes lasting up to a few days are usually terminated by extreme physical exhaustion and/or exhaustion of the drug supply. The binges may occur one or more times per week and, depending on the available supply, the abuser may consume 10–20 g or more. When money is readily available, as much as $250,000/yr may be spent on cocaine.

The propensity of stimulants to produce a compulsive abuse pattern appears to be due to at least three factors: (1) their powerful reinforcing property; (2) the short half-lives of stimulants, which is especially true for cocaine (<90 min); and (3) development of tolerance–tachyphylaxis to the pharmacological euphoria. In addition, high doses and rapid administration routes increase the intensity leading to a high-dose transition to abuse. [*See* NONNARCOTIC DRUG USE AND ABUSE.]

C. Withdrawal

The "crush" phase, the initial phase of stimulant withdrawal, immediately follows a binging episode. Initially marked depressive dysphoria, anxiety, and agitation are noted followed by craving for sleep over the next few hours. Often, the individual uses a wide variety of sedative-anxiolytic drugs such as alcohol to overcome the early agitated state and to induce sleep. Prolonged hypersomnia, often lasting 24–36 or more hr is not unusual during this phase. Notably, addicts report minimal desire for the abused drug during this immediate phase of withdrawal.

As the individual recovers from the crash phase, a period of anhedonia, dysphoria, and decreased mental and physical energy ensues (intermediate withdrawal phase). This phase can last from several days to weeks. Emerging from this state of ennui, mood and energy dysfunction, a stimulant craving returns, frequently initiating recidivism. Unfortunately, this craving is exquisitely sensitive to various environmental stimuli (often learned during binging episodes through their association with the stimulant high during the binging episodes); for example, individuals in this withdrawal phase can experience a sudden onset of stimulant craving by merely returning to a place that is associated with drug use (e.g., a hotel room used for purchasing crack). With continued drug availability, it is not unusual to observe repetitious cycles of binging with intervening crash and intermediate withdrawal phases over a period of months or even years with a devastating result.

With more sustained abstinence through the intermediate withdrawal phase (e.g., with help from a treatment program), a more "natural" baseline state returns (long-term withdrawal phase). However, although decreased in frequency and intensity, urges to return to stimulant use can recur after months or years of abstinence, most frequently triggered, again, by environmental stimuli. Moreover, a single "taste" of stimulant can induce a full set of behavioral responses, which may have originally taken weeks or months of chronic stimulant use to evolve (termed grease slice return). Thus, the individual may become psychotic (usually a sequela of chronic binging episodes) within minutes to hours of return to stimulant use. Changes observed during the long-term withdrawal phase suggest that chronic stimulant abuse may produce long-lasting (permanent?) residual changes in the brain, a hypothesis that is being actively tested.

V. Cocaine-Related Morbidity and Mortality

A. Medical Complications

The cocaine-related deaths of famous athletes and entertainers, such as those of Len Bias and John Belushi, have focused attention on several types of stimulant-related toxicity. These examples demonstrate that stimulant-related complications occur even in healthy young individuals and directly contradict various claims of relative safety. Important medical and psychiatric complications that are ascribable to stimulant abuse have been observed for almost a century.

Many of the important stimulants' toxicity is mediated by their actions on the cardiovascular system. Reported cardiovascular toxic pathology associated with stimulant use include spasm of blood vessels (leading to decreased oxygen supply); hypertensive crisis, in which the person's blood pressure suddenly rises to very high levels causing rupture of large arteries and multiple hemorrhaging of smaller arteries and arterioles; heart attack (acute myocardial infarction); irregular heart rhythms (arrhythmias); inflammation of the heart muscle cells (myocarditis); rupture of large arteries; and multiple hemorrhaging of smaller arteries and of arterioles. Some of these are fatal complications, occurring very rapidly, just minutes after use, precluding treatment; others, if not properly treated, can also lead to death. Importantly, these complications can occur in individuals with no known prior history of medical problems.

Many of the above cardiovascular complications have been ascribed to sympathomimetic effects of stimulants (i.e., mimicking the actions of the sympathetic nervous system by enhancing their function; see above). For example, because increased heart rate and blood pressure generally increase the heart's oxygen consumption, such changes following stimulants may overwhelm the normal oxygen supply. This oxygen insufficiency, in turn, may ultimately lead to death of heart muscle cells (heart attack). Among other mechanisms that may contribute to cardiac toxicity, the apparent higher propensity of cocaine to produce cardiac arrhythmia (compared with amphetamine or other related compounds) may be secondary to the drug's local anesthetic effect. Local anesthetics in high doses are known to interfere with the heart's conduction of electrical activity leading to severe arrhythmias.

Stimulant-induced hypertensive episodes have been also associated with bleeding in the brain, which leads to permanent neurological deficit or even death. Exact factors predisposing individuals to this type of complication are not well known. Not infrequently, this complication occurs in individuals with blood vessel abnormalities in the brain such as local dilation (called cerebral aneurysm) secondary to a weakened vessel wall from congenital defect, disease (e.g., chronic high blood pressure) or injury. The sudden hypertension caused by a stimulant becomes too great for the weakened vessel wall to accommodate, leading to rupture and bleeding. Unfortunately, most cases of this structural abnormality are discovered *a posteriori* during, for example, an autopsy.

In addition to directly causing cerebral bleeding, stimulants can indirectly produce neurological symptoms via their actions on the heart. For example, heart attack or irregular heart beat can reduce blood flow to the brain, depriving the brain of oxygen. Depending on various factors (e.g., length of deprivation or simultaneous occurrence of seizures), this hypoxia (lack of oxygen) can induce a reversible injury or death of brain cells.

Other forms of stimulant-induced toxicity include the induction of seizure and/or uncontrollable high body temperature. In contrast to conditions discussed above that can occur at moderately high doses, these two conditions tend to be more associated with very high doses of stimulants. One of the most difficult stimulant overdose cases to treat is the one that occurs in the drug smugglers known as "body packers" who swallow cocaine-filled condoms. One or more of these condoms may rupture, leading to extreme high blood concentrations of cocaine and subsequent induction of fatal convulsions and uncontrollable high body temperatures. Many, if not most, of these cases are fatal despite heroic treatment effort.

In addition to the well-documented cardiovascular and neurological toxicity, other types of medical complications associated with stimulant use include extensive liver or skeletal muscle damages and induction of asthma attacks by crack smoking.

B. Psychiatric Complications

There are significant (more so than for physiological effects) individual differences in psychiatric toxicity of stimulants. In addition, even a single individual's response to the drugs can vary over time; for example, previous history of stimulant abuse is a critical factor determining behavioral responses to a single dose of stimulant. Illustrative of the time dependency is the grease slide phenomenon discussed above.

Serious psychiatric complications of stimulant use are not common in occasional, low-dose users; on the other hand, many adverse psychiatric reactions have been described in high-dose users, especially during binging episodes. Thus, an alarming trend toward frequent association of cocaine use with serious psychiatric complications has been shown following the advent of the cheap rapid dose form of crack cocaine.

As described previously, high-dose and/or rapid

routes of administration (e.g., crack smoking) are associated with intense euphoria. More dangerously, the exaggerated effect of stimulants is also manifested by impaired judgment, grandiosity, combativeness, and extreme psychomotor stimulation, leading not infrequently to accidents, atypical sexual behavior, or illegal acts.

In contrast to the euphoria associated with single stimulant doses, prolonged binging episodes can be associated with anxiety, irritability, transient panic reaction with terror, and psychosis. Generally, two types of psychotic behaviors are observed: either with or without confusion. Compared with the type without confusion, the one with confusion is more frequently characterized by higher frequency and doses of stimulant and a higher propensity toward violence and hallucinations. The psychosis without confusion tends to last longer than the other type and is characterized by delusions of persecution (false belief that one is being persecuted) and hallucinations in the background of a clear sensorium (e.g., the patient knows who he is, where he is, what the date and year is). This has generated much interest in the scientific community because it can be similar in appearance to paranoid schizophrenia—so similar that at times even experienced psychiatrists have not been able to distinguish between the two. This marked similarity in the two has led to the use of paranoid psychosis as an experimental stimulant model in animals for studying the mechanism(s) involved in human psychoses. [*See* SCHIZOPRENIC DISORDERS.]

VI. Treatment for Cocaine Abuse

The primary goal of treatment for cocaine abuse is the initiation and maintenance of abstinence from the drug with subsequent development of personal strategies for relapse prevention. Because of the intense conditioning of cocaine craving that occurs with various environmental stimuli (see above), initial avoidance of situations and people associated with stimulant abuse is critical. Avoidance is initially maintained by hospitalization or a move out of the conditioned stimulus-rich environment for a period of time. Subsequently, behavior therapy, involving identification of such risk situations including "mine sweeping the environment for cue triggers" and training in development of specific strategies for avoiding them, is becoming a more prominent feature of treatment. A variety of techniques including frequent contacts with the patient, peer-support groups, family or couples therapy, urine monitoring, education sessions, and individual psychotherapy are used to further facilitate successful treatment. In addition to these modes of treatment, supplemental drug therapies are being investigated; so far, they appear to be useful in facilitating cocaine abstinence.

VII. Summary

Central stimulants have been used from prehistoric time for their medicinal value as well as their ability to produce mental alertness and a sense of well-being, and to heighten energy, self-esteem, and emotions aroused by interpersonal interactions. Many of these mental effects are thought to be mediated by dopamine; other catecholamines are more responsible for some of the toxicity on the heart. Because of their powerful euphorogenic effects, the stimulants have been abused repeatedly. One specific stimulant dominates each epidemic (e.g., amphetamine in the 1960s, cocaine in the 1980s); cyclic pattern of the abuse is characterized by an initial enthusiasm about the drugs' pharmacological effects despite lessons from the past, followed by widespread abuse and an eventual decline secondary to increasing awareness of their toxicity. Recent data suggest that the current cocaine epidemic may be in its declining stage. One wonders, then, whether or not history will repeat itself in the near future with another stimulant, "ice," which is a smokable form of methamphetamine, an amphetamine-type stimulant.

Bibliography

Castellani, S., and Ellinwood, E. H., Jr. (1985). Cocaine: Mechanisms underlying behavioral effects. *In* "Psychopharmacology 2, Part I: Preclinical Psychopharmacology" (D. G. Grahame-Smith, ed.). Elsevier Science Publishers, New York.

Ellinwood, E. H., Jr., and Rockwell, W. J. K. (1988). Central nervous system stimulants and anorectic agents. *In* "Meyler's Side Effects of Drugs," 11th ed.

(M. N. G. Dukes, ed.). Elsevier Biomedical Publishers, New York.

Gawin, F. H., and Ellinwood, E. H., Jr. (1988). Cocaine and other stimulants: Actions, abuse and treatment. *N. Eng. J. Med.* **318,** 1173–1182.

Post, R. M., Weiss, S. R. B., Pert, A., and Uhde, T. W. (1987). Chronic cocaine administration: Sensitization and kindling effects. *In* "Cocaine: Clinical and Biobehavioral Aspects" (S. Fisher, A. Raskin, and E. H. Uhlenhuth, eds.). Oxford University Press, New York.

Weiss, R. D., and Mirin, S. M. (1987). "Cocaine." American Psychiatric Press, Inc., Washington, D.C.

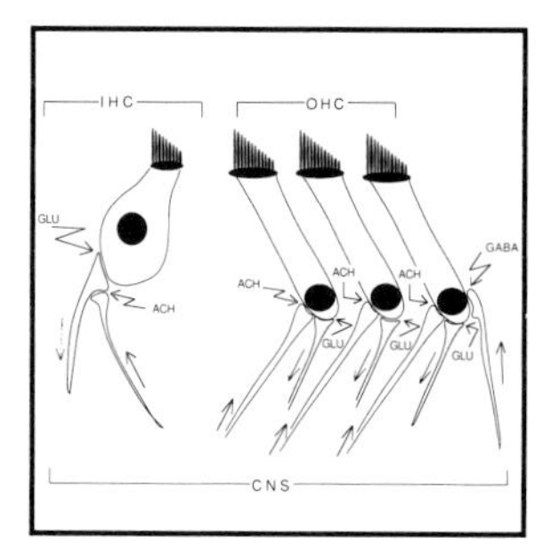

Cochlear Chemical Neurotransmission

RICHARD P. BOBBIN, *Louisiana State University Medical School*

Glossary

Afferent nerve fibers Nerve fibers that conduct action potentials toward the brain

Depolarization Change in the resting membrane potential of a cell in the positive direction, less negative

Efferent nerve fibers Nerve fibers that conduct action potentials away from the brain toward the cochlea

Hyperpolarization Change in the resting membrane potential of a cell in the negative direction, less positive

Neurotransmitter Chemical released from a sensory receptor cell or nerve cell on depolarization of that cell. On release the chemical diffuses across the gap between the releasing cell and an adjoining cell to act on the adjoining cell to induce a change in its electrical properties

Receptor protein Protein in the membrane of a cell that accepts a chemical messenger such as the neurotransmitter, changes its configuration on accepting the neurotransmitter, and so induces subsequent reactions in the cell, such as opening of an ion channel to allow for diffusion of that ion down its concentration gradient

Synapse Place where neurotransmission occurs between two cells

"CHEMICAL NEUROTRANSMISSION" is the term used to describe the process whereby nerve cells use chemicals to transmit information from one cell to another at a synapse. In the cochlea, this involves not only the transfer of information between nerve cells, but in addition the transfer of information from the sensory receptor cells to nerve fibers and from nerve fibers to the sensory receptor cells. Usually, in any given cell, only one primary chemical, the neurotransmitter, is used for this function. Modulators, secondary chemicals that are also involved, modify the action of the primary chemical.

A chemical must satisfy several criteria before it is considered the neurotransmitter or a modulator at a synapse. These include (1) when the candidate is applied to the synapse it must mimic the endogenous compound; (2) the candidate must be present in the presynaptic structure; (3) drugs that antagonize the endogenous compound must also block the exogenously applied candidate; and (4) the endogenous compound must be detected in the extracellular fluid when the synapse is activated. At some synapses these criteria are much more difficult to fulfill than at others. Then the criteria are only guides as to the identity of the transmitter, and proof remains an elusive goal.

I. Overview of Neurotransmission in the Cochlea

Neurotransmission in the cochlea occurs at four types of synapses (Fig. 1): (1) outer hair cell (OHC) to afferent nerve fiber; (2) inner hair cell (IHC) to afferent nerve fiber; (3) medial efferent nerve fiber to OHC; and (4) lateral efferent nerve fiber to afferent nerve fiber under the IHCs. Available evidence indicates that the OHCs and IHCs use one transmitter chemical [i.e., L-glutamate (GLU)], and most efferents use another transmitter chemical [i.e., acetylcholine (ACH)].

The neurotransmitters at all the synapses in the

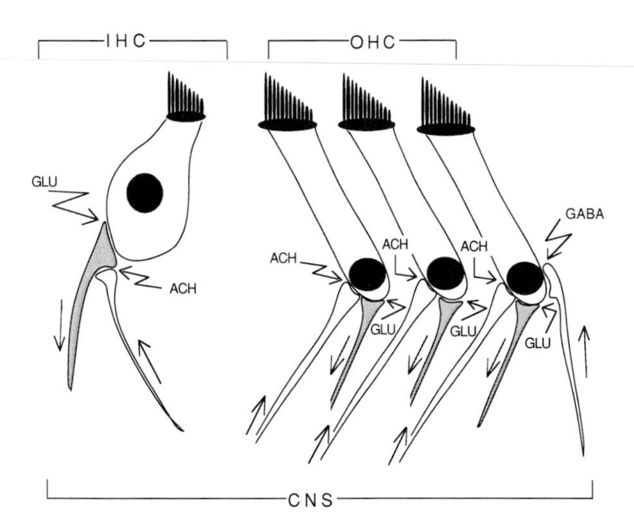

FIGURE 1 Schematic diagram showing principal neurotransmitters and their location in the cochlea. Neuromodulators have been omitted for the sake of clarity. Straight arrows indicate the direction of nerve conduction, afferent and efferent. IHC, inner hair cell; OHC, outer hair cells; CNS, central nervous system; ACH, acetylcholine; GLU, glutamic acid; GABA, gamma-aminobutyric acid.

cochlea appear to be released from storage packets, called *synaptic vesicles,* as in other neural structures. The synaptic vesicles in the hair cells (HCs) are unusual in that they surround a presynaptic rod-like structure, the function of which is unknown. A depolarization of the presynaptic cell at the synapse induces an opening of voltage-activated calcium channels, which allows calcium to enter the cell. The rise in intracellular calcium initiates a series of chemical reactions, which results in the release of the contents of the synaptic vesicles (neurotransmitter) into the synaptic space between the cells. The neurotransmitter then diffuses to the postsynaptic membrane where it interacts with a receptor protein. This chemical interaction alters the shape of the receptor protein, which changes either the membrane properties of the cell or chemical constituent in the cell. This in turn affects the electrical properties of the postsynaptic cell. The action of neurotransmitters are terminated either by enzymatic destruction or by diffusion and uptake into the cells surrounding the synapse.

II. Neurotransmitters of the Efferent Nerve Fibers

The medial system of efferent fibers synapse on the OHCs, whereas the lateral system synapses on the afferent nerve endings under the IHCs (Fig. 1). The neurotransmitter at the synapses of these efferents is ACH, the first neurotransmitter identified in the cochlea and in other HC systems. This conclusion was initially based on the localization of cholinesterase [the enzyme that terminates the action of ACH by converting it to acetate and choline] to the efferent nerve fibers in the cochlea. Subsequent studies have strengthened this hypothesis at both the medial and lateral systems of efferent nerve endings in the cochlea. Gamma-aminobutyric acid (GABA) appears to be a neurotransmitter at a much smaller population of efferents in both the medial and lateral systems (Fig. 1). Little is known about the chemistry or physiology of the GABA synapse.

The postsynaptic receptor for ACH is unusual and has not been definitively categorized. In other systems, the ACH receptor is identified as either muscarinic (e.g., glands, cardiac) or nicotinic (e.g., neuromuscular junction, autonomic ganglia). The receptor in the cochlea appears to be a nicotinic subtype. The reason for classifying it as nicotinic is that the action of ACH at the receptor is more sensitive to antagonism by nicotonic antagonists (decamethonium, hexamethonium, curare, alpha-bungarotoxin) than by muscarinic antagonists (atropine). However, converting tertiary atropine [with dominant muscarinic action] into the quaternary atropine [with nicotinic action dominant] enhances the action of atropine only about 10 times. More unusual is the fact that curare and decamethonium, the most potent antagonists, are about as potent as the glycine antagonist strychnine in antagonizing the action of these efferent fibers. Thus ACH appears to act on a receptor protein, which has a configuration that combines similarities of the glycine receptor and the nicotinic receptor, yet is not very different from the muscarinic receptor.

To date, only the crossed medial efferents have been studied in depth. (All that is known about the lateral efferents is that ACH is most likely the neurotransmitter; the effects are unknown.) The overall effect of activation of the medial efferents is to reduce the output of information from the IHCs to the brain. This set of efferents synapse predominantly on the OHCs, and how their action at these cells changes what is occurring at the IHCs is unknown. Some authors speculate that the OHCs change length (and width) both in response to sound and to stimulation of the efferents. This change in length then is thought to alter the response of the IHCs, probably through a mechanical and ionic interaction. The resting membrane potential of the OHCs is one factor in determining the length of these cells. Sound energy is thought to induce a change in the membrane potential of the OHCs either via stretch- activated ion channels located in the walls of the OHCs and/or via a mechanical ion channel located in the stereocilia. The efferents may fine tune the length of the OHCs by adjusting their membrane potential by about 5 mV. Information to date indicates that this is most likely accomplished by ACH altering an ion channel. The identity of the ion channel and the direction of change in the membrane potential (depolarization or hyperpolarization) are unknown. Physiologically, the change results in about a plus or minus 10 dB adjustment in the function of the cochlea. [*See* Ion Pumps.]

Several neuroactive substances have been colocalized to these efferents. The opioid neuropeptides such as enkephalins, dynorphin B, and alpha neoendorphin are present within the lateral subdivision of the efferents and appear to be absent from the medial efferents. In addition, dopamine and calcitonin gene-related peptide (CGRP) have been shown in the efferents. These chemicals are most likely released together with ACH or with GABA. It appears that ACH and GABA open ligand-activated ion channels to change the membrane potential of the postsynaptic structure. The other colocalized substances may act to modulate the action of the ion channels by regulating the levels of second-messenger chemicals within the postsynaptic structure.

III. Neurotransmitters of the Hair Cells

The other major category of neurotransmitters in the cochlea are those released by the two sets of HCs: the OHCs and the IHCs (Fig. 1). GLU is probably the neurotransmitter released by both sets of HCs. It is not the "proven" candidate because of the difficulty in achieving sufficient evidence for all the criteria mentioned above. For instance, GLU cannot be said to be solely localized to neurons where it is the transmitter because it is ubiquitous, having dominate roles in amino acid, energy, and nitrogen metabolism. Another disturbing facet of the problem in obtaining "proof" is that many endogenous compounds resemble GLU structurally and so activate the neurotransmitter receptors for the "endogenous compound." This is not unique to the cochlea or to HC systems, but is a problem throughout the nervous system. Wherever GLU may appear to be the transmitter in the nervous system, GLU is just the leading candidate, followed by other candidates (e.g., L-aspartic acid, L-homocysteic acid, *N*-acetylaspartylglutamate). All the chemical compounds in the group of candidates are excitatory amino acids.

As the HC transmitter, GLU may act on at least four receptor types belonging to the excitatory amino acid category (only four have been suggested to date but more are sure to follow). The four have been named after nonendogenous chemicals that activate them: *N*-methyl-D-aspartate (NMDA), 2-amino-4-phosphonobutyrate (APB), kainate, and quisqualate. However, only the kainate and quisqualate excitatory amino acid receptors have been functionally identified in the cochlea. Kainate receptors are present on the afferent nerve endings, which synapse on the IHCs during development, and they remain at maturity. However, at the OHCs, the afferents having kainate receptors appear to degenerate during development, leaving a type of afferent that is devoid of kainate receptors. Because GLU is probably the transmitter released by the OHCs onto the kainate receptor, it must also be the transmitter released onto the afferents that remain on the OHCs in the mature cochlea. However, the type of excitatory amino acid receptor on the remaining afferents is unidentified. GLU activation of the kainate receptor opens an ion channel that allows both sodium and potassium ions to diffuse down their gradients. In the cochlea, this results in depolarization of the postsynaptic, afferent cochlear nerve fiber endings, triggering postsynaptic potentials, and then action potentials at the axon hillock. Quisqualate receptors are also functionally present at the afferent synapse of the IHC. They may modulate the action of the kainate receptor by means of a second messenger; for example, in other systems, a type of quisqualate receptor has been found that is linked to the phosphoinositide cascade second-messenger system.

GLU has a weak affinity for the kainate and quisqualate receptors. This suggests that a high concentration of GLU (>60 mM) is in the presynaptic storage vesicles, allowing a concentrated solution of GLU to be released onto the receptors. In turn this low affinity increases the effectiveness of diffusion alone to terminate rapidly the action of GLU. The GLU that diffuses away from the receptors is thought to be taken up into neighboring cells through a transport system that is dependent on the presence of sodium ions. After being taken up into a

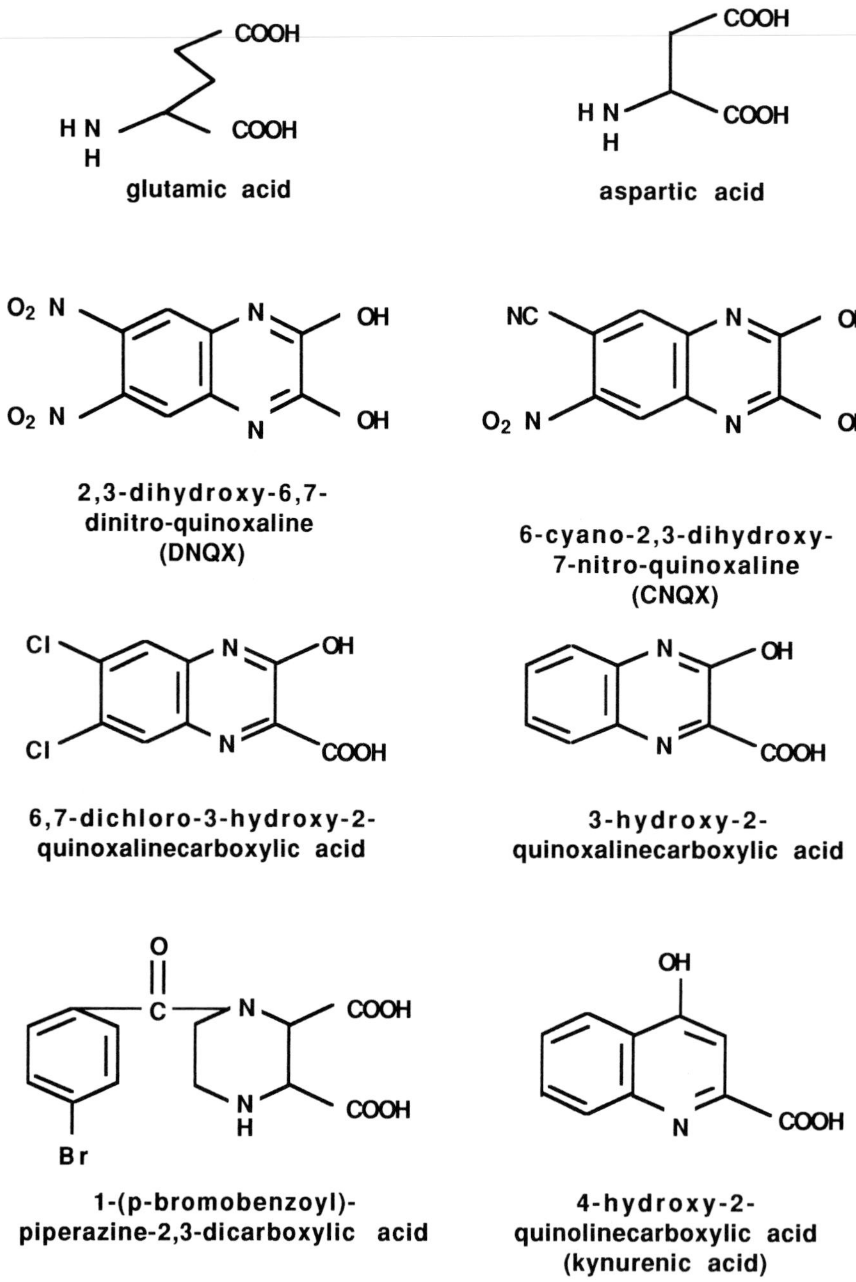

glutamic acid

aspartic acid

2,3-dihydroxy-6,7-dinitro-quinoxaline (DNQX)

6-cyano-2,3-dihydroxy-7-nitro-quinoxaline (CNQX)

6,7-dichloro-3-hydroxy-2-quinoxalinecarboxylic acid

3-hydroxy-2-quinoxalinecarboxylic acid

1-(p-bromobenzoyl)-piperazine-2,3-dicarboxylic acid

4-hydroxy-2-quinolinecarboxylic acid (kynurenic acid)

FIGURE 2 Chemical structures of neurotransmitter candidates for hair cells, glutamate and aspartate, and chemical antagonists of these candidates at kainate/quisqualate receptors at the hair cell to afferent nerve fiber junction.

cell, the GLU is converted to glutamine, which is then deposited into the extracellular space for uptake into the HCs. In the HCs, the glutamine is converted to GLU and repackaged into vesicles for release as HC neurotransmitter.

The third and fourth type of excitatory amino acid receptors, the NMDA and APB types, may be present, yet they have not been functionally detected in the cochlea. In other words, neither the agonists nor the antagonists active at the NMDA (D-alpha-aminoadipate, 2-amino-5-phosphonovalerate) and APB receptors in other nervous structures have any action in the cochlea. This may be a problem of experimental design. In contrast, in other HC systems such as the lateral line and vestibular systems, NMDA receptors have been found to be functionally present. Part of the problem could be that the

NMDA and/or APB receptors are the excitatory amino acid receptors on the afferents that synapse on the OHCs. At this synapse, GLU released by the OHC's interacts with afferents that comprise only a small number (5%) of the total afferent nerve fiber population (95% at the IHCs) in the cochlea. This may be such a small population that they are difficult to isolate and study functionally. However, some investigators believe that the OHC-afferent nerve fiber synapse may be nonfunctional in the mature cochlea.

The action of the neurotransmitter of the IHCs on the afferent nerve endings is antagonized by those chemicals found to block kainate and quisqualate receptors in other systems. The structural similarity of these antagonists to both GLU and aspartate is shown in Fig. 2. To date, no blockers adequately distinguish between kainate and quisqualate receptors. The first blocker tested with selective action in the cochlea was kynurenic acid (Fig. 2), which occurs naturally in the body. Up to that time (1987), blockers were either not very active or demonstrated activity on structures other than the synapse. Other compounds tested and active with about the same potency as kynurenic acid include 1-(*p*-bromobenzoyl)-piperazine-2,3-dicarboxylic acid and 3-hydroxy-2-quinoxalinecarboxylic acid. A 30-fold increase in potency was found with the antagonists 6,7-dichloro-3-hydroxy-2-quinoaxalinecarboxylic acid and 6-cyano-7-nitro-quinoxaline-2,3-dione (CNQX). An additional threefold (100-fold greater than kynurenic acid) was found with 6,7-dintro-quinoxaline-2,3-dione (DNQX). The latter four compounds are new and were just recently tested in the cochlea (1989). Given the success with these new compounds, it appears that in the near future compounds that are more selective and powerful will certainly become available.

IV. Summary

Our knowledge of neurocommunication between cells in the cochlea is rapidly advancing. ACH and GABA are efferent neurotransmitters, but their receptor type remains to be classified. Various neuromodulators at the efferent synapses are being identified. The transmitter released by both IHCs and OHCs is probably GLU. At the IHCs, GLU acts on kainate and quisqualate receptors, but the identity of the GLU receptor at the OHCs is unidentified. The chemical reactions initiated by these neurotransmitters and modulators at the membrane of the postsynaptic cell awaits definition. Overall, many questions remain.

Acknowledgment

Thanks to Maureen Fallon for preparing the illustrations and to Cindy Frazier for help in preparing this manuscript. The author's research cited in this manuscript is supported by grants from NIH (NIH-DC00379), NSF (BNS-84-19241), DRF, Kresge, and the Louisiana Lions Eye Foundation.

Bibliography

Altschuler, R. A., and Fex, J. (1986). Efferent neurotransmitters. In: "Neurobiology of Hearing: The Cochlea" (R. A. Altschuler, R. P. Bobbin, and D. W. Hoffman, eds.), pp. 383–396. Raven Press, New York.

Bledsoe, S. C., Jr., Bobbin, R. P., and Puel, J. -L. (1988). Neurotransmission in the inner ear. In: "Physiology of Hearing" A. F. Jahn, and J. R. Santos-Sacchi, eds.), pp. 385–406. Raven Press, New York.

Hoffman, D. W. (1986). Opioid mechanisms in the inner ear. In: "Neurobiology of Hearing: The Cochlea" (R. A. Altschuler, R. P. Bobbin, and D. W. Hoffman, eds.), pp. 371–382. Raven Press, New York.

Littman, T., Bobbin, R. P., Fallon, M., and Puel, J. -L. (1989). The quinoxalinediones DNQX, CNQX and two related congeners suppress hair cell-to-auditory nerve transmission. *Hear. Res.* **40,** 45–54.

Pujol, R., and Lenoir, M. (1986). The four types of synapses in the organ of Corti. In: "Neurobiology of Hearing: The Cochlea" (R. A. Altschuler, R. P. Bobbin, and D. W. Hoffman, eds.), pp. 161–172. Raven Press, New York.

Schwartz, I. R., and Ryan, A. F. (1986). Uptake of amino acids in the gerbil cochlea. In: "Neurobiology of Hearing: The cochlea" (R. A. Altschuler, R. P. Bobbin, and D. W. Hoffman, eds.), pp. 173–190. Raven Press, New York.

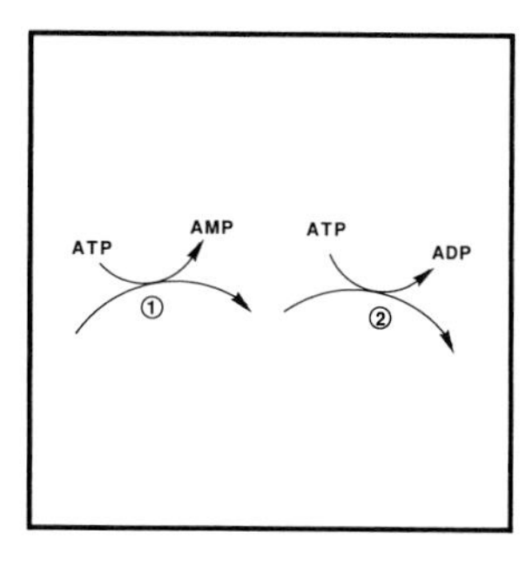

Coenzymes, Biochemistry

DONALD B. McCORMICK, *Emory University*

Glossary

Apoenzyme Protein moiety of an enzyme that requires a coenzyme

Coenzyme Natural, organic molecule that functions in a catalytic, enzyme system

Cofactor Natural reactant, usually either a metal ion or coenzyme, required in an enzyme-catalyzed reaction

Holoenzyme Catalytically active enzyme constituted by coenzyme bound to apoenzyme

Vitamin Essential organic micronutrient that must be supplied exogenously and in many cases is the precursor to a metabolically derived coenzyme

COENZYMES ARE ORGANIC MOLECULES that are bound to those enzymes that require their function to catalyze certain biochemical reactions. Though some simple proteins (unconjugated polypeptides) are able to function as enzymes competent to catalyze a modest range of reactions (e.g., hydrolyses), the limited chemical properties of amino acid side chains within proteins do not permit catalyses of many of the numerous essential reactions that operate by diverse mechanisms. Hence, additional reagents more broadly considered as cofactors serve with protein enzymes for reactions that would be difficult or impossible using only simple acid-base catalysis. Among such cofactors are inorganic materials such as metal ions as well as the organic compounds called coenzymes. Coenzymes bind to apoenzymes (proteins) to generate functional holoenzymes. Tightly bound coenzymes are sometimes referred to as prosthetic groups.

As major portions of some coenzymes cannot be biosynthesized by certain organisms, these precursors, normally made by other organisms, must be supplied exogenously. Many coenzymes are the metabolic result of converting an ingested vitamin, especially those of the B complex, to a form suitable for binding and function in an enzyme system. Of the 13 vitamins presently known to be required in the diet of humans, at least eight (thiamin, riboflavin, niacin, vitamin B_6, vitamin B_{12}, folacin, biotin, and pantothenate) are simply the essential precursors for coenzymic forms made in our bodies. Because coenzymes are indispensable coreactants in many enzyme-catalyzed reactions involved in the formation, metabolism, and degradation of almost all body components, it follows that the vitaminic precursors for most coenzymes are essential for normal growth and function. [*See* VITAMIN A.]

I. Thiamin Pyrophosphate

A. Chemistry

Thiamin pyrophosphate (TPP) is the principal if not sole coenzyme derived from thiamin (vitamin B_1).

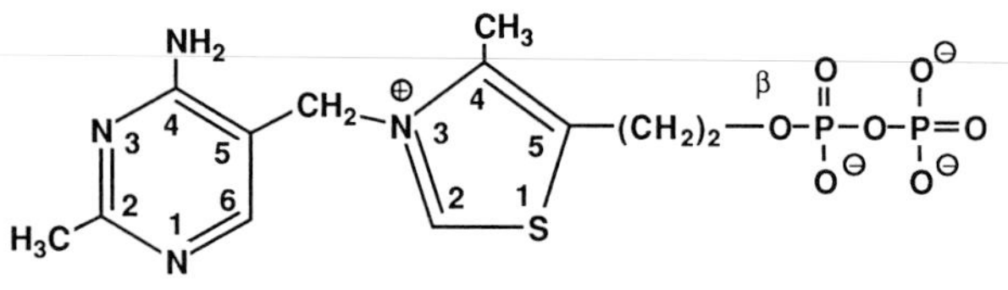

FIGURE 1 Structure with numbering for thiamin pyrophosphate.

The structure of TPP (Fig. 1) indicates that the physiologically active form is the ionized pyrophosphate ester formed at the β-hydroxyethyl substituent of the thiazole moiety of thiamin, which can be systematically named as 3-(2′-methyl-4′-amino-5′-pyrimidylmethyl)-4-methyl-5-(β-hydroxyethyl) thiazole.

TPP, as is the case with the parent vitamin, is relatively stable in acidic aqueous solutions, but sulfite causes cleavage at the methylene bridge to release the substituted pyrimidylsulfonate and the thiazole pyrophosphate. TPP is unstable in alkaline medium because the thiazole portion is subject to base attack at carbon 2. With hydroxyl ion, this leads to a pseudobase, thiazole ring opening and some disulfide under mild oxidizing conditions. Also, the 4-amino on the pyrimidyl portion can attack as an intramolecular base to form the tricyclic amino adduct, which can be oxidized with ferricyanide to yield thiochrome pyrophosphate. The thiochrome-level compound is fluorescent (λ excit = 385 nm; λ emit = 440 nm) and easily detected.

B. Metabolism

TPP is formed in a number of tissues from thiamin, and a fraction is subsequently converted, especially in brain, to the triphosphate. Hydrolysis of the latter to the monophosphate and some further release of free vitamin complete the interconversions shown in Fig. 2. Among the forms involved, TPP predominates in cells. Approximately 30 mg of thiamin-level compounds are found in an adult with 80% as pyrophosphate, 10% as triphosphate, and the rest as thiamin and its monophosphate. About half of the body stores are found in skeletal muscles with much of the remainder in heart, liver, kidneys, and nervous tissue, including brain, which contains much of the triphosphate. As concerns enzymes catalyzing the interconversions, thiaminokinase is widespread, but the phosphoryl transferase and

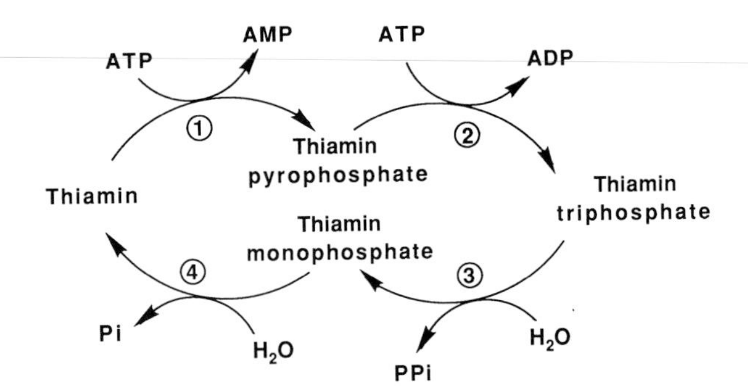

FIGURE 2 Interconversions of thiamin and its phosphates catalyzed by (1) thiaminokinase, (2) TPP-ATP phosphoryl-transferase, (3) thiamin triphosphatase, and (4) thiamine monophosphatase. Pi, inorganic phosphate; PPi, inorganic pyrophosphase.

membrane-associated triphosphatase are mainly in nervous tissue.

C. Functions

There are two general types of reactions where TPP functions as the Mg^{2+}-coordinated coenzyme for so-called active aldehyde transfers. First, in decarboxylation of α-keto acids, the condensation of the thiazole moiety of TPP with the α-carbonyl carbon on the acid leads to loss of CO_2 and production of a resonance-stabilized carbanion (Fig. 3). Protonation and release of aldehyde occur in fermentative organisms such as yeast, which have only the TPP-dependent dicarboxylase, but reaction of the α-hydroxalkyl-TPP with lipoyl residues and ultimate conversion to acyl-CoA occurs in higher eukaryotes including humans with multienzymic dehydrogenase complexes, as described in Section IX. The other general reaction involving TPP is the transformation of α-ketols (ketose phosphates). Although specialized phosphoketolases in certain bacteria and higher plants can split ketose phosphates to simpler, released products, the reaction of importance to humans and most animals is a transketolation, as mechanistically illustrated in Fig. 4. Transketolase is a TPP-dependent enzyme found in the cytosol of many tissues, especially liver and blood cells, where principal carbohydrate pathways exist. This enzyme catalyzes the reversible transfer of a glycoaldehyde moiety (α,β-dihydroxyethyl-TPP) from the first two carbons of a donor ketose phosphate to the aldehyde carbon of an aldose phosphate of the pentose phosphate pathway, which also supplies nicotinamide adenine dinucleotide phosphate, reduced (NADPH) needed for biosynthetic reactions.

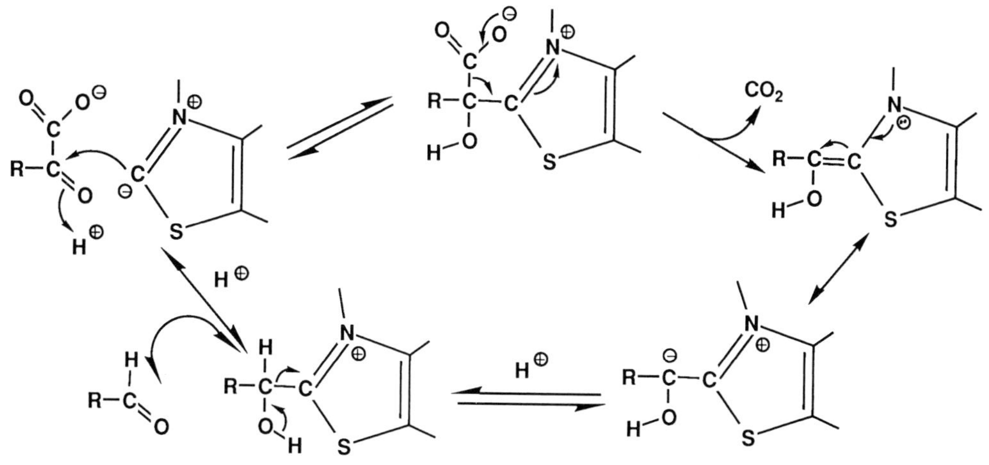

FIGURE 3 Function of the thiazole moiety of thiamine pyrophosphate in α-keto acid decarboxylations.

II. Flavocoenzymes

A. Chemistry

Flavin adenine dinucleotide (FAD) and its immediate precursor flavin mononucleotide (FMN) are two commonly encountered flavocoenzymes. The structure of these (Fig. 5) indicates a common tricyclic isoalloxazine nucleus with a D-ribityl side chain, as found in riboflavin [7,8-dimethyl-10-(1′-D-ribityl)-isoalloxazine]. FMN is riboflavin 5′-phosphate, whereas FAD is extended to include a pyrophosphoryl-linked 5′-adenosine monophosphate (AMP) moiety. In some less commonly encountered, but essential, forms of flavocoenzymes, there is covalent attachment of a peptidyl residue through an electronegative atom, usually in the 8α-position.

Flavocoenzymes are relatively more stable in acid than base and are photodecomposed, largely to lumichrome and lumiflavin, by cleavages at the ribityl side chain. In the natural oxidized (quinoid) forms, FMN and FAD are fluorescent (λ excit = 450 nm; λ emit = 530 nm) and readily detected when unbound by protein; however, FAD is significantly quenched (80%) by its intramolecular complex in solution. The observed oxidation-reduction potential near −0.2 V poises these coenzymes for electron transport usually after the more negative pyridine nucleotides and before cytochromes.

B. Metabolism

Biosynthesis of flavocoenzymes occurs within the cellular cytoplasm of most tissues, but particularly in the small intestine, liver, heart, and kidney. The obligatory first step is the adenosine triphosphate (ATP)-dependent phosphorylation of riboflavin catalyzed by Zn^{2+}-preferring flavokinase. The FMN product can be complexed with specific apoenzymes to form several functional flavoproteins, but the larger quantity is further converted to FAD in a second ATP-dependent reaction catalyzed by Mg^{2+}-preferring FAD synthetase. These coenzyme-forming steps (Fig. 6) are tightly regulated. FAD is the predominant flavocoenzyme present in tissues where it is mainly complexed with numerous flavoprotein dehydrogenases and oxidases. Less than 10% of the FAD can also become covalently attached to specific amino acid residues of a few important apoenzymes. Examples include the 8α-*N*(3)-histidyl-FAD within succinate dehydrogenase and 8α-*S*-cysteinyl-FAD within monoamine oxidase, both of mitochondrial localization. Turnover of covalently attached flavocoenzymes requires intracellular proteolysis, and further degradation of the coenzymes involves a pyrophosphatase cleavage of FAD to FMN and action by nonspecific phosphatases on the latter (Fig. 6).

C. Functions

Flavocoenzymes participate in oxidation-reduction reactions in numerous metabolic pathways and in energy production via the respiratory chain. The redox functions of a flavocoenzyme (Fig. 7) include one-electron transfers during which the biologically encountered, neutral, oxidized quinone level of flavin is half reduced to the radical semiquinone, which can exist within natural pH ranges as neutral or anionic species. A further electron can lead to a

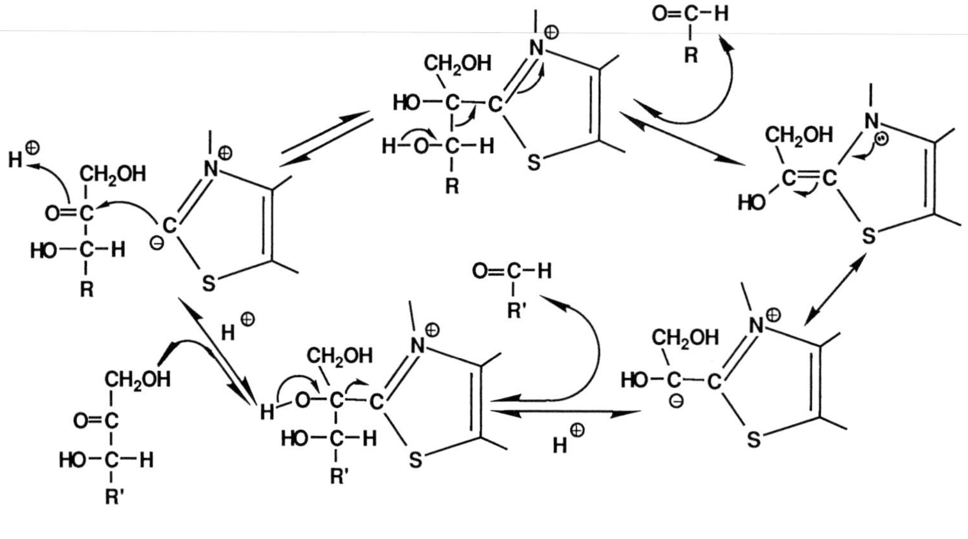

FIGURE 4 Function of the thiazole moiety of thiamine pyrophosphate in transketolations.

fully reduced hydroquinone. Additionally, a single-step, two-electron transfer from substrate to flavin can occur (Fig. 8). Such cases as hydride ion transfer from reduced pyridine nucleotide or the carbanion generated by base abstraction of a substrate proton may lead to attack at the flavin N-5 position; some nucleophiles such as the hydrogen peroxide anion add at the C-4a position.

There are flavoprotein-catalyzed dehydrogenations that are both pyridine nucleotide-dependent and independent, reactions with sulfur-containing compounds, hydroxylations, oxidative decarboxylations, dioxygenations, and reduction of O_2 to hydrogen peroxide. The intrinsic abilities of flavins to be varyingly potentiated as redox carriers upon differential binding to proteins, to participate in both one- and two-electron transfers, and in reduced (1,5-dihydro) form to react rapidly with oxygen permits wide scope in their operation.

III. Pyridine Nucleotide Coenzymes

A. Chemistry

Nicotinamide adenine dinucleotide (NAD) and its phosphate (NADP) (Fig. 9) are the two natural coenzymes derived from niacin. Both contain an N-1 substituted pyridine 3-carboxamide that is essential to function in redox reactions with a potential near −0.32 V. The oxidized coenzymes are labile to alkali including nucleophilic addition at the para (4) position, whereas the reduced (1,4-dihydro) coenzymes are labile to acid. Nicotinamide adenine dinucleotide, reduced (NADH) and NADPH (but not NAD and NADP) characteristically absorb light in the near ultraviolet (340 nm).

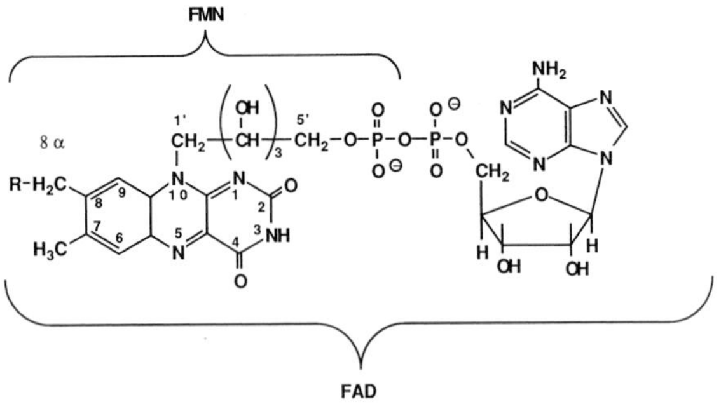

FIGURE 5 Structure with numbering for principal flavocoenzymes where R = H for FAD or an electronegative atom (N, O, S) with peptide for covalently attached species.

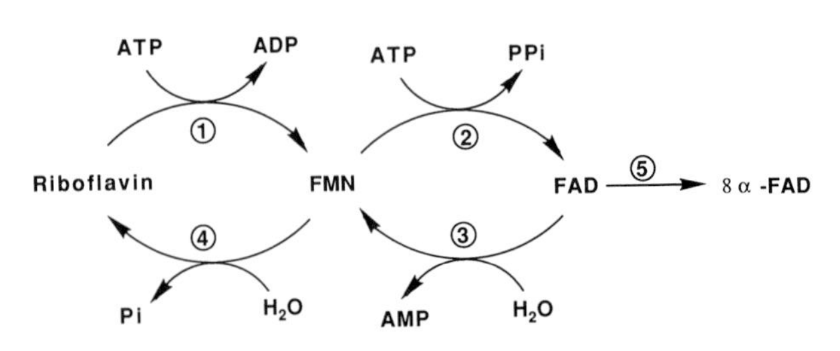

FIGURE 6 Interconversions of riboflavin and its coenzymes catalyzed by (1) flavokinase, (2) FAD synthetase, (3) FAD pyrophosphatase, (4) FMN phosphatase, and (5) posttranslational modification.

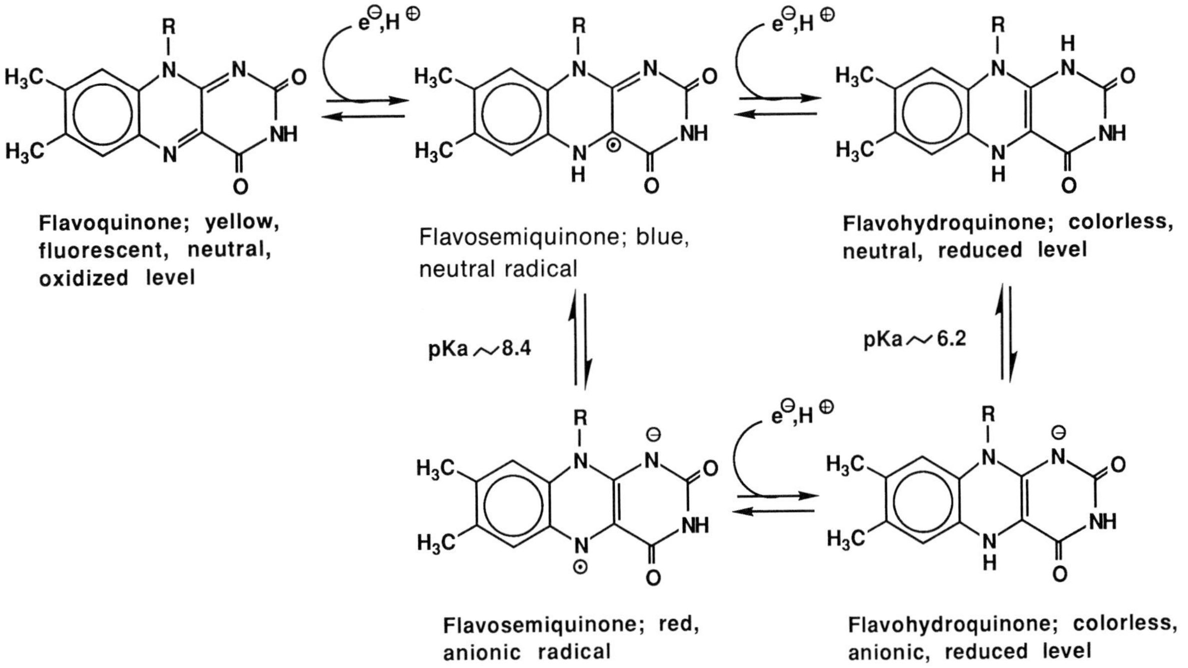

Flavoquinone; yellow, fluorescent, neutral, oxidized level

Flavosemiquinone; blue, neutral radical

Flavohydroquinone; colorless, neutral, reduced level

pKa ~8.4

pKa ~6.2

Flavosemiquinone; red, anionic radical

Flavohydroquinone; colorless, anionic, reduced level

FIGURE 7 Physiologically relevant redox states of flavocoenzymes with pKa values estimated for interconversion of the free species.

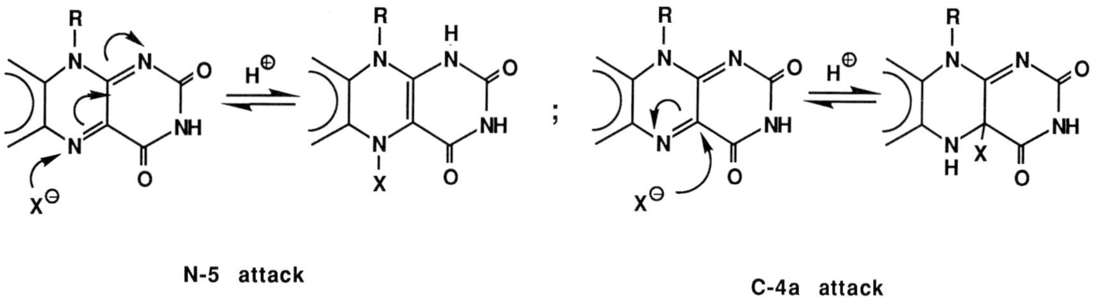

N-5 attack

C-4a attack

FIGURE 8 Reaction types encountered with flavoquinone coenzymes and natural nucleophiles (x^-).

B. Metabolism

Converging pathways lead to the formation of NAD (Fig. 10), a fraction of which is phosphorylated to NADP. Vitaminic precursors are converted to the coenzymes in blood cells, kidney, brain, and liver. Nicotinate and nicotinamide react with 5-phosphoribosyl-1α-pyrophosphate (PRPP) and nicotinic acid mononucleotide (NaMN) or nicotinamide mononucleotide (NMN), respectively. Additionally

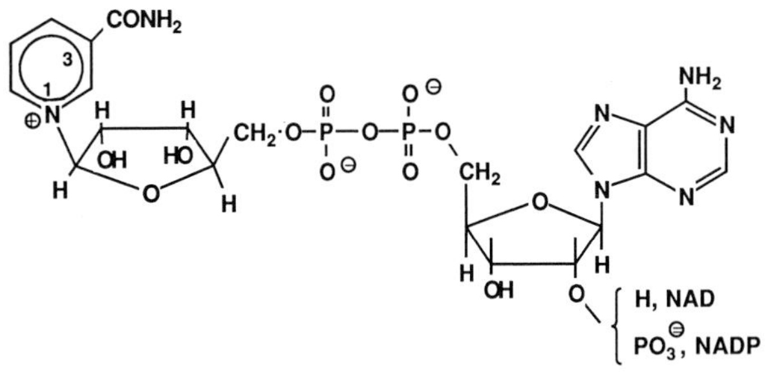

H, NAD

$PO_3^{\ominus}$, NADP

FIGURE 9 Structures with numbering for pyridine nucleotide coenzymes.

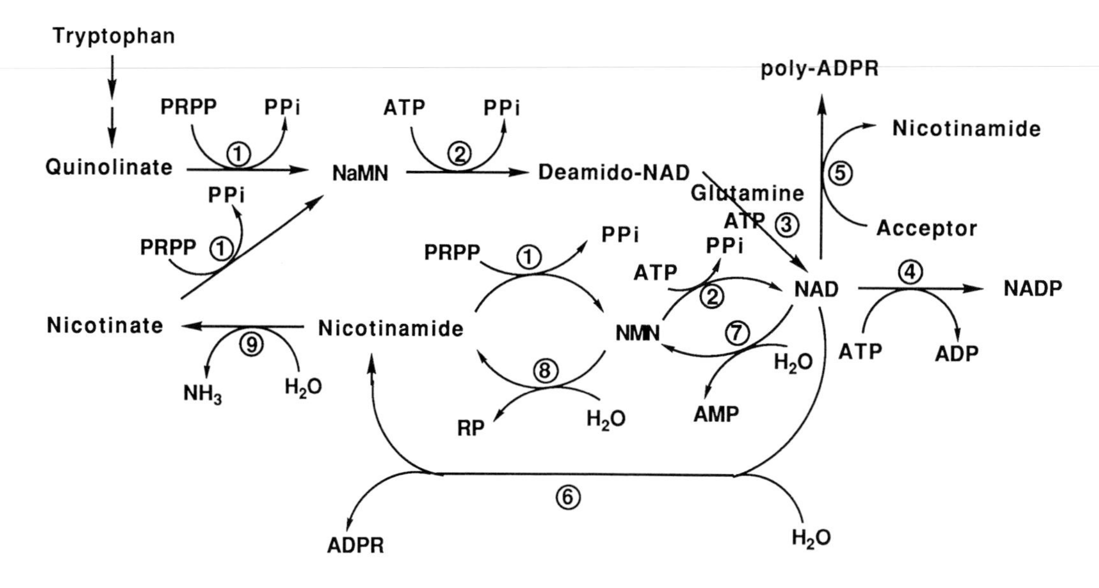

FIGURE 10 Interconversions of niacin-level precursors and its coenzymes catalyzed by (1) phosphoribosyl-transferases, (2) adenylyl-transferases, (3) synthetase, (4) kinase, (5) poly-ADPR synthetase (polymerase), (6) NAD glycohydrolase, (7) NAD pyrophosphatase, (8) NMN hydrolase, and (9) nicotinamide deamidase. RP is ribose 5-phosphate.

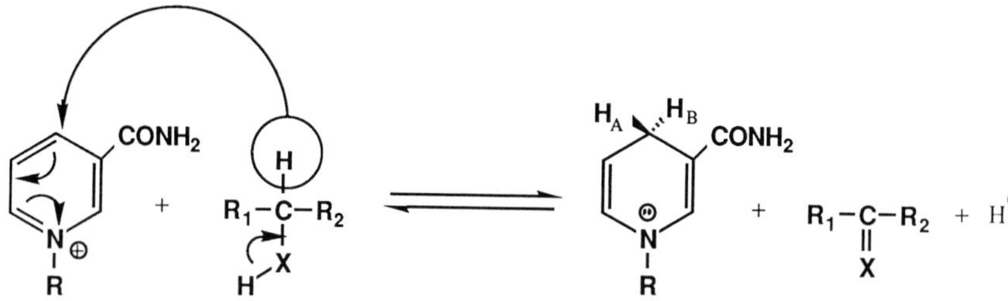

FIGURE 11 The hydride ion transfer for operation of pyridine nucleotide coenzymes.

in liver, quinolinate from catabolism of tryptophan is similarly converted with concomitant decarboxylation to NaMN. Subsequent reactions with ATP to incorporate the 5′-adenylate portion yield NAD from NMN and deamido-NAD from NaMN. The deamido compound reacts with glutamine in a cytosolic ATP-dependent step to yield NAD, glutamate, AMP, and pyrophosphate. In breakdown, NAD is hydrolyzed to NMN and the latter to nicotinamide, which, in turn, can be converted to nicotinate by a rather widespread microsomal deamidase. An NAD glycohydrolase (NADase) catalyzes hydrolysis of NAD to nicotinamide plus adenosine 5′-pyrophospho-5-ribose (ADPR). Some NAD glycohydrolases have the ability to transglycosidate, i.e., transfer the ADPR moiety of NAD to acceptor macromolecules.

C. Functions

Numerous enzymes require the nicotinamide moiety within either NAD or NADP. Most of these oxidoreductases function as dehydrogenases and catalyze such diverse reactions as the conversion of alcohols (often sugars and polyols) to aldehydes or ketones, hemiacetals to lactones, aldehydes to acids, and certain amino acids to keto acids. The common mechanism of operation (generalized in Fig. 11) involves the stereospecific abstraction of a hydride ion from substrate, with para addition to one or the other side of carbon 4 in the pyridine ring of the nucleotide coenzyme. The second hydrogen of the substrate group oxidized is concomitantly removed as a proton and ultimately exchanges as a hydronium ion.

Most dehydrogenases utilizing NAD or NADP function reversibly. Glutamate dehydrogenase, for example, favors the oxidative direction, whereas others, such as glutathione reductase, catalyze preferential reduction. A further generality is that most NAD-dependent enzymes are involved in catabolic reactions, whereas NADP systems are more common to biosynthetic reactions. The additional function of NAD as a substrate for providing the ADPR moiety to modify macromolecules has been recently more appreciated. ADP-ribosyl transferase catalyzes such a modification of the prokaryote elongation factor 2, thereby blocking translocation on ribosomes. Poly-(ADPR) synthetases (polymerases) in eukaryotes catalyze a multiple addition of ADPR from NAD to form (ADPR)n-acceptor plus nicotinamide and hydrogen ion. This activity is found in mitochondria and bound to microsomes as well as in nuclei where it affects operation of DNA. This nonredox function of NAD probably accounts for the rapid turnover of NAD in human cells.

IV. Pyridoxal 5′-Phosphate and Pyridoxamine 5′-Phosphate

A. Chemistry

Two of the three natural forms of vitamin B_6 (pyridoxine, pyridoxal, and pyridoxamine) can be phosphorylated to directly yield functional coenzymes, i.e., pyridoxal 5′-phosphate (PLP) and pyridoxamine 5′-phosphate (PMP). The structures of these 4-substituted 2-methyl-3-hydroxy-5-hydroxymethylpyridine 5′-phosphates are shown in Figure 12. PLP is the predominant and more diversely functional coenzymic form, although PMP interconverts as coenzyme during transaminations.

At physiological pH, the dianionic phosphates of these coenzymes exist as zwitterionic meta-phenolate pyridinium compounds. They are very water-soluble, absorb light in the ultraviolet region, exhibit fluroescence, and, in general, are sensitive to light, particularly at alkaline pH. Both natural and synthetic carbonyl reagents (e.g., hydrazines and hydroxylamines) form Schiff bases with the 4-formyl function of PLP (and pyridoxal), thereby removing the coenzyme and inhibiting PLP-dependent reactions.

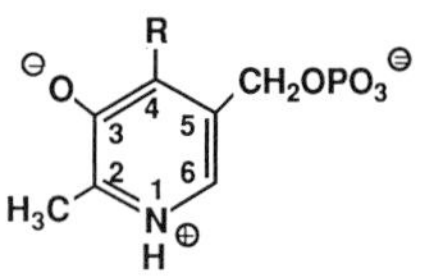

R = CHO, Pyridoxal 5'-phosphate;
CH_2NH_2 , Pyridoxamine 5'-phosphate

FIGURE 12 Structure with numbering for coenzyme forms of vitamin B_6.

B. Metabolism

The metabolic interconversions of vitamin and coenzymic forms of B_6 are shown in Fig. 13. Each of the three vitamin-level compounds is phosphorylated in the cytosol by ATP-utilizing pyridoxal kinase, which, in mammalian tissues, prefers Zn^{2+}. Most cells of facultative and aerobic organisms contain a cytosolic FMN-dependent, pyridoxine (pyridoxamine) 5′-phosphate oxidase responsible for catalyzing the O_2-dependent conversion of pyridoxine 5′-phosphate (PNP) and PMP to PLP. During aminotransferase (transaminase)-catalyzed reactions, PLP and PMP interconvert with amino and keto functions of substrate–product participants. Release of free vitamin, mainly pyridoxal when physiological nonsaturating levels of vitamin are absorbed, occurs when the phosphates are hydrolyzed by nonspecific alkaline phosphatase located on the plasma membrane of cells.

C. Functions

PLP functions in numerous reactions that embrace the metabolism of proteins, carbohydrates, and lipids. Especially diverse are PLP-dependent enzymes that are involved in amino acid metabolism. By virtue of the ability of PLP to condense its 4-formyl substituent with an amine, usually the α-amino group of an amino acid, to form an azomethine (Schiff base) linkage, a conjugated double-bond system extending from the α-carbon of the amine (amino acid) to the pyridinium nitrogen in PLP results in reduced electron density around the α-carbon. This potentially weakens each of the bonds from the amine (amino acid) carbon to the adjoined functions (hydrogen, carboxyl, or side chain). A given apoenzyme then locks in a particular configuration of the coenzyme-substrate compound such that maximal overlap of the bond to be broken will occur with the resonant, coplanar, electron-with-

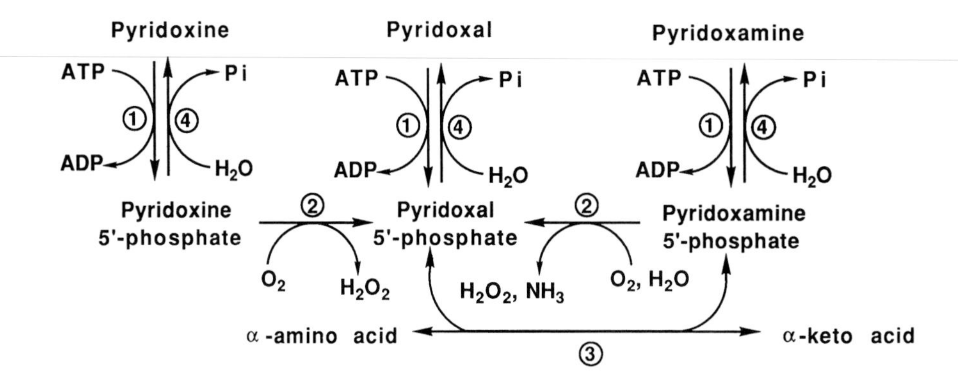

FIGURE 13 Interconversions of the vitamin B_6 group with coenzymes catalyzed by (1) pyridoxal kinase, (2) pyridoxine (pyridoxamine) 5′-phosphate oxidase, (3) aminotransferases, and (4) phosphatases.

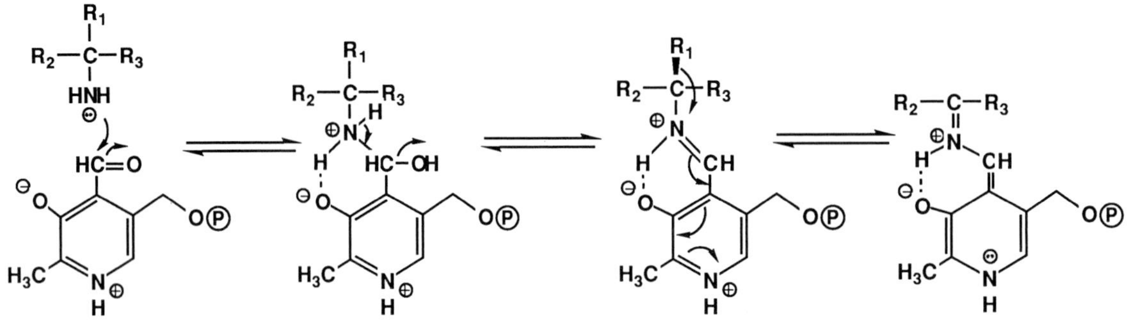

PLP + amine — Carbinolamine — Aldimine — Ketimine

FIGURE 14 Operation of PLP with a generalized amine.

drawing system of the coenzyme complex. These events are depicted in Fig. 14.

Aminotransferases effect rupture of the α-hydrogen bond of an amino acid with ultimate formation of an α-keto acid and PMP; this reversible reaction provides an interface between amino acid metabolism and that for ketogenic and glucogenic reactions. Amino acid decarboxylases catalyze breakage of the α-carboxyl bond and lead to irreversible formation of amines, including several that are functional in nervous tissue (e.g., epinephrine, norepinephrine, serotonin, and γ-aminobutyrate). The biosynthesis of heme depends on the early formation of δ-aminolevulinate from PLP-dependent condensation of glycine and succinyl-CoA followed by decarboxylation. There are many examples of enzymes, such as cysteine desulfhydrase and serine hydroxymethyltransferase, that affect the loss or transfer of amino acid side chains. PLP is the essential coenzyme for phosphorylase that catalyzes phosphorolysis of the α-1,4-linkages of glycogen. An important role in lipid metabolism is the PLP-dependent condensation of L-serine with palmitoyl-CoA to form 3-dehydrosphinganine, a precursor of sphingolipids. [*See* SPHINGOLIPID METABOLISM AND BIOLOGY.]

A simpler, but less frequently encountered, variation on the way in which PLP functions is provided by the pyruvoyl terminus of some enzymes. In these electrophilic centers [CH_3-C(β)O-C(α)O-NH-R′], the amino function condenses with the β-carbonyl, while the α-carbonyl enhances electron withdrawal from the resulting ketimine. For example, 5-adenosylmethionine decarboxylase from mammals as well as *Escherichia coli* uses such a system to form spermidine from putrescine and methionine.

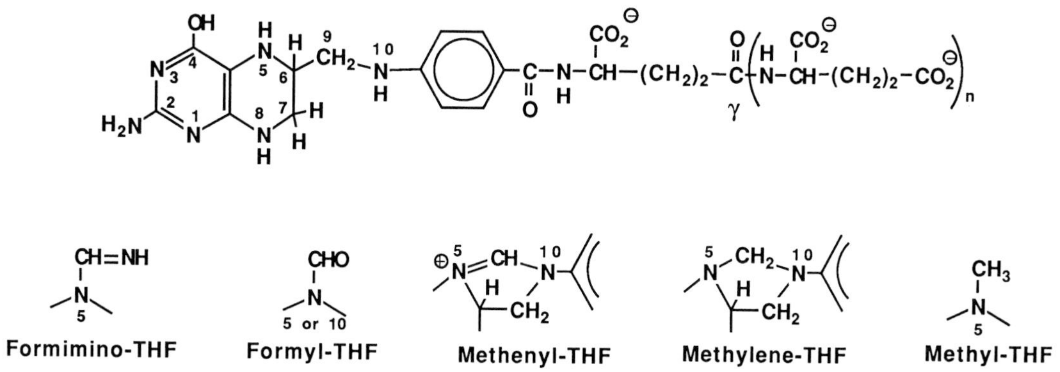

Formimino-THF   Formyl-THF   Methenyl-THF   Methylene-THF   Methyl-THF

FIGURE 15 Structures with numbering for tetrahydropteroyl-L-glutamates including the formimino, formyl, methenyl, methylene, and methyl derivatives.

V. Pterin Coenzymes

A. Chemistry

Among natural compounds with a pteridine nucleus, those most commonly encountered are derivatives of 2-amino-4-hydroxypteridines, which are trivially named pterins. Although a number of pterins when reduced to the 5,6,7,8-tetrahydro level function as coenzymes, the most generally utilized are poly-γ-glutamates of tetrahydrofolate (THF) (Fig. 15). The natural derivatives of tetrahydropteroylglutamates responsible for vectoring 1-carbon units in different enzymic reactions are also shown in abbreviated form in Fig. 15. All of these bear the substitutent for transfer at nitrogen 5 or 10 or are bridged between these basic centers. The number of glutamate residues varies, usually from one to seven, but a few to several glutamyls optimize binding of tetrahydrofolyl coenzymes to most enzymes requiring their function.

Less commonly encountered, but essential for some coenzymic roles of pterins, are those compounds shown in Fig. 16. Tetrahydrobiopterin cycles with its quinoid 7,8-dihydro form during O_2-dependent hydroxylation of such aromatic amino acids as in the conversion of phenylalanine to tyrosine. The most recently elucidated pterin cofactor in some Mo/Fe flavoproteins (e.g., xanthine dehydrogenase) is given in the right-hand side of Fig. 16.

Pterin coenzymes, most at the tetrahydro level, are sensitive to oxidation and have characteristic absorbance in ultraviolet light. Upon heating in aqueous media below pH 4, the pterin portion of the folyl type coenzyme tends to split from the para-aminobenzoyl glutamate portion. The xanthine dehydrogenase pterin during isolation loses hydrogen from the side chain to generate a double bond between sulfur-bearing carbons.

B. Metabolism

The interconversions of folate with the initial coenzymic relatives, the tetrahydrofolyl polyglutamates, are shown in Fig. 17. The dihydrofolate reductase necessary for reducing the vitamin-level compound through 7,8-dihydro to 5,6,7,8-tetrahydro levels is the target of such inhibitory drugs as aminopterin and amethopterin (methotrexate). A similar dihydropterin reductase catalyzes reduction of dihydro- to tetrahydrobiopterin. Tetrahydrofolate is intracellularly trapped and extended to polyglutamate forms that operate with THF-dependent systems. In some cases (e.g., thymidylate synthetase), there is a redox change in tetrahydro to dihydro coenzyme, which is recycled by the NADPH-dependent reductase. Turnover of coenzyme to folate at the monoglutamate level requires hydrolytic cleavage of the extra glutamyl residues. Cells of the small intestinal mucosa are especially rich in the γ-glutamyl peptidase ("conjugase") that degrades ingested natural folyl polyglutamates to the more readily absorbed folate.

C. Functions

An overview of some of the major interconnections among the 1-carbon-bearing THF coenzymes and their metabolic origins and roles are given in Fig. 18. Reactions include (1) generation and utilization of formate; (2) *de novo* purine biosynthesis wherein glycinamide ribonucleotide and 5-amino-4-imidazole carboxamide ribonucleotide are transformylated by 5,10-methenyl-THF and 10-formyl-THF, re-

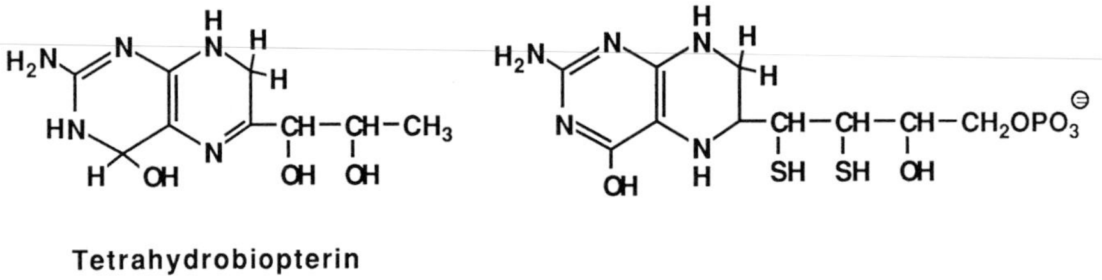

Tetrahydrobiopterin

FIGURE 16 Other pterin coenzymes: tetrahydrobiopterin of phenylalanine hydroxylase and the pterin cofactor of certain Mo/Fe flavoproteins.

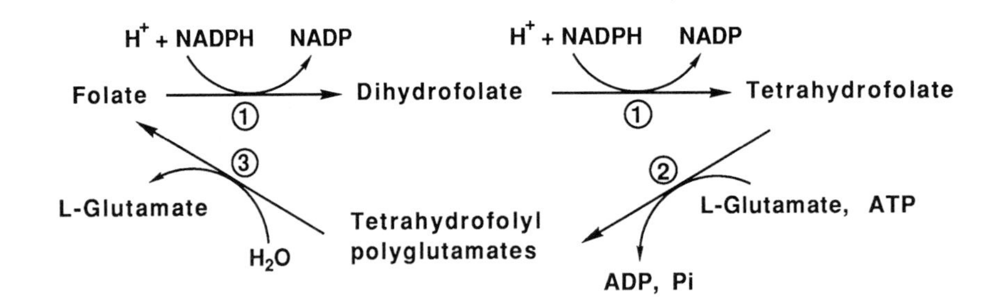

FIGURE 17 Interconversions of folate and its tetrahydro, polyglutamate coenzymes involving (1) dihydrofolate reductase, (2) folylpolyglutamate synthetase, and (3) pteroylpolyglutamate hydrolase.

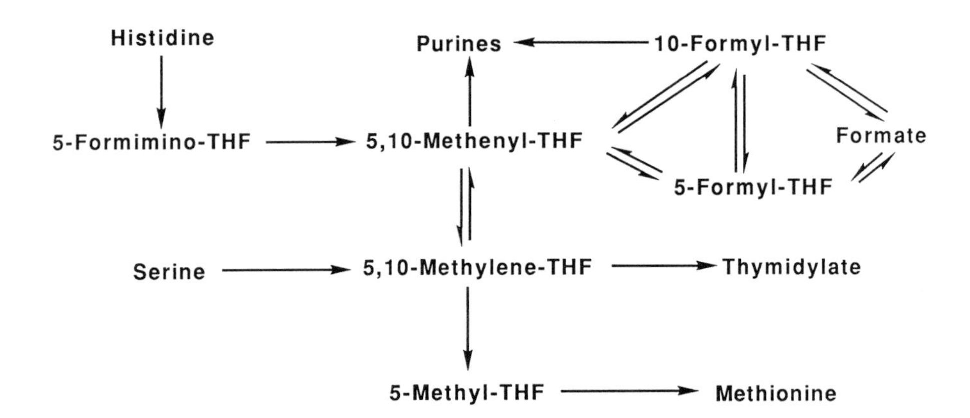

FIGURE 18 Origins, interconversions, and functions of tetrahydrofolyl coenzymes in 1-carbon transfers.

spectively; (3) pyrimidine nucleotide biosynthesis, wherein deoxyuridylate and 5,10-methylene THF form thymidylate and dihydrofolyl coenzyme; (4) conversions of some amino acids, namely N-formamino-L-glutamate (from histidine catabolism) with THF to L-glutamate and 5,10-methenyl-THF (via 5-formanino-THF), L-serine with THF to glycine and 5,10-methylene-THF, and L-homocysteine with 5-methyl-THF to L-methionine and regenerated THF.

VI. B_{12} Coenzymes

A. Chemistry

The coenzyme B_{12} (CoB_{12}) known to function in most organisms, including humans, is 5′-deoxyadenosylcobalamin (Fig. 19). A second important

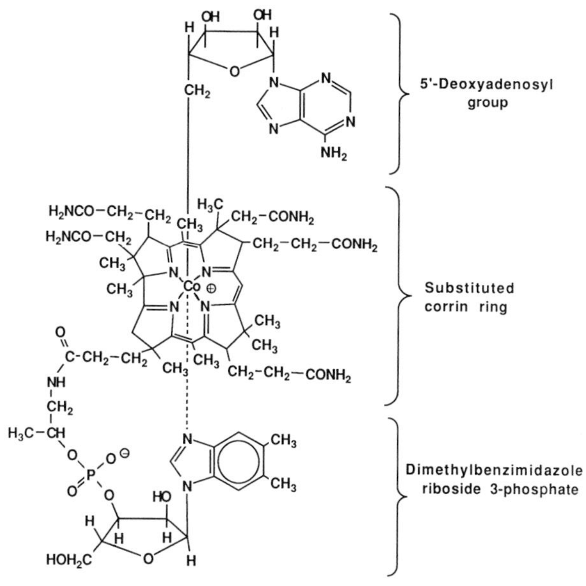

FIGURE 19 Structure with identification of principal components for coenzyme B_{12}.

coenzyme form is methylcobalamin (methyl-B_{12}), in which the methyl group replaces the deoxyadenosyl moiety of CoB_{12}. Some prokaryotes utilize other bases (e.g., adenine) in this position originally occupied by the cobalt-coordinated cyanide anion in cyanocobalamin, the initially isolated form of vitamin B_{12}.

Coenzyme forms of B_{12} are light-absorbing, photo-labile compounds that readily undergo photolysis to yield aquocobalamin (B_{12b}), in which H_2O is coordinated to cobalt in the corrin ring. Acid hydrolysis of CoB_{12} yields hydroxocobalamin (B_{12a}) with a coordinated hydroxyl ion.

B. Metabolism

The metabolic interconversions of vitamin B_{12} as the naturally occurring hydroxocobalamin (B_{12a}) with other vitamin- and coenzyme-level forms and the two B_{12} coenzyme-dependent systems in mammals are given in Fig. 20. As outlined, B_{12a} is sequentially reduced to the paramagnetic or radical B_{12r} and further to the very reactive B_{12s}. The latter reacts in enzyme-catalyzed nucleophilic displacements of tripolyphosphate from ATP to generate CoB_{12}, or of THF from 5-methyl-THF to generate methyl-B_{12}.

C. Functions

Seemingly all CoB_{12}-dependent reactions react through a radical mechanism, and all but one (*Lactobacillus leichmanii* ribonucleotide reductase) involve a rearrangement of a vicinal group (X) and a hydrogen atom. This general mechanism is illustrated in Fig. 21. For the CoB_{12}-dependent mammalian enzyme, L-methylmalonyl-CoA mutase, X is the CoA-S-CO- group, which moves with retention of configuration from the carboxyl-bearing carbon of L(*R*)-methylmalonyl-CoA to the carbon β to the carboxyl group in succinyl-CoA. This reaction is essential for funneling propionate to the tricarboxylic acid cycle. Without CoB_{12} (from vitamin B_{12}), more methylmalonate is excreted, but also the CoA ester competes with malonyl-CoA in normal fatty-acid elongation to form instead abnormal, branched-chain fatty acids. As indicated earlier (cf. Fig. 20), methyl-B_{12} is necessary in the transmethylase-catalyzed formation of L-methionine and regeneration of THF. Without this role, there would not only be no biosynthesis of the essential amino acid, but increased exogenous supply of folate would be necessary to replenish THF, which would not otherwise be recovered from its 5-methyl derivative.

VII. Biotinyl Functions

A. Chemistry

The coenzymic form of biotin only occurs naturally as the vitamin (*cis*-tetrahydro-2-oxothieno[3,4-*d*]-imidazoline-4-valeric acid), which has become amide linked to the ε-amino group of specific lysyl residues in carboxylases and transcarboxylase (Fig. 22).

Though the two ureido nitrogens are essentially isoelectronic in the biotinyl moiety, the steric crowding of the thiolane side chain near N-3′ essentially prevents chemical additions, which therefore occur at the N-1′ position.

B. Metabolism

Biotin is inserted into enzymes dependent on its operation by a holoenzyme synthetase that forms biotinyl 5′-adenylate as an intermediate from the vitamin and ATP. Usual proteolytic turnover of biotinylated enzymes release biocytin (ε-N-biotinyl-

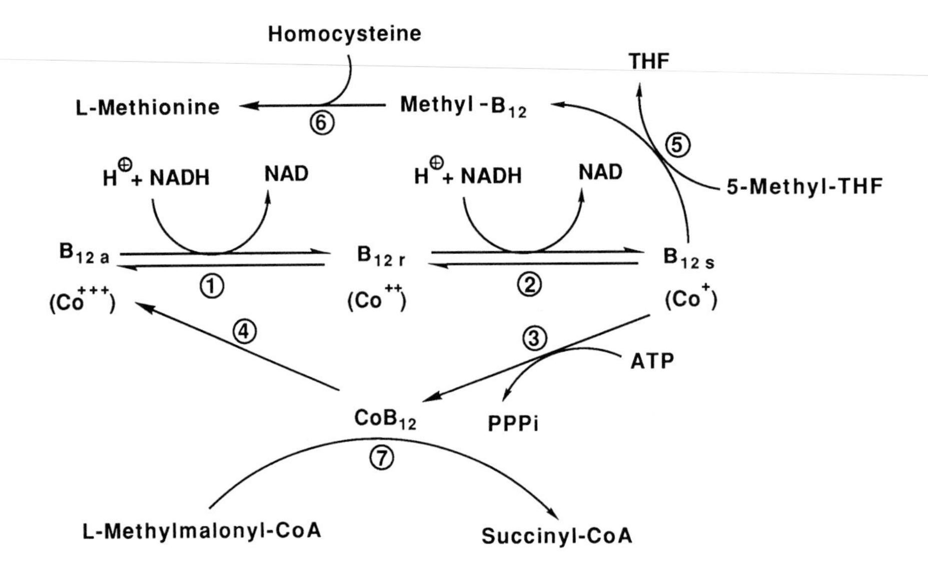

FIGURE 20 Interconversions of B_{12}, methyl-B_{12}, and Co-B_{12} in principal metabolic reactions include (1, 2) B_{12} reductases, (3) deoxyadenosyl transferase, (4) oxidations, (5, 6) methyltetrahydrofolate-homocysteine transmethylase system, and (7) L-methylmalonyl-CoA mutase.

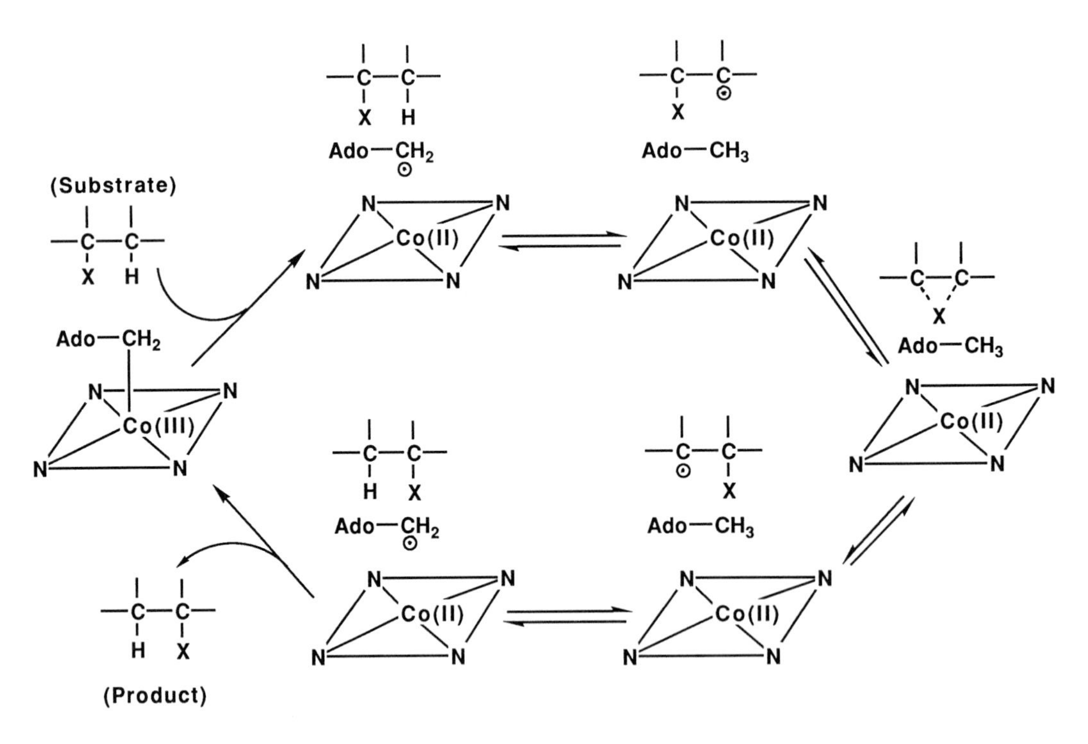

FIGURE 21 Radical intermediates in Co-B_{12} enzyme reactions.

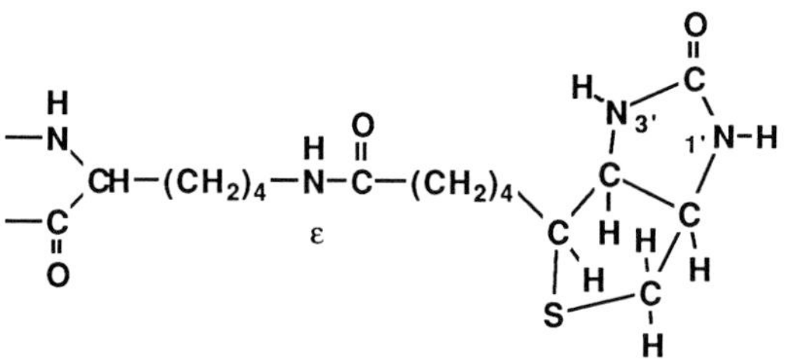

FIGURE 22 Structure with numbering for d-biotinyl-L-lysyl coenzyme moiety.

L-lysine), which is further hydrolyzed to release the vitamin and amino acid in a reaction catalyzed by biotinidase (biocytinase, biotin amidohydrolase).

C. Functions

There are nine known biotin-dependent enzymes: six carboxylases, two decarboxylases, and a transcarboxylase. Of these, only four carboxylases are found in tissues of humans and other mammals. These carboxylases, named for their substrates, are (1) a cytosolic enzyme that converts acetyl-CoA to malonyl-CoA for fatty-acid biosynthesis, (2) a mitochondrial enzyme that converts pyruvate to oxalacetate for citrate formation, (3) the enzyme for converting propionyl-CoA to D-methylmalonyl-CoA, and (4) the enzyme that carboxylates β-methylcrotonyl-CoA from L-leucine catabolism to form β-methylglutaconyl-CoA.

Biotin-dependent carboxylases operate by a common mechanism (Fig. 23). This involves phosphorylation of bicarbonate by ATP to form carbonyl phosphate, followed by transfer of the carboxyl group from this electrophilic mixed-acid anhydride to the sterically less hindered and nucleophilically enhanced N-1′ of the biotinyl moiety. The resulting N(1′)-carboxybiotinyl enzyme can then exchange the carboxylate function with a reactive center in a substrate, typically at a carbon with incipient carbanion character.

VII. Phosphopantetheinyl Coenzymes

A. Chemistry

The 4′-phosphopantetheinyl moiety, derived from the vitamin pantothenate [D-N-(2,4-dihydroxy-3,3-dimethyl-butyryl)-β-alanine] and β-mercaptoethylamine, serves as a functional component within the structure of coenzyme A (CoA) (Fig. 24) and as a prosthetic group covalently attached to a seryl residue of acyl carrier protein (ACP).

Because of the thiol terminus with a pKa near 9, phosphopantetheine and its coenzymic forms are readily oxidized to the catalytically inactive disulfides. CoA has a strong absorption maximum at 260 nm attributable to the adenine moiety.

B. Metabolism

Within cells, synthesis of CoA and the 4′-phosphopantetheinyl moiety of ACP occurs via successive enzyme-catalyzed conversions (Fig. 25). Following the ATP-driven steps in biosynthesis of CoA, formation of ACP occurs by transfer of the 4′-phosphopantetheinyl moiety of CoA, which binds via a phosphodiester link to apo-ACP in a reaction catalyzed by ACP holoprotein synthase. About 80% of pantothenate in animal tissues is in CoA form, much of it as thioesters. The rest exists mainly as phosphopantetheine and phosphopantothenate. Cleavage enzymes catalyzing hydrolysis of the phosphate moieties (CoA → dephospho-CoA → 4′-phosphopantetheine → pantetheine) and release of β-mercaptoethylamine (cysteamine) and pantothenate from pantetheine operate during turnover and release of the vitamin, some of which is excreted in urine.

C. Functions

The myriad acyl thioesters of CoA are central to the metabolism of numerous compounds, especially lipids and the ultimate catabolic disposition of carbohydrates and ketogenic amino acids. The chemical properties of the thioester, which has a high group-transfer potential, permits facile acylations and hydrolysis; the ready formation of enolate ions and the carbanionlike character of the carbon α to the carbonyl facilitate condensation reactions. For example, acetyl-CoA, which is formed during metabolism of carbohydrates, fats, and some amino acids, can acetylate compounds such as choline and hexosamines to produce essential biochemicals; it can also condense with other metabolites such as oxalacetate to supply citrate, and it can lead to cholesterol. The reactive sulfhydryl termini of ACP provide exchange points for acetyl-CoA and malonyl-CoA. The ACP-S-malonyl thioester can chain-elongate during fatty-acid biosynthesis in a synthase complex.

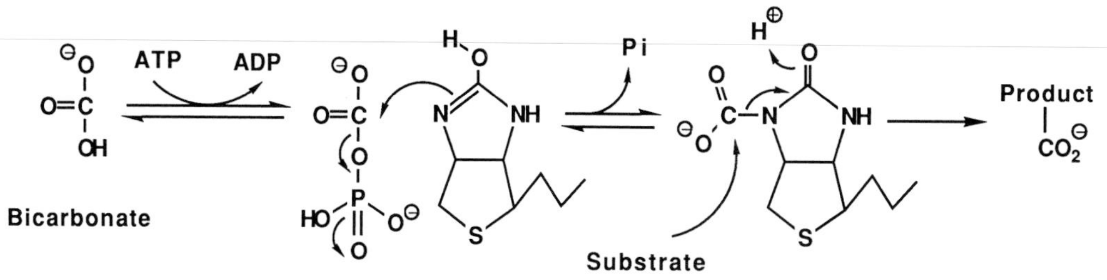

FIGURE 23 Mechanism of carboxylations by biotin-dependent enzymes with the intermediacy of the putative carbonyl phosphate.

IX. Lipoyl Functions

A. Chemistry

The coenzyme form of α-lipoic acid (thioctic acid) (Fig. 26) occurs in amide linkage to the ε-amino group of lysyl residues within transacylases. Hence, the functional dithiolane ring is on an extended, flexible arm.

Natural *d*-lipoic acid (1,2-dithiolane-3-pentanoic acid) is more soluble in organic than aqueous media and has a relatively weak absorption maximum at 333 nm. Because of the considerable ring strain, the sulfurs are especially subject to chemical and photochemical attack. The reduction potential (E_0' at pH 7, 25°C) is -0.3 V and near that of pyridine nucleotide coenzymes.

B. Metabolism

Lipoic acid is attached to appropriate lysyl residues of acceptor apoproteins via the 5′-adenylate in an ATP-dependent reaction catalyzed by a holoenzyme synthetase. Proteolytic turnover of lipoyl-enzymes involves release of the ε-N-amide of lysine and further hydrolysis to release lipoate.

C. Functions

Operation of lipoyl residues within transacylase subunits of multienzyme complexes that function as α-keto acid dehydrogenases for facultative and aerobic organisms is shown in Fig. 27. The lipoyl function mediates the transfer of electrons and activated acyl groups from the thiazole-attached α-hydroxyalkyl-TPP (cf. Section I). In this process, the disulfide bond is broken and a dihydrolipoyl residue is transiently generated; reoxidation requires coupling to a flavoprotein system. All α-keto acid dehydrogenases are comprised of varying numbers of the three basic subunits, i.e., TPP-containing α-keto acid decarboxylases, lipoyl-containing transacylases, and the FAD-containing dihydrolipoyl dehydrogenase. In humans and other animals, the three important and distinct mitochondrial α-keto acid dehydrogenases are for pyruvate to generate acetyl-CoA utilizing a transacetylase, α-ketoglutarate which forms succinyl-CoA using a transsuccinylase, and branched-chain keto acids from certain amino acids (valine, leucine, isoleucine) on which yet another transacylase is at work to yield the acyl-CoA products.

X. Pyrroloquinoline Quinone

A. Chemistry

Pyrroloquinoline quinone (PQQ) (Fig. 28) is a more recently studied cofactor that was originally isolated from bacteria possessing a methanol dehydrogenase. It is now recognized to be a functional co-

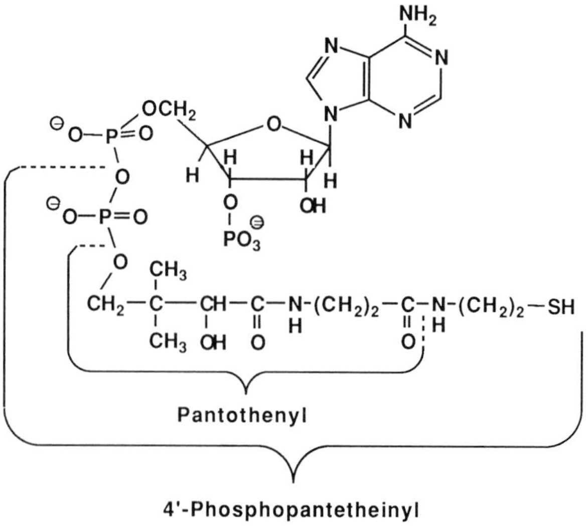

FIGURE 24 Structure for coenzyme A with pantothenyl and 4′-phosphopantetheine as components.

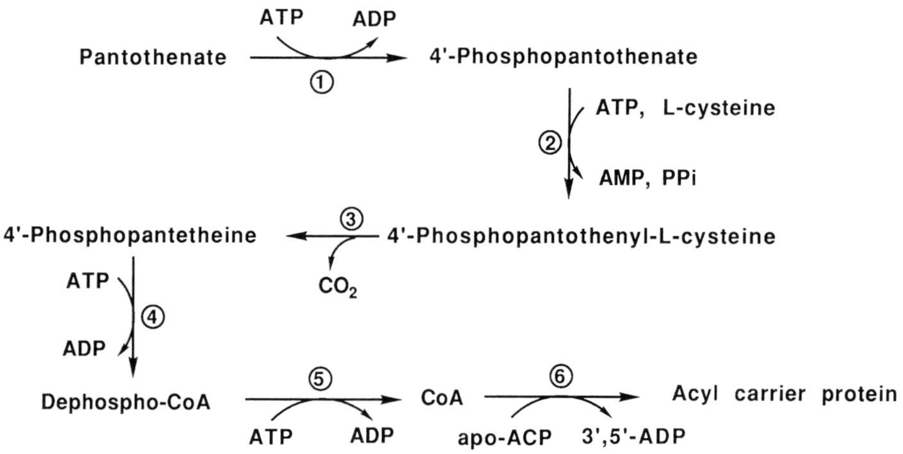

Pantothenate → 4'-Phosphopantothenate (① ATP, ADP)
4'-Phosphopantothenate → 4'-Phosphopantothenyl-L-cysteine (② ATP, L-cysteine; AMP, PPi)
4'-Phosphopantothenyl-L-cysteine → 4'-Phosphopantetheine (③ CO_2)
4'-Phosphopantetheine → Dephospho-CoA (④ ATP, ADP)
Dephospho-CoA → CoA (⑤ ATP, ADP)
CoA → Acyl carrier protein (⑥ apo-ACP, 3',5'-ADP)

FIGURE 25 Biosynthesis of CoA and the 4′-phosphopantetheine of ACP from pantothenate involving (1) pantothenate kinase, (2, 3) synthase and decarboxylase for 4′-phosphopantothenyl-L-cysteine, (4) a so-called pyrophosphorylase, (5) dephospho-CoA kinase, and (6) ACP synthase.

enzyme for many quinoproteins that occur in many organisms including humans and catalyze a range of oxidation-reduction reactions that replace or supplement other longer-known redox coenzymes. PQQ is usually tightly bound to its apoprotein partners; some of these associations probably involve esterification of the carboxyl groups to amino acid residues in the proteins.

PQQ is a reddish compound (λ max = 249, 325, and 475 nm at pH 7), which is fluorescent. Although fairly resistant to acidic and basic conditions, the quinone carbonyls are readily attacked by nucleophiles. PQQ is easily reduced to the nonfluorescent quinol with absorbance at 302 nm (pH 7). The redox potential is fairly high at an E_0' (pH 7) near +0.08 V.

B. Metabolism

Stringent experiments with rats suggest that PQQ may join the list of vitamins required by mammals. Microflora in the gastrointestinal tract, however, may symbiotically supply some of this cofactor to the host. Two amino acids have been implicated as the biosynthetic precursors in a *Methylobacterium*. That portion of the structure including ring atoms 6, 7, 8, and 9 and the two carboxyl groups at 7′ and 9′ is derived intact from glutamate; the remainder is from a symmetric product of the shikimate pathway, probably the L-dopaquinone produced from tyrosine in which the pyrrolic N-1 of PQQ was the α-amino group.

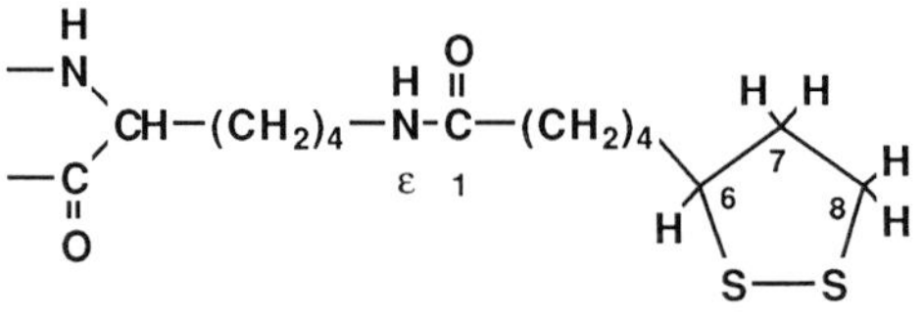

$$-NH-CH(-C(=O)-)-(CH_2)_4-NH-C(=O)-(CH_2)_4-$$ (lipoyl dithiolane ring, positions ε, 1, 6, 7, 8; S—S)

FIGURE 26 Structure with numbering for *d*-lipoyl-L-lysyl coenzyme moiety.

C. Functions

PQQ is a redox component of holoenzymes in which the coenzyme can be reduced by successive one-electron steps to the radical PQQH˙ and further to $PQQH_2$. In some enzymes, however, no radical intermediate is detectable, so two-electron processes seem to operate in which addition reactions may accompany hydrogen abstractions.

An important mammalian enzyme that contains PQQ is lysyl oxidase needed for collagen cross-linking. Dopamine-β-hydroxylase, which is essential for production of the neurotransmitter norepinephrine, contains a somewhat related trihydroxyphenylamine derivative as a cofactor.

XI. Ubiquinones

A. Chemistry

The ubiquinones (coenzyme Q's) are a group of "ubiquitous" 2,3-dimethoxy-5-methylbenzoquinones substituted at position 6 with variable-length terpenoid chains. The reduction of these to

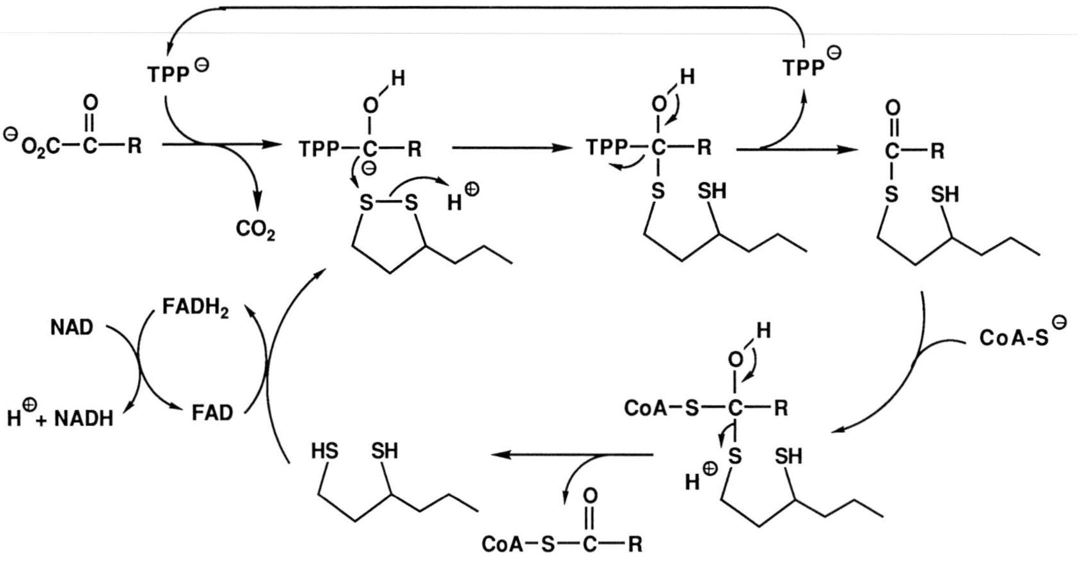

FIGURE 27 Function of the lipoyl-dependent enzymes involved in transacylations following α-keto acid decarboxylations (see Fig. 3). The transfer of an acyl moiety from an α-hydroxyalkyl-TPP to the lipoyl group and thence to CoA results in formation of the dihydrolipoyl group, which is cyclically reoxidized by dihydrolipoyl dehydrogenase.

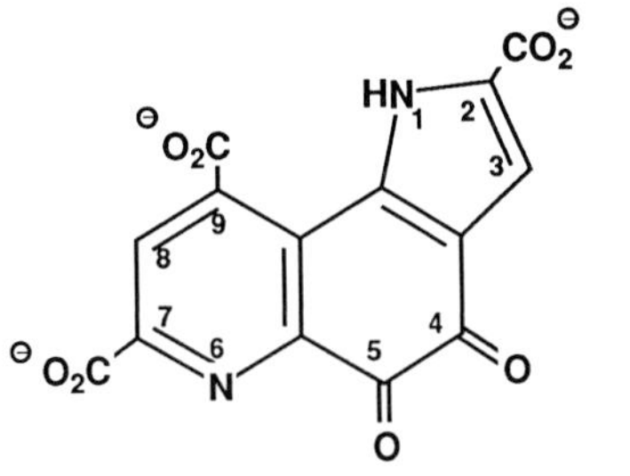

FIGURE 28 Structure with numbering for pyrroloquinoline quinone (PQQ; methoxatin; 4,5-dihydro-4,5-dioxo-1H-[2.3 f] quinoline-2,7,9-tricarboxylate).

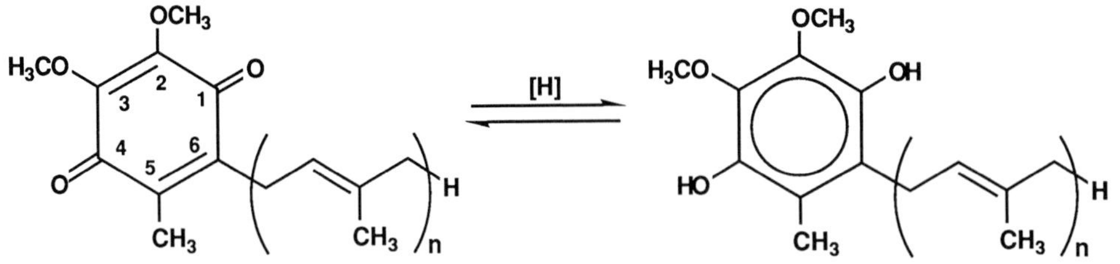

FIGURE 29 Structure with numbering of oxidized and reduced forms of ubiquinones (coenzyme Q) (n = 6–10 isoprenoid units).

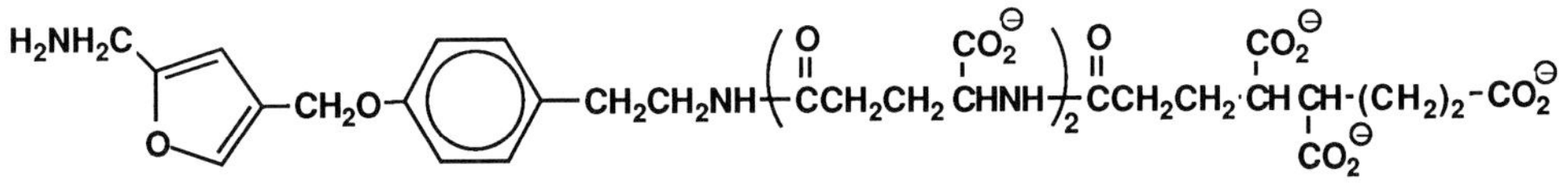

H_2NH_2C – CH_2O – CH_2CH_2NH ($\overset{O}{\|}CCH_2CH_2\overset{CO_2^{\ominus}}{|}CHNH$)$_2$ $\overset{O}{\|}CCH_2CH_2 \cdot CH\ CH$-$(CH_2)_2$-$CO_2^{\ominus}$, $CO_2^{\ominus}$, $CO_2^{\ominus}$

Methanofuran (MFR)

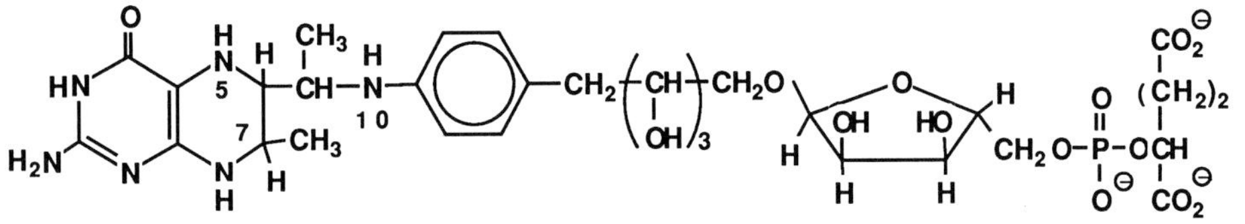

O, HN, H_2N, N, H, N, 5, H, CH_3, CH—N, H, 10, 7, CH_3, N, H, H, CH_2 (CH (OH)$_3$) CH_2O, O, H, OH, HO, H, H, H, CH_2O-P-OCH, O, $O^{\ominus}$, $CO_2^{\ominus}$, $(CH_2)_2$, $CO_2^{\ominus}$

Tetrahydromethanopterin (H_4MPT)

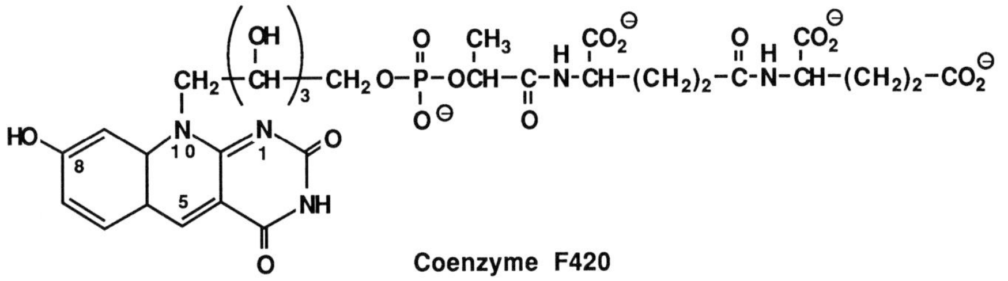

CH_2 (CH (OH))$_3$ CH_2O-P-OCH-C-N-CH-$(CH_2)_2$-C-N-CH-$(CH_2)_2$-$CO_2^{\ominus}$, O, $O^{\ominus}$, CH_3, O, H, $CO_2^{\ominus}$, O, H, $CO_2^{\ominus}$

HO, 8, N, 10, N, 1, O, 5, NH, O

Coenzyme F420

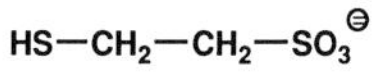

$HS—CH_2—CH_2—SO_3^{\ominus}$

Coenzyme M

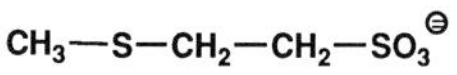

$HS—(CH_2)_6—\overset{O}{\|}C—\overset{H}{|}N—CH—\overset{CH_3}{|}CH—O—P{=}O$, $CO_2^{\ominus}$, $O^{\ominus}$, $O^{\ominus}$

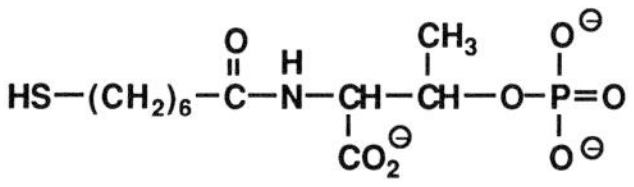

7-mercaptoheptanoylthreonine phosphate (HS-THP)

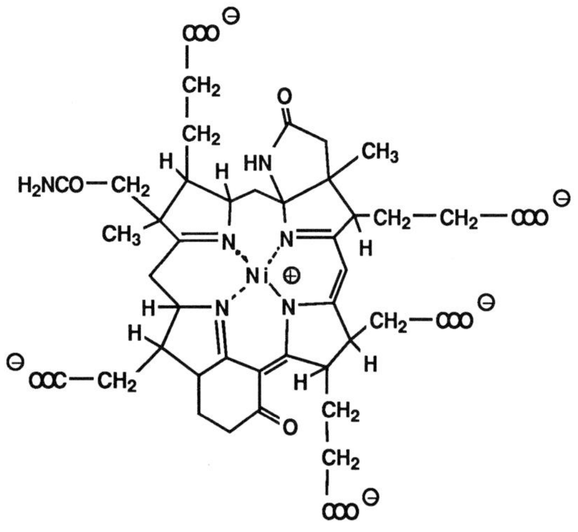

$CH_3—S—CH_2—CH_2—SO_3^{\ominus}$

S-methyl-CoM

$COO^{\ominus}$, CH_2, CH_2, O, H, H, HN, CH_3, $H_2NCO—CH_2$, CH_3, N, N, Ni ⊕, N, N, $CH_2—CH_2—COO^{\ominus}$, H, H, H, $CH_2—COO^{\ominus}$, $^{\ominus}OOC—CH_2$, H, H, O, CH_2, CH_2, $COO^{\ominus}$

FIGURE 30 Structures of coenzymes that participate in methanogenesis by certain archaebacteria.

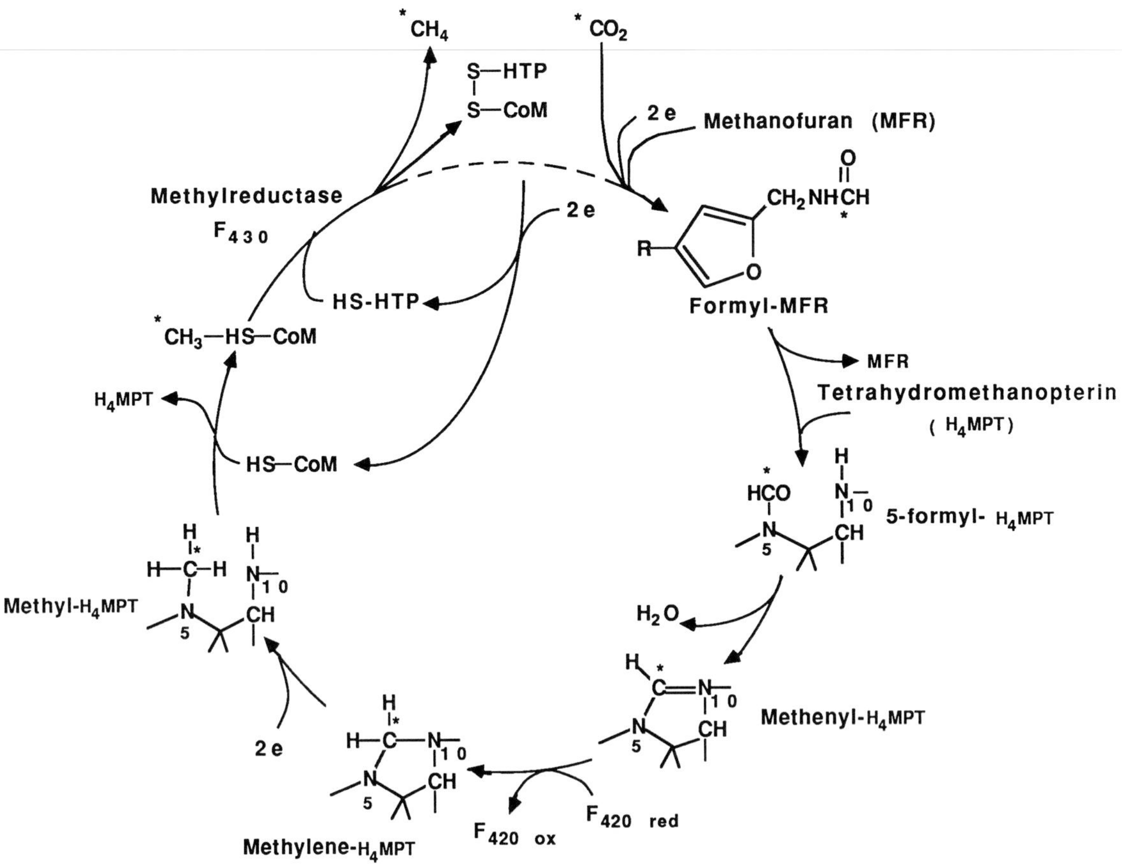

FIGURE 31 Proposed cycle for the reduction of carbon dioxide to methane. The carbon originating from CO_2 is tagged with an asterisk to follow its sequential fate in attachment to coenzymes during ultimate conversion to CH_4.

the 1,4-hydroquinone forms is illustrated in Fig. 29.

The ubiquinones are strongly hydrophobic and soluble only in organic solvents. The quinone form has an absorption band at 275 nm, which disappears upon reduction to the quinol (dihydroquinone). As with similar quinones, reduction can proceed by a two-electron process or by two single-electron transfers through a semiquinone intermediate, which can be neutral ($QH^{\cdot}$) or anionic ($Q^{\bar{\cdot}}$) depending on pH. The redox potential is +0.11 V.

B. Metabolism

The shikimate pathway is involved in biosynthesis of ubiquinones as well as plastoquinones, tocopherols (vitamin E), and nathoquinones (vitamin K). Because this pathway, also used in formation of phenylalanine and tyrosine, does not occur in mammals, the basic ring structures derive from bacteria and plants. Addition of the side chains, however, can be accomplished by microsomal prenylation in most organisms including humans.

C. Functions

In eukaryotes, ubiquinones are found mainly in the mitochondrial inner membrane where they function to accept electrons from several different dehydrogenases and relay them to the cytochrome system. Among substrates serving as electron donors are succinate and glycerol 3-phosphate, which are initially oxidized by flavoprotein systems, and malate, α-ketoglutarate, and β-hydroxybutyrate, which transfer their electrons initially to NAD. Hence, the positioning of ubiquinones in the electron–oxygen transfer process is near cytochrome b, after flavoproteins that couple in some cases with pyridine nucleotide-dependent enzymes, and before cytochromes c_1, c, a, and a_3. The relatively fluid movement of most of the ubiquinone within the phospholipid bilayer of the membrane serves well to provide a more mobile single-electron carrier between the rather large and rigid complexes that are constituted by other redox participants. In this connection, the relatively higher concentration of ubiquinone facilitates contact and the efficiency of electron flow. In heart mitochondria, for example, there is about seven times more ubiquinone than cytochrome a_3.

XII. Methanogen Coenzymes

A. Chemistry

The diverse coenzymes required to convert CO_2 to CH_4 are structurally elaborated and given in Fig. 30. These rather novel compounds vary from a complex nickel-porphyrin-like F_{430} to the phenyl ether of methanofuran, the unusual pterin of tetrahydromethanopterin, the deazaflavin of F_{420}, to the somewhat simpler sulfur-containing mercaptoheptanoylthreonine and coenzyme M.

B. Function

The proposed cycle in which the various coenzymes function in the methyl-coenzyme M methylreductase system is given in Fig. 31. The highly specialized archaebacteria (known as methanogens) that can accomplish such reduction of carbon dioxide through the formyl, methenyl, methylene, and methyl stages to methane are widespread in nature. The importance of the coenzymes that participate can be appreciated by the quantitative and global aspects of carbon cycling through methane-forming systems, which are not only in free-living microbes, but also in microbes in the intestinal tract of the human.

Bibliography

Chytil, F., and McCormick, D. B. (eds.) (1986). Vitamins and coenzymes, Parts G and H. *in* "Methods in Enzymology," Vols. 122, 123. Academic Press, New York.

Duine, J. A., Frank Jzn, J., and Jongejan, J. A. (1987). Enzymology of quinoproteins. *In* "Advances in Enzymology" (A. Meister, ed.), **59,** 170–212. John Wiley & Sons, New York.

Edmondson, D. E., and McCormick, D. B. (eds.) (1987). "Flavins and Flavoproteins." W. de Gruyter, New York.

Frey, P. A. (1988). Structure and function of coenzymes. *In* "Biochemistry" (G. Zubay, coord. au.). Macmillan Publishing Co., New York.

Korpela, T., and Christen, P. (eds.) (1987). "Biochemistry of Vitamin B_6." Birkhauser Verlag, Boston.

Rouvière, P. E., and Wolfe, R. E. (1988). Novel biochemistry of methanogenesis (Minireview). *J. Biol. Chem.* **263,** 7913–7916.

Cognitive Representation in the Brain

1
+1
?

MARK H. BICKHARD, *University of Texas at Austin*

Glossary

Connectionism Approach to representation as being distributed in the patterns of activations of nodes within a network. Inspired in part by the high interconnectivity of the nervous system

Encodingism View that representation is fundamentally constituted as elements or structures that are in known correspondences with what they represent

Ensemble Population of active elements in which the statistical properties of single elements over time is equal to the statistical properties of the population of elements at a single time

Incoherence problem Impossibility of specifying what an encoding is supposed to represent except in terms of some other already available representation, and the incoherence that results when that regress is supposed to halt with foundational encodings

Information processing View of cognition as consisting of the processing—the manipulation, combination, and generation—of symbolic encodings

Interactivism Functional and emergent approach to representation. Representation emerges as interactive differentiations of environments and consequent indications of possible further system activity in the service of goal-directed interactions

Skepticism Argument that it is impossible to check the accuracy of our representations of the world because it is impossible to know anything about the world except in terms of those representations themselves. Any purported check, then, is checking the representations against themselves—it is circular.

Transduction Transformation of form of energy. Also, a supposed generation of sensory encodings from encounters with environmental energy

STUDIES OF THE COGNITIVE ASPECTS of brain functioning cannot proceed without assumptions concerning the nature of representation. Since the demise of classical associationism, those assumptions, both in cognitive neuroscience and in cognitive psychology in general, have been dominated by the computer-inspired information processing model. In this model, representations are taken as being constituted as symbolic encodings, which are generated, processed, and transmitted by the central nervous system. This model has dominated so long and so thoroughly that it at times has seemed to attain the level of unquestionable common sense—the way things obviously must be.

More recently, however, several competing alternatives to this standard position, and criticisms of these standard assumptions, have arisen. As a result, future studies will increasingly be forced to take into explicit account their conceptual assumptions concerning representation as well as the neurophysiological and psychological results against which their models are tested.

One important alternative to information processing views is that of connectionism or parallel distributed processing (PDP). Another position has recently emerged in robotics but has not yet had much impact in brain studies. This approach attempts to eschew representation altogether in favor of systems without data structures that can nevertheless accomplish their goals.

All three of these positions, however—information processing and its two alternatives—*share* one basic assumption concerning the nature of representation, in spite of their differences in how that assumption is developed. The assumption in common among them is that representation is constituted as encodings, whether or not these are taken to be symbolic. That assumption is itself subject to severe criticism, which, in turn, leads to a fourth alternative to all three positions: an interactive conception of representation.

This article is primarily a review of the four positions concerning the nature of representation and some of the arguments among them. With respect to the information processing position especially, this will also involve some illustrative examples of how that approach can be applied to brain functioning. I will be arguing in favor of the interactivist position.

I. Information Processing

The backbone of the information processing perspective is the presumed flow of information from environment to perception to cognition to language. Information originates in the environment and is processed through the senses into the brain or mind, where further cognitive processing occurs and where new encodings of resultant mental contents can be generated and transmitted as language utterances. Those utterances, in turn, will be received by an audience and decoded in accordance with their semantics into cognitive contents for that audience. The basic flow of information, then, is from the environment, through the senses and cognition, and into the environment again as language—from which it will in general reenter the nervous system via the perception and understanding of language.

The various steps of this sequence are of three general kinds: the transduction of new encoding elements in the primary perceptual organs in response to encounters with environmental information, the generation of new encodings on the basis of already present encodings in the various processing steps (this is a form of heuristic and perhaps implicit inference of new encodings on the "premise" of extant encodings), and the emission of encodings via language. All three of these processes are being investigated, and knowledge of differentiations and specializations within the central nervous system by sensory modality and form of cognition is growing rapidly.

The sensory nervous system is generally considered to provide only two possible forms of basic encoding. The transduction process must result in signals being transmitted along various *axons* carrying some *frequency* of impulse. Basic sensory encodings, then, must be implemented in some combination of line (axon or spatial) and frequency (temporal) encoding.

For example, human color vision is based on three different types of receptors, each attuned to transduce a differing range of electromagnetic wavelengths. The transduction sensitivities of these types of receptors are maximal in, respectively, blue, green, and red ranges of color. This gives rise to a *line encoding* of *color,* which undergoes several stages of further processing in the retina, the visual pathways, and the visual cortex. Both color receptors, cones, and more general light intensity receptors, rods, are distributed spatically over the retina (with cones concentrated in the fovea), and this gives rise to a *line encoding* of relative *spatial position* of light reception. *Light intensity* itself receives a *frequency encoding*. The topology of these spatial relationships tends to be maintained through the several layers of further processing; thus, the spatial encoding of relative position of reception tends to be maintained. The auditory system for another example, yields primarily a *line encoding* of *frequency,* although lower frequencies seem to involve some degree of *frequency encoding*. [*See* COLOR VISION; EARS AND HEARING.]

These relatively simple correspondences between properties of the stimulus, on the one hand, and lines or frequencies of neural activity, on the other, become more complex and less well understood with progressive steps of processing. Some of the more complicated and well-known examples are the apparent motion and "feature" detectors of the visual system. The feature detectors seem to be sensitive to such features as edges and orientations—important properties of visual boundaries. [*See* VISUAL SYSTEM.]

Such models of perceptual encoding are based on single neurons and their activities as the elements of the presumed encodings. However much insight these models may provide for perceptual representation, there are strong reasons to think that single neurons are not the locus of representation in the central nervous system. One consideration is simply that the limiting case of such single neuron representation yields single neuron "encodings" of all of our concepts and representations—a common *reductio ad absurdum* of this is the infamous "grandmother neuron" that represents our grandmother. Since there is a continual loss of neurons to cell death, we would experience random and total losses of representation of whatever those neurons represented, including, potentially, our grandmothers. Such unitized representational losses are not found with neural death; thus single neurons cannot be the microanatomical locus of concept representation. Furthermore, activities of single cortical neurons in response to repeated stimulus instances are found in general to be highly unreliable, also not giving a foundation for a single neuron locus of representational encoding.

One design solution to this unreliability problem (cell death is itself a version of unreliability) is redundancy: if many neurons are serving the same representational function redundantly, then the loss of one or more, or the unreliability of all of them, can in principle be compensated for by the activity of the whole redundant set. A more powerful hypothesis, however, is that the functional unit is not the single neuron at all but, rather, local populations of neurons that function as statistical ensembles. The relevant properties of such ensembles would be the temporally and perhaps spatially organized patterns of oscillations within the ensembles, which would modulate the similar activities of other ensembles. Cognitive activity would consist of such modulatory processes disseminating and interacting throughout the brain.

This is a very different notion from the classical "switchboard" model of brain activity in which impulses are generated at perceptual surfaces and then switched to various output neurons within the central nervous system (CNS). In this long outdated switchboard model, even the notion of frequency encoding is distorted in that the switchboard metaphor emphasizes on and off relationships, not frequencies. The ensemble model builds on the spatial and frequency characteristics of neural functioning, provides a redundancy with respect to single level neurons in that the ensembles will exhibit reliable statistical properties of their oscillations in spite of single neuron unreliability (this is in effect a reduction of noise or error variance via a larger sample), and, for the first time among the models discussed, acknowledges the endogenous activity of the central nervous system.

This later point is potentially quite important. Neurons in general are not quiescent until stimulated by synaptic transmissions or sensory input. Neurons exhibit baseline frequencies of axonal impulses, varying from neural type to neural type, and varying from zero to high frequency. This intrinsic neural activity is continuous. Sensory inputs do not switch this on or off so much as they modulate the frequencies and patterns in which this ongoing activity takes place. The ensemble notion is a population level version of this basic point concerning even single neurons.

Perception as a modulation of internal endogenous activity is a quite different notion from that suggested by the simple information processing flow from environment to perception to cognition. Modulation is not the same relationship as simple encoded input. Even the information processing models, however, acknowledge the necessity of contributions to perception and cognition from previously learned or innate sources—the sensory information is not adequate to cognition, nor, most models hold, even to perception. Memories, for example, might be postulated as involved in the inferences from basic sensory reception to full perceptions of objects located and moving in space and time. Modulation of ongoing activity, then, is not at such deep variance with such versions of the information processing approach. [*See* PERCEPTION.]

Even these forms of information processing models, however, retain the presupposition that all cognition is ultimately input from the environment (or perhaps provided innately). If not present in current sensory input, relevant information must have been provided in earlier experiences and be available in memory. That is, such models, except for the "out" of innatism, force an empiricist epistemology, in which the senses are the source of all knowledge. Empiricist epistemologies have not fared well in epistemology or the philosophy of science. Understanding the necessity of mathematical relationships, such as $1 + 1 = 2$, for example, has been a classical counter to empiricisms—it might be conceivable that we could learn *that* $1 + 1 = 2$ simply from experience, but no amount of experience will ever provide knowledge that this relationship is logically necessary. Mathematics would remain on a par with, say, astronomy, in which the number of planets also remains consistent, no matter how many times you look, but that number is *not* necessary. For that matter, there does not seem to be any perceptual realm at all for mathematics—we can see pebbles but not numbers. Many other cognitions, such as of virtues and vices, are not presentable in sensory form. Such considerations suggest that the endogenous activity of the CNS is not simply bringing to bear previously perceived information but that it is involved in some way in *emergent* representational phenomena.

A convergent consideration for the notion of endogenously active ensembles as units of functional activity is the acknowledgment that the organism is fundamentally engaged in physical activity in the environment. Such interaction with the world requires in most cases *correct timing* of the organism's side of the interactions. By correct timing is meant neither too fast nor too slow nor in the wrong phase. Walking, for example, is not so much a matter of pushing the legs back and forth as it is a matter of exciting and modulating an intrinsic oscillation in the spinal cord and of the skeletal–musculatory system itself. Most activities require such timing—driving a car, catching a ball, running, and so on—and timing is fundamentally based on oscillatory phenomena. We would expect, therefore, that oscillations and modulations would be fundamental to the operating design principles of the nervous system. Ensembles, then, not only provide an answer to the problem of unreliability and pose the problem and the promise of intrinsic endogenous activity, they are also endowed with the basic solution to the problem of timing in action and interaction.

The information processing approach has no intrinsic place for timing. The sequence of operations on symbolic encodings has all the same *formal* properties no matter what the timing may be of the steps involved in that sequence. It is clear that results may be obtained too late to be of any good, and, thus, that *speed* is necessary, but in this view, timing is irrelevant to the nature of cognitive activity per se. Cognition abstracts away from the timing considerations that are essential to action, according to this view, but even so we could expect the timing properties of oscillatory phenomena to dominate the functioning of the CNS. I will argue later that timing is in fact *not* irrelevant to cognition in general.

II. Connectionism

Considerations of the highly parallel manner of functioning of neurons and neural circuitry, and of the enormously complex interconnectedness of neural circuitry, contributed to the inspiration of one major alternative to standard information processing approaches—connectionism or parallel distributed processing. The underlying metaphor for the information processing approach is the von Neumann computer, which has only one locus of processing. Parallel processing in computers can be introduced by multiplying the number of simultaneously active processing units, but something more than this seemed to be taking place in the brain.

One highly persuasive consideration is that the brain accomplishes many tasks, such as various forms of pattern recognition, in a short amount of time and, therefore, in a small number of strictly sequential "steps" of processing. One potential solution was to posit that the many steps that seemed to be logically required were carried out simultaneously and in parallel in multiple processing units. This solution, however, retained the basic assumptions of the information processing approach and simply introduced multiple processors, and it was a conceivable approach only when the basic task did not involve internal dependencies that required sequential processing steps—only when the processing could in fact be broken down into multiple parallel streams.

Another strong consideration was that standard information processing approaches had failed miserably at modeling the phenomena of learning. Systems could be designed that succeeded in "learning" things that were close to what they had been designed to solve, but any significant generalization beyond the problem for which they were designed seemed unattainable. One perspective on this failure comes from noting that the information processing approach construes all processing in terms of the generation and elimination of *instances* of various *types* of encoding elements, but the basic types of encodings themselves must be designed in from the beginning—there is no way to generate new types of symbolic encoding representations.

The major excitement of the connectionist or PDP approaches is that they seem to solve this problem of learning, of the generation of new representations. PDP systems can undergo training with respect to sets of problems and generate solutions to them that then generalize beyond the training set. At times, the form of the generalizations found seems tantalizingly similar to human solutions to those same problems.

A PDP system engages in two levels of dynamic activity. The primary level is that activity by which a categorization, the representation, of an input array is settled upon. The secondary or meta-level of activity is that by which the system "learns" to correctly categorize such input patterns.

A PDP system is a set of functional nodes, each of which is capable of varying levels of "activation," interconnected by a fixed topology of paths. Each connection between nodes has a weight, which can be positive or negative. The nodes and interconnections are frequently organized into layers, perhaps with feedback among layers. Some subset of nodes are connected directly to the environment, from which they receive "activation," and they in turn activate the nodes to which they are connected in accordance with the weights of the connection paths. These nodes activate still further nodes in accordance with their connectivities and weights, and so on. The system eventually settles down into a fixed pattern of levels of activation in the nodes, or in some selected subset of the nodes, that is specific to that particular pattern of inputs. The first key to the power and appeal of PDP models is that that pattern of resultant activations can be taken as a classification of the input pattern: it will classify *together* all such input patterns that yield that same resultant activation pattern and will classify as *different* all input patterns that result in some different final activation pattern. The class of possible final activation patterns, then, forms the class of classification *categories* for the possible input patterns. Note that the processes of settling of the node activations is massively parallel among all the nodes and their weighted interconnectivities.

The second, and most important, key to the appeal of PDP models is that various adjustment rules can be used to adjust the weights of the connections among the nodes, in accordance with various training "experiences," in order to "learn" input pattern classifications. The organization of connections remains fixed in such training, but the changes in the *weights* of those connections can change the entire first level—classifying—dynamics of the overall system. In particular, it can result in differing resultant classifications of the input patterns. The system, in other words, can adjust to, can be trained to, new and desired classification schemes. Insofar as the input pattern-classifying activation patterns are taken to be *representations* of those categories of input patterns, the system can be construed as generating new representations of novel input categories, something that is impossible in the standard information processing approach.

With proper design and appropriate shifts in interpretation, PDP systems can manifest still other characteristics that are simultaneously powerful, exciting, and reminiscent of the way in which the brain functions. One important example derives from the possibility of the *input* activation pattern being any of several *sub*patterns of the overall *resultant* activation pattern—so that any piece of the overall pattern as input results in the activation of the whole pattern—in which case we have a model of content addressable memory. Content addressable memory is a form of memory that permits memory representations to be accessed directly in terms of their representational *contents,* rather than just in terms of their *location* in the memory organization. Human memory, in particular, manifests this phenomenon.

A different shift in interpretation considers the input patterns themselves to be whole patterns, but the resultant activation pattern to be a composite of the permitted input patterns. Under this interpretation, the system manifests an *association* between the various input patterns—an associative memory, again manifested in human memory.

Connectionist approaches, then, capture a parallelism at least reminiscent of the functioning of the brain, model the emergence of new categorizations, model a form of content-addressable memory and associative memory, and other properties of human memory. They are an exciting alternative to information processing approaches for these and additional reasons and are being pursued eagerly.

They are not without their critics, however. One of the most powerful criticisms of the potentialities of PDP approaches turns on what from another perspective is one of their greatest strengths—the singularity and lack of internal structure of the categorizing patterns of activation. This is an aspect of their greatest strength in that emergent novel representations would be expected to be singular and without internal representational structure. To simply put together already available representations in some new structure is what information processing approaches already do and does not constitute emergent novel representation. On the other hand, it is precisely the ability to construct new *structures* of already available representations and, thus, to implicitly capture not only the resultant representation but also its internal representational structure that is the *forte* of symbolic encoding information processing approaches. It is argued that for both language and cognition alike, this componentiality of representation is necessary and is not provided by PDP approaches. For example, any genuine cognitive system, so the argument goes, that is capable of thinking "John loves Mary" is also capable of thinking "Mary loves John." This generalization of ability is natural in a symbolic encoding framework but not in the more holistic PDP framework. Needless to say, the arguments and explorations continue.

One obvious notion, for example, is the possibility of hybrid systems in which a basic PDP-type layer provides the representational categories that can then be operated on and processed in a more conventional information processing manner. Whether or not such is feasible, and what might be gained, remains to be explored.

Connectionism boasts a natural manifestation of several inherent properties of CNS functioning: parallelism, emergentism, content addressability and associativity, and so on. Nevertheless, there are a number of inadequacies, or at lest disanalogies, of connectionist approaches with respect to this comparison. For example, PDP networks "represent" by virtue of static patterns of activation of the nodes, once settled into, while the CNS is engaged in continuous ongoing activity. It is at least plausible, and even likely, that cognition in the brain is a function of that activity and cannot be captured in such static models. In this respect, among others, PDP networks are *not* akin to neural ensembles. A similar observation is that the CNS is engaged in interaction with an environment (internal or external), while a PDP network has no comparable outputs at all. Furthermore, although PDP networks do manifest a kind of emergence of categorization abilities, the "learning" rules by which this is accomplished are relatively inefficient and are, in general, *not* plausible as models of learning in the brain.

III. Robotics

The information processing approach encounters many problems of interpretation. One important example is what is known as the empty symbol problem. The basic notion underlying this problem involves the sense in which encoded symbols in the information processing approach are supposed to represent various events and objects and facts in the environment by virtue of being in correspondence with those events and objects and facts. Transduction, for example, is fundamentally a change in form of energy from some environmental form to some form of neural activity. There is nothing epistemic or representational in this energy-change level of consideration—such changes in form of energy or activity occur ubiquitously in the physical world, without being confused with representation. Transduction in sensory systems, however, is taken not only to be constituted by such energy changes and their resulting correspondences, but those correspondences, in turn, are taken as representing whatever those correspondences are with, whatever was in fact transduced. The empty symbol problem arises, among other ways, from consideration of how those factual correspondences could represent what has been transduced or some other relevant aspect of the correspondence. Specifically, how does the system know what the correspondences are with? Or, among the multitude of things that are in fact in correspondence—light patterns, quantum electron processes in the surfaces of objects, chemical reactions in the retina, and so on—how does the system know which are being represented? The scientist–observer can analyze these correspondences and analyze as well which of those correspondences seem to be ecologically relevant, but this only establishes which correspondences do occur and which of those would be ecologically desirable to represent. There does not seem to be any way for the system itself to have epistemic contact with whatever it is in causal contact with, to obtain transduced *encodings,* not just transduced *energies*. The internal symbols, in other words, seem doomed to be empty. They differ in shape or size or some other formal properties that allow them to be differentiated and operated upon, but they carry no representational content, or, at least, it is not understood how they can carry any representational content *for the system itself*.

Because of this and related difficulties in the information processing approach, some researchers in robotics have proposed that representation be avoided altogether and, furthermore, that robots can function quite well without any representation at all. Insofar as that is correct, it may be that symbols and data structures are superfluous for robotics, that representation is the wrong level or the wrong sort of abstraction for robots.

In place of notions of representation communication and processing, design is in terms of minimally coordinated activity systems, each of which is competent to its own perceptions and actions. An energy transducer, for example, doesn't have to establish an encoding of a wall, as long as it controls the locomotion of the system to avoid bumping into walls.

In important ways, this is a return to the roots of cybernetics and control theory out of which computer information models evolved. The control of effective interaction with an environment has largely been lost from information processing approaches as being of any interest beyond that of robotic engineering. In particular, it is not generally understood to have any basic relevance to foundational issues of representation or cognitive science in general. Roboticists, clearly, have not been able to ignore such concerns quite so readily.

This antirepresentationalism of some roboticists emphasizes the interactive and hierarchically organized control structure aspects of the nervous system—aspects which are absent in both the information processing and the PDP approaches. I will be arguing that these aspects are not only of practical design relevance but that they are intrinsic to the nature of cognition and representation as well.

IV. The Encodingism Commonality

There is a common notion of representation among the three approaches. It is that representation is constituted as encodings of what is to be represented. In the information processing approach, this is a direct assumption, with the basic atomic encoding types designed directly into the system. In the PDP approach, these basic encodings are presumed to emerge in the learning process of the PDP network, but what is learned is still a correspondence between patterns that is presumed to constitute a representation—an encoding. The antirepresentationalism position accepts the same notions of encoding representation but concludes, not that they are wrong, but that robotics can proceed without them. I will argue that all three approaches are in error in this common assumption.

V. What's Wrong with Encodingism?

A prototypic encoding is a representational stand-in. It is some element that is specialized to serve a representational function, to carry a representational content, that is determined by some *other* representation—which may also be an encoding—thus, it "stands-in" for that other representation. In Morse code, for example, ". . ." stands-in for "S," while bit patterns serve the stand-in function in computers. In this sense, encodings most certainly exist and are quite powerful and useful. They specialize and differentiate representational functions and change the form of representational elements in such ways as to allow processing to occur that would otherwise be difficult or impossible: ". . ." can be sent over a telegraph wire, while "S" cannot.

The term "encoding," however, is also used in a variety of derivative ways that do not comport with these paradigmatic cases. Genes, for example, are often called encodings of the proteins whose construction they control, but they are not *representations* in any legitimate sense: instead, DNA base pair triples and strings of such are elements of a complex control organization that builds proteins. The selectivity of those DNA triples for certain amino acids in the control process is what motivates their being deemed encodings—there is a correspondence involved. This generalization of the paradigmatic encoding notion makes quite clear the seductive power of the correspondence notion, even though, in this case, there is no epistemic or knowing or representing agent at all. It is only the stand-in notion of encoding that is a representation, and, therefore, it is only this notion that I will analyze in terms of its sufficiency for the general notion of representation.

The existence and importance of encodings is not at issue. What is at issue is the assumption that representation is fully characterized by encodings. There is a complex of related criticisms and arguments against strictly encodingist conceptions of representations, some of which are of ancient provenance. Several of the core components of this complex of arguments will be summarized. I submit that these arguments render any simple encodingism logically incoherent: strict encodingisms cannot make logical sense, and certainly cannot ground models of cognition, at the neural level or more abstractly. That is, strict encodingism is not merely factually false, it logically cannot be true.

The first argument is that of skepticism. The basic skeptical argument first notes that in order to check whether or not our representations are correct we would have to compare those encodings against the world that they are taken to represent. But since we can know the world which is supposedly being represented only via those representations themselves, we cannot ever get independent epistemic access to the world to check our encoded representations of it. The conclusion, then, is the classical skeptical contention that we cannot have genuine knowledge of the world since we cannot ever check the accuracy of our representations of it.

The second argument asks not about the accuracy of our representations but about the construction of them in the first place. The point is that in order to construct elements that are in correspondence with the world, we must already know what those correspondences are to be with, but that cannot occur until the correspondent encodings are constructed. Therefore, there is no way to get started; no way to know what encodings to construct. We must already represent the world before we can construct our representations of it.

A closely related consideration is to note that the *factual* correspondences found in sensory transduction between neural activity and environmental events do *not* establish *epistemic* correspondences. The problem concerning how to know which encodings to construct leads in this context to the question of how the system, the CNS, could possibly know what those sensory correspondences are with and, therefore, what those sensory elements are taken to be encodings of. In other words, how can the system turn nonrepresentational energy transductions into representational encodings? It would have to know what the internal neural activity was in correspondence with in the world in order to know what the representational content of that neural activity should be taken to be. It would have to already have its representation of the world in order to construct its representation of the world.

A quick apparent answer to these questions is to mention evolution and assume that they have been solved there. But the problem is logical, and evolution has no more power to escape them than does maturation, learning, or development. Note that the power of evolution to construct active systems that successfully interact with their environments, along the lines of the antirepresentational robotics position, is *not* questioned by these arguments. What *is* put into question is the ability to construe those transductions as more than useful control signals, to construe them as encodings. Where does their representational power, their representational content, come from, and how does it come into being? The factual correspondences that obtain between the environment and neural activity help explain how and why that neural activity is in fact useful, but that useful functioning does not require any encoded representations. How does, or could, evolution, maturation, learning, development, human design, or any other constructive process, construct representations out of control organizations, or out of anything else, that is not representational already?

An additional level of consideration derives not from the question of the accuracy of encoded representations, nor from the question of which ones to construct, but from the question of how the system could possibly know what any encodings in a strict encodingist system were even supposed to represent, before such questions of accuracy or rational construction. For actual encodings, the answer to this question is provided by the representation for which the encoding is a stand-in: the encoding represents the same thing as that for which it stands in. But this introduces a regress into the origin of the representational contents involved: they might be provided for encoding X in terms of encoding Y, and for Y in terms of Z, and so on, but this regress must end in a finite number of steps. If we consider a supposed grounding level of encodings, that are not stand-ins for any other representations, that are logically independent encoding representations, we find that there is no way to provide the necessary representational content. For some purported grounding level encoding "X," we might attempt to define it in terms of some other representations, in which case it would not be at the ground level, contrary to hypothesis. But this leaves us at best with "'X' represents whatever it is that 'X' represents," and this does not establish "X" as a representation of anything. The requirement for logically independent encodings in order to ground any strict encodingist model encounters a logical incoherence: logically independent encodings cannot exist.

The underlying reason for all of these problems with encodingisms is that the notion of encodings focuses on, and is enormously powerful for, change of form of representations *that already exist,* and for combinations of representations *that already exist*. There is nothing in the notion of encodings that can explain the *origin* of representation, the *emergence* of representation out of a ground that is itself not already representational. This is so whether that emergence is evolutionary, maturational, learning, developmental, or by design. Encodingisms model things that can be done with representations that are already available, so to assume that representation in the broad sense can be fully captured in a strictly encodingist model is intrinsically incoherent. It encounters a requirement that encodings cannot serve—the requirement to explain representational emergence.

The closest attempt at such explanation within the encodingist framework is that of transduction, or its closely related notion of induction—a temporally extended transduction. But, as we have seen, these do not work. In both cases, the knower must already have the representation—the cognitive category or sensory encoding element—before it can notice or detect or transduce or postulate that "corresponded-with" element of category for its environment. Something more is needed, something that can account for representational emergence.

VI. Interactivism

Consider a system, or subsystem, in interaction with its environment. The internal course of that interaction will, in general, depend both on the internal organization of the system and on the particulars of the environment being interacted with. At the end of the interaction, the system will be left in some internal state, say state *A* or state *B*. If *A* and *B* are the only two possible final states of this system or subsystem, then those internal states will serve a function of *differentiating* possible environments into two classes—those environments that leave the system in *A*, and those that leave the system in *B*. A simple version of this differentiation involves systems or subsystems that have no outputs; they passively arrive at final internal states, e.g., energy transductions.

Note that at this point, the typical encodingist move would be to note that such differentiation establishes correspondences with the differentiated environments, and, therefore, *A* and *B* *encode* their respective correspondent categories of environments. The first step in this move is correct: the differentiations do establish factual correspondences with whatever it is that is differentiated. The second step is invalid: those factual correspondences do not in themselves constitute encoded representations of what the correspondences are with, of what the differentiations are differentiations of. The differentiations are more primitive than encodings, yet they do involve factual correspondences. It should be clear that the correspondences noted in actual sensory systems are in fact useful as differentiations and are epistemically nothing more than differentiations—all the system has functional access to is that state *A* is different from state *B* and that it is currently in state *A*. They are not encodings.

At this point, we are roughly in the position of the antirepresentationalist roboticists—potentially useful signals in a control structure—but with the recognition that something more is necessary. In particular, the antirepresentationalism of the roboticist position *accepts* the basic encodingist notions of representation, and there is independent reason to conclude that those notions are false and incoherent. The emergence of representation must itself be accounted for, and encodingism cannot do that.

We already have the emergence of a representational sort of function, differentiation, out of system organization that is not itself representational. What is yet missing is the emergence of representational content. Purely differentiating states *A* and *B* are truly empty; they differentiate, but, in standard senses, they represent nothing. The next task is to account for the emergence of their having representational content.

Suppose that the system with final states *A* and *B* is a subsystem of a larger goal-directed system. In this larger system, in general, various alternative strategies and heuristics will be available for attempting to reach various possible goals. In such a case, given some particular goal at a particular time, some selection among possible strategies and heuristics must be made. It may be that the system makes a choice of strategies or heuristics in part on the basis of whether the differentiating subsystem has reached final state *A* or *B*. If so—if, when attempting to reach goal *G72* and final state *A* obtains, try strategy *S17*, while if final state *B* obtains, try strategy *S46*—then such functional connections constitute implicit predications concerning the environments differentiated by *A* and *B*. In particular, the predications involved are "state *A* environments are appropriate to strategy *S17*; state *B* environments are appropriate to strategy *S46*." Such implicit predications are *about* the environments and can be true or false about those environments. They constitute *representations* of supposed environmental properties. They constitute *representational contents* attached to final states *A* and *B*.

Most importantly, they constitute *emergent* representational contents: there is nothing in the system organizations involved that is itself a representation nor that is representational in a more general functional sense. The function of representation emerges in the further selection of system activity on the basis of initial environmental differentiations.

These considerations suffice to show: (1) that the interactive model of representation suffices to account for at least one form, a functional and implicit form, of representation, (2) that this form is capable of emergence from nonrepresentational ground, and (3) that this form is not itself an encoding form. There remain, of course, many questions concerning the interactive approach to representation. Among the most important are those concerned with the adequacy of interactive representation to the many representational phenomena. One important version of the adequacy question focuses on abstract knowledge, such as mathematics; another focuses on language. Essentially, the adequacy questions lead into the basic programmatic adequacy of interactivism in general, and the answers constitute a general model of cognition, perception, representation, and language. These programmatic issues will not be pursued here. What is critical for current purposes is that we have found a conception of representation that is not an encoding, not subject to the many logical incoherencies of encodingism.

The fundamental new property is that interactive representation is functionally emergent in organizations of interactive systems. That is, it constitutes a model of the emergence of representation out of action. As such, it does not fall to the incoherence problem because the representational content is emergent in the strategies and heuristics that are selected, and they do not require any prior representations to come into being or to be used. It does not fall to the origins problem because the differentiation into *A* or *B* does not require the prior knowledge of what *A* or *B* environments are, nor the prior knowledge of which sort of environment the system is currently in. It does not fall to the skeptical problem because the functional information that the system is in, say, an *A* type environment is tautologically certain, while knowledge of *properties* of *A* type environments in the strategies and heuristics are defeasible and can in fact be tested—checked by *using* those strategies and heuristics to actually engage the environment, and checking to see if they work, if they succeed. The fact that representation is *emergent* from action has as one critically important consequence that representation can be checked *via* action without encountering the circularities of skepticism. Without that emergence, checking representation via action gets nowhere because there are no determinate representational interpretations of the actions or of their outcomes: there is no determinate crossing from action back to representation.

Further, interactive representation can serve as the ground for encodings: stand-ins for indicator states like *A* and *B* can be constructed and processed and can be useful for exactly the reasons encodings are useful. A simple differentiator might function strictly passively, although that is intrinsically of reduced power relative to interactive versions, and one form of such a passive differentiator would be a simple sensory transducer, another would be a PDP network. In this perspective, both the potential power of the connectionist approach and the limitations from their intrinsic passivity and non-goal-directedness, are evident: connectionist systems are passive differentiators, and as such, they cannot differentiate what would require *interaction* to differentiate, and they have no representational content—their activation patterns constitute empty "symbols." Interactive representation intrinsically and necessarily involves *open, interactive, goal-directed systems* of exactly the sort discussed by the antirepresentational roboticists: such robots, in the interactive view, *do* in fact involve representation—representation in its most fundamental form as a functional aspect of successful goal-directed interaction—and, therefore, there is a natural bridge to more standard encoding representations *in those circumstances in which encodings would be useful.* The general claim, clearly, is that the interactive model of representation captures the strengths of all three alternative approaches, without encountering their limitations and logical incoherences.

VII. Some Connections and Implications

Interactivism accommodates the correspondences involved in sensory neural activity but with a distinctly nonencoding interpretation: they constitute differentiations that are useful to the further functioning of the overall system in many and various ways, but they do not in any legitimate sense constitute representations of the light patterns, and so on, that they in fact differentiate. The level of analysis concerned with what is in fact differentiated is important to understanding *how* those sensory differentiations manage to be useful to the organism but do not constitute analyses of what the organism knows or represents.

Intermodulations of neural ensemble oscillatory activity *are* control relationships. They are precisely what an interactive control system necessarily involves: a *control* system because that is the level at which notions of interactive system and goal-directed system are constituted and an *oscillatory and modulatory* control system because successful action—thus, successful representation in most cases—requires correct timing at all levels. Interactive representation is emergent not from abstractly sequenced action but from correctly timed *inter*action.

In the interactive view, representation emerges in the organization of the ongoing oscillatory and modulatory activities of the CNS—differentiations of environments and subsequent differentiations, selections, of further activity. There is no need—in fact, it is *in general* quite inappropriate—to attempt to interpret those oscillations and modulations as *themselves* constituting representations (with the caveat of derivative, secondary encodings specialized for and based on that emergent representational function).

In particular, there is no need to attempt to understand language activity in terms of various encodings being transmitted from homunculus to homunculus in the brain for various processings and understandings. In the context of the currently dominant encoding understanding of the nature of language, it has been difficult to avoid this form of analysis of language phenomena at the level of neural functioning. In fact, language *cannot* be fundamentally an encoding phenomena for exactly the same reasons that perception cannot be: encodings cannot provide representational content that is not already there, including knowledge of what an utterance or a word is supposed to represent. Thus, it would be impossible to either learn or to understand utterances if language were in fact merely encodings. Wittgenstein, among others, made essentially this point some time ago, but, in the absence of alternative conceptions and the dominance of the information processing approach in general, it has had limited impact. [*See* LANGUAGE.]

VIII. Summary

Studies of cognitive phenomena in the brain have tended to maintain the same presuppositions concerning the fundamental nature of cognition and representation as has cognitive psychology in general. For some decades, those presuppositions were massively dominated by the information processing view, so much so that it began to take on a sense of taken-for-granted obviousness. More recently, several alternatives to the information processing view, and critiques of that view, have emerged. An unexamined taken-for-grantedness concerning cognition and prepresentation thus no longer suffices.

I have reviewed four of these views and argued that three of them—information processing, connectionism, and a version of robotics—although all interesting and different from each other in important ways, nevertheless share an underlying assumption concerning the nature of representation—an encodingist assumption. Furthermore, this encodingist assumption is wrong and logically incoherent in its foundations.

An alternative interactive model of representation is outlined, and it is argued that it captures the strengths of each of the other three appraoches and avoids their limitations and logical weaknesses. This approach introduces new understandings of the nature and significance of sensory and CNS activity and gives rise to novel questions concerning, and approaches to, such phenomena as language.

Bibliography

Baars, B. J. (1986). "The Cognitive Revolution in Psychology." Guilford, New York.

Bickhard, M. H. (1980). "Cognition, Convention, and Communication." Praeger, New York.

Bickhard, M. H. (1987). The social nature of the functional nature of language, *in* "Social and Functional Approaches to Language and Thought (M. Hickmann, ed), pp. 39–65. Academic Press, New York.

Bickhard, M. H. (in press). The import of Fodor's anticonstructivist arguments, *in* "Epistemological Foundations of Mathematical Experience" (L. Steffe, ed.). Springer-Verlag, New York.

Bickhard, M. H., and Richie, D. M. (1983). "On the Nature of Representation: A Case Study of James J. Gibson's Theory of Perception." Praeger, New York.

Brooks, R. A. (1987). "Intelligence without Representation." MIT Artificial Intelligence Laboratory, manuscript.

Burnyeat, M. (1983). "The Skeptical Tradition." University of California Press, Berkeley.

Campbell, R. L., and Bickhard, M. H. (1986). "Knowing Levels and Developmental Stages." Karger, Basel.

Carlson, N. R. (1986). "Physiology of Behavior." Allyn and Bacon, Boston.

Chapman, D., and Agre, P. (1986). Abstract reasoning as emergent from concrete activity, *in* "Reasoning about Actions and Plans, Proceedings of the 1986 Workshop" (M. P. Georgeff and A. L. Lansky, eds.). Timberline, Oregon, 411–424.

Gardner, H. (1987). "The Mind's New Science." Basic Books, New York.

Glass, A. L., Holyoak, K. J., and Santa, J. L. (1979). "Cognition." Addison-Wesley, Reading Mass.

Kenny, A. (1973). "Wittgenstein." Harvard Univ. Press, Cambridge.

McClelland, J., and Rumelhart, D. (1986). "Parallel Distributed Processing: Explorations in the Microstructure of Cognition, Vol. 2. Psychological and Biological Models." MIT Press, Cambridge.

Neisser, U. (1967). "Cognitive Psychology." Appleton-Century-Crofts, New York.

Palmer, S. E. (1978). Fundamental aspects of cognitive representation, *in* "Cognition and Categorization" (E. Rosch and B. B. Lloyd, eds.). Erlbaum, Hillsdale, N.J.

Pinker, S., and Mehler, J. (1988). "Connections and Symbols." MIT Press, Cambridge.

Rumelhart, D., and McClelland, J. (1986). "Parallel Distributed Processing: Explorations in the Microstructure of Cognition, Vol. 1. Foundations." MIT Press, Cambridge.

Thatcher, R. W., and John, E. R. (1977). "Functional Neuroscience, Vol. 1. Foundations of Cognitive Processes." Erlbaum, Hillsdale, N.J.

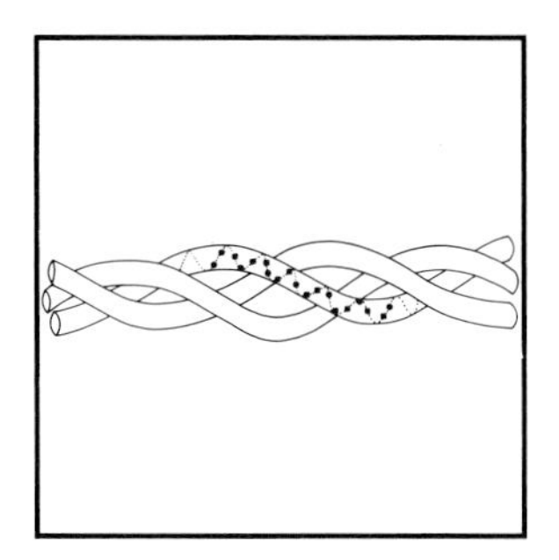

Collagen, Structure and Function

MARCEL E. NIMNI, *University of Southern California School of Medicine, Laboratory of Connective Tissue Biochemistry*

Glossary

α Chains Individual polypeptides which coil up around each other to form a collagen molecule
Collagen fibers Bundles of fibrils which can be visualized with the scanning electron microscope
Collagen molecule Triple-helical molecule that assembles spontaneously into fibrils, which are visible with the electron microscope and show a characteristic periodicity
Procollagen Intracellular precursor of collagen

COLLAGEN IS the single most abundant animal protein in mammals, accounting for up to 30% of all proteins. The collagen molecules, after being secreted by the cells, assemble into characteristic fibers responsible for the functional integrity of tissues such as bone, cartilage, skin, and tendon (Fig. 1). They contribute a structural framework to other tissues, such as blood vessels and most organs. Crosslinks between adjacent molecules are a prerequisite for the collagen fibers to withstand the physical stresses to which they are exposed. A variety of human conditions, normal and pathological, involve the ability of tissues to repair and regenerate their collagenous framework. In human tissues 13 collagen types have been identified.

I. Collagen Molecule

The arrangement of amino acids in the collagen molecule is shown schematically in Fig. 2. Every third amino acid is glycine. Proline and OH-proline follow each other relatively frequently, and the Gly–Pro–Hyp sequence makes up about 10% of the molecule. This triple-helical structure generates a symmetrical pattern of three left-handed helical chains that are, in turn, slightly displaced to the right, superimposing an additional "supercoil" with a pitch of approximately 8.6 nm.

These chains are known as α chains, and for the interstitial collagens (types I–III, Fig. 11) they show a molecular mass of approximately 100 kDa and contain approximately 1000 amino acids. The amino acids within each chain are displaced by a distance $h = 0.201$ nm with a relative twist of $-110°$, making the number of residues per turn 3.27 and the distance between each third glycine 0.87 nm. The individual residues are nearly fully extended in the collagen structure, since the maximum displacement within a fully stretched chain would be approximately 0.36 nm. This separation is such that is will not allow intrachain bonds to form (as does occur in the α helix), and only interchain hydrogen bonds are possible. The exact number of hydrogen bonds that stabilize the triple-helical structure has not been determined. One model describes two hydrogen bonds for every three amino acids, whereas another assumes one.

In addition to these intramolecular conformational patterns, there seems to exist a supermolecular coiling. Microfibrils, possibly representing intermediate stages of packing, have been described.

The process of self-assembly that causes the col-

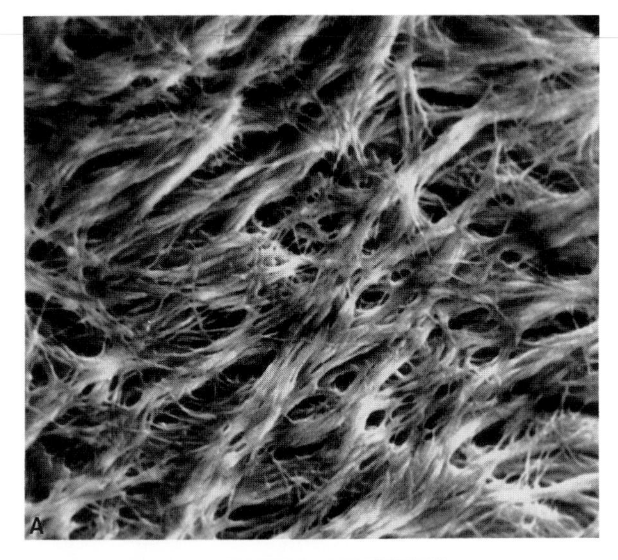

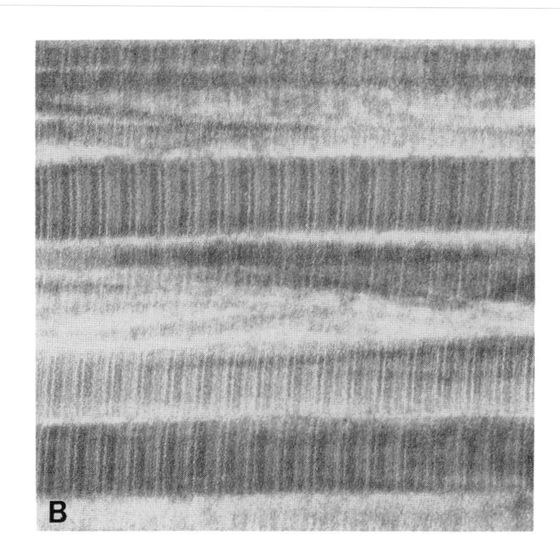

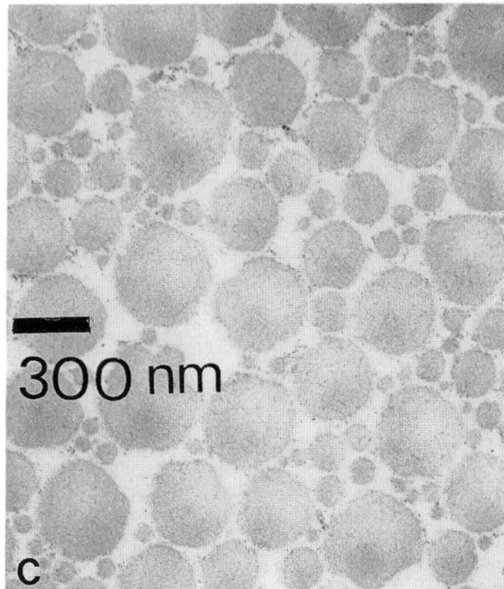

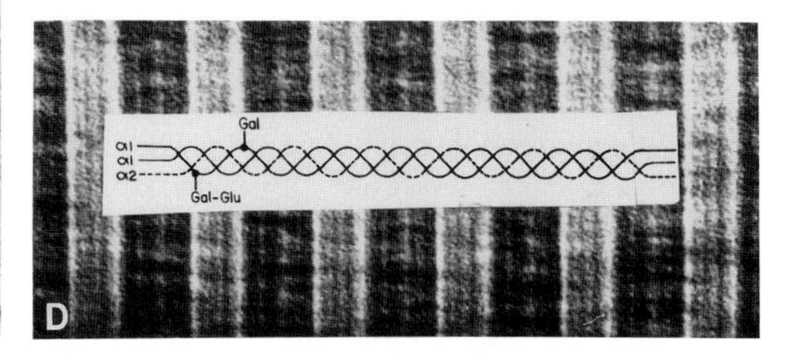

FIGURE 1 A, Bundles of collagen fibers in the osteoid (i.e., noncalcified bone matrix) lining of a haversian canal (i.e., a vascular tunnel) in a human long bone. The specimen was prepared by cleaning in a solution containing enzyme detergent for 6 hours and was then observed under a scanning electron microscope. Original magnification ×5400 (Courtesy of Dr. A. Boyde). B, Tendon containing collagen fibrils aligned parallel to each other. C, Cross-section of tendon showing fibrils of various diameters (dark bar = 300 nm.). D, Portion of a native collagen fibril stained with phosphotungstic acid displaying the characteristic 68 nm periodicity, overlaid by a diagram of a collagen molecule (type I) measuring approximately 300 nm.

lagen molecules to organize into fibers is shown schematically in Fig. 3. The thermodynamics of such a system involve changes in the state of the water molecules, many of which are associated with nonpolar regions of the collagen molecule.

II. Biosynthesis of Procollagen

In order for the organism to develop an extracellular network of collagen fibers, the cells involved in the biosynthetic process must first synthesize a precursor known as procollagen (Fig. 4). This molecule is later enzymatically trimmed of its nonhelical ends, giving rise to a collagen molecule that spontaneously assembles into fibers in the extracellular space. Procollagen molecules have been identified as precursors of the three interstitial collagens (i.e., types I–III). Several of the amino- and carboxy-terminal peptides (i.e., propeptides) have been characterized and their primary sequences have been determined.

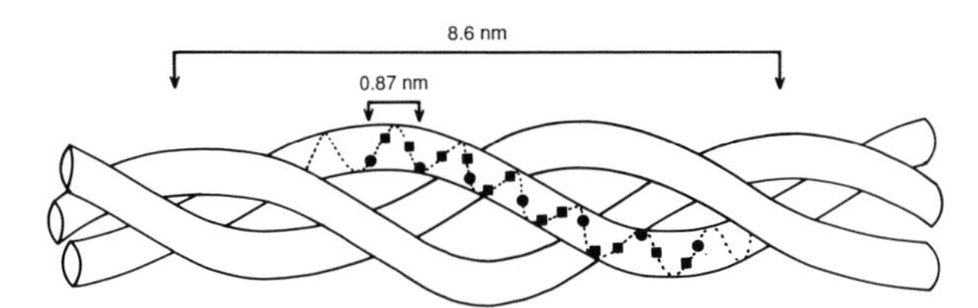

FIGURE 2 The collagen triple helix. The individual α chains are left-handed helices with approximately three residues per turn. The chains are in turn coiled around each other following a right-handed twist. The hydrogen bonds which stabilize the triple helix (not shown) form between opposing residues in different chains (interpeptide hydrogen bonding) and are therefore quite different from α helices which occur between amino acids located within the same polypeptide. ●, Glycine; ■, predominantly imino acids.

The carboxy-terminal propeptides of both pro $\alpha 1$ and pro $\alpha 2$ chains have molecular masses of 30–35 kDa and globular conformations without any collagenlike domain. These peptides contain asparagine-linked oligosaccharide units composed of *N*-acetylglucosamine and mannose. Once the molecule is completed and translocated to the cell surface, the extensions are enzymatically removed from those collagens which form fibrils. Enzymes that selectively remove these extensions can be found in a variety of connective tissues and in the culture media derived from collagen-secreting cells.

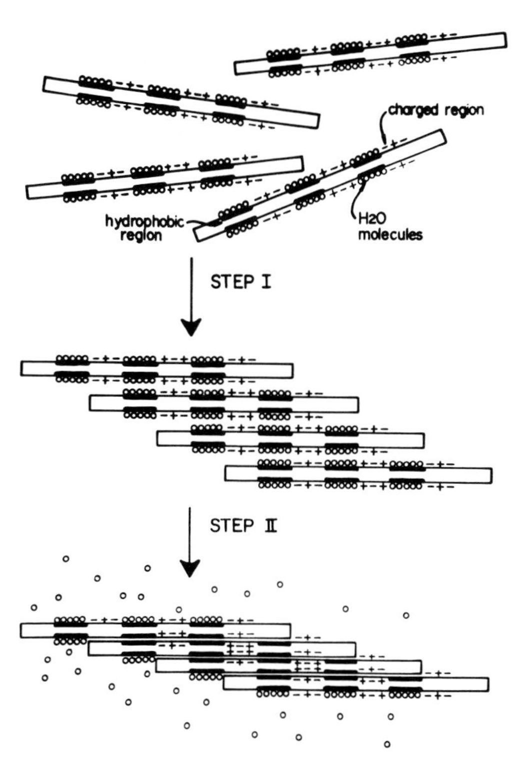

FIGURE 3 The nature of the forces associated with the early interactions of the helical regions of collagen via a mechanism of endothermic polymerization. This occurs *in vivo* and *in vitro* when the molecules still retain remnants of their telopeptides (i.e., nonhelical extensions), but can also occur *in vitro* with molecules devoid of such extensions. Further insight into the role of such telopeptides and modalities of assembly is provided in Figs. 4 and 9. Soluble collagen can be extracted from most tissues by cold neutral salt solutions. If these solutions are warmed to 37°C, the collagen molecules reassembe into native fibers. The upper part of the drawing represents molecules that aligned in a quarter-staggered overlap. The alignment is primarily due to interactions of opposing electrostatic charges, as depicted by + and − . As the temperature approaches 37°C, the hydrogen-bonded water molecules (○), clustered around the hydrophobic regions of collagen, "melt" and expose these nonpolar surfaces. Exclusion of water allows these surfaces to interact with each other, giving rise to hydrophobic bonds that greatly enhance the stability of the fiber. The driving force results from an increase of entropy of the system, since the release of "organized" water from initial sites and its transformation into "random" water increase the disorder of the system. Part of the aging process which leads to a gradual insolubilization of collagen can be associated with this phenomenon, which continues to operate through the life span of an individual.

A. Gene Expression

Since the discovery about 20 years ago of a distinct form of collagen in cartilage, now known as type II collagen, many other unique molecular species have been observed. Types I–III, V, and XI collagens are categorized as fiber-forming collagens. They all exhibit lengthy uninterrupted collagenous domains and are first synthesized as biosynthetic precursors (i.e., procollagens). Gene cloning experiments have demonstrated that the type I collagen genes are evolutionarily related, for they share a common ancestral gene structure. Human chromosome number 17 contains the coding information for the $\alpha 1$ chain of type I collagen, while chromosome 7 codes for its complementary $\alpha 2$ chain. A comparison of the five fibrillar collagens described shows that, with one exception [types III and $\alpha 2$(V) located on chromosome 2], all other genes are located on different chromosomes.

The collagen genes for fiber-forming collagens are large, about 10 times the size of the functional mRNA. Many of the exons (i.e., coding sequences) are 54 base pairs (bp) in length and are separated from each other by large intervening sequences (i.e., introns) that range in size from about 80 to 2000 bp. The gene itself contains 38,000 bp and is complex. The finding that most exons of these genes have identical lengths suggests that the ances-

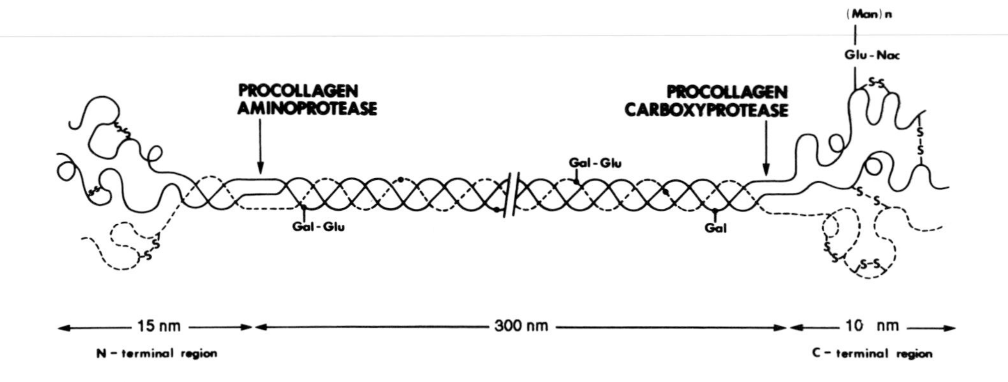

FIGURE 4 Procollagen molecule showing the nonhelical terminal extensions. The amino (N) terminus contains a small helical domian, and the carboxy (C) terminus is stabilized by interchain disulfide bonds. The sites of cleavage by procollagen peptidases are indicated by arrows.

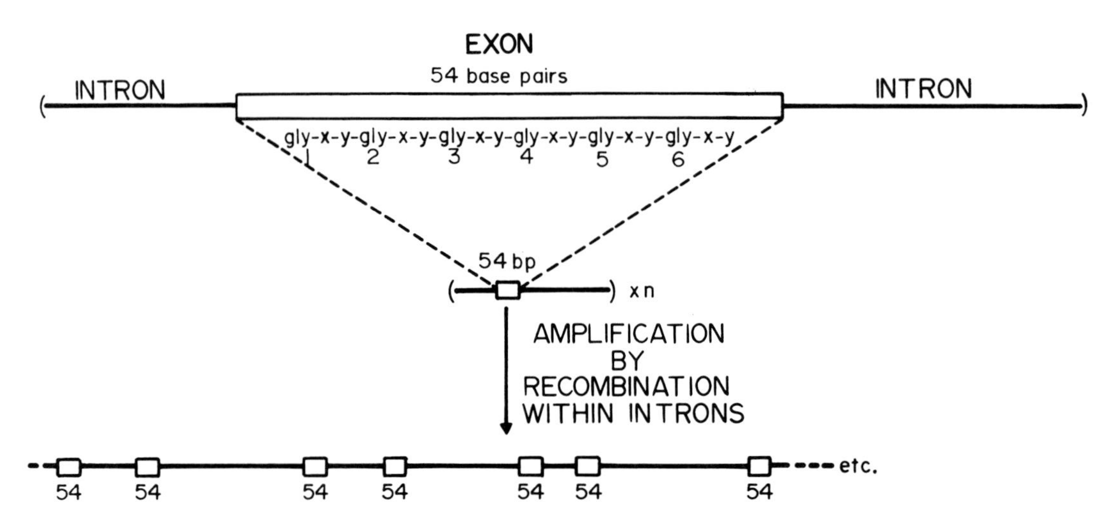

FIGURE 5 The collagen gene is made up of multiple units containing 54 base pairs, each of which corresponds to sequences of 18 amino acids. The conservation of this minimum sequence and the fact that it is repeated in such an exacting fashion provide valuable information to investigators interested in the process of evolution of proteins (Redrawn from Dr. DeCrombrugghe).

tral gene for collagen was assembled by multiple duplications of single genetic units containing an exon of 54 bp (Fig. 5). It is likely that a primordial exon this size could have encoded for a Gly–Pro–Pro tripeptide repeated six times ($3 \times 3 \times 6$). Such a polypeptide of 18 amino acids probably had the minimum length needed to form a stable triple-helical structure.

B. Translational, Cotranslational, and Early Posttranslational Events

After the gene is transcribed, it is spliced to remove introns and to yield a functional mRNA that contains about 3000 bases. Specific mRNAs for each chain and collagen type are translocated to the cytoplasm and translated into proteins in the rough endoplasmic reticulum on membrane-bound polysomes (Fig. 6). As the collagen polypeptide is synthesized in this region, it is modified in important ways. Two major constituents of collagen are the modified amino acids hydroxyproline and hydroxylysine, but neither of them can be directly incorporated into proteins. Instead, proline and lysine are incorporated and then modified by two hydroxylating enzymes, prolyl and lysyl hydroxylases. These enzymes require for their activity ferrous iron, ascorbate, and α-ketoglutarate. The degree of hy-

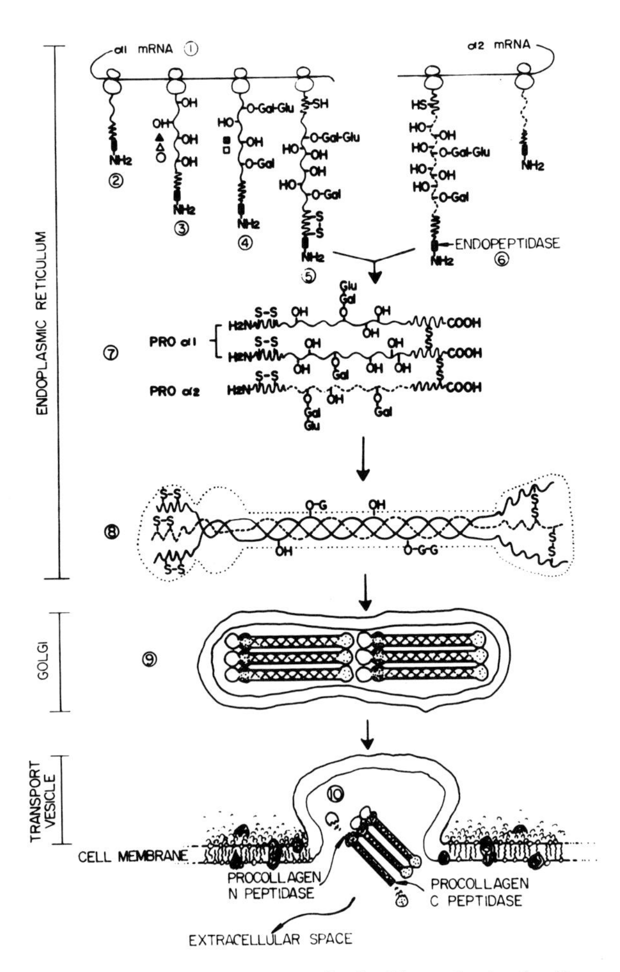

FIGURE 6 Sequence of events in the biosynthesis of collagen. (1) Synthesis of specific mRNAs for the different procollagen chains. (2) Translation of the message on polysomes of the rough endoplasmic reticulum. (3) Hydroxylation of specific proline residues by 3-proline hydroxylase (▲) and 4-proline hydroxylase (△) and of lysine by lysyl hydroxylase (○). (4) Glycosylation of hydroxylysine by galactosyltransferase (■) and addition of glucose by a glucosyltransferase (□). (5) Removal of the amino (N) terminal signal peptide. (6) Release of completed α chains from ribosomes. (7) Recognition of three α chains through the carboxy (C)-terminal prepeptide and formation of disulfide crosslinks. (8) Folding of the molecule and formation of a triple helix. (9) Intracellular translocation of the procollagen molecules and packaging into vesicles. (10) Fusion of vesicles with the cell membrane and extrusion of the molecule accompanied y the removal of the carboxy-terminal nonhelical extensions and part of the amino-terminal nonhelical extensions by specific peptidases.

droxylation differs from tissue to tissue and probably with the availability of substrate, rates of synthesis, and turnover, as well as the time during which the molecule remains in the presence of the hydroxylating enzymes. The time required for the synthesis of a complete pro α chain is about 6.7 minutes.

As lysyl residues in the newly synthesized pro α chains are hydroxylated, sugar residues are added to the resulting hydroxylysyl groups. Glycosylation is catalyzed by two specific enzymes: a galactosyltransferase and glucosyltransferase. Once the translation, modifications, and additions are completed, the individual pro α chains become properly aligned for the triple helix to form.

C. Intracellular Translocation of Procollagen and Extrusion into the Extracellular Space

The procollagen molecule, now detached from the ribosome, emerges from the endoplasmic reticulum and moves toward the Golgi apparatus through the microsomal lumen. In the Golgi apparatus the carboxy-terminal mannose-rich carbohydrate extension is remodeled, and the molecules are packaged in vesicles and carried toward the cellular membrane (Fig. 7). [*See* GOLGI APPARATUS.]

The small aggregates of oriented procollagen molecules are probably trimmed of their nonhelical amino and carboxyl extensions by specific peptidases when they reach the extracellular space. In the case of type I collagen, the first peptidase to act seems to be the aminoprotease; this is followed by a carboxyprotease. In type III collagen the sequence of removal might be reversed.

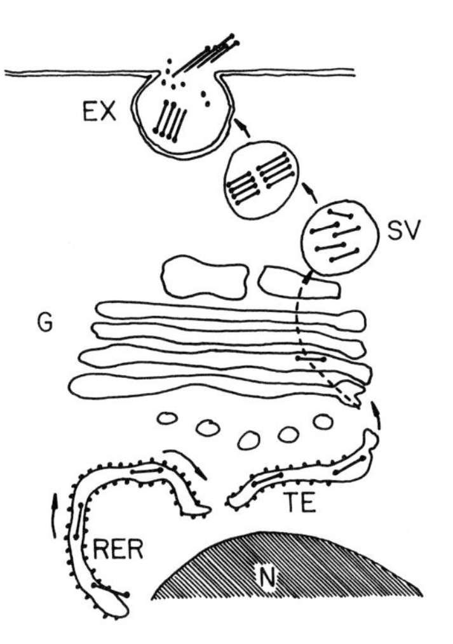

FIGURE 7 Movement of procollagen through the cisternae of the rough endoplasmic reticulum (RER) and through a transitional endoplasm (TE) to the Golgi apparatus (G), where it is packaged into secretory vesicle (SV) prior to extrusion (EX) by exocytosis. N, Nucleus.

R
|
C=O
|
C—$(CH_2)_4$—NH_2 —LYSYL OXIDASE→ C—$(CH_2)_3$—C(H)=O
|
NH
|
R

PEPTIDE-BOUND LYSINE α-AMINOADIPIC δ-SEMIALDEHYDE

FIGURE 8 The oxidative deamination of peptide-bound lysine by the enzyme lysyl oxidase generates the aldehydes associated with the collagen molecule.

D. Lysyl Oxidase

Recently formed microfibrils seem to be recognized by the enzyme lysyl oxidase, which converts certain peptide-bound lysines and hydroxylysines to aldehydes (Fig. 8). The enzyme is an extracellular amine oxidase, which has been purified from a variety of connective tissues. It requires Cu^{2+} and probably pyridoxal as cofactors, and molecular oxygen seems to be the cosubstrate and hydrogen acceptor. It is irreversibly inhibited by the lathyrogen β-amino propionitrile (βAPN), a substance found in the flowering sweet pea, *Lathyrus odoratus*. This enzyme exhibits maximal activity when acting on collagen fibrils rather than on monomeric collagen.

E. Fibrillogenesis

The tendency of collagen molecules to form macromolecular aggregates is well known. This tendency is common with most fibrous proteins that form filaments with helical symmetry and which occupy equivalent or quasiequivalent positions.

The exact mode in which the collagen molecules pack into microfibrils (the precursor of the larger fibrils) still remains a subject for speculation. A five-stranded microfibril was first suggested to account for such a substructure, which would satisfy the condition that adjacent molecules were equivalently related by a quarter stagger (overlaps staggered at intervals equal to approximate one-quarter the length of the molecule) (Fig. 9).

When monomeric collagen is heated to 37°C, it progressively polymerizes, generating a turbidity curve that reflects the presence of intermediate aggregates. The lag phase (persistence of monomers), the nucleation and appearance of turbidity (microfibrils), and the rapid increase in turbidity (fiber formation) have been equated to the way in which the cell might handle this process (Fig. 10).

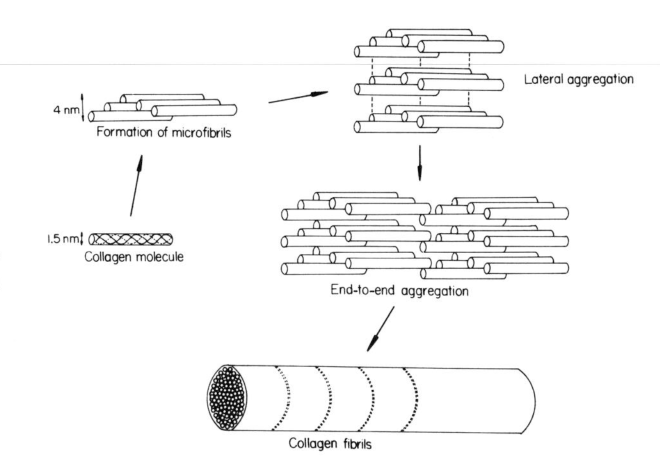

FIGURE 9 Formation of the five-membered microfibril and its potential for lateral and end-to-end aggregation to form fibrils.

III. Types of Collagen

Almost two decades have passed since we first realized that all collagen fibers within a particular organism are not made up of identical molecules. The different collagen types are usually identified using Roman numerals, assigned as they are purified and characterized. Figure 11 summarizes the molecular characteristics of most of the collagens identified to date.

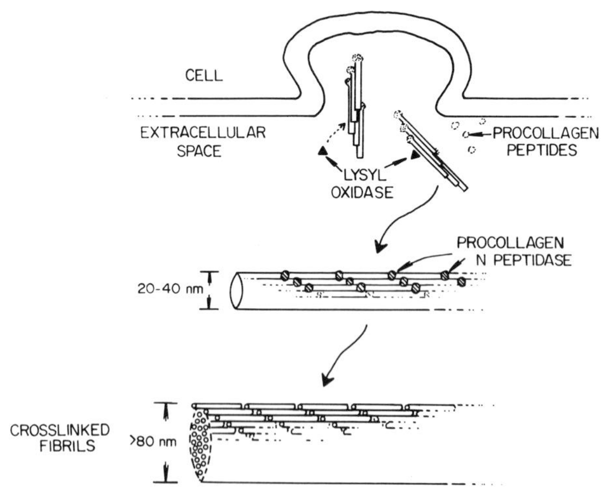

FIGURE 10 Fibrillogenesis: Microfibrils in a quarter-staggered configuration have lost their carboxy-terminal nonhelical extensions and part of their amino (N)-terminal extensions. In this form they seem to organize readily into small-diameter fibrils which retain part of the amino-terminal nonhelical extensions. After being relieved of these peptides by a procollagen peptidase, fibrils are able to grow in diameter by apposition of microfibrils or by merger with other small-diameter fibrils.

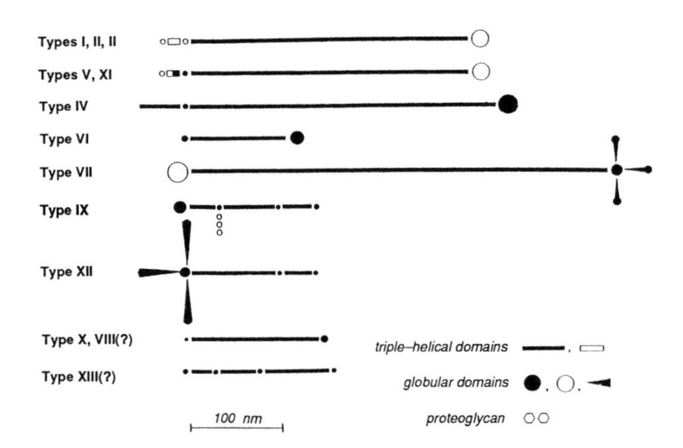

FIGURE 11 Collagens contain both triple-helical (solid and open rods) and globular domains (open and filled circles). Portions of the initially synthesized molecules are removed prior to their incorporation into insoluble matrices (open rods and circles) while the rest of the molecule remains intact in the matrix (closed circles and rods). The domains and their distributions are drawn approximately to scale. (Courtesy of Dr. Robert Burgeson.)

A. Type I

Before 1969 type I collagen was the only mammalian collagen known. It is composed of three chains, two identical $\alpha 1$ chains and one $\alpha 2$ chain. Type I collagen is most abundant in skin, tendon, ligament, bone, and cornea, where it makes up 80–99% of the total collagen. Bone matrix is essentially all type I collagen. The most common technique used to isolate this molecule and distinguish it both qualitatively and quantitatively from the other collagens involves the use of solvents of different ionic strengths and pH levels, followed by differential "salting out."

The amino acid compositions of some of the better characterized and more abundant human collagen types are shown in Table I. The amino acids that are present in a quantity significantly different from that of type I collagen are indicated in bold numbers. On many occasions these differences have been used to identify collagen types or their mixtures and to suspect the presence of new or abnormal collagen species. A simplified diagram illustrating type I collagen and its relationship to the other interstitial collagens, types II and III, is shown in Fig. 12.

B. Type II

Isolation and analysis of collagens from a variety of cartilaginous structures show that type II collagen is made up primarily of molecules containing three

TABLE I Amino Acid Composition of the Human Collagen Chains[a]

Amino acid	$\alpha 1$(I)	$\alpha 2$(I)	$\alpha 1$(II)	$\alpha 1$(III)	$\alpha 1$(IV)	$\alpha 2$(V)	$\alpha 1$(V)	$\alpha 3$(V)	$\alpha 1$(XI)	$\alpha 2$(XI)[b]
3-Hydroxyproline	1	1	2	0	**7**	**3**	**5**	1	···	···
4-Hydroxyproline	108	93	97	125	**133**	106	110	91	98	93
Aspartic acid	42	44	43	42	51	50	49	42	46	50
Threonine	16	19	23	13	20	29	21	19	17	25
Serine	34	30	25	39	37	34	23	34	25	28
Glutamic acid	73	68	89	71	79	89	100	97	107	98
Proline	124	113	120	107	**65**	107	130	98	109	119
Glycine	333	338	333	350	328	331	332	330	334	327
Alanine	115	102	103	96	**37**	**54**	**39**	**49**	**54**	**49**
Half-cystine	0	0	0	**2**	0–1	0	0	1	0	0
Valine	21	35	18	**14**	28	27	17	29	28	18
Methionine	7	5	10	8	13	11	9	8	10	9
Isoleucine	6	14	9	13	**29**	15	17	20	15	16
Leucine	19	30	26	22	**52**	37	36	**56**	35	39
Tyrosine	1	4	2	3	2	2	4	2	2	3
Phenylalanine	12	12	13	8	**29**	11	12	9	11	11
Hydroxylysine	9	12	**20**	5	**49**	**23**	**36**	**43**	**38**	**40**
Lysine	26	18	15	30	9	13	14	15	19	15
Histidine	3	12	2	6	6	10	6	14	6	11
Arginine	50	50	50	46	**26**	48	40	42	45	48
Gal-Hydroxylysine	1	1	4	···	2	3	5	7	···	···
Glc-Gal-Hydroxylysine	1	2	12	···	30	5	29	17	28	34

[a] Amounts are expressed as residues per 1000 total residues. Boldface numbers represent quantities significantly different from those for type I. Ellipses (···) indicate either trace amounts or nondetectable.

[b] $\alpha 3$(XI) appears to be almost identical to $\alpha 1$(II), either a similar gene product or a close genetic variant which has been altered post-translationally.

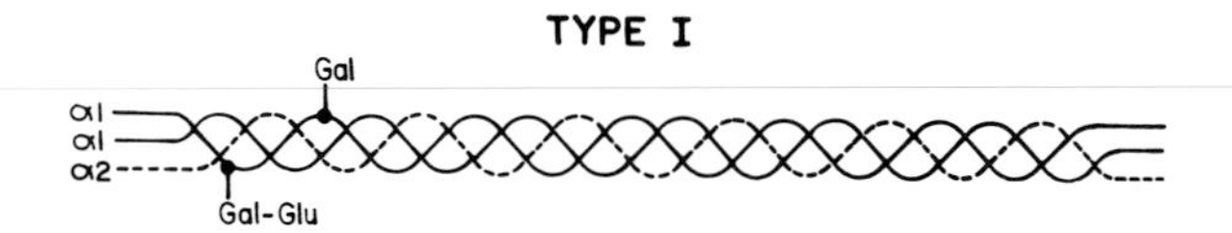

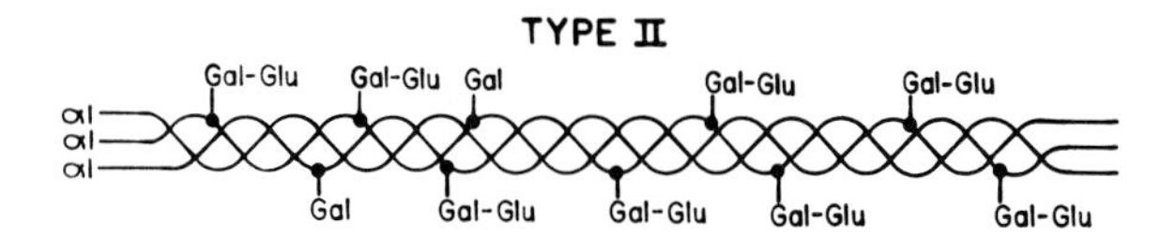

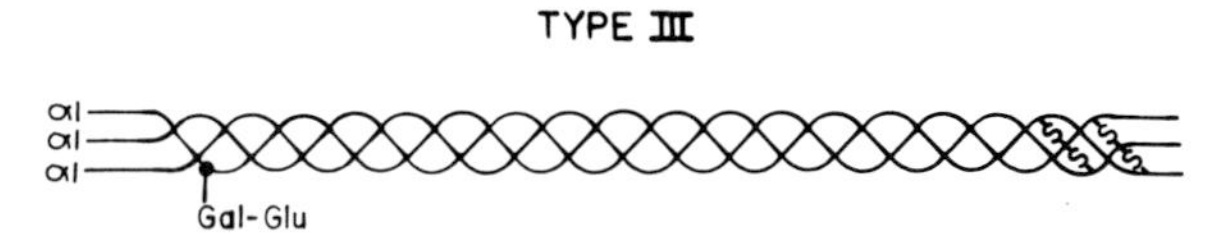

FIGURE 12 Diagram of the three interstitial types of collagen. Type I is present in skin, bone, tendon, etc.; type II is present in cartilage; and type III is present in blood vessels and developing tissues and as a minor component in skin and other tissues. There are differences in the chain composition and degrees of glycosylation. Disulfide crosslinks are only seen in type III collagen.

identical α chains. The most significant features of cartilage collagen are its high content of hydroxylysine and glycosidically bound carbohydrates. Type II collagen is also present in the intervertebral disk, which contains a central gellike region, the nucleus pulposus surrounded by concentric rings of highly ordered dense collagen fibers known as the annulus fibrosus. Another tissue that contains appreciable amounts of type II collagen is the vitreous body of the eye. [*See* Articular Cartilage and the Intervertebral Disc; Cartilage.]

C. Type III

When human dermis is digested with pepsin under conditions in which the collagen molecules retain their helical conformation, type I molecules can be separated from type III molecules by differential salt precipitation at pH 7.5. The type III molecules are composed of three identical chains. Characteristic of this collagen is the presence of intramolecular disulfide bonds involving two cysteine residues close to the carboxy-terminal region of the triple helix. Because the ratio of type I to type III collagen changes with age, type III being predominant in fetal skin, this type of collagen is many times referred to as fetal, or embryonic, collagen. Formation of intermolecular disulfide bridges by type III collagen could be of great advantage during early development and would healing, when collagen is deposited at a rapid rate in order to fill a gap.

Normal bone matrix might be the only tissue containing type I collagen that lacks type III. Blood vessels are particularly rich in type III collagen.

D. Type IV

Type IV collagen is the major component of basement membranes and is generally regarded as the most characteristic of a large number of macromolecules present within these specialized connective tissue structures. These macromolecules include, in addition to collagen, glycoproteins such as laminin, fibronectin, and proteoglycans. Basement membranes underlie epithelia and endothelia; are involved in cell differentiation and orientation, membrane polarization, and selective permeability to macromolecules; and are a target for a large number of diseases.

Type IV collagen differs from the interstitial collagens in its amino acid composition (Table I) and by the fact that it does not organize into a fibrillar structure. This molecule consists of a large triple-helical domain and noncollagenous extensions that make it resemble procollagen. Carbohydrate accounts for 10% of the mass.

It seems that type IV procollagen (with a molecular mass of approximately 180 kDa per chain) is incorporated as such into basement membranes without processing. The globular extensions at the end of the molecule associate to form a network. Thus, association of four molecules at their amino termini gives rise to tetramers that can aggregate further into a regular tetragonal or irregular polygonal meshwork. The pore size of these networks could differ in such a way as to account for the specific functions of basement membranes in different tissues.

E. Type V

Type V collagen was discovered in pepsin digests of placental membranes and other tissues. Three chains—α1(V), α2(V), and α3(V)—were initially isolated from this collagen. They are similar in size to the α chains of interstitial collagens, except that the α1(V) chains are larger. Two related collagen chains were isolated from cartilage.

Type V collagen is particularly abundant in vascular tissues, where it appears to be synthesized by smooth muscle cells, although it is also present in relatively large amounts within the corneal stroma, which lacks blood vessels. Type V collagen seems to be a unique form of collagen that contributes to the shape of cells by localizing on their surface and participates in the formation of an exocytoskeleton and binding to other connective tissue components.

F. Type VI

The type VI collagen molecule is a heterotrimeric assembly of three genetically distinct chains: $\alpha 1$(VI), $\alpha 2$(VI), and $\alpha 3$(VI).

Antibodies to type VI collagen indicate that the molecule has a ubiquitous distribution throughout connective tissues. It is shared between cartilage and noncartilaginous tissues. In skin, type VI filaments are highly concentrated around endothelial basement membranes, forming a loose sheath around blood vessels, nerves, and fat cells, separating these dermal and subdermal elements from the surrounding fibrillar network.

G. Type VII

Type VII collagen, also known as the anchoring fibril network, is the largest collagen thus far described, with a total molecular mass of approximately 1050 kDa for the precursor form of the molecule. The molecule is a homotrimeric assembly of $\alpha 1$(VII) chains.

The tissue distribution of type VII exactly correlates with that of anchoring fibrils. Anchoring fibrils are specialized fibrous structures found within the subbasal lamina of the basement membranes, which secure the lamina densa to the underlying connective tissue matrix by physically trapping collagen fibrils between the lamina densa and the anchoring fibril.

Individuals lacking anchoring fibrils suffer extensive cutaneous and mucosal blistering resulting from separation of the dermis from the epidermis along the subbasal lamina.

H. Type VIII

The peptides now known as type VIII collagen were first observed as cell culture products of bovine endothelial cells and rabbit corneal endothelial cells. The two current models of the molecular structure predict that type VIII is largely triple helical and, unlike most other collagens, contains little nonhelical structure.

I. Type IX

The first fragment of type IX collagen was isolated from pig hyaline cartilage after pepsin digestion. The intact molecule is unusual in possessing three collagenous domains, each of which is interspersed by a short, noncollagenous, highly flexible domain.

One unusual characteristic of type IX collagen is the presence of a covalently attached chondroitin sulfate on the $\alpha 2$ chain. In chick cartilage type IX is associated with type II on the surface of the banded fibrils.

J. Type X

A number of unique characteristics distinguish type X collagen from the other collagens. The molecule is 138 nm long, approximately one-half the length of the interstitial collagens. It has a restricted tissue distribution and is synthesized predominantly by hypertrophic chondrocytes during the process of endochondral ossification (i.e., bone formation preceded by cartilage). In the fetal and adolescent skeletons the molecule is a transient intermediate in cartilage, which is later replaced by bone. In the adult skeleton, type X collagen persists in the zone of calcified cartilage that separates hyaline cartilage from the subchondral bone.

K. Type XI

Type XI collagen was first discovered during differential salt fractionation of pepsin digests of hyaline cartilage. The protein is best recognized by the presence of three distinct α chains, smaller than $\alpha 1$(II) chains. This collagen resembles type V collagen in some of its properties. It might regulate the diameter of type II collagen fibrils, act as a specific building block of the basket of fibril surrounding chondrocytes, and mediate collagen–proteoglycan interactions.

L. Type XII

The gene for type XII collagen was recently cloned from tendon fibroblasts. The amino acid sequence predicted from its base sequence is homologous with, but not identical to, that of type IX collagen.

A schematic diagram illustrating the relative dimensions and fundamental features of the collagens described is shown in Fig. 11.

IV. Collagen Metabolism

Collagen is the most abundant of all body proteins. Tissues (e.g., bone) which are involved in active remodeling are responsible for the major turnover, while other less dynamic tissues in the full-grown individual, (e.g., skin and tendons) might exhibit slow, almost negligible, turnover. The collagen-synthesizing activity of cells is usually assessed by their ability to synthesize hydroxyproline or by the activities of specific enzymes such as the proline and lysyl hydroxylases.

A. Degradation by Bacterial and Mammalian Collagenases

Because of its triple-helical structure stabilized by hydrogen bonds, the collagen molecules are quite resistant to enzymatic degradation in their native configuration. They can be degraded by collagenases. The first molecules to be isolated were of bacterial origin, specifically *Clostridium histolyticum.* These enzymes are quite specific for collagen, but will also degrade gelatin, which is denatured collagen. They are inhibited by cysteine and other sulfhydryl compounds and by ethylenediaminetetraacetic acid (EDTA), a chelator for divalent cations. These enzymes are specific for peptide bonds involving glycine in a collagen helix conformation. Because of the abundance of this amino acid in collagen (i.e., every third residue), this enzyme generates a large number of small peptides.

The first enzyme derived from animal tissue capable of degrading collagen at neutral pH was isolated from the culture fluid of tadpole tissue. It cleaves the native molecule into two fragments in a highly specific fashion at a temperature below that of denaturation of the substrate. Since this discovery collagenolytic enzymes have been obtained from a wide range of tissues from animal species. In general, these enzymes have a number of fundamental properties in common: They all have a neutral pH optimum, and they are not stored within the cell, but, rather, are secreted either in an inactive form or bound to inhibitors. Figure 13 schematically summarizes the fundamental aspects of these enzymes and their modes of action. They appear to be zinc metalloenzymes requiring calcium, and are not inhibited by agents that block serine or sulfhydryl-type proteinases. Nearly all of the collagenases studied so far have a molecular mass that ranges

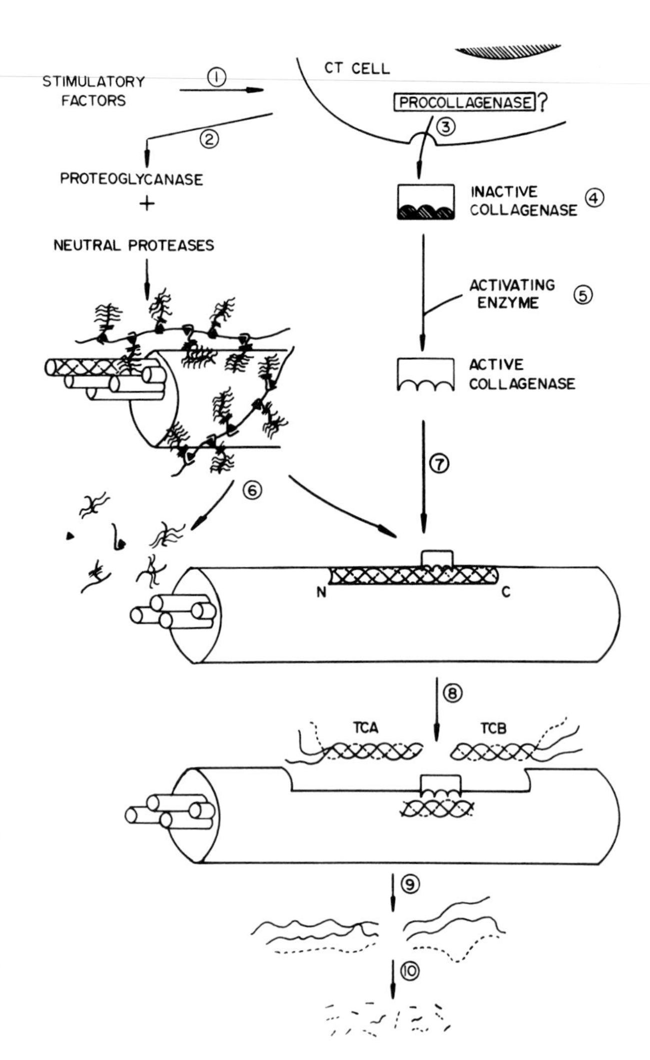

FIGURE 13 Sequence of events leading to the degradation of collagen fibers by the enzyme collagenase. (1) A variety of factors stimulate connective tissue (CT) cells to synthesize collagenase, glycosidases, and neutral proteases. (2) The proteoglycan-degrading enzymes remove the mucopolysaccharides which surround collagen fibers and expose them to collagenase. (3) Inactive collagenase is secreted. (4) The enzyme is usually found in the extracellular space bound to an inhibitor. (5) An activating enzyme removes the inhibitor. (6) Glycosidases complete the degradation of the proteoglycans. (7) The active collagenase binds to fibrillar collagen. (8) Collagenase splits the first collagen molecule into two fragments (TCA and TCB), which denature and begin to unfold at body temperature. The enzyme now moves on to an adjacent molecule. (9) The denatured collagen fragments are now susceptible to other proteases. (10) Nonspecific neutral proteases degrade the collagen polypeptides.

from 25 to 60 kDa. Mammalian collagenases display a great deal of specificity, cleaving the bond Gly–Leu or Gly–Ile. There are slight differences in the amino acid sequences surrounding the scission site; these could account for the differences in the rates at which various collagens are degraded.

The enzyme interacts tightly with the collagen fiber and appears to remain bound to the macromolecular aggregate during the degradation process. Approximately 10% of the collagen molecules in reconstituted collagen fibrils appear to be accessible for binding, in close agreement with the theoretical number of molecules estimated to be present near the surface of the fiber. The *in vitro* data obtained seem to indicate that digestion proceeds until completion, with hopping from one molecule to another without returning to the solution. Collagen from individuals of increasing age becomes more resistant to enzymatic digestion.

B. Urinary Excretion of Collagen Degradation Products

Because of the relatively large amounts of hydroxyproline (i.e., 12–14%) present in collagen, the assay of this amino acid in body fluids has raised considerable interest. More than 95% of the hydroxyproline is excreted in a peptide-bound form.

Altered urinary hydroxyproline values are found in conditions affecting collagen metabolism systemically, such as endocrine disorders, particularly those conditions accompanied by relatively extensive involvement of bone. The glycosides, galactosylhydroxylysine and glucosylgalactosylhydroxylysine, characteristic moieties of the collagen molecule, have been identified in urine and are also used as criteria of collagen degradation.

A significant amount of the collagen synthesized by cells can be degraded intracellularly before secretion, and such degradation products also appear in the urine. It has been estimated that between 10% and 60% of the newly synthesized collagen can be degraded by this route.

V. Crosslinking

A. Intramolecular and Intermolecular Crosslinks

Crosslinking renders the collagen fibers stable and provides them with a degree of tensile strength and viscoelasticity adequate to perform their structural role. The degree of crosslinking, the number and density of the fibers in a particular tissue, and their orientation and diameter combine to provide this function. Crosslinking begins with the oxidative deamination of the ε carbon of lysin or hydroxylysine to yield the corresponding semialdehydes and is mediated by the enzyme lysyl oxidase (Fig. 8). Enzymatic activity is inhibited by β-aminoproprionitrile, chelating agents, isonicotinic acid hydrazide, and other carbonyl reagents. Lysyl oxidase exhibits particular affinity for the lysines and hydroxylysines present in the nonhelical extensions of collagen, but can, at a slower pace, also alter residues located in the helical region of the molecule (Fig. 14).

In general, lysine-derived crosslinks seem to predominate in soft connective tissues such as skin and tendon, whereas hydroxylysine-derived crosslinks are prevalent in the harder connective tissues, such as bone, cartilage, and dentine, which are less prone to yield soluble collagens.

Several other crosslinks have been identified and their locations have been established. These more complex polyfunctional crosslinks can contain histidine (Fig. 15) or can result in the formation of naturally fluorescent pyridinium ring structures (Fig. 16).

The study of collagen crosslinking has advanced steadily, and even though hindered by the difficulty in dealing with an insoluble three-dimensional matrix composed of quarter-staggered molecules, many crosslinking regions, primarily those involv-

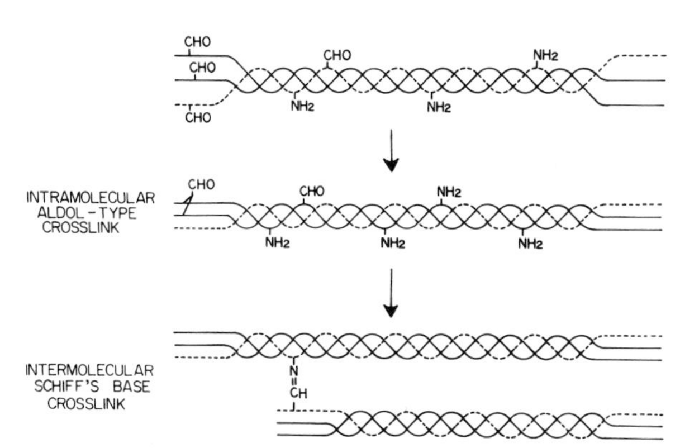

FIGURE 14 Formation of intramolecular and intermolecular crosslinks in type I collagen. Intramolecular crosslinks occur in the nonhelical regions and involve a condensation reaction between lysine- or hydroxylysine-derived aldehydes within a single molecule. Intermolecular crosslinks, on the other hand, involve aldehydes and ε-amino groups of lysine present in different molecules.

HISTIDINO-HYDROXYMERODESMOSINE

FIGURE 15 Histadinohydroxymerodesmosine: a more complex tetrafunctional covalent crosslink which bridges three different molecules. Two of the residues are part of an aldol condensation product (intramolecular crosslink) and therefore are associated with one single molecule.

ing the nonhelical extension peptides, have been identified (Fig. 17). It is interesting in this connection that covalent crosslinks between types I and III molecules have been recently identified, reflecting the presence of heterogeneous fibrils.

B. Collagen and Aging

The physical as well as chemical properties of collagen change with age. These changes in physicochemical properties have been attributed to the formation of both covalent and noncovalent

PYRIDINOLINE

FIGURE 16 Pyridinoline: this trifunctional pyridinium crosslink, which joins three adjacent collagen molecules, can be generated by one hydroxylysine residue and two hydroxylysine-derived aldehydes or by spontaneous interaction of two hydroxylysine-5-keto-norleucine residues formed from a hydroxylysine and a hydroxylysine aldehyde.

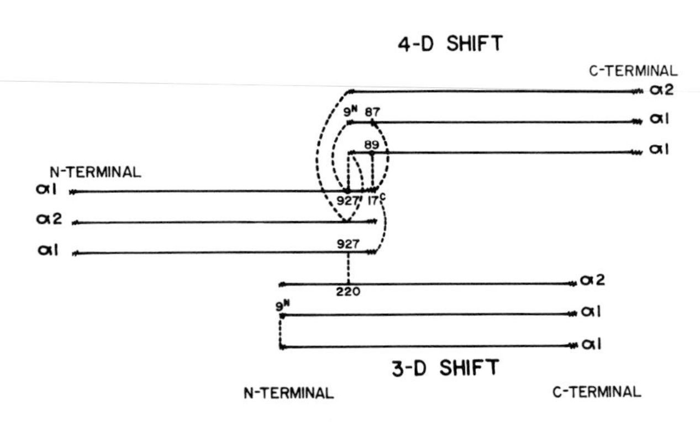

FIGURE 17 Schematic representation of type I collagen molecules aligned in three-dimensional (3-D) and four-dimensional (4-D) staggered positions. The known crosslinking sites are indicated by dashed lines. The intermolecular crosslink formed between hydroxylysine residues 9^N from the amino (N)-terminal region of $\alpha 1$ chains is among the first crosslinking sites to be recognized. The carboxy (C)-terminal hydroxylysyl residue 927 of $\alpha 1$ chain can crosslink to one or two $\alpha 1$ chains through residue 9^N. Hydroxylysyl residue 927 crosslinks to 9^N of an $\alpha 1$ chain or to 9^N of an $\alpha 2$ chain. The carboxy-terminal 17^C hydroxylysyl residue of the $\alpha 1$ chain crosslinks to hydroxylysyl residue 87 from an $\alpha 1$ chain of another molecule. The residue 17^C might also form an aldol-type intramolecular crosslink with a similar residue of the other $\alpha 1$ chain. Histidine 89 from the carboxy-terminal region of an adjacent molecule adds to this crosslink via a Michael addition. Recently, an intermolecular crosslink between hydroxylysyl residues 927 and 220 in the helical region was found in dentin and bone, supporting the observation that aldehydes do form in the collagen helix.

crosslinks. Neutral salt-soluble collagen, which has a low concentration of β components, will generate intramolecular bonds if gelled at 37°C. These intramolecular bonds seem to precede the formation of stable intermolecular crosslinks, since these gels can redissolve when cooled, to yield a soluble collagen with a higher content of β components (i.e., dimers of α chains) of intramolecular origin.

Although it is attractive, the crosslinking theory of aging awaits further experimental support. Obviously, stabilization of collagen fibers might occur by other means than by the formation of new covalent crosslinks. An increase in the number of weaker forces that stabilize macromolecules and their aggregates (e.g., van der Waals bonds, ionic interactions, hydrophobic bonds, and combinations of such forces) could account for changes in the physicochemical properties of the collagen fibers. For instance, a slow time-dependent exclusion of intermolecular water could not only lead to an increase in hydrophobic contacts, but could strengthen existing ionic bonds by placing them in an environment of decreased dielectric constant (Fig. 3).

C. Inhibition of Collagen Crosslinking: Aminonitriles and D-Penicillamine

Lathyrism is a connective tissue disorder associated with the ingestion or injection of βAPN and its chemical analogs or extracts of the sweet pea or other members of the Lathyrus family usually consumed during famine. The skeletal changes observed differ from species to species and vary with age, being much more pronounced in younger animals. The epiphyseal plate, the growth zone of the ends of long bone, is a prime target.

The connective tissue abnormalities are associated with crosslinking defects in collagen and elastin. They are revealed by an increased solubility in hypertonic neutral salt solutions, due to an inhibition of lysyl oxidase activity. Since copper deficiency also inhibits this enzymatic activity, the similarities of the defects induced by these two mechanisms are readily explainable.

Administration of D-penicillamine to animals and humans also causes an accumulation of collagen soluble in neutral salt in skin and various soft tissues. Two of the more characteristic properties of D-penicillamine—namely, the ability to trap carbonyl compounds and to chelate heavy metals—are of primary significance in impairing collagen crosslinking. The former property manifests itself in all effective dose ranges, whereas the latter occurs only at high doses, far beyond those administered to humans.

The collagen extracted from tissues of animals treated with D-penicillamine is able to form stable fibers *in vitro* and is not deficient in aldehydes, as is that from βAPN-treated animals. In fact, its aldehyde content is even higher than normal, suggesting that the mechanisms of action of βAPN and D-penicillamine are different (Fig. 18).

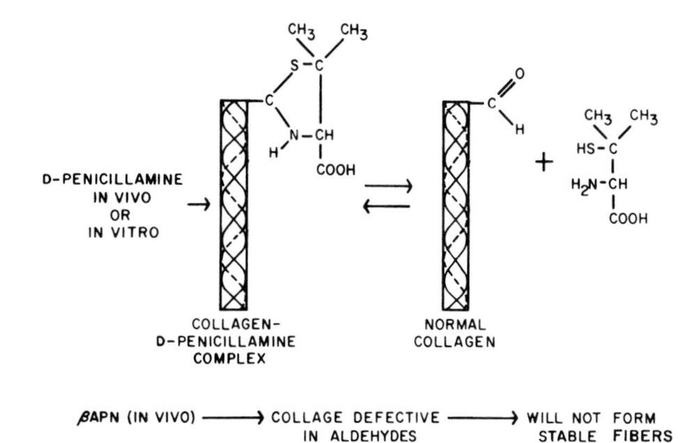

FIGURE 18 The modes of action of D-penicillamine and aminonitriles, inhibitors of collagen crosslinks. D-Penicillamine *in vivo*, as well as *in vitro*, interacts with aldehydes on collagen, preventing them from subsequently participating in the formation of intramolecular and intermolecular crosslinks. The reversibility of the collagen defect seen when D-penicillamine therapy is discontinued can be explained by the instability of the thiazolidine complex formed between D-penicillamine and the peptide-bound aldehydes. On the other hand, βAPN, a lathyrogen, inhibits aldehyde formation by irreversibly binding to the enzyme lysyl oxidase.

D. Wound Healing and Vitamin C

Collagen synthesis increases in wounds. There is an increase in the synthesis of type III collagen 10 hours after wounding, but by 24 hours its synthesis returns to a normal value. The early type III collagen deposited could be important in establishing the initial would structure and providing a basic lattice for subsequent healing events.

Well-healed mature scars in humans can reopen because of ascorbic acid deficiency. This phenomenon was recognized in ancient times, when sailors on extended sea voyages living on diets devoid of fresh fruits and vegetables noticed a breakdown in skin scars that had been healed for years. This is because metabolic activity is present in a scar, even after the healing process is completed, and ascorbic acid (i.e., vitamin C) is required for the hydroxylation of peptide-bound proline.

VI. Collagen Synthesis by Cells in Culture

Various cell types have been shown to synthesize collagen. Human fibroblasts in culture synthesize both types I and III collagen, type I accounting for 70–90% of the total. Cell density in the culture and nutrient availability can alter the ratio of collagen types synthesized: Human lung fibroblasts synthesize 26–68% more type III collagen at confluency than at low cell density.

Chondrocytes, which are highly differentiated mesenchymal cells, synthesize a unique blend of collagen (mostly type II) and proteoglycans. The normal pattern of proteoglycan and collagen syntheses can be altered by a variety of factors; these changes occur concomitantly with changes in cell shape. Although chondrocytes in culture are metabolically active, in adult cartilage their metabolic activity is relatively low and mitosis is rarely seen. When degenerative or trauma-related changes occur in the cartilage from joints, chondrocytes recover their ability to divide. The state of differentiation of chondrocytes can be monitored in a quantitative fashion by determining the collagen types they produce.

VII. Biomechanical Properties of Connective Tissues

The biomechanical properties of collagen fibers have mostly been studied primarily in tissues where these fibers are oriented in a parallel fashion (e.g., tendons and ligaments). The response to stretching is, however difficult to interpret because many variables significantly affect behavior. The toe region (Fig. 19a) of the stress–strain curve reflects the slack associated with elimination of the crimping of the fibers; during this phase the elastic modulus increases steadily. The linear part of the curve (Fig. 19b) is where stress increases rapidly. This is the area from which the modulus is usually calculated and reflects the actual contribution of collagen molecules aligned parallel to each other and stretching of the crystalline network. When the point of maximum stress (Fig. 19c) is reached, the tissue breaks; this is the failure point, which can occur suddenly in bone or can be preceded by a leveling off toward the strain axis, as in skin, ligaments, and tendons.

Denaturation of collagen by heat or chemicals can produce a collapse of the helical structure and a shrinkage in the direction of the longitudinal axis of the fiber. Whereas the collagen molecule (i.e., monomer) in a disperse solution melts at around 37°C, when assembled into fibrils it does not do so until it reaches a temperature of around 60°C. The tension developed during thermal shrinkage and the swelling that accompanies this event have been used to evaluate the physical properties of the collagen networks. Such an approach, particularly measurements of hydrothermal swelling, allow us to quantitate differences in the degrees of crosslinking of collagen that occur in skin with aging.

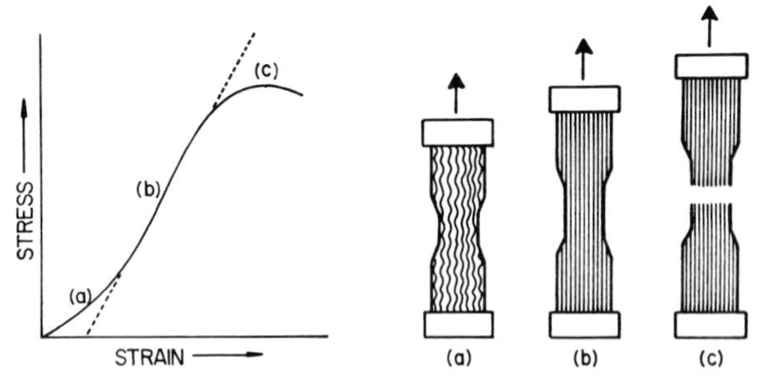

FIGURE 19 Changes in the internal structure of collagenous tissues during the stretching of a tendon or a ligament. (a) Elimination of crimpling. (b) Resistance displayed by the aligned fibers. (c) Failure point.

The cartilage of joints, a hard tissue characterized by a high degree of hydration, owes many of its properties to its ability to shift water molecules from one domain to another. Water is not evenly distributed through the extracellular space; rather, the upper 25% of that cartilage contains approximately 85% water, but as the depth increases the degree of hydration decreases in a linear manner to about 70%. It is hypothesized that one of the major contributors to the destruction of cartilage in osteoarthritis is hydration. The rupture of the collagenous framework seems to occur as a result of osmotic pressure imbalance within the tissue.

When cartilaginous tissue is in contact with a saline solution or a physiological fluid, the proteoglycans within the matrix exert an osmotic pressure which depends on their concentration. This osmotic pressure is able to counteract an externally applied force and enables cartilage to remain hydrated under load. Healthy cartilage is in thermodynamic equilibrium with the synovial fluid and is therefore balanced by a combination of tensile stresses exerted by the collagen network of the tissue and those coming from outside of the network, or applied pressure.

Whereas the molecules in type I collagen are separated by a distance of 14 Å, type II collagen molecules are separated by a distance of 16–17 Å. Measurements made on wet rat tail tendon (mostly type I collagen) show that 55% of the volume of a fibril is occupied by collagen molecules and 45% by water. From these estimates it can be calculated that a wet fibril of type II collagen could contain 50–100% more water than does a fiber of type I collagen that has the same number of molecules. One can easily envision how additional water in type II fibers can be advantageous to tissues such as cartilage and the nucleus pulposus of the intervertebral disks, whose main functions are to absorb or distribute compressive loads, in contrast to type I fibers present in tendon and skin, which transmit tension.

The normal mechanical stresses that act on the connective tissues seem to be essential for maintenance of the cellular activities required for the syn-

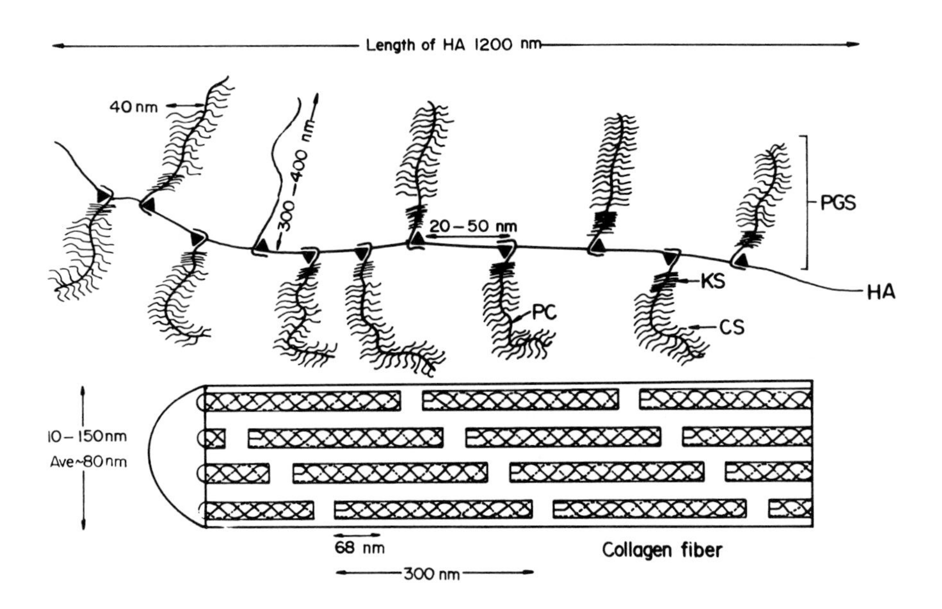

FIGURE 20 Collagen fibers do not exist in a vacuum. They are usually closely associated with the proteoglycans of the ground substance. A collagen molecule is depicted in cartilage, adjacent to a proteoglycan aggregate containing hyaluronic acid (HA), proteoglycan subunits (PGS), and link proteins (▲) which help to stabilize the structure. The PGSs consist of a protein core (PC) from which the negatively charged glycosaminoglycan chains of chondroitin sulfate (CS) and keratan sulfate (KS) radiate.

thesis, turnover, and organization of macromolecules of the extracellular matrix. Reduction of such stresses by inactivity or immobilization which results in a loss of muscle mass (atrophy of disuse) can also cause a loss of bone (i.e., osteoporosis) and cartilage. Immobility on the periarticular connective tissues often leads to rigid contractures, a major complication of fracture treatment using casts. A larger number of crosslinks could cause stiffness and be responsible for the contracture.

VIII. Collagen–Proteoglycan Interactions

To understand the physical properties of connective tissues such as cartilage and intervertebral disks, it is important to have an understanding of the salient features of the proteoglycan molecules (Fig. 20). There are essentially two types of glycosaminoglycans: those with weak negative changes (e.g., hyaluronic acid) and those with strong negative charges (e.g., the chondroitin sulfates, heparins, and dermatan sulfate, the latter comprising the largest bulk of the proteoglycans). Their distribution and physiochemical characteristics, which contribute to distinct functions, are also unique. Hyaluronic acid, with weak negative charges associated with the carboxylic acid residues present in glucuronic acid, has a tendency to form hydrated gels and can therefore contribute significantly to the viscoelastic fluidity of synovial fluid and to the turgency of the skin of an infant. [*See* PROTEOGLYCANS.]

On the other hand, the negatively charged polysaccharides that contain sulfonic acid residues are able to develop strong ionic bonds with the positively charged amino acids on the surface of the collagen fibers, particularly lysine, hydroxylysine, and arginine. Such tissues are more compact, resilient, and less hydrated and exhibit the viscoelastic behavior typified by hyaline articular cartilage. They are also more collagenous than their fluid counterparts. Synovial fluid has no collagen, the vitreous body of the eye has only small amounts of type II collagen, and the skin of a newborn rabbit has less than 2% collagen; this in contrast to a three-month-old rabbit, which has more than 15% collagen. Fetal skin has so little collagen that its wounds can heal without generating scars.

The physiochemical properties of the various types of connective tissues, their viscoelastic properties, the diffusion of macromolecules and of small ions through their midst, and the exclusion of molecules of various molecular masses (e.g., immunoglobulins) are, understandably, different.

IX. Mineralization of Collagen

A major proportion of human collagen is found in bone. Demineralized bone is almost exclusively type I collagen, with small amounts of characteristic noncollagenous proteins. Animal bones are a primary source of commercial gelatin in which, through a process of boiling, the collagen has been denatured.

The mechanism by which collagen mineralizes is a complex one. Essentially, it involves the nucleation of calcium and phosphate ions around functional groups in collagen and the subsequent formation of calcium phosphates, which mature into highly insoluble hydroxyapatites. This occurs on the surface as well as in the interstices of the fibril. It has been estimated that approximately one-half of the mineral is on the surface of the fibril and the remainder is inside, to a great extent in spaces left during the quarter-stagger packing of the collagen molecules (i.e., holes). Why some collagens mineralize and others do not is not clearly understood. It could be related to the intermolecular distances which result from the lateral packing, which in some cases could restrict the access of Ca^{2+} and PO_4, or from chemical modifications that occur subsequent to collagen deposition in the extracellular space (e.g., crosslinking and phosphorylation).

In addition to bone collagen, occasionally other tissue collagens calcify, usually as a result of pathological events (e.g., diseased heart valves, sclerotic blood vessels, and collagen implants). The mechanisms underlying these two forms of calcification are quite different: Whereas bone formation is a cell-mediated event, dystrophic calcification is not.

X. Fibrosis and Tissue Repair

Accumulation of collagen in excessive amounts is a major pathological event that underlies several clinical conditions, including pulmonary fibrosis, liver cirrhosis, and retrocorneal fibrous membrane formation, as well as various forms of dermal fibrosis, such as scleroderma, keloids, and hypertrophic scars. Although in many of these diseases the terminal fibrotic lesion is considered to be the sequela of cellular injury, the cell populations injured and the endogenous mediators responsible for the postinjury fibrotic response vary from organ to organ. In many instances we seem to be dealing with an uncontrolled repair mechanism, in which less organized and less specific connective tissue replaces a previously functional and carefully constructed matrix. In other instances we see an imbalance in the homeostasis of the extracellular matrix, in which synthesis of macromolecules exceeds breakdown, the end result being an excessive accumulation of collagen.

Healthy tissues contain optimum amounts of a well-organized extracellular matrix, of which collagen is a major component. The collagen types present seem to be closely associated with particular structures. Fibril diameter, orientation density, and mode of packing are optimal for the function of diverse tissues such as tendons, cornea, liver parenchyma, bone matrix, and glomeruli. The ability of mesenchymal cells to generate and maintain such structures is paramount to health. An inbalance can severely damage the ability of a tissue or organ to function and can lead to many of the problems briefly mentioned here, causing many degenerative or fibrotic diseases to occur.

In understanding the process of collagen biosynthesis at all levels, the nature and structure of the many types of collagen present in the human body, their morphology, physicochemical characteristics, and mechanisms of degradation greatly contribute to unraveling the mysteries of a fundamental aspect of our structural anatomy.

Bibliography

Cunningham, L. W., and Frederiksen, D. W. (eds.) (1982). "Methods in Enzymology," Vol. 82. Academic Press, New York.

Fleischmajer, R., Olsen, B. R., and Kuhn, K. (eds.) (1990). "Structure, Moleculer Biology and Pathology of Collagen." The New York Academy of Science, New York.

Glimcher, M. E., and Lian, K. B. (eds.) (1989). "The Chemistry and Biology of Mineralized Tissues." Gordon & Breach, New York.

Mayne, R., and Burgeson, R. E. (eds.) (1987). "Structure and Function of Collagen Types." Academic Press, Orlando, Florida.

Mecham, R. P. (ed.) (1986). "Regulation of Matrix Accumulation." Academic Press, Orlando, Florida.

Nimni, M. E. (1983). Collagen: Structure, function, and metabolism in normal and fibrotic tissues. *Semin. Arthritis Rheum.* **13,** 1–86.

Nimni, M. E. (ed.) (1988). "Collagen: Biochemistry, Biomechanics and Biotechnology," Vols. I–III. CRC Press, Boca Raton, Florida.

Olsen, B. R., and Nimni, M. E. (eds.) (1989). "Collagen: Molecular Biology," Vol. IV. CRC Press, Boca Raton, Florida.

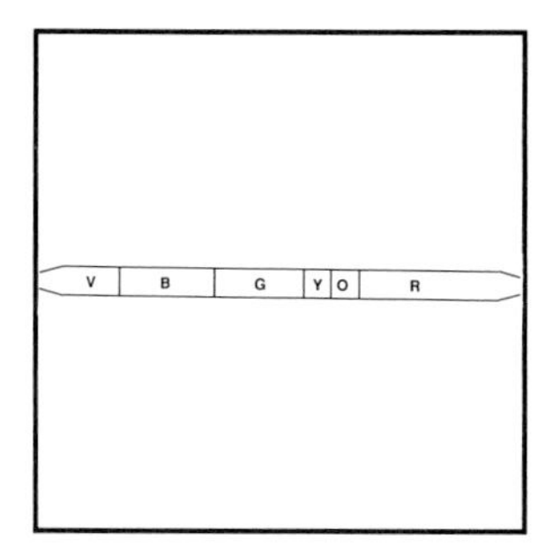

Color Vision

DAVIDA Y. TELLER, *University of Washington*

Glossary

Brightness One of the three dimensions of perceived color; perceptual quality that ordinarily varies with the intensity of light; lights of all colors can vary in brightness; also used to refer to white–grey–black variations of the colors of objects
Cones Photoreceptors that serve daytime (photopic) vision and make color vision possible; there are three kinds of cones: short-, mid-, and long-wavelength-sensitive, with maximal sensitivity at approximately 435, 535, and 565 nm, respectively; these terms replace the older nomenclature of blue, green, and red cones, respectively
Hue One of the three dimensions of perceived color (e.g., blue, green, yellow, orange, red, purple)
Isoluminant stimulus Pattern made up of purely chromatic variations, without variations in luminance (e.g., a set of red and green stripes, in which the red and green are matched in luminance)
Luminance Scientifically refined measure for specifying the efficiency (effective intensity) of lights of different wavelengths for human vision; lights that match in luminance will, in general, differ somewhat in (perceived) brightness; luminances of lights of different wavelengths are additive
Metamers Lights of different wavelength composition that are identical in appearance
Photopic vision Vision at daylight illumination levels; roughly, cone-mediated vision, including color vision
Rods Photoreceptors that serve nighttime (scotopic) vision
Saturation One of the three dimensions of perceived color; the variations white–pink–red and white–light blue–deep blue are variations of saturation
Scotopic vision Night vision; vision at dim illumination levels with the eyes adjusted to darkness; scotopic vision is colorless

IN ORDINARY LANGUAGE, the term color vision refers to one's capacity to see colors, or to tell objects or lights apart on the basis of color differences. The term color refers to a particular subset of the variations we perceive in the qualities of lights or objects. We use color terms (e.g., red, yellow, blue, green, purple, pink, lime green) to refer to this group of perceived variations.

It is not immediately obvious why color variations group themselves together as a natural perceptual category. In addition, the category boundary is ambivalently defined. Interestingly, in ordinary language brightness or lightness variations—the perceptual qualities of white, grey, and black—are sometimes included and sometimes excluded from the list of colors. For example, the following two exchanges are both acceptable usage: What color is your dress? —It's black and white. Is that a color TV? —No, it's black and white.

As a scientific discipline, color science encompasses two main domains. The first is psychophysics: the quantitative description of perceptions and their relations to the physical stimuli that give rise to them. In describing color perception quantitatively, color scientists ask such questions as: Can human subjects arrange colors consistently into groups or series on the basis of perceived similarity? In what ways and along how many dimensions do color perceptions vary? How can these and other color data be best represented in geometrical or mathematical terms? And in relating perceived colors to physical stimuli, we ask: What is the relationship between lights, objects, and perceived colors? What factors other than the wavelength of light influence perceived colors? These questions are treated in Section I of the present article. [*See* VISION, PSYCHOPHYSICS.]

The second main domain of color science is the attempt to explain color vision, on the basis of neural and mathematical models of information processing in the eye and brain. In this domain, we ask such questions as; How does information about color reach the eye? In what ways do the initial information encoding processes in the eye constrain the colors we see? How is information about color recoded and processed by the eye, and by the brain? Is it eventually recoded into a form that bears a recognizable correspondence to the characteristics of perceived colors? Questions of this kind are treated in Section II.

In Section III we consider the genetic basis of color vision and several kinds of naturally occurring variations in color vision. These include color vision deficiencies or color "blindness," the development of color vision in infants, and the color vision of different species of animals. We close with a comment (Section IV) on the intellectual value of the study of color vision and a list of recommended readings (Bibliography).

I. Perceptual Aspects of Color

A. Light

Electromagnetic radiation is one of the basic forms of physical energy. The range of wavelengths of electromagnetic radiation that is visible to the human eye is called light; these wavelengths range from about 400 to about 700 nm (1 nm = 10^{-9}m).

When a beam of white light (e.g., from a tungsten bulb) passes through a prism, it is spread out to reveal its different wavelengths. Under such conditions, a rainbow, or spectrum, of colors can be displayed. By testing human subjects with light of each wavelength in turn, one can examine the sensitivity of the human eye to each wavelength, and the variation of perceived color with wavelength.

B. Spectral Sensitivity

The human eye is differentially sensitive to lights of different wavelengths. The exact shape of the spectral sensitivity (luminous efficiency) curve, and the wavelength to which the eye is maximally sensitive, vary somewhat with light levels and methods of measurement. Under scotopic conditions—when the eye is adjusted to dim illumination levels, and dim lights are used for testing—sensitivity is maximal at about 500 nm, whereas under photopic conditions—when lights are at normal room illumination or higher—sensitivity is maximal at about 550 nm. Standard scotopic and photopic curves are shown in Fig. 1A. This shift in spectral sensitivity can sometimes be observed in daily life by noticing that red objects, which send mostly longer wavelengths of light to the eye, become quite suddenly relatively darker compared with blue or green objects as twilight deepens.

C. Spectral and Extraspectral Hues

1. Scotopic Conditions

Under scotopic conditions, all lights, regardless of wavelength composition, look whitish in color, and spots of light of all wavelengths can be made identical in appearance by varying their intensities to set them equal in brightness. This striking absence of color perception at dim illumination levels can be demonstrated by laying out a set of differently colored articles of clothing in the evening. When one wakes up at night in a very dimly illuminated room, the clothing appears colorless and varies only in shades of grey.

2. Photopic Conditions: The Spectral Hues

When higher light levels are used, the expected colors reappear. As wavelength increases from 400 to 700 nm, colors ranging from violets, through blues, greens, yellows, and oranges, to reds are seen, as shown in Fig. 1B.

Interestingly, at the perceptual level, the linear ordering of colors in Fig. 1B does not capture all of the facts of color similarity. In particular, the colors seen at the two spectral extremes—reds and violets—share a reddish component. To be true to an ordering of colors by perceptual similarity, the linear array would have to be bent into a segment of a circle, so that the two spectral extremes approach each other, as shown in Fig. 2A.

3. The Extraspectral Purples

The color purple is extraspectral; i.e., there is no wavelength of light that looks purple. Extraspectral colors are produced by superimposing lights of dif-

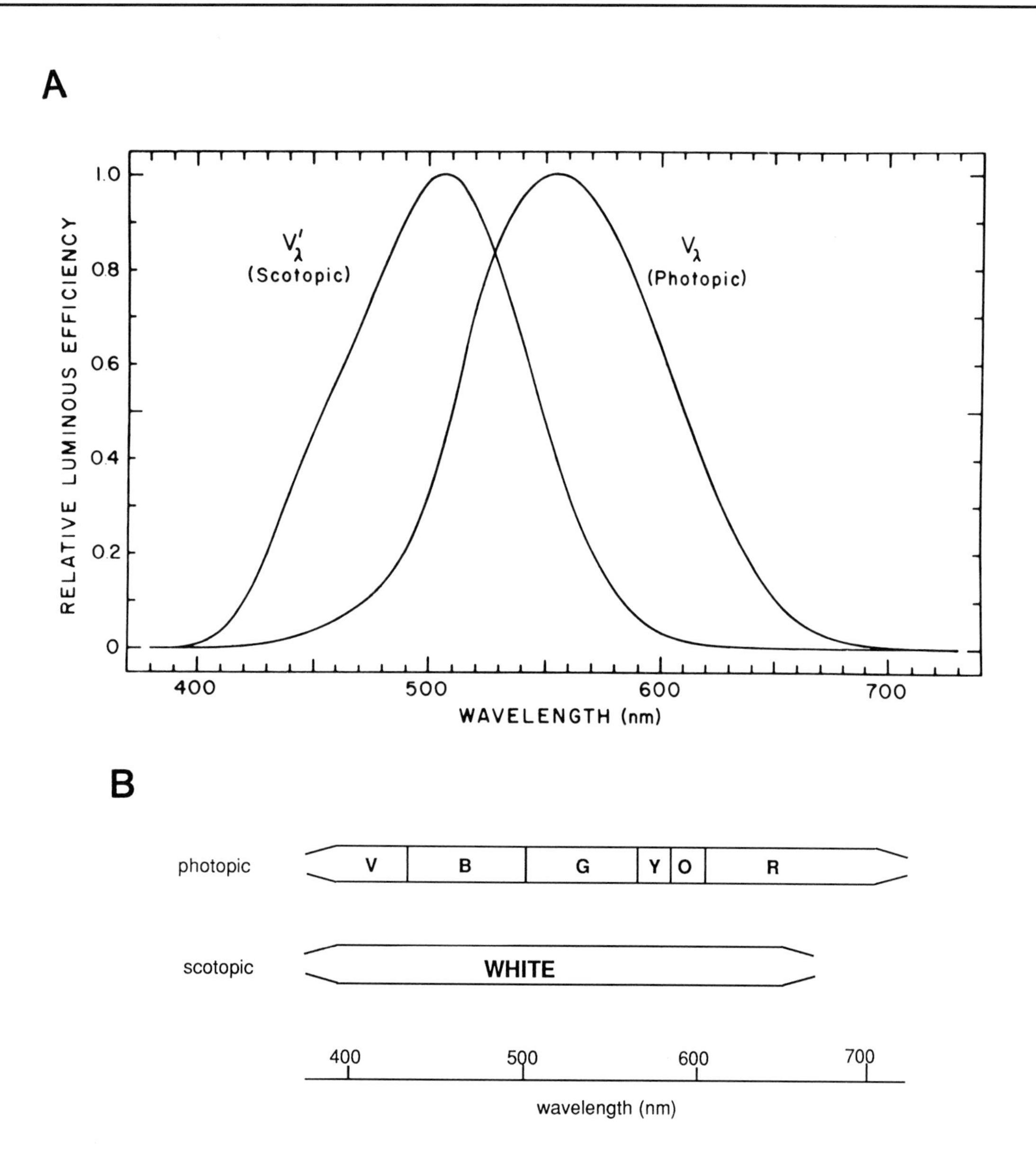

FIGURE 1 Light and color. A. Variations in the sensitivity (luminous efficiency) of the human eye with variations in the wavelength of light. Under scotopic conditions (very dim illumination levels) the sensitivity maximum occurs at about 500 nm. Under photopic (daylight) conditions, the sensitivity maximum occurs at about 550 nm. The curves V_λ and V'_λ are internationally adopted standard photopic and scotopic curves respectively. [Reprinted, with permission, from J. Pokorny et al. (eds.), 1979, "Congenital and Acquired Color Vision Defects," Grune & Stratton, New York.] B. The approximate colors perceived by a human subject viewing various wavelengths of light. Under photopic conditions, colors are seen; under scotopic conditions, all wavelengths of light give rise to the perception of whiteness. V, violets; B, blues; G, greens; Y, yellows; O, oranges; R, reds.

ferent wavelengths. In particular, if a very short wavelength (violet-appearing) and very long wavelength (red-appearing) light are superimposed, and their relative intensities varied, a continuous series of hues varying from violet through mid-purple, to reddish purple and red can be produced.

D. The Three Dimensions of Perceived Color

1. The Hue Circle

When ordered by perceptual similarity, brightness-matched patches of spectral and purple lights form a complete, continuous circle, called the hue circle. An example of a hue circle is shown in Figure 2A.

2. Unique and Mutually Exclusive (Opponent) Hues

Beyond the characteristics already discussed, the hues in the hue circle differ from each other in important qualitative respects. The key is that some colors appear to be analyzable into perceptual combinations of other colors, while others are not. The unitary, or unique, hues—red, yellow, blue, and green—are so-called because subjects report that these colors cannot be perceptually analyzed into subparts. In contrast, binary hues, such as purples, oranges, yellow-greens, and blue-greens, can be so analyzed. For example, most subjects agree that orange can be described as a reddish yellow, while red cannot be described as an orangish purple. Furthermore, of the unitary hues, there are two perceptually mutually exclusive, or opponent, pairs: red versus green and blue versus yellow, so-called because they cannot coexist as perceptual components (most people draw a blank when asked to imagine a reddish-green or a yellowish-blue).

Under ordinary viewing conditions, unique blue, green, and yellow occur at approximately 470, 520, and 575 nm, respectively; these values change somewhat with illumination level and other viewing conditions. Unique red is extraspectral, occurring when a small amount of short wavelength light is added to a patch of predominantly long wavelength light. It must be emphasized that there is nothing physically unusual about electromagnetic radiation of these wavelengths. The uniqueness and mutual exclusiveness of particular hues is perceptually, rather than physically, based.

In geometrical terms, the unique hue pairs are often used to define two cardinal axes—red–green and yellow–blue—for the hue circle; this rule was applied in construction of Fig. 2A.

3. Saturation

Besides the purples, other wavelength mixtures yield other extraspectral colors. The white-appearing light from the sun contains all wavelengths in relatively equal amounts. Mixture of this white with increasing proportions of a spectral color (such as red) again yields a continuous series of color variations, from white through pale pink and deeper pinks to red. Such color variations are called variations in saturation. Ordered by perceptual similarity, these colors fit within the hue circle, with white at the center and increasing saturations of each hue arranged on a line from white to the corresponding hue on the circle.

4. Brightness

The perceptual qualities of lights also vary along a third perceptual dimension, that of brightness. In general, variations of the intensity of light, without variations in wavelength, yield variations in perceived brightness. Addition of the brightness dimension to the two-dimensional hue circle yields a three-dimensional color solid, as shown in Fig. 2B. As noted in our original definition of the term color, the brightness dimension can be included or excluded from color terms depending on context.

5. The Color Solid: A Geometrical Representation of Color Vision

In summary, one of the most fundamental and theoretically enticing characteristics of color vision is its three-dimensionality: in mathematical terms, a closed two-dimensional surface, or plane, is both necessary and sufficient to display all of the hues and saturations that can be produced by spots of light, and a third dimension is required for brightness.

E. Wavelength Mixture

1. Complementary Wavelengths

The appearance of whiteness is not constrained to an equal energy mixture of all wavelengths. There is an indefinitely large number of pairs of wavelengths that, mixed together (i.e., superimposed) in the appropriate ratio, produce the perception of white. These pairs of lights are called complementary wavelengths. The placing of complementary wavelengths on the opposite ends of axes through white produces another major constraint on the geometry of the hue solid. Because the originally perceived colors of the two wavelengths are utterly lost from the perceived white of the mixture, the phenomenon of complementarity is perhaps the most dramatically counterintuitive aspect of color perception.

2. Metamers

Metamers are stimuli that differ in wavelength composition but are identical in hue, in saturation, and in brightness, i.e., identical in appearance and thus indiscriminable from each other. Perceptual whiteness provides a good example, because unlimited numbers of complementary pairs exist, and beyond these, there are an infinite number of mixtures of three, four, or any number of wavelengths, including the mixture of all wavelengths in equal proportions, that yield the perception of white. These lights can all be adjusted in intensity to form a set of indiscriminable lights, i.e., a metamer set.

Moreover, white is not the only color for which

A

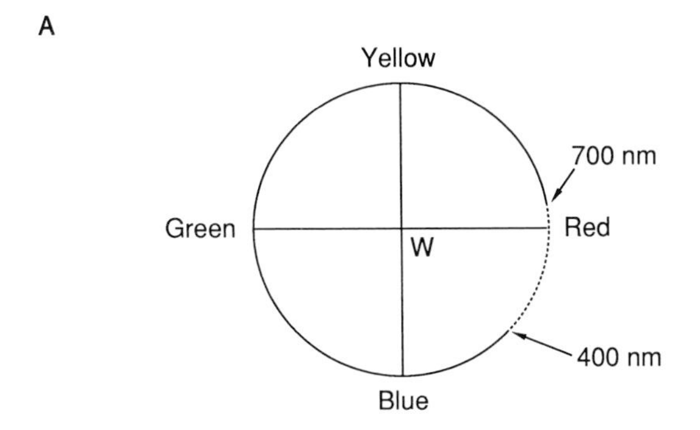

B

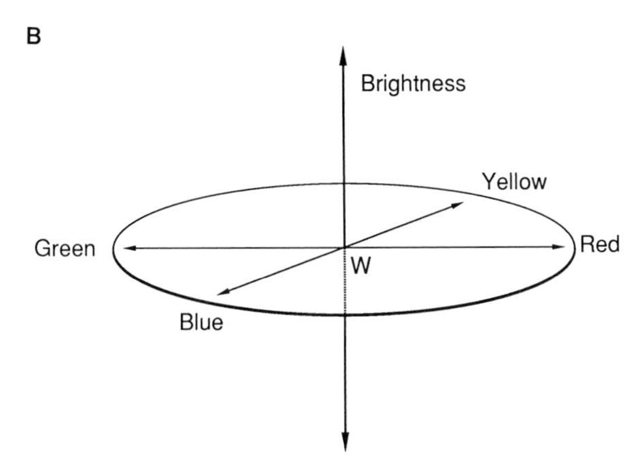

FIGURE 2 Perceptual color spaces. A. Hue circle. The solid line represents the physical (wavelength) spectrum. The cardinal axes are chosen to correspond to the two pairs of unique and mutually exclusive hues; red–green and yellow–blue. The dotted line represents the purples. White is located at the center, with variations of saturation (e.g., white–pink–red) represented on radii of the circle. A two-dimensional figure is sufficient to represent all of the perceived variations of the colors (hues and saturations) of lights, ordered by similarity. B. Three-dimensional color space, with brightness represented on the vertical axis. A three-dimensional figure is required to represent all of the perceived variations of the colors of lights when brightness is included.

metamers exist. The same phenomenon occurs for any other color, although the metamer sets will in general be larger the less saturated the color. Because the members of a metamer set differ in wavelength composition but cannot be told apart visually, the phenomenon of metamerism provides an important example of loss of information by the visual system.

3. Trichromacy

Metamer sets are not haphazard. The form and extent of metamerism is summarized by a general rule called the law of trichromacy. In inexact but intuitively useful form, the law of trichromacy is illustrated in Color Plate 12 and in Fig. 3.

Let A and B be two patches of light. Let A be composed of three superimposed lights of broadly separated wavelengths: λ_1, λ_2, and λ_3 (e.g., 430, 530, and 650 nm respectively). Let B be composed of any other light, of any intensity and wavelength composition (i.e., of any hue, saturation, and/or brightness). Then it is possible, simply by variation of the intensities of the three wavelengths in patch A, to make the two patches of light match each other exactly.

$$A \equiv B,$$

where the symbol ≡ is used to denote a metameric match. Therefore, roughly speaking, lights of three well-chosen wavelengths, mixed in differing combinations, are sufficient to generate a patch of light of *any* perceived hue, saturation, and brightness.

Two modifications are required to state the law of trichromacy rigorously. First, very saturated lights in patch B cannot be matched by leaving all three wavelengths in patch A; however, if one of the three is moved and combined with the saturated light in patch B, the two spots of light can always be made to match. And second, the wavelengths of the lights in patch A need not be those specified, nor need they be composed of only single wavelengths. Any set of four lights will do, provided that no two of them can be mixed to match a third. In its most general form, the law of trichromacy states that any set of four different lights can be arranged into two patches, such that a metameric match can be produced by variation of the intensities of three of them.

Thus, we arrive by wavelength mixture experiments at the same number we found in the perceptual ordering of colors. Color–brightness vision is a system with three and only three perceptual dimensions, and three and only three degrees of freedom.

F. Luminance

1. The Distinction between Brightness and Luminance

Patches of light that vary in hue and saturation can be equated in brightness by several different methods, but a problem arises in that different techniques and instructions yield systematically different results; therefore, color scientists distinguish between two concepts: brightness (with a new and more precise definition) and luminance.

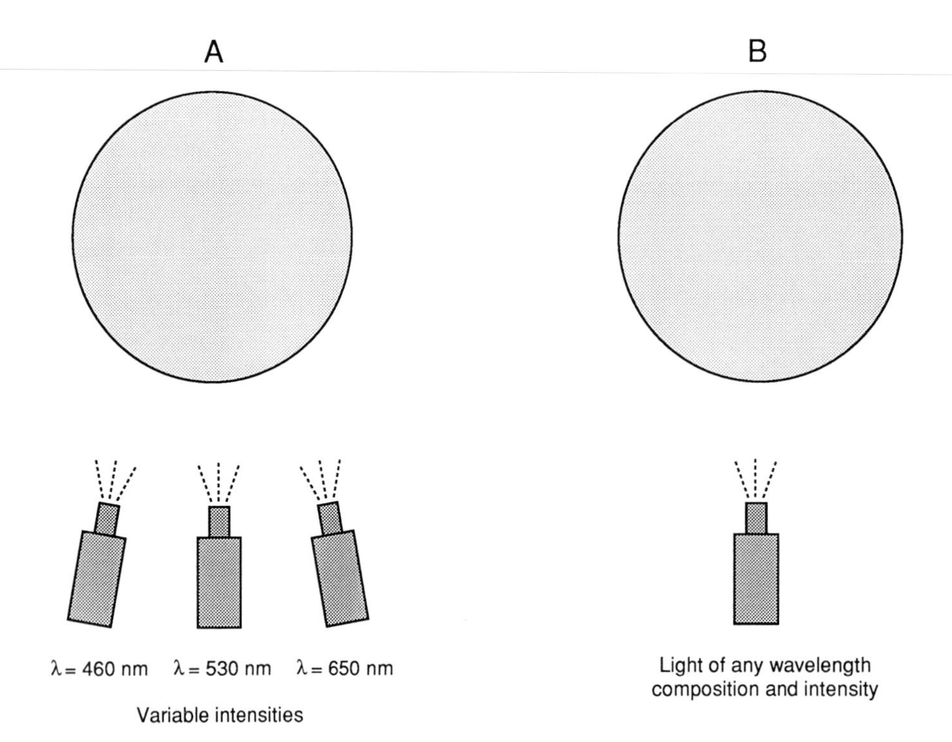

FIGURE 3 Trichromatic matching experiment. Two patches of light—A and B—are set up. Patch A is illuminated by lights of three wavelengths (e.g., 460, 530, and 650 nm); their intensities can be varied. Patch B is a light of any chosen wavelength composition and intensity. The law of trichromacy states that by varying the intensities of λ_1, λ_2, and λ_3, patches A and B can be made to match exactly. Lights of different wavelength composition that match exactly are called metamers.

In formal measurements of brightness, subjects are instructed to match the brightness of each spectral light to the brightness of a white standard light (direct heterochromatic brightness matching), or to the brightness of immediately neighboring wavelengths (step-by-step brightness matching). The results of such an experiment are shown in Fig. 4. Unfortunately, brightness values defined by such techniques suffer from the difficulty that they are not strictly additive; that is, if two spectral lights are each matched in brightness to the same white standard, and then each is reduced to half intensity and the two are superimposed, the combined patch is judged to be brighter than the original white light.

The inelegance of a nonadditive metric led color scientists to invent another, more satisfactory metric, called luminance. In flicker photometry, each spectral test light is alternated (flickered) with a standard white light at a rate of 10–15 cycles per second, and the subject's task is to vary the intensity of the spectral light to minimize the perceived flicker. In the minimally distinct border technique, test and standard lights are presented in two exactly contiguous patches, and the subject's task is to vary the intensity of each spectral light to make the edge, or border, between them appear minimally distinct. Although these measures have no particular face validity as quantifications of the brightness dimension, both measures yield highly similar spectral sensitivity curves, and the resulting values are additive in the sense defined above. For this reason, luminance rather than brightness is used to specify the visual effectiveness of lights for most scientific purposes. A world standard luminance curve, V_λ, has been established; this is the function shown in Fig. 1.

2. Vision with Isoluminant Stimuli

Isoluminant stimuli are patterns composed purely of color variations without variations in luminance (e.g., a field of luminance-matched red and green stripes). While black and white striped patterns as fine as 60 cycles per degree (c/d) of visual angle can be resolved, spatial resolution is much more limited (about 10 c/d) for isoluminant chromatic patterns.

In addition to the resolution of spatial detail, many other visual functions are poorly sustained when isoluminant stimulus patterns are used. If stimuli are composed of spatial variations in hue

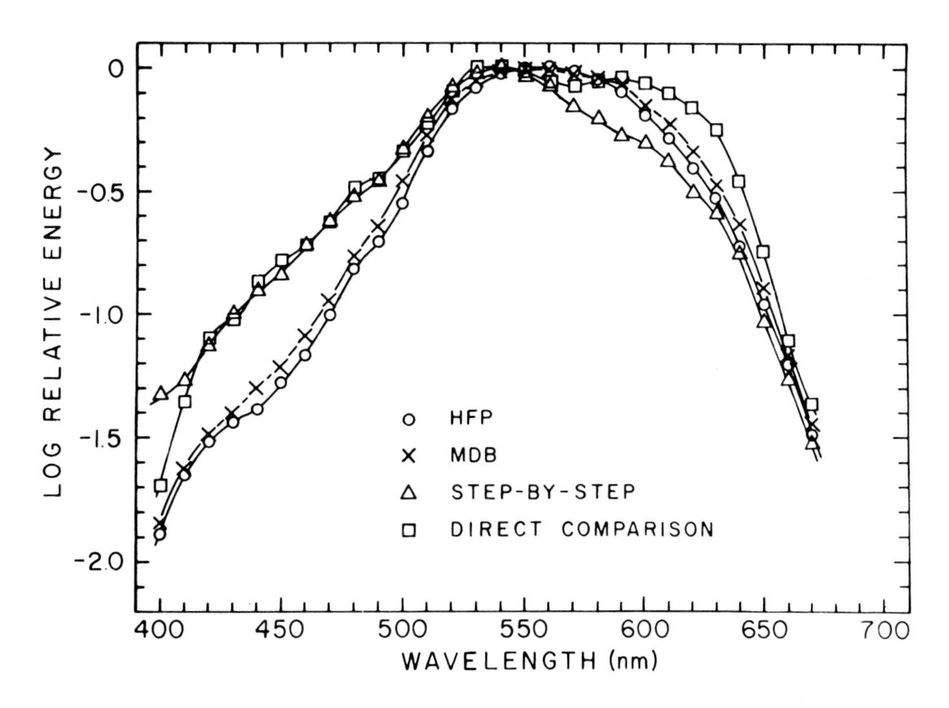

FIGURE 4 Brightness versus luminance. Different methods of equating the "brightness" of lights of different wavelengths lead to different spectral sensitivity curves. Color scientists distinguish between brightness, defined by direct comparisons of brightness between spots of light of different wavelength composition, and luminance, defined largely by a technique called heterochromatic flicker photometry (HFP). Minimally distinct border (MDB) matches agree with HFP matches, while step-by-step comparisons agree with direct brightness matches. [Reprinted, with permission, from J. Pokorny *et al.* (eds.), 1979, "Congenital and Acquired Color Vision Defects," Grune & Stratton, New York.]

and/or saturation without variations in luminance, the borders separating the colors become perceptually fuzzy and indistinct. There are losses of precision in stereoscopic vision (the ability to perceive depth and distance on the basis of combining inputs from the two eyes), accommodation (the ability to focus the lens of the eye), vernier acuity (the ability to see small spatial offsets), perception of the speed and direction of motion, and other visual functions. These visual losses occur when colors are matched in luminance (not brightness), and this fact, together with the characteristic of additivity, suggests that luminance measures might be revealing the more fundamental property of visual coding.

G. Alternative Axes for Color Space

Finally, we return to the question of the choice of cardinal axes for describing the three-dimensional nature of color perception. The axes originally shown in Fig. 2 were based on directly observable, qualitative perceptual dimensions. But other cardinal axes, based on other perceptual characteristics, are also useful in representing other, more sophisticated characteristics of color vision.

1. The Luminance Axis and the Isoluminant Plane

The additivity property of luminance suggests that luminance be substituted for brightness as the vertical axis of the color solid. The two-dimensional color plane, perpendicular to the luminance axis, then becomes a plane in which luminance does not vary (i.e., an isoluminant plane). Spatial patterns made up of stimuli selected from this plane will show the perceptual losses described above.

2. The Tritan Axis

When isoluminant patterns are used, there are in some cases more dramatic losses of vision for some color axes than for others. In particular, there is an axis called the tritan axis, which runs from yellow-green to mid-purple in Fig. 2A. This axis is defined in more detail in sections II.C.3 and III.A.3 below. Isoluminant chromatic stimuli composed of lights selected from along the tritan axis show particularly large perceptual losses. Borders, while indistinct on all chromatic axes, "melt" into invisibility for such tritan stimuli.

3. The Independence of Red–Green and Tritan Axes

In addition, certain perceptual phenomena transfer among most chromatic axes but do not transfer between red–green and tritan stimuli. For example, a subject is exposed to a light alternating in time (flickering) between isoluminant red and green. After adapting to this light for several seconds, the subject is asked to detect flicker along the other chromatic axes. He or she will be found to have lost sensitivity for flicker on most axes, including the yellow–blue axis, but little if any loss of sensitivity is found along the tritan axis. Similarly, adaptation to flicker along the tritan axis leads to a minimal sensitivity loss on the red–green axis. This and similar phenomena suggest that these two axes represent stimuli that are processed very nearly independently of each other. To capture the fact of this independence, red–green and tritan axes can be used as the two cardinal chromatic dimensions in color space.

4. An Alternative Color Space

A color space with luminance, red–green, and tritan axes is shown in Fig. 6B below. This version of color space departs somewhat from direct description of the perceptual aspects of color (Fig. 2B), particularly in that the tritan axis departs considerably from the old yellow–blue axis. The two color spaces represent different constellations of facts about color vision. We will suggest below that a series of recodings of color–brightness information occurs within the visual system, and that different color phenomena reveal the marks left by different stages of neural processing.

H. Contrast Effects

If a disk-shaped test field consisting of a fixed intensity of white light is surrounded by a white ring, or annulus, the test disk can be made to appear any shade of grey or black by variation of the luminance of the annulus. This perceptual phenomenon is called simultaneous contrast. A ratio of about 20 : 1 between annulus and disk yields the perception of black. Similarly, test disks of various wavelengths will appear increasingly darkened by the annulus; for example, a 475 nm light will change from blue through dark blue through navy blue to black, and a 600 nm disk will change from orange to brown to black.

Greys, blacks, browns, and other dark colors do not occur with single patches of light; they are added to the realm of color perception by simultaneous contrast. Color spaces intending to represent the dark colors often use the negative half of the brightness (or "lightness") axis to represent "darkness."

Annuli of different wavelengths will induce approximately complementary hues. For example, a red annulus surrounding a white disk induces a greenish hue in the disk. In combination with luminance variations, chromatic annuli can yield a surprisingly large range of color variations in the disk. Similar phenomena (called successive contrast) occur in the time domain; for example, a white test field following a green one will appear reddish.

I. The Colors of Objects

1. The Complexity of Object Colors

Up to this point, we have discussed mainly the perceived colors of lights. The colors of objects are more complex, for several reasons. First, most objects do not emit light; instead, they are visible because they reflect some of the light that falls upon them. The spectral reflectance function of an object describes the fraction of the incident light of each wavelength reflected by the object. Most objects have broad spectral reflectance functions. Different objects reflect light best in different spectral regions, and the perceived color of an object will depend on the particular mixture of wavelengths which that object sends to the eye.

Second, the spectral characteristics of the illumination that falls on an object, as well as the object's reflectance function, will influence the spectral mixture, and hence one should expect that the illumination under which an object is viewed will influence its perceived color. And third, objects are usually surrounded by other objects, and seen after other objects, so that simultaneous and successive contrast effects also routinely influence the perceived colors of objects.

2. Color Constancy

The concept of color constancy refers to the tendency of the perceived color of an object to remain constant across variations of the spectral composition of the incident light. Despite the difficulties discussed above, a fair degree of color constancy does occur in complex natural scenes, across changes among broad-band illuminants, such as direct versus indirect sunlight, and lights produced by tungsten versus fluorescent bulbs.

However, even with broad-band illuminants, color constancy can be imperfect when one is interested in exact color matches. Changes from daylight to tungsten light can upset the metameric match between one's tie and one's suit, or one's blouse and one's lipstick. Major failures of constancy occur, and major changes of the perceived hues of objects are seen, if the illuminant changes too much or becomes too narrow-band. For example, the low-pressure sodium lamps sometimes used for street lighting provide narrow-band illumination of about 590 nm and yield odd color perceptions.

Because the source of information is not obvious for separating the spectral reflectance functions of objects from the spectral characteristics of the incident illumination, the perceived constancy of object colors has been historically difficult to explain. Computational approaches to this problem are discussed below.

J. Summary

Much is known about the perceptual properties of color vision, and their dependencies on the properties of physical stimuli. But there is much about color perception that is specifically not directly predictable from the properties of light. Nothing about the physics of light leads us to expect the closed color circle, the three-dimensionality of color perceptions, the uniqueness and mutual exclusiveness of certain hues, the occurrence of complementary wavelengths, the trichromacy of color mixture, the reductions of visual function at isoluminance, or the fact that a surrounding annulus can completely change the perceptual qualities of a disk of light. We know a lot about which stimuli map to which perceptions, but the mappings are complex and not predictable simply from the physics of light. Their causes lie in the details of information processing within the eye and brain.

II. Physiological Aspects of Color

A. A Brief Sketch of the Visual System

Light from a visual stimulus is imaged by the optics of the eye onto the retina, a thin layer of neural tissue that lines the back of the eyeball. There it is absorbed by the photoreceptors, neurons (neural cells) specialized to absorb light. The resulting signals are processed through several layers of neurons, before exiting the eye via the neural processes (parts of neurons) that make up the optic nerve. The optic nerve ends at a way-station, the lateral geniculate nucleus, which in turn sends neural processes to the primary visual cortex. From there, the visual signals spread in two or more parallel pathways to other more distant parts of the cortex for further processing. [*See* VISUAL SYSTEM.]

Modern neuroscientific techniques make it possible to use microelectrodes to record the electrical activity of individual neurons within many different parts of the visual system and brain. Thus, one can select a neuron in, for instance, the lateral geniculate nucleus and listen to its response as lights of different wavelengths and intensities are shone upon the retina. Other modern techniques make it possible to stain the cells from which one records and to trace the pathways that run between the different areas and subareas of the brain. Much of what we know about color processing comes from experiments of this kind, carried out on primates or other animals.

B. Photoreceptors and Photopigments

1. Light and Photopigments

A photopigment is a substance that absorbs light in some portion of the visible spectrum. The normal human eye contains four kinds of photoreceptors, each containing its own unique photopigment. The four photoreceptor types are the long-wavelength-sensitive (LWS), mid-wavelength-sensitive (MWS), and short-wavelength-sensitive (SWS) cones, which have maximal sensitivities at about 435, 535, and 565 nm, respectively, and the rods, which have maximal sensitivity at about 500 nm. As shown in Fig. 5, the four sensitivity ranges overlap, and together cover (and define) the visible spectrum. The properties of photopigments and photoreceptors provide well-accepted explanations for spectral sensitivity, metamerism, and trichromacy.

To discuss the absorption of light by photopigments, we shift to the description of light in terms of packets of energy, called quanta. Lights of different wavelengths have different amounts of energy per quantum, and different probabilities of being absorbed by any given photopigment. It is the differential probability of quantal absorption that gives rise to the spectral sensitivity curves of the four photoreceptors.

The absorption of a quantum of light results in a

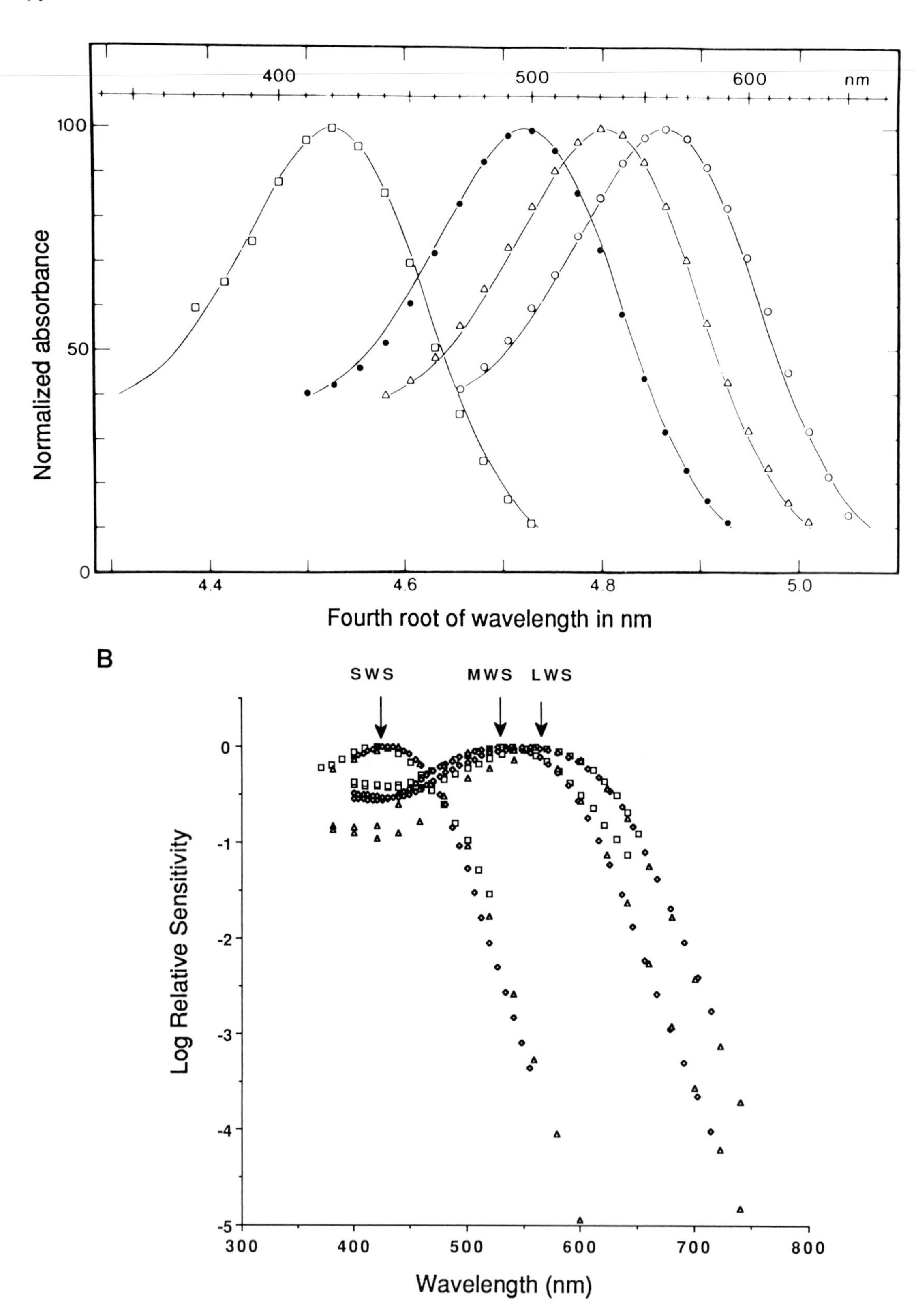

FIGURE 5 The spectral sensitivity curves of the human photoreceptors. A. The spectral sensitivities of the rods (●) and the short-wavelength-sensitive (SWS), mid-wavelength-sensitive (MWS), and long-wavelength-sensitive (LWS) cones plotted on a linear ordinate normalized to the maximum of each curve. These data were obtained by direct measurements (microspectrophotometry) of single photoreceptors. [Reprinted, with permission, from Mollon and Sharpe (1983). B. Spectral sensitivities for the three cone types, obtained by means of three very different techniques—psychophysics (◆), microspectrophotometry (□), and electrophysiological recordings from individual cones (▲). The logarithmic plot illustrates the overlap of the different cone spectra. This overlap is required for the encoding of wavelength information. Different wavelengths of light cause different ratios of quantal absorptions among the three cone types. [Figure provided by Peter Lennie; Adapted from P. Lennie and M. D'Zmura, 1988, Mechanisms of color vision. *CRC Crit. Rev. Neurobiol.* **3,** 333–402.]

particular change in shape—an isomerization—of the photopigment molecule that absorbs the quantum. This change of shape, when it occurs, is identical regardless of the wavelength of the quantum. Thus, wavelength information is lost at the instant of quantal absorption. This photochemical fact has profound implications for vision, because it means that individual photoreceptors can only signal the number of quanta they absorb. They cannot signal the wavelength of light, and a matrix of photoreceptors containing only a single pigment would have no means of preserving wavelength information.

2. Properties of Scotopic Vision

At very low illumination levels, rods initiate detectable signals while cones do not. The scotopic spectral sensitivity curve, V'_λ in Fig. 1, follows the spectral sensitivity curve of the rod photopigment rhodopsin (modified to allow for the absorption of light in the lens and other optical elements of the eye). Moreover, test patches of all wavelengths, weighted by this absorption spectrum to match them in brightness, are perceptually indistinguishable. The colorlessness of scotopic vision comes about because only a single photoreceptor type is active, and no single photoreceptor type can preserve wavelength information on its own.

3. Photopic Spectral Sensitivity and Luminance

The standard photopic spectral sensitivity curve, V_λ in Fig. 1, can be well fitted by a weighted sum of inputs from LWS and MWS cones. For this and other reasons it is believed that SWS cones contribute little if at all to luminance. They do contribute to perceived brightness, and their signals help to account for the difference between brightness and luminance in the short-wavelength region of the spectrum (Fig. 4).

4. Trichromacy

At photopic illumination levels, cones are functional, while rods reach the upper limit of their signaling range and cease to contribute meaningful signals. Logic insists that light of any particular wavelength composition and intensity can do nothing more than to produce a set of three quantum catches: L, M, and S, in the LWS, MWS, and SWS cone types, respectively. If two patches of light A and B produce differing quantum catches in any one or more of the three cone types, then patches A and B are potentially discriminable from each other. But if patches A and B were to produce identical quantum catches in all three cone types, they must be indistinguishable; i.e., metamers: if $L_A = L_B$ *and* $M_A = M_B$ *and* $S_A = S_B$, then $A \equiv B$.

The total quantum catch in any photoreceptor is simply the sum of the quantum catches at each wavelength. So, for patch A, composed of the three wavelengths λ_1, λ_2, and λ_3, the quantum catches L_A, M_A, and S_A in the LWS, MWS, and SWS cones, respectively, are

$$L_A = Q_1 l_1 + Q_2 l_2 + Q_3 l_3,$$

$$M_A = Q_1 m_1 + Q_2 m_2 + Q_3 m_3, \text{ and}$$

$$S_A = Q_1 s_1 + Q_2 s_2 + Q_3 s_3,$$

where Q_1, Q_2, and Q_3 are the incident numbers of quanta of wavelengths λ_1, λ_2, and λ_3, respectively; l_1, l_2, and l_3 are the probabilities of quantal catch at λ_1, λ_2, and λ_3, respectively (i.e., the spectral sensitivity of L at each respective wavelength); and the m's and s's are similarly defined. Thus the situation is described by a set of three equations in three unknowns, the unknowns being Q_1, Q_2, and Q_3, the intensities of the three wavelengths of light in patch A.

Light of any wavelength composition in patch B produces a characteristic set of quantum catch values L_B, M_B, and S_B in the LWS, MWS, and SWS cones, respectively. These values can be plugged into the above equations, and, because three equations in three unknowns are guaranteed solution, we know that the equations can be solved for the three light intensity values. The solution yields the intensity values of λ_1, λ_2, and λ_3 needed to make the metameric match.

It follows from these considerations, regardless of the wavelength composition of patch B, that patch A can always be made metameric to patch B by variation of the radiances of its three component lights alone. A negative value of Q corresponds to the physical operation of moving the corresponding light from patch A to patch B. The reader may readily generalize from the special case to the general one of matching any four lights by means of three intensity adjustments.

Thus, the behaviorally described law of trichromacy, as schematized in Color Plate 12 and Fig. 3, is explained exactly in terms of the properties of light and the properties of the photopigments, at the very first stage of retinal processing, the absorption of quanta by photopigments. The particular photopigments we have determine our particular metamer sets; if one or more of the pigments were shifted along the wavelength axis, the metamer sets would all change, but the property of trichromacy would remain. The spectral sensitivities of our photopigments, and the fact that there are three of them, leave irreversible marks on our perception of colors.

From a computational standpoint, information about wavelength composition, to the degree that it is encoded, is available in the relative quantum catches in the three photoreceptors, as these will be invariant across intensity variations. Information concerning the intensity of light is, in principle, available in the absolute levels of quantal catches in the three cone types.

5. A Three-Dimensional Photoreceptor Space

The information encoded by the three cone types can be represented quantitatively in the simplest of all color spaces, with the quantum catches of the three cone types represented on the three axes, as shown in Fig. 6A. Any individual wavelength of light (or wavelength mixture) of fixed intensity is represented as a point in this space. Variations of intensity of a light of fixed wavelength composition occupy a ray extending outward from the origin; variations in intensity produce variations along the ray.

This three-dimensional space, therefore, represents the information about wavelength composition and intensity that is available for visual processing, in the form in which it is available after the absorption of quanta by the SWS, MWS, and LWS cones. Because metamers plot to the same point, this space elegantly represents the phenomena of color mixture and the reasons for metamerism and trichromacy. But aside from its three-dimensionality, it does not look much like the perceptually derived color spaces of Fig. 2. To find the causes of other perceptual aspects of color, one must therefore look to recodings of these cone-generated signals at later levels of visual processing.

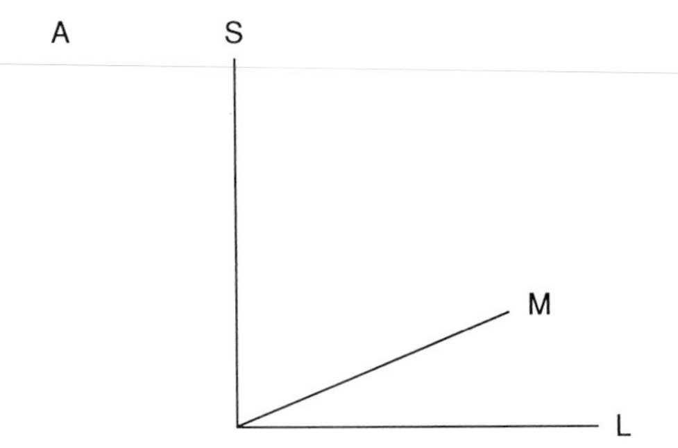

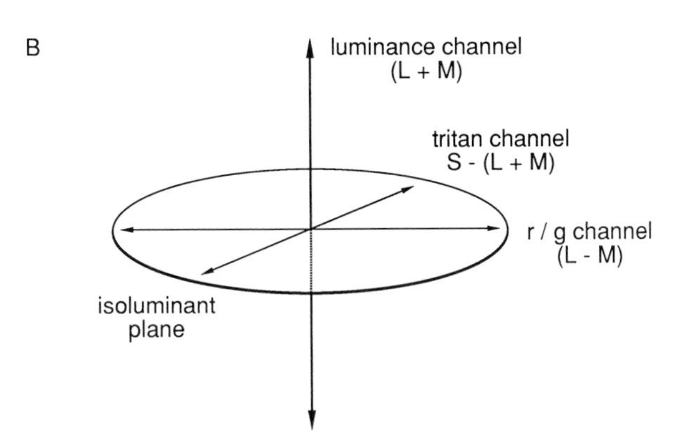

FIGURE 6 Physiologically based three-dimensional color spaces. A. Photoreceptor space. The quantum catches L, M, and S of the LWS, MWS, and SWS cones are plotted on the three axes. Any given light will cause a particular set of quantum catches in the three photoreceptors and, thus, can be represented as a point in this space. All members of a metamer set cause the same pattern of quantum catches, and thereby map to the same point. This space represents the reduced color–brightness information available after the absorption of quanta in the three kinds of cones. B. Three-dimensional space representing the alternate opponent model described in the text. The model suggests that the photoreceptor signals L, M, and S combine in particular sums and differences to make signals in three postreceptional channels. These are luminance channel (L + M), an r/g channel (L − M), and a tritan channel [S − (L + M)]. The plane perpendicular to the luminance axis is the isoluminant plane. A recoding of this general nature, and perhaps of this specific form, occurs in the early processing stages of the human visual system.

C. Recodings at the Level of Single Neurons

1. The Classical Color-Opponent Model

It has been argued by opponent process theorists for more than a century that the properties of the necessary neural codes for color and brightness can be deduced directly from the characteristics of color perception. The marked perceptual tridimensionality of brightness–color perceptions (Fig. 2) has been taken to imply that brightness–color information must be carried by signals in three separate and relatively independent cell types, or channels, corresponding to and signaling the three perceptual dimensions. These are a putative brightness (or white–black) channel and two putative chromatic channels, the latter having the specific characteristics required by the unique and mutually exclusive hues—red–green and yellow–blue.

The mutual exclusiveness of red and green, and yellow and blue, has been taken to reveal the presence of opponent coding in the two chromatic channels. That is, it is argued that the signals in these channels can deviate in either of two opposite and mutually exclusive directions (such as hyperpolarizations and depolarizations of the cell membrane) from a neutral state, and that the two members of a mutually exclusive hue pair are mutually exclusive precisely because they are coded by mutually exclusive deviations of the corresponding neural unit, in its two opposite directions from the neutral value. Unique hues (e.g., yellow–blue) are taken to occur when only the relevant chromatic channel is active, and the other chromatic channel (e.g., red–green) is in its neutral state. Binary hues occur when both chromatic channels are active. Perceived saturation depends on the relative strengths of signals in the brightness versus chromatic channels, and neutral colors (whites, greys, and blacks) occur when both chromatic channels are at their neutral values. In summary, this classical model suggests that three precisely defined classes of cells exist in the visual system: one that responds to luminance variations, one to blue–yellow variations, and one to red–green variations, with the latter two in opponent codes, with neutral values precisely predicted by the perceptually unique hues.

Of course, one need not accept this argument at face value, for two reasons. First, as a theory of visual processing it is incomplete. The early levels of the visual system must preserve and recode sufficient information to allow all aspects of vision, not just color vision, to occur. In particular, the visual system must encode and process information about spatial patterns if we are to recognize objects, and one should expect a code that begins both analyses, probably in some intertwined form. And second, even at more central levels, the requirement of a neural code with such simplistic correspondences to color perception may be more a matter of our own cognitive convenience than of logical necessity.

2. Horizontal Cells

Despite the reservations stated above, direct physiological evidence indicates that an immediate opponent recoding of the outputs from the different cone types does occur. Horizontal cells are retinal neurons that receive direct input from the photoreceptors. In fishes and other nonmammalian species, some classes of horizontal cells indeed exhibit one response (hyperpolarization) to lights of some wavelengths, the opposite response (depolarization) to lights of other wavelengths, and no response (a neutral value) at the transitional wavelength. Thus, the qualitative characteristics required for an opponent chromatic code are found at the very earliest stages of neural processing in some species.

However, for primates, neither the location nor the detailed form of the earliest retinal recodings are yet known. For the reasons described above, several stages of recoding will probably occur, and one would expect that a code corresponding to the perceptual dimensions would occur late rather than early in neural processing. To describe the earliest stages of neural recoding, we here adopt a single, relatively simple and current model, which we call the alternative opponent model.

3. An Alternative Opponent Model

The alternative opponent model suggests that three neural channels, which differ in detail from those of the classical opponent model, occur in the human visual system. The alternative model suggests that LWS and MWS cone signals are summed to produce the signal in a luminance channel (L + M). It also suggests two chromatic channels, a r/g channel constituted from the difference signal between LWS and MWS cones (L − M), and a tritan channel, constituted from the difference signal between SWS cones and the sum of LWS and MWS cones [S − (L + M)]. The tritan axis referred to briefly in Section I.F above corresponds to the tritan axis in this theoretical coding scheme. The largest difference between this model and the classical model is the substitution of a tritan for a blue–yellow axis. A three-dimensional stimulus space with the putative signals in these three channels represented on the three axes is shown in Fig. 6B. This model suggests, then, that three classes of cells, responsive to luminance, tritan, and red–green stimulus variations, respectively, should be found.

But visual neurons must analyze spatial as well as spectral information, and we here digress to consider the question of spatial coding.

D. Spatial Aspects of Neural Coding

1. Receptive Fields

Suppose one were to shine a tiny spot of light on the retina and record the response from a single neuron, somewhere within the visual system, as the location of the light is varied. The receptive field of a neuron is that retinal region, or set of photoreceptors, that when illuminated causes a response in the neuron. Most visual neurons receive inputs from many photoreceptors, not just one; i.e., they have extended receptive fields. Moreover, visual system neurons exhibit spatial opponency; i.e., they respond with an increase of activity to light in part of the receptive field, and with a decrease of activity to light in the rest of the receptive field. These two parts tend to be concentrically arranged, in a so-called center-surround configuration, as shown in the lower part of Fig. 7. Because the maximal response of the neuron will occur when a particular spatial pattern of light—for instance, light covering all of the center, but none of the surround—falls on its receptive field, this receptive field structure begins the analysis of spatial pattern.

2. Combined Spatial and Spectral Opponency

Many neurons in the early stages of visual processing are both spatially and spectrally opponent, with the center of the receptive field receiving inputs predominantly from cones of one type and the surround predominantly from comes of the second (or second and third) type. Examples of such spatially and spectrally opponent neurons, combining all possible combinations of inputs from LWS and MWS cones, are shown schematically in Fig. 7.

Cells with this kind of spectral–spatial coding have been seen by many researchers in the early processing stages of the visual system. There is little current consensus as to which combinations of cone types occur, particularly with regard to neurons carrying inputs from SWS cones. In the absence of consensus we again adopt the alternative opponent model and suggest that two basic kinds of opponent cells occur, namely r/g, with opposed L and M inputs, and tritan, with opposed S and (L + M) inputs. Each basic type can occur in several spatial configurations, such as those shown in Fig. 7, and with different weightings of cone inputs. Neurons of these two types have been seen clearly in one of the most recent studies of cells in the lateral geniculate nucleus (lgn).

Neurons such as these, with both spatially and spectrally opponent receptive fields, obviously carry both spatial and spectral information. Because they respond to both luminance patterns and chromatic patterns, they do not provide a clean separation of luminance from chromatic signals, and the notion that separate luminance and chromatic channels exist at the earliest stages of neural processing may have to be discarded.

3. Emergence of a Luminance Channel

One advantage of the alternate opponent model is that simple recombinations of pairs of such cells at a later stage could, in principle, allow the separation of red–green from luminance channels, as shown in the upper part of Fig. 7. Thus, a recombination scheme of this kind could yield cells corresponding closely to the three channels of the alternative opponent model. Although again there is no consensus on this point, some recent evidence indicates that a recoding of this type occurs in the early stages of cortical processing.

4. Psychophysical Considerations

Because it is easy to imagine that information carried in different cells can be processed independently, the existence of a neural processing stage of this kind provides a ready mechanism for modelling psychophysical observations such as the "melting" borders seen with tritan stimuli and the independence of adaptation effects between r/g and tritan axes, as discussed in Section I.G. If such a model is adopted, it also suggests that important aspects of border perception and flicker adaptation occur at a physiological level, at which the chromatic code remains in this particular form.

To return to our starting point: The earliest retinal recoding schemes for human color vision are not yet known. The best currently available evidence suggests that at the level of the lgn and in early cortical processing, the neural color–brightness code corresponds more closely to the alternate color space of Fig. 6B than to the more perceptually derived color space of Fig. 2B. If the uniqueness and mutual exclusiveness of red, green, yellow, and blue are taken to imply the existence of corresponding cells or physiological channels, one must expect further shifts of the chromatic code in individual cells at higher levels of visual processing. The number and kind of later transformations of chromatic axes, and the existence or nonexistence of a stage corresponding to that suggested by the unique and mutually exclusive hues, remain open questions at this time.

E. Parallel Processing Streams

We turn now from consideration of the coding of information in single neurons to the question of par-

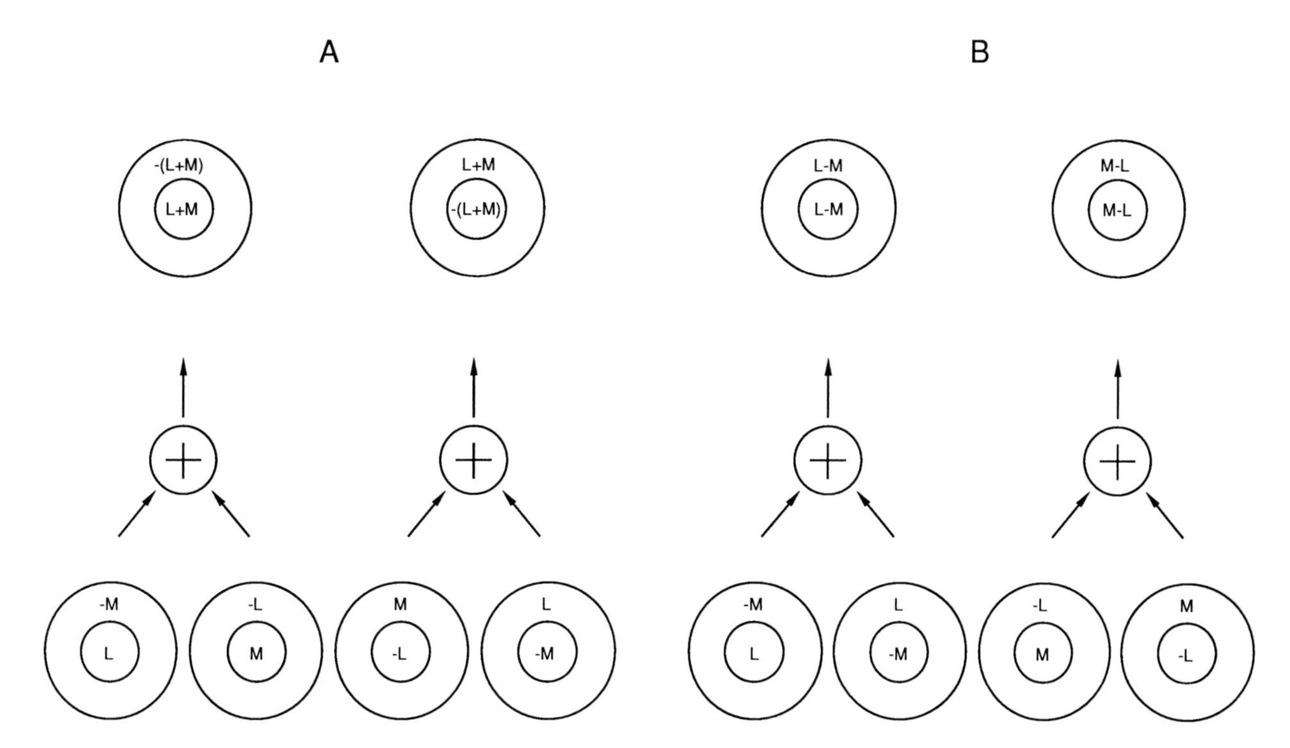

FIGURE 7 Spatial and spectral antagonism between L and M cone inputs. The bottom line of the diagram schematizes the receptive fields of individual postreceptoral visual neurons. These neurons exhibit center-surround antagonism, with one cone type (L or M) providing the predominant input (either positive or negative) to the receptive field center, and the other cone type (M or L) providing the predominant input (either negative or positive) to the surround. Many neurons of this kind occur in the retina and lateral geniculate nucleus of primate visual systems. The upper part of the diagram shows a scheme by which pairs of such spatially and spectrally opponent cells could be combined to yield A, spatially but not spectrally opponent luminance channels, with (L + M) in the center and −(L + M) in the surround, or vice versa; and B, spectrally but not spatially opponent r/g channels, with either (L − M) or (M − L) throughout both center and surround. A recoding of this general nature, and perhaps of this specific form, may occur in the early stages of cortical processing. [Adapted from P. Lennie and M. D'Zmura (1988) Mechanisms of color vision. *CRC Crit. Rev. Neurobiol.* **3,** 333–402.]

allel processing by larger populations of neurons. It is generally agreed that there are at least two major subpathways in early visual processing. These subpathways originate in the retina and diverge in the cortex to form two or more parallel and largely independent information processing "streams." Different aspects of visual processing occur rather separately and in parallel in these separate streams. Virtually all current theorists agree on the principle of parallel processing, but the exact numbers and specific functions of the different streams remain matters of controversy and speculation.

Parallel processing schemes provide a ready, if perhaps oversimplified, explanation for losses of visual function at isoluminance. If isoluminant stimuli create neural signals confined to a chromatic stream, and if this stream fails to access the neural machinery used for other specific aspects of visual processing—such as border perception, stereopsis, vernier acuity, and motion—then isoluminant stimuli would be incapable of supporting these aspects of visual function.

F. Algorithms for Color Constancy

Recent progress in models of color constancy has taken the form of computational schemes, or algorithms. That is, it has been shown that both the reflectance functions of natural objects and the spectral characteristics of natural illuminants conform to certain simplifying rules (technically, can be approximated by small numbers of basis functions). In principle, a system that had these rules stored in memory, confronted with a complex visual scene, could factor out the spectral reflectance functions of objects from the illumination spectrum with a fair degree of accuracy, and thus provide approximate color constancy, at least over a limited range of conditions. With this view, constancy would fail when either the illuminant or the reflectance function could not be approximated by the basis functions stored in memory.

Some evidence indicates that a particular area in visual cortex, called cortical area V4, contains cells whose response properties correlate more closely with perceived hue than with wavelength composition. However, there is no broad consensus on a neural model or a neural locus for the carrying out of a color constancy algorithm at this time.

G. Summary

Much is known about the processing of color and brightness information within the primate visual system. The facts of color mixture, including metamerism and trichromacy, can be attributed with confidence to the characteristics of photopigments, whose spectral sensitivity curves are now well established. Beyond the photoreceptors, a series of opponent coding stages is believed to occur. The initial stages, up to and including early levels of cortical processing, appear to form a code resembling, if not corresponding exactly to, the alternate opponent code. The processing of luminance and chromatic information in parallel streams provides a possible explanation for the dramatic losses of visual function seen with purely chromatic stimuli. Transformations of the chromatic code within the chromatic stream have not yet been studied in detail, and the specific neural bases for the more perceptual characteristics of color, including the uniqueness and mutual exclusiveness of specific hues and the partial color constancy exhibited by human subjects, remain elusive at the present time.

III. Variations of Color Vision

A. Color Vision Deficiencies (Color Blindness)

There are many different forms of color vision loss, both genetic and acquired. The simplest and most common are the "red–green" dichromacies and anomalies. These color deficiencies are surprisingly common (about 8% of Caucasian males are red–green color deficient). The red–green deficiencies and one other category, tritanopia, are discussed here because they provide interesting examples of reduced and altered forms of the normal trichromatic system and test one's understanding of that system. Other types of color vision deficiencies, although equally interesting, are beyond the scope of this chapter.

1. Red–Green Dichromacies

There are two types of red–green dichromacy: protanopia and deuteranopia, involving, respectively, the functional loss of LWS or MWS cones. With the loss of one cone type, color vision is reduced from a three-dimensional (trichromatic) to a two-dimensional (dichromatic) system. If, for example, the LWS cones were lost, the L axis in Figure 6A would collapse onto the M, S plane, and all stimuli that were originally represented along any line parallel to the LWS cone axis (i.e., discriminable by means of the L signal alone) would become metamers. A dimension would also be lost from Figure 6B, because without an L signal the luminance ($L + M$) and r/g ($L - M$) channels would both carry only the M signal and would be completely redundant. The equation for L would be dropped from the color mixture equations. Mixtures of only two wavelengths in patch A (Fig. 3) would be sufficient to match light of any wavelength composition in patch B. Metamer sets are larger for the dichromat, fewer color discriminations can be made, and two socks that obviously differ in color for the color-normal may match for the dichromat.

2. Red–Green Anomalies

There are also milder red–green deficiencies, called color anomalies. The anomalies probably stem from the replacement of either the LWS pigment (in protanomaly) or the MWS pigment (in deuteranomaly) by a different pigment whose spectral sensitivity (Fig. 5) is shifted along the wavelength axis. If, for instance, the MWS pigment were shifted, the values of m_1, m_2, and m_3 in the color mixture equations would change. Consequently, the intensities Q_1, Q_2, and Q_3 required to match patch A to patch B (Fig. 3) would change. An anomalous subject is still a trichromat, but he or she has different metamer sets than does the color-normal subject. A suit and tie that match for the color normal may not match for the anomalous subject, and vice versa.

3. Tritanopia

There is a third kind of dichromacy, called tritanopia. Tritanopes lack functional SWS cones. Like protanopes and deuteranopes, the color vision of tritanopes is reduced to two dimensions. The SWS cone axis (Fig. 6A) is lost. The tritan axis (Fig.

6B) is also lost, because the tritan channel [S − (L + M)] is the only channel that carries SWS cone signals; if they are eliminated, this channel becomes redundant with the luminance (L + M) channel. In fact, the tritan axis derives its name from its dependence on the SWS cone signal, and from the fact that it is effectively lost in the reduced vision of tritanopic observers.

B. Genetics

The red–green color deficiencies are transmitted as X-linked recessives; i.e., expressed in a male, carried without expression in all of his daughters, inherited by half of her children, and expressed in the males who inherit it. Other forms of genetic color vision deficiencies have other forms of inheritance.

Recently, it has become possible to locate the genetic structures responsible for the rod and cone photopigments in the human genome. The rod photopigment gene is located on chromosome 3, and the SWS cone pigment gene on chromosome 7. The LWS and MWS cone pigment genes are located in tandem on the X chromosome, as expected from the X-linked recessive mode of inheritance.

Surprisingly, neither the genetic structures for the LWS and MWS pigments nor the genotype–phenotype relationships are as simple as might have been expected. Color-normal individuals often have two or more MWS pigment genes, and a variety of hybrid genes (genes having parts of both the LWS and the MWS pigment genes) also occur. The genotypes of dichromats and anomalous trichromats often show the expected gene deletions and variations; however, genotypes vary even within color vision categories, similar genotypes sometimes occur in subjects with different color vision deficiencies, and fusion genes are often seen in red–green color-normal subjects. The further explication of phenotype–genotype relationships in red–green color vision is an area of intense research effort at present.

C. Development

Newborn infants can certainly see, because they will stare fixedly at bold black and white patterns. They have an adultlike scotopic spectral sensitivity curve and broad, variable photopic curves indicative of the presence of multiple cone types; however, they respond poorly if at all to isoluminant chromatic stimuli. The earliest color discrimination made is probably between red and other colors; this ability is present by 1–2 mo postnatal. Discriminations among many stimuli displaying large color differences have been well documented by 2–3 mo. Little is known about the progress of color vision from these primitive beginnings to its adult form.

Strong evidence indicates that young infants have functional rods and LWS, MWS, and SWS cones. Their failures to demonstrate full-fledged color vision may indicate a specific immaturity of postreceptoral chromatic pathways, or may simply be one additional manifestation of a broad and general immaturity of both achromatic and chromatic processing.

D. Other Species

Both rods and cones are present in the retinas of most vertebrates. Behavioral studies have provided evidence of at least some color vision in such disparate species as frogs, goldfish, pigeons, cats, dogs, ground squirrels, and many species of primates. The color vision of ground squirrels is dichromatic. The color vision of macaque (Old World) monkeys is trichromatic and highly similar to that of humans. New World monkeys show color vision that is variable, both among species and, in the case of squirrel monkeys, within individuals of the same species.

IV. Epilogue

In conclusion, the study of color vision provides a textbook example of a problem in modern systems neurobiology. We have learned much about the ways in which sensory information loss and sensory recodings leave their marks on our perceptual world. Color vision also provides an example of the profitable interplay of concepts from behavioral, mathematical, physiological, genetic, and anatomical sciences, in the accumulation of human knowledge.

Bibliography

Boynton, R. M. (1979). "Human Color Vision." Holt, Rinehart, and Winston, New York.

Cornsweet, Tom N. (1970). "Visual Perception." Academic Press, New York.

Jacobs, G. H. (1981). "Comparative Color Vision." Academic Press, New York.

Lennie, P., and D'Zmura, M. (1988). Mechanisms of color vision. *CRC Crit. Rev. Neurobiol.* **3,** 333–402.

Mollon, J. D. & Sharpe, L. T. (eds.). (1983). *Colour Vision: Physiology and Psychophysics.* Academic Press, London.

Piantanida, T. (1988). The molecular genetics of color vision and color blindness. *Trends Genet.* **4,** 319–323.

Pokorny, J., Smith, V. C., Verriest, G., and Pinckers, J. L. G. (eds.) (1979). "Congenital and Acquired Color Vision Defects." Grune & Stratton, New York.

Teller, D. Y., and Bornstein, M. H. (1987). Infant color vision and color perception. *In* "Handbook of Infant Perception, I. From Sensation to Perception" (P. Salapatek, and L. Cohen, eds.). Academic Press, Orlando, FL.

Teller, D. Y., and Pugh, E. N., Jr. (1983). Linking propositions in color vision. *In* "Colour Vision" (J. D. Mollon, and L. T. Sharpe, eds.). Academic Press, London.

Wyszecki, G., and Stiles, W. S. (1982). "Color Science: Concepts and Methods, Quantitative Data and Formulae," 2nd ed. John Wiley & Sons, New York.

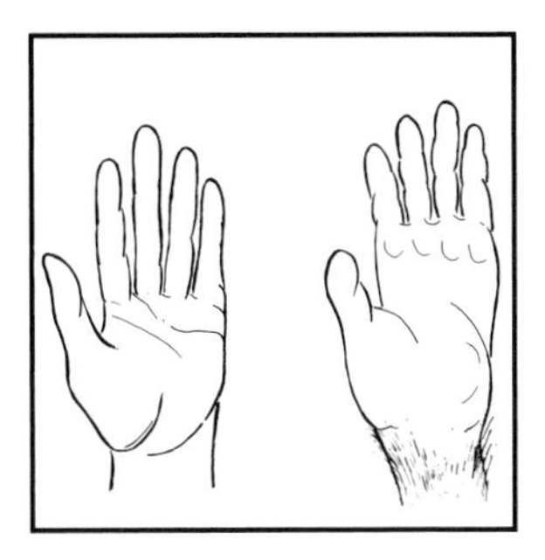

Comparative Anatomy

FRIDERUN ANKEL-SIMONS, *Duke University Primate Center*

Glossary

Dermatoglyphics Epidermal ridges with openings of sweat glands, arranged parallel in curved lines
Evolution Study of the historical development of the diversity of life on earth
Locomotion Power of moving from one place to another
Morphology Science of form and structure of animals and plants
Omnivore Organism that eats everything
Pentadactyl Five-fingered
Primates Order of mammals that contains lemurs, lorises, bushbabies, tarsiers, monkeys of the Old and New Worlds, lesser and great apes and humans
Saimiri Genus name for the New World squirrel monkey

ANATOMY STUDIES ALL ORGANISMS, their structure inside and out, and their similarity to each other. Thus, anatomy to a great extent is comparative. Comparative anatomy is the study of homologous structures in various organisms, their similarities or dissimilarities, ultimately the tool of taxomony, taxonomy being the attempt to classify all organisms into groups according to likeness, to define their relationship to each other, and even to address their possible historical ties with each other—when history is also called evolution.

The comparative anatomy of humans distinguishes us from the rest of the animal kingdom. Within the vertebrate class mammalia and the order primates—including lemurs, lorises, bushbabies, tarsiers, monkeys, apes, and humans—we are taxonomically placed alone in our own family: Hominidae. We are physically different in some ways from the nonhuman primates, but are we really different enough to be in our own taxonomic family? The answer to this question must be left open.

It is the comparison of *Homo sapiens sapiens*, such as you and me, with our closest relatives like the gorilla, the chimpanzee, even the macaque and all the nonhuman primates in general that physically defines our place in nature.

I. Introduction

For human biology, comparative anatomy provides the tool to understand what is truly human structurally. In the endeavor to answer the ever present human inquiry about what makes us different from all other animals—and how to define the place of humankind within the animate world as far as our anatomy is concerned—we must use comparison. Human biologists regard humans as primates. Primates are the group of mammals that are most similar to us and thus most closely related to us. [*See* PRIMATES.]

In this article, human structure is contrasted with that of one relative, our primate cousin the macaque. The macaque has been chosen for this purpose because it is less advanced structurally and less specialized than, for example, the apes or many other monkeys. The macaque appears therefore, to be well suited for this comparison.

A. Posture

Among all primates only humans are habitually bipedal. All the other primates are quadrupedal or exhibit variations of quadrupedality. Bipedality is one of the pivotal human features, and in the following comparison we will start at the feet, going upward through the human body and forward in the body of the macaque, culminating with the head.

B. Variability

Biochemically, all mammal tissues are alike. Even though we discuss "the human" and "the macaque," it must always be remembered that all primates and certainly humans exhibit an incredible degree of variability in all their physical features. This is equally true for monkeys. We must, however, ignore these variations in our brief attempt to comparatively define human structure. "The human" and "the macaque" are abstractions and simplifications.

II. Comparative Anatomy

A. Feet

The feet of the macaque are nimble grasping feet (Fig. 1). As such, they are more similar to human hands than to human feet. The great toe is long and robust in both. Human toes 2 through 5, however, are short, while these toes are long and nimble in the macaque foot. The entire macaque foot is loose and flexible like the human hand. The human foot, in contrast, is tightly bound by tendons and arched on the inward plantar aspect. Human toes are aligned parallel to each other and have only restricted mobility. Indeed, the big toe of humans appears to be large and strong but only toes 2 through 5 are shortened compared to those of the macaque. The big toe in unison with the second toe of the human is, infact, the strong counterbalancing axis of the relatively short striding foot that has to provide support for the comparatively enormous length of the upright body (10 in. balancing about 55 in., a ratio of approximately 1:5.5, while in the quadrupedal macaque the ratio is only about 1:2.7). The stability of the quadrupedally based body is doubtless much greater than that of the bipedal body, with the trunk erect and the entire weight based on only two feet. Even though most primates show a tendency to occasionally adopt an upright posture, they do not usually do so for any length of time. The bony elements of human and macaque feet are essentially the same, except for the weight-bearing talus and calcaneus; these are much more robust and strongly developed in the human foot than in the macaque foot. The phalanges of the human toes 2 through 5 are reduced in length. Also the heel (posterior extension of the calcaneus) is much larger and longer in human feet than in those of macaques, a fact closely related to the strong development of the calcaneus tendon (Achilles tendon) in humans, the tendon of the gastrocnemius muscle that is not nearly as well developed in macaques.

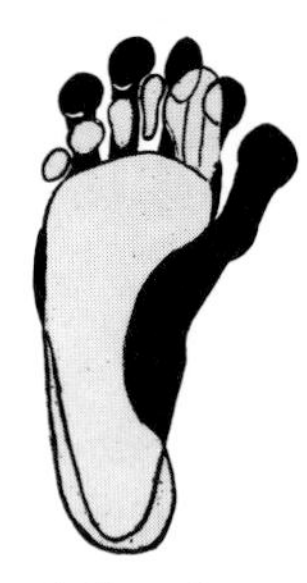

FIGURE 1 Human (shaded) and macaque (black) footprints (brought to approximately the same length). Human footprint superimposed over macaque footprint.

In combination with the alignment of the hallux (big toe) more or less in line with the other toes in humans, the abductor hallucis muscle is less strongly developed in humans than in the monkeys. The same is true for the other intrinsic foot muscles that provide mobility for the toes.

The plantar aspect of primate feet is covered by connective tissue pads and a friction skin that has characteristic dermatoglyphic patterns. All toes in both hand and foot have nails. The primitive toe length pattern as found in macaques is $3 > 4 > 2 > 5 > 1$, with the third toes being the longest. The third toe also represents the functional axis in the grasping foot. The toe length formula in humans is commonly $2 > 1 > 3 > 4 > 5$ or $1 > 2 > 3 > 4 > 5$, the functional axis thus either passing through the second or the first toe (hallux).

The nervous and vascular supply of the foot and leg is basically the same in both primates. Nerves and blood vessels, however, do vary individually to some extent in all primates.

B. Legs

The tibia and fibula are more rubust in humans than in macaques, with the muscles of the lower leg bulkier and the muscles themselves comparatively shorter in humans than the equivalent in macaques. These bulky muscles of humans extend into large, strongly developed tendons that are much less powerful in the quadrupedal monkey. Some minor differences in the way the muscles originate and insert in the two primates are closely related to the functional differences of the two different foot types: stability in the human foot, grasping mobility in that of the macaques. The tibia, being the major weight-bearing member of the lower leg, is especially adapted to this task by being comparatively robust in bipedal humans. The upper member of the leg, the femur, is also more robust in humans than in macaques, while that of macaques is more defined in its relief. The femoral head and neck are proportionally larger in humans than in macaques. The latter fact is again closely related to the weight-bearing demands of the femur in bipeds. A significant positional and size rearrangement of the hip musculature in the human biped, compared with quadrupedal macaques, results in a unique human pattern of this muculature: these rearrangements are intrinsically tied to the profound differences in the pelvic morphology.

Also, the human legs are proportionally much longer than the legs of the macaque. The great length of the human legs allows a much more efficient stride than short legs could.

C. Pelvis

The pelvis of the macaque has a long and narrow iliac blade that is bent outward. The distal and upper end of the macaque pelvis (the ischium) is covered by cartilaginous padding. These enlarged ends are known as "ischial callosities." When sitting, the monkeys are resting on these callosities rather than on muscles, as we do (Fig. 2).

Our pelvis, in contrast, is short and wide, and the iliac blades are bent inward in such a manner that the human pelvis is bowl shaped. The large hip muscle, gluteus maximus, is an abductor of the thigh, positioned at the side of the body in macaques, while in humans it is very bulky and positioned at the back of the pelvis rather than at its side. Thus, it functions as a powerful extender of the leg in the biped. Humans do sit on this muscle and on the muscles of the back of the upper thigh also, the so-called "hamstrings." While the muscles, blood vessels and nerves of the leg are indeed similar in both, macaque and human, the morphology of the pelvis and the concomitant differences of the musculature are closely related to their different ways of locomotion. Thus, for example, the human "hamstrings," gluteus medius and minimus, are involved in the stabilization of the bipedal body when standing. They also abduct and medially rotate the thigh in humans. The gluteus maximus provides power when striding and climbing in the biped and, together with the posterior portion of the medius, rotates the human thigh laterally. The gluteus maximus of macaques covers the side of the femoral articulation of the hip and covers the thick gluteus medius that occupies most of the pelvic fossa. The gluteus minimus also is positioned beneath the medius in the quadrupedal macaque, and these muscles act together in strong retraction of the leg and here are major abductors.

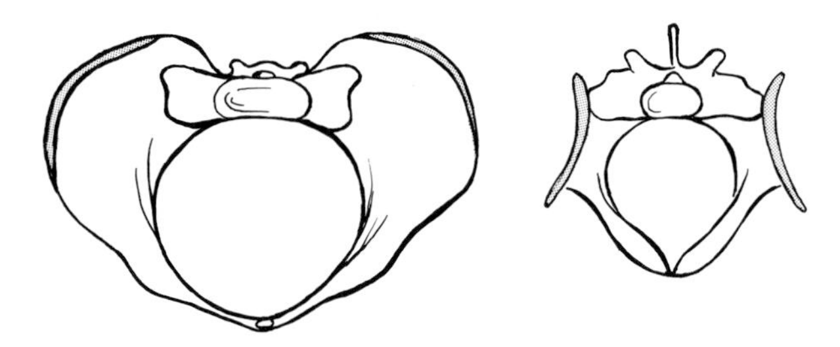

FIGURE 2 Human (left) and macaque (right) pelvis (brought to about the same size, seen from above (human) and the front (macaque). Iliac rim shaded to show the striking difference in the way they are curved.

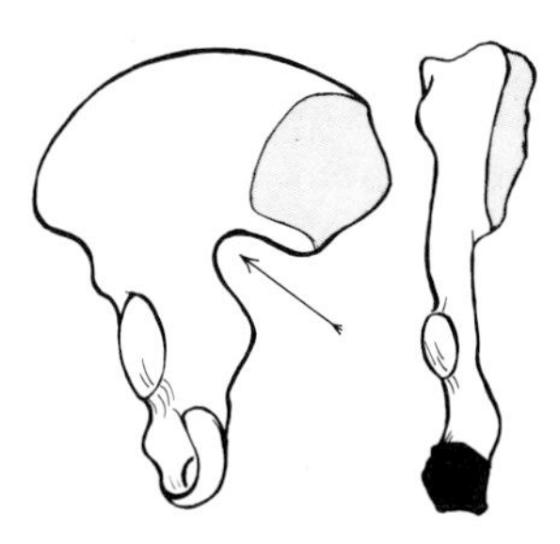

FIGURE 3 Hip bones seen from inside. Human (left), macaque (right). Ischial callosities in the macaque shaded dark. Arrow indicates the characteristically human ischiadic notch.

The pelvis is connected to the skeleton of the upper body by an element of the vertebral column, where several vertebrae are solidly joined to each other forming the sacrum. The sacrum is combined of three vertebrae in monkeys and of five or six in humans. It is the ilium of the pelvis that articulates with the sacrum. Together the two hip bones and the sacrum shape the pelvic outlet. The macaque pelvis is long and slender, and that of humans wide, short, and shaped like a bowl, and the back of the iliac blade is extended backward and downward. This extension of the iliac blade is the area of articulation with the sacrum (Fig. 3). The sacro-iliac articulation is thus situated much closer to and almost above the articulation of the femoral head with the pelvis. The closeness of these two major weight-transmitting articulations within the human pelvis is one of the crucial morphological adaptations within the human body that make true bipedal stance and walking possible. These two articulations transmit the entire weight of the upper body to the legs. In quadrupeds such as the macaque these two articulations between the vertebral column and the pelvis on one hand, and the pelvis with the head of the femur on the other, are positioned behind each other. The articulation of the femur with the pelvis of the macaque is positioned below and behind the sacro-iliac connection. The iliac flange of the human pelvis that extends backward and downward encloses an angle with the ischium that is also known as the incisura ischiadica, or the greater ischiadic notch. The greater ischiadic notch is uniquely characteristic of *Homo sapiens* and as such is one of the crucial clues that indicate bipedality.

D. Pelvis as Birth Canal

The pelvis, however does not only function in locomotion and weight bearing in all primates; it also shapes the canal through which in females the full term baby passes during birth. The necessity for the fetus to pass through this bottleneck puts a different restraint on pelvic morphology than locomotion. Primates are called primates because their well-developed brains are comparatively large, even in the newborn. Thus, in primates with large brains and usually single, large offspring, the bottleneck of the pelvis can be crucial. A birth problem exists in several primate genera, where the pelvic outlet and the size of the term fetus head are critically close (*Macaca, Saimiri*, and *Homo*, to name only three examples).

In the wild this situation is taken care of by nature's selective forces: if birth is difficult, frequently neither mother nor offspring survive. This used to be true for humans also, but medical intervention (Caesarean section) is now counteracting the selective force of birth difficulties caused by the large fetal brain. The sacrum in humans is unusually wide—thus enlarging the birth canal and also providing extensive surfaces for the articulation of the vertebral column with the pelvis and lower limbs. The pelvis contains parts of the urogenital tract. The bottom of the human pelvis is strengthened by muscles that are functionally different from those in macaques because macaques have tails and humans do not. The human sacrum is characteristically bent ventrally, while in macaques the last of the three sacral vertebrae is often tilted slightly upward. The width of the primate sacrum is ultimately correlated with the size of the lumbar vertebrae. They are narrow and high and comparatively long in the macaque, and wide, high, and stout in humans. In humans the lumbar vertebrae arise in a sharp angle upward from the last sacral vertebra. The body of the last sacral vertebra is enlarged ventrally and forms part of a promontory that is caused by the sacro-lumbar angulation of the human vertebral column. This promontory is an expression of the upright vertebral column. It is also a point of great stress and frequent injury in humans, an inherent weakness of our upright posture. [*See* VERTEBRA.]

E. Trunk

The width of the human pelvis is also reflected in the shape of the rib cage above. The lumbar region is short, however, being combined of only five vertebrae in humans, while it is long (seven lumbar vertebrae) in macaques. Also, the thorax of macaques is long, deep, and narrow, while that of humans is comparatively short, shallow, and wide (Fig. 4). The position of the transverse processes of the thoracic vertebrae is different in the macaque and in humans: they are positioned almost at a right angle to the midsagittal plane of the body (about 75° to 80°) in the former and tilted upward in an angle of about 45° in the latter. As the upper ribs articulate with these processes, the differences of angulation are important. The upper ribs articulate with the thoracic vertebrae in two places: between two adjacent vertebral bodies on the upper ends of the vertebral bodies and on the intervertebral disc with the ends of the ribs (capitulum costae) and at the end of the transverse processes of the adjoining thoracic vertebrae with the tuberculum costae. The length of the thoracic vertebral bodies and the angle of rib

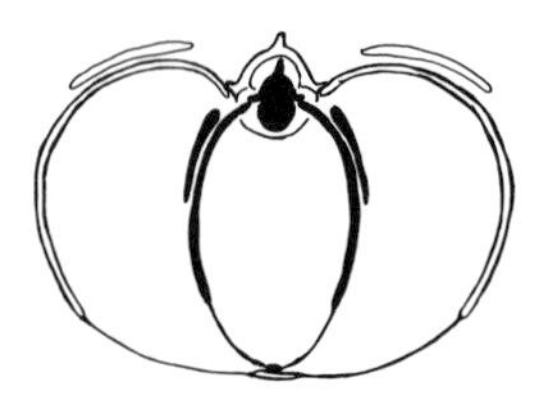

FIGURE 4 Cross sections of the human (white) and macaque (black) rib cage with shoulder blade position (Brought to the same depth).

insertion upon the transverse process both influence the angle of insertion of the ribs. The more dorsal the transverse process of the thoracic vertebrae is angled and the shorter the vertebral bodies of the region are, the steeper is the angulation of the rib. Rib angulation is steeper in humans than in macaques. Also, the way the rib itself is bent at its vertebral end determines the shape of the rib cage. In humans ribs are bent almost to a quarter circle at the vertebral end, while those of macaques are bent only slightly. These arrangements of the rib angulation and position in fact cause the vertebral column to be positioned rather inside the thorax in humans, while it is positioned at the back in macaques. By this means also the center of gravity lies more central in the human trunk. In consequence of these features, the human rib cage is barrel shaped and wider transversely than deep dorso-ventrally. It is narrow in the monkey. In concordance with this overall shape difference of the thorax, the sternum also is shaped quite differently: it is wide and comparatively short in humans and narrow and long in macaques, where it is also segmented, not fused like the sternum of humans. These differences in the shape of the thorax also have a crucial influence upon the position of the shoulder blades and thus the upper arm articulation in the two forms. This articulation is positioned high and somewhat backward on the rib cage in humans and forward and at the side of the thorax in macaques. This, in turn, causes the action radius of the arm to be considerably less restricted in humans than it is in macaques. A rearrangement of the soft tissues accompanies all these differences in the skeletal architecture of the trunk of macaques in accordance with the narrow trunk.

The presacral human vertebral column exhibits a unique and characteristic series of three curves when seen from the side. There is a lordosis (i.e., concave toward the back) in the neck region a kyphosis (i.e., convex backward) within the thoracic region, and a lordosis in the lumbar region. In fact, the human vertebral column is more similar to an elastic spring than to a true column. Even though it is rigid in specific ways in the different regions, it is also mobile within limits. The possible movements of the vertebrae are channeled by the position of the articulations between single vertebrae that are characteristically different in different regions. All vertebral centers are joined with each other by means of cartilaginous intervertebral discs with the exception of the first and second neck vertebrae (atlas and axis) and the vertebrae of the sacral region that are fused to each other. The characteristic triple curvature of the human presacral vertebral column, the promontory angle between lumbar region and sacrum, and the distinctive ventral curvature of the sacrum itself are all typical only of humans.

G. Shoulder

The human shoulder blade is triangular with a long medial border; that of macaques is rather more bladelike with long, almost parallel fore and hind margins and a short dorsal border. The two differently shaped shoulder blades allow different leverage for the enveloping musculature; that in turn is closely related to the very different use and reach of the forearms. Also the collar bones are long and angular in humans in conjunction with the barrel-shaped thorax. They are comparatively short and straight in macaques in accordance with the narrow trunk.

H. Inner Organs

The differences in the inner organs of humans and macaques are mostly differences in proportion and slight differences in position, but they are not characteristic or unique to either one of the two primates compared here.

In the human male, the testicles descend regularly before the infant is born. They do descend at birth in the macaque but are retracted back into the inguinal canal just afterward and only permanently descend into the scrotum at about 6 years of age.

The human penis is comparatively much larger than that of macaques. It is in fact almost the largest among all primates (only the chimpanzee rivals human males in size of the genitals).

It appears that only human females among primates exhibit permanent prominent breasts, which remain large even when not lactating.

I. Arms

As one example of the differences of the musculature of the forearm we will mention here that the deltoid muscle is a powerful pro- and retractor of the arm in the quadrupedal macaque. It has a larger clavicular insertion in humans than in the monkey and functions as a powerful abductor of the arm and also extends the arm in humans. Also in humans its most ventral portion rotates the arm medially; the most dorsal portion rotates it laterally.

In accordance with the barrel-shaped trunk and lateral position of the shoulder joint, the humeral head in humans is rotated inward dorsomedially (about 45°) compared with the position of the elbow joint of this bone. This torsion of the humeral shaft is also a distinctive human characteristic. The humeral head of macaques faces straight backward and is also proportionally smaller in its diameter than that of humans. Without this rotation of the human humerus, our arms and hands would face outward rather than toward the body in a relaxed position. Also the upper arm of humans is straight rather than curved parallel to the trunk as it is in macaques. In the ulna the olecranon is less prominent in *Homo* than it is in *Macaca*, and at its distal end the styloid process is considerably shorter and rounded in the former, rather than pointed as in the latter. Both these differences can be attributed to the difference in use: quadrupedal locomotion rather than free manipulation. Differences in the way the muscles are attached to the two bony elements of the lower arm also reflect the differences in function, which is much more restricted in the macaque. The ulna and radius in humans are positioned parallel to each other when the palm faces up (supination) and then cross over each other (radius over ulna) when the palm faces down (pronation). Macaques are not able to rotate their forearms much, and radius and ulna are positioned close to each other and are tightly bound.

J. Hands

All primate hands are basically built according to the primitive pentadactyl vertebrate plan (Fig. 5).

The human hand is characterized by comparatively greater overall width and a long pollex in contrast to that of the macaque. The hand of macaques is tightly bound and much less mobile and flexible than the human hand that, for instance, can be cupped. As in almost all higher primates, both human and macaque hands have nails on all five fingers. Even though the pollex of the macaque hand is opposable to digits II through V, this ability is much less efficient than the grasping abilities of the human hand. Human hands are not specialized, and they are built according to an old plan—they are structurally primitive. Lined on the palmar surface with a specialized, highly sensitive friction skin with individually characteristic dermatoglyphics, these hands are functionally omnipotent and doubtlessly the executing instruments of human civilization and, ultimately, culture. But this is only possible because these simple hands are doing what a highly evolved brain leads them to do.

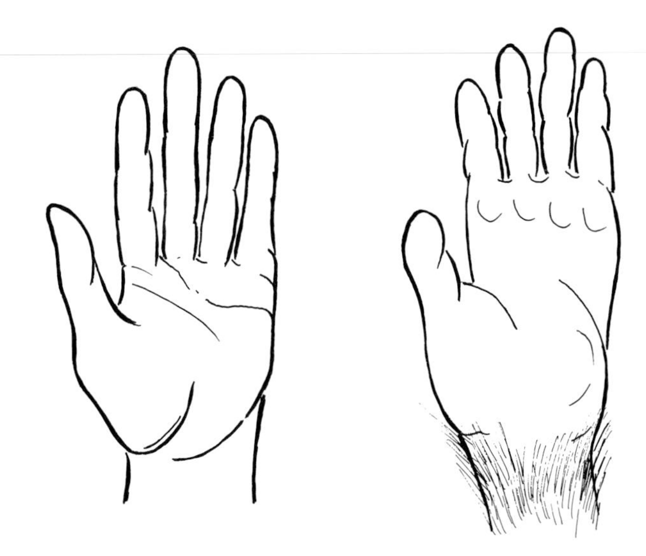

FIGURE 5 Human hand (left) and the hand of a macaque (right). (Brought to about the same length).

K. Neck

As we proceed up the neck we get to the pharyngeal region. Even though the anatomy of this region is structurally basically the same in humans and macaques, there is one fundamental difference: humans have language. However, none of the morphological features of the pharyngeal region or the upper airways in humans can be attributed with certainty as the crucial feature that makes language possible. Attempts to identify an anatomical feature that is the key to human language always fail. Even though, for example, the cantilever position of the human epiglottis is a mechanical prerequisite for the possibility of human speech, it can be found in many mammals that cannot talk. Also, the spina mentalis of the human chin is not an indicator for the presence of the ability to have language. Not even the representation of the motor control area of language in Broca's speech area of the brain (inferior frontal gyrus) is anatomical proof of the presence or absence of language. Broca's speech center coordinates the muscles involved in speech. Other cortex areas, however, are necessary for articulate and abstract speech formulation. In essence, the ability to voice abstract ideas cannot be documented anatomically.

L. Head and Face

The head is large and rather spherical in humans due to the dominating size and shape of the brain and braincase. The head of macaques is shaped more like that of a nonprimate mammal—with a long snout and small eyes, each covered above by a frontal torus, no forehead, and small braincase. The vertebral column meets the human head centrally from underneath, but it attaches under the back of the macaque's head. The human foramen magnum is tucked forward under the head and located almost in the middle of the basicranium (Fig. 6). It is directed backward in the macaque. The facial musculature is highly differentiated in humans compared with that in macaques. The face itself appears to be much larger in humans, which is mainly a function of the missing snout. The monkey face is dominated by this snout and impressive canines, and it is incapable of the varied expressions that are so telling and characteristic of humans. The human maxilla and mandible are tucked under the head. The teeth—even though the same in number and kind as in macaques—are relatively smaller in humans, especially the canines that are incorporated into the rounded arcade of front teeth. Also, the human molars have flattened occlusal reliefs compared with the rather prominent occlusal cusps of macaques. The lower first premolars in macaques are enlarged to an enormous bladelike honing tool for the dagger-shaped upper canine, while they are equally small and shaped more or less like incisors in humans, as are the canines.

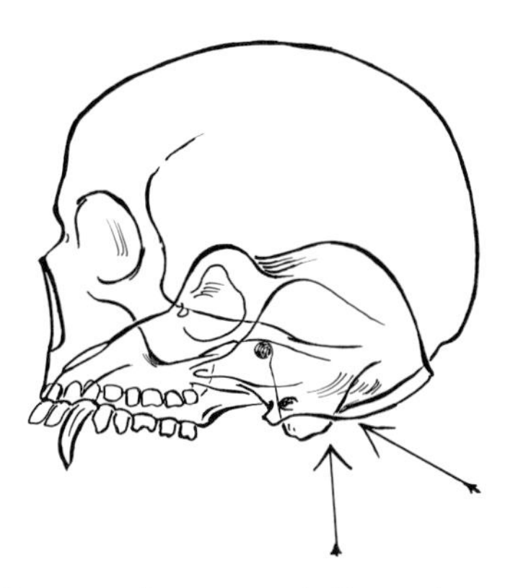

FIGURE 6 Human and macaque skull superimposed. (Brought to about to the same length.)

The human face is also characterized by the delicately chiseled lips—another feature that is uniquely human.

The use of the ear muscles is obsolete in humans; in contrast, the ears of macaques are quite mobile and play a part in social interactions among them. A prominent nose is another feature that is missing in macaques but always found in humans.

The eyes of humans are somewhat closer together than they are in the monkeys and directed forward.

M. Brain and Reproduction

Most striking is the difference in the brain. We can only highlight a few points in our brief comparison. It is the cortex that is absolutely enlarged in humans compared with macaques. The cortex of the human brain is folded into multiple gyri and sulci and also has a higher density of neurons in some areas than does the cortex of the macaque brain. The temporal and frontal lobes especially are enlarged and functionally perfected. It is not the volume of the brain alone that tells about its functional abilities. The sensory–motor cortex is not only larger than in monkeys, but certain areas like that for our hands are enormously enlarged. In humans we find cortex areas that are engaged with the development of skills and that govern foresight and memory as well as language and abstract thinking. A considerable part of our brain is devoted to learning. [*See* CORTEX.]

The large brain of humans also is instrumental in important changes concerning human reproduction. It involves strong mother–child dependency. Even though macaque and human females undergo similar reproductive cycles, there are important differences. Macaques have seasonality in their breeding and birthing times; humans do not. Humans are able to breed regardless of seasons. Humans engage in sex without the goal of progeny. Macaque females have color changes that advertise times of highest sexual receptivity to their males: the skin around the head and the anogenital area turns bright red or purple. Humans do not have such signals.

N. Hair

Humans are different from macaques in yet another striking way: humans are naked while macaques have fur. Humans usually have only remnants of fur on their scalps, under their arms, in their genital area, over their eyes, and around the mouth of males. Newborn macaques can cling to the mother's fur. Newborn humans are totally helpless, have no fur to cling to, and have to be carried. [*See* HAIR.]

III. Conclusion: What Makes Us Human?

What do we discover when we now look back at this comparison (Fig. 7)?

Anatomically, humans are in many ways not very different from macaques. Both are primates. What, then, makes us human and macaques monkeys? What makes humans the most successful primate on earth? Macaques are curious, manipulative, inquisitive, omnivorous, and social. Humans are curious, manipulative, inquisitive, omnivorous, and social. But they are more. Humans are intelligent, and stupid. They are compassionate, and aloof. They are loving, and full of hatred. Humans are imaginative and dull, industrious and lazy, authoritarian and submissive. They are generous and greedy. Humans are immensely creative and unbelievably destructive. How can it be that humans are so overpowering and successful? Is it reproductive prowess? Or is it just that humans are unlimited in being cunning? It seems that these qualities have arisen because of a highly refined, large brain, two specialized feet that have taken over locomotion, two able but unspecialized hands, and because of the ability to communicate through language.

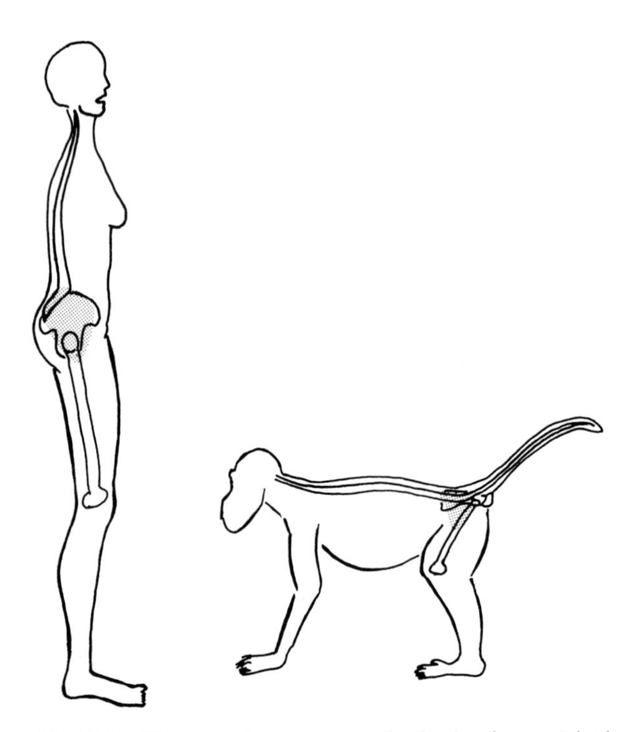

FIGURE 7 Human and macaque outlined, showing vertebral column, hip bone, and upper leg bone. Hip musculature shaded (not to scale).

Bibliography

Cartmill, M., Hylander, W. L., and Shafland, J. (1987) "Human Structure." Harvard University Press, Cambridge, Mass. and London, England.

Fleagle, J. G. (1988). "Primate Adaptation and Evolution." Academic Press, San Diego.

Jungers, W. L. (ed.) (1985). "Size and Scaling in Primate Biology." *In* "Advances in Primatology," Vol. VIII, Plenum Press, New York and London.

Kinzey, W. G. (ed.) (1987). "The Evolution of Human Behavior." State University of New York Press, Albany.

Shipman, P., Walker, A., and Bichell, D. (1985). "The Human Skeleton." Harvard University Press, Cambridge, Mass. and London, England.

Simons, E. L. (1989). Human evolution, *Science* **245,** 1343.

(This is Duke University Primate Center Publication # 479)

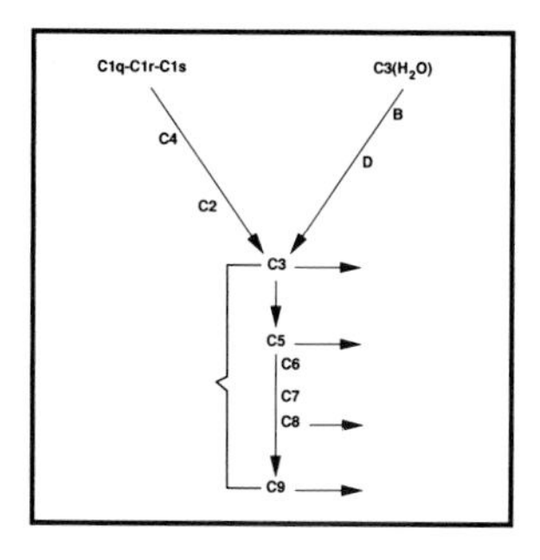

Complement System

GORDON D. ROSS, *University of Louisville*

Glossary

Complement Frequently referring to the entire complement system

C1 inhibitor Protease inhibitor in the plasma that irreversibly inhibits the activity of C1r and C1s, and also other proteases such as plasmin, kallikrein, and trypsin

CR_1, CR_2, CR_3, and CR_4 Complement receptor types one, two, three, and four; receptors for different parts of C3-complement molecules deposited onto activating surfaces

Decay-accelerating factor Membrane protein that protects host cells from their own C by dissociating C3-convertase enzymes deposited onto host tissue

Membrane attack complex C-protein complex of C5b, C6, C7, C8, and polymerized C9

Opsonization Coating of particles (e.g., bacteria) by serum proteins that facilitates phagocytosis

Paroxysmal nocturnal hemoglobinuria Acquired disorder in which red blood cells are lysed because normal control proteins (e.g., DAF) are missing from the red cell surface

Phagocytosis Particle (e.g., bacteria) ingestion by phagocytic white blood cells (neutrophils, monocytes, and macrophages)

Systemic lupus erythematosus Autoimmune disease characterized by a wide spectrum of circulating autoantibodies, some of which form immune complexes and activate C

THE COMPLEMENT SYSTEM is an important part of host defense, which functions together with the immune response to provide the effector mechanisms necessary to initiate inflammation, kill bacteria and other pathogens, and facilitate the clearance of bacteria and immune complexes. It is made up of 20 distinct plasma proteins and 12 different membrane proteins. Bacteria or immune complexes trigger activation of complement, resulting in a sequence of reactions in which one component activates another component in a cascade fashion. Along this cascade, inflammation and phagocytosis are initiated, and the terminal event is the generation of cytocidal (cell-killing) activity in the form of membrane-penetrating lesions. Because of the importance of complement, an inherited or acquired deficiency in any component of the system is frequently associated with either an increased susceptibility to infection or, as will be seen below, a lupus-like syndrome thought to result from diminished clearance of immune complexes. [*See* IMMUNE SYSTEM.]

I. Introduction

The primary function of complement may be defined as the destruction of both foreign organisms and immune complexes. This activity is carried out by two mechanisms. First, coating the particles (e.g., bacteria) with the proteins C3 and C4 results in particle phagocytosis through attachment of the

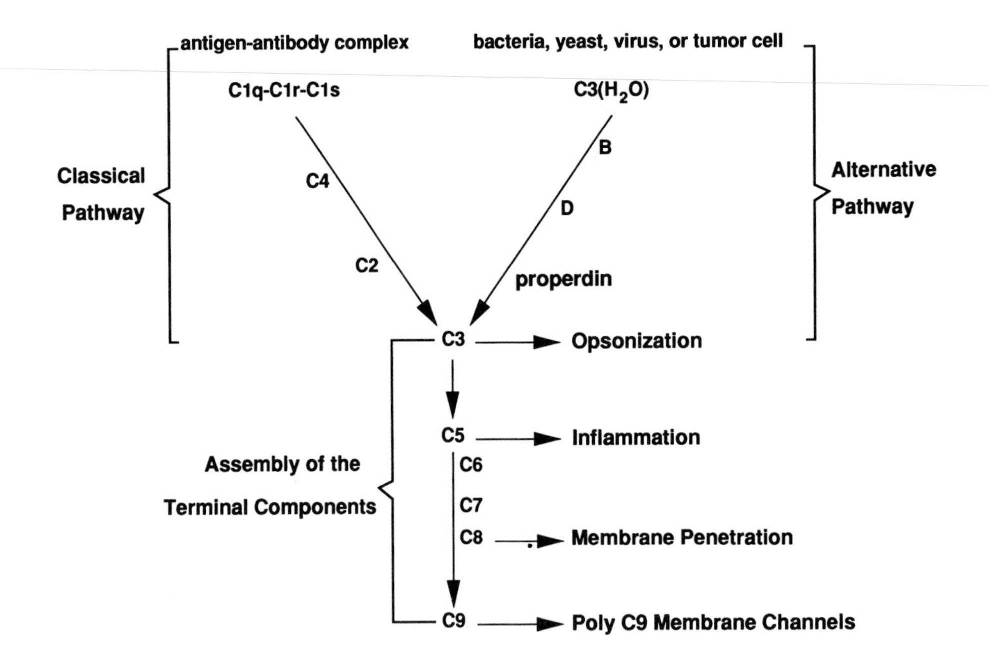

FIGURE 1 The complement system.

C3/C4 coating to receptors for C3/C4 on macrophages. Second, complement lyses organisms through insertion of hollow tubular structures composed of polymerized C9 molecules through cell membranes. There are two distinct pathways that lead to the deposition of C3, namely, the "classical" and "alternative" pathways (Fig. 1). [*See* MACROPHAGES.]

II. Classical Pathway of Complement Activation

The classical pathway[1] consists of five proteins (i.e., C1, C2, C3, C4, and C5), all present in plasma in inactive form. C3 is also a component of the alternative pathway, and C5 is part of the terminal component pathway that forms the membrane attack complex. C2, C3, C4, and C5 are present as single molecules, whereas C1 is made up of three noncovalently linked subcomponents (i.e., C1q, C1r, and C1s). Unfortunately, the components were named before their functional properties were elucidated, and it was ultimately found that C4 was misplaced in the sequence of activation, the order being C1, C4, C2, C3, and C5 (Fig. 2). Activation consists of the enzymic splitting of components. After activation, C1r, C1s, and C2 become enzymes capable of splitting proteins (proteases), whereas the activated C3 and C4 molecules become capable of binding covalently to immune complexes and cell surfaces.

1. This section on the classical pathway of complement activation was excerpted with permission from Hughes-Jones, N. C. (1986). The classical pathway. *In* "Immunobiology of the Complement System. An Introduction for Research and Clinical Medicine" (G. D. Ross, ed.), pp. 21–44. Academic Press, Orlando, Florida.

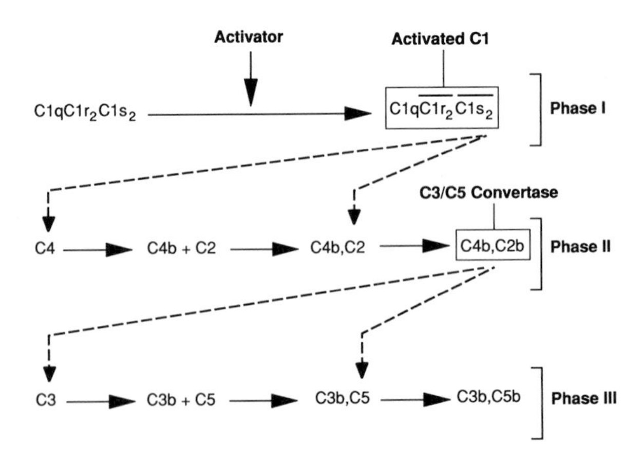

FIGURE 2 Activation of the classical pathway. Two protein-splitting enzymes are generated: activated C1 and C3/C5 convertase. The *arrows with dashed lines* indicate the substrates of these two enzymes: activated C1 splits both C4 and C2, and C3/C5 convertase splits both C3 and C5.

A. Nomenclature of the Classical Pathway

The nomenclature of the components and of their activated products has evolved as the molecular events were elucidated.

1. The native inactive components are named C1, C2, C3, C4, and C5.
2. Activation of C1 and its subcomponents C1r and C1s is signified by an overline (i.e., $\overline{\text{C1}}$, $\overline{\text{C1r}}$, $\overline{\text{C1s}}$). Activation of C2, C3, C4, and C5 occurs via proteolysis, with the resulting fragments being named by the suffix "a" and "b." In each case, the larger fragment is b (i.e., C2b, C3b, C4b, and C5b); these fragments interact with target membranes.
3. C3b and C4b are also further degraded into "c" and "d" fragments. In each case, the d fragment remains attached to target membranes, whereas the c fragment is released into the fluid phase.

B. Overall View of the Molecular Events

The classical cascade is divided into three phases: (1) the formation of activated C1, (2) the formation of the C3/C5 convertase, and (3) the splitting of C3 and C5 to their active forms (Fig. 2). Activated C1 and the C3/C5 convertase are the only important enzymes in the classical cascade.

1. Phase I: Formation of Activated C1

The first component, C1, contains three subcomponents: C1q, whose function is the binding of C1 to immune complexes, and C1r and C1s, which are proenzymes. Phase I (Fig. 2) consists of the binding of C1 via C1q to antibody on the target surface. The binding of C1 is followed by the autocatalytic conversion of C1r to an active protease, which then converts C1s to a similar active enzyme. $\overline{\text{C1s}}$ is the active enzyme used in phase II.

Activation of C1 occurs when a surface contains either closely paired IgG antibody doublets or IgM antibody. IgA, IgE, and IgD do not bind C1q and do not activate the classical pathway.

2. Phase II: Formation of the C3/C5 Convertase

The product of phase II is the C3/C5 convertase. The sequence of events is (1) antibody-bound $\overline{\text{C1s}}$ activated in the first phase cleaves and activates plasma C4; (2) activated C4 molecules (C4b) diffuse to the target surface and become attached close to $\overline{\text{C1}}$; and (3) C2 combines with the bound C4b, and this C2 in turn is also cleaved and activated by the neighboring $\overline{\text{C1s}}$). Phase II thus ends with the formation of a cell-bound C4b,C2b complex (the C3/C5 convertase), which has specificity for splitting both C3 and C5 (Fig. 2).

3. Phase III: Splitting of C3 and C5

This phase has two functions: (1) attachment of large numbers of C3 molecules to the target surface to opsonize the particle for phagocytosis, and (2) cleavage of C5 to initiate the assembly of the membrane attack complex. The sequence is as follows: (1) bound C3/C5 convertase (C4b,C2b) cleaves plasma C3; (2) activated C3 (C3b) attaches ("fixes") close to the C4b,C2b enzyme; and (3) C5 combines with the attached C3b, and as a result of this combination, a modification takes place in the C5 so that it is also susceptible to cleavage by the neighboring C4b,C2b complex. The C4b,C2b enzyme is thus used for the cleavage of both C3 and C5.

Phase III completes the events that generate activated C5 on the target surface. The action of the C3/C5 convertase on C5 is the last enzymatic event; the formation of the membrane attack complex proceeds via the polymerization of C5 to C9 initiated by activated C5.

C. Covalent Fixation of C4 and C3

Cleavage of C4 and C3 at their N-termini results in disruption of internal thioester bonds, producing C4b and C3b fragments that can covalently bind to cell surface sugars or proteins via ester or amide bonds, respectively (Figs. 3 and 4).

D. Mechanisms Confining Classical Pathway Activation to Target Membranes

Three mechanisms restrict the classical pathway to target membranes: (1) Activation of C1 requires binding of specific antibody to an antigen surface; (2) only C2 that has been modulated by combination with surface-bound C4b can form the C3/C5 convertase after cleavage by $\overline{\text{C1s}}$. Similarly, C5 can only be cleaved by the C3/C5 convertase after it has been modulated by binding to surface-bound C3b. (3) The extremely short life of the C3b and C4b combining sites restricts attachment of these molecules to a circular area of 40 nm in radius, centered on the activating C4b,C2b complex.

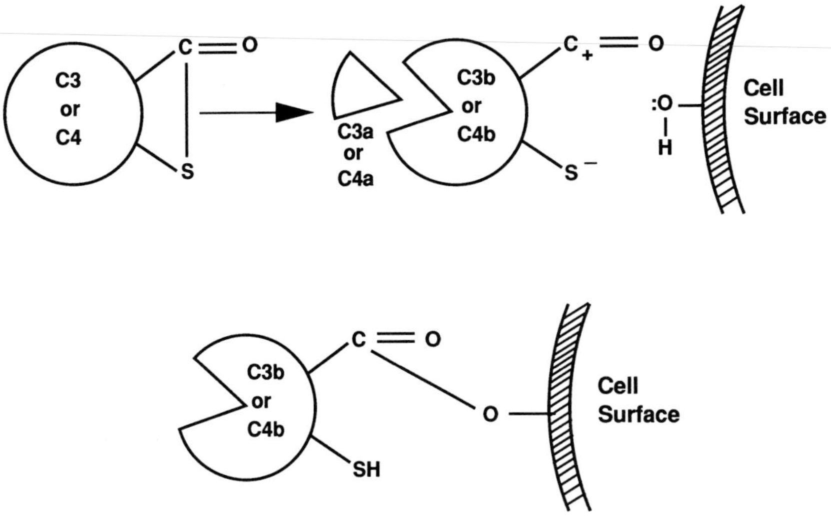

FIGURE 3 Activation and deposition of C3 and C4.

E. Regulatory Factors

Apart from confining activation to the target, regulatory proteins bring about the rapid destruction of the activated factors at each stage of the classical pathway. These regulatory proteins are required to prevent the complete consumption of plasma C4 and C2 in the fluid phase.

1. Control of C1 Activation

The rapid inactivation of the C1 enzyme to prevent uncontrolled activation of complement is brought about by C1 inhibitor (C1-INH). C1-INH functions by combining irreversibly with the active sites on both $\overline{C1r}$ and $\overline{C1s}$.

2. Inactivation of C3/C5 Convertase

The C3/C5 convertase (C4b,C2b enzyme) is inactivated in two stages. First, functional activity is lost by the spontaneous dissociation of C2b from the bound C4b. Two factors promote C2b dissociation (i.e., C4-binding protein (C4BP) and decay-accelerating factor (DAF). C4BP is a plasma protein that combines with C4b, preventing further association with C2b. DAF is a protein present within the membranes of many different host cell types, which appears to have a similar action to C4BP in bringing about a functional dissociation of C2b from C4b. The second stage of C4b,C2b inactivation is the degradation of the C4b molecule into C4c, which dissociates from the target surface, leaving only the small C4d fragment attached. This degradation is brought about by the enzyme, factor I (I for "inactivator"). Two proteins are known to act as cofactors for factor I proteolysis of C4b: C4BP and the red cell complement receptor CR_1.

3. Inactivation of C3b

Factor I is also the normal enzyme that cleaves bound C3b. This inactivation is of the greatest importance in the control of the feedback loop of the alternative pathway (see Section III, M). It also plays a part in the control of the classical pathway, as once C3b is inactivated, it can no longer bind C5 and hence further production of the membrane attack complex is prevented. Cleavage of C3b by factor I can only occur in the presence of a cofactor, and either factor H or CR_1 can act in this respect. Factor H is a plasma protein that binds to the C3b molecule and acts as a cofactor for the enzyme activity of factor I. Factor I splits the α chain of C3b, inactivating the C3b molecule, so that is termed *iC3b*. In the case of fixed iC3b exposed to red cells, further cleavage by factor I takes place, but on this occasion the sole cofactor is red cell CR_1. This latter cleavage also occurs in the α chain, with the result that the bulk of the molecule dissociates from the target surface as C3c, leaving the small C3dg fragment attached to the red cell surface. Other proteases (plasmin, trypsin, or elastase) can split off the C3g fragment from C3dg, leaving the C3d fragment bound to the red cell surface (Fig. 4).

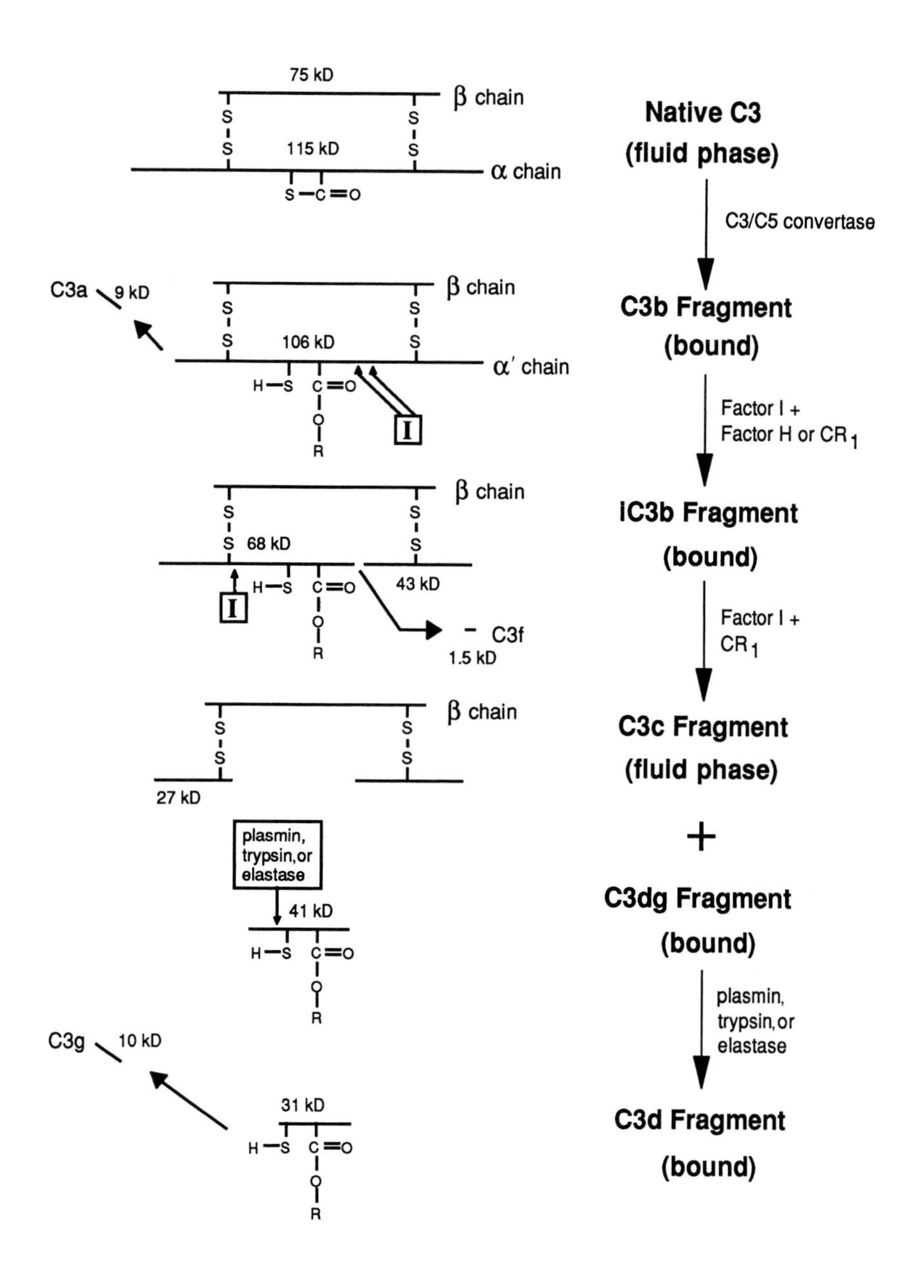

FIGURE 4 Structure of C3 and its physiologic fragments. Native C3 in plasma consists of two disulfide-linked α and β subunits. C3 convertase (C4b,C2b or C3b,Bb) proteolyses C3, splitting off the C3a fragment from the N-terminal of the α subunit, causing disruption of an internal thioester bond between a glutamate residue and a cysteine residue on the α subunit (indicated O═C—S). The broken thioester becomes a binding site capable to form a bond with any nearby hydroxyl group or amino group. C3b fragment then forms either an ester bond with exposed hydroxyls (such as those of sugar-containing bacterial cell walls) or an amide bond with protein amino groups. O═C—O—R indicates this site of ester linkage of C3b to an activating surface. The other end of the broken thioester, the cysteine residue, becomes a free sulfhydryl group (S—H in the figure). The C3b is cleaved rapidly by factor I at two closely spaced sites on the α subunit, releasing the small C3f peptide and causing a major conformational rearrangement of the C3 molecule that is now called iC3b. The sites of factor I proteolysis are shown with *arrows* emanating from the boxed letter I. The binding of fixed iC3b to the receptor CR_1 exposes a site in the α' subunit where factor I produces a third cleavage, releasing most of the fixed C3 molecule into the fluid phase as the C3c fragment and leaving only the small C3dg piece bound to the surface. The C3dg fragment is not broken down further in normal blood but is sensitive to a variety of serine proteases that may be present at inflammatory sites (e.g., trypsin, plasmin, or elastase). Proteolysis by one of these enzymes removes the C3g fragment and leaves only the C3d fragment bound to substrates.

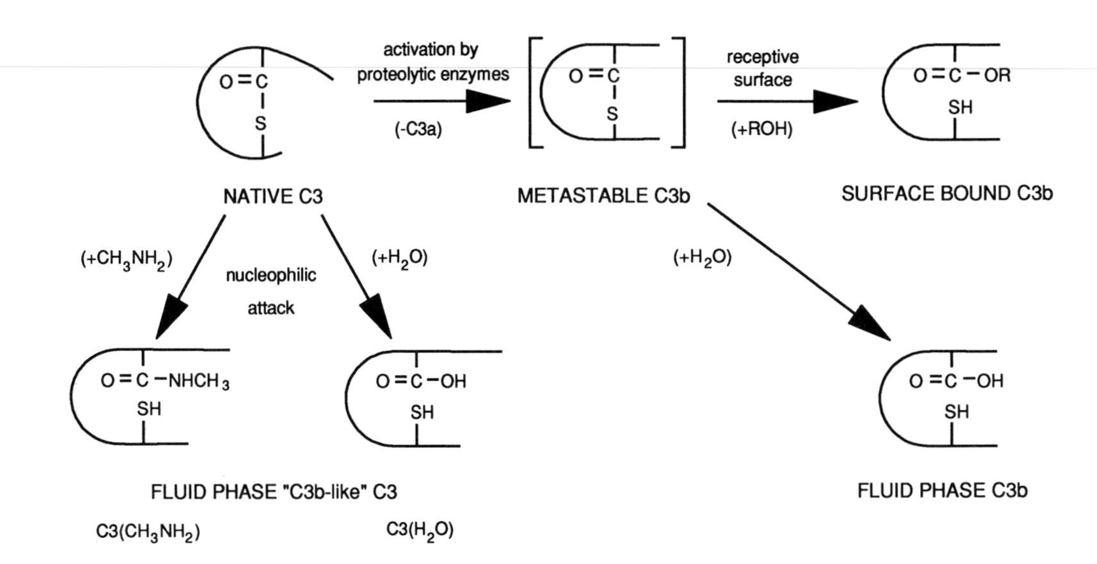

FIGURE 5 C3 activation or denaturation through disruption of its internal thioester bond.

III. Alternative Pathway of Complement Activation

The alternative pathway[2] provides a natural defense against infectious agents because it is capable of neutralizing a variety of potential pathogens in the absence of specific antibodies. The alternative pathway thus differs from the classical pathway in that it provides an immediately available line of defense that does not require prior immunization. It resembles the classical pathway in that both systems result in fixation of opsonizing C3 and initiation of membrane attack via the terminal pathway of complement. The six plasma proteins of the alternative pathway involved in activation perform a continuous surveillance function. They recognize a wide variety of potential pathogens within minutes after such organisms come in contact with plasma. Organisms sensitive to the alternative pathway include certain bacteria and fungi, a number of viruses, virus-infected cells, certain tumor cell lines, parasites such as trypanosomes, and erythrocytes from patients with paroxysmal nocturnal hemoglobinuria (PNH).

Activation of the pathway involves a unique amplification process, which produces the covalent attachment of large numbers of C3b molecules to the surface of the activating particle. The activation process is the result of a dynamic balance between the amplification process and regulation of this process. Specific regulatory components control this chain reaction–like process, allowing only minimal consumption of the native components.

2. This section on the alternative pathway of complement activation was excerpted with permission from Pangburn, M. K. (1986). The alternative pathway. *In* "Immunobiology of the Complement System. An Introduction for Research and Clinical Medicine" (G. D. Ross, ed.), pp. 45–62. Academic Press, Orlando, Florida.

A. Nomenclature

The five proteins unique to the alternative pathway are named factors B, D, H, I, and P (or properdin). C3 is an essential component of both pathways, and its numerical designation is retained in the alternative pathway. Two of the proteins (C3 and B) undergo proteolytic cleavage during activation and the fragments are assigned lower case letters: C3a, C3b, Ba, and Bb. Complexes formed by the noncovalent association of two or more proteins are written using a comma between the symbols (e.g., C3b,Bb).

B. C3

C3 plays a central role in the alternative pathway. Its activated forms (1) participate in initiation of the pathway in the fluid phase, (2) attach covalently in large numbers to biological particles during activation, (3) provide binding sites for C3 receptors on phagocytic cells, and (4) allow activation of C5, which leads to cytolysis. As in the classical pathway, cleavage of C3 produces C3b, which can bind to surfaces via the activated thioester site. Simulta-

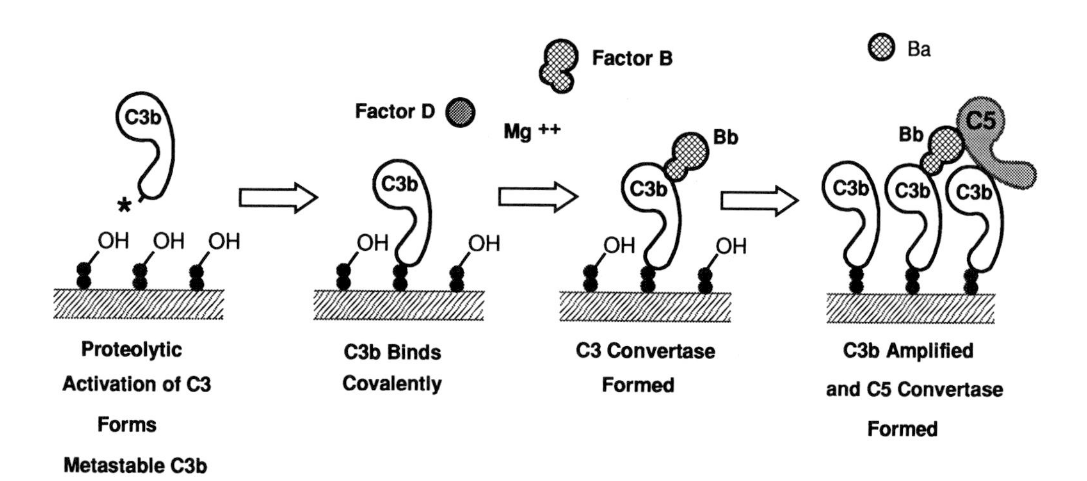

FIGURE 6 Initiation of the alternative pathway of complement activation.

neously, other binding sites appear on C3b for factors B, H, I, P, and C5.

Spontaneous low-level hydrolysis of the thioester bond in native C3 gives rise to C3(H_2O). The C3(H_2O) behaves as a C3b-like molecule with all properties of C3b except that it lacks a thioester binding site and thus is only found in the fluid phase (Fig. 5).

C3 and C3b perform a number of functions in the alternative pathway. C3 initiates the pathway through the spontaneous formation of C3(H_2O). In the classical pathway, many proteins (IgG, C1, C4b, C2b) bind to the activating particle surface before the attachment of C3b. In the alternative pathway, C3b is the first protein to attach to the target particle. Bound C3b participates in recognizing the particle as an activator or nonactivator and serves as a subunit of the cell-bound C3/C5 convertase on activators. Activation of C3 by the C3/C5 convertase deposits additional C3b molecules on the surface, simultaneously creating new convertase sites and allowing these enzymes to cleave C5 molecules that become attached to the C3b (Fig. 6).

C. Factor B

Factor B (or B) plays a key role in the alternative pathway. It binds to C3b and forms a proteolytic enzyme that cleaves and activates more C3. Factor B is cleaved only when it is bound to C3b, and this cleavage yields the Bb and Ba fragments. The Bb fragment expresses serine protease activity as long as it remains bound to C3b. Factor B is similar to C2 of the classical pathway. Both proteins form the catalytic subunit of C3 convertases, and both must be bound to be active. C2 is structurally similar to factor B, and the C2 and factor B genes are linked within the major histocompatibility locus where they are designated MHC Class III genes.

D. Factor D

Factor D (or D) is the enzyme that activates B to form the C3 convertase of the alternative pathway. D is a serine protease that circulates in active form. It is a highly specific enzyme that splits factor B only when B is bound to C3b. The action of D on the complex C3b,B releases the Ba fragment, leaving the C3 convertase C3b,Bb bound to the activating surface (Fig. 6).

E. Factor H

This protein is a regulator of alternative pathway activation. H binds to C3b and competes with the binding of B. H also binds to the C3b portion of C3 convertases (C3b,Bb) and accelerates dissociation of Bb from these complexes, thus inactivating them (Fig. 7). As in the classical pathway, H also competes with C5 for binding to C3b. These properties allow factor H to inactivate the enzymatic activity of the C3/C5 convertase and to regulate the use of C5 by the enzymes of both the classical and alternative pathways.

F. Factor I

Factor I (or I) not only cleaves and inactivates C3b, but also cleaves C3(H_2O). The role of factor I is to

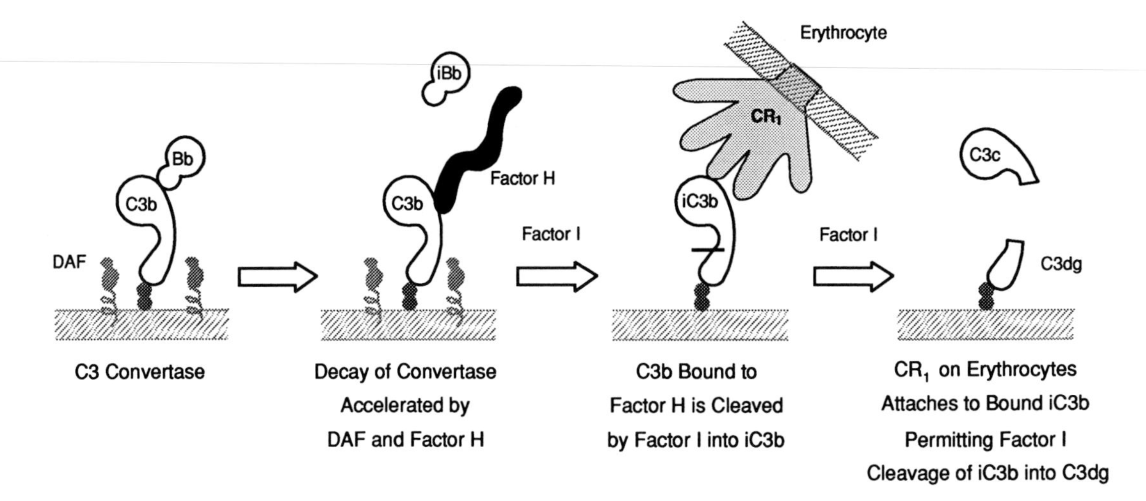

FIGURE 7 Regulation of the alternative pathway on normal tissue surfaces.

prevent formation of the C3 convertase by inactivating C3b permanently (Fig. 7). Failure to block convertase formation leads to consumption of C3 and B through a positive feedback process that is a unique feature of the alternative pathway.

G. Properdin

Properdin (or P) was the first component of the alternative pathway to be identified. Its function is to bind to the C3 convertase (C3b,Bb) and to increase the stability of the complex.

H. Decay-Accelerating Factor

DAF is an important control protein that is an intrinsic component of host cell membranes and functions to prevent assembly of the classical or alternative pathway C3 convertases on normal tissue. DAF is linked to membranes via a phosphatidyl inositol glycolipid, and its absence from red blood cells has been shown to be a major causative factor in the disease PNH.

I. Activation Process

Activation of the alternative pathway involves both reversible and irreversible interactions and can be divided into four phases: initiation, deposition of C3b, recognition, and amplification (Fig. 8). Activation begins in plasma with the formation of enzymes that cleave C3 and generate C3b. Through its metastable thioester site, C3b may attach randomly to nearby particles, to other proteins, or react with water. All these forms of C3b are rapidly inactivated unless the particle to which they are attached is recognized as an activator. Recognition involves bound C3b and factor H, as well as DAF. The exact process by which H distinguishes bound C3b on host tissue is not yet understood, but it results in H having a considerably higher affinity for C3b bound to host tissue as compared with C3b bound to bacteria. On the surface of activators, an amplification process rapidly deposits large numbers of C3b around the initial C3b (Fig. 8). This is followed by the activation of C5 and the membranolytic proteins.

J. Initiation

Initiation (Fig. 8) involves the spontaneous formation of the chemically and conformationally altered form of C3 [i.e., C3(H_2O)]. C3 altered by hydrolysis of the thioester bond, C3(H_2O), is formed continuously at a slow rate (0.005%/min) in aqueous solutions. This molecule has all the functional properties of C3b except that for a brief period after formation, the molecule is resistant to inactivation by H and I. C3(H_2O) forms a complex with factor B in the presence of Mg^{2+}. As illustrated in Fig. 8, the C3(H_2O),B complex is activated by factor D forming a fluid phase C3 convertase. C3(H_2O),Bb is the first enzyme of the pathway capable of generating C3b. The enzyme itself is confined to the fluid phase, but metastable C3b can diffuse as much as 300 Å to find and attach to nearby particle surfaces (abbreviated "S" in Fig. 8). It should be noted that this initiating process does not rely on specific initiators, but is spontaneous.

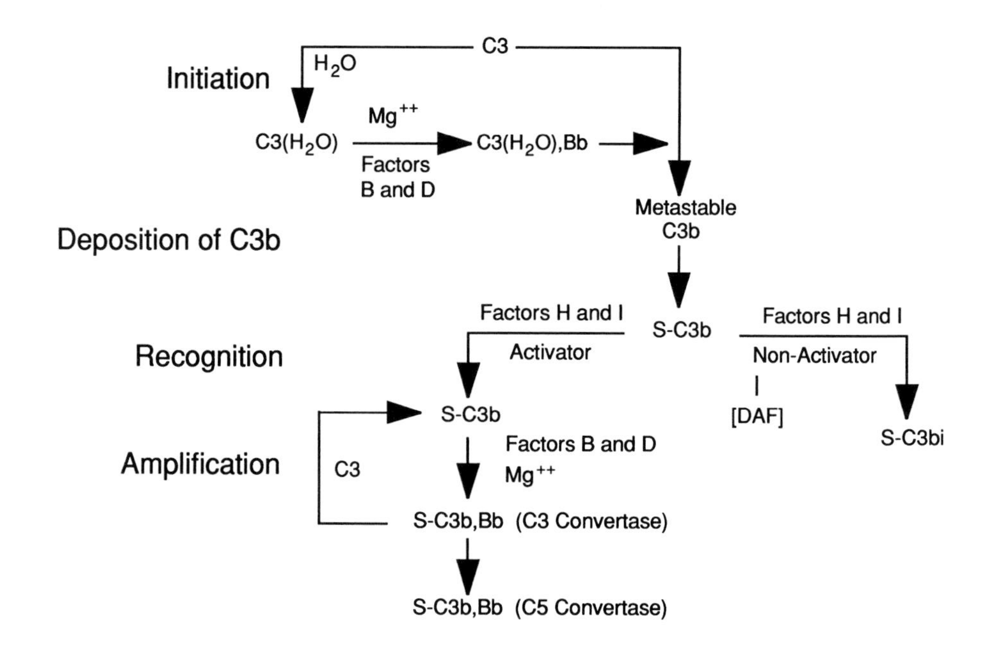

FIGURE 8 Summary of the activation and regulation steps of the alternative pathway of complement activation.

K. Deposition of the First C3b

The ability of the activated thioester in C3b to react with a wide variety of carbohydrates enables the pathway to deposit C3b onto a broad spectrum of organisms. This is important, because in the absence of specific antibodies, C3b may be the first host molecule to encounter and initiate a challenge to an invading pathogen. A unique feature of alternative pathway activation is that initial C3b attachment appears to be continuous and indiscriminate, occurring on host cells as well as on foreign particles.

L. Recognition

Discrimination between activators and nonactivators occurs soon after initial deposition of C3b. Discrimination is manifested by a reduction in the effectiveness of regulatory factors to control the amplification process when the initial C3b molecules are bound to activator surfaces (Fig. 8). Fluid phase C3b and C3b bound to host particles are rapidly inactivated by factors H and I. By contrast, when bound to activating particles, both C3b and C3 convertase are relatively protected from destruction by the fluid phase regulatory proteins. This appears to be determined by how effectively factor H can interact with activator-bound C3b, as well as by membrane DAF that functions at the host cell surface. C3b bound to activators exhibits a reduced affinity for factor H, whereas the binding of factors B, I, and properdin to C3b is unaffected by the type of particle to which the C3b is attached. This suggests that alternative pathway recognition resides in C3b or factor H or is expressed jointly by these two proteins at the surface of the particle. It is not yet clear what structures are recognized by the alternative pathway. Activators include many pure polysaccharides, lipopolysaccharides, certain immunoglobulins, viruses, fungi, bacteria, tumor cells, and parasites.

M. Amplification of Particle-Bound C3b

The C3b-dependent positive feedback process is a unique feature of the alternative pathway. C3b with B, D, and Mg^{2+} forms a C3 convertase capable of generating more C3b (Fig. 9). In the presence of B and D, each subsequent C3b has the potential of repeating this process. The initial C3b is thus amplified in number. H and I limit this process both in the fluid phase and on nonactivating particles. C3b deposited on an activator is relatively resistant to H and I, and the binding of B to C3b and its cleavage by D is unaffected by the nature of the surface to which the C3b is bound. The enzyme responsible for amplification is C3b,Bb or the properdin-stabilized form of this enzyme (see Section III, N). The enzyme C3b,Bb is labile, and spontaneous dissociation of Bb from the complex results in an irrevers-

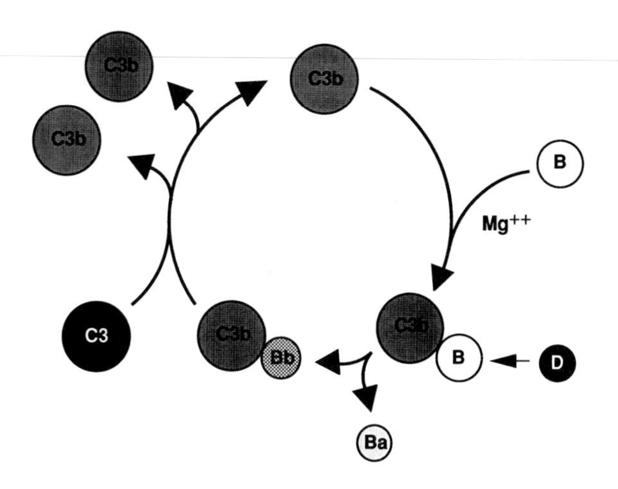

FIGURE 9 Positive feedback loop of the alternative pathway of complement activation.

ible loss of enzymatic activity. Factor H accelerates the dissociation of the C3b,Bb complex, resulting in a short half-life (3 min).

N. Role of Properdin

Properdin binds to the cell-bound alternative pathway C3 convertase, forming the trimolecular complex C3b,Bb,P. Binding of properdin stabilizes the C3b,Bb complex, slowing its dissociation. Both spontaneous and factor H–accelerated dissociation of the Bb subunit are slowed 5–10-fold.

O. Activation of C5 and the Terminal Pathway

The C3 convertase C3b,Bb can function as a C5 convertase provided that sufficient bound C3b molecules are present. The role of these C3b molecules is the same as in the classical pathway (i.e., to bind C5 and present it to the C3-convertase enzyme).

IV. Complement Cytotoxic Activity and the Terminal Components of Complement

Terminal complement protein activity[3] is directed toward the destruction of invading organisms. This occurs via the assembly on target membranes of five proteins (C5b, C6, C7, C8, C9) into the membrane attack complex (MAC). The MAC forms transmembrane channels, displaces lipids and other membrane constituents, and causes reorganization of lipids in the phospholipid bilayer. [*See* LIPIDS.]

A. Assembly of the MAC

1. Activation of C5

Cleavage of C5 by the classical or alternative pathway C3/C5 convertase liberates C5a, leaving the C5b fragment bound to the C3b unit of the C5 convertase. For a limited time, this C3b,C5b complex serves as the acceptor for C6.

2. Formation of the C5b–6 Complex

C5b binds stoichiometric amounts of C6. C5b–6 remains bound to the C3b subunit of the C5 convertase and serves as acceptor for C7 (Fig. 10).

3. Formation and Membrane Insertion of the C5b–7 Complex

The reactions described so far occur on the hydrophilic surface of membranes or particles, and the proteins involved retain their hydrophilic properties even after complex formation. Binding of C7 to C5b–6 causes an irreversible transition of the hydrophilic proteins to the amphiphilic (i.e., both hydrophobic and lipophilic) C5b–7 complex. In the complex of these three proteins (C5b,C6,C7), a site is exposed that is capable of binding to membranes. Insertion of C5b–7 into membranes bearing C3b is highly efficient and approaches 100%. if, however, the activating surface is not a phospholipid membrane, such as the surface of immune complexes or the carbohydrate of yeast cell walls, the C5b–7 has no substrate for hydrophobic insertion, and the complex is released into the fluid phase.

4. Binding of C8 and Formation of Small Transmembrane Channels

On binding of C8 to C5b–7, the C8 portion of the complex inserts itself into the hydrocarbon core of the membrane. Functionally, C5b–8 creates small membrane pores with an effective diameter of approximately 10 Å.

5. Binding of C9, C9 Polymerization, and Formation of Membrane Lesions

Interaction of C9 with C5b–8 complex causes polymerization of C9, forming "poly C9." Poly C9 is a hollow tubule formed by 12–18 molecules of C9

3. This section on complement cytotoxic activity and the terminal components of complement was excerpted with permission from Podack, E. R. (1986). Assembly and functions of the terminal components. *In* "Immunobiology of the Complement System. An Introduction for Research and Clinical Medicine" (G. D. Ross, ed.), pp. 115–137. Academic Press, Orlando, Florida.

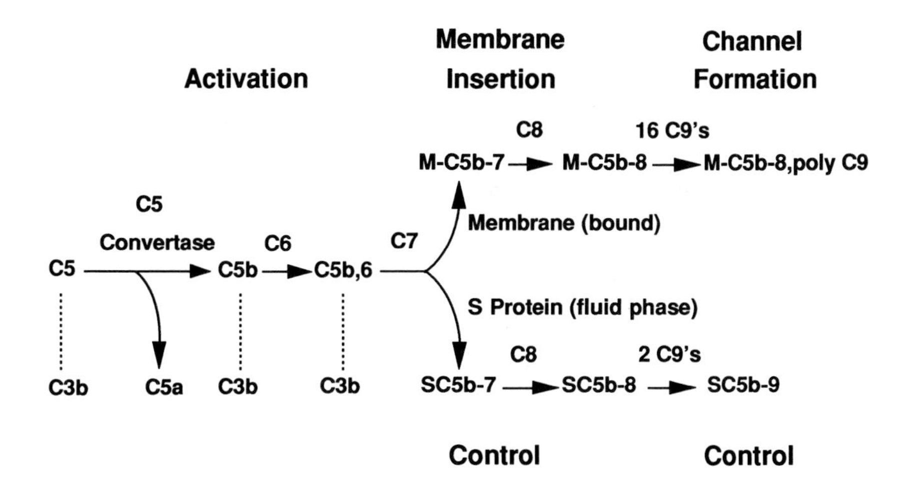

FIGURE 10 Activation of the terminal components and formation of the membrane attack complex.

(Fig. 11). This hollow tubular structure (tMAC in Fig. 11) is responsible for poly C9 cytolytic action. Incomplete C9 tubular structures (nontubular MAC or nt-MAC in Fig. 11) are also observed in electron microscopy of MAC-lysed cells and appear to consist of only 4–8 C9 molecules.

C5b–8 facilitates insertion of polymerizing C9 into membranes. The ultrastructure of the MAC corresponds to poly C9 with the exception that in MAC the C5b–8 complex is detectable as a long appendage attached to the torus of poly C9 (Figs. 10 and 11). The poly C9 is the channel through the membrane, and the subunits of C5b and C8 form the long appendage on the MAC-poly C9 torus.

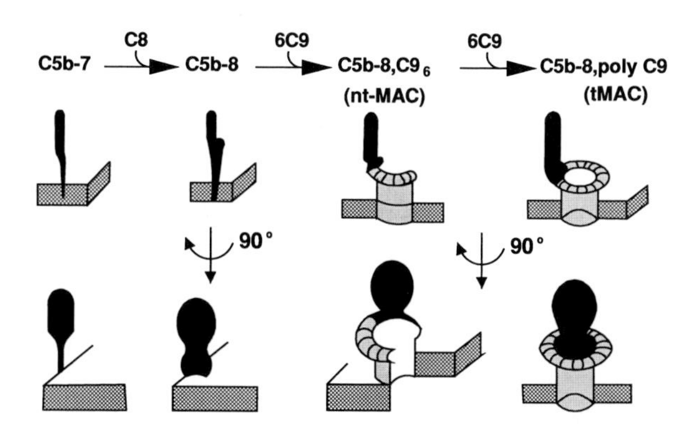

FIGURE 11 Appearance of the membrane attack complex at each stage of formation. These drawings were derived from photographs taken with an electron microscope. Components that make up each stage of the developing MAC complex are shown in the *top row*. In the *bottom row*, the appearance of the forming MAC complex is shown from a different angle produced by a 90° clockwise rotation of the MAC shown in the *middle row*. nt-Mac and t-Mac refer to nontubular and tubular MAC respectively.

B. Regulation of the MAC in the Fluid Phase and Host Cell Membranes

1. Control of the MAC by the S Protein

Transfer of the amphiphilic C5b–7 complex between cells is prevented by the plasma S protein and lipoproteins that form a complex with the fluid phase C5b–7. Two to three molecules of S protein bind to released but not to membrane-inserted C5b–7, giving rise to the SC5b–7 complex. This complex is fully water-soluble and hence has lost its capability for membrane insertion. The SC5b–7 complex still reacts with C8 and C9; C9 polymerization, however, is inhibited, and only two to three C9 molecules are incorporated into SC5b–9 (Fig. 10). S protein has been shown to be the same as vitronectin.

2. Protection of Host Cells by the Homologous Restriction Factor and CD59

The homologous restriction factor (HRF; also referred to as C8-binding protein) and CD59 are normal components of host cells that inhibit membrane insertion of the MAC. HRF shows homologous species specificity, such that HRF only protects human cells from human MAC; the HRF of nonhuman cells does not protect them from human MAC. Human cells are lysed efficiently by rabbit MAC, whereas rabbit cells are lysed efficiently by human MAC. HRF acts at the stage of C8 insertion into membranes, whereas CD59 acts somewhat earlier at the stage of insertion of the C5b–7 complex. Similar to DAF, both the HRF and CD59 are attached to membranes via a phosphatidyl inositol glycolipid and are deficient on red cells from patients with PNH. Red cells from patients with type III PNH are missing DAF, CD59, and HRF, leading to spontaneous hemolysis via the alternative pathway.

C. Functional Effects of the MAC on Target Membranes

1. Cell Death Caused by Formation of Transmembrane Pores

The MAC forms a pore through membranes of cells that may or may not penetrate through the entire membrane bilayer. MAC complexes that do penetrate the lipid bilayer cause leakage of small salt ions and the rapid uptake of water by the cell in an attempt to balance the higher osmotic pressure of the cellular cytoplasmic constituents that are too large to pass through the MAC pore. With red blood cells, this results in a rapid swelling of cells and hemolysis. Bacteria and nucleated cells may also be killed by the leakage of cellular salt ions without lysis.

2. Functional Effects of the MAC Independent of Pore Formation

The reorganization of lipid bilayers by MAC may adversely affect the stability of bilayer membranes. A second effect of MAC is caused by the displacement of membrane constituents by insertion of large numbers of MAC. Because MAC occupies a relatively large area in the membrane, the insertion (e.g., into bacterial membranes) increases the total membrane surface area by more than twofold, causing loss of structural integrity. This displacement of membrane constituents and the consequent physical alteration and surface expansion of attacked membranes may cause cell death independent of the effects caused by formation of pores.

3. Secondary Effects Contributing to Cell Death by the MAC

Pores created by the MAC in membranes of bacteria permit access to and degradation of the peptidoglycan layer of the bacterial cell wall by a lytic enzyme, lysozyme. Membrane pores also allow the entry of Ca^{2+} into the intracellular space, triggering a variety of cellular reactions. Pore formation is accompanied by the breakdown of the voltage difference at the two sides of the membrane (membrane potential) and by an efflux of K^+ and entry of Na^+. Compensatory ion pumping along with cell activation by Ca^{2+} entry may be responsible for the rapid depletion of ATP and high-energy phosphates observed in target cells. These effects may contribute to cell death. [*See* ION PUMPS.]

V. Inflammatory Function of Complement: The Anaphylatoxins

Anaphylatoxins[4] are small fragments derived from the complement components C3, C4, and C5 during complement activation. As the larger b fragments are generated that participate in the cascade reaction, the smaller a fragments formed simultaneously from the N-termini of each component's α chain (C3a, C4a, and C5a) have the function of causing inflammation. These small polypeptides share a common biological activity termed *spasmogenicity* (i.e., they promote smooth muscle contraction and induce increased vascular permeability).

One particular anaphylatoxin, C5a, plays a unique physiologic role as a mediator of the inflammatory responses. This glycopolypeptide differs from C3a and C4a in three important regards. First, C5a is a considerably more potent biological effector than C3a or C4a. Second, C5a, but not C3a or C4a, retains significant biological activity in serum. Third, C5a exerts a series of unique effects on human granulocytes (blood white cells) and thus promotes their participation in the inflammatory process. [*See* INFLAMMATION.]

A. Analysis of Anaphylatoxin Production as a Measure of Complement Activation

The extent of complement activation as well as the pathway responsible for the activation phenomena may be defined by quantitating anaphylatoxin production. Specific radioimmunoassay (RIA) procedures have been developed that permit selective quantitation of each anaphylatoxin (C3a, C4a, and C5a). Employing these assays, it has been shown that detection of elevated levels of C4a can be considered as diagnostic of classical pathway activation events. By contrast, the appearance of increased levels of either C3a or C5a, without evidence of increased C4a, is evidence for activation of the alternative pathway without activation of the classical pathway. Initial studies have suggested that the monitoring of C3a levels in patients with autoimmune disease or rheumatoid arthritis may be one of

4. This section on the inflammatory function of complement and the anaphylatoxins was excerpted with permission from Chenoweth, D. E. (1986). Complement mediators of inflammation. *In* "Immunobiology of the Complement System. An Introduction for Research and Clinical Medicine" (G. D. Ross, ed.), pp. 63–86. Academic Press, Orlando, Florida.

the best methods of assessing disease activity status. [*See* RADIOIMMUNOASSAYS.]

B. Control Mechanisms

1. Enzymatic Inactivation of the Anaphylatoxins

Once anaphylatoxins are formed, their spasmogenic activities are rapidly abrogated by a normal plasma enzyme [i.e., carboxypeptidase N (SCPN)] that acts as an anaphylatoxin inactivator. This enzyme removes the COOH-terminal arginine from each of the anaphylatoxin molecules and converts them to their "des-Arg" derivatives (Fig. 12). SCPN rapidly destroys all the activity of the C3a and C4a anaphylatoxins. C5a is less affected because it is converted to its des-Arg derivative at a slower rate; in addition, ~5% of the C5a formed during complement activation is resistant to cleavage by SCPN. Moreover, although enzymatically degraded C3a and C4a are biologically inert, the degraded form of C5a (des-Arg-C5a) retains considerable activity as an inflammatory mediator.

2. Breakdown of C5a

A second type of unique control mechanism exists for regulation of C5a activity. Both C5a and des-Arg-C5a bind to specific receptors on peripheral blood granulocytes. Once bound, the C5a is rapidly internalized by these cells and completely degraded with loss of activity. Thus, the cells that are activated by C5a during the inflammatory response play the major role in inactivating the molecule.

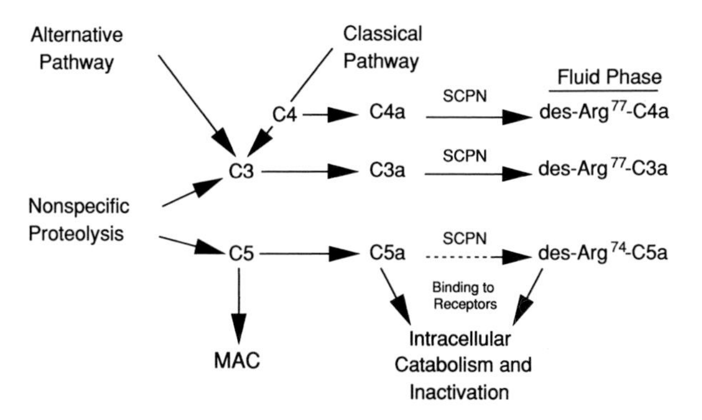

FIGURE 12 Generation and control of anaphylatoxins from the complement system. SCPN (serum carboxypeptidase N) is a plasma enzyme inhibitor of the anaphylatoxins that cleaves off the C-terminal arginine from C3a, C4a, and C5a.

C. Biological Activities of the Anaphylatoxins

1. Spasmogenic Properties

The spasmogenic activities normally ascribed to the anaphylatoxins include the ability to induce smooth muscle contraction, promote increased vascular permeability, and cause release of histamine from mast cells and basophils. Des-Arg-C5a also exhibits a lower level of spasmogenic activity.

The ability to induce smooth muscle contraction is usually defined by measuring the contraction of guinea pig ileal or uterine tissues, which are rich in smooth muscle. Increased vascular permeability may be assessed after intradermal injection of the anaphylatoxins into guinea pigs previously perfused with Evan's blue dye, which produces skin bluing at the site of injection, or into humans, where it causes a wheal and flair response. [*See* SMOOTH MUSCLE.]

The spasmogenic properties of the anaphylatoxins may be manifest during the initial phases of the acute inflammatory response, when increased vascular permeability and tissue edema are readily apparent.

2. Granulocyte-Related Activities of Human C5a

C5a is the most potent complement-derived mediator of granulocyte responses thought to be critical to the inflammatory process. The main cellular responses include (1) chemotactic migration; (2) augmented adherence to cells; (3) degranulation (i.e., the release of internal granules full of lytic enzymes); and (4) production of toxic oxygen derivatives. These properties of C5a, rather than spasmogenic activity, account for this molecule's importance as an inflammatory mediator. In addition, both C5a and des-Arg-C5a promote the chemotactic migration of neutrophils (a type of granulocyte) and monocytes (another type of white blood cell). With neutrophils, C5a expresses measurable chemotactic activity at low concentrations (10^{-10} M). Des-Arg-C5a is 10–50-fold less active than C5a.

C5a and des-Arg-C5a also augment adherence and/or aggregation of neutrophils and monocytes. These phenomena are manifest as a profound transient loss of granulocytes from blood and accumulation in the lung vessels.

Conceptually, production of even extremely low quantities (picomolar) of C5a at a localized site could promote adherence of granulocytes to the endothelium, induce their chemotactic migration into the site, and prime them to destroy the eliciting agent. All these events are observed in inflammatory foci and are important for host defense. However, both C5a and des-Arg-C5a may also act throughout the body and trigger similar types of granulocyte responses in blood or distant organs. When this occurs, C5a-activated granulocytes release cytotoxic substances that destroy normal tissues. In this case, normal host defense mechanisms may actually contribute to the causation of specific diseases (e.g., the adult respiratory distress syndrome or rheumatoid arthritis).

VI. Opsonization and Membrane Complement Receptors

Opsonization[5] is the process by which particles are made readily ingestible by phagocytic cells. Serum proteins (opsonins) coat particles and cause them to bind avidly to phagocytes and trigger ingestion. The complement system plays a major role in opsonization by coating particles such as bacteria with C3. The bacteria then bind to the C3 receptors at the phagocyte surface, with clearance of the bacteria. Viruses, soluble antigen–antibody complexes, and tumor cells are opsonized and removed by a similar mechanism. Other serum proteins, particularly IgG antibacterial antibody and fibronectin, may also opsonize bacteria. For each type of opsonin, phagocytes have an opsonin-specific membrane receptor responsible for binding particles coated with that opsonin. In the blood, C3-coated particles and immune complexes are bound first to red blood cells, which transport the bound complexes to macrophages in the liver. The complexes are stripped from the surface of the red cells, which then return to the circulation. At sites of infection, C activation generates C5a, which attracts phagocytic cells via C5a receptors. Once at the site of C activation, phagocytes use receptors for fragments of C3 to bind to bacteria or soluble complexes. This facilitates phagocytosis and the release of bactericidal substances and other mediators of inflammation. [*See* ANTIBODY–ANTIGEN COMPLEXES: BIOLOGICAL CONSEQUENCES.]

5. This section on opsonization and membrane complement receptors was excerpted with permission from Ross, G. D. (1986). Opsonization and membrane complement receptors. *In* "Immunobiology of the Complement System. An Introduction for Research and Clinical Medicine" (G. D. Ross, ed.), pp. 87–114. Academic Press, Orlando, Florida.

Nine types of C receptors are believed to exist, and structural data are now available on all except one.

A. Opsonization by Complement

Fixed C3 and IgG antibodies are the most important opsonins. On particles that activate the classical but not the alternative pathway, fixed C3b is not protected from the control proteins of the alternative pathway (factors H and I) and therefore is rapidly converted into fixed iC3b. This fixed iC3b is an important opsonin, recognized by three types of phagocyte receptors (CR_1, CR_3, and CR_4). Fixed iC3b that binds to CR_1 is subsequently broken down into fixed C3dg by factor I; if plasmin or leukocyte elastase is present, the fixed C3dg may be then cleaved to fixed C3d (see Fig. 4). Fixed C3dg and C3d are poor opsonins *in vitro,* and in patients with cold agglutinin disease, red cells may circulate with as many as 20,000 fixed C3dg molecules per cell without being being cleared by the macrophage phagocytic system.

B. Membrane Complement Receptors

Membrane C receptors of phagocytic cells have several important functions in mediating chemotaxis, phagocytosis, and release of cytotoxic and inflammatory substances. The functions of C receptors on nonphagocytic cell types, particularly lymphocytes, are less well-defined.

The receptors for fixed C3 fragments have overlapping C3 fragment specificities and have been named according to their order of discovery rather than for their specificity. The reader should refer to the diagram of C3 fragment structure (see Fig. 4). Table I below lists the major features of the four types of receptors for fixed C3 fragments.

1. Complement Receptor Type 1

Complement receptor type 1 (CR_1) binds to fixed C3b and with lower affinity to fixed C4b and iC3b. Proteolysis of iC3b into C3dg destroys CR_1 activity. Several types of cells express CR_1, including red blood cells [erythrocytes (E)], phagocytic cells, lymphocytes (B cells and some T cells), and kidney podocytes. CR_1 along with CR_3 and CR_4 (see Section VI,B,3,4) are the major opsonin receptors on phagocytic cells that promote the ingestion and killing of microorganisms. The CR_1 of erythrocytes (E CR_1) has two important functions in the clearance

TABLE I Receptors for Bound Fragments of C3

Type	Specificity	Structure	Cell-type expression
CR_1	C3b > C4b > iC3b	Four allotypes: A = 190 kDa B = 220 kDa C = 250 kDa D = 160 kDa	Erythrocytes—very low B cells—high; T cells—low Neutrophils, monocytes—high Kidney podocytes—high Macrophages—very low
CR_2	iC3b = C3dg > C3d > C3b Epstein-Barr virus	140 kDa	B cells—high Follicular dendritic cells—high Pharyngeal epithelial cells—low Thymocytes—very low
CR_3	iC3b, C3dg, LPS,[a] β-glucan, collagen, fibrinogen	165 kDa α chain 95 kDa β chain	Neutrophils—high Monocytes—high Macrophages—low
CR_4	iC3b, C3dg	150 kDa α chain 95 kDa β chain	Neutrophils—very low Monocytes—very low Macrophages—high

[a] Lipopolysaccharide or endotoxin.

of circulating immune complexes. First, circulating immune complexes and particles that activate C are bound rapidly to E via CR_1. Erythrocytes then serve as vehicles that transport the complexes to liver and splenic macrophages. Second, E CR_1 serve as the cofactor for factor I cleavage of fixed iC3b (or fixed iC4b) into fixed C3dg (or fixed C4d) and fluid phase C3c (or C4c). This is an important function in immune complex clearance, as fixed iC3b that is not degraded by this mechanism can attach immune complexes to neutrophils (via CR_1 or CR_3), triggering activation of neutrophils and release of cytotoxic enzymes, leukotrienes, and toxic oxygen metabolites. Although this is an important mechanism for recognition and destruction of bacteria with fixed iC3b on their surface, it can also be turned against the host and result in neutrophil-mediated tissue damage if the fixed iC3b is present on small immune complexes that become trapped in normal tissues [e.g., it is an important mechanism of autoimmune diseases such as systemic lupus erythematosus (SLE)]. [*See* NEUTROPHILS.]

The function of B-cell CR_1 has not been completely defined, but several studies have shown that its triggering can enhance activation of the B cells stimulated by other factors. CR_1 also aids in antigen recognition by attaching C3b- or iC3b-bearing antigens onto the B-cell surface. No data are available on possible functions of T-cell CR_1. Only small subsets of both helper and suppressor cells express CR_1, and the amount of CR_1 expressed per cell is approximately 10% of that expressed by B cells.

CR_1 exists in four forms that vary in molecular weight from 160 kDa to 250 kDa. These differences result from the presence of variable numbers of homologous 60–70-amino-acid repeating sequence motifs [''short consensus repeats'' (SCR)] that make the extracellular domain. Homologous SCR structures are found in six other C3/C4-binding proteins (i.e., CR_2, factor B, factor H, C4BP, C2, and DAF). The genes for factor H, C4BP, CR_2, and DAF have been shown to map to a region on chromosome 1 that has been named the regulator of complement activation (RCA) gene locus.

2. Complement Receptor Type 2

Complement receptor type 2 (CR_2) is expressed by B lymphocytes, lymph node follicular dendritic cells, pharyngeal epithelial cells, and thymocytes but is absent from mature T lymphocytes. CR_2 is not only a receptor for C3 fragments but also has a separate attachment site for Epstein-Barr virus (EBV). CR_2 is specific for the C3d portion of iC3b, C3dg, and C3d. The virus or large complexes containing these C3 fragments bind to CR2 and cause activation of B cells. EBV also uses CR_2 to gain entry into B cells, thereby causing either a virus infection (the disease infectious mononucleosis) or a rare form of cancer (Burkitt's lymphoma).

CR_2 consists of a single polypeptide chain of 140 kDa M_r whose extracellular domain is made up of 15–16 SCR units, with a high degree of homology to the SCR units of CR_1. The CR_2 gene is linked to the CR_1 gene.

3. Complement Receptor Type 3

Complement receptor type 3 (CR_3) is a major opsonin receptor on neutrophils and monocyte/macrophages involved in the clearance of bacteria. CR_3 binds with high affinity to fixed iC3b and with much lower affinity to fixed C3dg. CR_3 also contains two additional and separate binding sites for lipopolysaccharide (LPS), a component of bacterial surfaces, and for yeast cell wall β-glucans. These secondary binding sites appear to be of major importance in granulocyte recognition of bacteria and yeast cell walls. In addition, CR_3 functions, along with CR_4, (see Section VI, B, 4) to mediate neutrophil adherence to endothelial cells during inflammatory reactions. Under certain conditions *in vitro,* CR_3 may also express binding sites for collagen and fibrinogen, but it is unclear how these sites may be involved in cell matrix adherence reactions *in vivo*. Neutrophils stimulated by C5a express greatly increased amounts of CR_3 and CR_4, which attach the neutrophils to the vascular endothelium, allowing the neutrophils to become adherent and migrate out of blood vessels to sites of infection. Once at the site of infection, CR_3 allows attachment of the phagocytes to bacteria and yeast that bear fixed iC3b and/or β-glucans or LPS, promoting phagocytosis and a respiratory burst. Patients with a rare inherited deficiency of CR_3 and CR_4 have repeated life-threatening bacterial infections.

CR_3 consists of two glycoprotein chains known as α (165 kDa M_r, the CD11b antigen) and β (95 kDa M_r, the CD18 antigen) that are noncovalently associated.

4. Complement Receptor Type 4

Complement receptor type 4 (CR_4) is closely related to CR_3, sharing the same β chain, and with 87% homologous α chain. CR_4 appears to have the same C3 fragment specificity as CR_3 but is expressed preferentially on tissue macrophages. CR_3 and CR_4 molecules expressed by phagocytic cells differ in their linkage to the cytoskeleton. CR_3 is less restrained than CR_4 by cytoskeletal connections and consequently is more mobile than CR_4. As a result, relatively small amounts of CR_3 can aggregate rapidly at the site of particle contact, promoting particle attachment to the cell. Cytoskeleton-linked CR_4 initiates ingestion of particles that are trapped on the membrane via CR_3.

5. C5a Receptor

The C5a receptor of neutrophils, a glycoprotein of 40–47 kDa M_r, is responsible for mediating inflammatory responses to C5a and des-Arg-C5a.

VII. Human Diseases in Which Complement Is a Significant Factor

A. Autoimmune Diseases

Autoimmune diseases frequently involve the development of autoantibodies to normal tissue or blood cells. The immune complexes generated activate the classical pathway, causing major organ damage from complement-mediated cytotoxicity and complement-mediated recruitment of inflammatory cells, which release damaging cytotoxic substances. The major diseases in this class include autoimmune hemolytic anemia, SLE, and rheumatoid arthritis. [*See* AUTOIMMUNE DISEASE.]

B. Inherited Deficiencies of the Complement System

These diseases are rare, and many physicians may never see one in their clinical practices (Table II). Patients who have a genetic deficiency of C3 are unable to use any complement pathway or function in host defense and usually have a long history of repeated and life-threatening bacterial infections.

TABLE II Genetic Deficiency Diseases of the Complement System

Deficient component	Clinical characteristics
C1q	SLE, nephritis, recurrent infections, hypogammaglobulinemia, high mortality rate
C1r or C1s	SLE, recurrent infections, arthritis
C1-INH	Hereditary angioedema (HAE)
C4 or C2	SLE
C3, factor H, or factor I	Recurrent infections with pygenic bacteria
Factor D or properdin	Recurrent bacterial infections
C5, C6, C7, or C8	Disseminated gonococcal or meningococcal (neisseria bacteria) infections; SLE
C9	No apparent disease
CR_3 and CR_4	Recurrent bacterial infections of the skin and gingiva

By contrast, patients with deficiencies of one of the early classical pathway components (C1q, C4, or C2) have an illness resembling SLE, resulting from a diminished ability to clear circulating antigen–antibody (immune) complexes. In these patients, the normal immune complex clearance mechanism is absent because the classical pathway does not progress to the C3 stage at which immune complexes are normally attached to erythrocyte CR_1 and transported to the macrophage phagocytic system. Patients with deficiencies of one of the control proteins of the alternative pathway (factor H or factor I) are unable to control normal spontaneous activation of the alternative pathway. Continuous consumption of complement in these patients results in low levels C3, factor B, and all the terminal components. Such patients thus have an acquired C3 deficiency because their C3 is continuously consumed by the uncontrolled alternative pathway. As might be expected, they have the same types and frequencies of infections as patients with a primary genetic deficiency of C3. Patients with a deficiency of one of the terminal components (C5, C6, C7, or C8) have recurrent systemic infections with *Neisseria* bacteria but do not have problems with other types of bacterial infections. Apparently in patients with neisserial infection, the cytotoxic function of complement in host defense cannot be replaced by any other function.

C. Deficiency of C1 Inhibitor; Hereditary Angioedema

The most common genetic deficiency of the complement system among Caucasians is deficiency of C1-INH, which results in the disease called hereditary angioedema (HAE). Absence of C1-INH results in consumption of plasma C4 and C2 because any spontaneous activation of C1 in the fluid phase is uncontrolled. Cleavage of C4 and C2 in the fluid phase does not result in C4b2 complex formation and thus does not generate a C3 convertase. As a result, only C4 and C2 are consumed without C3 consumption. A diagnosis of HAE is suggested if patients have low levels of C4 but normal levels of C3, because all other disease processes that consume C4 also consume C3. Because patients with HAE have an acquired deficiency of C4 and C2, many develop a SLE-like illness resembling patients within inherited deficiency of C4 or C2.

Patients with HAE have occasional attacks of severe swelling in the epiglottis, with danger of asphyxiation, or in the extremities. Attacks may be years apart, the first one occurring sometimes as late as 30 years of age. Most patients have low levels (20–50% of normal) of a functionally active C1-INH, whereas 20% have normal levels of a functionally inactive C1-INH. Both types of patients have been treated successfully with androgens, which increase plasma C1-INH levels. The disease is autosomal dominant, so that patients have one normal and one abnormal C1-INH gene, and androgens are thought to increase synthesis of C1-INH by the normal gene. Treatment with normal C1-INH protein may become possible as a result of the cloning of C1-INH cDNA.

Bibliography

Campbell, R. D., Law, S. K. A., Reid, K. B. M., and Sim, R. B. (1988). Structure, organization, and regulation of the complement genes. *Annu. Rev. Immunol.* **6,** 161–195.

Esser, A. F., and Sodetz, J. M. (1988). Membrane attack complex proteins C5b–6, C7, C8, and C9 of human complement. *Methods Enzymol.* **162,** 551–578.

Gigli, I., and Tausk, F. A. (1988). C1, C4, and C2 components of the classical pathway of complement and regulatory proteins. *Methods Enzymol.* **162,** 626–638.

Janatova, J. (1988). C3, C5 components and C3a, C4a, and C5a fragments of the complement system. *Methods Enzymol.* **162,** 579–625.

Müller-Eberhard, H. J. (1988). Molecular organization and function of the complement system. *Annu. Rev. Biochem.* **57,** 321–347.

Pangburn, M. K. (1988). Alternative pathway of complement. *Methods Enzymol.* **162,** 639–652.

Ross, G. D. (ed.) (1986). "Immunobiology of the Complement System: An Introduction for Research and Clinical Medicine." Academic Press, Orlando, Florida.

Ross, G. D., and Medof, M. E. (1985). Membrane complement receptors specific for bound fragments of C3. *Adv. Immunol.* **37,** 217–267.

if this, then that.

Conditional Reasoning, Development

DAVID P. O'BRIEN, *Baruch College of the City University of New York*

Glossary

Inference schema Logical rule that specifies the propositions that can be derived from premises of a particular form

Modus ponens Inference schema inferring the consequent of a conditional when that conditional and its antecedent are true

Modus tollens Inference schema inferring the falsity of the antecedent of a conditional proposition when that conditional is true and its consequent false

Reductio ad absurdum Argument inferring the falsity of a proposition when the supposition of that proposition taken together with other propositions assumed true leads to a contradiction.

Schema for conditional proof Inference schema inferring a conditional proposition when its consequent is derived from the supposition of its antecedent taken together with other propositions assumed true.

Supposition Proposition that is treated as true, even though it may not be, in order to see what follows from it

CONDITIONAL REASONING occurs when someone supposes an event or state of affairs and infers outcomes under the supposition. Such a line of reasoning can be used to make a conditional assertion, e.g., of the form *if p then q,* to reason from alternative possibilities, or to disprove a supposition. Arthur Conan Doyle's Sherlock Holmes presents a paradigm of deductive reasoning, and Holmes's skills include sophisticated conditional reasoning. Consider the well-known case of the dog that failed to bark, in which Holmes explains to Dr. Watson that the murderer could not have been a stranger. The murderer had entered the barn at night, and there was a watchdog in the barn. If the murderer had been a stranger, the dog would have barked. However, the dog never barked; hence, the murderer was not a stranger.

Conditional reasoning is not limited to highly trained reasoners such as Sherlock Holmes; rather, reasoning under a supposition is as characteristic of human reasoning as Holmes believed barking at a stranger to be for a watchdog. Consider an ordinary problem-solving situation: Carol's motorcycle won't start. She knows the problem is neither the battery nor the starter because the engine turns over. "Maybe it's simply out of gas," she thinks, "and if this is the problem, I can add some gas to the tank and it will start." So Carol adds some gas, yet the motorcycle still fails to start. So she concludes that the problem is not caused by an empty gas tank, and she starts to look at the ignition system. Note that the logic of Carol's reasoning is similar to Holmes's. Both begin by adding a supposition to a set of assumptions and thinking through what then follows, both assert a conditional, and both use the discovery of a contradiction under the supposition to conclude that the supposition must be wrong.

I. What a Model of Conditional Reasoning Should Accomplish

A model of conditional reasoning should describe the logic that allows for arguments such as those of Holmes and Carol. However, in addition to logic abilities, some of the abilities required for conditional reasoning are extralogical, i.e., beyond the scope of logic. Holmes uses his knowledge about the characteristic behavior of watchdogs, and Carol uses her knowledge about how internal-combustion engines work: No system of logic can account for inferences based on world knowledge. Because younger children have less knowledge of the world, they are not able to make some inferences that are available to older children or adults; however, such limitations do not pertain to their logical abilities. Also outside the scope of logic is the pragmatic nature of ordinary reasoning: problems require solutions; goals or ends need to be met and things accomplished. Holmes wants to solve a crime, and Carol wants her motorcycle to run. Such pragmatic concerns help reasoners decide which facts and which possible suppositions are of interest. Also extralogical are the abilities needed to keep track of the information in an argument; human working memory is known to be fairly small, and for all but the simplest lines of thought, some skills in using one's memory are required. Researchers in child memory have found that young children typically demonstrate few such skills. Thus, there are apt to be age differences in which suppositional arguments are produced or comprehended that do not result from any differences in logical abilities.

Although these extralogical skills all need to be addressed by a more general theory of reasoning, and a real-time model for conditional reasoning would need to include such skills, they are beyond the scope of the present article. The present account is concerned with those logic abilities and reasoning strategies that pertain specifically to suppositions. A model of conditional reasoning should account for the logic in arguments such as those made by Holmes and Carol, and an account of the development of conditional reasoning should address when and how such logic abilities are available.

II. Logic of Conditionals

A. Truth and Inference Schemas

The logic used in ordinary human reasoning need not be identical to what is found in standard-logic textbooks, but standard logic is a good place to find direction in knowing how such abilities can be described. Logic is concerned with the truth or falsity of propositions and with the validity of inferences. It has been argued that truth is a cognitive primitive that makes language and reasoning possible, and even 1-year-olds make judgments of truth. All assertions require tacit judgments about truth, and interpreting an assertion is judging what would make it a true proposition. Truth and falsity are two parts of a single concept, and you can't have one without the other. Hence, making or interpreting an asserting presupposes an understanding of both truth and falsity and the knowledge that an assertion cannot be both true and false. There are no empirical grounds for thinking this thesis wrong, and we should begin by granting that a notion of truth need not be learned, but is a part of the basic abilities from which language and thought arise. Note that the thesis just presented does not require that young children know that they are making judgments of truth, any more than they know that *doggy* is a noun and *throw* is a verb when they produce an utterance with these words.

Compound propositions contain logical particles, such as the English words *or, and, not,* and *if.* Reasoning from suppositions is referred to as conditional reasoning because the governing logical operator is the conditional, expressed most clearly in English by the word *if.* Standard logic provides two general approaches to how compound propositions can be judged true or false. One is called truth–functional, e.g., defining a conditional, *if p then q,* as true either when both its antecedent, p, and its consequent, q, are true or when p is false and q is true or when both p and q are false, and as false only when p is true and q is false. This interpretation of the conditional is known as material implication, and basing a theory of conditional reasoning on it would be problematic. In particular, according to this interpretation, *if p then q* is true whenever its antecedent, p, is false. Such an outcome is sufficiently counterintuitive that it is referred to as one of the paradoxes of material implication. This suggests that the truth–functional approach will not provide an adequate theory of conditional reasoning. Indeed, a consensus has developed that human deductive reasoning does not proceed by making truth judgments of this sort.

An alternative, and psychologically more plausible, approach is suggested when one views propositions containing logic particles as inferences. Beginning with the work of Gerhard Gentzen in the 1930s, logicians have constructed logic systems in which valid inferences result from the application of inference schemas, i.e., rules that govern when one can infer statements of certain forms in a line of reasoning. Sound inference schemas are truth preserving, i.e., when applied to true propositions they yield true conclusions. For example, given that *p or q* and *not-p* are true, one can assert *q*, for any *p* and any *q*. Inference schemas describe ordinary logical-reasoning processes and can be used to model human deductive reasoning.

Two inference schemas have been viewed by logicians as essential to conditionals. One, *modus ponens*, holds that given *if p then q* and *p*, one can assert *q*. Thus, when both a conditional proposition and its antecedent are assumed true, one can assert the truth of its consequent. A second, the *schema for conditional proof*, holds that to derive or evaluate *if p then . . .* , first suppose *p*; for any proposition *q* that follows from the supposition *p* taken together with other information given, it is true that *if p then q*. Thus, when one can derive the consequent of a conditional from a set of assumptions taken together with the antecedent of the conditional as a supposition, one can assert the truth of the conditional on the truth of the assumptions alone. Holmes's reasoning illustrates the schema for conditional proof: The supposition that the murderer was a stranger is added to what he knows about the facts of the crime and about the characteristic behavior of watchdogs, and he infers that the dog would have barked. Thus, if the murderer had been a stranger, then the dog would have barked. This version of the schema for conditional proof differs from what is found in standard-logic systems in that it places no restriction on how the consequent can be derived. In standard logic the consequent must follow from application of logical inference schemas; as we saw above, no such restriction applies to ordinary reasoning. Modus ponens and the schema for conditional proof are basic schemas for conditionals in standard logic, and anyone who cannot apply these two schemas does not understand conditionals.

A third inference schema that pertains to conditionals is *modus tollens*, which holds that given *if p then q* and *not-q*, one can infer *not-p*. Thus, given the truth of a conditional and the falsity of its consequent, one can assert the falsity of its antecedent. Modus tollens is not part of most inference-schema models in standard logic; rather, the same inference is usually accomplished in such systems with an argument form known as the *reductio ad absurdum*, which holds that to infer *not-p*, first suppose *p*, and if any contradiction is derived from *p* taken together with other information given, one can infer *not-p*. Thus, whenever a supposition, taken together with propositions assumed true, leads to a contradiction, one can assert the falsity of that supposition. Holmes's reasoning illustrates the reductio: The supposition that the murder was a stranger leads to the inference that the dog would have barked, yet the dog did not bark. Hence, the supposition must be false. Note that to assert *not-p* through a reductio argument, one must first suppose *p*, and no system of logic proposes a set of instructions determining when such suppositions should be made. Successful use of reductio arguments depends on some strategic reasoning skills; thus, a person who might appreciate that a supposition must be abandoned when it leads to a contradiction might not think to make a supposition in order to find a contradiction that would falsify it.

Another argument form that relies on supposition can be applied when a disjunction, *p or q*, is assumed true. To assert *r*, set up two lines of suppositional reasoning, one under the supposition of *p* and the other under the supposition of *q*. If *r* is entailed under each of the suppositional lines of reasoning, where the supposition is taken together with premise propositions, then the truth of *r* can be asserted on the truth of the original premises alone. As with the reductio ad absurdum, this form of argument requires some strategic reasoning skills to know when it would be useful to consider a supposition.

In sum, truth-preserving inference schemas provide a model for the logic of conditional reasoning. The model defines the valid inferences that can be made but does not direct a reasoner as to when any particular supposition should be made. Thus, in addition to logic skills, a skilled reasoner also relies on pragmatic interests and goal setting, and on extralogical strategic skills.

B. Supposition and Truth

It should be noted that in ordinary argument to suppose something is to proceed as though it were true, and inferences under a supposition are judged on the truth of the supposition. There are two ways in which this makes ordinary conditional reasoning different from standard logic. First, supposition of the truth of a proposition sets up a context of argument in which a proposition incompatible with the supposition becomes inadmissible for use (even when it is true) for so long as the supposition continues to be made. Thus, a proposition whose truth depends on the falsity of the supposition cannot be used in the argument under the supposition, and to admit such a proposition requires abandoning the supposition. Particular care must be taken when the supposition is counterfactual. In evaluating *If Napoleon had won at Waterloo, he would not have been exiled to Elba* one is not permitted to appeal to any proposition whose truth depends on Napoleon's having lost, e.g., that following Waterloo he had neither army nor money. In standard logic any premise is permitted under any supposition, leading to such odd inferences as *If the American flag is red, green, and gold, then it is red, white, and blue.*

A second, and related, difference from standard logic concerns cases in which a conditional is to be evaluated when the supposition taken together with premise information entails the falsity of the consequent. In standard logic one could conclude either that the conditional is false or that its antecedent is false. However, in ordinary reasoning, which proceeds from the truth of the supposition (*if p is true then so is . . .*), such conditionals would be judged false.

C. Quantified Conditionals

Logicians distinguish between simple and quantified propositions. The following arguments illustrate the difference. The first concerns a simple conditional premise. Suppose you know that *If Marcus Aurelius wrote that book, he wrote it in Latin*, and you are told that *Marcus Aurelius wrote that book* and asked whether or not he wrote it in Latin. Applying modus ponens to these two premises would lead you directly to the conclusion that he wrote the book in Latin, and you would answer "yes." The second argument concerns a universally quantified conditional. For instance, you know that *If a Roman wrote that book, he wrote it in Latin*, and you are told that *Marcus Aurelius wrote that book* and asked the same question. You again would answer "yes," but before applying modus ponens, you would have had to *instantiate* the quantified conditional, i.e., you needed to know that Marcus Aurelius was a Roman, thus satisfying the quantified proposition. This additional reasoning step for the quantified conditional is made clear if you are told instead that *Hamilcar wrote that book.* Hamilcar was a Carthaginian, and the universally quantified conditional is not satisfied. The need to instantiate quantified propositions is not limited to conditionals; thus, difficulties in judging conditionals that result from the demands of processing quantification do not pertain to conditional reasoning per se.

D. Comprehension of Conditionals

Work in linguistics and in cognitive psychology usually makes a distinction between the surface structure and the meaning of an utterance, with surface structure being the utterance as expressed. Many surface-structure forms are ambiguous, i.e., they have more than one plausible interpretation. Consider universally quantified conditional sentences with the surface-structure form *If a thing is P then it is Q*. In this surface-structure form, universality is expressed indirectly through the indefinite article *a*, and recognizing the universality requires some linguistic sophistication. Indeed, the indefinite article reasonably could be construed as an existential marker, meaning *some* rather than *all*. Thus, although the conditionality is marked explicitly with the word *if*, the universality is marked only indirectly. Compare the surface-structure form *All P are Q*, which in standard logic is true in exactly the same conditions. In this case the universality is explicitly marked with *all*, but the conditionality is implicit.

If is the most explicit marker of conditionality in English, but a speaker is not limited in English to *if* to express conditionality. The simple conditional *If you mow the lawn I'll give you five dollars* also could be expressed as *Should you mow the lawn I'll give you five dollars* or as *Mow the lawn and I'll give you five dollars*. However, each of these surface-structure forms has a different emphasis, and there is no assurance that a subject in an experiment will interpret a surface-structure stimulus in the way that a researcher has in mind. Reasoning is applied to propositions as interpreted—not to sentences as expressed. Hence, a child who fails to appreciate either conditionality or universality when they are marked implicitly might not fail when the logic is marked explicitly.

E. Availability of Logic in Reasoning

The question of whether a logical ability is available in conditional reasoning is ambiguous. An ability might be available on all occasions, or only under some special conditions. An ability can be evaluated only by interpreting the assertions and judgments people make as revealing their available inferences. I refer to a *basic inference* as one that is always made for an explictly marked conditional, e.g., in English by the word *if*. A basic inference is incontrovertible, and any attempt to countermand it would be perceived as contradictory. Further, a basic inference should be available to young children as well as to adults.

An *invited inference* is made when the content, context, or surface-structure cues of a proposition make the inference seem sensible, i.e., it is available only for some construals of a conditionally marked surface structure. An invited inference is not incontrovertible and can be countermanded. For example, *If you mow the lawn, I'll give you five dollars* invites *If you don't mow the lawn, I'll not give you five dollars*. However, the invited inference of *if not-p then not-q* is not essential to *if p then q* and is not logically valid, although some invited inferences are valid.

A *specialized inference* is made only by some people or on some occasions, e.g., mathematical induction—with which one asserts a property for all members of an ordered set when it has been established that if a member of the set has that property then so does its successor, and that the first member of the set has that property. Although few laypersons are apt to have available a schema for mathematical induction, it is available to many mathematical logicians.

III. Empirical Findings

A. Suppositions and Conditional Assertions in Preschool Children

The core of conditional reasoning is to make inferences under a supposition. At what age do children typically show evidence of making inferences, when do they begin making suppositions, and when do they make their first conditional assertions? In this section I draw on two review articles that are concerned with first assertions of conditionals, by Judy Reilly and by Melissa Bowerman, both published in E. C. Traugott *et al.* (eds.), *On Conditionals*, and an article by Alan Leslie in *Psychological Review* in 1987 that is concerned with children's pretense.

Pretense occurs when someone acts "as if" some nonactual situation were true, and it requires acting on the inferences made under a counterfactual supposition. Typically, by about 18 months of age, children begin acting "as if," and early pretense includes understanding pretend roles by others, e.g., wiping up "spilled water" when another person has pretended that an empty cup contains water. Also occurring at this age are verbal indications of inferences from known information: Bowerman reports a child whose mother puts on a dry shirt before going out to play in water; the child says "Mommy wet shirt," thus expressing an inference about the anticipated consequence of playing in water.

At about 2 years of age, children begin using mental-state verbs, e.g., *know, think, remember, pretend, dream, wonder, believe,* and *wish*. Such verbs are of interest here because they do not entail truth assumptions about their constituents. For example, *The cup is red* is judged true or false by looking at the cup; however, *Jane thinks the cup is red* may be true even if the cup is blue. Also beginning about this time is the use of contingency markers such as *when, because, so, almost*. From the beginning of usage the clauses in these utterances are marked appropriately. At this age, however, antecedent clauses of conditionals are not yet marked explicitly and are often elliptical. For example, Bowerman reports a child saying "I don't want go out 'cause I get wet," but children do not yet produce expressions like "if I go out then I get wet."

At about $2\frac{1}{2}$ years of age children begin asserting their first explicitly marked conditionals, e.g., with *if* in English. These early conditional assertions include interrogatives (e.g., *If somebody tore off Michal's hands, then what?*), thus setting up a supposition and asking someone else to produce the inferences under it. From the beginning, when they assert conditionals, children show no difficulty in marking the antecedent clause appropriately. For instance, in languages that mark differences between hypotheticals and contingencies, children always mark them appropriately; in English this difference is seen with the word *when*, which is used in cases for which the even is expected to occur or known to have occurred, and the word *if*, which makes no claim that the event has occurred or will occur.

In sum, reasoning under a supposition is characteristic of 2-year-olds, and from the beginning of suppositional reasoning children make and appreciate counterfactual suppositions: Not only do children suppose *p* as true when its truth is unknown but also when *p* is known to be false. Further, 2-year-olds typically are able to make assertions whose truth depends on conceptual, rather than empirical, grounds. However, the assertion of clearly marked conditionals lags by about 6 months but clearly is exhibited by $2\frac{1}{2}$-year-olds. Thus, making and reasoning under suppositions is an early conditional-reasoning ability, as is making conditional assertions.

B. Comprehension Difficulties in Children

Although from the beginning of the time when children produce utterances, most show no difficulty in appropriately marking which clause is antecedent and which consequent, young children often fail in comprehension to differentiate *if p then q* from *p if q*, for example confusing *I put up my umbrella if it starts to rain* with *If I put up my umbrella it starts to rain*. Similarly *All P are Q* often is confused with *All Q are P*. Such difficulties decrease between 5 and 10 years of age. Apparently, children understand the surface-structure cues that mark conditionality or universality, but in determining which clause is antecedent and which is consequent they rely on semantic plausibility rather than surface-structure syntactic cues.

C. Evaluating a Simple Conditional

In two experiments young school-age children through college students were required to judge the truth or falsity of simple conditionals presented as conclusions. The problems had neutral content about boxes containing toy animals and fruits. Responses at all ages followed a similar pattern: For problems in which adding the antecedent of the conditional to the premises entails the consequent, the conditional conclusion was judged true; for problems in which adding the antecedent to the premises entails the falsity of the consequent, the conditional conclusion was judged false. These data do not correspond to the judgments that would be made in standard logic, but they are consistent with judgments that would be made through application of the schema for conditional proof when adding a supposition to the premises sets up a reasoning context in which the conditional means *if p is true then so is . . . .*

D. Evaluating a Quantified Conditional

When preadolescent school-age children are asked to judge the truth or falsity of sentences of the form *If a thing is P then it is Q*, they typically, and erroneously, respond "true" even when there is a case of *P and not-Q* as long as there is also a case of *P and Q*, thus responding as though the sentences meant *There is a thing that is P and Q*. Some college students also make this error; however, at no age does this error tendency stem from a lack of conditional-reasoning ability; rather it reflects some difficulty in appreciating the surface-structure cues to quantification. For second and fifth graders the error tendency even survives to similar problems that do not concern conditionals at all, e.g., *This thing is P; is it Q?* Further, the error is eliminated for adults and reduced for fifth graders when the consequent is marked as necessary (*then it has to be Q*), and presenting the surface-structure form *All P are Q* leads a large majority at all ages to the logically appropriate responses. Thus the error tendency stems not from an inability to appreciate the conditionality of the sentence but from confusion about quantification: The error is made most frequently on problems that mark universality with the indefinite article, *a*, and least frequent with the word *all*.

E. Reasoning from a Simple Conditional

The most common procedure for assessing reasoning from a simple conditional requires subjects, typically school-age children and young adults, to make judgments about conditional syllogisms. There are four forms of conditional syllogisms, each presenting an assertion of the form *if p then q* as a major premise, and one of *p, not-p, q,* or *not-q* as a minor premise. If modus ponens is a basic inference schema, then arguments with *p* as a minor premise lead directly to the judgment that *q* follows, and across many studies few subjects of any age tested fail to make this judgment. Children as young as 6 years old have no difficulty with modus ponens arguments.

Some investigators have argued that modus ponens may not be a basic inference because in some instances it can be countermanded. For example, given the premises *If it's Saturday we'll have a picnic, If it's not raining we'll have a picnic,* and *It's Saturday*, and asked whether *We'll have a picnic*, even adults respond "can't tell." However, given these premises, subjects are likely to interpret the problem as having the conditional premise *If it's Saturday and it's not raining, we'll have a picnic*, in which case modus ponens cannot be applied. No instances have been reported of a single simple conditional on which either children or adults fail to make modus ponens inferences, and modus ponens seems to be a basic inference.

Two of the conditional-syllogism forms do not lead to any valid inferences (the correct answer should be that nothing follows, or "can't tell"); however, the denial of the antecedent, with *not-p* as a minor premise, and the affirmation of the consequent, with *q* as a minor premise, often are judged erroneously by children, and sometimes even by adults, as leading to *not-q* and to *p*, respectively. Many investigators during the 1960s and 1970s were influenced by the Piagetian notion that adequate logical-reasoning abilities are lacking prior to adolescence, and they interpreted these errors by children as reflecting an inability to interpret *if* sentences appropriately. From this perspective, children are seen as limited to understanding *p* and *q* as corresponding to one another—each occurring only with the other, as though *if* meant *if, and only if.* If children are limited to the *if and only if*, or biconditional-like interpretation of conditionals, then the addition of a clause to the major premise, such as *but if not-p then maybe q or maybe not-q*, would make the conditional premise contradictory (and even modus ponens inferences should not be made). However, if this interpretation is merely invited, and the appropriate conditional interpretation is available, then the additional clause should countermand the invited inference and lead to correct performance. In fact, few erroneous judgments are made, either for the two fallacies or for modus ponens, at any age in this countermanding condition, suggesting that these errors do not result from a fundamental limitation in children's understanding but reflect an invited inference. Further evidence that these biconditional-like responses reflect an invited inference is their increased frequency for certain types of content, for example, when there is a causal relation between the two clauses.

The fourth form of conditional syllogism has *not-q* as its minor premise, and if modus tollens is a basic inference schema, these premises lead directly to the conclusion *not-p*. Younger school-age children usually accept the correct conclusion of *not-p* and the proportion of such correct responses increases with increasing age until about 15 years of age; however, older adolescents and young adults often respond incorrectly that nothing follows. The correct responses of children seem to be made for the same reason as the fallacies on the denial-of-the-antecedent and the affirmation-of-the-consequent arguments, i.e, the interpretation that either *p* and *q* both occur or neither occurs. When the biconditional-like interpretation is countermanded, children switch to the "can't tell" response on modus tollens problems. Adolescents and adults usually seem to realize that the biconditional-like interpretation is wrong, even without a countermand, but are unable to think through the correct conclusion, and they answer "can't tell."

Modus tollens inferences are made more often with certain types of content or situations, e.g., there is improved performance by adults on problems in which a universally quantified conditional concerns a social obligation or permission. Further, modus tollens inferences are made more often when the surface-structure form marks the consequent rather than the antecedent, as in *p only if q*. Generally, modus tollens seems to be an invited, rather than a basic, inference schema.

Apparently, some adults can come to the correct conclusion on modus tollens problems through a reductio ad absurdum argument, i.e., by supposing *p* and using modus ponens to derive *q*. The resulting contradiction with the minor premise, *not-q*, entails the falsity of the supposition. A study was carried out using a training procedure that presented a universally quantified conditional: *If a worker is __ years of age, or older, then that worker makes at least $350 a week.* Subjects were provided exemplars, e.g., a 60-year-old who makes $200, and asked what they could infer about the missing age in the rule. Typically, even adults erroneously judge the age in the rule as less than the age of the exemplar when the monetary amount in the exemplar exceeds that in the rule (analogous to the affirmation-of-the consequent fallacy). When they are subsequently given an exemplar that contradicts their erroneous inference, many adults correct their judgment. The training provides two possible sources for insight. First, subjects can be alerted to cases of *not-p and q*, countermanding the biconditional-like interpretation; second, they are provided experience in rejecting a proposition because of a subsequent contradiction. When presented conditional syllogisms after training, adult subjects were more likely to give correct modus tollens judgments; further, compared with untrained subjects, they rated modus tollens arguments as relatively difficult. Such difficulty ratings correlate highly with the number of steps required to solve a deductive-reasoning problem. Thus, we have reason to believe that some adults solve these problems through a reductio ad absurdum argument; however, the strategy is rarely undertaken spontaneously at any age, and many adults fail to benefit from the training. Thus the reductio ad absurdum is a specialized inference that is not readily available even to many adults.

F. Reasoning from a Universally Quantified Conditional

Universally quantified versions of the four conditional syllogisms have been presented by many investigators at ages ranging from 4-year-old children through college students, and the findings mirror those for simple-conditional syllogisms. Modus ponens inferences are available by the time children enter school and are available earlier when the major premise is implicit in general knowledge. For example, children believe that *All birds have wings;* when they are told that *a wog is a bird* they assert that *a wog has wings*. Even 4-year-olds make correct responses, both when the correct answer is "yes," and when the correct answer is "no," e.g., *A bongo is a kind of animal, but not a dog; is a bongo a poodle*? Thus, modus ponens is a basic inference for universally quantified, as well as simple, conditionals.

Children tend to accept the fallacies on universally quantified assertion-of-the-consequent and denial-of-the-antecedent problems. As with simple-conditional problems, these responses appear to result from invited inferences, and they occur more often when problem content makes them seem plausible and are reduced when problem content makes them less inviting. Further, the occurrence of such inferences is diminished even for 4-year-olds when the consequent clause is marked with necessity, e.g., . . . *have to be Q*.

As with simple-conditional problems, universally quantified modus tollens problems lead to correct responses by young children, although probably because of the invited biconditional-like inference, and the frequency of "can't tell" responses increases with increasing age. Thus, both for simple and for universally quantified conditionals, modus tollens seems not be a basic inference.

G. Complex Conditional Reasoning

The selection task is the best known complex reasoning problem for universally quantified conditionals. Subjects are presented four cards showing, e.g., *A, D, 4,* and *7*, respectively. Each card is known to have a letter on one side and a number on the other. A rule governing the four cards is presented: *If a card has a vowel on one side, then it has an even number on the other side*. Subjects are told that the rule may or may not be true and are required to select for inspection those cards, and only those cards, that can provide a test of whether the rule is true. Because only the cards showing *A* and *7* can lead to any useful (potentially falsifying) information, these are the only cards that should be selected.

Correct solution requires one to consider, for each card, what might be found on the hidden side, and to compare these possibilities with what should be there if the rule is true. Consider the possibility of turning over the card showing the letter *A*. Assuming that the rule is true, by universal instantiation and modus ponens, an even number must be on the other side. However, suppose that one were to find an odd number; then by reductio ad absurdum the conditional is false. The reasoning for the card showing *7* is more complex, requiring an additional reductio argument to realize that if it has a vowel on the other side the rule must be false. Not surprisingly, few people of any age solve standard versions of this task, and in some studies up to 40% of adults fail to choose the card showing *A*. It can be questioned whether modus ponens is a basic inference when so many adults fail to select the card showing *A*. However, such failure could stem from difficulty with other required inferences and does not necessarily reflect on the cognitive availability of modus ponens. Moreover, I know of no reasoning problems that require only modus ponens that are not solved by school-age children or by adults.

When the selection task is presented with realistic content, particularly referring to permissions or obligations, both school-age children and adults perform at levels well above what would be expected by chance. Why such materials frequently lead to success is not yet known. Permissions and obligations may convey some "pragmatic schemas," and one of these corresponds to modus tollens; from this perspective modus tollens can be viewed as an inference that is invited by such materials. A second possibility is that permissions and obligations directly mark which types of exemplars are permitted, obliged, or excluded. A third possibility is that certain types of content enable coding the problem so that a reductio argument is more likely to be undertaken. Some adults do seem able to carry out a complex line of reasoning using a reductio: Performance on the selection task was marginally better for adults after the contradiction training described in Section III,E, but even in this condition the majority of adults failed to solve the task appropriately, and it is likely that superior performance with realistic materials stems from invited inferences.

The THOG task is a complex reasoning problem requiring derivation of alternative suppositions, reasoning under each supposition, and comparing the outcomes under each to arrive at a final conclusion. In its standard version the task presents four designs constructed from two shapes and two colors. Subjects are told that one of the shapes and one of the colors was written down and are given the rule that if, and only if, any design includes either the shape or the color written down, but not both, then it is a THOG. Finally, they are provided an exemplar and asked to decide for each of the remaining three designs whether it is a THOG, is not a THOG, or cannot be determined. Although college students are able to figure out which two combinations of properties could have been written down, few use them to undertake the suppositional reasoning that could lead to solution. Apparently, subjects are discouraged from undertaking such a line of reasoning because the problem has some plausible and simpler solutions, e.g., thinking that the properties of the exemplar are those written down. When the simpler solution is discouraged, e.g., by separating the category labels of the rule from those of the exemplar, the number of correct solutions increases, but half of college students fail in all versions.

In sum, when faced with unfamiliar situations, few people at any age are skilled at setting up suppositions as the basis of a reductio argument, or at constructing parallel lines of reasoning comparing outcomes under multiple suppositions. We are not surprised at Sherlock Holmes constructing a reductio ad absurdum argument—Holmes is a legend of deductive reasoning—but, given how poorly people perform on these complex laboratory tasks, why are we not surprised at Carol's suppositional reasoning as she figures out why her motorcycle won't start? Carol probably would do poorly on the standard versions of the THOG or selection tasks. Ordinary reasoning is intentional, however; that is, people think about things and events, and they have little practice at reasoning when the only cues are words like *if, or,* and *not*. Thus, Carol likely sets up her suppositions and knows to test them because her knowledge of how a motorcycle engine works, rather than her logic skills, suggests interesting, and potentially fruitful, suppositions. These complex conditional-reasoning strategies are specialized; unlike modus ponens or the schema for conditional proof, they are not made automatically but usually are undertaken only when the appropriate suppositions and lines of reasoning are apparent on extralogical grounds.

IV. Summary and Discussion

Some kinds of inferences are made consistently by adults and are evident early in childhood; thus, these are the basic inferences for conditional reasoning. These basic inferences include judgments of truth or falsity—there is a theoretical reason to accept, and no empirical reason to reject, that truth, and its counterpart, falsity, are cognitive primitives that make discourse possible. Making suppositions and reasoning under them is also basic, and 18-month-old children show evidence of using suppositions in their pretend play. Making conditional assertions that are consistent with the schema for conditional proof, and making modus-ponens inferences are also characteristic of children, and these are also valid and basic inferences.

Those problems on which children have difficulty are the same ones that sometime give adults difficulty, and the errors made by children generally are the same as those made by adults. One source of erroneous judgment concerns misinterpretation of premise information. For example, most children and some adults accept the syllogistic fallacies because of a biconditional-like misinterpretation, and they misinterpret *If a thing is P then it is Q* as existential rather than universal. However, even young schoolchildren are able to provide correct responses on these problems, and adults almost never make errors, when the surface-structure cues discourage inappropriate interpretation and invited inference.

Modus tollens is not a basic inference but is invited easily by certain sorts of materials. Note that invited inferences cut two ways: They can lead to correct modus tollens responses but also to acceptance of the syllogistic fallacies. Whereas the basic inferences are made even by schoolchildren no matter what content is used, invited inferences are sensitive to context or content, and they can be countermanded. Children need explicit markers more than adults do, and children are less able to set aside invited inferences.

Construction of reductio ad absurdum arguments and reasoning under multiple suppositions are special abilities, i.e., available only to some people, or only on some occasions. These more complex suppositional-argument forms require some extralogical (or metalogical) skill in setting up appropriate suppositions, and other than premise misinterpretation, the major source of erroneous judgment on conditional-reasoning problems stems from strategic failure to construct an appropriate line of reasoning. Generally, strategic skills are poor at all ages, although some people do undertake appropriate complex strategies. Suppositions usually are made when they seem interesting but rarely are made in order to set up an argument that can falsify the supposition.

Reasoning skills, like other skills, require practice and often require some tuition. It may be that many apparent age differences in reasoning have more to do with schooling than with maturation; the extent to which that applies to the use of suppositions in reasoning strategies is not yet known. Children probably have fewer strategic skills than do adults, or their skills are less well established, although this has not yet been studied systematically.

Basic conditional reasoning skills enter into the behavioral repertory with the same sort of regularity that one finds with acquisition of language structures. All languages that have been surveyed have some way of marking conditionals, and production of conditionals comes in at about the same age across languages. To acquire a natural language expression, the mind must be able to represent it; thus, the mind must have an unlearned preexisting system for representing logical constructs. If the twin concept of truth and falsity is an unlearned cognitive primitive, then the supposition of truth in a line of thought also should be an unlearned ability. Humans are biologically disposed to acquire a natural language that expresses conditionals; thus, reasoning from suppositions is part of our biological inheritance.

Why would suppositional reasoning be a basic part of our bioevolutionary endowment? One would think that evolution would have provided a cognitive system that represents and makes inferences about the way the world actually is, rather than how it would be under some hypothetical or counterfactual supposition. An answer is provided by a thought experiment: Suppose that people could not reason suppositionally but were limited to reasoning from true premises. Planning of future events, as is characteristic of humans, would not be possible; suppositional reasoning allows us to do in thought what otherwise would have to be done in fact. Learning from past events would be crippled. Although it might seem a blessing not to think "if only I had . . . ," it would not be adaptive. Further, a thought experiment such as this would not be possible. Conditional reasoning is adaptive because it allows us to go beyond the information given.

Bibliography

Braine, M. D. S., and O'Brien, D. P. (in press). A theory of *if:* A lexical entry, reasoning program, and pragmatic principles. *Psychol. Rev.*

Braine, M. D. S., and Rumain, B. (1983). Logical Reasoning. *In* "Carmichael's Manual of Child Psychology. Vol. 3. Cognitive Development" (J. H. Flavell and E. M. Markman, eds.), (pp. 263–340). Wiley, New York.

Leslie, A. M. (1987). Pretense and representation: The origins of "theory of mind." *Psychol. Rev.,* **94,** 412–426.

Macnamara, J. (1986) "A border dispute: The Place of Logic in Psychology" The MIT Press, Cambridge, Mass.

O'Brien, D. P. (1987). Development of conditional reasoning: An iffy proposition. *In* "Advances in Child Development and Behavior." Vol. 20 (H. Reese, ed.) (pp. 66–91). Academic Press, New York.

O'Brien, D. P., Braine, M. D. S., Connell, J. W., Noveck, I. A., Fisch, S. M., and Fun, E. (1989). Reasoning about conditional sentences: Development of understanding of cues to quantification. *J. Exp. Child Psychol.,* **48,** 90–113.

Rumain, B., Connell, J. W., and Braine, M. D. S. (1983). Conversational comprehension processes are responsible for reasoning fallacies in children as well as adults: *If* is not the biconditional. *Dev. Psychol.* **19,** 471–481.

Traugott, E. C., ter Meulen, A., Reilly, J. S., and Ferguson, C. A. (eds.). (1986). "On Conditionals." Cambridge University Press, Cambridge.

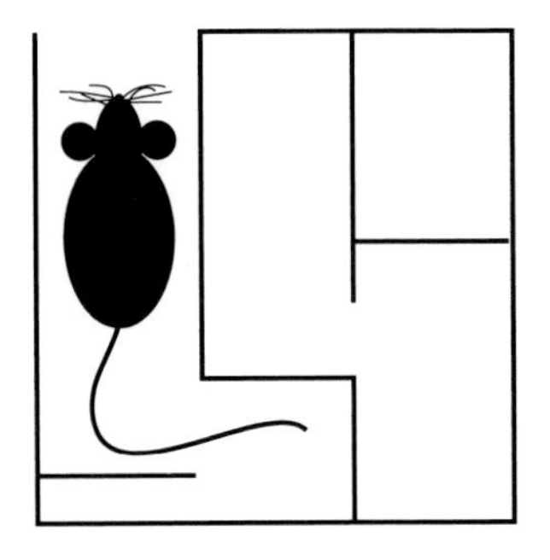

Conditioning

W. MILES COX, *North Chicago VA Medical Center/The Chicago Medical School*

ERICA A. COHEN, *The Chicago Medical School*

KENNETH H. KESSLER, *North Chicago VA Medical Center/The Chicago Medical School*

Glossary

Classical conditioning Simple form of learning in which a stimulus, initially incapable of evoking a response, acquires the ability to do so through repeated pairing with a stimulus that already elicits the response

Extinction In classical conditioning, the weakening or elimination of a conditioned response when the conditioned stimulus is no longer paired with the unconditioned stimulus. In instrumental, or operant, conditioning, the weakening or elimination of a conditioned response when the response is no longer followed by reinforcement

Instrumental (or operant) conditioning Form of conditioning in which the consequences of a response (i.e., reward or punishment) determine the likelihood of that response's occurring again

Learning Relatively permanent change in behavior that often results from reinforced practice. Conditioning is the simplest form of learning

Motivation Internal state of an organism that corresponds to observable behavior that is directed toward achieving a goal—either to get something that the organism wants or to get rid of something that it does not want

Schedule of reinforcement In operant conditioning, the plan according to which reinforcements are delivered, either in terms of the number of responses that have been emitted or an interval of time that has elapsed

Social learning Process by which a person learns new behavior simply by observing another person performing the behavior

Spontaneous recovery Following a period of rest, the reappearance of a conditioned response that has been extinguished

Stimulus generalization Phenomenon by which stimuli that are similar to a conditioned stimulus also elicit the conditioned response

CONDITIONING is the simplest most basic form of learning. It is the type of learning that psychologists study in the psychological laboratory to identify principles underlying more complex forms of behavior. There are two major types of conditioning. Classical conditioning occurs when a stimulus (the conditioned stimulus) that is initially neutral acquires the ability to elicit a response because it is presented with another stimulus (the unconditioned stimulus) that already elicits a response. Instrumental (or operant) conditioning occurs when organisms emit responses that are followed closely by reinforcement; such responses become more likely to occur again. Social learning (another type of learning that is studied experimentally) occurs when people acquire a new behavior merely by observing other people performing the behavior and imitating them in doing so. The principles of learning and conditioning have been useful in explaining why people acquire new behaviors (both desirable and undesirable), and therapists have developed conditioning techniques to help people change their behavior.

I. Introduction

Behavior is malleable. It can undergo distinct changes with time as a result of people's experi-

Encyclopedia of Human Biology, Volume 2.

ences and their interactions with the environment. Sometimes these changes are for the better. People can acquire new desirable behaviors, or they can rid themselves of undesirable behaviors. At other times changes in behavior are undesirable. People might acquire new undesirable habits, or they might stop performing desirable behaviors that they once performed regularly. Regardless of which type of behavior change is being considered, conditioning is the mechanism that is often responsible for the change.

Our purposes in this article are to examine the various forms of conditioning and the principles by which each of them operates and to demonstrate how the principles of conditioning can be used to understand, predict, and change people's behavior. There are two major types of conditioning that we consider here: classical and instrumental (or operant). In addition, we discuss social learning, another type of learning that is often studied in the laboratory.

Learning versus Conditioning

To psychologists, the term "learning" does not mean what it does in everyday language. According to its common usage, "learning" refers to the acquisition of knowledge and is considered to be desirable. By contrast, "learning" in a psychological sense refers to a change in behavior that might be either desirable or undesirable. More specifically, learning is a relatively permanent change in behavior that often results from reinforced practice. Learning, therefore, is distinguished from (1) temporary changes in behavior, such as those that occur when a person is under the influence of a psychoactive drug, and (2) changes that are permanent, but do not result from reinforced practice, such as changes resulting from injury to the body. Conditioning is the simplest form of learning, of which there are two major types: classical and instrumental (or operant). Psychologists often study conditioning in simple laboratory situations and in organisms simpler than humans because doing so allows them to understand the principles underlying more complex forms of human behavior. [*See* LEARNING AND MEMORY.]

II. Classical Conditioning

For centuries people have known that when two stimuli (i.e., objects or events) are closely associated with each other, they become linked in people's minds. In fact, associationism, a philosophical doctrine that was popular in the 18th and 19th centuries, held that people form concepts through principles of association. Two of these are the principle of similarity (i.e., ideas are formed when objects or events are similar to each other) and the principle of contiguity (i.e., ideas are formed when objects or events occur closely together in time or space). Classical conditioning extends associationism to include connections between stimuli and responses. This form of conditioning occurs when a stimulus that originally elicits no response comes to do so regularly because it has been associated with another stimulus that already elicits a response. A connection, or bond, is said to be formed between the old stimulus and the new response.

The original experiments on classical conditioning were performed by Ivan P. Pavlov, a famous Russian physiologist who was awarded a Nobel prize in 1904 for his work on digestion, which was the basis for his work in classical conditioning. While examining the neural mechanisms responsible for food-elicited salivation in dogs, Pavlov became curious about why the dogs in his experiments began to salivate as soon as they saw the food (or even when the person feeding them entered the room), rather than wait until food was actually placed into their mouths. Pavlov believed that the anticipatory salivation that he observed could not be explained physiologically, and he referred to it as a "psychic secretion." He began, therefore, to redirect his research to investigate the mechanisms controlling this psychological phenomenon.

Using dogs as subjects, Pavlov performed his experiments by pairing neutral auditory or visual stimuli with meat. He referred to the originally neutral stimulus (e.g., a tone or a light) as the conditioned stimulus (CS). He found that when a CS was repeatedly presented just before a dog was given food, the CS itself—when later presented alone—would gradually come to elicit salivation. He called the salivation elicited by the CS the conditioned response (CR), because the ability of the CS to elicit the response was dependent on its being repeatedly paired with food. Pavlov referred to the food itself as the unconditioned stimulus (UCS), because its ability to elicit salivation occurred automatically and was not dependent on the dog's training. The salivation that occurred in response to the food was called the unconditional response (UCR). Although the terms "conditioned" and "unconditioned" have come to be used in English to refer to the two types of stimuli and responses observed in classical conditioning, these terms actually are erroneous translations of the Russian words meaning "condition*al*" and "uncondition*al*." The exact translation would more accurately convey the intended meaning of these terms.

It was no revelation that dogs salivate in anticipation of receiving food; people had probably known that for centuries. What was ingenious about Pavlov's work, however, is that he identified the basic principles, or laws, by which organisms learn to make new responses to old stimuli. Because Pavlov was one of the original scientists to study conditioning, the type of conditioning that he studied has come to be called classical, or Pavlovian, conditioning. Classical conditioning experiments have three major features: acquisition, experimental extinction, and spontaneous recovery.

The acquisition phase of a classical conditioning experiment is the phase during which the CR is being acquired. Each time that the CS is paired with the UCS is referred to as a trial, and across successive trials of the experiment, there is a gradual increase in the ability of the CS to elicit the CR. During the course of the experiment, the experimenter measures the strength of the CR by occasionally administering a test trial (i.e., a trial during which the CS is presented alone without the UCS). When the results of a classical conditioning experiment are displayed on a graph, successive conditioning trials are plotted on the horizontal axis, and the strength of the CR (e.g., the number of drops of saliva) is plotted on the vertical axis. On such a graph we usually see a negatively accelerated learning curve. That is, across successive trials, the strength of the CR increases, but the magnitude of the increase becomes less and less, until eventually there is no further increase. At that point the learning curve is said to have reached asymptote.

Extinction, the next phase of a classical conditioning experiment, is when the experimenter repeatedly presents the CS in the absence of the UCS. As this occurs, the strength of the CR gradually extinguishes. For example, if an animal has been conditioned to salivate each time that a tone is presented, and the tone is then repeatedly presented in the absence of food, salivation as a CR to the tone will gradually decline. Even after a CR has become completely extinguished, however, it might regain some of its original strength without any reconditioning by the experimenter. This phenomenon is called spontaneous recovery and is observed following a period of rest after the extinction phase of a classical conditioning experiment. In fact, the amount of recovery of the CR, as a general rule, increases as the length of the rest increases.

Four additional classical conditioning phenomena that Pavlov identified are of interest: stimulus generalization, conditioned discrimination, experimental neurosis, and higher-order conditioning. Stimulus generalization is observed when stimuli that are similar, but not identical, to the original CR also elicit that responses. A gradient of stimulus generalization is observed, such that the greater the similarity between the original CS and the new test stimulus, the greater is the response to the test stimulus. For example, if the original CS were a tone of a certain frequency [e.g., 1000 cycles per second (cps)], tones of either a lower or a higher frequency (e.g., 800 or 1200 cps) would still elicit a CR, although the response would be weaker than one elicited by the CS itself. Tones still further removed from the original CS (e.g., 600 or 1400 cps) would elicit still weaker CRs. Stimulus generalization can be overcome through conditioned discrimination. That is, if the experimenter continues to consistently pair the original CS (called an S+) with the UCS, while repeatedly presenting a generalized stimulus (called an S−) alone, the organism will learn to discriminate between the S+ and the S−, consistently responding to the S+, while failing to respond to the S−.

While investigating stimulus generalization and conditioned discrimination, Pavlov discovered another interesting phenomenon, called experimental neurosis. He first conditioned dogs to consistently salivate in the presence of a circle (a S+), and not to salivate in the presence of an ellipse. On successive trials he made the circle more and more like an ellipse, and the ellipse more and more like a circle, until eventually the dog could no longer distinguish between the two stimuli. At this point the dogs became emotional and acted "neurotic." They were unable to stand quietly in the restraining harness of the conditioning apparatus, barking and struggling to get away. In fact, the dogs eventually became so disturbed that they resisted being taken into the conditioning laboratory.

Higher-order conditioning is another interesting feature of classical conditioning experiments. Once a CS comes consistently to elicit a CR, that CS can function as a UCS for the organism to acquire still another CR. For instance, if a tone has been repeatedly paired with food, so that the tone consistently elicits salivation, the tone could now be paired with a new neutral stimulus (e.g., a light) until the light also comes to elicit salivation. The original pairing of the CS (tone) and UCS (food) is called first-order conditioning. The pairing of the new CS (light) with the UCS is called second-order conditioning. If the new light were now paired with yet another CS, third-order conditioning would have occurred. Even second-order conditioning, however, can be difficult to establish, because at the same time the original CS is functioning as a UCS, it is no longer being paired with the original UCS (food). Hence, the CR to the original CS begins to undergo extinction.

Using Classical Conditioning to Understand and Change Human Behavior

The principles of classical conditioning discovered in the laboratory have direct applications to human behavior. These principles, for example, have helped psychologists understand how people acquire irrational fears and how they can be helped to rid themselves of such fears. People often become fearful of an object or event that was originally neutral, because it occurred in the presence of a noxious stimulus. For example, after being involved in an automobile accident, a person might develop a strong fear of driving. Similarly, a person might develop a phobia for snakes, bugs, heights, or almost any other object or event that occurs in close proximity with an aversive stimulus.

In the same way that people acquire irrational fears through classical conditioning, they can be helped to rid themselves of such fears with the aid of conditioning techniques. One such technique is called flooding. It requires that the phobic individual be exposed to (i.e., "flooded with") the stimulus that elicits his or her fear for an extended period of time in the absence of any actual danger. As exposure continues, with no actual danger occurring, the fear response undergoes extinction.

Another technique for eliminating fears is called systematic desensitization, the goal of which is to systematically desensitize a person to stimuli that he or she fears by conditioning a response to the fear-eliciting stimuli that is incompatible with fear. Systematic desensitization involves several steps. First, the therapist and client together devise a hierarchy of stimuli that elicit the client's fear, ranging from the least to the most fear-arousing stimulus. Second, the therapist teaches the client how to relax, usually by using progressive muscle relaxation techniques. Next, while the client is completely relaxed, the therapist introduces a fear-arousing stimulus that is very low on the patient's hierarchy of fear-arousing stimuli. Once the client is able to completely relax in the presence of the stimulus that once elicited a minimal amount of ear, the therapist then introduces the next stimulus in the hierarchy and the process is repeated. Eventually, the client will be able to relax in the presence of the most anxiety-arousing stimulus in the hierarchy. In other words, at that point the client's fear will have become extinguished. It should be noted that both systematic desensitization and flooding are typically conducted using imagined, rather than the actual, feared stimuli. Nevertheless, both techniques have been useful in helping clients eliminate unwanted fears.

Another classical conditioning procedure for helping people overcome undesirable behaviors is called aversive conditioning, which has been used with some success with alcoholics. The procedure involves administering a drug that induces nausea (i.e., an emetic) to alcoholics at the same time that they are given their favorite alcoholic beverage to drink. After repeatedly pairing the alcoholic beverage with nausea, the beverage itself becomes aversive, and the alcoholic avoids it. One difficulty, however, in using aversive conditioning with alcoholics is that the aversive properties that a particular alcoholic beverage has acquired might not generalize to other alcoholic beverages. Moreover, aversive conditioning does not correct the underlying problems that caused an alcoholic to overindulge in the first place. Thus, aversive conditioning in the treatment of alcoholism should be viewed as an adjunct to other treatment techniques.

III. Instrumental Conditioning

Instrumental conditioning generally involves a different category of responses than those that are classically conditioned. Whereas a CR or UCR occurs automatically when an organism encounters a CS or UCS, responses that are instrumentally conditioned are voluntary, because the organism can choose whether or not to emit them. (It should be noted, however, that there are some exceptions to this generalization. Under certain circumstances, involuntary responses can also be instrumentally conditioned.) A response becomes instrumentally conditioned (i.e., stronger or more likely to occur again) when it leads to reinforcement. If the response does not lead to reinforcement, it becomes weaker or less likely to occur. In other words, the response itself is instrumental in producing the consequences that will determine whether or not it is learned.

The study of instrumental conditioning began with the work of Edward L. Thorndike, an American psychologist who was a contemporary of Pavlov. The original purpose of Thorndike's work was to devise experiments to study animal intelligence. A well-known series of experiments that Thorndike designed for this purpose involved placing a hungry animal (e.g., a cat) inside a cage that had been specially designed so that, if the animal made a correct response (e.g., pulled a latch), the cage door would open and the animal would gain access to food outside. When the animal was initially placed inside this "puzzle box," it made various random movements, but eventually made the response that allowed the door to open. Thorndike believed that the animal's successful escape from the box caused an associative bond to be formed between the stimuli inside the box and the response that led to escape. Furthermore, each time that the correct response was made in the presence of these stimuli, the association between the two was strengthened. What this meant in terms of the hungry animal's behavior was that with each successive trial of being placed inside the puzzle box, it emitted fewer incorrect responses, and the correct response was made within a shorter period of time.

On the basis of these findings, Thorndike formulated his famous law of effect, which states that if a response is made in the presence of a stimulus and is followed by a "satisfying state of affairs," the association between the stimulus and the response will be strengthened. If the response is followed by an "unsatisfying state of affairs," the association will be weakened. The law of effect is considered the cornerstone of instrumental conditioning; if a response is instrumental in producing satisfying consequences, that response becomes conditioned.

Subsequent Research on Instrumental Conditioning

Thorndike's work paved the way for experimental psychologists' dominant interest in instrumental conditioning during the first half of the 20th century. However, the experimental procedure that they favored for studying instrumental conditioning was to train white rats to traverse a maze in order to obtain a food reward. In these experiments various parameters of the food reward were systematically manipulated by the experimenter to study the resulting changes in the rats' behavior. Such variations included (1) the size of the reward that was placed in the goal box at the end of the maze, (2) the length of the delay between the rats' entry into the goal box and the presentation of the food reward, and (3) the consistency with which reward was provided (i.e., whether it was given at the end of every trial or only on some portion of them).

The foremost learning theorist during this era was Clark L. Hull. He devised an elaborate mathematical theory of learning consisting of formal theorems, postulates, and corollaries from which he deduced various hypotheses to test empirically by conditioning white rats to traverse a maze for a food reward. On the basis of his experiments, Hull accepted certain aspects of his theory as experimentally confirmed, and he modified other aspects of the theory to take contradictory evidence into account. In Hull's view his theory would not only help to identify the laws that govern the behavior of white rats, but also would be useful in understanding the behavior of humans.

One important way in which Hull changed his theory was to assign an increasingly important role to motivational factors as determinants of behavior. Hull came to realize that whether or not an organism performs a given behavior is determined not simply by whether or not it has been instrumentally conditioned to do so. The occurrence of the behavior also vitally depends on the organism's motivation to perform it. One key determinant of motivation, in turn, is the value of the incentive that the organism receives for performing the behavior. As a general rule, the more valuable, or attractive, an incentive is to an organism, the more highly motivated it will be to perform the behavior that is necessary to obtain it.

IV. Operant Conditioning

The experimental paradigm for studying operant conditioning is the same as the one for instrumental conditioning. Operant conditioning occurs when an operant response (i.e., a voluntary response that operates on the environment) is followed closely by a reinforcing stimulus. When this occurs, the response becomes more likely to occur again. If the response is not reinforced, or is punished, it becomes less likely to recur. These are the same circumstances under which instrumental conditioning occurs. In fact, the similarity between instrumental and operant conditioning is so close that some authors make no distinction between them; they use the two terms interchangeably.

According to our use of the terms, however, there are two key differences between instrumental and operant conditioning. These include (1) the procedure for performing a conditioning experiment and (2) the measure of conditioning that is taken. An instrumental conditioning experiment involves discrete trials. Whether a cat, for example, is being conditioned to open a puzzle box or a rat is being trained to run through a maze for a food reward, the experimenter administers individual trials to the subject in succession. The strength of the conditioning can be measured by whether or not the animal makes a correct response and how rapidly it makes the response. By contrast, in an operant conditioning experiment the experimenter places the animal in a situation in which it is free to make an unlimited number of responses (e.g., press a lever for a food reward) in a given period of time. The rate of responding (or number of responses per unit of time) is the measure of conditioning in the free-responding operant-conditioning experiment.

The study of operant conditioning was begun by influential American psychologist B. F. Skinner. In large part Skinner devoted his work to the functional analysis of behavior, a simple laboratory analog of the way rewards and punishments control people's behavior in everyday life. He was especially careful to describe the variables controlling behavior as objectively as possible, assiduously avoiding theoretical and mentalistic concepts to account for the behavior that he observed. Hence, Skinner's atheoretical approach was in marked contrast to that of Hull.

To functionally analyze behavior, Skinner devised an apparatus called the operant conditioning chamber (or, as it is often referred to by other people, a "Skinner box"). This apparatus is simply a soundproof box that contains (1) a manipulandum (e.g., a lever for a rat to press or a lighted key for a pigeon to peck) and (2) a mechanism for delivering reinforcers (usually a dispenser for food pellets). Depending on the particular organism that is being studied, the experimenter decides on a particular response (e.g., lever-pressing for rats and key-pecking for pigeons) to reinforce or not to reinforce. The equipment attached to the operant conditioning chamber automatically records each time the designated response is made and each time a reinforcer is delivered. From the record that is produced, the experimenter can identify relationships between the number of responses that the animal made (and the pattern in which they were made) and the plan, or schedule, according to which the reinforcers were delivered.

Skinner's definition of a reinforcer differs markedly from that of Thorndike and Hull. In Skinner's view a reinforcer is simply any stimulus that follows a response and increases its probability of occurring again. He does not specify that a reinforcer be "satisfying" or "pleasurable" to the organism that receives it. Nevertheless, Skinner divides reinforcers into two major classes. A positive reinforcer is a stimulus (e.g., food or water) whose presentation increases the probability of the response that preceded it. A negative reinforcer is a stimulus (e.g., electric shock, a bright light, or a loud noise) whose removal increases the likelihood of the response that preceded it. Punishment, on the other hand, is the presentation of an aversive stimulus (or removal of an attractive one) that reduces the likelihood of the response that preceded it. (It should be noted that the term "negative reinforcer" is frequently used incorrectly to mean punishment.) From his research Skinner concluded that using positive and negative reinforcers to control behavior is more effective than using punishment. Reinforcers can be used to bring about enduring changes in behavior. Punishment, on the other hand, might temporarily suppress a response, but will not permanently extinguish it.

One of the best-substantiated findings from research on operant conditioning is that the rate of responding depends on the schedule according to which the reinforcement is delivered. If an organism is reinforced after every response, it is said to be on a continuous schedule of reinforcement. If an organism is never reinforced, it is said to be on an extinction schedule of reinforcement. When the organism is reinforced, but not after every response, it is on an intermittent schedule of reinforcement. There are four basic schedules of intermittent reinforcement, each of which is associated with a characteristic manner of responding. These schedules vary according to whether delivery of the reinforcement is based on (1) the number of responses that have been emitted or (2) the interval of time that has elapsed since the last reinforcement was delivered. Moreover, within each of these, the schedule can be either fixed or variable.

First, let us consider the fixed-ratio (FR) schedule of reinforcement. On this schedule the organism is reinforced after having made a fixed number of responses. For example, a rat being conditioned to press the lever in an operant-conditioning chamber might be on an FR15 schedule, meaning that whenever it presses the lever 15 times it will receive a reinforcement. This schedule produces a steady high rate of responding, because it is to the animal's advantage to make the required number of responses as rapidly as possible in order to receive its reinforcement. In everyday life a factory worker who is paid according to the number of units of work that he or she completes would be on an FR schedule.

Next, consider the variable-ratio (VR) schedule. On this schedule the organism again is reinforced for having made a certain number of responses, but in this case the number of responses that is required varies around a constant value. For example, if an animal is on a VR10 schedule, only part of the time will it be reinforced for having made exactly 10 responses. At other times it might be reinforced for having made say, five or 15 responses. By the end of the conditioning session, however, the average number of responses for which it will have been reinforced will equal 10. In other words, on this schedule the exact number of responses that is required to receive a reinforcement is unpredictable. As might be expected, therefore, VR schedules result in a very high rates of responding. In fact, animals have sometimes been observed to respond so rapidly on VR schedules that the amount of food that they receive as reinforcement is not sufficient to compensate for the great energy that they expend in responding. In the real world gambling devices (e.g., slot machines in Atlantic City and Las Vegas) are programmed to "pay off" (i.e., to deliver monetary reinforcements) on VR schedules. Thus, always expecting that their very next response might lead to the desired reinforcer (i.e., money), gamblers respond at very high rates, and often their behavior is difficult to extinguish.

A third schedule of reinforcement is the fixed-interval (FI) schedule. On this schedule the organism is reinforced for the first response that it emits after a fixed interval of time (e.g., 1 minute) has elapsed. Regardless of how rapidly the animal might respond, it is impossible for it to receive a reinforcer earlier. Hence, when on an FI schedule, the animal responds at a very low rate—or stops responding altogether—just after a reinforcer has been delivered. This phenomenon is known as the post-reinforcement pause. As the waiting interval draws to a close, however, the rate of responding becomes more rapid and continues at a rapid rate until the next reinforcer is delivered. Because of the erratic rate of responding, FI schedules are associated with the lowest overall rate of responding of any of the schedules that we discuss here. In real-life situations studying for examinations in order to get a good grade is an example of behavior that operates according to an FI schedule. As the day of an examination draws near, many students' studying behavior (i.e., their rate of responding) increases markedly. Then, immediately after the examination, many students study considerably less, or not at all. The same pattern repeats itself throughout the school term.

The final schedule that we consider is the variable-interval (VI) schedule. On this schedule a reinforcer again becomes available after an interval of time has elapsed, but, as with the VR schedule, the interval varies around a constant value. Thus, exactly when a reinforcer will be available is unpredictable. As might be expected, therefore, VI schedules result in more steady rates of responding than do FI schedules. In real life mail delivery is an example of reinforcement (or perhaps punishment, if the mail contains unwanted bills) that is delivered according to a VI schedule. Mail is delivered at an approximate (or average) time each day (e.g., 2:00 p.m.). On some days the carrier arrives early, whereas on other days he or she arrives late. As 2:00 p.m. approaches, a person might begin to check the mailbox, although probably not as early as 10:00 a.m. Thus, although VI schedules eliminate the exact predictability associated with FI schedules, they still allow for a certain degree of predictability.

The principles of operant conditioning are used to help people eliminate undesirable behaviors (e.g., a bad habit, such as smoking) or to acquire desirable behaviors (e.g., social skills) in which they are deficient. Operant techniques are often used in institutions both to help maintain order and to help patients acquire desirable behaviors. One example of such an operant technique is the token economy. This is a system that allows patients to earn tokens for emitting desirable behaviors (e.g., making one's bed, arriving at group therapy sessions on time, or making appropriate social responses) and to lose tokens for emitting undesirable behaviors. Tokens earned can later be exchanged for desired commodities or special privileges. The token economy, which at first might appear somewhat artificial, in many respects resembles the everyday world in which most people live and work. For example, by regularly going to work and performing adequately on the job, people earn money (i.e., tokens) with which they can later purchase items they want. On the other hand, if one violates a law (e.g., exceeds the speed limit while driving) a portion of his or her money (i.e., tokens) might have to be relinquished.

V. Social Learning

Although recognizing the value of classical and instrumental conditioning in explaining and controlling behavior, psychologists who have studied social learning believe that the laws of conditioning are incapable of fully accounting for human behavior. Social psychologists, as this group of scientists is sometimes called, emphasize that human behaviors is often learned from social interactions among people, and is not merely the result of trial and error. According to social learning theorists, people's cognitions (i.e., the thought processes that occur in their minds) are a strong determinant of which stimuli they attend to, the meaning that they attribute to these stimuli, and how they respond to them.

Social psychologists have devised experiments to demonstrate that people can acquire a new behavior by observing other people performing the behavior. A series of these experiments stimulated the initial interest in social learning. In one experiment one group of young children observed an adult model displaying aggressive behavior toward an inflated doll (a "Bobo" doll), whereas a control group observed the adult displaying innocuous behavior. Later, when the children themselves were allowed to play in the room with the Bobo doll, they imitated the behavior of the adult model whom they had observed; that is, only those children who had observed aggressive behavior were aggressive themselves. In another experiment each of two groups of children observed a model being aggressive, but one group saw the model being rewarded for being aggressive, and the second group saw the model being punished for being aggressive. Later, the children who had observed the model being rewarded for being aggressive were themselves more aggressive than those who saw the model being punished. However, when the children were themselves then offered a reward for imitating the behavior of the model, both groups displayed aggressive responses toward the Bobo doll. These experiments illustrate two important principles of social learning. First, reinforcement is not necessary for a person to acquire new behavior. It can be acquired merely by observing someone else performing the behavior. Second, reinforcement and punishment often determine whether a particular behavior is performed and not whether or not it is learned. [*See* BEHAVIORAL EFFECTS OF OBSERVING VIOLENCE.]

In everyday life a person is often influenced by the actual or perceived behavior of other people. For example, when people enter an ambiguous social situation (e.g., a gathering at their new employer's house) they tend to proceed cautiously until they have had a chance to observe others and gain an understanding of the dynamics of the situation. In the situation just described, the newcomer might look to others to determine if it is appropriate to drink alcohol or call the boss by his or her first name. In fact, a new employee might well have asked a senior employee about the appropriate dress code for the party in advance of the event. Thus, by modeling his or her behavior after that of others, the new employee would be able to avoid the embarrassment of arriving at the party dressed differently from the other party-goers.

Advertisers often use the principles of social learning in their attempts to influence consumers' behavior. For example, to change consumers' attitudes about a particular product (and thus their decision to buy the product), the advertiser might depict someone (i.e., a model) using the product who epitomizes the everyday-life role (e.g., career woman or mother) of persons at whom the advertisement is directed. The goal of the advertiser is to get consumers to change their attitude about the product by imitating the behavior of the model. In the event that a consumer already uses the product, the advertisement is designed to reaffirm his or her decision to do so.

Advertisements aimed at preventing drug abuse are another example of attempts to change people's behavior by having them imitate the behavior of a model. The goal of these advertisements is to convey vicariously to young people information about the negative consequences of drug use, without their having to acquire the information through their own experiences. In order for these approaches to succeed, however, the person hearing such an advertisement must be willing to accept the information communicated by the model in the advertisement. For this reason models whom young people feel positively about and whom they trust are recruited to discuss the negative aspects of using drugs. The models include individuals of high status (e.g., actors or rock stars) and people who have recovered from drug abuse through treatment.

VI. Conclusions

The study of conditioning in the laboratory has enabled psychologists to identify the principles according to which animals and humans acquire new behavior or change existing behavior. Through classical conditioning organisms learn to make new responses to neutral stimuli or to extinguish responses that they have previously learned. Through instrumental or operant conditioning the probability of organisms' emitting particular responses either increases or decreases, depending on whether or not a particular response is followed by reinforcement. Therapists have used the principles of conditioning identified in the laboratory to help people either rid themselves of an unwanted behavior or acquire a new desirable behavior. In short, classical, instrumental, and operant conditioning have all been useful in helping us understand, predict, and control behavior. Nevertheless, as the research on social learning indicates, conditioning principles alone do not account for the distinction between organisms' learning a particular behavior and their motivation to perform that behavior. For this reason psychologists have grown increasingly aware that people's behavior is strongly influenced by motivational factors and is not determined simply by the stimuli to which people have been classically conditioned to respond to or the responses that they have been instrumentally or operantly conditioned to emit.

Bibliography

Bootzin, R. R., Loftus, E. F., and Zajonc, R. B. (1983). Learning. *In* "Psychology Today: An Introduction" (R. R. Bootzin, E. F. Loftus, and R. B. Zajonc, eds.), 5th ed., pp. 177–197. Random House, New York.

Bower, G. H., and Hilgard, E. R. (1981). "Theories of Learning," 5th ed. Prentice Hall, Englewood Cliffs, New Jersey.

Domjan, M., and Burkhard, B. (1986). "The Principles of Learning and Behavior," 2nd ed. Brooks/Cole, Monterey, California.

Hill, W. F. (1982). "Principles of Learning: A Handbook of Applications." Mayfield, Palo Alto, California.

Hill, W. F. (1985). "Learning: A Survey of Psychological Interpretations." Harper & Row, New York.

Hull, C. L. (1943). "Principles of Behavior." Appleton-Century-Crofts, New York.

Hull, C. L. (1952). "A Behavior System." Yale Univ. Press, New Haven, Connecticut.

Kantowitz, B. H., and Roediger, H. L. (1984). Conditioning and learning. *In* "Experimental Psychology: Understanding Psychological Research" (B. H. Kantawitz and H. L. Roediger, eds.), 2nd ed., pp. 201–226. West Publ., St. Paul, Minnesota.

Klein, S. B. (1987). "Learning: Principles and Applications." McGraw-Hill, New York.

Klinger, E. (1977). "Meaning and Void." Univ. of Minnesota Press, Minneapolis.

Pavlov, I. (1927). "Conditioned Reflexes." Oxford Univ. Press, Oxford, England.

Petri, H. (1981). "Motivation: Theory and Research," 2nd ed. Wadsworth, Belmont, California.

Skinner, B. F. (1938). "The Behavior of Organisms. An Experimental Analysis." Appleton-Century-Crofts, New York.

Thorndike, E. L. (1898). Animal intelligence: An experimental study of the associative processes in animals. *Psychol. Rev. Monogr., Suppl.* **2,** 1–109.

Connective Tissue

KENNETH P. H. PRITZKER, *University of Toronto*

Glossary

Basement membrane Connective tissue layer between histologically dissimilar tissues

Collagen Major group of structural proteins that form connective fibers

Elastin Insoluble structural protein; elastin and fibrillin molecules become organized to form elastic fibers

Extracellular matrix Material in connective tissue outside of cells

Fibronectin Structural glycoprotein that can bind specifically to collagen, proteoglycans, and cell membranes

Glycosaminoglycan Structural sugar polymer containing repeating disaccharide units

Parenchymal cells Cells intrinsic to the functions of an organ, excluding connective tissue cells

Proteoglycan Complex structural macromolecule composed of one or more glycosaminoglycan chains attached to a protein core

CONNECTIVE TISSUE can be defined as the tissue that occupies the space between parenchymal organs and within organs, i.e., the tissue occupying space between parenchymal cells. The name connective tissue itself and the above topological definition imply functions of structural support and physiological integration of cells to organs and of organs to the body. Tissues are collections of cells and intracellular substances forming definite visible structures. Accordingly, connective tissues are classified by the distinctive, naked-eye appearance (color, texture, and consistency) of their predominant components (Table I). The density of capillary blood vessels in the reddish brown color of muscle, dissolved lipid pigments in yellow fat, chromophore molecules in orange-yellow elastic fibers, and the dense arrays of collagen fibers in white fascia or ligament demonstrate the variety of factors that contribute to connective tissue color. The relative concentration of nonfibrillar macromolecules to fibrillar macromolecules and the orientation of fibrillar macromolecules provides textural appearances as varied as the smooth, clear structure of the cornea, the glasslike hyaline cartilage, and the fibrous cable of tendon. The consistency of connective tissues may range from soft gelatinous or myxoid tissues and fatty tissues to firmer tissue such as fibrous, elastic, and cartilaginous tissue to the mineralized tissue of bone. Each tissue has a characteristic structure adapted to specific functions.

Connective tissues are usually complex in nature. For example, vascular tissue is normally present in most connective tissues except cartilage and cornea. Elastic fibers may be present in fibrous tissue or cartilage. Adipocytes (fat cells) may be present in fibrous tissue, in muscle tissue, or within bone. The soft connective tissue between organs, which contains a mixture of fibrous and gelatinous tissues, is often termed areolar connective tissue. Moreover, connective tissues themselves may be arranged into distinct organs, discrete tissues with specific architecture composed of multiple connective tissues arranged for specific functions. For example, a bone is an organ composed of bone, cartilage, adipose, and often hematopoietic (blood) tissues; a muscle contains muscle and vascular and fibrous connective tissues. As specific tissues and organs are considered elsewhere, this chapter will provide an

TABLE I Connective Tissue Classification

Type	Examples
Myxoid	Nucleus pulposus (intervertebral disc); vitreous humor (eye); synovium, areolar connective tissue
Fibrous	Tendon, ligament, fascia
Elastic	Ligamentum nuchae, artery
Adipose tissue	White fat, brown fat
Muscle	Striated muscle, smooth muscle
Vascular tissue	Artery, capillary, vein
Cartilage	Hyaline cartilage, fibrocartilage, elastic cartilage
Bone	Compact bone, trabecular bone

overview of the biological principles common to the structure and function of all connective tissues.

I. Structure and Composition

Connective tissue is derived embryonically from mesenchyme and ultimately from mesoderm. Connective tissues are the predominant bulk component of body mass and accommodate many diverse functions. This tissue is extremely heterologous both in its cellular population and in the composition of the intercellular material substance commonly called extracellular matrix. [*See* EXTRACELLULAR MATRIX.]

Connective tissue cells are of two general types. First, some cells produce, resorb, and maintain extracellular matrix. Cells of fibrous tissue (fibroblasts), joint lining (synovial) cells, cartilage cells (chondrocytes), and bone cells (osteoblasts, osteocytes, osteoblasts) have some or all of these activities. Second, other cells synthesize and maintain specialized internal structures or functions. These cells include myocytes (muscle cells), with their specialized contractile apparatus and adipocytes, which are adapted for vacuolar cytoplasmic storage of fat. Vascular cells (myofibroblasts and endothelial cells) share characteristics of both cell types. These cells have active contractile functions as well as actively produce extracellular matrix.

Each connective tissue contains extracellular matrix with distinct compositional properties. This matrix is a complex composite material, which varies not only from one matrix to another but also from domain to domain within each matrix. In connective tissues where intracellular functions predominate (e.g., muscle, adipose tissue), extracellular matrix is a minor component occupying <5% of the tissue. In connective tissues where most cells are specialized for matrix synthesis (e.g., tendon, cartilage, bone tissues), the extracellular matrix may occupy up to 95% of tissue volume. Most connective tissue matrices are intimate composites of three phases. First, a freely diffusible fluid phase consists of water and low molecular weight solutes. Less-diffusible medium-weight solutes and suspended noncollagenous proteins and lipids comprise the second phase. Both of these phases exist within and are constrained by a structural macromolecular phase composed of fibers such as collagen and elastin as well as nonfibrillar soluble macromolecules such as nonfibrillar collagens, glycosaminoglycans, and proteoglycans. Whereas some macromolecules such as elastin of elastic fibers are a single gene product, most macromolecules, including the collagens and proteoglycans, are products of multiple genes. Moreover, there is extensive heterogeneity among these macromolecules. Currently, over 13 different collagen types are recognized. Collagen types I, II, and III are the major types that form fibers. Type IV collagen is an important component of basement membranes; Type V-XIII collagens are termed minor collagens. Of these molecules, Types V, VI, and XI form fibrils. In bone, teeth, calcified tissue, and pathologic calcifications, a fourth solid phase is present. This phase is composed of the mineral calcium apatite $[(Ca)_{10}(PO_4)_6(OH)_2]$. The components of connective tissue matrices are derived variably from synthesis by cells within the matrix and from diffusion from the plasma of the circulating blood (Table II).

The structure of each connective tissue matrix is determined predominantly by the composition and arrangement of the fibrillar and amorphous macromolecules. This structure is highly related to mechanical function. For example, in tendon, parallel arrays of type I collagen fibers facilitate its tensile function. Cartilage, a tissue adapted to resist compressive forces, is composed of relatively more hydrated type II collagen and amorphous proteoglycan macromolecules arranged in a dense hydrogel. In arterial blood vessels, where the wall must modulate pulsatile pressure, elastic fibers are characteristic. In bone, where rigidity and strength are crucial, the matrix is a complex composite material composed mainly of type I collagen and the mineral calcium apatite. [*See* CARTILAGE.]

The arrangements and interactions among extracellular matrix molecules are highly controlled and

TABLE II Connective Tissue Composition

Component	Examples
Cells	Fibroblast, adipocyte, chondrocyte, osteocyte
Extracellular matrix	
Water	
Freely diffusible solutes	Gases: O_2, CO_2, N
	Ions
	Hormones
	Vitamins
	Cytokines
	Cell metabolites
Less-diffusible substances	Immunoglobulins
	Structural glycoproteins
	Fibronectin, laminin
	Enzymes, enzyme inhibitors
	Lipids
Structural macromolecules	Fibrillar molecules
	Collagens I, II, III, V, and VI
	Elastin
	Amorphous soluble polymers
	Proteoglycans
	Glycosaminoglycans
	Nonfibrillar collagens
Mineral	Calcium apatite (bone, teeth)

specify the functional capacity of the tissue. Some major intermolecular arrangements, discussed briefly below, are listed in Table III. Fibrillar macromolecules have specific orientations that are adapted to the ambient mechanical forces. For example, collagen fibers in tendon are aligned in parallel groups or fascicles to provide an arrangement that maximizes tensile strength. Collagen fibers in dermis (skin) are arranged in a matted or reticular network. This pattern becomes oriented to applied forces yet retains proteoglycans and other amorphous components within the tissue. In articular cartilage, the collagen fibers, while oriented in a nonrandom fashion, do not become aligned in a single direction with application of forces. This arrangement provides enhanced resistance to compression. [*See* ARTICULAR CARTILAGE AND THE INTERVERTEBRAL DISC; SKIN.]

TABLE III Connective Tissue Matrix Macromolecules: Arrangements and Interactions

Arrangement	Examples
Molecular orientation	Collagen in tendon or bone
Molecular polymerization and aggregation	Collagen fibril aggregation
	Collagen intermolecular cross-linkages
	Proteoglycan aggregation
Heteromolecular interaction	Collagen–proteoglycan binding
	Elastin–fibrillin binding
	Collagen–fibronectin–cell binding
	Proteoglycan–laminin–cell membrane binding
	Collagen and calcium apatite binding
Macromolecule–ion interaction	Ca^{++}, and Mg^{++} and glycosaminoglycans

Collagen is secreted from the connective tissue cells as a large molecule, procollagen. Within the matrix, procollagen molecules are cleaved by an enzyme to form collagen monomers. These molecules align themselves in a quarter-staggered fashion to form fibrils, which are then stabilized by cross-links. To a large extent, fibrillogenesis and the radial growth of collagen fibers is controlled by electrostatic binding of proteoglycans to specific sites on the collagen molecules.

Type IV collagen, present in basement membranes adjacent to epithelial and endothelial cells is a nonfibrillar collagen that binds noncollagenous proteins to proteoglycans. This matrix integration function is an important property of the minor collagens. Minor collagens such as Type V collagen associate cells to matrix. Type VI collagen, found particularly in blood vessels and skin forms microfibrils; Type VII collagen forms anchoring fibrils that facilitate adhesion of skin basement membranes to the underlying dermis. Type X collagen appears specific to hypertrophic growth cartilage and may assist in growth plate organization. Type IX and Type XII collagens are multidomain molecules that bind together Type II and Type I collagen fibers, respectively. Because of their properties, these substances are classified as Fibril Associated Collagens with Interrupted Triple Helices (FACIT) molecules. [*See* COLLAGEN, STRUCTURE AND FUNCTION.]

Elastic fibers are a composite material formed by the intimate interaction of an amorphous protein, elastin, with a fibrillar glycoprotein component, fibrillin. Elastic fibers, as the name implies, are deformable tissue elements. These fibers permit reversible absorption of energy by connective tissues without loss of tissue shape. [*See* ELASTIN.]

Glycosaminoglycans are long-chain sugars with repeating disaccharide units. Glycosaminoglycans such as hyaluronan contribute to the permeability and the visco–elastic properties of connective tissue matrix. By binding to cell membranes, these molecules help regulate cell adhesion and cell motion. Glycosaminoglycans link covalently to core polypeptides to form proteoglycans. Currently, five classes of proteoglycans, each with distinct properties are recognized: Major structural proteoglycans, interstitial proteoglycans, cell surface associated proteoglycans, basement membrane proteoglycans, and small serine- and glycine-rich proteoglycans. The major structural proteoglycans and proteoglycan aggregates determine many characteristics of connective tissue including stiffness, compliance, and viscosity responses to compressive forces as well as diffusion of molecules through the matrix. Interstitial proteoglycans bind avidly to gap regions in collagen fibers facilitating collagen fiber binding as well as a cell–collagen adhesion. By promoting organization of microfibrils, the cell surface associated proteoglycans assist binding of cells to the matrix. These proteoglycans also serve to modulate cell receptor interactions. Basement membrane proteoglycans assist in determining the organization of parenchymal tissues and provide selective permeability of molecules from epithelium or endothelium to matrix. These molecules possess antithrombogenic activity that facilitates blood flow as well as cell migration. The small serine- and glycine-rich proteoglycans are present in secretory granules in a variety of cells. By their anionic properties these proteoglycans may help regulate secretion of cationic peptides, nucleotides, and amines, substances that are active as neurotransmitters, paracrine factors, and inflammatory mediators. [*See* PROTEOGLYCANS.]

An important class of matrix interactions involves molecules that bind cells to structural macromolecules. These molecules contribute to the organization of connective tissue macromolecules around the cell. Fibronectin, a molecule that has separate binding sites for collagen, proteoglycan, and cell membranes, is an important molecule of this type. Laminin, a molecule that binds proteoglycan and cell membranes, serves this function in basement membrane structures that are present between parenchymal cells and connective tissues. The relationship of calcium apatite mineral to bone collagen fibers provides an excellent illustration of the exquisite regulation of connective tissue matrix. First, calcium apatite crystals that form within bone are very similar to each other in size and crystallinity; however, the density in which the apatite crystals are packed within the matrix varies considerably within bone domains and between bones. Moreover, the calcium apatite crystals are deposited anisotropically, i.e., with the axis of each crystal specifically oriented along the collagen fibril.

Some critical interactions in connective tissue involve the relations among elemental ions, small charged molecules, and proteoglycans. Proteoglycan molecules can be thought of as negatively charged ion exchange resins with charges usually counterbalanced by positive sodium ions and small molecular weight proteins. Ions diffusing through connective tissue are hydrated, i.e., associated with water molecules. Highly charged ions such as calcium ions will preferentially bind to glycosaminoglycans. This binding may change the conformation of the glycosaminoglycans as well as release water. Both of these activities will alter the local properties of the matrix including molecular diffusion.

Molecular diffusion is dependent quantitatively on the binding characteristics of the matrix molecules and the volume of water within the matrix. Matrix water content varies from tissue to tissue. More than 99% of the vitreous within the eye is water compared with $<$10% in dentin. Matrix water itself can be considered as a two-phase system. There is a free phase in equilibrium with interstitial fluid elsewhere and with blood plasma as well as a bound or restricted phase associated with macromolecules and ions.

II. Connective Tissue Functions

The functions of connective tissue are dependent on the matrix as a composite material. Connective tissue function depends not only on the quantity of the various components but also on the qualities of their orientation and molecular interaction. Perhaps the most important ideas concerning the varied functions of connective tissue are that each tissue has multiple functions and that the functions are nested within each tissue. Matrix domains of specific composition have specific characteristics for each function. Variation in matrix composition results in variation of each functional capacity of the matrix. Control of matrix structure and, hence, control of connective tissue function resides with the regulation of connective tissue cell metabolism. Common to connective tissues are functions that can be grouped as follows: structural support and mechanical adaptation, transport and barrier, storage, cell modulation, and tissue morphogenesis and repair (Table IV).

A. Structural and Mechanical Functions

Like other solid materials, connective tissue resists three types of stress forces: compression, tension, and shear. Connective tissue mechanical properties can be described in a similar manner as other materials, namely, by their quantitative change with the application of mechanical forces. Terms used include *strain*, the change in dimension of tissue compared with initial dimension; *extensibility*, the maximal strain at which the tissue fails (tears); *stress*, the force–unit area; *stiffness*, the change in stress/change in strain; and *toughness*, the strain energy absorption/area beneath the stress/strain curve. A characteristic of connective tissues is that these relationships change over the time that the force is applied. For tensile forces, this

TABLE IV Connective Tissue Functions

Function	Examples
Structural and mechanical	Support and protection of cells and organ arrangements
	Adaption to external forces (tension, compression, shear, etc.)
	Lubrication
	Cell shape determination
Transport barrier functions	Nutrients and wastes
	Hormones and paracrine molecules
	Immunoglobulins
	Microorganisms
Storage	Intracellular
	Carbohydrates
	Amino acids
	Lipids
	Lipid-soluble substances
	Extracellular
	Water
	Ions, Ca^{++}, PO_4^{-3}
	Cationic proteins
	Enzymes and inhibitors
	Immunoglobulins
	Growth-stimulating molecules and inhibitors
	Matrix degradation products
	Cell wastes
Cell modulation	Growth, poliferation, and differentiation
	Adhesion and migration
	Matrix synthesis and degradation
Tissue morphogenesis and repair	Parenchymal tissue organization
	Parenchymal tissue repair
	Connective tissue repair and regeneration

is termed viscoelastic behavior. The change in properties of connective tissues over time and with increasing load results from realignment of collagen and elastic fibers under load. The low resistance of the matrix to such rearrangements depends on the fluid properties of the proteoglycan hydrogel.

Connective tissues can be grouped by mechanical characteristics as tensile or ropelike materials, pliant or deformable materials, and rigid materials, which resist stress with little deformation. As applied to connective tissues, tendon is adapted to high-tensile strength and stiffness. The dermis of skin has moderate tensile strength but can be repeatedly stretched and relaxed without loss of shape. Arterial vessel wall stretches easily at low loads and becomes increasingly stiff at higher loads. Cartilage is moderately strong and stiff and has high extensibility, whereas bone is very strong, stiff, and inextensible.

Lubrication is a subset of connective tissue mechanical properties. Lubrication is the ability of tissues to accommodate to shear forces. Lubrication occurs not only at joints but at all surfaces where one tissue moves against another, whether this be at a macroscopic or microscopic level. The mechanisms and molecules involved in lubrication vary depending on the forces and the deformability of the tissues. For soft tissues, the glycosaminoglycan hyaluronic acid appears as the critically important molecule, whereas in cartilage specific glycoproteins or lipids may be implicated in this function.

Connective tissues maintain parenchymal cells in functionally effective conformations. These arrangements are adapted to support and protect the parenchymal cells to resist ambient forces including those induced by gravity and locomotion. Less obvious, but equally important, the interaction of cells with specific cell matrix molecules including proteoglycans and glycoproteins determines the shape and frequently the functional state of cells.

As materials, connective tissues are extremely complex and can degrade rapidly when studied out of the body. Therefore, while mechanical behavior at macroscopic and microscopic levels can be described in material science and mathematical terms, the functional understanding of connective tissue mechanical properties remains incomplete.

B. Transportation and Barrier Functions

The transport and barrier functions of connective tissue are associated mainly with the hydrogel macromolecular structure. This structure consists of water constrained by negatively charged (anionic) macromolecules counterbalanced by small molecular weight, positively charged cations. In this gel, gases and small molecular weight molecules diffuse more rapidly than larger molecules; cationic molecules diffuse faster than uncharged or anionic substances. Conversely, molecules bind with different affinities and capacities to connective tissue matrix components. An example of the complexity of the diffusion process is that attachment to negatively charged solutes may enhance the permeability and decrease binding of some molecules to connective tissue macromolecules. The diffusion of particular molecules varies from tissue domain to tissue domain related primarily to the charge density and hydration of tissues. These factors are, to some extent, regulated by the connective tissue cells. Functionally, this means that connective tissues are permeable to cell nutrients and wastes but are relatively impermeable to large molecular weight proteins and microorganisms. In tissues with a very high density of anionic proteoglycans such as cartilage, proteins secreted by the cells such as enzyme inhibitors diffuse slowly and bind avidly. This leads to their accumulation within the matrix. This property makes cartilage resistant to enzymatic degradation from external cells. One consequence of this property is the resistance of cartilage to penetration by blood vessels. Many processes in connective tissues may be regulated by the relative diffusion of molecules through the matrix. Cell activity is modulated by the relative diffusion of hormones compared with diffusion of hormone-degrading enzymes. Collagen and elastic fiber formation is dependent on the constrained diffusion and orientation of protein monomers. As another example, mineralization of connective tissues depends on relative diffusion and binding to achieve the critical concentrations of calcium and phosphate necessary to initiate calcium apatite crystal deposition.

C. Storage Functions

Connective tissues have important storage functions for cell nutrients, major and trace elements, water, and molecules with special functions such as immunoglobulins. For many substances, connective tissues are the major storage sites because of the mass and relative stability of these tissues. The storage function may be general, for the organism as a whole, or local, adapted to the requirements of cells and matrices within connective tissue domains. Most cell nutrients are stored intracellularly following anabolic reactions, which simultaneously store energy while other substances are stored extracellularly. Adipocytes (fat cells), for example, store large molecules of neutral lipid in cytoplasmic vacuoles; however, within these vacuoles, these cells also store lipid-soluble substances, both physiological substances such as fat-soluble vitamins and nitrogen and pathological compounds such as fat-soluble pesticides. Within muscle cells, amino acids are stored as contractile proteins, and carbohydrate as glycogen. All of these substances may be released during periods of exercise or undernutrition, where catabolic activity exceeds anabolism. [*See* ADIPOSE CELL.]

Extracellularly within the hydrogel of soft tissues, water, ions, and specific molecules such as immunoglobulins, growth factors and enzyme inhibitors accumulate. At these sites, stored molecules may have roles in local modulation of cells and modification of matrix properties. In the extracellular compartment of bone, many substances are extensively stored. Not only is 99% of the calcium phosphate stored but also large amounts of magnesium, carbonate, and trace elements such as fluoride. This is another example of nested functions. Bone tissue serves as a reservoir for mineral; bone mineral contributes to bone strength and rigidity. Many organic substances adhere to the bone mineral calcium apatite. Further deposition of apatite results in long-term retention of these substances. A marker of this storage capacity is the fluorescent antibiotic tetracycline, which can be seen years later in bone that formed at the time of drug administration.

D. Cell Modulation

Connective tissues provide much more than a passive material support for parenchymal cells. Extracellular matrix molecules are active in regulating such essential processes as cell adhesion–migration and cell proliferation–differentiation. Two classes of molecules appear to be involved: glycoproteins such as fibronectin and specialized proteoglycans. In general, regulation is mediated by charge-dependent binding of glycosaminoglycans to cell-adhesion molecules and growth factors or by binding of specific glycosaminoglycan sugar sequences to glycoproteins. These intermolecular linkages, which include linkages to the cell membrane, enhance adhesion by promoting organization of actin-intermediate filaments within the cell. Similarly, by blocking the binding sites on either the glycosaminoglycans or the cell membrane, soluble proteoglycans can promote cell mobility. Glycosaminoglycans such as heparan sulfate inhibit cell proliferation possibly by binding growth factors such as fibroblast growth factor and transforming growth factor β. This binding may in fact be a storage mechanism for growth factors. Matrix degradation, occurring in processes such as remodeling or inflammation, may release intact growth factors, which in turn stimulate cell proliferation.

E. Tissue Morphogenesis and Repair

From the formation of mesoderm, an embryonic tissue consisting of primitive fibroblast cells that produce predominantly proteoglycans and glycosaminoglycans, the connective tissue extracellular matrix modulates the form of organogensis. Three-dimensional form and the organization of other tissue elements appears to be controlled by the secretion of specific proteoglycans and specific glycoproteins at specific intervals of development. These molecules regulate the rate of cell proliferation, cell differentiation, and cell migration as well as the development of collagen and elastic fibers within the matrix. The exact recapitulation of this process following injury is termed regeneration, a process that occurs under very limited circumstances within human connective tissues. It is possible to obtain complete regeneration in the adult human, the foremost example being bone regeneration following fracture. More commonly following injury, connective tissue reacts in a process termed repair, which leads to a less functional result than regeneration. The repair process is a program in which, following resorption of tissue by the inflammatory response, there is a timed sequence of restorative cellular events. The repair process includes the orderly elaboration of proteoglycans, vascular invasion, collagen fibrillogenesis, collagen fiber polymerization and cross-linking, and elastic fiber formation, followed by dehydration and contraction of the tissue to form a dense fibrous scar. This repair tissue, while very effective in uniting tissues separated by injury, may be less functional than the original tissue for several reasons. To illustrate, in muscle reparative connective tissues, being noncontractile, lessen muscle strength. In skin, the less hydrated reparative tissue inhibits diffusion of nutrients and waste and contains poorly oriented macromolecules thereby being less adapted to mechanical forces. Furthermore, the rearrangement of connective tissue fibers in repair may prevent regeneration of structures including the arrangement of parenchymal cells within internal organs, and tissues such as muscle fibers and nerve axons. The processes or morphogenesis, regeneration and repair appear to share a similar program of cellular activity and matrix biomolecular synthesis. However, repair is distinguished from morphogenesis by the lesser degree of regulation. This is because the repair process of tissue is proceeding in domains with much greater separation between connective tissue cells and with abundant pre-existing matrix molecules, some partially degraded. These molecules are capable of storing and releasing various stimulatory and inhibition factors accumulated during normal growth and maturation of the tissue under conditions less fully controlled than in the pristine environment of embyronic development.

III. Connective Tissue Regulation

Although connective tissues are highly complex composite materials, which can vary considerably from domain to domain, each domain retains its overall structure and exquisite adaptation to multiple functions over long periods. Furthermore, with growth, repair, and aging, connective tissues undergo orderly changes in structure and function. Except for muscle and endothelial cells, tissue regulation occurs among cells with very limited intercellular contact. Mature connective tissue cells can produce differing amounts and types of extracellular matrix components depending on external stimuli. The most versatile cell is the fibroblast, which, depending on circumstances, can synthesize type I or type III collagens, elastic fibers, or proteoglycans. All of these products share with other cell products the highly regulated processes of protein formation through gene activation, transcription, and translation. This is followed by intracellular protein modification (e.g., hydroxylation, glycosylation) and translocation through the Golgi apparatus and the cell membrane to the extracellular space. For most other kinds of cells, metabolism is modulated by endocrine and neural stimuli mediated by receptors and signal transduction mechanisms within the cell membrane. The regulation of connective tissue cells appears more complex. This complexity is related to the further modification of connective tissue macromolecules in the extracellular space and their integration by orientation and bonding into a composite material. Moreover, connective tissue cells are shielded from short-term fluctuations in plasma hormones by the differential diffusion properties of the extracellular matrix. As well, connective tissue cells are capable of transducing mechanical and electrical signals present at the matrix–cell interface to modulate cell metabolism. Transduction of mechanical and electrical signals by individual connective tissue cells surrounded by matrix is not yet well understood. However, in fibroblasts, chondrocytes, and osteocytes, the solitary cilium with its prominent microtubular organization is thought to be implicated as the cell receptor organelle.

Hormones coming from the blood and locally produced paracrine regulators can permeate the extracellular matrix differentially and are selectively bound and released by its components. Similarly, cell metabolites in a feedback loop must pass through the matrix to other cells or to the bloodstream. In articular cartilage and intervertebral disc, the total diffusion distance may be up to several centimeters in length. The difference in connective tissue composition induced by hormones can be striking, the most visible example being the gender differences between women and men.

Adaptation of connective tissues to changing environments (i.e., growth, aging, traumatic injury, or disease) leads to changes in matrix composition, a process regulated by the connective tissue cells. In these circumstances, the cells secrete matrix-degrading enzymes such as glycosaminoglycanases, collagenases, and other proteases. The activity of these enzymes is controlled by ambient ions, cofactors, and inhibitors. The controlled degradation of matrix by pericellular proteolysis releases growth factors, which stimulate cell proliferation, cell differentiation, and manufacture of new matrix macromolecules. These macromolecules polymerize in the matrix in orientations constrained by pre-existing molecules and ambient mechanical forces.

IV. Aging

Aging is reflected in connective tissues by structural changes in bulk, color, texture, and consistency as well as functional changes. Tissues decrease in size; translucent tissues become opaque; white tissues develop yellow pigmentation; smooth tissues become rougher; and gelatinous tissues become firm. From the resiliency of the bouncing baby to the structural brittleness of the aged, connective tissue changes trend from a gel state toward a crystalline phase. These changes reflect the accumulation of extraneous substances within the matrix, loss of hydration, and increased intermolecular binding, in particular collagen cross-linking. All these changes are interrelated and must be considered as failure of the connective tissue cells to regulate effectively their matrix domains.

To place the sequence of aging events in perspective, we must recall that solutes bind differentially within the matrix. Among the solutes are degradation products of matrix components and cell membranes produced during growth and remodeling as well as from reactions to injury and repair. Ideally, the proteolytic enzymes elaborated by connective tissue cells together with the ability of these cells to synthesize and reconstitute matrix components would restore matrices to an optimal functional state. In aging, this ability decreases over time. Various molecules accumulate within the matrix, the most visible of which is the yellow, oxidized, lipid pigment lipofuscin. The changes in molecular composition decrease the capacity of connective tissue to retain water. Dehydration in turn brings connective tissue macromolecules closer together. This facilitates molecular bonding including collagen cross-linking and heterotopic mineralization such as calcium phosphate crystal deposits in vessel walls or calcium pyrophosphate dihydrate crystal deposition in cartilage.

Hormones and growth factors are substances that can accumulate in matrix. With aging, secretion and activation of proteolytic enzymes by connective tissue cells degrades macromolecules such as elastic fibers, fibronectin, and cell-associated proteoglycans releasing the bound growth factors. The altered concentration of cell growth factors in turn stimulates matrix macromolecule production by cells, modulated toward the production of fibrous molecules such as collagen rather than water-retaining molecules such as proteoglycans. The net effect is loss of tissue bulk and loss of functional capacity. This is reflected in visible manifestations of aging such as thinning and decreased elasticity of dermis, decreased vessel elasticity, loss of muscle bulk and strength, serous atrophy of fat, and osteoporosis.

Driving the rate of these changes in connective tissues is the biological clock of circadian hormonal and nutritional rhythms and their effects on connective tissue cells. Less well understood are the consequences of changes in the connective tissues including the permeability of hormones and feedback regulating molecules on the oscillations of hormone production by endocrine cells. It is likely that a functional understanding of connective tissue aging will involve the intimate interaction between the endocrine effects on connective tissue function and the connective tissue influence on endocrine function. In this paradigm, processes that allow or promote matrix degradation and resorption with accumulation of insoluble matrix molecule degradation products will accelerate aging; processes that promote matrix hydration and clearance of solutes may retard the aging process. [*See* CIRCADIAN RHYTHMS AND PERIODIC PROCESSES; ENDOCRINE SYSTEM.]

V. Disease

Given the ubiquity and mass of connective tissue, disease affecting these tissues can be classified as primary diseases affecting connective tissue molecules and diseases secondary to other disease processes that occur within connective tissue. Furthermore, connective tissue disease can be cross-divided topographically into generalized disease and localized disease. Generalized disease affects all connective tissue in a particular organ or particular connective tissues in the body as a whole. Localized disease involves selected regions in particular tissues (Table V).

Primary generalized connective tissue diseases affect the biosynthesis, structure, or degradation of a single or multiple classes of connective tissue molecules. A relatively well-studied example is the group of diseases known as osteogenesis imperfecta. These diseases are characterized by inherited defects in the structure of type I collagen resulting in brittle bones, thin sclera and dermis, and weak ligaments. Within the four clinical variants of this disorder, several single defects of amino acid deletions, insertions, or substitutions have been found in both primary polypeptide chains, which comprise type I collagen fibers. What must be emphasized is that while there are similar patterns of molecular defects, the potential heterogeneity of defects in these large molecules is immense and only partially correlates with the severity of the clinical syndrome. This reflects the observations that the de-

TABLE V Connective Tissue Diseases

Type/topology	Generalized	Localized
Primary	Primary/generalized	Primary/localized
Secondary	Secondary/generalized	Secondary/localized

fects may be variably expressed in the molecules and that the connective tissue cells may modify the structure of the collagen macromolecular organization to compensate for the weakness of fibrils that are defective at a single site. Another example of generalized connective tissue disease is scurvy, which is related to vitamin C (ascorbic acid) deficiency. Ascorbic acid is a factor required to promote extracellular maturation and cross-linking of collagen fibers and the formation of elastic fibers. Ascorbic acid deficiency results in defective fibril formation and maintenance resulting in increased fragility, particularly of structures that involve both components such as small blood vessels. Yet another example is myxedema, where deficiency of thyroid hormone results in the presence of excess proteoglycan in fibrous and areolar connective tissues. [*See* ASCORBIC ACID.]

The secondary generalized involvement of connective tissues in diseases is observed dramatically in disorders of interstitial fluid volume regulation in which dehydration or edema may ensue. Another example is diabetes mellitus where the increase of glucose within the extracellular matrix of connective tissues alters the permeability and binding characteristics of other solutes adversely.

Primary localized connective tissue diseases affect one or more classes of connective tissue molecules at local specific sites. An illustration of these diseases is the degenerative change in dermal collagen and elastic fibers following prolonged excessive exposure to sunlight. A second common example is the degenerative change in articular cartilage matrix following repeated impact injuries such as occurs with excessive athletic activity.

Secondary localized diseases of connective tissue embrace an immense spectrum of chronic disease involving both connective tissue structures and connective tissue within parenchymal organs. First, the disease process may be localized to a single site typified by a scar following local injury. As noted previously, the scar may be dysfunctional because it is not as permeable or as contractile as normal tissue. For this group of disorders, the implication is that the disease process involves cells within or adjacent to connective tissues. The reaction in connective tissue leads to local changes in structural organization, which disrupt the functional capacity of the tissue. These changes are also observed in the invasive process of malignancy where tumor cells may elaborate proteolytic enzymes that facilitate tumor cell invasion. Alternatively, tumor cells may induce dense collagen fiber formation or excessive proteoglycan production, features that inhibit the access of antitumor substances to the tumor cells.

Secondary diseases may involve an entire class of sites. Certain autoimmune diseases provide illustration of this phenomenon. Antigen–antibody complexes localizing in vessel walls induce inflammation that increases the permeability of vessels leading to mucinous edema, a noted feature of "collagen diseases." Antibodies directed against connective tissue basement membrane antigens may give rise to inflammation, which, depending on circumstances, can provoke blistering of skin or inflammation of kidney glomeruli. Third, connective tissue repair at multiple sites within a parenchymal organ may contribute to parenchymal organ failure. Two examples are offered. The lack of contractility in the scar within the heart with expansion of the scar as an aneurysm can contribute to heart failure. Scarring of kidney glomeruli or tubules following scattered inflammatory events may lead to decreased kidney function. For most secondary processes involving parenchymal organs, the connective tissue reaction is appropriate to the stimuli inducing the change in the connective tissue organization. Disease ensues because the connective tissue organization or location is inappropriate for the normal functioning of the parenchymal tissue. [*See* ANTIBODY–ANTIGEN COMPLEXES: BIOLOGICAL CONSEQUENCES.]

Bibiliography

Evered, D., and Whelan, J. (eds.) (1986). "Functions of Proteoglycans." Ciba Foundation Symposium, 124. John Wiley and Sons, New York, Toronto.

Glimcher, M. J., and Lian, J. B. (1989). "The Chemistry and Biology of Mineralized Tissues." Gordon and Breach Science Publishers, New York.

Nimni, M. (1988). "Collagen," Vols. I, II, III. C. R. C. Press, Boca Raton, Florida.

Pritzker, K. P. H. (1991). Articular skeletal system, muscle, fat and other connective tissues. *In* "Functional Endocrine Pathology" (S. L. Asa and K. Kovacs, eds.). Blackwell, Boston.

Silver, F. H. (1987). "Biological Materials: Structure, Mechanical Properties and Modeling of Soft Tissue." New York University Press, New York.

Vitto, J., and Pereuda, A. J. (1987). "Connective Tissue Disease." Marcel Dekker Inc., New York.

Vogel, S. (1988). "Life's Devices." Princeton University Press, Princeton, New Jersey.

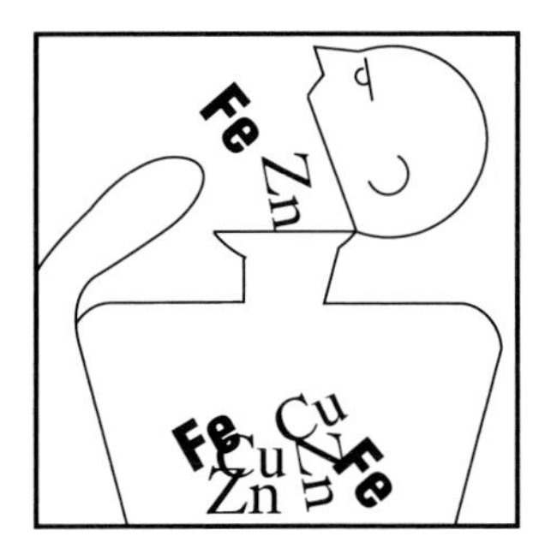

Copper, Iron, and Zinc in Human Metabolism

ANANDA S. PRASAD *Wayne State University*

Glossary

Ataxia Unsteady gait
Dysarthria Difficulty in speech
Erythropoeisis Production of red blood cells
Geophagia Clay- or dirt-eating
Hypogeusia Decreased taste acuity
Interleukin-2 Cytokine essential for proliferation of T helper cells
Neutropenia Decreased number of granulocytes in peripheral blood
Thymulin Zinc-dependent thymic hormone
Total parenteral nutrition All essential nutrients are supplied by intravenous feeding

THE IMPORTANCE OF IRON and iodine for human health has been known for over a century. The importance of other trace elements such as zinc, copper, maganese, selenium, and chromium, however, has been realized only in the past two to three decades. Prior to 1961, it was considered unlikely that zinc deficiency in humans could pose a significant clinical problem, inasmuch as zinc is ubiquitous in the environment. Reports from the Middle East established that zinc deficiency could occur under practical dietary conditions. Zinc deficiency is now recognized to occur in both developing and developed countries under various clinical conditions. Some investigators believe that a deficiency of zinc may be one of the most common nutritional problems in the world.

During the past two decades, considerable progress has been made in the field of trace elements. Impressive advances have been observed in the clinical, biochemical, immunological, and molecular biological areas as related to trace elements.

Three events appear to have contributed to these advances. The first, and perhaps the most important, event was the documentation that deficiency of trace elements in human subjects can occur under practical dietary situations and may be present in many disease states as well. The second event was the development and availability of the atomic absorption spectrophotometer in the 1960s, which simplified the assessment of trace element status in humans. The third event was the recognition of the many biochemical and immunological functions of trace elements that have been observed since 1960. For instance, zinc is now known to regulate the activities of at least 200 enzymes. It is known to be required for many functions of the lymphocytes and, as zinc finger proteins, it is known to be involved in gene expression. These are indeed exciting developments.

Of all the trace elements known to be essential for humans, the knowledge of the functions for human health is most complete for iodine, copper, iron, and zinc. The clinical importance and essentiality of copper, iron, and zinc for human subjects are well established. Distinct deficiency disorders of chromium and selenium have yet to be characterized, and clinical deficiency of manganese has been described on only one occasion.

In this chapter, clinical, biochemical, and metabolic effects of copper, iron, and zinc are addressed. A general overview of trace elements has been provided in another chapter in this book.

I. Copper

The presence of copper in plant and animal tissues was first recognized almost 170 years ago. In 1847, copper was shown to be present in the blood proteins of snails, and a copper-containing pigment (hemocyanin) was recognized to function as a respiratory compound in 1878. The first evidence of the dietary importance of copper in rats was reported in 1925 and, in 1928, it was observed that copper, in addition to iron, was necessary for blood formation in rats. Later, naturally occurring copper deficiency in livestock was recognized. Subsequent studies showed that copper-deficient pigs are unable to absorb iron, mobilize it from the tissues, and utilize it for hemoglobin synthesis.

A. Copper Deficiency in Humans

Copper deficiency has been implicated in the etiology of three clinical syndromes in the human infant. In the first of these, hypochromic microcytic anemia, hypoproteinemia, hypoferremia, and hypocupremia are present, and combined therapy with both iron and copper is required for complete recovery. The second syndrome is known to affect malnourished infants being rehabilitated on high-caloric and low copper-containing diets who exhibit hypochromic microcytic anemia, neutropenia, diarrhea, osteopenia, and hypocupremia. The third clinical condition in infants is Menke's kinky hair syndrome, which is a genetic disorder. In addition to these syndromes, copper deficiency resulting from prolonged administration of total parenteral nutrition (TPN) without copper supplementation has been reported. In adults, copper deficiency due to TPN and secondary to prolonged zinc administration in high dosage has been observed. Copper deficiency in humans has been induced experimentally by dietary manipulation (0.83 mg/day dietary intake). Increase in plasma cholesterol level and abnormal glucose metabolism and a decrease in the activity of superoxide dismutase (a copper- and zinc containing enzyme) in the red cells were observed. These studies demonstrate the importance of copper in clinical medicine.

1. Manifestations of Copper Deficiency

In animals, besides hypochromic microcytic anemia and leukopenia, hypopigmentation, defective wool keratinization, and abnormal hair texture, ataxia from abnormal defective myelination in newborn lambs, abnormal bone formation, reproductive failure, heart failure, spontaneous fractures, and arterial rupture due to aneurysm caused by abnormal elastin formation have been reported as a result of copper deficiency.

In humans, hypochromic anemia and neutropenia are consistent features of copper deficiency. The anemia appears to be related to defects in iron utilization due to a deficiency of copper. The pathogenesis of neutropenia remains poorly understood. In infants with copper deficiency and in patients with Menke's disease, osteoporosis and pathologic bone fractures as well as "scurvylike" bone abnormalities have been reported. Arterial tortuosity and aneurysms have reportedly occurred in patients with Menke's disease. The bony and vascular abnormalities in copper deficiency appear to be related to defective cross-linking in collagen associated with reduced activity of copper-dependent enzyme lysyl oxidase. Hypercholesterolemia and abnormal glucose metabolism were observed in humans in whom a copper deficiency was induced by experimental means.

In patients with Menke's disease, hypopigmentation of hair has been reported. This may be due to decreased activity of the copper-dependent enzyme tyrosinase, which is needed for normal melanin synthesis. The other abnormality, namely kinky hair, observed in Menke's disease may be related to defective formation of disulfide bonds.

Neurologic manifestations have not been described in cases of nutritional copper deficiency in humans, although a number of neurologic symptoms have been observed in patients with Menke's disease. These include hypotonia, hypothermia, episodic apnea, seizures, and mental retardation. These manifestations may be due to either decreased cytochrome oxidase activity or impaired catecholamine synthesis in the central nervous system. Whether or not alterations of lipid composition or some other mechanisms are involved in this syndrome remains to be elucidated.

Other abnormalities in copper deficiency include hypoproteinemia and hypercholesterolemia in experimental animals. The mechanism of these alterations are not well understood.

2. Menke's Disease

Menke's disease was first described in 1962 as a syndrome characterized by growth retardation, focal cerebral and cerebellar degeneration associated with mental retardation, and white hair with peculiar twisting, brittleness, and breakage. Later studies showed that intestinal copper absorption was impaired and hepatic copper content was reduced but that red cell copper was normal and intravenous copper was handled normally. Anemia and neutropenia, which are consistently present in cases of nutritional deficiency of copper, are not seen in Menke's disease. Also the use of a wide range of copper preparations by various routes was found to be ineffective therapy although serum copper and ceruloplasmin levels showed an increase.

Copper content of duodenal mucosal cells has been shown to be increased, and copper appeared to be localized in the mucosal brush border in Menke's disease. In cultured fibroblasts, copper accumulation at the plasma membrane of cells has been observed. Other studies with ^{64}Cu have shown that efflux of copper from cultured fibroblasts is also abnormal in Menke's disease. Whether or not copper is bound abnormally to intracellular protein such as metallothionein or another ligand remains to be settled.

B. Copper Excess

The signs and symptoms of acute copper poisoning are the result of direct irritation to gastrointestinal tract by copper. Pancreatitis resulting from acute copper toxicity has also been observed. Intravascular hemolysis with resultant jaundice, hemoglobinuria, and acute renal failure has been observed in cases of acute copper toxicity. The precise mechanism of acute hemolysis is not known, but it is believed to be associated with decreased glycolysis and decreased activity of glucose-6-phosphate dehydrogenase. Increased osmotic fragility of red cells incubated with copper has been observed; however, it is not known if this mechanism plays a role *in vivo* in clinical hemolysis.

The clinical manifestations of chronic copper toxicity are less well defined. In those patients in whom copper is retained in the liver, sequestration of copper in lysosomes apparently results in protection from toxicity. Later, when this mechanism is overwhelmed, accumulation of cytosol copper results in hepatitis, which may subsequently lead to cirrhosis of the liver. Copper may also accumulate in renal tubules, cornea, brain, and other organs, and later damage to these structures takes place.

1. Wilson's Disease (Hepato-Lenticular Degeneration)

Wilson described a series of patients with the autosomal recessive genetic disorder in the 1912 volume of "Brain"; however, only in 1948 was its association with increased copper recognized. It is believed that in this disease the accumulation of copper in the liver begins at birth (Stage I). The subject remains asymptomatic in the early stages and later the only physical finding may be the presence of Kayser-Fleischer ring (copper deposition in the cornea). In most patients at about the age of adolescence, the capacity of the liver to store copper may be exceeded, and there is sudden hepatic copper release and redistribution (Stage II). At this time, the patient may exhibit a hemolytic episode or experience hepatic failure with findings suggestive of chronic aggressive hepatitis or even hepatic necrosis. Subsequently, macronodular cirrhosis develops. If the patient survives, he/she enters Stage III, in which cerebral copper accumulation begins until in Stage IV, when the patient becomes symptomatic with neurologic disease including dysarthria, gait disturbance, tremor, and loss of fine motor coordination. These cerebral changes due to excessive copper deposition are associated with degeneration and cavitation of the putamen as well as other cerebral structures.

Laboratory features of this disease include hypoceruloplasminemia, increased nonceruloplasmin-bound serum copper, impaired incorporation of radio copper into ceruloplasmin, and increased urinary copper excretion.

It has been suggested that ceruloplasmin probably plays a central role in the pathophysiology of this disease; however, this is obscured by the observation of patients with familial hypoceruloplasmimeia who have no other evidence of abnormal copper metabolism, whereas other patients with Wilson's disease have been noted to have normal ceruloplasmin levels. Extensive biochemical studies from many laboratories have failed to demonstrate any structural, immunochemical, or enzymatic defect of ceruloplasmin in patients with Wilson's disease. Cultured fibroblasts from patients with Wilson's disease have elevated copper concentration, but decreased cytoplasmic copper–protein ratio in comparison to the normal controls, the significance of which remains unknown.

Recent studies indicate that the Wilson's disease gene is on chromosome 13, while the ceruloplasmin gene is on chromosome 3. It has been hypothesized that the Wilson's disease gene produces a product that is involved in the synthesis, processing, handling, or excretion of ceruloplasmin or ceruloplasminlike proteins. Apparently biliary excretion of copper is decreased in Wilson's disease patients, thus leading to copper accumulation throughout life. Recently, it has been demonstrated that fractionated biliary secretions from normal subjects contained copper in two main peaks, a high molecular weight protein peak and a low molecular weight protein peak. When normal subjects were given a high-copper diet, biliary copper increased, including the high molecular weight protein peak. Patients with Wilson's disease lacked the high molecular weight protein peak of copper in their bile. On further study, this protein appeared to be a ceruloplasminlike protein. Thus it appears that in Wilson's disease patients, there is a failure to secrete excess copper via this ceruloplasminlike protein into the bile. This protein is also resistant to proteolytic degradation and, thus, its presumed role is to bind copper and block its intestinal absorption.

The usual treatment of Wilson's disease is penicillamine, a copper chelating agent, which is not well tolerated by all subjects and results in several side effects, some of which may be serious. Research over the past decade has shown that zinc therapy produces a negative copper balance in patients with Wilson's disease. Zinc absorbed by intestinal mucosal cells induces synthesis of the metal-binding cysteine-rich protein metallothionein in the intestinal cells and then exchanges the zinc for copper, which it binds firmly and prevents its movement through the cell and into the body. Zinc also blocks the intestinal uptake of oral ^{64}Cu effectively. Oral zinc therapy thus appears to be a good alternative treatment for Wilson's disease on a long-term basis, inasmuch as zinc is considerably less toxic than penicillamine. Zinc therapy may be an excellent approach for prevention of copper damage to liver, brain, and other organs in subjects who may be considered to be potentially affected by this disorder.

C. Biochemistry and Metabolism

1. Biochemistry

Neonatal ataxia and other neurological signs are commonly observed among several species that have been subjected to copper deficiency. Lack of myelination has been observed in both nutritional deficiency and in Menke's disease. Some observers have suggested that the reduction in myelinated axons is consistent with the observed degree of nerve cell death. A marked decrease in the activity of the myelin markers 2′, 3′, cyclic nucleotide-3′-phosphohydrolase in weanling rats whose mothers were deprived of copper was observed. Whether the reduction in myelin is the result of a specific metabolic defect or simply failure of nerve cell production and survival remains unknown.

A more specific metabolic defect in the brain relates to the metabolism of catecholamines. Nutritional copper deficiency results in depressed levels of norepinephrine in lambs and rats, but the serotonin levels are not affected. The low norepinephrine level can be explained by impaired activity of the copper-dependent enzymes dopamine-hydroxylase. On the other hand, the depressed dopamine levels are not reversed by copper repletion.

The decreased activity of another enzyme, cytosolic superoxide dismutase (SOD), in copper deficiency may also contribute to the low norepinephrine levels. The copper- and zinc-dependent enzyme SOD catalyzes the dismutation of the superoxide anions and forms hydrogen peroxide, which is subsequently transformed to water by the enzyme glutathione peroxidase.

The activity of cytochrome oxidase, a copper-dependent enzyme, is lowered in most tissues by copper deficiency in animals. The depressed activity has been postulated to play a role in the genesis of central nervous system pathology, but evidence for a specific effect is lacking.

Cardiovascular disorders are evident in almost all species subject to copper deficiency of either genetic or nutritional origin. The metabolic defect responsible for most, if not all, of the pathology, such as dissecting aneurysms and angiorrhexis, is failure of cross-link formation in the connective tissue proteins collagen and elastin. The cross-links in the proteins are formed from lysine or hydroxylysine after they are incorporated into the soluble precursor proteins. Lysyl oxidase, a copper-dependent enzyme, catalyzes the oxidation with oxygen serving as the electron acceptor. In most species, when copper is limiting, lysyl oxidase activity is decreased and cross-linking fails, which lead to loss of rubberlike elasticity of elastin and decreased tensile strength of collagen.

The emphysemalike lung that occurs during development under condition of copper deficiency appears to result from failure of cross-link formation. Other factors such as low SOD activity in copper deficiency may also contribute to peroxidative damage of lung tissue.

The metabolic defect in genetic copper deficiency (Menke's disease) remains poorly understood. The defect may involve abnormal intracellular binding of copper or failure to transport it out of the cell. One might postulate that the basic defect in the genetic copper-deficiency disease relates to the rate of synthesis and/or degradation of a thioneinlike protein in at least some cells. Unfortunately, no information exists relative to copper-binding proteins in the brain.

2. Metabolism

It has been estimated that a normal 70-kg human contains 80 mg of total copper; however, many factors may affect this estimate. Differences in dietary and environmental copper exposure have been noted; however, their effect on total body copper content is not well documented. Liver and brain are especially rich in copper and represent about one-third of total body copper. Skeletal muscle also represents one-third of total body copper, although the concentration of copper in muscle is relatively low. In adults, homeostasis is apparently maintained by the absorption of approximately one-third or more of the estimated daily intake of <2.0 mg/day dietary copper. Analysis of copper in 15 American diets has revealed a daily intake ranging from 0.2 to 3.48 mg/day, with 8 diets containing <1.0 mg/day. Of total daily copper losses, from 0.5 to 1.3 mg is excreted in the bile, 0.1–0.3 mg passes directly into the bowel, 0.01–0.06 mg appears in the urine, and small amounts are lost in sweat, skin, and hair. In copper balance studies, a minimum of 1.3 mg of copper intake was required to maintain balance. These estimates were minimal because surface loss of copper (i.e., for hair, desquamated skin, and sweat) was not included in the balance equation. The recommended dietary allowance (RDA) (United States National Academy of Sciences, Food and Nutrition Board) for copper has been set at 2.0 mg/day. Clearly many individuals in the United States do not reach this level of intake. Whether or not this poses any health problems remains to be determined, although, as mentioned earlier, under experimental conditions, human copper deficiency showed an increase in plasma cholesterol level and abnormal glucose tolerance.

II. Iron

Iron is essential for many biochemical processes. It exists in both ferric and ferrous states, and its importance in the oxygen and electron transport systems concerned with cellular energy production is well established. Iron was considered to be of celestial origin in ancient civilizations of the Eastern Mediterranean region. The "metal of heaven" was used in Egypt and Mesopotamia for therapeutic purposes, and its use has been described in the Ebers Papyrus, an Egyptian pharmacopea dating around 1500 B.C. Therapeutic use of iron has also been mentioned in ancient Indian (Hindu) history dating around 500 B.C.

Iron became an accepted method of treatment of a disease known as chlorosis following Sydenham's studies published in 1681, although the mechanism of its action was not understood and its use remained controversial. In 1893, it was demonstrated convincingly that iron administration to women with chlorosis resulted in an increase in the hemoglobin level. In 1938, it was unequivocally demonstrated that inorganic iron is incorporated quantitatively into hemoglobin. Biochemical studies later showed that iron is intimately involved in oxygen utilization by the tissues as well as in oxygen transport as part of the hemoglobin molecule.

A. Deficiency of Iron

Iron is essential for many biochemical processes. Although the Earth's crust consists of almost 4% iron and the human diet may contain plenty of iron, ferric iron is insoluble, thus its availability for human metabolism is low. The body is limited in the adjustments it can make to loss of iron due to hemorrhage. Indeed, iron deficiency is the most common cause of anemia throughout the world.

When the body is in a state of negative iron balance, the iron is mobilized from the body storage pool for the synthesis of hemoglobin. Iron absorption is increased when stores are reduced. Serum ferritin is decreased before anemia develops. Serum ferritin is an iron storage compound which correlates with total iron stored in the body. When the stores are entirely depleted, the stainable iron in the bone marrow is no longer present. The next parameter to be affected is the serum transferrin saturation, which falls to <15%. Transferrin, a glycoprotein is an iron-binding protein responsible for transport of iron to precursors of erythroid cells for hemoglobin synthesis. Recent studies indicate that as a result of iron deficiency, even though the subjects may not be anemic, there may be impaired exercise tolerance and physical work capacity and reduced cognitive function in children.

If the negative iron balance continues, anemia develops and the patient may show breathlessness, pallor, and tachycardia. The red cells become hypochromic and microcytic. The serum total iron-binding capacity increases, and the serum iron decreases. The reticulocyte count is low. Partial villous atrophy, with minor degrees of malabsorption of xylose and fat, reversible by iron therapy, has been reported in infants suffering from iron deficiency. This has not been observed in iron-deficient adults. Iron-containing enzymes such as the cytochromes, catalase, and trypotophan pyrrolase are usually better preserved in the tissues than other iron-containing compounds. In severe iron deficiency, however, the activities of various iron enzymes are decreased. Poor activation of lymphocytes transformation, diminished cell-mediated immunity, and impaired intracellular killing of bacteria (white blood cells) by neutrophils have been reported by some investigators, but detailed immunological studies are not available.

Iron deficiency is seen in subjects with chronic blood loss, malabsorption syndrome, and diets from which iron is poorly *bioavailable* and in subjects with increased requirement of iron. Iron-deficiency anemia is easily correctable by oral or parenteral supplementation of iron.

1. Sideroblastic Anemias

Sideroblastic anemias comprise a group of refractory anemias in which, although the peripheral blood may show the presence of cells with reduced hemoglobin content, their precursor, the normoblasts in the bone marrow contain excess of iron. In a significant number of normoblasts, the iron accumulates around the nucleus and these are called ringed sideroblasts.

The congenital sideroblastic anemia is a rare X-linked disorder affecting mainly males in childhood or adolescence. The females are heterozygotes with only one altered gene in their X chromosomes. In advanced cases, spleen and liver are enlarged; the serum iron is very high, and iron-binding capacity is decreased. The anemia is severe.

Primary acquired sideroblastic anemia affects usually middle-aged and elderly subjects and is considered to be a variety of altered development of white blood cells. Many of these subjects represent examples of prelymphoma, premyeloma, or preleukemia. The clinical symptomatology of primary sideroblastic anemias are related to severe anemia and iron overload. The treatment is not very effective.

Secondary sideroblastic anemias may be seen in conditions with abnormalities of vitamin B_6 metabolism, leading to disorder in heme synthesis. These include antituberculous chemotherapy, celiac disease, hemolytic anemia, and alcoholism. Secondary sideroblastic anemia seen in lead poisoning, alcoholism, and use of chloramphenicol is also due to disturbance of heme synthesis or mitochondrial function. Secondary sideroblastic anemia in patients with collagen diseases, malignancy, and megaloblastic anemias is due to ineffective iron utilization by the normaoblasts.

2. Anemia of Chronic Disorders

One of the most common types of anemia is anemia of chronic disorders. This is due to chronic infections such as tuberculosis, malignant diseases, chronic inflammatory diseases, and renal failure. The pathogenesis of this disorder is not well understood. The marrow fails to mount appropriate erythropoiesis in spite of anemia. Whether the production of erythropoietin is decreased or erythroblast response to erythropoietin is reduced in this disease is not known. The changes in iron metabolism are consistent with a block in the release of macrophage iron. The decreased serum iron level is believed to be caused by increased lactoferrin production from granulocytes and increased apoferritin synthesis by macrophages. A mild shortening of red cell life span has also been observed in these cases. There is an imbalance between plasma iron supply and the erythroid marrow requirements in these conditions.

B. Iron Excess

In idiopathic hemochromatosis, there is an excessive parenchymal iron storage, particularly in the liver leading to tissue damage. Clinically the disease is manifested by cirrhosis of the liver, endocrine disorders, skin pigmentation, and cardiac failure. The excess iron store is believed to be due to an unknown genetic defect of iron metabolism leading to an increased absorption of iron. The incidence of the disorder has been estimated to be 1 in 10,000 births.

Other causes of iron overload include ineffective red blood cell development such as thalassemias, sideroblastic anemias, and congenital dyserythropoietic anemias. Excessive intake of iron such as that seen in certain African groups from drinking native beers prepared in iron drums and excessive blood transfusion result in iron overload. There are other disorders such as alcoholic cirrhosis, idiopathic pulmonary hemosiderosis, rheumatoid arthritis, and paroxysmal nocturnal hemoglobinuria, which may be associated with iron overload in tissues.

Excess iron in tissues leads to peroxidation of cell membrane lipids due to iron-catalyzed hydroxyl radical formation and oxidation of intracellular proteins, which ultimately result in tissue damage. Damage to lysosomal membranes by hemosiderin has also been shown to occur, with release of lysosomal enzymes into other cell compartments. Liver damage in transfused patients may also be due to hepatitis B virus or non-A, non-B hepatitis.

The main treatment is iron chelation to solubilize iron by use of iron chelators such as desferrioxamine or repeated blood drainage which ultimately leads to removal of iron from the body.

C. Biochemistry and Metabolism

1. Biochemistry

Hemoglobin is one of the most important functional iron-containing proteins. Hemoglobin (molecular weight 64,500) contains four iron-containing heme groups linked to four globin chains and can bind four molecules of oxygen. The total amount of iron in the hemoglobin pool is 2,500 mg. Myoglobin present in muscles (molecular weight 17,000) contains approximately 10% of body iron as a single heme group attached to its one polypeptide chain. The total amount of iron in this pool is 300 mg. Myoglobin has higher affinity for oxygen than hemoglobin. The mitochondria contain a series of heme and nonheme iron proteins. These include the cytochromes a, b, and c, succinate dehydrogenase, and cytochrome oxidase, which form an electron transport pathway responsible for the oxidation of intracellular substrates and the simultaneous production of adenosine triphosphate, the energy currency of cells. [*See* HEMOGLOBIN.]

Cytochrome P_{450} is present in the endoplasmic reticulum. It is involved in detoxification of various chemicals by the liver. The iron-containing enzymes catalase and lactoperoxidase are involved in peroxide breakdown. Tryptophan pyrrolase, an iron enzyme, is needed for the oxidation or tyrptophan to formylkynurenine. Xanthine oxidase, aconitase, and nicotinamide adenine dinucleotide, dehydrogenase are iron-reduced, and sulphur-containing proteins. Iron is also necessary for the enzyme ribonucleotide reductase, an important enzyme for DNA synthesis. [*See* CYTOCHROME P-450; DNA SYNTHESIS.]

Heme consists of a protoporphyrin ring with an iron atom at its center. The porphyrin ring consists of four pyrrole groups united by methene bridges. The hydrogen atoms in the pyrrole groups are replaced by four methyl, two vinyl, and two propionic acid groups. Heme is synthesized from the precursors succinic acid and glycine. The enzyme ferrochelatase and glutathione are required for the incorporation of iron into the protoporphyrin molecule. Once the heme is synthesized, it combines with globin chains to make hemoglobin.

Transferrin (molecular weight 79,500) is the vital iron transport protein. It accounts for only 4 mg of body iron. It is a β-globulin glycoprotein. It is present in plasma and extravascular spaces, and its plasma half-life is 8–11 days. The transferrin gene is on chromosome 3, and the protein is synthesized in the liver. Synthesis is inversely related to iron stores. Plasma level is usually 1.8–2.6 g/liter. Two atoms of ferric iron may be attached to each molecule. The binding sites most likely contain three tyrosine, two histidine, and an arginine group. Binding of iron to transferrin also involves attachment of an anion, usually bircarbonate.

Lactoferrin (molecular weight 77,000) is a related glycoprotein and also binds two atoms of iron per molecule. It is present in milk and other secretions. Its presence in neutrophils is believed to have a bacteriostatic effect by depriving the offending organisms of the iron needed for their growth.

Ferritin (molecular weight 480,000) is the primary storage compound for iron. It is made up of a spherical shell enclosing a core of ferric hydroxyphosphate (up to 4,000 iron atoms). Human ferritin has 24 subunits of two immunologically distinct chain types H and L. The small amount of ferritin present in human serum contains little iron and consists almost exclusively of L subunits. It has been suggested that serum ferritin may be secreted by macrophages and/or hepatocytes, which have been stimulated by iron to synthesize ferritin.

Hemosiderin is a noncrystalline and water-insoluble iron storage compound. In normal subjects, storage iron is about two-thirds ferritin and one-third hemosiderin, but in iron overload the proportion of hemosiderin increases considerably.

2. Metabolism

Iron absorption depends on the dietary content of iron, its bioavailability, and the body's need for iron. A normal Western diet should provide approximately 15 mg/day iron. Only 5–10% of dietary iron is absorbed. Heme iron is better absorbed than nonheme iron. The absorption of heme iron is 20–30%, whereas only 1–5% of nonheme iron is absorbed in normal subjects. Ferric iron salts are less absorbed than ferrous compounds. Phytates, phosphates, tannates, oxalates, phosphorpeptides, and products of Maillard browning decrease the availability of iron for absorption. Iron absorption is increased by the following conditions: iron deficiency, increased erythropoiesis, ineffective erythropoiesis, pregnancy, and anorexia. Ascorbic acid increases iron absorption by reducing ferric to ferrous form. Iron absorption is decreased when the body is overloaded with iron and in acute and chronic infections.

Iron absorption may be regulated both at the stage of mucosal uptake (possibly by varying the number of brush-border iron receptors) and at the stage of transfer to the blood. Iron is maximally absorbed from the duodenum.

Iron uptake by mucosal cells appear to involve binding to specific receptors on the brush-borders followed by an energy-dependent transfer across the cell membrane. At higher doses, there may be a passive diffusion as well. Heme enters the mucosal cells intact. Heme oxygenase breaks up heme and releases iron intracellularly. Iron then may be transferred to the portal circulation or may enter ferritin to be eventually lost with the exfoliation of the mucosal cell. The nature of the intracellular iron pool and transport across the cell membrane remains unknown.

Iron from the mucosal cell is transferred to the plasma transferrin, which transports iron to bone marrow for synthesis of heme by the erythroblasts. The binding of iron to transferrin is facilitated by ceruloplasmin, a ferrooxidase. Transferrin picks up iron not only from the intestinal cells but also from macrophages and liver. Senescent red cells are phagocytized by macrophages where iron is released by the action of heme oxygenase. As ferrous iron, it can then either enter ferritin (where it is oxidized to ferric iron by the ferritin protein) or be released into plasma where it binds to transferrin and then is transported to the bone marrow. The mechanism of iron donation to transferrin is poorly understood.

Uptake of iron from transferrin by developing red cell or other cells requires presence of specific receptor sites. It has been reported that some of the receptors are shed into the plasma and their level correlates with the overall activity of the erythron. These receptors can be assayed by immunoassay techniques. The transferrin receptor is a transmembrane protein consisting of two monomers linked by a disulfide bridge; each subunit is able to bind one transferrin molecule. The human transferrin is identified by the monoclonal antibody OK T9. It has a much higher affinity for fully saturated, diferric transferrin than for monoferric transferrin. Some investigators believe that the transferrin receptor complex remains on the cell surface during iron release. Some evidence suggests that in both reticulocytes and hepatocytes a receptor-mediated endocytosis may occur, i.e., the transferrin–receptor complex is engulfed into the cells in small vesicles. The iron is released inside the cell perhaps by a fall in pH within the vesicle. Whether or not a specific intracellular carrier, either a protein or low molecular weight chelate, is then involved in transport of iron to the mitochondria or to ferritin remains uncertain.

About 90% of the iron used by the bone marrow cells each day is derived from red cells that are lyzed by the reticuloendothelial cells, and only 5–10% is derived from the slow turnover of the storage pool and from absorption from the gut.

III. Zinc

The importance of zinc for human health has been recognized only in the past two and a half decades. During the past 25 years, remarkable progress has been made in the clinical, biochemical, and immunological aspects of the role of zinc in humans. Although the essentiality of zinc for plants and animals was known, its ubiquity made it seem improbable that human deficiency occurred or that alteration in zinc metabolism could lead to significant problems in clinical medicine. Research over the last 25 years has shown that prediction to be in error. The first documentation of human zinc deficiency was reported by A. Prasad and his associates from the Middle East in 1963.

Deficiency of zinc was suspected to occur for the first time in 1958, in an Iranian 21-yr-old male who looked like a 10-yr-old boy. In addition to severe growth retardation and anemia, he had hypogonadism, hepatosplenomegaly, rough and dry skin, mental lethargy, and geophagia. He ate only bread (wheat flour) without any intake of animal protein and consumed approximately 500g of clay every day. The habit of geophagia is fairly common among the villages in that part of the world. There was no evidence of blood loss. The anemia was due to iron deficiency. Inasmuch as this syndrome was fairly prevalent in Iranian villages, hypopituitaism as an explanation for growth retardation and hypogonadism was ruled out.

It was difficult to explain all of the clinical features solely on the basis of tissue iron deficiency. Because growth retardation and testicular atrophy are not seen in iron-deficient experimental animals, the possibility that zinc deficiency may have been present was considered. Zinc deficiency was known to produce retardation of growth and testicular atrophy in animals. As such, it was speculated that some factors responsible for decreased availability of iron in these patients with geophagia may also have decreased the availability of zinc. Many other cases followed. Zinc deficiency was recognized based on the following: the zinc concentration in plasma, red cells, and hair was decreased and ^{65}Zn studies revealed that the plasma zinc turnover rate was greater, the 24-hr exchangeable pool was smaller, and the excretion of ^{65}Zn in stool and urine was less in patients than in control subjects.

Further studies showed that the rate of growth was greater in patients who received zinc as compared with those who received iron instead and those receiving only an adequate animal protein diet. Pubic hair appeared in all cases within 7–12 wk after zinc supplementation was initiated. Genitalia size became normal and secondary sexual characteristics developed within 12–24 wk in all patients receiving zinc. No such changes occurred in a comparable length of time in iron-supplemented groups or in the group on an animal protein diet. Thus, the growth retardation and gonadal hypofunction in these subjects was related to a deficiency of zinc.

Nutritional deficiency of zinc affecting growth in children and adolescents is fairly widespread throughout the world. Zinc deficiency is expected to occur in countries where cereal proteins are primary in local diet.

It is currently believed that the risk of suboptimal zinc nutrition may pose a problem for a substantial section of the United States and the world population.

A. Clinical Spectrum of Human Zinc Deficiency

During the past two decades, a spectrum of clinical deficiency of zinc in human subjects has been recognized. On the one hand, the manifestations of zinc deficiency may be severe; on the other, zinc deficiency may be mild or marginal.

A severe deficiency of zinc has been reported to occur in patients with acrodermatitis enteropathica (AE), following TPN without zinc, following excessive use of alcohol, and following penicillamine therapy.

The manifestations of severe zinc deficiency in humans include bullous pustular dermatitis, alopecia, diarrhea, emotional disorder, weight loss, intercurrent infections due to cell-mediated immune dysfunctions, hypogonadism in males, neurosensory disorders, and problems with healing of ulcers. If this condition is unrecognized and untreated, it becomes fatal.

A moderate level of zinc deficiency has been reported in a variety of conditions. These include nutritional deficiency due to dietary factors, malabsorption syndrome, alcoholic liver disease, chronic renal disease, sickle-cell disease, and chronically debilitated conditions.

The manifestations of a moderate deficiency of zinc include growth retardation and male hypogonadism in adolescents, rough skin, poor appetite, mental lethargy, delayed wound healing, cell-mediated immune dysfunctions, and abnormal neurosensory changes.

Although the clinical, biochemical, and diagnostic aspects of severe and moderate levels of zinc deficiency in humans are fairly well defined, the recognition of mild levels of zinc deficiency has been difficult. Zinc assay in plasma, urine, and hair have been proposed as potential indicators of body zinc status. Currently, plasma zinc appears to be the most widely used parameter for assessment of human zinc status, and it is known to be decreased in cases of severe and moderate deficiency of zinc. However, several physiologic and pathologic conditions also may affect zinc levels in the plasma and urine, thus a reduced plasma or urine zinc level cannot be taken necessarily as an indicator of low body zinc status. Zinc in hair and erythrocytes do not reflect active or recent status of body zinc, inasmuch as these tissues are slowly turning over. Furthermore, in the cases of mild deficiency of zinc in humans, the plasma levels of zinc may remain normal, and, clinically, overt evidences of zinc deficiency may not exist, thus creating a difficult diagnostic problem.

Therefore, assay of zinc in more rapidly turning over blood cells such as lymphocytes, granulocytes, and platelets as indicators of zinc staus in human subjects has been utilized. With the use of these data, mild deficiency of zinc in humans was recognized.

A mild deficiency of zinc was induced in human volunteers experimentally by dietary means. Measurable effects occurred on zinc concentration of cells such as lymphocytes, neutrophils, and platelets. Some subjects showed abnormal dark adaptation, decreased lean body mass, and hypogeusia, which were corrected with zinc supplementation. Natural killer cells as well as interleukin-2 activity decreased. Sperm count declined slightly during zinc restriction. Clearly, these observations show that a mild zinc deficiency in humans affects clinical, biochemical, and immunological functions adversely.

B. Biochemistry and Metabolism of Zinc

1. Biochemistry

Several studies have shown that zinc is required for protein synthesis. Zinc supplementation has been found to be beneficial for normal healing in deficient subjects. Collagen is the main fibrous protein of the connective tissue and is largely responsible for the development of tensile strength in tissue and for the development of tensile strength in the healing wound. In zinc-deficient rats, a significant reduction in total collagen in sponge connective tissue (bone) was observed, apparently caused by a generalized effect on protein synthesis and nucleic acid metabolism, rather than specifically on collagen synthesis, possibly through decreased proliferation of fibroblasts. Also, glucose tolerance of zinc-deficient animals appears to be impaired, possibly through a reduction of the rate of insulin secretion in response to a glucose stimulation. [*See* COLLAGEN, STRUCTURE AND FUNCTION.]

Total insulinlike activity and immunoreactive insulin have been reported to be decreased in zinc-deficient animals. One possible explanation is that insulin destruction is increased in zinc deficiency. [*See* INSULIN AND GLUCAGON.]

The role of zinc in gonadal function was investigated in rats. Body weight gain, zinc content, and weights of testes were significantly lower in the zinc-deficient rats than in the controls. The serum luteinizing-hormone (LH) responses of the pituitary gonadotropins and follicle-stimulating hormone (FSH) to gonadotropin-releasing hormone (GN-RH) administration were higher in the zinc-deficient rats, but serum testosterone response was lower than in the restricted-fed controls, suggesting an impairment of gonadal function through alteration of testicular steroidogenesis. Similar results have now been reported in experimentally induced zinc-deficient human subjects. Supplementation with zinc resulted in reversal of testicular failure in such cases. [*See* TESTICULAR FUNCTION.]

Zinc has been shown to improve filterability through a 3.0-μm nucleopore filter of sickle cells, red blood cells containing an altered form of hemoglobin, probably through an effect of zinc on red cell membrane. Recent studies show that the process of formation of irreversibly sickled cells involves the cell membrane. Calcium and/or hemoglobin binding may promote the formation of irreversibly sickled cells, thus hindering their filterability. Zinc may block the calcium and/or hemoglobin binding to the membrane.

2. Zinc and Immunity

Recent studies clearly indicate that zinc is required for lymphocyte transformation. The effect of zinc appears to be that of a mitogen, and the kinetics of these influences most closely approximate the effects of antigen stimulation on lymphocyte culture. Currently available data suggest a direct stimulatory influence of zinc on DNA replication. Direct cell-surface effects of zinc cannot be ruled out, however. It is conceivable that zinc could be operating at several different levels. [*See* LYMPHOCYTES.]

Assessment of the role of zinc in the development and functions of different lymphoid cell populations strongly indicates that this element has an effect predominantly on T lymphocytes. Recent studies have shown that thymulin, a thymic hormone, is zinc dependent. In zinc deficiency the serum level of active thymulin is decreased that may affect adversely the functions of T lymphocytes. In AE, zinc deficiency exerts a profound and apparently specific effect on the thymus, thymocytes, and cellular immune functions, which are reversible with zinc repletion.

Granulocytes from chronic uremics who are zinc-deficient show significantly impaired mobility, both chemotactic and chemokinetic, in comparison with subjects who are supplemented with zinc. Furthermore, a significant correlation between granulocyte chemotaxis and both plasma and granulocyte zinc concentrations among all patients supports a pathophysiologic relationship between the severity of impaired granulocytic chemotactic response and zinc deficiency in these patients. Abnormal granulocyte chemotaxis, corrected by zinc supplementation, has also been observed by others in nonuremic patients with AE.

3. Zinc and Metallothionein

Highly purified metallothionein isolated from equine and human kidney contains 26 SH groups per mole, and the protein binds cadmium and zinc as well as copper. In the liver, kidneys, and possibly other organs, metallothionein apparently functions in detoxification and storage of trace elements. In the intestine, metallothionein apparently mediates the absorption of copper and other trace elements. The antogonism among copper, cadmium, and zinc may result from competition for binding sites on metallothionein.

It is believed that zinc induces the synthesis of metallothionein *de novo*, although one cannot exclude the possibility that zinc stabilizes the apoprotein that is being continually synthesized but normally has a sufficiently short turnover time to prevent its accumulation in the liver. An overall homeostatic mechanism is proposed where metallothionein synthesis is controlled at the transcriptional level by body zinc status.

4. Zinc in Gene Expression

Recently, the importance of zinc-binding finger-loop domains in DNA-binding proteins as regulators of gene expression has been recognized. The first zinc finger protein to be recognized was the transcription factor-III A of *Xenopus laevis*, which contained tandem repeats of segments with 30 amino acid residues, including pairs of cysteines and histidines. The presence of zinc in these proteins is essential for site-specific binding to DNA and gene expression. The zinc ion apparently serves as a strut that stabilizes folding of the domain into a finger-loop, which is then capable of site-specific binding to double-stranded DNA. The zinc finger-loop proteins provide one of the fundamental mechanisms for regulating gene expression of many proteins. In humans, the steroid hormones (and related compounds, such as thyroid hormones, vitamin D_3, and retinoic acid) enter cells by facilitated diffusion and combine with respective receptors (which contain DNA-binding domain of two zinc finger-loops), either before or after entering the nucleus. Complexation of a hormone by its specific receptor evidently initiates a conformational change that exposes the zinc finger-loops, so that they bind to high-affinity sites on DNA and regulate gene expression.

5. Metabolism of Zinc

The zinc content of a normal 70-kg male is estimated as approximately 1.5–2.0 g. Liver, kidney, bone, retina, prostate, and muscle appear to be rich in zinc. In humans, zinc content of testes and skin has not been determined accurately, although clinically it appears that these tissues are sensitive to zinc depletion.

Zinc in the plasma is mostly present as bound to albumin, but other proteins such as α_2 macroglobulin, transferrin, ceruloplasmin, haptoglobin, and γ-globulins also bind significant amounts of zinc. Besides protein-bound fraction, a small proportion of zinc (2–3% of overall zinc) in the plasma also exists as ultrafilterable fractions, mostly bound to amino acids, but a smaller fraction is present as ionic form. Histidine, glutamine, threonine, cysteine, and lysine appear to have significant zinc-binding affinity. Whereas amino acids competed effectively with albumin, haptoglobin, transferrin, and IgG for binding of zinc, a similar phenomenon was not observed with respect to ceruloplasmin and α_2 macroglobulin, suggesting that the latter two proteins exhibited a stronger binding affinity for zinc.

Approximately 10–30% of ingested dietary zinc is absorbed. Data on both the site(s) of absorption in humans and the mechanism(s) of absorption, whether it be active, passive, or faculative transport, are meager. Zinc absorption is variable in extent and is highly dependent on a variety of factors. Zinc is more available for absorption from meat and meat products. Zinc is poorly available from cow's milk, but the availability from human milk is very good. Among other factors that might affect zinc absorption are body size, level of zinc in the diet, and presence in the diet of other potentially interfering substances, such as phosphate, phytate, hemicellulose, products of Maillard browning, lignin, phosphopeptides, casein, and other binding substrates.

According to a proposed model, a portion of the dietary zinc entering the lumen of the small intestine is transported across the mucosal brush-border membrane by a process probably requiring adenosine triphosphate. Within the intestinal cells, newly acquired cytoplasmic zinc equilibrates with "zinc pool" and is either transferred to high molecular weight proteins and/or metallothionein or to the plasma.

In animals, if zinc status is adequate, a significant amount of zinc is transferred to the plasma. If dietary zinc is high, the plasma zinc concentration and the *de novo* synthesis of metallothionein are concomitantly increased. A reduction in zinc absorption is directly correlated with the uptake of orally administered zinc into newly formed thionein polypeptides. In view of the interactions that dietary zinc undergoes during transit through the intestinal cells, it is conceivable that information regarding zinc status programs the rate and extent of zinc absorption, in part via changes in the concentration of inducible metallothionein.

Normal zinc intake in a well-balanced American diet with animal protein is approximately 10–12 mg/day, although the RDA for zinc is 15 mg/day for men and 12 mg/day for women. Urinary zinc loss is approximately 0.5 mg/day. Loss of zinc by sweat may be considerable under certain climatic conditions. Under normal conditions, approximately 0.5 mg of zinc may be lost daily by sweating. Endogenous zinc loss in the gastrointestinal tract may amount to 3–5 mg/day. [*See* ZINC METABOLISM.]

IV. Concluding Remarks

Copper, iron, and zinc are essential elements for human health. Copper is required for many biochemical functions including its vital role in iron utilization, collagen cross-linking, free radical reactions, and neurotransmitters. Deficiency of copper in humans is thought to be uncommon.

Iron is essential for hemoglobin synthesis, and iron deficiency anemia is very prevalent throughout the world. Iron is also required for many enzymatic functions. Defective utilization of iron results in sideroblastic anemia and anemia of chronic disorders. Hemochromatosis is a genetic disorder, characterized by excessive iron absorption and accumulation of excess iron ultimately leading to organ damage.

Zinc regulates the activities of over 200 enzymes and is also essential for gene expression of various proteins. Zinc is a growth factor, and several hormones appear to be zinc-dependent. Growth retardation is seen in zinc-deficient infants, children, and adolescents. Zinc is essential for cell-mediated immune functions. Deficiency of zinc is prevalent throughout the world.

Bibliography

Brewer, G. J., Yuzbasiyan, V. A., Iyengar, V., Hill, G. M., Dick, R. D., and Prasad, A. S. (1988). Regulation of copper balance and its failure in humans, *In* "Essential and Toxic Trace Elements in Human Health and Disease" (A. S. Prasad, ed.), pp. 95–104. Alan R. Liss, New York.

O'Dell, B. L. (1982). Biochemical basis of the clinical effects of copper deficiency. *In* "Clinical, Biochemical, and Nutritional Aspects of Trace Elements" (A. S. Prasad, ed.), pp. 301–314. Alan R. Liss, New York.

Pippard, M. J., and Hoffbrand, A. V. (1989). "Iron in Postgraduate Hematology," 3rd ed. (A. V. Hoffbrand and S. M. Lewis, eds.), pp. 26–54. Heinemann Professional Publishing, Oxford, England.

Prasad, A. S. (1978). "Trace Elements and Iron in Human Metabolism." Plenum Publishing Corporation, New York.

Prasad, A. S. (1982). Clinical and biochemical spectrum of zinc deficiency in human subjects. *In* "Clinical, Biochemical, and Nutritional Aspects of Trace Elements" (A. S. Prasad, ed.), pp. 3–62. Alan R. Liss, New York.

Prasad A. S. (1988). Clinical spectrum and diagnostic aspects of human zinc deficiency. *In* "Essential and Toxic Trace Elements in Human Health and Disease" (A. S. Prasad, ed.), pp. 301–314. Alan R. Liss, New York.

Williams, D. M. (1982). Clinical significance of copper deficiency and toxicity in the world population. *In* "Clinical, Biochemical, and Nutritional Aspects of Trace Elements" (A. S. Prasad, ed.), pp. 277–399. Alan R. Liss, New York.

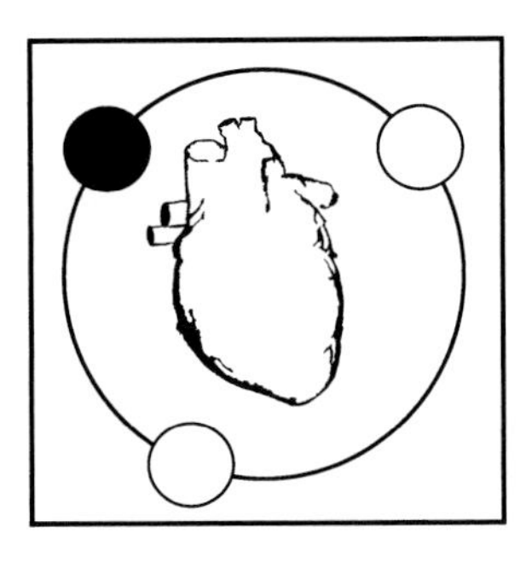

Coronary Heart Disease, Molecular Genetics

SAMIR S. DEEB, University of Washington

Glossary

Apolipoproteins Protein components of plasma lipoprotein particles
Atherosclerosis Disorder of arterial walls characterized by deposition of lipids and proliferation of smooth muscle cells; as these processes progress, the lumen of the artery narrows and artery walls harden, both of which decrease blood flow and may lead to a heart attack
Dyslipoproteinemia Abnormally high or low levels of plasma lipoproteins and associated lipids; hyperlipoproteinemia indicates an elevated level and hypolipoproteinemia indicates diminished concentration
Familial combined hyperlipidemia Phenotype characterized by individuals in a family displaying elevated levels of either cholesterol, triglycerides, or both; believed to be specified mainly by a dominant gene
Lipoprotein Spherical particle made up of a core of nonpolar lipids (cholesteryl ester and triglycerides) and a surface monolayer of cholesterol, phospholipids, and proteins
Lipoprotein lipase Enzyme attached to the endothelium of blood capillaries that catalyzes the hydrolysis of triglycerides in lipoprotein cores to generate fatty acids
Restriction endonucleases Enzymes that cleave DNA at specific deoxyribonucleotide sequences
Restriction fragment-length polymorphism Presence or absence of restriction endonuclease site on DNA results in the generation of short or long DNA fragments, respectively, when genomic DNA is cleaved by the endonuclease

CORONARY HEART DISEASE (CHD) remains the major cause of death in the United States and western Europe. Atherosclerosis accounts for the majority of these deaths. Epidemiological findings indicate that CHD is caused by the interaction of both environmental factors, such as diet, cigarette smoking, alcohol consumption, exercise, etc., and genetic determinants. Hyperlipidemia, hypertension, and diabetes are some of the risk factors for CHD that have definite genetic bases. The dietary components that contribute to the development of hyperlipidemia and CHD are cholesterol and saturated fatty acids. Individuals in the population differ genetically in their ability to "protect" themselves against long-term exposure to diets rich in these components.

The biochemistry and molecular genetics of hyperlipidemias have progressed dramatically during the past few years, offering the opportunity to delineate some of the genes that, when present in dysfunctional forms, predispose persons to premature hyperlipidemia and CHD. Several rare inborn errors of metabolism that affect blood lipid levels have recently been described and each assigned to a single gene. However, understanding the genetic basis of the common types of hyperlipidemia and

CHD will be the topic of intensive investigation during the next decade.

I. Plasma Lipoproteins: Structure and Metabolism

Results from a large number of epidemiological studies point to an association between elevated plasma cholesterol levels and an increased risk of CHD. Cholesterol is carried in plasma in several types of lipoprotein particles, which appear as spherical microemulsions of various sizes. They contain a central core of nonpolar lipids such as triglycerides and cholesteryl esters and a surface monolayer composed of unesterified cholesterol, phospholipids, and proteins called apolipoproteins (Apo) (Fig. 1). Based on density, composition, and function, the lipoproteins are classified into five major classes (Table I). The inverse relationship between size and density indicates that the larger particles have a high ratio of the low-density core lipids to the high-density surface proteins. The two largest particles, chylomicrons and very low-density lipoproteins (VLDL), which are triglyceride-rich, are formed in the small intestine and liver, respectively, and are then secreted into the blood. The triglycerides in these particles are progressively hydrolyzed by lipoprotein lipase, an enzyme attached to the luminal surface of capillaries of many tissues in the body, transforming them into the smaller and more cholesteryl ester-rich particles called chylomicron remnants, intermediate-density lipoproteins (IDL) and low-density lipoproteins (LDL) (see Figs. 2 and 3). The released fatty acids are taken up by fat cells for storage or by other cells for oxidation (used as an energy source). The remaining particles are taken up by liver and peripheral tissues where cholesterol is either stored or used mainly as a component of cell membranes, or for steroid hormone production, as is the case of certain tissues. Excess cholesterol can be secreted by the liver into the bile. Cholesterol can also be synthesized by all types of cells in the body from simple compounds. The uptake of chylomicron remnants, IDL, and LDL particles by cells is receptor-mediated. Apos E and B-100 function as ligands in the specific binding of lipoproteins to receptors found on most types of cells in the body. [*See* CHOLESTEROL; FATTY ACID UPTAKE BY CELLS; PLASMA LIPOPROTEINS.]

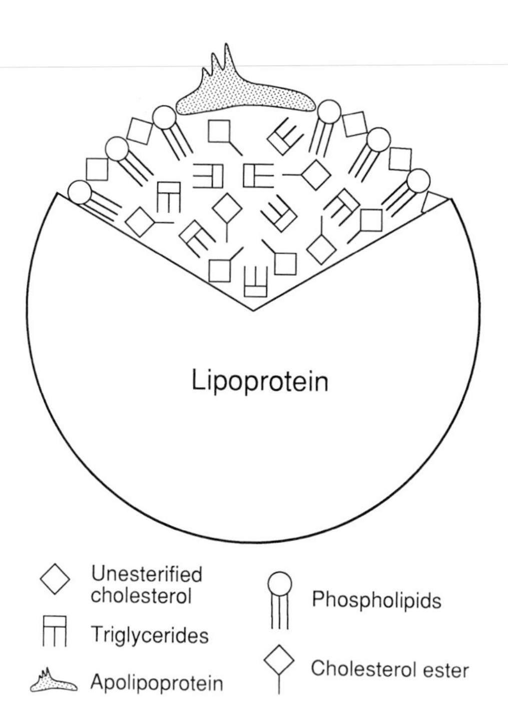

FIGURE 1 Schematic diagram of a plasma lipoprotein particle. The composition and structure of the spherical particle is shown. The surface monolayer is made up of unesterified cholesterol and phospholipids into which apolipoprotein molecules are embedded. The neutral lipid core is made up of cholesteryl esters and triglycerides.

TABLE I Properties and Composition of Human Plasma Lipoproteins

Major class[a]	Density (g/ml)	Diameter (nm)	Associated apolipoproteins
Chylomicrons	0.930	75–1,200	B-48, AI, AII, AIV
VLDL	0.930–1.006	30–80	CII, CIII, E
IDL	1.006–1.019	25–35	B-100, E, CII, CIII
LDL	1.019–1.063	18–25	B-100
HDL	1.063–1.210	5–12	AI, AII, CI, CII, CIII, E, D

[a] VLDL, IDL, LDL, and HDL stand for very low-, intermediate-, low-, and high-density lipoproteins, respectively.

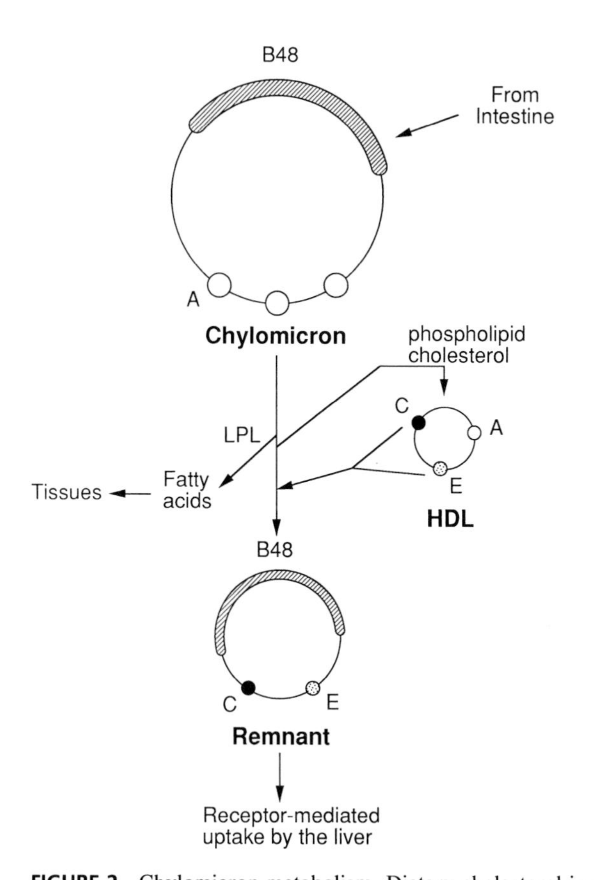

FIGURE 2 Chylomicron metabolism. Dietary cholesterol is absorbed by enterocyte cells that line up in the small intestine where it is esterified. Dietary triglycerides are hydrolyzed into fatty acids and monoglycerides in the small intestine, absorbed by the enterocytes, and re-esterified to form triglycerides. These lipids, together with lipoproteins (indicated as B48 and A) are assembled into chylomicrons and secreted into the blood. Once in circulation, apos C and E are transferred to chylomicrons from high-density lipoprotein (HDL). Most of the core triglycerides of chylomicrons are hydrolyzed by lipoprotein lipase, which is attached to the luminal surface of capillaries. At the same time, some of the surface lipids (phospholipids and cholesterol) are transferred to other particles such as HDL. The remaining particles, called chylomicron remnants, bind via the binding domain on apo E to specific receptor molecules on the surface of liver cells and are then taken up by these cells and hydrolyzed. The cholesterol content of these particles is either used by hepatocytes or secreted into the bile. LPL, lipoprotein lipase.

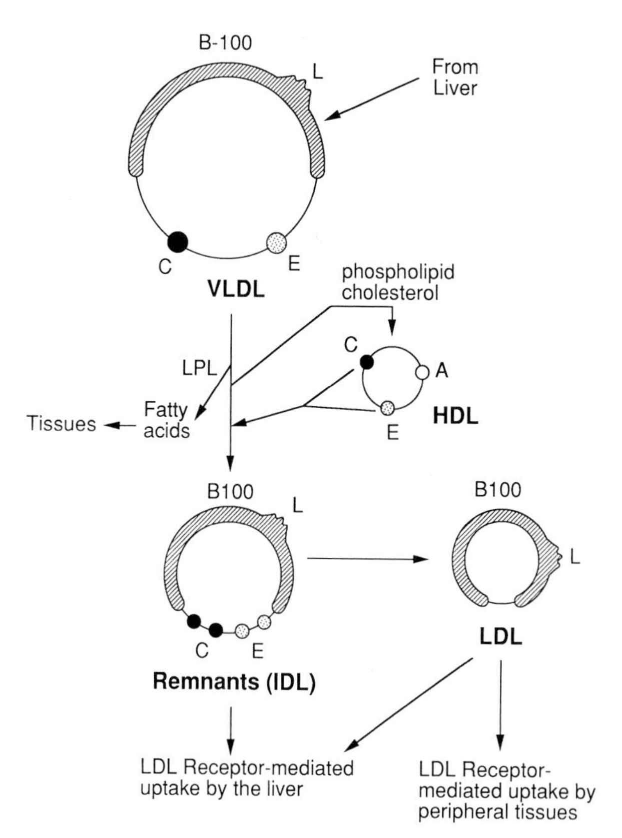

FIGURE 3 Very low-density lipoprotein (VLDL) metabolism. VLDL particles originate in the liver (hepatocytes) and provide the means by which excess fatty acids (in the form of triglycerides and cholesterol esters) are removed from this organ to be distributed to other tissues. VLDL particles are assembled from these lipids together with apos B-100 (a larger form of B-48 synthesized in the intestine), C, and E and are secreted into the blood. VLDL triglycerides are hydrolyzed by lipoprotein lipase, which is attached to the luminal surface of blood capillaries, into fatty acids which are taken up by tissues for either storage or oxidation to generate adenosine triphosphate. During this lipolysis, transfer of surface lipids (cholesterol and phospholipids) from VLDL to HDL particles and of cholesteryl esters in the reverse direction takes place. The VLDL remnants (intermediate density lipoprotein [IDL]) are then either taken up by the liver, as was the case in chylomicron remnants or are further catabolized by the hepatic lipase, which is attached to the luminal surface of capillaries in the liver, to form LDL particles. LDL particles circulate in the blood for a few days and eventually find their way into peripheral tissues by a LDL receptor-mediated uptake using apo B-100 as a ligand (L).

Receptor-mediated uptake of LDL accounts normally for the removal from plasma of approximately 70% of these particles. Unlike VLDL and IDL, LDL has a relatively long residence time in circulation (half-life of 2–3 days). During this time, it may be modified (oxidized or acetylated) to forms that are poorly recognized by the LDL receptor. Instead, these modified particles are taken up by the so-called scavenger receptors found on the surfaces of macrophages and endothelial cells. Accumulation of cholesterol in these two types of cells is one of the early events leading to the formation of atherosclerotic plaques on walls of coronary and other arteries. [*See* ATHEROSCLEROSIS.]

A small but variable proportion of LDL particles is found covalently linked to a large glycoprotein called apo(a), forming a lipoprotein particle referred to as Lp(a). This particle has attracted a great deal of attention recently as it seems to represent an independent risk factor for CHD.

Except for the relatively insoluble apo B, other apolipoproteins as well as surface cholesterol and phospholipids can move from one particle to another. Some of this transfer is mediated by lipid transfer proteins. This is how mature high-density lipoprotein (HDL) particles are formed. During hydrolysis of triglycerides from VLDL and chylomicrons, cholesterol, excess surface phospholipids, and certain apolipoproteins (C and A) are transferred to HDL precursor particles to form mature HDL (Figs. 2 and 3). HDL plays an important role in removing excess cholesterol from peripheral tissues and delivering it to the liver for secretion. High levels of plasma HDL are indicative of a highly efficient process of removal of excess cholesterol from tissues.

In addition to their function as structural components of the lipoprotein particles and as ligands for receptors, some apolipoproteins act as activators or inhibitors of enzymes involved in lipid and lipoprotein metabolism. The known functions of apolipoproteins are summarized in Table II.

II. Candidate Genes

The apolipoproteins, lipid transfer proteins, lipoprotein receptors, and the key enzymes that participate in lipoprotein metabolism (lipoprotein and hepatic lipases, lecithin-cholesterol acyl transferase) determine to a large extent the distribution of cholesterol and triglycerides among various tissues including blood. The genes that encode these proteins (about 15 total) are therefore potential candidates in determining genetic susceptibility to premature atherosclerosis and CHD. The majority of these genes have recently been cloned and assigned to specific regions of the human chromosomes (Table II). Some of these genes occur in clusters of three such as AI, CIII, and AIV on the long arm of chromosome 11 and CII, CI, and E on the long arm of chromosome 19.

The apolipoprotein genes exhibit similarities both in structure and sequence and, therefore, most

TABLE II Proteins Involved in the Metabolism of Plasma Lipids and Lipoproteins

Protein	Plasma concentration (mg/ml)	Chromosomal localization of the gene	Function
Apolipoproteins			
AI	1.0–1.2	11	HDL formation Activation of LCAT
AII	0.3–0.5	1	Unknown
AIV	0.15	11	Unknown
(a)		6	Formation of Lp(a) particles
B-100	0.7–1.0	2	Formation of VLDL and LDL Ligand for LDL receptor
B-48		2	Formation of chylomicrons
CI	0.04–0.06	19	Activation of LCAT
CII	0.03–0.05	19	Activation of lipoprotein lipase
CIII	0.12–0.14	11	Unknown
E	0.025–0.05	19	Ligand for LDL receptor
Other proteins			
LDL receptor	—	11	Cellular uptake of LDL particles
Lipoprotein lipase	—	8	Hydrolysis of lipoprotein triglycerides
Hepatic lipase	—	15	Hydrolysis of lipoprotein triglycerides
LCAT	—	16	Cholesterol esterification
Lipid transfer protein-I	—	16	Facilitates transfer of cholesteryl ester and phospholipids among plasma lipoproteins

LCAT, lecithin cholesterol acyl transferase.

likely belong to a multigene family. Except for that of apo B, the coding regions of these genes contain tandem repeats of 11 codons, suggesting that they were formed by tandem duplication of a primordial unit.

Abnormal plasma lipoprotein levels are fairly good predictors of the incidence of atherosclerosis leading to CHD. High LDL cholesterol, low HDL cholesterol, and high Lp(a) levels are associated with higher risk for atherosclerosis. Variations in genes that control lipoprotein production, processing, and catabolism or clearance from plasma could therefore underlie the genetic differences in atherosclerosis susceptibility that one observes among individuals in the population. Recent advances made in the understanding of lipoprotein metabolism and molecular genetics of apolipoproteins have now made it feasible to examine any patient's DNA to investigate the genetic variable responsible for a particular phenotype. As discussed in the next section, some progress has already been made in defining genetic lesions underlying certain disorders of lipoprotein metabolism.

III. Molecular Genetics of Disorders of Lipoprotein Metabolism

Most of our present knowledge on the genetic basis of atherosclerosis and CHD is derived from analysis of individuals and their families with markedly abnormal lipid levels. Although these cases are relatively rare and may not represent the more common and moderately abnormal phenotypes, the insights gained from these studies provide valuable clues in searching for the more common susceptibility gene variants.

A. Disorders of Lipoprotein Receptors: Familial Hypercholesterolemia

Familial hypercholesterolemia (FH) is characterized clinically by extremely elevated levels of plasma cholesterol, deposition of LDL-derived cholesterol in tendons, skin, and arteries, and premature CHD. FH is inherited as an autosomal dominant trait with a dosage effect such that heterozygote individuals (those who carry one normal and one defective gene) are less severely affected than homozygotes (i.e., those who carry two defective genes). Homozygotes have plasma cholesterol levels of 650–1,000 mg/dl and occur at a frequency of 1/1 million. Heterozygotes (1/500 persons in the general population) have plasma cholesterol levels of 350–500 mg/dl (normal level is 175 mg/dl). The excess cholesterol in these patients is almost entirely found in LDL particles.

Unusually high frequencies of FH have been observed in two parts of the world: Lebanon and Johannesburg, South Africa, with heterozygote frequencies of 1/171 and 1/100, respectively. This most likely occurrence is due to what is referred to as a "founder effect," where a small number of immigrants (among whom were individuals carrying the FH mutation) have founded a large population in a certain location.

The primary genetic defect in this disorder is a mutation in the gene encoding the receptor for plasma LDL particles. LDL receptor molecules on the surface of cells in the liver and other tissues bind LDL particles and promote their uptake and clearance from plasma. When LDL receptors are either defective or diminished in number, the rate of removal of LDL decreases and the level in plasma increases. Excess LDL becomes deposited in tissues such as tendons and skin and is taken up by macrophages, which are progenitors of cholesterol laden cells in atheromas. FH heterozygotes produce only half the number of functional receptors, hence their cells take up LDL at about half the normal rate.

The LDL receptor gene is on the short arm of chromosome 19 and encodes a protein made up of 839 amino acids. Four classes of mutations in the LDL receptor gene have been identified. In class 1, the most common, no LDL receptor protein is produced. Class 2 mutants produce a receptor that fails to be transported from its site of synthesis, the endoplasmic reticulum, to the Golgi complex on its way to the cell surface. Class 3 mutants result in the formation of a protein that reaches the cell surface but fails to bind LDL particles. Finally, class 4 mutants make a protein that binds LDL but fails to promote its uptake by the cell.

An animal model for FH exists in the Watanabe heritable hyperlipidemic rabbit. These rabbits have a mutation in the LDL receptor gene that is similar to the human class 2 mutants. They have extremely high-plasma LDL cholesterol and develop atherosclerosis in early life.

B. Disorders of Lipoprotein Receptor Ligands

1. Type III Hyperlipoproteinemia

This disorder is characterized by elevated concentrations of both cholesterol and triglycerides in plasma. This is due to the accumulation in plasma of cholesterol-rich VLDL and chylomicron remnants. Remnants are the product of action of lipoprotein lipase on these particles after secretion from the liver and intestine. Remnant particles are normally removed from blood by the liver via the LDL receptor. Remnants of type III patients are poorly taken up by liver cells. As in FH, excess remnants are deposited in peripheral tissues and in macrophages resulting in atherosclerosis.

The primary defect in this phenotype is not the LDL receptor but rather in apo E, one of the protein ligands for the LDL receptor found on remnant particles. Remnant particles with mutant apo E have diminished affinity for the LDL receptor.

Apo E is composed of 299 amino acids and the LDL receptor-binding domain has been localized to about 20 amino acids in the middle of the molecule. This domain is rich in basic amino acids (arginine and lysine) with positively charged groups that interact with negatively charged groups on the LDL receptor molecule. The apo E mutants (designated E-2), associated with the type III phenotype involve single amino acid substitutions in the receptor-binding domain. The most common mutation involves a substitution of the amino acid cysteine for the positively charged amino acid arginine at position 158 of the protein. This mutant binds to the LDL receptor with <2% of the activity of the normal protein.

With few exceptions, type III individuals have inherited two E-2 mutant alleles. Approximately 1% of the North American and northern European populations are of the E-2/E-2 phenotype, but only 1–5/5,000 individuals have type III hyperlipoproteinemia. Therefore, the mode of inheritance of this disorder is more complex than that of FH. Having two E-2 mutant genes is only a prerequisite for the type III phenotype. The coexistence of other genetic or dietary factors is also required. Evidence suggests that the coexistence of genes that cause other types of hyperlipidemia are required for the development of the type III phenotype. One such gene was shown to be responsible for what is called familial combined hyperlipidemia. In a study of a large family, only those individuals who have inherited both the E-2/E-2 phenotype and the dominant gene for the hyperlipidemia developed the type III phenotype. Body weight and caloric intake seem to have an important impact on type III expression. This is a good example where the phenotype is determined by both genetic and environmental factors.

2. Familial Defective apo B-100

Recently, another type of hyperlipidemia, familial defective apo B-100, in which plasma LDL cholesterol levels are elevated was shown to be caused by a mutation in the apo B-100 gene. The mutation results in a single amino acid substitution (glutamine for arginine) at position 3500 of the protein. This is analogous to the E-2 mutation in that a positively charged amino acid within the LDL receptor-binding domain of apo B-100 was changed to an uncharged residue. This mutation results in complete loss of affinity of LDL to its receptor resulting in its accumulation in plasma. The frequency of heterozygotes for this mutation in the general population is estimated to be approximately 1/500.

IV. Disorders of HDL Metabolism

HDL particles are assembled in plasma from lipids and apoliproteins. The major apoliproteins of HDL are AI and, to a lesser extent, AII. As mentioned previously, a large number of studies have shown that HDL cholesterol levels in plasma are inversely correlated with the incidence of CHD in the general population. Plasma HDL levels are determined by genetic, environmental, and hormonal factors. For example, excessive dietary saturated fats and alcohol elevate HDL cholesterol levels, whereas cigarette smoking and weight gain depress it. Androgens decrease HDL cholesterol levels, and estrogens increase it. The extent of genetic contribution to the variability in HDL levels in the population has been estimated to be between 30 and 45% of the total variance. Results of complex segregation analysis on a large number of families suggest that HDL levels are not influenced by any one major gene but by a number of genes with relatively equal contribution.

A number of rare inborn errors of metabolism affecting HDL levels are known. They are useful in defining some of the steps in HDL formation and metabolism. The following are some examples.

A. Defects in Apo AI and AII Synthesis

Two mutations in the apo AI gene and a third involving both the AI and CIII genes are known to result in substantial reductions in plasma HDL levels. A number of other electrophoretic variants of apo AI (resulting from single amino acid substitutions) that do not alter plasma HDL levels also exist in the population.

Apo AI Milano has been described in a large Italian kindred. The basic defect is due to the substitution of the amino acid cysteine for arginine at position 173 of the protein. Heterozygotes for this mutation have 33% of normal HDL cholesterol levels and approximately 60% of normal AI protein. Despite these low HDL levels, heterozygotes show no increase in the incidence of atherosclerosis.

Another mutation recently discovered in the author's laboratory is due to a small deletion (45 bases of DNA) and results in the formation of AI that is missing amino acids 140–164. An individual heterozygous for this mutation has approximately 10% of HDL and AI levels in plasma and an elevated plasma triglyceride level. This mutation therefore has a dominant effect on plasma HDL levels.

A third mutation, which results in the loss of both apos AI and CIII, was described in two sisters who had very low levels of HDL, extensive atherosclerosis, and CHD at an early age. The mutation involves an inversion of approximately 5.5 kilobase (kb) pairs DNA containing portions of the AI and CIII genes (the two genes, located on the long arm of chromosome 11, are separated by only 2.5 kb). The inversion inactivates both genes by disrupting their coding sequences. Although the above-mentioned mutations are rare in the population, they point to the importance of apo AI in the maintenance of normal HDL levels and in susceptibility to atherosclerosis.

Recently, a kindred from Japan was shown to have two sisters who had undetectable levels of apo AII in their plasma. AII is the other major protein component of HDL. These two individuals were asymptomatic and their HDL and LDL cholesterol and triglyceride levels fell within normal limits. The mutation was shown to be due to a single base-pair substitution in the AII gene. This substitution blocks the normal splicing of the primary RNA transcript to form functional mRNA. Therefore, AII seems to play at best a minor role in the metabolism of HDL particles.

B. Defects in HDL Processing

A few rare inborn errors interfere with HDL processing and affect plasma HDL levels. Examples are lipoprotein lipase deficiency, apo CII deficiency (activator of lipoprotein lipase), lecithin cholesterol acyl transferase deficiency, fish eye, and Tangier diseases, all of which result in very low plasma HDL cholesterol levels. These enzymes are important in the processing or formation of HDL particles. Unlike apo AI deficiency, however, low HDL levels resulting from the above processing defects are more likely to be associated with the development of premature atherosclerosis. This indicates that a low HDL cholesterol level *per se* is not atherogenic. Therefore, it seems that only a subset of metabolic defects associated with low HDL levels predispose to CHD.

V. Lipoprotein(a) and CHD

Lipoprotein(a) [Lp(a)] is a plasma particle that closely resembles LDL in lipid composition and the presence of apo B-100, the ligand for the LDL receptor. In addition to apo B-100, Lp(a) particles contain a high molecular weight glycoprotein named apolipoprotein(a) [apo(a)], which is linked to apo B-100 by a disulfide bond. Lp(a) has recently attracted considerable attention because several studies have shown that its concentration in plasma is significantly correlated with the risk of heart disease. Lp(a) was originally described as an antigen present in about one-third of a northeastern European population. It is now established that essentially all individuals have this lipoprotein and that plasma levels are distributed continuously in the population with a range of approximately 1–100 mg/dl. A high level of Lp(a) is considered to be an independent risk factor for the development of atherosclerosis and the incidence of myocardial infarctions.

Analysis of apo(a) from different individuals by electrophoresis indicates that it exists in different forms that range in size from approximately 400 to 700 kD. Generally, however, any one individual carries either one or two of these forms. A highly significant correlation exists between the size of the apo(a) and its concentration in plasma.

Recent genetic studies show that the level of Lp(a) in plasma and the size polymorphism are under strict genetic control, apparently by a single locus. This locus codes for apo(a) itself and is located on the long arm of chromosome 6. There are at least 11 known alleles (forms) of this gene that code for 11 apo(a) size forms. Only four of these alleles occur at a frequency of more than 10% in the general population. The different size forms are inherited in families in a codominant manner; i.e., if an individual carries two different size alleles then both are expressed equally to produce the corresponding two protein forms in plasma.

The apo(a) gene is highly homologous in sequence to that of plasminogen (it becomes a protease when activated) and, in fact, is located quite close to it on chromosome 6. These two genes probably evolved from a common ancestor by gene duplication followed by divergence. The apo(a) gene is remarkable in that it contains a large number of repeated coding units corresponding to those that code for one of the domains in plasminogen called Kringle IV. It is believed that alleles of the apo(a) gene that code for different size forms of apo(a) differ from each other by the number of Kringle IV type repeats they possess.

The striking homology between apo(a) and plasminogen led to the proposal that Lp(a) could interfere with the fibrinolytic process involved in lysis of a blood clot in coronary arteries. Unlike plasminogen, Lp(a) cannot be activated to become a protease upon exposure to tissue plasminogen activator (TPA). However, Lp(a) binds tightly and specifically to fibrin. By competing with plasminogen and TPA for binding to fibrin, Lp(a) could interfere with clot lysis by inhibiting the conversion of plasminogen to the active form plasmin. Efforts are now underway to test this hypothesis.

VI. Association of Restriction Fragment-Length Polymorphism DNA Markers at Apolipoprotein Gene Loci with Abnormal Plasma Lipid Levels and CHD

A. DNA Polymorphisms

The preceding discussions show that significant progress has been made in defining the genetic basis, at the molecular level, of relatively rare phenotypes characterized by markedly abnormal levels of lipoproteins; however, not much is known about the genetic factors that contribute to the more common and less severe types that constitute approximately 90% of dyslipoproteinemic patients seen in a typical clinic. The challenge of the coming few years will be in the identification of the relatively common gene variants that cause abnormal lipid levels. It is likely to be a difficult task given the complexities of lipoprotein metabolism and the patterns of interaction between environmental and genetic factors.

Strategies aimed at identifying genetic factors predisposing to hyperlipidemia and CHD rely on the prevalence of DNA sequence polymorphisms present in the general population that can be used to mark specific regions of the human genome. The obvious regions of the genome to investigate first are those that code for the apolipoproteins, enzymes, and receptors involved in lipoprotein metabolism. Many of these candidate genes have been cloned and mapped on the human chromosomes (Table II). Some of these genes are clustered (e.g., AI-CIII-AIV on chromosome 11, CI-CII-E on chromosome 19) and exhibit structural homology, suggesting that they evolved from a common ancestral gene by duplication and divergence.

Several polymorphisms in DNA sequence within or in close proximity to the above candidate genes are known. The majority of these polymorphisms are recognized because they involve a change in the sequences of nucleotides that are recognized by DNA cutting enzymes called restriction endonucleases (restriction fragment-length polymorphism [RFLP]). Over 150 restriction endonucleases are known, each capable of cutting double-stranded DNA at a specific sequence of 4–10 bases. These enzymes are normally synthesized by bacteria species to protect them against invasion by foreign DNA, and their names are derived from the scientific names of bacteria. For example, the enzyme *Eco*RI is derived from *Escherichia coli* (RI is for restriction enzyme I). Recognition sites for restriction enzymes occur with varying frequency in human DNA depending on the number of bases in the recognition sequence (for example, a recognition sequence of 4 bases is expected to occur 16 times more frequently on a random basis than one with 6 bases).

Alterations in DNA base sequence could either abolish or create recognition sequences for these enzymes resulting in alteration in the size of DNA fragments (called restriction fragments) generated by the enzyme. In most cases they do not alter the coding sequence of this gene. Consider, for example, a certain small region (at the apo B locus) on chromosome 2 that has three recognition sites for the restriction endonuclease *Xba*I (designated X), which are separated by 5.0 and 3.5 kb pairs, as shown in Figure 4.

If the central *Xba*I site is polymorphic in the population (i.e., some chromosomes have the site and others do not due to a mutation), then two restriction fragments would be generated from chromosomes that have the site (3.5 and 5.0 kb), and only one fragment (8.5 kb) from those that do not. Such DNA sequence polymorphisms are referred to as RFLPs. Analysis of size of DNA fragments generated by restriction enzymes is done by electrophoretic separation. To determine the genotype of an individual with respect to the presence or absence of the central *Xba*I site at the apo B locus, total genomic DNA is digested to completion with *Xba*I, and the product's size is determined by electrophoresis on an agarose gel. Because genomic DNA contains many other *XBa*I sites, a large number of fragments is generated. To distinguish the apo B-specific fragments from the rest, all DNA fragments are blotted from agarose onto filter paper and hybridized to radiolabelled DNA probes from the apo B locus. Such probes will form stable hybrids with only the *Xba*I fragments generated from the apo B locus (this is called Southern blot analysis). Thus, as shown in Figure 4, the DNA from one individual will yield a single 5.0-kb fragment (the 3.5-kb fragment is undetectable with this probe) if both chromosomes contain the restriction site, one 8.5-kb fragment if both chromosomes lack the restriction site, or two fragments (5.0 kb and 8.5 kb) if one of each type of chromosome is present.

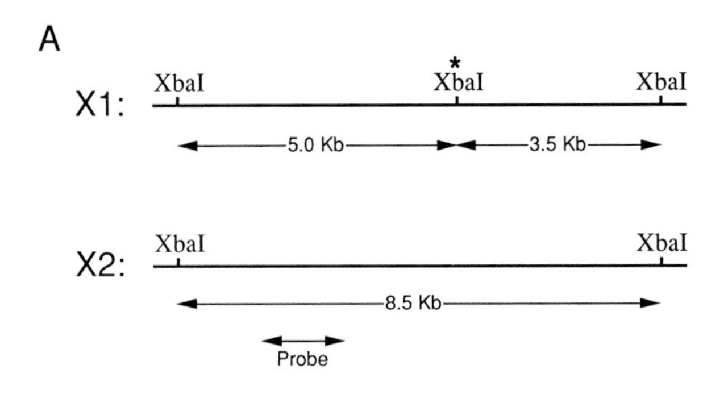

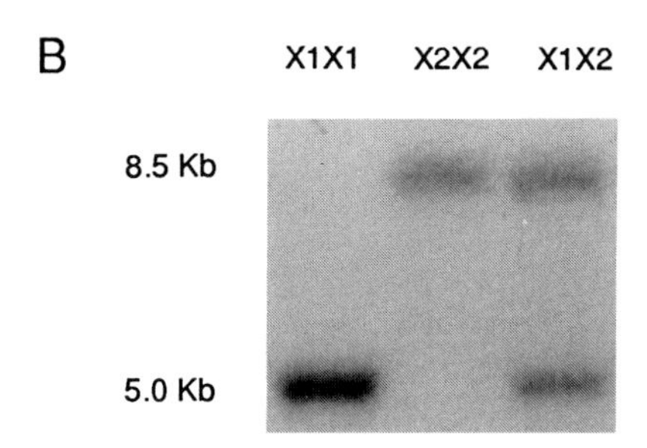

FIGURE 4 Restriction fragment-length polymorphism (RFLP). This figure illustrates a common type of variation in nucleotide sequence of DNA among individuals in the population. This polymorphism happens to occur within the recognition site of the restriction enzyme *Xba*I located in the gene-encoding plasma apo B on chromosome 2. A. The middle *Xba*I site indicated with an asterisk is the polymorphic one because some chromosomes have it (X1) and others do not (X2) due to a single base-pair difference. This RFLP could be used to tag the apo B gene in families and populations. B. An autoradiograph of a Southern blot generated by digesting DNA from three individuals with the enzyme *Xba*I, separating the different size fragments by electrophoresis on an agarose gel, transferring to filter paper, hybridization to radiolabeled DNA probe (indicated in A) followed by exposure of the hybridized DNA to an X-ray film. Electrophoresis is from top to bottom. Lane X1X1 shows the presence of a double dose (from the two homologous chromosomes) of the 5.0-kb fragment (the 3.5-kb fragment in A is undetectable using this probe because it does not hybridize with it). The pattern in the middle shows that the individual has two chromosomes of the X2 type (absence of restriction site). The third individual has one of each of the X1 and X2 types.

Other types of polymorphisms that are even more useful than the simple RFLPs are due to variation in the number of tandemly repeated DNA sequences in the population. These are called variable number of tandem repeats (VNTRs) and can also be recognized by examining the size of DNA fragments generated after digestion with restriction enzymes that recognize DNA sequence sites flanking the clusters of repeats (the larger the number of repeats, the longer the fragment). Usually, several VNTR alleles at a single locus exist in the population, offering a much greater degree of polymorphism and chance for specifically identifying a certain chromosome. [*See* DNA MARKERS AS DIAGNOSTIC TOOLS.]

B. Association with Disease

These polymorphic markers are extremely useful as tags in identifying certain forms of genes or chromosomes in populations and thus trace their inheritance in families. One type of approach used to investigate the possible involvement of a candidate gene (e.g., apo B in high LDL levels) is to test for co-inheritance in families of RFLP markers at the apo B locus and the high LDL phenotype. A high degree of co-inheritance would suggest that a cer-

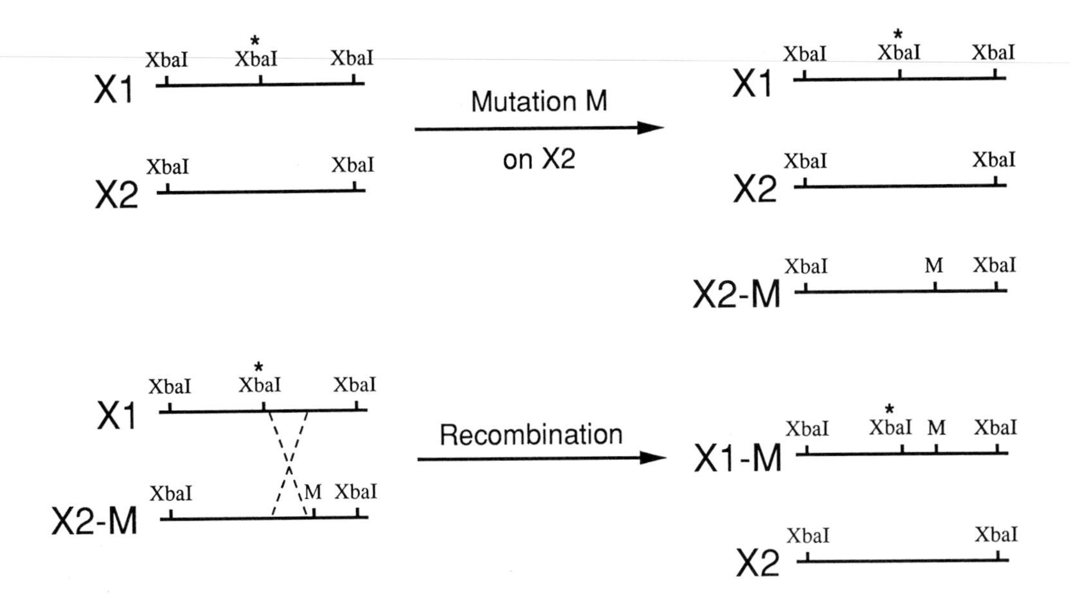

FIGURE 5 Association of RFLP markers with disease-causing mutations in the population. Studies of association between RFLP markers and the incidence of certain diseases have been used to implicate certain candidate genes in the causation of disease. This figure illustrates how an *Xba**I RFLP marker at the apo B gene locus (see Fig. 4.) could become associated with a mutation (M) causing, for example, high levels of plasma cholesterol and CHD. Assume that one can distinguish between two types of chromosomes that differ at the apo B *Xba*I site—X1 and X2 (Fig. 4); if mutation M had occurred on type X2 close to the *Xba*I site, the three types of chromosomes that one finds in the population are shown on the right. If *Xba*I* and M are so close that recombination between them is quite rare, then the mutant M (or the incidence of CHD) would much more likely be found in individuals who carry the X2 type of chromosome than those who carry the X1 type. If the RFLP marker and the disease-causing mutation are so far apart that recombination between them occurs frequently to generate X1-M, then after several generations M would be equally likely to exist in individuals carrying X1 or X2.

tain form of the apo B gene is responsible for elevated plasma LDL.

Another approach aims at detecting an association between an RFLP market at a candidate gene locus and a certain phenotype in the population at large. The following example illustrates the basis for this approach. As shown in the *Xba*I RFLP described above, there are two forms or alleles of an *Xba*I polymorphic site at the apo B locus in the population (X1 and X2). If a mutation (causing high plasma apo B levels) in the apo B gene occurred at site M and by chance on the chromosome of the X2 type, a third chromosome, X2-M, is formed (Fig. 5). With the passage of time, however, recombination between types X1 and X2M would generate a fourth type of chromosome, X1-M. If the polymorphic *Xba*I site and M are far apart, recombination between them would occur frequently, and eventually the mutation M would be equally distributed between the X1 and X2 alleles (i.e., linkage equilibrium). If, on the other hand, M and *Xba*I are very close, then recombination events would be rare and the X2 form would be more likely to be found associated with M and high apo B level. The mutant phenotype and RFLP markers are said to be in linkage disequilibrium and are inherited as a pair.

The possible contribution of a certain candidate gene to a certain phenotype is therefore determined by measuring the frequency of the two or more forms (alleles) of a polymorphism in individuals with, for example, low, intermediate and high plasma LDL levels. A significant correction between frequency of either allele and the level of LDL in plasma is an indication that this particular candidate gene plays a role in determining LDL levels. Furthermore, if a particular candidate gene plays a role in causing premature CHD, then the frequency of alleles of an RFLP marker at that gene locus would be different in a population with CHD than in that of normal controls. Clearly, careful matching of disease and control populations is extremely important in this approach in order to minimize the influences of socioeconomic status, ethnic background, country of origin, etc. It is extremely difficult to match disease and control groups, especially in as heterogeneous population as that of the United States, consequently increasing the complexities of interpretation of the results. To detect an association between a certain RFLP and a disease, the mutant allele that is closely linked to that RFLP would have to be common in the population.

Numerous results of association studies have been reported by a number of laboratories in Europe, Japan, and the United States. Results of some of these studies are given in Table III. Most of these results showed a positive association between RFLP markers and either CHD or dyslipoproteinemia. Other studies, however, failed to observe some of these associations. A number of these studies involved sampling biases from the population, used a small number of individuals, or were not subjected to proper statistical analysis. Furthermore, RFLP allele frequency at some of these loci differs in different populations. These may be some of the underlying reasons for the conflicting results reported by various laboratories. Another reason for the discrepancies may be that these diseases are genetically heterogeneous (i.e., they are caused by more than one genetic predisposing factor); therefore, one of these factors may be prevalent in one population and a different one in another population. Furthermore, not all RFLP markers at a single locus would necessarily be expected to show strong association with a certain phenotype; only those that are physically close enough to the site of the mutation to be in linkage disequilibrium. One of the associations that is most consistently observed in different populations is that of RFLP markers (*Xba*I, *Eco*RI) at the apo B locus and high levels of apo B and LDL cholesterol. RFLPs at two loci, the apo AI-CIII-AIV cluster of genes, and the lipoprotein lipase gene locus seem to be associated with high-plasma triglyceride levels and low HDL levels, the latter predisposing to CHD.

TABLE III Association of RFLP Markers at Apolipoprotein Gene Loci with Plasma Lipid Phenotypes and CHD

Locus	Phenotype
AI-CII-AIV	Hypertriglyceridemia
	Low apo AI and HDL levels
	Coronary heart disease
Apo B	LDL cholesterol
	Apo B levels
	CHD
Apo E-CI-CII	Type III hyperlipoproteinemia
	Levels of apo B, apo E, and cholesterol
	Coronary artery disease

C. Family Studies

Very few studies so far have addressed the question of linkage of RFLP markers with dyslipoproteinemia or heart disease in families. Preliminary results from one family study indicate coinheritance of apo B gene polymorphisms and high apo B levels. In another family study, a significant association occurs between RFLPs at the AI-CII-AIV locus and low apo AI and HDL levels. These results support the population-based studies in suggesting that variants at the apo B locus may underlie some of the variance in apo B and LDL cholesterol levels, whereas those at the AI-CII-AIV locus could contribute significantly to the variance in apo AI, HDL cholesterol, and possibly plasma triglyceride levels. It seems essential to include family studies to avoid the various artifacts inherent in population studies.

The results available at the present time are insufficient to allow use of these RFLP markers in predicting predisposition to dyslipoproteinemia or premature CHD, which ultimately is the goal of these studies. The apo B and AI-CIII-AIV loci seem to be two good candidates for predisposition to some of the disorders of lipid metabolism. At this time, additional family studies are necessary to place the previously observed associations on more firm ground and establish new associations. Once this is done, characterization of the mutant alleles at various loci that are co-inherited with these diseases would be necessary before they can be used reliably to assess potential risks for atherosclerosis and CHD.

VII. Future Trends

Two developments are crucial in the hunt for genes predisposing to atherosclerosis and CHD. The first is the ability to selectively amplify by the polymerase chain reaction (PCR) any desired DNA segment of the human genome and directly determining its nucleotide sequence. The automation of these procedures will make it feasible to examine sequences of candidate genes or parts thereof in families with dyslipoproteinemias and CHD. Common variants of the candidate genes that may be responsible for the common predisposition to high LDL cholesterol or low HDL cholesterol should not be too difficult to detect using these methods.

The second development, which is rapidly being realized, is the identification and mapping of enough

DNA polymorphic markers to span the entire human genome. In the next few years, when this becomes a reality, one should be able to test these markers for close association or linkage to any phenotype in family studies. It is estimated that a minimum of approximately 200 markers (preferably of the highly polymorphic VNTR type) uniformly distributed over the genome would be necessary. So far, about 60 such VNTR loci have been mapped on the human genome.

Finally, the mouse model offers a great opportunity to dissect the genetics of lipoprotein metabolism and atherosclerosis. Identification of inbred strains of mice that differ in their susceptibility to atherosclerosis when fed diets high in saturated fat and cholesterol would allow mapping of the responsible genes on the mouse genome. Mapping the gene in the mouse usually allows prediction of the map position of the homologous human gene. One such gene has already been identified on mouse chromosome 1 near the apo A-II locus. This gene influences susceptibility to low HDL cholesterol levels and atherosclerosis. In addition to identifying new genes, the mouse offers the opportunity to investigate the function of proteins known to participate in lipid metabolism. This could be achieved by the generation of transgenic mice in which such genes could be introduced and expressed at high levels or could be mutated by gene targeting.

The potential ability to diagnose disease predisposition antenatally or preclinically is already attracting considerable industrial interest. It is important at this time to address the complex legal and ethical questions involved in genetic diagnosis. Ethical and legal guidelines for genetic counseling, prenatal diagnosis, and screening individuals for alleles predisposing to disease are urgently needed. In the absence of such guidelines, discrimination in terms of employment and health insurance could reach serious proportions.

Bibliography

Cooper, D. N., and Clayton, J. F. (1988). DNA polymorphism and the study of disease associations. *Hum. Genet.* **78,** 299.

Deeb, S., Failor, A., Brown, B. G., Brunzell, J. D., Albers, I. I., and Motulsky, A. G. (1986). Molecular genetics of apolipoproteins and coronary heart disease. Cold Spring Harbor Symposium on Quantitative Biology, Vol. LI, p. 403.

Goldstein, J. L., and Brown, M. S. (1989). Familial hypercholesterolemia. *In* "The Metabolic Basis of Inherited Disease," 6th ed., Vol. I. (C. R. Scriver, A. L. Beaudet, W. S. Sly, and D. Valle, eds.), p. 1215. McGraw-Hill, New York.

Havel, R. J., and Kane, J. P. (1989). Structure and metabolism of plasma lipoproteins. *In* "The Metabolic Basis of Inherited Disease," 6th ed., Vol. I. (C. R. Scriver, A. L. Beaudet, W. S. Sly, and D. Valle, eds.), p. 1129. McGraw-Hill, New York.

Hegele, R. A., and Breslow, J. L. (1988). Apolipoprotein genetic variation in the assessment of atherosclerosis susceptibility. *Genet. Epidem.* **4,** 163.

Lusis, A. J. (1988). Genetic factors affecting blood lipoproteins: The candidate gene approach. *J. Lipid Res.* **29,** 397.

Mahley, R. W., and Rall, S. C., Jr. (1989). Type III hyperlipoproteinemia: The role of apolipoprotein E in normal and abnormal lipoprotein metabolism. *In* "The Metabolic Basis of Inherited Disease," 6th ed., Vol. I. (C. R. Sciver, A. L. Beaudet, W. S. Sly, and D. Valle, eds.), p. 1195. McGraw-Hill, New York.

Utterman, G. (1989). The mysteries of lipoprotein(a). *Science* **246,** 904.

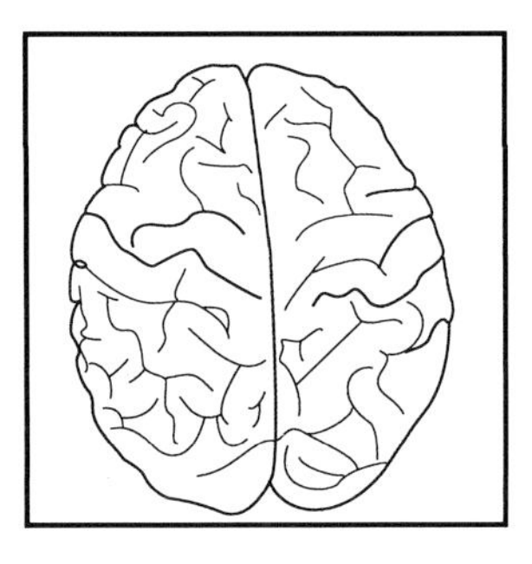

Cortex

KARL ZILLES, *University of Cologne*

Glossary

Brain stem Lowest part of the brain immediately rostral to the spinal cord

Cytoarchitecture Differences in cell densities and cell types among different areas, nuclei, or layers of the brain

Neurotransmitter Chemical compound which can be released at synapses and which effects neurotransmission

Ontogeny Growth and differentiation of organs during embryonic, fetal, and early postnatal periods

Synapse Special contact between two nerve cells, where information is transmitted from one cell to another

Telencephalon Most anterior part of the brain, which comprises the cortex, underlying white matter, and part of the basal ganglia

Thalamus Part of the diencephalon (i.e., the most caudal part of the forebrain), which consists of modality-specific (e.g., vision, hearing, pain, tactile sensations, and taste) and nonspecific nuclei

THE CEREBRAL CORTEX of all mammalian species consists of nerve cells and support (glial) cells arranged in a stratified manner and represents the superficially exposed part of the telencephalon. The human cortex is folded, causing its typical macroscopic appearance with gyri (ridges) separated by sulci (furrows). The microscopically visible strata (cortical layers) are defined by variations in the packing densities of nerve cell bodies and by differences in predominating cell types. Although the largest part of the human cerebral cortex consists of six layers (the isocortex), a minor part has a different lamination (the allocortex). Moreover, there are further subdivisions of the isocortex and the allocortex into numerous areas. These architectonic studies lead to a cortical map with areas which differ in connection and function from other brain regions. The most pronounced increase in volume during evolution of the human brain is found within the isocortex. Since this part of the cortex is of relatively recent evolution, it is also called the neocortex.

The neocortex comprises the areas where the sensory input from the receptors of the skin, taste buds, inner ear, and eye terminates, as well as motor control areas and association areas (the largest part of the human neocortex). Association areas receive major input from numerous other cortical areas. Higher functions of the central nervous system (e.g., understanding of verbal or written information and memory) are localized in these association areas.

The allocortex comprises target areas of olfactory input, which are collectively called the rhinencephalon, or paleocortex, and areas involved in diverse functions such as emotional activities, learning, and memory. The latter part of the allocortex is called the archicortex and is often collectively called the cortical part of the limbic system.

A cortical area receives direct input from the ascending sensory systems via the thalamus, from other cortical areas of the same or the opposite hemisphere, and is also connected with the basal ganglia and regions in the lower brain stem and the spinal cord.

The cortex is not only stratified by layers parallel

to the surface, but is also subdivided by small vertically oriented cell clusters, which are restricted to one layer or cross all layers. These periodically arranged units are called modules, columns, stripes, or blobs.

This article ends with a short review of the distribution and localization of some neurotransmitters and their receptors.

I. Macroscopic Anatomy

A. Outer Appearance

The cerebral hemispheres consist of the cortex, the underlying white matter, parts of the deeply located basal ganglia, and the lateral ventricles, which contain the cerebrospinal fluid. Cortex and white matter constitute the pallium (i.e., the mantle of the brain). The hemispheres are partially separated from each other by the median fissure (Fig. 1). A large connection between both hemispheres, called the corpus callosum, is found in the central part. Two smaller interhemispheric connections, the rostral (anterior) and the hippocampal commissures, are located beneath the most anterior and posterior ends of the corpus callosum, respectively. [*See* HEMISPHERIC INTERACTIONS.]

Some of the sulci of the human cortex can be used as landmarks for the division of the hemispheres into six lobes. The frontal lobe is demarcated from the parietal lobe by the central sulcus. The parietooccipital sulcus separates the parietal from the occipital lobe. The temporal lobe is bordered by the lateral sulcus from both the frontal and parietal lobes. Only the border between the occipital and temporal lobes is not defined by a sulcus and remains arbitrary. The insular lobe is completely covered by parts of the frontal, parietal, and temporal lobes and can be found only after an artificial dilatation of the lateral sulcus. The limbic lobe is not well defined. It encircles the rostral part of the brain stem and consists of small portions of the frontal, parietal, temporal, and occipital lobes. This subdivision of the hemispheres also reflects a functional aspect. Besides other functions (see below) the frontal lobe comprises the motor centers, the parietal lobe comprises the somatosensory centers, the temporal lobe comprises the auditory centers, and the occipital lobe comprises the visual centers. Some other sulci are also consistent features in all human brains (Fig. 1). Most of the sulci, however,

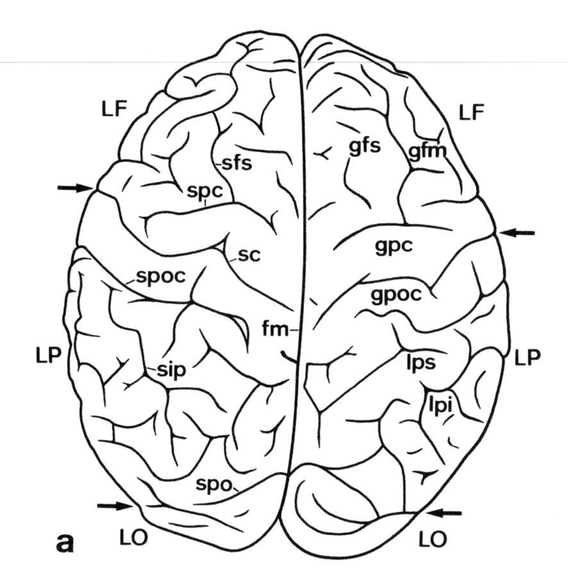

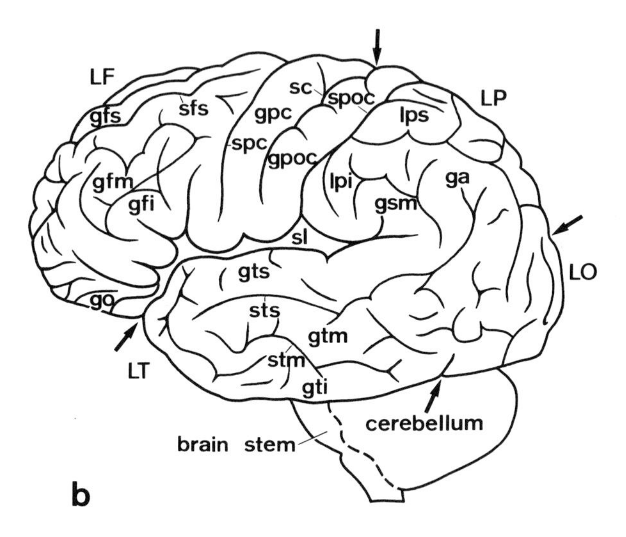

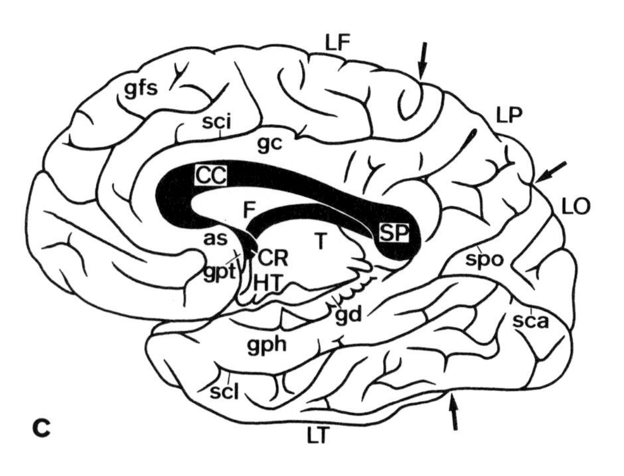

vary greatly among the brains of different individuals.

B. Quantitative Data

The volume of the cortex of one hemisphere shows a considerable variability among individuals, ranging from 190 to 360 cm^3 depending on sex, body size, age, and various other reasons. The mean volume of the cortex in males is larger by about 30 cm^3 than in females. The difference can be explained at least partially by the larger average body size of the males. In both sexes the cortex of the two hemispheres represents 46% of the total brain volume. The volume of the white matter is about 10% smaller than that of the cortex. The mean volume of the cortex is 6% smaller in people over 80 years of age compared with individuals younger than 50 years. The most extensive reduction is found in the frontal lobe (−12%).

The total surface of the cortex amounts to 1600–1700 cm^2 (i.e., both hemispheres), but nearly two-thirds of the surface are buried in the depth of the sulci. Comparative quantitative data corroborate that the human cortex has the highest degree of folding compared with other primates and most of the other mammals. The degree of folding reaches the highest values over the association regions of the brain. The estimated number of nerve cells in the cortex of one hemisphere varies from 7 to 9 $\times 10^9$ cells. The number of glial cells is about 10 times larger. Although exact data on the number of synaptic contacts are lacking for the human cortex, measurements derived from cortices of other mammals give an estimate of 2×10^{15} synapses in the human cortex of one hemisphere.

FIGURE 1 Human forebrain in dorsal (a), lateral (b), and medial (c) views. CC, Corpus callosum; CR, rostral commissure; F, fornix; HT, hypothalamus; LF, frontal lobe; LO, occipital lobe; LP, parietal lobe; LT, temporal lobe; SP, splenium; T, thalamus; as, subgenual area; fm, median fissure; ga, angular gyrus; gc, cingulate gyrus; gd, dentate gyrus; gfi, inferior frontal gyrus; gfm, medial frontal gyrus; gfs, superior frontal gyrus; go, orbital gyri; gpc, precentral gyrus; gph, parahippocampal gyrus; gpoc, postcentral gyrus; gpt, paraterminal gyrus; gsm, supramarginal gyrus; gti, inferior temporal gyrus; gtm, medial temporal gyrus; gts, superior temporal gyrus; lpi, inferior parietal lobule; lps, superior parietal lobule; sc, central sulcus; sca, calcarine sulcus; sci, cingulate sulcus; scl, collateral sulcus; sfs, superior frontal sulcus; sip, intraparietal sulcus; sl, lateral sulcus; spc, precentral sulcus; spo, parietooccipital sulcus; spoc, postcentral sulcus; stm, medial temporal sulcus; sts, superior temporal sulcus. Arrows indicate the borders between the different lobes of the hemisphere.

It is generally believed that the number of nerve cells decreases with aging. However, recent observations show that the total number of cortical nerve cells is fairly constant between the 20th and 100th years. A reason for the believed decrease could be the effect of the slow increase in brain size during the last 100 years. This means that in a sample of brains from individuals of a wide age range, older persons have smaller brains and, therefore, a lower number of cortical cells.

II. Microscopic Anatomy

A. Laminar Pattern

The stratification of the cortex into cell layers parallel to the surface is the most important architectonic feature. The largest part (approximately 95% of the total cortical volume) of the cortex has a six-layered structure (the isocortex). The smaller part of the human cortex (the allocortex) shows a greatly varying laminar structure (mostly less than six layers).

The six layers of the isocortex can be delineated in histological sections in which the nerve cell bodies are visualized with Nissl stain (Fig. 2). The

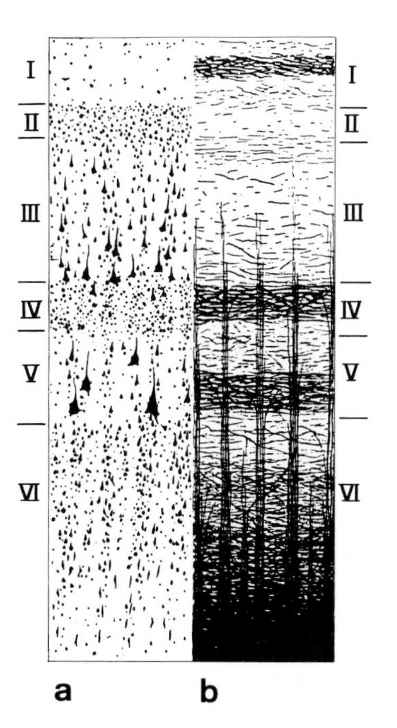

FIGURE 2 The laminar pattern of the human isocortex with the delineation of six layers (I–VI). (a) Nissl stain for the demonstration of cell bodies. (b) Myelin stain for the demonstration of myelinated nerve fibers.

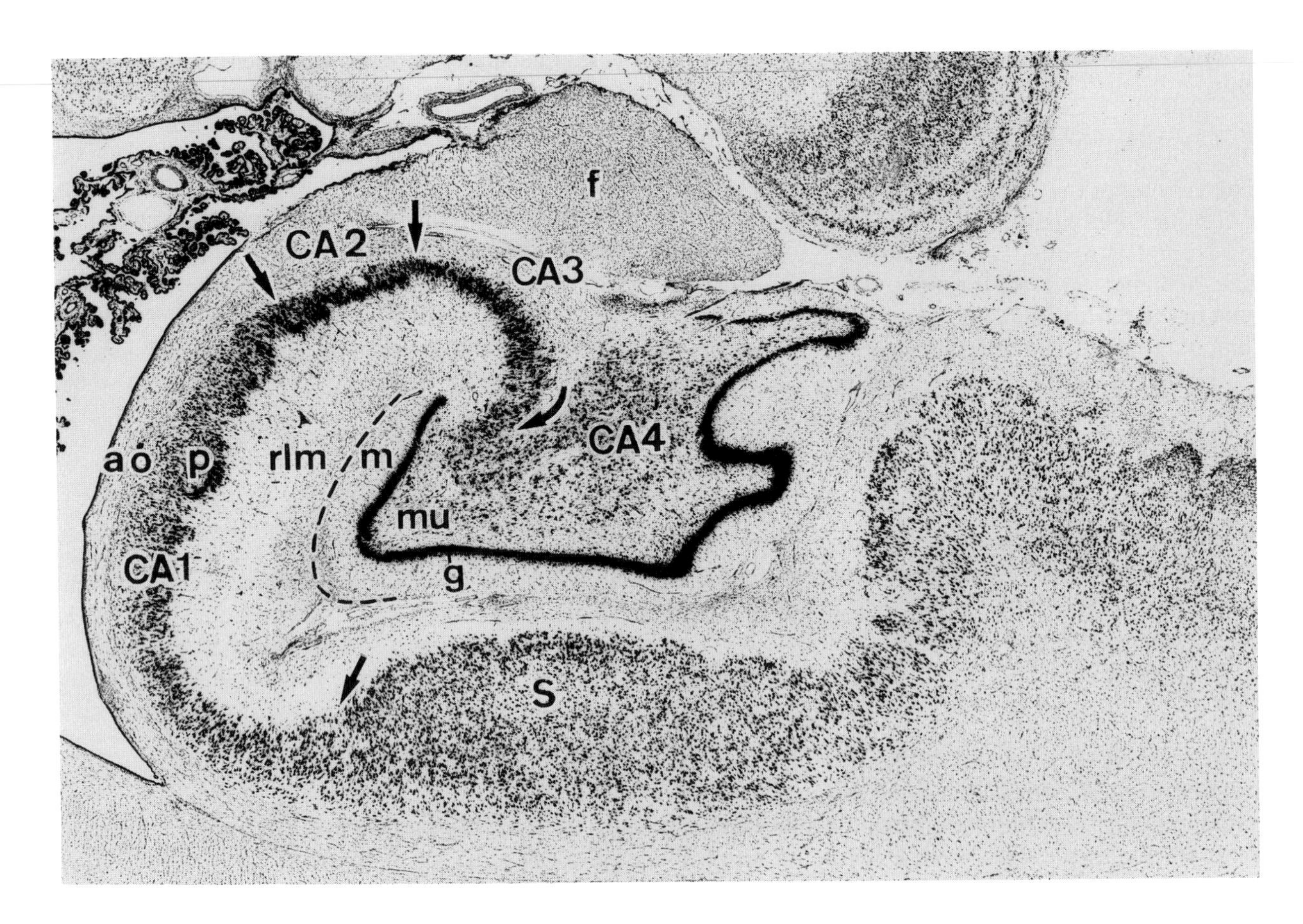

most superficial layer is lamina I (the molecular layer), which contains only a low number of nerve cells, or neurons. The second layer (the outer granular layer) is densely packed with small nerve cell bodies (granular cells). Layer III (the outer pyramidal layer) has a lower packing density. The cell bodies are medium sized and many of them display a triangular (i.e., pyramidal) shape. The fourth layer (the inner granular layer) has the highest packing density of all layers. Small and mostly round cell bodies are found here. The fifth layer (the inner pyramidal layer) contains the largest cell bodies, and most of the cells have a pyramidal shape. The packing density is low in this layer. The innermost, sixth, layer displays an increasing cell packing density with bodies of greatly varying shapes (the polymorphic layer).

The myelin stain demonstrates cellular sheaths wrapped around the axons of nerve cells. The sheaths are formed by glial cells and contain a high concentration of lipids (i.e., myelin). Myelin-stained sections of the isocortex demonstrate its stratified structure, in addition to many vertically oriented bundles of myelinated axons (Fig. 2).

FIGURE 3 Nissl stain showing the cytoarchitecture of the human hippocampus magnification ×10. a, Alveus; CA1–CA4, regions of Ammon's horn; f, fimbria hippocampi; g, granular layer; m, molecular layer; mu, multiform layer; o, oriens layer; p, pyramidal layer; rlm, radiatum–lacunosum–moleculare layer; S, subiculum. The arrows indicate the borders of the subdivisions (CA1–CA4) of Ammon's horn.

A silver impregnation procedure called the Golgi method reveals neuronal cell bodies together with their cell processes. Since this method is capricious and allows the impregnation of only a minor portion of the total cell population, it visualizes selected aspects of the cortical structure.

A description of the widely differing laminar patterns of all allocortical areas is not within the scope of this article. The cytoarchitecture of the hippocampus, which represents the major part of the archicortex, is illustrated as one example (Fig. 3). This simpler three-layered structure contrasts clearly with the isocortical lamination pattern. [*See* HIPPOCAMPAL FORMATION.]

B. Cell Types

The cortex is composed of many different cell types. A fundamental difference is found when the dimensions of the efferent cell processes (i.e., axons) are analyzed. The axon of a cortical neuron can be short and terminates in the vicinity of its cell body, or is long and travels to target areas far from the place of origin. The short axon cells are, therefore, the elements of the local cortical circuitry, whereas the neurons with the long axons are the cellular basis for far-reaching projection systems. The former cells are called interneurons; the latter, projection neurons.

All pyramidal cells (Fig. 4) belong to the class of projection neurons. They have a roughly triangular cell body with a thick dendrite leaving at the apical part of the cell body and numerous dendrites originating at the basis of the cell body. The dendrites are arborizing in different layers of the cortex. They bear some thousands of so-called spines, which are the postsynaptic parts of synapses. The dendrites offer the major portion of postsynaptic contacts on which the presynaptic terminals of numerous other neurons end. Most of the synapses on dendrites are excitatory, but inhibitory synapses are also found. Pyramidal cells have exclusively inhibitory synapses on the cell body and the initial axon segment.

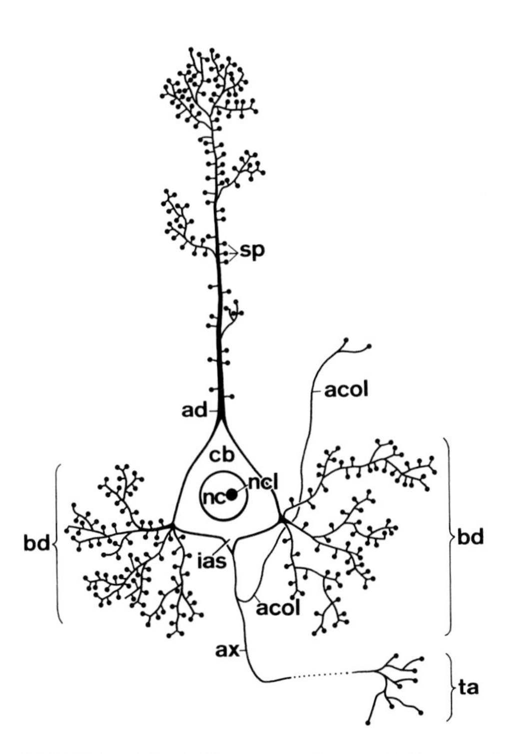

FIGURE 4 A Golgi-impregnated pyramidal neuron in the isocortex. acol, Axonal collateral; ad, apical dendrite; ax, main stem of the axon; bd, basal dendrites; cb, cell body; ias, initial axonal segment; nc, cell nucleus; ncl, nucleolus; sp, spines; ta, terminal arborization of the axon.

The class of interneurons comprises many different cell types, all of which are characterized by dendrites with few or no spines at all. The cell bodies are of greatly varying size and shape. They can be larger than small or medium-sized pyramidal cells and can be round, oval, or irregularly shaped. The dendrites and axons form different, but for each cell type characteristic, territories.

III. Ontogeny

The ontogeny of the human cortex starts with bilateral evaginations of the dorsolateral walls of the forebrain during the fifth embryonic week (Fig. 5). A cerebral vesicle comprises the primordia for the pallium (neo-, archi-, and paleopallium) and parts of the basal ganglia.

The primordial pallium is a thin neuroepithelial layer. Each epithelial cell spans the whole thickness of the wall of the vesicle. The first afferent nerve fibers from lower brain regions arrive in the primitive paleopallium and extend during the following week into the neopallial primordium. These generate a three-layered structure with an abundance of nerve fibers and some immature neurons in a superficial layer, the marginal plexiform lamina. This layer is separated from the densely packed neuroepithelium (the matrix or ventricular zone) by the primordial white matter (the intermediate zone) (Fig. 5).

The number of mitoses in the ventricular zone increases during the seventh and eighth weeks, and the first descending corticofugal fibers are detectable in the primordial white matter. The newly generated immature neurons of the ventricular zone move in an opposite direction into the marginal zone. These cells migrate in a highly ordered manner along processes of radial glial fibers, which span the whole width of the primordial pallium. The immature neurons stop their migration in the center of the marginal zone, forming an initially thin, but rapidly thickening, new layer, the cortical plate (Fig. 5). The marginal zone is divided by this developmental process into a superficial layer (the later layer I of the cortex) and a subplate zone (the later layer VIB). Other observers have defined the sub-

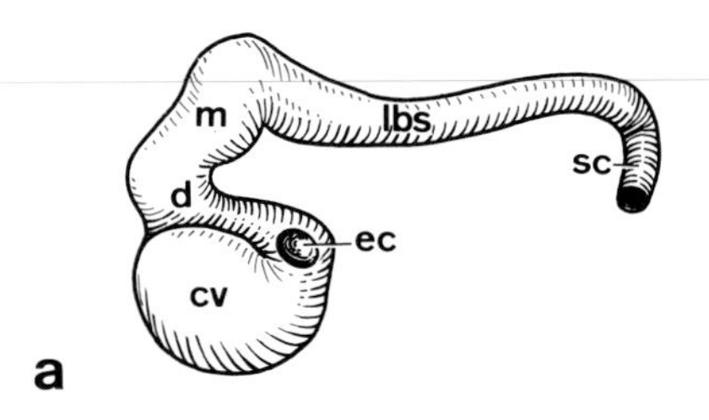

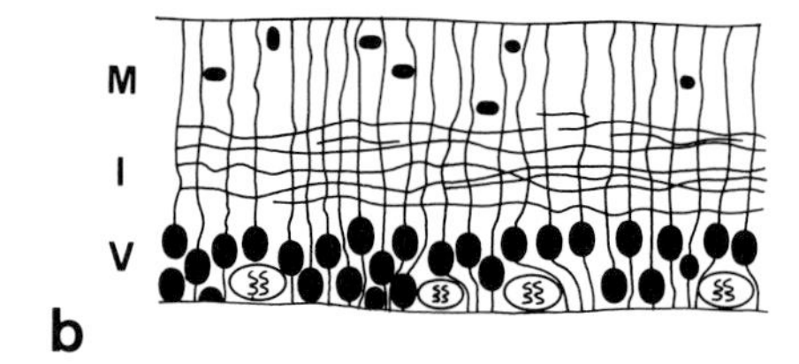

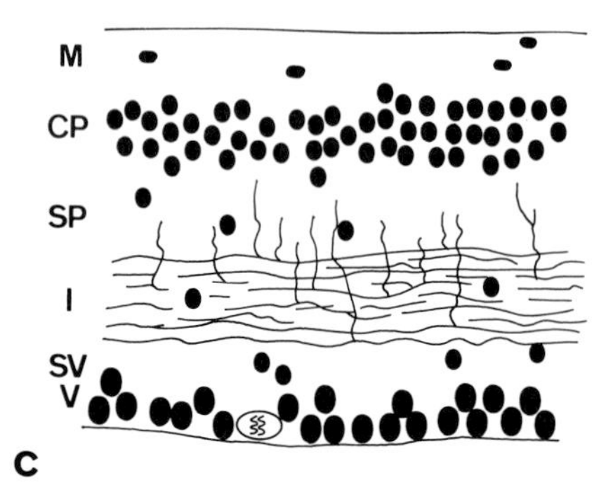

FIGURE 5 Ontogeny of the human cortex. (a) Cerebral vesicle (cv) of the brain during the fifth embryonic week. d, Diencephalon; ec, eye cup; m, mesencephalon; lbs, lower brain stem; sc, spinal cord. (b) Three-layered structure of the primordial pallium with the marginal (M), intermediate (I), and ventricular (V) zones. (c) Cortical plate (CP) and cortical stratification during fetal stages with the marginal (M), intermediate (I), subventricular (SV), and ventricular (V) zones and the subplate layer (SP).

plate zone as a superficial part of the intermediate zone.

The cortical plate gives rise to adult cortical layers II–VIA. The first cells arriving in the cortical plate form the lamina VIA of the adult cortex. The following waves of migrating neurons traverse this layer and are located more superficially; the result is an inside-out layering, with the latest wave of migrating neurons forming the adult layer II. The first synapses are found in layers I and VIB at the time of the formation of the cortical plate. The first afferent fibers to the cortical plate arrive around the 15th week. During the following week the arrival of thalamocortical, callosal, and other corticocortical fibers and the differentiation of nerve cells proceed from the lower to the more superficial parts of the cortical plate.

The generation of neurons in the ventricular zone stops around the time of birth. Since these neurons migrate into the cortical plate, the ventricular zone disappears. Only a single layer of epithelial cells, the ependyma, separates the adult white matter from the ventricular cavity.

The adult volume of the primary visual cortex (area 17) is reached at the beginning of the fifth postnatal month. At about the same time the volume proportion of the neuropil—the space between the cell bodies of nerve cells and glial cells containing cell processes of both types and most of the synaptic connections—reaches adult values.

The cortical surface is smooth during the first 23 weeks of ontogeny. At this time the enlargement of the surface leads to the appearance of the first cortical sulci. The adult degree of folding is acquired at the time of birth (Fig. 6).

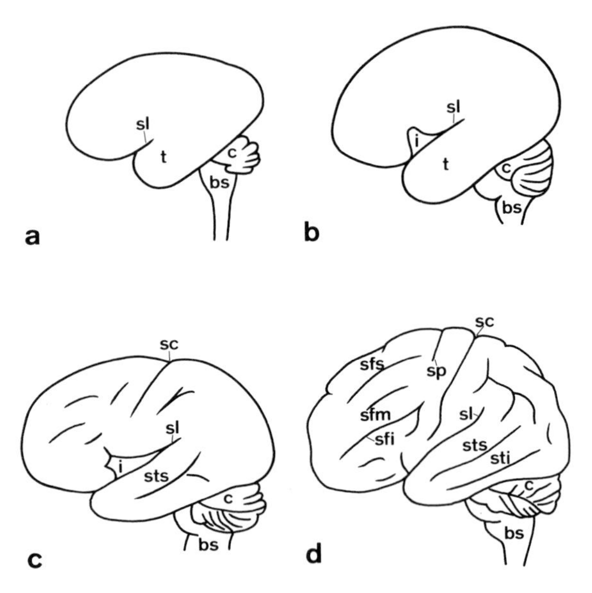

FIGURE 6 Ontogeny of the sulcal pattern of the lateral cortical surface. (a) Smooth cortical surface during the 16th embryonic week. The temporal lobe and the lateral sulcus are visible. (b) The temporal lobe and the lateral sulcus are enlarged, together with the frontal and pareital lobes, during the 25th embryonic week. The surface of the prospective insular lobe is not growing as fast as the other lobes, leading to a retrusion of the insular region. (c) A group of typical sulci is recognizable during the 33rd embryonic week, and (d) the adult pattern is visible around the time of birth. bs, Brain stem; c, cerebellum; i, insular lobe; sc, central sulcus; sfi, inferior frontal sulcus; sfm, medial frontal sulcus; sfs, superior frontal sulcus; sl, lateral sulcus; sp, precentral sulcus; sti, inferior temporal sulcus; sts, superior temporal sulcus; t, temporal lobe.

IV. Comparative Anatomy and Evolution

A comparison of the surface areas of the cortices between different mammals reveals the progressive development of the human cortex. Its surface is not only absolutely larger than that of most nonprimates, but surpasses the cortical surface of all living primates. Since total brain and cortex volumes are associated with body size, an analysis of the cortical volume independent of body size can prove these findings. The volume of the paleocortex of a hypothetical insectivore scaled to human body size is more than three times that of the human paleocortex. The volume of the human archicortex, however, surpasses that of the hypothetical insectivore four times and that of a rhesus monkey of equal body size almost twice. Even more pronounced is the evolution of the neocortex. The human neocortical volume is 156 times larger than in the insectivore and 45 times larger than that of the rhesus monkey in relation to body size. The progressive evolution of the human cortex is, therefore, characterized mainly by the growth of the neocortex (i.e., neocorticalization). [*See* COMPARATIVE ANATOMY.]

The size of the association regions of the neocortex dominates the primary sensory and motor regions, indicating that human cortical evolution favors higher levels of information processing. The association regions of the frontal, parietal, and temporal lobes comprise more than two-thirds of the neocortical surface. Within the occipital lobe only 25% of the cortical surface is occupied by the primary visual area, whereas 75% is represented by association areas.

V. Areal Pattern, Functional Parcellation, and Connectivity

A. Paleocortex

The human paleocortex comprises the olfactory bulb, retrobulbar region, olfactory tubercle, piriform cortex, septum, amygdala, and parts of the insular cortex (Fig. 7).

The olfactory bulb is less differentiated than in any other primate. The most important input comes from the olfactory epithelium through the fila olfactoria of the first cranial nerve. The olfactory bulb projects to the other paleocortical areas via an olfactory fiber tract at the basal surface of the frontal

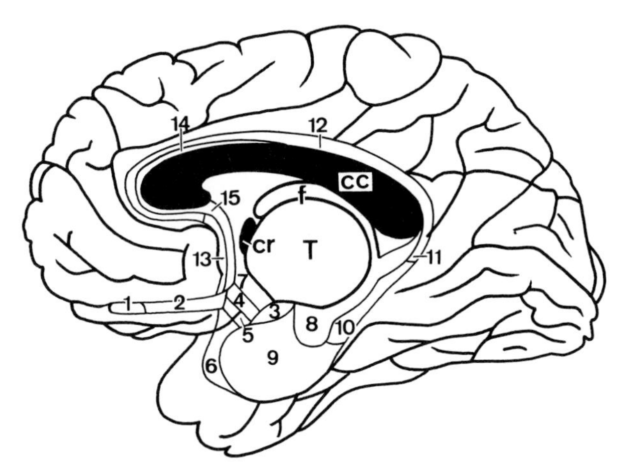

FIGURE 7 Medial view of the human telencephalon with paleocortical and archicortical areas. cc, Corpus callosum; cr, rostral commissure; f, fornix; T, thalamus; 1, olfactory bulb; 2, retrobulbar region and olfactory tract; 3, amygdala; 4, olfactory tubercle; 5, piriform cortex; 6, peripaleocortical region; 7, septum; 8, retrocommissural hippocampus; 9, entorhinal area; 10, pre- and parasubiculum; 11, retrosplenial cortex; 12, periarchicortical cingulate areas; 13, subgenual area; 14, supracommissural hippocampus; 15, precommissural hippocampus.

lobe. In contrast to other mammals, an accessory olfactory bulb, which is connected with the vomeronasal organ, cannot be found in the adult human brain. Destruction of the olfactory bulb leads to loss of the sense of smell. [*See* OLFACTORY INFORMATION PROCESSING.]

The retrobulbar region has often been termed an anterior olfactory nucleus. Despite the low degree of architectonic differentiation of this area in the human brain, comparative anatomical studies have corroborated the cortical character of this rhinencephalic region. The retrobulbar region is found at the transition of the olfactory tract into the basal frontal cortex. Its main afferent fibers originate in the olfactory bulb. The retrobulbar region is an important relay station for olfactory information.

The olfactory tubercle is located in the anterior perforate substance, which borders the retrobulbar region. Although heavily reduced in the human brain, the cortical structure of this area can be recognized. The olfactory tubercle has reciprocal connections with the other paleocortical areas.

The piriform cortex is also located in the anterior perforate substance of the basal forebrain. A three-layered cortical structure is visible. The most superficial layer contains the lateral part of the olfactory tract. The piriform cortex is reciprocally connected with the other paleocortical regions and projects to the hippocampus and the hypothalamus. Besides its olfactory function, the piriform cortex plays a role in sexual behavior.

The septum is found on the medial side of the olfactory tubercle and extends onto the medial surface of the hemisphere. It is classified as part of the paleocortex, because comparative anatomical observations have shown that at least parts of the human septum are equivalent to paleocortical areas in other primates. The septum receives input from paleocortical areas, the hippocampus, and isocortical and subcortical regions. A strong projection leaves the medial part of the septum via a fiber bundle beneath the corpus callosum, called the fornix, and terminates in the hippocampus and surrounding archicortical areas, where the axonal terminals release acetylcholine as a neurotransmitter. This renders the septum part of the magnocellular basal forebrain regions, the source of cholinergic innervation in cerebral cortex. Alzheimer's disease is associated with a remarkable degeneration of acetylcholine-containing neurons in the basal forebrain, including the septum.

The amygdala is a huge agglomeration of neurons in the temporal lobe just in front of the lower horn of the lateral ventricle. The hippocampus caudally borders the amygdala. One part of the amygdala shows a clear cortical structure, whereas the major part represents a subcortical nuclear formation. Reciprocal connections exist to paleo-, archi-, and isocortical regions and the hypothalamus. The cortical parts of the amygdala influence nutritional and sexual behavior, whereas the subcortical parts are involved in anxiety and aggression.

B. Archicortex

The archicortex comprises the hippocampus and the periarchicortical entorhinal, pre- and parasubicular, retrosplenial, and cingulate areas.

The major (i.e., retrocommissural) part of the hippocampus extends from the medial wall of the lower horn of the lateral ventricle to the caudal end of the corpus callosum (called the splenium corporis callosi). The hippocampus continues around the splenium to the dorsal surface of the corpus callosum (the supracommissural part) and follows this structure to its rostral end, where the precommissural hippocampus is located. The supracommissural and precommissural parts are small structures which do not show the typical lamination pattern nor regional subdivisions of the retrocommissural hippocampus. Further descriptions are relevant for the retrocommissural hippocampus.

This archicortical region can be subdivided into the Ammon's horn, dentate gyrus, and subiculum (Fig. 3). Ammon's horn and the dentate gyrus shows a basic three-layered cortical pattern with the strata oriens, pyramidale, and radiatum–lacunosum–moleculare in the former and the strata moleculare, granulosum, and multiform in the latter subregion. The hippocampus is covered with a thin layer of myelinated nerve fibers on the ventricular surface, called the alveus. The alveus contains afferent and efferent fibers. The alvear fibers converge in the fimbria hippocampi, which continues as the fornix at the level of the splenium. The width and cell packing density of the pyramidal layer are the basis for further parcellation of Ammon's horn into CA1–4 regions. The CA4 region is often considered together with the multiform layer of the dentate gyrus as the hilus. The main cell type of the pyramidal layer of CA1–4 is the pyramidal cell. The main cell type of the dentate gyrus is the granule cell. All of the other neurons of the hippocampus are interneurons. The subiculum is a transitory area between Ammon's horn and the adjoining periarchicortical regions.

Three major systems of afferent fibers originating in cortical areas reach the hippocampus: One system comes from the septum (see above) and terminates mainly in the stratum radiatum of Ammon's horn; the other two systems (i.e., perforant path and the alvear tract) originate in the entorhinal area, the most important entrance to the hippocampus, and terminate mainly in the molecular layer of the dentate gyrus. The granule cells give rise to a bundle of axons (mossy fibers), which form synapses with the apical dendrites of the pyramidal cells in CA3. These cells send their axons via the fornix to the septum, the hippocampus of the contralateral side or other target areas. Collaterals of these axons (i.e., Schaffer collaterals) travel to the apical dendrites of the pyramidal cells in CA1. These cells again form a major efferent pathway to the subiculum or via the fornix to the septum.

The activity in the intrahippocampal pathways is highest during slow-wave sleep and is strongly inhibited during rapid eye movement sleep or waking states. Although an association of the hippocampus with memory seems to be an important functional aspect, we are presently far from understanding sufficiently the functional impact of the human hippocampus.

The presubicular, parasubicular, and entorhinal areas are characterized by a cytoarchitectonic peculiarity, a nearly cell-free layer in the central part of the cortical width. Therefore, the term "schizocortex" has been coined for these periarchicortical areas. They are located on the parahippocampal gyrus between the subiculum and the laterally adjoining neocortex. The entorhinal area receives input from the total neocortex and gives a summary of all of this information into the hippocampus.

The retrosplenial areas are found in the caudal parts of the cingulate gyrus. The cingulate areas are located in the rostral parts of the cingulate gyrus. Both areas have lamination patterns in the regions near the archicortex, which allow classification as the periarchicortex. The laminar patterns of those parts of the retrosplenial and cingulate areas, which are nearer the neocortex, show, however, an increasing differentiation. Therefore, classification as the proisocortex seems legitimate. The rest of the cingulate cortex near the cingulate sulcus is characterized by the typical six layers of the isocortex. The retrosplenial and anterior cingulate areas are part of the Papez circuit, which connects the anterior thalamus via the cingulate gyrus and the schizocortex with the hippocampus. Since the Papez circuit has been claimed to be important for memory, the retrosplenial and anterior cingulate areas could serve this complex function. Additionally, changes in affectivity and lowering of spontaneous activity are observed after lesions of the cingulate gyrus.

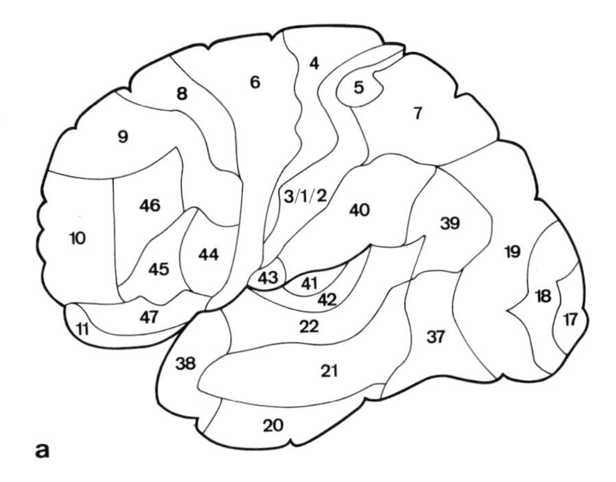

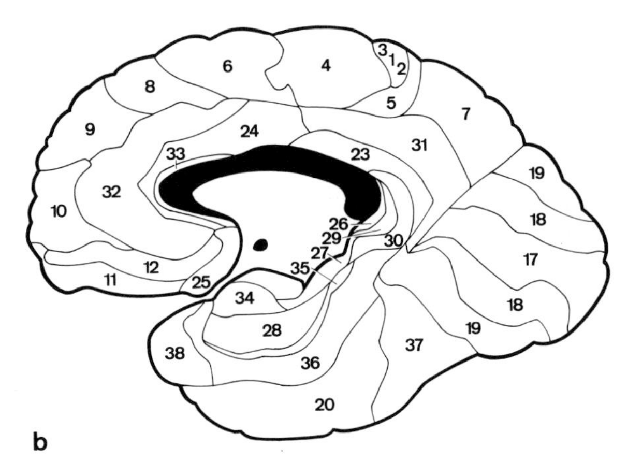

FIGURE 8 Areal map of the human cortex, according to Brodmann (1909). (a) Lateral and (b) medial views of the hemisphere.

C. Neocortex

The most widely accepted map of the human neocortex into areas of specific architecture and function has been proposed by neuroanatomist Korbinian Brodmann in his classical monography from 1909. Although numerous corrections of details seem necessary, the more general aspects of his cortical map are of great value even today. His map (Fig. 8) is therefore the basis for the following descriptions of the areal pattern of the human neocortex. [*See* NEOCORTEX.]

The areas in the neocortex of the frontal lobe can be subdivided architectonically into a group with a clearly visible inner granular layer (layer IV) and a group lacking this layer. The agranular part shows the six-layered structure of the typical isocortex only during the fetal period, but loses layer IV during the early postnatal period. This subdivision coincides roughly with the definition of a prefrontal (granular) and a motor (agranular) cortex in the frontal lobe. The prefrontal cortex comprises Brodmann's areas 8–11 and 44–47. The motor cortex comprises primary motor area 4 and premotor area 6.

The prefrontal areas are reciprocally connected with nearly all brain regions. The connection with the mediodorsal thalamic nucleus in the diencephalon is of special interest, because this system defines and delineates all prefrontal areas. A strong efferent projection of the prefrontal areas terminates in the basal ganglia.

Lesions of the prefrontal cortex in humans lead to attention disorders connected with cognitive deficits and preceptual distortions. Humans with destructions in the dorsolateral part of the prefrontal cortex show apathy and a lack of spontaneous movement. Lesions in the basal (orbital) part lead to hyperkinesia, euphoria, and disinhibition. Patients with frontal lesions cannot organize new and deliberate sequential behavior, owing to deficits in short-term memory, planning, and interference control. These deficits show that the prefrontal cortex is clearly an association region of the highest order. It might be a crucial structure for the high adaptive capacity of humans (i.e., their ability to develop new behavioral strategies).

Area 8 is at least partially equivalent to the frontal eye field found in other primates. This area contains cells responsive to visual stimuli, which are active before eye movement. Dominance of the right frontal eye field is found when nonverbal signals are analyzed.

Area 44 and, eventually, area 45 represent Broca's motorical speech region. Lesions in this area lead to motorical aphasia, which is characterized by a slow and effortful speech delivery. The speech lacks normal fluidity and continuity, and the articulation of words is disturbed. It is interesting that this area is functionally active only in the left hemisphere of most humans.

The primary motor cortex (area 4) is found in the rostral wall of the central sulcus. Brodmann has defined area 4 by the presence of huge pyramidal cells (Betz cells) in layer V. Area 4 receives input from thalamic nuclei and many cortical areas. The main efferent pathway of area 4 is part of the pyramidal tract. These axons terminate in the brain stem and the spinal cord, where they control motor neurons, which give rise to cranial and spinal nerves. Area 4 is therefore the primary motor center for voluntary movements. This area shows a conspicuous somatotopy; that is, the muscles of the body are represented at distinct sites in the cortex. These sites have the same topological relationships as the muscles of the body. However, the extents of their cortical representation areas are correlated not with the dimensions of the muscles, but with their functional importance. In fact, there is an overproportionate representation of the hand muscles, especially the muscles of the thumb, and a small cortical representation of the buttock muscles (Fig. 9). Lesions of area 4 cause disruptions of motor functions on the contralateral side of the body. In an initial stage the tonus and the reflexes of muscles are lost, but after some time spasticity and hyperreflexia appear.

Area 6 is located between the primary motor and the prefrontal cortices and can be subdivided into a supplementary motor area and a premotor area. The small supplementary motor area is found on the medial side of the hemisphere. The premotor cortex covers the rest of area 6. Both subdivisions have a similar cytoarchitecture (i.e., agranular isocortex), but are connected to different parts of the ventrolateral thalamic nucleus in the diencephalon. They receive afferent fibers from nearly all cortical areas and many subcortical areas. The efferent projections of the supplementary motor area reach the spinal cord directly, whereas the projections of the premotor cortex arrive via a synaptic connection in the brain stem. Both subfields project to the basal ganglia, cerebellum, and some other brain stem nuclei. A bilateral activation of these subfields is found, when a person is imagining, but not executing, complex voluntary movements. A destruction of the premotor area impairs mainly the coordination of proximal muscles of the limbs. A lesion of the supplementary motor area leads to forced grasping, a reduction in spontaneous movements of the contralateral hand, and impairment of bilaterally coordinated activities. A bilateral destruction of this area results in a permanent akinesia.

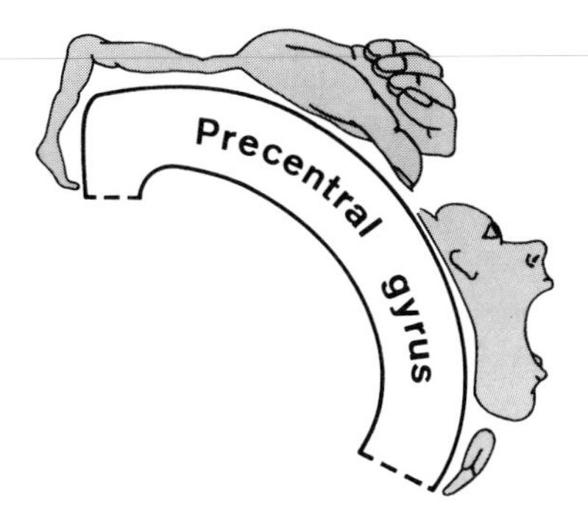

FIGURE 9 Representation of the different parts of the body in area 4.

The isocortical areas of the parietal lobe can be subdivided into primary and association areas. The anterior part of the parietal lobe roughly coincides with the postcentral gyrus, where the primary somatosensory cortex (i.e., areas 3, 1, and 2) is found. The posterior part consists of superior and inferior parietal lobules separated by the intraparietal sulcus (Fig. 1). The superior lobule is composed of areas 5 and 7, whereas the inferior parietal lobule includes areas 39 and 40. The whole parietal cortex shows the basic six-layered structure of the isocortex.

The primary somatosensory cortex has a conspicious layer IV. Area 3 is bordered by area 4 rostrally and area 1 caudally. Area 3a is found at the transition from area 3 to area 4 and has a broad layer V and a small layer IV. This area receives its main input from the muscle spindles informing about the actual degree of muscle tonus. Information from slowly adapting mechanoreceptors of the skin arrive in area 3. Area 1 is informed by rapidly adapting mechanoreceptors of the skin, whereas the respective peripheral organs of area 2 are the mechanoreceptors of joints. As in area 4, the body is represented in a somatotopic manner in the primary somatosensory cortex.

Recent observations have shown that there are multiple cortical representations of one part of the body. The major input of areas 3, 1, and 2 comes from specific thalamic nuclei. Additionally, afferent fiber systems from other cortical areas of the same and the opposite hemisphere (through the corpus callosum) terminate in the primary somatosensory cortex. Efferent fibers project to thalamic nuclei, brain stem regions, and the primary motor and prefrontal cortices. A lesion of the primary somatosensory cortex leads to paresthesia, tactile agnosia, and impairment of tactile discrimination.

Areas 5 and 7 of the superior parietal lobule are association areas. They are reciprocally connected with the prefrontal cortex and the inferior parietal lobe. Afferent fibers arrive from most of the isocortical areas. The efferent projections of this part of the parietal cortex terminate in many cortical areas, as well as in the basal ganglia, the thalamus, and regions of the lower brain stem.

Areas 39 and 40 of the inferior parietal lobule are reciprocally connected with various areas of the frontal, temporal, and occipital cortices. Additional input arrives from the hippocampus, retrosplenial cortex, and brain stem regions. The efferent projections terminate in various cortical areas, basal ganglia, thalamus, and further brain stem regions.

Lesions in the posterior parietal lobe lead to impairment in visual perception of space, size, and distance. Fixation of gaze and an inability to move the eye voluntarily toward a target are also found, together with losses of the capacity to write, right–left orientation, and calculation. The maintenance of a spatial reference system for goal-directed movements seems to be the main function of this association region.

The isocortical part of the temporal lobe comprises Brodmann's areas 41, 42, 20–22, 37, and 38. The primary auditory cortex and its belt region are represented by areas 41 and 42, whereas the other areas form the auditory association (area 22) and nonauditory association cortices. All of these areas reveal the basic isocortical lamination pattern, with the highest cell density in layer IV of the primary cortex.

Areas 41 and 42 (auditory cortex) are located on the dorsal plane of the temporal lobe (Fig. 10). Their position is recognizable by the appearance of a transverse gyrus (i.e., Heschl's gyrus). A massive fiber bundle ascending from the medial geniculate body of the thalamus is the most important input to the primary auditory cortex. Efferent projections terminate in the auditory association area, other isocortical regions, the medial geniculate body, and lower brain stem nuclei. Observations with positron emission tomography, which detects functionally active areas in a living subject, have revealed a tonotopic (i.e., different sites for different tones) organization in the primary auditory area.

The auditory association cortex (area 22) is found on the lateral surface of the superior temporal gyrus and on the dorsal surface of the temporal lobe posterior to Heschl's gyrus. The latter region is called the planum temporale. The most posterior part of laterally exposed area 22, together with the planum temporale, resembles a cortical field in which the sensory speech region of Wernicke is located. It receives input from the auditory and visual association cortices and gives rise to a strong fiber bundle, which terminates in Broca's area. Some authors include the posterior parts of area 40 and the whole of

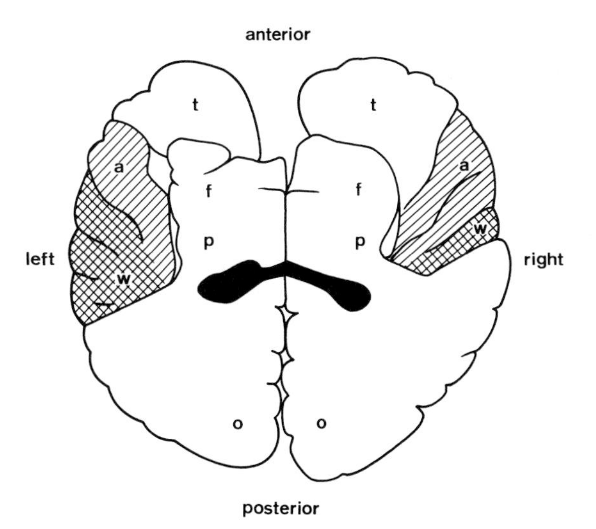

FIGURE 10 Dorsal plane of the temporal lobe with the primary auditory cortex (a) and part of Wernicke's area (w). The frontal (f), parietal (p), and occipital (o) lobes have been partially removed to give a view of the surface of the temporal (t) lobe from above. Wernicke's area of the left side is much larger than that of the right side (representing the dominance of the left hemisphere for the speech function).

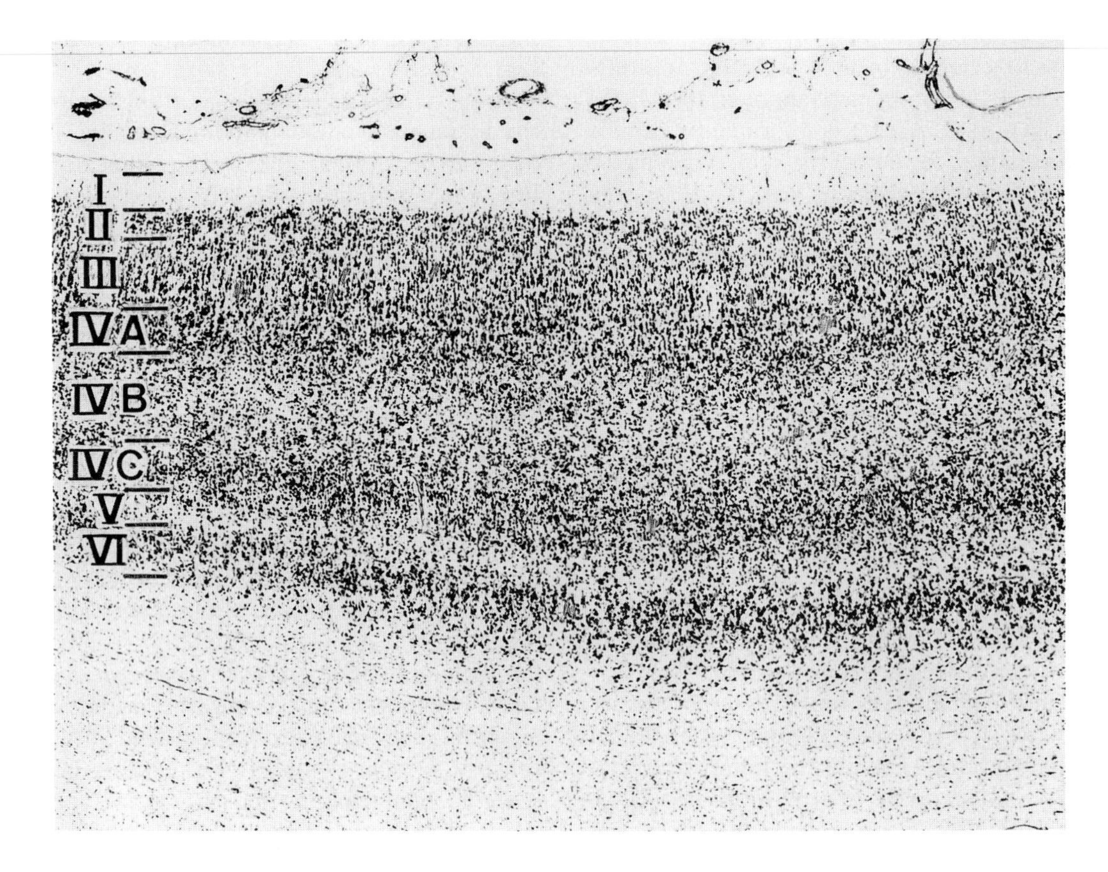

FIGURE 11 Nissl stain showing the lamination pattern of the primary visual cortex. Original magnification ×13. Roman numerals indicate the different isocortical layers in this highly differentiated area.

area 39 in their definition of Wernicke's area. A destruction of this area leads to a lost understanding of verbal information and an inability to write. This defect was described in the last century as sensoric aphasia and agraphia. It has been shown that the planum temporale of the left hemisphere is much larger in most human brains than that of the right side (Fig. 10). This lateralization of an anatomical structure is associated with the same lateral preference of function.

The nonauditory association cortex comprises all other isocortical areas of the temporal lobe. Efferent connections of this region lead to the prefrontal and premotor areas and to the hippocampal region. The afferent fibers originate in sensory association cortices. Lesions of this area correlate with memory deficits for more complex and abstract visual patterns and deficits in verbal learning.

The occipital lobe includes areas 17–19. The human primary visual cortex (area 17) shows the most differentiated laminar structure of all isocortical areas (Fig. 11). A conspicuous structure of this area is a stripe of myelinated fibers in layer IVB, which runs parallel to the cortical surface (i.e., Gennari's stripe). This stripe can be seen with the naked eye on unstained sections. Therefore, area 17 is also called the striate area. It extends from the occipital pole over the total length of the calcarine sulcus. The major part of this area is buried in the sulcus. The largest portion of area 17 receives input from both eyes (binocular part). The monocular part, which receives information only from the contralateral eye, is small and located in the most rostral extension of area 17. The whole cortical area is the target of information from the entire contralateral visual hemifield and a small part of the ipsilateral hemifield.

The strongest afferent fiber bundle to area 17 is the visual radiation, which originates in the lateral geniculate body of the thalamus of the same side. Since input from both eyes is already present in one lateral geniculate body, the information arriving in the area 17 of one side is also from both eyes. The optic radiation retains a clear separation of the input from both eyes. This is also conserved in the visual cortex, because the visual input from both eyes terminate separately in alternating stripes in layer IVC. The stripes consist of densely packed small neurons, synapses between these neurons, and the geniculocortical axon terminals. They have been demonstrated in many mammals, including primates, by axonal tracing techniques (Fig. 12). These stripes appear as columns in cross-sections through the visual cortex and have been termed ocular dominance columns. Additional types of vertically oriented periodical structures (i.e., modules) can be found in area 17 (modules of orientation-selective or color-specific cells). Efferent projections of the primary visual cortex terminate in the secondary visual area (area 18), tertiary visual area (area 19), and other isocortical areas, the lateral geniculate body, the pulvinar complex of the thalamus, and regions in the brain stem. Removal of the striate area leads to a loss of visual perception.

Area 18 is bordering the striate area. The strongest input to this area originates in the pulvinar and the primary visual cortex. The efferent projections terminate in area 19 and other isocortical areas; the pulvinar and regions, in the lower brain stem.

Area 19 is a single cortical area in Brodmann's map, but more recent observations have demonstrated that it contains numerous separate areas, each representing a complete visual half-field. Therefore, the term "third visual tier" has been introduced. All subdivisions of this region are strongly visually influenced. Areas 18 and 19 can be summarized as the extrastriate visual association cortex. Electrical stimulation of the extrastriate areas leads to complex optical hallucinations, in contrast to simple visual sensations after stimulation of area 17.

Despite areal differences in laminar structure and input–output specificities, some general layer-specific aspects of connectivity in the isocortex can be

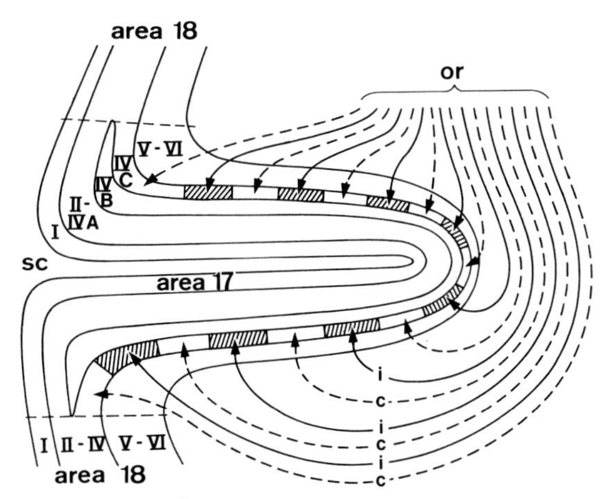

FIGURE 12 A coronal section through areas 17 and 18 with the ocular dominance columns (hatched and open areas in layer IVC) in the primary visual cortex. The ocular dominance columns are areas of alternating input from both eyes. The geniculocortical afferent fibers [optic radiation (or)] contain fibers from the ipsilateral (i) or contralateral (c) eye, which terminate in separate areas of layer IVC. Layer IVB is the stripe of Gennari, which disappears at the border between areas 17 and 18. Roman numerals indicate isocortical layers. sc, Calcarine sulcus.

stated. Table I summarizes these findings for the most important connections.

VI. Neurochemistry

A. Neurotransmitters and Neuropeptides

All classical neurotransmitters and neuropeptides as well as their receptors can be demonstrated in the cortex. Their regional and laminar distribution varies greatly between different cortical areas and layers. This indicates a highly differentiated neurochemical organization, which is summarized below, considering only the major afferent systems of the

TABLE I Isocortical Layers as Origins or Targets of Afferent and Efferent Fiber Systems, Respectively[a]

Origin of afferent fibers	Target of efferent fiber	Layer
Thalamus and cortex	—	I
Cortex	Cortex	II
Thalamus and **cortex**	**Cortex**	III
Thalamus and cortex	Cortex and basal ganglia	IV
Thalamus and cortex	**Spinal cord**, **brain stem**, thalamus, **Basal ganglia**, **cortex**	V
Thalamus and cortex	Thalamus, **claustrum,** cortex	VI

[a] The most important origins or targets are shown in boldface type.

neocortex and the neurotransmitter receptor distribution in the striate area.

The human neocortex has no acetylcholine-containing neurons, but receives a cholinergic innervation from the basal nucleus of Meynert, which is an area in the basal forebrain region at the transition between the telencephalon and the diencephalon. Over 90% of the neurons in the basal nucleus of Meynert contain acetylcholine. The number of these neurons and the acetylcholine content of the cortex are markedly reduced in patients with Alzheimer's disease. Noradrenaline-containing fibers originate in the locus coeruleus, which is a nuclear configuration in the brain stem near the floor of the fourth ventricle. They are most densely packed in the primary somatosensory area and are found in all cortical layers, with the lowest density in layer I. Dopamine-containing fibers arrive from the ventral tegmental area of the midbrain. They reach the highest densities in the primary motor and premotor cortices. Layers I and V–VI of the neocortex show maximal innervation densities. The axons of neurons in the raphe nuclei of the midbrain provide the serotonin innervation of the neocortex. A relative uniform distribution is described over all areas, with peak values in layer IV.

γ-Aminobutyric acid (GABA) is the major inhibitory neurotransmitter of the cortex. GABA-containing neurons are found in great numbers within the cortex. Nearly all of them are local circuit neurons. Glutamate is the major excitatory transmitter of the cortex, and is produced by a vast number of cortical neurons, especially pyramidal cells. These cells project to the basal ganglia, the thalamus, and brain stem nuclei. The vasoactive intestinal polypeptide, cholecystokinin, and corticotropin-releasing factor show the highest concentrations of neuropeptides in the cortex. Many of the peptides are colocalized in one neuron. Peptides might also be colocalized with classical neurotransmitters in the human cortex. A colocalization of acetylcholine, for example, with vasoactive intestinal polypeptide has been described for the rat neocortex, but comparable studies in the human cortex are still lacking. [*See* PEPTIDES.]

B. Receptors

All of these transmitters act on pre- or postsynaptically localized receptors during neurotransmission. The receptors are protein molecules traversing the cell membrane. Some of these directly regulate the opening or closing of ion channels; others are connected with secondary messenger systems. Drugs have been developed which are active at these receptors and can, therefore, influence neurotransmission. Table II gives a short summary of the laminar distribution of some neurotransmitter receptors in the human striate area. This table and results from many other cortical areas reveal that the single receptor shows a considerable laminar specificity (i.e., distribution pattern). [*See* NEUROTRANSMITTER AND NEUROPEPTIDE RECEPTORS IN THE BRAIN.]

A comparison of the laminar distribution patterns of different receptors in the human primary visual cortex demonstrates that some receptors have similar laminar distributions (i.e., M1 receptors with glutamate binding sites; glutamate binding site with $GABA_A$, $5\text{-}HT_1$, and $5\text{-}HT_2$ receptors; $GABA_A$ with $5\text{-}HT_1$ and $5\text{-}HT_2$ receptors; $5\text{-}HT_1$ with $5\text{-}HT_2$ receptors), but other receptors (e.g., M2 and α_1) do not show this laminar codistribution. The codistribution is a structural prerequisite for functional interaction of different transmitter systems within a cortical area.

Presynaptically localized receptors (e.g., M2 receptor) reduce the transmitter release, whereas postsynaptic receptors influence the neuronal excitability. It has been shown that M2 receptors are found on the axonal terminals of basal forebrain neurons projecting into the cortex. Consequently, the density of M2 receptors is lowered in the cortices of brains from Alzheimer's disease patients, which frequently show a massive neuronal degeneration in the basal forebrain complex. Since certain drugs can mimic the effects of natural transmitters

TABLE II Laminar Distribution of Some Neurotransmitter Receptors in the Human Primary Visual Cortex[a]

Layers	Receptor types
I	D1, α_1
II	M1, Glut, $GABA_A$, $5\text{-}HT_1$, **D1**, α_1
III	**M1**, **Glut**, $GABA_A$, $5\text{-}HT_1$, $5\text{-}HT_2$
IV	M1, M2, **Glut**, **$GABA_A$**, **$5\text{-}HT_1$**, **$5\text{-}HT_2$**, D1
V	**M2**, D1
VI	—

[a] Only the layers with high receptor densities are listed. Boldface type indicates the cortical layers with the highest densities of a specific receptor. M1, M2, Muscarinic cholinergic receptors; Glut, glutamate receptors; $GABA_A$, GABA receptor; $5\text{-}HT_1$, $5\text{-}HT_2$, serotonin receptors; D1, dopamine receptor; α_1, noradrenaline receptor.

at specific receptor sites, a more detailed analysis of the receptor distributions in normal and pathologically changed human cortices is necessary for the development of new therapeutic strategies in neurological disorders.

Bibliography

Brodmann K. (1909). "Vergleichende Lokalisationslehre der Grosshirnrinde." Barth Verlag, Leipzig, Germany.

Carpenter, M. B. (1985). "Core Text of Neuroanatomy," 3rd ed. Williams & Wilkins, Baltimore, Maryland.

Fuster J. M. (1989). "The Prefrontal Cortex," 2nd ed. Raven, New York.

Kandel, E. R., and Schwartz, J. H. (1985). "Principles of Neural Science," 2nd ed. Elsevier, New York.

Nieuwenhuys, R. (1985). "Chemoarchitecture of the Brain." Springer-Verlag, Berlin.

Peters, A., and Jones, E. G. (eds.) (1984–1988). "Cerebral Cortex," Vols. 1–7. Plenum, New York.

Schmitt, F. O., Worden, F. G., Adelman, G., and Dennis, S. G. (eds.) (1981). "The Organization of the Cerebral Cortex." MIT Press, Cambridge, Massachusetts.

Zilles, K. (1990). Cortex. *In* "The Human Nervous System" (G. Paxinos, ed.). Academic Press, San Diego, California.

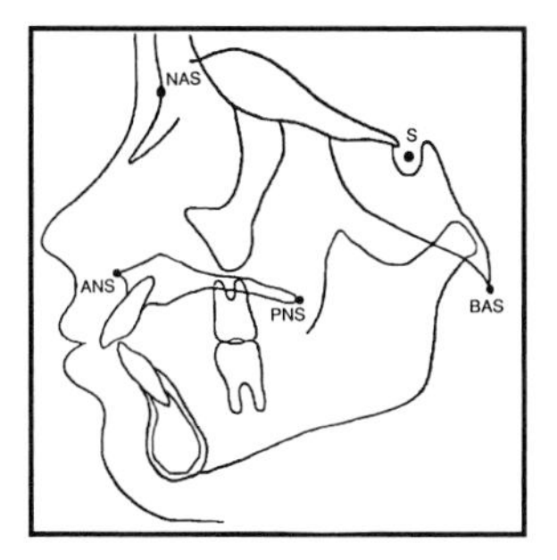

Craniofacial Growth, Genetics

LUCI ANN P. KOHN, *Washington University School of Medicine*

Glossary

Craniofacial complex Pertaining to structures of the head, including the face and braincase

Finite element methods Method from continuum mechanics, recently applied to morphology, which allows the estimation of form change local to landmarks on the skull, as the deformation necessary to produce a resultant form from an initial form

Form change Sum of the size and shape change necessary to explain the difference between two forms

Genotype Collection of genes possessed by an individual

Growth Result of form change between two ages; can be summarized by the sum of the size and shape changes necessary to develop a resultant form from an original form

Landmarks Homologous anatomical structure that can be reliably identified in all individuals of a sample

Phenotype Observable properties of a trait, produced by a combination of the genetic make-up of the individual, as well as nongenetic factors (environment)

Roentgenographic cephalometry Method of estimating form change from X-rays in which an individual's head is in a standardized position and a constant distance from the X-ray film and source; size change is estimated by the differences between linear measurements between landmarks on two or more forms, whereas shape change is estimated as the difference (or change in angle) between intersecting lines through landmarks

THE HEAD is an immensely complex structure that performs numerous diverse functions. The brain and its closely associated special sense organs (i.e., the eyes, ears, and olfactory region) receive and interpret information from the outside world. Mastication, respiration, and vocalization are performed by the oral, nasal, pharyngeal, and laryngeal portions of the head. The factors that determine normal craniofacial growth are of interest to individuals in a number of scientific disciplines. The orthodontist and plastic surgeon are concerned with recognition and treatment of abnormal dental and facial growth patterns. The anatomist and the biological anthropologist are interested in normal human variation, inferences on human evolution, and further understanding of the underlying "mechanisms" of growth events. A geneticist is interested in the inheritance of craniofacial form for purposes of developmental and evolutionary studies and for genetic counseling.

I. Prenatal Formation of the Craniofacial Complex

The bony head and face begin to form during the embryonic period, at approximately 22 days after conception, after the formation of other neural tissues such as the brain, eyes, and peripheral nerves. The tissues that will develop into the bones and cartilages of the craniofacial complex are from two tissue sources: mesenchyme derived from neural crest cells, and mesenchyme from the cranial somites. Neural crest mesenchyme covers the head region and invades the branchial arches, with the first two branchial arches contributing to the face. Many of the neural crest cells have their origins distant to the face and must migrate into their proper location and merge with the comparable bilateral structures. Imperfection in cell proliferation and fusion result in clefting of the affected structures. Facial proportions further develop during the fetal period (after 10–14 weeks), when they may still be influenced by numerous substances, including hormones, vitamins, growth factors, and placental and maternal factors (e.g., nutrition, drugs, intrauterine space). [*See* EMBRYO, BODY PATTERN FORMATION.]

Growth of the bones of the face are by either endochondral (formation of bone from a cartilage model, e.g., portions of the sphenoid, occipital, temporal bones) or intramembranous (formation of bone within membrane) ossification, as seen in the frontal, parietals, maxillae, and mandible. While the bones grow, they undergo remodeling on both their internal and external surfaces, ensuring that the cranial cavities also enlarge and that proper bone thickness is maintained. Although this process begins prenatally, it is not completed until well after birth. [*See* BONE, EMBRYONIC DEVELOPMENT; BONE REMODELING, HUMAN.]

II. Controlling Factors

The control of craniofacial growth has been discussed by van Limborgh as being divided into the three main categories of intrinsic genetic, epigenetic, and environmental factors. Intrinsic genetic factors are those that are encoded for by genes active within a cell and have their entire influence on the cell or the tissue containing the cell. Intrinsic genetic factors are involved in the differentiation of cranial tissues, as well as in the processes of intermembranous and endochondral ossification. Epigenetic factors are factors that have their influence on adjacent or distant cells or tissues, producing either local or general (distant) influences. Epigenetic factors can be determined genetically or environmentally. Local epigenetic factors from various head structures (e.g., muscles, organs) influence osseous cranial growth at sutures, as well as at the periosteum (the membrane covering the external surface of bone) and cartilaginous regions.

Intramembranous and endochondral ossification are also influenced by general epigenetic factors, with intramembranous ossification having additional control by local epigenetic factors, particularly muscle tension. Nongenetic extrinsic influences, also termed *environmental factors,* include such effects as food and nutrition and abiotic pressures. These factors influence growth throughout the face and cranial vault.

The concept of the skull as being composed of functional matrices proposes that the skull is composed of numerous units, or functional cranial components, and the size, shape, and position of each unit is determined by its function. Each functional cranial component is composed of a functional matrix, which performs the function, and a skeletal unit, which supports and protects the functional matrix.

The skeletal unit is composed of one or several bones and may not directly correspond to the named individual bones of the cranial skeleton. The growth of the skeletal unit is dependent on the growth and function of the functional matrix.

There are two types of functional matrices: capsular and periosteal matrices. The capsular matrix surrounds tissues, such as the brain or eyes, or the functional spaces of the nose, mouth, or pharynx. The skeletal units of these functional cranial components are embedded in the capsule surrounding the tissues or spaces. Five functional cranial components include capsular matrices—neural, orbital, otic (ear), nasal, and oral. The neural functional component is composed of the neural mass (brain, meninges, and cerebrospinal fluid) and its surrounding cranial vault (including the parietal bones and portions of the frontal, temporal, sphenoid, and occipital bones). A periosteal matrix is a skeletal muscle that generally attaches to its skeletal unit by periosteum. For example, a functional cranial component is created on the mandible at the temporomandibular joint by the lateral pterygoid muscle (the functional matrix) and the condyloid process (the skeletal unit).

The size, shape, and position of each functional cranial component may not necessarily dependent on any other functional cranial component. Additionally, genetic activity is responsible only for the initial osteogenesis of bone, and subsequent activity relating to size, shape, growth, and maintenance of skeletal tissue is accounted for by genetic and environmental activity of the related functional matrices (i.e., epigenetic factors). These epigenetic factors evolve as the structure and function of the components become more complex.

III. Inheritance of Continuous Traits

There is much evidence that factors influencing craniofacial growth and morphology are inherited as continuous traits. That is, these factors are determined by numerous genes, each with a small effect on any trait. Genes may be pleiotropic, affecting more than one trait, and different loci may be activated during different phases of development. With quantitatively inherited traits, the appearance of a trait in an individual, or its phenotype (P), is a sum of the individual's genotypic (G) and environmental (E) values. This is typically written in the form of a linear equation:

$$P = G + E \tag{1}$$

and it is generally assumed that there is no interaction between genotypic and environmental values. Similarly, the variability observed among individuals in the trait's phenotype (V_P) is the sum of the genetic variance (V_G) and the environmental variance (V_E). This can also be expressed mathematically as

$$V_P = V_G + V_E \tag{2}$$

with the assumption of no covariation between genetic and environmental effects.

The genotype is the collection of genes possessed by an individual and is composed of an additive genetic value (A), dominance deviation (D), and epistatic effect (I), or

$$G = A + D + I \tag{3}$$

where the additive genetic value is the sum of the separate effects of all genes influencing a trait over all loci, and the dominance deviations are the result of dominance interactions among alleles at a locus, by one allele masking the effect of another. The epistatic effects arise from the interaction of genes at different loci. The variability among individuals is summed as

$$V_G = V_A + V_D + V_I \tag{4}$$

The environmental factors can be partitioned into more specific influences [e.g., nutrition, climate, maternal effects (prenatal and postnatal)], as well as unknown factors.

The additive component of the genetic variance, V_A, is the principal determinant of the resemblance between relatives. The level of V_A also determines the response of a trait to natural or artificial selection. It measures the extent of inherited differences between individuals. The proportion of the phenotypic variance attributable to additive genetic variance is known as narrow-sense heritability, h^2,

$$h^2 = V_A/V_P \tag{5}$$

and ranges in value from 0 (no heritable variation) to 1.0 (all observed variation is due to genetic factors). Broad-sense heritability estimates, H^2, presents the proportion of the phenotypic variance that is attributable to total genotypic variance,

$$H^2 = V_G/V_P \tag{6}$$

Narrow-sense heritability and broad-sense heritability are equal if it is assumed that there are no dominance or epistatic effects on a trait, a common assumption in quantitative genetic studies.

Analyses of traits in related individuals is requisite for the estimate of the genetic variance or heritability of a trait. For traits that are inherited, individuals who share genes are expected to be phenotypically more similar to each other than to unrelated individuals. The degree of inheritance of a trait (e.g., the genetic variance of a trait) can be estimated from a population of related individuals. The degree of relationship between individuals prescribes the degree to which their similarity is expected to be caused by genetic factors. For example, parents and offspring share half of their genetic material, and the covariance between them for a trait estimates half of the genetic variance of the trait. Table I presents a summary of the proportion of the genetic and dominance variance determined by various relationships.

Estimates of phenotypic and additive genetic variance are specific to the population from which they are estimated, as they are a function of the gene frequencies within the population.

TABLE I Estimation of Additive Genetic Variance (V_A) and Dominance Variance (V_D) by Covariance Between Relatives

Relationship	Estimate
Parent–offspring	$(1/2)V_A$
Full siblings	$(1/2)V_A + (1/4)V_D$
Half siblings	$(1/4)V_A$
Identical twins	$V_A + V_D$
Fraternal twins	$(1/2)V_A + (1/4)V_D$

IV. Methods of Analyses of Normal Human Craniofacial Growth and Morphology

Numerous analyses of normal human craniofacial morphology and growth have been conducted since the late 1930s and the introduction of standardized cranial X-rays, or roentgenographic cephalograms (RCM). The X-rays are recorded with the head in a standard position a fixed distance from the X-ray anode and the X-ray film. Bony landmarks are then located on the cephalograms, and their locations are recorded. Figure 1 presents a cephalogram and five commonly identified landmarks whose definitions are found in Table II. With our recent respect for decreased exposure to radiation, additional collection by these methods is unlikely.

Growth studies may be performed on either cross-sectional or longitudinal data. In cross-sectional growth studies, individuals are observed only once. Cross-sectional growth studies can provide information on the average size of individuals at a given age or, in a sample of related individuals, the influence of genetic parameters on craniofacial morphology. Although large samples of cross-sectional data are relatively easy to collect, individual patterns of variation are obscured. Such studies concentrate on the influence of genetic parameters on craniofacial morphology and the results of craniofacial growth. However, these studies ignore the developmental mechanisms that produce the observed patterns. The heterogeneous age structure of cross-sectional family studies is usually handled in one of two ways. These analyses are generally limited to adults, or individuals of various ages are included but are statistically adjusted to a single age before analysis. Both of these methods control for the effect of age on craniofacial morphology.

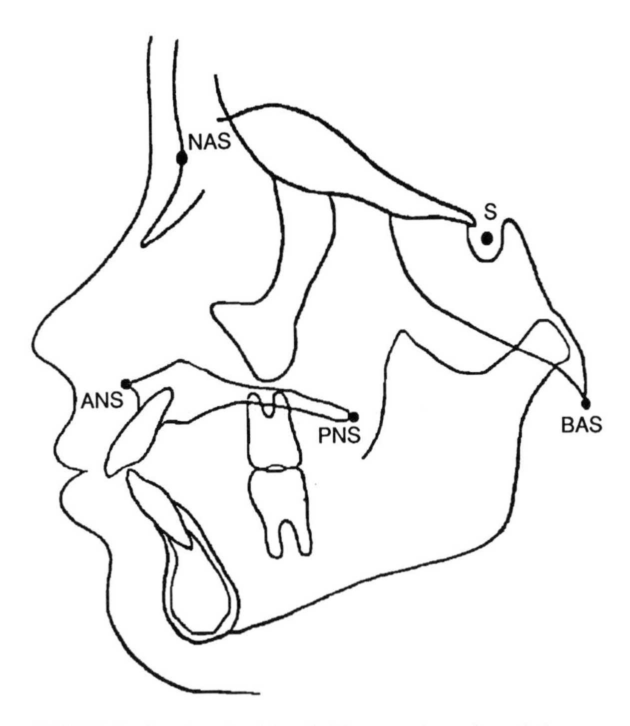

FIGURE 1 Landmarks identifiable on a lateral cephalogram. ANS, Anterior Nasal Spine; PNS, Posterior Nasal Spine; S, Sella; NAS, Nasion; BAS, Basion. Landmark locations are defined in Table I. [From Enlow, D. H. (1990) "Handbook of Facial Growth." W. B. Saunders, Philadelphia.]

Alternatively, participants are observed repeatedly during an extended period of time in longitudinal growth studies. This provides the advantage of information on individual variation and the change of variation with age. However, particularly with human studies, such analyses take a great deal of time and money and are greatly dependent on participant loyalty to the project. Therefore longitudinal analyses of human growth, especially those that include family members, are rare.

In traditional roentgenographic cephalometry, craniofacial morphology has been characterized by linear distances between landmarks, with size change estimated by the change in the linear distances with age. The angle formed by the intersection of these lines is identified as a measure of shape, which may also be estimated at different ages. It is assumed that little or no growth occurs at one landmark, usually sella, and that only linear growth occurs between two given landmarks, usually along the sella-nasion line. Measures are then expressed with reference to this landmark and line, a practice also known as registration.

Objections to RCM concern the representation of

TABLE II Landmark Names, Abbreviations, and Definitions

Landmark name	Abbreviation	Definition
Anterior nasal spine	ANS	tip of anterior nasal spine on maxilla
Posterior nasal spine	PNS	tip of posterior nasal spine on palatine
Sella	S	midpoint of sella turcica
Nasion	N	most anterior point on nasofrontal suture
Basion	B	median point of anterior margin of foramen magnum

craniofacial growth and form by linear segments and angles, and the registration of cephalograms during analyses. The assumptions surrounding the registration of cephalograms have been questioned. There are no landmarks of the skull where growth does not occur. Therefore, no landmark, including sella, can be considered to be an origin or location of no change, because its growth forces a change in its relation to other landmarks in the craniofacial complex. In addition, growth along the sella-nasion line is not simply a linear increase, but rather bends during growth.

An additional problem occurs because of the estimation of growth as the change in linear measurement between two landmarks, which best estimates the response of the two landmarks to growth. However, many traditional measures involve landmarks on separate regions of the craniofacial complex, forcing the line segment to cross more than one region of growth and remodeling. RCM analyses do not allow us to distinguish where growth is occurring or the magnitude of form change local to each landmark.

By representing craniofacial growth and form as change in linear segments and angles between linear segments and by analyzing cephalograms within registered systems, RCM may not adequately model the biological processes of craniofacial growth, obscuring the growth processes or underlying genetic control of growth. RCM is not inappropriate for all models of craniofacial growth, as it may appropriately be used for the description of generalized differences between two forms at a single point in time.

Few alternative methods have been proposed for use in analyzing morphological change, and many of these have their own associated errors. One of the earliest proposals of morphometric techniques came from D'Arcy Thompson, who suggested that form change between two morphs could be modeled by the transformation of cartesian coordinates of one form, the initial form, into the target form. Forms were registered on all landmarks, and no landmark was left unaltered in the transformation. Unfortunately, the methods were incompletely described and difficult to quantify, and they remained inaccessible until the methods of biorthogonal grids and finite element scaling were introduced.

Before finite element analyses, homologous landmarks are identified on all individuals. Lines connect landmarks at vertices to form finite elements, which may be triangular, quadrilateral, hexahedral, or other forms. A complex biological form may be divided into a number of elements, with each element representing that part of the organism included within the bounded landmarks (Fig. 2). Registration occurs simultaneously on all landmarks, making the analyses free of registration bias. Growth is then estimated local to each landmark, as the size and shape change necessary to deform an individual at one age into a subsequent age. Size and shape are estimated independently by the finite element methods. Thus the estimates of form change are inherent to the biological form and can be localized.

V. Genetics of Craniofacial Growth

Although numerous studies have found that adult craniofacial morphology has a large genetic component to observed variation, only three analyses have focused on genetic influences on craniofacial growth. The earliest longitudinal genetic analysis was carried out on a sample of 12 monozygotic and 10 dizygotic twin pairs observed during a 10-year

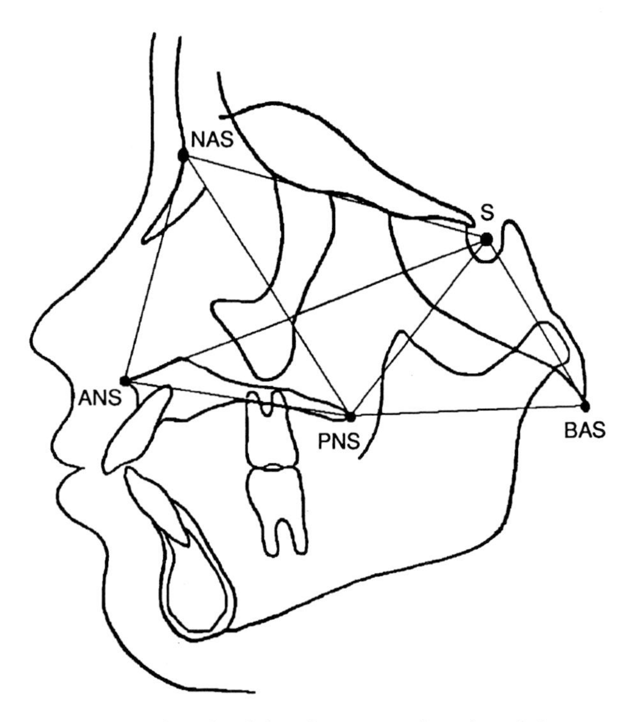

FIGURE 2 Triangular finite elements on lateral cephalogram. ANS, Anterior Nasal Spine; PNS, Posterior Nasal Spine; S, Sella; NAS, Nasion; BAS, Basion. Landmark locations are defined in Table I. [From Enlow, D. H. (1990) "Handbook of Facial Growth." W. B. Saunders, Philadelphia.]

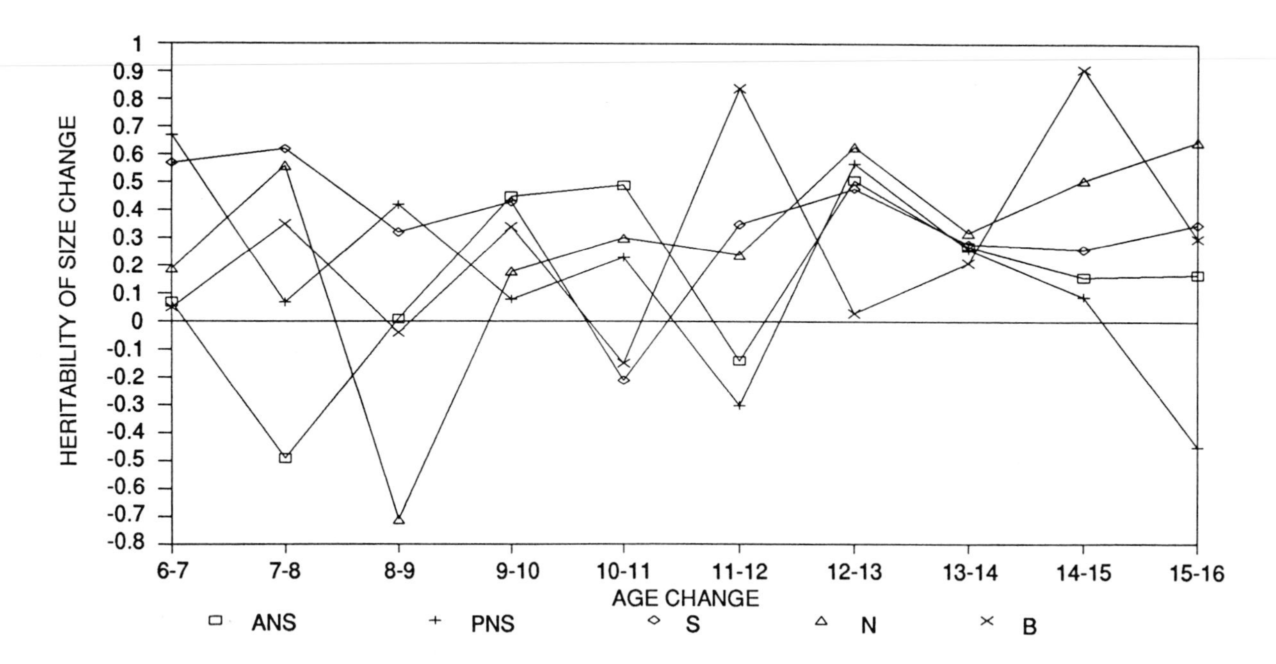

FIGURE 3 Heritability estimates of annual size change estimates at five landmarks based on analysis of variance. The sample represents sex-adjusted data from 114 twin pairs aged 6 to 11 years (58 monozygotic and 56 dizygotic twin pairs) and 117 twin pairs aged 11 to 16 years (54 monozygotic and 63 dizygotic). Estimates greater than 0.60 are significantly different from zero. [From Kohn, L. A. P. (1989). Ph.D. Thesis, Department of Anthropology, University of Wisconsin, Madison, Wisconsin.]

period. This analysis showed that only four of 15 linear and angular dimensions exhibited significant genetic influence on growth. A later analysis of sibling pairs for similarity of annual growth rates through the growth period (ages 0.5–17.5) found no consistent trend in sibling heritability during maturation. Correlations among siblings for some traits exceeded the expected correlation of 0.50, suggesting that shared sibling environment in addition to genetic factors determines craniofacial size and shape.

Finite element scaling methods were recently used to examine genetic variance and heritability of annual growth rates at five landmarks in a sample of 185 monozygotic and dizygotic twin pairs who were participants in the Forsyth Longitudinal Twin Study. The landmarks included in the study allow estimation of growth in the midface (anterior and posterior nasal spines) and the cranial base (nasion, sella, basion). Before genetic analyses, regression was used to adjust estimates of size and shape for differences caused by sex. Genetic variance and heritability were estimated from analysis of variation between (or among) twin pairs as compared with variation within twin pairs for each twin type. Because the genetic variation and heritability estimates followed similar patterns through the growth period, only the heritability estimates are presented in Figs. 3 and 4.

Little genetic variation or heritability of growth rates was observed in these data. There were, however, three distinct patterns of variation, which may provide insight into the underlying genetic and nongenetic control of craniofacial growth (Figs. 3 and 4). The first pattern, exemplified by size change at sella and nasion, is one of little stable genetic variation. Another pattern consists of large fluctuations of heritable variation throughout the growth period, as in size change at basion and shape change at anterior and posterior nasal spines and nasion. The third pattern is one of oscillations before age 12 or 13 and larger and more stable variation for the remainder of the growth period. The observed patterns of age-to-age changes in growth were significantly influenced by genetic factors for size change at sella before age 11, and after age 11 for size change at nasion and basion. However, the majority of the observed patterns in growth rates at these landmarks was due to environmental factors.

There are several possible explanations for these results. First, it may be expected that twins or siblings would exhibit greater genetic variation at the time of growth events at a landmark. Once a growth

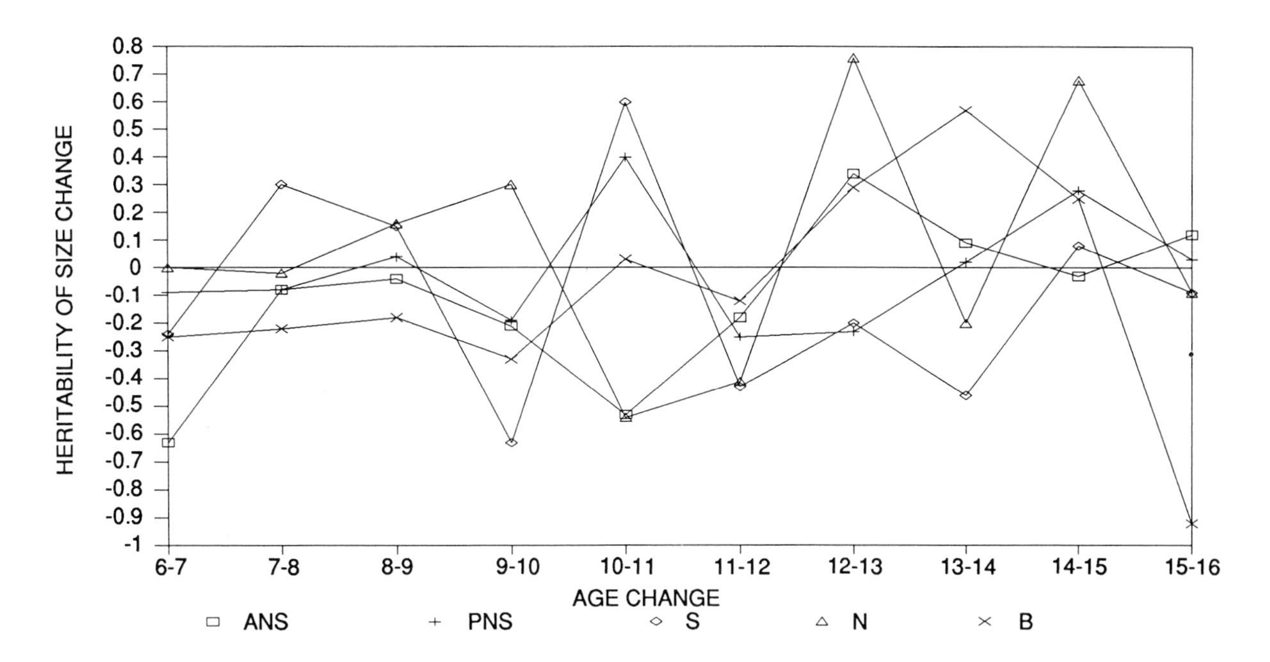

FIGURE 4 Heritability estimates of annual shape change estimates at five landmarks based on analysis of variance. The sample represents sex-adjusted data from 114 twin pairs aged 6 to 11 years (58 monozygotic and 56 dizygotic twin pairs) and 117 twin pairs aged 11 to 16 years (54 monozygotic and 63 dizygotic). Estimates greater than 0.60 are significantly different from zero. [From Kohn, L. A. P. (1989). Ph.D. Thesis, Department of Anthropology, University of Wisconsin, Madison, Wisconsin.]

event has occurred, similarity of a pair may decrease with an accompanying increase in environmental variation. The negative heritability estimates, although impossible in reality, indicate ages for which estimated environmental variance exceed phenotypic variance. Peaks observed in fluctuations of genetic variation and heritability patterns may correspond to the timing of growth events that cannot be identified with cephalometric data. That is, they may represent hormonal influences on bone or soft tissue associated with growth at these landmarks. It is conceivable that a number of such events may occur during the growth period. Stabilization of the genetic variation and heritability patterns between ages 12 and 13 correspond to the approximate timing of the male peak growth velocity and the end of the female peak growth velocity at these landmarks. Genetic variation may be expected to increase for the events of cessation of growth at the beginning of adolescence.

A second explanation of observed patterns of genetic variation is that critical growth events may occur early (i.e., either prenatally, in infancy, or in early childhood). High genetic variation and heritability of growth rates would be expected to be observed at these times. If this is the case, observed variation in growth in childhood may be largely environmental in origin. Again, cessation of growth at the end of adolescence may represent growth events for which genetic variation may increase.

With similar inheritance patterns of growth rates observed in three studies, it seems unlikely that these results are simply statistical artifacts. Adult morphology can be considered the sum of growth events and may therefore be expected to exhibit greater variability than its component parts. That genetic factors contribute significantly to adult variation and little to growth rates suggests that there is strong natural selection for adult morphology and not for the rates of attaining that morphology. This may explain the patterns of "fits-and-starts" observed during the growth period.

Acknowledgments

The helpful comments of Drs. James M. Cheverud and Allen J. Moore (Washington University) and the generosity of Dr. Coenraad F. A. Moorrees (Harvard University) are gratefully acknowledged. A revised version of this paper was presented at the 58th annual meeting of the American Association of Physical Anthropology.

Bibliography

Bookstein, F. L. (1978). "The measurement of biological shape and shape change." Lecture notes in biomathematics, vol. 24. Springer-Verlag, New York.

Byard, P. J., Lewis, A. B., Ohtsuki, F., Siervogel, R. M., and Roche, A. F. (1984). Sibling correlations for craniofacial measurements from serial radiographs. *J. Craniofac. Genet. Dev. Biol.* **4,** 265–269.

Cheverud, J. M., Lewis, J. L., Bachrach, W., and Lew, W. D. (1983). The measurement of form and variation in form: An application of three-dimensional quantitative morphology by finite-element methods. *Am. J. Phys. Anthropol.* **62,** 151–165.

Dudas, M., and Sassouni, V. (1973). The hereditary components of mandibular growth, a longitudinal twin study. *Angle Orthod.* **43,** 314–323.

Enlow, D. H. (1990). "Handbook of Facial Growth." W. B. Saunders, Philadelphia.

Falconer, D. S. (1981). "Introduction to Quantitative Genetics." Longman, New York.

Kohn, L. A. P. (1989). "A Genetic Analysis of Craniofacial Growth Using Finite Element Methods." Ph.D. Thesis, Department of Anthropology, University of Wisconsin, Madison, Wisconsin.

Moore, W. J. (1981). "The Mammalian Skull." Cambridge University Press, Cambridge.

Moss, M. L., and Salentijn, L. (1969). The primary role of functional matrices in facial growth. *Am. J. Orthod.* **55,** 566–577.

Moyers, R. E., and Bookstein, F. L. (1979). The inappropriateness of conventional cephalometrics. *Am. J. Orthod.* **75,** 599–617.

Nakata, M. (1985). Twins in craniofacial genetics: A review. *Acta Genet. Med. Gemello.* **34,** 1–14.

Richtsmeier, J. T., and Cheverud, J. M. (1986). Finite element scaling analysis of normal growth of the craniofacial complex. *J. Craniofac. Genet. Dev. Biol.* **6,** 289–323.

Thompson, D. (1961). "On Growth and Form." Cambridge University Press, Cambridge.

Van Limborgh, J. (1982). Factors controlling skeletal morphogenesis. *In* "Factors and Mechanisms Influencing Bone Growth" (A. D. Dixon and B. G. Sarnat, eds.). Alan R. Liss, New York. 1–17.

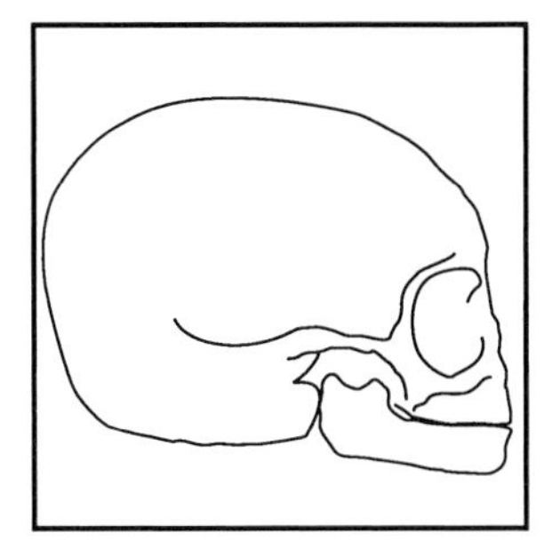

Craniofacial Growth, Postnatal

BERNARD G. SARNAT, *University of California–Los Angeles and Cedars–Sinai Medical Center*

Glossary

Alveolar process Bone surrounding and supporting a tooth

Ectodermal dysplasia Clinical condition wherein a number of structures (e.g., teeth, sweat and sebaceous glands, and hair) of ectodermal origin are deficient

Endosteum Fibrous connective tissue lining the marrow cavities of bone

Periosteum Fibrous connective tissue covering the outer surfaces of bone

Ramus Posterior vertical part of the mandible. The superior portion, the condyle, articulates with the temporal bone (see Fig. 5).

Vital marker Substance which, when administered to a living animal, marks its tissues

THE BLUEPRINT of a bone is inherent. Postnatal craniofacial growth is but a continuation of prenatal craniofacial growth interrupted by the event of birth. *In utero* the fetus is subjected to the vicissitudes of the maternal environment. After birth the individual is subjected to the effects of the general environment, thereby altering the external form and the internal architecture of a bone or a complex of bones. Changes in craniofacial form are related to the synchronous coordination of three-dimensional, multiple, differential skeletal growth sites, activities, and associated structures.

I. Bone Growth

Although significant reports in regard to bone growth appeared in the literature 200 years ago, many questions are still unanswered. What are some of the problems in need of study? What are the inherent difficulties? Any determination of bone growth must concern itself with one or more of the following questions: What are the sites? What are the amounts? What are the rates? Do they vary? When? What are the directions? What are the changes in proportion? What factors are influential?

Growth of bone occurs in three ways: cartilaginous, sutural, and appositional and resorptive (i.e., remodeling) (Table I). Growth is change. A basic physiological concept is that, throughout life, bone is in a continuous state of apposition and resorption. Consequently, skeletal size and shape are always subject to change. When skeletal mass increases, as in children, apposition is more active than resorption. Cartilaginous and sutural growth are active (i.e., positive growth). When skeletal mass is constant, as in the adult, apposition and resorption, although active, are in equilibrium (i.e., neutral growth). Cartilaginous and sutural growth have ceased. When the skeletal mass decreases, as in old age, resorption is more active than apposition (i.e., negative growth). [*See* BONE DENSITY AND FRAGILITY, AGE-RELATED CHANGES.]

The physiological stability of the bony components is the result of many interrelated factors, normal functional use being a prominent one. Well recognized are the effects of either excessive use, with hypertrophy (i.e., an increase in the mass of bone), or disuse, with atrophy (i.e., a decrease in the mass of bone). Thus, modifications in the functions of a part are reflected in alterations in the form of the part.

TABLE I Bone Growth and Repair[a]

	Growth					
	Cartilaginous	Sutural	Remodeling	Skeletal mass	Repair	Clinical considerations
Infancy and childhood	Active	Active	Apposition greater than resorption	Increasing (positive growth)	Active	Growth deformities
Adulthood	Inactive	Inactive	Apposition equal to resorption	In equilibrium (neutral growth)	Active	Acromegaly
Old age	Inactive	Inactive	Apposition less than resorption	Decreasing (negative growth)	Active	Senile osteoporosis
Clinical considerations	Conditions affecting cartilage: achondroplasia, rickets, etc.	Conditions affecting sutural growth: synostosis, etc.	Skeletal adjustments to various conditions	Changes in size and shape	Fracture, osteotomy, ostectomy, bone grafting	

[a] From B. G. Sarnat, *J. Am. Dent. Assoc.* **82,** 876–889 (1971) by permission of the American Dental Association.

Bony tissue, despite its hard, semirigid, supporting, mineralized nature, by virtue of the highly sensitive periosteal and endosteal membranes, is dynamic and ever-changing, adaptable to every nuance of tension and pressure. The basic and dual response of resorption and apposition is evident in the reaction of bone to growth, healing of fractures, alteration in muscular balance, orthopedic therapy, change in the position of bones after osteotomy and/or ostectomy, and other intrinsic and extrinsic factors.

II. Some Methods of Assessing Bone Growth

A variety of methods have been used in the study and measure of bone growth (Table II). Each, however, has its limitations. One might yield information about the sites of growth, another about the rate, while still another about direction. A combination of methods, however, potentially will yield more information and, in certain instances, more accurately than one alone. Although some of these methods lend themselves primarily to experimental work on animals, they nevertheless will contribute to our fundamental knowledge of the subject. Only a few methods are presented here.

A. Direct Measurements

1. Vital Staining

a. Madder Feeding Madder is a plant that possesses a deep-red root. For hundreds of years, madder was used for dyeing cloth. It was noted that the dye would be retained better if the cloth were previously soaked in lime water, a calcium compound. When fed to animals, madder causes bone to be stained red. Alternate layers of red and white bone correspond to periods when madder was fed and withheld. Since all animals will not accept diets containing madder, this certainly should not be considered to be a normal diet and could cause abnormal bone growth.

b. Alizarin Red S Injections Alizarin is one of the principal tinctorial agents found in madder and is available in synthetic form. In contrast to the method of prolonged madder feeding with diffuse staining, a sharp vital staining of calcifying substances might be obtained by a single intraperitoneal or intravenous injection of a 2% solution of alizarin red S (color index 1034, Coleman & Bell Company, Norwood, Ohio).

For microscopic studies ground sections are prepared, since the staining effects are lost by the action of the acid used in preparing decalcified sections. Ground sections (25–50 μm thick) show sharp red lines which may be studied with a dissecting microscope under reflected light. Under high magnification and strong transmitted illumination, the red lines (5–20 μm in width) are readily counted, and the distance between them can be accurately measured with a micrometer.

Injections of alizarin have been used to determine the rate of calcification of bone under both control and experimental conditions. They also have been used in studies of calcification in other conditions, such as the healing of fractures, kidney casts, calcified plaques of an atheromatous aorta, and calcified scars. The egg shell, the shell of the turtle, and the dentin of teeth are also stained by alizarin red S.

Other agents, such as the procion dyes, tetracycline, fluorochrome, lead acetate, trypan blue, and sodium fluoride, have been shown to be of value, with their advantages and disadvantages. One or more different vital markers have been used in the same animal.

2. Implants

Implants as reference markers have been used in the study of bone growth (Table III). John Hunter inserted two pellets along the length of the shaft of the tarsus (a foot bone) of a young pig and measured the distance between the pellets. When the tarsus was fully grown, he found that the distance between the pellets had remained the same and that the bone had increased in length at the ends. Thus, this experiment demonstrated that there was no interstitial growth of bone. Implantation of gold, silver, dental silver amalgam, stainless steel, Vitallium, or tantalum in the form of screws, pegs, pins, clips, or wires within a single bone can be used for studying the total amount of bone growth by measuring the increase in distance between the implants and the outer borders of the bone. This direct study, however, does not yield serial data without reoperation or killing of the animal. In addition, this method has been used to determine sutural growth by placing implants on either side of the suture (see Section II,B,2).

TABLE II Approximate Information Obtained from Various Methods Used to Study Bone Growth[a]

Method	Growth				Type of study	Limitations
	Site	Amount	Rate	Direction		
Direct measurements						
Osteometry						
Skeletal remains	0	X	X	X	Cross-sectional	Material of unknown history, posthumous distortion
Living	0	X	X	X	Longitudinal	Soft tissues restrict accurate measurement
Vital staining	XXX	XXX	XX	X	Longitudinal	Toxicity, method requires refinement
Implant markers	XX	XXX	X	X	Longitudinal	Local reaction to implants, requires reoperation
Histological and histochemical methods	XXXX	X	0	X	Cross-sectional	Sections show conditions at time of death of tissue only
Indirect measurements						
Impressions and casts	0	XX	XX	XX	Longitudinal	Soft tissues restrict accuracy of impression
Photographs	0	X	X	X	Longitudinal	Two-dimensional study
Serial radiography	0	XXX	XX	XX	Longitudinal	Must obtain stable landmarks, three-dimensional information not entirely accurate, radiation exposure
Measurements in combination						
Serial radiography and implantation	XXX	XXX	XXX	XXX	Longitudinal	Three-dimensional information not entirely accurate, radiation exposure
Serial radiography and metaphyseal bands	XXX	XXX	XXX	XXX	Longitudinal	Record of a toxic process, rate of growth not normal, radiation exposure
Radioautographs	XXX	0	0	0	Cross-sectional	Primarily of qualitative value

[a] 0, Gives no information; X, shows trends; XX, relatively accurate; XXX, grossly accurate; XXXX, microscopically accurate. [Modified from B. G. Sarnat and B. J. Gans, *Plast. Reconstr. Surg.* **9,** 152 (1952).]

TABLE III A Brief Historical Review of Implant Markers Used in the Longitudinal Study of Bone Growth[a]

Investigator	Year	Material used	Bone studied	Animal
Gross (direct) studies				
Hales	1727	Holes	Tibia	Chicken
Duhamel	1742	Silver stylets	Long bone	Pigeon, dog
Hunter	1770	Lead shot	Tibia, tarsometatarsal	Pig, chicken
Humphry	1863	Wires	Mandible	Pig
Gudden	1874	Holes	Parietal, frontal	Rabbit
Wolff	1885	Metal	Frontal, nasal	Rabbit
Giblin and Alley	1942	Wax	Parietal, frontal, etc.	Dog
Roy and Sarnat	1956	Stainless steel wire, black silk suture	Rib	Rabbit
Gross (direct) and/or serial radiographic (indirect) studies				
Dubreuil	1913	Metal	Tibia	Rabbit
Gatewood and Mullen	1927	Shot	Femur	Rabbit
Troitzky	1932	Silver wires	Skull	Dog
Levine	1948	Dental silver amalgam	Frontal, nasal	Rabbit
Gans and Sarnat	1951	Dental silver amalgam	Various facial	Monkey
Sissons	1953	Metal	Femur	Rabbit
Selman and Sarnat	1955	Dental silver amalgam	Frontal, nasal	Rabbit
Robinson and Sarnat	1955	Dental silver amalgam	Mandible	Pig
Björk	1955	Tantalum	Various facial	Human
Elgoyhen *et al.*	1972	Tantalum	Various facial	Monkey
Sarnat and Selman	1978	Dental silver amalgam	Nasal	Rabbit
Sarnat and McNabb	1981	Tantalum	Plastron	Turtle

[a] From B. G. Sarnat, *Am. J. Orthodont. Dent. Orthop.* **90,** 221–233 (1986).

B. Indirect Methods

1. Radiographs

Cephalometric radiography is an indirect method of studying growth of the skull (Fig. 4). A refinement in the cross-sectional method was made by the use of radiography and the superimposing of tracings of the serial radiographs over various supposedly stable bone landmarks, to obtain the pattern of growth. The accuracy of this method depends on standardization of the technique. Selection of a stable anatomical base, however, for superimposing the radiographic tracings is the key to reliable findings, since any shift of the baseline distorts the true direction of growth.

This method eliminated the most serious deficiencies of the anthropological technique. It permits a dynamic study of the growing child, that is, the increase in size and the change in proportion of the same growing bone (e.g., the mandible) or a group of bones forming a complex (e.g., the middle third of the face and the neurocranium). It reveals the rate, amount, and relative direction of bone growth. It does not, however, reveal either the sites or the mode of bone growth. This, of course, is a two-dimensional study of a three-dimensional subject.

2. Serial Cephalometric Radiography and Implantation

Use of a combination of serial radiography and radiopaque implants is a more accurate and reliable approach for a dynamic longitudinal study of bone growth (Tables II and III). This method has been used in the growth study of the face in both lower animals (Fig. 5) and humans. The serial radiographs demonstrate the increase in size and the change in proportion. In addition, a stable base for superimposing the serial radiographic tracings is obtained by inserting two or more radiopaque implants. Thus, the ensuing growth can be accurately determined and measured by superimposing tracings of radiographs over the images of the metallic implants. To avoid foreshortening, implants must lie in a plane parallel to the X-ray film. This approach is particularly useful in studying the growth pattern of the mandible or other single bones if they are clearly outlined on the radiograph.

Another advantage of this combined method is the ability to measure the amount of new bone formation and resorption that occurred from one period to another without reoperation or killing of the subject. There is also no interference with the normal diet, such as occurs in madder-fed animals. A

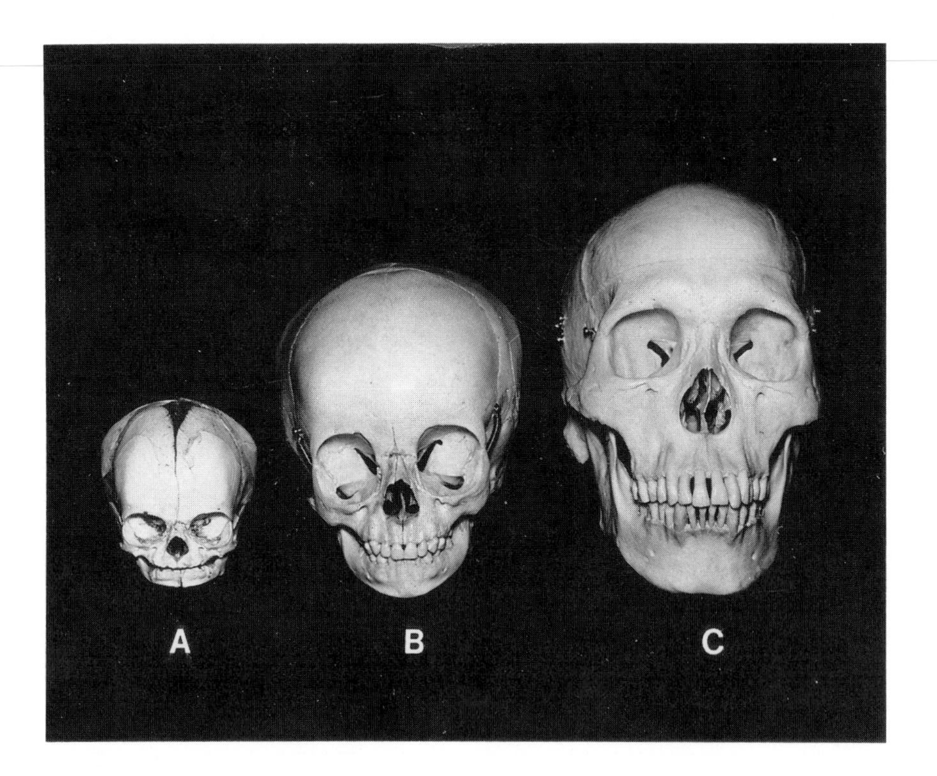

FIGURE 1 Normal growth of the human skull. (A) A clinically edentulous skull at about birth; (B) the skull of a child with completely erupted deciduous primary dentition; (C) the skull of an adult with completely erupted permanent secondary dentition. Note that in the infant the neurocranium is prominent, while the face is much less so, representing a lesser amount of the total skull size. Also note that the orbit makes up a large part of the face. In the adult the face is prominent and represents a large part of the total skull size. The orbit makes up a considerably less part of the total face in the adult than in the infant. Differential growth takes place at different times and rates in various parts of the skull. [From B. G. Sarnat, *Angle Orthodont.* **53,** 263–289 (1983).]

disadvantage is that the radiograph demonstrates the sum total of apposition and resorption at that particular time, without the detailed intervening changes shown with vital markers and histological sections. In addition to the study of the growth pattern of a single bone, this combined method is valuable in the study of sutural growth by placement of implants on either side of a suture.

To the above armamentarium, digital subtraction radiography, computerized tomography, and magnetic resonance imaging can offer more detailed and accurate information. Furthermore, newer methods for describing craniofacial growth, which are relatively independent of registration and superimposition (e.g., biorthogonal grids, elliptical Fourier functions, and finite elements) might be of future value.

III. Normal Craniofacial Growth

The craniofacial skeleton changes in size and shape in all three planes: height, width, and depth. It grows, however, in these three dimensions of space differentially in time and rate (Figs. 1–4). The dynamics of normal postnatal growth, change, and nonchange of the craniofacial skeletal system in both the young person and the adult is a fascinating, complex, and incompletely understood problem in the field of biology.

Craniofacial bones grow in three principal ways (Table I). One is cartilaginous at the nasal septum

FIGURE 2 Lateral cephalometric radiographs of the skulls shown in Fig. 1. (A) Note in the infant skull the presence of unerupted teeth in the jaws. (B) In the child skull the primary dentition is fully erupted and the permanent teeth are forming within the jaws. (C) In the adult skull the permanent teeth are fully erupted. The maxillary (m) and frontal (f) sinuses are not evident in the infant skull, are in early development in the child skull, and are fully developed in the adult skull. Note the open actively growing suture (s) in the infant neurocranium in contrast to the closed inactive suture (s) in the adult neurocranium. st, Sella turcica; o, orbit; h, head holder apparatus. (With assistance of Stuart C. White).

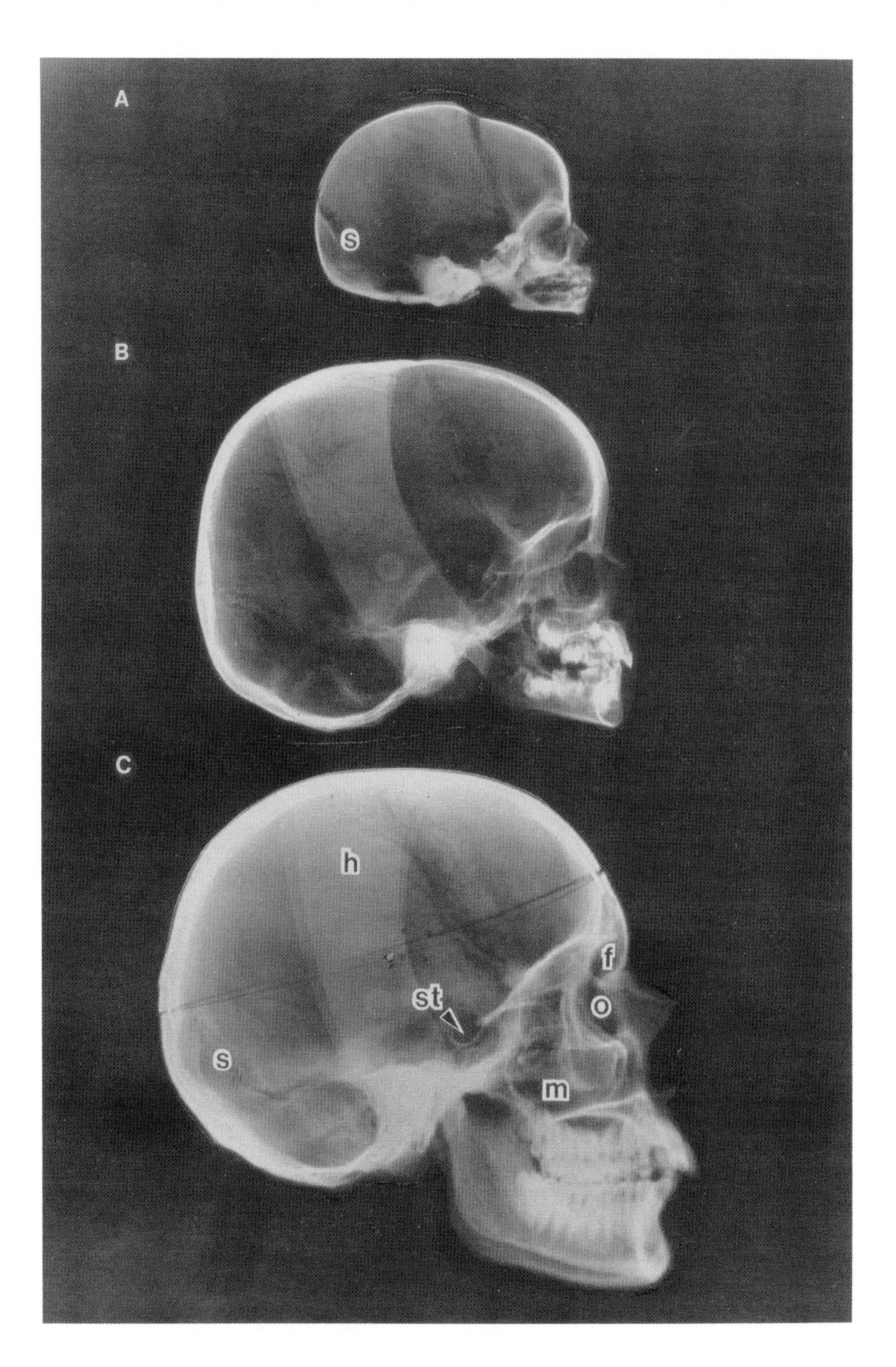

A
s
B
C
h
f
st
o
s
m

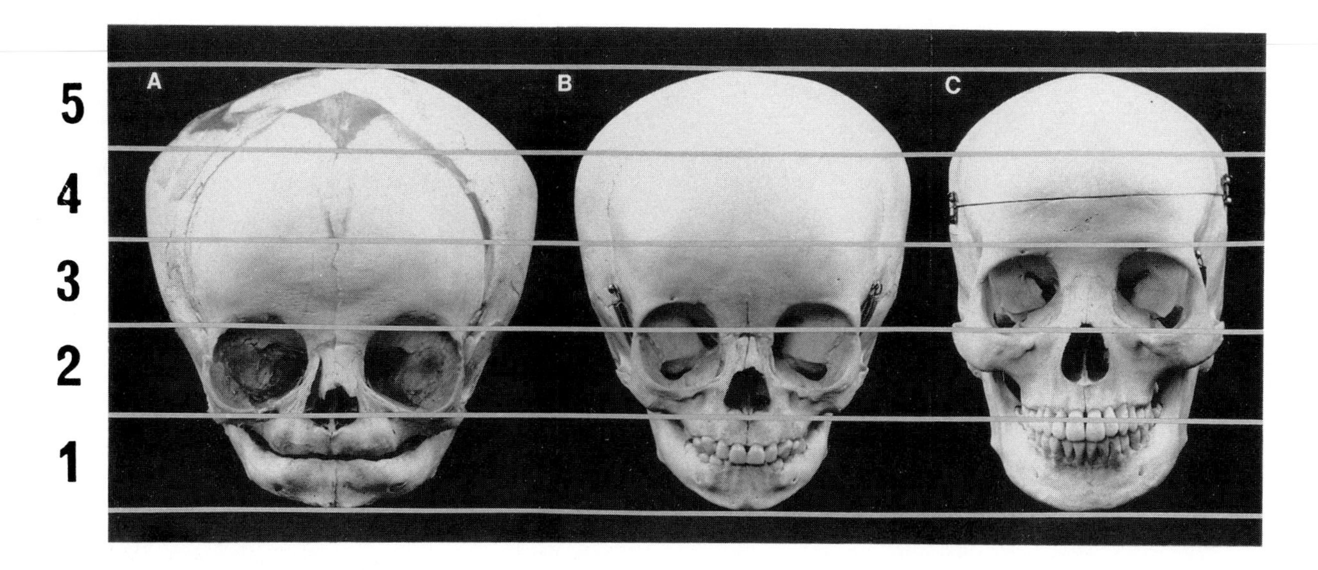

FIGURE 3 Frontal view of the skulls shown in Figs. 1 and 2, enlarged here to about the same skull height and oriented in the Frankfurt horizontal plane. Note the differences in form and proportions of the total skull and its components. The distance between the lower border of the mandible to the superior border of the orbit represents about 40% of the skull height in the infant and 60% in the adult. Orbital height is nearly the same in all three skulls. Neurocranial height represents about 60% of the skull height in the infant and 40% in the adult. Skull height is divided into fifths. [From B. G. Sarnat, *Angle Orthodont.* **53,** 263–289 (1983).]

and *endochondral growth* (i.e., the replacement of cartilage by bone) at the base of the skull at the spheno-occipital and sphenoethmoidal junctions. In addition, endochondral growth of bone occurs at the septopresphenoid joint and the mandibular condyle. These bones are joined by cartilage (synchondroses). A second way is by *sutural growth,* in which bones are united by connective tissue (synarthroses). This is found only in the skull. Sutures grow differentially. The amount of growth can vary on either side of a suture, the rate varies for different sutures at a particular time, and the same suture grows differentially at different times. These sites, as well as the endochondral, are of limited growth and usually cease activity upon reaching adulthood. A third type is *appositional and resorptive growth* (i.e., remodeling), which occurs on the outer (periosteal) or inner (endosteal) surfaces of bone. The differential responses and interrelationships of these processes are important.

The size and shape of the skull are determined not only by the growth of bone but also by its cavities. Increases in the contents of the neurocranial and orbital cavities of the skull influence the growth of adjoining bones and sutures. This occurs by a combination of resorption and deposition of bone on the surfaces and adjustments at the sutures. The maxillary, frontal, ethmoid, and sphenoid sinuses, although air containing, also increase in size (Fig. 2).

Thus, the skull, a complex of bones, has proved to be a rich source of study, particularly since the combination of different types of bone growth and increase in size of various cavities are not found elsewhere in the body. Serial cephalometric radiography has been a great aid (Fig. 4). A number of excellent references are available which describe

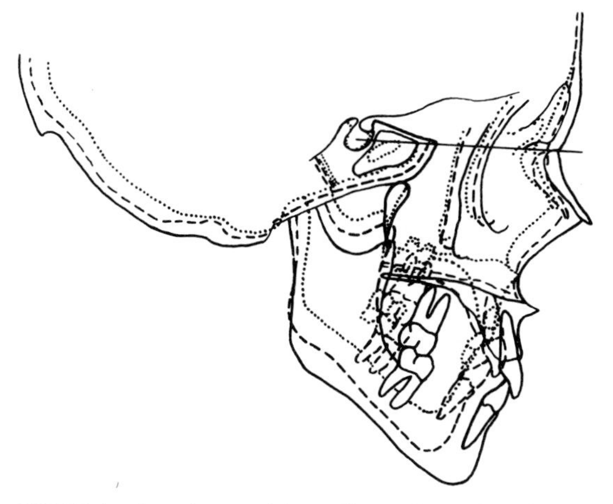

FIGURE 4 Superimposed (on sella turcica and nasion) tracings of cephalometric radiographs of a normal person at 5 years, 5 months (····); 10 years, 5 months (----); and 15 years, 7 months (—). Note in particular the downward and forward movement of the face and jaws. [From B. G. Sarnat, A. G. Brodie, and W. H. Kubackl. *Am. J. Dis. Child.* **86,** 162–169 (1953).]

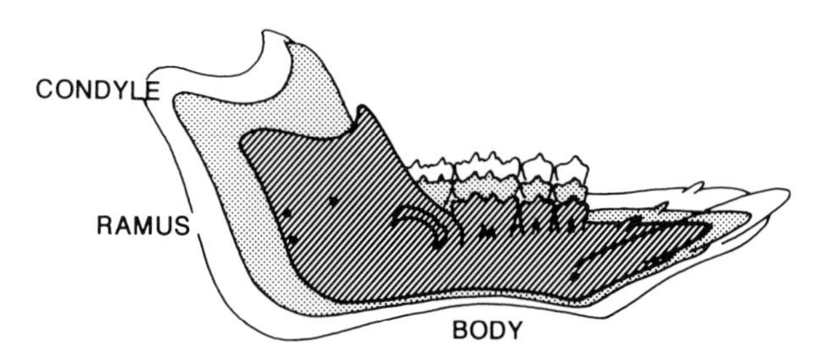

FIGURE 5 Serial tracings of lateral cephalometric radiographs of a pig superimposed on the outlines of four implants. Note the direction and differential growth that have occurred between various periods along all borders except the anterior border of the ramus, which has been resorbed. ▨, 8 Weeks; ▩, 16 weeks; □, 20 weeks. [From I. B. Robinson and B. G. Sarnat, *Am. J. Anat.* **96,** 50 (1955).]

the details of craniofacial growth and movement and the anatomical structures.

A. Lower Face

Mandibular growth occurs in two ways. One is appositional and resorptive, with differential bone remodeling at various periosteal and endosteal surfaces. In the ramus the posterior border is a particularly active site of bone apposition, whereas the anterior border is a particularly active site of bone resorption (Fig. 5). The second type of mandibular bone growth is endochondral at the condyle. Growth of the condyle and the ramus, the most prolific sites, is in a superior and posterior direction. Because the condyle articulates with the mandibular fossa of the temporal bone, the final result is a downward and forward movement of the mandible. Another consideration is that the mandible is distracted by the inframandibular muscles, with secondary growth at the condyle.

The condyloid process of the mandible grows by the replacement of cartilage by bone (i.e., endochondral bone growth). Microscopic examination of a growing condyle reveals the presence of three zones: chondrogenic, cartilaginous, and osteogenic (Fig. 6). The condyle is capped not by hyaline cartilage, as are nearly all articular surfaces, but by a narrow layer of avascular fibrous tissue that contains a few cartilage cells. The inner layer of this covering is chondrogenic, giving rise to the hyaline cartilage cells that constitute the second zone. In the third zone destruction of the cartilage and formation of bone around the cartilage scaffolding occur. [*See* CARTILAGE.]

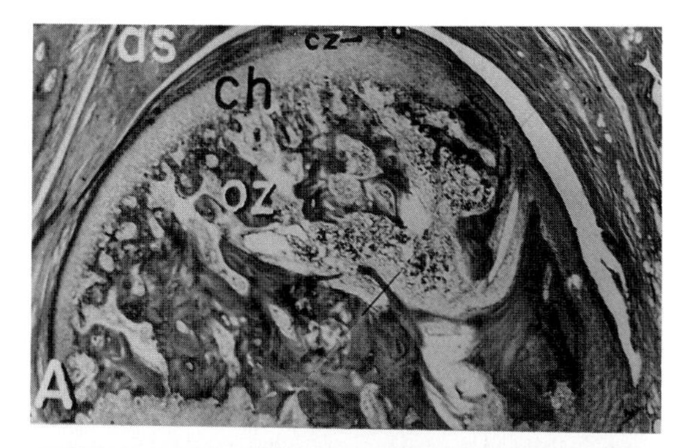

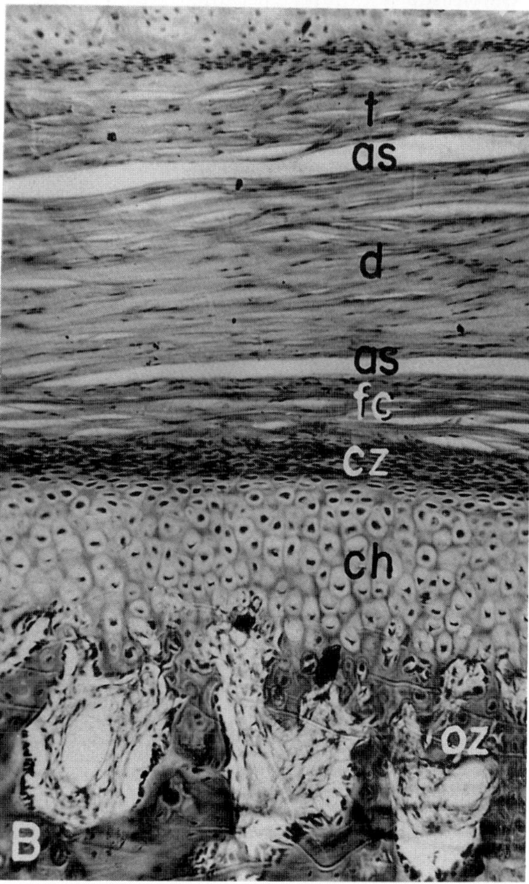

FIGURE 6 Sagittal hematoxylin–eosin-stained section through the left temporomandibular joint of a growing monkey. t, Temporal bone; as, articular space; d, disk; fc, fibrous connective tissue covering of the condyle; cz, chondrogenic zone; ch, cartilaginous zone; oz, osteogenic zone. (A) Original magnification ×20; (B) original magnification ×120. [From B. G. Sarnat, *Am. J. Surg.* **94,** 19–30 (1957).]

This cartilage is not a derivative of a cartilaginous model, as are epiphyseal and articular cartilages in long bones. While the latter are derived from the primary cartilaginous skeleton, the condylar cartilage of the mandible can be classified as a secondary growth cartilage. Appositional and interstitial growth of the cartilaginous portion of the condyle contributes directly to the increases in mandibular height and length. The cartilage, however, is not homologous to an epiphyseal cartilage because it is not interposed between two bony parts, and it is not homologous to articular cartilage because the free surface bounding the articular space is covered by fibrous tissue, rather than hyaline cartilage. Thus, defects in the condylar articular surface are repaired more rapidly and completely than similar injuries to other articular surfaces, such as those in long bones.

The growth pattern of the pig mandible has been studied with serial cephalometric radiographs in combination with metallic implants (Fig. 5). The condylar region contributed about 80% to the total ramus height of the mandible, while the posterior border contributed as much as 80% to total length of the mandible. Since the amount of apposition at the posterior border was about twice the amount of resorption at the anterior border, ramus width increased. Resorption of the anterior border of the ramus played an important role in lengthening the body of the mandible in the molar region. Thus, it could be stated that in the growing mandible the ramus of today will be the body of tomorrow.

As the anterior border of the ramus continued to be resorbed, it created not only more body, but also simultaneously permitted exposure of the crowns of erupting permanent molars. There is usually a harmonious relationship between dental development and growth of the mandible. This growth, however, was not dependent on the developing teeth. Normal eruption of the permanent molars, however, was dependent in part on the normal growth of the mandibular ramus.

The roles played by the alveolar processes and mandibular bone proper were quite different. In growth of the mandible, the only part subservient to erupting teeth was alveolar bone. The contribution of the growth of the alveolar bone to the increase in the height of the mandibular body was about 60%. The important feature of alveolar bone was its ability to serve the needs of the ever-changing requirements of the teeth. In contrast, however, the formed mandibular bone remained relatively constant, except for changes in proportion.

B. Upper Face

Many sites contribute to growth of the upper facial skeleton, a three-dimensional mosaic of bones and cavities. It is influenced by endochondral growth at the cranial base by the spheno-occipital and the sphenoethmoidal synchondroses. Sites of growth for the maxillary complex are the maxillary tuberosities and three sutures on each side: the frontomaxillary, zygomaticomaxillary, and pterygomaxillary. This maxillary complex moves downward and forward. Transverse growth at the median palatine suture, which is affected by the downward and divergent growth of the pterygoid processes, is both simultaneous and correlated with the widening of the downward-shifting maxillary complex. Anteroposterior growth occurs along the transverse palatine suture between the horizontal plate of the maxilla and the palatine bone and along the posterior margin of the palatine bone.

The downward shift of the hard palate, which increases the size of the nasal cavity, is by resorption of bone on its nasal surface and apposition on its oral surface. This shift could also be related to growth of the cartilaginous nasal septum. The downward, forward, and lateral growth of the subnasal part of the maxillary body is accompanied by the eruption of the teeth and apposition of bone at the free borders of the alveolar process. This contributes to the increases in height and width of the upper facial skeleton.

The upper facial skeleton also enlarges with the increased size of the orbital cavities, the nasal cavity, and the air-containing maxillary, frontal, and ethmoid sinuses (Fig. 2). Concurrent with all of these, growth and movement processes of the upper face are continuous differential apposition and resorption of bone. According to the functional matrix hypothesis, the above are secondary sites of growth dependent on the relationship to the adjoining soft tissue.

C. Orbit

Facial growth is related to orbital growth. Shape and size of the orbit result from the balance of a number of genetic and epigenetic factors which may function on a systemic, regional, and local basis. Is there a key factor(s) which influences orbital growth? Is there a correlation between orbital size

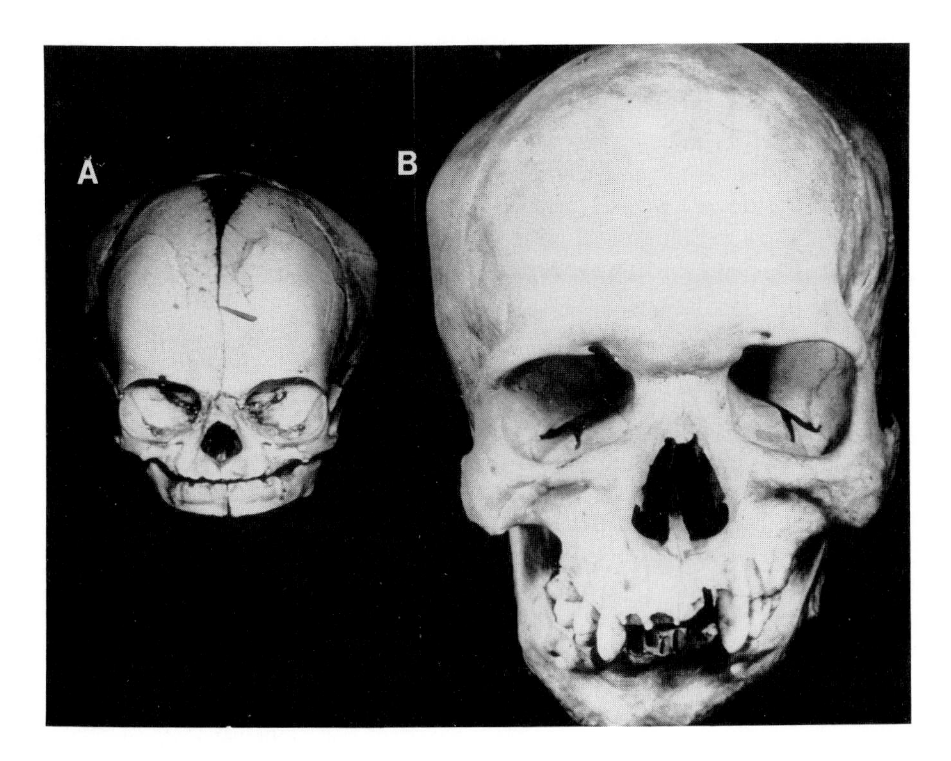

FIGURE 7 Comparison of (A) a normal symmetrical infant human skull with (B) an abnormal asymmetrical adult human skull. Note undergrowth primarily of the right condylar region, mandible, and face as well as an ankylosis of the right temporomandibular joint and a malocclusion. Trauma to, or infection of, the temporomandibular region during infancy or early childhood could have led to this deformity. [From B. G. Sarnat. *Surg., Gynecol. Obstet.* **148,** 659–669 (1979) by permission of *Surgery, Gynecology & Obstetrics*.]

and intraorbital mass? If so, what roles do the muscles and other extraocular structures (e.g., the globe, the vitreous body, the aqueous humor, and the lens) play?

The relative capacity of the orbit and the size of the eye diminish with increase in body weight. In the fetus and the newborn the eyeball is so large in relation to the socket that it projects considerably beyond the orbital rim so that a normal fetal exophthalmos exists. In infants the eyes are not only larger in proportion to body weight than in the adult, but also in proportion to the size of the orbit. In humans orbital height at birth is about 55% of the adult and 79% of the total growth at 3 years of age (Fig. 4). At 7 years of age, orbital height is about 94% of the adult, while facial height is still only 80% (Figs. 1–4).

D. Nasal Bones

The growth pattern of the nasal bone region in the rabbit was studied using the same method for studying the growth pattern of the mandible, also a single bone. Two radiopaque implants were inserted into each left and right nasal bone. Ventrodorsal cephalometric radiographs were obtained in a specially designed head holder. From these radiographs separate tracings were made on matte acetate paper of the left and right nasal bone regions, including the radiopaque implants for two different periods. Since the implants maintained the same relationship to each other in each nasal bone, they served as stable reference sites. The differences in the two established outlines represented the changes in size and shape in two dimensions. Growth at the proximal and distal borders was approximately equal and about twice that of the lateral border and about five times that of the medial border. What factors are active to produce this differential in growth?

E. Relationship of Teeth to the Jaws and the Face

The importance of the deciduous and permanent teeth in the growth of the jaws and the face has been a much-debated question. For example, the anteater, born without teeth, has long jaws. A patient with complete anodontia and ectodermal dysplasia presented an unusual opportunity to study this problem. Serial cephalometric radiographs were taken from 21 months to 16 years of age, a period of growth and eruption of the deciduous and permanent dentitions. Study of the superimposed tracings of the radiographs indicated that growth was within small normal limits and that the complete absence of teeth did not significantly impair development of the face and the jaws, with the exception of the alveolar bone. Five sets of full upper and lower artificial dentures were designed, constructed, and delivered during this time. Each successive denture was larger and contained more and larger teeth to accommodate the increase in the size of the jaws. Thus, in instances of surgical procedures on the jaws, disturbance of the teeth or tooth buds could lead to loss, malformation, or malposition of the teeth, with changes in alveolar bone. The general size and shape of the jaws and the face, however, are not affected because of damage to the teeth. [*See* Dental and Oral Biology, Anatomy.]

IV. Environment and Growth

Throughout our lives we are constantly reacting to our environment. Variations in temperature, light, humidity, atmospheric pressure, terrestial and extraterrestrial radiation, and gravity affect us. In addition, the vast number of noxious chemical, physical, or biological agents, intentionally or unintentionally ingested in our food and water and inhaled in our air, determine our destinies. Deficiencies in essential nutrients might also play a part. Consider the effect of our environment on the skeletal growth sites and the resulting changes in size and shape of the skull (Fig. 7) and the body.

Young rats exposed to cold stresses had a smaller skull, a longer face in relation to the cranial vault, a narrower nose, a rounder neurocranium, and a shorter femur. Populations living at high altitudes and exposed to the environmental stresses of hypoxia and cold have a slower postnatal growth and less of an adolescent growth spurt than do other groups. On Earth gravity is considered normal, or 1.0 g. What skeletal and other changes might occur in an environment of hypogravity (the moon, 0.18 g) or hypergravity (Jupiter, 2.65 g)? These and other factors are of great interest to the field of cosmic biology and should be of both interest and concern to us.

Bibliography

Björk, A., and Skieller, V. (1983). Normal and abnormal growth of the mandible. A synthesis of longitudinal cephalometric implant studies over a period of 25 years. *Eur. J. Orthodont.* **5,** 1–46.

DuBrul E. (1980). "Sicher's Oral Anatomy," 7th ed. Mosby, St. Louis, Missouri.

Enlow, D. H. (1982). "Handbook of Facial Growth," 2nd ed. Saunders, Philadelphia, Pennsylvania.

Johnston, L. E. (1976). The functional matrix hypothesis: Reflections in a jaundiced eye. *In* "Factors Affecting Growth of the Midface" (J. A. McNamara, Jr., ed.), Craniofacial Growth Ser., Monogr. 6, pp. 131–168. Univ. of Michigan, Ann Arbor.

Krogman, W. M. (1974). Craniofacial growth and development: An appraisal. *Yearb. Phys. Anthropol.* **18,** 31–64.

Sarnat, B. G. (1983). Normal and abnormal craniofacial growth: Some experimental and clinical considerations. *Angle Orthodont.* **53,** 263–289.

Sarnat, B. G. (1984). Differential growth and healing of bones and teeth. *Clin. Orthop. Relat. Res.* **183,** 219–237.

Sarnat, B. G. (1986). Growth pattern of the mandible: Some reflections. *Am. J. Orthodont. Dent. Orthop.* **90,** 221–233.

Sarnat, B. G., and Laskin, D. M. (eds.) (1991). "The Temporomandibular Joint: A Biological Basis for Clinical Practice," 4th ed. Saunders, Philadelphia, Pennsylvania.

Crime, Delinquency, and Psychopathy

RONALD BLACKBURN, *Ashworth Hospital North–Liverpool*

Glossary

Electroencephalogram Electrical record of brain rhythms which are conventionally divided into four frequency bands of delta (0.5–3 Hz), theta (4–7 Hz), alpha (8–13 Hz), and beta (14–30 Hz) activity

Socialization Process by which social values and moral prohibitions become internalized in the form of conscience, or superego

Temperament Emotional and expressive characteristics distinguishing an individual

Wechsler tests Commonly used measures of intellectual ability, or intelligence quotient (IQ), divided into verbal and nonverbal or performance components

CRIMINAL BEHAVIOR is action prohibited by law, but since laws vary over time and among societies, no act is inherently criminal. Delinquency refers to acts of juveniles which would be criminal if committed by an adult, but also misbehavior which derives from the status of being a minor. A criminal or a delinquent is strictly someone who has been subject to the official sanctions of the criminal justice system, but much criminal behavior remains undetected. Modern criminological research therefore also includes as criminals those who admit to criminal acts in self-report measures. Criminality hence can be regarded as a tendency or a disposition to engage in criminal behavior, which varies in degree throughout the population at large. Psychopathy, in contrast, is a psychological concept denoting abnormality of personality. The term lacks a universally agreed meaning, but relates to a disregard for the rights and feelings of others.

Psychopathic personalities are identified by deficiencies in emotional experiences which normally restrain harmful behavior and, hence, by characteristics such as callousness, lack of empathy and guilt, and impulsive, often criminal, actions. Common synonyms for psychopathic personality are sociopathic or antisocial personality disorder, the latter being the currently preferred term in American psychiatry. Those described as psychopaths by clinicians are not homogeneous, and a distinction is therefore frequently made between primary and secondary psychopaths, the former characterized by assertiveness and an absence of guilt and anxiety, the latter, by social anxiety and poor self-esteem. Psychopathy, like criminality, can be construed as a continuum, but not all criminals display psychopathic personality traits and not all of those showing such traits violate the law. However, more extreme criminality overlaps with psychopathy, and theories of psychopathy are relevant to the explanation of serious and persistent criminal behavior.

I. BIOLOGY AND CRIMINOLOGY

A. Study of Crime

Criminology is of interest to many disciplines, but has been dominated by sociology, psychiatry, and psychology. Sociologists are concerned with identifying which sections of society are more likely to be criminal. Crime rates are higher, for example, among males, in urban centers, in inner city slums, and in the United States among black minorities.

Also, the peak rates of crime occur among those aged 15 to 18, although only a minority of delinquents become adult criminals. Sociological theories account for these variations in terms of unequal economic opportunities, the effects of different subcultures, and the influence of the controlling social forces within the family, school, and community which discourage violation of social rules. Sociologists view crime as a creation of society and seek the formal causes of crime in terms of the activities of powerful groups which determine what behavior is to be defined as criminal.

Psychiatry and psychology, on the other hand, aim to identify which individuals are more likely to violate laws, and hence emphasize the efficient, or antecedent, causes of criminal behavior. These disciplines assume that social forces are insufficient to explain criminal acts, since, among sections of society with high crime rates, only a minority become criminal. Variations between individuals are therefore invoked as causal factors. Psychiatry focuses on abnormal mental states which might be correlated with the commission of crimes, although only a small minority of criminals show serious mental disorders. While psychological theories of crime overlap with psychiatric perspectives, they are more typically concerned with criminal acts as variations of normal behavior and seek explanations in terms of personal attributes and behavioral tendencies and the processes of development and learning through which such attributes arise, rather than in terms of discrete abnormalities.

The notion that biological endowment could contribute to crime has always been represented in psychiatric and psychological theories. A biological approach inquiries about variations in physiological and neurochemical functioning which might differentiate those who violate laws. The most extreme view was that of Italian physician and anthropologist Cesar Lombroso, who in 1876 proposed the concept of the "born criminal," a remnant of early human ancestry, characterized by primitive physical and psychological development. This was discredited by early anthropometric studies of prisoners, and biological perspectives on crime have subsequently been viewed with disfavor. Such perspectives are criticized for disregarding the apparently overwhelming role of social institutions in defining what is criminal and in creating conditions conducive to crime.

However, current biological approaches do not postulate a gene for crime, and they focus instead on the contribution of normal biological variation to adaptive learning, including the learning of society's rules. They view criminal behavior in terms of a tendency to break rules rather than the commission of specific criminal acts. Since learning depends on a social environment, biological approaches are also concerned with biosocial interaction and do not assert a unidirectional biological determinism. This view therefore recognizes the influence of the family, school, or social group in shaping rule-breaking behavior, but challenges an environmental determinism which assumes that individuals are merely passive recipients of these influences.

B. Theories of Criminality

Psychological theories of criminality traditionally focus on the way in which individuals are socialized through interactions with significant others, and child-rearing practices clearly play a significant role in this process. Parents of antisocial children commonly fail to punish deviant behavior with consistency and moderation, do not reward socially acceptable behavior, and do not themselves behave as models of conformity and achievement which the child can imitate. However, the child's own contribution to this process remains debated. Since there is evidence from behavior genetics studies that ability and temperament variables have a genetic origin, the child's personal attributes might well influence the causal chain leading to undersocialization and the subsequent drift to delinquency. For example, infant temperament might significantly affect the response of the parents.

Not all theories take into account the potential influence of biological variation. Psychoanalysis sees crime in terms of a failure to tame biologically determined erotic and aggressive instincts during the early years of life through the formation of the superego. The psychopath fails to develop a superego and, hence, engages in antisocial and often violent acts without moral qualms. The "neurotic" delinquent, in contrast, has too strict a superego and acts out repressed instinctual wishes in symbolic form as antisocial behaviors which invite punishment. Criminal acts are thus the expression of biological forces, but the psychoanalytic theory emphasizes biological similarities among people in the form of universal human instincts. [*See* PSYCHOANALYTIC THEORY.]

Learning theorists reject psychodynamic instinct theory. Social learning theorists propose that much criminal behavior is learned in the same way as other behavior, through imitation and the rewarding influences of peers. Others, however, notably Hans Eysenck, retain the notion of a generalized conscience which restrains antisocial acts, but attribute failure of socialization to biological differences among people.

Eysenck's theory does not propose that crime is biologically determined, but rather that biological variation influences the capacity to learn control of self-gratification. Variations in human temperament are attributed to three independent personality dimensions of neuroticism (N), extroversion (E), and psychoticism (P), each of which has a biological basis. N is held to reflect greater reactivity in the limbic and autonomic nervous systems and determines emotional responses to stress. Underlying E is the level of cortical arousal or arousability, governed by activity in circuits linking the reticular system of the brain stem and the cerebral cortex. Extroverts have low arousal, relative to introverts, and are predicted to form conditioned responses less readily, and to require more intense stimulation from the environment. More tentatively, P is related to circulating androgens.

Conscience formation depends on the classical conditioning of anxiety reactions. Punishment of antisocial behavior by parents, teachers, or peers generates anxiety in the child, and cues associated with punishment, including internal stimuli, become conditioned stimuli which themselves arouse anxiety. The child avoids anxiety by resisting the temptation to indulge in the punished behavior, and through stimulus generalization a generalized conscience is acquired. Since extroverts form conditioned responses slowly, they will be less well socialized than introverts and, hence, more likely to engage in antisocial behavior. Their need for stimulation also leads to impulsive "thrill-seeking," which might take the form of prohibited acts. Additionally, however, the traits of the P dimension (i.e., hostility, insensitivity, and cruelty) make antisocial behavior more likely. Eysenck, therefore, proposed that criminals in general, and psychopaths in particular, show the more extreme characteristics associated with E and P.

Research has established a significant association between personality measures of P and criminality, and the suggested greater need of criminals for stimulation also receives support in studies of sensation-seeking. The link with E is less consistently supported, and while self-reported delinquency does seem to be correlated with E, officially defined criminals have not been found to be significantly more extroverted than nonoffenders. The postulated links between E and arousal and conditionability, and between P and androgen level, remain to be substantiated. [*See* SENSATION-SEEKING TRAIT.]

This theory has been criticized for assuming that different physiological response systems condition in a uniform fashion. One alternative view is that much criminal behavior arises primarily from deficiences in child-rearing, but that some individuals (e.g., psychopaths) are relatively resistant to such training as a result of deficits in passive avoidance learning. Passive avoidance involves learning to inhibit punished behavior, but it is dependent on the conditioning of anxiety responses specifically, rather than on conditionability in general. [*See* CONDITIONING.]

It has been proposed that individual differences are based on specific forebrain systems, notably the behavioral inhibition system (BIS), a septohippocampal system responsive to conditioned stimuli for punishment or the absence of an anticipated reward, which mediates passive avoidance. The BIS interacts with the behavioral activation system (BAS), which mediates responsiveness to conditioned stimuli for reward or nonpunishment, although the biological substrate for this is less clear. Psychopaths are believed to have a weakly reactive BIS, leading them to be insensitive to threat stimuli. No dysfunction of the BAS is suggested, but reward-seeking might be disinhibited in psychopaths through the failure of punishment stimuli to inhibit rewarding behavior.

II. Genetics and Criminality

Much research has focused on the general proposition that there are hereditary influences on criminality, rather than on the specific characteristics of the nervous system which might mediate antisocial behavior. This does not assume that complex human actions are "preprogrammed," since genetic effects on behavior are polygenic and probabilistic. Genotypes give an initial direction to development by providing basic elements of behavior, which are incorporated into larger adaptive units through learning. They influence phenotypes through the combination of genes and environments supplied by parents, through differential reactions from others to biologically different individuals, and through differential selection of environments by these individuals. Genetic studies of criminality attempt to isolate innate influences on these complex pathways through several research designs.

A. Family Studies

Family studies compare the distribution of antisocial behavior in the biological relatives of offenders and nonoffenders. Criminal parents tend to be more likely to have criminal children. Family studies of delinquent females also suggest that they have more socially deviant relatives than do other females and have more familial pathology than do antisocial males, although familial factors seem to be of equal importance for the development of criminality in both sexes. However, family studies do not permit any clear separation of genetic and environmental influences.

B. Twin Studies

Monozygotic (MZ) twins have the same genotypes, whereas dizygotic (DZ) twins are no more alike, genetically, than other siblings. Studies of twins therefore use the logic that phenotypic differences between same-sex MZ and DZ twin pairs are likely to reflect genetic influences, assuming similar rearing conditions. The relevant differences are most commonly expressed as the percentage of criminals with twins whose twin is criminal (i.e., pairwise concordance).

About 12 twin studies of varying sample sizes have been reported. Earlier studies found that, on average, MZ twins showed 60% concordance for criminal history, and DZ, 30%, but differences are most apparent for adults, and twin studies of officially defined juvenile delinquents suggest minimal genetic influence. However, recent research reports greater similarity of self-reported delinquency for MZ than for DZ twins, while also indicating significant gene–environment interaction.

Earlier studies used selected samples of twins and unreliable methods of determining zygosity. Lower concordance rates emerge from recent Scandinavian studies, which relied on unselected samples of twins from national registers and determined zygosity from blood samples. For example, among Danish males who became criminal, pairwise concordance rates were 35% (MZ) and 13% (DZ), while for females, they were 21% (MZ) and 8% (DZ). The high discordance rates indicate substantial nongenetic effects, but the MZ concordance rate is nevertheless substantially higher than the DZ rate. However, similar research in Norway found male concordances of only 26% (MZ) and 15% (DZ). The difference is not significant and raises doubts about hereditary effects on crime. Nevertheless, all studies find differences between MZ and DZ concordance rates in the same direction, and the difference between the Norwegian and Danish studies remains unexplained.

One common argument is that phenotypic similarities of MZ twins represent a greater similarity of treatment by parents. Although it is equally plausible that any experience of environmental similarity by MZ twins might be an effect of their genetic similarity, this remains a possible confound in all studies of twins reared together. Without data on the criminality of twins reared apart, twin studies remain suggestive, rather than conclusive.

C. Adoption Studies

If children adopted shortly after birth resemble biological more than adoptive parents in some attribute, this is strong evidence for genetic influence. Research on criminality in adoptees uses two designs. The first identifies criminal parents who have given up their offspring for adoption, and compares their children with the adopted offspring of noncriminal biological parents. Studies using this design provide support for a genetic contribution to crime. For example, more of the adopted offspring of female offenders had acquired a criminal record in comparison with controls in one American study, while in another, adoptees whose biological parents were diagnosed as antisocial personalities showed a greater probability of receiving this diagnosis as adults.

The second design begins with a heterogeneous sample of adoptees and compares criminal and noncriminal adoptees in terms of criminality in biological and adoptive parents. Several studies have been reported during the past two decades from Scandinavia, and it has been found that the biological fathers of adoptees who are psychopathic show a higher incidence of psychopathy than do the adoptive fathers. It has also been shown that the likelihood of an adoptee's becoming criminal increases significantly if the biological father was also criminal, but that it increases further if both biological and adoptive fathers have criminal records. These data point to a genetic influence on criminality, but also to a genotype–environment interaction. However, recent work emphasizes the heterogeneity of criminals. Research in Stockholm suggests a genetic involvement for petty criminality, but not for violent crime. For some criminals, however, violent behavior is symptomatic of alcohol abuse, which could itself have a genetic basis.

In none of these studies has more than a minority of the adopted children of criminal parents become criminal or antisocial, and what is implied by the results is a genetic contribution to crime, given certain environments, rather than any overwhelming genetic determination. However, preadoption influences are not solely genetic, but include perinatal complications determined by the living conditions of the mother. Whether these might account for the relationship between criminality in biological parents and their adopted children remains unclear.

D. Chromosome Anomalies

Variations from the normal complement of 23 chromosome pairs usually arise from errors in cell division, and represent genetic factors which are innate, but not inherited. These rare anomalies have been linked with behavior disorder, and particular interest has been shown in sex chromosome patterns which depart from the usual configurations of 46,XY in males, and 46,XX in females. Most research has focused on males showing a 47,XYY complement, or the 47,XXY configuration (Klinefelter's syndrome). [*See* CHROMOSOME ANOMALIES.]

The XXY pattern has long been known to be more frequent among the mentally retarded, but research in the 1960s also indicated elevated frequencies of XYY males among prisoners and mentally disordered offenders. Most were of above average height, had a low intelligence quotient (IQ), and had personality disorders. Subsequent research has confirmed the higher prevalance of the XYY complement in institutionalized antisocial populations, particularly mentally disordered offenders, but has also shown an incidence in the general population of about 0.1%, which is higher than was previously thought.

The initial studies appeared to illustrate a genetic determination of criminal behavior and were also interpreted in terms of what might be contributed to crime by the normal Y chromosome. It was suggested that the sex difference in crime reflected the masculine traits contributed by the Y chromosome, and that the extra Y exaggerated masculinity and violence. However, possession of the extra X chromosome also correlates with criminality, and characteristics of the XYY male have typically been inferred from small samples in institutions.

A study reported from Copenhagen in 1976 avoided the problem of sampling bias and identified 12 XYY and 16 XXY individuals in a large birth cohort. While 42% of the XYYs, 19% of the XXYs, and 9% of controls had criminal records, confirming an association of the extra Y with criminality, the offenses of the former groups were mainly petty and nonviolent, and criminal behavior was confined to a minority. Since both the XYYs and the XXYs were also of lower intelligence, and the former showed electroencephalogram (EEG) abnormalities, it would appear that chromosome abnormality makes a nonspecific contribution to criminality through the medium of genetic disorganization and developmental failure. Interest in this rare phenomenon has subsequently declined, although there have been some speculations about the length of the Y chromosome in criminals.

E. Physique and Appearance

Although body constitution is influenced by the environment, it is fixed by the genotype. Lombroso claimed that criminals showed primitive bodily features, but this rested on statistically unsound observations. More modern research has nevertheless indicated a correlation between unattractive physical appearance and antisocial behavior. The association could, however, reflect a self-fulfilling prophecy, in that those who are facially unattractive are judged negatively, and react accordingly.

There is a long European tradition of attempting to link physique to temperament and psychiatric disorder, which has focused on body build, or somatotype. Modern studies of body build in delinquents derive from the American research by Sheldon in the 1940s. Sheldon assessed somatotype in terms of three embryologically derived concepts: endomorphy (i.e., fat and circular), mesomorphy (i.e., muscular and triangular), and ectomorphy (i.e., thin and linear). These were proposed to relate to specific temperament components. Rating somatotype components from photographs, Sheldon found that delinquents were significantly more mesomorphic and less ectomorphic than were students. A 30-year follow-up of Sheldon's delinquents has extended these findings, and has also shown that more serious criminals had higher ratings on andromorphy, a measure of masculinity of secondary sex characteristics.

Sheldon's research was criticized for the subjectivity of his ratings and his imprecise criteria of delinquency, and some argue that somatotype might be more appropriately assessed by height and width of body build. Nevertheless, subsequent studies confirm that those who violate legal rules are more likely to be muscular and less fragile in physique. This relationship might be mediated by associated temperament factors, since mesomorphs have been found to describe themselves as significantly more active and aggressive, and ectomorphs, as more socially avoidant. The combination of mesomorphy and andromorphy also suggests higher testosterone levels. Alternatively, delinquent peer groups might differentially reward toughness, while criminal justice agents might react more negatively to tough appearance.

III. Psychophysiological and Biochemical Correlates

A. Theoretical Issues

Genotypes set limits on phenotypic variation through the intermediaries of enzymes, hormones, and neurons, and probably contribute to rule-breaking tendencies through the medium of neurochemical functions associated with learning and temperament. Processes associated with stimulation-seeking, passive avoidance, conditionability, and emotional responsiveness have provided the main potential links between genetically determined characteristics of the nervous system and antisocial behavior, and correlates of criminality have therefore been sought in peripheral recordings of cortical and autonomic activity or biochemical assays. The concept of arousal has been prominent in this research. Although this term refers broadly to the level of physiological activation of the nervous system, there is no single psychophysiological measure of general arousal. The EEG provides the closest approximation, but the arousal level is often inferred from activity in particular autonomic systems, which might not reflect what is going on in other parts of the nervous system.

In Eysenck's theory extroverts have low levels of arousal and, hence, a higher optimal level of stimulation. The notion that psychopaths and delinquents generally have a suboptimal level of arousal has been popular, since much antisocial behavior appears to represent risk-taking consequent on boredom. Arousal, nevertheless, remains an ambiguous concept, and recent research on sensation-seeking suggests that the optimal level of stimulation might have more to do with activity in brain catecholamine systems involved in information processing than with the nonspecific arousal level. However, more specific psychophysiological processes have also been implicated in undersocialized behavior. For example, recent proposals suggest that deficient activity in Gray's BIS is manifest in hyporesponsiveness of the electrodermal system. Since such variations should be most clearly shown by those who are least socialized, research has focused particularly on psychopaths.

B. Electrocortical Correlates

Clinical interest in the EEG centers on abnormalities in the wave form, but the notion of "abnormality" is somewhat arbitrary. It includes unusual discharges, but more predominantly refers to diffuse slow wave activity or excesses of theta activity in temporal areas. However, these features are apparent in about 15% of normal adults, and more than one-quarter of young children. Although they are interpreted variously as indicating developmental delay, limbic system abnormalities, or cortical arousal level, their functional significance remains unclear.

There is an extensive literature on EEG correlates of behavior disorder, and high frequencies of abnormalities have been reported in aggressive and psychopathic samples. In one early report of psychiatrically disordered combat troops, for example, EEG abnormalities were displayed in the records of 65% of aggressive psychopaths, 32% of inadequate psychopaths, 26% of neurotics, and 15% of nonpatient controls. There are, nevertheless, many inconsistencies in the literature, which is generally marred by unreliability in EEG analysis and subject classification.

Recent research using replicable measurement of psychopathy and EEG quantification has raised doubts about the extent of abnormalities among antisocial personalities, some studies finding no differences between psychopaths and nonpsychopaths. However, when psychopaths have been divided into primary and secondary subgroups, it is the latter group which is distinguished by more theta activity and lower alpha frequency. EEG abnormalities and, by implication, low cortical arousal are not, then, uniformly characteristic of psychopaths. While there is slightly more consistent evidence of abnormalities in violent adults and hyperactive children, it is not unequivocal.

Although EEG abnormalities have also been reported in samples of incarcerated delinquents, the evidence is again inconsistent. However, prospective studies of Scandinavian boys have found that those who became delinquent were more likely than nondelinquents to show an excess of slow alpha activity in records taken in early adolescence. These delinquents were predominantly property offenders, and the results therefore indicate that EEG abnormalities are not specific to violent offenders. However, they are also not specific to offenders, and the presence of high-amplitude slow waves in the EEG might be associated with an increased risk for social maladjustment generally, rather than antisocial behavior specifically.

Recent attention has focused on event-related potentials not readily analyzable from the raw EEG trace. Of particular interest is the question of whether psychopaths show idiosyncrasies in information processing. A few studies have examined the contingent negative variation (CNV), or "expectancy wave," a slow negative potential elicited in forewarned reaction time experiments. Early suggestions of smaller-amplitude CNVs in psychopaths have not been confirmed, and there is probably no direct relationship between psychopathy and CNV amplitude. While there have been several investigations of sensory-evoked potentials in psychopaths, the diversity of methods, subjects, and components examined precludes any firm conclusions. However, there is some indication in this research that psychopaths allocate more attentional resources to events of immediate interest.

Individual differences in evoked potentials are also observed in the modulation of stimulus intensity, or augmenting–reducing. With increasing stimulus intensity some individuals show an increase in the amplitude of the early wave component (augmenters), while others show a decrease (reducers). Augmenting is thought to reflect a less "sensitive" nervous system, and there is some evidence that sensation-seekers and male delinquents show an augmenting response. However, this area has not been extensively investigated.

C. Autonomic Nervous System Activity in Psychopaths

In experimental tasks involving avoidance learning, psychopaths do not show any deficit in active avoidance of punishment stimuli, but their postulated deficit in passive avoidance learning has been found when electric shock was used as a punishment paradigm. In contrast, under punishment conditions such as social disapproval or loss of money, psychopaths perform similarly to nonpsychopaths and do not, therefore, appear to have any generalized deficit in passive avoidance. However, it has recently been suggested that the passive avoidance deficit might be confined to situations in which there are competing cues of both reward and punishment, and experiments have confirmed that psychopaths overfocus on reward, which interferes with attention to punishment cues. Since this failure to alter a dominant response set parallels the effects of septal lesions in rats, these findings are consistent with deficient functioning of the BIS.

Research on electrodermal (ED) and cardiac responses of psychopaths prior to noxious stimulation confirms a deficiency in anticipating punishment. For example, when forewarned that a particular stimulus in a series will be followed by a shock stimulus, psychopaths produce less anticipatory ED arousal. They also show weaker acquisition of conditioned ED responses when shock is the unconditioned stimulus, and poorer vicarious ED conditioning has been found for both primary and secondary psychopaths observing aversive stimulation delivered to others. However, a study which involved both noxious and pleasant conditioned stimuli, and which recorded both ED and heart rate responses indicated that deficient conditionability in psychopaths is confined to anticipatory responses to noxious stimuli and to the ED system. These results contradict Eysenck's hypothesis of a generalized deficit in conditionability in psychopaths, but are consistent with a deficit in the BIS.

However, threatening aversive stimuli which elicit smaller ED reactions in psychopaths simultaneously produce larger heart rate responses. It has been suggested that this dissociation of cardiac and ED responding reflects a "cortical tuning" mechanism involving a cardiovascular-induced reduction of cortical arousal via baroreceptors in the carotid sinus. This model of a protective mechanism which reduces the emotional impact of aversive stimulation has similarities to the concept of augmenting–reducing, but remains to be investigated further.

There is no consistent evidence that psychopaths in general are hyporesponsive in their orienting reactions (ORs) to simple nonaversive stimuli, but this has been observed in secondary psychopaths and schizoid offenders and could relate to attentional inefficiency. The rate of habituation of the ORs in normal ED records is correlated with the appearance of nonspecific fluctuations, and higher rates of these fluctuations and slower habituation of the ED ORs are associated with superior vigilance and more effective allocation of attentional resources. Some research suggests that psychopaths habituate more slowly in cardiac ORs to auditory stimuli, but more rapidly in ED ORs, and also show lower rates of NSF. This dissociation of the ED and cardiac systems is related to drowsiness, and these findings point to lower cortical arousal. However, there is some evidence that this pattern is more prominent in secondary than in primary psychopaths.

One other ED variable of interest is response recovery time. Slow autonomic recovery could lead to a failure of fear reduction to reinforce passive avoidance responses, and slower recovery has been demonstrated in criminals and psychopathic delinquents. However, the functional significance of recovery time remains debated.

Psychopaths do not exhibit generally lower autonomic arousal levels, but some prospective research suggests lower cardiac rates in children who subsequently become delinquent. Children of criminal fathers have also been found to have lower heart rates than do children of noncriminals, although this does not characterize children of psychopaths. If, as some suggest, low tonic heart rate is associated with low cortical arousal, these findings imply that lower arousal in general might be a correlate of petty delinquency, but not psychopathy.

The above research supports the view that psychopaths are underresponsive in the ED system and, hence, in the BIS. There is less clear evidence of low arousal in psychopaths, and recent EEG research contradicts this proposal. However, any conclusions must be tempered by inconsistencies in the research to date, which could reflect the heterogeneity of "psychopaths." There is, for example, some indication that lower arousal might be characteristic of secondary psychopaths.

D. Biochemical Correlates

Hormones secreted by the endocrine glands influence behavior through their role in normal development as well their effects on the temporary state. Correlates of criminal behavior and related temperment variables have been sought in variations in the production and level of hormones secreted by the gonads (i.e., androgens and estrogens), the adrenal glands (i.e., adrenaline and noradrenaline), and the pancreas (i.e., insulin). [*See* HORMONAL INFLUENCES ON BEHAVIOR.]

Androgens are crucial in sexual differentiation and the appearance of secondary sexual characteristics at puberty, and testoserone level has therefore been considered a possible factor for the universal correlation of criminal behavior with gender and age and for the greater aggressiveness of males. Although initial studies found that both the rate of production and the level of testosterone correlated positively with hostility in younger nonoffender subjects, subsequent studies have failed to confirm this. Results from offenders have been equally inconclusive. Higher testosterone levels have been reported in aggressive prisoners and extremely violent rapists, but there are several negative findings. The evidence to date is therefore not sufficient to indicate a direct influence of testosterone.

Testosterone might, however, have an indirect effect on behavior through the activity of brain neurotransmitters. It inhibits the activity of the enzyme monoamine oxidase (MAO), which metabolizes several neurotransmitters, and this might allow monoamines (e.g., noradrenaline) to accumulate to higher levels in the brain. Blood platelet MAO activity is lower in human males than in females, and lower MAO levels have been found to be associated with "disinhibitory" temperament variables (e.g., impulsivity, sensation-seeking, and undersocialization) in nonoffenders. Low MAO activity is also correlated with low concentrations of 5-hydroxyindoleacetic acid (5-HIAA), a derivative of the neurotransmitter serotinin, in the cerebrospinal fluid, and low 5-HIAA has also been reported to be correlated with impulsivity and sensation-seeking in nonoffenders and with habitually aggressive offending.

Female crime has long been thought to relate to hormonal changes occurring in the menstrual cycle, particularly premenstrual tension (PMT). PMT symptoms include depressed mood, lethargy, and headaches, and PMT might make females more vulnerable to deviant behavior as a consequence of irritability. Studies have found that female prisoners are more likely to have committed their crimes during the 8 days preceding or during menstruation (i.e., paramenstruum), this being most marked for theft, and among those experiencing PMT. Disciplinary problems among female prisoners and mentally disordered females have also been found to occur more frequently in the premenstrual week. However, the peaking of antisocial behavior during the paramenstruum is not consistently associated with PMT, and it is not specific to violent behavior. Also, since severe PMT symptoms occur in up to 40% of women, they are obviously neither necessary nor sufficient to account for female offending.

Criminal acts have been reported to occur in states of hypoglycemia, and recent research indicates an association of violence with dysfunction in glucose metabolism. Hypoglycemia is related to increased insulin secretion which could follow from starvation or alcohol ingestion, and impairs cerebral function. Aggressive behavior in psychiatric patients has been found to be associated with glucose dysfunction and EEG abnormalities, and slower recovery from experimentally induced hypoglycemia has similarly been observed in habitually violent adult offenders diagnosed as antisocial personalities, particularly those with a history of violence under the influence of alcohol. Hypoglycemia proneness also correlates with self-reported aggressiveness among nonoffenders.

Secretion of the catecholamines adrenaline and noradrenaline has been of interest because of earlier hypotheses linking adrenaline increase with fear and noradrenaline increase with aggression. These hypotheses are now seen as too simple, but a few studies have examined urinary catecholamine in relation to stress responses in antisocial individuals. Among offenders awaiting trial, for example, psychopaths show less increase in adrenaline immediately prior to trial, suggesting less stress responsiveness. Lower adrenaline secretion under stress has also been reported in studies of persistent bullies, and adolescents who later obtained criminal convictions. However, adrenaline increase might be more related to cortical alertness than specifically to stress. [*See* CATECHOLAMINES AND BEHAVIOR.]

These findings might be integrated in terms of autonomic balance. Insulin secretion is normally opposed by adrenomedullary hormones, while testosterone increases catecholamine output by lowering MAO activity. Low cortical arousal, lower heart rate, hypoglycemia proneness, and low adrenaline output would all be consistent with a dominance of the vagoinsulin system, or parasympathetic balance. However, the data are not wholly consistent, since hypoglycemia proneness is associated with violence, but other features are more characteristic of nonviolent delinquents.

IV. Higher Nervous Functions

A. Brain Pathology and Crime

Damage to the brain can result from perinatal complications, head injuries, tumors, infections, or exposure to toxic substances (e.g., atmospheric lead), but such conditions do not invariably lead to structural damage. Brain damage must be severe before significant psychological disturbance ensues, and unsocialized behavior is not a necessary consequence. For example, a survey in Finland found that less than 6% of 507 veterans who had sustained head wounds received a criminal conviction leading to imprisonment in the 30 years following injury.

In the absence of known cerebral insult, brain pathology remains difficult to detect. Most research relies on indirect signs, such as medical history, "soft" neurological signs, EEG records, or neuropsychological tests, which detect brain dysfunction, and not necessarily tissue damage. Moreover, causal implications are not always clear. While antisocial acts might be positive symptoms of brain disorganization, representing the release of subcortical activity from inhibitory control, they might alternatively by a negative symptom of brain pathology, in that cerebral impairment results in a deficiency of functions necessary for cognitive development and socialization.

Some 0.5% of the population suffer from epilepsy, a symptom of brain disorganization. Although confined largely to incarcerated samples, the evidence indicates significantly higher rates among offenders. However, this does not necessarily indicate a direct biological effect on behavior, since social problems of sufferers from epilepsy often reflect the experience of the stigma resulting from societal reactions. [*See* EPILEPSY.]

Although there is legal interest in the question of whether crimes can be committed automatically and without awareness (i.e., automatism), automatism rarely accounts for the offences of criminals with epilepsy, and the evidence indicates that criminal behavior which goes beyond the fragmentary character of automatisms is unlikely to reflect a seizure. It is widely believed that temporal lobe epilepsy increases the likelihood of violence, but the evidence remains controversial. Rage reactions or aggressive outbursts have been reported in more than one-third of adult and child patients with this form of epilepsy, but these estimates might reflect referral biases. In some studies prisoners with epilepsy were no more likely to have committed violent crimes than were those without epilepsy, although other research has found a correlation between violent history and epileptic symptoms among delinquents. Disparate findings could reflect differing criteria of both violence and epilepsy.

Also controversial is the concept of the dyscontrol syndrome, in which brain lesions are assumed to be causes of violence, even though observable seizures are absent. Dyscontrol is characterized by explosive outbursts of violent behavior with minimal provocation, with evidence of adequate adjustment between episodes, and such behavior is believed to reflect epilepticlike discharges from sites of focal abnormality in the limbic system. Evidence for these abnormalities, however, rests largely on such indirect indicators as EEG abnormalities or soft neurological signs, which are common in patients referred for neurological or psychiatric evaluation following recurring episodes of unprovoked rage. However, dyscontrol simply describes a correlation between a vaguely defined behavioral syndrome and signs of brain pathology whose validity is not clear established, and histories of childhood deprivation and family violence might equally account for the violence of these patients. Specific neural triggers for aggression have not, in fact, been clearly established in laboratory research, and recent studies of neurological patients suggest that the involvement of brain dysfunction in violence is indirect and nonspecific.

B. Minimal Brain Dysfunction, Hyperactivity, and Learning Disabilities

Many child behavior disorders are assumed to be symptomatic of brain dysfunction resulting from cerebral trauma sustained in the perinatal period or early infancy, and histories of head and facial injuries are found with some frequency in serious delinquents. Such histories have encouraged use of the concept of minimal brain dysfunction (MBD), which refers to a symptom complex of motor restlessness, impulsivity, deficits in attention and learning, and soft neurological signs. ''Hyperactivity'' has been used interchangeably with ''MBD'' to denote a similar symptom complex, currently referred to as attention deficit hyperactivity disorder. However, the validity of these concepts remains controversial, and firm evidence that hyperactivity or MBD is symptomatic of brain damage has not been forthcoming.

Although hyperactivity has long been associated with child conduct problems (e.g., aggression or stealing), recent studies suggest that children displaying hyperactive behavior are not uniformly antisocial. However, follow-up studies indicate that childhood hyperactivity is associated with antisocial behavior in adolescence and adulthood and that hyperactive children are more likely to meet criteria for antisocial personality disorder later in life. Apart from the problems of social training presented by a child who is restless and inattentive, the nature of the contribution of hyperactivity to criminality remains unclear. Delinquent outcomes in hyperactive children seem to be related to both early biological aspects and later social factors, including the reactions of parents and others.

Diagnoses of MBD and hyperactivity often overlap with learning disabilities (LDs), which refer to a discrepancy between what is expected of a child on the basis of established ability and actual educational achievements, and which are assumed to have a constitutional basis. The role of LDs in delinquency is of interest in view of the established correlation between poor school performance and delinquency. Retrospective prevalence estimates of LDs in delinquents have ranged from 26% to 73%, but the link remains tenuous. Although learning problems are probably common among delinquents, ''LD'' subsumes heterogeneous disorders which are unlikely to have a single etiology. Recent attempts to distinguish among LDs indicate varying forms which are differentially related to both academic achievement and behavior disorder, and one recent study found that only 14% of a delinquent sample showed a pattern overlapping with that predominating in a nondeliquent LD group.

C. Neuropsychological Dysfunctions

Performance of offenders on neuropsychological tests has been cited as evidence for brain dysfunction in a number of studies. Consistent evidence of deficits comes from findings with intelligence tests, notably the Wechsler scales, on which delinquents not only score lower than nondelinquents, but also tend to produce larger discrepancies between performance (P) and verbal (V) IQ in favor of the former. The PIQ > VIQ sign is generally interpreted in terms of deficient verbal skills, rather than superior nonverbal abilities, and might be a correlate of LD. VIQ and PIQ are also thought to have some association with the differential functions of the left (i.e., linguistic processing and sequential analysis) and right (i.e., spatial and qualitative analyses) cerebral hemispheres, respectively. The PIQ dominance has therefore been interpreted in terms of reduced left hemispheric lateralization in delinquents. [*See* CEREBRAL SPECIALIZATION.]

Recent studies of both officially defined and self-reported delinquency have found that delinquents produce not only a larger PIQ > VIQ sign, but also poorer scores on a variety of neuropsychological tests. The pattern of deficits shown by delinquents suggests only minimal dysfunction in motor skills or gross sensory functioning, but significant deficiencies in problem-solving abilities requiring verbal, perceptual, and nonverbal conceptual skills, although female delinquents tend to differ from males.

Other studies, however, have suggested that lateralized deficits in the left (i.e., dominant) hemisphere, particularly in frontal and temporal areas, are characteristic of violent and psychopathic offenders. Frontal lobe damage has long been associated with impulsive poorly planned behavior, and the frontal lobes are thought to be involved in the integration and direction of voluntary behavior. Frontolimbic connections are also part of Gray's BIS. Recent findings point to less specialized lateralization of language functions in psychopaths, suggesting that their cerebral organization of language might be marked by poorer integration of affective and other components linking cognition and behavior. This might account for deficiencies in emotional experiences.

Although neuropsychological studies of offenders have so far been limited by an overemphasis on institutionalized populations, impairment of language-related skills and regulative functions controlled by the frontal lobes have been found with some consistency. Current views associate these with development failure, rather than neurological damage, and suggest that delinquents, particularly those identified as violent and impulsive, might have a relative inability to use inner speech to modulate attention, affect, thought, and behavior under conditions of stress.

V. CONCLUSIONS

Children and adults who are more prone to violate social rules differ from those who are more socialized in a number of bodily functions and processes which are likely to reflect innate or congenital influences. Inconsistencies of results in several areas might be attributed to the relatively modest and indirect role played by biological processes in socialization and to the heterogeneity of antisocial groups, and it is unlikely that biological factors are equally relevant for all classes of deviant individuals. Nevertheless, the evidence to date indicates that criminology cannot ignore the relevance of a biological level of analysis. While no one would now claim that there are "born criminals" or deny that crime is critically dependent on social environments, biological factors appear to increase the likelihood that some individuals will become criminals in some environments.

Bibliography

Blackburn, R. (1991). "Psychology and Criminal Conduct: Theory, Research, and Practice." Wiley, Chichester, England.

Ellis, L. (1987). Relationships of criminality and psychopathy with eight other apparent behavioral manifestations of suboptimal arousal. *Person. Individ. Differ.* **8,** 905.

Eysenck, H. J., and Gudjonsson, G. H. (1989). "The Causes and Cures of Criminality." Plenum, New York.

Miller, L. (1988). Neuropsychological perspectives on delinquency. *Behav. Sci. Law* **6,** 409.

Moffitt, T. E., and Mednick, S. A. (eds.) (1988). "Biological Contributions to Crime Causation." Nijhoff, Dordrecht, The Netherlands.

Rowe, D. C., and Osgood, D. W. (1984). Heredity and sociological theories of delinquency: A reconsideration. *Am. Sociol. Rev.* **49,** 526.

Zuckerman, M. (1984). Sensation seeking: A comparative approach to a human trait. *Behav. Brain Sci.* **7,** 413.

Cryotechniques in Biological Electron Microscopy

MARTIN MÜLLER, *Laboratory for EMI, Institute of Cell Biology, ETH-Zürich*

Glossary

Cryofixation Stabilization of cyroimmobilized, biological structures at low temperatures by crosslinking agents (Aldehydes, OsO_4, Embedding-resins) The term "Cryofixation" is frequently misused in place of "Cryoimmobilization"

Cryoimmobilization The near immediate arrest of cellular processes and solidification of a biological specimen by very rapid cooling. (Cryoimmobilized samples are stable only at very low temperatures (e.g., 100–100 K).

Cyrotechniques Biological specimen preparation procedures for electron microscopy utilizing the solid state of water at low temperature

Segregation pattern Compartments formed during ice crystal growth by exclusion of solutes from the crystal lattice. Segregation patterns are the most obvious artifact of microscopical cryotechniques, reflecting insufficient cooling rates

Vitrification Solidification of specimen water in a glass-like, amorphous state

CRYOTECHNIQUES have become very useful tools in modern biology; macromolecules, cells, and even entire organs can be stored under conditions that preserve viability. Cryotechniques can also contribute to microscopical ultrastructural studies. Procedures required to maintain the vital functions of biological matter differ in many ways from those needed to preserve the structural integrity necessary for ultrastructural studies. This article examines the role of cryotechniques applied to biological electron microscopy.

I. Introduction

One of the important tasks of biological electron microscopy is to provide structural information with which one may correlate structure and function. It is currently the only methodology with the inherent potential to observe structures down to molecular dimensions within the context of complex biological systems. Specimen preparation techniques should be directed towards preserving the smallest significant details in order to fully exploit this unique, integrating feature of biological electron microscopy, thereby complementing the progress of the techniques used in cell biology, biochemistry, and molecular biology.

Problems encountered during preparation of biological specimens for electron microscopy arise from the necessity to transform the hydrated biological sample into a solid state in which it can resist the physical impact of the electron microscope (e.g., high vacuum, electron beam irradiation). Ideally one would like specimen preparation procedures that simultaneously guarantee absolute preservation of the dimensions and spatial distribution of diffusible elements. Antigens, receptors, lectin-binding sites, etc. should become demonstrable through immuno- and other cytochemical techniques. A universal specimen preparation procedure will, perhaps, remain a dream. One must, nevertheless, attempt to realize it so that the integrating power of electron microscopy can further develop into a complementary tool in modern biological research. [*See* ELECTRON MICROSCOPY.]

The immobilization of biological material kept under optimally controlled physiological conditions is the first and most critical step in attempting to preserve the complex interactions of organelles, macromolecules, ions, and water in a close relationship to the living state. Immobilization must be sufficiently rapid; trapping dynamic events at membranes, (e.g., membrane fusion and exocytosis, which occur on a milisecond time scale) and preventing the lateral displacement of lipids and proteins within the membranes. A lateral diffusion coefficient of fluorescent lipid analogs of 3×10^{-11} m^2sec^{-1} would allow a lateral displacement of approximately 300 nm within 1 msec (Knoll *et al.*, 1987) as well as the displacement of ions, which, depending on their interaction with macromolecules and water molecules, have a lateral diffusion coefficient of up to 10^{-9} m^2sec^{-1}.

In this respect, techniques based on chemical immobilization (aldehyde fixation) seem to approach their limits, although their use will remain indispensable in solving many relevant biological problems. Chemical fixatives react relatively slowly and cannot preserve all cellular components. Most of the diffusible ions are lost or redistributed during sample preparation. Fixation influences the diffusion properties of the membranes and results, therefore, in alterations of shape, volume, and content of the cell and its components. It becomes evident that the initial potential of electron microscopy to serve as an integrating source of primary information on the structural complexity at cellular to macromolecular dimensions, can hardly be approached by techniques based on chemical fixation. Immunocytochemical methods represent an important exception because for many questions it is irrelevant whether the antigen is detected, for example, in a structurally preserved or a distorted organelle. Electrophysiological experiments of Hülser *et al.* (1989), demonstrated the effect of 0.2% glutardialdehyde on electrically coupled homocaryons of BICR/MIR_k-cells (a permanently growing tumor cell line in monolayer culture derived from a spontaneous mammary tumor of the Marshall rat). A collapse of membrane potentials was found after approximately three min. The authors concluded from their measurements that glutardialdehyde cannot preserve gap junction channels in their open state.

Cryoimmobilization represents a farther reaching alternative. High-cooling rates (10^4K sec^{-1}–10^6K sec^{-1}) are required to prevent the formation and growth of ice crystals by which the structural integrity would be affected. The high-cooling rates at the same time bring along a rapid arrest of the physiological events (i.e., a high time resolution for dynamic processes in the cell and consequently structural immobilization closely related to the living state).

II. Cryoimmobilization

The structural integrity of biological material is guaranteed only if cryoimmobilization brings about solidification of water or solutes in a vitreous or microcrystalline state in the absence of any chemical pretreatment. Heat can be extracted only through the surface of the sample. Heat transfer from deeper within the specimen is limited by the low thermal conductivity of water and the developing solid layer. Extremely high-cooling rates can be achieved at the surface of the sample. This may lead to the immobilization of a thin-surface layer in the vitreous state. Insufficient cooling rates allow ice crystals to form deeper in the sample. More heat is produced by ice crystal formation with increasing depth than is transferred through the ice to the cooled surface. This progressively reduces the cooling rate and results in increasing ice crystal dimensions.

A. Vitrified Samples

Freezing cellular water in the amorphous state (vitrification) would be ideal because the basic nature of the liquid phase would be preserved. Vitrification of cellular water, however, requires very high-cooling rates, present only in thin layers at the specimen surface. These layers are usually too thin (100 nm) to encompass the bulk of the sample, and further processing for electron microscopy entails some difficulties. True vitrification of biological solutions has been demonstrated by low-temperature electron diffraction using the "bare grid" technique. Thin aqueous layers (<100 nm) of suspensions of viruses, phages, liposomes, and macromolecules are formed by this technique within the meshes of an electron microscope grid, either bare or covered with a perforated carbon foil, vitrified by immersion into liquid ethane, and observed in the microscope at low temperatures using a cold stage.

B. Rapid Freezing Techniques

Aqueous layers of suspensions of cells, microorganisms, organelles, and tissue cultures up to a thickness of approximately 10–20 μm are adequately cryoimmobilized by several rapid freezing techniques. These techniques include: plunge freezing, wherein the thin sample layer is sandwiched between two metal platelets and plunged into liquid coolants (propane and ethane are preferentially used); propane-jet freezing, wherein two jets of liquid propane are directed onto the specimen sandwich from opposite directions simultaneously; and spray freezing, wherein small droplets of the sample are sprayed into the liquid coolant. Impact freezing allows adequate cryoimmobilization of a thin layer at the surface of suspension droplets or tissue pieces by forced contact with a polished metal surface kept at liquid-helium or liquid-nitrogen temperature.

The freezing rate achievable in aqueous layers of 10–20 μm may be too low for vitrification but high enough to prevent growth of damaging ice crystals. Solutes, however, are excluded from the crystal lattice when ice crystals grow and are concentrated between neighboring ice crystals to form a eutectic. The eutectic appears in electron micrographs as a network of segregation compartments in freeze-substituted or freeze-fractured biological material that has been cryofixed with insufficient cooling rates. In practice, the absence of segregation patterns is a generally accepted indication of adequate cryoimmobilization, reflecting high-cooling rates but not necessarily vitrification. The concentration of solutes by crystallization leads to drastic local changes in osmolality and pH, which may induce conformational changes (e.g., of proteins). Structural details may, therefore, be altered even if no segregation patterns are visible.

The appropriate rapid freezing procedure can reproducibly yield adequate freezing of very thin samples, immobilizing the specimens in a state most closely related to the living situation, and thus allowing one to investigate significant structural details as a function of the physiological state. Rapid-freezing procedures are, however, generally not suitable for cryoimmobilization of more complex samples (e.g., animal and plant tissues). Useful structural information in a thin superficial zone at the natural or cut surface of tissue samples can be obtained by impact freezing. The thickness of the zone, in which no segregation patterns are visible, depends on the concentration of cellular components that exhibit cryoprotective properties and may often reach 20 μm. This is generally, however, still insufficient to analyze intact tissue cells.

C. High-Pressure Freezing

Thicker, more complex systems can be studied by cryofixation-based electron microscopy only if the physical properties of the cellular water are manipulated in a way that adequate cryoimmobilization is achieved with much slower cooling rates. The impregnation of larger samples with cryoprotectants, usually in combination with aldehyde-fixation, is frequently employed. Numerous artifacts, however, can be introduced by these procedures.

The sole aim of cryofixation is to physically immobilize the specimen. The development of an alternative method based on the application of high-hydrostatic pressure was started 20 years ago by Moor and co-workers and adequate instrumentation subsequently became commercially available. Freezing increases the volume of water. This expansion is hindered by high pressure thereby increasing the viscosity. This effect is demonstrated by a lowering of the freezing point and reduced rates of nucleation and ice crystal growth. Consequently, less heat is produced by crystallization, and less heat has to be extracted per unit time by cooling. This means that adequate immobilization can be achieved with reduced cooling rates. At a pressure of 2045 bar, the melting point of water is lowered to 251K, and the temperature of homogeneous nucleation reduced to 181K, as deduced from the phase diagram of water.

High-pressure freezing is the only practical way of cryofixing larger unpretreated samples up to a thickness of 500 μm. It permits, at least in theory, structural analysis of more complex systems i.e., fungus–host interactions, tissue culture cells grown on established substrates (e.g., glass coverslips), or, in the center of a tissue sample, of cells that have not suffered from traumatic excision. These advantages are somewhat reduced by the relatively slow-cooling rates (approximately 500K s^{-1}) achieved in the center of the sample. These rates may be too slow to catch dynamic events at membranes or to prevent structural alterations due to lipid phase-transition and segregation phenomena (Knoll *et al.*, 1987). On the other hand, the transition temperature of membrane lipids is raised by approximately 20 K $kbar^{-1}$. This means that by applying a pressure of more than 2 kbar (attained in approximately 15 ms), the membrane lipids may be immobilized very quickly purely by the action of high pressure. Experimental data supporting this assumption as well as on other short-lived high-pressure effects on biological material are not yet available. Possible reactions of biological specimens exposed to high pressures for periods in excess of 1 min have been observed.

Unsatisfactorily low yields of adequately frozen biological samples frequently hamper attempts to correlate structure and function. This has been especially true for high-pressure freezing until Studer *et al.* (1989) improved the transfer of pressure and cold to the biological specimen sandwiched between two metal supports that contain an appropriate cavity. This method gives high yields by quickly immersing the samples in 1-hexadecene prior to freezing. We believe that 1-hexadecene removes the free water surrounding the specimens, thereby reducing the danger of extraspecimen ice nucleation. It also completely fills the space unoccupied by the specimen in the cavity of the metal sandwich. This assures a good thermal contact with the metal platelets.

Exision of tissue specimens, as well as preparation of tissue samples suitably sized to fit into the metal supports for high-pressure freezing, or to be mounted on special holders for plunge freezing or impact freezing, requires some time, during which physiological changes and structural alterations undoubtedly occur. These changes may reach an extent that prohibits any attempt to correlate structure and function at high resolution. This problem, while inherent to most ultrastructural studies, may only be overcome by *in situ* freezing approaches, which are as yet unavailable.

III. Follow-Up Procedures

Successfully cryoimmobilized samples have to be further processed for electron microscopic analysis by various follow-up procedures, each of which yields different information and poses different technical problems. Subsequent processing has to be performed at sufficiently low temperatures at which devitrification and secondary ice crystal growth are avoided. The devitrification range for vitreous, amorpheous water was determined to be approximately 140K through means of low-temperature electron diffraction. Vitreous water recrystallizes into cubic ice at higher temperatures. In this state no effects of phase separation due to ice crystal formation are visible when employing the most frequently used follow-up procedures (i.e., freeze-fracturing, freeze-substitution). At still higher temperatures, however, (e.g., above 190K) cubic ice may be transformed into hexagonal ice in which modification ice crystals may rapidly grow and alter the specimen.

A. Physical Procedures

Low-temperature electron microscopy of vitrified, thin aqueous layers and of cryosections, as well as freeze-fracturing, and freeze–drying, are considered to be direct, purely physical procedures. They provide reliable structural information most closely related to the living state. Cryosectioning of untreated, cryofixed biological material is a very demanding technique, and the sections are often too thick to provide structural information at high resolution. Cryosections observed in the microscope at low temperatures, however, either frozen hydrated, or freeze–dried, represent the best, if not the only, means of obtaining qualitative and quantitative information on spatial distribution of diffusible ions by x-ray microanalysis.

Freeze-fracturing is the easiest and best established physical technique to obtain a "safe" representation of structural details (down to a resolution of approximately 5nm). Many reviews treat this technique in detail (e.g., Robards and Sleytr, 1985). Freeze-fracturing exposes specific structural aspects depending on the fracturing behavior of the sample and its components. It is especially suited to characterization of membranes, because for energetic reasons, the fracture plane proceeds through the hydrophobic interior of the membranes, thus providing information about size and distribution of intramembraneous particles (IMP). The pattern formed by the IMPs is characteristic for each specific membrane fracture face. Alterations of these specific patterns may reflect dynamic processes at membranes, or if the sample is cooled too slowly, the occurrence of phase transitions and segregation phenomena of the lipid phase. The nature of IMPs is still under discussion. They may indicate the positions of transmembrane- or intramembraneous proteins but may also be lipid in nature.

None of the aforementioned direct physical procedures is suitable for immunocytochemical work unless mild chemical fixation and cryoprotection precede cryofixation because immunocytochemical reactions must be performed at room temperature.

B. Hybrid Techniques

Freeze-fracturing is a straight forward, easily handled technique that provides reliable structural information. Its major disadvantages are that it can be used for hardly anything other than structural description, and that the fracture plane proceeds at will. These problems can be partially overcome by hybrid techniques, which combine the advantages of cryofixation with those of the conventional plastic embedding and thin-sectioning procedures. Freeze-substitution and freeze–drying are frequently employed. Both procedures are essentially dehydration processes; freeze-substitution dissolves the ice in a cryoimmobilized specimen with an organic solvent and freeze–drying eliminates the frozen water by sublimation in a vacuum chamber. Freeze–drying and freeze-substitution must be executed well above the devitrification range of amorphous water (~140K). Due to the low vapor pressure, freeze–drying at temperatures below approximately 150K would lead to impractically long drying times. The temperature limits during freeze substitution are set by the melting point of the solvent used and the amount of water the solvent can take up at low temperatures. Generally, temperatures of 180–190K are considered to be "safe" for cryoimmobilized biological samples because of the rather high natural cryoprotective activity of many cellular components. After completion of the dehydration process, the samples are warmed to room temperature, infiltrated with the embedding resin, and heat polymerized. With respect to the preservation of structural integrity, these hybrid techniques are much more obscure than the purely physical follow-up procedures such as cryosectioning or freeze-fracturing (discussed above), since effects of the organic solvents (e.g., lipid extraction) and the embedding chemistry cannot be excluded. The structural description provided by the physical procedures, in which the water remains in the specimen represents, therefore, the standard by which all the other procedures have to be measured. Freeze-substitution and freeze–drying may allow an accurate control of the dehydration process, but due to our incomplete knowledge of cellular water they are still insufficiently understood. Our present knowledge about the role of water in the cell and the effects of its removal are summarized by the following statements:

1. Water in the cell exhibits different physicochemical properties and is classified into two major groups: "bulk" or "free water," in contrast to anomalous water referred to as "bound water" or "nonfreezable water." This anomalous water is supposed to be closely associated with surfaces (e.g., of macromolecules, membranes, and ions) and is sometimes termed "hydration shell," "surface modified water," or "vicinal water."
2. This surface modified water is extremely important for all metabolism as well as for the maintenance of the structural integrity of proteins and other cell constituents.
3. One may conclude that bulk water is more easily removed and effects preservation of structural integrity less than the water of hydration shells during the dehydration process in biological electron microscopy.

These assumptions are supported by the nonlinear shrinkage behavior of cells and tissues during conventional dehydration of chemically fixed material by organic solvents at room temperature. Cells start to shrink when approximately 70% of cellular water is replaced by an organic solvent. Fully dehydrated, they shrink up to 30–70% of their initial volume. This is a primary indication that a specific fraction of the cellular water can be removed that does not introduce gross dimensional changes. There is, however, water closely associated with the cellular structures. Removing this residual water may lead to conformational changes of cellular components (collapse) and aggregation. The temperatures above which different types of macromolecules collapse when exposed to dehydrating agents such as organic solvents and vacuum have been determined. These temperatures range from 215 to 263K and appear to depend solely on the temperature and the polarity of the dehydrating agent. Experimenters have used an ultra-high-vacuum apparatus to study the freeze–drying of test specimens containing deuterium oxide (D_2O) instead of H_2O and followed its escape with a mass spectrometer. They observed that a first peak of D_2O evaporating in the temperature range of 180 to 190K approached zero after 2 hr at 190K. A second peak of D_2O was observed only after further heating the specimen and reached a maximum between 220 and 230K. Thus, the water is held in the tissue by different forces, and it may be concluded that some of the specimen water is bound, therefore, needing a higher energy for evaporation than the free water. The temperature at which the second peak of D_2O was observed is within the range of the collapse temperature, and it may be speculated that it corresponds to the water of the hydration shells.

Ideally, freeze–drying and freeze-substitution could be used to control the residual water content (i.e., how much water must remain in order that cells maintain their structural and functional integrity, and how much water must be removed to permit successful plastic embedding). Experiments have shown, however, that the hydration shells can prevent an efficient copolymerization between biological material and resin (Humbel and Müller, 1986). Strongly hydrated organelles may therefore not become embedded at all, and resin and biological material may separate very easily along membranes. Freeze–drying, in contrast to freeze-substitution, has not yet found wide application for dehydrating cryofixed samples for subsequent plastic embedding. It is, however, used very successfully in combination with metal shadowing in transmission as well as in scanning electron microscopy. The organic solvents used to freeze-substitute cryofixed biological samples frequently contain fixatives (e.g., OsO4, uranylions, aldehydes). One assumes these stabilize the biological structures at the ice solvent interphase, or during the gradual or stepwise increase of the temperature, but little is known about their reactivity at these high-subzero temperatures. There is experimental evidence, however, that uranylions react and prevent the extraction of phospholipid by solvent at even the lowest temperatures (180K). A reaction of OsO_4 with the double bonds of unsaturated fatty acids has been reported to occur at 203K. Glutaraldehyde starts to effectively cross-link proteins at 223K. Fixatives may be necessary to reduce solvent effects (e.g., the loss of lipid and other low molecular constituents). They might also help to reduce the effects of conformational changes and aggregations of macromolecules and supramolecular structures, which inevitably occur as the hydration shells are removed at higher temperatures (*cf.* collapse temperatures). The much more homogeneous finer-grained appearance of the cytoplasm after freeze-substitution, when compared to conventionally dehydrated samples, supports this assumption. The presence of fixatives in the substitution medium is essential if freeze-substitution is followed by conventional embedding at room temperature and heat polymerization at 335K. Samples prepared in this way exhibit excellent structural detail with conservation of the dimensions comparable to freeze-fracturing, yielding valuable new ultrastructural information, and identifying many artifacts associated with the conventional procedures based on chemical fixation and dehydration. All membranes possess smooth contours, which, after chemical fixation and dehydration, frequently appear undulated. Plant vacuoles appear fully turgescent. Mitochondria exhibit a close apposition of the two membranes, which is reflected by a frequent deflection of the fracture plane between the inner- and the outer membrane after freeze-fracturing and a multilaminar appearance after freeze-substitution. Figure 1 illustrates the different appearance of the ultrastructure of rat liver after chemical fixation with glutaldehyde and osmium tetroxide (OsO_4) (Fig. 1A), and after high pressure freezing followed by freeze-substitution in acetone/osmium tetroxide (Fig. 1B). Both samples were embedded in Araldite/Epon at room temperature and heat polymerized at 333K. Excellent structural preservation is obtained by freeze-substitution in acetone or methanol containing fixatives like osmiumtetroxide, uranylions, or aldehydes either alone or combined, followed by conventional embedding and heat polymerization. Samples prepared in this way, however, are hardly useful for cytochemical studies. Specimens that permit both an optimal structural description and

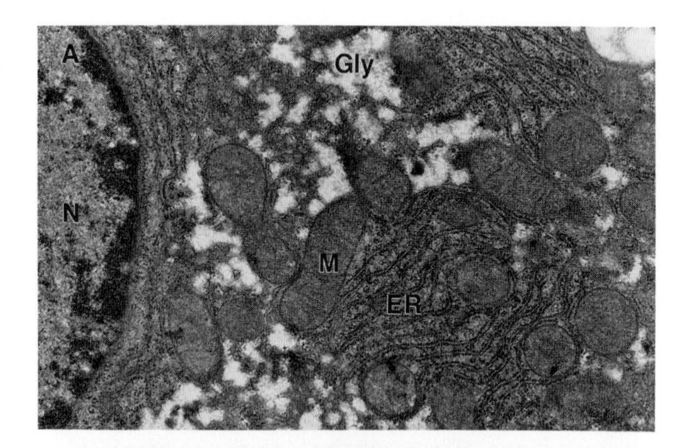

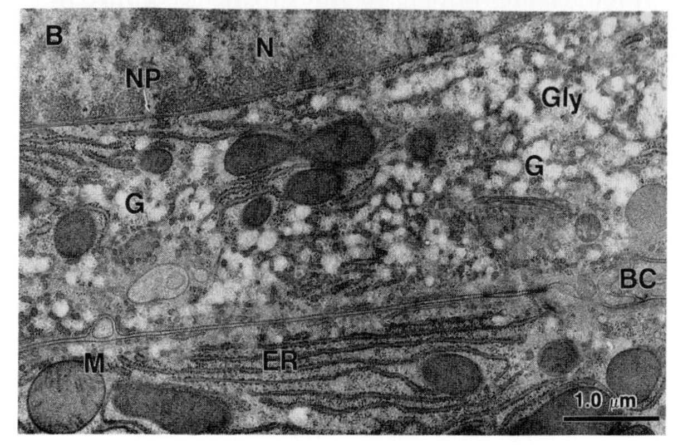

FIGURE 1 The appearance of rat liver ultrastructure after chemical fixation and dehydration (A) and that of high pressure cryo-immobilized, followed by freeze-substitution (B). Note the various aspects (e.g., membranes). N, Nucleus; NP, Nuclear Pore; M, Mitochondrion; ER, Endoplasmatic Reticulum; G, Golgi Complex; Gly, Glycogen; BC, Bile Caniculus.

the labeling of intracellular antigens are obtained by combining freeze-substitution with low-temperature embedding in Lowicryl (Humbel and Müller, 1986). Freeze-substitution followed by low-temperature embedding and polymerization is currently under study in many laboratories and will undoubtedly find more applications, due to the possibility of high-pressure cryoimmobilization of larger and more complex samples.

The label efficiency of freeze-substituted and low-temperature embedded samples is mainly affected by osmiumtetroxide as a stabilizing additive. It should therefore be avoided. Uranylions and aldehydes at low concentration often show no untoward effects with respect to label efficiency; they again help to minimize effects of the organic solvent and the Lowicryls. Furthermore, they may improve the stainability of the biological structures. Figure 2 presents an example. A primary culture of adult rat heart muscle cells, grown for seven days, on a carbon coated, 50 μm thick glass substrate was high pressure frozen and freeze-substituted in ethanol containing 0.5% uranylacetate. Freeze-substitution was performed at 183K for 8h followed by 3h at 223K. Infiltration with Lowicryl HM 23 and polymerization by UV-light was executed at the same temperature. Ultrathin sections were cut parallel to the glass substrate. Figure 2A shows an overview and illustrates adequate structural preservation and identification. The thin section is treated with an antibody against a gap-junction protein. The primary antibody is visualized by 8nm gold colloids, coupled to protein A from Staphylococcus aureus. The internalized gap–junction–vesicle, shown in Fig. 2B at higher magnification is clearly identified by the gold colloids. Freeze-substitution in a pure solvent can be combined with low-temperature embedding at low temperatures under carefully controlled conditions.

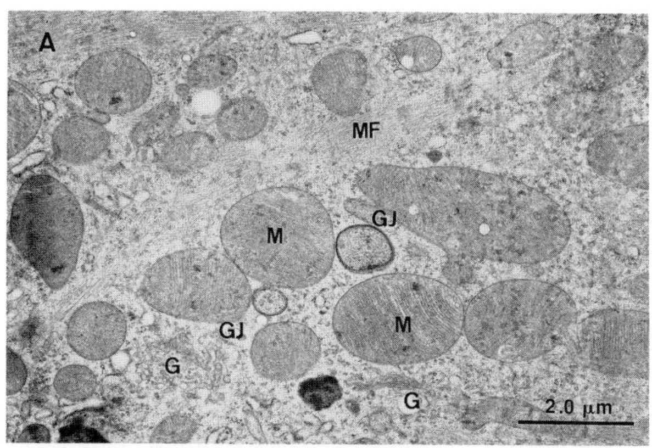

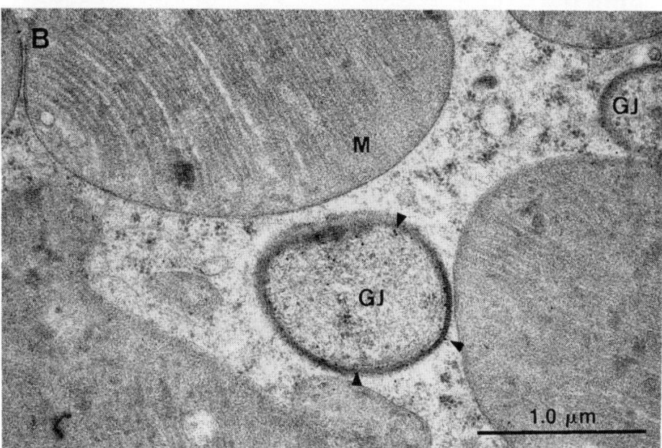

FIGURE 2 A portion of a high pressure cryoimmobilized rat heart muscle cell in culture, freeze-substituted in absence of fixatives, low temperature embedded, and polymerized at 223K. Thin sections of material prepared in this way frequently permit successful immunolabeling and display adequate structural preservation (A). An internalized gap-junction vesicle (GJ), labeled with an antibody against a gap-junction protein is shown in (B) at higher magnification. The arrowheads point to 8 nm colloidal gold particles used to localize the primary antibody. GJ, Internalized Gap-Junction vesicle; G, Golgi Complex; M, Mitochondria; MF, Myofilament.

IV. Cryotechniques in Scanning Electron Microscopy

The factors effecting the preservation of the structural integrity are identical in both scanning- and transmission electron microscopy. Shrinkage due to complete dehydration and drying may be even more pronounced because the removed water is usually not replaced by an embedding resin. Techniques based on cryofixation help to overcome the major problems.

A. Low-Temperature Scanning Electron Microscopy

Low-temperature scanning electron microscopy (LTSEM) embodies the direct, physical approach to structural studies. Modern LTSEM-equipment consists of a high-vacuum preparation chamber attached directly to the scanning microscope. The specimen can be kept in the preparation chamber at controlled temperatures. It may be retained intact, fractured, or dissected and kept either fully frozen hydrated, partially freeze–dried ("etched"), or fully freeze–dried. The samples may be coated for subsequent observation in the SEM. The gate valve between preparation chamber and microscope is then opened and the sample is transferred onto a temperature controlled stage in the scanning electron microscope where it is examined at low temperatures (e.g., 100K). Uncoated samples may be repeatedly dissected or "etched" if necessary. They may also be partially freeze–dried under visual control in the microscope. LTSEM experiments illustrate various dimensional and structural changes that occur in partially or completely freeze–dried specimens. These suggest that the traditional classification of "free" or "bulk" water as opposed to "bound water" or water of hydration shells might provide a much simplified view of cellular water. Models that more comprehensively describe the complex interactions of macromolecules, membrane surfaces, ions, and water are supported (e.g., Clegg 1979, Negendank 1986). LTSEM is currently bound to lower magnifications (e.g., up to 10,000 ×), mainly due to technical limitations of the equipment (e.g., stability of cold stages, moderate resolving power of conventional SEM-instruments). A safe representation of structural facts is, however, always more valuable than the high-resolution detection of insignificant structural details. LTSEM will undoubtedly further develop towards improved resolution. Reasonably priced, easily handled high-resolution SEM-instruments, equipped with reliable field emission guns as well as more sophisticated cryopreparation attachments are now commercially available. Furthermore, high-pressure freezing offers adequate cryofixation of samples of significant size. [*See* SCANNING ELECTRON MICROSCOPY.]

B. Freeze-Drying and Freeze Substitution for SEM Observation at Room Temperature

Freeze-drying of cryofixed samples can be performed under well controlled conditions only if the temperature is homogeneous throughout the sample. This is achieved only in very thin samples that remain in perfect thermal contact with the stage of the freeze drier (e.g., membranes, frozen hydrated cryosections, macromolecular solutions). There is evidence that such specimens can be kept partially freeze-dried at 193K without apparent shrinkage. A certain amount of water is removed from the specimen at this temperature. More water is released only when the specimen is warmed to the temperature range between 223 and 243K and is referred to as the water of hydration shells. Shrinking now occurs. This temperature range is well in accordance with the collapse temperatures (263–223K) for freeze-dried model solutions. Removal of the hydration shells most certainly leads to conformational changes of proteins and macromolecules. Whether or to what extent it is the main factor responsible for dimensional alterations (e.g., shrinkage) is not yet clear. In a cell, water may play a much more complex role than in the above mentioned model experiments and its controlled removal may be very difficult.

Only fully freeze-dried samples can be examined in a SEM at room temperature. Specimens, therefore, have always suffered from shrinkage and collapse. Nevertheless fully freeze–dried samples can provide useful information, because in contrast to the conventional critical point drying procedure, any interaction with organic solvents is avoided. Such a solvent effect is illustrated by the different appearance of the yeast cell surface after conventional and cryofixation based preparation for electron microscopy. The cells reveal a smooth surface after chemical fixation and dehydration in graded ethanol series followed by critical point drying. Hairlike structures, "fimbriae," some of which are identified as acid phosphatase—a highly glycosylated glycoprotein—are detected in SEM after cryoimmobilization followed by complete freeze-drying and in TEM after partial freeze–drying ("deep–etching") and replication or freeze-substitution.

Recently, ultra-high-resolution scanning electron microscopes of the "in lens" type became commercially available. Equipped with a field emission gun, they allow one to examine the samples with a diameter of the electron probe of <1 nm. Adequate structural preservation is required to make full use of this resolving potential. Experiments, using the T_4-polyhead mutant as a model specimen, showed that only rapid freezing followed by freeze–drying revealed the ringlike structure of the capsomers of 8nm diameter composed of six subunits (Fig. 3). Conventional critical point drying after chemical dehydration of the aldehyde fixed samples by ethanol as well as after freeze-substitution, failed to preserve structural details at this level. Figure 3 shows a secondary electron image of a rapidly frozen, freeze-dried T_4-polyhead preparation. It was freeze-

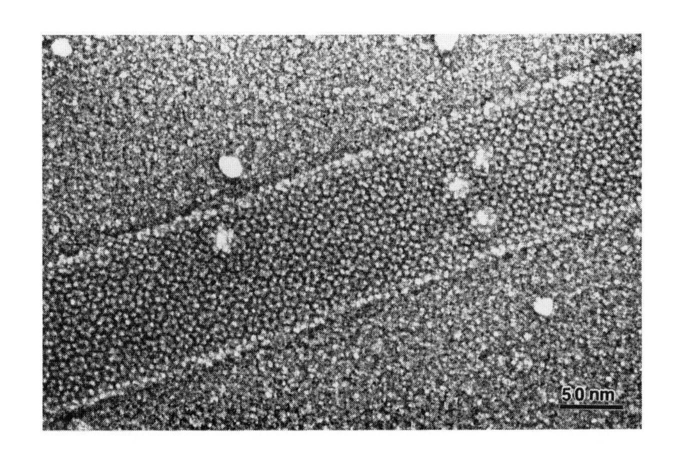

FIGURE 3 The resolution that can be obtained with modern "in-lens" scanning electron microscopes, provided that adequate specimen preparation procedures are available. T_4-Polyheads were adsorbed to a carbon foil and rapidly frozen by plunging into liquid ethane. Freeze-dried at 188K, the samples were coated with a thin (1–2 nm), continuous film of chromium. The capsomers consisting of six subunits are readily identified.

dried at 188K for 30 min, coated with a thin (1–2 nm) continuous film of chromium, then warmed to room temperature and examined in a ultra-high-resolution scanning electron microscope (Hitachi S-900) at a primary magnification of 200,000 ×. Figure 3 illustrates the potential that scanning electron microscopy, optimized with respect to both structural preservation and instrumentation, might achieve in future.

Freeze-drying should be performed with a clean solvent (e.g., distilled water) in order to avoid the deposition of solutes onto the specimen surface during drying. Aldehyde prefixation, therefore, is frequently required to render the specimen resistant against treatment with distilled water. Freeze-substitution followed by critical point drying helps to partially overcome this problem in the medium resolution range. Barlow and Sleigh (1979) systematically tested various freeze-substitution media for stabilizing the metachronal wave of the cilia of paramecium in the SEM.

VI. Conclusions

Cryofixation alone can immobilize biological structures close to the living state. The application of cryofixation based preparative procedures is mandatory for the correlation of structure and function at the ultrastructural level. Purely physical techniques provide safe structural description and preserve the spatial distribution of diffusible elements. Hybrid techniques (e.g., freeze substitution) faciliate thin sectioning and, in combination with low temperature embedding, immunocytochemical studies of intracellular antigens. The quality of structural preservation, however, has to be gauged by purely physical procedures such as cryosectioning and freeze-fracturing. Adequate procedures to cryofix suspensions and thin superficial layers of tissues are available. High-pressure freezing offers the chance to cryofix samples up to a thickness of approximately 0.5 mm. Samples that can be prepared in thin layers are successfully cryoimmobilized by the rapid-freezing techniques under controlled physiological conditions. The relatively long time needed to excise tissue samples and to size them suitable for cryofixation frequently prohibits the correlation of structure and function at high resolution.

Acknowledgment

I wish to thank my colleagues D. Studer, M. Michel, R. Hermann, and P. Schwarb for providing the micrographs and M. L. Yaffee and M. Lampert for their help with the manuscript.

Bibliography

Beckett, A, and Read, N. D. (1986). Low temperature scanning electron microscopy. *In* "Ultrastructure Techniques for Microorganisms" (H. C. Alrich and W. J. Todd, eds.) pp. 45–86. Plenum Press, New York/London.

Clegg, J. S. (1979). Metabolism and the intracellular environment: The vicinal-water network model. *In* "Cell-Associated Water" (W. Drost-Hansen and J. S. Clegg, eds.) pp. 363–413. Academic Press, London/New York.

Dubochet, J., Adrian, M., Chang, J. J., Homo, J. C., Lepault, J., McDowall, A. W., and Schultz, P. (1988). Cryo electron microscopy of vitrified specimens. *In* "Quarterly Review of Biophysics" 21/2: 129–228.

Gross, H. (1987). High resolution metal replication of freeze-dried specimens. *In* "Cryotechniques in Biological Electron Microscopy" (R. A. Steinbrecht and K. Zierold, eds.) pp. 205–215. Springer, Berlin.

Knoll, G., Verkleij, A. J., and Plattner, H. (1987). Cryofixation of dynamic processes in cells and organelles. *In* "Cryotechniques in Biological Electron Microscopy'' (R. A. Steinbrecht and K. Zierold, eds.) pp. 258–271. Springer, Berlin.

Huelser, D. F., Paschke, D., and Greule, J. (1989). Gap junctions: Correlated electrophysiological recordings and ultrastructural analysis by fast freezing and freeze-fracturing. *In* ''Electron Microscopy of Subcellular Dynamics'' (H. Plattner, ed.) pp. 33–49. CRC Press, Boca Raton.

Plattner, H. (1989). ''Electron microscopy of subcellular dynamics.'' CRC Press, Boca Raton.

Negendank, W. (1986). The state of water in the cell. *In* ''The Science of Biological Specimen Preparation'' (M. Müller, R. P. Becker, A. Boyde, and J. J. Wolosewick, eds.) pp. 21–32. SEM, AMF O'Hare, Illinois.

Pinto da Silva, and P., Kan, F. W. K. (1984). Label fracture: A method for high resolution labelling of cell-surfaces. *J. Cell Biol.* **99,** 1156–1161.

Plattner, H., and Bachmann, L. (1982). Cryofixation: a tool in biological ultrastructural research. *Int. Rev. Cytol.* **79,** 237–304.

Kellenberger, E. (1987). The response of biological macromolecules and supramolecular structures to the physics of specimen cryopreparation. *In* ''Cryotechniques in Biological Electron Microscopy'' (R. A. Steinbrecht and K. Zierold, eds.) pp. 35–63. Springer, Berlin.

Moor, H. (1987). Theory and practice of high pressure freezing. *In* ''Cryotechniques in Biological Electron Microscopy'' (R. A. Steinbrecht and K. Zierold, eds.) pp. 175–191. Springer, Berlin.

Robards, A. W., and Sleytr, U. B. (1985). ''Low temperature methods in biological electron microscopy.'' Elsevier, Amsterdam.

Sitte, H., Edelmann, L., and Neumann, K. (1987). Cryofixation without pretreatment at ambient pressure. *In* ''Cryotechniques in Biological Electron Microscopy'' (R. A. Steinbrecht and K. Zierold, eds.) pp. 87–113. Springer, Berlin.

Steinbrecht, R. A., and Müller, M. (1987). Freeze-substitution and freeze-drying. ''Cryotechniques in Biological Electron Microscopy'' (R. A. Steinbrecht and K. Zierold, eds.) pp. 149–172. Springer, Berlin.

Steinbrecht, R. A., and Zierold, K. (1987). ''Cryotechniques in Biological Electron Microscopy'' Springer, Berlin.

Studer, D., Michel, M., and Müller, M. (1989). High pressure freezing comes of age. Scanning Microscopy Supplement **3,** 253–269.

Tanford, C. (1980). ''The Hydrophobic Effect: Formation of Micelles and Biological Membranes.'' 2nd ed. John Wiley & Sons, New York.

Tokuyasu, K. T. (1984). Immunocryoultramicrotomy in immunolabelling for electron microscopy. *In* ''Immunolabeling for Electron Microscopy'' (J. M. Polak and I. M. Varndell, eds.) pp. 71–82. Elsevier, Amsterdam.

Wildhaber, I., Gross, H., and Moor, H. (1982). The control of freeze-drying with deuterium oxide (D2O). *J. Ultrastruct. Res.* **80,** 367–373.

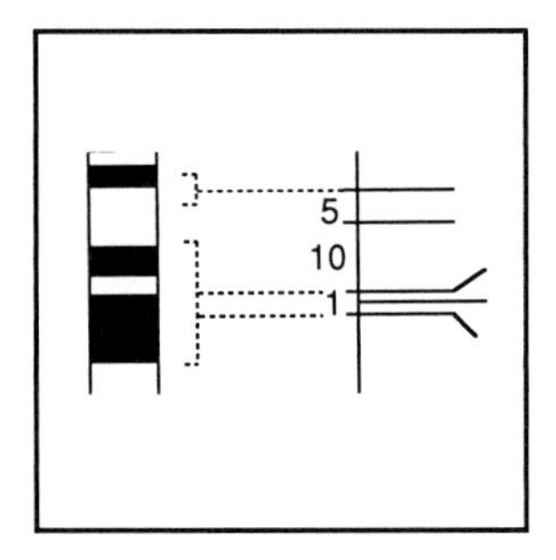

Cystic Fibrosis, Molecular Genetics

LAP-CHEE TSUI, *The Hospital for Sick Children, Toronto; University of Toronto*

Glossary

Adenosine-3′-triphosphate High energy-storing compound used as substrate in energy-dependent biochemical reactions
CentiMorgan (cM) Unit of distance on chromosomes measured by genetic recombination events (1 cM = 1% recombination)
Complementary DNA From a messenger RNA template rather than a DNA template
DNA marker Segment of DNA, usually defined by its associated sequence polymorphisms recognizable by restriction endonuclease cleavage (thus, also termed restriction fragment-length polymorphisms)
Hybridization Annealing of complementary DNA strands or DNA and RNA strands
Linkage analysis Statistical method of mapping genetic loci with respect to each other by following their inheritance in families
Secretory epithelium Cell layer lining the exterior surface of body organs

CYSTIC FIBROSIS (CF) is the most common severe recessive genetic disorder in the Caucasian population. The frequency of the disease is about 1 in 2,500 livebirths, and it varies among geographical locations. The major clinical symptoms of CF include chronic obstructive pulmonary disease, pancreatic enzyme insufficiency, and elevated sweat electrolyte levels. If untreated, affected children usually die at an early age because of severe lung infection and malnutrition; however, due to advances in clinical management, the life span of patients has increased markedly, and many of them now live to adulthood. Electrophysiology studies show that the excessive mucus accumulation in CF patients is probably related to the abnormal regulation of ion transport in the secretory epithelia of the affected organs. More recently, the gene responsible for this disease has been isolated through molecular cloning studies, and the major mutation has been defined at the DNA sequence level.

I. Background

A. The Disease

In 1938, "cystic fibrosis of the pancreas" was comprehensively described. Patients with CF suffer from excessive mucus accumulation resulting in severe clinical consequences in the respiratory, gastrointestinal, and genitourinary tracts, namely chronic obstructive lung disease, pancreatic enzyme insufficiency, and infertility, respectively. All these symptoms are consistent with defects of exocrine glands, as first suggested in 1945 when the mucovisidosis was introduced, a name still popular in parts of continental Europe. CF patients also have elevated electrolyte levels in their sweat, an observation that became the basis of a popular, reliable clinical diagnostic test for CF.

B. Ion Transport

The basic defect in CF appears to reside in the secretory epithelia, in the transport of water, electrolytes, and other solutes across these membranes. A decrease in fluid and electrolyte secretion in CF epithelia may eventually lead to blockage of exocrine outflow in the affected organs. In the early 1980s, altered electrical properties were recognized in CF respiratory epithelium, and it was suggested that the basic defect was associated with abnormalities in sodium and chloride ion transport. Chloride ion impermeability in CF sweat gland ducts was also shown. More recent electrophysiological (patch-clamp) studies suggested that the chloride ion transport defect was due to a failure of a specific anion channel to respond to regulatory signals, presumably mediated through a pathway involving phosphorylation by protein kinase A or C. Progress has been made in the isolation of polypeptide components of epithelial chloride channels, but their relationship to the kinase-activated pathway and CF has yet to be established.

II. The Gene

The mode of inheritance of CF in families is autosomal recessive. On this basis, it was possible to identify the genetic locus (the presumed gene) responsible for this disease by following and trying to match its inheritance in families with other known genetic markers (i.e., linkage analysis).

A. Gene Mapping

In 1985, the CF locus was linked to a genetic marker called *PON* (which determines the activity of paraoxonase, an enzyme made in the liver). The distance between the two was estimated to be about 10 cM, but the chromosome location of *PON* itself was not known. A DNA marker (officially known as *D7S15*) was identified at about 15 cM from the CF locus, suggesting that the CF locus was on chromosome 7, where the marker was located. This led to the discovery of two additional RFLP markers, *MET* (the met protooncogene) and *D7S8* (another arbitrarily selected DNA marker), respectively, both estimated to be about 1 cM from CF and known to be on the long arm of chromosome 7. The result of a collaborative study suggested that the CF locus was flanked by *MET* and *D7S8* (Fig. 1).

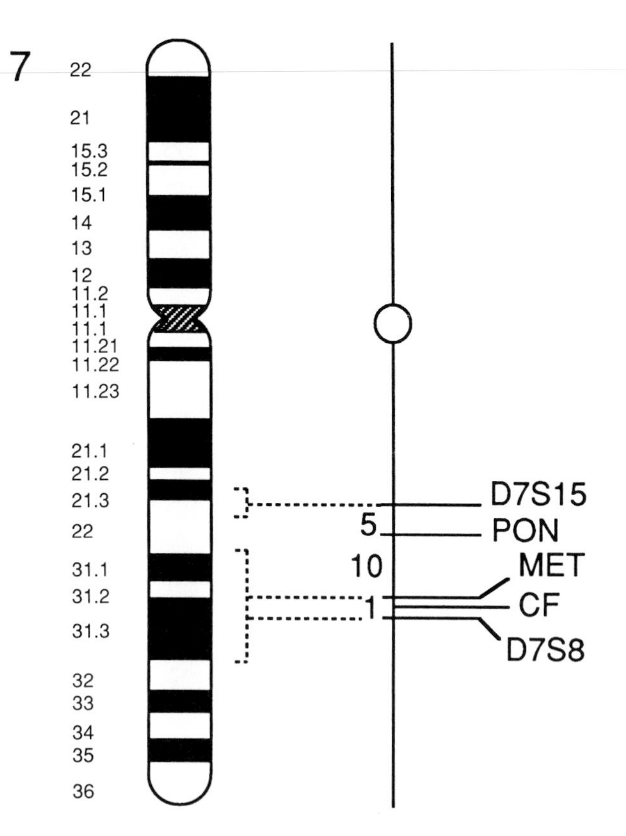

FIGURE 1 Genetic markers on the long arm of human chromosome 7 surrounding the CF locus. A schematic drawing of the chromosome with Giemsa staining is shown on the left; the numbers indicate cytogenetic banding assignment. The corresponding genetic map is shown on the right with distance between genetic markers indicated in centiMorgans.

Among the many different approaches, the chromosome-mediated gene transfer technique was used to introduce a small region of human chromosome 7 presumed to contain the CF locus into a mouse cell line. Because the human DNA could be easily distinguished from that of the mouse by means of detecting human-specific repetitive DNA sequences, a DNA segment (*D7S23*) thought to contain the CF gene (Fig. 2) was isolated from this cell line. Genetic studies showed that *D7S23* was extremely close to the disease locus. The physical distance between *MET* and *D7S8* was also determined with *D7S23* between them.

In 1988, over 250 DNA segments were isolated from chromosome 7 and >50 of them were assigned to the CF region. Genetic and physical mapping studies showed that two of these segments (*D7S122* and *D7S340*) were between *MET* and *D7S8* (Fig. 2). The distance between *D7S122* and *D7S340* was found to be only about 10 kilobase pairs (kb), and

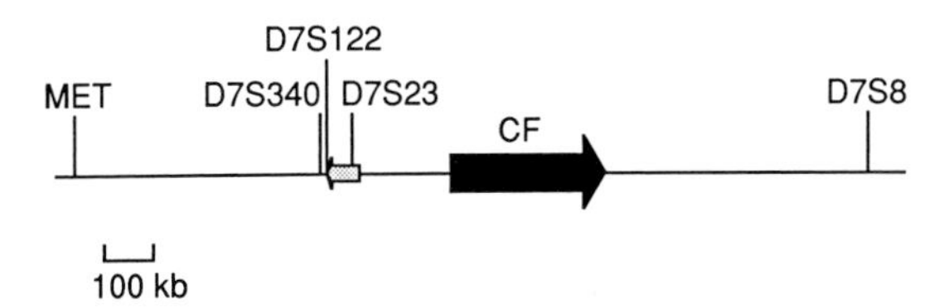

FIGURE 2 A physical map of the CF gene and its adjacent genetic markers. The distances between markers are based on long-range restriction endonuclease mapping; the arrows indicate the transcription direction of two of the genes in this region.

the two are about 500 kb from *MET* and 950 kb from *D7S8*. Further studies showed that *D7S23* was between *D7S340* and *D7S8*, about 80 kb from the former. In addition, family studies showed that the CF locus was between *D7S23* and *D7S8*. [*See* GENETIC MAPS.]

B. Gene Isolation

In 1989, approximately 280 kb of a continguous DNA segment were isolated by chromosome walking and jumping initiated from *D7S122* and *D7S340*. Small portions of DNA segments from this region were examined individually for their ability to detect possible conserved sequences in other animal species by DNA hybridization, possible messenger RNA (mRNA) products or their complementary DNA (cDNA) clones from affected tissues, and possible protein-coding regions by direct sequence analysis. As a result of these intensive investigations, several genes were identified from this chromosome region, and one of them was proved to be the gene responsible for CF on the basis of predicted biochemical properties and genetic analysis.

The CF gene spans approximately 250 kb of DNA sequence (Fig. 3). Transcription of this gene yields a mature mRNA of about 6,500 nucleotides in length, which is the spliced product of at least 27 exons (regions corresponding to the coding sequences, which are retained after the removal of intervening sequences contained in the original transcript of the 250-kb gene). The CF gene product can be found in a variety of body tissues including lung, pancreas, sweat gland, liver, nasal polyps, salivary gland, and colon, all of which are tissues affected in patients with CF. While this tissue distribution pattern is in good agreement with the disease pathology, there is no qualitative or quantitative difference detectable between the mRNA isolated from tissues of CF and control subjects, consistent

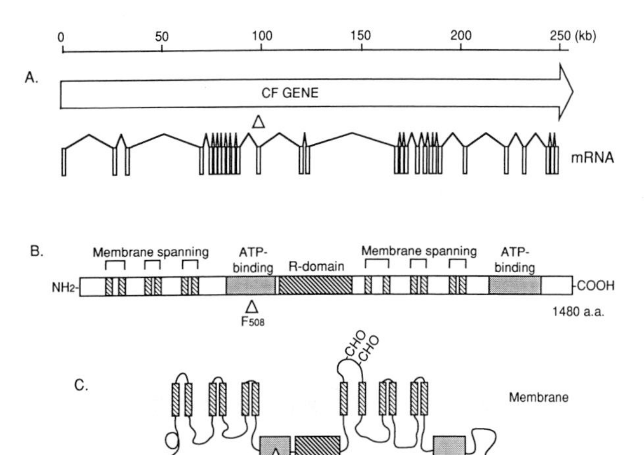

FIGURE 3 The CF gene and the predicted gene product. A. Schematic diagram of the gene together with its processed mRNA; at least 27 exons have been discovered. B, C. Putative polypeptide and a model of CFTR, respectively, with the presumed functional domains highlighted. The amino(NH_2-) and carboxyl (-COOH) termini of the polypeptide and the two possible glycosylation sites (-CHO) are indicated. a.a., amino acids.

with the assumption that CF mutations are subtle changes at the nucleotide level.

III. The Protein

Computer analysis of the nucleotide sequences derived from the cDNA clones predicted a coding region capable of producing a polypeptide of 1,480 amino acid residues with a molecular mass of 168,138. This putative CF gene product was named the cystic fibrosis transmembrane conductance regulator (CFTR) protein, to avoid confusion with the previously named, unrelated CF protein, CF antigen, and CF factor.

A. Predicted Properties

The most characteristic feature of the predicted CFTR protein is the presence of two repeated motifs, each of which consists of a domain capable of spanning the membrane several times and sequences resembling consensus nucleotide adenosine-3′-triphosphate (ATP)-binding domains (Fig. 3). Because there is no apparent signal-peptide sequence at the amino-terminus of CFTR, the highly charged hydrophilic segment preceding the first transmembrane sequence is probably oriented in the cytoplasm. Little of the polypeptide is predicted to be exposed to the exterior surface, except the region between transmembrane segments 7 and 8, where two potential sites for attachment of carbohydrate side chains (N-linked glycosylation) are located. These characteristics are remarkably similar to those of the mammalian P-glycoprotein, which confers multidrug resistance in cancer cells by removing the anticancer drugs from the cells, and a number of other membrane-associated transport proteins in prokaryotes and lower eukaryotes.

A highly charged cytoplasmic domain can be identified in the middle of the predicted CFTR polypeptide, linking the two halves of the protein (Fig. 3). This domain is thought to have a regulatory function (thus named the R-domain). It contains >25% of polar residues arranged in alternating clusters of positive and negative charges, as well as a large number of amino acid sequences that are potential sites for phosphosphoylation by the protein kinases A and C.

Although CFTR must be directly responsible for CF, how this protein is involved in the regulation of ion conductance across the membrane of epithelial cells remains unclear. The presence of ATP-binding domains in CFTR suggests that ATP hydrolysis is involved and required for its function. While many of the biochemical properties described above are consistent with CFTR being a transport protein, it is possible that CFTR is not an ion channel itself but only serves to regulate ion channel activities. A better knowledge of CFTR requires detailed structural and functional analysis of the protein. Useful information may also be derived from characterization of mutations that cause the disease.

IV. Mutations

A. Location

The major mutation, which accounts for about 70% of all CF chromosomes, is a 3-bp deletion, which results in the absence of a phenylalanine residue (position 508) in the predicted CF polypeptide (Fig. 3). This single amino acid deletion (ΔF_{508}) is located in the first ATP-binding domain, and the mutation is thought to come from a single origin. A second single amino acid deletion is also detected in the same region, removing either the isoleucine residue at position 506 or 507 (ΔI_{507}). Many more (>60) mutations have been discovered subsequently as the result of a worldwide effort in identifying additional mutations; they include small deletions and insertions causing frameshifts in translation, nonsense, and missense mutations, and point mutations affecting mRNA processing. Except for ΔF_{508}, most other CF mutations are relatively rare in the population as they are only represented by a small number of examples.

B. Functional Implications

Some of the missense mutations involve amino acid residues highly conserved among different ATP-binding proteins, suggesting that these mutations affects ATP-binding which is essential for the function of CFTR. The ΔF_{508} and ΔI_{507} mutations suggest that the phenylalanine residue at position 508 and isoleucine at position 507 (or 506) are also essential but the function of these residuals is less apparent. Deletion of either of these residues may prevent proper binding of ATP or the conformational change required for normal CFTR activity. Since the length of the polypeptide in this region varies among different ATP-binding proteins and the amino acid sequences are less conserved, the functional importance of Ile507 and Phe508 is probably unique to CFTR. The length of this region of the polypeptide is, however, probably more important than the actual amino acid residues in CFTR, as amino acid substitutions in this region have been found in rare variants of normal chromosomes. Moreover, the large number of nonsense as well as frameshift mutations due to small deletions or insertions occurring at different regions of the predicted polypeptide suggests that CFTR is perhaps dispensable for most body functions.

C. Correlation between Genotype and Phenotype

Genetic analysis suggests that there are two groups of mutations in the CF gene, one termed *severe* and the other *mild*. About 90% of the CF chromosomes belong to the *severe* group and the rest of the *mild* group. Patients with both CF chromosomes carrying a *severe* mutation should show the pancreatic insufficient phenotype, which is found in 85% of the patients with CF. The other 15% of CF patients who have sufficient pancreatic functions are expected to carry one or both chromosomes with *mild* mutations. The ΔF_{508} mutation is considered to be in the *severe* group and, as predicted by the hypothesis, patients homozygous for ΔF_{508} are almost exclusively pancreatic insufficient. Since a small number of these latter patients are initially diagnosed as pancreatic sufficient but develop pancreatic insufficiency at a later time, the genotype information may prove to be useful in disease prognosis. Individuals with 2 copies of the *severe* mutations are also thought to be more likely to develop meconium ileus, a condition which occurs in about 10% of newborns with CF and which is generally ascribed to failure of pancreatic enzyme secretion and digestion of intraluminal contents in the uterus.

D. Heterozygote Advantage?

The overall gene frequency of CF mutations in the Caucasian population of European origin is about 2% and that in other races is considerably lower (<0.3%). Various hypotheses have been proposed to explain this relatively high frequency in a specific population; the difference mechanisms include high mutation rate, multiple loci, heterozygote advantage, genetic drift, increased fertility, and reproductive compensation. While it is difficult to test any of these hypothesis without a good knowledge of the basic defect, some of them (e.g., the possibility of multiple genes involved in CF) could be eliminated on the basis of the genetic data. Increased resistance to diseases (e.g., tuberculosis or cholera) in CF heterozygotes due to presumed alterations in secretion has been suggested as a possible mechanism.

V. Clinical Applications

A. Genetic Diagnosis

Genetic diagnosis for CF has been in practice since the discovery of the closely linked DNA markers. In families with affected children, one can identify the DNA marker alleles that are associated with the mutant genes by analyzing the patients and both of their parents, and then use the information to predict the status of the unknowns (confirming disease status, carrier detection, and prenatal diagnosis). Because of the fortuitous association of the major CF mutation with an uncommon combination of the DNA marker alleles (allelic and haplotype association, or linkage disequilibrium), this type of diagnosis with linked DNA markers is highly informative—an accurate diagnosis is possible for almost all families at risk. Some prediction can even be made on the basis of the strong allelic association, where there is no DNA information from any affected family member. [*See* DNA MARKERS AS DIAGNOSTIC TOOLS.]

With the CF mutations characterized at the DNA sequence level, it is now possible to perform genetic testing for the disease status for any random individual. Detection of ΔF_{508} may be achieved by specific oligonucleotide hybridization or by heteroduplex formation in a polymerase chain reaction. In North America, about 50% of CF patients without a previous family history can be accurately diagnosed by DNA analysis, and 70% of the CF carriers can be identified with the ΔF_{508} mutation. The proportion of this mutation varies among different geographic locations, with higher numbers (80–85%) in northern Europe and lower numbers in the south (35–40%).

B. Better Treatment

While most of the current treatments for patients are based on the symptoms of the disease, discovering the basic defect in CF is the first step toward the development of more effective means of treatment. Both specific pharmacological reagents or gene therapy are realistic possibilities. In addition, because of the detected correlation between genotype and phenotype, CF clinicians may be able to consider different treatment plans according to the requirements in the future.

Bibliography

Beaudet, A., Bowcock, A., Buchwald, M., Cavalli-Sforza, L., Farrall, M., King, M.-C., Klinger, K., Lalouel, J.-M., Lathrop, M., Naylor, S., Ott, J., Tsui, L.-C., Wainwright, B., Watkins, P., White, R., and Williamson, R. (1986). Linkage of cystic fibrosis to two tightly linked DNA markers: Joint report from a collaborative study. *Am. J. Hum. Genet.* **39,** 681.

Boat, T. F., Welsh, M. J., and Beaudet, A. L. (1989). Cystic fibrosis. *In* "The Metabolic Basis of Inherited Disease," 6th ed. (C. L. Scriver, A. L. Beaudet, W. S. Sly, and D. Valle, eds.). McGraw-Hill, New York.

Estivill, X., Farrall, M., Scambler, P. J., Bell, G. M. Hawley, K. M. F., Lench, N. J., Bates, G. P., Kruyer, H. C., Frederick, P. A., Stanier, P., Watson, E. K., Williamson, R., and Wainwright, B. J. (1987). A candidate for the cystic fibrosis locus isolated by selection for methylation-free islands. *Nature* **326,** 840.

Kerem, B., Rommens, J. R., Buchanan, J. A., Markiewicz, D., Cox, T. K., Chakravarti, A., Buchwald, M., and Tsui, L.-C. (1989). Identification of the cystic fibrosis gene: Genetic analysis. *Science* **245,** 1073.

Mastella, G., and Quinton, P. M. (eds.) (1988). "Cellular and Molecular Basis of Cystic Fibrosis." San Francisco Press, San Francisco.

Riordan, J. R., Rommens, J. M., Kerem, B., Alon, N., Rozmahel, R., Grzelchak, Z., Zielenski, J., Lok, S., Plavsic, N., Chou, J.-L., Drumm, M. L., Iannuzzi, M. C., Collin, F. S., and Tsui, L.-C. (1989). Identification of the cystic fibrosis gene: Cloning and characterization of complementary DNA. *Science* **245,** 1066.

Rommens, J. M., Iannuzzi, M. C., Kerem, B., Drumm, M. L., Melmer, G., Dean, M., Rozmahel, R., Cole, J. L., Kennedy, D., Hidaka, N., Zsiga, M., Buchwald, M., Riordan, J. R., Tsui, L.-C., and Collins, F. S. (1989). Identification of the cystic fibrosis gene: Chromosome walking and jumping. *Science* **245,** 1059.

Tsui, L.-C., and Buchwald, M. (1991). Molecular genetics on cystic fibrosis. *In* "Advances in Human Genetics" (H. Harris and K. Hirschhorn, eds.). Plenum, New York. (in preparation)

Tsui, L.-C., Buchwald, B., Barker, D., Braman, J. C., Knowlton, R., Schumm, J., Eiberg, H., Mohr, J., Kennedy, D., Plavsic, N., Zsiga, M., Markiewicz, D., Akots, G., Brown, V., Helms, C., Gravius, T., Parker, C., Rediker, K., and Donis-Keller, H. (1985). Cystic fibrosis locus defined by a genetically linked polymorphic DNA marker. *Science* **230,** 1054.

Wainwright, B. J., Scambler, P. J., Schmidtke, J., Watson, E. A., Law, H.-Y., Farrall, M., Cooke, H. J., Eiberg, H., and Williamson, R. (1985). Localization of cystic fibrosis locus to human chromosome 7cen-q22. *Nature* **318,** 382.

White, R., Woodward, S., Leppert, M., O'Connell, P., Nakamura, Y., Hoff, M., Herbst, J., Lalouel, J.-M., Dean, M., and Vande Woude, G. (1985). A closely linked genetic marker for cystic fibrosis. *Nature* **318,** 382.

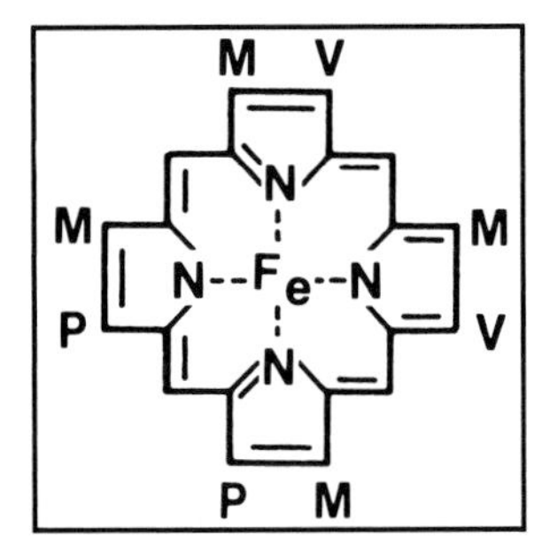

Cytochrome *P*-450

FRANK J. GONZALEZ, *National Cancer Institute, National Institutes of Health*

Glossary

Allele One of two or more alternating forms of a gene occupying corresponding sites on homologous chromosomes; multiple alleles for a single gene can exist in a population, and these can be recognized by DNA changes within or flanking a gene or by differing properties of a gene product; DNA changes of various types can give rise to mutations that inactivate a gene—these are known as mutant alleles

Complementary DNA In this chapter, refers to a double-stranded DNA that is copied from a messenger RNA (mRNA); the primary amino acid sequence of the mRNA-encoded protein can be deduced from the sequence of the complementary DNA (cDNA)

Concerted evolution Following a gene duplication, two or more related genes diverge, as during divergent evolution, and then become more similar to each other by exchanging DNA or genetic information. This phenomenon is recognized when a subfamily is examined in two species (A and B) and the two genes (1 and 2) in a single species are known to have existed prior to the divergence of the two species; as a result of concerted evolution, gene 1 will be more similar in sequence to gene 2 in species A than to gene 1 in species B

Divergent evolution Following a gene duplication, the two daughter genes begin to diverge by accumulating random base substitutions and, hence, become less similar to each other

Gene conversion Generally thought to be the physical mechanism of convergent evolution, gene conversion is a nonreciprocal recombination event in which a segment of one gene replaces the corresponding segment of a related gene

Genetic polymorphism Occurrence together in a population of two or more genetically determined phenotypes in such proportions that the rarest of them cannot be maintained merely by recurrent mutation

Ligand (1) Organic molecule that donates electrons to form coordinate covalent bonds with metallic ions such as iron; (2) compound that specifically binds to a protein receptor

Orthologous Gene or protein in one species is said to be orthologous to a gene or protein in another species if both are believed to have evolved from a single gene present at the time of divergence of the species

Xenobiotic Compound that is foreign to the body, such as a drug or an environmental pollutant (*xeno-*, foreign; *-biotic,* of biological origin)

THE TERM CYTOCHROME *P*-450 refers literally to a colored substance in the cell that absorbs light at around 450 nm, within the visible spectrum (*cyto-*, cell; *-chrome,* color; *P,* pigment; *450,* wavelength). In general, a cytochrome is a cellular heme-containing protein (hemoprotein) whose principal function is electron transport. The heme moiety is a molecule called protoporphyrin IX complexed with an iron ion, the latter of which can undergo reversible changes in valency. The protein environment surrounding the heme governs the selective properties of hemoproteins in the organism or cell. For example, hemoglobin serves to transport O_2 within the body, whereas *P*-450 is a monooxygenase using one atom from O_2 and two electrons to oxidize chemical substrates. The *P*-450 s are ubiquitous enzymes that have probably been present since very early in biological evolution and are found in bacteria, in fungi, and throughout the plant and animal kingdoms. *P*-450s exist in multiple forms and are classified into a superfamily of enzymes, all of which include a conserved cysteine-containing region near their carboxy termini, which serves to bind the heme molecule. These enzymes use electrons from nicotinamide adenine dinucleotide (NADH) and/or nicotinamide adenine dinucleotide phosphate (NADPH) and O_2 to oxidize their substrates. In humans, *P*-450s perform three critical functions: (1) five are involved as key enzymes in pathways leading to steroid synthesis; (2) some *P*-450s convert arachidonic acid, in the epoxygenase pathway, to signal metabolites that mediate a wide range of biological processes; and (3) many other *P*-450s catalyze the oxidation of foreign compounds or xenobiotics ingested in the form of plant metabolites, drugs, and environmental contaminants. These enzymes are the primary interface between humans and the chemical environment. Human deficiencies of both the steroidogenic and xenobiotic-metabolizing *P*-450s are responsible for a severely debilitating and sometimes fatal disease called congenital adrenal hyperplasia. Deficiencies in xenobiotic-metabolizing *P*-450s are associated with clinical drug oxidation polymorphisms that result in toxic reactions to some prescription drugs.

I. Biochemistry of *P*-450s

A. Physical Properties

Cytochrome *P*-450s are a large group of enzymes that have several common features. They all contain a single iron ion that is bound to a protoporphyrin IX molecule in a complex called heme. This green pigment is noncovalently-bound to the enzyme and found in many other proteins including hemoglobin, myoglobin, and cytochrome *c*, and in enzymes such as thyroid peroxidase, xanthine oxidase, and myeloperoxidase. Collectively, these enzymes can all bind and/or metabolize oxygen. Thus, *P*-450s also metabolize oxygen in the form of atmospheric O_2. Generally, the concentration of O_2 that is in equilibrium with water in living cells is more than sufficient to support the catalytic activity of *P*-450s. Every *P*-450 found in bacteria, yeast, plants, and animals possesses heme and metabolizes oxygen in the same manner.

The heme-binding region can be readily identified in any *P*-450 sequence. This region is found within the carboxy-terminal third of the *P*-450 polypeptide chain and centers around the amino acid cysteine (Fig. 1). The sulfur or thiolate portion of this amino acid serves as the fifth ligand to the heme iron and is absolutely required for *P*-450 enzymatic activity. The precise positioning of this thiol group determines the spectral properties of the *P*-450 in the presence of electrons (which reduce the iron from a valency of Fe^{3+} [ferric form] to Fe^{2+} [ferrous form]). In the presence of the strong ligand carbon monoxide (CO) (which binds to the heme in the sixth coordinate position to the iron, in the same way as O_2, but is not metabolized), the enzyme dis-

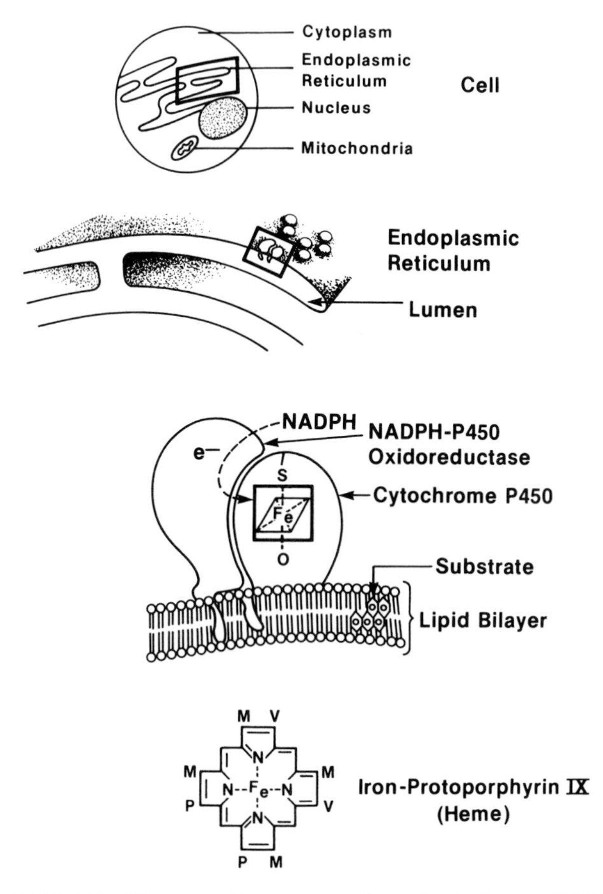

FIGURE 1 Diagrammatic representation of cytochrome *P*-450 and its relationship to other components of the cell. The elements enclosed by rectangles are expanded in more detail in the lower panels.

plays a unique absorption of light at around 450 nm of the visible spectrum. Hemoglobin, cytochrome b_5, cytochrome *c*, and other heme proteins absorb at a lower wavelength of about 420 nm. This difference in absorption properties reflects the polypeptide sequence that surrounds the heme molecule and, more importantly, the positioning of the cysteine residue. Other hemoproteins appear to have basic amino acids in the fifth position, which in *P*-450 is occupied by cysteine. In fact, with denaturation, *P*-450 no longer absorbs light at 450 nm but, instead, at 420 nm, suggesting that the heme is displaced from the cysteine and that another, perhaps basic amino acid, is the fifth ligand to the iron. [*See* HEMOGLOBIN.]

All *P*-450s are similar in size, ranging from 48,000 to 60,000 kDal, and all contain approximately 500 amino acids, except the smaller bacterial $P\text{-}450_{cam}$, found in *Pseudomonas putida,* and one unusually large enzyme found in the bacterium *Bacillus megaterium* (see below). The bacterial *P*-450s isolated to date are soluble in water, whereas the yeast, plant, and animal enzymes are hydrophobic molecules that are bound to cell membranes. In animals, most *P*-450s are part of an intracellular membrane network called the endoplasmic reticulum (Fig. 1). This membrane network extends throughout the cell and consists of large sheets or tubes containing an inner space called the lumen. The outer region of the membrane faces the cytoplasm, and this is where *P*-450s are found. In fact, *P*-450s are frequently referred to as microsomal enzymes. Microsomes are actually fragments of endoplasmic reticulum that form vesicles when a cell is disrupted. These vesicles can be separated from soluble cellular components and other macromolecular organelles by differential high-speed centrifugation.

A few of the *P*-450s involved in steroid biosynthetic and metabolic pathways are found in the inner membranes of mitochondria. These enzymes are encoded by genes located in the cell nucleus. The endoplasmic reticulum-bound and mitochondrial *P*-450s are inserted into their respective cellular compartments by distinctly different mechanisms. The mitochondrial enzyme must be transported through the outer membrane of this organelle to be embedded ultimately in the inner membrane. This transport process is mediated by the system that is responsible for the incorporation of many inner mitochondrial enzymes. The endoplasmic reticulum-bound *P*-450s are inserted directly into the endoplasmic reticulum lipid bilayer by a complex pathway that involves recognition of a hydrophobic peptide (the signal sequence) located at the amino terminus of the *P*-450. [*See* CELL MEMBRANE TRANSPORT.]

The predicted structure of an endoplasmic reticulum-bound *P*-450 is shown in Fig. 1. The enzyme is anchored to the phospholipid membrane bilayer of this membrane system by the amino-terminal portion of its polypeptide chain. The bulk of the enzyme faces the cytoplasm of the cell or the outside of the endoplasmic reticulum network. This cellular location is ideally suited for the *P*-450 to bind to and metabolize hydrophobic substrates such as the polycyclic aromatic hydrocarbon that dissolves in lipid (Fig. 1). The endoplasmic reticulum also contains other enzymes, such as the conjugating enzymes or transferases, which are crucial in the further metabolism and elimination of *P*-450 metabolites resulting from *P*-450-catalyzed oxidations of these types of substrates.

B. Enzymology

The *P*-450s have been called mixed-function monooxygenases because only one atom from O_2 is inserted into the substrate, in contrast to dioxygenases such as cyclooxygenase and lipoxygenase, which insert both atoms from O_2 into their substrates. *Mixed function* refers to the fact that these enzymes are capable of incorporating one oxygen atom into water and the other into the substrate, the latter of which includes numerous structurally diverse chemicals. *P*-450s use O_2 and electrons, supplied by the cofactor NADPH, to add a single oxygen atom to the substrate, usually in the form of a hydroxyl group or an epoxide. This process can be seen in the diagram of the *P*-450 catalytic cycle (Fig. 2). The substrate "RH" binds to the enzyme near the heme residue. The heme iron is then reduced from Fe^{3+} to Fe^{2+} by the addition of an electron from NADPH, which, in microsomal *P*-450s, is donated by the flavin-containing enzyme NADPH-*P*-450 oxidoreductase (note that the electron is transferred as a hydride ion, H^-). After reduction, an O_2 molecule binds to the heme, and a second electron, usually donated by another molecule of NADPH (again through NADPH-*P*-450 oxidoreductase), is added. However, with certain substrates, the second electron may come from NADH via cytochrome b_5 and NADH-cytochrome b_5 oxidoreductase. The mitochondrial enzymes receive both electrons from NADPH using the iron-sulfur pro-

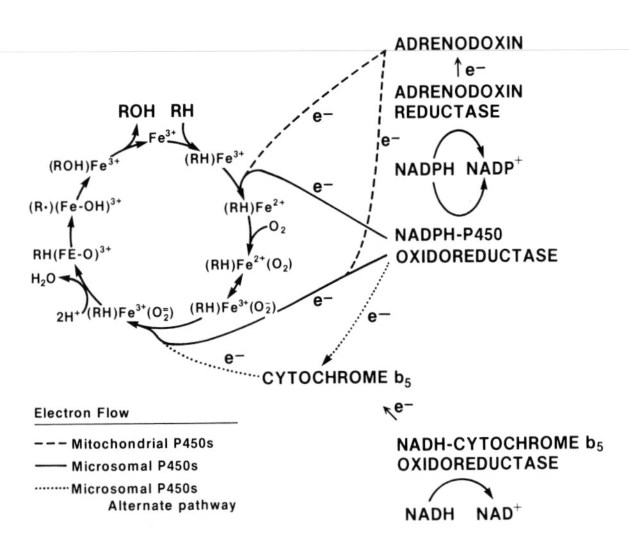

FIGURE 2 Catalytic cycle of cytochrome *P*-450s. Two electrons are individually introduced into the cycle via participation of other enzymes.

tein adrenodoxin and adrenodoxin reductase. Like the NADPH-*P*-450 oxidoreductase, adrenodoxin reductase is also a flavin-containing enzyme.

The addition of a second electron results in the splitting of oxygen and the release of one oxygen atom as water. The second oxygen atom is then in such a state that it can be inserted into the substrate to form an epoxide or alcohol "ROH." Prior to the release of substrate, an active intermediate "R·" is sometimes formed; the rearrangement of this intermediate may result in the loss of stereospecificity and the formation of a number of other products. In many cases, rearrangement results in the removal of a hydrogen rather than the addition of an oxygen.

Many activities are associated with *P*-450-catalyzed reactions, depending on the substrate and the nature of the radical intermediate. Some of these are shown in Fig. 3, where R_1 and R_2 represent alkyl and aryl side chains. The high-energy intermediates of some *P*-450-catalyzed reactions are the principal basis for cell toxicity and cell transformation mediated by chemical carcinogens. In general, these chemicals yield high-energy intermediates that cannot undergo intramolecular rearrangement. In these cases, the intermediates, usually elec-

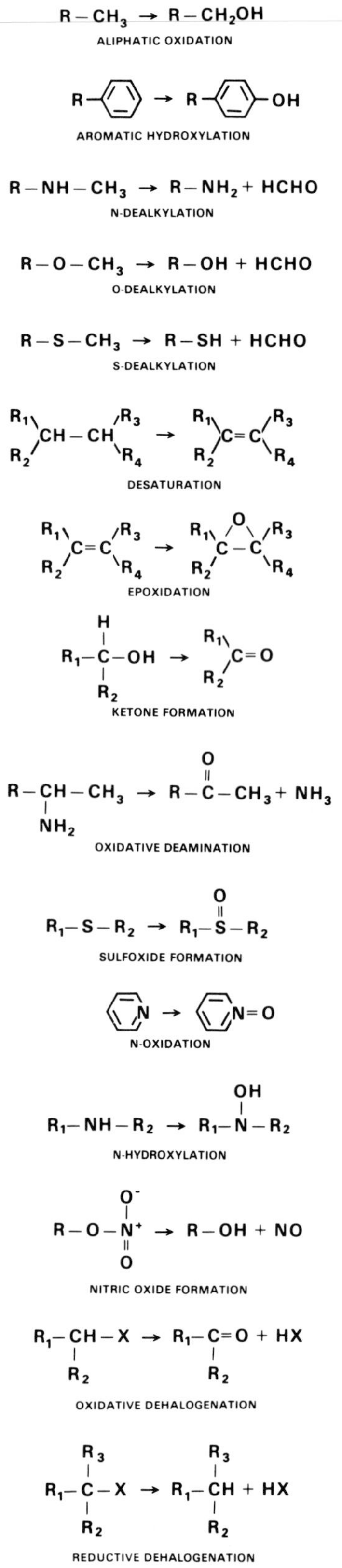

$$R-CH_3 \rightarrow R-CH_2OH$$
ALIPHATIC OXIDATION

$$R-C_6H_5 \rightarrow R-C_6H_4-OH$$
AROMATIC HYDROXYLATION

$$R-NH-CH_3 \rightarrow R-NH_2 + HCHO$$
N-DEALKYLATION

$$R-O-CH_3 \rightarrow R-OH + HCHO$$
O-DEALKYLATION

$$R-S-CH_3 \rightarrow R-SH + HCHO$$
S-DEALKYLATION

$$(R_1)(R_2)CH-CH(R_3)(R_4) \rightarrow (R_1)(R_2)C=C(R_3)(R_4)$$
DESATURATION

$$(R_1)(R_2)C=C(R_3)(R_4) \rightarrow (R_1)(R_2)C\overset{O}{-}C(R_3)(R_4)$$
EPOXIDATION

$$R_1-CH(R_2)-OH \rightarrow (R_1)(R_2)C=O$$
KETONE FORMATION

$$R-CH(NH_2)-CH_3 \rightarrow R-C(=O)-CH_3 + NH_3$$
OXIDATIVE DEAMINATION

$$R_1-S-R_2 \rightarrow R_1-S(=O)-R_2$$
SULFOXIDE FORMATION

$$C_5H_5N \rightarrow C_5H_5N=O$$
N-OXIDATION

$$R_1-NH-R_2 \rightarrow R_1-N(OH)-R_2$$
N-HYDROXYLATION

$$R-O-N^{+}(O^{-})=O \rightarrow R-OH + NO$$
NITRIC OXIDE FORMATION

$$R_1-CH(R_2)-X \rightarrow R_1-C(R_2)=O + HX$$
OXIDATIVE DEHALOGENATION

$$R_1-C(R_2)(R_3)-X \rightarrow R_1-CH(R_2)(R_3) + HX$$
REDUCTIVE DEHALOGENATION

FIGURE 3 Reactions catalyzed by cytochrome *P*-450s. Note that no attempt has been made to balance these reactions.

trophiles, can exit the *P*-450 active site and react with other cellular compounds, particularly endogenous chemicals with nucleophilic properties. Among the most nucleophilic cellular constituents are ribonucleic acid (RNA) and deoxyribonucleic acid (DNA), the genetic information source of the cell. When DNA or specific genes such as oncogenes or tumor-supressor genes become damaged or mutate, cell transformation and cancer can result. Thus, *P*-450-mediated carcinogen activation is the first step in chemically induced cancer. Cigarette smoking-associated lung cancer is one example of a cancer that results from exposure to carcinogens.

II. *P*-450 Nomenclature

The nomenclature system for *P*-450s is based solely on primary amino acid sequence relatedness among *P*-450 forms. Computer programs capable of paired comparisons of multiple primary amino acid sequences are used to generate percent sequence similarities between *P*-450s. By examining sequence similarities for multiple *P*-450s, arbitrary percentage cutoff values have been assigned to distinguish among gene families and gene subfamilies. By definition, *P*-450 from one gene family displays ≤40% sequence similarity to a *P*-450 in any other gene family. With only a couple of minor exceptions, *P*-450s within individual families are >40% identical in sequence. Within these families, *P*-450s that display >59% sequence similarity have been assigned to subfamilies.

The individual human *P*-450 genes are named using the root symbol *CYP* (cytochrome *P*-450), an Arabic number to denote the gene family, a capital letter to designate the subfamily, and another Arabic number to represent the individual gene. For example, the gene encoding a *P*-450 that oxidizes polycyclic aromatic hydrocarbons is *CYP1A1*. The *CYP1A1* gene product or enzyme is also designated CYP1A1, except without italics. *P*-450s are sometimes identified by their trivial names (see Fig. 4). In some cases, the numbering of *P*-450 gene families was based on certain reactions carried out by their encoded enzymes. For example, the *P*-450 involved in the synthesis of 11-deoxycorticosterone from progesterone oxidizes the latter steroid at the 11-carbon position, and hence its family was assigned the name *CYP11*. In fact, all of the five families of

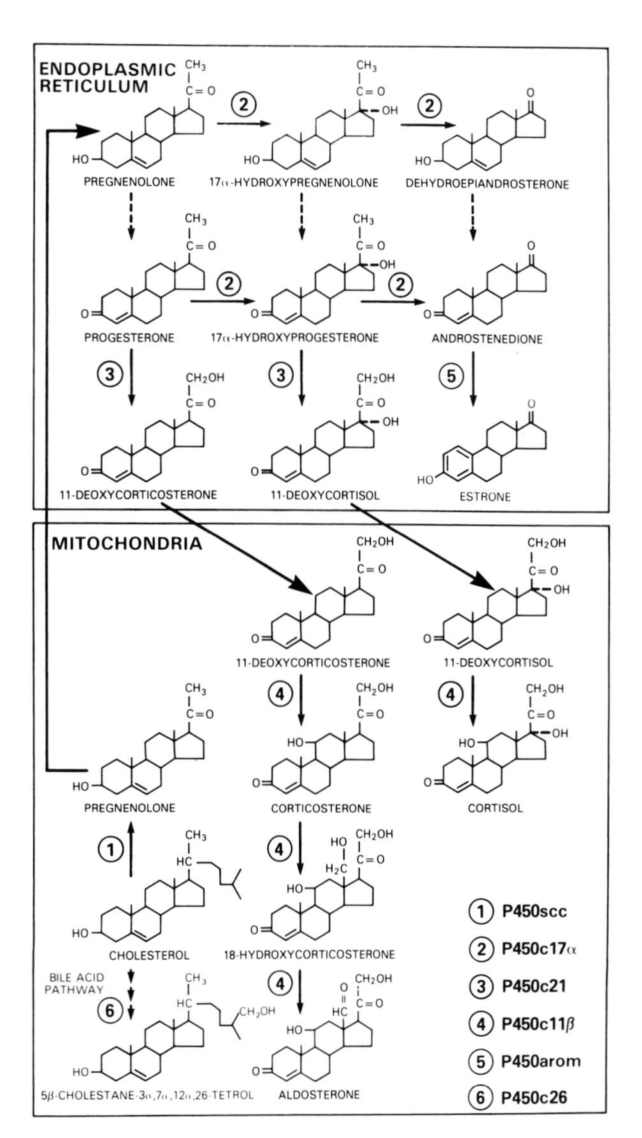

FIGURE 4 List of those cytochrome *P*-450s and their cellular compartments that carry out reactions in steroidogenic pathways. These enzymes are identified by their trivial names.

steroidogenic *P*-450s were named in this fashion. The four families containing the *P*-450s involved in drug and fatty-acid oxidations were named 1–4. To allow for newly discovered families in higher eukaryotes, the numbering scheme began at 51 for yeast *P*-450 families, 71 for plant *P*-450s and at 101 bacterial *P*-450s.

The *P*-450s that are known to be present in humans are listed in Table I. Surely, others are yet to be discovered. In some cases, as with the steroidogenic *P*-450s, the human enzyme, complementary DNA (cDNA), or gene has not been sequenced but

TABLE I List of Human *P*-450s, Tissues in Which They Are Expressed, and Some of Their Substrates

Gene[a]	Tissue[b]	Substrates[c]
CYP1A1	Liver	Benzo(a)pyrene (C)
	Lung	7,12-Dimethylbenz(a)-anthracene (C)
	Placenta	
	Other	
CYP1A2	Liver	Acetaminophen (D)
		2-Acetylaminofluorene (C)
		Aflatoxin B_1 (C)
		Heterocyclic arylamines (C)
		Phenacetin (D)
CYP2A3	Liver (P)	Coumarin
	Lung	*N*-nitrosodimethylamine (C)
	Nasal epithelium	
	Other	
CYP2B7	Liver (P)	7-Ethoxycoumarin
	Lung	Aflatoxin B_1 (C)
	Other	
CYP2C8	Liver	Tolbutamide (D)
		R-mephenytoin (D)
CYP2C9	Liver	Tolbutamide (D)
		R-mephenytoin (D)
		Warfarin (D)
CYP2D6	Liver (P)	Bufuralol (D)
	Kidney	Debrisoquine (D)
		Dextromethorphan (D)
		Nortriptyline (D)
		Propranolol (D)
CYP2E1	Liver	Acetoacetate (E)
	Other	Acetol (E)
		Acetaminophen (D)
		Ethanol (D)
		Halothane (D)
		N-nitrosodimethylamine (C)
CYP2F1	Lung	7-Ethoxycoumarin
	Liver	Testosterone (E)
	Other	
CYP3A3	Liver	Aflatoxin B_1 (C)
	Other	Cortisol (E)
		Cyclosporine (D)
		Erythromycin (D)
		Midazolam (D)
		Nifedipine (D)
		Warfarin (D)
		Testosterone (E)
CYP3A4	Liver	Similar to IIIA3
	Other	
CYP3A5	Liver (P)	Cyclosporine (D)
	Other	Testosterone (E)
CYP3A6	Liver (fetal)	Dehydroepiandrosterone-3-sulfate (E)
CYP4A1	Liver	Lauric acid
	Other	Palmitic acid (E)
		Arachidonic acid (E)
CYP4A2	Liver	Similar to IVA1
	Other	
CYP4B1	Lung	Testosterone (E)
	Other	
CYP7A1	Liver	Cholesterol (E)
CYP11A1	Adrenal gland	Cholesterol (E)
	Ovary	
	Testis	
CYP11B1	Adrenal gland	17α-hydroxyprogesterone (E)
	Ovary	Progesterone (E)
	Testis	
CYP17A1	Adrenal	Dehydroepiandrosterone (E)
	Ovary	17α-hydroxyprogesterone (E)
	Testis	17α-hydroxypregnenolone (E)
		Progesterone (E)
CYP19A1	Ovary	Androstanedione (E)
	Placenta	
CYP21A2	Adrenal gland	Progesterone (E)
	Ovary	17α-hydroxyprogesterone (E)
	Testis	
CYP26A1[d]	Liver	Cholesterol (E)

[a] The amino acid sequences of all *P*-450s listed have been determined using human cDNA libraries, except for *ZYP26A1*, which encodes the cholesterol 26-hydroxylase. Because this enzyme is a member of a cascade of enzymes involved in bile acid formation, it is presumed to exist in humans. The *CYP4A1* and *CYP4A2* cDNAs have not been completely sequenced at the time of writing, but the available data suggest that they correspond to the rat *CYP4A1* and *CYP4A2* *P*-450s, which encode fatty-acid hydroxylases.

[b] The tissues listed are known to express *P*-450s; however, not every human tissue has been carefully examined. On the basis of studies in rodents, some *P*-450s are believed to be expressed in other tissues. Some genes are polymorphically expressed (P).

[c] The substrates listed fall into the classes of carcinogens (C), drugs (D), and endogenous compounds (E). The unmarked substrates are chemicals that happen to be substrates but are not drugs or carcinogens. The carcinogenicity of these compounds in rodents varies considerably from the potent aflatoxin B_1 to the weak heterocyclic arylamines. In most cases, the carcinogenic potency in humans is unknown. This list is not inclusive. Several of these *P*-450s are known or presumed to metabolize many other compounds. It should also be noted that a single substrate can be metabolized by multiple *P*-450 forms (e.g., aflatoxin B_1, 7-ethoxycoumarin).

[d] Present in rodents and believed to be present in humans.

is known to exist. These cases are indicated by asterisks. The cDNA for all human *P*-450s listed in the table have been sequenced to deduce their protein sequences. Substrates are also listed; however, as will be discussed below, the *P*-450s within families 1–3 can metabolize many different compounds. Therefore, only partial lists of the best-studied substrates are included.

The most confusing aspect of the nomenclature system is the naming of individual genes. The complexity of this system reflects difficulties in assigning orthologous *P*-450s between species and is particularly evident in families 2 and 3. As will be discussed below, *P*-450s are evolving quite rapidly and in a species-specific manner. That is, *P*-450s within one subfamily in rat are evolving quite differently than those within the same subfamily in humans. This evolutionary diversity is probably a consequence of species-specific adaptation as a result of environmental and dietary exposures.

The *CYP2D* subfamily can be used to illustrate species differences in *P*-450 genes. There are five genes in the rat *CYP2D* subfamily, and these have been designated *CYP2D1* through *CYP2D5*. Two genes and an inactive pseudogene, designated *CYP2D6, CYP2D7,* and *CYP2D8P* (*P* stands for pseudogene), exist in humans. None of the human *CYP2D P*-450s has a high degree of amino acid sequence similarity to any of the five rat *CYP2D P*-450s. Practically, therefore, it is impossible to determine which rat *P*-450 gene is the evolutionary equivalent of the human *CYP2D6* gene. In these cases, *P*-450s have been named as separate entities, even though it is believed that all *CYP2D* genes evolved from the same gene or genes in the predecessor of rats and humans more than 80 million years ago. In some instances, however, the assignment of orthologues is more straightforward. For example, the human *CYP2A3* gene product displays a high level of sequence similarity to one of the three rat *CYP2A* genes. Hence, both rats and humans have *CYP2A3,* although rats also possess two other genes, designated *CYP2A1* and *CYP2A2,* and humans possess *CYP2A4*.

It is important to note that individual *P*-450s are *forms* of *P*-450, not isozymes. These enzymes cannot formally be considered isozymes because, even though they have similar catalytic cycles and all metabolize O_2, they do not necessarily carry out the same reactions on the same substrates.

III. Evolution of *P*-450s

Primary amino acid sequences have been determined from individual *P*-450s found in bacteria, yeast, plants, and animals. More than 150 different *P*-450 sequences representing 28 families were known at the time this chapter was written, and it is presumed that many more sequences will be determined, particularly those of plant *P*-450s. These sequence data can be analyzed by computer programs that align multiple sequences simultaneously, yielding percentage differences between a single *P*-450 and all others. The resultant percentages are then converted into accepted point mutations (PAMs), using an evolution table developed from M. O. Dayhoff's protein sequence atlas. Phylogenetic trees are constructed using these data and methods such as the unweighted pair group method of analysis (UPGMA), in which an arbitrary unit, termed the evolutionary distance, is used to plot divergence times between *P*-450s. The evolutionary distance can be converted to a real time scale using data derived from fossil evidence of divergence times between, for example, birds and mammals or, for more recent divergence times, the mammalian radiation that occurred approximately 75 million years ago. An example of a *P*-450 phylogenetic tree and further details concerning the derivation of such trees can be found within readings in the bibliography.

By examining the *P*-450 phylogenetic tree it becomes apparent which *P*-450s existed before the divergence of eukaryotes and before the mammalian radiation. By using these observations in conjunction with the known catalytic activities of *P*-450s, it is possible to speculate about the evolution of these enzymes.

The early *P*-450s appear to have been more clearly related to the modern *P*-450 enzymes that are known to metabolize cholesterol and fatty acids. These may have been required to maintain membrane integrity in primordial organisms. Two types of eukaryotic *P*-450s emerged, and these differed in their intracellular location and electron source. The mitochondrial *P*-450s, involved in steroid and bile acid biosynthetic pathways, receive electrons from the iron-sulfur protein adrenodoxin via the flavoprotein adrenodoxin reductase. These seem to be the most closely related to the well-studied bacterial *P*-450_{cam}, which metabolizes camphor and also receives electrons from an iron-sulfur protein. The eukaryotic and prokaryotic iron-sulfur proteins are both reduced by NADPH using an evolution-related, flavin-containing enzyme. Therefore, the mitochondrial *P*-450s appear to have evolved from bacterial *P*-450_{cam}. A second bacterial *P*-450, expressed in *Bacillus megaterium, CYP102,* is a *P*-450 hemoprotein that is fused with an electron transport flavoprotein related to the NADPH-*P*-450 oxidoreductase. This bacterial enzyme may be the precursor of the microsomal *P*-450s and the flavoprotein NADPH-*P*-450 oxidoreductase, which transfers electrons directly to the *P*-450 (without the aid of an iron-sulfur protein). Therefore, the mitochondrial and microsomal *P*-450s and their electron carriers appear to have evolved from separate origins.

While all mitochondrial *P*-450s catalyze reaction steps in critical biosynthetic pathways, only four microsomal *P*-450s carry out reactions in steroid biosynthesis. The majority of the microsomal *P*-450s are involved in xenobiotic metabolism. It is in these enzymes that significant species differences and intraspecies polymorphisms are evident (see below).

All of the known mammalian *P*-450 families were in existence 400 million years ago. An explosion in the number of *P*-450 genes, especially those involved in xenobiotic metabolism has occurred within the past 75–100 million years. What caused this large diversification in *P*-450s? Current thinking suggests that these enzymes were needed as a defense against plants, which typically produce chemicals that are highly toxic to animals. The *P*-450s may have evolved to enable land animals to feed on plants. (Aquatic animals also have *P*-450s but, in contrast to terrestrial animals, many lipid-soluble toxic compounds can be eliminated through their gills.) Plants would then have developed new toxins and animals would have evolved or made use of new *P*-450 genes. This animal–plant *warfare* is an excellent example of coevolution and is probably the major reason for the large number of *P*-450s and the diverse substrate specificities of these enzymes.

IV. Steroidogenic *P*-450s

Among the most critical *P*-450s in humans are those involved in steroid biosynthetic pathways. Three endoplasmic reticulum-bound *P*-450s and two mitochondrial *P*-450s participate as key enzymes in the synthesis of steroid hormones, including aldosterone, androstanedione, cortisol, estrogen, and progesterone (Fig. 4). These enzymes are encoded by genes within five separate families and are expressed in various tissues of the body that produce steroids, such as the adrenal gland, which manufactures aldosterone and cortisol, and the gonads, with make sex steroids. Two P450s, CYP7A1 and CYP26A1, are expressed in liver and are involved in the pathway by which cholesterol is converted to bile acids, the major route of cholesterol elimination from the body. [*See* BILE ACIDS; CHOLESTEROL; STEROID HORMONE SYNTHESIS.]

The steroidogenic enzymes are of crucial medical importance. Approximately one in 7,000–15,000 livebirths results in a condition called congenital adrenal hyperplasia, in which the child is deficient in the production of aldosterone, cortisol, and other steroids. A lack of these steroids can cause severe debilitation, including growth and developmental abnormalities and a loss of salt from the body, which can result in death within the first 2 wk of life. In cases involving other, milder conditions or partial deficiencies, symptoms do not occur until later in life. [*See* STEROIDS.]

Congenital adrenal hyperplasia results, in most cases, from mutations in genes encoding the steroid-synthesizing *P*-450s shown in Fig. 4. By far the most common deficiency is a lack of functional *P*-450c21 (*CYP21A2*), commonly called steroid 21-hydroxylase. However, mutations have also been found in genes encoding $P\text{-}450_{scc}$ (*CYP11A1*), *P*-450c17α (*CYP17A1*), and *P*-450c11β (*CYP11B1*), albeit at much lower frequencies than the mutation affecting *P*-450c21 (*CYP21A2*). The symptoms of this deficiency result not only from a lack of steroid hormone production but also from an accumulation of intermediates in the steroid-synthesizing pathways. These intermediates are shunted into other biochemical pathways (e.g., androgen synthesis, resulting in the masculinization of females). The accumulation of intermediates in the steroid pathway results from loss of negative feedback regulation of the pituitary resulting in excess production of the peptide hormone adrenocorticotropic hormone (ACTH). Excess ACTH production in the pituitary is accompanied by an increase in other peptide hormones, such as melanocyte-stimulating hormone.

The genetic defects responsible for congenital adrenal hyperplasia can sometimes be detected using cloned gene probes for *CYP21A2*. More importantly, it is possible to detect prenatally the presence of two copies of a defective gene. The precise diagnosis of all mutant genes is difficult because the genetic lesions responsible for the disruption of the *CYP21A2* gene are quite diverse. For example, mutations in one gene may be caused by a deletion while another gene becomes inactivated as a result of an amino acid substitution.

$P\text{-}450_{arom}$, the key enzyme in estrogen synthesis (Fig. 4), may also of importance in the treatment of breast cancer. Because many breast cancer cells are estrogen-dependent, their growth can be arrested by inhibiting the synthesis of estrogen. This can be accomplished by treatment with agents such as 4-hydroxyandrostein-3,17-dione and 10-propargylestr-4-ene-3,17-one and related drugs that inhibit the activity of $P\text{-}450_{arom}$, resulting in a cessation of estrogen synthesis and a halt in the growth of estrogen-dependent tumors. These agents are also effec-

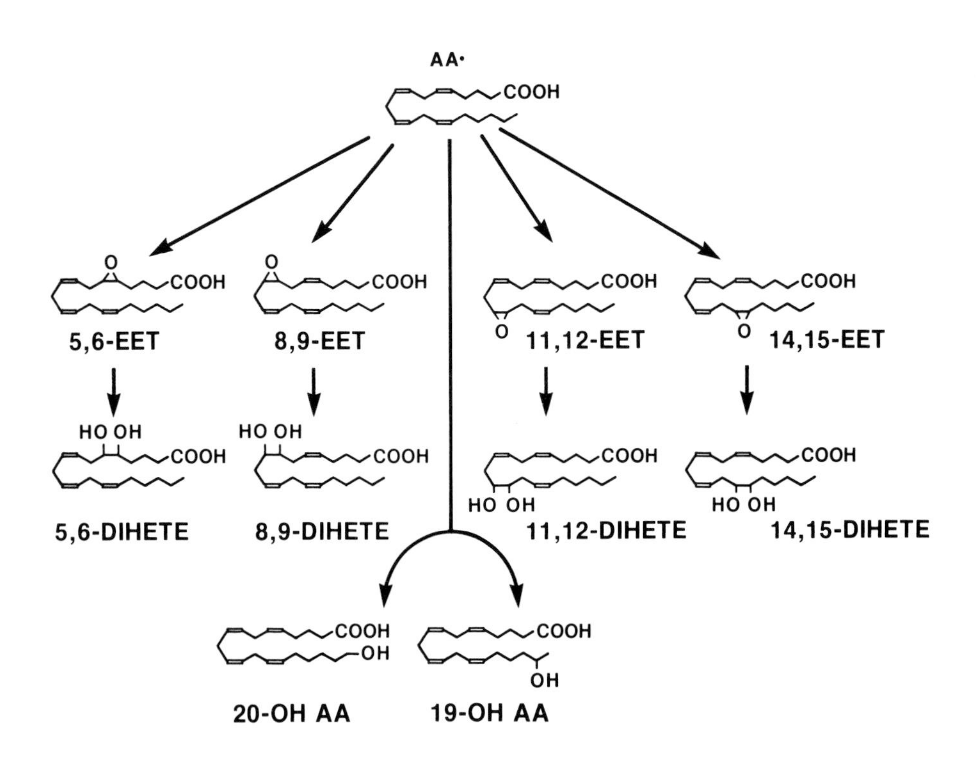

FIGURE 5 Metabolism of arachidonic acid by cytochrome *P*-450s, commonly known as the epoxygenase pathway of arachidonic acid metabolism. AA, arachidonic acid; DIHETE, dihydroxyeicosatrienoic acid; EET, epoxyecosatrienoic acid. [Reproduced, with permission, from F. A. Fitzpatrick and R. C. Murphy Cytochrome *P*-450 metabolism of arachidonic acid: formation and biological actions of epoxygenase-derived eisosanoids. 1988, *Pharmacol. Rev.* **40,** 229–241.]

tive in treatment of androgen-dependent prostratic cancers. [*See* BREAST CANCER BIOLOGY.]

Other *P*-450 enzymes (e.g., *CYP26*, Fig. 4) are of critical importance because of their role in the degradation of cholesterol, which has been shown to contribute to arteriosclerosis and heart disease. The cholesterol 7α-hydroxylase *P*-450 (CYP7) may also be important in lowering cholesterol levels.

A mitochondrial *P*-450 that catalyzes the 1′-hydroxylation of vitamin D_3 is found in the kidney. This enzyme, which is known to exist but has not yet been purified, is critical in the production of active hormone.

Finally, a yeast *P*-450 called CYP51, or lanosterol 14α-demethylase, is involved in the biosynthetic pathway leading to the production of the steroid ergosterol. When production of this enzyme is inhibited, lanosterol and other 14-methylsterol metabolites accumulate in the cells, resulting in growth inhibition. Some potent antifungal agents including diniconazole and ketoconazole, inhibit CYP51.

V. Fatty Acid-Metabolizing *P*-450s

A group of *P*-450s in the *CYP4A* subfamily metabolize fatty acids. These enzymes catalyze the oxidation of fatty acids, including palmitic acid and arachidonic acid. The *P*-450s within the *CYP4A* subfamily are expressed in many tissues, a finding that is consistent with their possible role in fatty-acid metabolism. Other, yet uncharacterized, *P*-450s are involved in the synthesis of arachidonic acid epoxides or epoxyeicosatrienoic acid (EET), many of which are biologically active (Fig. 5). Among the many known biologic activities of these compounds are (1) to promote the release of peptide hormones from brain tissue and insulin from pancreas, (2) to regulate ion transport in a variety of tissues, (3) to stimulate vasodilation, and (4) to stimulate platelet aggregation. An CYP4A or related *P*-450 is also expressed in polymorphonuclear leukocytes, where it converts the potent chemotactant leukotriene B_4 to a hydroxylated derivative that is biologically inactive.

VI. Xenobiotic-Metabolizing *P*-450s

A. General Characteristics

The majority of *P*-450s are involved in the metabolism of foreign compound's (xenobiotics) such as drugs, environmental pollutants, and plant metabolites. These enzymes appear to exist solely for this purpose. If an individual did not ingest drugs, plants containing toxic chemicals, or other foreign chemicals, he or she could probably survive without xenobiotic-metabolizing *P*-450s.

The xenobiotic-metabolizing *P*-450s are found in families 1, 2, and 3. In humans, more than 30 *P*-450s may be expressed in family 2 and perhaps six in family 3. In general, the oxidation of drugs by these enzymes results in the abolition of the compounds therapeutic activity, although in some instances a drug may be activated to a therapeutically active metabolite. Oxidation of lipid-soluble drugs by *P*-450s can render them more water-soluble, aiding in their elimination through the kidney by reducing reabsorption. Without *P*-450s, some drugs would be retained by the body for long periods.

The xenobiotic-metabolizing *P*-450s are noteworthy for their capacity to metabolize numerous structurally diverse compounds. A single *P*-450 can oxidize a large number of drugs. CYP2D6, for example, can metabolize compounds with a wide range of structures and clinical uses (Fig. 6), including β-blocking agents, tricyclic antidepressants, cardiovascular drugs, and other miscellaneous drugs, including dextromethorphan, the active component of over-the-counter cough medicine. This enzyme is also capable of converting the inactive analgesic codeine to its active derivative, morphine (Fig. 6).

P-450s also have overlapping substrate specificities. A single compound can be metabolized at the same position by multiple *P*-450s. The affinity of different *P*-450 forms can vary significantly, such that one form may be active against a compound at low concentrations, whereas another *P*-450 may require high substrate concentrations. When a drug is first ingested, many *P*-450s may contribute to its metabolism, but later, when the drug is present at lower concentrations, only a single *P*-450 may be able to inactivate the substance.

A single compound or drug can also be metabolized at different positions by one or more *P*-450s, and this is particularly true of large compounds with complex structures. For example, note the positions of hydroxylation of the cardiovascular drugs guanoxan and encainide, shown in Fig. 6. In these cases, the *P*-450 can bind and hydroxylate these compounds at different positions. Frequently, oxidation at one site inhibits substrate binding and oxidation at another site on the same molecule. It is also possible that a single compound can be metabolized at different sites by different *P*-450 forms.

Finally, it is fairly common to find that a single substrate or drug is metabolized primarily by a single form of *P*-450. Therefore, when this *P*-450 is not expressed, the drug cannot be metabolized. This condition can lead to a drug oxidation polymorphism.

B. *P*-450 Polymorphisms

Genetic polymorphisms in *P*-450-mediated reactions occur in humans. These have been identified clinically as drug oxidation polymorphisms. For example, the debrisoquine polymorphism affects about 10% of European and North American Caucasians, who cannot metabolize the cardiovascular drug debrisoquine and numerous other compounds that are specific substrates for a single *P*-450 form, CYP2D6. However, such clinical polymorphisms do not affect the metabolism of all drugs that act as substrates for CYP2D6 (Fig. 6), possibly because other *P*-450s can also metabolize these compounds when CYP2D6 is not expressed.

The debrisoquine polymorphism results from mutations in the structural gene encoding CYP2D6, the *CYP2D6* gene. Several mutant alleles of the *CYP2D6* gene exist that produce CYP2D6 primary RNA transcripts that are incorrectly spliced, and consequently *P*-450 protein synthesis does not occur. Other mutations have also been detected, including these insertions that disrupt the coding region of the mRNA and amino acid changes that result in an enzymatically inactive CYP2D6 protein.

The clinical consequences of *P*-450 polymorphisms can be quite severe. When debrisoquine was used in Europe in the early 1970s many patients suffered exaggerated side effects, including dramatic decreases in blood pressure, because they were unable to metabolize and inactivate the drug. This variability in clinical response led to the discontinuation of debrisoquine use. Fortunately, other drugs could be substituted for debrisoquine. The metabolism of some other drugs (e.g., bufuralol) has also been found to be affected by the polymorphism in CYP2D6 and, as a consequence, these agents have not been widely used.

Several other *P*-450s are also known to be polymorphically expressed in humans but do not complicate the clinical use of drugs, in part, because many drugs can also be metabolized by other

β-ADRENERGIC BLOCKING AGENTS

Bufuralol **Metoprolol** **Propanolol** **Timolol**

TRICYCLIC ANTIDEPRESSANTS

Amitriptyline **Nortriptyline** **Desmethylimipramine**

CARDIOVASCULAR DRUGS

Debrisoquine **Sparteine** **Guanoxan** **Propafenone**

Encainide **Perhexiline** **N-Propylajmaline**

MISCELLANEOUS

Dextrometorphan **Methoxyamphetamine** **Phenformin**

Codeine **Morphine**

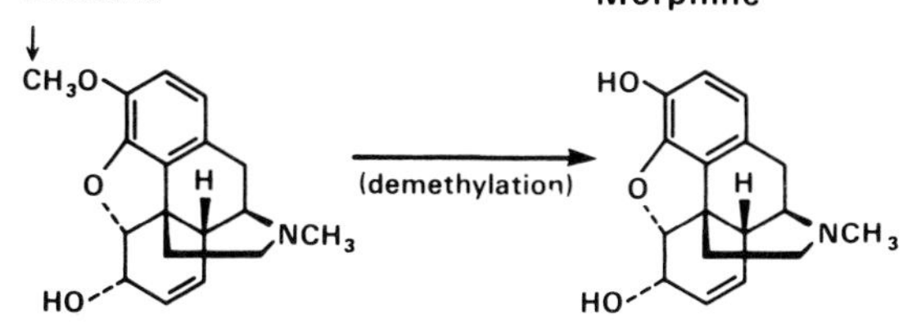

FIGURE 6 Substrates for CYP2D6. The hydroxylation positions of each substrate are indicated by arrows. The conversion of the analgesic codeine to its active derivative morphine is shown at the bottom of the figure. The author thanks Dr. Urs A. Meyer for use of this figure.

P-450s. It should also be recognized that some potential drugs that are subjected to polymorphic metabolism in humans are never taken beyond the development or preclinical testing stage. It is possible, therefore, that many potentially useful compounds are eliminated because of *P*-450 polymorphisms. Uncertainties regarding drug metabolism will decrease considerably with the development of tests to detect mutant *P*-450 genes and, hence, identify individuals who are incapable of metabolizing a certain drug. *P*-450 gene testing could be undertaken using DNAs derived from small blood samples. These tests will allow physicians to tailor drug prescriptions to an individual's *P*-450 phenotype.

Certain rare toxicities are sometimes associated with the use of drugs. For example, acetaminophen hepatotoxicity occurs very rarely and is usually fatal. This common analgesic compound is known to be metabolized by a *P*-450, and it is therefore possible that a *P*-450 defect is responsible for its toxicity. Such a rare defect would not be considered a polymorphism but might be a spontaneous germ line mutation in a gene encoding a specific *P*-450 form. Alternatively, a regulatory defect in a gene may result in increased levels of a certain *P*-450 that produces toxic metabolites.

The fact that *P*-450s metabolize and activate carcinogens suggests that they may be involved in the susceptibility or resistance to human cancer. For example, it is well established that cigarette smoking produces a 10-fold increase in the relative risk for lung cancer. On the average, about one in five heavy smokers develop lung cancer. Because the primary cancer-causing agents in tobacco smoke are chemical carcinogens that must be activated by *P*-450s, it is possible that susceptibility to lung cancer may be associated with the presence in some individuals of a *P*-450 that activates a chemical carcinogen to a DNA-damaging and mutagenic metabolite. Conversely, a *P*-450 that is polymorphically expressed may be able to convert a chemical carcinogen to an inactive metabolite. The role of these enzymes as risk factors in susceptibility to cancers of the lung and other organs is still unknown and is under investigation. [*See* TOBACCO SMOKING, IMPACT ON HEALTH.]

Each year the drug and chemical industries produce tens of thousands of new chemicals during the development of drugs, insecticides, plastics, polymers, and other products. Many of these are tested for toxicity and carcinogenicity in animal-based systems. For example, rats and mice will be given a specific chemical and observed for toxic reactions. If a chemical harms a rodent, it is usually regarded as harmful to humans. However, as discussed earlier, rodents can have *P*-450s and enzyme activities that are distinctly different from those of humans. In these cases, the development of potentially useful chemicals will be discontinued even before it has been determined that they are actually harmful to humans. Today, cell culture systems are being developed in which *P*-450s derived from cloned human cDNAs can be used to test chemicals. It is also possible to produce transgenic animals that have human *P*-450 genes. [*See* CARCINOGENIC CHEMICALS.]

C. *P*-450 Regulation

P-450s are regulated by a number of xenobiotics. In many cases, the cellular content of a *P*-450 is elevated by the same chemical that it metabolizes. The induction of *P*-450s by certain chemicals is known to proceed via a receptor-mediated mechanism. This type of regulation is common in biologic systems involving hormones. The chemical or ligand enters the cell and binds to a receptor protein. This binding confers on the receptor an ability to specifically interact with regulatory regions of the *P*-450 gene, resulting in the activation of gene transcription and a subsequent increase in mRNA and protein levels. *P*-450 then rapidly degrades the chemical.

The induction of other *P*-450s is governed by a substrate-induced stabilization of *P*-450 levels. In the presence of substrate, *P*-450 levels are stabilized, or the rate of *P*-450 degradation is decreased. The cellular level of enzyme is then increased until the chemical or substrate is metabolized and its concentration lowered. In some cases, the mRNA encoding a *P*-450 becomes stabilized against degradation, resulting in an increase in mRNA level and a subsequent increase in *P*-450 synthesis.

Induction of *P*-450s has important implications for clinically significant drug interactions. For example, oral contraceptives have been found to fail in patients taking the antibiotic rifampicin. This antibiotic is known to induced *P*-450s that can metabolize 17α-ethynylestradiol, the principal component of birth control pills. The synthetic steroid dexamethasone and the anticonvulsant agent phenobarbital can also induce *P*-450s and impairs usage of other drugs.

Bibliography

Fitzpatrick, F. A., and Murphy, R. C. (1988). Cytochrome *P*-450 metabolism of arachidonic acid: Formation and biological actions of epoxygenase-derived eicosanoids. *Pharmacol. Rev.* **40,** 229.

Gonzalez, F. J. (1988). The molecular biology of cytochrome *P*-450s. *Pharmacol. Rev.* **40,** 243.

Gonzalez, F. J., Nebert, D. W. (1990). Evolution of the P450 gene superfamily: animal-plant "warfare," molecular drive and human genetic differences in drug oxidation. *Trends Genet.* **6,** 182.

Gonzalez, F. J., Skoda, R. C., Kimura, S., Umeno, M., Zanger, U. M., Nebert D. W., and Gelboin, H. V., Hardwick, J. P., and Meyer, U. A. (1988). Characterization of the common genetic defect in humans deficient in debrisoquine metabolism. *Nature* **331,** 442.

Guengerich, F. P. (ed.) (1987). "Mammalian Cytochromes *P*-450," Vols. I and II. CRC Press, Boca Raton, Florida.

Miller, W. L. (1988). Molecular biology of steroid hormone synthesis. *Endocr. Rev.* **9,** 295.

Nebert, D. W., Nelson, D. R., Coon, M. J., Estabrook, R. W., Feyereisen, R., Fujii-Kuriyama, Y., Gonzalez, F. J., Guengerich, F. P., Gunsalus, I. C., Johnson, E. F., Loper, J. C. Sato, R., Waterman, M. R., Waxman, D. J. (1991). The P450 superfamily: update on new sequences, gene mapping and recommended nomenclature. *DNA Cell Biol.* **10,** 1.

Nelson, P. R., and Strobel, H. W. (1987). Evolution of *P*-450 proteins. *Mol. Biol. Evol.* **4,** 572.

Ortiz de Montellano, P. R. (1989). Cytochrome P450 catalysis: Radical intermediates and dehydrogenation reactions. *Trends Pharmacol. Sci.* **10,** 354.

Schuster, I. (ed.) (1989). "Cytochrome *P*-450: Biochemistry and Biophysics." Taylor & Francis, New York.

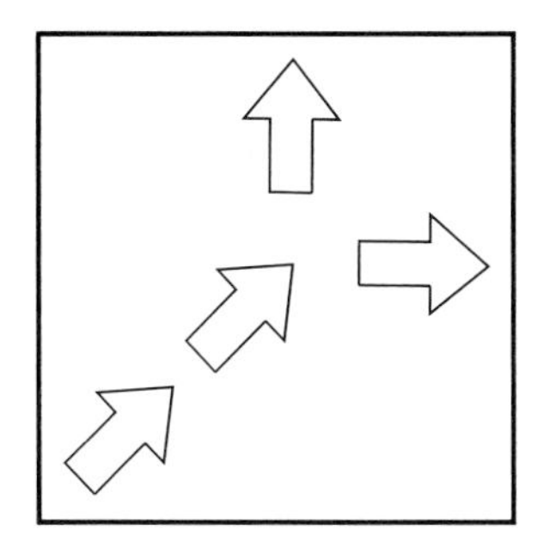

Cytokines in the Immune Response

LYNN A. BRISTOL, JOOST J. OPPENHEIM, SCOTT K. DURUM,
National Cancer Institute, Frederick Cancer Research Facility

Glossary

Antigen Foreign substance with which a component of the immune system reacts

B lymphocyte Bone marrow-derived cells in mammals that are distinguished by their surface expression of immunoglobulins and that produce antibodies after exposure to antigens

Cytokines Hormonelike peptides (e.g., interleukins, interferons, tumor necrosis factors, and transforming growth factor β) that are released in the course of immunological and inflammatory reactions by both immune and nonimmune cells

T lymphocyte Bone marrow-derived, thymus-dependent cells that mediate delayed-type hypersensitivity, graft rejection, and cellular cytotoxicity

CYTOKINES ARE SOLUBLE MEDIATORS produced by both immune and nonimmune cells of the host in response to foreign or injurious agents. Cytokines can modulate the specific immune response by regulating (1) immature B- and T-lymphocyte differentiation and maturation and (2) immune B- and T-lymphocyte activation. Consequently, cytokines can also upregulate the effector functions of these cells such as lymphokine production, cytotoxicity, and antibody production. Also, a number of cytokines are not growth factors for lymphoid cells but do exhibit antiproliferative activities. Thus, B- and T-lymphocyte cell-cycle progression is influenced by the stimulatory and inhibitory actions of various cytokines.

I. Introduction

A consequence of inflammatory and immunological reactions to foreign agents or antigens is the production of highly active polypeptides called cytokines. Virtually all normal cell types, including T and B lymphocytes and monocytes or macrophages, upon stimulation produce cytokines. Only some transformed neoplastic cell lines spontaneously secrete cytokines. Cytokines function as intercellular signals that regulate local (such as chemotactic factors) and systemic (such as on the fever center of the brain) inflammatory responses of the host to foreign antigens. Cytokines also regulate the growth and differentiation of immature lymphoid cells (T and B lymphocytes). The proliferative clonal expansion of more mature immunocompetent populations of T and B lymphocytes is also regulated by cytokines. In addition, cytokines contribute to the binding of antigen to distinct antigen-specific receptors on the surface of T and B lymphocytes. Thus, cytokines participate not only in the inflammatory "efferent limb" of the immunological and nonimmunological host responses, but also are vital participants in the initiation of the specific immune responses, the so-called "afferent limb" of immunity. [*See* ANTIBODY-ANTIGEN COMPLEXES: BIOLOGICAL CONSEQUENCES; INFLAMMATION; LYMPHOCYTES; MACROPHAGES.]

Many of the cytokines have overlapping biological functions. In the case of interleukin (IL)-1, tumor necrosis factor (TNF)-α, and IL-6, for example, each can trigger the fever response, the acute phase response of the liver, and can also promote immune T-cell activation. Despite overlapping actions however, each cytokine interacts with a target cell by selectively binding to a distinct complementary high-affinity receptor.

Cytokines are a heterogeneous group of peptides with molecular weights ranging from 5 to 30 kDa. They include the interleukins (ILs), the interferons (IFNs), and a number of other soluble peptide mediators. While many of the systemic consequences of cytokines have been recognized for many years (e.g., fever), in the 1980s dramatic progress has been made in cytokine research with the isolation, identification, and cloning of a number of these agents. Cloning makes it possible to acquire a pure cytokine in sufficient quantities to permit detailed studies of its structure, synthesis, and biological action. [*See* INTERFERONS.]

The following discussion will first outline how cytokines induce T- and B-lymphocyte differentiation and maturation and promote immune T- and B-lymphocyte activation. Then, the immunological properties of the cytokines that are involved in the afferent limb of the immune response will be addressed.

II. Effects of Cytokines on T Lymphocytes: Overview

A. T-Cell Development

T-cell precursors originate from hematopoietic stem cells primarily in bone marrow. How the numbers of pre-T cells produced from stem cells are regulated at this stage is unknown, and this could involve some as yet unidentified cytokine. T-cell precursors migrate to the thymic cortex, and a complex process of growth, differentiation, and selection occurs. A number of cytokines have been implicated in these intrathymic processes, primarily based on *in vitro* experiments (Fig. 1).

The earliest thymic stem cell lacks both surface markers CD4 (helper T cells) and CD8 (cytotoxic–supressor T cells), and thus is termed double negative. Many cytokines have been shown to promote growth of double-negative thymocytes *in vitro*, including IL-1, IL-2, IL-3, IL-4, IL-5, IL-6, IL-7, and granulocyte macrophage colony-stimulating factor GM-CSF). These cytokines probably also play a role *in vivo*, because they are produced by thymic stroma, macrophages, dendritic cells, and other thymic T cells. At the double-negative stage, thymocytes rearrange the genes that encode the various subunits of the T-cell receptor. Four genes can undergo such rearrangment: alpha, beta, gamma, and delta. T cells can be classified into two subsets on the basis of those genes: those that express subunits for the alpha and beta receptor and those expressing the gamma and delta receptor. Evidence from *in vitro* studies suggest that IL-1 serves as a growth and differentiation factor for alpha and beta cells, whereas IL-2 induces clonal expansion of gamma and delta cells. [*See* CD8 AND CD4: STRUCTURE, FUNCTION, AND MOLECULAR BIOLOGY; T-CELL RECEPTORS.]

T cells enter a stage of development termed double positive, during which both CD4 and CD8 are expressed on the membrane. Apparently, this is the stage at which "tolerance" develops: T cells that are reactive against self-antigens are triggered by those antigens, and the cell self-destructs by an autolytic process termed apoptosis. T cells that survive the double-positive stage may later still be positively selected by a low-affinity reaction to self major histocompatibility antigens. Whether or not cytokines participate in any of the events at the double-positive or single-positive stage is still unclear.

In the next stage of T-cell development, T cells lose either the CD4 or CD8 markers, becoming single positive, and enter the thymic medulla. Such CD4- or CD8-positive T cells are mature thymic T cells. Again, whether or not cytokines participate in their maturation, migration, or survival is not known. Mature T cells finally migrate from the thymus and maintain a continuous recirculation pattern from the lymphoid organs, through the lymphatics, into the blood, and back to the lymphoid organs until they encounter their specific antigen. Clearly, from the marked lymphoid hypoplasia seen in germfree animals, exogenous antigenic stimulants are responsible for the cytokine-mediated generation of peripheral lymphoid tissues. Although to date no homeostatic regulators of lymphopoiesis have been identified, cytokines produced during development may be responsible for lymphopoiesis.

B. T-Cell Activation by Antigen

Mature T cells are activated by antigen presented on the surface of antigen-presenting cells (APCs). In the case of helper T lymphocytes responding to peptide antigens, the types of cells that can serve as

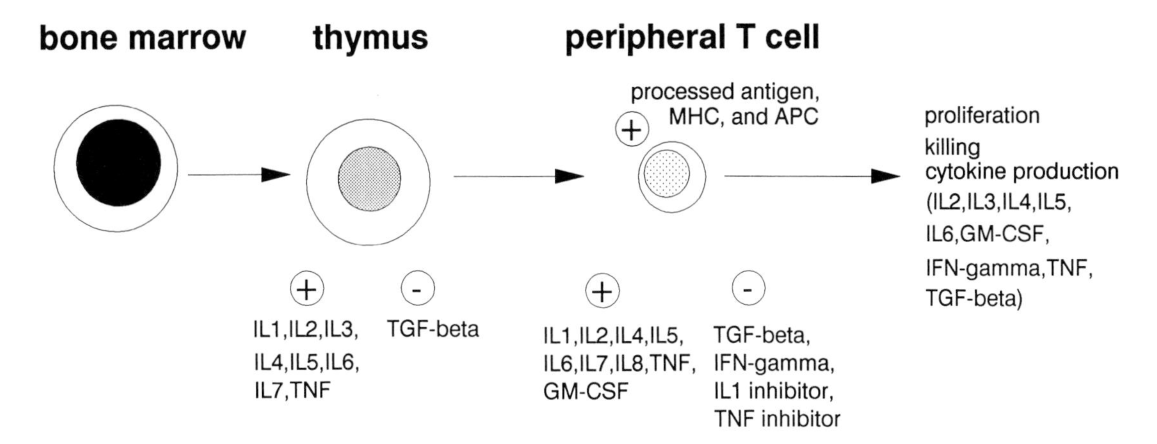

FIGURE 1 Cytokines and the T lymphocyte.

APCs include macrophages, endothelial cells, dendritic cells, and B lymphocytes. The APC performs several functions in the T-cell activation process. Peptide antigens are taken up by the APC, proteolytically fragmented, and then re-expressed on the outer cell membrane of the APC as a complex with type II histocompatibility antigen (HLA-DR in humans and Ia in mice). If a T cell bearing an antigen receptor recognizes an antigen-Ia complex on a neighboring APC, the receptor delivers a signal to partially activate the T-lymphocyte (other signals come from cytokines; see below). The binding of the T cell to the APC is stabilized further by several different adhesion molecules on the T cell including LFA-1, CD2, and CD4, which bind to their corresponding ligands on the accessory cell, ICAM-1, CD3, and Ia, respectively.

The APC also secretes cytokines that further activate the T cell (Fig. 1). In fact, the APC can be entirely replaced with cytokines if the T cell receives a strong artificial stimulus through the T-cell antigen receptor (such as antibodies or lectins that cross-link the receptors). The APC-derived cytokines, including IL-1, TNF-α, and IL-6 are released from the APC in response to exogenous or endogenous stimulants. The APC-derived cytokines act as cofactors, together with the signals generated from antigen binding to the T-cell receptor, and the T cell is driven from G_0 into the G_1 stage of the cell cycle. These cytokines also induce the production of the T-cell growth factors IL-2 and IL-4, which are synthesized during the G_1 phase of the cell cycle. IL-7 can also act as a T-cell growth factor *in vitro*, but it is not yet clear whether IL-7 actually participates in immune responses *in vivo*.

In addition to APC, other cell types (such as keratinocytes and fibroblasts) can cooperate in the activation of T cells through their production of cytokines such as IL-1, IL-6, and TNFα. Although such cells may be ineffective in presenting antigens, they can potentially act in concert with APCs to promote T-cell-dependent immune responses by secreting cytokines.

There is considerable overlap in the actions of IL-1, IL-6, TNF, and IL-7 for the promotion of T-cell activation. There can also be strong synergy of cytokines involved in T-cell activation: IL-1 with IL-6, IL-4 with IL-6, and IFN-γ with TNF.

Inhibitory cytokines may also play an important role in the negative feedback of immune response pathways. IFN-α, -β, -γ, and TGFβ exert powerful antiproliferative effects on some T cells. Also, specific endogenous inhibitors for some cytokines (IL-1 and TNF) have been identified that competitively inhibit the respective cytokine. Interestingly, both growth-promoting and growth-inhibitory cytokines can be simultaneously produced by some accessory cells in response to a given stimulus.

Cytokines can indirectly affect T-cell activation and proliferation through the action of the cytokine on nonlymphocytic cells. IFN-γ can augment the expression of Ia on many cell types, including macrophages and endothelial cells, thus enhancing the ability of such cells to present antigen to T cells. GM-CSF can also stimulate macrophages to express more Ia, and IL-4 can similarly stimulate B cells. IL-1 and TNF can promote the expression of adhesion proteins such as ICAM-1 on APC and LFA-1 on T cells, thus promoting the cell–cell interaction, which can lead to T-cell activation.

Another indirect effect of cytokines on T cells is fever. IL-1, TNF, IL-6, or IFN-α can stimulate an increase in temperature, which can result in augmentation of the T-cell activation process. Some cytokines can induce prostaglandin production (in many cell types) and/or glucocorticoid production by the adrenal medulla, both of which are inhibitors of T-cell proliferation.

III. Effects of Cytokines on B Lymphocytes: Overview

A. B-Cell Development

B lymphocytes develop from uncommitted hematopoietic progenitor cells located primarily in bone marrow. The cytokines that regulate the differentiation of hematopoietic stem cells to B-lymphoid stem cells have not been identified; however, several stages in precursor B-cell development are controlled by cytokines (Fig. 2). Immunoglobulin (Ig) gene rearrangement characterizes maturation of B cells. The first identifiable cell of the B lineage, the pro-B cell, is without Ig gene rearrangement. IL-7 stimulates the proliferation of pro-B cells without inducing rearrangement of the Ig genes or other evidence of differentiation. The next stage in B-cell development is the pre-B cell, which has gene rearrangements for both heavy and light Ig chains. These cells express the message for the heavy chain and the corresponding heavy-chain protein, both of which are found only in the cytoplasm. IL-1 can stimulate pre-B cells to synthesize light-chain mRNA and protein. When both Ig genes are expressed, heavy and light chains can be assembled into complete Ig molecules, followed by their expression as antibodies on the cell surface. Proliferation of pre-B cell clones is also stimulated by IL-7. While IL-1 and IL-7 have biological effects on cells of B-cell lineage, the list of cytokines involved in the differentiation and development of precursor B cells is probably incomplete.

B. B-Cell Activation by Antigen

Mature B cells migrate to lymphoid tissue, and homing may be influenced by cytokines. In the peripheral lymphatic system, each clone will survive only a few days unless it is activated by specific antigen interaction coupled with lymphokine signals from T cells. The B-cell activation process consists of several stages including (1) entry from G_0 into G_1 of the cell cycle, (2) clonal expansion, (3) secretion of IgM, (4) switching to other Ig isotypes, (5) terminal differentiation and death, or (6) survival as memory cells. Several stages in B-cell activation are controlled by cytokines (Fig. 2).

Entry into G_1 of the cell cycle is driven by a combination of signals and takes approximately 24 hr. Specific antigen binds to Ig receptors on the B-cell surface. If sufficient numbers of B-cell membrane Ig receptors are engaged and cross-linked, antigens can deliver a partial activation signal. The antigen is then internalized, processed, and re-expressed on the membrane in association with Ia. T cells bind to this antigen presenting B cell, recognize the antigen-Ia complex, and deliver several types of stimulatory signals to the B cell. Some of these signals from the T cell are cytokines, but others seem to require cell contact between T and B cells. During cell contact, cytokines may also be locally released by T cells into the tight synapse formed between the T- and B-cell membranes. The best understood T-cell signal is IL-4, a cytokine that is released from T cells either diffusely or into the synapse. IL-4 acts on B cells in G_0 and in association with T-cell contact or extensive membrane Ig cross-linking and drives the B cell into the G_1 phase of the cell cycle. IL-4 also induces (1) increased expression of Ia on B cells and macrophages, (2) receptors for IgE on B cells, and (3) immunoglobulin class switching in B cells, as will be discussed.

Clonal expansion of B cells is induced by various growth stimuli such as IL-5. Further exposure to IL-2 or IL-4, or even to further Ig cross-linking, can also drive replication.

Differentiation of B cells to secrete IgM is promoted by IL-6, but other cytokines can also induce this secretory phase, including IL-2, IL-4, IL-6, and IFN-γ.

Switching from IgM to other classes of Ig involves the gene rearrangement of a VDJ cluster from C_μ to another downstream C gene. IL-4 promotes switching to $C_{\gamma 1}$, C_ε and $C\alpha$ and impedes switching to $C_{\gamma 2b}$ and $C_{\gamma 3}$. IL-4 does not directly induce the switch, but apparently it prepares B cells to subsequently switch to these C genes. This pre-switch action of IL-4 may involve unmasking the switch regions on these C genes as well as increasing their transcriptional activity. In contrast, IFN-γ promotes switching to $C_{\gamma 2a}$ and inhibits the IL-4 induced switch to $C_{\gamma 1}$ and C_ε, as well as other IL-4-induced effects on B cells. IL-5 is reported to promote switching to C_α gene expression.

The mature B-cell response to antigenic stimulation can also be downregulated by cytokines such as

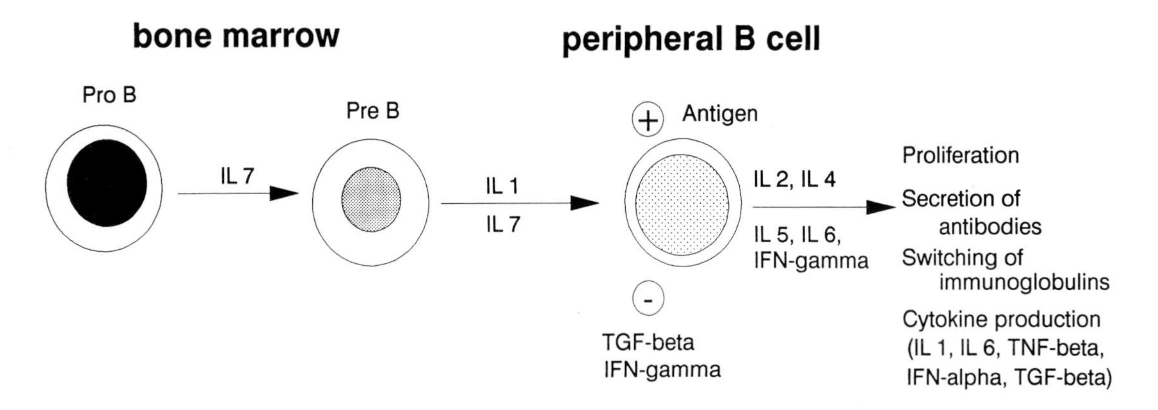

FIGURE 2 Cytokines and the B lymphocyte.

IFN-γ and TGFβ. Other as yet unidentified growth-promoting or -prohibiting factors probably also function in the B-lymphocyte cell cycle. [*See* B-CELL ACTIVATION, IMMUNOLOGY.]

IV. Immunologic Properties of the Cytokines

In the following sections each cytokine will be discussed to provide further detail concerning their structure and actions in the immune system. The characteristics of the relevant cytokines are summarized in Tables I and II.

A. Interleukin 1 (IL-1)

As early as 1972, it was known that a soluble mediator was released by adherent peripheral blood cells. This mediator, termed lymphocyte-activating factor, was mitogenic for murine thymocytes and could enhance the proliferative response of murine thymocytes to lectins. This factor, later renamed interleukin 1, was shown to be identical to endogenous pyrogen, also derived from macrophages. Although some cells constitutively express IL-1, most normal cells must be stimulated with insoluble materials (urate or silica crystals, or microbial organisms) or soluble stimuli (LPS, MDP, TNF, CSF, IFN-γ, or T cell-derived IL-1-inducing factor) to induce the synthesis of IL-1. A variety of cells are capable of IL-1 expression, but macrophages produce by far the most IL-1. Two distinct types of IL-1 molecules exist (IL-1α and IL-1β), yet they resemble each other in their potency, biologic activities, and binding affinity to the IL-1 receptor.

Extensive work has been done on the characterization of the role of IL-1 in T-cell activation and replication. Most evidence suggests that IL-1 is a costimulus with antigens or mitogens, and that by inducing production of IL-2 and receptors for IL-2, the T cells will ultimately proliferate. From studies of T-cell clones, it is clear that one subtype of T helper cells (termed "Th_2") responds vigorously to IL-1 together with antigen whereas Th_1 is virtually unresponsive. Some reports contradict the idea that IL-1 is an important costimulus based on the inability of anti-IL-1 antibodies to inhibit proliferation of antigen- or mitogen-stimulated spleens and lymph nodes. This failure may, however, be due to the presence of other cytokines (IL-4, IL-6, and TNFα), which can be produced in culture under these same stimulatory conditions, and may substitute for IL-1 in coactivating T cells. Several conflicting reports have also emerged concerning the ability of IL-1 (and mitogen or anti-T-cell receptor antibody) to promote the proliferation of T cells depleted of accessory or adherent antigen-presenting T cells. These results suggest that surface-receptor interaction between T cells and accessory cells in addition to IL-1 may be required for triggering of T-cell activation.

B. Interleukin 2 (IL-2)

T-cell growth factor, later renamed interleukin 2, was initially discovered as a signal for T cells to proliferate as well as for the sustained growth of T cells in culture.

IL-2 is produced after activation of either mature helper T cells or a medullary subpopulation of thymocytes, and to a lesser degree by the suppressor–cytotoxic subset of T cells and large granular lymphocyte clones (LGL). Activation of T cells by APC or T-cell-receptor cross-linking agents along with cytokines (IL-1, IL-6, or TNF) results in *de*

TABLE I Properties of Interleukins

	Biochemistry		Cell source	Activity
IL-1	α	β	Macrophages, keratinocytes, endothelial cells, fibroblasts, astrocytes, T and B lymphocytes	Enhances antibody production Promotes B-cell proliferation Enhances IL-2 receptor expression Induces lymphokine production T-cell comitogen
Proform	31 kDa	31 kDa		
Mature	17 kDa	17 kDa		
IL-2	15 kDa		Peripheral T cells, thymocytes, large granular lymphocytes	Induces proliferation of activated T-cell subsets, B cells, and large granular lymphocytes Stimulates lymphokine secretion Augments lymphokine production
IL-3	14–28 kDa		T cells	Hematopoietic differentiation
IL-4	20 kDa		T helper cells, mast cells	Comitogenic for B cells Enhanced Ia and FcR_E expression in B cells Promotes switching to IgG, IgE, and IgA Comitogenic for thymocytes and mature T cells

novo synthesis of IL-2 followed by secretion. T-cell division requires IL-2 binding to receptors, which are also synthesized *de novo* following activation. The growth of both cytotoxic T cells and T helper cells is supported by IL-2.

In addition to growth, other reported effects of IL-2 on activated T cells *in vitro* include the production of IFN-γ, IL-3, -4, -5, and -6, and TGFβ. IL-2 also enhances the cytotoxicity of T cells. IL-2 can augment natural killer (NK) cell responses and T-cell alloantigen responses *in vivo*. Administration of IL-2 *in vivo* has also been shown to increase serum levels of ACTH and cortisol, which lead to immunosuppressive effects, suggesting a possible mechanism for downregulation of the initial IL-2 response.

B lymphocytes are also stimulated by IL-2. Using B cells that are antigen-stimulated to induce the expression of IL-2 receptors, IL-2 can induce B cells to increase antibody production, to proliferate, and to switch isotypes. IL-4 can inhibit IL-2-enhanced specific IgM antibody switching in antigen-primed cultures of human B cells.

In addition to being produced by LGL, IL-2 promotes the proliferation of, the production of other cytokines by, and enhances the NK activity of LGL. In fact, the chemotherapeutic lymphokine activated killer (LAK) cells which are generated in vitro from circulating lymphocytes treated 3-5 days with IL-2, consist predominantly of natural killer (NK) cells. [*See* INTERLEUKIN-2 AND THE IL-2 RECEPTOR.]

C. Interleukin 3 (IL-3)

Activated T cells and mast cell lines have been the only cells reported to secrete IL-3. IL-3 stimulates hematopoiesis but does not appear to act on lymphopoiesis *in vivo*. IL-3 does, however, show modest costimulatory activity for proliferation of immature human thymocytes *in vitro*. Other studies on murine cells have shown that IL-3 induces expression of the enzyme 20-α-hydroxysteroid dehydrogenase in cultures of splenic lymphocytes, and promotes the proliferation of NK lymphocytes in culture.

D. Interleukin 4 (IL-4)

IL-4 was originally called B-cell stimulatory factor based on its ability to induce proliferation of anti-IgM-activated B cells. IL-4 can, however, also promote the differentiation and growth of mature T cells, certain thymocyte populations, and monomyelocytic and mast cells. Cellular sources of IL-4

TABLE II Properties of Tumor Necrosis Factors, Interferons, and Transforming Growth Factors

	Biochemistry			Cell source		Activity
		α	β	α		
TNFα, -β						
Proform		25 kDa	25 kDa	Macrophages, keratinocytes, fibroblasts, astrocytes T and B lymphocytes, endothelial cells β T lymphocytes (T helper 1)		Enhances antibody production Promotes B-cell proliferation Enhances IL-2 receptor expression Induces lymphokine production Thymocyte comitogen
Mature form		17 kDa	17 kDa			
	α	β	γ	α, β	γ	
IFNα, -β, -γ	18–20 kDa	23 kDa	20–25 kDa	Leukocytes, fibroblasts	T lymphocytes, large granular lymphocytes	Enhancement of CTL generation Inhibition of T helper 2 subset Promotes proliferation of B cells and controls immunoglobulin switching Enhancement of CTL generation Inhibition of T helper 2 subset Promotes proliferation of B cells and controls immunoglobulin switching
IL-5	18 kDa			T helper cells		Costimulant of B-cell proliferation and differentiation Enhancement of IL-2 receptor expression on B cells Promotes isotype switching to IgA in B cells Comitogen for thymocytes and mature T lymphocytes
IL-6	22–30 kDa			T and B lymphocytes, monocytes, endothelial cells, epithelial cells, fibroblasts		Cofactor for Ig secretion by B cells Comitogen for thymocytes and T cells
IL-7	25 kDa			Bone marrow stromal cells, thymus		Proliferation of T and B precursor cells
IL-8	8 kDa			Macrophages–monocytes, T lymphocytes, fibroblasts, endothelial cells		T lymphocytes: chemotaxis, lymph node homing, peripheral infiltration
TGFβ	β_1	β_2	β_3			
Proform	55 kDa	55 kDa	55 kDa	Kidney, bone, T and B lymphocytes		Inhibits proliferation of T and B lymphocytes, early hematopoietic stem cells, MLR, and generation of CTL
Mature	25 kDa	25 kDa	25 kDa			

include peripheral T cells, murine Th_2 clones, and mast cells.

Mature (CD3+) thymocyte populations exhibit greater responsiveness to IL-4 than CD3-depleted populations (see Section II. A). Phenotypic analysis of unfractionated thymocytes stimulated with IL-4, PHA, and phorbol esters show increased numbers of single-positive and double-negative cells, with concomitant decreases in double-positive cells. Under the same stimulatory conditions, fractionated immature (CD3−) thymocytes generated greater numbers of double-negative cells. Thus, the ability of IL-4 to costimulate immature thymocyte proliferation suggests that it may play a role in T-cell ontogeny.

Numerous studies suggest that IL-4 also stimulates mature T lymphocytes. IL-4 is comitogenic in the activation of normal resting T cells from lymph nodes of mice and peripheral blood of humans. IL-4 is the autocrine growth factor that sustains the growth of murine Th_2 helper clones, and IL-4 also supports the growth of cytotoxic T-cell and NK-cell clones. Both IL-1 and IL-2 enhance IL-4-supported proliferation of mitogen-stimulated T cells.

The stimulatory activity of IL-4 on T cells is apparently not mediated by IL-2 production in T cells, because neither IL-2 message nor biological activity are detected in such T cells, and antibodies to the IL-2 receptor do not block the activity of IL-4.

E. Interleukin 5 (IL-5)

IL-5 is a T cell-derived B-cell growth factor (formerly BCGF II). Before the cloning of IL-5, it was distinguished from IL-4 on the basis of its ability to promote proliferation of dextran-stimulated B cells and the inability to costimulate resting B cells with anti-IgM. IL-5 is produced by T cells activated by mitogens, antigens, alloantigens, or lectins and by murine Th_2 clones. Although a receptor for IL-5 has not been cloned, IL-5-binding proteins have been identified on IL-5-responsive cell lines by chemical cross-linking.

Subsequent studies on IL-5 revealed the ability of the cytokine to promote IgM secretion by activated (but not resting) B cells, to promote B-cell differentiation into IgA-secreting cells, and to synergize with IL-4 by increasing the expression of IgA from splenic B cells over that for IL-5 alone. Recently, it was shown that IL-5 increases the induction of IgE synthesis mediated by IL-4 on normal B cells, and that the effect is inhibited by IFN-γ. IL-5 synergizes with IL-2 in inducing IL-2 receptor expression on splenic B cells and a mature subset of murine thymocytes. In the presence of IL-2, IL-5 can also induce the differentiation of mature thymocytes into cytotoxic T cells.

F. Interleukin 6 (IL-6)

IL-6 is produced by many cells (monocytes, endothelial cells, epithelial cells, plasma cells, fibroblasts, and T cells) in response to a variety of stimuli such as TNF, IL-1, platelet-derived growth factor, antigens, mitogens, and LPS. IL-6 exhibits a broad spectrum of activities such as induction of B-cell differentiation, enhancement of plasmacytoma growth and T-cell proliferation, and hepatocyte activation. These activities were confirmed using purified recombinant products obtained from cloning of the IL-6 gene.

IL-6 can act independently as a costimulus for proliferation of rigorously purified mature T cells and thymocytes. IL-6 synergizes with IL-1 or IL-2 and lectins to enhance thymocyte proliferation, and in the course of proliferation, more IL-6 is produced by the responding cells. While mature thymocytes proliferate in response to IL-6 and lectin, it was demonstrated that immature thymocytes can only respond to IL-6 when IL-4 and PMA are also present. The primary *in vitro* generation of cytolytic T cells in a mixed lymphocyte culture was also enhanced by the addition of IL-6 to the reaction, indicating that IL-6 can promote differentiation of antigen-activated CTL precursors. Although freshly isolated T-cell subsets expressing a helper or cytotoxic–suppressor phenotype appear to be equally responsive to IL-6 and lectin, T-cell lines and clones are virtually unresponsive to direct stimulation by IL-6. Perhaps the long-term culture of these cells contributes to the loss of IL-6 sensitivity, or the cells are only responsive during a critically defined period of the cell cycle. For one murine Th_2 clone, IL-6 synergized with IL-1 in the presence of mitogens to enhance proliferation, but not with mitogens alone or with antibody cross-linking of the antigen receptor.

IL-6 does not enhance the expression of IL-2 or the IL-2 receptor for normal human T cells, and coculturing human T cells with antibodies to the IL-2 receptor has little or no effect on IL-6 activity. These findings are in contrast with the IL-2 and IL-2 receptor-inducing effects of IL-6 on murine T-cell proliferation. Thus, for human T cells, IL-6 may regulate T-cell proliferation through IL-2-independent mechanisms, whereas murine T cells may proliferate in response to IL-6 through both IL-2-dependent and independent pathways.

IL-6 is a late-acting factor in B-cell activation, promoting differentiation to IgM secretion. In transformed B cells, as represented by human multiple myeloma cells *in vitro*, IL-6 appears to act as an autocrine growth factor. IL-6 can also regulate growth of murine plasmacytomas *in vitro*. These plasmacytomas, induced by intraperitoneal oil injections *in vivo*, require IL-6 for *in vitro* growth until the cells eventually become growth-factor independent.

IL-6 also enhances NK cell activity, but unlike IL-1 and TNF, IL-6 does not induce the production of a cascade of lymphokines.

G. Interleukin 7 (IL7)

IL-7 is produced by stromal cells of the bone marrow and converts bone marrow precursor cells into early pre-B cells. Cloning of IL-7 has permitted analyses *in vivo*, which have identified IL-7 mRNA expression in the murine thymus. Thus, IL-7 may also play a role as a lymphopoietin in the thymic microenvironment.

IL-7 is a potent comitogen for unfractionated thymocytes stimulated with lectin, and synergizes with IL-1 and lectin to enhance thymocyte proliferation. IL-7-mediated thymocyte proliferation is less susceptible to inhibition by TGFβ than IL-1, suggesting that these mediators use different pathways of activation or act on different subsets of thymocytes. Double-negative thymocytes have been shown to proliferate in response to IL-7 in the absence of comitogen. Finally, both subsets (CD4+ and CD8+) of mature T cells respond equally to IL-7 and mitogen.

IL-7 plus lectin increases IL-2 and IL-2 receptor expression on mature peripheral T cells, resulting in proliferation. In contrast, recent reports indicate that IL-7 and PMA induce proliferation of purified T cells through an IL-2-independent mechanism.

H. Interleukin 8 (IL-8)

The recently cloned neutrophil-activating peptide or IL-8 is expressed by IL-1 or TNF-stimulated peripheral blood mononuclear cells, fibroblasts, and keratinocytes. IL-8 is an *in vitro* chemoattractant for T lymphocytes as well as neutrophils. Lymphocyte infiltration occurs in rat ears injected subcutaneously with low doses of IL-8, while higher doses elicit neutrophils. Injection of IL-8 into the lymphatic drainage areas also causes increased lymph node weight and selective lymphocyte accumulation. IL-8 may stimulate the expression of cell-surface adhesion molecules on T lymphocytes and neutrophils to increase their adhesiveness for vascular endothelium.

I. Tumor Necrosis Factor (TNF-α and -β)

TNFα was first identified as a factor in serum from BCG and endotoxin-treated animals that was able to elicit hemorrhagic necrosis of tumors in recipient animals. Since then, it has been shown that TNFα is produced by cells of various lineages in response to bacteria and bacterial cell wall products (BCG, mycoplasma, LPS, and MDP), PMA, and tumor cells as well as endogenous mediators (IL-1, IL-3, TNF, colony-stimulating factors, leukotrienes, platelet-activating factor, and IFN-γ with LPS).

T lymphocytes are among the many cell types that respond to TNFα. TNFα has been shown to stimulate thymocyte proliferation and cytolytic T-cell generation and to upregulate the mixed lymphocyte response. Although resting T cells do not express TNFα receptors, TNFα receptors are expressed upon activation by T-cell receptor cross-linking or mitogens, thus enabling TNFα to enhance mitogen-induced T-cell proliferation and to stimulate increased expression of class II MHC antigen and IL-2 receptor. Activated thymocytes and mature T cells can also produce TNFα. TNFα synergizes with IL-1 and IL-2 in thymocyte proliferation but does not enhance the signal of IL-4. In mature T cells, TNFα synergizes with IL-1 and IL-2 to enhance proliferative activity. TNFα-stimulated thymocytes and T cells produce IL-2 and express IL-2 receptor, suggesting that IL-2 is an intermediate lymphokine in the process of TNFα-mediated replication.

Lymphotoxin or TNFβ, while distinct from TNFα, binds to the same receptor and shares many biological properties with TNFα. TNFβ was first identified as a cytotoxic factor produced by activated T cells. IL-2 and mitogens or T-cell receptor cross-linking agents induce the expression and secretion of TNFβ by peripheral T cells, and the effect of IL-2 is enhanced by IFN-γ.

B lymphocytes and B-lymphocyte lines are also reported to express both TNFβ message and activity in response to mitogens and phorbol esters. Conversely, TNFα promotes antibody production by B cells.

J. Interferon (IFN-α, -β, and -γ)

IFN-α and -β (also termed type I IFN) and IFN-γ (type II IFN) represent a class soluble mediators that have been classically defined as having antiviral activity. IFN-α is primarily a product of activated macrophages, whereas IFN-β is produced by many cell types (such as fibroblasts) following induction by viral infections or artificial stimuli such as double-stranded RNA. IFN-γ is produced exclusively by T cells in response to antigens or mitogens and by LGL.

IFN-α, -β, and -γ exert both positive and negative effects on delayed hypersensitivity, graft-versus-host response, and mixed leukocyte reactions. The inhibitory and stimulatory effects that occur both *in vitro* and *in vivo* are determined by the concentration and time of administration of the IFN's. In general, low concentrations and late administration stimulate whereas at high concentrations and early administration, prior to antigen, suppress immune responsiveness *in vivo*. IFN-γ is inhibitory for IL-4-mediated immature thymocyte proliferation. However, IFN-γ is a potent inducer of certain cell surface structures on T lymphocytes, thymocytes and bone marrow cells, which are implicated in the process(es) of activation: such as Ia and adhesion proteins. IFN-γ is also inhibitory for proliferation of lymphokine and antigen-stimulated murine Th_2 clones but not for Th_1 clones.

Both IFN-α and -γ affect B-lymphocyte responses to antigens. In activated human B cells, IFN-γ enhances proliferation, differentiation to immunoglobulin production, and isotype switching. IFN-γ controls the expression of IgG_{2a} production, while suppressing both IgG_1 and IgE. In contrast, IFN-γ has inhibitory effects on murine B cells. IFN-α has also been shown to induce proliferation and differentiation of human B cells.

K. Transforming Growth Factor β (TGF β-1, -2 and -3)

First identified as a factor able to promote anchorage-independent growth of normal rat kidney fibroblasts in soft agar assays, TGFβ is a pleoitropic agent that is made by most cells (including macrophages and T cells) implying that it has broad biological significance. To date, five molecular forms of TGFβ (TGFβ1–5) have been identified and cloned. Almost all normal cells, including resting T cells, express receptors for TGFβ, and it appears that the receptors also bind TGFβ-1, -2, and -3 equally.

TGFβ is a potent antiproliferative factor in T-cell immunity. At picomolar concentrations, TGFβ antagonizes the effects of IL-1 induction of thymocyte and mature lymphocyte proliferation, in addition to IL-2-dependent T-cell proliferation. TGFβ also inhibits proliferation of IL-2- and IL-4-dependent T-cell clones. The generation of cytotoxic T lymphocytes in a mixed lymphocyte reaction was also inhibited by TGFβ *in vitro*. Because T cells also produce TGFβ, TGFβ possibly can function as an autocrine signal in T-cell proliferation.

The mechanism(s) by which TGFβ inhibits immune responses are not well characterized. TGFβ does not affect the transcription or translation of IL-2 mRNA by human peripheral blood T cells, however, its effects on IL-2 receptor expression are variable and may reflect the response of different normal human T-cell populations (peripheral blood vs. tonsillar T cells) to TGFβ. TGFβ has been shown to downregulate transferrin receptors and can also inhibit transcription of c-*myc* in T-cell lines, a nuclear factor that is important in T-cell replication.

B lymphocytes also synthesize and secrete TGFβ and increase their level of expression of TGFβ receptors in response to mitogen. In addition, TGFβ inhibits IL-2-induced proliferation and immunoglobulin secretion (IgM and IgG) by B lymphocytes *in vitro*. The inhibition of NK cell activity by TGFβ is mediated by TGFβ binding to high-affinity receptors. Increased NK cell cytolysis in response to IFN-α is inhibited by TGFβ, suggesting that TGFβ downregulates IFN-α receptors in NK cells. In contrast, TGFβ does not effect IL-2 induced enhancement of NK activity. [*See* TRANSFORMING GROWTH FACTOR-β.]

L. Cytokine Inhibitors

In recent years, endogenous inhibitors of cytokines have been identified and characterized for their mode of action. Two inhibitors that are derived from the urine of febrile patients and that specifically compete with the binding of cytokines to their respective receptors have been described for IL-1 and TNFα. The IL-1 inhibitor has recently been cloned and has been shown to ablate the comitogenic effect of IL-1 on T cells. Whether or not these inhibitors will have activity *in vivo* has not yet been shown.

V. Conclusions

Cytokines are a heterogenous group of peptides released from cells in response to a wide variety of exogenous and endogenous stimuli. Cytokines serve as crucial regulators of cell growth, differentiation, and maturation. Cytokines enable a small number of immune cells to signal and elicit numerous reactions from both neighboring and distant cells. A number of features of the immune response are controlled by cytokines, including early development of T and B lymphocytes, the antigen-driven proliferative clonal expansion of T and B lymphocytes, and the expression of T- and B-lymphocyte genes responsible for the effector functions of these cells such as lymphokine production, cytotoxicity, and antibody production. Some cytokines also exert negative feedback effects on T and B cells during the immune response. Thus, it is likely that the regulation of the T- and B-lymphocyte cell cycle consists of a complex network of agonistic and antagonistic interactions among various cytokines.

We have discussed the role of cytokines that are relevant to the development and activation of T and B lymphocytes, but the data suggest that several other, as yet unidentified, signals may be involved in lymphocyte regulation. The contribution of cytokines to homeostasis remains uncertain. The factors that control (1) the commitment of hematopoietic progenitor cells to T- and B-cell lineages, (2) the homing of T cells from the bone marrow to the thymus as well as their eventual migration to the peripheral lymphatics, and (3) the homing of B cells from the bone marrow to the lymphoid tissue need to be ascertained. Cytokines may be involved in the regulation of mature T- and B-lymphocyte activities, such as endogenous cytokine inhibitors or cytokines involved in Ig class-switching. Thus, the cytokine research begun in the 1980s will continue with the identification of new cytokines, new insights as to their mechanism of action, and the development of therapeatic trials to explore their potential clinical applications.

Acknowledgement

We are grateful for the critical reading of this chapter to Drs. Ira Green, Pat Latham, Dan Longo, and Mark Smith and to the editorial assistance of Bobbie Unger.

Bibliography

Interleukin 1

Dinarello, C. A. (1988). Biology of interleukin 1. *FASEB J.* **2,** 108–115.

Mizel, S. B. (1987). Interleukin 1 and T-cell activation. *Immunol. Today* **8,** 330–332.

Oppenheim, J. J., Kovacs, E. J., Matsushima, K., and Durum, S. K. (1986). There is more than one interleukin 1. *Immunol. Today* **7,** 45–56.

Interleukin 2

Greene, W. C., and Leonard, W. J. (1986). The human IL2 receptor. *Annu. Rev. Immunol.* **6,** 69–96.

Smith, K. A. (1988). Interleukin-2: Inception, impact and implications. *Science* **240,** 1169–1176.

Interleukin 3

Andreef, M., and Welte, K. (1989). Hematopoietic colony-stimulating factors. *Sem. Oncol.* **16,** 211–229.

Kimoto, M., Kindler, V., Higaki, M., Ody, C., Izui, S., and Vassalli, P. (1988). Recombinant murine IL-3 fails to stimulate T or B lymphopoiesis in vivo, but enhances immune responses to T cell-dependent antigens. *J. Immunol.* **140,** 1889–1894.

Interleukin 4

Paul, W. E., and Ohara, J. (1987). B cell stimulatory factor-1/interleukin 4. *Ann. Rev. Immunol.* **5,** 429–59.

Yokata, T., Arai, N., deVries, J. Spits, H., Banchereau, J., Zlotnick, A., Rennick, D., Howard, M., Takebe, Y., Miytake, S. *et al.* (1988). Molecular biology of interleukin 4 and interleukin 5 genes and biology of their products that stimulate B cells, T cells, and hemopoietic cells. *Immunol. Rev.* **102,** 137–187.

Interleukin 5

Harriman, G. R., and Strober, W. (1989). The immunobiology of interleukin-5. *In* "The year in Immunology 1988," Vol. 5. (J. M. Cruse and R. E. Lewis, eds.), pp. 160–177. Karger, Basel.

Yokata, T., Arai, N., deVries, J. Spits, H., Banchereau, J., Zlotnick, A., Rennick, D., Howard, M., Takebe, Y., Miytake, S. *et al.* (1988). Molecular biology of interleukin 4 and interleukin 5 genes and biology of their products that stimulate B cells, T cells, and hemopoietic cells. *Immunol. Rev.* **102,** 137–187.

Interleukin 6

Revel, M. (1988). Interleukin 6. *In* "Monokines and Other Non-Lymphocytic Cytokines" (M. Powanda, J. J. Oppenheim, M. J. Kluger, and C. A. Dinarello, eds.). Alan R. Liss, New York.

Wong, G. C., and Clark, S. C. (1988). Multiple actions of IL-6 within a cytokine network. *Immunol. Today* **9,** 137–139.

Interleukin 7

Chantry, D., Turner, M., and Feldmann, M. (1989). Interleukin 7 (murine pre-B cell growth factor/lymphopoietin 1) stimulates thymocyte growth: Regulation by

transforming growth factor beta. *Eur. J. Immunol.* **19,** 783–786.
Henney, C. S. (1989). Interleukin 7: Effects on early events in lymphopoiesis. *Immunol. Today* **10,** 170–173.

Interleukin 8

Larsen, C. G., Anderson, A. O. Apella, E., Oppenheim, J. J., and Matsushima, K. The neutrophil-activating protein (NAP-1) is also chemotactic for T lymphocytes. *Science* **243,** 1464–1466.
Matsushima, K., and Oppenheim, J. J. (in press). Interleukin 8 and MCAF: Novel inflammatory cytokines inducible by IL-1 and TNF. *Cytokine.*

Tumor necrosis factor

Beutler, B., and Cerami, A. (1989). The biology of cachectin/TNF—A primary mediator of the host response. *Ann. Rev. Immunol.* **7,** 625–655.
Le, J., and Vilcek, J. (1987). Biology of disease. Tumor necrosis factor and interleukin 1: Cytokines with multiple overlapping biological activities. *Lab. Invest.* **56,** 234–248.

Interferons

Gajewski, T. F., and Fitch, F. W. (1988). Anti-proliferative effect of INF-gamma in immune regulation. 1. INF-gamma inhibits the proliferation of Th2 but not Th1 murine helper T lymphocyte clones. *J. Immunol.* **140,** 4245–4252.
Pestka, S. (1987). Interferons and their actions. *Ann. Rev. Biochem.* **56,** 727–777.

Transforming growth factors

Sporn, M. B., and Roberts, A. B. (1988). Transforming growth factor-beta: New chemical forms and new biological roles. *BioFactors* **1,** 89–93.
Wahl, S. M., McCartney-Francis, N. and Mergenhagen, S. E. (1989). Inflammatory and immunomodulatory roles of TGF-β. *Immunol. Today* **10,** 258–261.

Cytokine inhibitors

Larrick, J. W. (1989). Native interleukin 1 inhibitors. *Immunol. Today* **10,** 61–66.
Seckinger, P., Isaaz, S., and Dayer, J.-M. (1988). A human inhibitor of tumor necrosis factor α. *J. Exp. Med.* **167,** 1511–1516.

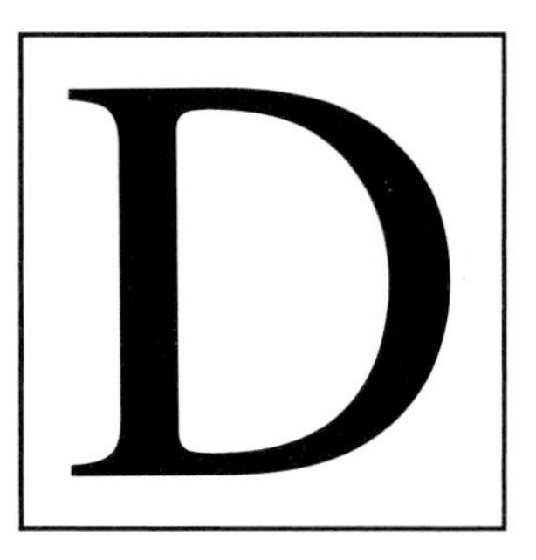

Dementia in the Elderly

ALICIA OSIMANI, MORRIS FREEDMAN, *Baycrest Centre for Geriatric Care and Mount Sinai Hospital*

Glossary

Aphasia Disorder of language and/or comprehension caused by damage to the language zone of the dominant cerebral hemisphere (usually the left)

Apraxia Ideomotor apraxia refers to an impairment in the performance of pantomimed movements to command and possibly to imitate, but not when using the real objects. Ideational apraxia is the inability to perform sequential movements correctly with the actual object

Dysarthria Disorder of speech caused by neurogenic impairment of the structures involved in respiration, phonation, and/or articulation

Huntington's disease Familial disease affecting the basal ganglia, characterized clinically by involuntary movements, changes in muscular tone, and dementia

Hypophonia Decreased volume in speech

Paraphasia Substitution of one word for another with a similar meaning or a similar sound (verbal paraphasia), or omission, addition, and/or substitution of letters within one word (literal paraphasia)

Pick's disease Degenerative disease of the brain affecting primarily the frontal and temporal lobes and producing progressive dementia

Procedural memory Learning of tasks or skills that is not dependent on conscious awareness (e.g., learning how to ride a bicycle). It has been defined as ''learning how'' as opposed to ''learning what''

Progressive supranuclear palsy Degenerative disease of unknown cause affecting several structures of the upper brainstem. The clinical features consist mainly of impaired ocular movements, impaired gait, frequent falls, dysarthria, and dementia

Wilson's disease Familial disease affecting the basal banglia of the brain and the liver, caused by a disorder of copper metabolism. It produces a progressive impairment of liver function, involuntary movements, rigidity, and dementia

DEMENTIA has been defined as an acquired impairment of intellectual function with compromise in at least three of the following spheres of mental activity: language, memory, visuospatial skills, emotion, personality, and manipulation of acquired knowledge (abstraction, calculation, judgment).

In the United States it has been estimated that approximately 10 to 15% of the population over age 65 are demented. Ten percent are mildly to moderately impaired, whereas 5% are severely demented. The prevalence of dementia is quite homogeneous throughout the world in areas where studies have been done. Any variation in prevalence rates is largely due to differences in the age of distributions of the populations. Because dementia predominantly affects older individuals, prevalence rates are higher in populations with a greater proportion of elderly.

Women are more susceptible than men to some dementing illnesses [e.g., Alzheimer's disease (AD) and Pick's disease], whereas multi-infarct dementia (MID) and Parkinson's disease (PD) with dementia affect men more than women.

I. Etiology

Dementia may affect the brain as a primary disorder, or it may be a secondary manifestation of general medical disease. In the elderly, primary dementia represents 80% of the total cases, while the remainder are due to a miscellaneous group of metabolic, endocrine, and systemic diseases.

The most common cause of primary dementia is AD. The second most common cause is MID. Other less frequent disorders producing dementia include degenerative disease (e.g., Pick's disease), progressive supranuclear palsy, and PD, as well as other disorders including chronic alcoholism, depression, drug intoxication, and normal pressure hydrocephalus. Most of these conditions are of unknown cause. Some are familial (e.g., Huntington's disease and Wilson's disease). Some cases of AD and Pick's disease are also familial. [*See* ALZHEIMER'S DISEASE; HUNTINGTON'S DISEASE; PARKINSON'S DISEASE.]

Parkinson's disease is a degenerative disease that primarily produces a motor disorder, but it may produce dementia. Although some authors report that the incidence of dementia is as much as 93% in Parkinson's patients, others claim that cognitive impairment is seen in only 4% of these patients.

Among the causes of secondary dementia, depression and drug intoxication must always be taken into consideration in the elderly, as they are reversible. The elderly tend to be sensitive to many drugs. Doses that may be appropriate for younger people may be toxic for the elderly. Two mechanisms account for this exaggerated sensitivity: altered pharmacokinetics in the elderly, with prolonged times of metabolism or excretion, and an enhanced response to certain drugs, caused by involutional changes in the central nervous system. Combinations of drugs may have deleterious effects on mental function in the elderly, causing confusion and behavioral changes that are often interpreted as dementia, but which gradually disappear once the drug is discontinued. Alcoholic brain damage may also be arrested or partially reversed if diagnosed in the early stages. Chronic alcohol abuse may also produce Korsakoff's syndrome, which is an irreversible amnestic disorder with loss of past memories and impaired ability to learn new information. [*See* ALCOHOL TOXICOLOGY; PHARMACOKINETICS.]

Normal pressure hydrocephalus is a condition that may be secondary to a previous disease of the brain (e.g., subarachnoid bleeding, meningitis), but it may appear without any previous history of neurologic disease. The diagnosis of this condition is important, because it can be neurosurgically treated with a shunt and is thus one of the treatable causes of dementia.

II. Pathology

A wide variety of changes have been described in aging and in the different dementing processes seen in the elderly. Several changes may be found in the brain of normal elderly individuals without any clinical manifestation of neurologic disease during life (normal aging). The brain loses weight, there is a loss of neurons and myelin, and there is accumulation of amyloid in the vascular walls of the brain. All these phenomena are age-related, but they are more prominent in degenerative dementing processes.

Granulovacuolar degeneration, neurofibrillary tangles, senile plaques, and Hirano bodies may occur in small numbers and in relatively restricted areas of the normal aged brain; their density and distribution are much increased in degenerative dementia.

Although all degenerative processes affect the brain in a selective fashion, there are differences in the distribution of the lesions. In Pick's disease, for example, the degenerative process affects the anterior temporal lobes and the frontal lobes. The involvement is unilateral in 50% of the cases and bilateral in the other 50%. In AD, however, the degeneration affects the cortical association areas in the temporal, parietal, and frontal lobes, as well as the amygdala, hippocampus, and basal forebrain. The nucleus basalis of Meynert is part of the basal forebrain and is the source of 90% of the cortical cholinergic input to the cerebral cortex. Because the cholinergic system has been related to cognition, lesions in the nucleus basalis may be extremely important in the pathogenesis of dementia. In AD, the primary motor and sensory cortices as well as the cerebellum are usually intact. The nucleus basalis of Meynert may be affected in other degenerative diseases (e.g., PD and Pick's disease). This predilection for specific areas of the brain explains the characteristic symptomatology that is seen in the different forms of dementia. [*See* BRAIN.]

Each dementing disorder is characterized by specific neuropathologic features, although there may be some overlap between certain disorders. In AD, there are senile plaques, neurofibrillary tangles, and Hirano bodies; in Pick's disease, there are Pick's bodies, which are intracytoplasmatic inclusions within the never cells; in Creutzfeldt–Jacob disease, there is spongiform degeneration; in PD, there are Lewy's bodies, which are intracytoplasmatic inclusions within nerve cells in the substantia nigra and locus ceruleus of the brainstem.

III. Clinical Features

Dementia is a clinical neurobehavioral syndrome. The diagnosis of dementia can, therefore, only be made on the basis of a thorough clinical history and mental status examination.

A sudden onset and a stepwise deterioration are characteristic of MID, although in some cases, infarctions (i.e., ischaemic strokes) have been small so that neither the family nor the patient can pinpoint the onset or specific episodes of neurologic damage as manifested by sudden episodes of deterioration.

Degenerative diseases (e.g., AD) usually present insidiously and progress gradually over months or years. In some cases, an external cause that seems totally unrelated may precipitate the symptoms. An elderly patient with some mild memory problems may show a rapid deterioration after moving to a new house or after suffering a mild febrile disease or after an episode of confusion caused by drug overdose or dehydration. In other cases, the initial symptom is depression. The patients become apathetic, show no initiative, and display no interest in anything that surrounds them.

As defined above, dementia is a deterioration of cognitive function involving several areas. The following is a brief summary of the findings that may be seen in these areas.

A. Attention and Concentration

One of the early signs in most dementing disorders is a decrease in attention and concentration. Patients cannot focus their attention for more than a few seconds at a time. They are often distracted by external stimuli, losing track of their conversation or the tasks they are performing.

B. Memory

Memory is one of the most complex areas of cognition. It includes several processes, and it is also dependent on other functions. It is particularly dependent on attention and concentration. It is also related to the ability to develop strategies and to associate ideas. These processes are normally somewhat reduced in the elderly, but they are particularly altered in dementia.

Episodic memory refers to personal memory for events and their location in time. Remembering a face and placing it in the context of when it has been seen or remembering an event read in the newspaper some time before are part of the process of episodic memory. Semantic memory, however, refers to the fund of knowledge the individual has accumulated throughout life, such as knowing that an elephant is an animal or that water freezes at 32 degrees.

Patients with amnesia (i.e., patients with memory loss but with relative preservation of other cognitive areas) show a loss of episodic memory with intact semantic memory. Demented patients show impairment of both processes.

Another kind of memory that has lately been the focus of attention is procedural memory. Some amnestic patients have shown an intact ability to learn motor tasks and other tasks (e.g., mirror reading) without being consciously aware of this learning. Procedural learning is usually not affected in amnesia and in some dementing diseases (e.g., AD) and is affected in others (e.g., PD).

Although memory loss is considered an early symptom in dementia, this is true in most but not in all cases. Some patients with Pick's disease, for instance, may show severe language problems for a long time, with little or no changes in memory. [*See* LEARNING AND MEMORY.]

C. Language

The language examination is perhaps one of the most valuable diagnostic tools in dementia. In so-called subcortical dementia (e.g., in PD, Huntington's disease, and progressive supranuclear palsy), speech impairment is an early sign. There is frequently dysarthria and hypophonia, but there are seldom signs of linguistic impairment in the use of language itself. In other dementing diseases, however, language disorders are common. In AD, for instance, speech is neither dysarthric nor hypophonic, but there is usually an early impairment of language production manifested by word finding difficulty and later by frequent paraphasias. At this stage, comprehension and repetition are usually unimpaired. In the more advanced stages of this disease, comprehension is affected. In final stages, language production is poor, with many paraphasias, unfinished sentences, and poor comprehension.

In Pick's disease, language impairment may be the only salient sign of dementia during the early stages, or it may be accompanied by severe behavioral changes. It later evolves to other areas, affecting memory, attention, and concentration.

Reading and writing are also affected in dementia. In PD, reading is preserved until advanced stages. Writing is altered not only by the motor problems that produce a tremulous handwriting but also by micrographia. In AD and Pick's disease, reading and writing are altered at some stage in the disease. Misspellings are frequent, with omissions and substitutions of letters and frequent perseverations. [*See* SPEECH AND LANGUAGE PATHOLOGY.]

D. Praxis

Ideational apraxia may be an early sign in some dementing disorders (e.g., AD) although it is frequently absent in others. In daily life, patients have difficulty in activities requiring sequential use of objects (e.g., preparing a cup of coffee) or in the manipulation of certain objects (e.g., peeling an orange with a knife or unwrapping a package). Ideomotor apraxia is also frequent in some dementias, but it is more a diagnostic examination tool than an obstacle in daily activities. [*See* MOTOR CONTROL.]

E. Visuospatial Skills

Visuospatial skills have a tendency to decrease with age. These skills are especially altered in many dementing disorders (e.g., AD, PD, progressive supranuclear palsy, and Huntington's disease). There may be some perceptual difficulties, but there are also problems of performance. Altered perception may sometimes underlie some of the hallucinations of the demented patients.

F. Manipulation of Acquired Knowledge

This area includes the different aspects of cognition and thought, requiring logical thinking, abstraction, and calculation. Deterioration in this area is frequently an early sign of dementia, and it seriously interferes with normal activities. Patient's thoughts become concrete, with increasing difficulty in manipulating old knowledge. Judgment is also impaired. Because the dementing process is usually gradual, it often goes unnoticed in the course of daily routine by the family or the persons surrounding the patients. It may become evident, however, when abnormal circumstances arise (e.g., a fire, an accident, or a traffic problem).

G. Behavioral and Mood Changes

Depression may be associated with early dementia, especially in AD and PD. In other disorders, depression may not be prominent. In fact, patients may seem euphoric, without any concern for their problem. Mood may also be normal.

Behavioral changes may occur early in dementia. Some patients with Pick's disease may show only personality changes for some time until the full picture of dementia develops. Social graces are often lost early, and the changes are of such magnitude that they frequently require psychiatric hospitalization. In AD, on the contrary, social graces are usually preserved until late in the disease.

Delusions, hallucinations, and paranoid ideation are frequent in dementia. Patients may believe their family is trying to harm them or rob them, thus creating secondary problems in behavior and in their relations. [*See* DEPRESSION; MENTAL DISORDERS.]

H. Frontal Signs

A special section will be dedicated to frontal signs because they are so frequent and so prominent in most dementing disorders. The frontal lobes constitute almost two-thirds of the brain, and they are affected in most dementing disorders.

Personality changes are frequently related to frontal lobe dysfunction. Disinhibition, impulsivity and irritability are common manifestations. Patients may show perseveration (i.e., a tendency to persist on the same task, the same thought, or the same movement, with an inability to shift set). If they are drawing a square, for instance, and are asked to draw a flower, they may continue drawing squares. Fragments of previous sentences may sometimes contaminate the new sentences, or movements may be repeated again and again. A common sign is what has been called being "stimulus bound." The patients' attention is pulled by the stimulus perceived at that moment, so that they act in consequence and forget to focus on their previous plan of action. Patients may, for instance, leave their room to go to the dining room because they feel hungry, but seeing a door on the way they may enter another room and forget their initial plan. In some cases, apathy is the central feature. Difficulty in initiation of speech may also be part of the syndrome.

The patterns of impairment vary according to the etiology of dementia. In AD, there is usually an early impairment of memory, language, and manipulation of acquired knowledge. Progressive deterioration leads to impairment of most of the other areas without physical signs in the general neurological examination until late in the disease.

In Pick's disease, the changes of behavior and language are the prominent sign in the early stages of the disease. Other areas of cognition (e.g., memory) are often affected later in the process.

In PD, speech is altered, but the linguistic processes are usually intact; there is a loss of attention and concentration, some memory problems, and often visuospatial problems. Frontal symptoms are frequent. The most salient sign is the slowness of thought. Other areas of cognition are affected only late in the disease. This dementia occurs in a context of motor signs of PD.

In alcoholic dementia, frontal signs are prominent; poor attention and concentration, memory loss, and decreased manipulation of acquired knowledge are early features, as well as impulsivity and irritability. In all these cases, the course may be variable. Some patients develop dementia with a slow course, so that they are still functioning at home 8 or 10 years after onset, whereas others are bedridden and totally dependent within a year or 2. In degenerative diseases, the average survival after the diagnosis of dementia is from 3 to 6 years. Death is usually caused by intercurrent diseases (e.g., infections, aspiration pneumonia).

In MID, the symptomatology varies in relation with the localization of the strokes within the brain. There is usually dysarthria, attention and memory problems, and frontal signs, but the rest of the features vary. The process may be arrested if no new strokes occur.

IV. Diagnosis

The identification of dementia is based on clinical history and mental status examination. In most cases, the etiology is usually evident on the basis of the clinical examination.

Mental status examination can be further complemented by formal neuropsychological testing. Modern neuropsychological batteries include sensitive tests that explore different areas of cognition and compare results to norms standardized for age, sex, and education. They may detect subtle impairments and contribute to the diagnosis, and they may also be used as objective measures in the follow-up of a dementing process.

Laboratory tests and neuroimaging techniques are helpful in determining the cause of dementia. In the case of some of the degenerative diseases, the cause of dementia can be suspected but cannot be definitely proven, except by pathology studies.

Neuroimaging techniques, such as X-rays, are helpful tools in the differential diagnosis of etiology of dementia. Brain tumors, hemorrhages, strokes, and normal pressure hydrocephalus may be diagnosed by computerized tomography of the brain (CT) or magnetic resonance imaging. Modern techniques of nuclear medicine [e.g., PET and SPECT (single photon emission computerized tomography)] have become a promising tool for the understanding of dementia. PET is an expensive test, which has so far been used for experimental and research purposes but has not been used as a routine diagnostic tool. SPECT provides a simpler, less expensive means of studying brain function through the measurement of cerebral blood flow. It is based on the concept that cerebral blood flow is usually directly related to the brain metabolism, which, in turn, is related to brain function. Areas of lower perfusion are therefore indicative of lower function. In AD, SPECT classically shows a pattern of low perfusion affecting in particular the posterior parietotemporal areas and frontal lobes. In Pick's disease, the typical pattern is low perfusion of the anterior temporal lobes and frontal lobes. Areas of infarction (i.e., ischemic stroke) are seen in SPECT as areas without perfusion. The technique is not always sensitive, but it is helpful in confirming the clinical diagnosis or sometimes revealing strokes that were not seen on CT.

As to other laboratory tests, they are performed to rule out metabolic, endocrine, or systemic disease that may be the underlying cause of dementia.

In very early stages of degenerative dementia, the diagnosis may not be clear. Two circumstances pose special diagnostic difficulties: depression and normal aging. Depression in the elderly is often accompanied by cognitive impairment, and some dementing disorders may present with depression. Neuroimaging techniques may show no pathology in early stages of dementia, and both mental status examination and neuropsychological testing may show cognitive impairment. It may sometimes be difficult to distinguish between those two conditions on these grounds. In this case, a therapeutic trial with antidepressants may be helpful. Cognitive improvement after treatment suggests depression, whereas the opposite is indicative of early dementia.

The changes of normal aging in mental status examination and CT scan are variable and in studies of large series the findings overlap those of early stages of dementia. This is especially so in the ninth decade and beyond, for which standardized tests are lacking. The diagnosis of degenerative dementia must be made cautiously and after several follow-up examinations during periods ranging from 6 months to 1 year.

V. Management

Some dementing processes are reversible if treated in the early stages. Hence the diagnosis of dementia and the underlying process should be identified as early as possible. Nutritional and endocrine causes should be treated. Normal pressure hydrocephalus and chronic subdural hematoma can be surgically treated.

Multi-infarct dementia may be arrested by preventing new infarcts. This can often be achieved by managing high-risk factors (e.g., hypertension, diabetes) or treating the underlying cardiovascular disease.

Degenerative diseases have no currently curative medical treatment, and they represent a difficult problem for caregivers. A neurobehavioral assessment is helpful in detecting intact cognitive areas that may be used in the management of the patient. It should also provide a guide and support to the caregivers. Psychotropic drugs are often needed for treatment of depression, agitation, or paranoid ideation, but these drugs should be used very cautiously because they often have marked side effects in the elderly and, in particular, in the brain-damaged patient.

Bibliography

Albert, M. L. (1984). "Clinical Neurology of Aging." Oxford University Press, New York.

Brodal, A. (1969). "Neurological Anatomy." Oxford University Press, New York.

Cummings, J. L., and Benson, D. F. (1983). "Dementia: A Clinical Approach." Butterworth Publishers, Boston.

Heilman, K. M., and Valenstein, E. (1979). "Clinical Neuropsychology." Oxford University Press, New York.

Luria, A. R. (1980). "Higher Cortical Functions in Man." Basic Books, New York.

Mesulam, M. M. (1985). "Principles of Behavioral Neurology." F. A. Davis, Philadelphia.

Plum, F., and Geiger, S. R., eds. (1987) "Handbook of Physiology, vol. V, Higher Functions of the Brain, part 2." Mountcastle, VB, American Physiological Society, Bethesda, Maryland.

Reisberg, B. (1983). "Alzheimer's Disease." The Free Press, Collier Macmillan, New York.

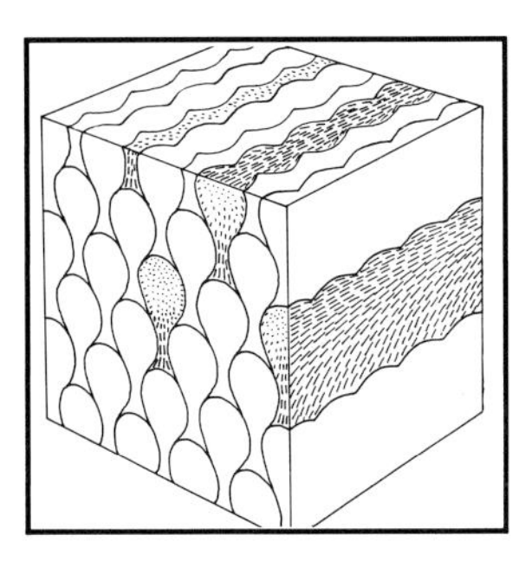

Dental and Oral Biology, Anatomy

A. DAVID BEYNON, *University of Newcastle upon Tyne*

Glossary

-Blast Denoting a formative cell (i.e., ameloblast, enamel-secreting cell; odontoblast, dentine-secreting cell)
Cusp Rounded or pointed elevation on or near to the masticatory surface of a tooth
Deciduous dentition Primary dentition consisting of 20 teeth, subsequently replaced by the permanent dentition
Distal Away from the midline along the dental arch
Fossa Depression or concavity on the masticatory surface of a tooth, into which cusps on opposing teeth usually fit
Mesial Toward the midline along the dental arch
Occlusion Functional contact relations between teeth in opposing jaws; anatomical alignment of teeth within jaws
Permanent dentition Secondary dentition consisting of 32 teeth, which replaces the deciduous dentition

DENTAL AND ORAL ANATOMY includes the structure, both gross and microscopic, development, and function of oral and dental tissues. Teeth are essential for initial food processing, and their early loss may have life-threatening consequences in primitive societies. They consist of durable mineralized tissues, including enamel and dentine, which are supported in tooth sockets by a fibrous sling (periodontal ligament) anchored on the tooth side by cementum and to the jaw bone by specialized bone of attachment. The interface between the tooth and gingiva is highly specialized, forming an attachment site that is vulnerable to damage from microorganisms in the oral cavity. Oral health depends on a healthy mucosal lining, which is maintained by the secretions of salivary glands. Salivary glands are differentiated on the basis of their types of secretions, which perform specific functions within the oral cavity.

I. Oral Cavity

The oral cavity is a space devoted to food preparation and quality control in the form of taste reception and to the modification (articulation) of sounds formed initially in the larynx (phonation). It is a space bounded for the most part by high mobile structures. The lips form the anterior boundary, forming an essential seal to permit chewing and swallowing. The cheeks lie laterally, and it is roofed by the palate. The floor of the mouth is occupied by the highly mobile tongue, and its posterior limit is the anterior tonsillar pillar containing the palatoglossus muscle, which acts as a posterior seal. The palate comprises a rigid anterior segment (i.e., the hard palate, supported by the palatal processes of the maxilla and the palatine bone) and a moveable posterior segment (i.e., the soft palate, which forms a seal between mouth and nose and between mouth and pharynx during swallowing and speech articulation). The teeth lie toward the periphery of the oral cavity, dividing it into a smaller peripheral vestibule bounded laterally by the lips and cheeks, with the large oral cavity proper internally.

The teeth are supported by the alveolar processes of the mandible and maxilla. The lips and cheek muscles are attached to the lateral surfaces of those bones and serve to bring contents in the vestibule between the teeth to allow efficient chewing. Inadequate neuromuscular control leads to cheek biting. The oral cavity is lined with mucosa, which is an epithelium moistened by salivary secretions. Mucosa is organized in three categories based on their microscopic anatomy and functions. Masticatory mucosa is found where the mucosa is abraded by food processing, typically around the necks of the teeth and on the hard palate, where mucosal ridges (or rugae) are found. Rugae may play a minor role in food breakdown before swallowing. Lining mucosa covers the mobile lips and cheeks in the vestibule, reflecting onto the alveolar process, and is present internally on the floor and on the underside of the tongue. Specialized or gustatory mucosa is present on the tongue and is characterized by the presence of papillae, which comprise four forms.

The nerve supply to the oral cavity is complicated, arising from five cranial nerves. The motor nerve supply to the cheeks and lips is through the facial (VII) nerve, the soft palate from the pharyngeal branch of the vagus (X), and tongue by the hypoglossal (XII) nerve. Common sensation from the oral cavity returns to the central nervous system via the trigeminal (V) nerve. Special visceral (taste) sensation from the anterior two-thirds of the tongue and the palate return with the facial nerve (nervous intermedius), while the posterior one-third of the tongue is innervated by the glossopharyngeal (IX) nerve. The blood supply to the oral cavity arises from branches of the external carotid artery, including the lingual, facial, and several branches of the maxillary artery, including the inferior and superior alveolar branches, palatine arteries, and infraorbital branches. Lymphatic drainage is to the submental, submandibular, and deep cervical lymph nodes.

II. Teeth

Humans possess two dentitions, with the primary or deciduous dentition being replaced during childhood by the secondary or permanent dentition (Fig. 1). The deciduous dentition consists of two incisors, a canine, and two molars in each quadrant of the two dental arches (upper and lower), each arch being split into two quadrants (left and right)

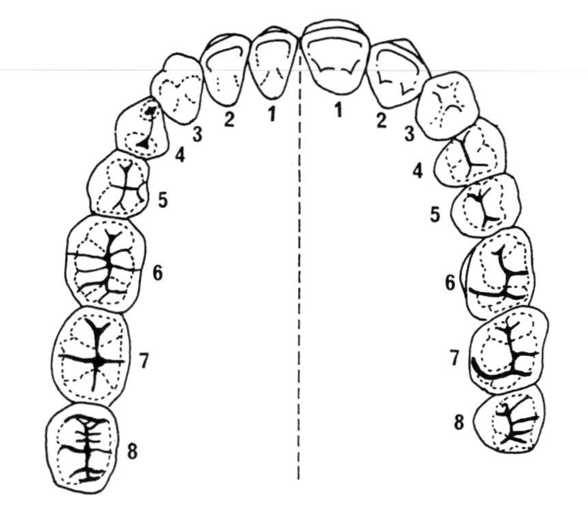

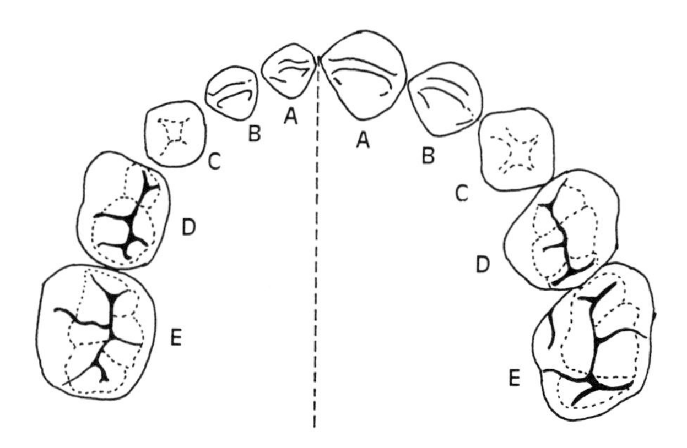

FIGURE 1 Diagram of human permanent dentition (*top*) and deciduous dentition (*bottom*). The mandibular crown morphology is indicated on the *left side*, maxillary morphology on the *right side*. Permanent teeth are numbered (1, central incisor; 2, lateral incisor; 3, canine; 4, first premolar; 5, second premolar; 6, first molar; 7, second molar; 8, third molar). Deciduous teeth are indicated in capital letters (A, central incisor; B, lateral incisor; C, canine; D, first molar; E, second molar).

(Table I). The permanent dentition contains two incisors, one canine, two premolars or bicuspids, and three molars in each quadrant (Table II). The successional secondary dentition replaces the primary dentition with exact correspondence of tooth number in the incisor, canine, and deciduous molar/premolar sectors. The three permanent molar teeth erupt behind the deciduous molars during childhood, with the first molar at ≃6 years, the second molar at ≃12 years, and the third molar (wisdom tooth) at ≃18 years, although this tooth is highly variable in its eruption times. The deciduous teeth are smaller, whiter, and have more bulbous crowns, and the incisors and canines have much smaller

TABLE I Deciduous Dentition

	Central incisor (upper/lower)	Lateral incisor (upper/lower)	Canine (upper/lower)	First molar (upper/lower)	Second molar (upper/lower)
First calcification (months I.U.)	4/4.5	4.4	5.2/5	5	6
Crown complete (months)	4	4/4.2	9	6	10–12
Eruption (mean) (months)	7.5/6.5	6.5/7	16–20	12–16	20–30
Root complete (years)	1.5–2	1.5–2	2.5–3	2.5–3	3

roots than their permanent successors. The deciduous molars have short but widely divergent roots to accommodate the developing successional premolar teeth. The enamel on deciduous teeth is relatively thinner and is less highly mineralized, allowing relatively rapid tooth wear to take place.

Teeth may be grouped into three functional sets. Incisors have a narrow elongated incisal edge, with uppers overlapping lower incisor teeth, giving rise to a cutting or shearing action suitable for nipping off fragments of food. Canine teeth are pointed and, although now smaller than in our earlier ancestors, still play a minor role in puncturing and tearing food held in the anterior part of the mouth. Premolars and molars have complicated crown morphologies, with elevations (cusps) on one tooth entering into depressions (fossa) on opposing tooth surfaces, similar to a mortar and pestle relation. These teeth are specialized for crushing and grinding of food in the posterior part of the mouth, where the biomechanical relations permit the generation of the greatest biting forces on the tooth surfaces.

In both permanent and deciduous dentitions the largest incisor is the upper central, the smallest being the lower central, with the upper and lower lateral incisors being more similar in size. The incisors are wedge-shaped and have a single conical root. Canine teeth have a single pointed cusp with sloping arms, of which the mesial is generally shorter relative to the distal arm. The permanent premolar teeth have two main cusps, a buccal and lingual cusp separated by the mesiodistal occlusal fissure. The upper premolars have a flattened oval morphology, and the two cusps are more equal in height, although the buccal cusp is always the highest. The lower premolars in contrast have a more rounded occlusal morphology, with relatively much reduced lingual cusps particularly in the first premolar. The second premolar may possess two lingual cusps. Premolars tend to have a single mesiodistally flattened root, although it is common for the upper first premolar to have two roots, buccal and lingual. The permanent molar teeth show the greatest differences between the upper and lower dentitions. The upper molars characteristically have a rhomboid-shaped crown with two cusps (mesiolingual and distobuccal) in broad contact, separating mesiobuccal and distolingual cusps widely, and also giving rise to an H-shaped fissure pattern. Upper molars have three roots: two buccal roots, which are usually flattened mesiodistally, and one larger rounded palatal root. There is a progressive tendency when passing distally along the tooth row from first to third molar for the distolingual cusp to be reduced, with the crown eventually becoming three cusped and with progressive merging and fusion of the three roots. Lower molars, in contrast, have a rectangular crown with four or five cusps separated by an approximately cross-shaped fissure pattern. They have two large roots, placed mesially and distally, and these are flattened in a mesiodistal direction. The first molar has five cusps, with a distal cusp usually being present. The second molar generally has only four cusps, with the loss of the distal cusp. The third molar shows a tendency to have five cusps, which may be supplemented by extra cusplets and fissures. The deciduous molar teeth show specific differences in crown morphology, most particularly in the first deciduous molar teeth. In both upper and lower first deciduous molars, the mesiobuccal corner is elongated (the molar tubercle) giving the crown an irregular quadrilateral shape. The buccal and lingual cusps tend to converge toward the midline of the tooth. The upper first deciduous molar has the simplest morphology, with a mesiodistal fissure dividing the occlusal surface into two cusps, buccal and lingual. The buccal cusp is long and narrow, forming a cutting edge, whereas

TABLE II Permanent Dentition

	Central incisor (upper/lower)	Lateral incisor (upper/lower)	Canine (upper/lower)	First premolar (upper/lower)	Second premolar (upper/lower)	First molar (upper/lower)	Second molar (upper/lower)	Third molar (upper/lower)
First calcification (months)	3–4	10/3–4	4–5	17/21	30	birth	30	≃9 yr
Crown complete (years)	4–5	4–5	8–9/6–7	5–6	6–7	2.5	7–9/7–8	≃12
Eruption (mean) (years)	7.2/6.5	8.5/7.5	11.2/10.5	10.5/10.5	11/11.2	6.7/6.5	12/12	≃18
Root complete (year)	10/9	11/10	13–15/12–14	12–13	12–14/13–14	9–10/9–10	15	≃21

the lingual cusp is more pointed. There is a prominent buccal cingulum extended mesially to form the molar tubercle. This tooth, together with the second deciduous molar, shares the upper molar characteristic of having three roots. The lower first deciduous molar is elongated mesiodistally and is divided into buccal and lingual halves by a mesiodistal fissure. The buccal half consists of two cusps, mesiobuccal and distobuccal, separated by a shallow buccal groove, of which the mesiobuccal cusp is larger. The lingual half is narrower and comprises two cone-shaped cusps, of which the mesiolingual cusp is again larger. The mesiolingual and mesiobuccal cusps are commonly connected by a transverse ridge dividing the fissure system into a mesial pit and distal fissure. This tooth has a mesial and distal root, like all lower molars. The upper and lower second deciduous molars closely resemble in morphology the first permanent molar tooth, which erupts behind them.

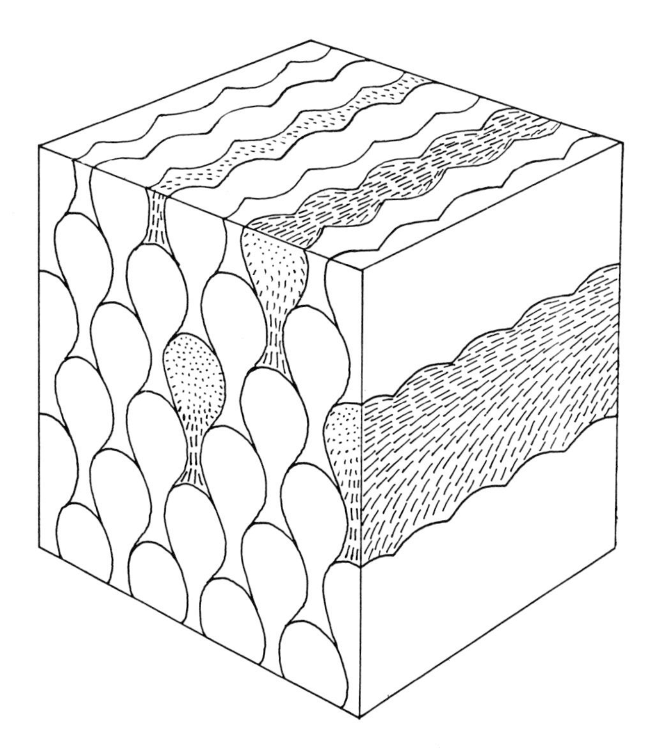

FIGURE 2 Diagram of human enamel prisms. Right-hand vertical face shows prisms cut in vertical longitudinal section along their lengths. Left-hand vertical face shows prisms cut in vertical cross section showing the characteristic keyhole shape. The top face shows a horizontal section passing through heads and tails of keyhole-shaped prisms. Crystal direction in four prisms is indicated. Crystals run in the long axis of the body of the prism, with crystals diverging from this longitudinal axis passing into the tail to an increasing extent, producing a marked change in crystal orientation at the boundary between prisms.

III. Dental Tissues

A. Enamel

Enamel is the hardest structure in the body, with a hardness value similar to that of mild steel. Its color ranges from white to gray-yellow, depending on its thickness, which allows the yellow color of underlying dentine to show through under thinner enamel. It consists of 96% by weight of inorganic material, largely composed of hydroxyapatite (HA), with the remainder consisting of less than 1% of organic material and water. The organic material consists principally of a specific enamel protein called *enamelin.* The mature tissue in the mouth is without cells, and therefore it is incapable of repair, unlike all other calcified tissues. Enamel structure is wholly dependent on the orientation of HA crystalites within it. These crystalites are packed in a highly ordered fashion, producing rod (or prism) structural units, each of which is about 5μm across (Fig. 2). Prisms viewed from their ends have a keyhole shape, with a relatively large body, tapering down to a constricted waist with a final minor enlargement in the tail, which is directed toward the neck of the tooth. Crystals running in the body of the prism are broadly parallel to its length, but passing into the tail region, the crystalites diverge progressively more from this direction, terminating at an acute angle in the base of the tail adjacent to the adjacent inferior prism. The prismatic structure of enamel is a consequence of changes in crystal orientation at each prism boundary. Individual crystalites are hexagonal in cross-striation and are relatively long, indeed their actual length is unknown in enamel.

Crystals in enamel are approximately 10 times larger than those in calcified connective tissues including dentine, cementum, and bone; this may be a consequence of differences in the organic matrix. Enamel proteins are mobile and labile and are partially removed during development, in contrast to the more rigid collagen scaffolding in other mineralized tissues. This rigid cross-linked structure of collagen possibly restrains crystalite growth during development in calcified tissues. The ultimate hardness of enamel is a function of both the relatively large size of enamel crystals and the relatively small amount of proteins remaining in the tissue. Enamel formation is a continuous process punctuated by repetitive fluctuations over short and longer periods. The short-period fluctuations are caused by the circadian rhythm, which affects the rate of secretion of enamel, giving rise to alternating varicosities and constrictions along the length of enamel prisms. Longer-period fluctuations in enamel formation give rise to striae of Retzius, which are orientated broadly perpendicular to the prism direction, running obliquely toward the enamel–dentine junction (EDJ). These accentuated markings represent the position of the whole secretory ameloblast sheet at the time of the disturbance. Striae manifest themselves in two different ways, illustrating differing aspects of enamel growth. Over the cusp tip, they run obliquely from one side of the tooth to the other, without reaching the enamel surface, and represent successive layers of enamel deposited during the phase of secretion of full-thickness enamel over the cusp tip. Once this cuspal thickness is achieved, ameloblasts continue to secrete enamel on the sides of the crown, often for relatively large distances over an extended period, to complete the formation of the full crown height. In this lateral enamel, striae run obliquely from the EDJ toward the surface, producing a tile-like arrangement of layers of enamel (imbricational enamel). These produce alternating transverse ridges (perikymata) and depressions. There is accumulating evidence that striae have a systemic origin and repeats over a circaseptan period (6–9 days, with a modal value of around 8).

Prisms do not follow a straight course from within outward, which would be prone to fracture. They follow a sinuous or undulating course, which is most marked over the cusp tips in the form of gnarled enamel. Here prisms undergo frequent and sharp changes in direction, giving rise to a complex interlocking structure. In the bulk of the lateral enamel, the prisms undulate two or three times from side to side in a horizontal plane, straightening in the outer one-third to one-quarter of the enamel. These sinuosites move progressively out of phase at differing horizontal levels, reaching the opposite phase approximately after 20 prism levels, approximately 50–100 μm apart. This decussating structure causes the dispersion of cracks as they propagate into the deeper enamel and acts as an energy absorbing device, limiting further crack extension.

The internal surface at the EDJ is scalloped, with long shallow undulations extending over a distance of approximately 50 μm. These depressions in the dentine surface are mirrored by shallow elevations on the enamel surface. This arrangement may represent a device to increase the area of enamel–dentine surface contact, which may help to prevent less elastic enamel shearing away from the elastic underlying dentine. Extensions of the odontoblastic processes commonly pass through the EDJ into the enamel, particularly over the dentine horn under the cusps. These extensions are called *enamel spindles*.

The external surface of lateral enamel is characterized by the presence of surface ridges, or perikymata, which reflect the imbricational enamel structure. Recent interest has focused on the possibility of counting the numbers of perikymata on lateral enamel in teeth to gain an estimate of how long crown formation takes to be completed, using nondestructive methods. Crystalites in subsurface enamel are commonly orientated parallel to one another, without prism boundaries. This aprismatic enamel layer is typically 20–30 μm thick.

B. Dentine

Dentine is the second most highly mineralized tissue in the mature tooth, 70% of which (by weight) is HA, and 20% organic material, principally collagen. Dentine, which is yellow in color, differs from bone and cementum in that cells are never found within the body of the tissue, although processes from the odontoblasts extend throughout the tissue. The odontoblast cell bodies are located internally at the periphery of the dental pulp. This specialized cell–tissue relation enables dentine to respond to injury and irritation by either laying down mineralized tissue within the tubule containing the processes (peritubular dentine) or to mount a pulpal repair response. Collagen fibers in peripheral dentine, particularly in the crown, are orientated perpendicular to the EDJ; this arrangement distinguishes this most peripheral (mantle) dentine from the remaining circumpulpal dentine, in which the collagen fibers are arranged broadly parallel to the tooth surface. In circumpulpal dentine, the collagen fibers form a meshwork at right angles to the dentinal tubules, which traverse the tissue. In the early development of dentine, the dentinal tubules are relatively large and filled by the odontoblastic process, which extends to the periphery of the dentine. With the passage of time there is deposition of hypermineralized peritubular dentine within the original tubule, associated with a retraction of the process. This peritubular dentine is distinguished from the intervening intertubular dentine by the absence of a collagen matrix and by its relatively high mineral level. The organic constituents of peritubular dentine appear to be glygosaminoglycans, although these have not been properly characterized. There are also differences in the mineral phase, peritubular dentine consisting of octocalcium phosphate. Earlier studies suggested that the odontoblastic process did not extend to the periphery in mature teeth, but according to recent evidence they extend through the breadth and length of the tubules where they remain patent. Some dentinal tubules are penetrated by nerve fibers for relatively short distances, although these nerves do not extend throughout dentine. [*See* COLLAGEN, STRUCTURE AND FUNCTION.]

Incremental lines are present in dentine and are broadly similar to those in enamel in their periodicity. Short period lines, equivalent to cross-striations and related to the circadian cycle, are lines of von Ebner. Longer period lines are represented by contour lines of Owen, which are variable in their spacing and expression and generally inconstant. Hypomineralized defects occur in dentine, the larger ones occurring in the coronal dentine in the form of interglobular dentine. These are caused by failure of fusion of globular calcification masses called calcospherites. Interglobular dentine is rare in the root, possibly because the mineralization of this part of the tooth occurs linearly. In root dentine, however, there is a characteristic defect called *Tomes granular layer,* just within the peripheral dentine. The individual defects in this layer resembles interglobular dentine but are much smaller. This granular layer is usually present throughout all root dentine and is bounded externally by a structureless (hyaline) layer, about 20 μm thick.

On the pulpal side of dentine is a uniform uncalcified layer called *predentine*. Its thickness varies between 20 and 40 μm and is equivalent to unmineralized bone (osteoid). The odontoblasts secrete collagen and ground substance into this layer, which then undergoes mineralization becoming dentine.

Dentine responds to irritation in two ways. Mild stimuli, including gradual dentine exposure during tooth wear, may result in the obliteration of tubules by peritubular dentine. This has the beneficial effect of sealing off the pulp from the oral cavity using a hypermineralized tissue, which will resist further tooth wear. Severe irritation, for example, arising from carious attack on the tissue, may cause the odontoblasts to die and loss of processes from the tubules. A secondary response beneath dead tracts cause the deposition of irregular or reparative secondary dentine, which seals off the pulp at the pulpal side. Another form of secondary dentine occurs with tooth aging and involves a continuous deposition of dentine throughout life. The pulp chamber becomes progressively smaller, until the original odontoblasts can no longer be accommodated at its periphery. A proportion of the odontoblasts die, and the remaining ones regroup and secrete irregular secondary dentine, which is characterized by a reduction in the number of dentinal tubules and by an abrupt change in their direction.

C. Pulp

The pulp is a specialized connective tissue whose function is to support and supply nutrients to dentine. The odontoblasts are located at the periphery of the pulp with processes extending into dentine. They are tall columnar cells, being approximately 30 μm in length and are polarized: the nucleus is located away from the predentine, whereas a well-developed rough endoplasmic reticulum is situated near the predentine in the distal part of the cell. Recent evidence suggests that the odontoblasts are joined to one another by a continuous zone of tight junctions at the secretory end of the cell. This junction may serve to restrain the free passage of mineral ions into forming dentine. Between the odontoblasts and the pulp there is first a cell-free zone and then a cell-rich zone. The majority of the pulp cells are fibroblasts, which secrete collagens and abundant ground substance. The rate of tissue turnover in pulp is high, particularly for the ground substance. With increasing age the fibroblasts tend to become flatter and to deposit more collagen fibers. Apart from fibroblasts, two other principal cell types are associated with neural and vascular elements.

Nerves are present in relatively large numbers, comprising both unmyelinated and myelinated nerve fibers. Nerves in pulp are either sensory, transmitting only the sensation of pain, or are efferent sympathetic fibers, responsible for altering blood flow in vessels. Sensory nerve fibers form a network below the odontoblasts, called the *plexus of Raschkow,* and terminate on and around the odontoblast cell bodies. In certain parts of the pulp, particularly toward the pulpal horns, sensory nerve fibers accompany the odontoblastic cell processes in the dentinal tubules for a distance of 300–400 μm. Although nerves are restricted to the inner part, the whole dentine, even the outer layer, is extremely sensitive to pain. Various theories have been proposed to account for this. The most plausible explanation (hydrodynamic theory) is that disturbance to outer dentine causes fluid flow through the numerous dentinal tubules present in dentine (approximately 40,000/mm^2 in coronal dentine). The resulting movement and displacement of odontoblasts and their processes may then trigger a response in the nerve plexus surrounding these cells.

Other cells of the dental pulp include macrophages and lymphocytes. Macrophages reside in perivascular locations and act as pulpal scavenger cells. Lymphocytes are defense cells, which are responsible for producing antibodies, thus playing a role in cellular immunity. [*See* LYMPHOCYTES.]

The vascular supply to the dental pulp is extensive. One or two larger vessels (arterioles) enter through the apical foramen and proceed upward centrally through the pulp, giving off a network of branches, which extends toward the subodontoblastic area, forming an extensive capillary network. Some capillaries penetrate a short distance into the odontoblast cell layer. Discontinuities (fenestrations) are present in the walls of these capillaries, allowing rapid transfer of nutrients to the extracellular space. Venules collect the returning blood in relatively large vessels. Arteriovenous shunts (i.e., anastomoses) within the pulp direct blood from the arterial inflow into the venous outflow, diverting it away from the capillary circulation. Lymphatic vessels arise as blind thin-walled vessels in the coronal part of the pulp and drain out through the apical foramen.

The pulp chamber is effectively a rigid closed box, with one or several narrow entry points from the apical foramen(ina). The pulp is therefore susceptible to damage when there is an inflammatory response secondary to infection, which commonly arises from dental caries. Inflammation causes an increase in tissue fluid pressure; when this exceeds the low pressure in the venous outflow, the blood supply to the pulp is cut off, ultimately leading to pulpal necrosis and death. [*See* DENTAL CARIES; INFLAMMATION.]

D. Periodontium

The periodontium comprises cementum, periodontal ligament, the alveolar bone of the tooth socket, and the gingiva applied to the tooth (dentogingival junction). The function of the periodontium is primarily to achieve a mobile but strong tooth supporting system, to enable teeth to move slightly during biting and mastication, and also to form a seal between the tooth-supporting tissues and the oral cavity. This sealing function is the primary responsibility of the dentogingival junction.

1. Cementum

Cementum is a mineralized connective tissue that closely resembles bone. It is mineralized to about 50% by weight, the remainder consisting of organic material principally collagen, and water. Cementum is avascular and depends for its nutrition on blood vessels in the periodontal ligament. It does not normally undergo remodeling. Successive layers are added onto preexisting layers to provide new functional attachment for periodontal ligament fibers during the life of the tooth. A recent classification of cementum divides it into afibrillar and fibrillar forms. Afibrillar cementum is only found, in small amounts, in the cervical part of the tooth. The bulk of cementum is fibrillar, consisting of collagen fibrils. Fibrillar cementum is further subdivided: one part, which is located predominantly in the cervical third of the root, consists primarily of extrinsic (or Sharpey's) fibers (i.e., is engaged in tooth attachment); the other part, located in the more apical parts of the root, also has intrinsic fibers. This classification is a functional one: a high proportion of extrinsic fiber content implies an exclusively attachment role; a lower proportion indicates a filler function.

An older classification involves two categories: acellular cementum (broadly equivalent to extrinsic fiber) and cellular cementum (mixed fiber). Acellular cementum is formed relatively slowly and is restricted to the neck region of the tooth. Cellular cementum is principally formed toward the root apex, and also in the bifurcation region of multirooted teeth. In both locations it serves to compensate for loss of crown material caused by attrition and/or jaw growth. With increasing age, teeth wear both occlusally (i.e., at the surface of contact with opposite teeth) and approximally between teeth. To avoid progressive spacing developing between adjacent teeth and between jaws, teeth migrate mesially and occlusally. This occlusal movement, into the mouth cavity, is compensated by elongation of the tooth root through deposition of rapidly formed cellular cement.

2. Alveolar Bone

The function of this tissue is tooth support, achieved by the anchoring of periodontal ligament fibers into its most superficial layers. The embedded parts of the ligament fibers are the Sharpey's fibers (see above). Bone containing such extrinsic fibers, called *bundle bone,* is relatively deficient in intrinsic fibers secreted by osteoblasts. The deeper layers of alveolar bone are lamellar and provide the greatest strength per unit mass of any bone tissue type. Alveolar bone forms the wall of the socket and is perforated by numerous canals (for further details see Section VI).

3. Periodontal Ligament

This tissue forms a fibrous sling, connecting the cementum to the alveolar bone. Its width is approximately 0.2 mm in juveniles and may become thinner with increasing age. Its primary function is to convert forces, placed on the teeth during biting of hard objects, into tensile forces on the alveolar bone. It also prevents excessive displacement of the tooth apically, which would damage the nerves and blood vessels entering the pulp. It serves an important proprioceptive function in that nerves present in the periodontal ligament monitor tooth contact positions and assist in the control of masticatory movements.

The ligament is unusual in several respects. It differs from other typical ligaments in that the collagen fibers do not form straight, closely appoximated bundles, but follow an undulating course and tend to be dispersed between the fibroblasts. It is also unique as a connective tissue in its high rate of metabolic activity and collagen turnover, with a half-life on the order of only 1–2 days, much shorter than in other tissues. This high rate of turnover is achieved through the activities of the periodontal fibroblasts, which constantly secrete collagen, at the same time digesting and degrading previously secreted fibers.

Collagen fibers are arranged in a series of groupings called *principal fibers*. These include gingival fibers, which are principally attached to cervical cementum and support the gingiva. Alveolar crest and horizontal fibers are attached to the alveolar bone crests and insert into cementum. The oblique group is the most abundant, and acts to prevent extreme intrusion of the tooth. Other minor groups include the apical group at the root apex and interradicular fibers found at the bifurcation in multirooted teeth. A system of unusual connective fibers, called *oxytalan fibers,* resembling preelastin fibers, follows a cervico-apical course. They originate either in the gingiva or in the cervical cementum and run downward and outward across the oblique collagen fibers to terminate at the periphery of the ligament, usually in association with blood vessels. Their function remains disputed, but it is likely that they play some role in the regulation of blood flow within the ligament. The principal blood supply comes from branches of the dental arteries. Vessels entering the pulp give off branches supplying the apical part of the ligament, and the remaining ligament is supplied by arteries that penetrate the alveolar cribriform (perforated) plate to supply the bulk of the periodontal ligament. Blood vessels running within the gingiva supplement this supply around the tooth neck and are responsible for supplying nutrients to the dentogingival tissues.

4. Dentogingival Junction

This is the specialized junction between the gingiva and the tooth, which is essential for the maintenance of the integrity of the tooth-supporting tissues. It consists of the mucosa facing toward the tooth and is divided into the coronally directed (or crownward) part, called *sulcular* or *crevicular epithelium,* lining the gingival sulcus, and inferiorly junctional epithelium, which is attached to both the underlying connective tissues and the enamel. The gingival sulcus is a shallow groove between the gingiva and the tooth, with a depth of approximately 1–2 mm in health. The sulcus provides a sheltered environment that is difficult to cleanse, in which oral microorganisms thrive, causing tissue irritation and damage with their products. These damaging effects are to some extent counteracted by the outward flow of crevicular fluid, which helps to wash out noxious substances. [*See* GUM (GINGIVA).]

The junctional epithelium forms an adherent cuff around the tooth, wider toward the sulcus and narrow toward the cervix. Junctional epithelium originally derives from enamel epithelium, which formed enamel, but is replaced by downgrowing oral epithelium approximately 3 years after tooth emergence. The epithelium does not contain fibrous keratin and the structural features of the cells reflect its functions of attachment and cell replacement. Epithelial cells are attached to each other by desmosomes, although these are much fewer in junctional epithelium than in typical oral epithelium, and in consequence the intercellular spaces are larger. The basal cells adjacent to the basal lamina at the border with the connective tissue are attached to it by hemidesmosomes. On the tooth side the epithelial cells are attached by hemidesmosomes, in a comparable manner, to a thin organic layer, or primary cuticle, on the surface of the enamel. Specialized junctions preventing fluid passage between cells are absent, and membrane coating granules, which are thought to play a role in "waterproofing" epithelia, are few. These features probably account for the leakiness of this tissue, which allows the outflow of crevicular fluid. This fluid crosses the basal lamina, which appears to act as a semipermeable filter to the passage of proteins outward through the junctional epithelium. Cell turnover rate in this tissue is high, on the order of 6 days. Daughter cells produced at all levels migrate toward the oral cavity and are shed into the gingival sulcus. Polymorphonuclear leukocytes and monocytes are also commonly observed migrating through the junctional epithelium, to be shed into the mouth. They may play some role in the defense against bacteria and their products in this local environment.

VI. Mandible, Maxilla, and Temporomandibular Joint

The mandible is a U-shaped bone consisting of two vertically orientated rami located posteriorly, linked by a stout elliptically shaped body. The ramus possesses an anterior coronoid process, which is an attachment for the temporalis muscle, separated by a mandibular notch from the posterior condylar process, which forms the inferior articulation of the temporomandibular joint. The posteroinferior angle of the mandible is the insertion site of the masseter muscle exteriorly and of the medial pterygoid muscle internally. The temporalis, masseter, and medial pterygoid are the principal jaw elevator muscles and are responsible for jaw closure. The lateral pterygoid muscle, which inserts into the neck of the condylar process, is responsible for pulling the condyle forward during jaw opening and is the principal muscle capable of jaw protrusion.

Muscles attached anteriorly to the body of the mandible include the digastric and mylohyoid muscles, as well as the inferior origins of the muscles of facial expression. The digastric is the principal jaw depressor muscle but may be assisted by the mylohyoid when the hyoid bone is fixed by contraction of infrahyoid muscles. Near the midline of the mandible, which is called the *symphysis,* there is the head of the genioglossus muscle, which runs backward into the body of the tongue, and is responsible for protrusion of that organ. Between this muscle and the mylohyoid is the geniohyoid, which also helps to protrude the tongue. Other muscles attached posteriorly to the body of the mandible include the buccinator muscle, which is responsible for maintaining the cheek in close approximation to the posterior teeth, and part of the origin of the superior constrictor muscle, which encloses the pharynx and assists in swallowing. The body of the mandible is a relatively dense bone, because considerable stresses are generated in it during biting. Unilateral biting on posterior teeth produces powerful shearing forces adjacent to the loaded teeth, as well as considerable torsional forces across the midline symphyseal region.

The mandible develops by intramembranous ossification at about 40 days of intrauterine development. The site of initial ossification is near the bifurcation of the inferior dental nerve into the incisive and mental branches, on both sides. Ossification spreads from both points forward toward the symphysis, where a midline joint persists until about the first year of postnatal life. Ossification backward from the initial centers causes the development of the ramus and formation of the coronoid and condylar processes. In these two latter sites, as well as in the symphysis and transiently in the angle, secondary cartilages appear, possibly to stabilize these structures initially. The coronoid cartilage disappears before birth, and the symphyseal one within the first years of life, leaving the coronoid process as the principal growth site in the mandible until adulthood. [*See* CRANIOFACIAL GROWTH, POSTNATAL; DENTAL EMBRYOLOGY AND HISTOLOGY.]

The maxilla is attached directly to the skull, with processes extending up toward the frontal and zygomatico-temporal regions. The maxilla abuts posteriorly against the pterygoid plates, and the palatal processes of the maxilla meet in the midline, across the roof of the palate, where they are supported above by the vomer and ethmoid bones. These wide maxillary attachments to the skull disperse forces generated during mastication, and accordingly, the construction of the maxilla is much less heavy than that of the mandible. It also lacks attachment of powerful muscles of mastication, although most of the small muscles of facial expression gain their superior origin from this bone. The maxilla develops, like all the facial bones, intramembranously and has minor contributions from secondary cartilages, which, if present, are obliterated before birth.

Alveolar processes develop on both the maxilla and the mandible initially as outgrowths at about 60 days of uterine life, forming primitive crypts enclosing tooth germs. True alveolar processes do not develop until the eruption of the teeth, which begins shortly after birth with the deciduous incisors and terminates in late adolescence with the eruption of the third permanent molar tooth. The alveolar process consists of an external continuation of the cortical plate of the two bones, which is reflected back into the socket. This internal alveolar plate is perforated by numerous canals called the *cribriform plate* and allows the passage of blood vessels and nerves into the periodontium. The bone lining the tooth socket consists of bundle bone, which is bone of attachment containing embedded Sharpey's fibers. In the mature jaw the underlying alveolar bone, including that of the outer cortex, is lamellar, with circumferential lamellae on the external surfaces.

Alveolar bone is dependent on tooth function for its proper development and maintenance. If teeth are congenitally absent it fails to develop, and if a tooth is lost, it undergoes resorption. During life, alveolar bone undergoes active remodeling, at a higher rate than any other bone in the adult. This active reworking is probably a consequence of the fact that the teeth undergo slow but continuous movement throughout life, to compensate for tooth wear on both occlusal and contact zones between teeth. The latter process is accentuated along the length of the dental arch in populations who chew particularly vigorously; for example, in Eskimos it may amount to as much as 10 mm during a lifetime. Tooth wear of this fashion, if uncompensated for, would leave spaces between the teeth, and damage to the gums from food impaction. Spacing is prevented by the process of mesial drift, where all the posterior teeth move toward the midline, maintaining close contact between individual teeth. This process is believed to be mediated through the activity of fibroblasts in the transeptal fiber group, which join the teeth into one continuous arch. [*See* BONE REMODELING, HUMAN.]

The temporomandibular joint is a complex articulation between the cranium and mandible. In a functional sense the joints of the two sides should be considered together, because movement in one causes movement in the other. The joints are unusual in that there is a fibrous disc between the cranial articulation on the temporal bone and the condylar process of the mandible. The disc separates the joint space into a superior compartment in which movements are largely of a translatory forward type, and in an inferior compartment they are predominantly rotory. The joint is usually described as being fibrocartilaginous in structure, although this is an inaccurate description. The tissue on all its articulating surfaces is fibrous, and there is normally no cartilage in the temporal bone or disc elements, except possibly in old age. On the condylar surface the fibrous articulating layer is separated from an underlying hyaline cartilage layer by a functionally distinctive cell layer, called the *intermediate cell zone*. This layer is the source of proliferating cells, of which the majority pass downward and differentiate into chondroblasts. They are responsible for most of the postnatal growth in face height. A minority of cells migrate upward to replace effete fibroblasts in the surface articulating layer. This pattern of appositional growth of the condylar cartilage is unique postnatally: in other cartilaginous joints it is usually interstitial. Growth of the condylar cartilages ceases at about 20 years of age, but some believe that it can be restarted in acromegaly, by excessive production of growth hormone; then proliferation of the condylar cartilage gives rise to the protrusive bulldog-like jaw characteristic of untreated cases. [*See* ARTICULATIONS, JOINTS BETWEEN BONES; CARTILAGE.]

The joint is enclosed in a capsule that is strengthened laterally by the temporomandibular ligament. This ligament consists of an oblique superficial component and horizontally orientated deep part. Its function is to prevent the posterior displacement of the mandible, and it is sufficiently strong to cause fracture of the condylar neck rather than backward displacement of the jaw as a result of force exerted on the chin. Also associated with the joint are the sphenomandibular and stylomandibular ligaments. The sphenomandibular ligament is the embryological remnant of the perichondrium surrounding the primary cartilage of the lower jaw, Meckel's cartilage. The stylomandibular ligament extends from the tip of the styloid process toward the angle of the mandible and is part of the fascia enclosing the posterior margins of the parotid gland. Neither ligament exerts a significant role in controlling and stabilizing jaw movements. The capsule is lined on its inner surface by a synovial membrane, which is absent over the articulating surfaces.

Jaw opening involves the depressor muscles in the floor of the mouth and also the lateral pterygoids, which pull the condyle forward onto the articular eminence. This jaw movement involves both a rotatory component and a forward translatory component, which take place in the inferior and superior joint spaces, respectively. Jaw closure is produced by the contraction of elevator muscles and by the controlled relaxation of the major inferior head of the lateral pterygoid muscle. The minor superior head of the pterygoid inserted into the articular disk contracts during this phase, showing that its movements are separate from those of the jaw. Protrusion is achieved by contraction of both lateral pterygoid muscles, and the jaw is pulled back by the horizontal posterior fibers of temporalis as well as digastric muscles. Unilateral jaw shifts, as occur in chewing, take place around an axis located posteriorly to the condyle on the displacement side, with the opposite condyle being shifted anteromedially by its lateral pterygoid muscle.

V. Oral Mucosa

Oral mucosa, which lines the oral cavity, has properties intermediate between skin and the lining of the rest of the digestive tract. It consists of a superficial epithelial layer and an underlying connective layer, the lamina propria. This mucosa forms an envelope, protecting the underlying connective tissues, and has sensory functions, including taste.

Oral epithelium is divided into keratinized (or para-keratinized) and nonkeratinized epithelium. In keratinized epithelium, the cells of a surface layer are completely filled with keratin, and intracellular organelles, including nuclei, are absent. In para-keratinized epithelium, the surface cells are filled with keratin, but some nuclei are still present. Nonkeratinized epithelium contains nuclei in superficial levels, and keratin is scarce. Keratinized cells are impermeable to the passage of materals. Diffusion of substance between cells is also prevented by glycoproteins secreted by immediate level cells. Epithelium in the floor of the mouth, however, is relatively permeable, allowing certain drugs to be administered parenterally via this route. Other nonepithelial cells include pigment cells (melanocytes) which contain the pigment melanin, the nonpigmented Langerhans cells, which are thought to be members of the macrophage series important in presentation of antigenic material to lymphocytes, and Merkel cells, round clear cells that are usually in close approximation to intraepithelial nerves and could have a sensory function.

Oral epithelium is divided into three main categories: masticatory, lining, and specialized or gustatory.

A. Masticatory

Masticatory mucosa, which is usually keratinized or para-keratinized, is found in sites subjected to abrasion, including the gingival tissues around the teeth and the hard palate. The interface between the epithelium and the lamina propria is deeply indented, producing epithelium ridges and troughs filled with prolongations (papillae) from the lamina propria. Masticatory mucosa is firmly bound to the underlying periosteum covering the alveolar processes around the teeth and on the hard palate. In the palate there is a submucosal layer containing fat cells anteriorly and mucous cells posteriorly.

B. Lining

In contrast to masticatory mucosa, lining mucosa is usually nonkeratinized and has a thicker epithelium and a less indented interface with the underlying connective tissues. It covers the lips and cheeks in the oral vestibule, reflecting onto the base of the alveolar process, and meets masticatory mucosa at the muco–gingival junction. Within the oral cavity proper, lining mucosa is restricted to the floor of the mouth.

C. Gustatory

Specialized or gustatory mucosa, which is found on the superior and the lateral aspects of the tongue, is characterized by the presence of papillae, of which four types are recognized. The two smaller types project from the surface of the tongue, one mushroom-shaped (fungiform papillae) with associated taste buds and the other conical keratinized (filiform papillae). Foliate papillae are occasionally found on the sides of the tongue and consist of alternating vertical clefts, lined with epithelium containing taste buds. Circumvallate papillae which are the most distinctive, have taste buds on their inner walls, and respond principally to bitter taste sensation. They are found at the boundary between the anterior two-thirds and posterior one-third of the tongue, anterior to the sulcus terminalis. They are typically 2–3 mm across and are surrounded by a circular ditch or vallum, into which serous glands (of von Ebner) drain.

VI. Salivary Glands

The oral cavity is kept moist by secretions from salivary glands. There are three pairs of major glands located external to the oral cavity, with long ductal systems. Small minor glands are present on the inner surface of the lips and cheek and also in the tongue. Saliva consists principally of water (>99%), and it contains large numbers of epithelial cells and polymorphonuclear leukocytes.

The total volume of saliva secreted in 24 hr is approximately 0.6–1.2 liters, of which the great majority is produced on stimulation from masticatory and gustatory processes associated with eating. The functions of saliva are numerous, including lubrication and protection, and act as a solvent for food substances to be transmitted to taste buds. The most characteristic components are mucins, which are predominantly carbohydrate in composition and are strongly hydrophilic. Saliva contains various proteins with antibacterial properties, including lysozyme, and other substances including lactoferrin, and antibodies including immunoglobulin A, which aggregates oral bacteria. Saliva has important buffering properties (due to bicarbonate, phosphate ions, and amino groups on salivary proteins), which reduce the effects of acid production by oral microorganisms. It is saturated with calcium and phosphate ions, which are prevented from precipitating by the presence of specific proteins.

Salivary glands are formed by initial secretory units and a ductal system. Serous or seromucous secretory units, secreting carbohydrate components, are organized into spherical acini, consisting of cells with a central nucleus and abundant rough endoplasmic reticulum (RER) arranged principally in the apical pole where the small Golgi apparatus is also located. Salivary proteins, including amylase, are synthesized on ribosomes, packed into secretory granules in the Golgi apparatus, and secreted apically. Mitochondria are dispersed through the cell to provide energy for this process. Mucous secretory units have cells arranged in elongated tubular structures called *alveoli*. They contain a flattened nucleus, pushed toward the cell base, around which are small amounts of RER and mitochondria. There is a prominent Golgi complex from which large secretory granules containing mucin are secreted apically. Both types of primary secretory unit drain into intercalated ducts, with a lining of low cuboidal epithelium, and contain the antibacterial proteins lysozyme and lactoferrin. These ducts lead into striated ducts, which have a lining of columnar cells with a centrally placed nucleus and highly infolded external membrane associated with numerous mitochondria. This greatly enlarged cell surface, which its associated ion pumps, plays an important role in ion transfer into and from primary salivary secretions. The primary secretion is isotonic but is made hypotonic by resorption of sodium ions, with some replacement by potassium ions. These cells are the probable source of the important buffering ion bicarbonate, which is secreted in large amounts at high flow rates. Striated ducts terminate in excretory ducts, which combine uniting the secretions of the different parts of the gland. [*See* SALIVARY GLANDS AND SALIVA.]

A. Major Glands

The parotid gland contains almost exclusively seromucous secretory units, relatively long intercalated ducts, draining into prominent striated and excretory ducts, which converge into a single duct entering the mouth opposite the second maxillary molar. The parotid secretion is the most watery (or serous) and contains relatively little mucin but abundant salivary amylase, which may help to digest starch dietary products from around the teeth. The submandibular gland is mixed with a predominantly seromucous product. It contains serous acini, mucous alveoli, and mucous alveoli capped with crescents of serous cells. Intercalated ducts are shorter and striated ducts longer than in the parotid. The main secretory duct terminates in the floor of the mouth, below the tip of the tongue. The sublingual gland consists predominantly of mucous alveoli, with relatively few serous cells usually present as crescentic caps. Both intercalated and striated ducts are short, and the gland drains by multiple ducts into the anterolateral floor of the mouth.

B. Minor Glands

Minor salivary glands in the cheek, lips, and anterior tongue are small mucous glands with short ductal systems. The glands of von Ebner are exceptional, being serous and draining into the ditch around the circumvallate pipillae.

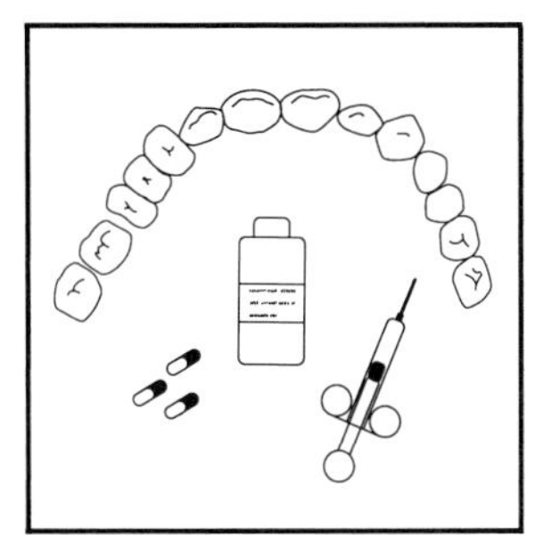

Dental and Oral Biology, Pharmacology

ARTHUR H. JESKE, University of Texas Dental Branch

Glossary

Amide local anesthetic Local anesthetic consisting of an aromatic ring linked to an alkyl chain with an amino terminus by an amide bond

Benzodiazepine One of a group of sedative–hypnotic drugs composed of a benzene ring coupled to a seven-member ring referred to as a diazepine ring

Broad spectrum Refers to an antibiotic that is equally effective against both gram-positive and gram-negative microorganisms

Ester local anesthetic Local anesthetic consisting of an aromatic ring linked to an alkyl chain with an amino terminus by an ester bond

Narrow spectrum Refers to an antibiotic that is limited to a relatively small range of microorganisms (e.g., only gram-positive ones)

Prostaglandin Twenty-carbon fatty acid containing three to five double bonds and a five-member, substituted ring; the several classes of prostaglandins serve a variety of roles as autacoids

Vasoconstrictor Drug used to elevate systemic blood pressure or to reduce bleeding by inducing contraction of vascular smooth muscle with a consequent reduction in blood vessel diameter

THE ADMINISTRATION AND PRESCRIPTION of drugs in dentistry has become considerably more complex over the years, due in large part to the development of a vast array of medically prescribed drugs that may interact with drugs prescribed by the dentist as well as to the introduction of many new drugs for dentistry. Concepts of the treatment of pain and anxiety in dental patients have undergone a virtual revolution in the past decade with the use of new analgesic and sedative drugs, as based on evolving research findings in neuropharmacology.

This same evolution has occurred in the therapy of periodontal disease, which is now increasingly being treated with nonsurgical methods, including antibiotics and other, topically applied antimicrobial agents.

This section will address the state-of-the-art in the most important drug groups used in dental practice (local anesthetics, antibiotics, analgesics, sedatives, and preventive agents).

I. Local Anesthetics

Modern dental local anesthesia utilizes several agents of the amide class of local anesthetics (Table I). They have almost entirely replaced the ester-type agents (e.g., procaine, or Novocain). Procaine is the prototypical drug in the ester class of local anesthetics. The other major class, amides, is typified by the drug lidocaine (Xylocaine and other tradenames). Local anesthetics are drugs that, when placed on a nerve, penetrate the nerve membrane and prevent the transient increase in sodium permeability, which ordinarily accompanies chemical or mechanical stimulation of the nerve. Because depolarization (the change of the electrical potential across the nerve membrane) is prevented in the affected portion of the nerve, impulse generation and conduction is locally blocked and pain impulses cannot be transmitted to the central nervous system. This mechanism of action differs from that of narcotic analgesics, for example, which affect the perception of impulses within the brain. [*See* PAIN.]

At present, local anesthetics are thought to affect nerve membrane sodium conductance through com-

TABLE I Dental Local Anesthetic Agents for Injection

Drug	Tradename	Concentration	Vasoconstrictor
Lidocaine	Xylocaine	2%	Epinephrine
Mepivacaine	Carbocaine	2%	Levonordefrin
		3%	—
Prilocaine	Citanest	4%	—
	Citanest Forte	4%	Epinephrine
Bupivacaine	Marcaine	0.5%	Epinephrine
Etidocaine	Duranest	1.5%	Epinephrine

bination with a receptor located on the internal aspect of the nerve membrane. Combination of the positively charged local anesthetic molecule with the receptor theoretically induces the inactivation of a gate controlling the internal opening of the sodium channel, stabilizing the inactivated, or nonconducting, gate configuration. This model of local anesthetic action also accounts for the phenomenon of use-dependence, in which more frequent nerve depolarizations enhance the rate of nerve block by the local anesthetics. This is presumably due to greater accessibility of the sodium channel and, therefore, the receptor during repeated opening of the channels.

Because of their ability to block inward sodium currents in electrically excitable tissues, local anesthetics can, at toxic blood levels, adversely affect the heart, blood vessels, and central nervous system and, most seriously, can cause central stimulation followed by depression and cardiorespiratory arrest. Toxic reactions to local anesthetics may be associated with overdosage, accidental intravascular injection, or an inability of the patient to effectively metabolize or excrete the local anesthetics. Correct dosages of local anesthetics are determined by the patient's body weight. For example, the maximum recommended dose of lidocaine is 4.4 mg/kg, or a total dose of no more than 300 mg in individuals weighing >70 kg.

In clinical usage, local anesthetics are generally employed in combination with a vasoconstrictor such as epinephrine. The rationale for adding a small amount of such a drug to a local anesthetic is to counteract the vasodilation that is ordinarily produced by the local anesthetic, as described above. This prolongs the duration of anesthesia as well as reduces the rate of absorption of the anesthetic, thereby reducing the overall potential for a toxic reaction.

Vasoconstrictors themselves can produce toxic reactions and may be especially dangerous in patients with cardiovascular disease. Because they constrict small arteries and arterioles, vasoconstrictors can elevate blood pressure, resulting in possible cerebrovascular hemorrhage and an increased work load on the heart, with the potential consequence of myocardial infarction in an already diseased myocardium. The doses of vasoconstrictors in cardiovascular patients must be reduced, and occasionally, if ordered by a physician, they cannot be used in these patients.

The side effects of vasoconstrictors include elevations in blood pressure, cardiac arrhythmias, and increases in cardiac work. In general, vasoconstrictors do not cause adverse effects in dental patients and greatly prolong the duration and increase the depth of local anesthesia. The types of vasoconstrictors used in local anesthetic solutions are found in Table I.

II. Antibiotics

A. Treatment of Dental Infections

In the past decade, the nature of dental infections has been further elucidated to establish the prominent role of obligate anaerobic bacteria (which can only exist in the absence of oxygen) as major infective agents, usually mixed with facultative organisms (which can live with or without oxygen). While up to 264 bacterial groups or species may exist within the oral cavity, several prominent genera (*Staphylococci, Streptococci, Bacteroides, Peptostreptococci,* and *Fusobacterium*) play a major role in dental infections and apparently produce pathological changes by working together through synergistic interactions. This synergism is apparently facilitated by some temporary breakdown in the immune defenses in the host.

B. Antibiotics Used in Dentistry

Penicillin remains the drug of choice for most odontogenic infections. This drug offers the advantages of being bactericidal and inexpensive, and its narrow spectrum (gram-positive) is well suited for application to dental infections. Penicillin kills bacteria by producing a defect in their cell walls, which ultimately allows osmotic forces to rupture the bacterium.

In most situations, antibiotic therapy is instituted

on an empirical basis with usual first-choice agents. Treatment of dental infections should be based on culture and sensitivity testing, although this is frequently not feasible and, in many cases, is unnecessary because mild infections of dental origin frequently respond to first- or second-choice agents when those antibiotics are administered in conjunction with drainage of the infected area and appropriate dental treatment (e.g., root canal therapy, tooth extraction). Culture and sensitivity testing serve as a basis for changing antibiotic therapy if initially selected agents prove to be ineffective.

Unfortunately, penicillin cannot be used in many patients because they are allergic to it, although in nonallergic patients it produces few, if any, side effects. Penicillin is available in oral and injectable forms. The potassium salt of phenoxymethyl penicillin (penicillin V) is usually selected as the oral form, with doses of 250–500 mg administered every 6 hr.

In cases of penicillin allergy, an alternative antibiotic must be selected for initial therapy. Erthyromycin is generally accepted as the best alternate antibiotic for use in routine outpatient infections. Its spectrum is narrow but very similar to that of the penicillins and is, therefore, useful in oral infections. Its major disadvantages include a bacteriostatic, rather than a bactericidal, mechanism of action, fairly rapid development of microbial resistance to the drug, and a relatively high incidence of gastrointestinal side effects, including nausea and vomiting. Several oral forms of erythromycin are available under numerous tradenames, including erythromycin base, erythromycin ethylsuccinate, erythromycin estolate, and erythromycin stearate. Other alternate antibiotics are summarized in Table II.

An important concept in the use of antibiotics in dentistry is that of infection prevention, or prophylaxis. Prophylactic use of antibiotics is generally directed at the prevention of bacterial endocarditis, which can result when microorganisms (especially *Streptococci*) infect diseased heart tissue, such as heart valves damaged by an episode of rheumatic fever, or prosthetic implant materials, such as artificial heart valves. The American Heart Association has developed prophylactic antibiotic regimens for the prevention of such problems. Dental patients with the aforementioned medical complications should receive antibiotic prophylaxis before any procedure that will introduce bacteria into the blood (bacteremia), including routine teeth cleaning (prophylaxis). [*See* ANTIBIOTICS.]

TABLE II Alternate Oral Antibiotics Used in Dentistry[a]

Drug	Usual adult dose
Erythromycin[b]	250–500 mg, q. 6 hr[c]
Cephalexin	250–500 mg, q. 6 hr
Ampicillin	500 mg, q. 6 hr
Tetracycline	250–500 mg, q. 6 hr

[a] Only the most common alternates are listed. Products are available under numerous tradenames.
[b] Base form. Like erythromycin, various salts of antibiotics are available but are not listed.
[c] Refers to dosage every 6 hr. q., "every"

III. Pharmacologic Control of Pain and Anxiety

A. Non-Narcotic Analgesics

Recognition of the role of prostaglandins in the genesis of pain and inflammation has resulted in the now widespread use of nonsteroidal, anti-inflammatory drugs (NSAIDs) for the treatment of dental pain. Aspirin is the prototypical drug in this class, but its use is limited to mild levels of pain in dentistry. When combined with 8, 15, 30, or 60 mg of codeine, it is regarded as a reliable analgesic for relief of moderate dental pain. The newer agents (Table III) are capable of relieving mild to moderate levels of pain, with side effects that, while qualitatively similar to those of aspirin, appear to have a lower incidence and/or reduced severity than those produced by aspirin. The side effects of these drugs include gastrointestinal irritation, inhibition of platelet aggregation (increased bleeding time), allergic reactions of varying severity, fluid retention and edema, and, occasionally, visual and hearing disturbances. The NSAIDs are cross-allergenic with aspi-

TABLE III Oral Nonsteroidal Anti-inflammatory Drugs Used in Dentistry

Drug	Usual adult dosage
Ibuprofen	400 mg, q. 6 hr p.r.n.[a]
Diflunisal	1,000 mg initially, then 500 mg, q. 12 hr p.r.n.
Naproxen[b]	550 mg initially, then 275 mg, q. 6–8 hr p.r.n.
Ketoprofen	25–50 mg, q. 6–8 hr p.r.n.

[a] p.r.n. indicates "as needed" (for pain).
[b] Dosage listed is for sodium salt. q., every.

rin and should not be used in aspirin-allergic individuals. All NSAIDs, including aspirin, can interact with other drugs that prolong bleeding time, irritate the gastrointestinal tract, or lower blood pressure. Protein-bound drugs, such as oral antidiabetic drugs, may be potentiated by drugs in this group.

Acetaminophen is a commonly used analgesic and, in dentistry, is regarded as an alternative to aspirin for the treatment of mild pain or, in combination with codeine, for the treatment of moderate pain. It is usually selected for aspirin-allergic patients or patients with gastrointestinal or bleeding disorders.

B. Narcotic Analgesics

The narcotic analgesics are generally reserved for use in treating moderate-to-severe levels of pain in dentistry. Their use for moderate pain has recently declined with the increasing use of NSAIDs, but they are still widely used for the treatment of severe levels of pain. Codeine is widely used in combination with aspirin or acetaminophen, as described above for moderate pain, as are pentazocine and hydrocodone. Oxycodone, meperidine, methadone, and dihydromorphone are prescribed for oral treatment of severe pain, whereas morphine is usually reserved for the parenteral treatment of the most severe types of pain.

The use of narcotic analgesics is limited by their side effects, the most notable of which in dental applications include nausea, vomiting, and depression of the central nervous system. All the narcotic analgesics mentioned thus far are classified as "controlled substances" by federal and state laws and, as such, have the potential to produce tolerance, physical dependence, and psychological dependence. As central depressants, these drugs can dangerously interact with other drugs, including alcohol, antihistamines, sedative–hypnotics, and others.

In addition to their use in controlling dental pain, narcotics are widely used in combination with sedative–hypnotics to produce enhanced analgesia and patient cooperation during intraoperative, intravenous sedative procedures. A newer group of drugs, including butorphanol and nalbuphine, are partial narcotic antagonists and are being used in intravenous sedation to limit the degree of respiratory depression encountered with classical narcotics, such as morphine or meperidine.

C. Sedative–Hypnotic Drugs

This group of drugs constitutes the major group of agents used in dentistry to control pre- and intraoperative anxiety. In the past, barbiturates such as pentobarbital saw widespread use for dental sedation, but because of their attendant problems (respiratory depression, larygnospasm and abuse potential), they have been replaced by the benzodiazepine-type agents (Table IV). These drugs represent a significant advantage in producing less respiratory depression than the barbiturates. They are also useful as orally administered, preoperative sedative–hypnotic drugs. Some drugs in this group (flurazepam, lorazepam, temazepam, and triazolam) are better for hypnosis (induction of sleep on the night before a dental appointment), whereas others (alprazolam, chlordiazepoxide, clorazepate, diazepam, and oxazepam) are preferred for their antianxiety effect during waking hours. Until recently, diazepam was the preferred benzodiazepine for intravenous sedation in dentistry; however, because it is not water-soluble and its vehicle (propylene glycol) produced thrombophlebitis in some patients, midazolam (a water-soluble benzodiazepine) is now gaining popularity for this application.

The benzodiazepines are classified as controlled substances, because long-term use can produce tolerance and dependence. Their principal side effects include drowsiness and other types of central depression, paradoxical central excitation, allergy, weight gain, nausea, vomiting, and xerostomia. They can, of course, interact adversely with other central depressant drugs. They are contraindicated in allergic individuals, pregnancy, some types of

TABLE IV Benzodiazepine Drugs

Official name	Tradename
Alprazolam	Xanax
Chlordiazepoxide	Librium
Clonazepam	Klonopin
Clorazepate	Tranxene
Diazepam	Valium
Flurazepam	Dalmane
Lorazepam	Ativan
Midazolam	Versed
Oxazepam	Serax
Przepam	Centrax
Temazepam	Restoril
Triazolam	Halcion

glaucoma, liver or kidney disease, and in chronic obstructive pulmonary disease.

In addition to barbiturates and benzodiazepines, a number of other drugs may be used for sedation and anxiety control in dentistry. In children, hydroxyzine, an antihistamine, and chloral hydrate are widely used and can be administered by the oral route. Other nonbarbiturate agents include carbamates (ethinamate and meprobamate), ethchlorvynol, paraldehyde, glutethimide, and methyprylon. The oral administration of a drug to produce sedation is limited because it relies on patient compliance, the absorption of the drug can be very erratic, the drug effect cannot be titrated, and the duration of action tends to be prolonged.

Nitrous oxide (laughing gas) is used frequently to produce relief of anxiety and control of gagging in dental patients. Nitrous oxide is administered by inhalation of 20–50% concentrations in combination with 80–50% pure oxygen, respectively. Nitrous oxide has a rapid onset of action and can be rapidly titrated to the level of sedation needed; it is very safe, nonallergenic, and not metabolized by the liver. Nitrous oxide is rapidly excreted by the lungs so that patient recovery is extremely rapid, unlike other forms of sedation. Because nitrous oxide lacks significant effects on the cardiovascular system and other organs, it is very useful in patients with systemic disease. It is contraindicated in patients with some types of respiratory diseases, pregnancy, and in uncooperative patients. Long-term exposure to nitrous oxide can produce a polyneuropathy and, in females, has been implicated in increasing the risk of spontaneous abortion and fetal malformations.

IV. Fluorides and Antiplaque Agents

The beneficial effects of fluoridation of drinking water on dental caries have been known since the 1950s. The systemic administration of optimal fluoride levels in infants, children, and young adolescents results in the incorporation of fluoride ion into the enamel crystals of developing teeth, rendering them significantly less soluble in the acids produced by dental plaque bacteria. Where drinking water concentrations are suboptimal (<1 part per million, or 1 mg/liter), systemic supplementation may be accomplished by the prescription of sodium fluoride (in liquid or tablet form). The recommended total dietary fluoride ion intake is 0.25 mg for infants (neonates to age 2 yr), 0.5 mg for ages 2–3 yr, and 1 mg/day for ages 3–13 yr.

Fluoride also confers reduced acid solubility when applied topically to teeth. Prescription-strength preparations for in-office or home application include 0.2% sodium fluoride; 0.5–1.23% acidulated phosphate fluoride gels, sodium fluoride, and phosphoric acid topical solutions and gels; and 0.4% stannous fluoride gels. Fluoride, as sodium fluoride, is available over-the-counter in a variety of dentifrices as well as in 0.05% concentrations in mouth rinses.

More recently, agents specifically targeted at dental plaque have become available. Of these, chlorhexidine is now available in prescription form as a 0.2% mouthrinse and acts as a broad-spectrum antibacterial. Chlorhexidine reduces the formation of plaque, and its side effects are limited to a bitter taste, reversible staining of teeth and restorations, and occasional irritation of the oral mucosa.

Other Drugs Used in Dental Practice

A. Corticosteroids and Antihistamines

Corticosteroids, such as prednisolone, are used to a limited degree in dentistry to control inflammation and as secondary agents in the treatment of immediate-type (anaphylactic) allergic reactions. Antihistamines, including diphenhydramine, are prescribed for the symptomatic treatment of mild, delayed-type allergic reactions.

B. Emergency Drugs

Epinephrine is recommended for the emergency treatment of acute, life-threatening allergic reactions and is administered subcutaneously or intravenously in this application. Other emergency drugs typically found in a dental office include nitroglycerin (antianginal), diazepam (anticonvulsant), spirits of ammonia (smelling salts), naloxone (narcotic antidote), glucose (for hypoglycemia), and oxygen with positive-pressure ventilation assist.

Bibliography

Cooper, S. A. (1983). New peripherally-acting oral analgesic agents. *Ann. Rev. Pharmacol. Toxicol.* **23**, 617–647.

Council on Dental Therapeutics (1984). "Accepted Dental Therapeutics," 40th ed. American Dental Association, Chicago.

Gangarosa, L. P., Ciarlone, A. E., and Jeske, A. H. (1983). "Pharmacotherapeutics in Dentistry." Appleton-Century-Crofts, Norwalk, Connecticut.

Gjermo, P. (1989). Chlorhexidine and related compounds. *J. Dent. Res.* **68,** 1602–1608.

Malamed, S. F. (1989). "Sedation–A Guide to Patient Management," 2nd ed. The C. V. Mosby Co., St. Louis.

Moenning, J. E., Nelson C. L., and Kohler, R. B. (1989). The microbiology and chemotherapy of odontogenic infections. *J. Oral Maxillofac. Surg.* **47,** 976–985.

Newbrun, E. (1989). "Cariology," 3rd ed. Quintessence Publishing Co., Inc., Chicago.

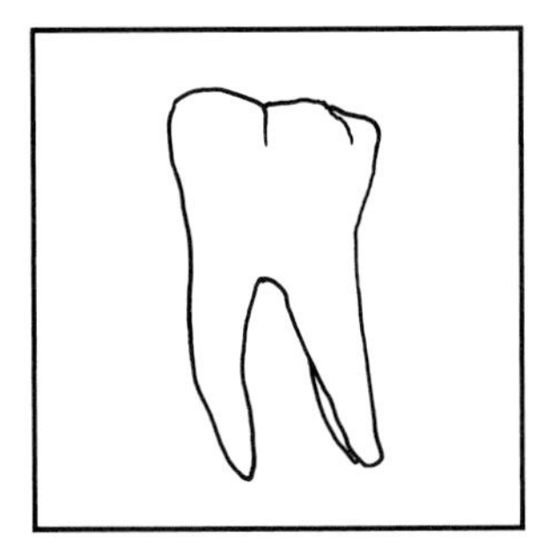

Dental Anthropology

G. RICHARD SCOTT, *University of Alaska, Fairbanks*

Glossary

Antimere Corresponding teeth in the left and right sides of a jaw (e.g., left and right upper first molars)
Carabelli's trait Morphological character derived from cingulum of mesiolingual cusp of upper molars
Cingulum Bulge or shelf passing around the base of the tooth crown
Cusp Pointed or rounded elevation on the occlusal (chewing) surface of a tooth crown
Isomere Corresponding teeth in the upper and lower jaws (e.g., left upper and lower first molars)
Labret Ornament worn in and projecting from a hole(s) pierced through the upper and lower lips and cheeks
Phenetics Classificatory method for adducing relationships among populations on the basis of phenotypic similarities
Quadrant One half of the upper or lower dentition
Shoveling Mesial and distal marginal ridges enclosing a central fossa on the lingual surface of the incisors

DENTAL ANTHROPOLOGY IS A field of inquiry that utilizes information obtained from the teeth of either skeletal or modern human populations to resolve anthropological problems. Given their nature and function, teeth are used to address several kinds of questions. First, teeth exhibit variables with a strong hereditary component that are useful in assessing population relationships and evolutionary dynamics. Given their role in chewing food, dental pathologies and patterns of tooth wear can indicate kinds of food eaten and other aspects of dietary behavior, including food preparation techniques. Teeth can also exhibit incidental or intentional modifications, which reflect patterns of cultural behavior. Finally, as the process of tooth formation is highly canalized (i.e., buffered from environmental perturbations) developmental defects provide a general measure of environmental stress on a population. Researchers in several disciplines, including physical anthropology, archeology, paleontology, dentistry, genetics, embryology, and forensic science conduct research that falls directly or indirectly within the province of dental anthropology.

I. The Human Dentition

A. Terms and Concepts

A tooth has two externally visible components, crown and root, and is made up of three distinct hard tissues, enamel, dentine, and cementum, and one soft tissue, the pulp, which provides the blood and nerve supply to the crown and root. Teeth are anchored in the bony alveoli of the upper and lower jaws by one or more roots and the periodontal membrane. Terms of orientation for teeth are mesial (toward the anatomical midline, or the point between the two central incisors); distal (away from the midline); buccal (toward the cheek); labial (toward the lip); lingual (toward the tongue); and occlusal (the chewing surface of a tooth). [*See* DENTAL AND ORAL BIOLOGY, ANATOMY.]

The reptilian dentition is homodont (generally uniform, single-cusped, conical teeth for grasping food objects) and polyphyodont (multiple generations of teeth). By contrast, the mammalian dentition is heterodont (four types of teeth, each performing different functions) and biphyodont (only two generations of teeth, primary and permanent). Thus, humans share with other mammals the presence of four distinctive tooth types (i.e., incisors, canines, premolars, and molars) and two successional dentitions. In the human primary dentition, there are two incisors, one canine, and two molars per quadrant. In the permanent dentition, there are two incisors, one canine, two premolars, and three molars per quadrant.

In the human dentition, the four tooth types are found on both the left and right sides of the upper jaw (maxilla) and lower jaw (mandible). Antimeres exhibit mirror imaging, but isomeres differ in both size and morphology. Incisors and canines, referred to collectively as anterior teeth, have one cusp and a single root. Premolars exhibit one buccal cusp, one lingual cusp, and a single root. Upper molars are characterized by four major cusps and three roots, while lower molars exhibit five cusps and two roots. The multicusped premolars and molars are referred to collectively as posterior or cheek teeth. Variations on these normative characterizations of cusp and root number fall within the area of dental morphological variation.

An important concept that relates to the different types of teeth in mammals is Butler's Field Theory. When this concept was adapted to the human dentition by A. A. Dahlberg, he used the phrase tooth districts to describe eight morphological classes corresponding to the four types of teeth in the two jaws. Within each tooth district, there is a "key" tooth, which shows the most developmental and evolutionary stability in terms of size, morphology, and number. For humans, the key tooth in a given tooth district is usually the most mesial element (e.g., upper central incisor, lower first molar); the only exception is in the lower incisor district where the lateral incisor is the key tooth. In a given tooth district, variation increases with distance from the key tooth. In the two molar districts, for example, the first molar is the most stable (least variable), while the third molar is highly variable (least stable); second molar variation falls between these extremes. The implication is that the key teeth best reflect the genetic-developmental program controlling tooth development, whereas the distal elements of a field are more susceptible to environmental effects. This may be related to the relatively protracted period of tooth development in humans; early-developing teeth (e.g., M1) are the most stable, while late-developing teeth (e.g., M3) exhibit more environmentally induced variability.

B. Why Study Teeth?

For the resolution of anthropological problems, a number of advantages are associated with the study of human dental variation.

1. Preservability

Teeth preserve exceptionally well in the archeological record (due in part to the chemical properties of enamel) and are frequently the best represented part of a skeletal sample. This is evident in both Holocene archeological series and Pleistocene hominid fossil remains.

2. Observability

For the most part, human biologists interested in biochemical polymorphisms, dermatoglyphics, and other anatomical and physiological variables are limited to the study of living populations. Likewise, most variables of interest to human osteologists can be observed only in prehistoric and protohistoric skeletal remains. Teeth, on the other hand, can be directly observed and studied in both skeletal and living populations (e.g., through intraoral examinations, permanent plaster casts, extracted teeth). Because teeth are observable in both extinct and extant human groups, they provide a valuable research tool for the analysis of short-term and long-term temporal trends.

3. Variability

Because teeth are critical in food-getting and food-processing behavior, their development is controlled by a relatively strict set of genetic-developmental programs. On the other hand, as the dentition interfaces directly with the environment, teeth are also modified postnatally by physical factors associated with mastication and disease factors related to the interplay of dietary elements and a complex oral microbiota. Thus, a wide variety of dental variables are available for analysis; some provide information on the genetic background of a population, while others reflect environmental and behavioral factors that impinge on individuals in a given population.

C. Teeth as Indicators of Age

An accurate determination of age and sex is fundamental to any inquiry relating to human skeletal remains in both archeological and forensic contexts. One characteristic of the dentition, which make teeth useful in aging individual skeletons, is a predictable sequence of developmental events, including crown and root formation, calcification, and eruption. As this genetically controlled sequence of events varies to only a limited extent among recent human populations, the principles of aging children by stages of dental development can be applied to all human groups. Before the age of 12 yr, teeth are the best and most readily available indicator of age. During adolescence, teeth provide a useful adjunct to patterns of epiphyseal fusion in age determination. After the permanent dentition is completed with the eruption of the third molars (ca. 18 yr of age), degree of crown wear and gradients of wear between the first, second, and third molars allow researchers to estimate adult age by decade or within broader age categories (e.g., young, middle, and old adult). Because tooth wear in adulthood has a strong cultural component, it is necessary to apply different standards to spatially and temporally circumscribed populations. For example, medieval Europeans exhibited much greater degrees and rates of crown wear than modern Europeans, so tooth wear standards for modern Europeans would not be applicable to their medieval forebearers.

II. Dental Phenetics and Phylogeny: Inferring History from Teeth

Traditionally, physical anthropologists have been interested in human population origins and relationships. Historical questions can be posed on a broad geographic scale (e.g., where did Native Americans come from?; who are they most closely related to?; when did they arrive in the New World?) or have more regional focus (e.g., what is the relationship between peoples of the Jomon culture and modern Japanese and Ainu populations?). Many biological traits have been employed to address such questions, including general morphology, body size, pigmentation, dermatoglyphic patterns, genetic markers of the blood, mitochondrial DNA, and metric and nonmetric skeletal traits. Although variation in human tooth size and morphology has long been recognized, historical inferences based on teeth have been limited until recently.

The derivation of historical relationships from dental data requires variables with a significant genetic component. Variables that meet this requirement fall under the broad headings of tooth size, crown and root morphology, hypodontia (missing teeth), hyperodontia (supernumerary teeth), and eruption sequence polymorphisms. As most historical analyses focus on tooth size and morphology, this discussion is limited to metric and morphologic variables.

A. Tooth Size

In studies of human tooth size variation (odontometrics), the measurements reported most commonly are maximum crown length (mesiodistal [MD] diameter) and maximum crown breadth (buccolingual [BL] diameter). In some instances, measurements are reported for crown height and intercuspal distances, but crown wear must be minimal or the landmarks used for measurement are obliterated. Recently developed techniques to measure the volume of individual teeth are promising, but little comparative volumetric data are currently available.

In human populations, there is a modest sex dimorphism in tooth dimensions. A comparison of individual crown diameters within a single population usually shows that male teeth are 2–6% larger than those of females. This dimorphism is most pronounced in canine dimensions. Because there is a slight but consistent dimorphism in tooth size, workers generally present data separately for male and female tooth diameters.

In the human dentition, a high degree of dimensional intercorrelation exists; i.e., the size of one tooth is not independent of the size of all other teeth. Correlation matrices among the 32 MD and BL diameters (not including antimeres) show that interclass correlation coefficients vary from about 0.30 to 0.60 between specific tooth dimensions. Principal component analyses usually reveal that three to seven underlying components account for 50–75% of the covariance among the 32 crown dimensions. Although different populations may not exhibit comparable component loadings, some of the primary latent structures involve (1) general size; (2) anterior vs. posterior teeth; (3) MD vs. BL

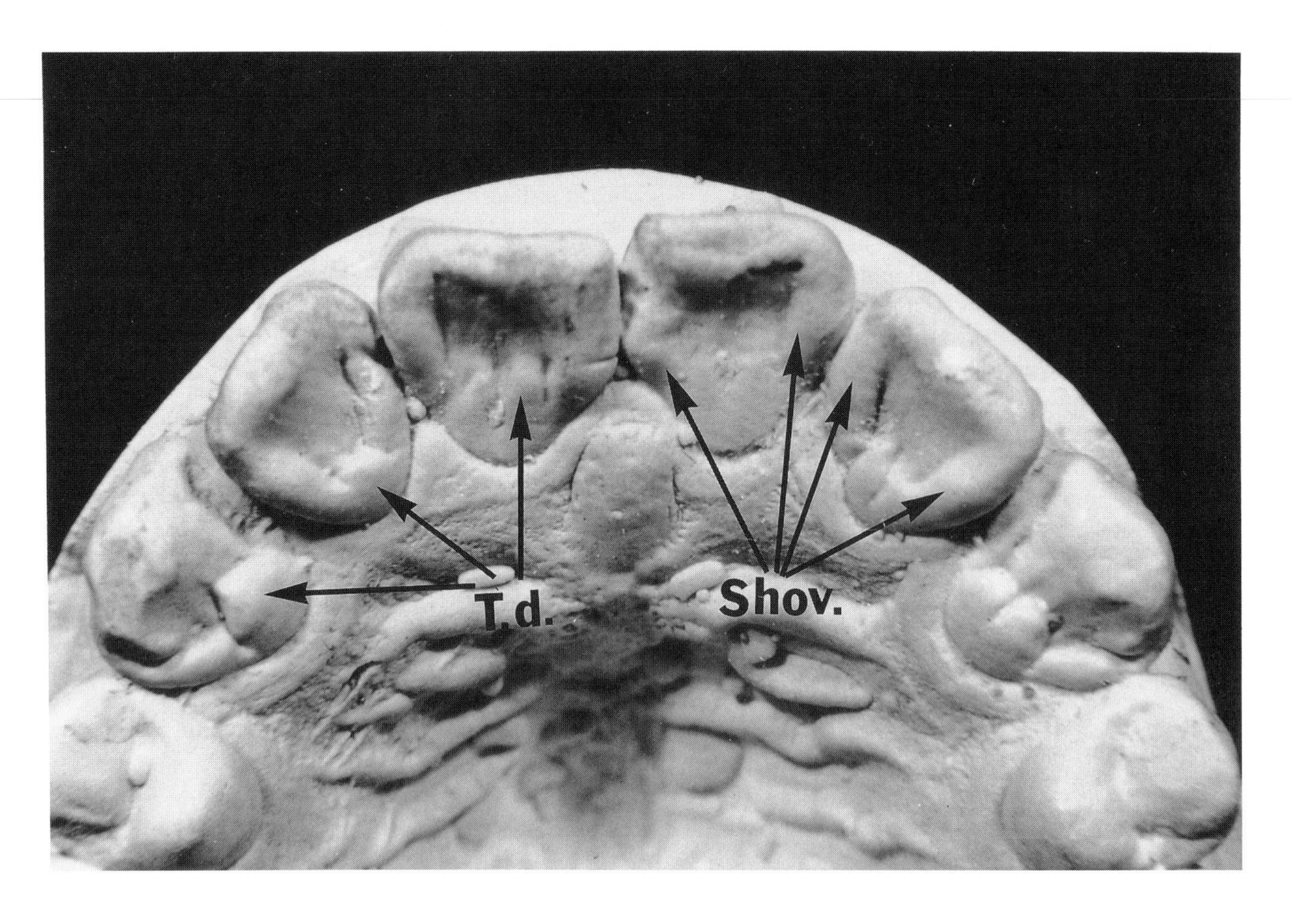

FIGURE 1 Upper anterior teeth exhibiting shoveling (Shov.; mesial and distal lingual marginal ridges) and *tuberculum dentale* (T.d.; cingular ridges and tubercles).

diameters; and (4) premolars vs. molars. Typically, component loadings for maxillary and mandibular isomeres are similar. Although the exact significance of the latent structures underlying the major tooth size components remains unknown, they probably reflect some aspect of the genetic programs that moderate tooth crown development.

In addition to interdimensional correlations, crown diameters are also associated with other dental variables, including hypodontia, hyperodontia, and, to some extent, crown morphology. Within European populations, large-toothed (megadont) individuals are more likely to have supernumerary teeth, while small-toothed (microdont) individuals are more likely to have missing teeth. There is also a detectable but weak relationship between crown size and the expression of certain morphologic traits (e.g., the hypocone and Carabelli's trait of the upper molars).

B. Crown and Root Morphology

Teeth exhibit two types of morphological variation. First, there is variation in the form of recurring structures (e.g., labial curvature of the upper central incisors). However, most morphological crown and root traits that have been operationally defined take the form of presence–absence variables. That is, within a population, some individuals exhibit a particular structure while others do not. For tooth crowns, such structures may be exhibited as accessory marginal or occlusal ridges (cf. Fig. 1, incisor shoveling; Fig. 2, deflecting wrinkle), cingular derivatives (cf. Fig. 1, *tuberculum dentale*), and supernumerary cusps (cf. Fig. 2, cusps 6 and 7). Morphological root traits are most often defined in terms of variation in root number; lower molars, for example, can exhibit one, two, or three roots. For most crown and root traits manifested as presence–absence variables, presence expressions vary in degree from slight to pronounced.

Although some morphological variables exhibit significant sex differences (e.g., the canine distal accessory ridge), the majority of these traits show similar frequencies and class frequency distribu-

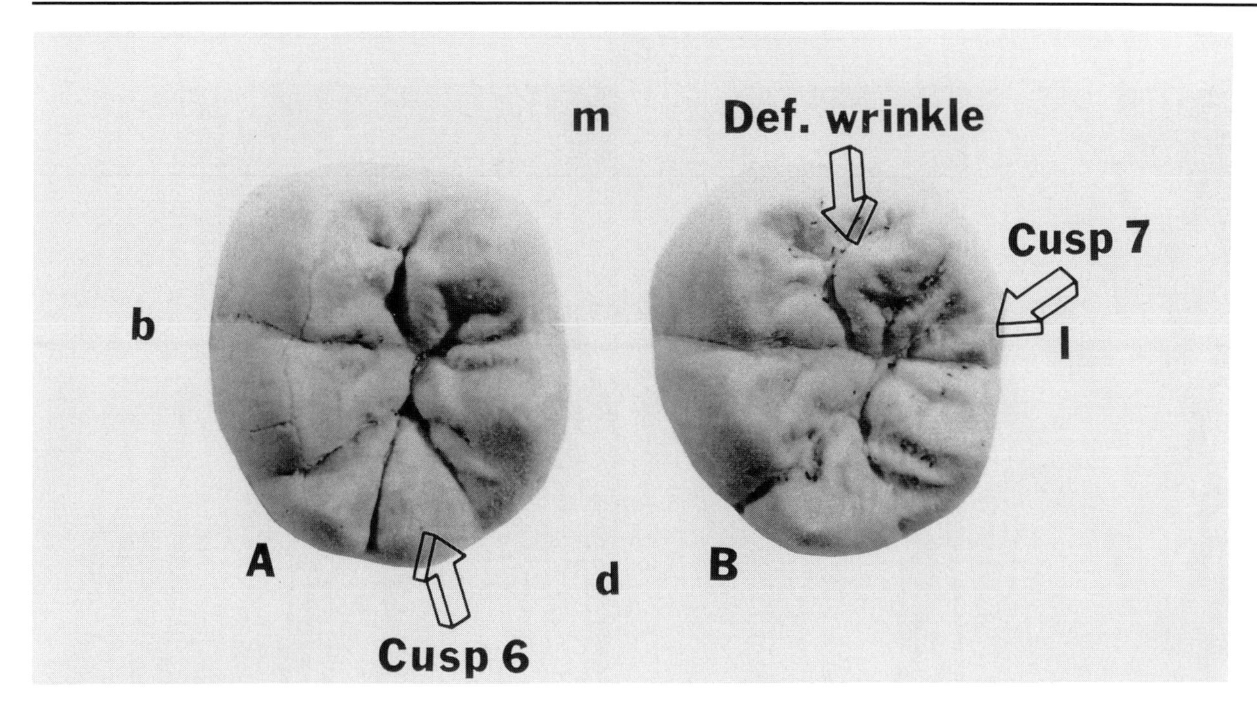

FIGURE 2 Two left lower first molars: tooth A exhibits five major cusps plus a supernumerary cusp (cusp 6); tooth B exhibits an occlusal ridge variant on the mesiolingual cusp (deflecting wrinkle) and a supernumerary cusp between the mesiolingual and distolingual cusps (cusp 7). Orientation: m, mesial; d, distal; b, buccal; l, lingual.

tions for males and females. For this reason, population frequencies are generally reported for combined data on males and females.

Crown morphology is characterized by high within-field correlations in trait expression (cf. Fig. 1; incisor shoveling on U11 and U12; *tuberculum dentale* on UI1, UI2, UC). Shoveling, which can be expressed on all four upper and four lower central incisors, should not be viewed as four or eight distinct traits. Rather, it is a single trait that may be expressed on all teeth of the upper and lower incisor districts. Although there are a few exceptions (e.g., Carabelli's trait and the protostylid), different morphological traits usually show little or no correlation among themselves.

C. Dental Genetics

Monozygotic (MZ) twins share identical genotypes, a fact widely exploited by geneticists interested in determining the relative genetic and environmental components of variance underlying the expression of biological traits. Although there are complicating factors, such as common prenatal and postnatal environments, traits with a strong genetic component should show similar expressions between MZ twin pairs. The dentitions of identical twins, in fact, often show remarkable parallels in crown size, the timing and sequence of tooth eruption, general crown form, and small morphological details.

Two matched MZ twin pairs, shown in Fig. 3, illustrate close similarity in tooth size, form, and morphology. However, the expression of one morphologic feature in these twin pairs can be used to illustrate a general point. Carabelli's trait, when present, is manifested on the mesiolingual cusp of the deciduous second molars and permanent molars. In Fig. 3, this cingular trait is expressed to a moderate degree on the permanent first molars of all four twins. However, in one set of twins (B1-B2), there is almost perfect concordance in the form and expression of this trait, while degree of expression differs between A1 and A2. As MZ twins have the same genotype, such differences in expression are environmental in origin. In general, discordance between MZ twins for any dental variable is attributable to environmental effects. The differences in tooth structure and form between MZ twins may be caused by environmental factors similar to those that produce asymmetry between the left and right sides of the dentition.

As a series of metric variables, the hereditary basis of human tooth size has been assessed through the methodology of quantitative genetics. Workers who have analyzed tooth size in twins and families estimate that the heritability of crown dimensions is relatively high (ca. 0.60–0.80). These estimates in-

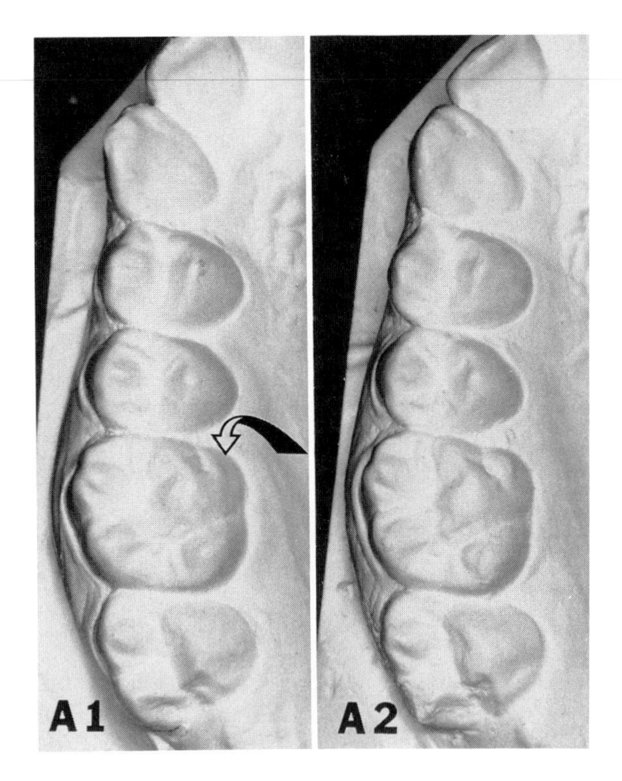

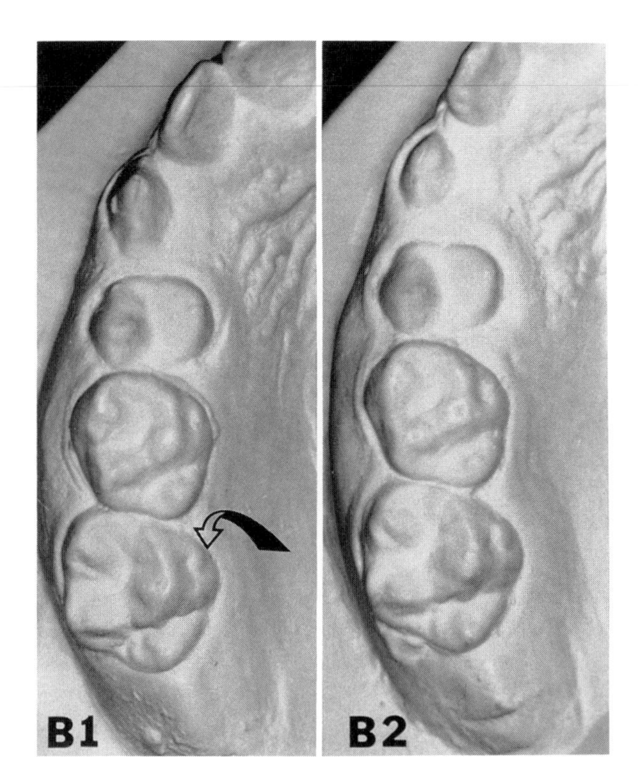

FIGURE 3 Right upper dentitions of two pairs of monozygotic twins; Carabelli's trait (indicated by arrows) differs in degree of expression between A twins (A1, A2) but is identical between B twins (B1, B2) on both the deciduous second molar and the permanent first molar.

dicate a strong genetic component in the development of tooth size, but environmental factors, including maternal effects, also have some influence. Studies showing significant differences in tooth size between generations also point to an environmental component in dental development.

Considering the presence–absence nature of crown and root traits, early workers thought these variables might follow simple dominant-recessive patterns of inheritance. It now seems likely that morphologic traits, like other threshold traits, have polygenic modes of inheritance similar to continuous variables. That is, the genotypic distribution underlying trait expression is continuous (with multiple loci and/or alleles), but there is an underlying scale (absence) and a visible scale (presence) associated with this distribution, with the two scales separated by the presence of a physiological threshold. Genotypically, individuals who fall below the threshold fail to express the trait, whereas those just beyond the threshold exhibit slight expressions. With increasing distance from the threshold along the visible scale, there is an associated increase in degree of expression.

Although complex segregation analysis suggests that major gene effects may be associated with the expression of some dental morphological variables, it is still not possible to reduce this variation to gene frequencies. Currently, dental morphological data are presented in the form of class frequency distributions (i.e., the frequency of absence and each degree of trait presence) or total trait frequencies.

D. Tooth Size and Population History

Teeth from many human populations, skeletal and living, have been measured for mesiodistal and buccolingual crown diameters. These basic tooth crown dimensions are often broken down into two components for between-group odontometric comparisons. First, there is a size component that centers on the absolute dimensions of the tooth crowns. The second component, shape, is a measure of among-tooth proportionality. Between-group analyses of odontometric variation often utilize multivariate distance statistics, which are designed to take into account both absolute dimensions and among-tooth proportions. In particular, the newly developed tooth crown apportionment method may prove to be a powerful method for utilizing crown size to assess affinities among human populations. However, for a general characterization of tooth size variation, two synthetic variables can serve to illustrate dimensional and proportional variation among the major subdivisions of humankind.

To summarize human tooth size variation on a global scale, summed cross-sectional crown areas of upper and lower premolars and molars (excluding M3) were calculated for males from 75 recent skeletal and living samples representing 12 geographic regions or population groupings. The derived means and ranges for posterior crown areas (see Fig. 4) show four general divisions of humankind: (1) <700 mm^2, this small-toothed grouping includes two broad geographic populations, India and the Middle East, and two relatively small groups from Europe (Saami) and South Africa (San); (2) 700–750 mm^2; this grouping includes Asian (East Asia, Southeast Asia) and Asian-derived (Eskimo-Aleut, American Indian) populations and Europeans (including European-derived groups); (3) 750–800 mm^2, this range includes subSaharan Africans (excluding San) and Melanesians; and (4) >800 mm^2, with a mean posterior tooth crown area of 864.3 mm^2 and a range of 835.9–912.8 mm^2, native Australians fall well beyond any other regional population in tooth size. It should be noted that the limited range for three of the regional populations (San, Saami, Melanesia) is due primarily to the small number of samples used in the calculations. Moreover, the distantly related yet small-toothed San and Saami populations also share small body size. The question of tooth size–body size scaling is an important avenue of inquiry that should be further explored in different geographic contexts.

Although absolute tooth dimensions provide useful information on relative population relationships, odontometric comparisons are even more discriminating when tooth shape is also taken into account. To illustrate how information on shape differs from that of size, the population variation of the incisor length index (UI1 MD diameter/UI2 MD diameter × 100) has been summarized on a global scale. A low index indicates that the upper lateral incisor is broad relative to the upper central incisor, whereas a high index signifies a relatively narrow lateral incisor.

The means and ranges of the UI1/UI2 index, derived from 105 samples, are shown in Fig. 5 for the same 12 geographic areas (or populations) compared for overall crown dimensions. The worldwide sample range for this index is 113.5–139.4, with group means between 119.3 and 130.0. Of the first five groups with the lowest mean indices, four represent Asian and Asian-derived groups (five if one includes Melanesians, but their historical status is less certain). At the other extreme, Asiatic Indians, Middle Easterners, and Europeans have the highest indices. SubSaharan Africans, the San, and native Australians fall in the middle of the global range.

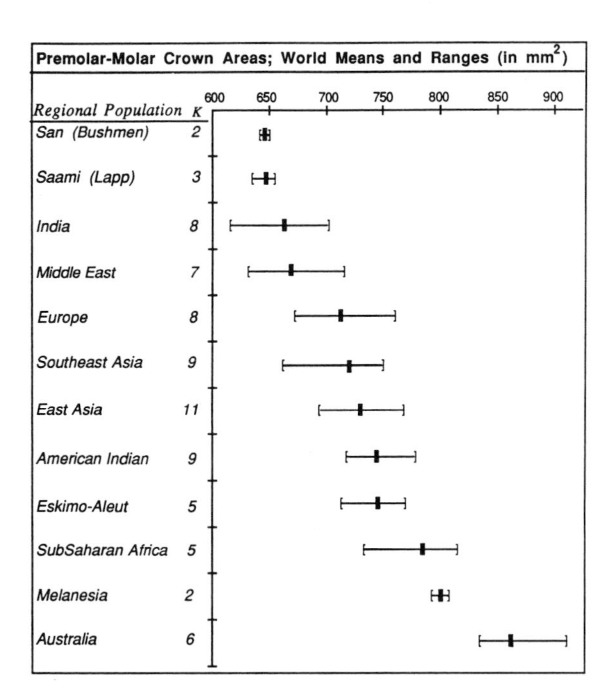

FIGURE 4 Global variation in cross-sectional crown areas of male upper and lower premolars and molars (excluding third molars). Vertical bars denote means, and horizontal lines show ranges within geographic regions. K, number of samples used in calculations.

When the major geographic subdivisions of humankind are analyzed on the basis of simple genetic markers, population geneticists find that (1) Africans are the most highly differentiated from all other regional populations; (2) Asiatic Indians, Middle Easterners, and Europeans form a coherent genogeographic grouping; and (3) mainland Asian and Asian-derived groups in the Americans and the Pacific cluster together at low to intermediate levels of differentiation. Australians remain the most enigmatic population from a genetic standpoint, with hints of distant historical ties to both Southeast Asia and Africa.

Geographic variation in absolute tooth dimensions and the UI1/UI2 index broadly parallel the relationships indicated by measures of genetic dis-

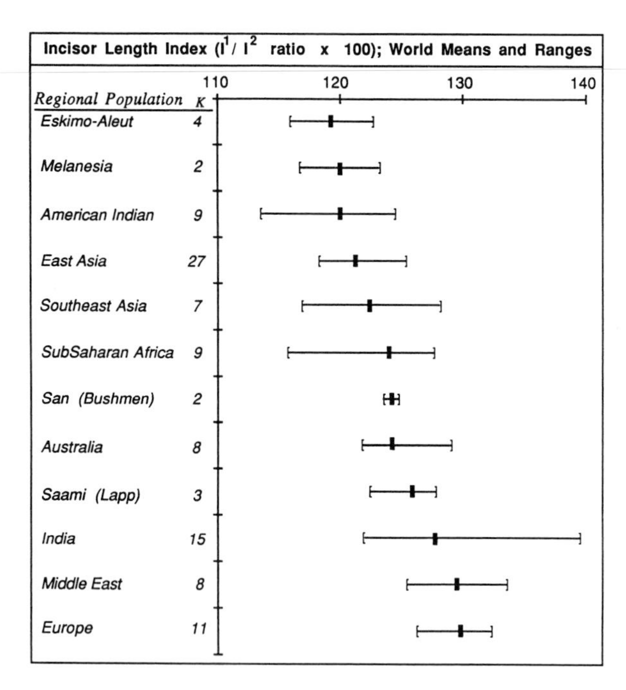

FIGURE 5 Global variation in incisor length index. Vertical bars represent means, and horizontal lines show ranges within geographic regions. K, number of samples used in calculations.

tance. East Asians, Southeast Asians, American Indians, and the Eskimo-Aleut clustered for both tooth size and the incisor length index. Europeans clustered with Asiatic Indians and Middle Easterners for the incisor length index, although tooth size is somewhat larger in Europe. The Saami, with exceptionally small teeth, fall between European and Asian groups for the incisor length index as they do for many other genetic and biological variables. Australians, who reflect a world extreme for tooth size, are in the middle of the range for the incisor length index, more closely aligned with African than with European or Asian groups. Africans are, however, less distinctive odontometrically than they are genetically. Based on this summary, they show more dental similarity to Melanesia and Australia than to either Europe or Asia.

In general, when both tooth size and shape are taken into account, odontometric data provide a useful tool for assessing population relationships. Of course, this discussion focused on the distributions of sample means for some of the major subdivisions of humankind. There is still some debate among researchers on the discriminatory power of odontometric data when comparisons are made among closely related populations. Improved analytical methods (e.g., tooth crown apportionment) and refinements in measuring tooth mass (e.g., volumetrically) should ultimately increase resolution in studies of tooth size microdifferentiation.

In addition to addressing questions of population affinities, tooth size variation has also been used to assess temporal trends in recent human evolution. In many different parts of the world, temporally divided Holocene skeletal collections show a significant decrease in tooth size over the past 12,000 years. Frequently, this decrease in tooth size parallels the origins and development of agriculture. Some workers have explained this trend in terms of natural selection (either reduced selection and concomitant reduction in tooth mass or positive selection for smaller and morphologically simpler teeth), but environmental factors may also play a role. Some domestic animals of the early Neolithic also tend to have smaller teeth than their wild ancestors. The correlates of sedentism (increased population density, endemic and epidemic diseases, between group warfare and other forms of general stress), along with the development of new food preparation techniques and the greater reliance on plant domesticates, may have produced a similar effect in the "domestication" of human populations. [*See* EVOLUTION, HUMAN.]

E. Dental Morphology and Population History

Although human populations exhibit a great deal of within- and between-group variation in the frequencies of various crown and root traits, this fact was not fully exploited in studies of population relationships and origins until recently. This can be attributed to (1) improved and expanded standards of observation, increasing intra- and interobserver reliability in scoring morphologic trait expression; (2) the ready availability of computers and statistical packages for calculating biological distance measures and performing cluster and components analysis; and (3) the demonstration that tooth morphology is useful in assessing population relationships at levels of differentiation below the major geographic race.

The utility of dental morphology in resolving questions of population history is well illustrated by a problem that has concerned anthropologists for decades: the question of Native American origins. Many years ago, A. Hrdlička argued that American Indians were most closely related to Asian populations; one biological feature used to support this position was incisor shoveling (Fig. 6). Asians and

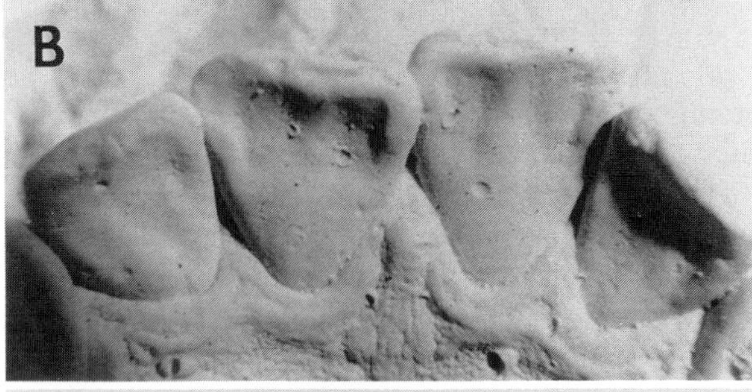

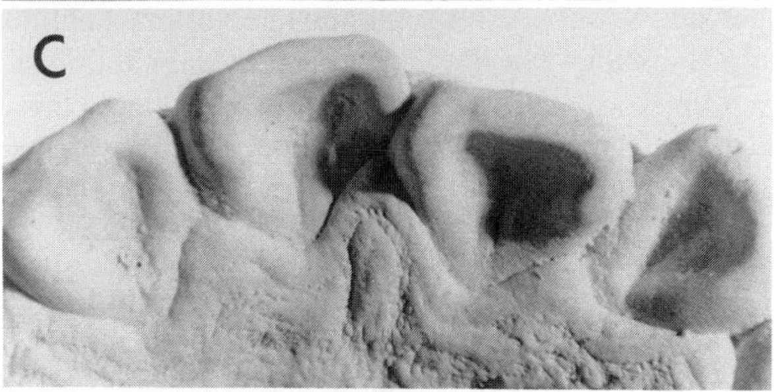

FIGURE 6 Variation in expression of upper incisor shoveling; slight shoveling shown in A and B characteristic of Europeans, while pronounced expressions in C and D limited largely to north Asians and Native Americans.

Native Americans expressed this trait in high frequencies and often pronounced degrees, whereas Europeans and Africans were characterized by lower frequencies and slight degrees of trait expression. Genetic markers and other biological traits eventually vindicated this position; the similarity between Asians and Native Americans in shoveling was an indication of homology, not analogy, and could reasonably be attributed to common ancestry.

The analysis of population variation in another trait, three-rooted lower first molars (3RM1), led to a further refinement of knowledge on Native American origins. After making observations on samples of Aleut, Eskimo, and American Indian skeletal remains, C. G. Turner II noted that the frequencies of 3RM1 fell into two distinct clusters: Eskimos and Aleuts were characterized by high frequencies (20–40%), whereas American Indians showed uniformly low frequencies (ca. 5%). The only exceptional group among American Indians was a small Navajo sample, representing the Na-Dene language family, with an intermediate frequency of 27%. On the basis of this single trait, Turner hypothesized that recent populations of the New World (i.e., American Indians, or Macro-Indians, south of the Canadian border; Na-Dene speakers of Alaska and Canada; Eskimo-Aleuts) represent the descendant groups of three major migrations out of Asia. Subsequent observations on 23 crown and root traits among dozens of samples representing North and South American native populations supported the initial model based on 3RM1 frequencies. One slight modification was to expand the Na-Dene grouping beyond this language family to include other Indian groups residing in the Pacific Northwest.

The peopling of the Pacific was also addressed by Turner through the use of dental morphologic data. However, to unravel the biological diversity of Oceanic populations, a data base was first established for mainland Asian populations. The examination of numerous Asian skeletal series showed a major dichotomy among populations combined traditionally under the general term of Mongoloid (or Asian). In contrast to the broad Mongoloid Dental Complex, defined primarily on the basis of Japanese dental morphology, Turner defined two complexes: Sinodont (North Asian) and Sundadont (Southeast Asian).

Several historical patterns emerged following the definitions of the Sinodont and Sundadont complexes. First, the suite of variables that characterized the Sinodont complex of north Asians also characterized all Native American populations. Although there is dental variation among New World populations, it appears that all were derived from ancestral populations in North Asia. Second, Polynesians and Micronesians exhibited crown and root trait frequencies in accord with the Sunadont pattern, so the historical inference is that these groups were ultimately derived from Southeast Asian populations. Moreover, the Ainu of Japan exhibit a Sundadont rather than Sinodont pattern, so it is possible that the range of Sundadont populations may have extended further to the north at one time, eventually to be displaced and/or replaced by north Asian groups. In this matter, comparisons of both tooth size and morphology support the position that the Ainu are the descendants of the aboriginal Jomon population of Japan, while the modern Japanese population, allied dentally with other Sinodont groups of the Asian mainland, are descended from relatively recent migrants to the Japanese archipelago.

In addition to assessing broad patterns of historical relationships, dental morphology has also been used to measure microdifferentiation among local populations within circumscribed geographic regions (e.g., Middle East, Solomon Islands, Yanomama, American Southwest). At these levels of divergence, trait frequency differences and the associated measures of biological distance are relatively small. Still, population relationships indicated by dental morphology show close correspondance to those suggested by simple genetic markers, other biological systems, and language.

As dentally based biological distance estimates among closely related populations are small when compared with distances between groups with a more remote common ancestry, it may be possible to estimate times of divergence from these values. Turner demonstrated the potential of this approach in a comparison of dental distances among 85 samples with a primary focus on Asian and Asian-derived groups. For the most part, there is close agreement with dentally derived times of divergence and independent estimates from archeology, geology, and linguistics. Further development of this method, labeled dentochronology appears warranted in light of current interest surrounding the origins and dispersal of modern *Homo sapiens*.

III. The Environmental Interface: Teeth and Behavior

Once a tooth crown is fully developed, enamel and dentine are subject only to physicochemical changes. Interest here is with alterations of the tooth crown, which indirectly reflect four classes of human behavior: (1) dietary; (2) implemental; (3) incidental cultural; and (4) intentional cultural.

A. Dietary Behavior

Over the past decade, isotope and trace element analyses of bone collagen and apatite have been widely used to infer general characteristics of the diet of earlier human populations. Inferring dietary constituents and their relative proportions is also a goal of those who study the natural processes that result in the cumulative loss of enamel and dentine from the occlusal surfaces of the teeth, referred to as crown wear.

Crown wear, which occurs on both the chewing surfaces of the teeth (occlusal wear) and at the contact points between adjacent teeth (approximal wear), is generated by the combined action of attrition and abrasion. The process of attrition results from direct tooth-on-tooth contact. Within- and between-group variation in attrition may reflect the nature of foodstuffs being consumed (how much chewing is required for particular foods), the amount of energy brought to bear between the upper and lower teeth by the muscles of mastication, the thickness and quality of crown enamel, and the nonmasticatory grinding of one's teeth (e.g., bruxism). Abrasion, the second component of wear, is caused by the introduction of foreign material (e.g., grit) into the foods being consumed. Abrasive elements may be inherent in foods (e.g., silicate phytoliths in plants, grit in shellfish), incidentally generated by certain food preparation techniques (e.g., a fine grit is added to the flour when seeds are processed by grinding stones) or introduced accidentally into foods by external sources (e.g., windborne silt, sand).

In all human groups, crown wear is produced by both attrition and abrasion. Regarding general subsistence levels, it appears that crown wear in earlier agricultural groups had a significant abrasive component because of the reliance on stone-ground grains and the greater amounts of windborne grit due to lands cleared for farming. Because meat is less abrasive than plant foods, crown wear in hunter-gatherer groups may have a relatively higher attrition component than in agricultural groups, due in part to a more powerful chewing musculature. However, the potential for introducing abrasive elements in the diet is present in all human populations, regardless of subsistence level. Both early hunter-gatherer and agricultural populations are characterized by rapid rates and pronounced degrees of crown wear although the relative contributions of attrition and abrasion to this wear was prob-

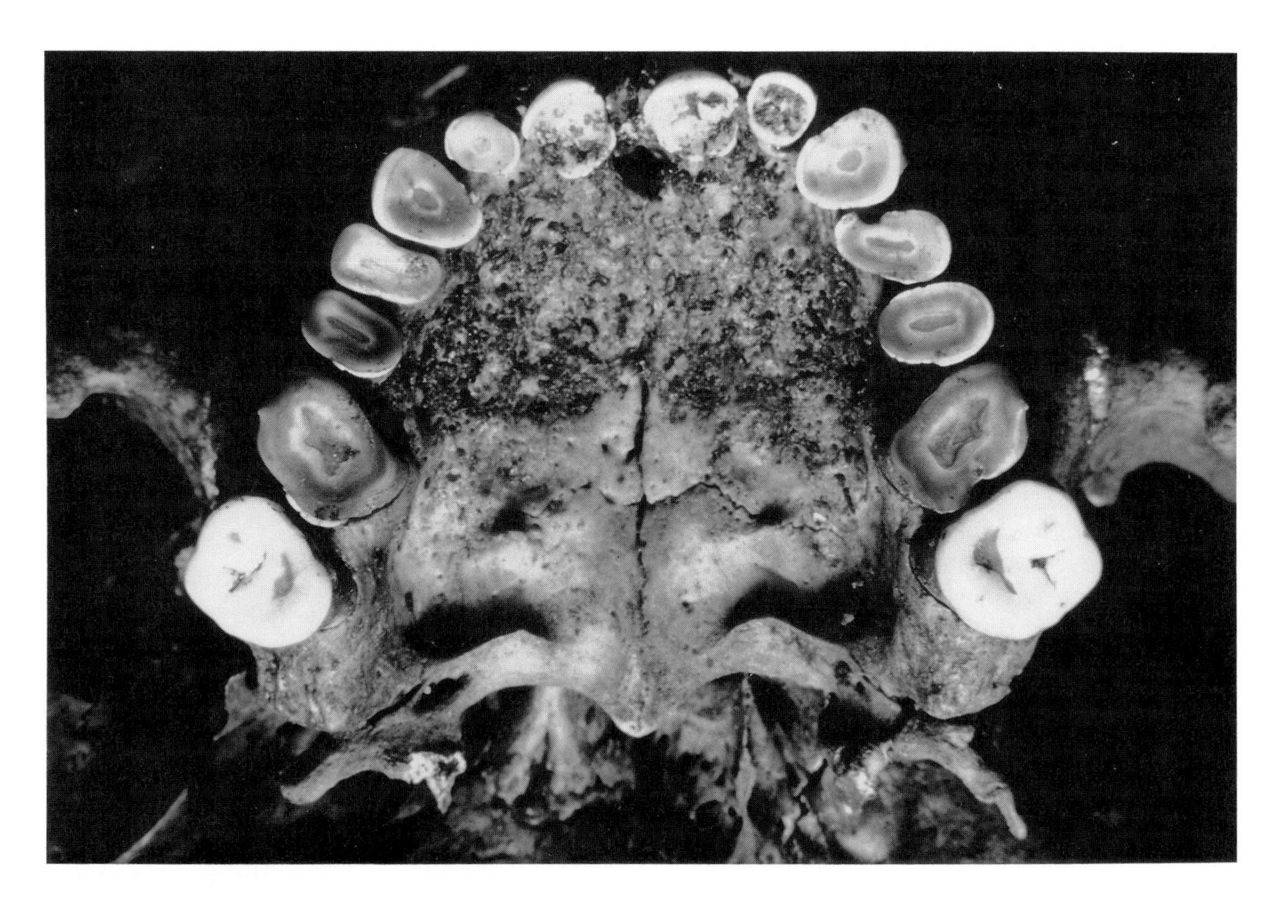

FIGURE 7 Pronounced crown wear in the upper dentition of a medieval Norwegian from St. Gregory's Church, Trondheim, Norway. The relative roles of attrition and abrasion in producing wear in this individual cannot be discerned.

ably highly variable (Fig. 7). It has been suggested that angle of crown wear, rather than absolute degree of wear, may distinguish groups practicing different subsistence economies; i.e., agriculturalists seem to exhibit a steeper angle of wear on the posterior teeth in contrast to the flatter wear plane of hunter-gatherers.

As many factors are involved in crown wear, it is difficult to generalize from comparisons among distantly related populations residing in contrasting environmental settings. The most fruitful lines of study are those that focus on closely related populations from similar environments or involve temporal comparisons among groups living in a circumscribed locale. As this approach controls for some of the variables contributing to crown wear (e.g., craniofacial morphology, tooth size, plant and animal resources), measurements of degree, rate, and angle of wear can reveal significant differences and/or changes in resource utilization and diet.

In addition to crown wear, certain dental pathologies can be utilized to make inferences about dietary and other cultural behavior. Dental caries in particular is useful for addressing dietary differences and/or changes among and within groups. Although the etiology of caries involves a complex interplay between the oral microbiota, dietary elements, dental microstructure, and saliva, the study of earlier human populations indicates that diet played a critical role in the formation of carious lesions. When the term diet is used in the context of caries, specific reference should be to carbohydrate consumption, in particular the ingestion of simple sugars (e.g., sucrose, glucose, fructose), which serve as the substrate for acidogenic bacteria. Fats and proteins do not promote carious lesions. [*See* DENTAL CARIES.]

As the emergence of food production (domestication of plants and animals) did not occur until the Holocene, an economy based on hunting and gathering wild foods characterized the first several million years of human evolution. As the constituents of a hunter-gatherer diet did not generally promote the formation of carious lesions, these groups are characterized by low caries frequencies. In fact, populations of the far north subsisting on high protein–high fat diets often had no caries at all.

Despite the positive effects associated with the rise of agriculture in the Old and New Worlds, it also had its adverse consequences. With an increased reliance on plant foods and food preparation techniques, which broke down complex carbohydrates into simpler sugars, caries rates increased. But, despite the fact that carious lesions increased in earlier agricultural populations, this increase was modest compared with the extremely high caries rates in modern populations. Today, the widespread usage of refined sugars processed from sugar cane and sugar beets has led to a major increase in caries frequencies.

The analysis of caries rates, like crown wear, is most informative when studied in the context of circumscribed geographic populations. Temporal comparisons among British populations, for example, show how caries rates can indicate the introduction of specific dietary components during particular historical periods. Comparisons between prehistoric and modern populations of the far north also show a dramatic rise in caries rates following the introduction of refined carbohydrates into native diets that had hitherto consisted primarily of animal products (protein and fat).

FIGURE 8 St. Lawrence Island Eskimo female using anterior teeth to soften and crimp walrus or bearded seal hide for mukluk sole [Used, with permission, from the Denver Museum of Natural History, photo archives, BA21-419K; all rights reserved.]

B. Implemental Behavior

Modern technology provides us with a tool for almost any mechanical task. Lacking such advantages, earlier human populations with relatively simple tool kits were often forced to rely on their own biological equipment. This is particularly true for teeth, which can serve functionally as pliers, vises, strippers, gravers, etc. In this sense, teeth can literally be used as a third hand with unique properties. While the use of teeth as tools is most commonly associated with populations of the far north, particularly Eskimos, this behavior is not limited in either time or space. Humans throughout history have taken advantage of the strength, form, and ready availability of their teeth to perform a variety of functions from carding wool to holding bobby pins.

Teeth do not record all instances of tool use, but they can reflect repetitive behaviors and traumatic episodes. In many hunter-gatherer populations, for example, the anterior teeth were used for skinworking (Fig. 8). The ultimate effect of this usage was to generate a distinctive pattern of labial rounding especially evident on the anterior teeth (Fig. 9). When manipulation centered on the manufacture of threadlike items from sinew or grasses, the pattern generated on the anterior teeth would take the form of notches or grooves rather than uniform surface wear.

In addition to patterns of uniform wear generated by attrition and abrasion, enamel and dentine can also be removed through traumatic fracturing. This process, called dental chipping, occurs when teeth are subjected to forces that exceed their load-bearing capacity. Commonly, such chipping takes the form of small enamel flakes removed around the margins of the teeth, much like the pressure-chipping evident along the margins of stone tools. Although chipping can be caused by such things as grit accidentally introduced into food, it is frequently attributed to using the teeth as tools, especially among Eskimos.

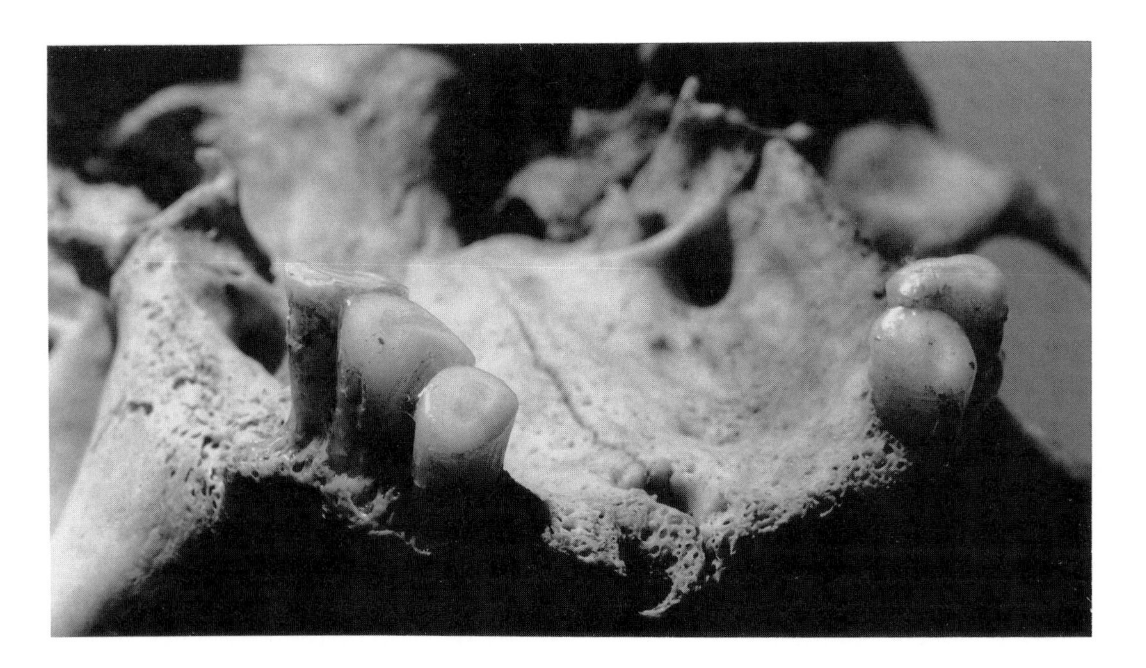

FIGURE 9 Upper dentition of prehistoric Alaskan Eskimo. Incisors lost antemortem, but canines and premolars exhibit distinct pattern of labial rounding indicative of implemental use (e.g., skin-working).

C. Incidental Cultural Behavior

Several patterned behaviors, which do not reflect either implemental use or intentional modification, leave an imprint on teeth. One such behavior is pipe smoking. Habitual pipe smokers commonly hold a pipe on either or both sides of the mouth in the region of the left or right canines. Pipes with highly abrasive clay stems are among the worst offenders for generating deep ovate notches that extend over several teeth. A more hygienic habit, the use of probes, or toothpicks, to remove food debris lodged between the teeth, may leave grooves on the interstitial surfaces of the tooth crowns. Another cultural practice that leaves unintended wear is labret usage. Labrets, which are inserted through the cheeks or lips, come in different shapes and sizes and are made from a variety of raw materials, including wood, ivory, bone, and stone (Fig. 10). The wear pattern produced by labret use is very distinctive; it is manifest as a polished facet on the labial or buccal surfaces of the anterior or posterior teeth, respectively (Fig. 11). A thorough perusal of the ethnographic literature would probably reveal many other cultural practices that leave unintentional marks on the teeth.

FIGURE 10 Early nineteenth century Eskimo male from Kotzebue Sound, Alaska, with paired composite labrets. The internal aspect of such labrets would contact the buccal surfaces of the lower anterior teeth, resulting in polished wear facets [Sketch by Tim Sczawinski, courtesy of Ernest S. Burch, Jr.; all rights reserved.]

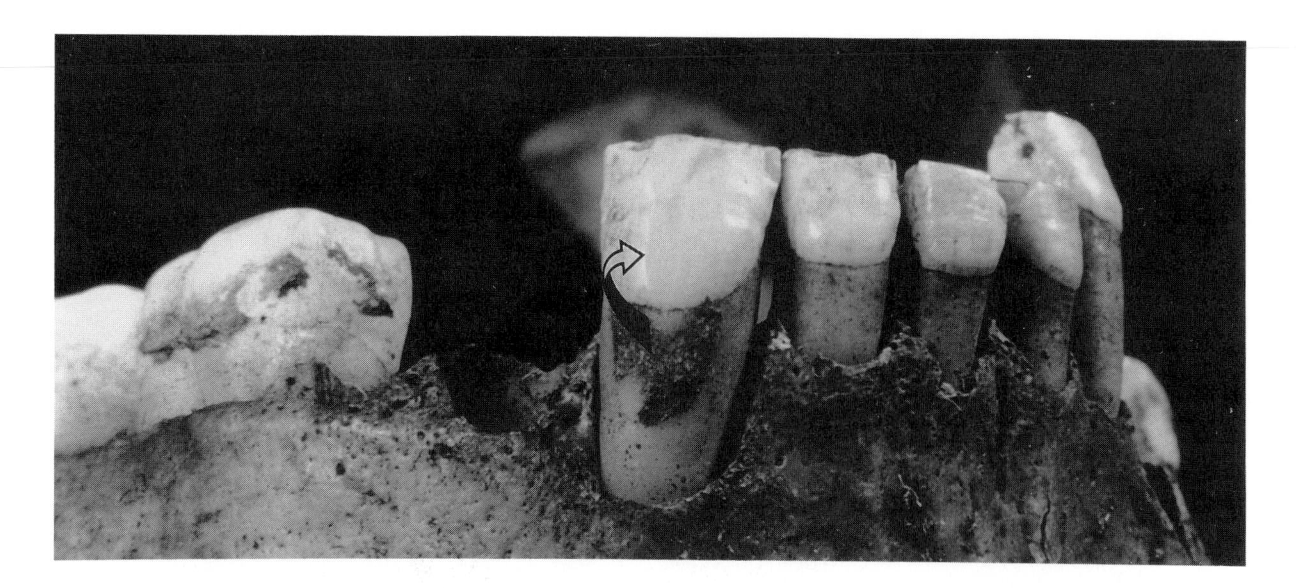

FIGURE 11 Labret facet on the lower right canine of a prehistoric Alaskan Eskimo. Arrow points to distal margin of facet, which extends from crown–root junction to occlusal surface.

D. Intentional Cultural Modification

Although the primary function of the oral cavity relates to the ingestion of food and water, the mouth also serves as a major social organ for many animals, including humans. Unlike other animals, however, which make do with the biological equipment they are provided with, humans can modify the appearance of their mouths in a variety of ways. In some cases, these modifications are external in nature, e.g., beards, tatoos, lip plugs, labrets, lipstick, etc.; In others, groups directly modify the appearance of their teeth, especially the more visible incisors and canines.

Culturally prescribed dental mutilation takes several forms. For example, individuals can chip or file their incisors and canines to produce notched or pointed teeth. Incising tools can be used to engrave patterned lines on the labial surfaces of the anterior teeth, or small holes can be drilled to serve as settings for precious metals (e.g., gold) or gemstones (e.g., jade, turquoise). Precious metals can also be inlayed as bands on the labial surface or around the entire crown. In addition to modifications in form, entire teeth can be removed traumatically through the practice of ablation.

The reasons for dental mutilation may be idiosyncratic or culturally prescribed. In the first instance, individuals may choose to modify their teeth to achieve a desired cosmetic effect (e.g., to enhance beauty or fierceness). On the other hand, some populations require some form of dental mutilation as a symbol of group membership. In some cases, mutilation is involved in rites of passage, especially those rites involving a transition in status (e.g., adolescence to adulthood, unmarried to married). An interesting and not yet fully exploited anthropological usage of dental mutilation would be to assess the diffusion of specific practices from one region to another.

IV. Dental Indicators of Environmental Stress

Anthropologists have long sought methods to estimate relative levels of environmental stress on earlier human populations. Growth arrest lines in long bones (i.e., Harris or transverse lines) provide one measure of this phenomenon. In addition, events that disrupt crown and root formation may also mirror episodes of environmentally induced stress. Although the course of dental development is, to a large extent, stable and predictable in terms of matrix formation, calcification, and eruption, the dentition is not immune from external influences. This is apparent in size asymmetry between antimeres and micro- and macrostructural defects in the enamel and dentine.

Antimeres exhibit mirror imagery because dental development is moderated by a common genetic control system for the teeth in the two sides of the jaw. When differences are evident between antimeres in size, morphology, and number, they are

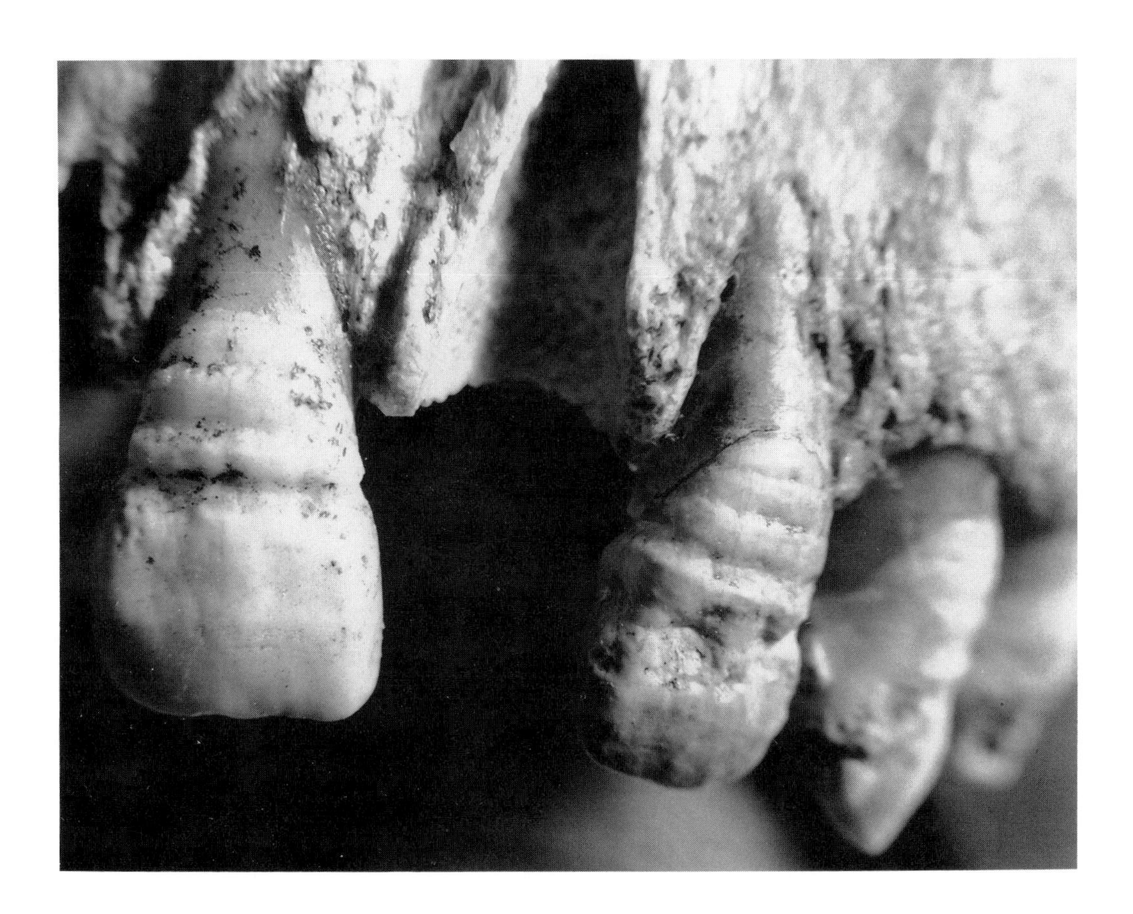

FIGURE 12 Pronounced linear enamel hypoplasia on anterior teeth of prehistoric St. Lawrence Island Eskimo. Three distinct bands can be observed on right central incisor, while four bands are visible on left lateral incisor.

environmental in origin. At one time, the varying levels of fluctuating asymmetry in tooth size among different populations were thought to provide some indication of relative levels of environmental stress. Although the logic of the argument remains sound, methodological problems pertaining to measurement error and small sample size have recently curtailed the use of dental asymmetry as a relative measure of stress.

As dental asymmetry appears to have certain limitations as a broad scale indicator of comparative stress levels, dental anthropologists have shifted their attention to the analysis of irregularities in the tooth crown that arise during amelogenesis (enamel formation) and dentinogenesis (dentine formation). The most readily observed manifestation of such growth irregularities is linear enamel hypoplasia (LEH), which takes the form of horizontal circumferential bands and/or pits on the tooth crown (Fig. 12). LEH can be observed on any tooth, although it is more common and pronounced on the anterior teeth. Experimental and clinical evidence shows that a wide range of phenomena can disrupt amelogenesis and stimulate hypoplastic banding/pitting. However, the key stimulus in earlier human populations probably involved some combination of nutritional deficiency and disease morbidity.

Because matrix formation and calcification for specific teeth occur during predictable time intervals, it is possible to estimate the ages at which specific LEH bands developed and the approximate duration of a particular episode of stress. For this reason, enamel defects are used to address a number of problems in the analysis of human skeletal remains. For example, in some early agricultural groups in North America, LEH banding appears to concentrate in the 2-4-yr-old age category, possibly marking the shift from mother's milk to a weanling diet of cereal gruel deficient in certain essential amino acids. The distances between bands, in some cases, may indicate seasonal patterns of stress, a common phenomenon in earlier populations. The numbers of bands and their degree of expression may also provide insights into the differential treatment of male and female children or differences in status within a population. Further experimental and comparative work on surface irregularities and dental histological indicators of growth disturbance (e.g., Wilson bands), provide a promising avenue of research for scholars interested in discerning differential stress levels among and within earlier human populations.

Bibliography

Brace, C. L., Rosenberg, K. R., and Hunt, K. D. (1987). Gradual change in human tooth size in the late Pleistocene and post-Pleistocene. *Evolution* **41,** 705–720.

Hillson, S. (1986). "Teeth." Cambridge University Press, Cambridge.

Kelley, M. A., and Larsen, C. S. (eds.) (1990). "Recent Advances in Dental Anthropology." Alan R. Liss, Inc., New York.

Lukacs, J. R. (1989). Dental paleopathology: Methods for reconstructing dietary patterns. *In* "Reconstruction of Life from the Skeleton" (M. Y. Iscan and K. A. R. Kennedy, eds.). Alan R. Liss, Inc., New York.

Rose, J. C., Condon, K. W., and Goodman, A. H. (1985). Diet and dentition: Development disturbances. *In* "The Analysis of Prehistoric Diets (R. I. Gilbert, Jr., and J. H. Mielke, eds.). Academic Press, New York.

Scott, G. R., and Turner, C. G., II (1988). Dental anthropology. *Ann. Rev. Anthrop.* **17,** 99–126.

Turner, C. G., II (1986). The first Americans: The dental evidence. *Nat. Geogr. Res.* **2,** 37–46.

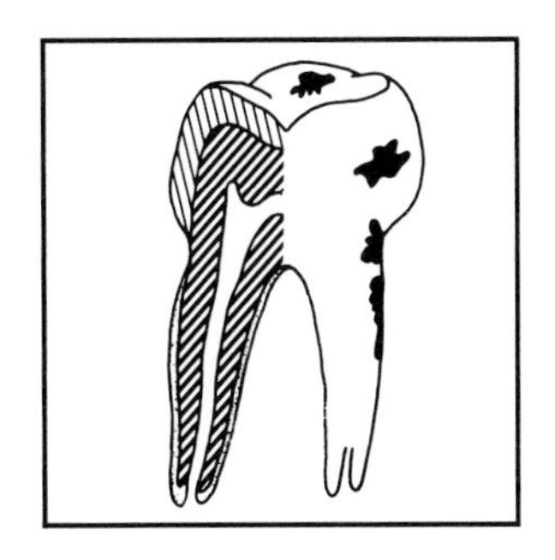

Dental Caries

RENATA J HENNEBERG, *University of Cape Town; University of the Witwatersrand*

Glossary

Dental calculus Mineralized microbial plaque covering the enamel or root surfaces firmly attached to them; supragingival and subgingival calculus types are distinguished with respect to where they form on the root relative to gingivial margin
Gingivitis Inflammation of the gingiva
Periodontal disease General term denoting diseases of tissues (periodontium) that support teeth in jaws; periodontium includes gingiva, cementum, the periodontal ligament, and the alveolar bone
Plaque Layer of microorganisms of various species and strains embedded in an extracellular matrix and covering tooth surfaces; structure of plaque layer is highly variable

TOOTH DECAY, OR DENTAL CARIES (from Latin *caries,* rottenness), is a pathological process of external origin and multifactorial nature in which localized destruction of the hard tissues of a tooth occurs. The disease begins with dissolution of the inorganic structures of the tooth surface (softening of the surface, decalcification, demineralization) and progresses to disintegration of the organic matrix of the tooth tissues, ultimately producing cavitation.

Several theories attempt to explain caries initiation and development. More prominent among them are the chemicoparasitic, or acidogenic, theory, the proteolytic theory, the proteolysis-chelation theory, the sucrose-chelation theory, and, most recent, the autoimmunity theory. Postulated in the nineteenth century, the chemicoparasitic theory is the oldest, and available evidence seems to provide it with the most support. According to this theory, microorganisms present in the oral cavity produce acids from food carbohydrates at or near the tooth surface. The acids dissolve apatite crystals in the tooth enamel.

I. Scoring Techniques Used for Studying Caries

A. Background

Several classifications of dental caries were used in dental practice until 1961; the same situation existed among authors describing archeological dental material. This resulted in difficulties for comparative studies. In 1961, the International Dental Federation (FDI [Federation dentaire internationale]) made the first attempt to develop standards and a unitary system for compiling statistics on caries. The results of the FDI Special Commission on Oral and Dental Statistics were later adopted by the World Health Organization (WHO) and, with further improvements, were used in epidemiological studies of caries worldwide. In archeological and anthropological studies, the WHO-approved standards are also used but with some modifications due to specificity of the human prehistoric material (Table I).

From the diagnostic point of view, caries may be classified as initial caries and clinical caries. Initial caries (synonyms: microscopic carious lesion, radiographic lesion, questionable caries) is described as a white, chalky or discolored rough spot on the enamel without visible surface breakdown. Clinical caries (synonyms: macroscopic carious lesion, untreated carious defect, cavity) is described as a visible cavity, which can be diagnosed by physical (clinical) examination. The main interest of dental

TABLE I Methods Used to Describe Occurrence of Caries in Living Human Populations and in Skeletal Samples

Living human populations	Skeletal samples
1. Caries frequency	
$\frac{\text{no. of individuals affected by caries}}{\text{total no. of individuals examined}}$	
or	
$\% = \frac{\text{no. of individuals affected by caries}}{\text{total no. of individuals examined}} \times 100$	
2. Caries intensity[a]	
a. $\frac{\text{no. of carious teeth}}{\text{no. of individuals with caries}}$	a. $\frac{\text{no. of carious teeth}}{\text{no. of individuals with caries}}$
b. $\frac{\text{no. of cavities}}{\text{no. of individuals with caries}}$	b. $\frac{\text{no. of carious teeth}}{\text{no. of teeth examined}}$
c. $\frac{\text{no. of carious surfaces}}{\text{no. of individuals with caries}}$	
d. $\frac{\text{no. of carious surfaces}}{\text{no. of carious teeth}}$	
e. $\frac{\text{no. of cavities}}{\text{no. of carious surfaces}}$	
3. DMF index (DMFT or DMFS)[b] recommended by WHO, average number of decayed (D), missing (M) (due to caries), and filled (F) permanent teeth (T) or tooth surfaces (S) per person.	**3. DM index (dmt or DMS)**[b] Average number of decayed (D) and missing (M) (due to caries) permanent teeth (T) or tooth surfaces (S) per individual.
4. dmf index (dmft or dmfs) Recommended by WHO, sum of decayed (d), missing (m), and filled (f) primary (deciduous) teeth, which should be present in the mouth at the time of examination, or tooth surfaces (s) per person.	**4. dm index (dmt or dms)** Sum of decayed (d) and missing (m) primary (deciduous) teeth (t) or surfaces (s) per person.
5. def index Sum of primary teeth decayed (d), beyond repair (e), and filled (f).	**5. —**
6. Root Caries Index (RCI)	
a. $\frac{\text{no. of root caries lesions} \times 100}{\text{no. of teeth or surfaces} \times \text{no. of individuals with gingival recession}}$	
b. $\frac{\text{no. of root caries lesions} \times 100}{\text{no. of carious teeth or surfaces}}$	

[a] Describes severity of caries experience.

[b] DMF or DM indices can be made age-specific, sex-specific, or population-specific. WHO recommends epidemiologic surveys to be made when children and youths are 5, 12, 15, and 18 yr old.

DM or dm indices for skeletal material are very rough estimates due to errors resulting from incompleteness of archaeological samples, and thus, their numerical values should be treated with caution.

clinicians, epidemiologists, and paleopathologists is concentrated on clinical caries. According to the WHO recommendations, the examination for dental caries should be conducted with a plane mouth mirror and a sharp, sickle-shaped dental explorer in natural light. Carious lesions found during such macroscopic examination may be classified as follows: (1) according to anatomical location on a tooth (Fig. 1), (2) according to the speed with which the lesion progresses, and (3) according to the tissue attacked (Fig. 2).

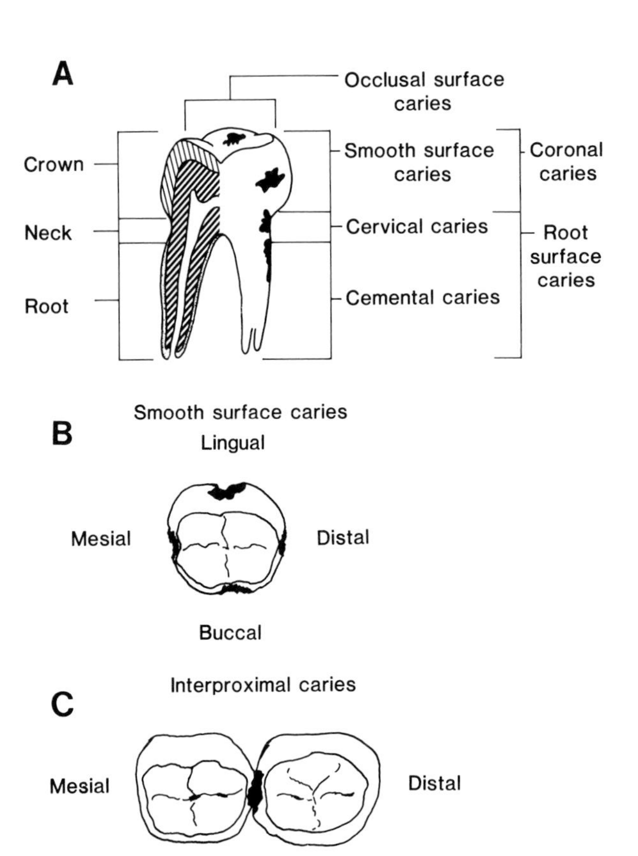

FIGURE 1 Classification of carious lesions according to anatomical location on a tooth. Coronal caries is a lesion localized on the crown of a tooth. Coronal caries may occur on the occlusal surface of the tooth (A), where it starts as pits or fissures on the surface, or on the smooth surface of the crown, where it can be localized on the buccal (or labial), lingual, mesial, or distal surface (B); a special case of smooth-surface caries is an interproximal lesion (C), in which adjacent surfaces of two neighboring teeth (proximal surfaces) are involved. Root surface caries (A), in which cementum covering the root of a tooth is affected, is a lesion that may be localized on the cemento enamel junction of the tooth (cervical caries, caries of the tooth neck) or lower down on the root (cemental caries). A separate category of carious lesions is ''circular caries,'' which occurs on the buccal surface of the hypoplastic enamel of the teeth.

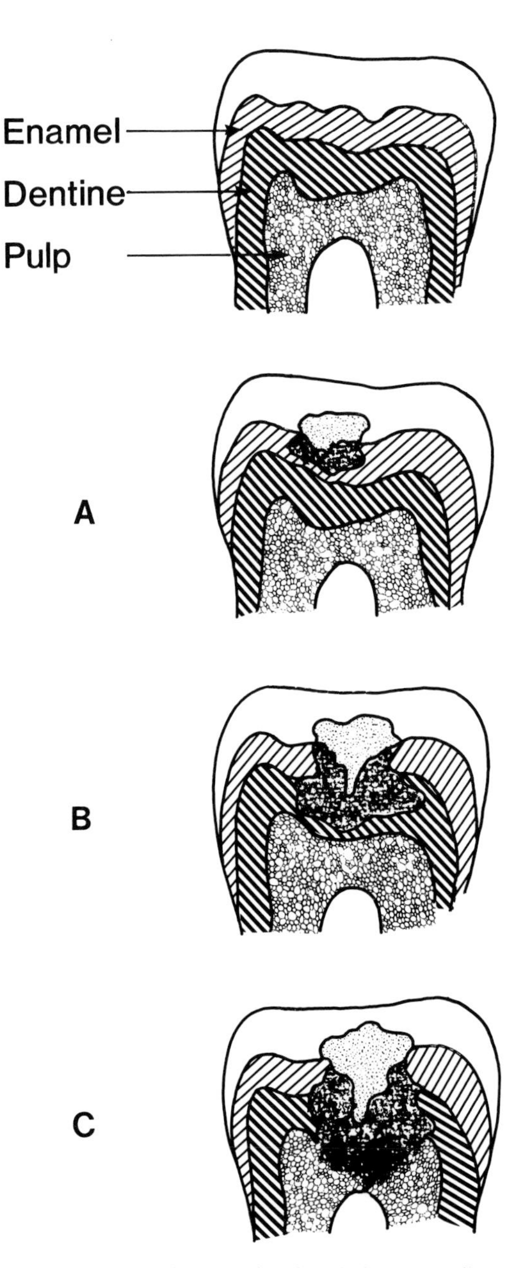

FIGURE 2 Classification of carious lesions according to the tissue affected. A. Enamel caries (first-degree caries, caries prima, caries superficialis). B. Caries of the dentine (second-degree caries, caries media). C. Caries of the pulp (third-degree caries, caries profunda). This classification also describes the depth of the lesion's penetration in the coronal part of the tooth. Cemental caries constitutes a separate category of the classification, but categories of root caries have not been fully established yet. The term fourth-degree caries is sometimes used to denote complete tooth destruction by the carious process (missing tooth). Another classification is that based on the progression of the disease: nonpenetrating caries, in which only enamel and dentine are affected, and penetrating caries, in which the pulp chamber is opened. This classification is most often used by paleopathologists rather than dental practitioners.

B. Scoring of Caries in Living People

The total amount of dentition destruction by caries, including tooth loss due to decay, is an individual's or a population's life caries experience. The rate at which caries disease occurs in a population is described as caries prevalence (i.e., a fraction of a population having any experience of caries). The caries prevalence is usually calculated separately for various age groups (age-specific prevalence). Caries incidence is the rate per unit of time (usually 1 yr) at which new carious lesions develop in an individual or a population. Occurrence of caries in a population is expressed by means of various indices (Table I).

In epidemiologic studies, assessment of dental caries may be desired for the entire dentition, in which case all teeth are examined, or by using the half-mouth method, in which observations are made and recorded for one-half of the upper dental arch and one-half of the lower arch of the opposite side (e.g., upper-left and lower-right quadrants).

The latter method gives a reasonably accurate assessment of dental caries in living populations, because caries is considered largely a bilateral phenomenon. In archeological material that is often fragmentary, recording of caries on all teeth available for examination is logical. Observations may be recorded in various ways. Recording it usually done using a simple code. The WHO recommended Basic Oral Health Assessment Form and the Treatment Assessment Form, which contain sections on caries, are widely used in both clinical practice and epidemiologic surveys. The individual's dentition is divided into quadrants denoted numerically (1, upper left; 2, upper right; 3, lower left; 4, lower right) and teeth numbered 1–8 from the central incisor to the third molar in each quadrant. On a chart, each tooth is given a box into which information is written with code (Table II). The chart also shows how many surfaces were restored on the tooth or the reason for eventual extraction.

C. Scoring of Caries in Archeological Samples

An example of a scoring chart used for human skeletal remains is presented in Table III.

D. Radiographic Techniques in Caries Identification

Modern radiographic techniques are a very useful tool for identification of carious lesions in dental practice and in epidemiologic studies. Caries appears on a radiograph as a radiolucent (dark) area surrounded by more radio-opaque (lighter) tissues of a tooth. Depending on its anatomical location, caries may be visible on a radiograph more or less readily. Some carious lesions are visible at such an early stage that they are undiagnosable by physical examination (with a dental probe).

Intraoral radiographs have proved to be extremely helpful as an aid in clinical examination. For instance, detection of interproximal caries with the use of X-rays improves more than twofold over detection by the mirror-probe method. Occlusal caries can be observed radiographically only after

TABLE II Codes Used for Tooth Assessment

	Dentition	
	Primary	Permanent
Sound tooth (no treated or untreated caries)	A	0
Decayed tooth (macroscopically visible caries)	B	1
Filled tooth with no active caries	C	2
Filled tooth with primary decay	D	3
Filled tooth with secondary decay	E	4
Deciduous teeth missing due to caries	M	—
Permanent teeth missing due to caries (only in individuals <30 yr old)	—	5
Permanent teeth missing for any reason other than caries (<30 yr old only)	—	6
Permanent teeth missing for any reason (>30 yrs old)	—	7
Unerupted tooth	—	8
Excluded tooth	X	9

TABLE III Scoring Chart for Caries in Skeletal Material[a]

		Tooth[b]							
Right side		8	7	6	5	4	3	2	1
Maxilla (upper)	Degree of penetration	u		2	3	4		a1	p1
	Surface attacked			o	o				
Mandible (lower)	Degree of penetration		1, 2	2		1		1	1
	Surface attacked		o, m	d		l		n, b	n, d

Degree of penetration (tissue attacked)	1, enamel caries; 2, dentinal caries; 3, pulpal caries; 4, crown completely decayed; a1, antemortem tooth loss; p1, postmortem tooth loss; u, impacted (unerupted) tooth (if visible).
Surface attacked	o, occlusal; b, buccal; l, lingual; d, distal; m, mesial (d + m = i, interproximal); n, neck (cervical caries, root surface caries)

[a] Only one-half of the chart (for the right side) is shown.
[b] Examples given on the chart: Right upper third molar (8) is impacted. Right upper first molar (6) shows occlusal surface dentinal caries. Right upper second premolar (5) shows occlusal surface caries that penetrate into the pulp. Right upper first premolar (4) displays completely decayed crown due to caries. Right upper lateral incisor (2) was lost during life, most probably due to caries. Right upper first incisor (1) was lost after death, most probably during excavations. On the lower second molar (7) there are two cavities;—one in the enamel only (first degree) on the occlusal surface and the other of the second degree (dentinal caries)—interproximal with the cavity on the distal surface of the first molar (6). Right lower first premolar (4) shows enamel caries on the lingual surface. Both right lower incisors (2, 1) show enamel caries located on buccal (2) and distal (1) surfaces of the neck of the tooth.

the disease has progressed to the dentinoenamel junction. At this stage the occlusal lesion is already clearly identifiable macroscopically. Detection of buccal, lingual, and cemental lesions can be done radiographically, but differentiation of these types with X-ray pictures is unnecessary because such lesions are clearly visible clinically. Panoramic radiographs (panorex) are an excellent survey tool. They are taken by various types of apparatus based on the principle of tomography (scanning). Orthopantomography produces one continuous image from the right to the left temporomandibular joint (Fig. 3). As a result of technological developments and education in radiology, use of radiographic techniques for caries diagnosis is increasing, especially in developed countries such as Sweden, Japan, Finland, and Denmark; it is also popular in the United States.

E. Caries in Forensic Medicine

Because of the progressive nature of the disease, caries as such has little value in the process of forensic identification. The pattern of treated carious lesions, however, and materials and methods used by various dental practitioners may play an important role in the elimination of false identifications (negative identification). The reliability of results from such examinations increases substantially when radiographs of the dentition of the person in question exist in dentist's files.

II. Caries as a Multifactorial Disease

A. Antiquity of Caries

Dental caries is a disease of the human species, and is one of the most common microbe-caused diseases in modern man. Although some carious lesions have been described in animals, they are uncommon and differ from those found in humans. Caries has been found in fossil reptiles, camel, mastodont, and European cave bears. Caries occurs in extant wild animals, especially among aged individuals of primates, horses, rabbits, rats, dogs, sheep, and cats. It can be induced in laboratory animals such as hamsters and rats.

Dental caries accompanied hominids during their evolution. The earliest examples of caries are found in South African australopithecines. The evidence of the first authentic dental caries of human type is

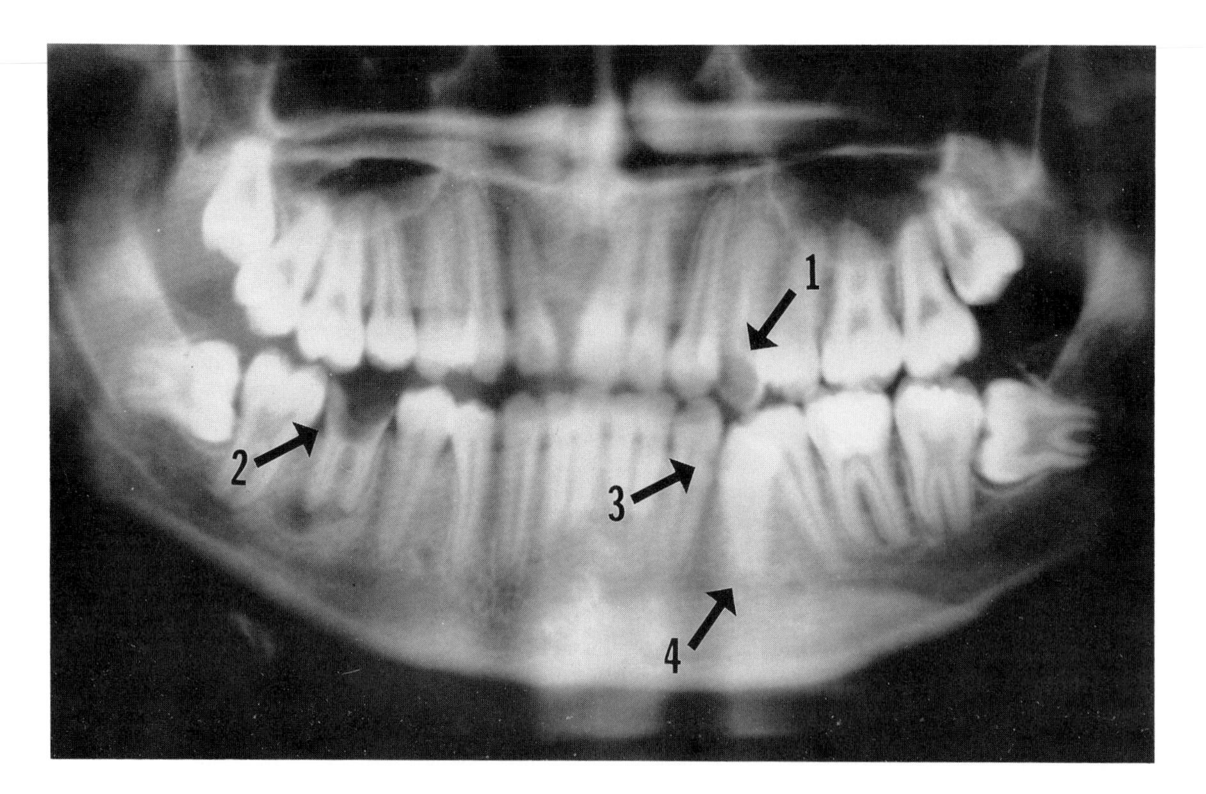

FIGURE 3 Orthopantomogram of jaws indicating presence of caries. 1. Caries of occlusal and buccal (labial) surface of the upper left first (third) premolar. 2. Penetrating caries of the right lower first molar. The crown is almost entirely destroyed by the disease. Radiolucency around the tip of the root of the tooth is due to the inflammation of surrounding tissues (periodontitis), which is a sequel to pulp involvement following penetration of the pulp chamber. 3. Interproximal caries on the cementoenamel junction of the distal surface of the left lower canine (cervical caries). 4. Rotated left lower second (fourth) premolar with periapical root resorption probably following trauma. [Courtesy of R. Hendricks, University of Western Cape.]

found in the dentition of the skull from Kabwe in Zambia (''Rhodesian Man,'' Broken Hill). Of the 13 teeth preserved in this specimen, 11 showed carious lesions. This is an unusually high caries intensity for this period (Middle Paleolithic; Table IV). Caries prevalence was very low in hunters and gatherers. It increased substantially during the first major economic transition of neolithization, during which time dietary conversion occurred due to the introduction of agriculture as a major food source, and further increased dramatically with the second dietary conversion when refined food, rich in carbohydrates, became the basic source of nutrients. Caries is an undoubted disease of civilization.

B. Heritability and Prevalence of Caries

Evidence that genetic factors influence the prevalence of caries is derived from (1) animal experiments, (2) human family and twin studies, and (3) studies of metabolic diseases and defects, which themselves are heritable. Selective breeding of rats has demonstrated that genetically caries-resistant and caries-susceptible strains of animals can be developed. Under the same conditions of a carbohydrate diet, the caries-resistant rats produced caries at a rate several times slower than rats from the caries-susceptible strain. In the investigation of a large number of families, caries frequency among siblings of caries-susceptible school children was approximately double the frequency among siblings of caries-resistant individuals. Children of highly caries-resistant parents had lower DMF scores than children of caries-susceptible parents, despite the same exposure to water containing fluoride.

Twin studies showed greater variation of caries experience among dizygotic (DZ) twins than among monozygotic (MZ; i.e., identical) ones. Recent studies of twins reared apart confirm earlier findings that MZ twins show a strong resemblance of caries status despite different dietary patterns, oral hygiene practices, professional dental care, etc. In DZ twins, resemblance in caries status is weaker.

TABLE IV Intensity and Prevalence of Caries in Various Periods of Prehistory and History

Period	Mode of subsistence	% Carious teeth	% Individuals with caries
Middle Paleolithic (100,000–40,000 B. C.)	hunting and gathering	0–?	0–?
Upper Paleolithic (35,000–10,000 B. C.)	mostly large game hunting and gathering	1.0–?	—
Mesolithic (10,000–4,000 B. C.)	small game hunting, fishing, and gathering	0.4–7.7	?–41.6
Neolithic (8,000–2,000 B. C.)	agriculture	1.4–12.0	0–60.0
Classic antiquity (1,000 B. C.–500 A. D.)	agriculture	5.0–14.6	?–67.0
Middle ages (500–1500 A. D.)	agriculture	2.8–40.0	24.8–65.5
Modern times[a]	industrialized agriculture	5.9–45.0	36.3–100.0

?, no data.
[a] Cane sugar became commercially available at the beginning of this period.

Individuals who could not taste the phenylthiocarbamide (PTC) were observed to have 25% lower def scores than those who could. It is believed that individuals with Down's syndrome, whether institutionalized or living at home, have lower-than-average caries experience. The complexity of caries indicates that its hereditary determination is polygenic.

Genetic factors, however obviously present, seem to be less important for the development of caries than environmental factors such as food, fluoride content of water, oral hygiene, etc. Further detailed studies of the genetic–environmental interaction in the development of caries are needed.

C. Factors Influencing Prevalence of Caries (Fig. 4)

Etiologic factors responsible for dental caries can be divided into two major groups: essential and modifying. Factors essential for the development of caries are host susceptibilty (i.e., natural susceptibility of the tooth), plaque, and food taken orally. Modifying factors are composition of the saliva, soil and water mineral contents, systemic diseases, sex and age of individuals, tooth anatomy, and also variations in all three essential factors such as composition of the plaque, variations in susceptibility or resistance, variations in food composition, and the frequency of food intake. These modifying factors may affect distribution of caries (site on a tooth, type of tooth), speed of the lesion's progression and remineralization of the lesion.

1. Host Susceptibility

Caries-resistant teeth have less complicated morphology and better enamel integrity. The simple evidence of this fact may be that molars and premolars, having a more complex structure and being generally bigger, are more prone to caries than anterior teeth. The first permanent molar is known to be the most carious tooth in the entire human dentition. The pattern of caries distribution by tooth type is usually as follows (in decreasing order of intensity): molars, premolars, incisors, and canines. This pattern has not changed much from ancient to modern populations. Teeth that do not erupt and are impacted in the jaws do not develop caries.

2. Plaque

Plaque formation begins with coating of the tooth surface with saliva. Salivary film is colonized by species of *Streptococcus, Actinomyces, Staphylococcus, Veillonella, Lactobacillus,* and a few others. There is a difference between the bacterial composition of the plaque from caries-active and caries-resistant individuals. Proof that microorganisms embedded in plaque are essential for the occurrence of carious lesions comes from animal experiments. Susceptible rodents do not develop caries on a highly cariogenic diet when kept under sterile, germ-free conditions. The same animals develop caries after infection with bacteria.

People who develop caries have a higher level of *Streptococcus mutans* and Lactobacilli in their plaque. Distribution of these organisms inside the mouths correlates to some extent with the distribu-

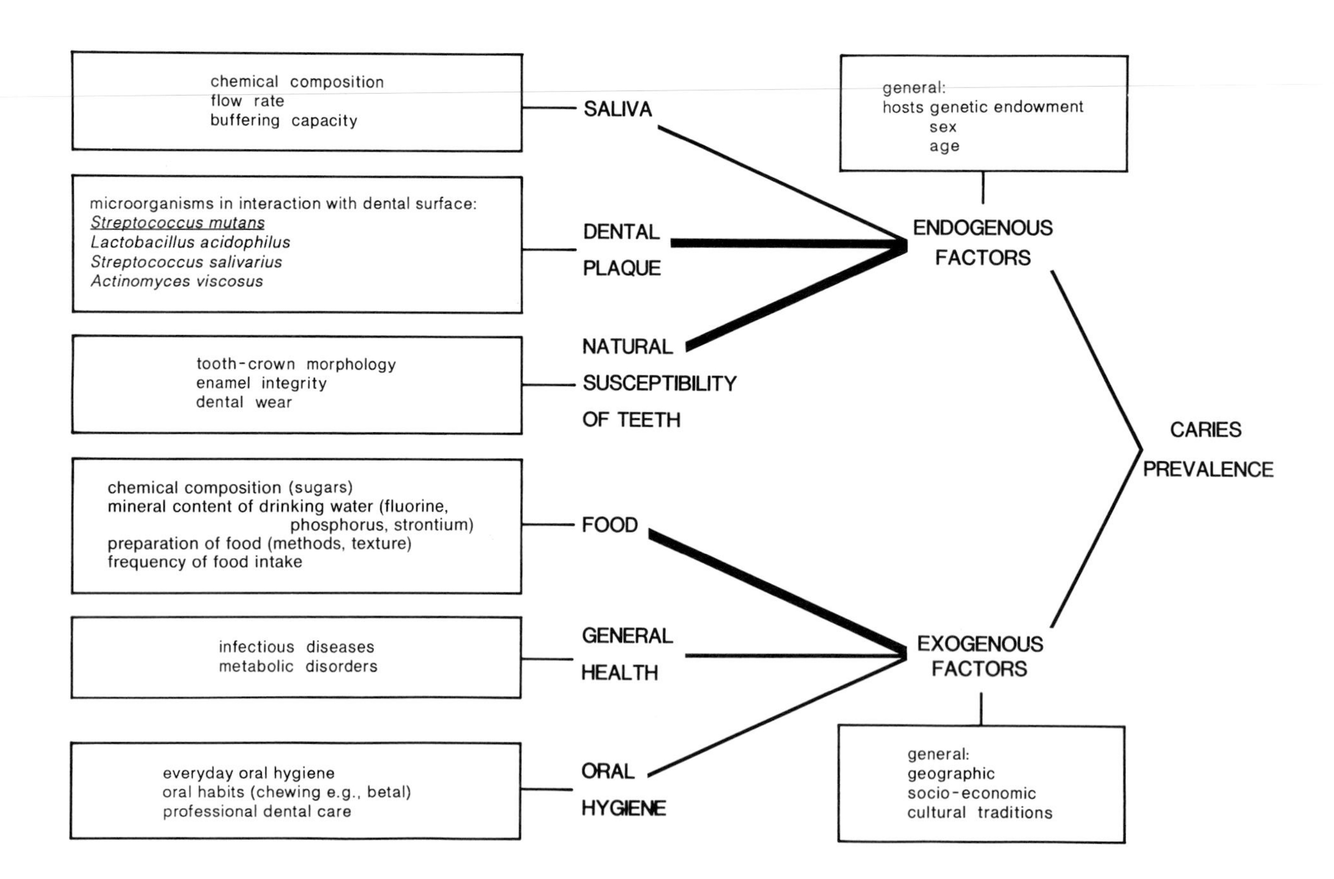

FIGURE 4 Schematic representation of factors influencing prevalence of caries.

tion of carious lesions. *S. mutans* is considered to be the most active etiologic agent in caries disease. Acids produced by organisms living in plaque can decrease pH at the tooth surface from neutral to <5.0, which seems to be a critical point for initiation of the carious process.

3. Food

Food has the most impact on the initiation and development of caries. The concept of food carbohydrates as a major cariogenic factor has been proven in studies of human and experimental animal diets.

The cariogenicity of various carbohydrates depends on the speed of their disintegration in the process of fermentation and on their ability to reduce the pH of plaque and saliva by production of acids. Simple sugars—mono- and disaccharides—ferment more rapidly than complex polysaccharides. Sucrose is the greatest cariogenic agent among simple sugars; it is closely followed by glucose. Fats and proteins are not metabolized by cariogenic bacteria; they do create rather high pH conditions, and their cariostatic properties are well documented. Cariogenicity of the diet is determined by (1) content of simple sugars, (2) texture of foodstuffs, and (3) frequency of food intake. Soft, sticky substances are more cariogenic than foods of rough, hard texture. Hard and rough food requires more chewing and stimulates flow of the saliva, thus providing better cleansing of teeth. Sticky foods remain on the tooth surface longer and constitute a continuous source of substrates for fermentation. It has been proven that snacking increases incidence of caries. Food intake seems to be the most flexible among factors influencing the multifactorial phenomenon of carious disease.

4. Saliva

Saliva contains proteins (enzymes: amylase, lactoperoxidase, lysozyme, lactoferrin, cationic glycoprotein, esterases, ribonucleases, and secretory IgA) and minerals (with ions of calcium, phosphate, magnesium, bicarbonate, chloride, sodium, potassium). Proportion of contents varies and depends on the type of salivary gland (parotid, submandibular, sublingual, minor glands), on intensity of stimulation, time of the day, diet, age, sex, a number of diseases, and pharmacological agents. Salivary pH is extremely variable, ranging from 5.8 to 7.7 under normal circumstances. By changing its contents and the rate of flow, saliva can affect caries by way of mechanical cleansing of the tooth surface; reducing enamel solubility by providing calcium, phosphate, and fluoride; buffering and neutralizing acids produced by cariogenic organisms and foods; and by antibacterial activity. [*See* SALIVARY GLANDS AND SALIVA.]

5. Mineral Content of Drinking Water

a. Fluoride People who live in areas where drinking water contains a high concentration of fluoride (F^-) develop fewer carious lesions than people whose water does not contain fluoride salts. The inhibitory effect of fluoride depends on the time of exposure to the mineral and its concentration. The best results are found in populations exposed to F^- from birth and where F^- concentration in drinking water is 1.0–1.5 ppm (parts per million).

Higher fluoride concentration in water and food results in increased concentration of the mineral in the tooth enamel and dentine. The surface of the enamel and outer layers of dentine contain more fluoride than inner layers. It is suggested that F^- incorporated chemically in the hard tissues of the tooth as fluorohydroxyapatite or fluoroapatite either during tooth formation or after its eruption reduces the solubility of the enamel in acids. A constant supply of F^- in water, food, and dental care treatment (topical application of fluoride, fluoride tablets, toothpastes, mouthwashes, etc.) provides a reservoir of F^- when demineralization of the enamel in low pH occurs. F^- can substitute for carbonate in the apatite structure of dental hard tissues and, thus, increases their resistance to dissolution of minerals. Recent studies support the view that fluoride is particularly effective in remineralization of the tooth. It inhibits the progression of caries and stimulates the flow of minerals back into the enamel structure by binding them in chemical forms more resistant to dissolution. F^- has little effect on initiation of the carious lesion.

b. Strontium Another trace element found in soil and water of which cariostatic properties were proved is strontium (Sr^{+2}). Similar to F^-, strontium is incorporated in the dental hard tissues. The element can substitute for calcium in the structure of the apatite in the tooth. A positive correlation has been found between concentrations of strontium in drinking water and concentrations of the mineral in the enamel. Unlike F^-, which has greater posteruptive effect on the resistance to caries, strontium is most effective against caries when the element is supplied in high concentrations during tooth development. No posteruptive effect of strontium on caries prevalence has been observed.

Other trace elements such as molybdenum, manganese, vanadium, lithium, and boron also show anticariogenic properties. On the other hand, the presence of selenium, lead, cadmium, and silicon in drinking water increases the incidence of caries.

6. Systemic Diseases

Certain systemic diseases are accompanied by an increase in dental caries; in others caries activity is lower than average. High dental caries experience in neurologic and renal diseases may be explained mainly by difficulties in carrying out dental treatment. In Sjorgen's disease with xerostomia (dry mouth symptom due to reduced or absent flow of saliva) caries develops rapidly unless special precautions are taken. Decrease of dental caries incidence was observed in hypothyroidism related directly to thyroxine deficiency and indirectly to deficient secretion of thyroid-stimulating hormone. Diabetics develop extensive caries despite diets containing very small amounts of carbohydrates. Individuals with congenital syphilis usually have increased caries levels. Their hypoplastic teeth can be more susceptible to caries because of thinner enamel and an increased number of pits and grooves on the crown.

7. Sexual Dimorphism in Caries Experience

Sex differences in the prevalence of caries were documented in almost every study of permanent teeth. Usually females exhibit more caries than males of the same age. No difference between sexes was found with respect to caries in deciduous teeth. The most popular theory emphasizing the earlier tooth eruption in females and their relatively longer exposure to cariogenic environment, does not fully explain the difference. Although the difference may be partly explained by diet differences between males and females (in some historical populations or in pregnancy) or by gender-specific chewing habits, the differential susceptibility to caries in males and females remains unexplained.

8. Age and Caries

In prehistoric and historic times, caries seemed to be associated with old age rather than childhood. Caries prevalence was observed to increase in adults from the age of 20 yr upwards. In modern humans, caries is typically a childhood disease. It develops rapidly up to 18 yr of age and then stabilizes. Almost 100% of adults of Western economies experience caries at least once in their lives (life caries experience). Distribution of various types of carious lesions changes with individual's age. Older people develop more root surface caries, because of gingival recession causing root exposure.

D. Caries and Socioeconomic Status

Many aspects of the socioeconomic status (SES) such as income per capita, education, age of parents, dietary customs, oral hygiene habits, availability of dental services, dental attendance patterns, and others were studied in relation to the incidence and prevalence of caries. Because of the complexity of the socioeconomic status as a study variable, results of various studies are inconsistent and vary with the country in which they were conducted.

In developed countries, low SES groups have DMF (or dmf) scores higher than high SES groups. This is generally due to poor availability of dental services and poor oral hygiene habits, while sugar consumption is very high. In contrast, in underdeveloped countries higher DMF scores are found among higher SES people. These are associated with increased sugar consumption as compared with low SES groups, while other SES factors play a minor role.

Socioeconomic status, because of its complexity, has only an indirect influence on dental caries prevalence. Thus incidence of the disease cannot be treated as a simple and universal indicator of living conditions.

III. Conclusion

The reasons for which dental practitioners and anthropologists study caries are different. The practitioners are interested in combating the disease, while anthropologists use it as an indicator of varying living conditions.

The prevalence of dental caries in developed countries declined rapidly by about 50% during the last two decades. This was due to improvements in prevention and dental care and to changing lifestyles of people. A further decline in DMF scores by another 60% is expected to occur by the year 2050. Developing countries will—at the same time—face the problem of increases in caries due to lack of efficient preventive action and increased sugar consumption. The changing patterns of caries throughout the world and a constant need for better preventive methods will remain the major subjects of investigation in the next decades.

Anthropologists have found the phenomenon of caries a useful tool to study living conditions of prehistoric and historic populations. Observations of caries experience and frequency can shed light on major changes in lifestyles such as subsistence strategies, techniques of food preparation, and dietary customs. Reconstruction of prehistoric diets helps to explain patterns of human evolution and microevolution.

Bibliography

Brothwell, D. R. (1963). The macroscopic dental pathology of some earlier human populations. *In* "Dental Anthropology" (D. R. Brothwell, ed.). Pergamon Press, London.

Driessens, F. C. M., and Woltgens, J. H. M. (eds.) (1986). "Tooth Development and Caries." CRC Press, Boca Raton, Florida.

Hefferren, J. J., and Koehler, H. M. (eds.) (1981). "Foods, Nutrition and Dental Health," Vol. 1. Pathotox Publishers, Park Forest South, Illinois.

Keene, H. J. (1981). History of dental caries in human populations: The first million years. *In* "Animal Models in Cariology," a special supplement to "Microbiology Abstracts" (J. M. Tanzer, ed.). Information Retrieval, Washington, D. C.

Nikiforuk, G. (1985). "Understanding Dental Caries." Vol. 1 and 2. Karger, Basel.

Ortner, D. J., and Putschar, W. G. J. (1985). "Identification of Pathological Conditions in Human Skeletal Remains." Smithsonian Institute Press, Washington, D. C. [see especially chapter "Lesions of Jaws and Teeth"].

Powell, M. L. (1985). The analysis of dental wear and

caries for dietary reconstruction. *In* "The Analysis of Prehistoric Diets" (R. I. Gilbert, Jr. and J. H. Mielke, eds.). Academic Press, Orlando, Florida.

Turner, C. G., II (1979). Dental anthropological indications of agriculture among the Jomon people of central Japan. *Am. J. Phys. Anthro.* **51,** 619.

World Health Organization (1977). "Oral Health Surveys. Basic Methods." World Health Organization. Geneva

Wuehermann, A. H., and Manson-Hing, L. R. (1981). "Dental Radiology," 5th ed. C. V. Mosby Co., St. Louis [see especially p. 305].

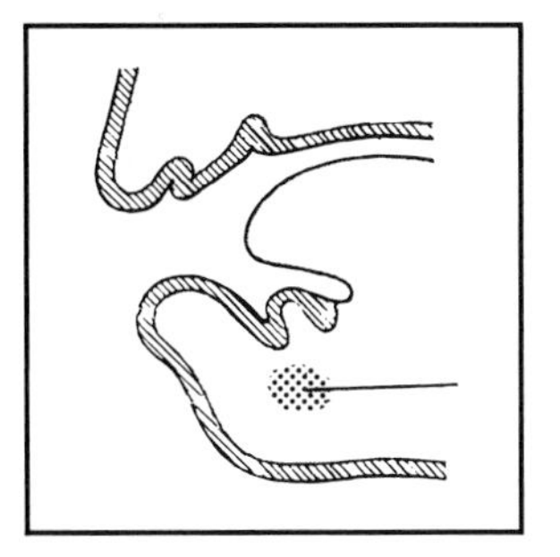

Dental Embryology and Histology

GEORGE W. BERNARD, *UCLA*

Glossary

Alveolar bone Supporting bone immediately surrounding the tooth
Amelogenesis Development of tooth enamel, which is the outer cover of the tooth crown
Cementogenesis Development of cementum, which is the outer cover of the tooth root
Dental (enamel) organ Crown-shaped epithelial masses growing from the dental, successional, and molar laminae from which the enamel develops
Dental lamina Localized epithelial ingrowths from the epithelial band where each of the 20 deciduous teeth will develop
Dental pulp Innermost soft tissue of the tooth in which the lymphatic, nervous, and vascular structures are found; develops from the dental papilla
Dentinogenesis Dentin formation, which is the inner hard structure of the tooth
Gingiva Outer coat of epithelium and adjacent connective tissue (mucosa) that surrounds the tooth in the mouth
Molar lamina Rear extensions of the dental lamina from which the 12 molar teeth will develop, three for each quadrant of the dental arches
Periodontal ligament Oriented connective tissue fibers (collagen) that connect the cementum of the tooth to the alveolar bone
Primary epithelial band Thickened horseshoe-shaped band of oral epithelium seen at the future site of the mandible and the maxilla
Successional lamina Lingual extensions from the dental lamina, which is the site where 20 permanent incisor, canine, and bicuspid teeth will develop
Tooth bud/germ Structure from which the tooth and adjacent structures develop; includes the dental organ, dental papilla, and dental follicle (sac)

FROM THE 27th to the 37th day of interuterine life two ectodermal epithelial bands are formed one in the presumptive upper jaw (the maxillary process), the other in the presumptive lower jaw (the mandibular process). From this band, a downgrowth of epithelial cells into the deeper layer of mesenchyme becomes the dental lamina from which 20 deciduous and 32 permanent tooth germs will develop. They become the baby or deciduous and the permanent teeth. Each tooth, whether deciduous or permanent, is made up of four different parts. Three are mineralized; the enamel, the dentin, and the cementum. The fourth is the soft tissue dental pulp, which occupies the central portion of the tooth.

Concomitant with the development of the teeth is the synchronous growth and development of adjacent tissues, which become an intrinsic part of the functioning tooth. The adjacent tissues are the periodontal ligament, the gingiva, and the alveolar bone. When the crowns and roots of the roots are mineralized, each tooth emerges into the mouth by breaking through the mucosa, ultimately meeting and articulating with its corresponding tooth in the opposite jaw.

Each tooth, whether deciduous or permanent, essentially follows the same pattern of development. Therefore the accompanying description of the development of *a* tooth is useful as a description of *all* teeth.

I. Development of the Primitive Mouth

The earliest indication that a presumptive mouth is forming is the appearance of the stomatodeum, which is formed by the anterior end of the embryo folding downward, the head fold. This creates a space bounded by the neural plate above and the cardiogenic area below. The bucco-pharyngeal membrane lies between the stomatodeum and the foregut. On either side (i.e., laterally), the stomatodeum becomes bounded by the first branchial arch, in which will ultimately develop the bony maxilla and the mandible. The boundaries of the stomatodeum consist of ectoderm on the outside and mesenchyme inside derived from the inner areas of the first bronchial arch and the head fold. Thus, the primitive mouth is a cavity lined by ectodermal epithelium, supported by mesenchymal connective tissue, and delineated internally by a temporary buccopharyngeal membrane. After the disappearance of this membrane, the mouth and the foregut are connected presaging a continuous gastrointestinal tract.

A. Primary Epithelial Band

The teeth begin when two bands of epithelial tissue thicken in the upper and lower arches of the stomatodeum at 37 days postfertilization. This is the primary epithelial band. From it, two subbands will form, the vestibular lamina, which will become the cleft or sulcus between the teeth on one side and the cheek and lips on the other, and the dental lamina, from which all the teeth will develop (Fig. 1). At 20 distinct areas, there will be downgrowth of the epithelium at the sites where the 20 deciduous (baby) teeth will grow. Surrounding these sites, there is increased mitosis of the ecto-mesenchyme, primitive connective tissue that is derived from neural crest ectoderm. It is the expansion, growth, and refinement of these two interacting tissues (epithelium and ecto-mesenchyme) that will result in the formation of 20 deciduous teeth followed by 32 permanent teeth.

II. Development of Teeth (Fig. 2)

A. Dental Lamina

There are 20 areas of growth of epithelium into the ectomesenchyme along the epithelial band, 10 in each presumptive jaw. These become string-like, remaining attached at one end to the oral ectoderm. On the end farthest away from the mouth, the epithelial cells of the dental lamina begin to divide at a greater rate, heralding the development of the deciduous tooth germ or bud.

B. Successional Lamina

On the lingual side of each of the 20 outgrowths of the dental lamina, an outcropping of epithelial cells indicates where permanent teeth will begin to develop. All the deciduous incisors and canine teeth will be replaced by permanent teeth of the same name from successional laminae, branches of the deciduous dental laminae. The successional laminae of the eight deciduous molars give rise to eight permanent bicuspids, two in each quadrant of the dental arches.

C. Molar Lamina

Posterior to the last or second deciduous molar on each side of each jaw, there is a backward extension of the dental lamina. These epithelial outgrowths on each side are the beginning sites of permanent molar tooth bud formation.

All the deciduous teeth begin to form between the sixth and eighth week of intrauterine life. The successional permanent teeth begins in the 20th week and ends in the 10th month after birth. The permanent molars are formed between the 20th week of fetal life and 5 years into childhood.

D. Dental or Enamel Organ

As the cells at the free ends of the dental laminae divide, they form knots of epithelial cells, which begin a pattern of morpho- and histo-differentiation for the 20 deciduous teeth.

1. Bud Stage

This is the first indication that the epithelia are forming into a defined structure. The surrounding ecto-mesenchyme also begins to condense around this knot of cells, an indication of increased mitosis.

2. Cap Stage

The name of this stage is descriptive of the shape that the dividing epithelia take as they multiply. At this stage, differentiation has established the primitive dental (enamel) organ. It is made of an internal

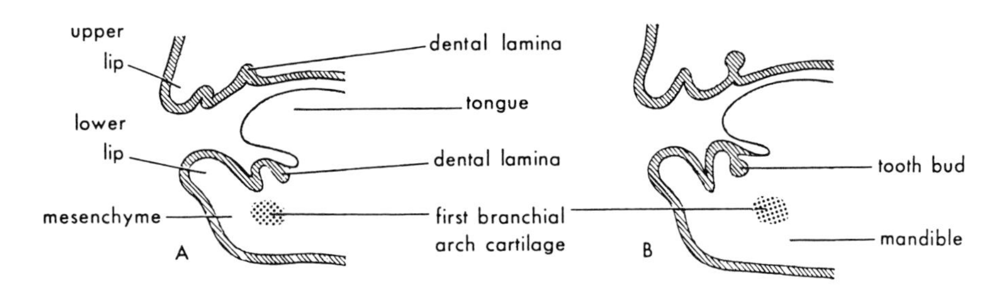

FIGURE 1 Diagrammatic sketches of sagittal sections through the developing jaws, illustrating early development of the teeth. A, Early in the sixth week, showing the dental lamina. B, Later in the sixth week, showing tooth buds arising from the dental laminae.

dental (enamel) epithelium, an external (enamel) epithelium, and an intermediate zone of less differentiated epithelial cells. The ecto-mesenchyme, in the meantime, has condensed around the cap as the dental follicle (sac) and under the cap as the dental papilla. The cells of the dental follicle will differentiate into the periodontal ligament, the cementum, and the inner portion of the alveolar bone. The cells of the dental papilla will become the dental pulp as the dentin is mineralized around it. The dental follicle, dental papilla, and the dental organ comprise the tissues of the tooth germ from which all the tooth and its adjacent structures will develop.

3. Bell Stage

The tooth germ continues to grow into the shape of a bell. Mitosis expands the size of the tooth germ. A layer of flattened differentiated cells has condensed in the dental organ side of the cuboidal inner-enamel epithelia. This is the stratum intermedium, which is a source of new cells for the internal enamel epithelium. Water inhibition into the center of the dental organ causes intercellular spaces to appear. Macula adherens attachments between cells at limited spots on the plasma membranes now cause stretching of the cells into star shapes. This middle zone of the dental organ is now called the *stellate (star-shaped) reticulum*. The linear area at the periphery of the bell, where the inner enamel epithelium meets the outer enamel epithelium, is called the *cervical loop*. All the cells of the dental organ are attached to each other by maculae adherens. The whole dental organ is separated from the ecto-mesenchyme of the dental papilla and the dental follicle by a basement membrane.

The original dental lamina begins to break up into small clusters of cells, which disappear leaving the deciduous tooth germ and the successional lamina as isolates. When this occurs, the dental organ gradually takes the shape of the future crown. The stellate reticulum is squeezed to the sides of the dental organ and is sparse at the tip of the presumptive tooth crown. The dental papilla is now the site of ingrowth of blood vessels and nerves. In contrast, the dental organ is avascular and must receive its nutrition by diffusion from blood vessels in the dental follicle. During the late bell stage, cells of the internal enamel epithelium change shape from cuboidal to tall columnar. Internally these cells develop organelles and the nucleus moves to the basal pole of the cell closer to the cells of the stratum intermedium. This cellular differentiation results in presecretory ameloblasts. Ecto-mesenchymal cells of the dental papilla close to the apical pole of the presecretory ameloblasts differentiate into elongate cells and line up along the underlying basement membrane. These cells are now called *odontoblasts*. They begin to secrete organic matrix, which will become calcified as dentin. The basement membrane disappears at this time. The ameloblasts will become secretory only after 2–3 μm of dentin has been laid down, but odontoblasts will not be differentiated without the inductive activity of the ameloblasts. This epithelial–mesenchymal interaction is called *reciprocal induction*.

4. Appositional Stage

The shape of the crown has been determined by the design of the dental organ. Dentin has formed at the cusp tip. Enamel begins to calcify immediately after organic enamel matrix is secreted by the secretory ameloblasts (Fig. 3). During this stage, differentiation of ameloblasts and odontoblasts continues down the sides of the crown and immediately, dentin followed by enamel, is mineralized. In the older areas, both dentin and enamel thicken as they mineralize in opposite directions from each other, the dentin toward the papilla, the enamel toward the external enamel epithelium. The stellate reticulum

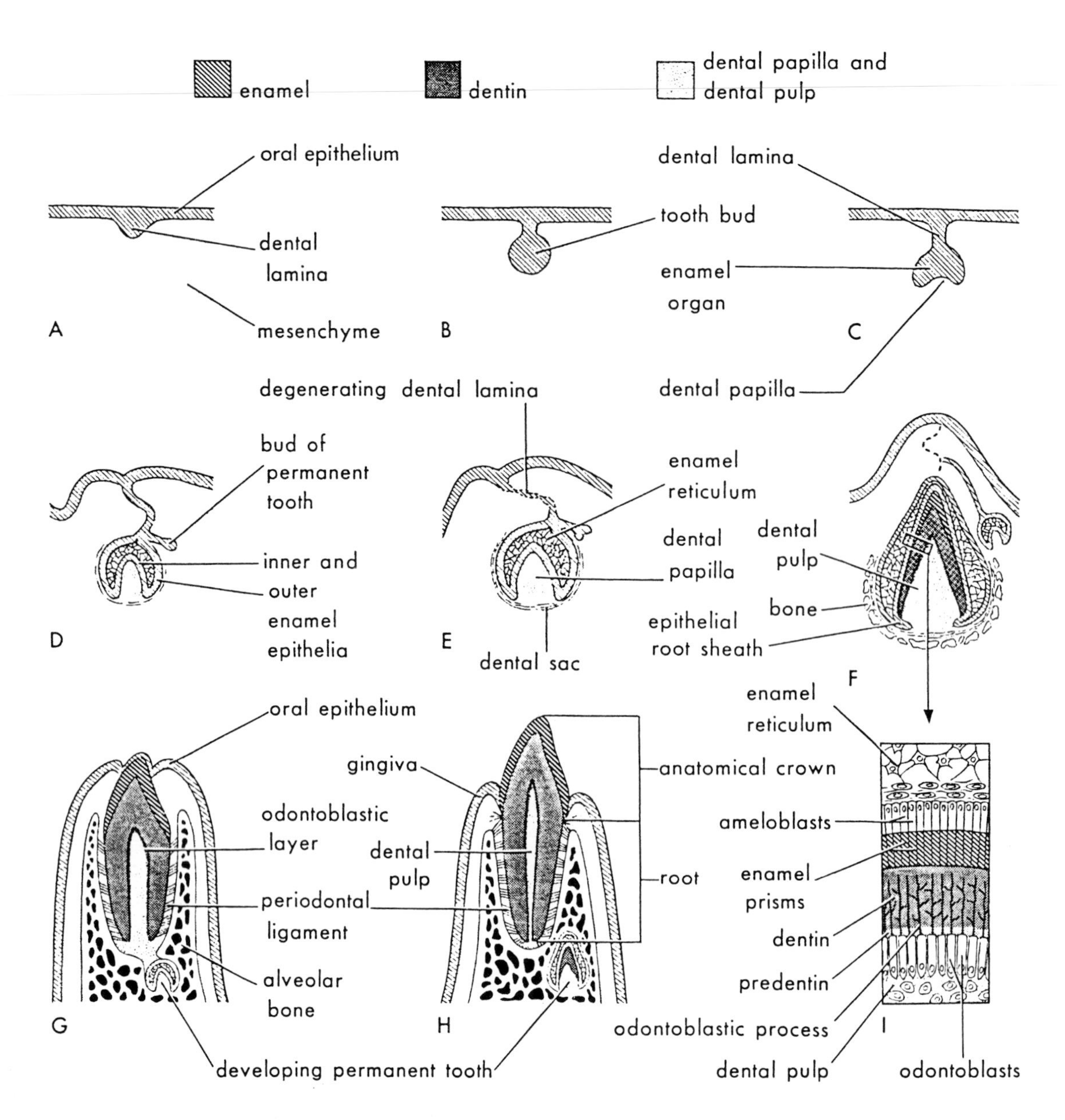

collapses at this stage, allowing the ameloblasts to come closer to their nutrient vascular supply from the dental follicle.

FIGURE 2 Schematic drawings of sagittal sections, showing successive stages in the development and eruption of an incisor tooth. A, Six weeks, showing the dental lamina. B, Seven weeks, showing the tooth bud developing from the dental lamina. C, Eight weeks, showing the cap stage of teeth development. D, Ten weeks showing the early bell stage of the deciduous tooth and the bud stage of the developing permanent tooth. E, Fourteen weeks, showing the advanced bell stage of the enamel organ. Note that the connection (dental lamina) of the tooth to the oral epithelium is degenerating. F, 28 weeks, showing the enamel and dentin layers. G, Six months postnatal, showing early tooth eruption. H, 18 months postnatal, showing a fully erupted deciduous incisor tooth. The permanent incisor tooth now has a well-developed crown. I, Section through a developing tooth, showing the ameloblasts (enamel producers) and the odontoblasts (dentin producers).

5. Root Formation

The cervical loop of the appositional stage is where the inner enamel epithelium ends. It is also where the enamel ends its formation, as the last cells of the inner enamel epithelium are differentiated into ameloblasts. At this time cells of the inner and outer enamel epithelium continue dividing and form a two cell–layered sheet of epithelial cells, which grow away from the crown. This is Hertwig's

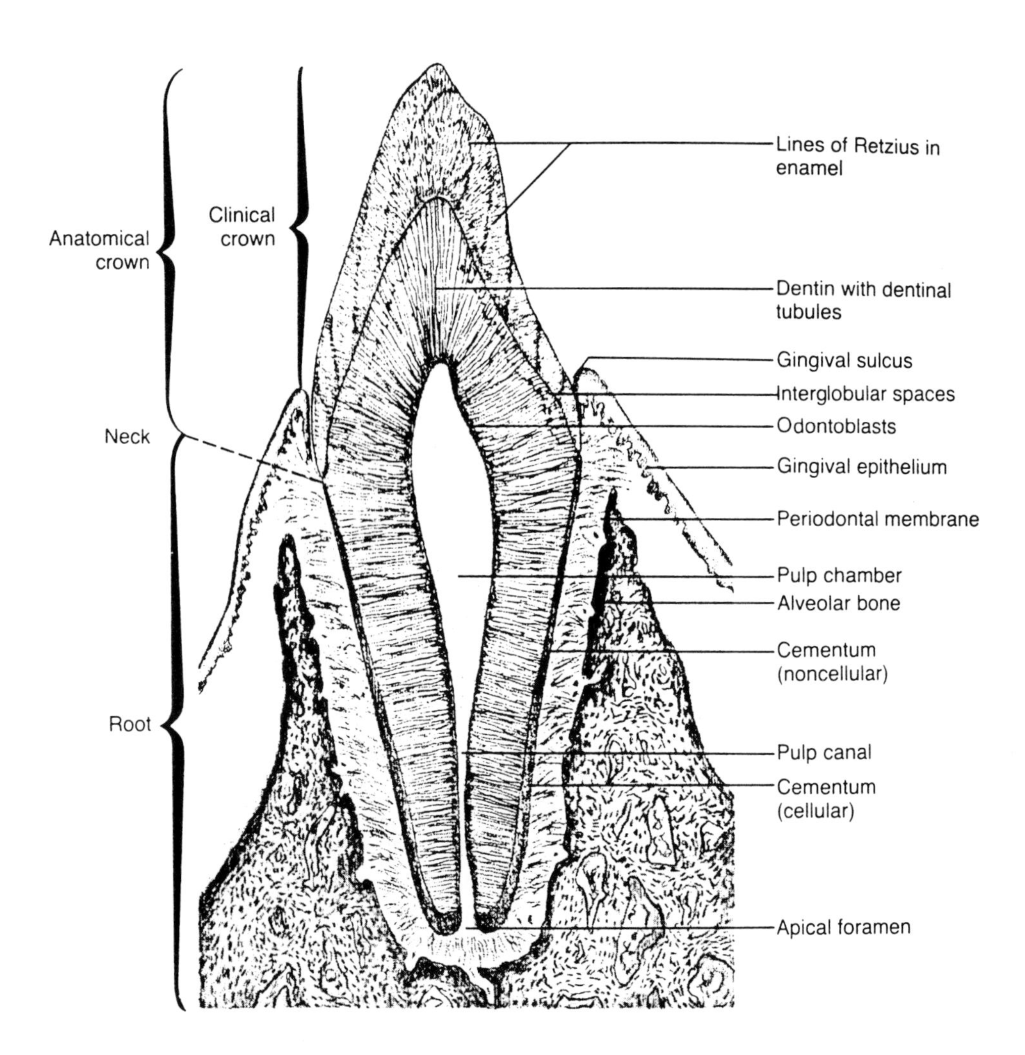

FIGURE 3 The fully formed functional tooth and adjoining structures. [Source: A. A. Paparo, T. S. Leeson, and C. R. Leeson. (1988). "Text/Atlas of Histology." p. 402. W. B. Saunders Co., Philadelphia.]

epithelial root sheath, which will provide a scaffolding for the development of the tooth root. Cells of Hertwig's sheath are similar in origin to the ameloblasts, which have induced odontoblasts to differentiate from the ecto-mesenchyme of the papilla. They repeat this process (i.e., papilla cells differentiate to form odontoblasts adjacent to Hertwig's sheath). These odontoblasts will now secrete matrix that will calcify as root dentin. Continued growth of Hertwig's sheath and subsequent dentin formation is how the root is formed. Multirooted teeth such as molars are developed from ingrowths of Hertwig's sheath, which allows each root to grow as a separate structure but attached to the crown of the tooth.

III. Development and Structure of Parts of Teeth

Odontogenesis means tooth embryogenesis or development. Amelogenesis is enamel development, dentinogenesis is dentin development, and cementogenesis is cementum development. Concomitant with tissue embryogenesis is the initiation and growth of the mineral within the tissues (mineralogenesis). As the mineralized tissues develop they surround the growing soft tissue in the dental pulp.

A. Amelogenesis

Enamel formation is composed of two processes: the growth and development of the dental organ, and the differentiation of its quintessential substructure, the inner enamel epithelium. As the dental organ matures from the cap through the bell, appositional and root formative stages, the inner enamel epithelium matures from a flattened, squamous cell to a cuboidal cell, to a columnar presecretory cell, to a specialized secretory ameloblast, to a mature columnar cell, and finally to a cuboidal protective covering cell, which ultimately will be lost from the mineralized enamel surface after the tooth erupts into the mouth.

As discussed above, the presecretory ameloblasts induce proximal ecto-mesenchymal cells of the dental papilla to become a layer of juxtaposed odontoblasts. These cells produce extracellular matrix, which will calcify in a process discussed in Section III,B. After 2–3 μm of dentin is mineralized, presecretory ameloblasts differentiate into secretory ameloblasts and begin to produce enamel matrix. Secretory ameloblasts differentiate into secretory ameloblasts and begin to produce enamel matrix. Secretory ameloblasts are distinguished by the trix (see Section III,A,1). This matrix calcifies almost immediately with mineralized crystals of calcium, phosphate, and hydroxyl moieties, called *hydroxyapatite* (HA). These grow from the HA crystals already mineralized in the dentin. As enamel continues to thicken, the thin plate-like crystals of HA grow in length, width, and height at the expense of a pH labile enamel protein, which is denatured and resorbed. Eventually less than 1% of organic matrix remains in mature enamel. Enamel protein is exquisitely sensitive to small changes in pH. In the presence of potassium, an abundant extracellular cation, the crystal surface pH is elevated, significantly influencing protein depolymerization. There are one-half as many crystals of HA at the surface of enamel as compared with the smaller circumference at the dentino-enamel interface or junction (DEJ), indicating fusion of crystals as they grow and mature. Enamel matrix has two major constituent proteins: amelogenin, which is 19 times more abundant than the second protein, enamelin. Amelogenins with a molecular weight of approximately 4,000 Daltons is unbound to HA and therefore is the protein that is resorbed during amelogenesis. Enamelin has a high molecular weight (67,000 Daltons), and because it is bound to HA, it mostly is retained in mature enamel. [*See* EXTRACELLULAR MATRIX.]

1. Structure of Enamel

The first enamel to form is relatively homogeneous. After 2 or 3 μm of this early enamel have been laid down, Tomes processes develop on the apical surfaces of the ameloblasts. Almost immediately the enamel becomes divided into rod and interrod structures, mediated by Tomes processes. The enamel rod or prism is the basic unit of enamel and is roughly hexagonal to round with a variable keyhole shape in cross section. It is filled with innumerable crystals of HA, oriented more or less perpendicular to the DEJ. Theoretically, HA crystals of enamel can extend the whole 2–3-mm width of enamel from the DEJ to the tooth surface. Rod diameter averages 4 μm, varying between 2.5 μm at the DEJ to 5 μm at the surface. There are 5 million rods in a lower incisor and 12 million rods in an upper molar. They run a wavy course and therefore are longer than the 2–3-mm thickness of the enamel of the cusp tip.

Interrod substance is filled with small HA crystals and contains a greater amount of organic material than does the rod itself. Each rod has 4-μm cross striations, indicating a diurnal rhythmicity of formation. Periodically, there are larger, more distinct rhythmic bands called *incremental lines* or the *stria of Retzius*, caused by long-acting metabolic alterations or minute changes of rod direction. Sometimes, when there are even more prolonged metabolic changes, the incremental lines are increased, as either in febrile diseases of early childhood or the neonatal line of deciduous teeth or in permanent teeth that are in the process of mineralization at the time of birth. Just as is the case in the first enamel formed without Tomes processes of the ameloblasts, surface enamel shows no rod outlines because the maturing ameloblasts, which secrete the final enamel matrix, also are devoid of Tomes processes.

2. Manifestations of Developmental Faults in Enamel

Enamel lamellae are faults or cracks extending from the DEJ to the enamel surface.

Enamel tufts are interwoven areas of poor mineralization looking life tufts of grass extending out from the DEJ up to one-fifth to one-third of the enamel thickness.

Enamel spindles are poorly mineralized beginnings of odontoblastic processes (see below) trapped in the early enamel.

B. Dentinogenesis

Dentin formation begins with the differentiation of secretory odontoblasts from undifferentiated ecto-mesenchymal cells. These cells line up as a single layer on the basement membrane, which was previously secreted by presecretory ameloblasts. The secretory odontoblasts have a basally placed nucleus, which is at the pulpal pole of the tall columnar cell. The large number of intracellular membranous organelles is indicative of a secretory cell. The odontoblasts now secrete extracellular matrix (predentin). They migrate away from the DEJ, leaving behind cellular extensions called *odontoblastic processes*, which remain as part of the odontoblasts throughout life. The precalcified extracellular matrix contains the primary protein of connective tissues (i.e., collagen), as well as glycoproteins, phosphoproteins, glycosaminoglycans, lipoproteins, and most importantly, matrix vesicles. The matrix vesicles are the primary calcification loci of dentin. It is within the matrix vesicle that the first crystals of HA are formed. Calcification that originates in matrix vesicles is called *primary* (initial calcification). As crystals continue to grow either related to subunits of the matrix vesicle membrane or by secondary crystallization from already crystallized HA, the matrix vesicle membrane breaks down and the crystals continue to grow in, on, and within ubiquitous collagen fibers. This in time separates dentin into mineralized dentin and unmineralized predentin. Predentin lies between the mineralized dentin and the secretory odontoblast. The odontoblastic process continues to grow in length as the odontoblast migrates toward the pulp. A highly calcified structure surrounds it in the mineralized dentin (*peritubular dentin*), forming a canal, the dentinal tubule, in which the odontoblastic process courses.

The first dentin that is formed is *mantle dentin,* which represents the coalescence of calcification nodules. These are outgrowths of matrix vesicles. After several microns of this first dentin has been formed, the odontoblasts stop secreting matrix vesicles and begin to secrete collagen in sheets, more or less oriented radially from the mantle dentin toward the pulp. When this is mineralized, it is called *circumpulpal dentin*.

If ameloblasts do not develop, dentin will not form. If dentin does not form, enamel will not develop because each is reciprocally related to the other in development.

1. Structure of Dentin

It is 70% inorganic, essentially HA, and 30% organic, which is primarily collagen with glycoproteins, phosphoproteins, lipoproteins, and glycosaminoglycans. Dentin underlies the enamel of the crown and the cementum of the root. As such, it comprises the bulk of the calcified structure of the tooth. It has a slightly yellowish color, which gives the tooth much of the background shade as light is reflected through the translucent enamel. All dentin, including the root, has mantle dentin beginning at the junctions with enamel or cementum and extending pulpward for about 10–20 μm. Continuing for the rest of the dentin is circumpulpal dentin, which is lined by a continuously viable layer of cells (i.e., the odontoblasts). This is the boundary for the dental pulp. Another way to subdivide the mineralized structure of dentin is: peritubular dentin, the heavily calcified zone surrounding the odontoblastic processes, which lie in the dentinal tubules, and intertubular dentin, the areas between the processes. Still another way is temporal. Primary dentin forms before the root has been completed. Secondary dentin forms afterward and is represented by a sharp change of direction of the dentinal tubules. Tertiary dentin (reparative dentin) results from a response to insult or injury, and the tubules are irregular in their arrangement. Dentinal tubules are straight in the cusps and the roots and S-shaped in the crowns. In cross section, there are 20,000 tubules mm^2 at the DEJ and 45,000 at the pulp, indicating crowding of the odontoblasts as the dentine grows. Whenever dentinal tubules continue to calcify and completely occlude, it is called *sclerotic dentin*. Daily growth lines are called *incremental lines of von Ebner*. These are 6 μm in the crown and 3.5 μm in the roots.

The contour lines of Owen are analogous to the striae of Retzius of the enamel. That is, they are structural bands resulting from long-lasting metabolic changes during dentinogenesis.

C. Cementogenesis or Cementum Formation

The extension of the inner enamel epithelial cells into Hertwig's epithelial root sheath forms the cellular basis for inducing root papillae ecto-mesenchymal cells to differentiate into root odontoblasts. Odontoblasts from several microns of mantle dentin of the root. Hertwig's sheath now disintegrates. Mesenchymal cells from the dental sac (follicle) invade through the spaces caused by the breakdown of Hertwig's sheath. When these mesenchymal cells reach the mantle dentin of the root, they differentiate into cementoblasts. Cementoblasts, which are similar to osteoblasts, secrete collagen, glycoproteins, and glycosaminoglycans. This matrix calcifies by *secondary (subsequential) calcification* as crystals of HA grow in, on, and between collagen fibers by epitaxy from the HA of the mantle dentin. Groups of sheath cells, which do not degenerate, may last into adult life as epithelial rests of Malassez. These sometimes become cystic.

Collagen oriented perpendicularly to the surface of the mantle dentine is secreted by fibroblasts lying on the outside of the cementoblasts, and this collagen forms between the cementoblasts. Matrix secreted by the cementoblasts engulfs this collagen. All these structures are calcified from the mantle dentin by secondary calcification. The oriented collagen bundles produced by the fibroblasts when they are embedded within calcified cementum are called *Sharpey's fibers* and are the tooth terminal end of the periodontal ligament fibers, which terminate at the other end in alveolar bone as Sharpey's fiber as well. [*See* COLLAGEN, STRUCTURE AND FUNCTION.]

1. Structure of Cementum

This tissue is less hard than dentin because it is only 50% inorganic. Like other mammalian mineralized tissues, the basic inorganic moiety is HA. The organic substructure is primarily the fibrous protein collagen with smaller amounts of glycoproteins and glycosaminoglycans.

There are two types of cementum: (1) acellular, in which there is no incorporation of cells in the mineralized structure. It is sometimes missing in the apical one-third of the tooth; and (2) cellular, in which cellular cementum is studded with spider-like cementocytes with canaliculi, mainly directed toward the outer root surface, which are mineralized into the cementum. It is found in the apical one-third and the *furcations* of teeth. Furcations are the areas where roots join at the body of the tooth.

Collagen is found as intrinsic fibers, which are secreted by cementoblasts, and extrinsic fibers, which are secreted by periodontal ligament fibroblasts as Sharpey's fibers. Cementum is thinnest at the cemento–enamel junction (20 μm–50 μm) and thickest at the apex (150 μm–200 μm).

Incremental lines are parallel to the surface of the tooth and are found in both cellular and acellular cementum.

The cemento-enamel junction is a butt joint 30% of the time, does not meet at all 10% of the time, and exhibits an overlap of the cementum over the enamel 60% of the time. When overlapping, the cementum is *afibrillar*.

Cementum is the tooth tissue to which the collagen bundles of the periodontal ligament attach. Cementum is incremental throughout life. Resorption is seen in deciduous teeth but less frequent in the adult. Sharpey's fibers attach only to the outside layer of cementum. Each incremental layer has a new attachment of fibers. Deposition of apical cementum compensates for occlusal wear, to maintain the vertical height of the tooth.

D. Dental Pulp

Its origin is in the dental papilla, which begins as a condensation of ectomesenchymal cells close to the dental organ when it reaches the cap stage of development. As dentin grows inward from the crown and downward to become the root of the tooth, nerves and vascular and lymphatic vessels grow in from the root side. The pulp is primarily a connective tissue with certain primitive qualities. The cells include odontoblasts, which have a peripheral location, fibroblasts, mesenchymal cells, macrophages, nerve cells, vascular cells, and lymphocytes. There are four zones in the pulp:

- odontoblastic cell zone,
- cell-free zone of Weil,
- cell-rich zone, and
- pulp core.

The extracellular matrix contains glycoproteins and glycosaminoglycans, but unusually, it has a high concentration of Type II embryonic collagen as well as the more common Type I collagen. When the pulp is completely contained by the crown and the root, it is said to reside in the pulp chamber. There is access into the pulp chamber only through the variably sized apertures at the root tip.

E. Vasculature and Lymphatics

One hundred fifty micron arterioles enter the pulp at the root tip, remain centrally located, shed some musculature, and give off lateral branches in the root that extend to the pulp periphery. Here branching develops into a subodontoblastic capillary network, which continues into the coronal area where the network is the most extensive. Venules of equivalent size accompany the arterial supply. Lymphatics function to relieve fluid pressure in the pulp.

F. Nerves

Nerves of different sizes follow the arborization of blood vessels. They form (like the vasculature) a subodontoblastic plexus of nerves called the *plexus of Raschkow* in the cell-free zone of Weil. Axons are primarily myelinated sensory afferents of the trigeminal nerve and nonmyelinated sympathetic branches from the superior cervical ganglion, which innervate blood vessels. Myelin is lost by the time the nerves reach the subodontoblastic plexus. Some nonmyelinated fibers enter dentinal tubules for a short distance after passing between the cell bodies of the odontoblasts.

G. Dental Pain

Pain is stimulated by diverse stimuli (e.g., hot, cold, mechanical stimulation, dessication). The tooth is most sensitive at the dentino–enamel junction. There is increased sensitivity with an inflamed pulp. The possible mechanisms of pain are:

- stimulation of free endings in the dentinal tubules,
- odontoblasts and their processes may serve as pain receptors, and/or
- fluid movement in the dentinal tubules may cause stimulation of free nerve endings in the dentin and pulp.

[*See* PAIN.]

H. Pulp Stones

Pulp stones are calcified masses in the pulp chamber. They may be free or attached. True pulp stones are surrounded by odontoblasts. False pulp stones are dystrophic calcifications without odontoblasts.

I. Age Changes

There is a gradual reduction in volume of the pulp chamber caused by the continual growth of secondary and tertiary dentin. Some dentinal tubules become sclerotic dentin as they calcify.

IV. Development and Structure of Dental Supporting Tissues

All the supporting tissues of the teeth develop from the dental follicle or sac. As described above, during cementogenesis, Hertwig's epithelial root sheath disintegrates, and cementoblasts differentiate from follicular mesenchymal cells to secrete cementum matrix. Some mesenchymal cells become fibroblasts, which secrete periodontal ligament collagen fibers, whereas others differentiate into osteoblasts and produce the adjacent alveolar bone.

A. Alveolar Bone

Alveolar bone develops by intramembranous osteogenesis in both the maxilla and the mandible. By the third fetal month there is a bony groove that contains the tooth germs, alveolar nerves, and blood vessels. In time, bony septae separate the tooth buds and the alveolar bone (which surrounds the teeth). Alveolar bone now becomes firmly attached to the body of the maxilla and the mandible without any distinct boundary between them.

The alveolar process is that part of the maxilla and mandible that forms the sockets and supports the teeth. Strictly speaking, it is defined only after the teeth have erupted and disappears when teeth are lost.

There are two parts to the alveolar process:

1) Alveolar bone proper, which gives attachment to the periodontal ligament fibers and therefore surrounds the tooth; and
2) Supporting alveolar bone, which is composed of two *cortical plates* on either side of *spongy bone*. This is attached to the outside of the alveolar bone proper.

In the anterior regions of both jaws, the supporting bone is thin so that the cortical plates merge with the alveolar bone proper.

In oral radiographs of the teeth and jaws, the alveolar bone proper is called the *lamina dura*. Because this bone is perforated by nutrient canals, it is sometimes called the *cribiform plate*.

Marrow spaces contain hemopoeitic bone marrow in the young but mostly only fatty marrow in the adult.

Alveolar bone begins as woven bone, and because there are no calcified structures present in the vicinity, calcification is initiated in matrix vesicles, produced by osteoblasts. It is only when several microns of woven bone have developed that matrix vesicles disappear and lamellar bone appears, calcification proceeding by secondary nucleation and growth of HA crystals.

A peculiar and unique type of bone is found only in the alveolar process and is called *bundle bone*. Bundle bone has incremental areas of lamellar bone with incorporated Sharpey's fibers, even in the deeper layers where no attachment to the periodontal ligament exists.

Alveolar bone, like all bone, is 65% inorganic, primarily calcium HA, and 35% organic, primarily collagen and glycoprotein.

B. Gingival Tissues

Gingival tissues are the pink mucosal soft tissues that surround the teeth in the mouth. They originate from ectodermal cells and underlying ecto-mesenchymal cells of the stomatodeum. Early in life, for example, at birth, there is only the primitive oral mucosa devoid of both gingiva and teeth. Gingiva can only be demonstrated after teeth have erupted into the mouth, because its structure and function is dependent on the presence of teeth. Its primary function is to seal the oral cavity from the connective tissue surrounding the tooth. It resists abrasive, hot, cold, and irritating foods. It forms a first line of defense against invasion of microorganisms. The importance of this function is underscored by the fact that bacteria are normal inhabitants of the oral cavity and obviously must be contained.

Gingival tissue is made up of epithelium and underlying connective tissue. There are three gingival epithelial subdivisions.

1) The outer gingival epithelium covers the outer surface of the gingiva. Its surface is stratified, squamous, and keratinized as the horny outer layer. It has predominant epithelial ingrowths, the rete pegs, between which are connective tissue outgrowths.

2) The oral sulcular epithelium lines the superficial half of the gingival crevice. It is stratified, squamous, and nonkeratinized with prominent rete pegs. The gingival crevice or sulcus is the space formed around the tooth between the epithelium and the tooth. It is bounded below by the epithelial attachment to the tooth and is open to the mouth above.

3) The junctional epithelium lines the deep half of the gingiva below the crevice with stratified squamous nonkeratinized epithelium. It has no rete pegs and is only one to five cells thick.

Between the junctional epithelial cells and the tooth, a glycoproteinacious cement substance attaches the cells to the tooth. This is the epithelial attachment. It is secreted by the epithelial cells, which in turn are attached to it by hemidesosomes. These cells are attached to each other with tight junctions, which acts as a further seal against bacterial invasion.

The attached connective tissue is organized as dense collagen fibers mainly perpendicular to the surface. There are no elastic fibers and no muscle fibers. The collagen fibers form a variety of structures that run from the alveolar bone to the gingiva, from the cementum to the gingiva, and around the tooth below the gingival crest.

C. Periodontal Ligament

The periodontal ligament is really a group of precisely oriented collagen fibers, which surrounds the tooth, cradles it, and attaches it through the cementum with the alveolar bone (Fig. 4). At one time it was called the *periodontal membrane* because in early light microscopy it appeared homogeneous rather than, as it turned out, fibrillar. The *periodontium* is the term used to describe the tissues that surround and attach the tooth in its bony socket. It consists of cementum, periodontal ligament, inner alveolar bone, and junctional epithelial gingiva. Peridontal ligament originates in the ecto-mesenchymal cells of the dental follicle. These cells differentiate into fibroblasts, which begin to secrete collagen and associated noncollagenous proteins as well as glycosaminoglycans. The first collagen fibers are seen after the initiation of root formation. The first fibers are oriented in an oblique direction passing upward from the cementum because the tooth develops inferior to the crest of alveolar bone. As the tooth erupts, it moves past the alveolar crest, and the fibers become horizontally oriented. When the tooth is completely erupted, the most coronal fibers again become oblique, but now with fibers directed from the cementum downward to the alveolar crest.

After the root has formed, the periodontal ligament becomes denser and the fibers assume particular orientations depending on position in the tooth socket. The fibers are categorized into the following groups: alveolar crest, horizontal, oblique, transitional, apical, interdental, circular, and tangential. The alveolar crest group are oblique fibers that run downward between the cementum, at the cemental–enamel junction (CEJ) and the crest of alveolar bone. Just apical to them are horizontal fibers followed by oblique fibers that run upward from the cementum to the alveolar bone. Next is a group of horizontal fibers that are transitional to the oblique and vertical apical fibers surrounding the apex of

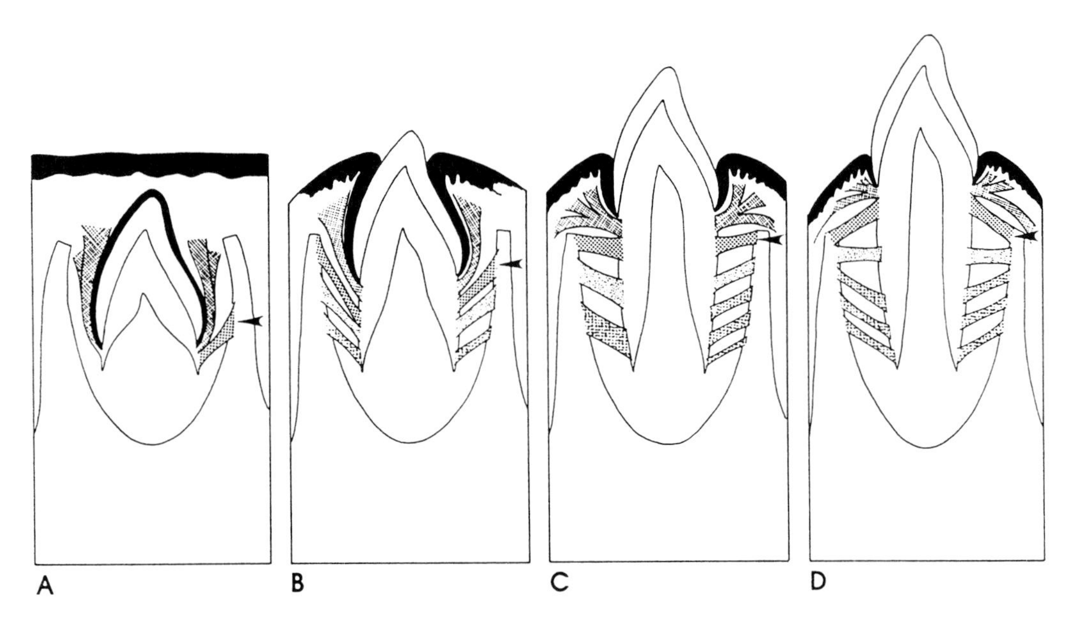

FIGURE 4 Development of the periodontal ligament. A: First fibers are oblique. B: With eruption of the tooth, new fibers are added. C: With continued eruption, the first fibers become horizontal. D: Later still, they become oblique again but in the opposite direction. New fibers are added as the root continues development.

the tooth. Interdental (or transseptal) fibers traverse the area above the gingival crest, attaching cementum of one tooth with the cementum of an adjoining tooth. Circular fibers surround the tooth, attaching, in part, the gingiva to the bone, and tangential fibers run from one side of the alveolar bone to the other. The collagen of periodontal ligament fibers are attached on either side by being calcified into bone and cementum. These partially calcified fibers are called Sharpey's fibers. The diversity of orientation of the fibers suspends the tooth in the alveolar socket. This permits a slight limited movement during mastication while holding the tooth firmly in place. In addition to collagen, there are some elastic fibers associated with blood vessels and a unique fiber, the *oxytalin* fiber, which could be an elastin precursor. The periodontal ligament is highly vascularized, the blood supply being derived from the inferior alveolar artery for the mandibular periodontium and teeth and from the superior alveolar artery for the maxillary structures. Collagen of the periodontal ligament is secreted and resorbed rapidly by the same cells, the fibroblasts. The ligament increases in thickness with increased force on the tooth.

Because all the components of the periodontium have the same origin, it is not surprising that the growth of bone, ligament, and cementum stem from the same cells. These precursors to the periodontal ligament fibroblast subsequently can differentiate into osteoblasts, cementoblasts, and fibroblasts. All these cells and the derived tissues function in unison under varying physiologic conditions. For example, if a tooth is moved orthodontically, the periodontal ligament is compressed against the alveolar bone. Osteoclasts are activated, causing the bone to resorb. On the other side of the tooth, where the periodontal ligament is under tension, the width of the ligament is increased, fibroblasts secrete oriented collagen, and new bone is deposited. Sharpey's fibers in both cementum and bone are either destroyed or remade on both sides of the tooth.

V. Eruption of Teeth into the Mouth

At the time that the enamel of any tooth, deciduous or permanent, has been mineralized and the crown is complete, the dental organ is compressed on the enamel surface as the reduced enamel epithelium. This two-layered structure is the remains of the dental organ. The layer in contact with the tooth consists of the nonsecretory protective phase ameloblasts and is firmly attached to the enamel. The outer layer is the remnant of all the other dental organ components. As the tooth erupts, the connective tissue that lies between the cusp of the tooth and the oral epithelium is resorbed. The outer cells of the reduced enamel epithelium and the inner cells of the oral epithelium merge. The cells now are induced to proliferate, particularly on the sides of the crown as the cusp tip pushes through the merged epithelium. The merged epithelial cells along the side of the tooth become the gingiva of the dentogingival junction. By continued growth of the tooth into the mouth and proliferation of the adjacent cells, the gingiva becomes the functional tissue described above. With age there is an apical migration of the dentogingival junction. In all this growth and development, the epithelium maintains an absolute separation of the gingiva and the crown from the underlying connective tissue.

1. Tooth Movement During Development

The deciduous teeth all begin development from the 6th to the 8th week postfertilization. The permanent teeth begin development from the successional laminae of the incisors and the molar laminae of the first molars during the 20th week. During this phase and the subsequent time period when the rest of the permanent teeth develop from the successional laminae, the jaws are enormously crowded. Fortunately, the jaws continue growth in all dimensions, permitting the teeth to continue their orchestrated growth. The permanent incisor and canine teeth begin growth lingual to (i.e., on the tongue side of) the deciduous teeth. As simultaneous growth of both deciduous and permanent teeth continues, each lies within its own bony crypt. The permanent premolars (bicuspids) grow between the roots of the deciduous molars. The permanent molars develop behind the deciduous molars from the molar lamina, at first in tight quarters. As the jaw gets larger, the tooth germs move into position in the greater space that growth has provided. As the tooth germs move forward and toward the midline, remodeling takes place in the bony crypts. There is resorption ahead of movement and deposition of bone behind the movement of the tooth germ. After root formation has begun and the tooth has begun to erupt, the overlying bone is resorbed, allowing the tooth to migrate through the epithelium into the mouth. The deciduous tooth, when erupted into the mouth, continues its root development until the apex is calcified. Meanwhile, the successional permanent tooth continues its growth. This pressures the deciduous tooth and starts the process of resorption of the roots. Osteoclast-like cells, the odontoclasts, mediate the resorption process. In time, the deciduous tooth roots are resorbed, and the successional permanent tooth crown is fully formed. The roots are partially complete. The deciduous tooth, which is now only a disembodied crown, is exfoliated into the mouth. Shortly afterward, the permanent tooth erupts into the mouth through a canal in the bone called the *gubernacular canal*. In the canal are the remains of the successional lamina.

Most evidence points to the pulling characteristics of the periodontal ligament fibers as the major force that causes the eruption process to continue. This is true for both deciduous and permanent teeth, although the deciduous teeth are eventually exfoliated because of the joint forces of root resorption and masticatory pressure. Exfoliation of deciduous teeth begins with the lower incisors at 6–7 years of age, then the upper incisors. The lower canines are followed by the upper canine and all the deciduous molars almost simultaneously at about 8–10 years of age. The successional permanent teeth erupt shortly after the deciduous teeth are exfoliated. The permanent first molars erupt at 6–7 years, the second molars at about 12 years, and the third molars between 17 and 22 years.

VI. Mineralogenesis—Primary and Secondary Calcification

Mineralization of tissues *in vivo* can be divided into two distinct phases, each giving rise to distinct anatomic entities. (Phase I is primary or initial calcification; phase II is secondary or subsequential calcification.) Initial calcification involves the matrix vesicle, which provides the locus for the nucleation of HA in woven bone, calcified cartilage, and mantle dentin. Subsequential calcification is the process by which lamellar bone, circumpulpal dentin, cementum, and enamel develops. It does not depend on matrix vesicles for initiation of mineralization; rather HA continues growing by direct extension and secondary nucleation of the preexistent crystaline structure formed during initial calcification. Although each adult calcified tissue develops from a different cell lineage, all share the matrix vesicle as the initial calcification locus.

Matrix vesicles are only found to be associated with mineralization of the primary dentin (mantle dentin) of the tooth. The tissue is unique in that mineralization develops segmentally, beginning at the top of the crown and extending to the apex of the root. Therefore, during development, initial mineralization may occur in one segment of the root while several segments of the crown are completely mineralized. In any one segment, mineralization of the tooth begins in matrix vesicles close to the presumptive interface with the enamel of the crown or the cementum of the root. As is true in woven bone and calcified cartilage, random growth of HA crystals from the initial nucleation site in the matrix vesicles of dentin determines the spheriodal structure of calcification nodules. These nodules coalesce as the HA crystals continue to grow into, onto, and between the collagen fibrils of the dentin extracellular matrix. When several micrometers of mantle dentin have calcified, HA crystals grow in two directions simultaneously. The first is directed toward the pulp of the tooth, as constantly receding odontoblasts secrete matrix glycoproteins and collagen precursors. The collagen fibers, particularly, are oriented in parallel array, more so than is apparent in the random orientation seen in the mantle dentin. This orientation of collagen signals the formation of circumpulpal dentin, which constitutes the major bulk of dentin. Structurally, HA crystals appear to grow from already formed crystals in the mantle dentin as the crystals infiltrate the collagen. The second direction in which the HA crystals grow begins at the enamel–mantle dentin or the cementum–mantle dentin interfaces. This growth is directed to the outside of the tooth, exactly opposite to the growth in the circumpulpal dentin. In cementum, HA crystals are related to collagen in an analogous pattern to the relation in circumpulpal dentin. Initiating enamel formation, ameloblasts secrete enamel matrix (enamel protein and glycoprotein) directly onto the surface of the mantle dentin HA. Characteristic long, thin crystals of enamel HA are initiated from the HA crystals in the mantle dentin, maturing and growing in all dimensions at the expense of labile enamel protein.

TABLE I Lineage of Mammalian Calcification

Initial or Primary	Subsequential or Secondary
Woven bone	Lamellar bone
Calcified cartilage	
Mantle dentin	Circumpulpal dentin
	Cementum
	Enamel

From these observations, a pattern of growth and development of the different calcified tissues has emerged. There are two subdivisions of mineralogensis: (1) primary or initial, and (2) secondary or subsequential. The criteria for the inclusion of a mineralizing tissue as primary calcification are that calcification begins in the matrix vesicle and that there is no preexisting calcification in the tissue before matrix vesicle emergence in the matrix. The criteria for inclusion as secondary calcification are that growth of HA crystals is a continuation by direct extension and secondary nucleation from a previously formed seam of calcified tissue and not from matrix vesicles. There are three tissues that have HA nucleation sites in matrix vesicles and therefore are in the primary calcification subdivision: woven bone, calcified cartilage, and dentin (see Table I). There are four tissues that are in the secondary calcification subdivision: lamellar bone, which develops subsequent to woven bone, and enamel, cementum, and circumpulpal dentin, all three of which develop subsequent to mantle dentin.

In endochondral osteogenesis, bone formation begins *de novo* on the surface of calcified cartilage. That is, matrix vesicles of woven bone initiate mineralogenesis as if there were no seam of calcified tissue in the region. This may be because direct extension of HA crystals is impeded by a carbohydrate-rich coating on the surface of calcified cartilaginous spicules.

Bibliography

Bernard, G. W., and Marvaso, V. (1982). Matrix vesicles as an assay for primary calcification *in vivo* and *in vitro*. *In* "Matrix Vesicles." (Ascenzi, A., Bounuci, E., de-Bernard B, eds.), pp. 5–11. Wichtig Editore, Milane.

Moss-Salentijn, L., and Hendricks-Klyvert, M. (1990). "Dental and Oral Tissues." Lea and Febiger, Philadelphia.

Provenza, D. V. (1988). "Fundamentals of Oral Histology and Embryology." Lea and Febiger, Philadelphia.

Ten Cate, A. R. (1989). "Oral Histology, Development Structure and Function." C. V. Mosby, St. Louis.

Depression

PAUL WILLNER, *City of London Polytechnic*

Glossary

Hypomania Form of mania (i.e., a pathological combination of elation and energy) mild enough not to require hospitalization

Hypothalamus Area at the base of the brain important in the regulation of the internal environment. Hormones released from the hypothalamus control the activity of the pituitary gland and, through this, the endocrine system.

Neurotransmitters Chemicals which, when released, achieve transmission between nerve cells [e.g., the catecholamines (noradrenaline and dopamine) and serotonin (collectively known as monoamines), and acetylcholine]

Synapse Space between two nerve cells across which neurotransmitters diffuse. The transmitter is released from the presynaptic terminal and acts on receptors in the membrane of the postsynaptic cell.

DEPRESSION IS A complex psychiatric syndrome involving changes in mood, thought, activity, social behavior, and vegetative functions. Mood is sad or empty, and responses to pleasurable events can be reduced or absent; thoughts center on themes of hopelessness, helplessness, worthlessness, and guilt; fatigue is prominent and normal levels of activity require great effort; social interaction is diminished; sexual behavior, appetite, and weight are decreased; and sleep is poor. Frequently, there are marked diurnal variations, usually with improvements toward the evening. No one symptom is invariably present; all symptoms can also occur in other conditions.

I. Diagnosis

In everyday use the term "depression" denotes a transient lowering of mood, a mixture of sadness and inertia. Depression of clinical severity is a complex psychiatric syndrome, encompassing a far broader range of symptoms. The central features are a profound lowering of mood (described by the patient as depressed, sad, hopeless, not caring, or unable to experience pleasure) and/or a loss of interest or pleasure in usual activities or pastimes, which is often more obvious to observers than to the patient. Also present are a reduced ability to concentrate, thoughts of suicide, and feelings of worthlessness and guilt, which can be so excessive and inappropriate as to reach delusional proportions (i.e., psychotic depression).

There are, in addition, a range of biological symptoms: loss of energy, decreased sex drive, sleep disturbances, loss of appetite (or, sometimes, overeating), and psychomotor retardation and/or agitation. The psychomotor changes are particularly striking: characteristic agitated behaviors include an inability to sit still, pacing, hand-wringing, pulling of hair or clothing, and outbursts of shouting, complaining, or crying; retardation refers to slowed body movements, a decrease in amount of speech, with protracted pauses, and a lack of facial animation. The prominence of biological symptoms in severe depression, together with the fact that depression can be treated physically, with drugs or electroconvulsive shock treatment, have led to much research attempting to characterize the biological basis of depression and clarify the relationships between biological and psychological processes in the control of mood.

The biological symptoms tend to be associated with the endogenous subtype of depression. The distinction between endogenous and reactive depressions is often thought to refer to whether the depression has an obvious precipitant (reactive) or not (endogenous). This view is mistaken and has been a source of much confusion. The distinction in fact refers to whether the patient responds with mood elevation to psychosocial interventions (reactive) or not (endogenous). In the diagnostic system of the American Psychiatric Association, (DSM-III), endogenous depression is known as melancholia, the defining feature being the inability to experience pleasure. Endogenous depression is usually, though not always, more severe than the nonendogenous form; the presence of pronounced psychomotor change almost invariably denotes an endogenous depression.

Another important distinction is between unipolar and bipolar depressions. Depression is a recurrent disorder, which is likely to recur after 4–5 years, and subsequently with increasingly shorter periods between episodes. In bipolar disorder, episodes of mania or hypomania are interspersed between the episodes of depression. The depressive pole of bipolar disorder tends to be of the endogenous type and thus is less variable than unipolar depression. Otherwise, unpolar and bipolar depressions do not differ in their symptomatology.

Depression will usually resolve, with or without treatment, within 6–12 months, but in around 15% of episodes, depression persists. A chronic depression may be a genuine residuum of an acute major depressive episode. More commonly, chronic depressions are either the product of a longstanding depressive character trait, which preexisted the acute episode, or secondary to some other chronic psychiatric or medical complaint. An important question is whether depressions of clinical severity differ quantitatively or qualitatively from normal low mood swings. Such as it is, the evidence indicates that the differences are quantitative: For example, some biological markers characteristic of severe depression (see Section IID, E) are also present in patients suffering chronic mild depressive or bipolar conditions, and normal persons who experience large mood swings tend to have bipolar patients among their first-degree relatives.

II. Pathogenesis

The likelihood of entering an episode of depression is increased five- or sixfold in the 6 months following a stressful life event. Life events also increase the likelihood of a variety of other psychiatric disorders. However, there is some evidence that loss events (e.g., bereavement or separation) might predispose more specifically to depression. Depression is also more likely in conditions of chronic stress; indeed, it might be more appropriate to consider life events as chronic stressors, since they exert their long-term effects by exacerbating existing minor stresses and undermining self-esteem. A second predisposition to depression arises from a complex of social factors. Chief among these is the absence of close confiding relationships and, more generally, of social support. Depressed people are also more likely to have inadequate social skills and to have experienced childhood loss events.

A number of personality traits predispose to depression. An introverted, gloomy, "depressive" character has been described, which shares some biological markers characteristic of endogenous depression. Depression is also seen in the context of an unstable histrionic personality, often in association with heavy drug and alcohol use, and particularly in the context of social inadequacy. There is also a complex interface between depression and anxiety: In many theoretical models anxiety is seen as a precursor to depression, but this issue is far from being resolved at present.

In addition to these psychosocial determinants of depression, a number of pharmacological and physiological precipitants have also been identified, which probably play a rather minor role numerically, but are of great theoretical interest. The most familiar pharmacological depressions are those induced by the antihypertensive agent reserpine and related agents, which act primarily by depleting the brain of catecholamine neurotransmitters, and by drugs that activate the muscarinic subtype of cholinergic receptors. Both classes of drugs exacerbate preexisting depressions, although in nondepressed subjects the primary effects are apathy and sedation. Depression has frequently been reported as a side effect of neuroleptic therapy in schizophrenia (and an episode of mania or hypomania might follow neuroleptic withdrawal). A variety of hormonal conditions can also precipitate or predispose to depression, including hypothyroidism, which can often be an unrecognized cause of resistance to antidepressant therapy, adrenocortical hyperactivity, and estrogen. Depression can also develop in the context of physical illnesses, particularly Parkinson's disease, and in the aftermath of viral infections such as glandular fever.

Unipolar, but not bipolar, depressions are two to three times more frequent among women than among men. It is not clear to what extent the sex difference is psychosocial or hormonal in origin; this issue is not resolved either by the recent finding that the incidence of depression is greater only in women who have been through childbirth or by the high incidence of depression shortly after childbirth (i.e., puerperal depression). However, the fact that childbirth can sometimes precipitate a manic-like episode (i.e., puerperal psychosis) suggests that physiological changes at childbirth can have powerful psychological effects.

III. Biological Features

It is clear that the distinction between psychological and biological features of depression is largely spurious: Physiological changes in the brain and, to a lesser extent, elsewhere in the body are likely to have psychological consequences, and all psychological events must have a basis in the physiological activity of the brain. Nevertheless, a number of features of severe depression are biological in the sense that they cannot be explained easily at the level of psychological processes; these features include both biological precipitants of psychological changes and biological markers indicative of the changes in brain function that underlie the experience of depression.

A. Bipolarity

As with depression (see Section II) individual episodes of mania may sometimes have clear precipitants. However, the alternation between periods of depression and mania in bipolar manic–depressive disorder is usually assumed to reflect an underlying instability in the nervous system. A small number of patients have been described who switch rapidly between depression and mania in a 48-hour cycle. Spontaneous fluctuations in brain activity and behavior, reminiscent of rapid mood cycling, have been described in animals subjected to repeated low-intensity electrical stimulation (i.e., "kindling") of the amygdala or frontal cortex. Kindling may be prevented by drugs such as lithium or carbamazepine, both of which are effective in bipolar disorder.

An infrequent but interesting side effect of tricyclic antidepressants is a switch from depression to hypomania. This response is seen only in bipolar patients, and its presence in an apparent unipolar patient indicates an underlying tendency to mania that will eventually emerge. Pharmacological hypomania is also seen with drugs (e.g., L-dopa) that activate dopamine receptors, but is relatively infrequent with antidepressants selective for the serotoninergic system (see below). [*See* ANTIDEPRESSANTS.]

B. Seasonal Affective Disorder

It has recently been recognized that some people suffer episodes of depression or mania at the same time each year. Winter depression, which is the best characterized of these seasonal affective disorders, appears to be triggered by a decrease in the length of the day. Accordingly, winter depression can be rapidly and effectively treated by exposure to bright light: broad-spectrum light which mimics natural sunlight is administered just before dawn and/or just after dusk, thereby artificially lengthening the light period. There is some evidence that these effects might be mediated by the pituitary hormone melatonin, which is produced during the hours of darkness and inhibited by bright light.

C. Genetic Markers

There is strong evidence for the genetic transmission of bipolar disorder: The concordance rate for monozygotic (i.e., identical) twins is in excess of 70%, compared with approximately 10% for dizygotic (i.e., fraternal) twins. The evidence for genetic transmission is much weaker in the case of unipolar depression. A number of genetic linkage studies have identified chromosomal markers associated with bipolar disorder within pedigrees from families with a high incidence of the disorder. Some studies, in predominantly bipolar pedigrees, have argued for an association between depression and a region of the X chromosome close to that responsible for red–green color blindness; others have been unable to replicate these findings. Recent studies have begun to use recombinant DNA techniques; however, as yet no two studies have identified the same DNA marker. It is not known exactly what characteristics are transmitted genetically in bipolar disorder. However, in some of the genetic linkage studies, the marker associated with bipolar disorder lies close to the area of the chromosome that controls the synthesis and metabolism of the catecholamine neurotransmitters. [*See* AFFECTIVE DISORDERS; GENETIC MARKERS.]

D. Endocrine Changes

A number of neuroendocrine abnormalities have been described in depressed patients. The most extensively studied of these is an increase in levels of circulating adrenal corticosteroids. The release of corticosteroids is controlled by adrenocorticotropic hormone (ACTH) released from the pituitary, which, in turn, is controlled by corticotropin-releasing factor released from the hypothalamus. Some studies have detected elevated levels of the latter in the cerebrospinal fluid (CSF) of depressed patients, which indicates that activity is elevated throughout the hypothalamus–pituitary–adrenal (HPA) system. This, in turn, points to changes in the brain as the cause of abnormally high blood corticosteroid levels. The release of corticotropin-releasing factor is largely under the control of three neurotransmitter systems, noradrenaline (NA), which is inhibitory, and acetylcholine and serotonin [i.e., 5-hydroxytryptamine (5-HT)], which are excitatory. Thus, an increase in HPA activity is compatible with either a decrease in noradrenergic inhibition or an increase in cholinergic and/or serotoninergic excitation. The release of ACTH and cortisol by drugs that stimulate cholinergic or serotoninergic receptors is enhanced in depressed patients. [*See* ADRENAL GLAND, HYPOTHALAMUS; PITUITARY.]

Changes in adrenocortical function in depression are most commonly studied by means of the dexamethasone suppression test (DST). Dexamethasone is a synthetic corticosteroid which inhibits the release of ACTH; if administered at night, dexamethasone suppresses the major peak of corticosteroid secretion, which normally occurs in the early morning. As a result corticosteroid levels are very low in blood samples taken the afternoon following dexamethasone administration. However, approximately 45% of endogenously depressed patients "escape" dexamethasone suppression and have corticosteroid levels within the normal range. Abnormal DST results are also seen in a number of other conditions, particularly those involving weight loss (although weight loss alone cannot explain the DST abnormality in depression). However, DST nonsuppression is rarely seen in cases of nonendogenous depression: Within the population of depressed patients, the DST specifically identifies the endogenous subtype. Abnormal DST results are not uncommon in mania; however, the DST discriminates well between psychotic (i.e., delusional) depression and schizophrenia. The DST usually normalizes upon recovery from depression, and failure to do so predicts a likely relapse.

A second major hormonal abnormality in depressed patients, which again is specific for the endogenous subtype, is a blunted growth hormone response to the drug clonidine. Clonidine causes growth hormone release by stimulating the α_2 subtype of adrenergic receptor. Thus, like DST nonsuppression, a reduced growth hormone response to clonidine is also compatible with a decrease in noradrenergic transmission in the brain. Responses to two other drugs that stimulate growth hormone through noradrenergic mechanisms, insulin and desmethylimipramine (a tricyclic antidepressant), are also blunted in depression; however, there is a normal growth hormone response to amphetamine, which is mediated through a different neurotransmitter system. Blunting of the growth hormone response to clonidine or insulin is also observed in nondepressed postmenopausal women and among heavy alcohol users; both of these groups are at high risk to develop depression. The growth hormone response to clonidine is blunted following the administration of high doses of corticosteroids, which suggests that abnormal growth hormone responses in depression might be secondary to overactivity of the HPA system.

A third endocrine abnormality, seen in approximately 30% of depressed patients (but also in other psychiatric conditions) is a reduction in the release of thyroid-stimulating hormone (TSH) from the pituitary gland in response to TSH-releasing hormone. The mechanism of this change is not known. As noted in Section II, hypothyroidism and borderline hypothyroid conditions can contribute to the etiology of depression. Although thyroid hormones are not themselves effective as antidepressants, supplementation of antidepressant therapy by the addition of thyroid hormones has beneficial effects in a high proportion of patients who are nonresponsive to antidepressants alone. [*See* THYROID GLAND AND ITS HORMONES.]

E. Sleep and EEG Abnormalities

It is well known that depression is often accompanied by insomnia. Recent electroencephalographic (EEG) studies have described a number of features in which the sleep patterns of depressed and nondepressed people differ, including a loss of slow-wave sleep, a decrease in total sleep time, with increased periods of waking, and early morning waking; 90% of patients have some form of sleep abnormality. The most characteristic abnormality, which is associated strongly with the endogenous subtype, is a pronounced shortening of the latency to enter the first period of rapid eye movement (REM) sleep. REM sleep, during which active narrative dreaming takes place, occurs in 20- to 30-minute episodes at 90-minute intervals throughout the night. The first REM period usually occurs some 45–60 minutes after falling asleep; in endogenously depressed patients this latency is typically halved. [*See* SLEEP.]

Like the HPA axis, the onset of REM sleep and the oscillation between REM and non-REM sleep phases are under the inhibitory control of noradrenergic neuronal systems and the excitatory control of cholinergic systems. The reduction in REM sleep latency is therefore compatible with a reduction in noradrenergic tone and/or an increase in cholinergic tone. The infusion of drugs that stimulate muscarinic cholinergic receptors during a period of non-REM sleep induces an early onset of the next REM period. This response is supersensitive in severely depressed patients, and there is evidence that the abnormality persists following recovery.

A high proportion (i.e., approximately 60%) of depressed patients show improvements in mood following a night without sleep; however, the benefit is temporary and does not usually survive a period of normal sleep. Similar but longer-lasting effects have been claimed in endogenous, but not nonendogenous, patients following selective deprivation of REM sleep. This effect is not well established; however, it is interesting that all effective antidepressant drugs suppress REM sleep to some extent, although electroconvulsive shock treatment does not.

A number of studies have reported abnormalities in the waking EEG in depressed patients. The most consistently reported observation is an increase in EEG activation over the right frontal lobe. An increase in right frontal activation is also suggested by the observation of an increase in leftward eye movements (which are generated in the right frontal lobe) in depressed patients. In most people the left cerebral hemisphere is dominant for speech, while the right hemisphere is specialized for visuospatial functions. In contrast to the EEG data, neuropsychological testing of depressed patients has consistently suggested the presence of nondominant (i.e., right) hemisphere pathology: Verbal performance is normal, but depressed patients typically perform poorly on visuospatial tests. These neuropsychological abnormalities normalize on recovery.

The discrepancy between the apparent increase in right frontal activity and poor performance on tests of right hemisphere function could reflect a specialization within the nondominant hemisphere. While severe endogenous depression is likely following damage to the dominant (i.e., left) frontal lobe, damage to the right frontal lobe typically results in indifference or even euphoria. However, depression is likely following more posterior damage in the right parietal or temporal lobe or in association with nondominant temporal lobe epilepsy. These depressions are typically associated with neurotic features (e.g., anxiety or irritability).

F. Neurochemical Changes

For most practical purposes direct methods are not available to study the neurochemistry of the living human brain. The evidence of neurochemical abnormalities therefore derives largely from postmortem studies and indirect sources (e.g., measurement of the levels of neurotransmitter metabolites in the blood or the CSF).

The best-established neurochemical abnormality is a decrease in CSF concentrations of the dopamine (DA) metabolite homovanillic acid in a group of depressed patients who display pronounced psychomotor retardation, suggesting a reduction in the release of DA in these patients. DA neurons are an important component of the motor output system of the brain, and it remains unclear whether decreased DA activity is causal in depression or simply reflects the decreased motor output of retarded patients. However, the high incidence of depression associated with Parkinson's disease (which results from the degeneration of DA neurons) suggests a causal role for DA depletion.

A second major abnormality is a decrease in CSF concentrations of the 5-HT metabolite 5-hydroxyindoleacetic acid (5-HIAA). Many studies have reported this abnormality; however, an equal number have not, and some have reported increases in CSF 5-HIAA. Chronic alterations in neurotransmitter release are often accompanied by compensatory changes in the number of receptors for the transmitter, and some postmortem studies have indeed found evidence of an increase in 5-HT receptor numbers in some brain areas; again, however, these findings are controversial. While there is disagreement over the status of 5-HIAA in depression, it is well established that 5-HIAA levels are reduced in people who have attempted suicide; some of the postmortem studies have confirmed low post-mortem levels of brain 5-HT and 5-HIAA levels in successful suicides. CSF 5-HIAA has been found to be low in murderers and other aggressive offenders and in patients with disorders such as bulimia nervosa; these studies, together with the lack of consensus in studies of depressed patients, suggest that this marker relates to poor impulse control, rather than depression per se. In general, studies that have identified a decrease in CSF 5-HIAA in depressed patients have also reported the same abnormality in mania and following recovery: Low CSF 5-HIAA appears to be a trait marker, rather than a marker for the depressed state. [*See* NEUROTRANSMITTER AND NEUROPEPTIDE RECEPTORS IN THE BRAIN.]

There are many biochemical similarities between 5-HT neurons and blood platelets, including the machinery by which 5-HT is transported into the cell from the extracellular fluid. The tricyclic antidepressant imipramine, which inhibits 5-HT uptake (see Section V,A) binds to a site adjacent to the 5-HT transporter. Both 5-HT uptake and imipramine binding are decreased in blood platelets from depressed patients. The abnormality in imipramine binding normalizes upon recovery, but the decrease in 5-HT uptake does not. While these are reliable findings, their significance for the operation of the 5-HT system is unclear, and the relationship between platelet abnormalities and 5-HT function in the brain is obscure. Some postmortem studies have reported decreases in imipramine binding in the brains of suicide victims, but other studies have not.

As noted in Section III,D,E, some of the hormonal and EEG markers of depression are consistent with a reduced level of NA function. The major biochemical marker of NA activity is the metabolite methoxyhydroxyphenylethyleneglycol (MHPG), which has been studied in CSF and urine. (While substantial amounts of NA are released peripherally in the autonomic nervous system and from the adrenal medulla, peripheral NA is metabolized differently, and a high proportion of MHPG in the urine originates in the brain.) Many studies have confirmed a decrease in urinary MHPG in bipolar patients; and in longitudinal studies of small numbers of bipolar patients, urinary MHPG levels were found to be higher during manic than depressed episodes. Studies in unipolar patients are less consistent: Decreased urinary MHPG has often been reported, but increases have also been seen. Studies of CSF levels of NA and MHPG have largely been negative; however, there is some evidence that these indices might also be decreased in bipolar depression, but increased in mania. CSF NA and MHPG are sometimes found to be elevated in unipolar depression, but these increases are more closely correlated with levels of anxiety than with the severity of depression. Increases in NA and MHPG levels in depressed patients are difficult to interpret, given the evidence for a reduction in the sensitivity of some postsynaptic receptors for NA.

IV. Psychobiology

As is apparent from their frequent mention in earlier sections, biological theories of depression have tended to focus on the monoamine neurotransmitters NA, 5-HT, and recently, DA. The monoamine hypotheses of depression, formulated in the 1960s, postulated that depression was due to a functional deficiency of one or another of these neurotransmitters. It is pertinent to consider briefly the functional characteristics of the monoamine systems and their role in normal behavior.

The cell bodies of the major pathways in all systems are located in a small number of compact nuclei in the hindbrain and the midbrain. Their axons run forward in well-defined pathways, but then diverge to terminate in wide areas of the forebrain. (The precise innervation differs between the three systems and is not discussed here.) The postsynaptic actions of monoamine transmitters are relatively long lasting, and although there is some topographical organization within each system, the cells within each nucleus tend to fire synchronously, resulting in the simultaneous release of transmitter throughout the area of innervation. This mode of functioning suggests that the monoamine systems act to bias or modulate activity states in the forebrain, rather than to carry precise information.

Of the three systems the DA system has characteristics most obviously relevant to the psychobiology of depression. It has been known for many years that the major DA pathway (i.e., the nigrostriatal system) is an important component of the motor control system. However, it is now also well established that the mesolimbic DA pathway, which terminates in the limbic forebrain and the prefrontal cortex, is crucially involved in the control of motivated behavior; if this pathway is damaged or inactivated pharmacologically, animals are unable to respond appropriately to rewards. These properties make the mesolimbic DA pathway a prime suspect in a disorder characterized by apathy and the inability to experience pleasure.

An early suggestion that NA also might function as a "reward transmitter" (similar to DA) has been largely discounted, although some recent evidence suggests that NA might play a more subtle role in the perception of reward. The major forebrain NA projection appears to be involved primarily in attention and the efficient processing of sensory information. A reduction in NA activity would lead to an inability to sustain attention or concentration, and this seems the most promising approach to understanding the role of NA in depression. It is of great interest that the NA system is inactivated by prolonged or uncontrollable stress.

The functions of the 5-HT system are less straightforward. Studies in animals are broadly consistent with the clinical data (see Section III,F) indicating that depletion of 5-HT causes aggressive behavior, as part of a more general inability to control impulsive behavior. However, depletion of 5-HT in animals also has antianxiety effects, leading to the hypothesis that 5-HT is involved in responses to aversive stimuli: According to this hypothesis, it is excessive 5-HT transmission, rather than a deficiency, that would be compatible with depression. It seems likely that this paradox will eventually be resolved by the recognition that different behavioral functions are mediated by different subtypes of the 5-HT receptor. (Further discussion of this important issue is, unfortunately, outside the scope of this article.)

V. Treatment

Although antidepressant drugs were introduced in the 1950s, we are only now beginning to understand fully their mechanism of action. More recently, cognitive therapies for depression have been introduced. It is not clear whether cognitive therapies are as effective as antidepressant drugs or whether they work in the same populations of patients. There seems to be little reason to see pharmacological and psychological treatments of depression as presenting any kind of philosophical contradiction. However, we do not yet know the degree of similarity in the psychological and physiological mechanisms by which these treatments achieve their effects.

A. Tricyclic Antidepressants

Tricyclic antidepressants are effective in approximately two-thirds of patients. However, they are rather unpleasant drugs, and compliance is poor; there is reason to believe that their success rate would be closer to 90% if all patients took their medicine and if sufficiently high doses were administered. Tricyclics take several weeks to work, and little improvement is seen in the first 2 weeks of treatment. They are widely believed not to elevate mood in nondepressed individuals, although the evidence to support this claim is surprisingly sparse.

It has traditionally been believed that tricyclic antidepressants work by potentiating the effects of NA and/or 5-HT. The major mechanism by which these transmitters are inactivated is reuptake into the presynaptic terminal: Tricyclics inhibit the reuptake mechanism, thereby increasing the concentration of transmitter in the synaptic cleft. However, the immediacy of this effect contrasts with the prolonged course of clinical action. More recent work has focused on the adaptive changes that take place in receptor systems during prolonged drug treatment. The best-known adaptation to prolonged antidepressant treatment, which is produced by most antidepressants, is a decrease in the number of β-adrenergic receptors; however, this is more likely to be a side effect than the mechanism of antidepressant action. Tricyclics also increase the responses to stimulation of the α_1 subtype of adrenergic receptor and of DA receptors in the mesolimbic system. There is evidence that these effects are responsible for the therapeutic action of tricyclic antidepressants in certain animal models of depression. These data are compatible with the clinical evidence of deficiencies in the functional activity of NA and DA systems in depressed patients (see Section III).

The effects of long-term antidepressant administration on 5-HT receptors is also complex, with different effects on different receptor subtypes: Antidepressants increase the responsiveness of 5-HT_1 receptors, but decrease the responsiveness of 5-HT_2 receptors. Both of these effects might be of clinical relevance; the evidence is limited, but arises both from animal models and from clinical studies.

In general, it seems likely that actions mediated by different transmitter and receptor systems will be reflected in different components of recovery from depression; some data from animal models of depression support this view, but there is as yet no clear clinical evidence.

B. Monoamine Oxidase Inhibitors

The second major class of antidepressant drug, the monoamine oxidase (MAO) inhibitors, are far less effective than are tricyclics in severe depression. However, like tricyclics, they are effective in mild depression in which anxiety is a prominent feature. The recognition that MAO exists in two forms led to the development of specific inhibitors of MAO-A and -B; the activity of MAO-A inhibitors might be more comparable to that of tricyclics.

MAO is an enzyme which destroys excess NA, DA, and 5-HT in nerve terminals; MAO inhibitors prevent this action, leading to an increase in concentrations of all three transmitters. The consequences of this effect, following acute administration, are functionally similar to the action of tricyclics: Both potentiate monoaminergic neurotransmission. The adaptive changes that follow prolonged MAO inhibitor administration are also broadly similar to the effects of tricyclics. MAO is present in red blood cells, where its actions include the breakdown of tyramine. Because excessive levels of tyramine can have untoward cardiovascular effects, tyramine-containing foods (e.g., cheese, chocolate, and red wine) must be avoided during MAO inhibitor therapy.

C. Electroconvulsive Shock Treatment

Electroconvulsive shock treatment (ECT), which is administered by passing an electric shock through the head at an intensity sufficient to induce convulsions, has been banned in several states in the United States. However, it remains the most effective antidepressant therapy, particularly in severe delusional depression, in which condition tricyclics are ineffective. Like antidepressant drugs, ECT is effective after a course of several treatments. The biochemical consequences of such a treatment regimen are broadly similar to those of tricyclics; the basis for the differential effectiveness of ECT and tricyclics in delusional depression is not understood, however.

Amnesia and confusion are important side effects of ECT. While ECT is usually administered bilaterally (i.e., to the whole head), there is evidence that restricting convulsions to the nondominant hemisphere is equally as effective, but minimizes the side effects. This observation might be related to the evidence that depression is associated with increased electrical activity over the right frontal lobe (see Section III,E).

D. Atypical Antidepressants

In recent years a large number of atypical antidepressants have been introduced which are neither tricyclics nor MAO inhibitors. Some of these compounds were developed to maximize one or another of the many biochemical properties of tricyclics; others have radically different mechanisms of action [e.g., γ-aminobutyric acid (GABA) agonists and opiate partial agonists]. Most of the latter group are still in the early stages of clinical testing, and it is too early to assess their effectiveness or their strengths and weaknesses relative to tricyclics.

From a theoretical point of view, the most interesting of these new drugs are those developed as selective inhibitors of the uptake of either NA or 5-HT (tricyclics having both of these effects). Drugs of both types appear to be effective as antidepressants (although whether they work as well as tricyclics is uncertain); the 5-HT uptake inhibitors are also effective for certain forms of anxiety (e.g., panic and obsessive–compulsive disorder) and as appetite suppressants.

E. Treatment of Resistant Depression

There is increasing evidence that patient who fail to respond to tricyclic antidepressants can recover following the addition of thyroid hormone (see III,D above) or lithium. Delusional depression is the exception. Such depression does, however, respond to ECT (see V,C) or to an antidepressant/neuroleptic drug cocktail.

VI. Evolutionary Antecedents

Depression is sometimes considered to be a uniquely human disorder. However, it is difficult to see any justification for this view. It is true that we are unable to ask animals about their mental state, but this is no more the case for depression than for any other psychological condition. The starting points for understanding the evolutionary antecedents of depression are depressive behavior and our increasing knowledge of the physiological substrates of depression.

Behavior reminiscent of retarded endogenous depression is readily induced in animals by exposure to prolonged or uncontrollable stress, which causes reductions in locomotor activity, appetite, sexual activity, motivation, and responsiveness to rewards. Frequently, animal experiments use levels of stress so high as to be unrealistic as analogs of the kinds of stressors implicated in the etiology of depression. However, mild stressors can also cause a similar behavioral pathology in animals. The physiological changes responsible for these stress responses are similar to those in the depressed brain, insofar as we can deduce these abnormalities from the indirect evidence summarized above in Section III.

The initial response to stress in animals is usually one of active opposition, which might, if successful, result in stress reduction. If active stress management fails, the passive behaviors that occur following prolonged or uncontrollable stress represent an adaptive response, which conserves energy. This process is part of a more general pattern of adaptive changes, which include an increase in adrenocortical activity and a predominance of the parasympathetic branch of the autonomic nervous system. If these parallels are appropriate, depression can be viewed to some extent as an adaptive response to uncontrollable stress; some psychological accounts have described depression in just these terms. However, the switch to a passive mode of coping with stress is only appropriate if the severity of stress is correctly appraised; part of the reason that depression appears to be maladaptive is that depressed people distort and magnify their problems. Cognitive therapy for depression aims to restore a proper sense of proportion and is clearly effective in some patients.

One manifestation of the passive coping syndrome that could have particular relevance for depression is the submissive behavior displayed by animals at the bottom of a social hierarchy. There are some compelling parallels between depressive behavior and that of people in the lowest social groupings (e.g., tramps); a similar parallel might be drawn between manic behavior and that of certain "top" individuals (e.g., corporate executives). These speculations, together with the evidence of social factors in the etiology of depression, suggest that depression and mania might, to some extent, be residues of behaviors appropriate to primitive social hierarchies.

Behaviors reminiscent of agitated or psychotic depression, including aggression, might be induced in animals by prolonged social isolation. These parallels are less convincing than the effects of stress, but are nonetheless intriguing, particularly considering the importance of social deprivation in the genesis of depressive pathology. However, while outwardly directed aggression has clear adaptive functions, suicide equally clearly does not, and there are no obvious evolutionary precursors for this shift in the object of aggressive behavior.

A number of diverse behavioral paradigms have been developed as animal models of depression. Their common feature is that abnormal behaviors, typically in rodents, are normalized by antidepressant treatment. In most of the early models, the behavioral abnormalities were induced pharmacologically. More recently, a range of models have been introduced in which behavioral abnormalities are induced by acute or chronic stress, social isolation, or brain damage. As with any model the ultimate test of animal models of depression is the extent to which they generate predictions about depression that are confirmed in humans. It is beyond the scope of this article to discuss these models in detail. Suffice it to say that they are of value for unraveling the physiological mechanisms underlying depressivelike behaviors and in the development of novel antidepressants.

Bibliography

Checkley, S. (ed.) (1990). Antidepressant drugs. *Int. Rev. Psychiatry,* 2, No. 1.

Deakin, J. F. W. (ed.) (1986). "The Biology of Depression." Gaskell, London.

Meltzer, H. Y. (ed.) (1987). "Psychopharmacology: The Third Generation of Progress." Raven, New York.

Willner, P. (1985). "Depression: A Psychobiological Synthesis." Wiley, New York.

Willner, P. (1989). Animal models of depression: An overview. *Pharmacol. Ther.* **45,** 425–455.

Zohar, J., and Belmaker, R. H. (eds.) (1987). "Treating Resistant Depression." PMA Publ., New York.

Depression, Neurotransmitters and Receptors

ROGER W. HORTON, *St. George's Hospital Medical School, University of London*

Glossary

Neurotransmitter receptors Specialized membrane macromolecules that detect neurotransmitters and mediate their cellular effects

Neurotransmitters Chemicals that mediate communication between nerves and between nerves and other organs

Receptor agonists Drugs that mimic the effects of a neurotransmitter

Receptor antagonists Drugs that prevent the effects of a neurotransmitter

DEPRESSION AND MANIA are termed affective disorders; the primary abnormality being in mood or "affect." Affective disorders are often recurrent, may require repeated periods in a hospital, and diminish the quality of life for the sufferers and their families. Various forms of treatment are available, but the development of more effective treatments requires a better understanding of the chemical changes in the brain in affective disorders.

I. Clinical Features of Depression

A. Symptoms

The clinical concept of depression embraces a wide range of symptoms. Patients with clinically depressed mood will describe themselves as feeling sad, miserable, hopeless, pessimistic, or some similar colloquial equivalent. Loss of interest in usual and pleasurable activities is always present. Guilt and a sense of worthlessness are exaggerated and often inappropriate. A decrease in energy level is invariably present and is experienced as sustained fatigue in the absence of physical effort. Difficulty in concentrating, slowed thinking, and indecisiveness are common. Delusions and hallucinations are sometimes present. Patients often feel they would be better off dead, and suicidal thoughts and acts are common. Depressed mood often lessens in severity as the day goes on (diurnal variation). Physical symptoms are also usually present. Appetite is frequently reduced with consequent weight loss; a minority of patients eat more and gain weight. Sleep is nearly always disturbed, often with early morning wakening and sometimes with difficulty getting to sleep. Some patients are unable to be still, constantly pacing or wringing their hands (psychomotor agitation); others may show slowed body movements and slowed and monotonous speech (psychomotor retardation). The presence or absence and severity of particular symptoms varies widely among patients. [*See* DEPRESSION.]

The distinction between depressive illness and the downward mood swings, which we all experience as part of daily living, lies in the severity and persistence of symptoms throughout nearly every day for at least 2-wk duration and by changes from previous social and occupational functioning.

Some patients experience only depressed mood (termed unipolar depression); others experience periods of euphoric mood (mania), intermingled with depressive episodes (termed bipolar disorder or manic depressive illness). Manic episodes are characterized by inflated self-esteem, ranging from uncritical self-confidence to delusional grandiosity, often resulting in reckless behaviors. Energy is increased with a decreased need for sleep and food and increased sexual interest. Thought and speech are rapid and copious, with abrupt and frequent changes from topic to topic. Frequently, manic patients have little insight into their illness and resist treatment.

Symptoms of depression usually develop over days to weeks, and, untreated, an episode may last 6 mo or longer. Manic episodes usually last from a few days to months and begin and end more abruptly than depressive episodes.

B. Frequency and Epidemiology

Depressive illness is among the most prevalent psychiatric disorders in adults. Estimates of point prevalence (the proportion of the population with the diagnosis at a point in time) for unipolar depression is 2–3% in adult males and 5–9% in adult females, with a lifetime risk of 8–12% for men and 20–26% for women. Bipolar disorder is much less common with a lifetime risk for men and women of about 1%. Risk factors for unipolar depression include being female between the ages of 35 and 45 yr, being of low socioeconomic class, having a family history of depression or alcoholism, parental loss before the age of 11 yr, living in a negative home environment, lacking a confiding relationship, and having a baby in the previous 6 mo. Bipolar disorder is more prevalent in higher socioeconomic classes, and mean age of onset is earlier than in unipolar depression. Although negative life events play a similar role to unipolar depression, being female or lacking a confiding relationship does not increase risk for bipolar disorder. Bipolar disorder is more strongly associated with a family history of affective disorder than unipolar depression, suggesting a stronger genetic component.

C. Classification and Measurement

The division into bipolar and unipolar depression is generally accepted. Further subdivisions of unipolar depression have been proposed but the terminology is confusing. Endogenous depression refers to a symptom cluster, including marked physical signs (early morning wakening, diurnal variation, weight loss), with a relative lack of external precipitant and a well-adjusted premorbid personality. Melancholia is an approximate equivalent term. Nonendogenous depression refers to a symptom cluster where physical signs are not prominent, mood varies with external circumstances, and delusions and hallucinations never occur. Other terms used include depressive neurosis and reactive depression; however, there is considerable disagreement as to whether these are distinct illnesses, which it is important to differentiate, or extremes of a continuous spectrum.

The inadequacies of these classifications are important for research investigations because of the need to identify relatively homogeneous populations. Consequently, a number of classification systems have been developed based on standardized structured interviews and operational definitions of symptoms. A high inter-rater reliability is achieved. The Research Diagnostic Criteria (RDC) and the Diagnostic and Statistical Manual of Mental Disorders (DSM III-R) are widely used for research purposes.

Standardized rating scales exist for the numerical assessment of the severity of depression. The observer-rated Hamilton depression rating scale and the self-rating Beck scale are widely used for research purposes. These measures are quite sensitive to changes in the intensity of the illness and eliminate subjective bias.

D. Treatment

Tricyclic antidepressants, or second-generation drugs (see Section II.B), are usually the first line of treatment for depression. They may require 3 wk or more to be effective and are usually continued for at least 6 mo to reduce the risk of relapse. Monoamine oxidase (MAO) inhibitors are less widely used because of dietary restrictions (foods containing tyramine can produce hypertensive crises in patients receiving MAO inhibitors); however, the development of isoenzyme-specific MAO inhibitors and of reversible inhibitors has rekindled interest in this area. Electroconvulsive therapy (every 2–3 days for 10 treatments) works quicker than drugs, is often the treatment of choice in suicidal patients, and is more effective than tricyclics if delusions are present. Psychological treatments (interpersonal therapy and cognitive therapy) appear to be as effective as drugs in mild depressions, particularly depressive neurosis. Psychosurgery (subcortical tractotomy) is occasionally performed in severe long-standing cases of depression, where other treatments have failed. [*See* ANTIDEPRESSANTS.]

Acute mania is often treated initially with neuroleptics (most often haloperidol) and lithium carbonate. Neuroleptics are withdrawn as soon as the effect of lithium is established. Long-term lithium treatment is effective in preventing further episodes. Carbamazepine has been advocated as effective treatment in rapid cycling bipolar patients (who experience four or more episodes of affective illness per year).

II. Theories Underlying the Biology of Depression

A. Historical Background

Attempts to explain depressive illness in terms of altered brain chemistry have been dominated for more than 30 yr by theories involving abnormalities in the functioning of the monoamine neurotransmitters noradrenaline (NA) and 5-hyroxytryptamine (5-HT, serotonin). The link between these neurotransmitters and affective state in humans arose from a combination of clinical observations and animal experimentation. In the 1950s, reserpine was widely used as an antihypertensive drug, and a proportion of patients developed symptoms of depression and attempted suicide. Clinical trials of the antituberculosis drug iproniazid produced mood elevation and reversal of depressive symptoms associated with chronic physical illness. Reserpine administered to animals caused profound depletion of brain monoamines by preventing their intraneuronal storage, allowing them to leak into the cytoplasm and be metabolized by the enzyme MAO. Iproniazid reversed the behavioral effects of reserpine in animals by inhibiting MAO and allowed leakage of monoamines, rather than their inactive metabolites, from nerve terminals. Imipramine also prevented the behavioral effects of reserpine, not by inhibiting MAO, but by blocking the reuptake mechanism of monoamines (the main mechanism of inactivation), thus increasing their synaptic lifetime.

B. Formulation and Evolution of the Monoamine Hypotheses of Depression

This, and other evidence, led to the formulation of the catecholamine hypothesis of depression, which stated that ". . . some, if not all, depressions are associated with an absolute or relative deficiency of catecholamines, particularly NA, at functionally important receptor sites in the brain. Elation conversely may be associated with an excess of such amines . . . ." Because much of the evidence was unable to distinguish between an involvement of NA and 5-HT, an analogous indoleamine hypothesis of depression quickly followed. It was argued that some patients may have a selective deficit in NA, and others in 5-HT.

A large number of MAO inhibitors and amine reuptake blocking drugs (tricyclics) were developed and found to be clinically effective antidepressants. Some tricyclics are more selective for inhibition of NA uptake; others for inhibiting 5-HT uptake. The therapeutic benefits of these classes of drugs added considerable support to the monoamine hypotheses of depression, but without significantly altering the NA versus 5-HT debate.

Although the monoamine hypotheses have undergone refinement over the years, further developments have stimulated a major reexamination. The introduction of second-generation, or atypical, antidepressants, such as mianserin and iprindole, which lack both MAO and amine reuptake blocking properties, and the lack of antidepressant action of established amine reuptake blockers, such as cocaine and amphetamine, cast doubts on the relationship between these biochemical events and antidepressant drug action. This is further emphasized by the rapid onset of uptake or MAO inhibition compared to the therapeutic response in patients, which is delayed for up to 3 wk despite daily medication. This temporal mismatch suggests that longer-term adaptive changes, which may be initiated by the acute effects, are necessary for the development of therapeutic antidepressant activity. This has prompted numerous studies on the effects of repeated administration of antidepressant drugs to animals. A range of techniques have been used including behavioral studies and electrophysiological responses of single nerve cells to direct application of neurotransmitters. The quantitation of neurotransmitter receptors *in vitro* by labeling with radioactive drugs (radioligand binding) has made an important contribution.

An important finding was that antidepressants reduce the number of β-adrenoceptors and diminish the response of NA-sensitive adenylate cyclase (an intracellular enzyme that catalyzes the conversion of adenosine triphosphate to cyclic adenosine monophosphate (cAMP), which acts as a second messenger system to mediate the effects of β-adrenoceptor stimulation) in rat cortex. This effect was not apparent after acute administration but was found to develop slowly over a 3-wk period of daily antidepressant administration. This property is shared by most antidepressant drugs (including tricyclic, MAO inhibitors, and atypical antidepressants), but not other psychoactive drugs, and by repeated but not single application of electroconvulsive shocks (ECS, a procedure in animals designed to mimic electroconvulsive therapy). The importance of this finding is that the time course of these effects is similar to that of clinical improvement in depressed patients and that the effects are shared by most antidepressant treatments (although the highly selective 5-HT uptake inhibitors such as fluoxetine appear to be an exception).

However, effects of repeated administration of antidepressants are not limited to the β-adrenoceptor system. Reductions in α_2-adrenoceptor number and function have been reported but are more variable depending on the individual drug, the dosage, and the duration of administration. Decreases in the number of 5-HT_2 receptors have also been reported following antidepressant drugs of many chemical classes. A similar time course is seen to the effects on the β-adrenoceptor system. However, repeated ECS, unlike drug treatment, increases 5-HT_2 receptor number and function.

Although the detailed mechanisms involved in these adaptive receptor changes in animals and their relationship to antidepressant action remains unclear, these findings raise the possibility that depression per se may be related to abnormalities in the number or function of neurotransmitter receptors. Such effects may be secondary to primary abnormalities in neurotransmitter availability or turnover.

III. Methods Used to Study the Biology of Depression

Animal studies of antidepressant drugs have certainly strongly influenced thinking about depression and the mechanisms involved in the therapeutic benefits of treatment. However, critical testing of the hypotheses generated require experiments in depressed patients. Such studies are limited by practical and ethical issues. If the objective is to identify biological abnormalities associated with depression, a relatively homogenous patient group, usually based on operational classification systems such as DSM III-R or RDC (see Section I.C), is compared with healthy volunteers. To exclude the confounding effects of recent or current drug treatment, depressed patients need to have never previously been treated with drugs (rarely achievable) or to be drug-free for a period (ideally 4–6 wk, in practice more often 1–2 wk) prior to measures being performed. To minimize other confounding variables, patients and controls need to be matched as closely as possible for other factors, such as age, sex, race, socioeconomic group, and, in females, menstrual status. Depressed patients are often studied when drug-free and throughout a standardized course of treatment. Such studies not only provide information on the effects of treatment but allow distinctions to be made between biological abnormalities that are related to mood (so called state-dependent markers) and those unrelated to current mood (so-called trait markers), which may reflect a predisposing abnormality. The following section describes some of the methods used, together with their advantages and disadvantages.

A. Measurements in Blood, Urine, and Cerebrospinal Fluid

Measurements of neurotransmitters and their metabolites in readily accessible body fluids, such as blood and urine, are poor substitutes for our inability to perform such measures in living human brain. Measurements in blood and urine are technically possible; the difficulty is in trying to distinguish that component derived from the brain from that derived from other body organs. For most neurotransmitters, this distinction is not possible. Measurements in cerebrospinal fluid (CSF) are a more useful, but still imperfect, measure of neurotransmitter turnover in the whole of the central neural axis; in the case of some neurotransmitters, there may be a significant contribution from the spinal cord. Concentration gradients within CSF necessitate matching patients and controls for height.

B. Peripheral Model Receptor Systems

Blood platelets possess a number of features in common with nerve cells, including a common embryonic origin. Platelets are able to take up 5-HT by a carrier-mediated mechanism and store it in intracellular organelles in a manner analogous to nerve cells. Platelets also possess a number of cell-surface receptors for neurotransmitters, including 5-HT and NA. Thus, platelets have been widely studied as readily accessible models of neurons. However, the accuracy of such models has not been convincingly established. [*See* PLATELET RECEPTORS.]

C. Neuroendocrine Challenge Tests

The basis of neuroendocrine challenge tests lies in the ability of drugs to interact with central neurotransmitter systems to influence circulating concentrations of pituitary hormones. Such tests have advantage over many procedures in that they provide an index of transmitter function, rather than a static measure of receptor numbers. There are some well-established neuroendocrine challenge tests (e.g., stimulation of growth hormone secretion by the α_2-noradrenergic agonist clonidine), but the number of tests has been limited by the availability of selective pharmacological agents suitable for administration in humans. [*See* NEUROENDOCRINOLOGY.]

D. Postmortem Studies

The opportunity to study brain tissue from depressed patients is obviously a great advantage over the more indirect approaches outlined above; however, it is rare that such tissue is available from living subjects. An alternative is to study brain tissue obtained after death (i.e., postmortem examination). However, great care is needed to ensure that apparent biological differences between depressed subjects and controls are due to depression and do not arise from some other cause. The greatest difficulty is in identifying suitable subjects. Two approaches have been used. One is to study subjects with an antemortem diagnosis of depression (based on established criteria) who die by natural causes in a hospital. Such subjects are invariably elderly. The second approach is to study suicide victims. The rationale for this approach is that the majority of people who commit suicide are suffering from psychiatric illness, of which depression is the most common. A limited number of studies have been performed in selected groups of suicides in whom there was sufficient documentary medical evidence to allow a firm retrospective diagnosis of depression. Such studies are likely to identify biological abnormalities more specifically associated with depression than studies performed in more heterogeneous groups of suicides, which are likely to include subjects with other psychiatric diagnoses (such as schizophrenia, personality disorder, or alcoholism). However, it is rarely possible to achieve the detailed subclassification of depression that is possible in living subjects. The choice of control subjects in postmortem studies is also important. Control subjects need to be free of psychiatric illness and to be closely matched to the index subjects, particularly for age, sex, and postmortem delay (the time from death to freezing of tissue). Subjects dying after protracted physical illness, particularly if drug treatment is involved, are clearly less than ideal. There is also a clear need in postmortem studies to establish that biological differences do not arise from artefactual causes related to instability of substances after death or upon storage. [*See* SUICIDE.]

IV. The Involvement of 5-HT

A. CSF Studies

Measurement of CSF 5-hydroxyindoleacetic acid (5-HIAA) concentration, the major metabolite of 5-HT, has been extensively studied as an index of central 5-HT function in depression. Although many studies have reported lower CSF 5-HIAA concentrations in drug-free depressed subjects than in controls, other studies have found no differences. Thus, an association between low CSF 5-HIAA concentration and depression does not appear to be very consistent. Stronger evidence is in favor of an association between low CSF 5-HIAA concentration and suicidal behavior, particularly in subjects who make violent suicide attempts. This relationship is not restricted to depressed patients; it is also seen in subjects with personality disorder, alcoholism, and schizophrenia. Low CSF 5-HIAA concentration has also been associated with various forms of impulsive and aggressive behavior. The important relationship appears to be between low CSF 5-HIAA concentration and impulsivity and aggression, which may be inwardly directed in the case of violent suicide, rather than with depression per se.

B. Peripheral Models

Most studies that have employed adequate methodology have reported platelet 5-HT uptake (measured *in vitro*) to be lower in drug-free depressed patients than in controls, although considerable overlap in values is evident in all studies. The reduction in uptake is due to a decrease in the maximum velocity of uptake (which is directly related to the number of uptake sites) rather than to a difference in the affinity of the uptake carrier for 5-HT.

The 5-HT uptake sites in platelets and brain can be quantitated by radiolabeling with the antidepressant drug ^{3}H-imipramine. A practical advantage of this approach is that the platelet ^{3}H-imipramine-binding sites are stable to repeated washing and freezing, whereas active platelet 5-HT uptake is unable to withstand such treatments and requires that experiments are completed within a few hours of blood sampling. Initial studies reported that the number of platelet ^{3}H-imipramine-binding sites was lower, by as much as 50%, in drug-free depressed patients than in controls, whereas binding affinity was unaltered. Many studies have replicated this finding, although the magnitude of the difference has generally been less than that initially reported. However, a significant number of studies have found no differences in the number of ^{3}H-imipramine-binding sites between depressed patients and controls. A number of factors have been proposed in an attempt to explain these discrepant findings, including differences in assay methodology and patient selection. However, to date no single factor has been convincingly demonstrated to account for these differences. It has been argued that reduced platelet ^{3}H-imipramine binding may be restricted to certain subgroups of depressed patients, such as those with a family history of depression or dexamethasone test nonsuppressors (see Section VII), or with specific clinical symptoms, such as retardation or psychosis; however, no such associations have been consistently replicated. In contrast to the association between low CSF 5-HIAA concentration and suicidal behavior, little evidence suggests such a relationship for ^{3}H-imipramine binding. One of the most consistent features of both measures of platelet 5-HT uptake is a lack of association with established measures of severity of depression. It is also interesting to note that in studies where platelet-active 5-HT uptake and ^{3}H-imipramine binding have been measured in the same groups of depressed patients, and both shown to be reduced, there is a surprising lack of correlation between these two measures. One tentative explanation is that ^{3}H-imipramine labels a modulator site associated with, but distinct from, the 5-HT recognition site. A putative endogenous ligand for the modulator site has been proposed, but the physiological function of this site and its possible role in depression remains to be established. A number of more potent and selective 5-HT uptake inhibitors (paroxetine, citalopram, indalpine) than imipramine have been radiolabeled and used to quantitate 5-HT uptake sites. However, to date the only study comparing platelet ^{3}H-paroxetine binding in depressed patients with controls found no difference in the number of binding sites.

Platelet 5-HT receptors, which appear to be of the 5-HT$_2$ subtype (see Section III.D), reportedly do not differ in number or affinity between drug-free depressed patients and controls. However, treatment of depressed patients with tricyclic antidepressants or administration of desmethylimipramine to healthy volunteers resulted in a marked increase in the number of platelet 5-HT$_2$ receptors. This is in sharp contrast to the reduction in 5-HT$_2$ receptors in rat cortex following repeated antidepressant treatment. Reportedly, platelet 5-HT-mediated aggregation is reduced and unaltered in drug-free depressed patients compared with that in controls.

C. Neuroendocrine Studies

The lack of 5-HT agonists suitable for administration in humans has limited 5-HT-mediated neuroendocrine challenge tests to metabolic precursors of 5-HT (tryptophan or 5-hydroxytryptophan) and fenfluramine (which releases 5-HT and inhibits its reuptake). Prolactin and growth hormone responses to tryptophan infusion have been reported in several studies to be lower in drug-free depressed patients than in controls. In one study, the growth hormone response was almost absent in endogenously depressed patients. Reduced prolactin response to fenfluramine has also been reported in depressed patients. This effect did not correlate with the state of depression or with the severity of depressed mood but was correlated with history of suicide attempts in patients with major depression. Although there are doubts concerning the involvement of 5-HT synapses in the growth hormone response to tryptophan, the similar results with tryptophan and fenfluramine suggest an overall reduced sensitivity of the 5-HT system, at least in some depressed patients. However, the ability of 5-hydroxytryptophan to increase plasma cortisol concentration is greater in drug-free depressed patients than in controls, suggesting an enhanced 5-HT-mediated response. The 5-HT receptors involved in the fenfluramine–prolactin and 5-hydroxytryptophan–cortisol responses are not entirely clear, and thus the possibility of subsensitivity and supersensitivity within different 5-HT receptor subtypes in depression cannot be discounted.

D. Postmortem Studies

There are several studies of 5-HT and 5-HIAA concentrations in brain tissue from psychiatrically undefined suicide victims. Although isolated differences have been reported, most studies found that concentrations were not significantly different from controls in cortical areas. The most consistent finding has been a reduction in 5-HT and/or 5-HIAA concentration in the hind brains of suicides. In suicides who were retrospectively diagnosed as depressed and had not been receiving antidepressant treatment prior to their death, 5-HT and 5-HIAA concentrations did not differ from controls in frontal or temporal cortex, but 5-HIAA concentration was higher in the amygdala of suicides. The evidence take together suggests that the lower CSF 5-HIAA concentrations seen in violent suicide attempters is likely to be a reflection of reduced turnover of 5-HT in hind brain, rather than in higher brain centers. 5-HIAA concentration in patients with an antemortem diagnosis of depression did not differ from controls in frontal or occipital cortex or hippocampus.

Studies of ^{3}H-imipramine-binding sites in suicide victims have yielded inconsistent results. In frontal cortex, increased, decreased, and unaltered binding has been reported. One detailed autoradiographical analysis found ^{3}H-imipramine binding to be significantly lower in the frontoparietal cortex but significantly higher in the hippocampus of suicides compared with that in controls. The lack of consensus in results for ^{3}H-imipramine binding in frontal cortex of suicides may in part be related to a reported marked asymmetry between left and right hemispheres. In control subjects, the number of binding sites was twofold higher in right than in left hemispheres, whereas in subjects with a variety of psychiatric disorders the converse was true. The significance of this observation is not clear; however, it may be important in view of the proposed modulator site labeled by ^{3}H-imipramine (see Section IV.B) that in a study using ^{3}H-paroxetine to label 5-HT uptake sites no hemispheric asymmetry was found in control subjects or depressed suicides, and no differences in the number of sites was found between depressed suicides and controls in any of the 10 brain regions studied, including frontal cortex and hippocampus. In patients with an antemortem diagnosis of depression, ^{3}H-imipramine reportedly is lower in occipital cortex and hippocampus, whereas ^{3}H-paroxetine binding in frontal cortex did not differ.

It is now generally accepted that 5-HT interacts with three distinct classes of receptors: 5-HT$_1$, 5-HT$_2$, and 5-HT$_3$. Also, substantial evidence indicates further heterogeneity within the 5-HT$_1$ class, with five subsets of binding sites: 5-HT$_{1A–E}$. Although not extensively studied, there are no reports of altered cortical 5-HT$_1$-binding sites either in psychiatrically unclassified suicides, depressed suicides, or subjects with an antemortem diagnosis of depression. There is one report of lower 5-HT$_1$ binding in the hippocampus of depressed suicides. The general lack of difference of total 5-HT$_1$ binding should be treated with caution since no studies to date have examined specific subsets of 5-HT$_1$ sites.

As with ^{3}H-imipramine binding, studies of 5-HT$_2$ receptors have produced conflicting results. Two independent groups have reported the number of 5-HT$_2$ binding sites to be higher in frontal cortex of psychiatrically undefined suicides than in that of controls. One study was limited largely to suicides who died by violent means; the second, with a much larger group of subjects, concluded that the higher number of 5-HT$_2$-binding sites was limited to those suicides who died by violent means. A further study reported no difference in 5-HT$_2$-binding sites in frontal or occipital cortex of a group of suicide victims or a subgroup with definite or possible clinical history or depression. Suicides were not divided according to violence of death. A significantly lower number of 5-HT$_2$-binding sites has been reported in the hippocampus of depressed suicides. The difference was restricted to those suicides who died by violent means. In the same study, the number of 5-HT_2-binding sites did not differ in frontal, temporal, or occipital cortex or amygdala, but the number of sites tended to be higher in violent than in nonviolent suicides. No significant difference in 5-HT_2 binding was found in frontal cortex of subjects with an antemortem diagnosis of depression. An association between increased frontal cortical 5-HT_2-binding sites and impulsive or violent behavior, rather than depression, seems a distinct possibility and is worthy of further study.

V. The Involvement of NA

A. CSF Studies

Studies of CSF NA and 3-methoxy-4-hydroxyphenylglycol (MHPG), the major central metabolite of NA, in affective disorder patients have not produced dramatic or consistent findings. Most studies report greater variation in both NA and MHPG concentrations in patients than in controls. In general, manic patients and some unipolar depressed patients have increased concentrations of NA and MHPG, whereas bipolar patients when depressed tend to have lower concentrations than controls.

B. Peripheral Models

Blood platelets possess cell-surface α-adrenoceptors of the α_2 type, which have a similar pharmacological profile to their central counterparts. Activation of these receptors *in vitro* induces platelet aggregation. This has been used as an *ex vivo* test of platelet α_2-adrenoceptor function.

Quantitation of human platelet α_2-adrenoceptors by radioligand binding is dependent on the pharmacological nature of the radioligand used. Selective α_2-adrenoceptor antagonists, such as yohimbine and rauwolscine, appear to label the total population of α_2-adrenoceptors with uniform affinity. In contrast, α_2-adrenoceptor agonists, such as clonidine and UK 14,304, label a proportion of sites with high affinity and the remainder of sites with lower affinity. The relative affinities for these two classes of sites varies from one agonist to another. High- and low-affinity conformation sites are in dynamic equilibrium, and the proportion of sites can be reversibly altered *in vitro* by changes in the ionic composition of the medium and the presence of guanine nucleotides.

Several studies have examined platelet α_2-adrenoceptors in drug-free depressed patients and in controls using antagonist radioligands. None of the studies found differences in the number of binding sites. In contrast, higher numbers of α_2-adrenoceptor-binding sites have been reported in drug-free depressed patients compared with controls using the agonist ligand ^{3}H-clonidine. The increase was initially reported to be confined to the high-affinity binding sites. One might conclude that while the total number of binding sites is unaltered in depression there is a selective increase in the proportion of sites that exist in high-affinity agonist conformation; however, not all the evidence is in favor of this conclusion. A subsequent study with the same ligand found the number of both high- and low-affinity sites to be increased in depressed subjects; other studies using restricted concentrations of other agonists, selected to label only the high-affinity conformation, have found the number of sites to be reduced or unaltered in depressed patients. Studies showing increased sensitivity to the aggregatory effects of adrenaline in platelets from depressed patients tend to support the findings of increased high-affinity binding sites. Other studies have shown that the aggregatory responses do not differ.

Although blood platelets possess a small number of β-adrenoceptors, these receptors have been most extensively studied in white blood cells. Two early studies provided evidence that lymphocyte β-adrenoceptor function was reduced in depression. Stimulation of adenylate cyclase with the β-adrenoceptor agonist isoprenaline resulted in lower cAMP formation in lymphocytes from drug-free depressed and manic patients than from control subjects. β-adrenoceptor binding was also lower in patients than in controls. A specific reduction in β-adrenoceptor mediated cAMP production was advocated since prostaglandin E_1 (PGE_1) stimulated cAMP production (via specific receptors distinct from β-adrenoceptors) did not differ between the groups of subjects. Similar independent findings of lower isoprenaline, but unaltered PGE_1-stimulated cAMP production were reported in unipolar depressed and bipolar manic patients. A subsequent study has confirmed the reduced isoprenaline-stimulated cAMP production, but in this case in the absence of any alteration in β-adrenoceptor binding. A further interesting approach has been the establishment of lymphoblastoid cell lines in culture, by transforming lymphocytes with Epstein–Barr virus, from manic depressed patients, their unaffected relatives, and controls. β-adrenoceptor binding was reduced to less than half the control values in 4 out of 6 cell lines from manic depressed patients but only 1 out of 18 cell lines from unaffected relatives. Other cell-surface markers did not differ between cell lines derived from manic depressive and control subjects.

The further classification of β-adrenoceptors into β_1- and β_2-subtypes is now well documented. In rodent and human brain, β_1-adrenoceptors are the predominant form in most brain regions (except cerebellum) and the reduction of β-adrenoceptors in rat brain following repeated antidepressant administration is limited to the β_1-subtype. Although the evidence is not convincing, lymphocyte β-adrenoceptors appear to be predominantly of the β_2-subtype. This casts doubts on the suitability of lymphocyte β-adrenoceptors as a suitable model of central β-adrenoceptors.

C. Neuroendocrine Studies

The intravenous infusion of clonidine in humans induces a release of pituitary growth hormone. This is mediated by α_2-adrenoceptors, thought to be located in the arcuate nucleus of the hypothalamus. The growth hormone release in response to clonidine has been reported by several groups to be lower in patients with endogenous depression than in sex- and age-matched controls. Since growth hormone release induced by dopamine agonists is unaltered, this suggests a selective reduction in the sensitivity of the α_2-adrenoceptors that mediate this response, rather than a generalized defect in growth hormone secretion in depressed patients. Most studies have found that other effects induced by clonidine, such as reduced blood pressure and increased sedation, which are also mediated by α_2-adrenoceptors, do not differ between endogenously depressed patients and controls. This suggests a selective reduction in the sensitivity of certain α_2-adrenoceptors in depression rather than a global change in all α_2-adrenoceptors. Preliminary experiments in drug-free recovered depressed patients indicate that the growth hormone response remains reduced, suggesting that α_2-adrenoceptor subsensitivity represents a state-independent biological abnormality in endogenous depression.

There are no well-established neuroendocrine challenge tests of β-adrenoceptor function in humans. The most promising area of study is the release of melatonin from the pineal gland, which is stimulated by β_1-adrenoceptor activation via a sympathetic innervation from the superior cervical ganglion. Unstimulated melatonin secretion appears to be reduced in depressed patients, suggesting that net noradrenergic activity within this system is reduced in depression. Treatment of depressed patients for up to 3 wk with desmethylimipramine consistently increased night-time plasma melatonin concentrations, suggesting that down-regulation of β-adrenoceptors is not the predominant effect of antidepressant treatment, at least within this system. [*See* PINEAL BODY.]

D. Postmortem Studies

Although few in number, studies in postmortem brain from suicides or depressed patients have generally found no differences in NA or MHPG concentrations compared with controls.

Two studies have reported markedly higher β-adrenoceptor binding in frontal cortex of suicide victims compared with that of controls. One study reported a mean increase of 50% in the number of sites in the suicide group, with increases in individual subjects ranging from 30 to 110% of the values in individually matched controls. The increase in β-adrenoceptor binding was largely restricted to the β_1-adrenoceptor subtype. The other study found a mean increase of 73% in β-adrenoceptor binding in suicides compared with controls. The suicides in both studies died largely by violent means. In contrast to these relatively large differences, at least two other studies found no difference between suicides and controls in β-adrenoceptor binding to frontal cortex. As with other studies in psychiatrically unclassified suicide victims, it is difficult to attribute the biological differences, or the lack of them, specifically to depression. The only study restricted to suicide victims in which there was sufficient medical evidence for a retrospective diagnosis of depression, and who had not been recently treated with antidepressant drugs, found decreases, rather than increases, in β-adrenoceptor binding. Decreased binding was restricted to cortical areas, and regional differences within cortical areas seemed to be related to the violence of death. In subjects with an antemortem diagnosis of depression, β-adrenoceptor binding in frontal cortex did not differ from controls.

Studies of α_1-adrenoceptors appear to be limited to subjects with an antemortem diagnosis of depression. No differences were found in frontal and occipital cortex, but in the hippocampus binding was almost 50% lower in depressed subjects.

It is rather surprising in view of the accumulating evidence from the clonidine growth hormone challenge test (see Section V.C) of a subsensitivity of α_2-adrenoceptors in depression that few studies have examined α_2-adrenoceptors in postmortem tissue. Preliminary results in a small, but well-defined, group of depressed suicides demonstrated a significant increase in the number of high-affinity agonist-labeled (^{3}H-clonidine) binding sites in frontal cortex, which was most marked in those subjects who had not received antidepressants prior to death. α_2-adrenoceptor-binding sites labeled with antagonist ligands reportedly do not differ in frontal or occipital cortex or hippocampus between subjects with an antemortem diagnosis of depression and controls.

VI. Involvement of Other Neurotransmitters

A. Dopamine

Homovanillic acid (HVA) concentration, the major metabolite of dopamine (DA), has been reported in several studies to be lower in drug-free depressed patients than in controls, whereas patients with delusional–psychotic depression and bipolar manic patients have higher HVA concentrations than nondelusional depressed patients and controls. Although the number of studies is rather limited, neuroendocrine responses to DA agonists have not been shown to be consistently different from controls in depressed or manic patients. Antidepressant effects have been reported with DA agonists, particularly in bipolar depressed patients. There are also numerous reports that high doses of neuroleptics, which block DA receptors, induce depressive symptoms, and rebound improvement in mood, and even hypomania, are seen following neuroleptic withdrawal. Low doses of neuroleptics in combination with antidepressants are a clinically acceptable treatment for delusional–psychotic depression, and substantial evidence indicates that the efficacy of neuroleptics in treating mania is due to blockade of DA receptors.

B. Acetylcholine

Evidence suggests that drugs that mimic or antagonize acetylcholine have effects on mood in humans. Direct-acting muscarinic agonists and inhibitors of acetylcholine esterase (the enzyme that catalyzes the inactivation of acetylcholine) can intensify depressive symptoms in unipolar patients, cause a depressed mood in euthymic bipolar patients, and reduce symptoms in manic patients. Also, ancedotal evidence indicates that muscarinic antagonists alleviate depression. Some tricyclic antidepressants are potent muscarinic antagonists. This property does not appear to be necessary for antidepressant activity since many clinically effective second-generation drugs lack this action. Antimuscarinic action is an important source of troublesome side effects such as dry mouth and blurred vision. While there are isolated reports of increased muscarinic binding in cultured fibroblasts from affective disorder patients and in frontal cortex of suicides compared with controls, other studies have not replicated these findings.

C. γ-Aminobutyric Acid

Several lines of evidence have recently emerged to suggest an involvement of γ-aminobutyric acid (GABA)-mediated neurotransmission in antidepressant drug action. GABA is a major inhibitory neurotransmitter in the mammalian central nervous system and exerts its action through two separate classes of receptors: bicuculline sensitive $GABA_A$ receptors, which include binding sites for benzodiazepines and barbiturates and are coupled to chloride ion channels, and bicuculline-insensitive $GABA_B$ receptors, which are linked to potassium and calcium ion channels. While selective agonists at $GABA_A$ or $GABA_B$ receptors are not antidepressant, mixed agonists at $GABA_A$ and $GABA_B$ receptors, such as progabide and fengabine, are active in animal behavioral models of antidepressant drug activity and have been shown to be clinically effective antidepressants in open and double-blind trials. Prolonged administration of classical antidepresssant drugs and mixed GABA agonists to rats have been reported in some, but not all, studies to increase the number of $GABA_B$-binding sites in frontal cortex and hippocampus.

However, there are relatively few studies of indices of GABA neurochemistry in depressed patients. CSF GABA and plasma GABA (which is thought to be largely of central origin) concentrations have generally been reported to be lower in drug-free unipolar depressed patients and, in some studies, higher in mania than in controls. The number of benzodiazepine-binding sites (a marker for $GABA_A$

receptors) has been reported to be higher in the frontal cortex of depressed suicides than of controls but did not differ in temporal cortex, amygdala, or hippocampus. $GABA_B$-binding sites did not differ in frontal or temporal cortex or hippocampus between drug-free depressed suicides and controls. Glutamic acid decarboxylase activity (the enzyme that catalyzes the synthesis of GABA from glutamic acid) has been reported to be lower in several brain regions of depressed patients, but no differences were found between depressed suicides and controls, when subjects who had died by carbon monoxide poisoning were excluded.

VII. Endocrine Abnormalities

Hyperactivity of the hypothalamic–pituitary–adrenal (HPA) axis is one the best-established biological findings in depression. HPA activity is manifest as increased plasma cortisol, increased urinary excretion of free cortisol and cortisol metabolites, and failure of the synthetic steroid dexamethasone to reduce plasma cortisol (the dexamethasone suppression test [DST]). The DST clearly distinguishes groups of depressed patients from controls; about half the patients with major depressive disorder were nonsuppressors in the DST (i.e., had midafternoon plasma cortisol concentrations >5 μg/dl on the day following dexamethasone) compared with only 7–8% of controls. However, comparable rates of nonsuppression to major depressive disorder are also found in acutely psychotic and demented patients. [*See* ENDOCRINE SYSTEM.]

It has been proposed that the hypercortisolemia of depression is due to a defect at or above the hypothalamus, which results in hypersection of corticotropin-releasing factor (CRF), leading to increased release of adrenocorticotrophic hormone (ACTH) from the pituitary and increased release of cortisol from the adrenals. The corticotrophic cells subsequently become less sensitive to CRF and ACTH secretion returns to normal, but the hypersensitivity of the adrenal cortex to ACTH continues to result in hypersecretion of cortisol. The neurosecretory cells of the hypothalamus receive numerous neuronal inputs from higher brain centers, acting via many different neurotransmitter systems. The involvement of these systems in the proposed hypersection of CRF remains unclear. It has also been suggested that alterations in neurotransmitter receptor number and function in depression may be the result of sustained increases in circulating cortisol. [*See* ADRENAL GLAND; HYPOTHALAMUS; PITUITARY.]

Increased secretion of CRF may not be limited to the hypothalamus. Strong evidence indicates that, in addition to its endocrine role, CRF acts as a neurotransmitter within the brain. CRF administered into the brain ventricles of animals activates the locus coeruleus (the nucleus containing the cell bodies of the ascending NA pathways) and decreases feeding and sexual behavior. CRF concentrations reportedly are increased in CSF of drug-free depressed patients compared with that of controls. CRF receptor-binding sites have also been reported to be lower in frontal cortex of suicides compared with that of controls. This finding is compatible with a decrease in CRF receptors in response to chronic hypersecretion of CRF.

VIII. Future Prospects

Despite considerable advances in recent years, the biological basis of depression still remains in doubt. Heterogeneity in the clinical presentation of the illness may indicate heterogeneity within the underlying chemical pathology and many account for the disappointing lack of replication of findings between studies. However, developing techniques provide exciting prospects for the future. The ability to image neurotransmitter receptors in the brains of living subjects is now only dependent on the development of suitable radioligands and should make a vital contribution to our understanding of the illness within a decade. Although affective disorders demonstrate complex inheritance, the powerful techniques of molecular biology have begun to make significant impacts, particularly in bipolar disorder.

Bibliography

Deakin, J. F. W. (ed.) (1986). "The Biology of Depression." Gaskell Psychiatry Series. Royal College of Psychiatrists, London.

Meltzer, H. Y. (ed.) (1987). "Psychopharmacology: The Third Generation of Progress." Raven Press, New York.

Porter, R., Bock, G., and Clark, S. (1986). "Antidepressants and Receptor Function." Ciba Foundation Symposium, 123. John Wiley & Sons, Chichester, United Kingdom.

Willner, P. (1985). "Depression: A Psychobiological Synthesis." John Wiley & Sons, Chichester, United Kingdom.

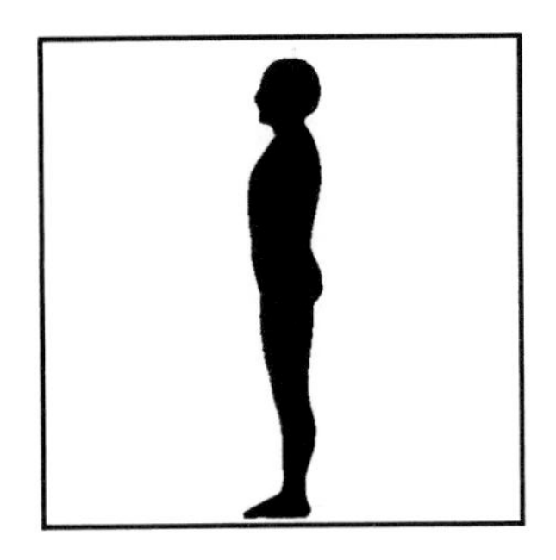

Development of the Self

ROBERT L. LEAHY, *Center for Cognitive Therapy, New York*

Glossary

Object-relations Process by which the individual forms attachment and subsequently individuates self–other

Schematic processing Bias in information processing resulting in greater attention and recall of information consistent with a prototype

Self-schema Representation of the self that forms the early prototype of self-understanding

Structural model Psychoanalytic model of the id–ego–superego of the self

ONE OF THE MOST IMPORTANT questions to be answered by any psychological theory of human behavior is how the self develops. The self is not distinctly "human," if we include in our definition of the self the ability to recognize one's own physical attributes, since monkeys and cats are capable of this. If we limit our definition to self-recognition—that is, the ability to recognize that a spot or cap on one's head does not belong there when seen in a mirror image—then we can say that this is clearly established in infancy. These clever experiments on nonverbal subjects, however, do not investigate what we generally refer to as "self." Any review of theory and research on self-development leads to a recognition of the often incompatible variation with which the self is defined and explained. Is the self that which controls anxiety and impulses, are we always conscious of the self, does the self change with age, is the self primarily emotional or cognitive, are there different contents to the self, can we describe psychopathology by reference to self, and what accounts for individual differences in the self? Each of these intriguing questions has special relevance to different theories of the self.

This article attempts a description of the self's development by examining how different theoretical systems approach this issue. It begins with a discussion of psychoanalytic theory which reflects the importance of how the understanding of the self has led to revolutionary developments in the theory itself such that contemporary object–relations theory often seems far removed from the original work of Freud. Next it turns to cognitive models of the self, which, in a sense, are more "general-process" approaches than "developmental" in that they do not describe qualitative stages of functioning which are related to age or development. Finally, it examines "cognitive–developmental" theories which attempt to relate self-functioning to qualitative changes in cognitive and social functioning associated with age.

I. Psychoanalytic Theories of Self

A. Freudian Theory of the Self

Freud's theory of the self is composed of two models—the *topographic model* and the *structural model*. According to the topographic model, consciousness of the self's emotions and experiences may be blocked by repression and denial. Freud distinguished between conscious, preconscious, and unconscious knowledge and motivation, such that preconscious thought could be accessed with some effort, but unconscious thought is generally unaccessible. According to Freud, thought remains unconscious because of its capacity to arouse anxi-

Encyclopedia of Human Biology, Volume 2.

ety. These unconscious qualities of the self are expressed through ego defenses, such as projection, introjection, displacement, and reaction formation.

According to the structural model, the self is composed of three often competing elements—id, ego, and super-ego. The id refers to innate libidinal energy—such as hunger, sex, and aggression—which seeks release and satisfaction. According to Freud's theory of anxiety, frustration during infancy is first experienced with the infant's separation from the mother. In order to reduce this frustration, Freud postulated, the infant forms a mental representation of the mother ("hallucinatory image") which allows the infant to internalize the mother's presence in her absence. [*See* PSYCHOANALYTIC THEORY.]

This early representation of the external world is the first source of the infant's differentiation of self from external world (i.e., mother) and marks the origin of the ego which attempts to control the id and negotiate adaptation to reality: "Where id was, ego shall be." This "libidinized cognition" was criticized by Heinz Hartmann, who proposed that perception and cognition are "preadapted" to reality and are not derived through frustration of drive. Hartmann's theory of *ego psychology* placed considerable emphasis on analyzing the functions of the ego which are independent of early infantile or oedipal conflict.

Freud outlined psychosexual stages of development—oral, anal, genital, latency, and phallic—which correspond to body modalities in which the self attaches libidinal energy ("cathects"). Thus, at the oral stage—during the first 2 years of life—the self experiences the world largely through oral preoccupations such as dependency, narcissism, and "oral rage" (when frustrated). Freud viewed the sequence of these stages as invariant due to biological maturation of body zones. Failure to resolve conflicts at any stage, or experience of excessive gratification at a stage, would result in negative and positive fixation in later development. For example, strict and punitive toilet training would be expected to result in fastidious and retentive patterns in later life.

Important during the genital stage is the development of the superego. The superego represents the internalization of the values of parents and others. According to Freud, the male child fears castration by the father for his sexual desires for the mother. In the Oedipal complex, he resolves this fear by "identifying with the aggressor"—that is, he renounces his desires for his mother, internalizes the father's identity (as his superego), and displaces his sexual and aggressive energy in more culturally acceptable behavior through sublimation during the latency period. Classic Freudian theory reflects a Victorian male bias in that it proposes that the female, lacking the threat of castration, has a less severe superego and, thus, is less capable of cultural achievement. Feminist critics have indicated that, to the contrary, females are equal and, in some cases, more advanced on moral and social–cognitive indices of development.

Freud's theory of development places the origins of most important qualities of the self in the "pre-Oedipal" phase, that is, between 4 and 6 years of age. Except for his discussion of the origins of the ego during the oral phase, there is little that is interpersonal in the self. The self, in Freud's theory, seems to be an interiorized self with conflicts raging between different layers and different structural unities.

B. Non-Freudian Psychoanalytic Theories of Self-Development

Harry Stack Sullivan advanced an *interpersonal theory* of the self, which he proposed as a psychoanalytic alternative to Freud's emphasis on sexuality in self-development and the emphasis on the "privacy" of Freud's ego. In Sullivan's view, the earliest source of anxiety for the child is *empathy* by the infant of the mother's emotions. The self arises as a means of reducing this anxiety. The infant splits the self into three components—good-me, bad-me, and not-me—where the not-me corresponds to those traumatic experiences that cannot be integrated into consciousness. Sullivan's terminology—dramatization, consensual validation, egocentricity—all point to the interpersonal construction of the self in his model.

Erik Erikson has outlined eight psychosocial stages which mark changes in self-development. In Erikson's model, each stage reflects a social, not just sexual, quality of self-construction. These "eight ages of man" reflect competing issues: trust vs. mistrust, autonomy vs. shame and doubt, initiative vs. guilt, industry vs. inferiority, identity vs. role confusion, intimacy vs. isolation, generativity vs. stagnation, and ego integrity vs. despair. In Erikson's system, failure to resolve an earlier conflict results in persistence of that conflict at later stages. For example, failure to positively resolve the conflict of trust vs. mistrust may result in mistrust as an impediment in identity achievement during late adolescence or early adulthood.

Of specific interest to us is Erikson's discussion of "ego identity," which he defines as the awareness of self-sameness and continuity of one's "style of individuality" where this sense corresponds to the meaning for significant others. Ego identity is defined interpersonally, just as all other stages in Erikson's system have interpersonal implications. Erikson claims that commitment and choice (e.g., career, religion, marriage) contribute to the formation of ego identity. According to Erikson, many individuals experience considerable anxiety (i.e., an "identity crisis") in the formation of their identity, resulting in uncertainty about the self's past, present, and future, as well as intense questioning of one's own values. Some individuals, however, who experience "identity foreclosure" never question their conventional values or identity, simply accepting the traditional stereotypes of their parents.

Jane Loevinger (1976) has proposed that development is characterized by successive levels in the functioning and structure of the ego which parallel intellectual, moral, and social functioning. Loevinger's levels span the life span, from birth to maturity: presocial, symbiotic, impulsive, self-protective, conformist, self-aware, conscientious, individualistic, autonomous, and integrated. These levels are characterized by differences in impulse control, interpersonal style, conscious preoccupations, and cognitive style. Ego-development theory originates with Sullivan's interpersonal theory. Loevinger's view of the ego goes beyond Freud's model of id–ego–superego: in Loevinger's system, the self not only controls anxiety, but it also implies the manner by which others are viewed and how relationships are judged. Because of the clear interpersonal parallels of self–other conceptualization that develop with each level, Loevinger's theory has had more direct impact than Freud's theory on developmentalists interested in social cognitive functioning.

Object–relations theory attempts to trace interpersonal developments of the self. It posits several levels of self-development, beginning in the first months of life as "symbiotic" relations, such that there is no distinction between infant and mother. This symbiotic stage gives way to phases of differentiation of self and mother, to "rapprochement" with (return to) mother, and final independence. An extension of the theory proposes that the representation of self and other follows parallel development in that the self is largely an internal representation, or mirror, of the image of the other. Primitive, and sometimes unresolved, images of self and other involve idealization and splitting, with idealization referring to a projected positive omnipotence and nurturance, often as a result of a defense against a projected mother-image which is negative, punitive, and destructive. Thus, during infancy the other is viewed as either meeting all needs or frustrating all needs. This "splitting" of the image of other is mirrored in the image of self, either all good or all bad. Normal development beyond infancy results in the modification and integration of positive and negative images, allowing for the direction and control of anxiety.

Freud's emphasis on repression of sexuality and his focus on hysterics and obsessive–compulsives reflected the typical Victorian patient of his day. However, many of the psychiatric disturbances seen today are better conceptualized in an object–relations model than in the Oedipal model of Freud's time. Object–relations theorists have been especially helpful in elucidating the complexities of serious personality disorders, such as the borderline and narcissistic personalities. These disorders are viewed as failures to resolve the earlier splitting which results in the inability to integrate the complex, competing emotions in the self. [*See* PERSONALITY DISORDERS (PSYCHIATRY).]

II. Cognitive Theories of Self-Development

Cognitive theories of the self place emphasis on the role of thought processes in the construction of the self. Unlike psychoanalytic theory which emphasizes regulation of anxiety in the development of the self, resulting in an emphasis on defense mechanisms in regulating the self, cognitive theories view the self as an information-processing system. There are a variety of cognitive theories of self-development, with no single unifying approach.

A. Information-Processing Models of the Self

According to Aaron Beck's cognitive model of depression, individuals differ in their underlying *self-schemas*. These self-schemas refer to labels of the self, established during early childhood, which direct information processing. For example, the individual who begins with a self-schema of being unworthy will differentially focus on information consistent with the idea that he is unworthy and filter out information inconsistent with that image. This process of *schematic-processing* results in cognitive distortions which are referred to as *automatic thoughts*. Typical automatic thought distortions include mind-reading ("He thinks I'm a failure"), discounting the positive ("That test was easy"), all or nothing thinking ("I fail at everything"), mislabeling ("I'm a loser"), and magnification ("That quiz was really important"). In addition, the negative self-schema is reinforced by *maladaptive assumptions* which are the "formulas" or rules by which information is evaluated. These include "if-then" statements ("If I fail one thing, then I'm a failure" or "If someone criticizes me, then I'm wrong") and "should-statements" ("I should be perfect at everything," "I should understand things immediately," "I should be better than everyone else"). The simultaneous operation of automatic thoughts, maladaptive assumptions, and negative self-schemas is schematic processing.

Although these concepts do not provide a truly developmental model of the self (that is, a model suggesting qualitative stages of development), they indicate that early experiences of loss, rejection, or punishment contribute to the formation and persistence of these self-schemas. Moving to personality disorders, cognitive therapists have attempted to speculate about the origins of early self-schemas.

Another information-processing model proposes that not everyone who encounters experiences of failure develops depression. The central factor affecting depression, according to this model, is the set of *attributions* or explanations the individual gives for his failure. For example, if one explains his failure by claiming that it was caused by an internal-stable cause (e.g., lack of ability), and that others would have done well ("personal helplessness"), and that this failure will generalize to many other situations ("global helplessness"), then self-critical depression will result. There are no "developmental" stages of the self in the attribution model. In fact, many of the attribution patterns characteristic of depressed adults are also characteristic of young children who demonstrate helplessness on difficult tasks. The depressive attribution pattern is related to early childhood experiences of exposure to punitive and rejecting parents and to early separation or loss of a parent.

In a somewhat different vein, *self-perception theorists* have proposed that the self-concept is developed by the individual "observing" his own behavior and, subsequently, drawing inferences about his motives, thoughts, or traits, for example, "I must like pasta, because I'm always eating it." An assumption of this model is that we make inferences about ourselves in a manner similar to the way we make inferences about others. If children are rewarded for behavior which was already intrinsically interesting to them, there is a subsequent decrement in their interest in that activity. The inferred information processing that "turned play into work" was that the child would say to himself, "It must not be that interesting if I have to be bribed to do it." Too great an emphasis on reinforcements in early education may cause children to lose interest in intrinsically interesting behavior.

III. Cognitive–Developmental Models of the Self

The foregoing models of self are not truly "developmental" in that there is no clear description of qualitative changes in self associated with age. In this section we shall review cognitive–developmental models of the self which propose that there are either parallels or structural similarities in self-cognition with other forms of cognitive development.

A. Social–Cognitive Developments Affecting Self-Image

The self is an object of cognition similar to other social objects. Consequently, it is reasonable to assume that there may be parallel developments in social and self-cognition. With increasing age, there is increasing emphasis on intention, distal causes (that is, causes which are not immediately present in the situation), complexity, qualification, past time perspective as well as future time perspective, and social competence in evaluating self and other. Between the ages of 5 and 10 there is a dramatic increase in the emphasis of "social comparison" information in evaluating the self's performance; that is, older children use the norms of other children's performance to evaluate their own performance. In addition, motivational systems also undergo change during this age period, with older children placing a greater emphasis on competence or mastery, with less reliance on either social approval or extrinsic reinforcement.

One model of the self emphasizes the *self-regulatory* aspects in self-development. It draws on the work of the Soviet psychologists Luria and Vygotski, who proposed that speech begins as social, or external, speech, but with development, speech becomes internal, or private. This internal speech serves a self-regulating function, such that the older child may rely on these internal representations to control and guide behavior. Children are taught to stay on task, examine all alternatives, and engage in self-reward for completion of the task. This cognitive model was first applied to children with behavioral deficits (e.g., hyperactive children), teaching these children to use self-instruction to control their impulsive behavior. Although successful in reducing impulsive behavior and in increasing "on-task" behavior, self-control seldom generalizes to situations other than the initial training situation. There is little that is "developmental" in the model and little that allows us to determine the content or developmental course of self-regulatory processes. The only development appears to be the overcoming of the *deficit* in self-regulation.

John Bowlby's model of attachment has been extended to include self-development. Contrary to traditional psychoanalytic theory, which proposes that attachment is a result of the learned association of the mother with reduction of drive (e.g., hunger or anxiety), this model advances an *ethological* theory according to which attachment is an innate behavior pattern which seeks completion. Self–other development reflects disturbances in these attachment patterns. According to the model the child develops *object–representations* of self–other interactions, where the self may be viewd as unlovable, alone, or helpless. The model also refers to the self as an information-processing system in which information threatening to the self is excluded from memory. This theory places considerable emphasis on the priority of early experience in affecting all future object representations. Thus, early loss of a parent (through death or divorce) or threats to attachment bonds result in lifelong vulnerabilities to depression, agoraphobia, or anxiety disorders.

This theory does not propose any qualitative stages in self-development and does not specify the content of the self at any level. It appears to have replaced Freud's structural model (id, ego, superego) with an ethological model emphasizing innate tendencies (rather than the id) and information processing (rather than ego and superego). It explicitly excludes "psychic energies" or "libido" in favor of information processing and innate attachment tendencies. Its consideration of the self is primarily focused on self–other interactions rather than on the private experiences of the self.

George Herbert Mead proposed that the self is a phenomenal object distinct from all other objects in that it is reflexive. The problem for the self would be how the individual is able to stand outside the self to view the self. The proposal is that the self becomes an object of experience by taking the role of others toward the self—for example, "How does the teacher (or parent) see me?" This theory of the self emphasizes the cognitive (for example, role taking) rather than the emotional aspects of the self and proposes that the self is a social construction—that is, the self is known only from the perspective of other selves.

In this theory, the epistemology of the self is based on an analogy to games. According to this view, the child comes to understand who he is and what individual differences are through playing social games with other children. The theory distinguishes several stages in this "self-understanding." The first stage is the *play stage,* during which the child plays at reciprocal roles (for example, teacher–student) in which the child acts like another toward himself (for example, praising himself). At the *game stage* the child coordinates the views of others in a game through which the "team perspective" is organized and constructed. Further, games entail cooperation by which the child internalizes the standards and expectations of others ("the generalized other").

This theory has been extended by Robert Selman who proposes that self-understanding changes with qualitative changes in role-taking ability. This model identifies a series of stages of role taking, which develop between childhood and adolescence. At the egocentric stage, the child is unable to recognize differences in knowledge and perspective between self and other, whereas at the subjective stage the child understands that these differences do exist. With self-reflective role taking the child or adolescent understands that he may be the object of someone else's thoughts. At the level of mutual role taking the individual understands that the self and other may be constructed from a third person's point of view. Finally, with conventional systems role taking, the individual recognizes that the views, needs, and dilemmas of self and other may be reconciled and unified through conventions.

Another proposal is that increased development is characterized by increasing differentiation and internalization of values. Numerous studies support this model, showing that higher self-image disparity (between the way one sees oneself and the way one wishes to be) is associated with greater chronological and mental age, IQ, and social competence, with emotional maladjustment unrelated to disparity.

An attempt to specify the cognitive factors accounting for this greater disparity, tries to identify social cognitive factors accounting for greater self-image disparity. As already discussed, it has been proposed that the self becomes an object of experience through the process of role taking, and that through role taking the child internalizes the values of his peers and parents. Moreover, levels of moral judgment may be conceptualized as increasingly more general and abstract forms of role taking. Consistent with this structural model, it was found that greater self-image is associated with higher levels of moral judgment and higher levels of role taking.

It has been suggested that the foregoing findings imply that there are costs of development: Increasing social cognitive development results in an increased capacity for self-critical depression and an increased probability of identity diffusion. This is implied by the fact that development results in greater internalization (higher ideal self-image), greater self-image disparity, increased uncertainty and qualification in describing the self, increased uncertainty regarding the "true" perspectives of others, and greater future-time perspective.

B. Cognitive-Development and Self-Understanding

A comprehensive model of the development of self-understanding that is the individual's conceptualization of different schemes of the self (e.g., physical, activity, social, and moral) has been advanced. Four *levels* of self-understanding of the "me" component are identified. This model stands in contrast to other developmental models which have proposed that self-understanding begins with descriptions of physical attributes and actions and later develops into concepts of social, psychological, and moral phenomena (which correspond to *self-schemes*). These self-schemes are operative at every level of development, but what changes is the manner in which they are conceptualized.

IV. Conclusions

The self is described and explained in a variety of ways. Psychoanalytic theory has undergone considerable modification from Freud's early description of a self with different layers of consciousness and competition among id–ego–superego. With the advent of object–relations theory, the self has gained primary importance in psychoanalysis, with its essential developmental milestones established in the first few years of life. The information-processing models of self—especially, the cognitive theory and the ethological theory—have been important in advancing our understanding of affective disorders, such as depression, which result from negative cognitive distortions in processing information and from disturbances in early attachment. Finally, the cognitive–developmental approaches to self have demonstrated that the self-concept undergoes qualitative change with increasing age.

Bibliography

Beck, A. T., Rush, A. J., Shaw, B. F., and Emery, G. (1979). "Cognitive Therapy of Depression." Guilford, New York.

Bowlby, J. (1980). "Attachment and Loss (Vol. 3): Loss, Sadness and Depression." Hogarth Press, London.

Damon, W., and Hart, D. (1988). "Self-Understanding in Childhood and Adolescence." Cambridge Univ. Press, New York.

Guidano, V. (1988). "The Complexity of the Self: A Developmental Approach to Psychopathology and Therapy." Guilford, New York.

Guidano, V., and Liotti, G. (1983). "Cognitive Processes and the Emotional Disorders." Guilford, New York.

Kernberg, O. (1975). "Borderline Conditions and Pathological Narcissism." Jason Aronson, New York.

Kohut, H. (1977). "The Restoration of the Self." International Univ. Press, New York.

Leahy, R. L. (ed.) (1985). "The Development of the Self." Academic Press, San Diego.

Masterson, J. F. (1983). "The Narcissistic and Borderline Disorders: An Integrated Developmental Approach." Bruner Mazel, New York.

Developmental Neuropsychology

JOHN E. OBRZUT AND ANNE UECKER, *University of Arizona*

Glossary

Affective disorders Group of disorders characterized by a disturbance of mood and not caused by any other physical or mental disorder: unipolar depression, disorder of individuals who have experienced episodes of depression but not of mania; bipolar disorder, disorder of people who have experienced episodes of both mania and depression or mania alone.

Aphasia Absence or impairment of the ability to communicate through speech, writing, or signs, due to dysfunction of brain centers. It is considered to be complete or total when both sensory and motor areas are involved.

Broca's area Area located in the inferior frontal gyrus of the brain that is responsible for expressive language. A lesion to this area results in expressive aphasia.

Cerebral asymmetries Specialization of the two cerebral hemispheres. The left hemisphere has superiority for functions such as language, whereas the right hemisphere specializes in behaviors such as music and the holistic perception of patterns and faces.

Corpus callosum Great commissure of the brain connecting the two cerebral hemispheres

Down's syndrome A form of moderate to severe mental retardation most generally caused by an extra 21st chromosome

Equipotentiality All areas of the cerebrum possessing an equal ability to assume behavioral functions. Specific areas in the brain do not govern specific aspects of behavior.

Hemidecortication Removal of one-half of the cortex. A hemidecorticate is an individual with only one hemisphere.

Intelligence quotient Standardized measure indicating how far an individual's raw score on an intelligence test falls away from the average raw score of his or her age group; abbreviated IQ

Lateralization Localization doctrine in respect to hemispheric specialization for behaviors such as language, spatial tasks, and visual perceptions. For example, in the majority of the population, language is assumed to be lateralized to the left hemisphere.

Localization Doctrine which suggests that certain brain areas are dedicated to specific behavioral or psychological functions

Plasticity Undamaged areas of the brain taking over the function of damaged areas

Schizophrenia Group of psychotic disorders characterized by disturbances in thought, emotion, and behavior. Individuals are usually impaired in daily functions such as work, social relations, and physical care.

Unilateral hemidecortication Removal of the cerebral cortex from one side of the brain. Removal of the left or right cerebral hemisphere has permitted further investigation into equipotentiality, lateralization, and localization.

Wernicke's area Area found in the posterior portion of the left superior temporal gyrus that is responsible for the comprehension of written or spoken words. Damage to this area results in receptive aphasia.

DEVELOPMENTAL neuropsychology is the study of brain–behavior relationships as they apply to the developing human organism. It can be said that developmental neuropsychology attempts to discover and understand the intricate neurological mechanisms involved in learning. Although child and adult neuropsychologies are similar in research and clinical issues, they differ in emphasis and conceptualization. However, a problem in developmental neuropsychology is a lack of theoretical direction and data-based studies. Scientific and clinical goals cannot be met without a clearly established direction. A common goal of neuropsychology in laboratories throughout the United States, Canada, and other countries is to learn more, in a descriptive way, about brain–behavior relationships. Once validity has been established, the ultimate goal is the development of remediation and rehabilitation procedures.

To attain an adequate description of the patient's problems and to learn more about brain–behavior relationships in children, the developmental neuropsychologist is faced with a variety of problems. A foremost issue in the study of neuropsychological dysfunction in children is the many developmental factors which must be considered. The age of the child at the time of injury, the type and size of lesion, the extent and location of damage, and the specific mental activity involved and its cognitive complexity are just a few of the factors which must be considered. The consideration of these factors affects the reliability and validity of research in developmental neuropsychology.

Although developmental neuropsychology has developed out of the larger field of clinical psychology, it is imperative that one does not generalize adult data to a child population. Data from a child population follow a set of rules not necessarily similar to that of the adult population. As suggested above, age of onset of the injury is a main factor involved in determining patterns of behavioral deficit. Also, the effects of children's injuries seem to be more generalized, whereas damage in adults tends to be more localized and specific.

There are several obstacles in interpreting data from a child population: (1) adult brain injuries are usually more apt to undergo diagnostic and surgical procedures, and therefore receive more attention than childhood cerebral dysfunction; (2) assessments of children are more difficult because of the complex nature of interaction between brain injury and the natural progressive changes due to development; (3) there is a limited range of dependent variables that are appropriate for use with children; and (4) it is difficult to obtain a representative sample of brain-damaged children to study.

I. Theoretical Issues and Research

Developmental neuropsychology has a long history. In the 1800s Gall postulated the brain to consist of numerous individual organs. Each organ, according to Gall, had a specific psychological function. An organ of the brain might be responsible for reading, writing, arithmetic, walking, talking, or friendliness, among other things. In addition, it was the size of the organ which determined the amount of function. A gifted reader would be said to possess a large reading organ, while a learning disabled reader would be said to possess a smaller reading organ. It was this belief about organ size that led to the study of skull configuration, or phrenology. [*See* BRAIN.]

Gall's theories serve as the basis for the localizationist doctrine, which, however, did not meet with unconditional acceptance. Another scientist, Flourens, found little support for it in his experiments. When he selectively removed portions of pigeon and chicken brain, he found that the area removed had little to do with the nature of the symptoms shown by the hen or the chicken. Flourens' observation that the mass of the lesion was the cause of the symptoms led to the equipotential theory of brain functioning. He stated that all areas of the brain are equipotential; there is no differentiation of brain tissue for psychological behavior, as was suggested by the localizationist. It is from this assumption that the name "equipotentialism" comes, the name indicating that all brain tissue is equivalent in terms of what it does or can do. A second, related, assumption is the postulate of mass action. Since all brain tissue is equal, the effects of brain injury are determined by the size of the injury, rather than its location.

Localizationist doctrine and equipotential theory have essentially remained the same since the time of the first studies and have been the two approaches that have generally dominated American psychology and education. Theories of brain function, rehabilitation, and assessment use the assumptions of one of these approaches in their formulations, although these underlying theoretical beliefs are not always recognized. For example, the classic description of the brain-damaged child is an equipotential explanation. The classic symptoms include attentional deficits, emotional lability, coordination difficulties, and poor academic functioning and are characteristic of such children. Although never stated, such a description implies that all brain-damaged children are alike, regardless of the localization of their injury, and that the brain is homogeneous in terms of function; the description is thus a reflection of equipotential thinking.

A. Developmental Neurolinguistics

Lenneberg is a well-known proponent of equipotentiality in the study of language. His three now well-known tenets include: (1) Hemispheric equipotentiality exists at birth for language mediation. (2) Lateralization of language processes (generally to the left hemisphere) is realized gradually. Influenced by both maturational and environmental factors (exposure to and use of language), the development of lateralization (a subsequent decrease in equipotentiality) proceeds most rapidly between 2 and 3 years of age and more slowly after that until puberty, when the process is felt to be complete. (3) Interhemispheric plasticity exists for language development. This plasticity (actually denoting the combined effects of tenets 1 and 2 above) is necessary for learning language naturally and completely (at the critical age for language learning) and also enables the right hemisphere to assume language mediation in cases of damage to the left, language-lateralized, hemisphere. [*See* HEMISPHERIC INTERACTIONS.]

These ideas generated much neurolinguistic research on the topics of hemispheric specialization and interhemispheric plasticity within the maturing child. Evidence to the contrary, however, suggests a left-hemisphere specialization for language, present from birth, which could limit the degree of interhemispheric plasticity available to the brain-damaged child.

B. Development and Measurement of Cerebral Lateralization

By 1977 there were enough data to suggest that newborns exhibit rudimentary hemisphere specialization by at least 2–3 months of age, and possibly even before full-term birth, analogous to the adult pattern of left-hemisphere superiority for language-related functions and right-hemisphere superiority for music and the holistic perception of patterns and faces. [*See* CEREBRAL SPECIALIZATION.]

Lateralization is an important issue in developmental neuropsychological research. For example, it has been suggested that deficiencies in cognitive tasks such as reading and language are due to an abnormal or weak pattern of lateralization. The study of lateralization, though, presents many problems. For instance, the validity of the method is often suspect. The wide applicability of these methods, however, as well as the lack of any clearly superior alternatives, provides incentives to investigators to continue using laterality methods while attempting to increase their validity. However, other evidence, such as that derived from computerized axial tomography (CAT) scans, neurological examination, and medical history, is used to confirm or support signs of lateralized dysfunction. In time other noninvasive measures (e.g., the recording of cortical evoked potentials) or minimally invasive procedures (e.g., positron emission tomography scanning) could prove to be far superior.

Lateralized brain damage or dysfunction is determined by comparing performance of the right and left hemispheres. The two hemispheres are contralaterally organized, so that the right side of the body is primarily controlled by the left hemisphere, and the left side of the body is primarily regulated by the right hemisphere. While the somatosensory, motor, and auditory systems are almost completely crossed, other pathways send impulses from the same side of the body to the same hemisphere (e.g., right ear to right hemisphere). The visual system is more complex than the other systems because the visual fields (not the eyes) are crossed in the hemispheres. Thus, part of the left visual field projects to the right visual cortex, and part of the right visual field projects to the left visual cortex. Several noninvasive methods exist with which to measure brain lateralization; these are reviewed below.

1. Dichotic Listening

This noninvasive procedure can be used on all subjects. In this test two lists of numbers are presented simultaneously, one to each ear, arranged in such a way that one number arrives at the left ear at the same time a different number arrives at the right ear. Subjects are asked to listen to the numbers and then to report as many as they could, in any order. Normal subjects are more accurate on the right ear than on the left, and clinical patients with known language lateralization are better on the ear contralateral to their language-dominant hemisphere. The results suggest that the contralateral pathway from each ear to the cerebral cortex is more efficient than the ipsilateral pathway, and with competitive dichotic stimulation there is a suppression of the ipsilateral input by the contralateral input. This right ear advantage is thought to reflect the left-hemisphere representation for language. Most right-handed children report right ear stimuli more accurately than left. However, children with learning problems do not show this same pattern of performance.

2. Manual Preference

A gross estimate of handedness is obtained through questionnaires. There is a significant relationship between handedness and speech lateralization. However, handedness is not an adequate basis for identifying speech lateralization, because it leads too frequently to misclassifications. In fact, although the majority of the population are left-hemispheric for both language and production and language perception, there are many exceptions, especially among left-handers.

3. Dichhaptic Stimulation

In this procedure stimuli are presented through the sense of touch. Subjects are given two different shapes to palpate simultaneously, one with each hand. Because the ascending somatosensory systems are crossed, information from the right hand is transmitted first to the left hemisphere, while the reverse is true for left-hand information. This gives the dichhaptic procedure a superficial advantage over the dichotic procedure, in that there are no ascending ipsilateral pathways from the hands.

4. Tachistoscopic Procedures

Visual asymmetry studies commonly used tachistoscopic procedures. Verbal or spatial stimuli are presented briefly, usually less than 180 msec, to the right and/or left visual field. The unilateral procedure involves random presentation to the right or left of a central fixation point; bilateral presentation, in contrast, involves different stimuli being presented simultaneously to the left and right of fixation. Stimuli perceived in the left visual half-field are processed in the right cerebral hemisphere; stimuli perceived in the right visual half-field are processed in the left cerebral hemisphere. It is common for children to perform poorly at brief exposure durations, and it is often necessary to increase the exposure duration into the range in which eye movements are possible in order to raise accuracy to acceptable levels. Also, to identify verbal material, the child must be able to read it; consequently, there is an inevitable confound between reading ability and accuracy that could influence observed laterality measures.

5. Dual-Task Performance

Dual-task performance can also be referred to as verbal–manual time sharing. Time sharing is a type of experiment that contrasts the subject's ability to perform concurrent activities when they are performed in the same hemisphere (e.g., speaking and right-hand manual activities) and when they are programmed in separate hemispheres (e.g., speaking and left-hand manual activities). The consequence of this effect of "hemisphere sharing" seems to be competition and "crosstalk" between incompatible timing mechanisms hierarchically organized in the brain. Basically, the idea is to occupy one hemisphere with a particular task and to see how this affects the performance of some other task. Thus, for example, reciting animal names or a familiar nursery rhyme should involve left-hemisphere activity and disrupt right-hand tapping more than left-hand tapping. Such an effect is found, but is not related to age between 3 and 12 years.

Although the methods of measurement of cerebral lateralization and asymmetry are imperfect, much has been learned from their studies. A pattern emerges that lateralization effects do not change significantly after 3 years. There might be meaningful changes in the first few years of life, but better techniques and more careful longitudinal studies are needed to establish their significance.

C. Anatomical Evidence of Lateralization

Anatomical studies could present a more solid case for the early lateralization of language, because the left temporal planum, a region of the brain important for receptive language, is larger than the right planum in about 88% of infants. The temporal planum is part of Wernicke's area and therefore is in the language territory. The existence of this asymmetry in newborns lends to the interpretation that language lateralization begins very early.

Another language-specialized area in the left hemisphere is Broca's area in the frontal lobe. The size of this crucial speech production area is paradoxically smaller in the left hemisphere than in the right in the majority of both adult and fetal brains, when measured as the visible surface area. However, it is more deeply fissurated in the left hemisphere. If the cortical surface is measured, including the cortex buried inside the folds, it is larger on the left than on the right in three-quarters of the cases.

D. Plasticity and Recovery of Function in the Central Nervous System

The issue of plasticity also presents an ongoing debate in the field of developmental neuropsychology, between structuralists, who believe that early brain damage is deleterious to the developmental potential, and those advocating plasticity, who believe that the young brain shows a greater restoration of function in comparison to an adult brain. Still other theorists suggest that there are critical periods for the successful transfer of function. Although there is agreement that the earlier the damage, the better the chances for transfer of function to occur, there is no agreement as to the optimal times when damage can be minimized. A majority of children between the ages of 6 and 15 years show substantial improvement in language functions 1 year after injury to the left hemisphere, and language impairment is less severe when left hemispheric damage occurs in infancy as opposed to later in life. [*See* PLASTICITY, NERVOUS SYSTEM.]

As to the period beyond which restoration cannot occur, some investigations put it at about age 14, whereas others put it at about 2 years. In general, the earlier the damage, the better the chance for transfer of function.

When unilateral hemidecortications are performed prior to the development of language (i.e., in the first year of life), transfer of function is possible. If the left hemisphere is removed, simple language tasks can be performed by the right hemisphere without a decrease in visuospatial abilities; and if the right hemisphere is removed, simple visuospatial abilities can be mediated by the left hemisphere without impairment to language functions. While simple tasks can be performed by the remaining hemisphere, more complex tasks cannot be. Thus, left hemidecortication results in a loss of complex language functions, whereas right hemidecortication is followed by a loss of complex visuospatial abilities. It seems that each hemisphere has the ability to mediate functions of the opposite hemisphere, but neither hemisphere is able to assume all of the functions of the other. These studies support the idea of plasticity in the young brain, but not equipotentiality.

II. Common Disorders

The most common childhood neuropsychological disorders include mental retardation, learning disabilities, epilepsy, closed-head injuries, and psychiatric disorders. Although each of these areas has been researched, the majority of neuropsychological research with exceptional children has focused primarily on the learning-disabled child. Neuropsychological research in the other areas has not progressed as rapidly, due to technological and psychometric limitations in evaluating such children. It is hoped that in future years continued research into the spectrum of neuropsychological disorders will help to reveal aspects of brain–behavior relationships of interest to the field of developmental neuropsychology.

A. Mental Retardation

The concept of mental retardation is, like the concepts of brain damage and epilepsy, clouded by attempts to identify a homogenous or unitary entity from what is, in reality, a widely varying assortment of neuropsychological conditions, often with little more in common than a poor intelligence quotient (IQ) on formal IQ tests. Brain dysfunction is currently viewed as being concomitant to mental retardation; in fact, neurostructural damage was found in necropsies on a large sample of institutionalized retarded individuals. A tendency was also noted for the less severe cases of retardation to have the less severe brain anomalies. Further evidence can be seen with electroencephalographic (EEG) procedures; specifically, with increasing severity of mental retardation in children and adolescents, there is an increase in EEG abnormalities. Down's syndrome subjects seem to be an exception, with significantly fewer EEG abnormalities.

Neuropsychological aspects of mental retardation have not received attention for several reasons: (1) many difficulties exist when administering neuropsychological tests to this population, (2) there are few specific neuroanatomical and neurophysiological correlates of behavior in this population, and (3) there is much uncertainty with regard to the normal neuropsychological data.

Little laterality research has been completed with severely language-impaired retarded children. The usual dichotic listening procedure requires that subjects comprehend and reliably follow verbal instructions, attend to and discriminate the dichotic stimuli, remember them, and report back what has been heard. Children with severely limited language lack some of the skills necessary to perform the task adequately. Recent research looking at ear advantage in Down's syndrome children using dichotic listening tests has indicated that these children have a left ear advantage for linguistic serially processed auditory stimuli (e.g., digits and common objects). This is in contrast to the right ear advantage commonly found. This left ear advantage seems to be related to the syndrome itself, and not to retardation in general, since a control group of non-Down's syndrome, but retarded, children were shown to have the usual right ear advantage. There is also an increase in the incidence of dichotic listening left ear advantage for autistic compared to normal children, suggesting that severe and pervasive language disabilities might be associated with an increased incidence of right-hemisphere specialization for language functioning.

B. Learning Disabilities

"Learning disability" is a generic term that refers to a heterogenous group of disorders manifested by significant difficulties in the acquisition and use of listening, speaking, reading, writing, reasoning, or mathematical abilities. Little is known about the etiology of learning disabilities. Neuropsychological studies have shown that children who have large discrepancies between their verbal and nonverbal abilities are often the most severely learning impaired.

It has been proposed that learning-disabled children have deficits in their ability to transfer information from one hemisphere to the other through the corpus callosum. The result of this deficit is that the two cerebral hemispheres in these children might function somewhat independently and without the contralateral interaction found in normal children.

The difficulty in studying learning disabilities is that most childhood disorders are rare and are also diffuse or indeterminate, making it difficult to state with any assurance that something is wrong in any particular area of the brain. For learning disabilities there is rarely any neurological evidence for characteristic lesions, and the evidence obtained from many children is contradictory.

C. Epilepsy in Children

Epilepsy is the most predominant neurological disorder of childhood. Epileptic children attending ordinary schools are at a greater risk of developing learning problems than other children. In one study of 85 school children with recurrent seizures, it was reported that 16% were regarded as falling seriously behind and 53% were functioning at a below-average educational level; 42% of the children were described as inattentive by their teachers and had poorer school performance. There are several factors affecting intelligence score outcomes in epilepsy. Low seizure frequency is associated with higher intelligence scores, whereas early-onset seizures tend to be indicative of lower intelligence scores (IQ). [*See* EPILEPSY.]

Children with minor motor and atypical absence (petit mal) seizure tend to have the lowest IQ, whereas those with generalized tonic–clonic and classic absence have the highest average full-scale IQs. With respect to other areas of cognitive functioning, no specific pattern of impairment has been identified.

D. Closed-Head Injury in Children

A head injury can range from a simple bump on the head to a more complex penetrating injury. Infants and young children are at a high risk for head injury, especially for nonpenetrating closed-head injuries. Closed-head injury can have no observable consequences or, in a very severe case, can result in death. Brain injury from trauma to the skull is one of the more common neurological disorders in children. The effects vary considerably with age, because the brain of the child is still developing postnatally.

Trauma at an early age affects an incompletely developed brain and an incomplete repertoire of behaviors. Early brain lesions have a greater effect, perhaps because of the smaller repertoire of skills and knowledge the young individual has to rely on. Around the age of 5 years, damage to the right hemisphere no longer has much of a disruptive effect on language. Between the ages of 5 and 12 years, a left-hemisphere injury produces aphasia, although it is generally milder and more transitory than that found with similar injury later in life. It is only after about age 16–18 that adultlike aphasia is seen with left-hemisphere injuries. In a group of children whose injuries in the left hemisphere occurred at various times, from infancy through the preschool years, deficits were evident in both verbal and nonverbal skills relative to healthy children. However, dysphasia, which is typically seen in adults with left-hemisphere damage, was not prevalent.

Perhaps even more important, the infant brain often shows a reduction in size following brain damage, suggesting that, from a structural point of view, early damage might be more disastrous than later damage. When the mature brain of the adolescent or adult is damaged, loss of function tends to be more highly localized and specific.

Intellectual impairment several years after head trauma is more frequent in children younger than 8 years of age at the time of injury than in those over 10 years of age. Thus, although younger children recover more rapidly, they do so less completely. A lesion occurring during the first year of life tends to be associated with intellectual deficits involving both verbal and nonverbal skills; lesions occurring after the first year have effects more dependent on the side of the lesion. Later left-hemisphere lesions were found to be related to decreased verbal and nonverbal test scores, whereas right-hemisphere lesions were found to be associated with impaired nonverbal skills.

E. Psychiatric Disorders

Psychiatric disorders present a continuing puzzle to developmental neuropsychologists. Defects in the lateralization of functions to the hemispheres of the brain have frequently been cited as an etiological factor in a number of unadjusted behaviors (e.g., learning disorders and emotional disturbance). Confused lateralization seems to be especially prominent in those with schizophrenia, a disorder that is more clearly related to an overall level of neuropsychological dysfunction than specific patterns of impairment. As a group, patients with affective disorders, both unipolar and bipolar, were relatively more deficient on right-hemisphere tasks than schizophrenics. Inconsistency in the patterns of peripheral activities in children might accompany emotional instability and compromised frustration tolerance. [*See* MENTAL DISORDERS; MOOD DISORDERS; SCHIZOPHRENIC DISORDERS.]

Left-hemisphere dysfunction has been repeatedly suggested to be related to schizophrenia; such a dysfunction is suggested by flat affect, speech disorder, and paranoia. In neuropsychological studies a reduced anterior left-hemisphere activation was found when ipsilateral skin conductance responsiveness and cerebral blood flow were used as measures of activation. In studies of the cognitive capacity of the left and right hemispheres, schizophrenics were found to be much less able to process both verbal and nonverbal information tachistoscopically presented to the left hemisphere.

Head injury is also likely to produce psychiatric symptoms, different in left- or right-hemisphere damage. Right-hemisphere damage is likely to result in clinical levels of anxiety, general denial, and inappropriate indifference to the medical condition. The emotional response to lesions of the left hemisphere seem to be most often expressed as depression–catastrophic reaction.

III. Perspectives and Issues in Assessment

Much remains to be learned regarding neuropsychological developmental disorders. Lateralization studies provide one avenue of investigation, but an important area not to be neglected is the standard neuropsychological test battery. The use of these tests can aid in further defining brain–behavior relationships in neuropsychological disorders. In addition, use of the tests can help provide answers as to how best to rehabilitate the patient in need of neuropsychological assistance.

A. Standard Neuropsychological Batteries for Children

Two common neuropsychological test batteries include the Halstead–Reitan Neuropsychological test batteries for children and the Luria–Nebraska Test Battery: Children's Revision. The Halstead–Reitan test is said to epitomize North American standardized assessment procedures. The focus in this test is on quantitative norms. The Luria approach, in contrast, is more qualitative in nature and uses a functional approach. The Luria–Nebraska Battery is an attempt to wed Luria's techniques with American clinical neuropsychology.

The Halstead–Reitan Neuropsychological Test Battery for Older Children (i.e., 9–14) and the Reitan–Indiana Neuropsychological Test Battery for Younger Children (i.e., 5–8) are two of the most commonly used neuropsychological test batteries for children. These batteries were developed by Ralph Reitan based on the adult version of the Halstead–Reitan Neuropsychological Test Battery. The major theoretical basis of the Halstead–Reitan and Reitan–Indiana tests is the propositions that behavior has an organic basis and that performance on behavioral measures can be used to assess brain functioning.

The Luria–Nebraska Neuropsychological Test Battery was developed by Charles Golden, who studied with A. Luria in the Soviet Union. Luria's test procedures have several advantages over techniques traditionally used in the United States. First, they comprehensively assess impaired neuropsychological functions from a clinical perspective. Functions are broken down into their most basic components. As a result, a qualitative evaluation of the patient is performed, allowing for the identification of the basic neuropsychological processes underlying overt behavior. A second major advantage is that the Luria–Nebraska Test Battery is somewhat speedier.

Finally, the Luria–Nebraska Test Battery was also designed for the development of rehabilitation programs directed toward the individual's deficits, using techniques that optimize recovery and minimize staff time. Considering the limitation of rehabilitation sources available for the brain-injured patient, this is an important and potentially powerful advantage.

Luria repeatedly stressed that variability and flexibility in the administration of neuropsychological measures were crucial. The emphasis here is on how problems are solved by the patient, what types of compensations are possible, and areas of deficit and strength. Within such an intraindividual model, the psychologist is encouraged to adapt neuropsychological assessment procedures in such a way that the maximum amount of information is gleaned. However, lack of standardization has limited the use of Luria's test procedures in the United States.

In summary, actuarial and clinical approaches to assessment in neuropsychology are adequately illustrated in the Halstead–Reitan and Luria–Nebraska test batteries. Reitan's test hallmark is the strict scientific standardization of the clinical approach. Luria's method, on the other hand, is actuarial and encourages the clinician to exercise his professional expertise. The clinical approach is more readily accepted in scientific circles, having conformed to the rigors of experimental investigation; actuarial approaches are not usually so well accepted. Nevertheless, it is acknowledged that both approaches are necessary in pediatric neuropsychological practice.

B. Prospectives of Developmental Neuropsychology

Clinical developmental neuropsychology is more than a search for tests of brain damage. Diagnostic information as an end result of neuropsychological assessment limits the effectiveness of the testing situation. The recognition by many clinical neuropsychologists of the need to complete assessments with more direct treatment implications is an important step forward. While the procedures have shown clinical validity for differentiating behavioral deficiencies resulting from brain dysfunction, the main objective in neuropsychological evaluation is to provide descriptive information about the behavioral consequences of neuropathology, which can be used to design relevant rehabilitation or remedial programs for individuals with brain-related disorders.

The therapeutic role of the neuropsychologist is only beginning to be defined. Traditionally, the neuropsychologist has worked as a diagnostician, with little emphasis on devising treatment programs. It is now a common argument in favor of a neuropsychological perspective that goes beyond the diagnosis of impaired neurological processes to the structuring of educational programs that maximize a child's assessed strengths.

Until recently, the brain and its relationship to behavior have been largely unrecognized in attempts to rehabilitate and educate the individual. Developmental neuropsychology can signify a great step forward in its promise to teach humans about humans. It is important to move away from diagnosis as a sole criterion of neuropsychological assessment and move toward ways in which the individual can best be helped. The study of children, as developing individuals, has the potential to teach us much about a subject we are most interested in: the human being.

Bibliography

Bryden, M. P. (1982). "Laterality: Functional Asymmetry in the Intact Brain." Academic Press, New York.

Hynd, G. W., and Obrzut, J. E. (eds.) (1981). "Neuropsychological Assessment and the School-Age Child: Issues and Procedures." Grune & Stratton, New York.

Hynd, G. W., and Willis, W. G. (1988). Pediatric Neuropsychology." Grune & Stratton, Philadelphia, Pennsylvania.

Kolb, B., and Whishaw, I. Q. (1980). "Fundamentals of

Human Neuropsychology.'' Freeman, San Francisco, California.
Lenneberg, E. H. (1967). ''Biological Foundations of Language.'' Wiley, New York.
Obrzut, J. E., and Hynd, G. W. (eds.) (1986). ''Child Neuropsychology,'' Vol. 1. Academic Press, Orlando, Florida.
Obrzut, J. E., and Hynd, G. W. (eds.) (1986). ''Child Neuropsychology,'' Vol. 2. Academic Press, Orlando, Florida.
Orton, S. T. (1937). ''Reading, Writing, and Speech Problems in Children.'' Norton, New York.
Rourke, B. P., Bakker, D. J., Fisk, J. L., and Strang, J. D. (1983). ''Child Neuropsychology.'' Guilford, New York.
Segalowitz, S. J., and Gruber, F. A. (eds.) (1977). ''Language Development and Neurological Theory.'' Academic Press, New York.

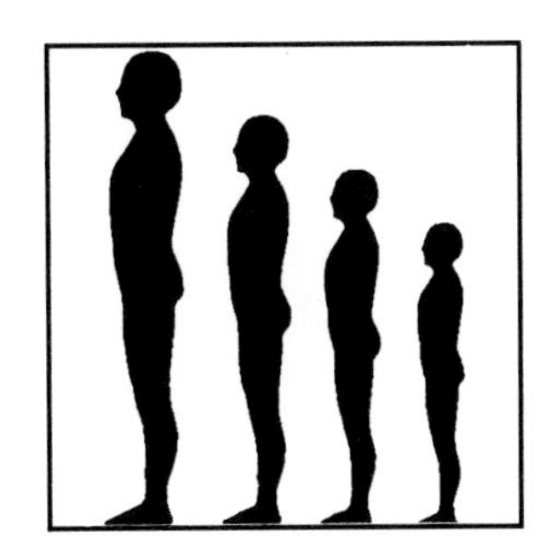

Development, Psychobiology

MYRON A. HOFER, *Columbia University*

Glossary

Critical or sensitive periods Limited time during development when a system is particularly open to modification of its characteristics in response to some external influence

Epigenesis Historically, the idea that complex structures and functions originate from formless material within the egg. Now used to refer to the influences on development that do not arise from the actions of genes. In particular, denotes the stepwise nature of behavior development, whereby new characteristics emerge serially as a result of the repeated interactions of the organism with its environment.

Heterochrony Evolutionary change in the onset or timing of development. Rates of development of features are not synchronized within each individual but are capable of separate variation. Particularly, the rate of development of a feature in descendants may be either accelerated or retarded in relation to its schedule in ancestors.

Ontogenetic adaptation Specific adaptive mechanism, peculiar to a given stage in development and its special environmental conditions, which is not present in the adult

Stage Period during the development of a system or behavior during which changes are taking place relatively slowly but which differ from earlier or later stages in important characteristics. Stages are bounded by transitions in characteristics of the organism (e.g., infant, juvenile) or its environment (e.g., prenatal, postnatal).

THE FIELD OF DEVELOPMENTAL PSYCHOBIOLOGY brings together research efforts aimed at understanding the development of behavior and of the physiological systems related to it. The goal is to discover the underlying processes that determine the course of development through experimental and observational studies of animals, including humans. An effort is made to integrate psychological with biological frames of reference, through an approach to behavior that is based on the principles of evolution and ecology. The development of physiological systems and of behavior are treated as parts of an integrated unit organized to promote adaptation of the developing organism to its changing environment. Research in this field has shown that these adaptive interactions provide complex neural and hormonal feedback to the developing nervous system that regulates the course of behavior development and ultimately determines the nature of the organism. This approach promises to resolve the ancient "nature vs. nurture" riddle by shifting the level of analysis to the psychological, physiological, cellular, and genetic mechanisms involved in the processes of development.

I. History

The relationship between the processes underlying inner experience, behavior, and bodily functions has puzzled humans since they first began to think. Once Darwin established the evolutionary continuity between animals and humans, the stage was set for a vast broadening and enrichment of scientific approaches to the problems of psychology and of development. The word *psychobiology* has been used since the end of the 19th century to denote an integrative approach that cuts across traditional academic disciplines to emphasize the fundamental unity of the processes underlying these various manifestations of human nature. Comparative psychology in North America and ethology in Europe emerged as early forerunners of contemporary psychobiology, and in the area of medicine, Adolph Meyer is generally credited with establishing the term in his "psychobiological life history" approach.

In the first half of the 20th century, however, these trends were almost buried in waves of extreme genetic determinism (e.g., eugenics) and of a polar opposite, extreme environmentalism (e.g., behaviorism). Furthermore, for most of this century, developmental psychology and embryology (or "developmental biology") were fields that maintained almost complete isolation from each other, turning away from the synthesis that had been emerging at the close of the 19th century. Finally, in the years after the second world war, research on displaced war refugees, followed by more controlled experimental work with animals, produced dramatic examples of the impact that early life experience can have on the development of behavior and of underlying neural structures. Embryologists (e.g., C. F. Waddington) as well as ethologists (e.g., Konrad Lorenz) were writing about critical or sensitive periods in development when experience could have unique effects. Developmental psychologists (e.g., Piaget) were discovering stages in the acquisition of knowledge by children during which information was processed in ways that were different from the adult. Moreover, research in endocrine influences on behavior (e.g., Frank Beach and Daniel Lehrman) demonstrated how biological regulatory systems, behavior, and experience interacted in orderly sequences to produce complex adaptive patterns such as courtship, mating, and nursing.

These research findings in a number of different fields appeared to have much to contribute to one another and created the need for an organization to foster the study of developmental processes that would draw not only from psychology and embryology but also from physiology, genetics, biochemistry, endocrinology, and neuroanatomy. The potential for novel applications of these approaches to clinical medicine attracted psychiatrists, pediatricians, and neurologists. In 1967, The International Society for Developmental Psychobiology was formed, and the first issue of the journal *Developmental Psychobiology* was published in March of 1968. Since that time, the Society for Neuroscience (in 1971) and the International Society for Developmental Neuroscience (in 1981) have come into being as forums for research on the nervous system, primarily at the cellular and molecular levels.

Currently, the study of development is being carried out in fields that are defined by levels of biological organization and the different methods that are applicable at each level: cellular and molecular neuroscience at one extreme, cognitive/developmental psychology at the other, and developmental psychobiology operating in the area between these two, at the level of integrative physiology and behavior.

II. Evolution and Development

Ultimately, the understanding of development derives from evolutionary theory in that developmental patterns are products of evolution. But attempts to integrate the two forms of biological change within a single explanatory principle have not yet been successful. In Darwin's day, adaptations that were acquired during the lifetime of an individual, like particular habits of behavior, were thought to gradually become part of the inherited characteristics of individuals after a number of successful generational repetitions. In this way, natural selection was thought to be capable of building a developmental schedule of adaptive behaviors from the history of experiences that occurred over a number of generations. For example, Douglas Spalding, probably the first developmental psychobiologist, discovered in 1872 that chicks raised without visual experience would peck at moving insects with "infallible accuracy" on their first experience after removal of the hoods he had fitted on them at hatching. He concluded that such highly complex behavior patterns ("instincts") were the result of the "accumulated experiences of past generations . . . and . . . may be conceived to be, like memory, a turning on of the nerve currents on already established tracks."

The use of memory as an analogy for the mechanisms of heredity and development was central also to the immensely influential "biogenetic law" of Ernst Haeckel, put forward in 1874. He portrayed the stages in the development of a human as "recapitulating" the adult forms of its ancestors progressively from single cells through cellular aggregates, invertebrate, vertebrate, and lower mammals in a stepwise march through our evolutionary history. This portrayal of human development as an accelerated version of the developmental patterns of all our ancestors linked together was captured in the phrase still taught in high school biology: "ontogeny recapitulates phylogeny."

The rediscovery of Mendel's experimental work on the mechanisms of inheritance in 1900 and the work of experimental embryologists in the first half of the century provided a mass of evidence that forced the abandonment of these appealing theories that so neatly linked together development and evolution. They are mentioned here because they have not yet disappeared from the psychological literature and clinical psychiatric works on development, and the student should be aware of their historical place.

Our best present evidence leads us to believe that the germ cells containing the genes are isolated and are not affected by the consequences of events in the life of the individual. And we have learned that developmental rates of different features may be delayed as well as accelerated in evolution. Thus, early features of an ancestor's development may be retained and appear later in the development of descendants as frequently as the accelerated appearance of ancestor's adult features described by Haeckel. Thus, developmental processes are clearly not offering us a "time capsule" in which we can trace the history of evolution in any direct manner. The ontogeny of a species is what has been retained *and subsequently modified* in the process of evolution. It embodies transformations that have come about through mutations in structural genes and through heritable alterations in regulatory genes that control the timing of developmental events in different cell lines and within the organism. New species can originate through the addition of new traits early in development (insertion) as well as at the end of a developmental sequence (terminal addition). And alterations in the rate of development of one system or characteristic in relation to others (heterochrony) can have profound long-term effects, as will be described at the end of this section. [*See* EVOLUTION, HUMAN.]

But, evolution has not created new forms with developmental patterns that are entirely new designs. Evolution has conserved in the human many basic developmental processes by which the long line of our ancestors developed, going back to single-cell organisms and beyond. The general similarity of early stages in development across widely divergent species is the result of our common ancestry. But in different species, similar structures and functions may develop on different schedules and through different developmental processes. Likewise, similar developmental processes can have different outcomes in different species. Within closely related groups such as the mammals, however, the conservative nature of evolution is predominant, with many developmental processes being closely similar and principles derived from one mammalian species often being found to generalize to others. This provides us with the basis for the use of animal models to understand human development.

In the patterns of human development we see the footprints of our evolution. The diversity of the complex patterns underlying our behavioral development appears to have the same origin as the diversity of successful life forms. For these patterns embody residues from phases of evolutionary history when the units of natural selection shifted from the simplest replicating nucleotides to more complex molecules, organelles, then to cells, and finally to multicellular organisms. The many different paths and processes by which the nervous system of even the simplest animals is assembled appear to represent the end result of the operation of a multitude of selection pressures in changing environments over a prodigious time scale.

Since regulation by feedback from the immediate environment and the effects of activity are present even in relatively simple biochemical processes (e.g., end-product inhibition of enzymes), the role of environmental interactions became an integral part of developmental processes from the beginning. Environments exist at various levels, ranging from that within cells to the ecology of species, and have become as important a part of what is inherited as the information carried in the genome. Developmental processes have made use of both sources of information at every stage in the transformation of germ cells into an adult organism in all species (see Section III, C).

What appears to be missing in our understanding of development is a simple principle by which the diversity of processes and schedules found in human development can be said to operate, something similar to the explanatory power that the principle of natural selection provides for evolution. And yet, if our present understanding (as outlined above) is correct, such a principle does not exist. Instead, developmental processes are the result of natural selection, and as such they represent diverse solutions to the problem of building a successful organism.

Thus far we have viewed developmental processes primarily as the results of evolution. But the recent discovery of genes that regulate the timing and occurrence of developmental events has suggested a mechanism by which evolution may have taken place through changes in schedules of development. New species appear to have evolved through relatively simple modifications in regulatory genes controlling the timing of development in different systems. A particularly intriguing example is the divergence of humans and chimpanzees from their common ancestor, the two species having in excess of 99.9% identical DNA. The retention of certain early bodily features into adulthood (''neoteny'') and in particular the large, slowly developing brain of the human are likely to be due to changes in a few genes regulating the relative timing of developmental events within different cellular systems. The prolonged period of responsiveness to experience in humans, resulting from this genetic mechanism, is thought to underlie our acquisition of complex language, abstract thought, and even civilization itself.

III. Principles

A. Stages and Transformations

Development is often defined in dictionaries as ''a gradual unfolding, a fuller working out of things'' (Oxford English Dictionary). But it has a very different character, which is the result of the forces considered in the previous section. Evolution did not invent wholly new developmental plans at each stage in the history of a species but utilized existing developmental processes, modifying them through terminal addition, insertion, and heterochrony as described above. The inheritance of these modifications, involving changes in sequences and timing of growth, proliferation, and differentiation in different cellular systems, gives development its complex programmatic character in which reasonably well-defined stages of relative stability alternate with periodic rapid transformations in multiple systems. The temporal organization of development in current species thus reflects a long evolutionary history of successful strategies, pieced together and modified in novel ways.

Stages are periods during which changes are taking place relatively slowly but which differ from subsequent or past stages in important characteristics of structure or function. Transitions mark the limits of stages. Structures and functions are generally continuous within stages and may be continuous or discontinuous across transition periods, depending on the extent of reorganization taking place during the transition. Thus embryos, fetuses, newborns, children, adolescents, adults, and the elderly are remarkably different and inhabit different environments. They differ more from each other than do adults of different species.

B. Adaptations and Precursors

One of the tasks of a developmental approach is to attempt to understand organisms at a particular developmental stage and their adaptation to the unique environment of that stage. For environments change as much during development as do the young themselves (for example, the transition from prenatal to postnatal life in mammals). Special characteristics that appear transiently and suit developing forms to a given stage-specific environment are known as ''ontogenetic adaptations.''

Two forms of selection pressure have acted together to determine the nature of the developing organism at any particular stage. One has selected features that prepare for more complex and more adaptive functioning in subsequent stages (e.g., capabilities for perception and for action are necessary antecedents for learning to occur). The results of this form of selection give development its ''progressive'' character. The other has selected features that favor adaptation to the unique environment of a given stage (for example, the following response of newly hatched birds that favors attachment to the parent). Features that maximize adaptation to such early environments as the uterus or egg have resulted in greater numbers of individuals surviving to a reproductive age.

Thus, traits of the young are usually understood both in terms of their contribution to future development, as precursors, and in terms of their role within the ecology of their stage in development, as adaptive traits. The interesting issue then becomes: how do these two processes interact? For variations in the nature of adaptive behavior at a given stage often feeds back upon developing systems (through the neurochemical and hormonal effects of these experiences) to alter the schedule and even the direction of their development in subsequent stages. For example, different patterns of mother–infant interaction have been shown to alter the behavioral and physiological responses of young as they develop into juveniles and adults.

A great deal of current research activity in the field is directed at understanding more precisely how these interactive effects take place, both in terms of what later changes are produced as a result of the early experiences and also in terms of the mechanisms by which earlier adaptive interactions alter developmental trajectories in specific systems. Sensitive or "critical" periods when certain experiences (e.g., sensory stimulation, learning, drug exposure) have maximal influence on subsequent development are of particular interest, along with the developmental events that form the boundaries of these sensitive periods. For example, the period for imprinting of attachment in certain species of birds is ended by the development of "fearful" avoidance responses to novel moving objects.

C. Dynamic Forces

The foregoing considerations help us to understand the stepwise form that development takes but do not account for its dynamic quality. What are the forces that propel the developing organism through this series of stages and transformations? How are these forces regulated?

It is difficult to single out certain elements as primary in a complex process like development that operates through a series of interactions between different kinds of events occurring at different levels of organization. But to state that everything depends on everything else does not provide a good basis for further understanding. I have provided a conceptual schema (see Table I) which proposes three classes of events as the most dynamic and pervasive ("driving forces") and others which act to modulate, guide, and shape the interplay of these forces ("regulatory processes"). This schema is designed to provide a basis for discussing mechanisms of development at the level of behavior and integrative physiology and is congruent with a similar schema proposed by Gerald Edelman for developmental processes at the level of groups of cells (brain cells in particular). In Edelman's classification, the driving forces are cell division, migration, and death; while regulatory processes are cell adhesion, differentiation and induction. These cellular events result in the formation of a diversity of tissues and organs (morphogenesis) from cells with identical genomes. The natural extension of morphogenesis is maturation, which I have listed as the first of three driving forces of development at the level of integrative physiology and behavior. Thus maturation can be considered to be a link across levels of biological organization in development.

Maturation denotes the growth and functional changes in the brain and other organs that result from the interactions of structural and regulatory genes as controlled by intracellular 'messenger' proteins linking the genome with the environment of the organism in a series of steps involving neural and hormonal pathways. The second driving force, behavior, begins to exert its effects when maturing nerve cells begin to generate and conduct impulses in early fetal life. Activity in motor and sensory nerve cells plays a major role in shaping the musculoskeletal system and the organization of brain networks for processing information. Motor activity propels the developing organism into new interactions with the environment after birth, and sensory function allows these interactions to be organized into novel patterns (e.g., through learning and memory). The third major force is environmental change. As development proceeds, the environment that the organism encounters changes radically. Some of these changes are thrust upon the young (e.g., birth), some are the result of its own increasing range of activity, and many are due to both operating together.

These three forces, operating together, propel the development of behavior along its course. But in order for development to follow a relatively predictable route and for the organism to adapt, within limits, to variation in its ecology, a number of regu-

TABLE I Components of Psychobiological Development

		Driving forces		
	MATURATION	BEHAVIOR	ENVIRONMENTAL CHANGE	
		Regulatory processes		
NUTRIENT	THERMAL	HORMONAL	SENSORI-MOTOR	INTEGRATIVE

latory influences exist which initiate and control the operation of these forces and act to integrate them in adaptive patterns. These regulatory processes exist in an extraordinary variety, and the analysis of specific instances of such regulation is the subject of much current research in the field. In outline, these regulatory processes can be categorized into five main areas: nutrient, thermal–metabolic, endocrine, sensori-motor, and higher integrative functions (see Table I).

I will give one or two brief examples from the wide range of regulatory influences in the five major categories. Nutrient intake regulates feeding behavior in different ways at different ages; early deprivation of nutrient can have irreversible long-term effects on brain structure and cognition in adulthood; nutrient has unexpected actions in regulating early cardiovascular function; and young raised on diets of different composition show characteristic behavioral differences. Core body temperature, together with oxygen-dependent metabolic processes, controls rates of maturation of cells and other physiological processes. Ambient temperature regulates the expression of a variety of behaviors in infants and influences the nature of the mother–infant interaction. Endocrine influences have their most profound effects on reproductive development. For example, low levels of testosterone in the late fetal period of male primates appear to be essential for masculine sexual behaviors to develop in puberty when activated by the greatly increased rates of secretion of testosterone in males at that later stage of development. Sensori-motor regulation is perhaps the largest category, involving influences in all sensory modalities as well as feedback from all behavioral activity. Motor activity takes place in human fetuses from the 7th week on, and sensory function develops at about the same time. These early sensori-motor activities constitute a major shaping influence on maturational processes. For example, the shapes of joints between bones and the growth of supporting cartilage structure is dependent on the existence of the jerky, uncoordinated activity of the early fetus. Later, the development of coordinated and adaptive movements is dependent on shaping by sensory feedback and by the temporal relationship between motor and sensory functioning. The fifth category, higher integrative functions, includes learning, memory, emotion, motivation, cognitive plans, and inner consciousness of these mental events. Much of developmental psychology is concerned with this last category, which nevertheless rests upon the operation of many less-apparent developmental processes within the other four categories.

D. Levels of Organization

The driving and regulatory forces just described clearly operate at a number of different levels of organization, ranging literally from molecules to mind. This is inevitable when considering human development which begins with single cells, the sperm and the egg, and progresses to complex psychological events. Bridging these different worlds are a number of properties which are of central interest. *Environments* exist for cells and molecules as well as for the adult organisms and provide similar opportunities for interactions or *experience*. Cells as well as organisms are changed as a result of that interaction, through molecular mechanisms. *Behavior* is a word that we use to describe molecules and cells as well as complex physiological systems and the actions of the whole organism. The use of this word in these different settings is not simply metaphorical. For the three basic properties of behavior as we know it (activity, receptivity, and integration) are found even within the chemical systems of cells. Some of the simplest of life processes have an organized and adaptive character. This principle underlying the organization of behavior across many levels is understandable in terms of evolutionary history: behavior has been extraordinarily useful.

These continuities across levels of organization help the developmental psychobiologist make sense out of the diverse events that characterize different stages of development. And in the adult human, a given behavior or psychological process can be analyzed at a number of different levels of organization using similar principles, although the methods of analysis may differ enormously between the levels of molecular genetics and cognitive psychology.

E. Emergent Properties

It is evident that as one advances into more complex levels of organization, new properties emerge. Psychological terms such as *learning* and *emotion* are used to deal with properties of complex levels of organization for which biological terms do not exist. Much of current research in developmental psychobiology centers on the general area of the transition from complex physiological systems to psychological processes. Behavioral terms which apply to both systems provide a conceptual bridge for moving back and forth between the historically (and academically) separate realms of psychology and biology.

Throughout the course of human development, new properties emerge at each stage. In its first emergence, we have an example of the property in a simple form when we are most likely to be able to analyze and understand it. Since evolution and development build up more complex organization out of earlier simpler structures, these are not abandoned but are used as building blocks. Thus this field hopes to understand such complex mental functions as motivation, emotion, and cognition by finding out how they are put together.

F. Nature and Nurture

It should be evident from this discussion of the forces at work during development that there can be no events which are solely the result of environmental influences or are simply "programmed" by the genes. No "blueprint" for structure or for behavior exists. Instead, a series of events takes place involving a number of participants or contributors. Genes are as influenced by their location in the developing embryo as children are by their location among the cultures of the world. The contribution of genetic and epigenetic processes can only be evaluated in terms of the processes underlying a given developmental event and not predicted by generalizations such as that genes determine form while experience determines function. (For example, the shape of a lobster's claw and the markings of a Siamese cat's fur are strongly influenced by experience at sensitive stages of development, while the ability of some nestling birds to fly does not require practice.)

One of the major discoveries of the past two decades in this field is the extent to which biological as well as psychological systems are shaped by environmental influences during development. Conversely, the extent to which psychological processes are influenced by genetic differences is rapidly becoming appreciated. These trends have led to the abandonment of notions that associated biology with nature and psychology with nurture. This unification has greatly expanded the horizons of biology while giving psychology a firm foundation upon which to launch comparative and developmental branches of the field.

IV. Events and Processes

It would be well beyond the scope of this article to summarize current knowledge in the field. It is possible, however, to select some of the major events in human development and use them to illustrate the operation of the principles outlined above.

The first phase of human development takes place in the form of germ cells within individuals of the previous generation, prospective parents. The ova have a long existence, being formed during the fetal period of the mother. The sperm are formed 60–90 days before fertilization in the prospective father and become motile when ejaculated with hundreds of millions of others into the fluid environment provided by the male and female at the time of sexual intercourse. The swimming behavior of the sperm is the first critically important behavior in a human life, for one of the 300–400 million sperm will be the first to make its way through the cervical canal, the cavity of the uterus, and the fallopian tube to fertilize the egg, blocking all other sperm from penetrating the egg membrane. Individual sperm vary greatly in shape, motility, and surface proteins, as well as in the genetic material within their nuclei. The competitive balance between different sperm can be shifted by their interaction with characteristics of the female genital tract and the egg. The mother's behavior and emotional state, through changing hormonal and autonomic balance, may alter the environment through which the sperm must move (antibodies, mucous flow, and uterine contractions) and thereby act to select which sperm will fertilize her egg. The genetic content of the sperm also may influence the selection process by affecting sperm surface proteins and motility patterns. (For example, more girl babies are born to women who have intercourse soon after ovulation.) The gene–environment interaction is nowhere more directly represented than in this prefertilization stage of human development. [*See* SPERM.]

Fertilization represents the first major transformation in human development and brings to an end the bicellular phase of human life, one that is barely recognized by us, although it is both long and in the end, eventful. The next phase lasts for only 9 months, but in that short time a whole human being is built from a single cell. Until recently, it has been viewed as a stage in our life devoted to the development of structure, while behavioral development was considered to begin at birth. The invention of ultrasound imaging, the ability to support life in younger premature babies, and advances in developmental neurobiology in the past decade have revealed for the first time the extent and nature of prenatal behavior. [*See* FERTILIZATION.]

The fetus begins to move early in gestation, during the 7th week after fertilization in the human. By 9 or 10 weeks, 15 different kinds of movements can be described, including startles, hiccups, breathing movements, and sucking, as well as movements of limbs, trunk, and head separately and together. Sensory function begins, independently, at about the same time with the first responses occurring to stimulation in the area of the mouth, at a stage when the growing sensory nerves have not yet reached the basement membrane of the skin. Early axonal outgrowths conduct impulses and transport trophic substances even before synapses are fully formed and before primitive muscle tissue becomes differentiated. [*See* FETUS.]

What is the "purpose" of this early behavior? It has become clear that neural activity and the transport of trophic substances are major components of the means by which the nervous system is constructed and by which its fine structure is organized. For example, it is the pattern of early neural activity that determines whether muscles will develop into "fast twitch" or "slow twitch" types, with their different biochemical structure and performance. Activity in sensory systems is needed for the formation of cerebral architecture such as "cortical columns." The early, frequent, and persistent movements of fetal limbs are critical for the shaping of joint cavities and the formation of intra-articular ligaments. This use of early behavior as a sculptor of the developing brain and limb is a good example of an ontogenetic adaptation. Early behaviors are not fully explained as precursors of later, more complex behaviors; they serve an adaptive function unique to the particular requirements of that stage in ontogeny. The dynamic forces of maturation and activity are clearly illustrated here as well.

The intrauterine space provides a controlled, predictable environment that is reliably heritable. High rates of nutrient and oxygen flow can be supplied, as well as other substances such as steroid hormones and neuroactive peptides, which can serve a regulatory function for the developing fetus. We are just beginning to learn about maternal influences on behavior development during this period. For example, in the last few years it has been discovered that the mother sets the biological clock of her fetus in the last trimester to her own circadian rhythm and day length. Also, recent evidence suggests that the fetus learns the special characteristics of its own mother's voice so that it can recognize her as distinct from others within hours of birth. These functions clearly serve an adaptive role as precursors of traits useful in a later stage after birth.

As the fetal period progresses, we can observe a series of transformations that represent the maturation of progressively higher neural systems which come to exert a modulating and organizing effect on the activity of earlier developing units. Motor activity of the fetus reaches a peak between the 15th and 17th weeks, declining gradually thereafter as movements become more patterned in time and are coordinated with each other into sequences. This gathering together of units from within the broad matrix of early diffuse activity tends to organize behavior into a hierarchial structure with different levels of organization. New properties emerge. For example, a cycling between activity and rest characterizes the behavior of a single neuron, but then assemblages of neurons begin to function together with their own particular patterns, and when cardiovascular and respiratory systems are joined with neuromuscular patterns, an integrated behavioral state is formed that can be specially suited to carrying out certain highly specialized functions. The appearance of a precursor of dreaming sleep (rapid eye movement, or REM, sleep) appears at 32 weeks of gestation. In this state there is a profound inhibition of muscle tone and movements while diffuse activating volleys project from the brain stem to higher cortical centers as well as to eye and ear muscles, blood vessels, and heart. This intense internal activation in an otherwise passive organism appears to represent an evolutionary solution to the paradoxical requirements for neural activation in preparation for life after birth and behavioral quiescence in the confined space of the uterus.

Birth represents the most abrupt regularly imposed environmental change in our normal development. It is clearly a most powerful force in the development of human behavior since it provides a whole new range of stimuli with which the infant can interact. This greatly increases the complexity of sensori-motor interactions, sets the stage for learning, and initiates the possibility of higher cognitive functioning. The range of regulatory or shaping forces is thus expanded to a great degree over those present during the prenatal period.

The behavioral capabilities of the newborn human, in contrast to its environment, change very little in the period immediately after birth. Continuity with the sleep–wake state organization of the late fetal period is the most striking characteristic, although episodes of clear-cut wakeful state continue to become more frequent and quiet sleep gradually begins to appear, alternating with the REM sleep previously described. The newborn's behavior when held in its mother's arms reveals that its visual and auditory systems have become predisposed to respond to the human face and voice and its sensori-motor systems predisposed to respond to the breast with a complex sequence of behaviors enabling it to find the nipple, attach, and suck.

The competence of the newborn and its ability to discriminate and learn have only become apparent as investigators have found ways of studying its behavioral capacities in its natural environment. This is because early adaptive behaviors have evolved in relation to the heritable aspects of the environment and not to other modalities, intensities, and patterns of stimuli, such as those present in laboratory tests of learning. The distinction that until recently was drawn between instinctual behaviors and those that are learned has now been abandoned since it has become clear that all animals are predisposed to learn certain things more easily than others and that even the learning of novel tasks is guided by preexisting hierarchies and patterns of response tendencies. In turn, predispositions and complex response sequences, like suckling, that appear to occur in newborns without any opportunity for learning, have been found to depend on nursing experience in order to be maintained by the infant. And with careful analysis, their first appearance can be found to depend on a number of prior interactions between activity and local environmental cues within developing sensory and motor systems in the prenatal period.

Of the many behavioral and physiological systems that have been studied experimentally in animal model systems by developmental psychobiologists, those mediating feeding have been the most extensively analyzed. One of the major surprises to come from this work is the finding that, in all species studied, independent feeding on solid food and fluid does not develop out of nursing as a precursor behavior but instead has its own developmental course that is for the most part independent of nursing experience. An infant's suckling thus is another example of ontogenetic adaptation.

Experimental studies of the mechanisms underlying nursing behavior in rats have revealed an interweaving of response predisposition, sensori-motor regulation through olfaction and touch, more complex associative learning, and dependence on feedback from active initiation of behavior, all of which illustrate the dynamic forces described in the previous section. Pups are born with sensory-guided response tendencies (present also prenatally) which bring them to their mother's ventrum. The scent of amniotic fluid deposited by the mother on her nipple lines by licking after birth guides them closer; then highly developed touch and sensory hairs of their snouts provide the most proximal sensory cues. The first few nursing bouts are the result of these response tendencies, but if some of the pups are artificially fed by stomach tube, these behaviors will gradually cease to be elicited by their dam, and nursing by the pups will cease in 2–3 days. Even if the pup is placed on the nipple each day, nursing will fade away at the same rate. But if the pups are merely placed near the ventrum and are allowed to search out and locate nipples for as little as 15 minutes each day, their capacity to suckle will be maintained. And this maintenance through active initiation does not depend on reinforcement by milk.

The next phase in nursing, its decline at weaning, is assumed in humans to be under control by the mother. But some infants seem to wean themselves, and little is known of the processes underlying this developmental transformation which does so much to define the stage of infancy from that of childhood. Some understanding has been gained by experimental analysis in laboratory rats. The decline in nursing by pups between days 15 and 21 has been shown to be mediated, at least in part, by maturation of a serotonergic receptor forebrain system. This is not to suggest that weaning takes place simply due to processes intrinsic to the pup. Dams of 21-day-old litters given to 14-day-old pups will hasten weaning, and dams of 7-day-old pups will retard it. In fact, placing a single pup of 15–17 days age with a dam and litter of 14-day-old pups and then replacing the dam and litter with a new 14-day-old "family" each week will result in rats continuing to nurse into young adulthood, well after sexual maturity (50–70 days), when they are often as large as the mother. This phenomenon has been shown to be the result of continued milk availability plus a lesser component contributed by the social facilitation of the younger litter mates.

It will be clear from these analyses of the processes underlying the development of early feeding that the infant mammal and its parent participate in a range of interactions in which they are linked by their sensitivity to subtle cues originating from the other. The course of nursing is initiated, maintained, and regulated by these interactions. But nursing is only one of a number of transactions taking place between the parent and the infant. Parents also keep infants warm, clean them, and engage in a wide range of communicative actions and reciprocal sequences referred to as "play." These activities have roles in the health of the infant, the development of its emotional life, its ability to communicate, and a wide variety of motor skills.

Recent experimental research in laboratory rats has revealed a number of inapparent processes, built into the many mother–infant interactions, by which the developing physiological and behavioral systems of the infant are regulated. Growth hormone and corticosterone levels, autonomic cardiovascular function, sleep–wake cycles, central catecholamine neurotransmitter levels, and behavioral reactivity have all been shown to be maintained at their characteristic levels in infants by various aspects of the intensity, frequency, and patterning of different interactions between the dam and her pups. Each system is controlled by a particular modality or aspect of the interaction. For example, behavioral reactivity is regulated by intermittent tactile and olfactory stimulation, cardiac rate by gastric interocepters sensitive to nutrient composition, and sleep–wake states by the periodicity and timing of nutrient and behavioral interaction.

These regulatory systems within early social relationships are an extension of the principles of regulatory forces in development that were present in prenatal life. Their existence is made possible by the predictable inheritance of the early mother–infant interaction secondary to the mammalian characteristic of nursing young. The behavioral attachment system provides the young with a powerful motivational system for maintaining close contact with early social companions. This system makes developmental regulation by the parents and feeding by nursing possible.

As a result of the existence of these systems, prolonged separation of young infants from their mothers can have widespread and even lethal consequences. In addition to the loss of nutrition from nursing and mobilization of the attachment system in prolonged but unsuccessful efforts to search out and reunite with the mother, all maternal regulatory systems are withdrawn at once, producing a pattern of widespread physiological changes. These separation responses give dramatic evidence of the existence of these several developmental processes that characterize this phase of our development.

As development proceeds and new functions mature, the infant gradually comes to increasing independence not only in feeding (as described above) but also in homeostatic regulation of its internal physiology and expansion of its range of behaviors beyond the confines of interaction with its immediate caretaker. But there continue to be dependences of development in a number of systems on social interactions. For example, children born congenitally deaf show similar vocalizations to hearing children during the first 3 months of life. They cry and murmur normally and show normal "gooing." But they do not, beginning in the 4th month, produce normal, resonant vowel-like sounds, and they lag months behind hearing infants in the production of mature syllable strings like "ba ba ba." Without special training, deaf babies' vocalizations do not progress beyond the early stages. This represents the importance of auditory experience in maintaining normal development of speech.

V. Range and Prospects of the Field

These few examples only hint at the range of topics currently under study in developmental psychobiology and focus on very early stages, although development continues throughout life into old age. There is a good deal of activity in the area of abnormal experiences, environmental hazards, and how these influences contribute to the production of pathological deviations in development. Animal model systems useful for the study of clinical conditions in psychiatry (such as anxiety disorders) and pediatrics (e.g., prenatal drug abuse, sudden infant death syndrome) have been developed. Altered susceptibility to stress as a result of differential early experience has been found in studies demonstrating marked influences on corticosteroid responses, the production of gastric erosions, and hypertension. The course of experimental diseases such as cancer, infections, and diabetes in animals has been found to be significantly influenced by modification of early experience such as premature separation, handling by the experimenter, or crowding of the pregnant dams. The physiological regulatory processes that originate in the early environmental interactions described in Sections III and IV account for these complex and wide-ranging early experience effects.

Future progress in the field should result in a deepening understanding of the behavioral and physiological mechanisms of these effects, operating both in normal and disturbed development. The advent of genetically engineered strains with specifiable discrete changes in DNA should lead to a new level of analysis of environmental and other epigenetic processes in gene expression.

Bibliography

Aslin, R. N., Alberts, J. R., and Petersen, M. R. (ed.) (1981). "Development of Perception. Psychobiological Perspectives. Volume 1: Audition, Somatic Perception and the Chemical Senses." Academic Press, New York.

Blass, E. M. (ed.) (1988). "Handbook of Behavioral Neurobiology. Volume 9: Developmental Psychobiology and Behavioral Ecology." Plenum Press, New York.

Edelman, G. M. (1988). "Topobiology. An Introduction to Molecular Embryology." Basic Books, New York.

Gould, S. J. (1977). "Ontogeny and Phylogeny." Belknap Press of Harvard University Press, Cambridge, Mass.

Greenough, W. T., and Juraska, J. M. (1986). "Developmental Neuropsychobiology." Academic Press, New York.

Hofer, M. A. (1981). "The Roots of Human Behavior. An Introduction to the Psychobiology of Early Development." Freeman, New York.

Krasnegor, N. A., Blass, E. M., Hofer, M. A., and Smotherman, W. P. (ed.) (1987). "Perinatal Development. A Psychobiological Perspective." Academic Press, New York.

Prechtl, H. R. F. (1984). "Continuity of Neural Functions from Prenatal to Postnatal Life." Lippincott, Philadelphia.

Raff, R. A., and Kaufman, T. C. (1983). "Embryos, Genes and Evolution." Macmillan, New York.

Shair, H. S., Barr, G. A., and Hofer, M. A. (ed.) (1991). "Developmental Psychobiology: New Methods and Changing Concepts." Oxford University Press, New York.

Smotherman, W. P., and Robinson, S. R. (ed.) (1988). "Behavior of the Fetus." Telford Press, Caldwell, N.J.